D1223374

Insects

Their Natural History and Diversity

With a photographic guide to insects of eastern North America

Insects
Their Natural History and Diversity

With a photographic guide to insects of eastern North America

STEPHEN A. MARSHALL

FIREFLY BOOKS

Published by Firefly Books Ltd. 2006

Second printing

Publisher Cataloging-in-Publication Data (U.S.)

Marshall, Stephen A.
Insects - their natural history and diversity : with a photographic guide to insects of eastern North America / Stephen A. Marshall.
[720] p. : col. photos. ; cm.
Includes bibliographical references and index.
Summary: An examination of the characteristics, habitat and behavior of insects, including comprehensive picture keys for insect identification.

ISBN-13: 978-1-55297-900-6
ISBN-10: 1-55297-900-8

1. Insects -- North America. I. Title.
595.7/097 22 QL473.M377 2006

Library and Archives Canada Cataloguing in Publication

Marshall, S. A. (Stephen Archer)
Insects : their natural history and diversity : with a photographic guide to insects of eastern North America / Stephen A. Marshall.
Includes bibliographical references and index.

ISBN-13: 978-1-55297-900-6
ISBN-10: 1-55297-900-8

1. Insects--North America--Identification. 2. Insects--East (U.S.)
3. Insects--Canada, Eastern. 4. Insects--East (U.S.)--Identification.
5. Insects--Canada, Eastern--Identification. I. Title.

QL473.M37 2006 595.7'0974 C2005-901179-3

Published in the United States by
Firefly Books (U.S.) Inc.
P.O. Box 1338, Ellicott Station
Buffalo, New York 14205

Published in Canada by
Firefly Books Ltd.
66 Leek Crescent
Richmond Hill, Ontario L4B 1H1

Illustration on page 14 by Richard Lewington
Insect Picture Keys designed by Dave Cheung
Cover design by Sari Naworynski
Interior design and layout by Tinge Design Studio

Printed in China

The publisher gratefully acknowledges the financial support for our publishing program by the Canada Council for the Arts, the Ontario Arts Council and the Government of Canada through the Book Publishing Industry Development Program.

TITLE PAGE
Far left: *Baccha elongata* (Syrphidae)
Middle left: *Sphex pennsylvanicus* (Sphecidae)
Middle right: *Lema daturaphila*, the Three-lined Potato Beetle (Chrysomelidae)
Far right: Damselfly adult emerging from its nymphal skin.

Table of Contents

Preface

This book is based on material originally gathered in support of a third-year entomology course – "The Natural History of Insects" – that I started teaching at University of Guelph in 1982. The text is based on the lecture notes for that course, the picture keys are based on the course manual and the photographs are part of a collection that was initiated to provide color to my lectures in several entomology courses over the past 20 years.

The text sections in this book provide an introduction to insect diversity and natural history, with basic information about all major insect families. The photos and captions provide a visual overview of the diversity of each family with discussions of common or especially interesting genera and species. Picture keys are provided to the orders and common families of most orders. The emphasis is on northeastern North America, loosely interpreted as anything east of the Mississippi River and north of the state of Georgia.

Insect identification

Although the focus of this book is on the common families of northeastern North American insects, the keys and photos should be useful for identifying orders and most families anywhere in the world. If you are trying to identify an insect to order, start with the illustrated keys (pages 615–666). When you think you have a match, turn to the appropriate section of the book and look over the full spectrum of photos for that order.

If you know the order and want to identify your insect to the family level you can either jump right to the photos and captions, or start with the illustrated keys. The illustrated keys may not take you right to the family level, but they will guide you to the correct part of the book to look for further information. The photos and captions themselves should serve as a practical field guide to the family or subfamily level for common insects from anywhere in North America, and will serve for positive identification of some eastern insects at the genus or species level.

Almost all the photographs are of northeastern species, although a few groups that do not occur in the east (honey ants, pollen wasps) were slipped in for interest, and a few groups are illustrated with photos from outside North America, as noted. The great majority of the photographs were taken in Ontario, Canada (mostly the Bruce Peninsula or southern Ontario), but a few are from Mountain Lake Biological Station, Virginia (where I teach a field entomology course), and elsewhere in the eastern United States.

The tail end of this Abbot's Sphinx (*Sphecodina abbottii*) caterpillar has an intimidating fake eye instead of the "tail" found on most sphingid larvae.

The illustrated family keys are designed to be as user-friendly as possible, with an emphasis on characteristics visible to the naked eye or easily discernable with a handheld magnifying glass. Most keys represent a compromise between ease of use and comprehensiveness. The keys in this book lean towards ease of use and should be treated as shortcuts rather than definitive roadmaps. The keys to families in the larger orders (Diptera, Hymenoptera, Coleoptera) are designed to aid in the identification of typical members of commonly encountered families, and the odd rarity or exception will not key out. Comprehensive keys to the families of these orders are listed in the references, but most require experience, patience and a good microscope to use. For example, the key to families of beetles in *American Beetles* (Arnett et al., 2002) is 185 complex couplets long; the key to families of flies in the *Manual of North American Diptera* (McAlpine et al., 1981) is 152 couplets long. Those keys will work for almost all North American beetles and flies; the simplified keys in this book will probably work for over 95 percent, including almost all routinely encountered taxa. I think it is a good compromise, but it is a compromise, and the serious student will want to check problematic identifications using more technical literature.

The "id" and "idae" of entomological jargon

This book is organized around insect orders and families. Names given to orders (big groups, like beetles and flies) do not have standard endings, but orders are divided into families, and the names of families always end in "–idae." Insect families are routinely mentioned by informally contracting the family name to end with "–id." Ground beetles, for example, are formally called the family Carabidae, but are informally referred to as "carabids." Sometimes it is useful to talk about a number of related families together, in which case we talk about superfamilies and the names always end in "–oidea." Some families, especially large families like the leaf beetles (Chrysomelidae), are usefully divided into subgroups called subfamilies (subfamily names end in "–inae").

The scientific name of a species always has two parts, the genus (always capitalized and in italics) and species (always in italics, never capitalized). For example, the Monarch Butterfly (*Danaus plexippus*) is a nymphalid butterfly in the family Nymphalidae and the subfamily Danainae. In scientific papers (but not in this book) the name of a species is usually followed by the name of the person who first described and named it (in parentheses if it was first described in another genus). The Monarch, then, would appear as *Danaus plexippus* (L.), with L. serving as a short form for its "discoverer" Linnaeus.

Taxonomic hierarchy	**Suffix**
Superfamily	–oidea
Family (formal)	–idae
Family (informal)	–id
Subfamily	–inae

These seed bugs feeding on a coneflower closely resemble the Small Milkweed Bug (*Lygaeus kalmii*), but the bright Y-shaped markings on their heads identify them as *Lygaeus turcicus*.

Many insect species, especially large or economically important species, have common names like the Monarch, but most do not. Where widely used common names exist they are included here. Common names are normally capitalized. Family names appear in bold face when they first appear in the text.

Classifications are in constant flux as we discover more about the relationships between groups of organisms, and the names used in this book may be different from those you are already familiar with. I have taken a conservative approach to higher classification, using the family concepts in current usage unless there are compelling reasons to accept a recent change. Significant changes are indicated in the text and captions, and generally follow a recent checklist, catalogue or monograph in the Selected References section at the back of this book.

An Overview of Six-legged Life

We live in a world of insects. They are our continual and closest neighbors, so much a part of day-to-day life that most of us hardly take notice of them unless they are particularly loud or obnoxious, or they stand accused of robbery or assault. It is easy to forget that human beings form a tiny two-legged minority in an overwhelmingly six-legged world – a world where a bit of knowledge about our dominant neighbors can unlock the door to a surprisingly diverse local environment. The key to seeing and understanding insect diversity is knowledge of the common orders and families of insects. Armed with that knowledge, a sizeable proportion of the multitude of walking, crawling and flying creatures you share your daily life with will become familiar neighbors, replete with predictable habits.

Insects are influential and interesting creatures well worth getting to know. It is self-evident that you can hardly step outdoors without exposure to a usually ignored infinity of insect types, but surprisingly few people are aware that these ubiquitous animals make up the staggering majority of all living things. Most named species of living things, including close to eighty percent of the approximately one and a half million named animal species, are insects. These numbers would be all the more impressive if they took into account the millions of insect species still awaiting discovery and formal naming. We can only guess that the total number of insect species is somewhere between five and ten million, and we have only the crudest idea of how many individual insects there are. According to one estimate, there are about 200,000 ants for every living human being, and we can safely assume that each of us is also matched by many thousand flies, fleas, bugs and beetles.

Compsobata univitta is the most common eastern species in the Micropezidae subfamily Calobatinae.

Perhaps you are a practical person, unimpressed by raw numbers and the mere fact that insects are the overwhelmingly dominant inhabitants of our planet. You might still find it worthwhile to gain some insight into a group of organisms that bite, carry diseases, cost us billions of dollars every year in crop losses and lead us to contaminate our environment with a frightening variety of toxic chemicals in our attempts to get the better of them. Insect-borne diseases, like plague and typhus, have periodically wiped out sizable portions of the Earth's human population, and have repeatedly turned the tides of war by killing far more people than guns and swords. Even today, half the world's population is at risk from mosquito-borne malaria, and another 90 million people in 76 countries suffer from insect-borne filariasis. Chagas' disease, caused by a bug-borne protozoan, affects another 16 million to 18 million people in South and Central America, and countless millions are affected daily by fly-borne food and water contamination.

The tiny *Poecilognathus unimaculatus* (until recently treated as *Phthiria*) is a common southeastern bee fly (see page 458).

Introduction

If insects are worth getting to know as an enemy of humankind, they are even more worthy of attention as our benefactors. We could probably get by without insect products such as honey and silk, but our food production systems would be wholly changed in the absence of insect pollinators, and entire ecosystems would collapse if there were no insects in the food chain. Furthermore, insect predators and parasitoids keep pest insect numbers down, plant-eating insects regulate plant density, and microbe-eating insects, such as maggots, regulate the decomposition phase that every living thing must go through.

Although our history, health and finances all pivot to a surprising degree on insects, the best reasons to be interested in insects lie elsewhere. One of the great appeals of these invertebrates is their remarkable combination of seemingly infinite variety with comforting predictability. Insects as a whole, and each of the major subgroups, exhibit several constant and predictive attributes despite their dazzling array of novelties and remarkable modifications. Therefore, one can easily learn to recognize the major insect groups and their basic biology, and make safe generalizations about most of the insects encountered day to day.

At the same time, it is always possible to find new things. Even after decades as a professional entomologist, I frequently come across insects I've never seen before or watch insects doing something new and remarkable to me, perhaps something I've read about with interest, or perhaps something never previously observed. Entomology – the study of insects – is both an absorbing hobby and a scientific frontier we can push forward through backyard observations. I hope that the following overview of insect diversity and natural history will provide both an introduction to the science of entomology, and a framework for the observation and appreciation of the small life that surrounds us all.

What is an insect?

Just what is an insect anyhow? To put that question in perspective, think of a tree representing all animal life. Make it a huge old tree with a bunch of spindly twigs growing off the lower part of its massive trunk. Those little branches represent relatively small groups such as sponges, the sea cucumber lineage, the vertebrate lineage, and the mollusk and worm lineage. The main tree trunk, above those spindly twigs, is made up entirely of animals that periodically shed their skins to allow growth (Ecdysozoa). These moulting animals include the nematodes and their relatives as well as a few very small groups such as the penis worms, but most Ecdysozoa, including around 90 percent of all animal species, belong to the phylum Arthropoda.

The Slender Meadow Katydid, *Conocephalus fasciatus*, is a widely distributed and common member of the Conocephalinae (Tettigoniidae).

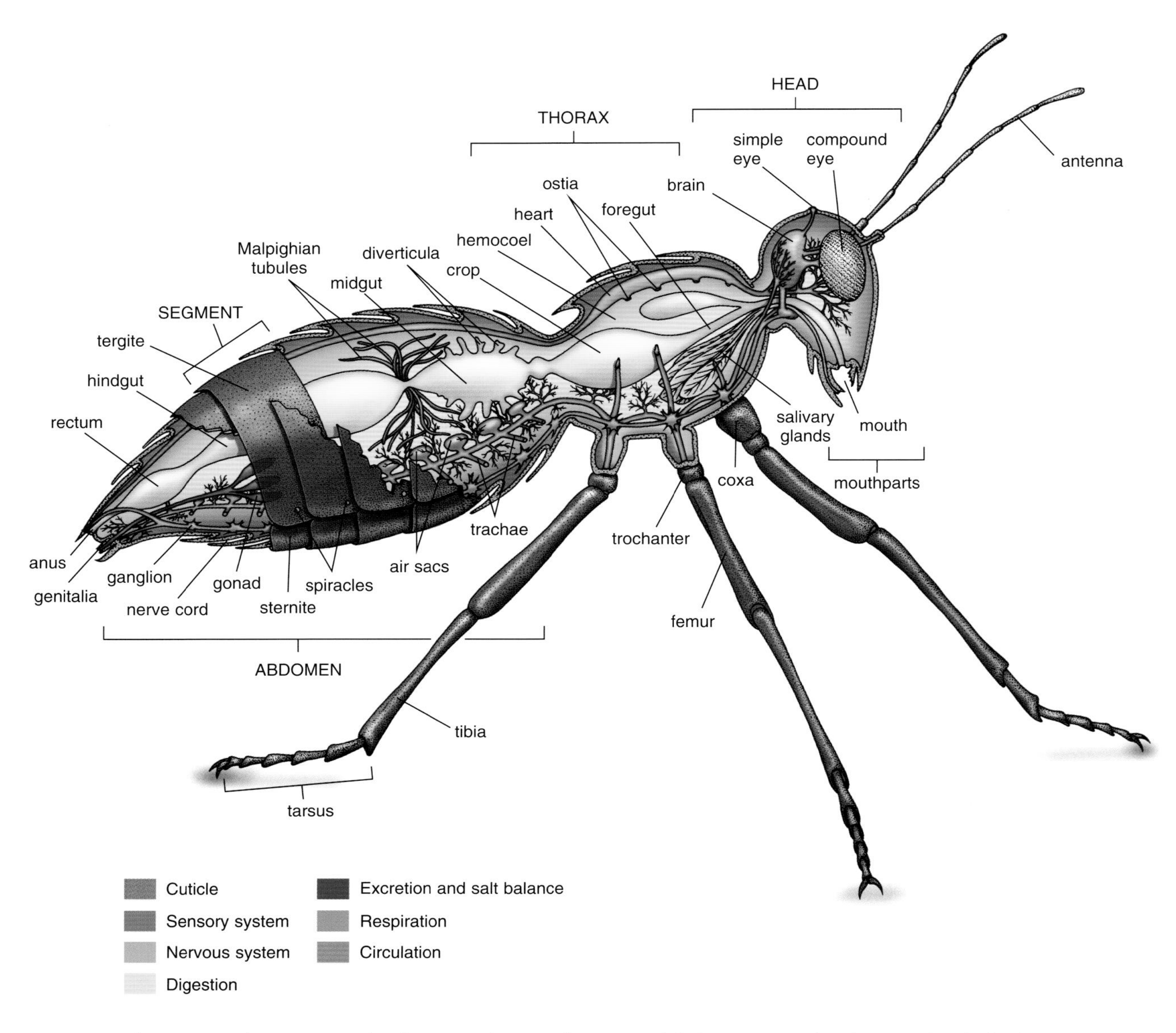

This cutaway shows a generalized insect with most of its external skeleton removed and its muscles, fat body tissue, haemolymph (blood), and most of its air circulation system (tracheae, air sacs) excluded. The remaining tracheae and their connections to spiracles are shown in pale green. The digestive system is shown in yellow, with the excretory system (malpighian tubules) shown as green tubules meeting the digestive system at the junction of the midgut and hindgut. The nervous system, shown in orange, runs along the bottom (ventral side) of the insect, looping up around the oesophagus (part of the digestive system) in the head to form a dorsal brain. There are no blood vessels, but a dorsal heart, shown in pink, circulates the haemolymph ("blood") around the body.

Arthropods can be thought of as "inside out" relative to vertebrates because their skeleton is an external shell. The term "arthropod" means "jointed feet," and refers to a conspicuous side-effect of having a hard external shell. Jointed appendages were an inevitable attribute for knights in armor, and the same goes for invertebrates with a hardened, or sclerotized, external skeleton.

The phylum Arthropoda is divided into several classes, including the arachnids (spiders and their relatives), myriapods (millipedes, centipedes and their relatives), crustaceans (crabs and their relatives), and insects. The class Insecta, which includes the overwhelming majority of all arthropod species, is defined by several unique attributes which might explain why insects are so successful. The most obvious of those attributes, and one reflected in the name "insect," is the division of the body into three sections: the head, thorax and abdomen.

It is useful to think of insects as a group in which different parts of the body were assigned different tasks, and given appropriate appendages for each those different tasks. The insect head is the result of the first few body appendages getting the jobs of scouting out the environment and handling food. In the basic insect body plan, the part of the head in front of the mouth has a pair of sensory antennae followed by an upper lip, or labrum. The appendages immediately behind the mouth are modified into unsegmented jaws called mandibles. A pair of food-handling and sensory appendages called maxillae follows the mandibles, and the maxillae are followed by a similar pair of appendages fused in the middle to form a lower lip, or labium. Both the maxillae and labium have distinctive segmented processes, called palpi (or palps).

Many insects, such as grasshoppers, cockroaches and beetles, retain simple chewing mouthparts in which you can easily examine these basic bits and pieces, but in some groups of insects, the mouthparts are strongly modified. It helps to be familiar with simple chewing mouthparts, like those in grasshoppers, in order to appreciate how the specialized mouthparts of butterflies, bugs, mosquitoes and many other groups can be derived by simple modification of those basic oral appendages.

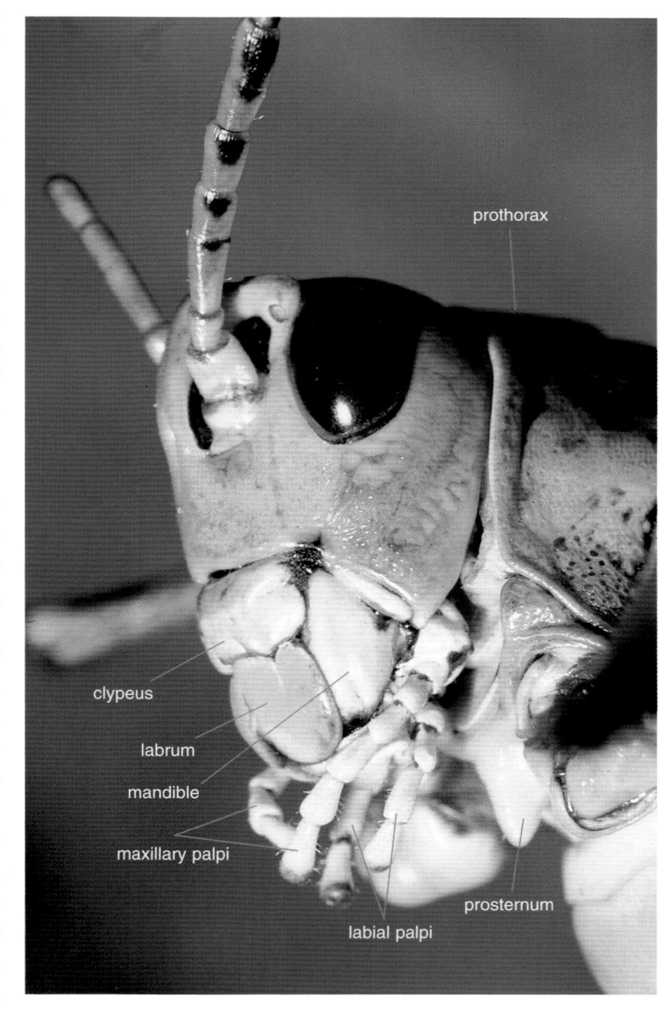

The main parts of the insect head are clearly visible on this grasshopper.

The second of the insect's three body sections is the thorax, which is usually a big muscle-packed box made up of three segments, each with appendages that have become specialized for locomotion. The thorax is the insect's transportation center, almost always bearing three pairs of legs and usually with two pairs of wings. Insects with wings make up not only the majority of the class Insecta, they make up the majority of *all* animals.

The third insect section is the abdomen, essentially a segmented sack of digestive and reproductive organs, usually devoid of any conspicuous appendages other than those involved with sex or laying eggs. Even a wasp's stinger, quite a conspicuous structure at times, is really just a modified egg-laying tube derived from appendages at the end of the abdomen.

Grasshoppers are great insects for a quick and casual study of insect parts, and you should have no problem getting your hands on one of these common insects. The grasshopper's body segments should be readily apparent once you realize that the wings usually project from the thorax right back over the abdomen. You might start by looking at the head appendages – this is easily done, as the big, triangular, chewing mandibles make it easy to get your bearings on the other mouthparts. Don't worry about the brown saliva (we called it tobacco juice when I was a kid) that living grasshoppers dribble over your probing fingers, just wipe it off and take note of the two pairs of segmented palpi following the mandibles. The first palpi arise from the maxillae; the second ones come off the labium.

Now, hold your grasshopper by the hard section just behind the head. This section is hard because its top and sides are enveloped by a big shield made from the top of the first of the three thoracic segments. The second and third thoracic segments both support wings, but the straight-edged, somewhat leathery forewings usually conceal the hind wings. By grasping a forewing and gently pulling it away from the body, you will expose the large, fanlike hind wing.

Now that you have checked out your hopper's wings, look closely at the sides of the abdomen for the abdominal spiracles, which look like a series of portholes. The spiracles, also found on the side of the thorax, open into a system of breathing tubes which form an air delivery system reminiscent of our own system of blood vessels and capillaries. Insects don't have blood vessels like ours, and insect blood, more properly called haemolymph, isn't used for respiration like we use blood. Insect blood, which fills the body cavity, serves a variety of other roles ranging from internal transport of nutrients and chemical messengers through to hydraulic operation of various body parts. Insect blood does circulate, but it isn't restricted to vessels like vertebrate blood.

Your grasshopper even has a heart of sorts, really just a perforated dorsal tube that pumps to circulate the blood toward the head. In a way, insects are not only "inside out" relative to those of us supported by internal skeletons, but they are also "upside down" because the pumping organ is not in a ventral (lower) position like the human heart, but in the dorsal (upper) position like the human spinal cord. The insect equivalent of a spinal cord is a nerve cord running along the ventral part of the body cavity. The insect nerve cord loops up around the esophagus to form a brain of sorts, in the form of a concentration of nervous tissue in the head.

When I suggested that you pick up a grasshopper to browse the major insect bits, I was making the safe assumption that most people know what a grasshopper is. In the following chapters, I will discuss hexapod highlights in the context of specific insect groups, starting with some primitive wingless types and moving up the evolutionary tree to ultimately arrive at a sort of insect treetop, somewhere among the ants, wasps and bees.

The winged insect orders comprise the mainstream of insect evolution. Most of this book goes with that flow, but chapter one dabbles in some of the interesting little trickles of six-legged life that ultimately give rise to the torrent of flying insects that inundate the planet, and chapter 13 leaves the hexapods for a quick look at some other arthropod orders.

This Atlantic Grasshopper (*Paroxya atlantica*) is feeding on a goldenrod flower (see page 84).

Insect Orders

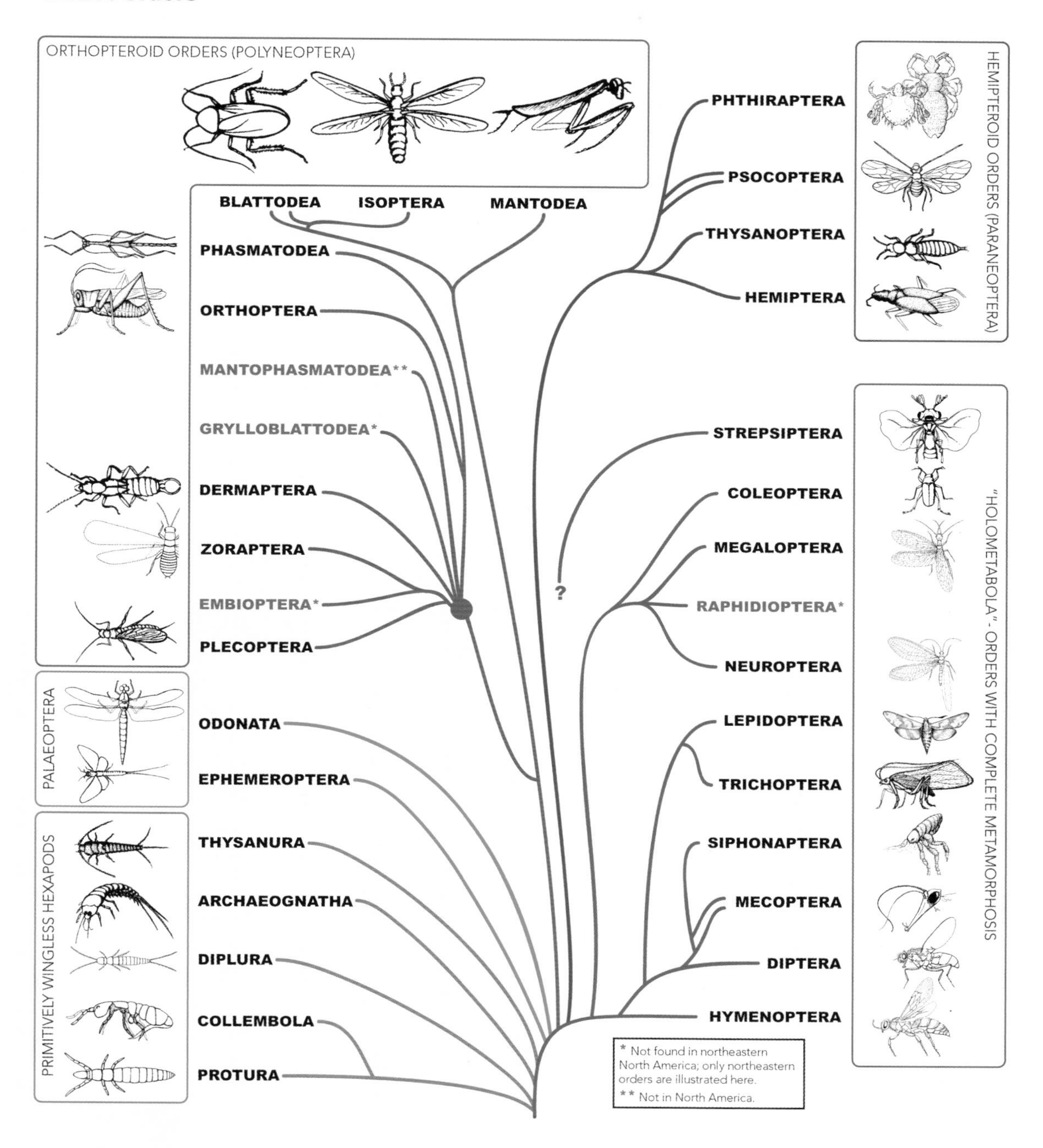

This insect "family tree" illustrates the probable relationships between the hexapod orders. Single branches comprise natural (monophyletic) groups (such as the hemipteroids); multiple separate lines indicate artificial (paraphyletic) groups (such as the primitively wingless insects).

CHAPTER 1

The Wingless Insects: Springtails, Diplurans and Bristletails

Ask an entomologist what an insect is and he or she will tell you that it is an arthropod with a number of special insectan features, the most obvious of which are a head with mandibles (jaws) and a labium (lower lip), a thorax with three pairs of legs and an abdomen that usually lacks conspicuous appendages. They might also mention that almost all adult insects have wings, a very important insect feature unlike anything else in the known universe. Adult insects that lack wings usually belong to groups that have lost their wings as an adaptation to some specialized way of life. Fleas and lice, for example, are descended from winged ancestors that became specialized as wingless parasites. Only a few kinds of wingless insects stem from the base of the evolutionary tree of the six-legged arthropods, below the point where the origin of wings set the stage for the great diversity of modern insects. These are the primitively wingless insects, an assemblage of odd groups truly outside the mainstream of insect evolution.

Despite the fact that older books used to lump all of the primitively wingless insects into one group (sometimes called the Apterygota because they are apterous, or without wings), these odd groups really don't have much in common with one another and they don't form a natural group. In other words, they do not form one branch of the tree of life, but instead are quite separate twigs arising independently from the base of an enormous single branch that includes all winged insects. The winged insect group includes all the descendants of a single common ancestor, and thus forms a natural group (the Pterygota) in contrast to the artificial grouping of wingless insects.

Older classifications often reflect our apparently innate tendency to organize things into those with some attribute on one hand, and those without that attribute on the other. Modern classification, however, is based on the premise that only natural groups – defined by the possession of some character inherited from a uniquely shared ancestor in which that character first appeared – are stable and predictive, and thus only natural groups should be named. In other words, whenever we consider a group, or taxon, like the Apterygota, we must ask whether the characters we use to recognize the group are novel characters that define the group, or merely the lack of novel characters that define some other group.

The Apterygota is a classic example of a taxon no longer recognized because it was based on the lack of one or more characters (wings), rather than the possession of novel or derived characters.

When a biologist describes something as primitive, the immediate image is of something inferior, awkward or inadequate, the way a Model T Ford is primitive compared to a brand new automobile. That is a bit misleading since primitive groups, belonging to lineages that never developed important new adaptations like wings, are often successful and highly advanced in their own, albeit different fashion. The diversity of winged insects is probably in a large part due to their "discovery" of flight, but some of the relatively few species of primitively winged insects are tremendously successful. For an illustration of that success, take a shovelful of soil or leaf litter, dump it on a white sheet and watch for strange little wingless hexapods bouncing themselves around by whacking the sheet with their springlike tails. Estimates of the abundance of springtails in North American soils range from tens of thousands up to millions per square meter, so it is a safe bet that there will be some in your shovelful. Springtails are also common on water, in moss, in fungi – in fact in almost every moist habitat from the Antarctic to the Arctic.

Springtails

Although springtails are ubiquitous and obvious by dint of their sheer abundance, it takes a bit of patience and a hand lens to get a look at one of these tiny (usually 1–3 mm), wingless hexapods. It is worth the effort, though, because they are amazing animals. Some are covered with scales, like those of a butterfly, many are brilliantly colored and all are morphologically bizarre, starting from the long, forked tail used to make Herculean leaps, and ending with the deeply pocketed mouth that makes springtails look like they have lost their dentures and then sucked on a bunch of lemons.

In between the tail and mouth, springtails sport a variety of strange structures. Right in the middle of the body, on the underside of the first abdominal segment, there is a broad, tubelike structure called a collophore, or glue peg. The primary roles of the collophore are probably water uptake, excretion, and grooming rather than adhesion, but the glue peg idea has some validity for springtails that frequent slick surfaces, and calm water surfaces often abound with water-repellent springtails that seem to stick to the surface using their non-water-repellent collophores. At least for these species the ordinal name **Collembola** (page 23), meaning "glue piston," is rather descriptive. Behind the collophore is a little latch (tenaculum) that the forked tail snaps into prior to springing down and launching the springtail into the air.

The biology of springtails is just as strange, relative to advanced insects, as their morphology. For one thing, they don't mate in the "advanced" way, with direct genital contact between the male and female. A typical male springtail deposits his sperm in packets, or spermatophores, away from the female, and the female must subsequently pick up one of those packets with her genital opening. Sometimes the male expedites this process by building a fence of stalked spermatophores around a female, a strategy described by one entomologist as putting the female's virginity "at stake"! Scattering spermatophores seems like a roundabout way to have safe sex, but this sort of indirect fertilization is the rule rather than the exception among primitively wingless insects, and is often coupled with some way of getting the right species of female in touch with the right sperm.

Despite the indirect nature of actual sperm transfer, some springtails have elaborate courtship behavior, and males are often strikingly modified to expedite sperm transfer. The male of one species grasps the female with his antennae, pulls her toward him and then transfers his sperm using his third pair of legs. In another species (a comically rotund globular springtail that often abounds on the near-shore surfaces of calm waters), the male is much smaller than his globose mate, yet he attaches himself to her antennae with hooklike structures on his modified antennae and, with considerable effort, drags or directs her to his sperm droplet. In yet another species, the male taps the female's head with a horn on his head, then turns around and presents his genitalia to the female. She responds by taking a drop of sperm in her mouth, setting it on the ground and then lowering her genitalia on it. Unlike other insects, springtails continue to molt after

Orchesella villosa (Entomobryidae), a common, introduced springtail.

reproducing, alternating reproductive instars with feeding instars. (An instar is a period between molts.)

Despite their intriguing biology, the small size of most of the world's 7,000 or so species of springtails makes field observations of individual organisms difficult. This does not mean that the average naturalist will never notice springtails, because at times their sheer abundance makes them impossible to ignore. Dense swarms of these individually minute creatures are a common phenomenon, never more conspicuous than when they occur against the backdrop of a contrasting surface such as white snow. What observant naturalist has not seen a living blotch of black or gold shimmering on the snow along a snowshoe track or cross-country ski trail? Some springtail species, commonly known as snow fleas, are active on the snow surface on sunny, mild winter days and often reach such concentrations that they blacken (or golden in the case of one western species) square meters of surface. These dense snow-surface aggregations probably expedite sexual encounters, but they might also have something to do with feeding, or perhaps the search for new habitats most easily reached across frozen and predator-free surfaces.

Predation must be a significant factor driving the evolution of springtails, as these minute hexapods seem to make up a large part of the diet of many other arthropods. Common shoreline rove beetles have sticky mouthparts specialized for nabbing springtails, long-legged flies envelop springtails in their spiked lower lips and some specialized ground beetles and ants have traplike jaws adapted to snap shut on Collembola. Collembola, in turn, offer a variety of defenses against predation. Most can make evasive leaps, some have little pores from which they can exude toxic blood, many are covered with scales that make them difficult to grab and some seem to time their activities to avoid predators.

Snow fleas and other insects that avoid predators by going about their business on the snow surface are able to tolerate subzero temperatures using a variety of strategies. It is no coincidence that winter active insects often have dark bodies that soak up the rays, but there are more elaborate mechanisms for either avoiding or tolerating freezing. Most insects avoid freezing by spending the winter in sheltered places and by eliminating substances, like stomach contents, that could provide the initial site or "nucleus" upon which ice crystals could get started on the path towards fatal fracturing of frozen cells. Winter-active insects, like snow fleas and winter stoneflies, can't keep an empty stomach as they graze upon fungal spores and algae, but are nonetheless protected by antifreeze compounds in the blood, just as our car radiators are protected from midwinter ruptures by antifreeze. The most common insect antifreeze, glycerol, is not that different from the ethylene glycol in your car radiator.

As you might guess, because of their sheer abundance, springtails are important organisms. Although a few feed on plants and a very few are even plant pests, springtails really come in to their own as a major component of healthy soil. A square meter or so of good dirt is likely to conceal tens of thousands of springtails, not to mention millions of little collembolan feces slowly releasing essential nutrients. Springtails disperse soil microorganisms and play a major role in soil respiration and decomposition by grazing on fungal hyphae. Springtail grazing can stimulate a kind of symbiotic association between fungi and certain plants, known as mycorhizal growth, which enhances plant growth, and their fungal munching may even prevent the outbreak of disease. It is a good bet that the myriad of springtails you see in that shovelful of soil constitute a good indicator of that soil's ability to provide nutrients in a form available to plants, and therefore ultimately in a form available to your dinner table.

The group of arthropods most similar to springtails is an odd little order that is unlikely to come to the average biologist's attention. Members of the tiny, soil-dwelling order **Protura**, like springtails but unlike other insects, have their mouthparts deep in a pouch, and for that reason Collembola and Protura are often referred to as the entognaths (*ento* = inside, *gnatha* = mouthparts). Some authorities currently exclude the entognaths from the class Insecta, putting them in their own class, Parainsecta, or putting them with the similarly entognathous Diplura in a class called Entognatha. Others treat Collembola as a class on its own. The fossil record provides some justification for this because the Collembola have been out there as a distinct lineage for about 400 million years – about 200 million years before the familiar classes of vertebrates appear in the fossil record!

Furthermore, some recent evidence suggests that springtails are not closely related to insects, and may be more closely related to crustaceans than to true insects. Whether or not to separate the springtails and proturans from the formal class Insecta makes for interesting discussion, but we will continue to use the common name "insect" for the six-legged arthropods, mouth-in-a-pouch or not.

Bristletails and other apterous insects

The springtails, although primitive in their lack of wings, form a highly specialized, successful lineage that hardly fits our image of a primitive insect. Some of the other primitively wingless groups have a more ancient appearance combined with some new characteristics lacking in springtails and proturans but found in all other insects. Three groups (diplurans, jumping bristletails and bristletails) seem to have branched off the insect evolutionary tree later than the springtails, but still before the origin of wings turned the insect mainstream into the great river of life we are all swimming in today.

Despite the fact that at least two of the "bristletail" orders show a relationship to the more advanced insects through specialization in the antennae, eyes and other features, these three wingless orders can all be thought of as "entomological antiques." To quote the great 19th-century scientist and statesman Sir John Lubbock: "In the very interesting genus *Campodea* we have a form closely resembling the parent stock, from which the various orders of insects have arisen."

The genus *Campodea* mentioned by Lubbock is the most abundant genus of the odd, eyeless order **Diplura** (page 25). Diplurans are tiny, obscure, unpigmented soil insects with mouthparts superficially similar to those of springtails. Some entomologists consider the Diplura to be closely related to springtails and proturans ("Entognatha"); others consider them to be more closely related to other insects ("Ectognatha" or "true insects"). The name – "Di-plura" – refers to the two long antennae and the two cerci, or tails, at the other end; the cerci are long and thin in *Campodea* and its relatives (Campodeidae), and short, stout and curved in Japygidae and Parajapygidae. The stout cerci of Japygidae are used to grab prey, and some species bury themselves head down in the soil, leaving the cerci exposed like a snap-trap for small insects. The more delicate Campodeidae are probably herbivorous or fungivorous.

These little-known insects are quite common in some situations, especially under rocks and logs in moist hardwood forests. Some diplurans lay eggs on stalks suspended from the roof of a special underground chamber where the female stays on guard until the eggs hatch. Diplurans are rather obscure animals, but the remaining two primitively wingless insect orders are larger, distinctive and sometimes abundant to the point of being pests.

One of the two wingless orders often called bristletails is the order **Archaeognatha** (page 25), or jumping bristletails.

Firebrat (*Thermobia domestica*), page 26, caption 2.

These primitive, robust insects are easiest to observe along rocky seashores. Inland species are a little harder to find, but jumping bristletails are such striking insects it is worth turning over a few rocks to see them. Good places to look are open areas, like the alvars of the Great Lakes region, and rock piles along old roads. An affinity for hard surfaces like alvars probably reflects the special needs of archaeognaths at molting time. In order to cast a skin to enable growth, some jumping bristletails use their own feces to anchor themselves down to a solid surface, such as a rock. They also glue their eggs to rocks. Mating in jumping bristletails is indirect, but still elaborate, as the male makes silken guides and steers the female onto the waiting sperm droplet. Males are relatively rare and many populations are parthogenetic (consisting of only females).

Jumping bristletails are bigger and much more cylindrical and robust than familiar household bristletails such as silverfish (**Thysanura**), and have huge eyes, a scale-covered body and well-developed appendages on each abdominal segment. The abdominal appendages are unusual among insects and reflect the multilegged ancestry of the whole class Insecta.

Bristletails, jumping and otherwise, are all elongate, scaly, wingless insects with long antennae and tails. Despite this superficial similarity, the silverfish-firebrat group of bristletails (**Thysanura**, page 26) is more closely related to winged insects than to jumping bristletails. Mandibles of jumping bristletails, like those of more primitive arthropods, swing on a single pivot, much like a leg. This is reflected in the order name (*Archaeo* – old, *gnatha* – jaw). In contrast, Thysanura and the winged insects have two similar ball and socket joints on the mandible, rendering the mandible a more specialized structure. If this similarity is inherited from a common ancestor shared between Thysanura and the winged insects, but not jumping bristletails, then it is evidence that Thysanura plus the winged insects form a natural group excluding the jumping bristletails. The jumping bristletails are relatively primitive insects, truly out of the mainstream.

The bristletails include a variety of native species, some of which live as guests in ant nests, but the only bristletails seen by most North Americans are widespread, introduced domestic species. These flattened, fast-running household pests, known as silverfish and firebrats, are common in warm houses where they dine on food scraps and starchy materials like book bindings and wallpaper paste.

Like other primitively wingless insects, silverfish males deposit sperm droplets rather than mate directly with the females. They then help the female make contact with the sperm by spinning a sort of silken snare line which they use to direct the female over the sperm droplet. Jumping bristletails have similar mating behavior, sometimes using a silken line as a carrier thread on which the male deposits his sperm droplets before nudging the female into position to contact his sperm with the tip of her abdomen.

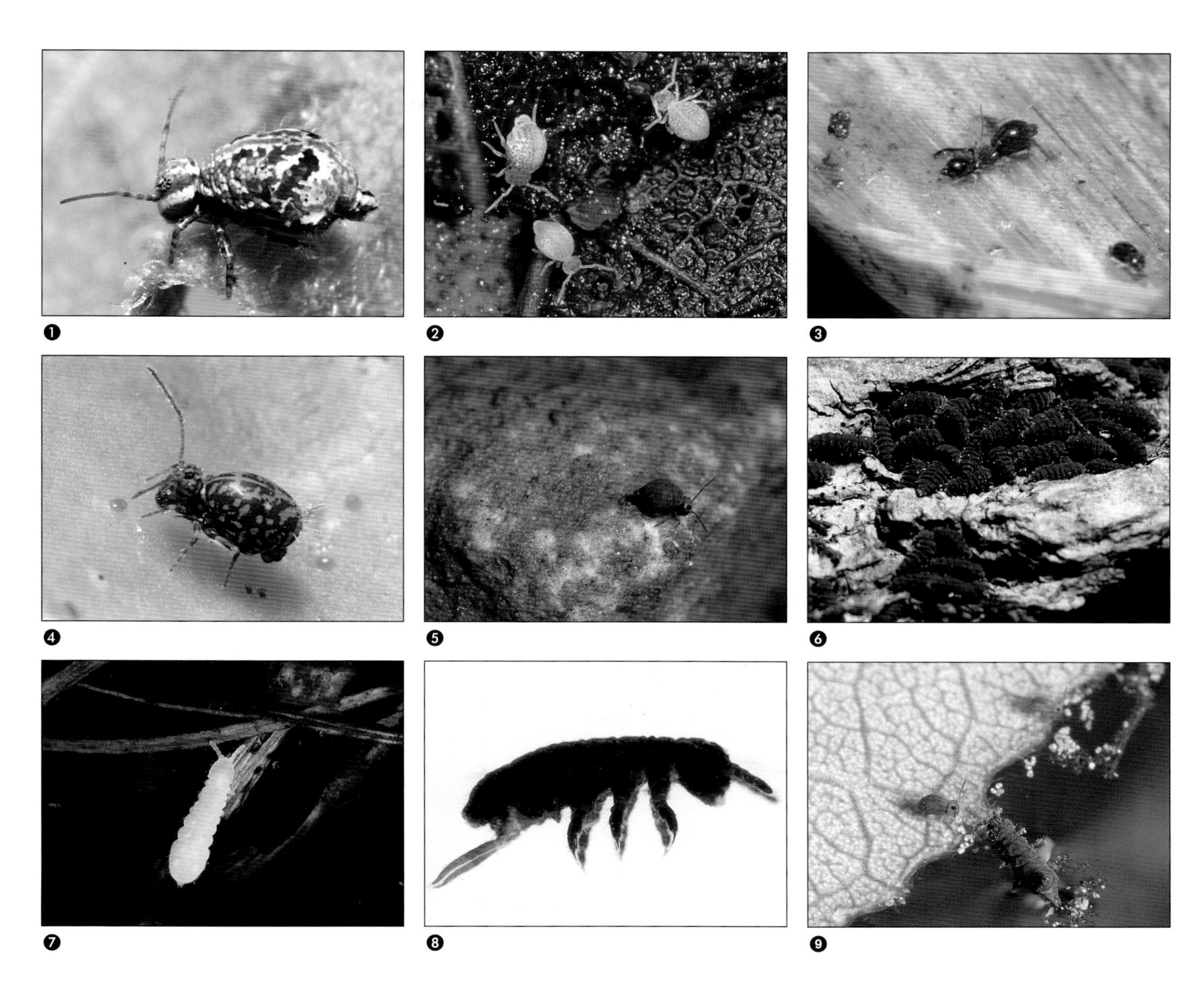

FAMILY SMINTHURIDAE ❶ Globular springtails form a distinctive suborder (Symphypleona), most species of which are usually placed in the family Sminthuridae. ❷ These colorful globular springtails (***Arrhopalites hirtus***) are grazing on the surface of a decaying leaf. ❸ Sex in the wingless insects is an indirect affair that does not involve genital contact. This minute male globular springtail (***Sminthurus*** sp.) is working to increase the chance that the larger female will make contact with his sperm droplet. ❹ This ***Ptenothrix marmorata*** (until recently known as *Dicyrtoma marmorata*), was photographed in a cup fungus. Some authors split up the Sminthuridae, putting *Ptenothrix* and its relatives in the **Dicyrtomidae**. ❺ The **Garden Springtail** (***Bourletiella hortensis***) is a widely distributed globular springtail that sometimes occurs in great numbers in gardens and tree nurseries, where they can cause significant damage to new seedlings. *Bourletiella* and relatives are now sometimes put in the family **Bourletiellidae**. **FAMILY HYPOGASTRURIDAE** ❻ Most Collembola other than the globular springtails are called "elongate-bodied springtails", and although they do not form a natural group there is no general agreement about their subordinal classification. Springtails that appear by the thousands on warm winter days, often on the surface of the snow, are called snow fleas. The **Common Snow Flea**, ***Hypogastrura nivicola***, belongs to the family Hypogastruridae, one of several families of elongate-bodied springtails. **FAMILY ONYCHIURIDAE** ❼ Members of the family Onychiuridae are usually very pale, don't have any eyes and don't have much of a tail. Although they are usually found in soil, this specimen was washed up in some riverside flood debris. **FAMILY PODURIDAE** ❽ ***Podura aquatica*** (just over 1mm), the only member of the family Poduridae, sometimes occurs in huge numbers on the surfaces of puddles and ponds. This species has a much longer tail than the similar Hypogastruridae that are also common on standing water. ❾ This *Podura aquatica* is eating pollen alongside a floating leaf with some immature globular springtails.

FAMILY TOMOCERIDAE ❶ All of our Tomoceridae are in the genus ***Tomocerus*** and were until recently included in the Entomobyridae. These elongate-bodied, hairy-tailed springtails occur in a variety of damp environments. **FAMILY ENTOMOBRYIDAE** ❷ ***Lepidocyrtus paradoxus*** is an introduced and widespread species with a remarkable mesonatal hump. ❸ ***Orchesella celsa***, seen here on a recently cut hemlock stump, is a distinctively colored and common springtail. ❹ ***Orchesella villosa*** is an introduced European species that has spread through southeastern Canada and the adjacent States. **FAMILY ISOTOMIDAE** ❺ ***Folsomia candida*** is a small, blind, pale springtail that abounds in agricultural soils throughout the world. This parthenogenetic (all female) species is easily reared, and has become a "standard" test organism for estimating the effects of pesticides and environmental pollutants on nontarget soil arthropods. ❻ The polychromatic springtails seen here amongst their recently shed skins belong to the genus ***Isotoma***. Unlike many other springtails, *Isotoma* do not eat their cast exuvia after molting (the tiny metallic springtail in this picture, taken under bark in mid winter, is an entomobryid). **FAMILY NEANURIDAE** ❼ **Seaside Springtails** (***Anurida maritima***) are common all along the Atlantic coast, where they are submerged at high tide and survive in air pockets. Although the Seaside Springtail lacks the forked tail (furcula, or furca) that gives most springtails their jumping ability, it can still pop on and off the water surface through interaction between its ventral tube (collophore) and the surface film. ❽ This small neanurid springtail is on a slime mould fruiting body. ❾ The bulky, tuberculate species of ***Morulina*** are springless springtails, without a furcula. This holarctic genus is mainly northern but this ***M. delicata*** was photographed in Virginia.

DIPLURANS
ORDER DIPLURA

❶ ❷ ❸

FAMILY CAMPODEIDAE ❶ Diplurans are rarely encountered, but ***Campodea*** species like this one are sometimes found under rocks or logs, usually in wet deciduous forests. Campodeids are only 5 or 6 mm in length. **FAMILY JAPYGIDAE** ❷ ❸ Diplurans in the families Japygidae and Parajapygidae differ from the more common Campodeidae in having pincerlike cerci. These are both in the most common eastern japygid genus, ***Metajapyx***.

JUMPING BRISTLETAILS
ORDER ARCHAEOGNATHA

❶ ❷ ❸

FAMILY MACHILIDAE ❶ Jumping bristletails are usually associated with rocky areas like alvars or escarpment cliffs where they graze on algae and lichen. This species (***Pedetontus saltator***, 12 mm) is known only from females. ❷ The shoe-sole-shaped ocelli under the massive eyes of this ***Pedetontus saltato***r distinguish it from the small triangular ocelli of *Trigoniophthalmus alternatus,* the other machilid occurring in inland northeastern North America. A third species of machilid *(Petrobius brevistylus)* occurs on rocky cliffs along the east coasts of Canada and New England. ❸ ***Trigoniophthalmus alternatus***, a species probably introduced from Europe, is sometimes common in disturbed areas. This was one of many exposed, along with several isopods, by peeling bark from a dead tree in Hamilton, Ontario. **FAMILY MEINERTELLIDAE** ❹ Species in the single North American genus of meinertellid, ***Machiloides,*** differ from the family *Machilidae* in lacking scales on the basal antennal segments. Machiloides is known only as far north as New Jersey.

❹

❶

❷

FAMILY LEPISMATIDAE ❶ **Silverfish** (***Lepisma saccharina***) are common in damp places where they scavenge on a variety of foods, including kitchen scraps and paper. This is the most common of the three species of silverfish often found in North American homes. ❷ **Firebrats** (***Thermobia domestica***) are especially common in warm places like furnace rooms. Silverfish and Firebrats are about 12 mm in length.

CHAPTER 2

Mayflies, Dragonflies and Damselflies: The "Old-Winged" Insects

Most insects – and thus most animals – belong to a tremendously successful group characterized by the possession of wings. Analogous structures appear in two or three relatively small groups of flying vertebrates (birds, bats, pterodactyls), but insect wings developed as novel structures a few hundred million years before the first birds or bats used mere modified limbs to take flight. Insect wings seem to have evolved from articulated appendages on the bases of ancestral insect legs, probably starting off as something similar to the winglike abdominal gills of mayfly nymphs. Most insects have the second and third thoracic segments developed into a muscle-packed box that supports and operates two pairs of wings that arise just above the mid and hind legs.

Although most living winged insects have a unique type of wing-folding mechanism that allows them to fold their wings neatly over the body (much the way you might fold up an umbrella), when the first insect took to the air it did so with permanently outstretched wings. The useful – and now widespread – ability to fold the wings out of the way did not arise until much later, with the origin of a complex mechanism used to fold and tuck the wings over the body. Since this folding ability arose sometime after the origin of the winged insects, insects that can fold their wings are called the Neoptera ("new-winged" insects), and the winged insect lineages that evolved before the origin of wing folding are called the palaeopterous insects ("old-winged" insects). All winged insects, including both the Neoptera and palaeopterous insects belong to a single lineage called the Pterygota.

Several groups of insects became established between the time that wings originated and the time that wing-folding appeared; however, all but two of those early lineages are now extinct. The two surviving old-winged (palaeopterous) lineages, Ephemeroptera (mayflies) and Odonata (dragonflies and damselflies), are ancient groups that have been "on the wing" for about 400 million years – four times as long as we mammals have been around.

Compared to major insect groups such as flies or beetles, the palaeopterous orders are small groups, with only about 1,000 species in North America. For observant naturalists, however, the fascination engendered by swarms of delicate mayflies and conspicuously common, brightly colored dragonflies is out of proportion to the relatively low diversity of these groups.

Ancient but ephemeral – The mayflies (Order Ephemeroptera)

An old joke has St. Peter welcoming a man to the gates of heaven saying, "Congratulations, Mr. Smith, you have been reincarnated – as a mayfly. Have a nice day," – reflecting the common knowledge that mayfly adults are indeed very short-lived. This generalization is more formally reflected in the scientific name for the order, Ephemeroptera, a name derived from the Greek words *ephemeros* ("lasting only a day") and *ptera* ("wings"). As the Greek name suggests, it is only winged mayflies that are ephemeral. A mayfly spends a long period as a wingless, aquatic, immature nymph before shedding its nymphal skin to emerge as an aerial, milky-winged preadult, called a subimago, which usually sheds its skin again to transform into the clear-winged adult stage. This in itself makes the mayflies a remarkable group, since no other living insect sheds its skin (molts) once it has reached the fully winged stage.

Fossil evidence shows that many extinct groups of winged insects developed gradually, and their life cycles included stages with intermediate-sized wings. Today's winged insects skip these intermediate stages and instead undergo a metamorphosis directly from a wingless or effectively wingless form to a fully winged form.

Dragonflies and mayflies, like cockroaches, bugs and grasshoppers, have wings that appear externally as inconspicuous wing buds in the immature stages, remaining undeveloped until they are ready to make the spectacular transformation from tiny wing buds directly to full-sized wings. This kind of metamorphosis, known as hemimetaboly, or incomplete metamorphosis, seems to have originated independently in most lineages of winged insects, so there must have been a good reason for winged insects to develop a special metamorphic transformation that allows them to jump from a wingless stage to a fully winged stage.

It is likely that stages with intermediate-sized wings – big enough to get in the way and prevent rapid escape from predators, but not big enough to fly with – have been rigorously selected against, and molts from winged stage to winged stage have therefore disappeared from all winged insects except the mayflies. The extra molt still found in mayflies is a sort of evolutionary appendix, serving to remind us of the way winged insects developed before metamorphosis became the rule. That is not to say that the subimago is a useless developmental vestige, since the relatively hairy, water-repellent subimago is well-equipped to escape the nymphal habitat before molting to the adult stage.

Adult mayflies vary widely in size and color, but are uniform in general appearance. Their most distinctive

Ephemera guttulata. This strikingly colored burrowing mayfly is sometimes referred to as a "coffinfly."

features are the large, almost triangular forewings held together above the body (fly-fishing enthusiasts call them "upwings"), the two or three long, thin tails and the big-eyed head with tiny antennae. The hind wings are small or even absent. As is the case among many insects that form male swarms, male mayflies often have strikingly large eyes. Males in the common family **Baetidae** (page 34) have especially huge, turbanlike upper eyes that seem to have been piled on top of the normal eyes as an afterthought. Many male mayflies have extraordinarily long front legs used to reach underneath the female and grip her wing bases during mating. Mating and laying eggs fully occupy adult mayflies during their short lives, which usually last two or three days, although some species fly for only a couple of hours. Adult mouthparts are nonfunctional and the stomach is little more than an air sac that aids in the rhythmic upward flying and downward drifting of swarming mayflies. Such swarms consist mostly of males waiting to mate with any female flying into the swarm, but some mayflies form mixed-sex swarms.

After leaving the swarm, some female mayflies crawl right into the water to lay their eggs, while some of our smallest mayflies (**Caenidae**, page 35) fly onto the water and literally explode on contact, releasing their eggs as the abdomen ruptures. Most mayflies fly over the water depositing their eggs through the water surface, either one by one or in clumps that spread out when they hit the water.

Spent females often end up dead or dying on the water surface, sometimes subsequently drifting along with their wings outstretched in a familiar pattern imitated by a trout-fishing fly called a "spent wing." Specific trout flies often imitate single species of mayfly, and are fished at the right time and place to capitalize on their similarity to that species. For example, the "Brown Hen Spinner" is similar to a particular species in the genus *Ephemerella*. The term "spinner" refers to a fishing fly that imitates an adult mayfly, whereas flies called "duns" imitate subimagos. Fly-fishers are well aware that the subimago stage is popular with trout because of the mayfly's vulnerability as it leaves the nymphal skin, and for the first few minutes after it emerges, before its wings have hardened enough to carry it away from the water for the final molt from the subimago to imago (adult) stage.

Many families of mayflies have swimming or crawling nymphs, but the conspicuously flattened mayflies found pressed against solid objects in running water are all in one family – **Heptageniidae** (page 35), or flat-headed mayflies. Pick a rock out of any reasonably clean and swift-moving stream and you are very likely to see strongly flattened mayfly nymphs with conspicuously long tails (usually three, but sometimes two). The tails act like vanes to keep the streamlined nymph oriented into the current. Also look at the leaflike gills that run down the sides of the abdomen, sometimes together forming a sort of suction-cup structure that helps the flat mayflies stick to rock surfaces in swift-moving water. These rows of abdominal gills, covered by flaps or a carapace on some kinds of mayfly nymphs, are found on all mayfly nymphs. Superficially similar stonefly nymphs never have rows of abdominal gills, although one kind of stonefly has tufts of gills only on the first two or three abdominal segments. If in doubt, look at the claws on the end of the hind leg. Mayflies have only a single claw where stoneflies have a pair.

Nymphs of some mayfly groups are strictly burrowers, living in burrows in the sand, gravel or mud at the bottoms of rivers, ponds or lakes. Most of our common burrowing mayflies are in the family **Ephemeridae** (page 36). Some of these form U-shaped burrows in the soft bottoms of lakes and streams, where they feed on particles filtered from water drawn through the burrows with the aid of characteristically large, feathery gills undulated along the top of the abdomen.

Enrichment of water through the addition of sewage and farm runoff increases the food available to these mayflies and allows them to build up incredible populations. Impressive piles of adult mayfly bodies often build up under streetlights near major lakes in late June or early July, and I recently saw a commercial dumpster at a highway gas station full to the brim with hundreds of thousands of burrowing mayfly bodies swept up from around the pump lights.

During the 1940s and 1950s, *Hexagenia limbata* (common burrowing mayflies sometimes known locally as "fishflies" or "shadflies") used to emerge from Lake Erie in such numbers that streetlights were blocked, snowplows had to be used to remove the mayflies from highway bridges and sidewalks were often clogged with rotting piles of bug bodies. People complained about the nuisance and the smell of rotting mayflies (like rotting fish), for the most part unaware of the wonderful job these mayflies were doing in removing tons of phosphorus and nitrogen from the "enriched" waters of Lake Erie.

Unfortunately, even for *Hexagenia* there is a fine line between enrichment and pollution, and the numbers of mayflies in parts of Lake Erie declined from several hundred per square meter around 1950 to almost none in the early 1960s. *Hexagenia* populations have rebounded since then, and warm July nights along Lake Erie are once again entomologically enlivened by snowstorm-like hordes of adult burrowing mayflies.

Although much less frequently encountered than our two genera of Ephemeridae (*Hexagenia* and *Ephemera*), burrowing mayflies in the family **Polymitarcyidae** (pale burrowers, page 37) occur in rivers throughout eastern North America. Adults of this family are exceptionally short-lived, and the females do not even bother to molt from the subimago to adult stage. Males do molt, but they do so quickly and in flight because their midlegs and hind legs are undeveloped (females have no functional legs at all). *Ephoron leukon*, the only common northeastern member of this group, is a striking, snow-white insect that you are unlikely to see on the wing because they emerge, mate, lay eggs and die all between dusk and dawn of a single night.

Mayfly nymphs occur in weed-choked ponds, raging torrents and every kind of aquatic environment between those extremes, so it is not surprising to find that they exhibit a correspondingly wide range of body forms, ranging from streamlined swimmers to stout crawlers. The largest group of streamlined swimming mayfly nymphs is the family Baetidae (small minnow mayflies), and the small nymphs of this family abound in a variety of aquatic habitats. Adult small minnow mayflies are distinctive for their tiny or completely lost hind wings, and for the huge, turbanlike eyes of the males. Most of the larger, streamlined mayfly nymphs belong to the small families **Siphlonuridae** (minnow mayflies, page 37) and **Metretopodidae** (cleftfooted minnow mayflies, page 37), some of which are significant predators on midge larvae and other small aquatic insects.

Most mayfly nymphs feed on dead plant material, but many are at least facultative predators and a few are filter feeders. Nymphs of the common brush-legged mayfly *Isonychia bicolor* have distinctively fringed front legs, used like a net to filter food out of flowing water. *Isonychia*, although superficially similar to the minnow mayflies in the family Siphlonuridae, is now usually put in its own family, **Isonychidae** (page 37). In contrast to the minnow mayflies and brush-legged mayflies, most other common mayfly nymphs are awkward swimmers at best. For example, **Leptophlebiidae** (page 37), or pronggill nymphs, swim with an apparently inefficient undulating motion and normally cling to debris in still or slow-moving waters. The name "pronggill" refers to the adorned abdominal gills that usually end in narrow, distinctively shaped processes.

Nymphs of several mayfly families are crawlers that rarely swim, such as the extremely common spiny crawlers (**Ephemerellidae**, page 38). Spiny crawlers are ubiquitous denizens of trout streams and are imitated by dozens of trout fly patterns, with such names as "Beaverkill," "Hendrickson," "Brown Hen Spinner," "Pale Sulfur Dun" and "Small Cream Variant."

The remaining families of mayfly are less often encountered, but can be locally abundant. Small squaregills (**Caenidae**) and little stout crawlers (**Leptohyphidae**, page 38), for example, are only a few millimeters long but often emerge in impressively large numbers. Armored mayflies (**Baetiscidae**, page 39) are relatively uncommon insects, the nymphs of which have an enormous thoracic shield that gives them an armored, tanklike appearance. Like many mayflies, armored mayflies seem to be associated with special – and often sensitive – habitats.

Emeralds, Rubyspots and other flying jewels – The dragonflies and damselflies (Order Odonata)

Brightly colored beetles, butterflies and other insects are often enthusiastically compared to beautiful jewels, but the finest jewels of the insect world could well be the dragonflies and damselflies. Living dragonflies exhibit a wealth of patterns and shimmering iridescence on their distinctively outstretched wings, brilliant colors on their huge, opalescent eyes and a rainbow of pigments on their familiar elongate abdomens and robust thoraces. Dragonflies and damselflies, some species of which are sensitive and easily observed indicators of water quality, are also jewel-like symbols of pristine natural habitat.

The term "dragonfly" normally refers only to robust, fast-flying Odonata with four broad, outstretched wings, as opposed to the more delicate, narrow-winged damselflies. Dragonflies and damselflies are each suborders of the order Odonata, a name derived from the Greek words for "toothed" and "jaws." A carelessly handled dragonfly can give you a nip with its big jaws (mandibles), but dragonflies are harmless insects despite this, and in spite of inappropriate common names like "horse stingers" or "devil's darning needles."

If you are fast enough to catch one, first observe the remarkable head, which looks like a giant pair of eyes mounted loosely on the boxlike thorax. Dragonflies and damselflies are somewhat unusual among insects in having only tiny antennae, leaving them almost entirely reliant on superb vision, complemented by an ability to rotate the head almost 360 degrees. Furthermore, the dragonfly thorax is rotated so that the modified legs are directed forward to form a spiny basket or net conveniently located under the head. Think of these voracious predators as highly maneuverable, well-aimed flying baskets continually filling up with mosquitoes and black flies! Dragonflies normally scoop up prey on the wing, either by hunting in flight, as the large darners do, or by darting out from perches, like many skimmers. Damselflies also usually hunt from perches, but snatch prey off surfaces like plant stems or leaves rather than scoop up flying victims.

Dragonflies and damselflies are not only "primitive winged" in that they can't fold their wings out of the way; they also differ from most insects in how they fly. The wings work independently of one another, and muscles inserted directly on the wing base power the downstroke. This is in contrast to moths, wasps and other more "advanced" flying insects that have the front and hind wings latched to one another, and power both the upstroke and downstroke with muscles that distort the thorax, indirectly causing wing movement. Wing speed is also slow in dragonflies, at about 30 flaps per second as opposed to about 1,000 flaps per second in a typical fly. If you have watched dragonflies darting after prey, or perhaps had one fly alongside your car on a rural road, you will see this as a good example of how "primitive" attributes are not necessarily inferior or inefficient. Dragonflies can take off vertically, hover, fly backwards and hit speeds of as much as 45 miles per hour (70 km/h). Some regularly migrate thousands of miles, and any insect collector can testify to their ability to outmaneuver the most deftly swung insect net.

Band-winged Meadowhawk (*Sympetrum semicinctum*), page 42, caption 2.

Pairs of dragonflies and damselflies are often seen flying around while joined together into loops, the tail of the male apparently connected behind the head of his mate, and the tail of the female looped forward to complete the ring. This behavior not only looks peculiar, it is peculiar from start to finish. It all starts when the male develops his sperm in the normal genital area at the tip of his abdomen, then arches his abdomen forward to deposit his sperm on the underside of the front part of his abdomen, just behind his hind legs. The abdominal sternites where he deposits his sperm are highly modified, with all sorts of scoops and hooks, as well as a penislike structure.

When it is time to mate, he seeks out a female and grabs her just behind the head using a pair of claspers at the tip of his abdomen. She then arches her long abdomen around to make contact with the specialized structures on the male's abdominal sternites, thus completing that loop. This is a pretty weird kind of sex, not like the wingless insects that deposit sperm droplets away from the female, or like other winged insects, in which the males insert their genital organs directly into the female. It seems even stranger when examined in detail.

Apparently some males are accomplished cuckolds, and use some of their specialized appendages to make sure that when they mate with a female, *their* sperm will be more successful than that of previous suitors. They even have tools to scoop out the sperm placed in the female by rival males! Of course, the same thing could happen to their sperm after they leave the female, so males of several species of dragonflies and damselflies hang onto the female and guard her until the eggs are deposited. One commonly sees damselflies in pairs, the male's abdomen gripping the female's "neck" as she lowers herself down the stem of an aquatic plant, inserting eggs in the submerged part of the stem. Other species guard their parental investment in different ways, in some cases releasing the female but remaining nearby just in case.

One thing any dragonfly watcher is sure to notice is the very specific territorial behavior exhibited by some species. Males often patrol their territories, mating with females that stray onto their turf, and behave aggressively toward invading males. Aerial battles are not infrequent but, as is often the case in macho encounters across the animal kingdom, there may be mechanisms whereby one of the boys can lose the fight without suffering real physical damage. In one of our commonest species of dragonflies, *Libellula lydia*, the Common Whitetail, sexually mature males are conspicuous in having a broad, somewhat flattened abdomen, the upper side of which is a brilliant bluish white. These conspicuous, banded-winged dragonflies fly slowly along the margins of ponds, often hovering or repeatedly perching on the same prominent spot. If another male enters his territory, the defender will raise his bright abdomen in a threat display and the intruder will usually depress his abdomen and flee until his own territorial boundary is approached. If a relatively dull-colored female appears on the scene the male approaches her with his abdomen down and his display colors concealed, his accessory genitalia loaded and his terminal claspers ready to grab her by the neck.

The Common Whitetail belongs to a large and common group of dragonflies called skimmers (**Libellulidae**, page 40)

Common Whitetail (*Libellula lydia*), page 40, caption 6.

because mated females characteristically skim along the surface of water, periodically dipping the tip of the abdomen into the water to deposit their eggs. The two other families in the superfamily Libelluloidea have similar habits. **Corduliidae** (page 44), called emeralds because of their jewel-like eyes, sometimes stick their eggs in mud or water using a spoutlike appendage, and sometimes extrude "baskets" of eggs at the tip of the abdomen before dipping them into water. **Macromiidae** (page 44), called cruisers because males in this small family are powerful fliers that cruise along shorelines, lay eggs while dipping or dragging the abdomen in the water.

In contrast to the egg-dropping behavior of female Libelluloidea, the large and spectacular darners (**Aeshnidae**, page 45) and spiketails (**Cordulegastridae**, page 46) have sharp ovipositors used to insert their eggs directly into plants or other substrates. Spiketails are so-called because they can be seen hovering over shallow stretches of running water, periodically dipping the abdomen straight down through the surface to stick an egg in the substrate using a spikelike abdominal tip. Darners inject their eggs into soft, submerged or partially submerged plants or decayed wood.

Darners and spiketails belong to the superfamily Aeshnoidea along with the petaltails (**Petaluridae**, one rare southern species in eastern North America) and clubtails (**Gomphidae**, page 47). Clubtails and petaltails are relatively primitive dragonflies with widely separated eyes like those of damselflies but unlike the contiguous eyes of darners and spiketails. Petaltails have a well-developed ovipositor used to lay eggs in mud or wet soil where the semiaquatic larvae live; clubtails have a reduced ovipositor and usually just drop either single eggs or egg bundles into clean streams or rivers.

Damselflies, like darners, have well-developed ovipositors with which they insert eggs, usually in submerged plant tissue, although some damselflies (**Lestidae**, page 48) lay their eggs on vegetation far above the water surface. The common name for the family Lestidae is "spreadwings" because they rest with their wings outstretched at an angle of about 45 degrees, in contrast to our other damselflies that rest with the wings clasped together above the body.

The vast majority of our damselflies belong to the narrow-winged damselfly family (**Coenagrionidae**, page 48), common little damselflies that usually hold their delicate, clear, petiolate wings clasped parallel to the abdomen. Eastern North America's only other damselflies are the few species in the broad-winged damselfly family (**Calopterygidae**, page 51). Broad-winged damselflies (called demoiselles elsewhere in the world) are distinctive for their red or black wing pigmentation as well as their relatively broad wings. Look for these conspicuous insects near running water, where clouds of glittering males can make a summer walk up a small stream a breathtaking experience. Female broad-winged damselflies usually insert their eggs into submerged plants, sometimes going entirely underwater for extended periods to do so.

Dragonfly eggs hatch into hydraulic monstrosities quite unlike the aerial adults. Try scooping some bottom debris from your local pond, then put your sample in a pan of water for observation. With luck, your sample will include some robust greenish or brownish dragonfly nymphs, torpedo-shaped if you have scooped up a darner nymph, perhaps broader and more sprawling if you have a skimmer nymph. Either way, if you disturb the nymph as it sits at the bottom of your pan of water you will see it fold its legs and shoot across the container as though by jet propulsion, blasting a clear path of debris with its jet exhaust. If you dropped a bit of ink behind the nymph first, you would see that it is propelling itself by taking water in through its anus, then using its muscular rectum to blast that water out its anus along with a conspicuous plume of ink. The same system is used, albeit at a slower pace, to ensure a continuous supply of fresh water to the gills, which are up the rectum. Damselflies don't follow this peculiar plan, and instead have broad, gill-like lamellae conspicuously displayed at the tip of the abdomen. Damselfly lamellae are used like paddles, and are probably as important for locomotion as for respiration.

Both dragonflies and damselflies have a hydraulic-powered lower lip (labium) reminiscent of, but much more interesting than, the tongue of a frog or chameleon, which can be shot forward to nab an item of prey. Of course it is best to have a good look at a dragonfly nymph in action, but you can get a good idea of how a dragonfly's lower lip works by holding your elbow against your chest and the palm of your hand cupped over your mouth. Now shoot your hand out and grab something and bring it back to your mouth in about a tenth of a second. Of course you are using an armful of muscle and bone rather than a lower lip shot out by a combination of hydraulic pressure and stored elastic energy, but there is similarity between your elbow and the main joint in the dragonfly's lower lip, between your hand and the part that does cover the dragonfly's mouth, and between your fingers and the two palpi that grab the prey. Dragonfly nymphs typically sit concealed among the debris at the bottom of a pond or river, waiting for a small fish or other organism to come within striking range of that deadly lower lip. Some dragonflies and damselflies climb on aquatic vegetation from which they spot potential prey, and some even stalk their prey from a distance.

Most dragonfly and damselfly nymphs spend at least a year before transforming to the aerial adult, although a few species adapted to life in temporary ponds, such as some common little *Sympetrum* skimmers and some spreadwings (*Lestes*), hatch from the egg in early spring and develop in a matter of weeks. The Wandering Glider (*Pantala flavescens*), a migratory species that spends the winter in the south then flies north to breed, is able to develop extremely quickly in a variety of ephemeral habitats. Other species, including some of our large darners, can take years to develop. We even have one species, the Common Green Darner (*Anax junius*), which has two alternative development strategies. Some individuals spend the winter as nymphs that transform into adults the following July; others emerge later in the year (August and September) and fly south for the winter along with Monarch butterflies and a few other more familiar migrants. The migrating Green Darners never return, but their progeny come back in early spring to lay eggs in our ponds and in doing so set the stage for next year's fall-emerging, migratory generation.

When mature, dragonfly and damselfly nymphs climb out of the water and the nymphal skin splits along the back; the delicate adult emerges from the split nymphal skin and slowly expands its wings. Immediately upon emerging, the adult dragonfly is soft, cannot fly and is quite vulnerable. If you are an early riser, you may have had the pleasure of watching this transformation, perhaps on the edge of your dock or perhaps on a reed at a local pond. If you have not already noticed them, look for the cast skins of dragonfly nymphs left clinging to the place where adult emergence took place. These beautiful insect skeletons are worth a careful look, as much of the life of their former occupants is reflected in their translucent structures.

Skimming Bluets (*Enallagma geminatum*), page 49, caption 6.

FAMILY BAETIDAE (SMALL MINNOW MAYFLIES) ❶ This adult mayfly has the large turbanlike eyes characteristic of males in the family Baetidae. The abdomen is mostly translucent – little more than an air sac. ❷ The green color of this female ***Procloeon*** is unusual among mayflies. ❸ The small but robust nymphs of Baetidae give rise to the name "small minnow mayflies." The platelike structures on the abdomen are the gills. This is ***Cloeon dipterum***, the only North American *Cloeon* and one of our most common pond and lake mayflies.
❹ As implied by the species name, adults of *Cloeon dipterum* completely lack hindwings (unlike similar *Callibaetis* species).
❺ ❻ ❼ ❽ ***Callibaetis*** species can be found in a variety of habitats ranging from sewage treatment ponds through to clean lakes and ponds.
❾ A milky-winged subimago female small minnow mayfly, probably a *Pseudocloeon* species.

FAMILY BAETIDAE (continued) ❶ This female small minnow mayfly is entering the water from a rock partially submerged in the waters of a rushing river. Her eggs will be laid while she is entirely underwater. (The eggs visible along the air-water edge are black fly eggs.) **FAMILY CAENIDAE** (SMALL SQUAREGILLS) ❷ ***Caenis*** nymphs are small (often less than 5 mm), but can be very common. The gills on the second abdominal segment are enlarged into two contiguous flaps that cover the gills on the following abdominal segments. ❸ ***Caenis*** species often emerge in such enormous numbers they are quite conspicuous despite their minute size. **FAMILY HEPTAGENIIDAE** (FLATHEADED MAYFLIES) ❹ Flatheaded mayfly nymphs are often found among leaves trapped under flowing water. ❺ Flatheaded mayfly nymphs are very common, and most rocks pulled out of flowing water will have some of these heavily flattened insects scurrying crablike out of the light. This is a ***Stenonema vicarium*** nymph. ❻ ***Epeorus*** nymphs are characteristic of clean, cold, fast-flowing water. ❼ The gills of some flatheaded mayfly nymphs, like this ***Epeorus***, are enlarged and together form a friction disk or "suction cup" to hold the nymph against the substrate in torrential waters. ❽ This photograph shows a male (the one with large eyes) and female subimago of the flat mayfly ***Stenacron interpunctatum*** on a window screen. Fishers call this species the "Light Cahill." ❾ Some flatheaded mayfly adults, like this ***Stenonema***, are strikingly pale.

FAMILY HEPTAGENIIDAE (continued) ❶ This male flatheaded mayfly has relatively large eyes and, like all mayflies, has minute, almost bristlelike antennae. ❷ This female flatheaded mayfly has relatively small eyes. Like all flatheaded mayflies, she has two tails and a relatively well-developed hind wing. ❸ Mayflies, like many other aquatic insects, are often attracted to artificial lights. This flatheaded mayfly, along with thousands of tiny midges, ended up in a spider web near my porch light. **FAMILY EPHEMERIDAE** (COMMON BURROWING MAYFLIES) ❹ This is the nymph of the burrowing mayfly ***Hexagenia limbata***, an insect that can occur in very high densities in larger rivers and lakes. ❺ During a burrowing mayfly hatch, near-shore waters are often awash with the shed skin, or exuviae, of mature nymphs. Even though this is just an empty shell, you can see the feathery gills, upturned tusks and rounded "forehead" that characterize ***Hexagenia*** nymphs. ❻ Nymphal mayflies hatch to a winged subadult stage, called a subimago, which is duller in color than the adult. In the subimago stage, as with this ***Hexagenia***, there are fine fringes on the wing margins; these are not found on adults. ❼ The subimago usually flies away from the water to molt to the adult stage. This ***Hexagenia*** adult is almost entirely out of the subimago skin. ❽ This ***Hexagenia limbata*** adult is easily recognized as a male by its paired penes and paired claspers (at the tip of the abdomen, just below the long tails). Fly-fishers sometimes call this species the "Olivewinged Drake." ❾ Spent adult mayflies can carpet the surface of the water following a mass emergence.

FAMILY EPHEMERIDAE (continued) ❶ The forked process on the front of the head, combined with the up-curved tusks, marks this burrowing mayfly nymph as a member of the genus ***Ephemera***. ❷ Adult *Ephemera* differ from the similar large burrowing mayflies in the genus *Hexagenia* in having marked wings. This is ***Ephemera simulans***, a widespread species known to fly-fishers as the "Brown Drake." **FAMILY POLYMITARCYIDAE** (PALE BURROWERS) ❸ ❹ ***Ephoron leukon*** mayflies emerge, mate, lay eggs and die during a single night, so few people see them on the wing. As you can see from these males (photographed in a streamside spider web), the middle and hind legs of pale burrowers are atrophied and apparently useless. **FAMILY METRETOPODIDAE** (CLEFTFOOTED MINNOW MAYFLIES) ❺ Members of the small family Metretopodidae, like this ***Siphloplecton*** nymph, usually occur in slow flowing waters. **FAMILY ISONYCHIIDAE** (BRUSH-LEGGED MAYFLIES) ❻ ***Isonychia bicolor*** nymphs are aptly called brush-legged mayflies because of the way they use the hairs on the inside of their front legs to make a net to filter food out of the current. ❼ Adults of ***Isonychia bicolor*** have distinctively contrasting black forelegs and white midlegs and hind legs. Fly-fishers know these common mayflies by various names, including "Whitegloved Howdy" for the adult and "Leadwing Coachman" for the subimago. **FAMILY SIPHLONURIDAE** (MINNOW MAYFLIES) ❽ ***Siphlonurus*** nymphs, like this one photographed in a spring pool in Ontario's Algonquin Park, often feed on midge (Chironomidae) larvae. **FAMILY LEPTOPHLEBIIDAE** (PRONGGILLS) ❾ Pronggills are so-called because the nymphs have distinctively forked gills with long fingerlike projections. In this genus (***Leptophlebia***) there are single projections, but in other genera the projections are often more elaborate.

FAMILY LEPTOPHLEBIIDAE (continued) ❶ This pronggill adult (***Leptophlebia intermedia***), emerging from its subimago skin, had flown some distance from a lake into the forest before molting. ❷ This striking, black male pronggill was picked out of an enormous midafternoon swarm over a small stream running into Lake Huron. **FAMILY EPHEMERELLIDAE** (SPINY CRAWLERS) ❸ ❹ These photographs show a spiny crawler subimago during and after its emergence at the surface of the water. The nymphal skin (exuvia) is visible in the water below the fully emerged subimago. ❺ Spiny crawler nymphs are the only mayflies with the wing pads largely fused and the abdominal gills exposed and obvious. Similarly stout mayflies in related families have most gills concealed, usually under a single pair of flaplike gills. ❻ Clear-winged adults and milky-winged subimago stages of the same species can often look quite different – these two mayflies are both ***Ephemerella subvaria***. This is the common mayfly species known to fly-fishers as the "Hendrickson" or "Bluewinged Hendrickson." ❼ ❽ These ***Ephemerella*** started mating in a large aerial swarm before dropping to the ground. The single female shown here has already started to extrude the mass of eggs that would normally be released into a river. **FAMILY LEPTOHYPHIDAE** (LITTLE STOUT CRAWLERS) ❾ Nymphs of ***Tricorythodes*** are inconspicuous because they usually bury themselves in the sediment (and they are often covered with sediment, like this one).

FAMILY LEPTOHYPHIDAE (continued) ❶ ❷ ***Tricorythodes*** subimagos and adults have a distinctive, somewhat humpbacked appearance because of the stout thorax (blackish in the adult). They also lack developed hind wings. **FAMILY BAETISCIDAE** (ARMORED MAYFLIES) ❸ Although this is a relatively uncommon group, nymphs of the aptly named armored mayflies are spectacular insects because of the way the thoracic shield covers much of the body, including the gills.

SUBORDER ANISOPTERA (DRAGONFLIES)

FAMILY LIBELLULIDAE (SKIMMERS OR CHASERS) ❶ Skimmer nymphs have a cuplike lower lip (labium) ending in large, toothed palpi that they use to grab their prey. The lower lip is hinged back under the head at rest, and envelops the lower part of the head. ❷ The cast skins of dragonfly and damselfly nymphs are common features on shoreline rocks and vegetation. In this image, the back of a skimmer nymph has split open to allow the adult dragonfly to emerge. ❸ The **Slaty Skimmer** (***Libellula incesta***) is, as the name suggests, slate-gray in color when mature. Newly emerged adults are yellowish with brown stripes. ❹ **Widow Skimmers** (***Libellula luctuosa***) are distinctive for the broad black patch at the bases of both front and hind wings. This female, like most skimmer females, will deposit eggs by flying along the water surface and repeatedly skimming the water to drop an egg. ❺ This juvenile male **Widow Skimmer** (***Libellula luctuosa***) differs from the female in having white wing patches; as it matures it will develop a powdery, bluish bloom on the abdomen. ❻ This **Common Whitetail** (***Libellula lydia***) male has a bright white abdomen used to warn other males to stay out of its territory. Females lack the white abdominal bloom, and have a different wing pattern more like that of Twelve-spotted Skimmer (*L. pulchella*) females. ❼ This male **Twelve-spotted Skimmer** (***Libellula pulchella***) is easily recognized by the 12 black wing spots interspersed with white spots. Female Twelve-spotted Skimmers lack the white spots and resemble females of the Common Whitetails except for their continuous light abdominal stripe – the abdominal stripe is broken on Common Whitetails. ❽ **Chalk-fronted Skimmers** (***Libellula julia***) are among our most common dragonflies. Both sexes have a similar chalky color on the front half, but juvenile adults are a dull red without a conspicuously chalky front. ❾ The **Four-spotted Skimmer** (***Libellula quadrimaculata***) is a common dragonfly, easily recognized by the yellow pigmentation along the leading edge of each wing.

FAMILY LIBELLULIDAE (continued) ❶ ❷ Juvenile male **Great Blue Skimmers** (***Libellula vibrans***) (1) do not have the intensely blue abdomen seen in adult males. Females start out yellow and black (2) and darken to brown with age. This is our largest skimmer, at about 6 cm in length. ❸ Male meadowhawks (*Sympetrum* spp.) are common, small and generally reddish in color. The **White-faced Meadowhawk** (***Sympetrum obtrusum***) is the most abundant meadowhawk species in the northeast. ❹ ❺ Male meadowhawks are usually red, but females are usually yellowish or brown as you can see from these **White-faced Meadowhawks** (***S. obtrusum***). ❻ These **Yellow-legged Meadowhawks** are flying in tandem, periodically dipping down so the female can lay eggs along the edge of this weed-choked pond in late fall. The eggs will hatch when the pond fills with water in spring. Yellow-legged Meadowhawks (***Sympetrum vicinum***) fly later in the year than other dragonflies, and this picture was taken in Ontario during October. ❼ The **Band-winged Meadowhawk** (***Sympetrum semicinctum***) is easily distinguished from other meadowhawks by the brown band along the wing bases. This male is perching in a typical position; in hot weather they will perch with their abdomen pointed directly toward the sun to minimize sun exposure.

FAMILY LIBELLULIDAE (continued) ❶ These copulating **Band-winged Meadowhawks** (***Sympetrum semicinctum***) show the color differences that sometimes exist between male and female meadowhawks. ❷ This male **Band-winged Meadowhawk** (***Sympetrum semicinctum***) is hanging on to his mate to protect his paternity. The couple will fly in tandem along the water's edge until the female has deposited her eggs. ❸ The **Wandering Glider** (***Pantala flavescens***) earns its common name by making long migratory flights north to breed. They lay their eggs in a wide variety of habitats, including temporary pools, where the larvae quickly develop into adults that fly south again. ❹ The **Black Saddlebags** (***Tramea lacerata***) is the most common northeastern member of the genus *Tramea*. Members of this genus are called saddlebags because of the saddlebag-like patch restricted to the hind wing. ❺ The **Carolina Saddlebags** (***Tramea carolina***) is a rare species in the northeast, although it occasionally occurs as far north as southern Ontario. ❻ Small skimmers in the genus ***Leucorrhinia*** are among our most common dragonflies, and most ponds are well populated by their broad-bodied nymphs. ❼ This juvenile female **Frosted Whiteface** (***Leucorrhinia frigida***) is eating a deer fly. Members of the genus *Leucorrhinia* are called whitefaces for obvious reasons. ❽ This male **Frosted Whiteface** (***Leucorrhinia frigida***) looks very little like the brightly marked juvenile female. Mature females are more similar to the male in color. ❾ This pair of **Red-waisted Whitefaces** (***Leucorrhinia proxima***) illustrates the sexual dimorphism common in this genus.

FAMILY LIBELLULIDAE (continued) ❶ This is a juvenile **Crimson-ringed Whiteface** (***Leucorrhinia glacialis***). As the common name suggests, adult males are differently colored and the yellow ring around the base of the abdomen turns red. ❷ The aptly named **Dot-tailed Whiteface** (***Leucorrhinia intacta***) is an easy species to recognize. ❸ The **Elfin Skimmer** (***Nannothemis bella***), our smallest dragonfly, is found in bogs and fens. ❹ In contrast to the usual pattern, female **Elfin Skimmers** (***Nannothemis bella***) are brightly colored and older males are a dull, dusty blue-black. Young males look like females. ❺ Females and juvenile males of the **Eastern Pondhawk** (***Erythemis simplicicollis***) are similarly colored; older males are pale blue. ❻ You can still see the ripples in the water where this green female **Eastern Pondhawk** (***Erythemis simplicicollis***) has been dipping her abdomen to lay eggs. The powder-blue male is guarding his mate from above. ❼ **Blue Dasher** (***Pachydiplax longipennis***) males frequently perch on twig tips and similar vantage points, threatening each other by raising their blue abdomens. Females and juveniles are relatively dull-colored. ❽ **Halloween Pennants** (***Celithemis eponina***) like to perch on top of tall plants, and will often return repeatedly to a good perch. ❾ Male **Calico Pennants** (***Celithemis elisa***) are easy to photograph, as they are reluctant to give up their perches, usually on relatively low vegetation, from which they scan for potential mates and prey.

FAMILY LIBELLULIDAE (continued) ❶ Female **Calico Pennants** (***Celithemis elisa***) look a bit like a washed out version of the male. **FAMILY CORDULIIDAE** (EMERALDS) ❷ This ***Cordulia shurtleffi***, one of our most common emeralds, is resting on an emergent stem in a small lake. ❸ Some emeralds, like the **Racket-tailed Emerald** (***Dorocordulia libera***), have broad abdominal tips like clubtails. Unlike clubtails (Gomphidae), emeralds have eyes that meet on top of the head. ❹ The spoutlike appendage below the abdominal tip of this female **Williamson's Emerald** (***Somatochlora williamsoni***) is used to lay eggs in mud. ❺ This male **Brush-tipped Emerald** (***Somatochlora walshii***) is easily recognized by the brushlike claspers at the tip of the abdomen. ❻ This is an older **Beaverpond Baskettail** (***Epitheca canis***) female with heavily browned wings; wings of this species are usually almost clear in contrast with most other members of the baskettail genus (*Epitheca*), which have dark wing patches. Baskettail females extrude a basketlike mass of eggs at the abdominal tip, then release the whole batch into the water with one dip of the abdomen. ❼ Female baskettails, like this **Spiny Baskettail** (***Epitheca spinigera***) lay eggs in gelatinous strands that spread out in the water to form strings of hundreds of eggs, sometimes added to by other females to form communal egg masses. **FAMILY MACROMIIDAE** (CRUISERS) ❽ **Stream Cruisers** (***Didymops transversa***) are usually found in shady, wooded areas. ❾ This is a typical pose for mating **Stream Cruisers** (***Didymops transversa***). After mating, the female will skim along the surface of a woodland stream, dragging the tip of her abdomen in the water to lay eggs as she flies along.

FAMILY MACROMIIDAE (continued) ❶ **Illinois River Cruisers** (***Macromia illinoiensis***) are large, powerful fliers that patrol high and far, usually following regular routes like a path or stream bank. **FAMILY AESHNIDAE** (DARNERS) ❷ The male of this pair of **Canada Darners** (***Aeshna canadensis***) has grasped the female behind the head using a pair of claspers at the tip of his abdomen; the tip of her abdomen is in touch with his secondary genitalia, under his second abdominal segment. ❸ Dragonfly heads seem to be all eyes and massive mandibles, here being used by a **Canada Darner** (***Aeshna canadensis***) to munch on a deer fly. Darners usually consume prey in flight. ❹ The large darners in the family Aeshnidae are undoubtedly the most familiar dragonflies. This **Shadow Darner** (***Aeshna umbrosa***) is obviously a harmless insect despite its formidable appearance. ❺ The **Shadow Darner** (***Aeshna umbrosa***) gets its name from its habit of flying in the shadows, often much later in the day than most dragonflies. ❻ ***Aeshna eremita***, the **Lake Darner**, is a common dragonfly in large marshes but rarely seen in other habitats. These large darners are wonderful to watch as they cruise amongst the reeds, periodically hovering motionless as they search for flying prey. ❼ *Aeshna* adults, like this female **Zigzag Darner** (***Aeshna sitchensis***) are often seen resting on branches and tree trunks near the shallow, slow or standing waters in which they develop. ❽ ❾ Darner nymphs have a relatively flat labium (lower lip), hinged under the head at rest but quickly extended when a small minnow or other potential prey passes by.

FAMILY AESHNIDAE (continued) ❶ **Common Green Darners** (***Anax junius***) are named for the uniformly green thorax. Mature males are usually more brightly colored than females, with a brilliant blue abdomen. ❷ This close-up of a **Common Green Darner** (***Anax junius***) head illustrates how small the antennae of dragonflies are in comparison with the massive eyes. ❸ A **Springtime Darner** (***Basiaeschna janata***) nymph seen from underneath, with the lower lip in resting position. ❹ This dead **Springtime Darner** (***Basiaeschna janata***) nymph has been posed with its lower lip in the extended position. ❺ This **Springtime Darner** (***Basiaeschna janata***) was swiftly patrolling up and down a small stream until briefly detained for this photograph. ❻ ❼ These photographs show the nymph and adult of the **Fawn Darner** (***Boyeria vinosa***). These pretty darners tend to patrol close to the surface of streams, usually inconspicuously flying along in the shadows of streamside vegetation. **FAMILY CORDULEGASTRIDAE** (SPIKETAILS) ❽ Cordulegastridae, like this **Arrowhead Spiketail** (***Cordulegaster obliqua***), are called spiketails because of the female's sharp abdominal tip, used to jab eggs into the sediment below shallow, flowing water. ❾ The **Twin-spotted Spiketail** (***Cordulegaster maculata***) is named for the pattern of spots on the abdomen. Spiketails look a bit like darners, but the eyes just barely touch instead of being broadly joined as in the darners.

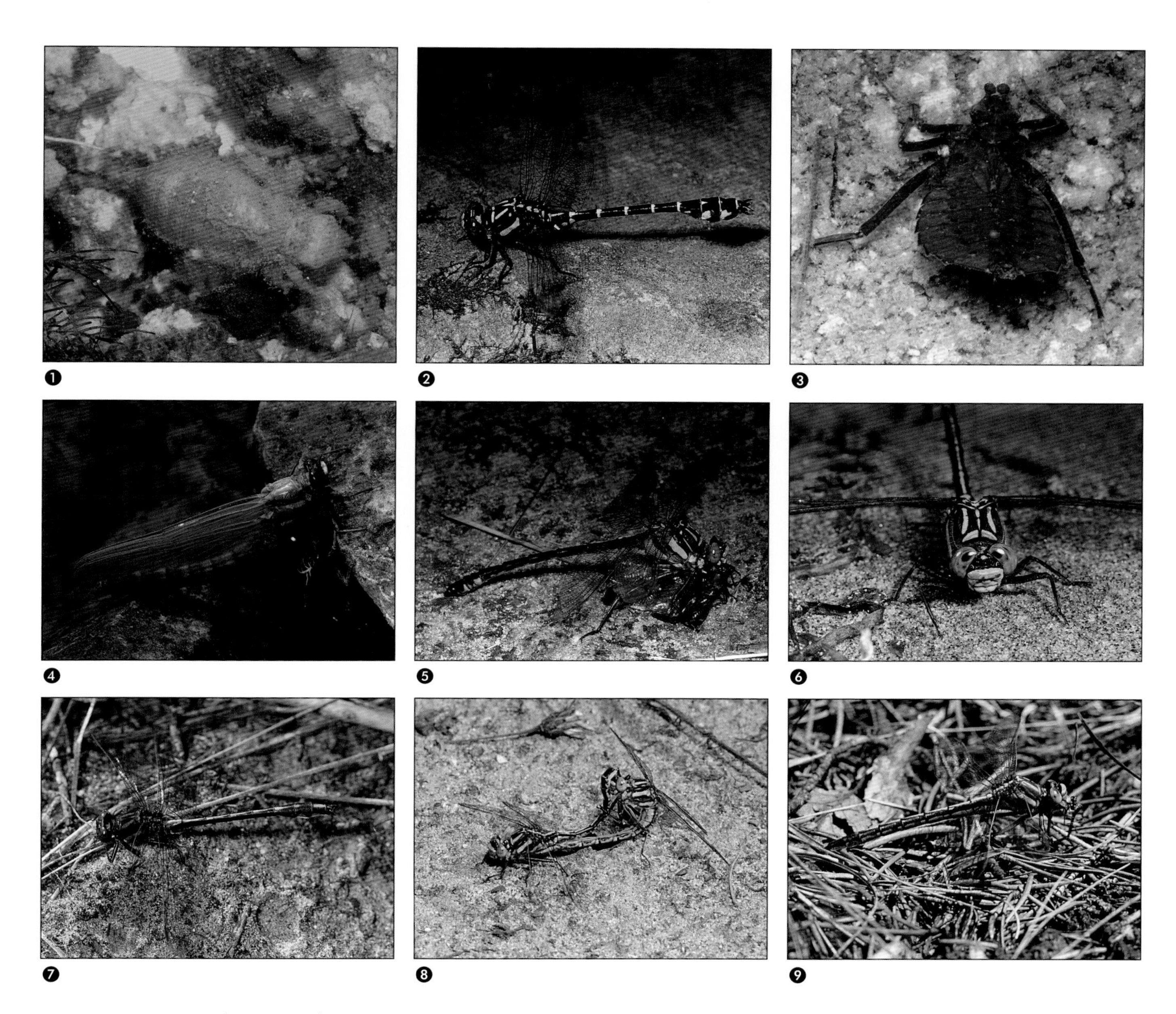

FAMILY GOMPHIDAE (CLUBTAILS) ❶ Clubtail nymphs, like this ***Ophigomphus***, often strike out at prey from concealed positions amongst silt, mud or gravel, and can be very well camouflaged. ❷ Gomphidae are called clubtails because in many species the tip of the abdomen is widened like a club. This is a **Zebra Clubtail** (***Stylurus scudderi***), an elusive species that patrols up and down trout streams in search of prey, which can include other dragonflies such as Fawn Darners (*Boyeria vinosa*). ❸ The distinctive, broadly oval nymphs of the **Dragonhunter** (***Hagenius brevistylus***) are exceptionally long-lived, leaving the water to transform into an adult after four or more years. Dragonhunter nymphs can be found along the edges of large lakes and in streams. ❹ This **Dragonhunter** (***Hagenius brevistylus***) has just emerged from the nymphal skin, and it cannot yet fly. When it has had time to pump blood into its wings and harden up it will be brilliant black and yellow. ❺ The **Dragonhunter** (***Hagenius brevistylus***) is the largest clubtail species, and one of the largest dragonflies in our skies. As the name suggests, these robust dragonflies often feed on other dragonflies. This one is eating a Calico Pennant (*Celithemis elisa*). ❻ Although large clubtails like this **Dragonhunter** (***Hagenius brevistylus***) look a bit like darners, their eyes do not meet at the top of the head as do darner eyes. ❼ The club is relatively inconspicuous in some clubtail species, like this **Dusky Clubtail** (***Gomphus spicatus***). ❽ When clubtails land, they usually do so on flat surfaces like the beach this mating pair of ***Gomphus*** has chosen. ❾ This clubtail is eating a damselfly. Our most common clubtails, including this one, belong to the genus ***Gomphus***.

SUBORDER ZYGOPTERA (DAMSELFLIES)

FAMILY LESTIDAE (SPREADWINGS) ❶ Spreadwing nymphs have a spectacularly long lower lip (labium) that can be shot out forward farther than the combined length of the nymph's head and thorax. ❷ The common name "spreadwing" refers to the way members of the family Lestidae hold their wings at rest. This is the **Slender Spreadwing** (***Lestes rectangularis***). ❸ This **Elegant Spreadwing** (***Lestes inaequalis***) is ingesting a smaller damselfly. ❹ A pair of **Emerald Spreadwings** (***Lestes dryas***) mating. **FAMILY COENAGRIONIDAE** (COMMON DAMSELFLIES) ❺ The three tapered, tail-like gills on the hind end of the slender damselfly nymph (foreground) stand in contrast to the robust dragonfly nymph (background), which has its gills tucked away up the rectum and out of sight. This damselfly nymph belongs to the genus ***Ischnura*** (forktails). ❻ This male **Eastern Forktail** (***Ischnura verticalis***), with its bright green thorax and blue abdominal tip, is one of our most common and familiar damselflies. Eastern Forktails are easily found from late spring to autumn. ❼ This bright orange young female **Eastern Forktail** (***Ischnura verticalis***) is eating a dance fly. Some females of this species are green like the males, possibly a strategy for avoiding harassment from males at high population densities. Older females of this species are usually neither green nor orange, but a dull blue-gray instead. ❽ A bright male **Eastern Forktail** (***Ischnura verticalis***) mating with a dull-colored female. The male is using appendages at the end of his abdomen to grasp the female at the front of her thorax, while she arches her abdomen forward to connect with his secondary genitalia. ❾ Damselflies, like this forktail female, usually insert their eggs in submerged or partly submerged vegetation.

FAMILY COENAGRIONIDAE (continued) ❶ The **Fragile Forktail** (***Ischnura posita***), easily recognized by the exclamation-mark-shaped stripe on the thorax, is found in shadier areas than the much more common Eastern Forktail (*I. verticalis*). ❷ This bluet (***Enallagma***) nymph has broader, less tapered gills than forktail nymphs. ❸ ***Enallagma boreale***, the **Boreal Bluet**, is one of the most common bluets in northern Ontario. ❹ Female bluets, like this one eating a small moth, are often duller in color than the males. ❺ Members of the genus *Enallagma* are called "bluets" because most but not all males are bright blue, like this common **Hagen's Bluet** (***Enallagma hageni***). *Enallagma* is a large genus and species identification can be difficult, although the external male genitalia are often distinctly different between species. ❻ These **Skimming Bluets** (***Enallagma geminatum***), given their common name for their habit of flying low over the water, are mating over a lily pad in a slow stream. ❼ The female in this mating pair of **Marsh Bluets** (***Enallagma ebrium***) is differently colored than the male. As is the case for many bluets, some females of this species are colored like the males. ❽ These **Tule Bluets** (***Enallagma carunculatum***) are mating as they hang from a vertical stem emerging from shallow water.

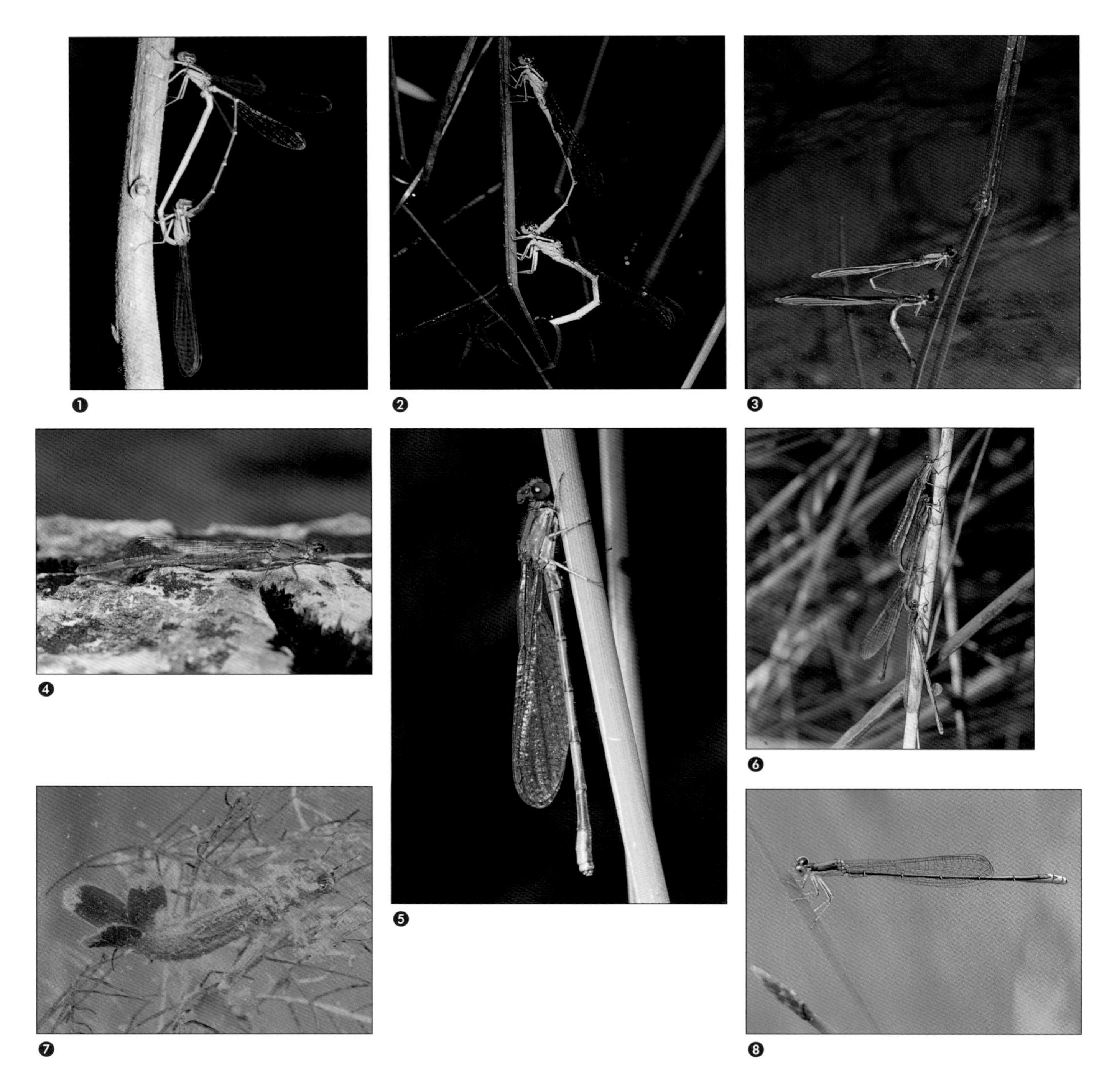

FAMILY COENAGRIONIDAE (continued) ❶ Although it sounds like an oxymoron, the common name of this species is **Orange Bluet** (***Enallagma signatum***), as it is a member of the bluet genus, but is marked with orange rather than blue. ❷ This **Taiga Bluet** (***Coenagrion resolutum***) male is guiding his mate into the water on an emergent stem, where she will insert her eggs. ❸ This pair of **Powdered Dancer** damselflies (***Argia moesta***) has backed down an emergent stem; they are several centimeters underwater, where the female will insert her eggs into the stem. ❹ A male **Powdered Dancer** (***Argia moesta***). ❺ Male **Variable Dancers** (***Argia fumipennis***) are usually bright violet or purple in color. ❻ Female **Variable Dancers** (***Argia fumipennis***) are much duller in color than their mates. Variable Dancers usually insert their eggs into wood submerged in slow-flowing water. ❼ Nymphs in the dancer genus (***Argia***) are distinctively stocky, with broadly oval gills. ❽ **Sedge Sprites** (***Nehalennia irene***) are distinctively colored, delicate damselflies that often abound in the low vegetation of bogs and fens.

FAMILY COENAGRIONIDAE (continued) ❶ The anomalous-looking yellow patch on the underside of the thorax of this damselfly identifies it as ***Chromagrion conditum***, a beautiful insect associated with small springs and spring pools. ❷ ***Amphiagrion saucium***, the only eastern species in the genus, is a distinctively colored inhabitant of boggy seeps. **FAMILY CALOPTERYGIDAE** (BROAD-WINGED DAMSELFLIES) ❸ Nymphs of broad-winged damsels, common in running water, have distinctively long antennae. ❹ Males of the **Ebony Jewelwing** (***Calopteryx maculata***) species have entirely black wings. ❺ Female **Ebony Jewelwings** (***Calopteryx maculata***) have a white spot near the end of each wing. This female is laying eggs in partially submerged vegetation. ❻ **Ebony Jewelwings** (***Calopteryx maculata***) are common around small streams with lots of vegetation. Two females, laying eggs, and one vigilant male can be seen in this photograph. ❼ The wings of **River Jewelwing** (***Calopteryx aequabile***) males are clear in the basal part, while the outer part is entirely dark. Females are less distinctly pigmented and have a white spot in the black wing apex. ❽ ❾ **American Rubyspots** (***Hetaerina americana***) tend to stay close to the streams in which they develop. Females, which are much duller in color than males, go right underwater to lay eggs, sometimes staying submerged for as long as an hour.

CHAPTER 3

Stoneflies

No insect order better epitomizes cold, clear, fresh water than Plecoptera: the stoneflies. Almost all develop as nymphs in clean, moving water, and then spend their somewhat reclusive adult lives within a stone's throw of the water's edge.

Adult stoneflies are usually inconspicuous because their wings – unlike the outstretched wings of dragonflies and damselflies – fold neatly against their backs. When they do unfurl their wings for short, fluttering flights, stoneflies expose broad hind wings that have pleated, or fanlike, trailing edges, an attribute reflected in the scientific name Plecoptera, which literally means "pleated wings." Plecoptera is a small order, with only about 650 North American species.

Stoneflies can be conveniently grouped as "summer stoneflies" or "winter stoneflies." The adult stoneflies you are most likely to notice are relatively large species that emerge during the spring and early summer, and which often have a cockroach-like, flattened appearance (although, unlike cockroaches, they have a fully exposed head). One family of stoneflies (**Peltoperlidae**, page 54) is even officially called the "roachlike stoneflies". Roachlike stonefly nymphs abound in mountain streams but the most commonly encountered stonefly nymphs throughout North America are in the family **Perlidae** (common stoneflies, page 54). These relatively large, flattened, brightly patterned, predaceous nymphs abound on sticks and stones in clean, swiftly flowing waters, usually along with superficially similar flat mayfly nymphs. Stonefly nymphs look a bit like mayflies, but never (at least in eastern North America) have rows of gills along the sides of the abdomen as mayflies do. Common stoneflies have tufts of gills on the thorax, which give them the appearance of having "hairy armpits."

Perlidae are the most frequently encountered stoneflies that emerge as adults during the warmer months, but the northeastern fauna does include a few other families of summer stoneflies. Larger streams and rivers, for example, often support huge stoneflies, often over 50 mm long, with gills on the first two abdominal segments (as well as the thorax). The gills are present even in the adult, albeit shriveled and nonfunctional. These big stoneflies belong to a family descriptively called the giant stoneflies or salmonflies (**Pteronarcyidae**, page 54). Yet another occasionally encountered family of summer stoneflies (**Chloroperlidae**, page 55) includes several small, attractive, lime-green species. Although there are also yellowish and brownish species in the family, both the common name (green stoneflies) and scientific name of the family reflect the striking appearance of common green species. There is yet another large family of summer stoneflies which includes green (as well as yellow or brown) adults, uninspiringly called perlodid stoneflies (**Perlodidae**, page 55). Perlodid nymphs usually resemble small summer stoneflies without the branched gills under the legs.

Contrasting with the predaceous nymphs and nonfeeding adults that characterize the familiar summer stoneflies, winter stoneflies generally eat algae or pollen as adults and detritus as nymphs. These are the small, often very narrow stoneflies you might find walking on snow on a late winter day or emerging from the water and molting to the adult stage in the space between the water surface and the ice. Not surprisingly, given the cold temperatures under which they emerge, many winter stoneflies don't fly, and some have very short wings or no wings at all.

The really abundant winter stoneflies, which occur by the hundreds on the snowy margins of clear streams on warmer days of February and March, are small (usually less than 10 mm), flightless stoneflies belonging to the family **Capniidae** (slender winter stoneflies, page 55). Slightly larger winter stoneflies emerging at the same time, but from streams of varying water quality, are likely to belong to another winter stonefly family, the taeniopterygid broadbacks (**Taeniopterygidae**, page 57). Two additional families of winter stoneflies, nemourid broadbacks (**Nemouridae**, page 56) and rolled-winged stoneflies (**Leuctridae**, page 57) normally occur in smaller streams. Look for them in accumulations of debris trapped against rocks or logs.

Some male stoneflies have conspicuous abdominal swellings or lobes, called hammers or vesicles, used to drum on logs or other surfaces. Each of the species that does this, and each sex, drums at a different frequency or pattern, establishing a unique communication system. Unlike the drumming that humans sometimes do to communicate with one another, stonefly drumming does not involve aerial sound waves. The vibrations are transmitted through the substratum, and send messages to the effect of, "I'm a male of species A, and I'm looking for a mate," or "I hear you, and I am an unmated female of species A." Communication systems of this nature are found throughout the insect world, but in most cases involve sound, vision or smell rather than "good vibrations."

Once the drumming results in a correct match, female stoneflies usually fly to the water surface and deposit masses of eggs that typically fall apart, leaving the individual eggs to sink to the bottom. Winter stonefly eggs adhere to the substrate with a sticky jellylike coat and those of summer stoneflies use anchor plates for the same purpose.

Once hatched from the eggs, nymphs usually complete their development within a year, but many take longer. The most common of the large summer stoneflies spend two or three years as nymphs before crawling out of the water to emerge as adults. The fact that they do usually crawl out of the water before transforming into adults, unlike mayflies, which usually emerge from the surface of the water, means that stonefly emergences don't trigger the same kind of fish-feeding frenzy that a mayfly "hatch" can induce. Nonetheless, imitation adult and nymphal stoneflies are popular in fly-fishing, and the first recorded artificial fly was based on a species of stonefly found in Britain.

Nymphal perlodid stonefly (*Clioperla clio*), page 55, caption 8.

FAMILY PELTOPERLIDAE (ROACHLIKE STONEFLIES) ❶ Stonefly nymphs, like this roachlike stonefly (Peltoperlidae, 8–15 mm) do not have rows of gills on the abdomen like superficially similar mayflies. Members of the small roachlike stonefly family (only 20 North American species) are often abundant among submerged dead leaves in small mountain streams. **FAMILY PERLIDAE** (COMMON STONEFLIES) ❷ Adult stoneflies are somewhat cockroach-shaped but, unlike cockroaches, the heads of stoneflies are not concealed by a pronotum. The characteristic fanlike hind wings (reflected in the name of the order – Plecoptera, meaning "pleated wings") are folded flat over the body and concealed by the front wings. ❸ Common stonefly nymphs, like this ***Acroneuria abnormis***, have tufts of branched gills near the base of each leg. ❹ ❺ The most frequently encountered big (usually 10–30 mm) stoneflies in larger streams and rivers belong to the common stonefly family, Perlidae. Many, like ***Agnetina capitata***, are brightly colored. ❻ ❼ Common stonefly nymphs, like this ***Paragnetina media***, are usually voracious predators of aquatic insects, such as mayflies and caddisflies. Adults are short-lived, and do not feed. **FAMILY PTERONARCYIDAE** (GIANT STONEFLIES) ❽ The largest stoneflies in the northeast are the aptly named giant stoneflies. ❾ Giant stonefly nymphs, like this ***Pteronarcys dorsata*** can exceed 50 mm in length (not counting the tails), and have gills on the first two abdominal segments as well as the thorax. Nymphal giant stoneflies eat detritus and plant material, sometimes developing for two or three years before emerging from the water as a nonfeeding adult.

FAMILY PTERONACYIDAE (continued) ❶ The ordinal name "Plecoptera" means "pleated wing," and refers to the expanded hind wing, normally concealed on living stoneflies but visible on this pinned ***Pteronarcys dorsata***. **FAMILY CHLOROPERLIDAE** (GREEN STONEFLIES) ❷ ❸ Most members of the green stonefly family are green or greenish yellow. A few are darker colored, and some members of the family Perlodidae are greenish or yellowish much like Chloroperlidae. ❹ Stoneflies usually have their wings folded flat over the body like cockroaches, but this ***Sweltsa*** has such an infestation of parasitic mites, its wings are forced into an abnormal position. ❺ This green stonefly nymph has started to molt – its back is split and its wing pads are pushed out (note: green stonefly nymphs usually have parallel wing pads unlike the otherwise similar Perlodidae). The short tails of this nymph are typical for the family. Nymphal green stoneflies are predators, and are usually only 5–12 mm in length. **FAMILY PERLODIDAE** (PERLODID STONEFLIES) ❻ ❼ Perlodid stoneflies are variable in size and color, as illustrated by this greenish ***Isoperla*** and a darker ***Cultus***. Some look like common stoneflies (Perlidae), but common stoneflies usually have remnants of their nymphal branched gills visible around the leg bases; perlodids never have branched gills. Others look very much like green stoneflies. ❽ Nymphal perlodid stoneflies, like this ***Clioperla clio***, lack branched thoracic gills unlike superficially similar common stoneflies, and have divergent wing pads unlike superficially similar green stoneflies. Nymphs of all three families are predators; they have a lower lip (labium) with a deep V-shaped cleft at the middle, giving the lip a bilobed appearance (unlike the winter stonefly families Capniidae, Nemouridae, Taeniopterygidae and Leuctridae, in which the labium has three shallower notches). Nymphal perlodids are usually between 8 mm and 16 mm in length, excluding tails. **FAMILY CAPNIIDAE** (SLENDER WINTER STONEFLIES) ❾ Although four families (Capniidae, Nemouridae, Taeniopterygidae and Leuctridae) are often called winter stoneflies, most of the stonefly adults walking on the snow surface in late winter are slender winter stoneflies in the genus ***Allocapnia***, which are usually about 5 mm in length. Adult slender winter stoneflies have short but multisegmented tails, unlike similar broadbacks and rolled-winged stoneflies, which have one-segmented tails.

FAMILY CAPNIIDAE (continued) ❶ ❷ Slender winter stoneflies often emerge well into the spring in colder waters. These nymphs, nymphal exuviae (cast-off skins) and adults of ***Allocapnia*** were photographed along Georgian Bay in late May. Nymphs feed on detritus, and adults often eat algae. ❸ ❹ Slender winter stoneflies are often short-winged, and males of ***Allocapnia vivipara*** lack wings entirely. **FAMILY NEMOURIDAE** (NEMOURID BROADBACKS) ❺ ❻ Nemourid broadbacks are small (usually 3–8 mm) stoneflies that often emerge in large numbers from clean, cold springs and small streams in early spring. They are easily recognized by the distinct "X" made by some of the wing veins near the front edge of the outer half of the front wing. ❼ Nemourid broadback nymphs have distinctively divergent wing pads and sometimes (like this ***Amphinemura***) have collarlike tufts of gills behind the head. Like most winter stoneflies, they are detritus feeders. ❽ Broadback nymphs in the genus ***Soyedina*** are common in springs and seeps.

FAMILY TAENIOPTERYGIDAE (TAENIOPTERYGID BROADBACKS) ❶ Taeniopterygid broadbacks look much like relatively large (usually 10–15 mm) nemourid broadbacks, but lack an "X" mark on the front wing. ❷ Nymphal taeniopterygid broadbacks, like this ***Strophopteryx fasciata***, have distinctively divergent wing pads and have the first and second tarsomere (tarsal segment) of each leg equal in length. ❸ ❹ ***Taeniopteryx nivalis***, unlike most stoneflies, is able to tolerate a wide range of water conditions. Adults appear in late winter or very early spring, and are often seen walking on the snow along with *Allocapnia* (Capniidae). FAMILY LEUCTRIDAE (ROLLED-WINGED STONEFLIES) ❺ ❻ Adult rolled-winged stoneflies look like little cigars because of the way the wings are rolled around the body. Nymphs of these small (usually 6–10 mm) stoneflies are very similar to those of slender winter stoneflies (Capniidae), and are rarely noticed because of their small size and their habit of staying deep in the substrate, where they feed on detritus.

CHAPTER 4

Cockroaches, Termites, Mantids and Other Orthopteroids

Thanks to the popularity of *Star Trek* and similar android-populated fiction, the suffix "-oid" should be familiar to everybody. "Android" is a convenient Greek compound word used to refer to man-like things, and similar "...oid" words are used to refer to convenient groupings of insect orders. All hemimetabolous insects (insects with incomplete metamorphosis) other than dragonflies, mayflies and stoneflies are usually placed in one of two such convenient groupings: the orthopteroids (Orthoptera-like insects), and the hemipteroids (Hemiptera-like insects). The hemipteroids form a natural group made up of true bugs and their close relatives, but orthopteroids form a heterogeneous assemblage including all or almost all hemimetabolous Neoptera (insects with incomplete metamorphosis and folding wings) that are not hemipteroids.

Up until recently, several orthopteroid lineages, including cockroaches, walkingsticks, mantids, grasshoppers and crickets, were all included in the order Orthoptera. Pooling this wide assortment of insects into one order was inappropriate, because some are more closely related to other orders than to crickets and grasshoppers, and the Orthoptera in the old sense was thus an artificial assemblage. Cockroaches, mantids and walkingsticks are now removed from the Orthoptera and each placed in separate orders, but the fact remains that they are Orthoptera-like in life cycle and in some morphological attributes such as their straight-edged, parchment-like forewings. These similarities can be reflected by informally grouping roaches, mantids, walkingsticks, grasshoppers and crickets together with similar orders (earwigs, termites) as the orthopteroid insects. Different authorities disagree on whether the orthopteroids form a natural group, and whether the small orders Zoraptera (zorapterans), Embioptera (webspinners) and perhaps even the Plecoptera (stoneflies) should be included with the orthopteroids. Some entomologists call the orthopteroid insects "Polyneoptera."

Within the orthopteroids, three orders stand out as particularly closely related. These three orders represent a well-established and well-known group you can easily remember using the familiar acronym IBM, which of course stands for Isoptera (termites), Blattodea (cockroaches), and Mantodea (mantids). In an evolutionary sense, the mantids and termites are really just specialized cockroaches, so let's look at the Blattodea first.

Cockroaches (Order Blattodea)

If you could jump in a time machine and move to a world a few hundred thousand years into the past or a few thousand years into the future, at least one of our planet's most familiar creatures would probably be there to greet you at both ends of the time scale. Fossil evidence suggests that roach-like insects scurried about in the primeval forests of the Carboniferous period (about 360 million to 290 million years ago) just as roaches scurry around apartment kitchens today, and the resilience and adaptability of modern cockroaches suggest that they will continue scurrying about long after apartments have ceased to exist. Of course, "roaches" comprise a whole order of primarily tropical insects (**Blattodea**, page 66), including several familiar species that live as scavengers in both temperate and tropical households. Cockroaches eat almost anything, although those that live in your house or apartment probably specialize on kitchen waste. One publication reported that 12 roaches can live on the glue of a postage stamp for a week!

Next time you spot a cockroach, try to suppress your reflexive "squash" response, and instead take a closer look. If it slows down enough, you will see a flattened, shiny brown insect with its hind four-fifths covered by overlapping, translucent, almost parchment-like wings. The front one-fifth is covered by an oval, hard, shining plate that is actually the top of the prothorax (the pronotum). You might see a bit of the head projecting in front of the pronotum, and you will certainly notice the long antennae borne by that usually hidden head. If you are fast enough to pick up a cockroach, take a whiff of its highly characteristic body odor, produced by special repugnatorial glands. Thanks to this odor, some pest control operators can walk through a door, sniff a couple of times and accurately guess the level of cockroach infestation.

Australian Cockroach (*Periplaneta australasiae*), page 67, caption 1.

If you can tolerate the distinctive odor of the roach long enough to continue your examination, pull the wings away from the body. You will see that the forewings form a somewhat leathery, tough shield over the broad, delicate hind wings. Roach hind wings are the flight wings but the household roaches of North America are not frequent fliers, and the most common domestic species do not fly at all.

Flightlessness is an especially obvious attribute in one of our four or five common household roaches, the Oriental Cockroach (*Blatta orientalis*). The short wings of the male, and the very short wings of the female, give this cockroach a very shiny appearance. This appearance may explain why Oriental cockroaches are often called "waterbugs"; however, "waterbug" is more likely to be a clever euphemism dreamed up by some pest control operator. After all, wouldn't you rather have a "waterbug problem" than a "cockroach infestation"?

My student days in southern Ontario exposed me to a variety of domestic cockroaches, especially that urban scourge we call the German Cockroach (*Blatella germanica*). These fully winged, yet flightless little roaches are easily identified by the two longitudinal stripes on the pronotum. Similar to other roaches, the female German Cockroach produces eggs in a shiny brown, purselike case, or ootheca, almost as big as her abdomen. It is easy to make this size comparison because, unlike some other common household roaches, the expectant *frau* German roach runs around with her egg case protruding almost entirely out the tip of her abdomen. She hangs on to this relatively huge case until hatching time, in contrast to the Oriental Cockroach or the American Cockroach (*Periplaneta americana*), which deposit their egg cases on the ground or elsewhere after carrying them for 24 hours or so. When the eggs hatch, around 20 to 30 soft, tiny, white cockroaches emerge. Their pale exoskeletons soon harden and turn brown through exposure to the air.

Roaches really get around. They are common on boats and aircraft from other countries and continents, and some of these new "immigrants" can be big news. Only a few years ago, for example, a species similar to the German roach arrived in central Florida. Originally from Japan and Southeast Asia, this little insect differed from the familiar German Cockroach in its ability to fly. Now familiar to Floridians as the all-too-abundant Asian Cockroach (*Blatella asahinai*), this prolific orthopteroid may yet become a boon to pest control operators over a wider area. North Americans already spend more than a billion dollars a year on insecticides for household cockroach control, a figure sure to increase as the Asian roach and other newly established pests spread.

The Asian roach probably did originate in Asia, but the American Cockroach, our big (about 30 mm) winged species, is not American, nor is the German Cockroach German. These great travelers seem to have been given their common names by people who wanted to blame their neighbors for their problems. In fact, most domestic cockroach species seem to share the human cradle of life to the point of having an African origin, and cockroaches as a whole are essentially equatorial insects. There are thousands of tropical species ranging in size from a few millimeters to over 10 centimeters, and ranging in color from the familiar greasy brown of our pest species to the white, green or even bright blue of some tropical species.

One well-known tropical roach is the 70–80 millimeter long Madagascar Hissing Cockroach (*Gromphadorhina portentosa*), a species that thrives in captivity and is often reared in university laboratories. These stout, wingless roaches blast air out their spiracles when disturbed, making quite a startling hissing sound.

Given that cockroaches originated in the steaming tropics, it is not surprising to find that here in the north they are primarily creatures of warm, humid microenvironments, such as that nice little space behind your kitchen cupboards. These greasy orthopteroids are so sensitive to desiccation that proper air circulation in a house will go a long way toward their control, and it has even been suggested that proper house design with air circulation everywhere, even behind cupboards, will eliminate the cockroach problem.

Not all North American cockroaches are denizens of our dens, kitchens and bathrooms, nor are they all pests. North America can claim some native cockroaches, which should not be tarred with the same brush as the introduced pest species. These interesting natives are to domestic cockroaches as the whooping crane is to the pigeon – species to be eagerly sought and observed by the same naturalist who would dust her baseboards with borax to annihilate lurking household roaches. Such a fascinating native cockroach is the Brown-hooded Cockroach (*Cryptocercus punctulatus*), a wingless cockroach that forms small colonies in decaying logs. The hooded roach family (Cryptocercidae) includes only a few species, found in widely separated areas of Asia, western U.S. and eastern U.S. Called hooded cockroaches because of the hoodlike pronotum, these shiny, dark insects can feed on wood because, like termites, they harbor symbiotic microorganisms that assist with digestion of cellulose. Microorganisms are passed from individual to individual by anal trophallaxis much like that found in closely related termites, which means that nymphs acquire the needed gut fauna by feeding on protist-rich fluids egested by their parents. Colonies usually include only an adult male, an adult female, and a couple of dozen slowly developing nymphs that feed on protist-rich maternal feces over a period of year. The native wood roaches of southeastern Canada and the northeastern States belong to another genus, *Parcoblatta*, and another family, Blatellidae. Wood roaches usually occur in rotting wood or under bark, although they are also sometimes very abundant in rocky, open areas such as alvars and they are occasionally minor nuisances in cottages and other rural dwellings.

Termites (Order Isoptera)

It shouldn't come as too much of a surprise to find that cockroaches have a lot in common with termites because termites (**Isoptera**, page 67), praying mantids and cockroaches are close relatives, descended from something similar to today's cockroaches that scuttled about on the earth 300 million years ago. Let's call that ancestor the "primal termite" (or primal cockroach or primal mantis) to take advantage of an Ogden Nash poem: "Some primal termite knocked on wood/ And tasted it, and found it good!/ And that is why your cousin May/ Fell through the parlor floor today."

That cockroach-like primal termite developed some interesting attributes in order to become today's well-known wood eaters. First, as is the case with the cockroach *Cryptocercus*, today's termites digest wood and other high-cellulose material only with the aid of symbiotic micro-organisms that break down the cellulose for them. Furthermore, the termites chomping on your parlor floor are social insects. They are not just eating your floor as individuals; they are doing it as a diabolically well-organized superorganism, the development of which probably had something to do with termite dependence on gut micro-organisms. Wood-feeding hooded cockroaches, like termites, are born without symbiotic microorganisms, and furthermore they lose these essential partners when they molt. Both termites and wood-feeding cockroaches get around this problem by an activity called "proctodeal trophallaxis," which is a nice, scientific way of saying they eat each other's fresh feces. The exchange of partially digested food between individuals, both orally and anally, distributes the needed symbiotic microorganisms among the social group, or colony.

The essential difference between termites and their cockroach relatives is the way in which feces-eating termites obtain more than just cellulose-crunching microorganisms. Some termites produce and distribute compounds that regulate the development of the recipients, and ultimately regulate the development of the social structure for which termites are justly famed. The founding female of a colony, the queen, produces pheromones that are spread through the colony by proctodeal trophallaxis, suppressing the development of other reproductives and relegating her sons and daughters to roles as workers or specialized soldiers. Some males, of course, are allowed to become sexually mature because a queen needs a king, and some individuals are allowed to develop into secondary or "spare" reproductives that can take over if the queen dies.

New termite colonies are usually founded by massive reproductive swarms of winged males and winged females that periodically emerge from well-established colonies. The males and females get together during these flights, settle down together, break their wings off and work together to start a new nest as queen and king. The king fertilizes his queen in the nest, then stays and repeatedly fertilizes her throughout his life. This is in marked contrast to the social bees, wasps and ants in which the male fertilizes the female once and then dies. The resulting termite colonies are highly organized, but quite unlike colonies formed by the familiar ants, wasps and bees whose workers and soldiers are all female, with the males serving only as one-time sperm donors. Furthermore, wasps belong to a group with complete metamorphosis, so their young are immobile, usually helpless, larvae. Termites have gradual metamorphosis, so the young nymphs are very similar to the adults and play an active role as colony workers. Workers, including males and females (unlike the all-female worker caste in wasps, ants and bees) tend the eggs, forage for food, feed and clean the reproductives, and work on nest repair and construction. Soldiers, which can also be male or female, are specialized for colony defense and usually have some sort of conspicuous soldierly attribute such as armored heads with huge mandibles, or long gunlike snouts that shoot sticky streams of gluey material at their adversaries. All termites are social, whereas sociality developed independently in several groups of Hymenoptera.

North America has only a few of the world's 2,000 termite species. The common termite species of northeastern North America, the Eastern Subterranean Termite (*Reticulitermes flavipes*), is normally seen as wingless, white, antlike individuals that spell bad news when they issue forth from some rotting corner of your porch, stairs or foundation. I recently saw evidence of a thriving colony in a child's sandbox, the wooden walls of which were sunk into sandy soil as a perfect invitation to termite activity. Stumps and buried wood scraps left around construction sites also make great snacks for subterranean termites.

Termite sightings are not restricted to those of us curious enough to poke around underground. Subterranean termites

Eastern Subterranean Termites (*Reticulitermes flavipes*), page 67, caption 1.

sometimes make conspicuous covered runways, like brown tunnels, running from the soil to a good piece of wood. Similar to other termites, they also form large and conspicuous swarms of long-winged male and female reproductives. The wings are long, clear and equally sized, as the scientific name for termites (*iso* = equal, *ptera* = wings) suggests. Winged termites are rare where I live (Ontario), as colonies along the northern fringe of *Reticulitermes'* range slowly spread underground, budding off portions of the population to found new colonies instead of swarming.

If you do see an Eastern Subterranean Termite swarm it is likely to be in very early spring, and you had better hope it is not coming from your house foundation. If it is, you will be glad to know that the search is on for alternative ways to control termites now that some formerly popular termite poisons are off the market. According to the U.S. Environmental Protection Agency, before chlordane and related chemicals were banned in 1987 they were used to treat approximately 30 million structures for termite control. One of the more interesting alternatives is a bait block that kills the termite's intestinal symbiotic microorganisms, preventing the digestion of cellulose. Bait blocks are also used to deliver slow-acting poisons that are distributed around the colony by grooming and trophallaxis.

Under natural conditions, termites are beneficial insects, accelerating the breakdown of cellulose, and aerating and mixing soil. Termites that abound in tropical areas serve much the same role as earthworms in temperate soils, but they do it on an impressive scale, with a total mass about three times the mass of all human beings on the planet. In some areas around one-third of the soil surface is tied up in termite colonies, and in other areas about one-third of all the wood, grass and leaves produced annually gets cycled through termite guts. Their global role is enormous, but easily forgotten until part of your house gets tied up in one of those colonies.

Mantids (Order Mantodea)

Everyone with enough interest in insects to be reading this is familiar with praying mantids (**Mantodea**, page 68). These large (50–80 mm), conspicuous orthopteroids, with their long pronotum and characteristic grasping forelegs, seem as much a part of the northeastern autumn as goldenrod and Monarch butterflies. Much as termites are, in an evolutionary sense, social cockroaches, mantids are predacious cockroaches. Give a roach large, spiny, grasping forelegs by enlarging its fore coxae, then stretch its prothorax out to keep the mobile head and spiny forelegs far forward of the vulnerable abdomen, and you have a well-designed predator called a mantid.

Mantids even lay eggs in cases, as do cockroaches, although instead of being shiny and smooth like roach egg cases, common mantid egg cases look like blobs of sponge toffee or foam insulation. The foamy egg cases of our familiar European Mantis (*Mantis religiosa*) are quite common on twigs and stems from late fall to spring, at which time they hatch into dozens of voracious, tiny praying mantids. Indiscriminate eaters, these insects are as aptly called "preying" mantids as "praying" mantids, because the same enlarged forelegs that give them a praying appearance are used to grasp any insect that strays within ambush range.

Despite their large eyes and comically mobile head (mantids are the only insects that can look over their shoulders), female mantids seem unable to distinguish males from meals. A male mantid is lucky to get into a copulatory position, and even then he could end up as his mate's dinner. It is common for a male to get his head bitten off during copulation; this serves to sever a nerve that normally inhibits the male's sexual movements, so even with his head gone his abdomen just keeps on moving.

European Mantis (*Mantis religiosa*), page 68, caption 1.

The European Mantis is by far the most common of the three species of mantid in northeastern North America, although the larger Chinese Mantis (*Tenodera aridifolia sinensis*) is abundant in some areas. Both of these beneficial predators were introduced to North America around the turn of the century, and egg cases of the Chinese Mantis are now sold in many garden stores and supermarkets. Another introduced species, the Narrow Winged Mantid (*Tenodera angustipennis*), occurs in the southeast along with nine native species. Like their close relatives, the termites and cockroaches, mantids are essentially tropical insects. Before the introduction of European and Chinese mantids, only one of the almost 2,000 species of Mantodea occurred in the northeast (the Carolina Mantis reaches as far north as southern Pennsylvania and Illinois), and the whole continent only had 18 mantid species.

Earwigs (Order Dermaptera)

"Tis vain to talk of hopes and fears; And hope the least reply to win; From any maid that stops her ears; In dread of earwigs creeping in!" – T. Hood

This little bit of prose from the early 1800s sounds like good advice to me. When I was a single man I quickly learned that I generally had little in common with women who feared the stings of dragonflies, reacted negatively to beautiful beetles or thought that earwigs would get into their ears. A more enlightened woman (like the one I eventually married) would be interested to know that our common earwigs (**Dermaptera**, page 69) spend their daylight hours in a variety of dark places other than ears, then emerge during the night to exhibit outstanding omnivory, distinctive dimorphism and a very interesting life history.

Earwigs are diverse in tropical parts of the world, as is typical for orthopteroid orders, but only a handful of the world's 1,800 or so species of Dermaptera occur in North America. Of these, the only one seen by most people is the overwhelmingly abundant, omnivorous, introduced European Earwig (*Forficula auricularia*). The European Earwig eats almost everything, from caterpillars to cauliflower, but seems to prefer the finest flowers, best heads of lettuce and your favorite foliage plants. If you have a garden or yard, you certainly know, and probably dislike, these elongate, short-winged, shiny brown insects with their threatening forked tails. The tails, or cerci, differ between the sexes, with the male cerci strongly curved like a pair of tongs and the female cerci long and relatively straight.

European Earwigs respond to the onset of winter by nesting underground in pairs, but the female evicts the male from the nest after she lays a batch of eggs in late winter. She

Spine-tailed Earwig (*Doru aculeata*), page 69, caption 2.

then broods over her eggs until they hatch, and she continues to care for the young nymphs for some time thereafter. The new earwig family tends to stay together for some weeks, so if you find a dense cluster of whitish, elongate insects in your house foundation don't jump to the conclusion that you have found a termite colony. Look again for the distinctive cerci of the abundant and relatively innocuous European Earwig.

Only one native earwig, the Spine-tailed Earwig (*Doru aculeata*) occurs as far north as Canada, where it hides in the leaf axils of emergent plants in southern Ontario wetlands. Both *Doru* and *Forficula* are in the family Forficulidae, but two or three other families of earwigs occur in northeastern North America. The small (about 5 mm) introduced species *Labia minor* (Labiidae) is often seen on the wing, unlike the larger European Earwig. The Seaside Earwig (*Anisolabis maritima*, family Carcinophoridae) is a large (over 20 mm) wingless species found under debris along seashores. The Seaside Earwig, like *Labia minor* and the European Earwig, is introduced, as are the striped earwigs (Labiduridae) that range up the coast north to North Carolina. Several other earwigs occur in the southeast, and Florida now has a dozen dermapteran species.

Stick insects (Order Phasmatodea)

Like most other orthopteroids, the 2,500 or so species of stick insects (**Phasmatodea**, page 70) are primarily tropical, like the huge species called the "spiny devil" sold in some North American pet stores. Despite their formidable appearance, these skinny giants, native to New Guinea, are strictly vegetarian and are relatively safe for students to handle. Some of the big tropical walking sticks can inflict a painful stab with their leg spines, but none bite.

The Common Walkingstick (*Diapheromera femorata*), page 70, caption 2.

The common stick insect of northeastern North America, a wingless species called *Diapheromera femorata*, is rarely seen, although it is periodically abundant in areas of extensive oak forest. The large eggs of these stick insects, which are simply dropped to the ground without any evidence of maternal care, can be found in oak litter during the autumn months. The eggs of some stick insects are routinely gathered by ants, which do not kill the eggs but instead store them in the parasite-free environment of underground ant nests. The ants are "paid" for this service by the edible outgrowths on the stick insect's egg.

Stick insect nymphs and adults are protected from predators by a combination of woody camouflage and their twiglike stances, but some species can also change color to match a background color, and others have outgrowths similar in appearance to moss or lichen to further enhance their "disappearing acts." In the unlikely event that one of these carefully concealed orthopteroids is spotted and grabbed by a would-be predator, stick insects still have a few tricks up their skinny sleeves. Some species, including one common in the southeastern United States, have a second line of defense in the form of a powerful chemical spray. This unusually fat and cheeky stick insect can direct a spray, which is similar to tear gas in its composition and effect, at approaching birds. A third and more usual line of defense is to voluntarily shed appendages. This in itself is not an unexpected escape mechanism among insects, but young stick insects have the unusual capability of regenerating the appendages at the next molt. Adult stick insects, like all adult insects, do not molt, and can not regenerate appendages the way their younger counterparts can.

Zorapterans (Order Zoraptera)

Zorapterans (**Zoraptera**, page 70) are like mysterious small termites. As is the case with termites, they live in groups in cellulose-rich environments, such as sawdust piles or old logs, and occur in groups that may include winged and wingless forms. Unlike termites, however, zorapterans are not social (although they live in groups), nor do they eat and digest wood with the aid of symbiotic organisms. Most zorapterans appear to be fungus-feeders. These small (under 4 mm) insects occur in winged and wingless forms; the latter include primarily wingless individuals that are pale and eyeless, and secondarily wingless individuals (individuals that have shed their wings after a dispersal flight) that are dark and possess compound eyes. The only northeastern species in this small, obscure order is *Zorotypus hubbardi*, which occurs under bark and in sawdust piles as far north as southern Pennsylvania.

Rock crawlers (Order Grylloblattodea)

I once stopped in the mountains en route from San Francisco, California, to Reno, Nevada, on a sunny December day, and scraped some leaves off the frozen soil to reveal several active, entirely wingless insects that looked a bit like earwigs with thin cerci, and a bit like skinny cockroaches. It only took a few moments to come to the exciting realization that these cold-loving orthopteroids were icebugs, or rock crawlers, rarely encountered insects first discovered in western Canada in 1914 by the great Canadian entomologist E.M. Walker. This order of primitive orthopteroids does not occur in the east, and is restricted to mountains in western North America and Asia where they feed on other insects, presumably including less cold-tolerant species that have been blown onto glacier margins where rock crawlers are able to thrive. The emblem used by the Entomological Society of Canada has a stylized rock crawler in the middle, making it the closest thing Canada has to a national insect.

A superficially similar order of insects, Mantophasmatodea or African Rock Crawlers, was described in 2002, making it the most recently discovered order of insects. African Rock Crawlers are currently restricted to southern Africa, but the order is also known from Baltic amber fossils.

Webspinners (Order Embioptera)

Webspinners are distinctive insects, somewhat similar in shape and size to earwigs but with an enormously swollen tarsomere on each front leg. These conspicuous swellings hold silk glands which webspinners use to make silken galleries in which they normally remain hidden as they feed on lichens, moss and dead plant material. Females never have wings, but some males are fully winged and occasionally fly to lights at night. Webspinners are insects of warm habitats, and only one species (*Diradius vandykei*, family Teratembiidae) ranges as far north as North Carolina in eastern North America. A few other species and two other families occur farther south and west.

Webspinners are widely referred to as Embiidina as well as Embioptera; either name is correct since ordinal names are not governed by the same strict rules as names at the family level and below. Embioptera is to be preferred because the suffix "ina" is normally used for subtribes, not orders, and the ending "ptera" (wing) is a familiar ordinal suffix.

A Webspinner, page 70, caption 3.

❶ ❷ ❸ ❹ ❺ ❻

FAMILY BLATELLIDAE ❶ The **Pennsylvanian Wood Cockroach** (***Parcoblatta pennsylvanica***) is the most common northeastern wood roach, although there are several other *Parcoblatta* species. This is a mature nymph, photographed during the winter months. ❷ This **Pennsylvanian Wood Cockroach** has just molted from a nymphal to a fully winged adult form, and it has not yet hardened or obtained its adult color. Such soft, newly molted individuals are referred to as "teneral." ❸ This female adult **Pennsylvanian Wood Cockroach** has an egg case, or ootheca, projecting from her abdomen. ❹ ***Ischnoptera deropeltiformis*** is a common wood roach in the eastern United States, but it does not occur in Canada. ❺ The **Small Yellow Cockroach** (***Cariblatta lutea***) is common from North Carolina south to Florida and the Caribbean and occasionally turns up as far north as Ontario. ❻ The **German Cockroach** (***Blatella germanica***) remains one of the most common urban roaches, but it is much less abundant than in previous years. ❼ The **Brown-banded Cockroach** (***Supella longipalpa***) is an introduced species, native to India, and is an increasingly common household pest throughout our area. ❽ These pinned specimens are **Oriental Cockroaches** (***Blatta orientalis***, family Blattidae). Female Oriental Cockroaches are very short-winged, and this one has her egg case still projecting from her abdomen as she would have carried it in life.

❼

❽

❶

❷

❸

FAMILY BLATTIDAE ❶ The **Australian Cockroach** (***Periplaneta australasiae***) and American Cockroach (*Periplaneta americana*) are similar, large (around 30 mm) domestic roaches. The Australian Cockroach shown here has more distinct black spots on the pronotum and paler shoulders. Both species originated in the Old World tropics (probably Africa) and are now widespread household pests. **FAMILY CRYPTOCERCIDAE** ❷ The **Brown-hooded Cockroach** (***Cryptocercus punctulatus***) was until recently considered to be the only North American species in a family of 3 species (with the other two species in Asia), but one western and 4 eastern North American species of *Cryptocercus* are now recognized, with the latter ranging from Pennsylvania to Georgia. Furthermore, some authors now consider *Cryptocercus* to be part of the sand roach family, Polyphagidae. **FAMILY BLABERIDAE** ❸ The family Blaberidae is a mostly tropical group, but the beautiful green ***Panchlora nivea*** occurs in the southeast. This species is common in the Caribbean, where it is known as the **Green Banana Cockroach**.

TERMITES ORDER ISOPTERA

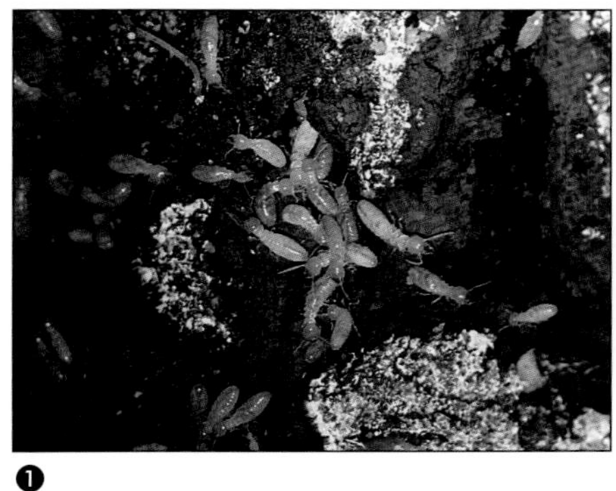

❶

❷

❸

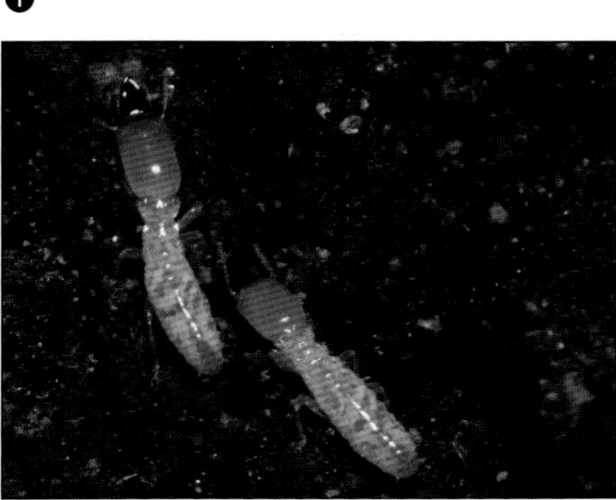

❹

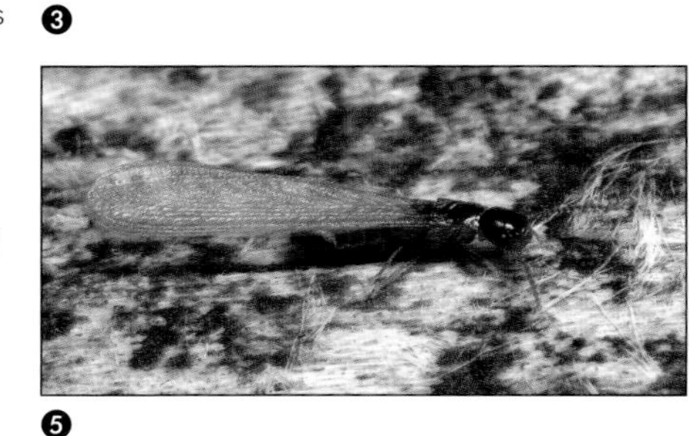

❺

FAMILY RHINOTERMITIDAE ❶ The **Eastern Subterranean Termite** (***Reticulitermes flavipes***) is the most important termite species of the eastern United States and Canada, and the only termite most northeastern residents are likely to see. Two other *Reticulitermes* species occur in the northeast. ❷ Most individuals in a ***Reticulitermes*** colony are pale, soft-bodied workers. ❸ Peeling bark off this poplar stump revealed foraging termites (***Reticulitermes flavipes***). ❹ ***Reticulitermes*** soldiers have enlarged mandibles used in colony defense. ❺ Winged termites like this ***Reticulitermes*** shed their wings soon after completing a mating flight.

❶ ❷ ❸ ❹ ❺ ❻ ❼

FAMILY MANTIDAE ❶ ❷ The **European Mantis** (***Mantis religiosa***) is the common mantid of the northeast, and one of two well-established introduced mantids in our area. These two photographs show the green and brown forms of this species. ❸ The **Chinese Mantis** (***Tenodera aridifolia***) is an introduced species that is well established as far north as southern Ontario, although its range does not extend as far north as the more abundant European Mantis (*Manits religiosa*). ❹ This **Chinese Mantis** (***Tenodera aridifolia***) egg case was purchased at a local garden supply store. Nymphs are just starting to emerge from the case. ❺ ❻ ❼ The only native mantid that might be found in the northeast is the **Carolina Mantis**, ***Stagmomantis carolina***, which ranges from southern Pennsylvania to Florida. Several other native mantid species occur in the southern states. These images show the green form, egg case, and brown form of the Carolina Mantis.

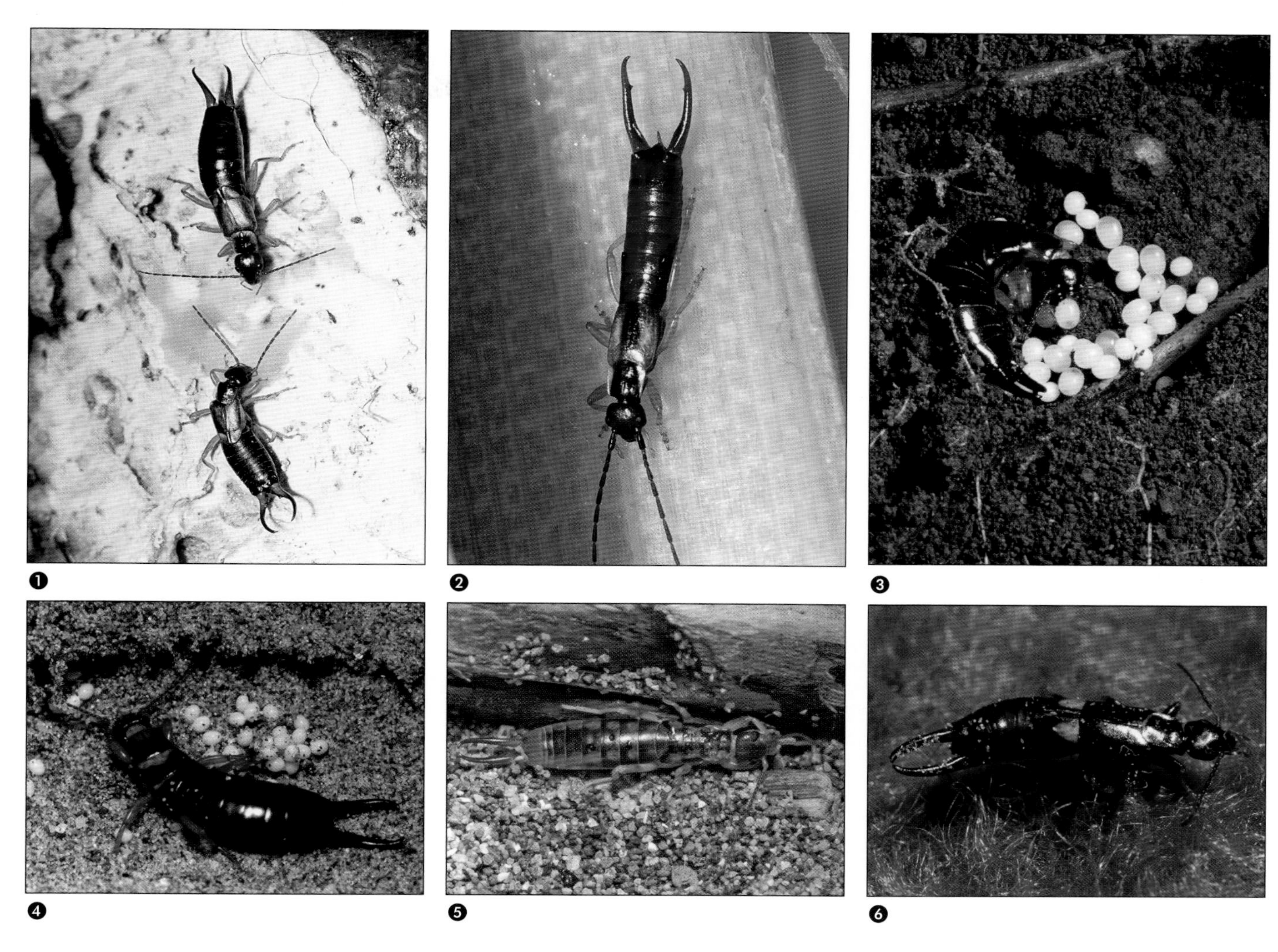

FAMILY FORFICULIDAE ❶ The **European Earwig** (***Forficula auricularia***) is an abundant omnivore throughout our area. This male (strongly curved cerci) and female (weakly curved cerci) are feeding on a damaged squash. ❷ This male **Spine-tailed Earwig** (***Doru aculeata***) was photographed in the base of a cattail (*Typha*) stem in southern Ontario. This rare species is the only native earwig in the northeast. FAMILY CARCINOPHORIDAE (SEASIDE AND RING-LEGGED EARWIGS) ❸ This female **Ring-legged Earwig**, ***Euboriella annulipes***, was exposed in her nest full of eggs by turning over a stone. This cosmopolitan species ranges as far north as North Carolina. ❹ This **Seaside Earwig** (***Anisolabis*** sp.) and her eggs were exposed by flipping a piece of driftwood. One introduced species of this cosmopolitan genus, *A. maritima*, ranges up the Atlantic coast to Canada. FAMILY LABIDURIDAE (STRIPED EARWIGS) ❺ Striped earwigs occur on seashores throughout the world. This is an endemic New Zealand species of ***Labidura***, but the very similarly pigmented, relatively large (20-30 mm), cosmopolitan *Labidura riparia* occurs in the United States from North Carolina on south to Florida. FAMILY LABIIDAE (LITTLE EARWIGS) ❻ The two eastern species in the genus *Marava* normally occur in the southeast (New Jersey on south), but this ***Marava arachidis*** was part of a thriving indoor population in an Ontario chicken barn. ❼ The only common member of this family found outdoors in the northeast is the tiny (about 5 mm) introduced species ***Labia minor***. ❽ ***Vostox brunneipennis*** is a common earwig found under the bark of dead pine trees in the eastern United States.

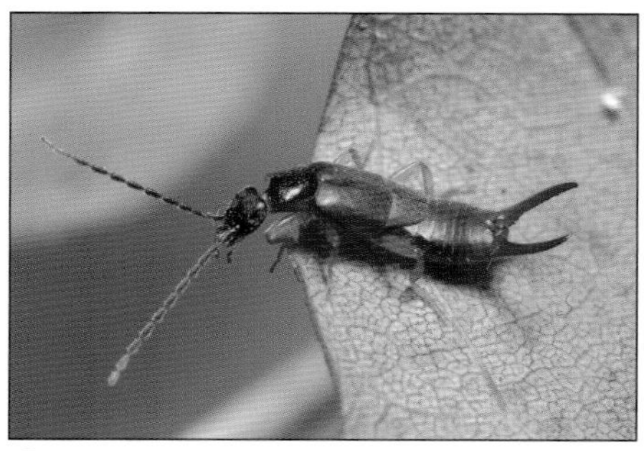

❼

❽

STICK INSECTS
ORDER PHASMATODEA

❶ ❷ ❸

FAMILY HETERONEMIIDAE ❶ ❷ The **Common Walkingstick** (***Diapheromera femorata***), the only northeastern walkingstick, can be found on oak foliage throughout our area as far north as Ottawa, Ontario. Walkingstick pairs often remain *in copula* for a very long time – the small male protects his paternity by remaining in position for days. **FAMILY PSEUDOPHASMATIDAE** (STRIPED WALKING STICKS) ❸ The two similar North American species of Pseudophasmatidae, both in the genus ***Anisomorpha***, are mostly southeastern, with one species (*A. ferruginea*) ranging north to Virginia and Illinois. The species shown here is common in Florida, and famous for its long copulations and habit of squirting milky, caustic defensive fluid at attackers (including insect collectors).

ZORAPTERANS
ORDER ZORAPTERA

WEBSPINNERS
ORDER EMBIOPTERA

❶ ❷ ❸

FAMILY ZOROTYPIDAE ❶ ***Zorotypus hubbardi*** is the only northeastern species in the small, rarely seen order Zoraptera. This small (about 2 mm) eyeless and wingless individual was one of many under the bark of a dead pine tree in North Carolina. Winged individuals (with eyes) also occur, but they only retain their wings for a short period before shedding them. The phylogenetic relationships of this odd order are a matter of debate, with some scientists considering them "orthopteroid" and others considering them "hemipteroid." Most evidence suggests that they are closely related to webspinners. **ORDER EMBIOPTERA** ❷ ❸ Webspinners use silk secreted by glands in the conspicuously swollen tarsi of their forelegs to spin retreats where they hide by day, coming out at night to feed on plant material of various sorts. Males are usually winged and short-lived; females are always wingless. This is a mostly tropical order but a few species occur in the southern United States (these images are of tropical species).

CHAPTER 5

Grasshoppers, Crickets and Katydids

Grasshoppers and crickets are among the most familiar of animals, and together make up an easily recognized order of hemimetabolous insects equipped for jumping with massive and muscular hind legs. The order name "Orthoptera" (*ortho* = straight, *ptera* = wings), however, does not refer to the conspicuous hind legs that most obviously characterize this order, but instead describes straight and parchment-like front wings that conceal more membranous hind wings. Similar wings are found in several other related orders and in the past the ordinal name Orthoptera was used in a broad sense to include mantids, cockroaches, stick insects, and various other groups now treated as separate orders (see chapter 4). Orthoptera in this older sense was not a natural group (some of its members had their closest relatives outside the group), so the definition of "Orthoptera" is now narrowed down to a single natural group characterized by fat hind femora and associated saltatorial skills.

The order Orthoptera, in its current and jumpy sense, is divided into two suborders, one including the long-horned (long antennae) forms, such as crickets and katydids, and the other made up of short-horned (short antennae) grasshoppers. These two groups differ from one another in how they sing, how they hear, how they lay eggs and even how they mate.

Long-horned Orthoptera (Suborder Ensifera)

Our most familiar crickets (**Gryllidae**, page 76) are the fat, black field crickets that abound under debris in meadows, roadsides and other relatively open places in spring and fall. Both the spring and fall forms were thought to be the same species until it was realized that those found in spring produced a different chirping "song" from those found in fall, and that the "Field Cricket" was actually two species. Both species occur across North America, with the spring-maturing species (*Gryllus veletis*) less often encountered than the familiar, gregarious, late summer and fall species (*G. pennsylvanicus*). Recent work on the phylogenetic relationships amongst all eleven North American species of *Gryllus* has shown that the similar spring and fall "field crickets" do not even belong to the same species group, and are more closely related to other *Gryllus* species than to each other.

Next time you spot one of these common crickets, take a closer look. Is it a male, with broad wings and two short tails (cerci), or a female with narrow wings and a long, sharp egg-laying tube? Like the males of most long-horned Orthoptera, male field crickets sing by rubbing their front wings together. These wings have large veinless areas that serve as resonating membranes, amplifying the songs produced when a basal scraper on the upper surface of one front wing is rubbed along a filelike ridge on the lower surface of the other. During song, the wings are elevated above the body and rubbed back and forth, with sound generally produced only on the closing stroke. Crickets are usually "right handed," with the right wing on top and doing the filing, whereas other long-horned Orthoptera tend to be "left handed," with the left wing on top.

The messages inherent in the wing-scraping songs of male crickets and other long-horned Orthoptera are picked up by both sexes using eardrums, or tympana, on the front legs. Crickets have songs for aggression, defense and territoriality, but most stridulating songs are amorous in intent. If a male's serenading is successful and a female is attracted, he switches to a softer song and may add other inducements. Melodious males of our small, pale green tree crickets, for example, take a "song and a box of candy" approach, secreting substances from glands on their backs which the females seem to find irresistible.

Say's Bush Cricket (*Anaxipha exigua*), page 76, caption 7.

One of the species of tree cricket common across North America (the Snowy Tree Cricket, *Oecanthus fultoni*) is known as the "thermometer cricket" because, like other crickets, it increases its rate of singing as the temperature increases. (Apparently, it is possible to obtain the current temperature in degrees Fahrenheit by counting the number of chirps in 13 seconds and adding 40.) Snowy Tree Crickets also often synchronize their chirps, so their song pattern can be easily distinguished from the more irregular, asynchronous chirping of field crickets or the continuous trill of most other tree crickets. Although some species of tree crickets are fairly habitat-specific (Pine Tree Crickets, for example, are found in pine trees, and Larch Tree Crickets are found in larch trees), our most common tree crickets abound among goldenrod and blackberry vines late in the season, when their melodious singing seems to mark the end of summer. Cricket females most obviously differ from males in having simple, straight wings and in having a well-developed egg-laying tube, or ovipositor. The long, sharp tree cricket ovipositor is used to insert eggs in twigs, sometimes killing the twigs in the process.

Despite doing occasional damage to nursery stock while laying eggs, tree crickets are generally beneficial insects that feed primarily on aphids and other small insects. Field crickets, on the other hand, are well known for their omnivorous natures, which lets them take the blame for chewing through binder twine on hay bales, damaging tomatoes and generally feeding on all sorts of things we would rather they left alone – although they do eat the eggs and pupae of insect pests. Another common cricket species, the introduced House Cricket (*Acheta domestica*), is generally considered a pest because of its tendency to chew up damp clothing and similar household items.

Woods, fields and wetlands abound in inconspicuous native crickets (mostly little brown ground crickets in *Allonemobius* and some less common related genera), but the field crickets, tree crickets and introduced House Crickets are the "true" crickets encountered most frequently. Aptly named for their domestic preferences, House Crickets have been associated with human dwellings for centuries. Keeping House Crickets and other long-horned Orthoptera for their song and for use in cricket fights is an ancient custom in parts of Asia, and elaborately constructed cricket cages can often be seen in displays of Chinese antiques. Here in North America, House Crickets are reared in huge numbers, but primarily as bait and food for vertebrate pets rather than as pets in their own right.

Mole crickets (**Gryllotalpidae**, page 79), so called because of their molelike burrowing lifestyle and molelike shape, comprise yet another group to which the name "cricket" is applied. Despite their subterranean life, mole crickets have fully developed wings, complete with file and scraper. These omnivorous burrowers are very common in the southern United States, but they are relatively rare in the north and they barely get across the Canadian border. Their tunnels are often visible as wormlike tracks near the surface of sandy soils near ponds and lakes, but you are more likely to hear their melodious nocturnal singing than to see mole crickets. Males construct "amplifying burrows," with elaborate entranceways that serve as resonant sound chambers to magnify their melodious chirps so that they can be heard over a mile away.

The "true" crickets all fall into the families Gryllidae (including several subfamilies) and Gryllotalpidae, but some other Orthoptera are loosely referred to as crickets even though they are more closely related to katydids than to true crickets. The wingless cave and camel crickets, for example, are in the family **Rhaphidophoridae** (page 78). These strongly humpbacked insects frequent dark, damp places like basements and woodpiles. The name "camel cricket" accurately describes their brown, humpbacked appearance, while the name "cave crickets" reflects the habitat preference of many species (if you venture into a cave during the day you will probably see the walls lined with these large, glistening, wingless hoppers). At night the foraging activities of cave crickets may take them outside the cave, where they often feed on dead or dying invertebrates. Members of a closely related family, **Gryllacrididae** (page 79) or leaf-rolling crickets, feed on aphids at night and hide in rolled leaves by day. The only North American leaf-rolling cricket occurs in the southeast.

Katydids belong to a group collectively known as long-horned grasshoppers because they are flattened side to side like grasshoppers, but have very long antennae like crickets. Orthoptera specialists sometimes split this group into a number of families, but it is easier to lump the whole lot (coneheads, katydids, meadow grasshoppers and shield-backed grasshoppers) into a single family called **Tettigoniidae** (page 79), and divide it into subfamilies if you are so inclined. Whatever you call them, long-horned grasshoppers are spectacular insects, including some common, large, bright green (or occasionally pink) species with spectacular swordlike ovipositors.

The big katydids most frequently seen on low vegetation in late summer and fall are bush katydids (subfamily Phaneropterinae, especially the very common genus *Scudderia*). Although singing is usually the prerogative of male Orthoptera, female Phaneropterinae respond to the male's song with clicklike call, and one prominent katydid specialist reports attracting singing males by clicking his

True Katydid *(Pterophylla camellifolia)*, page 79, caption 8.

fingernails in response to the males' short, relatively soft song.

The rasping song of bush katydids is a familiar sound on warm autumn nights, but the famous song that gives rise to the name "katydid" emanates from a distant relative of the bush katydids, called the True Katydid (*Pterophylla camellifolia*, subfamily Pseudophyllinae). Both males and females of the elusive True Katydid sing from high in the trees, but only the male belts out the distinctive "katy did, katy didn't" song that enriches well-treed suburbs of the northeast on warm late summer evenings. Unlike the more commonly seen bush katydids, True Katydids have strongly convex wings that entirely surround the abdomen. The True Katydid is the only member of the Pseudophyllinae in the northeast, but this group is common in the tropics where it includes a thousand or so species, some of which are spectacular leaf mimics.

Several cricket-like insects, such as the famous western North American pest species called the Mormon Cricket, belong to a group of long-horned grasshoppers known as the shield-backed grasshoppers. Shield-backed grasshoppers, traditionally treated as the subfamily Decticinae but now included with the subfamily Tettigoniinae, are diverse in the west but form a small group in northeastern North America. One of the most common northeastern species, *Metrioptera roeselii*, was accidentally introduced from Europe. These attractive hoppers normally mature without ever developing functional wings but when high population densities occur they produce a winged form that disperses to less crowded locales.

A short-winged male grasshopper, *Melanoplus pachycercus*.

The long-horned grasshoppers that contribute the most to late summer stridulatory symphonies are those you are least likely to see. If you have the patience, try tracing each different song to its skittish, well-camouflaged producer. Careful nocturnal listening in open meadows might lead you to the common but rarely seen coneheads (subfamily Copiphorinae), large long-horned grasshoppers with the head produced in front of the antennae in an outlandish conelike fashion and the ovipositor strikingly long and bladelike. Should your meadow be a wet one or near a stream or pond, you are even more likely to track down the very common but well-hidden meadow grasshoppers (subfamily Conocephalinae). Long-horned grasshoppers seen on sedge stems are usually small meadow grasshoppers (*Conocephalus*) or – much harder to find – large meadow grasshoppers (*Orchelimum*).

Northeastern North America is now home to two additional subfamilies of long-horned grasshoppers thanks to recent introductions of the Matriarchal Katydid (Saginae) and the Drumming Katydid (Meconematinae) from Europe. The former, a predacious species with raptorial front and middle legs, is apparently established only in Michigan; the latter is a day-active species introduced to New York State but now spreading throughout the northeastern States. Drumming Katydids (*Meconema thalassinum*) do not stridulate but the small (about 16 mm) males 'drum' by tapping leaf or other surfaces with a hind leg.

Short-horned Orthoptera (Suborder Caelifera)

Short-horned grasshoppers are almost exclusively day-active vegetarians devoid of the musical inclinations of their normally nocturnal, often omnivorous, long-horned cousins. If they sing, they do so in a comparatively crude fashion, usually by rubbing pegs on the hind leg against the wing covers. They also don't "hear" the way crickets and katydids do. Most long-horned Orthoptera have an elegant "ear" (called a tympanum) on the front leg at the base of the tibia, whereas grasshoppers hear using a tympanum on the first abdominal segment.

Long-horned and short-horned grasshoppers really show their differences when the combination of music made by the male and its perception by a receptive female leads to sexual encounters. In the coupling of long-horned Orthoptera, the female normally mounts the male, initiating a copulation that ultimately leaves her with a large, nutritious spermatophore hanging out the end of her abdomen where she can reach an edible, gelatinous, sperm-free part of it with her mouthparts. (A spermatophore is a capsule used as a sperm-carrier by male orthopterans and many other insects; the gelatinous sperm-free part found in long-horned Orthoptera is called a spermatophylax.) In contrast, the short-horned grasshopper male takes the top position before inserting his spermatophore fully inside the female. Male short-horned grasshoppers often ensure paternity by riding the female for long periods, so coupled grasshoppers are a common sight. The fertilized short-horned grasshopper female has no swordlike ovipositor to inject single eggs into twigs the way a cricket or katydid would. Instead, she typically works her whole abdomen deep into the soil, leaving behind a pod of eggs.

That pod of eggs, if deposited close to other pods of eggs, could develop into progeny quite unlike their mother. Some short-horned grasshoppers in the family **Acrididae** (page 83) have a Dr. Jekyll and Mr. Hyde propensity to lay eggs that develop into either solitary, innocuous individuals, or into the large, gregarious, devastatingly voracious forms we know as locusts. Development into the solitary form, as opposed to a gregarious form, depends on the circumstances under which the young individuals grow to maturity. In the case of locusts, crowding results in changes not only in behavior but also in color, physiology and proportions of the body parts. The noncrowded populations develop into the solitary phase, and are just ordinary grasshoppers. The crowded populations develop into the gregarious phase, and become the locusts of biblical fame.

The word "locust" justifiably strikes fear into the hearts of people in many parts of the world, where these migratory short-horned grasshoppers can form swarms of over a billion individuals that move over vast distances, consuming crops along the way. Locust swarms have been recorded throughout history as among the most overwhelming and devastating types of natural disasters. Although locusts remain among the most important insect pests internationally, especially in Asia and Africa, major locust swarms have been unknown in North America since the mid-1800s, when the Rocky Mountain Locust, *Melanoplus spretus*, was a scourge of major proportions. The last living *M. spretus* was collected in Manitoba in 1902, but vast masses of their frozen bodies can still be found in some glaciers in the American northwest, attesting to their previous abundance. If you are willing to make the daylong hike from the nearest road, you can still see locust-laced glaciers at a couple of sites in Montana (although this opportunity will pass soon if the glaciers continue to shrink at their current rapid rates). Several species of grasshoppers, including some *Melanoplus* species common in northeastern North America, still occasionally occur in outbreak numbers, but none exhibit the phase polymorphism that characterizes "true" locusts.

The vast majority of commonly encountered short-horned grasshoppers are in the genus *Melanoplus*. This massive genus is in a group called the spur-throated grasshoppers (subfamily Melanoplinae) because they have a spurlike process below and behind the head.

The large family Acrididae, which includes most short-horned grasshoppers, includes some other subfamilies that look quite different from the ordinary spur-throated grasshoppers. One such group, the band-winged grass-hoppers (subfamily Oedipodinae), contains several showy and familiar species with hind wings banded in colors including red, black and yellow. The most familiar of these is a large gray grasshopper characteristic of dry, disturbed areas such as roadsides, where it is commonly seen flying from spot to spot in a conspicuous flurry of black and yellow wings. This species is often called the Carolina Locust, though it is not really a locust because it has no gregarious phase. Two other subfamilies of grasshoppers, most species of which have strikingly receding faces, are aptly called the slant-faced grasshoppers (subfamilies Acridinae and Gomphocerinae). Yet another subfamily of Acrididae, the bird grasshoppers (Cyrtacanthacridinae) includes a few species that are almost birdlike both in size and flying ability.

Eastern North America is home to only a couple of other families of short-horned Orthoptera and they are small and rarely noticed. The **Tetrigidae** (page 89), or pygmy grasshoppers, are really bizarre-looking little grasshoppers, common along pond and stream margins where there is exposed, damp earth. They are hard to catch, but if you can nab one you will see that the pronotum (the top of the first segment of the thorax) extends like a camouflaged shield over the whole top of the body. These small (about 15 mm) grasshoppers are also often unusually polymorphic, and neighboring individuals of the same species can be very differently pigmented. Pygmy grasshoppers seem to get more abundant late in fall, but that might only be because they spend the winter as adults or nymphs. Most grasshoppers spend the winter in the egg stage, but another little group that spends the winter as adults is the peculiar family **Tridactylidae** (page 90), or pygmy mole crickets. These gregarious, nocturnal, rarely encountered insects are often found on sandy river banks where they can evade capture by flying, swimming and burrowing. They do look a bit like tiny (less than 10 mm) mole crickets, but are in the grasshopper group, not the cricket group.

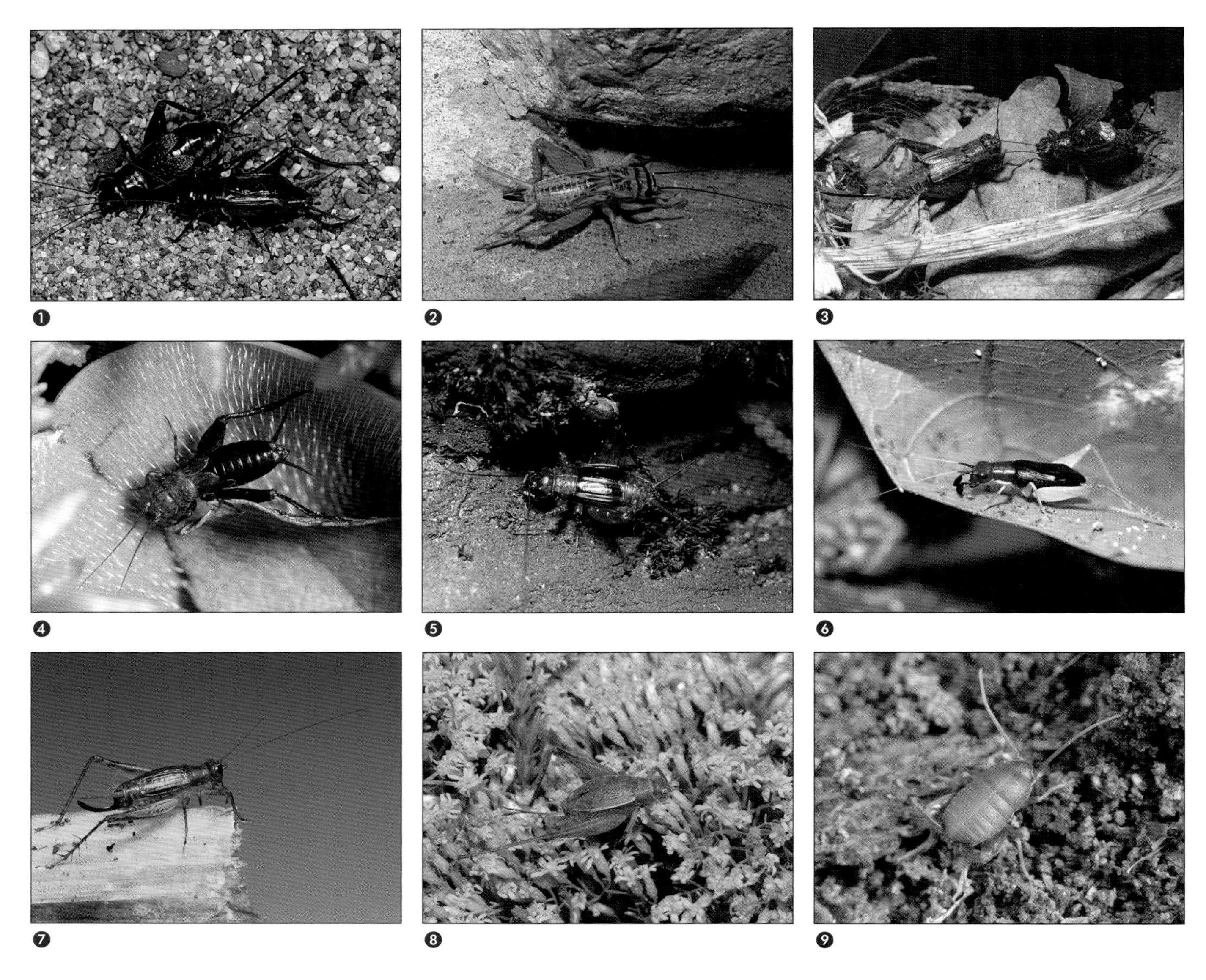

SUBORDER ENSIFERA

FAMILY GRYLLIDAE Subfamily Gryllinae (HOUSE AND FIELD CRICKETS) ❶ Male crickets, like the male of this pair of **Fall Field Crickets** (***Gryllus pennsylvanicus***) have broad wings and two short tails (cerci). Females have relatively narrow wings and a long, sharp egg laying tube (ovipositor) in addition to the cerci. ❷ The **House Cricket** (***Acheta domestica***) is a common, introduced insect, mostly restricted to indoor habitats in colder areas but widespread out of doors in the south. Rearing them for fish bait or for pet food is a big business, and they are widely available for sale. Subfamily Nemobiinae (GROUND CRICKETS) ❸ The male of this pair of **Striped Ground Crickets** (***Allonemobius fasciatus***, 9–11 mm) is singing by rubbing his elevated front wings together to make a series of short, buzzing chirps. This is the most common of several similar ground cricket species in the northeast. ❹ ***Neonemobius palustris***, known as the **Sphagnum Cricket** or **Marsh Ground Cricket**, feeds and lays eggs only on Sphagnum moss, although this one was photographed as it made a hazardous crossing over a pitcher plant leaf. These small (5–6 mm) crickets are usually shiny black, but specimens from some localities are partially brown to golden in color. ❺ ***Eunemobius carolinus***, the **Carolina Ground Cricket**, is common as far north as central Ontario and Quebec despite the southern implications of its name. Females, like this one, differ from other ground crickets in having a very short ovipositor. Subfamily Trigonidiinae (SWORD-TAIL CRICKETS) ❻ The distinctively colored Handsome Trig or **Red-headed Bush Cricket** (***Phyllopalpus pulchellus***) ranges north to Illinois and New Jersey. ❼ As the common name suggests, the little **Say's Bush Cricket** or **Say's Trig** (***Anaxipha exigua***; 5–7 mm) is usually found on foliage or twigs unlike the superficially similar ground crickets. This species is found throughout the eastern United States, but is restricted to extreme southern Ontario in the Canadian part of its range. The female is the one with the long ovipositor. Subfamily Eneopterinae ❽ The **Restless Bush Cricket** (***Hapithus agitator***) is one of two species of Eneopterinae in the northeast, neither of which ranges north to Canada. These crickets differ from other subfamilies in having teeth between the large spines on the hind tibia. Subfamily Myrmecophilinae (ANT-LOVING CRICKETS) ❾ ***Myrmecophilus pergandei*** (3 mm), the only eastern member of the subfamily Myrmecophilinae, is a myrmecophilous, or "ant loving," species. This one was found in a carpenter ant nest. Ant-loving crickets are sometimes treated as a separate family, Myrmecophilidae.

FAMILY GRYLLIDAE (continued) Subfamily Mogoplistinae (SCALY CRICKETS, 6–12 mm) ❶ The **Two-toothed Scaly Cricket** (***Cycloptilum bidens***), like other members of the subfamily, has wingless females and short-winged males. Scaly crickets are sometimes treated as a separate family, Mogoplistidae. ❷ ❸ Like other scaly crickets, **Slossons's Scaly Cricket** (***Cycloptilum slossonae***) is primarily southeastern in distribution. Scaly crickets have body scales much like those found on butterflies and moths. Females are totally wingless. Subfamily Oecanthinae (TREE CRICKETS, 12–15 mm) ❹ **Snowy Tree Crickets** (***Oecanthus fultoni***) are sometimes called thermometer crickets. ❺ The **Black-horned Tree Cricket** (***Oecanthus nigricornis***), named for the entirely black basal antennal segments, is a very common tree cricket in disturbed areas. Some individuals, like this male, are extensively black; others have the black restricted to the base of the antennae and top of the thorax. ❻ The **Four-spotted Tree Cricket** (***Oecanthus quadripunctatus***) gets both its common and scientific names from the black spots on the inner surface of the basal antennal segments. This is a common insect of open fields and weedy areas in the northeast. ❼ The **Narrow-winged Tree Cricket** (***Oecanthus niveus***) has a distinctive J-shaped black mark on the inside of the basal antennal segment (partially visible on the left antenna of this female). Live Narrow-winged Tree Crickets have a characteristic yellowish to orange patch on the top of the head, but this disappears on dead specimens. ❽ This female tree cricket is recognizable as a **Broad-winged Tree Cricket** (***Oecanthus latipennis***) by the unmarked basal antennal segments. The characteristic reddish color of the head and antennal base fades in dead specimens. This species occurs on shrubs and low trees.

FAMILY GRYLLIDAE Subfamily Oecanthinae **(continued)** ❶ ❷ The male of this pair of **Black-horned Tree Crickets** (***Oecanthus nigricornis***) is taking the "song and a box of candy" approach to courtship. In the first photograph, he is singing with his wings at right angles to the body, exposing glands on his back. In the second photograph the female is feeding on a substance produced in the male's glands. ❸ ❹ The similarly brown-pigmented **Larch Tree Cricket** (***Oecanthus laricis***) and **Pine Tree Cricket** (***Oecanthus pini***) are hard to distinguish from one another without host information. ❺ Like other long-horned Orthoptera, tree crickets have a long, sharp ovipositor, usually used to insert eggs into plant stems. In this photograph a sumac stem is broken open, revealing a row of tree cricket eggs. ❻ ***Neoxabea***, our only tree cricket genus other than *Oecanthus*, includes only the single species ***N. bipunctata***, the aptly named **Two-spotted Tree Cricket**. This photo was taken in extreme southern Ontario, the most northern point in this species' distribution. **FAMILY RHAPHIDOPHORIDAE** (CAVE AND CAMEL CRICKETS) ❼ Most northeastern cave and camel crickets are wingless insects in the genus ***Ceuthophilus***. The female of this pair is the one with the stout, bladelike ovipositor at the end of its abdomen. ❽ Although most North American Rhaphidophoridae belong to the genus *Ceuthophilus*, the introduced Asian **Greenhouse Stone Cricket** (***Tachycines asynamorus***) is found in greenhouses from New England north to Ontario and Quebec.

FAMILY GRYLLACRIDIDAE (LEAF-ROLLING CRICKETS) ❶ The **Carolina Leafroller** (***Camptonotus carolinensis***), New Jersey south to Florida, is the only North American species of gryallacridid. They use silk spun from the mouth to make leaf rolls in which to shelter during the day, and feed on aphids during the night. These wingless crickets are similar to camel crickets except for their widely separated antennae. **FAMILY GRYLLOTALPIDAE** (MOLE CRICKETS) ❷ The **Northern Mole Cricket** (***Gryllotalpa hexadactyla***) is well-equipped for its burrowing activities, with large fingerlike digging dactyls on its front tibiae. This is the only mole cricket in the northeastern United States and southern Canada. ❸ The **Southern Mole Cricket** (***Scapteriscus borellii***) has fewer spurs on the front tibia than the Northern Mole Cricket (*Gryllotalpa hexadactyla*), and is distinctly bicolored. This is a very common species in the southeast. **FAMILY TETTIGONIIDAE** Subfamily Tettigoniinae (SHIELD-BACKED GRASSHOPPERS AND RELATIVES) ❹ ***Atlanticus*** is an entirely eastern genus in North American with seven flightless species, three of which occur as far north as the Canadian border. This pair of ***A. monticola*** (**Davis' Shield Bearer**) was photographed in a sandy oak savanna in southern Ontario. ❺ A female **Davis' Shield Bearer.** ❻ ❼ The introduced shield-backed grasshopper ***Metrioptera roeselii*** (**Roesel's Katydid**) is now very common in eastern Canada and the adjacent states. These photos show *M. roeselii* in its more common short-winged form and the less common long-winged morph. Subfamily Pseudophyllinae (TRUE KATYDIDS) ❽ The **True Katydid** (***Pterophylla camellifolia***) occurs from Florida north to southern Ontario. The leaflike wings of True Katydids entirely encase the abdomen, but you are unlikely to see this attribute first-hand since True Katydids rarely descend from their treetop haunts, where females use a conspicuous ovipositor to lay eggs in bark. The male's distinctive (and loud) "katy did, katy didn't" song is often heard during late summer evenings. Females stridulate when disturbed, but with only a short, scraping call. Subfamily Phaneropterinae (LEAF KATYDIDS) ❾ The **Broad-winged Katydid** (***Microcentrum rhombifolium***), distributed from Florida to southern Ontario, is a large katydid with a short hind femur and widely separated antennae. Look for these well-camouflaged insects on low foliage of trees or shrubs.

FAMILY TETTIGONIIDAE Subfamily Phaneropterinae **(continued)** ❶ ❷ ❸ The **Oblong-winged Katydid** (***Amblycorypha oblongifolia***) is usually green, but occasionally occurs in a pink morph. These photos show adults of both morphs and a nymph of the pink morph. Similar color forms occur in the closely related *A. rotundifolia*, which also occurs in the northeast but which has more rounded front wings. ❹ ❺ ❻ Bush katydids (***Scudderia*** spp.), by far our most common katydids, differ from other leaf katydids in having the bases of the antennae right next to one another. These photos show an adult male, an adult female and a nymphal female of our most common katydid, the **Fork-tailed Bush Katydid** (***S. furcata***). The species of *Scudderia* can only be reliably distinguished from one another by examining reproductive structures. Subfamily Conocephalinae (MEADOW GRASSHOPPERS) ❼ ❽ Male and female **Black-legged Meadow Katydids** (***Orchelimum nigripes***). The male has his wings slightly elevated as he sings; the more narrow-winged female has a long, bladelike ovipositor. ❾ This male **Common Meadow Katydid** (***Orchelimum vulgare***) is singing by rubbing a basal scraper on the upper surface of the right front wing along a filelike ridge on the lower surface of the left front wing. The veinless area you can see near the base of the wing amplifies the song.

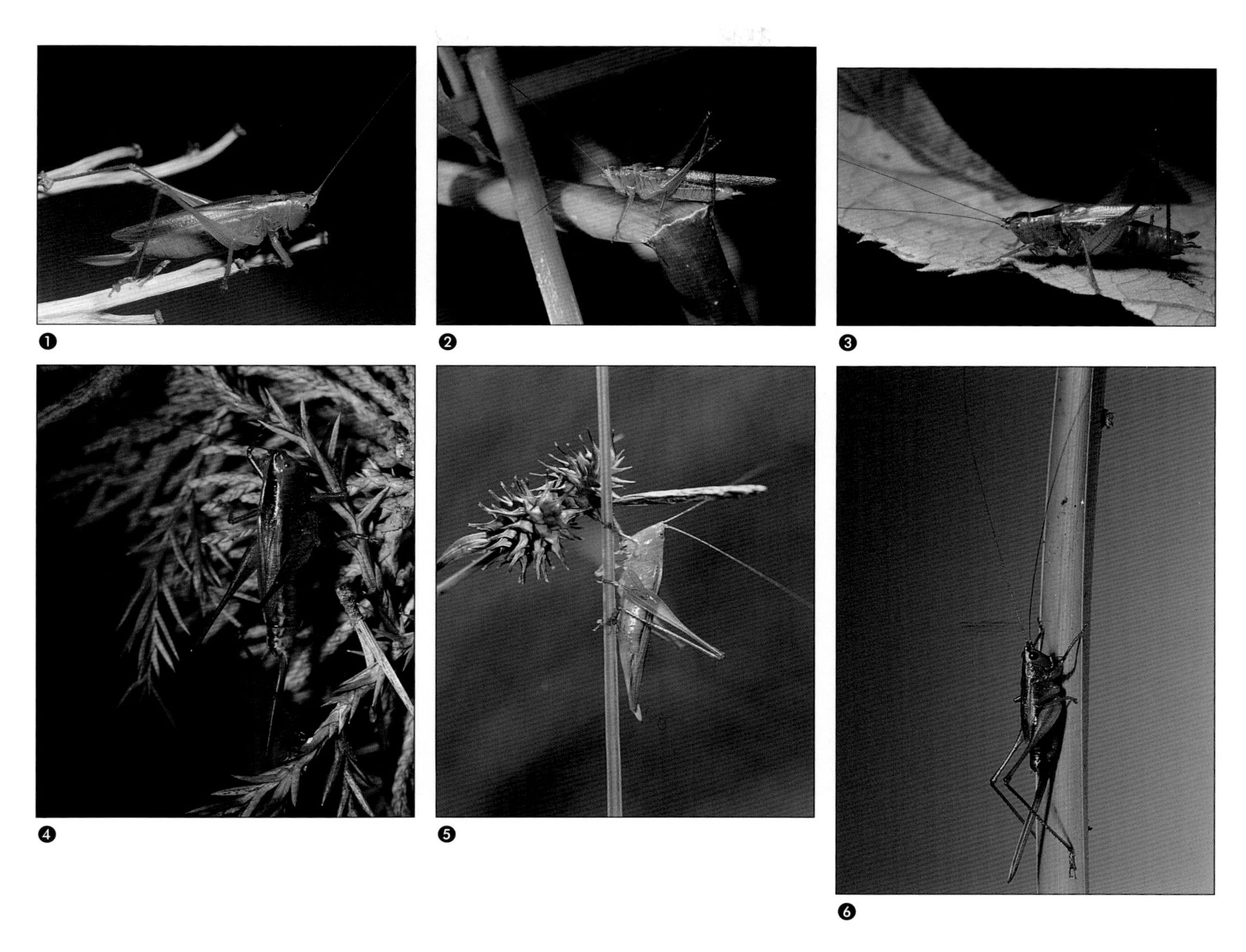

FAMILY TETTIGONIIDAE Subfamily Conocephalinae **(continued)** ❶ This female **Gladiator Meadow Katydid** (***Orchelimum gladiator***) has a strikingly straight, bladelike ovipositor. *Orchelimum* species are largely carnivorous, and the aptly named Gladiator Meadow Katydid has been observed attacking prey-laden paper wasps (*Polistes*) to expropriate their masticated prey. ❷ Male **Nimble Meadow Katydids** (***Orchelimum volantum***) have distinctively tapered cerci (the yellow processes at the tip of the abdomen). This relatively scarce northeastern species was photographed in a southern Ontario marsh, near the northern extent of its range. ❸ ❹ The **Short-winged Meadow Katydid** (***Conocephalus brevipennis***) is the most common short-winged *Conocephalus* in eastern North America. *Conocephalus* species in general are more slender and generally smaller (less than 17 mm, excluding the ovipositor) than the only other genus of meadow katydids (*Orchelimum*, more than 18 mm, excluding the ovipositor), and the Short-winged Meadow Katydid is smaller than similar species of *Conocephalus.* It is normally found near water, where it injects its eggs into grass stems. These photographs show a male singing, and a female with a long ovipositor. ❺ The **Slender Meadow Katydid** (***Conocephalus fasciatus***) is the most common of our fully winged *Conocephalus* species, and the only small, long-winged species. The maximum body size, excluding the bladelike ovipositor, is 14 mm. This species is common in a variety of habitats across the continent, from Mexico north to northern Ontario. ❻ The aptly named **Black-sided Meadow Katydid** (***Conocephalus nigropleurum***) is distinctive for the black-sided abdomen. This relatively large species occurs in marshes in the northeastern states, southern Ontario and Quebec. Females lay their eggs in the common willow "cone" galls induced by the midge *Rhabdophaga strobiloides.*

❶ ❷ ❸ ❹ ❺ ❻

❼ ❽

FAMILY TETTIGONIIDAE Subfamily Conocephalinae **(continued)** ❶ This female **Long-tailed Meadow Katydid** (***Conocephalus attenuatus***) has a spectacular ovipositor almost twice as long as her body. This relatively large species occurs in marshes in the northeastern states, southern Ontario and Quebec. Like some other *Conocephalus*, it occurs in both fully winged and short-winged forms. Subfamily Copiphorinae (CONE-HEADED GRASSHOPPERS) ❷ ❸ ❹ Although ***Neoconocephalus ensiger***, the aptly named **Sword-bearer Conehead** (look at the female's ovipositor) is our most common conehead, it is well camouflaged and hard to find. Listen for the distinctive, loud and rapid stridulation of male Sword-bearer Coneheads amongst tall grasses or weeds on late summer evenings. Like Oblong-winged Katydids (*Amblycorypha oblongifolia*), Sword-bearers sometimes occur as a striking pink morph. ❺ ***Neoconocephalus lyristes*** is a wetland species which has the underside of the "conehead" all black, in contrast with the partially black "conehead" of the common *N. ensiger*. The larger *N. robustus*, with little or no black on the cone, is the only other species of the genus to reach southern Canada, although several other *Neoconocephalus* occur in the eastern U.S.A. ❻ ***Belocephalus subapterus***, the **Half-winged Conehead,** is a southeastern species ranging north to North Carolina. Other short-winged coneheads (genus *Belocephalus*) occur in Florida and Georgia.

SUBORDER CAELIFERA

FAMILY ROMALEIDAE (LUBBER GRASSHOPPERS) ❼ **Eastern Lubber Grasshoppers** (***Romalea microptera = R. guttata***) are the familiar fat grasshoppers used in high school and university teaching laboratories. Although its native range extends only as far north as North Carolina, the Eastern Lubber Grasshopper sometimes shows up as an adventive (introduced but not established) as far north as Canada. The bright color pattern is a warning to predators, since lubber grasshoppers contain toxic substances and are able to exude a foamy defensive spray from the thorax. These are nymphs. ❽ Lubber grasshopper adults are large, short-winged insects that range in color from yellow to black. Like the common spur-throated Acrididae, lubber grasshoppers have a "spur" or tubercle (the pale, pointed structure behind and below the head) sticking down between the front legs.

FAMILY ACRIDIDAE Subfamily Melanoplinae (SPUR-THROATED GRASSHOPPERS) ❶ ❷ The **Grizzly Grasshopper** or **Pinetree Spurthroated Grasshopper** (***Melanoplus punctulatus***) is one of the few instantly recognizable species in the large and difficult genus *Melanoplus*. It occurs in isolated colonies in association with pine trees, and adults are usually seen on the tree trunks or in pine foliage. As can be seen in the second photograph, eggs are laid by inserting the entire abdomen into existing holes (usually beetle burrows) in dead wood. ❸ ***Melanoplus bivitattus***, one of the largest and most recognizable species of *Melanoplus*, is known in the United States as the **Two-striped Grasshopper** and in Canada as the **Yellow-striped Grasshopper**. This large (about 30 mm), brightly colored insect is common throughout our area and most of the rest of the continent, where it is a significant pest of various crops including corn and tobacco. ❹ The **Differential Grasshopper** (***Melanoplus differentialis***) is a very large (28–44 mm) insect usually recognizable by the yellowish hind femur marked by a distinctive "herringbone" pattern. This photograph was taken in southern Ontario, near the northern limit of the species range. Differential Grasshoppers feed on a variety of crops and ornamental plants, and are considered significant pests. ❺ The most common grasshopper through much of our area is the **Red-legged Grasshopper** (***Melanoplus femurrubrum***), so-called because the tibia is bright red. The grasshopper in this picture has just defecated, and is about to flick the pellet of frass away with a kick of a powerful hind leg. Red-legged grasshoppers eat a variety of plants, including some crops. ❻ Many ***Melanoplus*** are variable in color, and two of our most common species (***M. femurrubrum***, shown here, and *M. sanguinipes*) have an occasional pink form. ❼ The **Migratory Grasshopper** (***Melanoplus sanguinipes***) is abundant across the continent, and is one of our most important pest grasshoppers. This species can build up impressive densities during outbreak years, when they can be devastating to cereal crops. ❽ The **Broad-necked Grasshopper** (***Melanoplus keeleri***) is found in open sandy areas late in the season. ❾ **Dawson's Grasshopper** (***Melanoplus dawsoni***) is easily recognizable because of the black and white pattern on the abdomen.

FAMILY ACRIDIDAE Subfamily Melanoplinae **(continued)** ❶ ***Paroxya hoosieri*** (the **Hoosier Locust**) is a wetland grasshopper that occurs in small local populations in the midwestern states and into southern Ontario. Two other, longer winged, *Paroxya* species occur in the eastern United States. ❷ **Walsh's Grasshopper** (***Melanoplus walshii***) is an uncommon species with narrow, short wings. It occurs from Ontario south to Georgia and, at least in the northern part of its range, is associated with oak savanna or grassland habitats. ❸ The **Forest Grasshopper** (***Melanoplus islandicus***) is one of many short-winged species of *Melanoplus*. This copulating pair was photographed on Ontario's Bruce Peninsula. ❹ Many species of ***Melanoplus*** are brachypterous (short-winged) and their limited dispersal ability is often reflected in very small ranges. Most of the fourteen very similar species of the *M. viridipes* group occur in the Appalachian Mountains. This is ***M. pachycercus***. ❺ **Hebard's Green-legged Grasshopper** (***Melanoplus eurycercus***) is a short-winged grasshopper characteristic of open woodlands. This species, which occurs from southern Canada to the Carolinas, was previously treated as a subspecies of *M. viridipes*. ❻ **Scudder's Short-winged Grasshopper** (***Melanoplus scudderi***) is a short-winged species characteristic of grasslands. Subfamily Podisminae ❼ ❽ ❾ Our two very similar species in the genus ***Booneacris*** are entirely wingless. These photos show a brightly colored male, a duller green female and a mating pair of the **Wingless Mountain Grasshopper** (***B. glacialis***). Despite the common name, this species is found throughout much of eastern Canada and Maine, New Hampshire, Vermont, Massachusetts and New York. It can be quite common in blueberry heaths.

FAMILY ACRIDIDAE (continued) Subfamily Cyrtacanthacridinae (BIRD LOCUSTS) ❶ The genus ***Schistocerca***, which includes the famous desert locust of Africa, includes four northeastern species of medium to large-sized (up to 70 mm), strong-flying grasshoppers. ❷ The **American Bird Grasshopper** (***Schistocerca americana***) ranges from southern Canada all the way to Argentina, and is among our largest grasshoppers at 40-55mm. Subfamilies Gomphocerinae and Acridinae (SLANT-FACED OR TOOTH-LEGGED GRASSHOPPERS) ❸ ❹ Like many other grasshoppers, the **Short-winged Green Grasshopper** (***Dichromorpha viridis***) shows considerable variation in size and color. One of these photographs shows a green female; the other shows a brown female mating with a green male. These photos were taken in southern Ontario, near the northern limit of this species range. ❺ ***Mermiria intertexta*** is a distinctive slant-faced grasshopper found along the Atlantic and Gulf coasts of the United States. ❻ The **Pasture Grasshopper** (***Orphulella speciosa***) is another species that can vary from green to brown. ❼ The **Marsh Meadow Grasshopper** (***Chorthippus curtipennis***) is the most common slant-faced grasshopper in northeastern North America, and one of the most widespread of all North American grasshoppers. ❽ The **Short-winged Toothpick Grasshopper** (***Pseudopomala brachyptera***) is a distinctive species, easily recognized by its long, strongly slanted face. Look for them clinging to grass blades or stems, but if one sees you first it will quickly sidle around to the other side of the stem to avoid detection. ❾ The **Sprinkled Grasshopper** (***Chloealtis conspersa***) can usually be recognized by the distinctive dark patch on the side of the male's thorax. A grass feeder like most slant-faced grasshoppers, this is one of the few grasshoppers that lays eggs in rotting wood. It occurs across the continent.

FAMILY ACRIDIDAE (continued) ❶ The **Cow Grasshopper** (***Chloealtis abdominalis***), so called because it sometimes lays eggs in cattle droppings, is a relatively northern species that occurs across Canada and the northern United States where it feeds on sedges and grasses. **Subfamily Oedipodinae (BAND-WINGED GRASSHOPPERS)** ❷ ❸ These two photographs of the aptly named **Coral-winged Grasshopper** (***Pardalophora apiculata***) show how the most striking features of this and many other members of the band-winged grasshopper subfamily are hidden at rest. The brown goop seen on the fingers in this photograph is grasshopper vomit; many grasshoppers regurgitate foul fluids as a defensive strategy. Coral-winged Grasshoppers occur across the continent and throughout the northeast from northern Ontario to North Carolina. Adults occur in spring. ❹ The **Orange-winged Grasshopper** (***Pardalophora phoenicoptera***) is a large grasshopper with bright orange hind wings and a distinctive flash of color on the inner surface of the hind legs. It occurs in the eastern states from Pennsylvania to Florida. ❺ ***Psinidia fenestralis*** is sometimes called the **Long-horned Grasshopper**, since its antennae are longer than other Acrididae. Look for it in open sandy areas from southern Quebec to Florida. ❻ The **Dusky Grasshopper** (***Encoptolophus sordidus***) occurs from southern Canada to the Carolinas, and can be found in weedy fields, roadside ditches and similar grassy habitats. ❼ The **Mottled Sand Grasshopper** (***Spharagemon collare***) occurs across the continent, and abounds on sand dunes around the Great Lakes and the Atlantic coast. *Spharagemon* is distinguished from similar sand grasshoppers by the strong ridge, or carina, on top of the thorax, which is cut across by a single deep groove or notch, and by the red or orange hind tibia. The most similar band-winged grasshoppers common on eastern sand dunes belong to the genus *Trimerotropis*, in which the ridge on top of the thorax is weak and cut across by two grooves. ❽ **Boll's Grasshopper** (***Spharagemon bolli***) occurs in more vegetated habitats than the similar Mottled Sand Grasshopper, from which it differs in hind tibial coloration. ❾ The **Northern Mottled Grasshopper** (***Spharagemon marmorata***) occurs from southern Canada to the Carolinas, and favors open, sparsely vegetated soils where its mottled colors blend in with lichens and mosses.

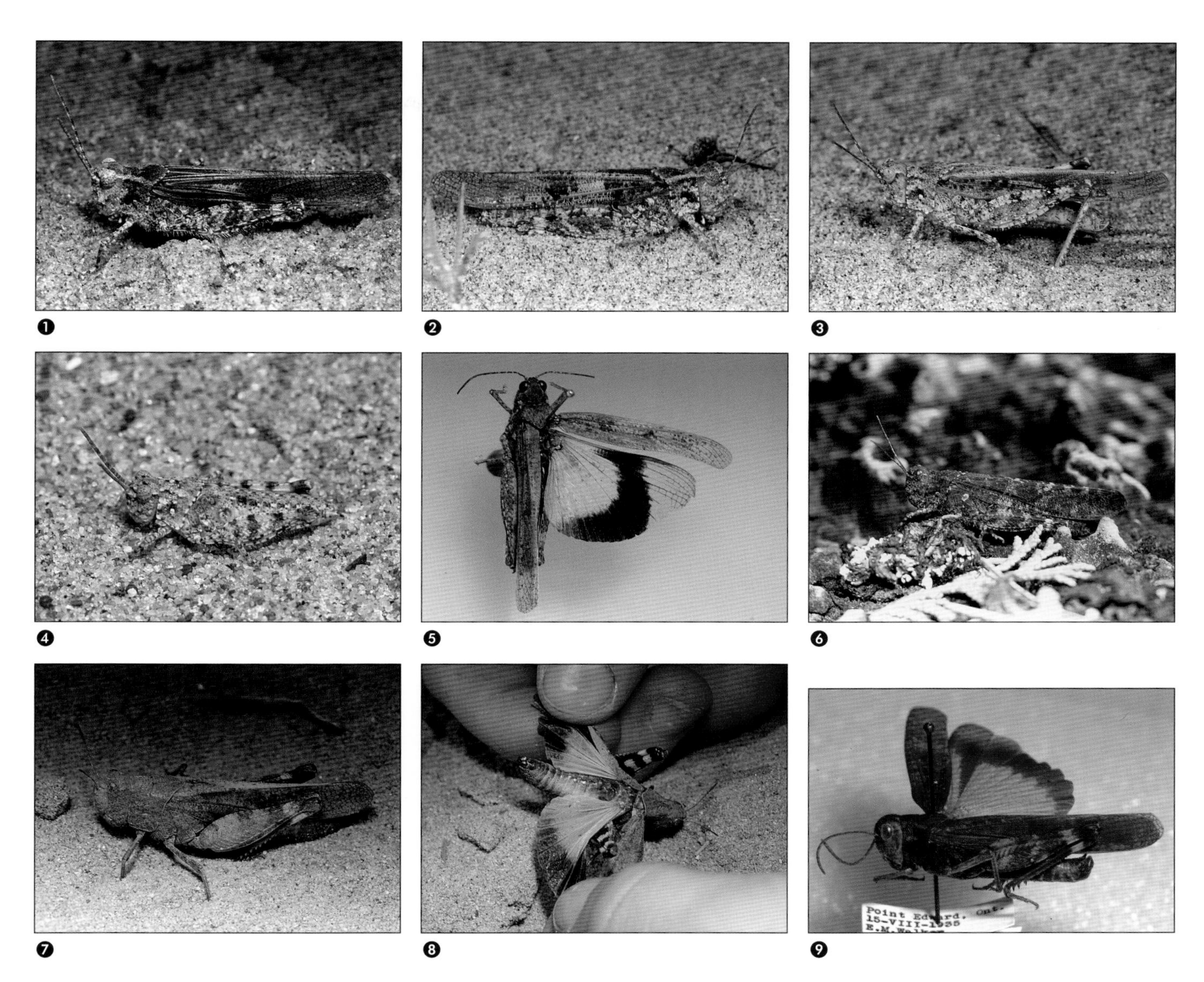

FAMILY ACRIDIDAE Subfamily Oedipodinae **(continued)** ❶ ❷ The **Lake Huron Locust** (***Trimerotropis huroniana***) is neither a locust nor a Lake Huron endemic, but it is a threatened species restricted to a relatively few high quality sand dunes along the shorelines of central Lake Huron and Lake Michigan. Along the southern shores of Lake Huron it seems to be replaced by the very similar Seaside Locust (*T. maritima*), a species that also occurs on the Atlantic coast. These two photographs show a female and a male (smaller and more mottled) Lake Huron Locust, photographed on Manitoulin Island. ❸ ❹ These two photographs show a female and a well-concealed nymph of the **Seaside Locust** (***Trimerotropis maritima***), a species (sometimes divided into two subspecies) that occurs both on Atlantic and Great Lakes shorelines. To distinguish this species from the less common Lake Huron Locust (*T. huroniana*) observe the groove along the bottom of the femur (the groove that accepts the tibia when the leg is folded). It is yellow with black bands in this species and mostly black in the Lake Huron Locust. ❺ Like most band-winged grasshoppers, the **Seaside Locust** (***Trimerotropis maritima***) has brightly banded hind wings. Those of the Lake Huron Locust (*T. huroniana*) have a paler yellow base. ❻ ***Trimerotropis*** is a mainly western genus, with only three northeastern species. This is the **Cracker Grasshopper** (***T. verruculatus***), so named because of the crackling noise called crepitation, made by expansion and contraction of the hind wing folds in flight. Males put on flight displays in which they hover, dart and hover again, all the while making a particularly loud snapping crepitation. These relatively dark grasshoppers are common on exposed, rocky places like alvars, in contrast to the sandy habitats preferred by other eastern *Trimerotropis* (which also crepitate). ❼ ❽ The **Spring Yellow-winged Grasshopper** (***Arphia sulphurea***), aptly named for its May–June period of activity and its bright yellow hind wings, is common in open woodlands or other sparsely treed habitats. Species of *Arphia* differ from other band-winged grasshoppers in that the ridge (median carina) on the top of the prothorax is simple and not cut by a notch, or sulcus. Our other common *Arphia* species, the Red-winged Grasshopper (*A. pseudonietana*), occurs later in the season. ❾ The **Red-winged Grasshopper** (***Arphia pseudonietana***) has a mostly western distribution, but also occurs in grasslands in Ontario, Michigan and Ohio.

FAMILY ACRIDIDAE Subfamily Oedipodinae **(continued)** ❶ ❷ Although it sounds like a contradiction of terms, the **Clear-winged Grasshopper** (***Camnula pellucida***) is a band-winged grasshopper that does not have banded hind wings. This small (17–29 mm) grasshopper is distinctively pigmented nonetheless, and the general coloration of the resting grasshopper, including a pale stripe along the side of the front wings, makes it fairly easy to recognize. This species is abundant across the continent, and is a major pest of cereal crops. ❸ ❹ ❺ These three photos show green and brown forms of one of the first grasshopper species to appear in spring. The **Northern Green-striped Grasshopper** (***Chortophaga viridifasciata***) spends the winter as a partly grown nymph, in contrast to most grasshoppers, which overwinter as eggs. This is a common species in grassy areas throughout most of the United States and southern Canada. ❻ ❼ By far the most common and familiar of the band-winged grasshoppers, the **Carolina Locust** (***Dissosteira carolina***) is neither a locust nor Carolinian, which may explain why the official (although less widely used) Canadian name for this species is the **Black-winged Grasshopper**. The hind wing is mostly black, with a narrow yellow border. Carolina Locusts abound on roadsides and waste places across the continent, even in urban playgrounds. ❽ The **Striped Sedge Grasshopper** (***Stethophyma lineatum***) is a large (26–36 mm) and strikingly colored grasshopper usually found in wet bogs and sedge meadows across Canada and in the northeastern states. ❾ The **Graceful Sedge Grasshopper** (***Stethophyma gracile***) is a sedge-feeder, usually found in fens. It occurs across Canada and in the northern United States. Some authors put this genus in the Acridinae.

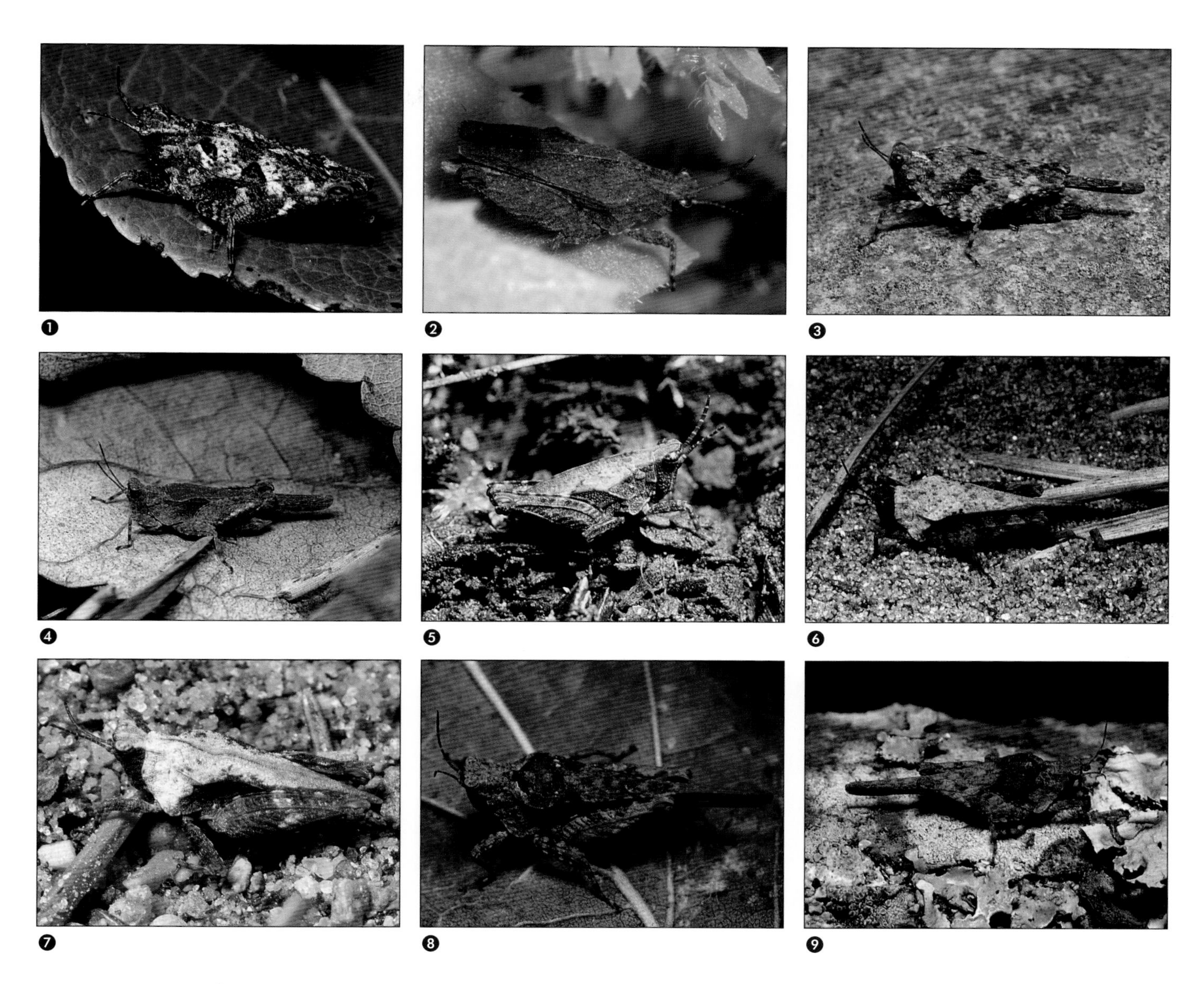

FAMILY TETRIGIDAE (PYGMY GRASSHOPPERS OR GROUSE LOCUSTS, 14–18 mm) ❶ ❷ The **Crested Grouse Locust** (***Nomotettix cristatus***), the only eastern *Nomotettix*, has an elevated keel along its back that give it a distinctive arched appearance. It is common in a variety of habitats in the eastern United States and southern Canada. ❸ ❹ The **Hooded Grouse Locust** (***Paratettix cucullatus***), our only *Paratettix*, is easily recognized by the flat-fronted head that does not extend in front of the eyes as in other grouse locusts. Like other Tetrigidae, it is variable in shape and color; two color forms are illustrated here. ❺ ❻ The four species of ***Tetrix*** found in the northeast, especially the very common **Ornate Grouse Locust** (***T. ornatus***), are almost infinitely variable in color, and are usually separated on the basis of small differences in the shape of the front of the head. ❼ ❽ ❾ ***Tetrix*** species exhibit a variety of color forms, as seen in ***T. arenosa***.

FAMILY TETRIGIDAE (continued) ❶ ❷ The **Slender Grouse Locust** (***Tetrix subulata***) is a common species found in a variety of habitats in Canada, the northern States, Europe and Asia. ❸ ❹ The **Black-sided Grouse Locust** (***Tettigidea lateralis***) is easily separated from other pygmy grasshoppers by the high number (over 14) of antennal segments, and of the way head seems pulled back into the thorax almost to the middle of the eyes. Some entomologists put *Tettigidea* in a separate family, the Batrachideidae, with two species in the northeast. They are common on a variety of soils. ❺ The **Armored Grouse Locust** (***Tettigidea armata***) has a more southerly distribution than the Black-sided Grouse Locust (*T. lateralis*), barely ranging north to the Canadian border. **FAMILY TRIDACTYLIDAE** (PYGMY MOLE CRICKETS) ❻ ***Ellipes minutus***, the **Minute Pygmy Mole Cricket** (4–5 mm), occurs throughout our area on wet sandy margins of ponds or slow-moving rivers. ❼ This **Larger Pygmy Mole Cricket**, ***Neotridactylus apicalis*** (5.5–6.5 mm), was found along a small stream flowing into a Lake Erie beach, where it was initially noticed with just the front part of its body sticking out from its burrow in the sand. Pygmy Mole Crickets normally stay in their burrows, even laying eggs in brood chambers at the ends of the burrows, but they can fly, swim and jump to avoid capture once flushed. This species ranges from southern Ontario and Quebec through to South America. ❽ As you can see on this pinned ***Neotridactylus***, pygmy mole crickets are well equipped for digging (forelegs) and swimming (hindlegs).

CHAPTER 6

True Bugs and Other Hemipteroids

The term "bug" is often used loosely to refer to any little invertebrate that creeps, crawls, or otherwise transports its spineless body around on any number of appendages, but the proper entomological use of the term refers only to a single group of insects equipped with syringe-like beaks used to suck the liquefied contents out of other organisms. The segmented beak you see when you look closely at a member of this group (the order Hemiptera) is the lower lip (labium), which is extended to form a long, trough or gutter-like, structure. Think of it as a sword-sheath containing not one blade but four. The four blades derive from the usual mouthparts (mandibles and maxillae) found between the upper lip (labrum) and lower lip (labium) of insects. In bugs, these mouthparts form four narrow blades (stylets) that fit into the long, sheath-like lower lip. One pair of blades (mandibles) is for cutting, and the other pair (maxillae) is for spitting and sucking. The maxillae combine to make a tube with two channels, one for sucking food up, and the other for spitting saliva back into the food. The slender, syringe-like stylets leave the bug's labial sheath when the mandibles and maxillae slide against each other to penetrate prey or plant tissue. The labium is sometimes bent back as the stylets snake their way into the host.

This kind of beak is characteristic of the order Hemiptera, an order divided into two easily recognized groups: (a) the true bugs (suborder Heteroptera; sometimes referred to using the term "Hemiptera" in a restricted sense) and (b) the homopterans (traditionally treated as the suborder Homoptera), including aphids, planthoppers, cicadas and others. The differences between these two groups are reflected in their traditional scientific names. "Homoptera" literally means "uniform wings"; these insects have homogeneous or uniformly membranous wings, usually angled alongside the body rather than flat above it. "Heteroptera," on the other hand, means "different wings," and refers to true bug forewings, which have basal, leathery halves and contrasting membranous tips. True bug wings fold flat against the top of the body, and the front wings usually overlap one another.

True bugs and homopterans also include many wingless species, but characters other than wings can be used to easily separate these groups. True bugs usually have a prominent beak that arises from the front of the head and hangs like a sheathed sword swung back under the head and body. In the homopterans, the long, thin beak usually arises from the back of the head and is sometimes held so tightly against the body it looks like a thread arising between the front legs.

New work on the phylogeny and classification of sucking

insects confirms that the true bugs and homopterans belong in a single order (Hemiptera), now broken into three or four suborders (plus another rare suborder found in the Southern Hemisphere). One of these suborders is the old familiar group of true bugs with half-leathery wings (the suborder Heteroptera); the other suborders are groups that used to be in the Homoptera. Almost everyone agrees that aphids, whiteflies, scales and their relatives form one of the homopteran suborders, called Sternorrhyncha. The remaining "homopterans" are treated here as the other suborder (Auchenorrhyncha) including the cicadas and relatives (superfamily Cicadoidea) and the planthoppers (superfamily Fulgoroidea), although some authors treat the cicada and planthopper superfamilies as separate suborders (Clypeorrhyncha and Archaeorrhyncha). No matter what you call them, the main groups of Hemiptera – true bugs, aphids and their relatives, cicadas and their relatives and planthoppers – are distinctive and common groups, well worth looking at in turn. Let's start with the true bugs (Heteroptera), dividing them for convenience into the land bugs, water bugs and shore or surface bugs.

True Bugs I: Land Bugs

The first entomological usage of the term "bug," probably from the 17th century, was for normally hidden insects that slipped on and off human hosts during the night, leaving itchy welts as evidence of their visits before returning to cracks and crevasses during the daylight hours.

Human Bed Bugs (*Cimex lectularius*) have been pestering us for a long time, and have probably tracked humankind's spread over the planet from the time we first holed up in some cozy cave. These flattened, little (about 5 mm) bloodsuckers have been relatively rare in Europe and North America for most of the last 50 years, but they were very common before that, and bed bugs are once again common in some surprising places. One recent survey found them in 43 percent of dwellings in old London (England) with as many as 4,000 to 5,000 bugs per household, and North American pest control companies from coast to coast have been reporting an impressive resurgence in bed bug problems, especially in hotels, since the late 1990s. Bed bugs probably first arrived in the New World on sailing ships with European colonists, and the new wave of bed bug problems likely reflects today's growth of international migration along with international movement of bedding, luggage and furniture that may harbor bugs or their eggs.

Remarkably, bed bugs are not proven carriers of disease organisms, although they are suspected of carrying leprosy, oriental sore and a variety of other diseases, and one researcher showed that HIV could survive for an hour in a bed bug. I would not lose any sleep worrying about getting diseases from bed bug bites, even though sleep deprivation in buggy places can be a problem, and frequent bites can lead to iron deficiencies.

Although a couple of species attack people, members of the bed bug family (**Cimicidae**, page 112) generally feed on bat or bird blood that they digest with the aid of microorganisms maintained in special pockets in the digestive tract. One species is common in swallow nests, and others inhabit caves where they feed on bat blood. Their disgusting (from the human perspective) habits extend beyond their culinary preferences and into their sex lives, characterized by apparently violent copulation known, somewhat anthropomorphically, as traumatic insemination. The bed bug male has a huge, swordlike penis with which he rips a hole in the abdomen of the female, injecting sperm and nutritive seminal fluid right into her body cavity. The female has a spongelike organ that soaks up the sperm, where it is stored until the female has imbibed the blood needed for her eggs to mature. Traumatic insemination is found elsewhere in the bug world (other Cimicoidea, including damsel bugs and minute pirate bugs), but few bugs are as well studied as bed bugs.

Stink Bugs and their Relatives (Pentatomoidea)

Shield-shaped stink bugs (**Pentatomidae**, page 112) are common sights on exposed surfaces such as stems and leaves, and if you pick up one of these conspicuous insects you will soon see, or rather smell, why such foul-smelling bugs have little need to conceal themselves from potential predators. The odoriferous aspect of most adult stink bugs can be attributed to scent glands that open on the bug's thorax, discharging nasty smelling compounds over an elaborately sculptured, wicklike surface near the hind legs. Immature bugs usually have different scent glands, which open on the abdomen. The scent glands produce compounds that often serve as a kind of chemical alarm for the insect's compatriots, as well as deterring would-be predators. Some bugs also use products from these characteristic glands to attract mates, and other gland secretions serve to inhibit harmful microorganisms, but the main function of bug glands seems to be the production of defensive chemicals. Many bugs stink, but stink bugs are particularly pungent and characteristically conspicuous. The Pentatomidae is the third largest family of true bugs after the Miridae and Reduviidae, although the 200 North American species make up only a tiny fraction of the over 4,100 world species.

About a third of our common stink bugs are predators (often beneficial) of caterpillars and other insects, and about two-thirds are plant feeders (sometimes pests), but some species feed on both plant and animal tissue or start life as plant feeders and later shift to predation. Predacious stink bugs belong to one subfamily (Asopinae), in which the first segment of the beak is fat and directed away from the head. Plant-sucking stink bugs have a relatively thin beak, pressed against the underside of the head.

Although the plant-sucking stink bugs include several significant pests, such as the Southern Green Stink Bug (*Nezara viridula*) and the Harlequin Bug (*Murgantia histrionica*), their impact as pests is more than counterbalanced by beneficial predacious species, such as the common spined soldier bugs (*Podisus* spp.) often seen sitting on flowers or leaves with impaled caterpillars suspended from their outstretched beaks. *Podisus* and a similar, recently introduced species (*Picromerus bidens*), with even bigger shoulder spines, are major predators of webworms and tent caterpillars, which they impale right through their silken shelters. Some predacious stink bugs also have gluttonous appetites for garden pests, and one pretty red and black species, the Two-spotted Stink Bug (*Perillus bioculatus*), is a major predator of Colorado Potato Beetles. Each Two-spotted Stink Bug nymph can suck the blood (hemolymph) out of hundreds of potato beetle eggs, and an adult stink bug can destroy 150 to 200 potato beetle larvae over the course of its life.

Stink bugs are part of a group of related families collectively known as the superfamily Pentatomoidea; the "penta" part of the name refers to the five-segmented antennae found in this group (most other terrestrial bugs have fewer antennal segments). Most pentatomoids are in the family Pentatomidae, and most pentatomids are stink bugs, but there are a few exceptions. Shield bugs (**Scutelleridae**, page 117) and negro bugs (**Thyreocoridae**, page 117), for example, have shell-like coverings because part of the thorax extends right back over the abdomen like a turtle's shell. These bugs sometimes look a bit like beetles, but they lack the beetle-like line where the wing covers meet. Burrower bugs (**Cydnidae**, page 118) look like black (or brown) stink bugs with conspicuously robust and spiny legs used to burrow deep into the ground where they feed on roots.

The eggs of stink bugs and their relatives are usually attractive, barrel-shaped structures laid in clusters stuck to each other and to flower heads or leaves by a sticky substance. A few species guard the eggs until they hatch, and even take care of the young nymphs. Care of the young is reflected in the common name "parent bugs" given to members of the small family **Acanthosomatidae** (page 117) that feed on birch catkins. Female parent bugs lay their eggs away from the catkins, then later lead the nymphs to the catkin food source, like a mother duck leading a column of ducklings to water.

Assassin Bugs, Ambush Bugs, Damsel Bugs, Lace Bugs and Flat Bugs

Assassin bugs (**Reduviidae**, page 118) are characterized by a wicked-looking beak conspicuously slung from the front of the head and partly sheathed in a special groove between the front legs. Depending on your inclination and the weather, a good place to start looking for assassin bugs would be in your kitchen light fixture. Unless you are much more fastidious than most, the odds are that among the crispy critters accumulated there you will find a large black assassin bug called the Masked Bed Bug Hunter (*Reduvius personatus*). The "hunter" part of its name refers to the propensity of this

Masked Bed Bug Hunter (*Reduvius personatus*), page 118, caption 6.

creature to seek and destroy a variety of little arthropods that end up in the corners of your household (more likely to be spiders and fruit flies than bed bugs), while the "masked" part of its name refers to the odd habits of the nymphs of this species. Nymphs exude a sticky substance that soon accumulates a body-masking layer of junk ranging from the dead bodies of victims to dismembered dust bunnies. Next time you spot some wandering lint in the corner of your living room, take a closer look to see if it's a Masked Bed Bug Hunter nymph.

Although most assassin bugs are beneficial predators with no inclination to bite people, the Masked Bed Bug Hunter, like some of its southern relatives, has to be handled with care lest it painfully impale you with its bug-sucking beak. A few species of Reduviidae inflict painless bites (in contrast with the bee-sting-like defensive bites of Masked Bed Bug Hunters) because they are specialized surreptitious bloodsuckers rather than blatant invertebrate assassins. Bugs in this group (subfamily Triatominae, the kissing bugs or bloodsucking conenoses) can be dangerous because they often harbor one-celled organisms called trypanosomes, similar to those that cause sleeping sickness in Africa.

American trypanosomiasis, or Chagas' disease, affects millions of people in South and Central America. This serious illness is sometimes known as Darwin's disease because Charles Darwin's infamous ill health may well have begun with an encounter with a kissing bug in South America. Darwin did write about the painless bite of the "giant black bug of the pampas," but if he contracted American trypanosomiasis it was from the feces rather than the bite of the bug. Infection takes place when trypanosome-filled bug feces are scratched into broken skin or rubbed into mucus membranes of the eye, nose or mouth. A couple of species of *Triatoma* found in houses as far north as Pennsylvania and Ohio do sometimes carry Chagas' disease and do sometimes bite people, but because they usually do not defecate on human hosts they rarely transmit the disease to people. Northeastern *Triatoma* can pick up the disease organism by feeding on wood rats, about 15 percent of which carry *Trypanosoma cruzi*.

The Reduviidae is divided into several subfamilies, most of which are either called assassin bugs (predacious species in several subfamilies) or kissing bugs (subfamily Triatominae). Members of the subfamily Emesinae are slender, inconspicuous bugs called thread-legged bugs. Although rarely noticed, thread-legged bugs are sometimes common among grass thatch, in old buildings and in trees, where they feed on small, soft-bodied insects.

Ambush bugs (**Phymatidae**, page 122) are bright black and yellow, squat, hourglass-shaped insects with a spiny thorax and remarkable forelegs with sickle-like tibiae, greatly reduced tarsi and huge swollen femora. If you look closely at fall flowers, especially goldenrod, these phenomenally well-camouflaged bugs are usually abundant, and are often easiest to spot by looking for the dead bees and flies they have recently impaled. Phymatidae is a small family, with only four northeastern species; it is very closely related to the Reduviidae and is sometimes treated as part of that family.

The common name "damsel bug" (family **Nabidae**, page 122) is an interesting choice for a family of aggressively predacious bugs since the term "damsel" hardly brings to mind images of effective predators. The name probably refers to the relatively delicate dimensions of these little insects, because just as damselflies are small predators relative to other Odonata (dragonflies), damsel bugs are small predators relative to the generally larger and more conspicuous, but somewhat similar, assassin bugs. The common damsel bug *Nabis americoferus* is described by Hemiptera specialists as one of the most common hemipterans throughout the United States, and careful examination of fall flowers will generally reveal these gray-brown bugs, often with half-consumed aphids hanging from their beaks. Another member of this small family encountered with regularity in the northeast is the black, antlike *Nabicula subcoleoptrata*. This shiny, usually short-winged, native bug occurs in meadows across the country, where it is said to feed mostly on a common introduced bug, the Meadow Plant Bug (family Miridae).

Lace bugs (**Tingidae**, page 123) are generally small,

A damsel bug, *Nabicula americolimbata*, page 122, caption 7.

elaborately sculptured bugs found in large numbers on the undersides of some kinds of leaves. Birch and basswood leaves, for example, are often speckled and deformed due to clusters of lace bugs, including adults with expanded backs and wings embossed with lacelike patterns, and nymphs bristling with needle-like defensive bristles. Charismatic clusters of multiple nymphal and adult bugs are common and remarkably free of parasites and predators, presumably in part due to the spines and defensive chemicals that characterize the nymphs. Eggs are placed in or on the undersides of leaves. The Tingidae is a fairly large family, with around 2,000 species worldwide, only around 140 of which occur in North America. Most of the strikingly "lacy" lace bugs in the northeast belong to the big genus *Corythucha*, which includes several host-specific species on trees including hawthorn, basswood, birch, walnut, willow, oak and sycamore; one species is a pest of chrysanthemums.

The small family **Piesmatidae** (ash-gray leaf bugs, page 123) is superficially similar to the Tingidae except for the presence of ocelli. Ash-gray leaf bugs are common on pigweed and related plants, and are sometimes serious pests of sugar beets.

The dimensions of the subcortical (under bark) habitat demand that all its denizens be greatly flattened, and members of the subcortical bug family **Aradidae** (flat bugs, page 124) are conspicuously shaped to meet that demand. Flat bugs usually feed on the higher fungi that thrive under the bark of recently killed trees, using long, thin, threadlike stylets that are normally held coiled up into a compact double spiral. During feeding, the stylets uncoil to as much as five or six times the length of the bug. Similar habits are found in the nymphs of some homopterans (planthoppers in the families Achilidae and Derbidae), but fungus feeding is in general an unusual habit for a sucking insect.

Flat bugs, Aradidae.

Although flat bugs occasionally damage pine trees by probing young tissues with their long beaks, aradids are of greatest interest because of their specialized habitat requirements. Because some species require particular types of trees, such as large and old trees, burned trees, particular tree species, particular stages of succession or trees with required fungal associates, many flat bugs are rare or habitat-restricted.

Seed Bugs, Stilt Bugs, Big-eyed Bugs and their Relatives (Superfamilies Lygaeoidea and Pyrrhocoroidea)

It seems that wherever there is something for them to suck on, be it bug blood, people blood, plant sap or even fungi, there are bugs. Some common species are even found in entomological laboratories where they are in popular use as experimental organisms. Milkweed bugs, for example, are frequently used as entomological "lab rats" in insect physiology research labs. These common red-and-black bugs can be easily maintained as long as they are supplied with milkweed seeds and adequate water. Bugs feed with a two-channeled straw, and an insect that sucks the contents out of a dry seed needs ample fluid to squirt into the food through one channel before slurping the resultant slop up the other beak channel.

Milkweed bugs and many other seed-feeding bugs belong to the family **Lygaeidae** (page 124), or seed bugs. Although the laboratory milkweed bug (properly called the Large Milkweed Bug, *Oncopeltus fasciatus*) is a species that doesn't survive the winter in the northern United States and Canada, a similar red-and-black milkweed bug (the Small Milkweed Bug, *Lygaeus kalmii*) is common both north and south of the border. Milkweed bugs, Monarch butterflies and milkweed beetles all wear similar bright colors as a warning to potential predators that they taste bad. All of these milkweed-eating insects sequester poisonous substances from their host plants, so once an animal has tasted one of these bright insects it is not too likely to eat another similar-looking bug.

The family Lygaeidae has been substantially redefined in recent years, and we now split the bugs traditionally included in the Lygaeidae into a number of different families, grouping these newly recognized families along with the stilt bug family (**Berytidae**, page 126) in the superfamily Lygaeoidea. This change is clearly justified on the basis of a close relationship between stilt bugs and some seed bugs, but it does make the identification of bug families considerably more difficult. Instead of one very large, easily recognized family Lygaeidae (characterized by the presence of simple eyes, the absence of a wing notch, and the presence of only four or five wing veins) we now have eight bug families to deal with.

Milkweed bugs and many other seed bugs are still in a more narrowly defined Lygaeidae, but some other familiar "seed bugs" are now in other families. For example, the Hairy Chinch Bug (*Blissus leucopterus hirtus*), a little black and white bug that sucks sap from grass crowns and stems, causing those brown patches on your lawn or at the golf course, is now in the small family **Blissidae** (page 125). (If you want to determine whether or not a lawn has chinch bugs, cut the bottom and top out of an old juice can to form a cylinder you then force into the lawn. Fill this cylinder with water, and then look at all the neat bugs that float to the surface as their grassy lair is flooded.)

Chinch bug adults are distinctive black bugs with reddish legs and characteristic white and black wings. True Chinch Bugs (same species as the Hairy Chinch Bug but a different subspecies) are major grain pests in the Midwest where overwintered adults lay eggs on wheat in spring and a second generation moves to other grains when the wheat matures or when harvest begins. Adults of the second generation fly to clumps of grass or some other sheltered place and hibernate during the winter. Hairy Chinch Bugs have a similar life cycle, but don't migrate from host to host.

Ischnodemus falicus, a grass-feeding bug in the family Blissidae.

Most of the small bugs formerly treated as Lygaeidae are now in the **Rhyparochromidae** (page 125). *Myodocha serripes*, a species with a shiny black head sitting on an impressively long and narrow neck, is perhaps the most distinctively shaped member of the family. This common phytophagous species sometimes damages strawberries. Big-eyed bugs (**Geocoridae**, page 125) are also distinctly shaped, and were also formerly in the Lygaeidae. These common, distinctively bulgy-eyed bugs eat other insects and insect eggs rather than seeds, and are important beneficial insects in various agro-ecosystems. The other families resulting from the break-up of the Lygaeidae are the **Cymidae** (page 125), with one small genus found amongst the seeds of *Carex* and similar plants, and the very small families **Artheneidae** (one introduced species, page 126), **Oxycarenidae** and **Pachygronthidae** (page 126).

The most common northeastern Lygaeoidea that were not formerly part of the family Lygaeidae are the stilt bugs (**Berytidae**, page 126), odd-looking insects that resemble small, spindly legged stick insects. Of course they have sucking mouth-parts like other bugs instead of chewing mouthparts like walkingsticks, and unlike stick insects they have odd antennae swollen in the middle like a spindle. Stilt bugs are often common on mullein plants where they apparently stab and suck the juices out of a variety of targets ranging from the plant itself through to thrips, butterfly eggs and leafhoppers.

The two North American families in the superfamily Pyrrhocoroidea are mostly southern. Red bugs or stainers (**Pyrrhocoridae**, page 129) are mostly tropical but can be common in the southern United States, where one species in the huge genus *Dysdercus* (the Cotton Stainer) is a significant pest. Largid bugs (**Largidae**, page 129) are similar to Pyrrhocoridae and are also mostly southern, but one species ranges as far north as New York State.

Leaf-footed Bugs, Broad-headed Bugs and Scentless Plant Bugs (Superfamily Coreoidea)

Although tropical leaf-footed bugs (**Coreidae**, page 127) often have spectacular flattened and expanded areas on their hind legs, most northeastern species have simple legs that render the common name "leaf-footed bugs" a bit of a misnomer. Some authors refer to Coreidae as the squash bug family, but that is misleading too, since coreids other than the Squash Bug (*Anasa tristis*) feed on a variety of plants. The northeastern coreid fauna is small, and most of our coreid species are dark-colored, robust and easily identified insects.

Leaf-footed bugs are much more abundant in the tropics, where it is not unusual to see colorful coreids with huge flattened tibiae hanging from the body like grossly oversized earrings. Some leaf-footed bugs, like the tropical Passion Vine Bug (*Diactor bilineatus*), have conspicuous, leaflike tibiae in both sexes, which probably serve to deflect predator attention from the bug's body. Amongst some leaf-footed bugs, expanded tibiae only occur on males, and they are probably used for sexual display.

One of the most common large coreids in our area today is the Western Conifer Seed Bug (*Leptoglossus occidentalis*), a western species which first appeared in eastern North America about 1980 and now frequently appears in alarming aggregations as it seeks adequate shelter, such as your porch or garage, for overwintering. The other common bug that often masses on sunny walls as it seeks a place to hibernate is the Box Elder Bug (*Boisea trivittata*), a bright red-and-black bug that also spread east from an originally western North American distribution. Box Elder Bugs feed on seeds of female trees of the Box Elder and sometimes invade houses in huge numbers in late fall in search of a place to spend the winter. Although closely related to leaf-footed bugs, Box Elder Bugs belong to a separate family, called the scentless plant bugs (**Rhopalidae**, page 128) because they lack the stink gland found in related bugs. Most other eastern Rhopalidae are relatively small and inconspicuous.

Broad-headed bugs (**Alydidae**, page 129) are among the many insects that presumably deter predators by pretending to be ants, a pretence especially well perpetrated by the strikingly antlike flightless nymphs. Both adults and nymphs of these large, dark bugs are also protected from predators by defensive chemicals, but the adults produce a particularly pungent stink from the conspicuous gland openings on the sides of the thorax. Broad-headed bugs, named for the long, broad head that is wider than the thorax, feed mostly on legumes, but are common on goldenrod and other flowers in weedy meadows late in the season.

The Rhopalidae, Coreidae and the closely related Alydidae are all plant-feeding bugs characterized by wing membranes crowded with longitudinal veins, in contrast with the superficially similar Lygaeidae, which have only a few wing veins.

Plant Bugs, Pirate Bugs and Unique-headed Bugs

If you make a collection of terrestrial bugs, the odds are that a good half of your smaller specimens will belong to the enormous plant bug family (**Miridae**, page 129). Seen from the side, plant bug wings have a characteristic angular bend where they are abruptly deflexed over the abdomen. The outside of the wing is notched at the bend, demarcating a part of the wing called the cuneus. That wing notch, plus the usual absence of ocelli and the presence of a couple of closed loops in the wing membrane, makes the Miridae easy to recognize as a family even though plant bugs are incredibly variable in color and shape. Miridae is the largest family of true bugs, with around 10,000 world species (about 1,800 in North America) making up almost one-third of all known Heteroptera species. Most plant bugs do indeed feed on plants as the name suggests, but some are scavengers, some are predators and many feed on both plants and other insects. Plant-eating mirids use their beaks plus a cocktail of enzymes to whip plant tissues into suckable soupy slurries, giving these plant-sucking bugs a much more nitrogen-rich diet than that imbibed by sap-sucking homopterans.

Tarnished Plant Bugs (*Lygus lineolaris*) are the most common mirids – and perhaps the most common bugs – in every state and province of North America. In contrast to most plant bugs, which are associated with a narrow range of host plants, these nondescript little (about 5 mm) brown

Tarnished Plant Bug, *Lygus lineolaris* (Miridae), page 130, caption 1.

bugs are recorded from over 300 plant species. Mirid feeding can cause all sorts of undesirable effects such as crinkling, shot-holing, tattering, lesions, wilting, undeveloped flowers, blemished fruit and stem die-back, so they are serious pests. Tarnished Plant Bugs are active in fall until well after the last frost, by which time it is sometimes hard to find a weed or flower that doesn't support a couple of adult bugs. After spending the winter as adults, Tarnished Plant Bugs feed on various plants including fruit trees, where they can cause premature fruit drop and malformed fruit. They are also pests of beans, pepper, canola and other crops.

Although mirids seem to get mostly bad press because of the impact of Tarnished Plant Bugs and other pests, most species are innocuous and some are even beneficial. Some species are specialized predators of insect eggs, lace bugs (Tingidae), thrips (Thysanoptera) or scale insects (Coccoidea). Mirids have even been deliberately moved from place to place as biological control agents, and some have great potential for control of aphids in greenhouses.

Minute pirate bugs (**Anthocoridae**, page 132) differ from seed bugs such as the superficially similar chinch bugs in having the wings notched and bent over the back part of the body and, more importantly, in sucking eggs and small insects rather than seeds. Members of this small family have a voracious appetite for small bundles of living but relatively defenseless proteins such as pollen grains or small insects, and many species are considered beneficial because of their role in eating pests and pest eggs in orchards and cornfields. One species, the Insidious Flower Bug (*Orius insidiosus*), is routinely mass-reared and can be purchased for release against thrips and other greenhouse pests – sort of a living insecticide. Insidious Flower Bugs are small black and white predators commonly found on a variety of flowers.

Members of the small families Lasiochilidae and **Lyctocoridae** (page 132), until recently treated as Anthocoridae, are small predators usually found on the ground. Some lyctocorids occur in bird and mammal nests, and occasionally dine on vertebrate blood.

Gnat bugs or unique-headed bugs (family **Enicocephalidae**, page 132) form another small, rarely encountered family. These odd-looking insects have a long head and raptorial (grasping) front legs used to prey on other insects under bark or under stones. The common name "gnat bug" refers to the propensity of these little (about 4 mm) bugs to form large swarms like some small flies (gnats). Swarming gnat bugs are commonplace in cloud forests of Central and South America, but are rarely seen in North America.

Gnat bugs and a couple of other obscure and rarely noticed group of tiny bugs, the **Ceratocombidae** (page 132), Dipsocoridae and Schizopteridae, are of special interest as the most basal (primitive) lineages amongst the North American true bugs, with many primitive features such as relatively uniform wings. Ceratocombidae are tiny (less than 2 mm) bugs that prey on small organisms in ground litter and decaying wood, with a single widespread eastern species (*Ceratocombus vagans*) sometimes seen on small fungi or in rotting wood. Dipsocorids and schizopterids are mostly tropical families of minute, ground-living bugs, but one species of beetle-like schizopterid ranges north to Maryland. One further bug family, the Microphysidae, includes three tiny ground-living species in eastern North America, of which two are introduced species recorded in North America only from the east coast of Canada.

True Bugs II: Water Bugs

The odds are good that your childhood memories include "pond aquaria" full of aquatic insects captured during exciting expeditions to neighborhood ditches and ponds. You might recall, then, how the dragonflies, mayflies and caddisflies confined to your jam-jar aquaria soon succumbed

Belostoma flumineum male with hatching eggs, page 133, caption 3.

due to depletion of oxygen in their confined quarters, leaving true bugs such as water scorpions as the most conspicuous survivors. This, of course, occurred because aquatic true bugs are not dependent on absorbing oxygen from the water. All North American aquatic Hemiptera (all of which are in the suborder Heteroptera) are able to obtain air at the surface using tubes (water scorpions), flaps (giant water bugs), the tip of the abdomen (backswimmers) or the thorax (water boatmen).

Unlike most other aquatic insects, aquatic Heteroptera feed, respire and swim in much the same fashion throughout their lives, although adult water bugs are usually winged and capable of an occasional aerial excursion. The ability to fly as an adult is important because it permits dispersal from pond to pond, and allows escape from deteriorating conditions such as those found in a dried up pond (or a neglected aquarium). Water bugs on the wing seem to be most common in early spring, right after they have spent the winter as inactive adults. Flying aquatic bugs are sometimes attracted to lights, and concerned citizens frequently show up at entomology laboratories bearing specimens of 50–60 mm long giant water bugs (usually a species aptly called the Giant Electric Light Bug, *Lethocerus americanus*) that they have found floundering around on brightly lit tennis courts or patios. Giant water bugs have huge, grasping forelegs and a stout beak, so people who bring in these intimidating bugs are relieved to find that they are not dangerous, except to small fish or large aquatic insects.

Superficially, giant water bugs live like most other water bugs. They take air at the surface using a pair of short, flaplike structures at the tip of the abdomen, they swim quite well and they are voracious predators. Water bugs are frequently seen with frogs, fish or other prey impaled upon the beak, which serves as both deadly syringe to inject toxic saliva and a drinking straw through which to imbibe the prey's dissolved body contents. **Belostomatidae** (page 132), the giant water bug family, includes two common genera in the northeast. The widespread Giant Electric Light Bug is typical of one genus, and lays masses of eggs on stones, plants, planks or logs in wet places near the ponds and slow streams in which the nymphs and adults live. Asian *Lethocerus* bugs are famous as food flavorings and have been described as being to Vietnamese cuisine what the truffle is to French cuisine. The characteristic flavor comes from glands found on the bug's third thoracic segment, which produce an apparently delicious chemical (trans-hex-2-enyl acetate). North American *Lethocerus* produce exactly the same chemical, thought to function in nature as a marker used by male bugs to relocate and protect their own eggs.

Other belostomatids (the widespread genus *Belostoma* and the southern and western genus *Abedus*) are about half the size of the Giant Electric Light Bug and have even more remarkable life histories. Instead of sticking her eggs on inanimate objects, the female glues them to the back of a male. The male is then stuck with this dense mat of eggs on his back, eggs that would die without his movements keeping them moist, ventilated and protected. One must suppose that the male benefits enough from improved survival of his offspring to offset the disadvantage of having to feed, hunt and dodge predators with a bulky egg mass on his back. This would be a bad deal indeed for a cuckolded male giant water bug, and the male mates repeatedly (sometimes more than a hundred times) to ensure that only his own sperm fertilizes the eggs to be laid on his back. The eggs hatch in about 10 days, with each nymph emerging from a flaplike lid at the top of the egg.

The only insects easily confused with giant water bugs are our few members of the primarily southern family **Naucoridae** (creeping water bugs, page 135), but creeping water bugs are smaller than the smallest giant water bugs (less than 15 mm). They are usually a bright greenish color rather than brown, and they lack tail-like appendages. They are also notoriously painful biters, and should be handled with care.

The water scorpion family (**Nepidae**, page 133) includes two genera in the northeast, one of which (*Nepa*) is uncommon and looks a bit like a giant water bug, and the other of which (*Ranatra*) is a very common, long (20–40 mm), skinny bug with the general dimensions of a stick insect. Water scorpions normally sit on aquatic vegetation while breathing through a tail-like air tube running to the surface. These ambush predators grab small organisms with their nutcracker-like front legs, impaling prey such as water boatmen on a short, needle-like beak.

Water boatmen (**Corixidae**, page 133) are distinctively shaped, soft-bodied aquatic bugs that are usually the most numerous adult aquatic insects in slow-moving and still water. Most of North America's 120 or so corixid species abound in ponds, pools or lake margins where they usually lay eggs in masses on the submerged parts of aquatic vegetation, although *Ramphocorixa* species preferentially lay eggs on crayfish carapaces. Boatmen eggs can be abundant enough to harvest for food use, and those of some Mexican species are dried and processed into flour, used as pet food or sold entire under the name "ahuauhtle" or "Mexican caviar." The fully winged adults, which often appear in huge numbers at porch lights as they make nocturnal flights from pond to pond, are easily recognized as boatmen by their oarlike hind legs, flat and mottled wings, and short, scooplike front legs. Front legs of males have a row of pegs which, when rubbed along the side of the head, make a sound comparable to the stridulation of male grasshoppers. The head overlaps the thorax, and the uniquely broad, short beak houses slender stylets small enough to puncture filamentous algae and suck out the chlorophyll. Water boatmen are grazers on algae and only rarely prey on other animals, in marked contrast to all other aquatic bugs including the superficially similar but entirely predacious backswimmers.

Backswimmers (**Notonectidae**, page 134), which do indeed swim about with their backs down, are hard, very convex, whitish insects which really bear only the most superficial similarity to the flattened, soft, mottled black water boatmen. The convex "back" of backswimmers is usually white or mottled, rendering these upside-down swimmers hard to see from below. The flat underside is darker, but often looks silvery due to bubbles of air held by rows of hairs under the body. Backswimmers and water boatmen do have some similarities, such as oarlike hind legs and the ability to stridulate, but the differences between them could be worth remembering if you have occasion to handle bugs that have colonized your swimming pool or backyard pond, as both these groups often do. Unlike water boatmen, backswimmers are voracious predators, and have a well-developed beak that they can use defensively with the same effect as a bee sting. The most common backswimmers, relatively large (10–16 mm) bugs in the genus *Notonecta*, are often seen literally hanging out at the water surface. They sit suspended at about a 45-degree angle, heads down and long hind legs stretched out, ready to escape if you approach too closely, or to attack if a small fish or invertebrate approaches. *Notonecta* species eat a variety of prey, including mosquito larvae. Our other common genus of Notonectidae, *Buenoa*, is smaller, doesn't hang out at the water surface and feeds mostly on smaller organisms.

The small family **Pleidae**, or pygmy backswimmers (page 134), includes a common species of backswimmer-like bug that abounds in many ponds but, with a body length of less than 3 mm, it often escapes notice until a pond sample is put in a jar or aquarium. These tiny, convex, brownish backswimming bugs are often mistaken for tiny beetles because they lack the overlapping forewings that characterize other Heteroptera, and instead have uniform forewings which are either fused or meet in a straight line down the back (as in beetles). Despite destroying diagnostic rules of thumb with their weird wings, pygmy backswimmers do have a well-developed, typical heteropteran beak, usually used for impaling small crustaceans.

True Bugs III: Shore and Surface Bugs

Even for the beginning bug person, a new family sighting can be a memorable event, and I can easily recall most of my unexpected first encounters with unusual insect families. One such encounter took place on a canoe trip down a southern Ontario river about 30 years ago when what appeared to be numerous baby toads appeared about my feet as I traversed a muddy portage. A closer look revealed these little "toads" to be bugs, in a family appropriately called toad bugs (**Gelastocoridae**, page 135). These extraordinarily toadlike bugs are characteristic of open, muddy shores where they pounce upon and eat other insects. I have seen them a few times since then, but they are not common north of the Great Lakes. Toad bugs have greatly reduced antennae, like aquatic bugs but unlike most shore and surface bugs, suggesting that they probably evolved from aquatic ancestors. The same is true for one other group of shore bugs with short antennae, the small, rarely encountered, velvety black bugs in the family **Ochteridae** (velvety shore bugs, page 135).

Unlike the remarkable toad bugs and rare velvety shore bugs, most bugs associated with shorelines and the water surface have conspicuous antennae, and are common and easy to find. Everyone has seen water striders (**Gerridae**, page 137) skating along the surface of ponds or lakes, their mid and hind feet forming dimples on the water surface, and their tiny front legs poised to grab small insects on or just below the surface. One large species, *Aquarius remigis*, is particularly abundant on the backwaters of rivers and streams of eastern North America, and on lakes and ponds in the west. Most other water striders occur on lakes, ponds and the calmer parts of rivers and streams. Some spend their lives at sea, even though the open ocean is the one frontier normally closed to insects. Out of millions of species of insects, only five species of marine water striders, or ocean skaters, have crossed that frontier. Ocean skaters occur in mid-ocean, but you are more likely to find these strikingly beautiful blue-gray striders on tide pools, or washed ashore among flotsam and jetsam. Places to lay eggs are scarce on the open ocean, and ocean skaters lay eggs on any floating object available, including living seabirds. Food is also scarce at sea, but ocean skaters obviously find enough floating fish eggs and other pierceable protein sources to survive.

Gerridae are assisted in their water walking by fine, water-repellent hairs on their feet. Like most other surface insects, they also have tarsal claws inserted up above the foot instead of at the end of the foot as in other insects. Some common relatives of the water striders have the claws even further modified to assist with surface survival. Riffle bugs (**Veliidae**, page 136), commonly seen swarming about on the surface below swift stretches of clean streams, have leaflike claws inserted well up a deeply cleft foot (tarsus). The claw is similar to the leading edge of a Chinese folding fan, an elegant swimming plume formed from a ray of feather-like hairs that fan out from the junction between the claw and base of the cleft foot. Veliidae includes not only the common riffle bugs (*Rhagovelia*) but also some tiny bugs (aptly named *Microvelia*) that abound on and along calmer waters. *Microvelia* are only a few millimeters long, and the majority of the tiny, blackish surface bugs encountered on a visit to a pond are likely to be wingless individuals of this genus. The similarly sized and colored velvet water bugs (**Hebridae**, page 136) have a distinctive black and white, velvety appearance, and have claws inserted in the normal apical position.

Some of the oddest looking surface bugs also have their tarsal claws inserted in the normal apical position. Look very closely at the surface near the edge of a calm pond for water measurers (**Hydrometridae**, page 137), which look much like minute walkingsticks, slowly and methodically walking along the water surface. The slow, careful tread might be due to the threat of puncturing the life-supporting water surface, but it might also be an adaptation for stalking the mosquito larvae which serve as a major menu item for water measurers. Water measurers are small bugs – not much longer and a whole lot thinner than the mosquitoes they eat.

Lampracanthia crassicornis, a peatland shore bug, page 136, caption 1.

Many groups of bugs are adapted to walk on calm waters, but perhaps the most abundant of all surface bugs are the small green water treaders (**Mesoveliidae**, page 137) that scurry around the surfaces of floating vegetation and nearby water surfaces in search of drowned insects. Water treaders, especially the common and widespread species *Mesovelia mulsanti*, normally lack wings and look like nondescript, greenish nymphs. Despite the relative rarity of winged forms of Mesoveliidae, most published insect keys diagnose this family on the basis of a distinctive veinless wing membrane. To make things more complicated, when winged adults do occur, they often mutilate their own wings after they have used them for a dispersal flight. Despite these problems, water treaders are easily identified in the field as the common greenish, slender, small (3–4 mm) bugs zipping along amongst floating vegetation. Wing polymorphism is a common occurrence in surface bugs, with wingless forms predominating and winged forms just turning up now and then to allow dispersal to new bodies of water.

If you try to catch shoreline bugs, you will quickly find that they usually don't have the wingless forms characteristic of their floating cousins. Common shore bugs (**Saldidae**, page 135) are particularly adept at making a quick escape when approached by other predators (they are predators themselves), or by insect collectors. Some collectors carry squirt guns full of alcohol used to stun these evasive little bugs. Shore bugs are almost invariably present on muddy, rocky or sandy shorelines, where they are easily recognized by short, quick flights, dark or black and white coloring and the four or five looplike closed cells in the wing margin. Our most unusual-looking shore bug, the beetle-like *Lampracanthia crassicornis*, occurs in Sphagnum bogs.

Homopterans

Homopterans are plant-sucking insects that differ from Heteroptera, or true bugs, in having uniformly membranous wings and a thin beak pressed so closely to the bottom of the head it looks like it comes out either between the front legs (suborder Sternorrhyncha) or at the very back of the head (suborder Auchenorrhyncha).

Unlike the Heteroptera, the homopterans do not form a natural group, so it is incorrect to include the very different leafhopper, planthopper and aphid groups in one order or suborder called Homoptera, as is sometimes done. Homopterans are an artificial assemblage made up of three very different component parts: (a) cicadas and their relatives, (b) aphids and their relatives and (c) planthoppers. Aphids and their relatives belong in one suborder (Sternorrhyncha) and other homopterans in another suborder (Auchenorrhyncha). The Auchenorrhyncha are in turn divided into two easily recognized superfamilies, the Cicadoidea (cicadas and their relatives) and the Fulgoroidea (planthoppers).

Cicadas, Leafhoppers and Spittlebugs (Suborder Auchenorrhyncha, superfamily Cicadoidea)

In one of the better known of Aesop's fables, there is an encounter between an industrious ant and a frivolous cicada. The cicada, sometimes depicted as a grasshopper in North American versions of this fable, has spent the summer making those familiar songs that most of us associate with long summer days. The onset of winter finds the cicada hungry and seeking food from a neighboring ant. When asked by the ant why he didn't gather his own food during the summer, the cicada replies that he was busy singing. "Singing, were you?" answers the ant. "Well, then, now you may dance!"

Seventeen-year Cicada nymph (*Magicicada septendecim*), page 139, caption 1.

In real life, the end of summer, and the end of the cicada's (**Cicadidae**, page 139) welcome song, leaves little opportunity for dancing because it means the end of the singing cicada's life. It has, however, been a long life – as much as 17 years for some species – culminating in the short adult stage regularly referred to in literature ranging from centuries-old Chinese poetry through to sexist doggerel like: "Happy are cicadas' lives, for they all have voiceless wives."

It is true that only the male cicada sings, but anthropocentric terms like "happy" are probably a bit off the mark, despite the fact that Fabre, the great 1800s naturalist and author, concluded his famous essay on cicadas with the suggestion that they sing for joy of their brief adult life. His was a nice suggestion, but it seems more likely that the shrill, loud, singing serves primarily to attract the opposite sex. Once that business is over with, the female uses a sharp ovipositor to insert eggs into tree twigs, where they later hatch into robust nymphs that fall to the ground. After burrowing into the ground and attaching to tree roots, cicada nymphs suck xylem sap and grow until the time comes to molt into an adult. When that time arrives, the stout-bodied, wingless nymphs crawl up the tree trunk, cling to the trunk with their stout, digging front legs, then split lengthways up the back to release the clear-winged adult cicadas. The nymphal shells are commonly seen on tree trunks, sometimes in huge numbers.

Our most common cicadas (*Tibicen canicularis*) are called Dog-day Cicadas because of their noisy abundance during the dog days of summer. Dog-day Cicadas spend several years as underground nymphs, but some emerge every year to join the serenade. Other cicadas, in a genus appropriately called *Magicicada*, are absent for years then "magically" appear in huge numbers. *Magicicada* species are called periodical cicadas because they have 13- or 17-year life cycles, resulting in spectacular periodic emergences of large, brightly colored, conspicuous, noisy adults. The numbers involved are incredible, with as many as 1.5 million cicadas emerging per acre during peak periods. All seven species of periodical cicada are eastern North American, with the four 13-year species mostly southern and the three 17-year species mostly northeastern. The 17-year cicadas occur in different broods with different distributions, with each brood normally including all three species. Brood X, which has an extensive distribution including much of Michigan, Ohio, Indiana and Kentucky, appeared in June 2004 but won't be seen again until 2021. The only 17-year cicada emergence anywhere between 2004 and 2012 will be the relatively small brood XIII emergence south and west of Lake Michigan in 2007; the next big emergence along the northeast coast will be the brood II emergence of 2013.

There seems to be some really serious pressure to go with the crowd here, probably because those cicadas that dare to be different get killed. If a few individuals of these noisy, bright insects emerge between mass emergence years, they are easily spotted by predators, and are likely to be eaten. In contrast, predators can hardly put a dent in a mass emergence of millions of individuals, unless of course they can predict and track mass emergences with their own life cycles. That is apparently very difficult to do with 13-year and 17-year life cycles.

Cicadas are such interesting bugs that you might want to have a look at how they make their loud song, probably the noisiest song of the insect world. In essence, the song is made in the same way you might make a sound by wobbling a saw blade – as the flat surface of the blade is distorted then popped back into shape, a loud sound is created. Instead of wobbling a saw blade, the cicada male uses drumskin-like membranes, called tymbals, inside the abdomen just behind the thorax. The tymbals are "wobbled" by muscular action, and the resultant sound is modified by adjacent structures and amplified by the hollow air chamber created by the air-filled abdomen. Flaps hide the tymbals and related structures including eardrum-like tympana in Dog-day Cicadas, but the flaps can be lifted to view the tymbal in action.

Next time you are outside listening to cicadas, sit down for a minute and look at the activity surrounding you at ground level. You will almost certainly be within view of a variety of little leafhoppers and other "hoppers" that are closely related to cicadas. Like cicadas, hoppers feed on plant sap and lay eggs with a bladelike ovipositor. They have similarly short, threadlike antennae, and even sing using similar sound-producing organs although their song is usually inaudible to the human ear. Three hopper families are common enough to be familiar to everybody – the leafhoppers, spittlebugs, and treehoppers. These three families, plus the Cicadidae, make up the superfamily Cicadoidea.

Have a good look at almost any plant, and the chances are good that you will see some small, active leafhoppers (**Cicadellidae**, page 140) scooting sideways or backward along the foliage, or perhaps disappearing in spectacular jumps as you move in for a closer look. Leafhoppers rival the diversity of butterfly colors and exceed the diversity of other sucking insect families with about 2,500 species in North America alone. The family is characterized by a fringe of bristles on the hind legs, but you can assume that most common, slender, brightly colored hoppers are cicadellids.

Like other homopterans, leafhoppers feed by sucking the contents of plants with their strawlike beaks. Stem-feeding leafhoppers usually imbibe phloem sap, and small, leaf-feeding species consume cell contents, causing local discoloration known as hopperburn. Similar damage is caused by several sap-sucking homopterans when they secrete a substance that hardens around the puncture made by the beak, forming a feeding sheath like a straw inserted permanently into the plant. Leafhopper beaks, like all bug beaks, have two channels, one for shooting in saliva and another for sucking up food. In many cases the saliva causes discoloration, spotting or even death of the plant. On top of this, leafhoppers and other homopterans frequently "use a dirty needle," as they stick a virus-infected beak into host plants and move viral diseases from plant to plant. Some hoppers further damage plants by inserting eggs into plant tissue with their sharp ovipositors.

Although some leafhoppers have a very wide host range, many are very host specific and some are even fussy about what part of the plant they live on. If you want to identify leafhoppers it is often useful to record the host plant. Some leafhoppers spend the winter as adults, so you can find these insects throughout the year. Most, however, spend the winter in the egg stage.

Paraphilaenus parallelus (spittle shelter removed from one of these two nymphs), page 144, caption 7.

The familiar glistening white gobs of phlegm-like bug "spittle" that abound in most meadows (and on many trees and shrubs) are each composed of multiple tiny bubbles whipped up into a froth by a nymphal bug called a spittlebug (**Cercopidae**, page 143). If you wipe away the bubbles to expose the spittlebug, beak inserted into the stem and head facing down, it will shortly show you how it constructs its bubble blob. The squat, short-winged hopper nymph will appear to blow sticky bubbles out its hind end, whipping them around with rotations of its tail and moving them forward with its legs to form a new foamy domicile.

Spittlebugs, like other kinds of homopterans, feed by inserting a strawlike beak into plants and sucking up massive amounts of sap but, unlike most other homopterans, spittlebugs tap the sap on its way up from the roots (xylem) rather than the sugary fluid flowing down from the leaves (phloem). This xylem sap contains enormous amounts of excess water that is pumped out about as fast as the sap is sucked in. While other homopterans deposit their extra juice in the form of droplets of sugar-rich honeydew, spittlebugs start out life by just dribbling excess fluids over their body. Newly hatched nymphs avoid drowning in their own dribble by having the underside of the abdomen rolled into an air-filled, tubelike canal (think of a rolled tongue) running along the underside of the body. They breathe through spiracles opening into this canal, and the canal opens out to the insect's hind end that can be extended, like a telescoping snorkel, out of the fluid.

Not until the spittlebug nymph has grown to its second stage, or second instar, does it start to use that air-filled ventral canal to blow those characteristic bubbles. First the tip of the abdomen, and thus the end of the canal, is withdrawn into the fluid covering. The bubbles are formed when the hopper contracts its muscles, squeezing the air canal like a tube and forcing air out the tip into the fluid. As that air is forced out the tip and through the fluid covering, glands in the side of the abdomen add materials that make the bubbles tough, sticky and perfect building blocks for a spittlebug shelter.

Only the nymphs of spittlebugs utilize bubble shelters. Once the flightless nymph has completed its growth, a process that usually takes up about six weeks, it molts to a preadult instar with conspicuous wing buds. Mature nymphs of our most common spittlebugs make new shelters of somewhat gelatinous bubbles when they are ready to transform to adults. The gelatinous bubbles dry up to form a protective dome in which the nymph molts to a fully winged adult. Some spittlebug species merely wander out of their spittle mass and molt to the adult stage while clinging to a nearby stem, and a few molt right in the nymphal spittle mass.

Adult spittlebugs (often called froghoppers) suck copious quantities of sap just like the nymphs, but expel excess fluid as droplets shot forward away from the body rather than just wallowing in waste products like the nymphs. Some observers have reported an audible "pop" associated with the adult spittlebug's habit of energetically ejecting globules of excess sap.

Female spittlebugs have a narrow, bladelike ovipositor for inserting eggs into plant tissue, usually late in the season. Eggs hatch the following spring, and by June spittle masses are common sights. The most commonly encountered spittlebug throughout much of North America is an enormously variable species that seems to feed on just about any juicy plant it can get its beak into. The Meadow Spittlebug (*Philaenus spumarius*) is a European insect that was accidentally introduced in both western and eastern North America, and odds are good that it abounds in your backyard. Occurring in almost every color from black to brown to mottled yellow, the Meadow Spittlebug is a great example of a polymorphic species and a sobering reminder that recognizing species by color alone is not always a reliable approach. Fortunately, the smooth wings and some other structural characters allow the sharp-eyed student of insects to recognize this tricky species. Most other spittlebugs are less variable, and the observant naturalist can easily add a dozen common species of these fascinating bubble blowers to his or her repertoire of familiar animals.

The treehopper family (**Membracidae**, page 145) is a common group of easily recognized bugs in which the pronotum is greatly enlarged and often extended back over much of the body. This big pronotum makes some species look like rose thorns, gives a strikingly antlike body form to some species and gives others a conspicuous armor in the form of bizarre sculptures projecting at all angles from the top of the thorax. Treehopper nymphs often hang out in conspicuous aggregations tended by honeydew-hungry ants, but the adults usually lead a more individualistic existence protected by their armor-like prothorax and their hopper heritage of a speedy escape mechanism.

Planthoppers (Suborder Auchenorrhyncha, superfamily Fulgoroidea)

Planthoppers (several families of Fulgoroidea) are often superficially similar to common froghoppers, leafhoppers or treehoppers, but planthopper antennae are usually inserted underneath the eyes and behind a characteristic ridge, or carina. Most planthoppers are small, rarely noticed insects, but every now and then some of the more conspicuous members of the superfamily, such as the relatively large, brilliant green, wedge-shaped **Issidae** (page 148), occur in eye-catching abundance. One of the two spectacular species of Issidae now common in northeastern North America used to be restricted to the southeast, and has only recently become common in the northeastern states and southern Canada. This large green issid (*Acanalonia conica*) joins an interesting assemblage of long-snouted **Dictyopharidae** (page 149), brilliant blue- or pink-striped **Derbidae** (page 149), beetle-like **Caliscelidae** (page 149), dusty purple **Flatidae** (page 148) and several other small families that make the fulgoroids favorites among informed insect watchers. Most planthoppers are sap-suckers like other homopterans, but nymphs of a few species feed on fungi. The Fulgoroidea of northeastern North America are only a pale shadow of the spectacular planthopper diversity one can see in both the New World and Old World tropics.

Aphids and Phylloxerans (Suborder Sternorrhyncha, superfamilies Aphidoidea and Phylloxeroidea)

Gardeners often refer to aphids as "plantlice," reflecting a correct perception that aphids are like plant parasites that live by sucking vascular fluids out of host plants, much as lice are associated with their vertebrate hosts. Another gardeners' term for aphids is "greenbugs," a less accurate description since these soft, pear-shaped insects come in a variety of colors. Whatever their hue, these slow-moving, plump homopterans are usually recognizable for their habit of feeding in dense groups. Aphid wings, if present, are membranous and are usually awkwardly held away from the body. Most aphids (those in the large family **Aphididae**, page 151), also have a pair of prominent tubes (siphunculi, or cornicles) projecting conspicuously from the abdomen, used to secrete a sticky substance that deters predators and may even trap and kill would-be parasitoids. Chemicals secreted by the cornicles also serve as alarm pheromones, sometimes triggering other aphids to run, jump or drop from the plant. The same chemicals may also attract ants to the defense of the threatened aphid, and other chemicals render some aphids toxic or distasteful.

Aphids (*Uroleucon helianthicola*) on a sunflower stem.

Chemical warfare notwithstanding, the main secret to aphid success is their incredible reproductive potential. Aphids regularly reach densities of over a million per acre. In the absence of predators, pathogens and parasites, and in the presence of good food and good weather, a single aphid could theoretically give rise to over 600 billion progeny in a single season. Fortunately, that awesome potential production is never achieved because aphid colonies are invariably infiltrated by a variety of killers. Blind, legless larvae of flower flies can usually be seen inching among aphids, periodically impaling victims on their mouth hooks, raising the doomed aphids off the surface and sucking them dry. Silver fly maggots and predacious midge larvae regularly join the carnage, along with brightly colored lady beetle larvae and typical red-and-black lady beetle adults. Brown and green lacewing adults and larvae are also major predators of aphids. One subfamily of stinging wasps (Crabronidae, Pemphredoninae) hunts aphids almost exclusively, some plant bugs (Miridae) are aphid predators and some significant groups of parasitic wasps are specialized internal parasitoids of aphids.

Like most other homopterans, aphids pierce the phloem tissue of plants, and feed by pumping huge amounts of sugar-rich and nitrogen-poor sap through their bodies. Homopterans must suck enormous quantities of sap to obtain sufficient protein, excreting the excess sugary fluid as copious quantities of honeydew. Far from the mere waste product it appears to be at first, honeydew is like a currency with which aphids pay for services provided by other organisms. The Woolly Ash Aphid (*Prociphilus fraxinifolii*) for example, has a root-feeding stage that occurs in association with a fungus that creates a hollow shelter for the aphids. In return for housing the aphids, the fungus benefits from the nutrient-rich honeydew.

More conspicuous mutually beneficial arrangements exist

between many aphids and ants that "tend" aphids and "milk" honeydew. Some ant-aphid interactions are remarkably pastoral in their detail. Cornfield ants (*Lasius* spp.) gather Corn Root Aphid (*Anuraphis maidiradicis*) eggs in fall and take them home for the winter, moving the eggs around the ant nest to maintain them at optimum temperature and humidity. In the spring, newly hatched aphids are solicitously placed on the roots of weeds, where the first generations of aphids feed. Later in the season, the ants pick up the aphids and transfer them to corn roots where they can do considerable damage. This damage is accentuated by the continuing activities of the ants in dispersing the aphids around the cornfield. Even in the absence of elaborate interactions like those between ants and Corn Root Aphids, honeydew-loving ants pay for their sweets by protecting aphids from parasitoids. Parasitoids that manage to get past the ant army and invade the bodies of aphids in the large superfamily Aphidoidea face yet another potent group of aphid defenders, in the form of aphid-specific symbiotic bacteria that kill the larvae of parasitoid wasps within their hosts.

Every garden has aphids, so it is easy for anyone to do a bit of aphid observation. Even in winter you can find little black aphid eggs on apple and other trees, usually near buds. In spring these eggs will hatch into wingless females, which will develop rapidly and give birth to more wingless females without bothering with sex or eggs. Later in the season, as things become a bit more crowded, winged females will appear and, depending on the species, move to other plants. Species that migrate from apple or other trees to weeds or other herbaceous plants during the summer produce winged males and females that move back to their host trees or shrubs when shortening days herald the end of summer. In temperate regions many aphids produce wingless sexual females in late fall; these mate with the winged males to produce eggs that withstand the winter and hatch to wingless females in the spring.

The above life cycles have several unusual features, each of which might be important in explaining why there are so many aphids. Parthenogenesis, or the production of young by virgin females, is a good trick for getting the jump on predators and competitors. Population numbers can be built up more quickly by not wasting time with immediately useless males. Viviparity, or the production of live young, bypasses the egg stage and offers another way to build up numbers quickly under good conditions. If you spend a few minutes watching a colony of aphids, you might witness a live birth. If the light is right, you might be able to see the developing young right through the translucent body of the mother. Winglessness also allows populations to build up quickly since the development of wings competes with developing embryos for limited protein. One scientist showed that short-winged forms of an aphid species had 32 percent more offspring than long-winged forms.

Winglessness, viviparity and parthenogenesis give aphids an advantage under optimum conditions, but winged forms are occasionally useful to escape suboptimum conditions, and eggs are sometimes useful as a resistant overwintering stage. Occasional sex assures a pool of genetic variability (males aren't entirely useless), and may also be important in lineage perpetuation through speciation. The occasional appearance of winged or sexual forms is regulated by internal chemical messages (hormones), which are sent in response to factors such as food quality, day length, temperature, crowding and external chemical messages (pheromones) received from other aphids.

Alternating between a winter woody host plant, such as an apple tree, and a summer herbaceous host, such as plantain, is another common and obviously adaptive aphid strategy. It allows aphids to exploit foliage that is either actively growing or old and senescent, during which times the phloem sap is relatively rich in amino acids. Host alternation might also have evolved to escape natural enemies. If they can establish new colonies on plants to which predators have not yet been attracted, aphids can build up large populations very quickly.

Soybean growers in northeastern North America were recently treated to a spectacular population explosion when Soybean Aphids (*Aphis glycines*) were accidentally introduced from Asia. This species multiplies on soybean plants as wingless females for much of the summer; it then produces winged females to disperse to new soybean fields, followed by winged males and females that move to buckthorn shrubs in the fall months. Soybean Aphids built up such huge numbers in southern Ontario in the summer of 2001 (their first summer in Canada) they were big news when midsummer dispersal flights blanketed some urban areas.

Aphids can do more to their host plants than merely rob them of some sap, since they are the major vectors of plant viruses. Even in the absence of viruses, the saliva that aphids pump into plants can cause injury or abnormal growth. Aphids feeding on the underside of a leaf, for example, can cause the leaf to curl around them, forming a convenient shelter. Other aphids induce distinctive swellings called galls.

Galls can be thought of as controlled tumor-like growths, built by plants to specifications in insect-delivered instructions. Many kinds of insects, including aphids, induce species-specific galls. Sumac trees, for example, are loaded every autumn with huge pink swellings full of Sumac Gall Aphids (*Melaphis rhois*, family **Pemphigidae**, page 153). Sumac Gall Aphids lay eggs on sumac leaves in spring, each egg hatching to a female that induces the sumac plant to make the hollow galls in which female aphids multiply asexually. The first few generations are wingless, but by fall the swollen, pink, conspicuous galls are packed with a mixture of winged and wingless aphids. The winged aphids fly to mosses where they spend the winter. The next spring the overwintered females produce both male and female offspring that mate before females move to sumac and lay eggs, starting the cycle anew. Galls, like sumac galls, not only give shelter, but also cause a local difference in metabolism that results in an improvement in the aphid's food supply. Even when they don't form galls or cause visible growth abnormalities, feeding aphids distort the metabolism of the plant in their favor.

Woolly aphids in the family Pemphigidae are common insects, usually with small males and asexual females, but sometimes with relatively large sexual forms that lack the conspicuous, tubelike cornicles seen on common aphids. Most people only notice woolly aphids during their conspicuous fall sex flights, but apple, maple and alder branches are often covered with these white, fuzzy insects earlier in the season. These groups of wax-covered aphids are often joined by predacious lacewing larvae covered with the empty bodies of their woolly victims, like wolves in sheep's clothing. Predacious caterpillars of harvester butterflies often graze the same aphid colonies.

The pine and spruce aphids, or **Adelgidae** (page 154), and the phylloxerans, or **Phylloxeridae** (page 154), are gall-making homopterans that induce characteristic swellings on leaves and roots of conifers, grapes and other plants. The Grape Phylloxera (*Daktulosphaira vitifoliae*) should be of particular interest to those who like wine. Phylloxeran galls can be found on the leaves and roots of most grapes, and wild North American grapes appear to have developed some resistance to this native eastern North American homopteran. European grapes, however, are highly susceptible, and the introduction of this pest from North America to Europe in the 19th century created a serious threat to the wine industry. This problem was solved, in part, by the grafting of European vines onto resistant North American root stalks. Big, single-variety vineyards in California have had problems recently with new varieties of phylloxerans, and many growers, like French growers of a hundred years ago, have had to replant using resistant root stalks.

Adelgidae and Phylloxeridae make up the superfamily Phylloxeroidea, while other aphids are in the superfamily Aphidoidea. In addition to the large family Aphididae and the familiar woolly aphids in the family Pemphigidae, the Aphidoidea includes several small families (**Lachnidae**, page 152; **Mindaridae**, page 153; **Drepanosiphidae**, page 152; **Hormaphididae**, page 154; and others) that used to be treated as part of the Aphididae.

Jumping Plantlice and Whiteflies (Suborder Sternorrhyncha, superfamily Phylloxeroidea)

Jumping plantlice, or psyllids (**Psyllidae**, page 154), look much like cicadas shrunk to only a few millimeters, with long antennae in contrast to the bristle-like appendages that pass for cicada antennae. Despite their hard-bodied, cicada-like appearance and jumping hind legs, jumping plantlice have much in common with aphids. Some cause galls and some, like the Pear Psylla (*Cacopsylla pyricola*) and Apple Sucker (*Cacopsylla mali*), are introduced pests. Look for newly hatched Apple Sucker nymphs on honeydew-spattered apple tree buds in spring, where their feeding can prevent fruit set. Pear Psylla nymphs excrete copious amounts of honeydew that gives pear trees (and fruit) a blackened appearance due to the growth of sooty mold.

The psyllids most likely to come to the attention of naturalists are those that make conspicuous products, like flocculent waxy masses or abundant and distinctive leaf swellings. Hackberry psyllids (*Pachypsylla* spp.), for example, can be found inside those small round galls that usually abound on hackberry leaves, and the bright white, cottony masses seen on alder branches early in the summer are waxy secretions produced by clusters of *Psylla floccosa*, or Cottony Alder Psyllids (similar waxy masses later in the summer are made by aphids). Cottony masses on the undersides of blackberry leaves are usually caused by feeding groups of the Blackberry Psyllid (*Trioza tripunctata*), a member of a small family (**Triozidae**, page 155) that is sometimes treated as part of the Psyllidae.

The smallest aphid-like suckers that may come to your attention as plant pests are the whiteflies, or **Aleyrodidae** (page 155). Greenhouse Whiteflies (*Trialeurodes vaporariorum*) are common pests on greenhouse foliage, where they often appear in clouds of tiny, mothlike insects covered with a fine, white, waxy powder. Greenhouse Whiteflies are ubiquitous pests in greenhouses, but do not survive out of doors in the northern United States and Canada. Other species can be found outdoors on a variety of plants, usually on the undersurfaces of leaves, and some are pests that occasionally occur in outbreak numbers. For example, the Sweet-potato Whitefly (*Bemisia tabaci*) is an insecticide-resistant pest species (or complex of species) that was dubbed a "superbug" in the national media when it appeared in outbreak numbers in California in the early 1990s, presumably after being accidentally introduced to the area on some poinsettias. That outbreak reportedly cost close to $200 million per year just in the Imperial Valley. Sweet-potato Whiteflies feed on an enormous variety of plants, and now occur in much of the world. In some areas, including northeastern North America, attempts have been made to control them with a tiny parasitic wasp (*Encarsia formosa*, Aphelinidae). The same wasp is routinely used to control whiteflies in greenhouses.

When whiteflies first hatch from their eggs they are flat, oval, active creatures that soon settle down and lose their legs and antennae. The resultant immobile insect sucks sap on the spot, often under a scalelike waxy secretion, until it is "ready" to become a waxy-winged adult whitefly, at which time it stops feeding as its appendages develop inside the motionless body. This nonfeeding stage is like the pupa of higher insects and, like a pupa, transforms to a completely different, fully winged adult. We usually think of the Holometabola (which includes flies, beetles, moths and so on) as the only insects with complete metamorphosis, but this kind of development, including completely wingless stages followed by a quiescent stage (pupa) that molts to the winged adult, has obviously developed independently in whiteflies and the Holometabola. Similar development shows up in the thrips and scales.

Pine Needle Scale (*Chionaspis pinifoliae*), page 156, caption 2.

Scale Insects and Mealy Bugs (Suborder Sternorrhyncha, superfamily Coccoidea)

When I was a graduate student at my first big Entomological Society meeting, I chanced to sit with some prominent entomologists at dinner one night. Upon being introduced to a famous expert on scale insects, whom I had never heard of, I naively asked, "What do you work on?" "Pussbags" was the curt answer. Since then, I've never been able to avoid the graphic image of a scale insect as an amorphous sac… not an inaccurate image of a wingless, legless, antennae-less, eyeless, adult female scale insect. Basal ("primitive") lineages of Coccoidea retain functional legs in both sexes, but legs are greatly reduced in most female scale insects and absent in female armored scales such as Oystershell Scales.

One of the common species in our area is the Oystershell Scale (*Lepidosaphes ulmi*). Each of the "oystershells" that often encrust the branches of fruit trees is a waxy covering produced by a degenerate adult female attached to the tree by her long, thin beak. Eggs are laid under the oystershell-like covering, where they spend the winter before hatching into tiny, active young, called crawlers, with legs and antennae. Female crawlers find themselves a good piece of branch where they settle down and degenerate, quickly losing appendages and secreting a wax house. Male crawlers develop into a small two-winged insect, a bit like a small fly but without functional mouthparts and with a thin tail.

Scale insects and their relatives (several families in the superfamily Coccoidea) are most important as tough plant pests, but they also provide several useful substances. Shellac, for example, is made from the waxy covering of a Southeast Asian member of this group. Branches covered with "lac" insects are harvested for the waxy material that is melted off and refined into the shellac we value so highly in various products ranging from furniture polish to glazing on chocolates.

Shellac production is still a significant business, but other homopteran products are more interesting from a historical perspective. One group of cactus-sucking scales native to the

New World can be used to produce a beautiful red-purple dye (Cochineal); another group of oak-sucking scales from the Old World is used in producing bright red dyes. Pigment-providing scale insects were big business until replaced by aniline dyes in the late 1800s, and many famous bright red products prior to that time – including the bright uniforms of British soldiers (Redcoats) – can be attributed to a bunch of squished bugs (there is currently a resurgence of interest in these beautiful, nontoxic natural dyes).

Scale insects provide an entomological explanation for the biblical story about the appearance of a food in the desert, called manna, which saved the fleeing Israelites from starvation. Exodus 16:31 records the appearance "on the face of the wilderness" of vast quantities of manna, described as rounded, gray masses of material tasting like "wafers made with honey." An abundant scale insect called the Tamarisk Manna Scale (*Gossyparia mannipara*) produces large quantities of honeydew that, in arid regions, solidifies to form a sweet material that fits the biblical description of manna. The concept of eating dried honeydew is not that unusual, since aboriginal Australians have gathered a similar homopteran product for hundreds of years, and the honeydew egested by scales and other phloem-feeding homopterans is an important food for a wide variety of other organisms. Many parasitic flies and wasps are dependant on honeydew as a source of energy for flight, suggesting that honeydew is an important contribution to the natural control of pest insects. On the other side of the coin, accumulations of honeydew on foliage can support the growth of damaging fungi such as sooty molds. Honeydew is produced by some but not all scale insects, and members of the largest family of Coccoidea (Diaspididae) lack an anus so produce no honeydew at all.

The most common families of Coccoidea in eastern North America are the soft, waxy, **Pseudococcidae** (mealy bugs, page 156) that unfortunately often abound on house plants; the hard-shelled **Diaspididae** (armored scales, page 156) such as Oystershell Scales; and the **Coccidae** (soft scales, wax scales and tortoise scales, page 155) such as the Magnolia Scale and Cottony Maple Scale. Other families you might encounter include the **Dactylopiidae** (cochineal insects, which occur only on *Opuntia* cacti, page 156), **Margarodidae** (giant scales, over 10 mm with well-developed legs), **Eriococcidae** (the European Elm Scale and other mealybug-like scales found on azalea, elm, and other plants), **Asterolecaniidae** (Golden Oak Scales and other pit scales, so called because some form pits in the bark of host trees) and **Cryptococcidae** (Beech Scales and relatives, found on trunks of maple and beech trees), and **Kermesidae** (Gall-like scales, page 157, conspicuously convex scales found on oak twigs, usually along with ants and other insects attracted to their abundant honeydew).

The Thrips (Order Thysanoptera)

Thrips are distinctively slender little insects equipped with peculiar tools designed to rip open food and imbibe the newly exposed fillings. The asymmetrical thrips' head supports a single large mandible (the left mandible) that punches through target tissues such as leaf surfaces, pollen or arthropod integument so the thrips' slender, needle-like maxillae can slurp up the insides.

Thrips are common, if inconspicuous, inhabitants of bark, fungi and various plant parts, and they are especially common in flowers. Look for numerous torpedo-shaped insects, about 2 mm long, adorning the center of the next daisy you encounter. If you have a hand lens, you can look closely and see that the mouthparts are by no means the only unusual attribute. The legs, for example, end in bladder-like structures that give the insect the appearance of a child's plastic toy bug that has been accidentally placed on a hot stove top, melting the ends of its legs into rounded blobs. Those blobs are actually blood-filled bladders that can be inflated or deflated according to the demands of the terrain, and generally function as "four wheel drive" for walking on slick leaf surfaces. The four wings, if present (winglessness is common in thrips), are unusually narrow and fringed, reflecting the ordinal name Thysanoptera, which literally means "fringed wing."

Even the life cycle of thrips is out of the ordinary. Fertilized eggs develop into females, and unfertilized eggs usually develop into males (much like the sex determination system in bees and other Hymenoptera). Many thrips species are rare or unknown as males, and normally produce only females (parthenogenesis). Eggs hatch to wingless thrips that feed for a couple of stages, then go through two or three resting, nonfeeding stages. Some thrips species remain wingless throughout their lives, but others evert their wings in these nonfeeding stages. The last nonfeeding stage is sometimes spent inside a silk cocoon before molting to the active, familiar adult. Thrips' life cycles have some striking similarities to life cycles found in the Holometabola (insects with complete metamorphosis), and the terms "larvae" and "pupae" are used for the wingless early stages and the last, nonfeeding stage of thrips.

The Thysanoptera is divided into two suborders. The tube-tailed thrips (Tubulifera) have the tip of the abdomen drawn out into a long tube that never carries a bladelike ovipositor; and the saw-tailed thrips (Terebrantia) have a shorter abdomen equipped (in females) with a bladelike ovipositor. Our only family of tube-tailed thrips (**Phlaeothripidae**, page 157) includes many common species, including one abundant on daisy flowers, some large predacious species found under bark and a variety of fungus-feeding species. Most pest thrips are in our four families of saw-tailed thrips, especially the common thrips family **Thripidae** (page 157). Plant-feeding common thrips are frequently problems in greenhouses, and they sometimes carry serious plant diseases. One species, called the Pear Thrips (*Taeniothrips inconsequens*) because it has been a pest of pears since its introduction from Europe in the 1830s, has recently become a major pest of maple trees. Reaching densities as great as 50 million thrips per acre in Vermont forests, the Pear Thrips' infestation of maple trees means less maple syrup and diminishment of brilliant fall foliage.

Booklice and Barklice (Order Psocoptera)

One doesn't hear much about barklice. Never mind that these are often beautiful little animals with striking body patterns and intricate pigmentation on their large, delicate, rooflike wings, and never mind that this is an ancient order with about 250 North American species, these inconspicuous insects have been neglected by most students of insects. Members of the barklouse order (Psocoptera) are common on bark or on the bare wood of standing dead trees, and also occur on foliage, in leaf litter, on lichen-encrusted rocks, in bird nests and all sorts of other odd places.

A good way to observe some attractive barklice is to put a white sheet under a pine branch, then whack the branch with a stick to dislodge the resident psocids. The odd-looking barklice that show up on your sheet are probably scavengers, as most psocids feed by chewing on lichens, fungi or other organic matter with their large, unequal mandibles.

Although the order Psocoptera is unfamiliar to most, its members are easily recognized by their large heads with a big bulge (a swollen clypeus) at the front, their long antennae and the way they hold their wings (if present) in a rooflike fashion. Their maxillae form rodlike picks, with which they scrape away at their food before grinding it in a sort of mortar and pestle apparatus derived from the tongue (hypopharynx). The salivary glands of barklice can produce silk, sometimes used to cover their eggs, and sometimes used to form tents to protect numerous barklice of all ages.

Although the 28 North American families of Psocoptera are generally small and difficult to identify, most of the common northeastern psocid species belong in the large family **Psocidae** or common barklice (page 158), some species of which graze gregariously as huge herds of lichen-munching adults and nymphs. Probably the most familiar Psocoptera are the "domestic" species that invade granaries, household foodstuffs and other household items such as books. Flightless domestic psocids, collectively called booklice, look remarkably like lice except for their decidedly unparasite-like long antennae. Completely wingless booklice in the family **Liposcelidae** (page 159) are commonly found feeding on mold or bindings in older books, and short-winged species in the family **Trogiidae** (page 159) are common household pests in rice and other foodstuffs. Some trogiids communicate with one another by tapping their abdomens on the substrate.

The comparison of booklice to parasitic lice is not entirely a spurious one, since the first parasitic lice were probably very much like booklice. A shift from chewing on dead organic matter to chewing on skin or feathers could easily have occurred in one lineage, which would later have become specialized for this new habitat. A lineage developing parasitic habits would be expected to evolve flattened forms with legs modified to grip the host, and to lose formerly adaptive features like long antennae and well-developed eyes, which have little role in a parasite's life. Such a lineage must have given rise to the lice, a group of entirely wingless little chewers and suckers we now know as the Phthiraptera.

Lice (Order Phthiraptera)

Lice are tiny, flattened parasites that spend their entire life cycles on their vertebrate hosts, and sometimes have an intimate relationship with human beings. Human lice are not the ever-present irritants they were in days gone by, but if you have school-age children you are probably familiar with the Head Lice and their associated "nits" (louse eggs glued to hairs) sometimes brought home by children who share hats, headphones or combs with other children at school. Head Lice outbreaks are routine in most public schools, so teachers are accustomed to telling parents how to use a special comb to get the nits out of their children's hair, and how to use louse shampoo to kill the adult lice.

Body Lice, sometimes called "seam squirrels" or "cooties," do not attach eggs to hairs like head lice, but instead stick eggs on the clothing of infested individuals. Body lice are associated with unsanitary conditions and dirty clothes where they can occur in horrifying densities. One patient researcher recorded 10,428 individual lice from a single shirt! Head Lice and Body Lice are usually considered to be different subspecies of the same species, *Pediculus humanus*. Like most lice, *P. humanus* is highly host specific, and this species only lives and breeds on humans ... and pigs.

The other commonly encountered human louse thrives only on the more private patches of human hair. If you are unlucky, you could pick up some Pubic Lice (*Phthirus pubis*, a.k.a. "crotch crickets" or "crabs") from infested bedding, toilet seats (honest!) or, more typically, a *very* close friend. These crab-shaped little suckers tend to settle amongst the coarse hairs of the armpit or genital area (or, more rarely, facial hair), where they tend to remain stationary, feeding and defecating in one spot. There is only one other species in the genus *Phthirus*, another primate-infesting louse called *P. gorillae*.

Human lice are now mercifully scarce in the sanitary societies of developed countries, despite past roles as vectors of some of the most important diseases in human history. There seems little doubt that lice, not bullets or swords, have been the big killers throughout much of humankind's war-torn history, and that louse-borne epidemic typhus dictated the course of every major war prior to World War II. One of the earliest and most historically justifiable uses of the now discredited insecticide DDT was in delousing campaigns to stop typhus epidemics during World War II.

Even today, when crowded, unsanitary conditions follow in the wake of a war, lice become common, and typhus outbreaks are regular events in war-torn regions. Typhus has been present here in North America since the Spanish introduced it in the 1500s, but our last serious typhus epidemic was in 1877. The only animal reservoir (other than man) for the disease is the Southern Flying Squirrel, but about the only way it can get from a squirrel to you is if you are unlucky enough to come into close contact with the feces of a Flying Squirrel Sucking Louse (*Neohaematopinus sciuropteri*), an insect that bites squirrels but not humans.

The microorganism that causes typhus (a kind of primitive bacterium called a rickettsia), also fatally infects Body Lice, and is usually transmitted to humans when the doomed, rickettsia-filled louse is crushed and scratched – along with louse feces – into the host's skin. It is best to avoid crowded, dirty conditions conducive to louse infestations, but if you do get bitten, don't scratch. Don't breathe either, since infective louse feces can easily become airborne as they dry out and find their way into your lungs.

Another disease you can get from scratching louse poop into your skin is called trench fever because great epidemics of this bacterial disease were associated with the crowded and dirty conditions of trench warfare. More recently, trench fever has become common among homeless people, even in North America and Europe. Fortunately, it is rarely fatal.

The lice that attack people are all "sucking lice" – lice with narrow heads and special mouthparts in which the maxillae, labium and the tongue form long, thin stylets used for piercing skin and sucking blood. The sucking lice make up a small group, with only a few hundred species, all of which attack mammals. Sucking lice are just a specialized offshoot from the middle of the order Phthiraptera, an order made up mostly of biting lice with broad heads and chewing mouthparts.

There are thousands of species of biting lice, none of which parasitize us. They live on both mammals and birds, where they usually either chew away at the bases of hair and feathers, or attack the skin itself. Some attack a variety of hosts, but others are highly specific not only with regard to what host they attack, but also in how they attack their hosts. One of our most extraordinary (and largest) chewing lice feeds only within pelican pouches; some others live only inside feather quills.

Chewing lice are much like the closely related booklice and barklice (Psocoptera), and probably evolved from something like a barklouse. It is not hard to imagine how skin and feather chewing could have evolved from a habit of munching detritus in bird's nests, and it is easy to see how blood sucking could have originated within the chewing lice. As Hans Zinsser put it, "I suppose given the habits of the first parasitic lice, it was inevitable that it should be discovered that under their feet ran an infinite supply of rich, red food."

The only lice inclined to tap your blood belong to the suborder Anoplura (sucking lice) families **Phthiridae** (Pubic Lice) and **Pediculidae** (Head Lice, Body Lice). Other families of North American Anoplura imbibe the blood of other animals, such as the pigs and horses preferred by the large sucking lice *Haematopinus suis* and *H. asini* in the family **Haematopinidae**. *Haematopinus* species also attack cattle, as do lice in the family **Linognathidae**. Other families of sucking lice have generally more esoteric habits, living only on seals, walruses and river otters (**Echinophthiriidae**), only on squirrels (**Enderleinellidae**) or rodents, hares, moles, and shrews (**Hoplopleuridae** and **Polyplacidae**).

Chewing lice belong to two other suborders, the Amblycera (antennae clubbed and hidden in grooves) and Ischnocera (antennae thin and visible). The three native eastern North American Amblycera families are found on birds, while the two Ischnocera families are found on either birds (**Philopteridae**) or mammals (**Trichodectidae**). Most lice belong to the diverse family Philopteridae.

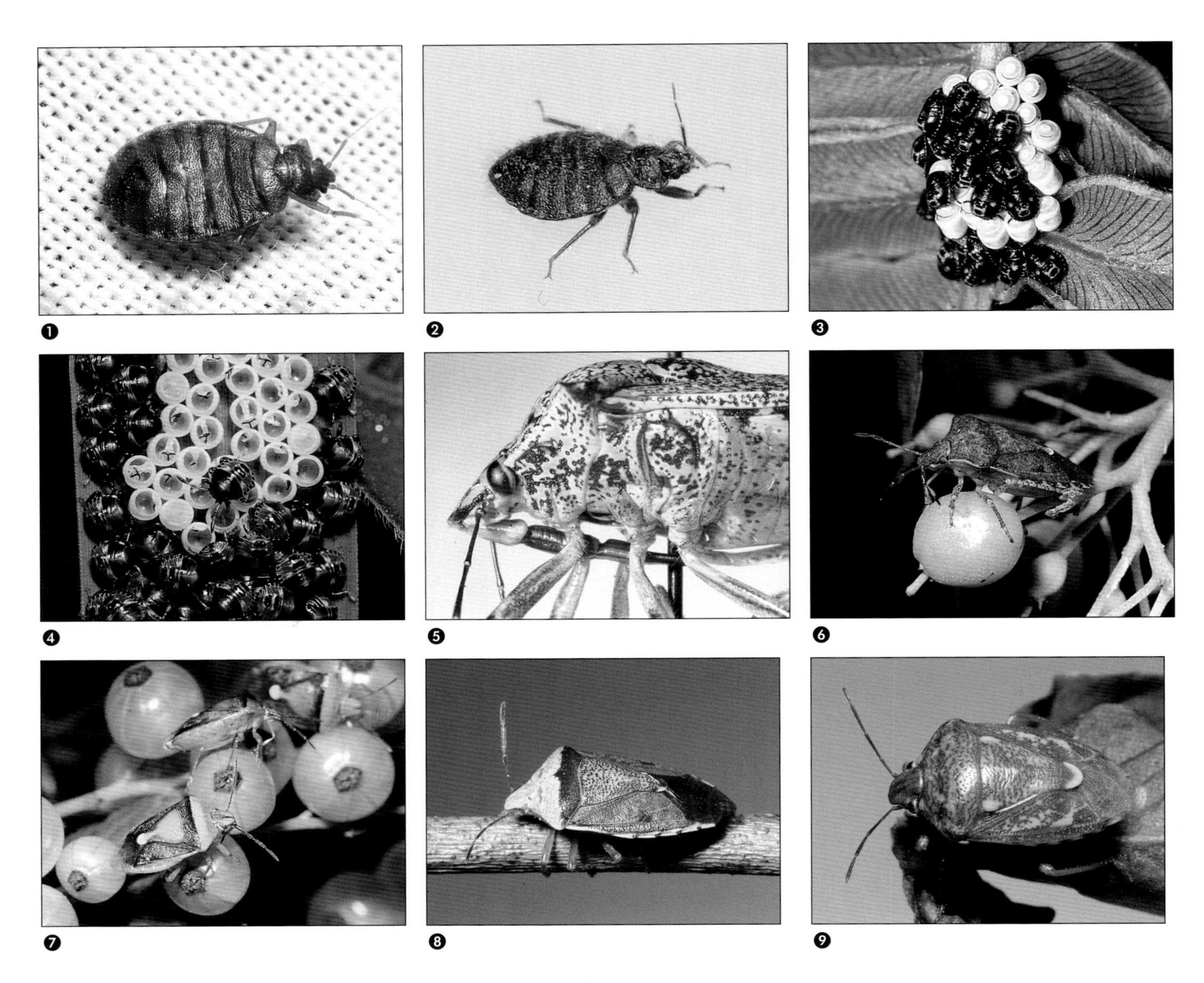

FAMILY CIMICIDAE (BED BUGS, SWALLOW BUGS AND BAT BUGS) ❶ **Human Bed Bugs** (***Cimex lectularius***) are not the ubiquitous household pests they once were, but they have become more common in North America over the past few years. Other species of *Cimex* occur in association with bats. ❷ **Swallow Bugs** (***Oeciacus vicarius***), which look like fuzzy bed bugs, occur in swallow nests. **FAMILY PENTATOMIDAE** (STINK BUGS) ❸ ❹ Stink bug eggs are usually distinctively barrel-shaped and glued in clusters to leaves. These clusters have just hatched, and several first instar nymphs are visible on top of the eggs. ❺ Predacious stink bugs in the subfamily Asopinae (like this ***Alcaeorrhynchus grandis***) have a relatively short and stout weapon-like beak directed away from the head. ❻ Plant-feeding stink bugs in the subfamily Pentatominae have a strawlike beak with its front part fixed in a groove under the head. This bug is inserting its beak into a berry, and you can see how the thick part of the beak (the labium) is just a sheath that bends out of the way while the thin stylets (mandibles and maxillae) enter the tissue. Subfamily Pentatominae (PLANT-FEEDING STINK BUGS) ❼ The **Banasa Stink Bug** (***Banasa dimidiata***) feeds on a variety of berries and seeds. This is probably our most common stink bug, and if you eat a handful of currants or gooseberries directly out of your garden, you stand a good chance of experiencing the pungent taste of this attractive, but smelly, green bug. ❽ ***Banasa calva*** has a distinctively two-tone pronotum like the more common Banasa Stink Bug (*B. dimidiata*), but it is usually more brown than green, and it has black spots on the side of its abdomen. ❾ The jade green ***Banasa euchlora*** occurs on cedar trees.

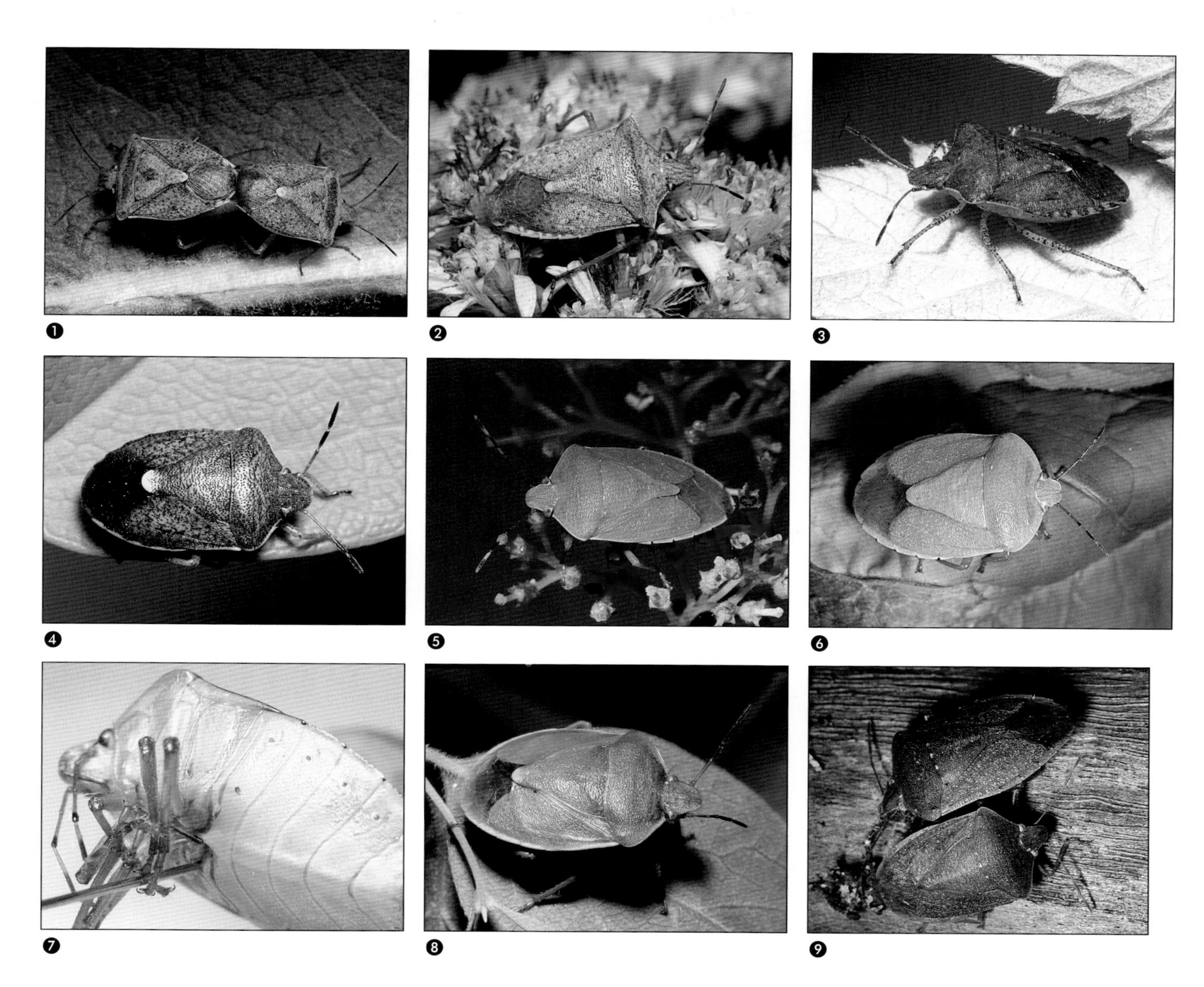

FAMILY PENTATOMIDAE Subfamily Pentatominae **(continued)** ❶ The **Brown Stink Bug** (***Euschistus servus***) is one of our most common stink bugs. This brown, round-shouldered stink bug, about 13 mm in length, can be found feeding on a variety of plants. ❷ The prominent shoulder spines of this **One-spotted Stink Bug** (***Euschistus variolarius***) distinguish it from related plant-feeding stink bugs. ❸ The **Dusky Stink Bug** (***Euschistus tristigmus***) is smaller than other common *Euschistus* (about 10 mm in length), and the underside of its abdomen has a row of black midspots (there are usually three spots, as suggested by the "*tri*" part of its species name). ❹ ***Holcostethus*** species (like this ***H. limbolarius***) are brownish yellow, plant-feeding stink bugs found in autumn fields where they feed on a variety of plants. The smaller size (7–8 mm) and the white tip on the scutellum (the white mark in the middle of the back) distinguish common *Holcostethus* from the much more common *Euschistus* species. ❺ A big (usually over 15 mm) green stink bug feeding on late summer or autumn plants and fruits is likely to be the **Green Stink Bug** (***Acrosternum hilare***). ❻ ***A. pensylvanicum*** has a shorter, broader head than the closely related *A. hilare*. ❼ ***Acrosternum*** species differ from similar green stink bugs in having a finger-like lobe projecting forward between the hind legs from the middle of the first abdominal segment. ❽ Some western species of ***Chlorochroa*** are destructive pests, but the only eastern species (***C. persimilis***) is rarely abundant enough to be a pest. It is most common in dry, sandy areas. ❾ The **Southern Green Stink Bug** (***Nezara viridula***) is a serious pest of soybean and other crops in the southern United States, but becomes rare northward and does not occur in Canada, although it has been intercepted or recorded as an adventive species in Ontario and Quebec. These overwintering adults were photographed under the bark of a tree in Florida.

FAMILY PENTATOMIDAE Subfamily Pentatominae **(continued)** ❶ ❷ The **Red-shouldered Stink Bug** (***Thyanta accerra***) is a small (about 10 mm), relatively rare stink bug that occurs in both green and brown forms, with the green forms predominating in summer and the brown forms found in fall. Some *Thyanta* individuals can change color from green to brown. Brown forms of *Thyanta* are distinctively covered with black dots; green forms are smaller than other green stink bugs. ❸ Our two species of ***Trichopepla*** (this is ***T. semivittata***) are distinctively hairy stink bugs found on a variety of weeds. ❹ Our one species of ***Cosmopepla***, ***C. bimaculata*** (sometimes called the Two-spotted Stink Bug, a common name more properly applied to *Perillus bioculatus*), occurs on a variety of host plants. This easily recognized, little (about 6 mm) red-and-black bug is common on mullein plants. ❺ Our one species of ***Mormidea***, ***M. lugens***, is a distinctively patterned little bug (about 6 mm) most easily found on native grasses. ❻ ❼ ***Menecles insertus*** (our only *Menecles*) is rarely seen, perhaps because this distinctive, medium-sized (12–14 mm), plant-feeding stink bug lives in trees and is active at night. These photos show a nymph and an adult of this distinctively shaped bug. ❽ The most common species of ***Neottiglossa*** found in eastern North America is our smallest (usually less than 5 mm) stink bug. ***N. undata*** occurs in open areas, where it feeds on grasses. ❾ Our only ***Coenus***, ***C. delius***, is a distinctively rounded, yellowish bug often found on native grasses. The big scutellum (shield) makes it look a bit like a shield bug, but note that the tips of the wings are exposed. About 5 mm.

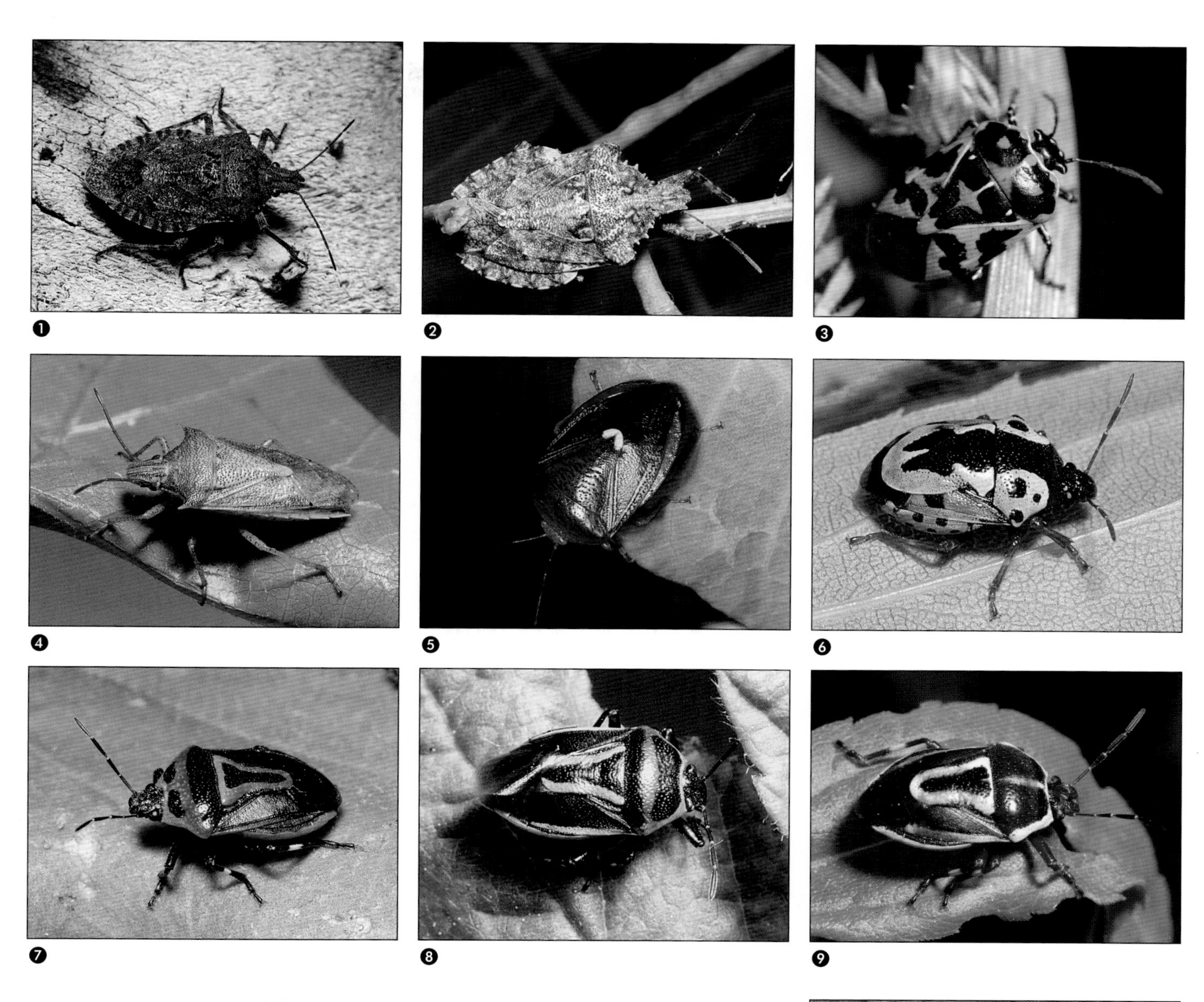

FAMILY PENTATOMIDAE Subfamily Pentatominae **(continued)** ❶ **The Four-humped Stink Bug**, or **Rough Stink Bug** (***Brochymena quadripustulata***), is usually found on tree trunks, where its barklike coloration and slow movements make it difficult to spot despite its large size (around 15 mm). Although they normally feed on trees, members of this species also eat other insects. ❷ ***Parabrochymena arborea*** has squarish, strongly toothed shoulders unlike those of the otherwise similar, but much more common Four-humped Stink Bug. ❸ The **Harlequin Bug** (***Murgantia histrionica***), a well-known consumer of cabbage and related plants, spread north from its native Mexican and Central American home about 100 years ago. It is now a pest as far north as the 40th parallel. ❹ The **Rice Stink Bug** (***Oebalus pugnax***) is a pest that feeds not only on rice, but also on wheat and other grasses as far north as Minnesota. Subfamily Edessinae ❺ ***Edessa florida*** is the only member of this mostly neotropical genus to reach northeastern North America. This one was photographed in North Carolina. Subfamily Asopinae (PREDACIOUS STINK BUGS) ❻ ***Stiretrus anchorago*** is a distinctive stink bug found in grassy areas as far north as southern Ontario. This species is a predator, especially of caterpillars and beetles. ❼ Our most common ***Perillus***, the **Two-spotted Stink Bug** (***P. bioculatus***). Nymphs are voracious consumers of potato beetle eggs, and adults destroy large numbers of beetle larvae. Two-spotted Stink Bugs are usually about 10 mm long, and although they vary widely in color, they usually differ from other *Perillus* species in having two black spots on a bright yellow, orange or red pronotum. ❽ ***Perillus exaptus*** is only about 6 mm, smaller than other *Perillus* species, and usually has a black bar across the pronotum. ❾ ***Perillus circumcinctus*** occurs from Ontario south to Missouri. ❿ The distinctively patterned ***Perillus strigipes*** is a predator like its more common congeners. This species is sometimes included in the genus *Mineus*.

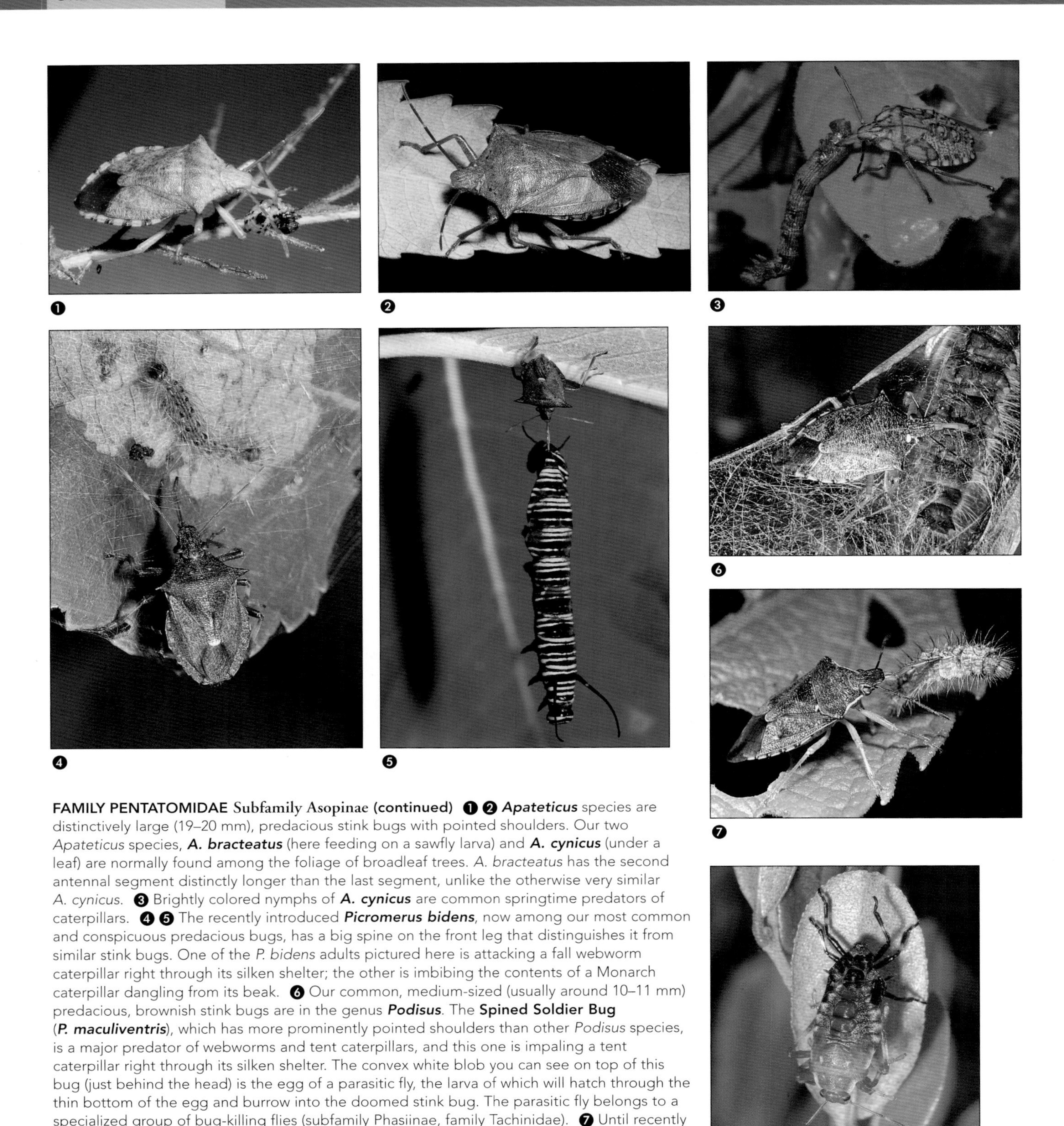

FAMILY PENTATOMIDAE Subfamily Asopinae **(continued)** ❶ ❷ ***Apateticus*** species are distinctively large (19–20 mm), predacious stink bugs with pointed shoulders. Our two *Apateticus* species, ***A. bracteatus*** (here feeding on a sawfly larva) and ***A. cynicus*** (under a leaf) are normally found among the foliage of broadleaf trees. *A. bracteatus* has the second antennal segment distinctly longer than the last segment, unlike the otherwise very similar *A. cynicus*. ❸ Brightly colored nymphs of ***A. cynicus*** are common springtime predators of caterpillars. ❹ ❺ The recently introduced ***Picromerus bidens***, now among our most common and conspicuous predacious bugs, has a big spine on the front leg that distinguishes it from similar stink bugs. One of the *P. bidens* adults pictured here is attacking a fall webworm caterpillar right through its silken shelter; the other is imbibing the contents of a Monarch caterpillar dangling from its beak. ❻ Our common, medium-sized (usually around 10–11 mm) predacious, brownish stink bugs are in the genus ***Podisus***. The **Spined Soldier Bug** (***P. maculiventris***), which has more prominently pointed shoulders than other *Podisus* species, is a major predator of webworms and tent caterpillars, and this one is impaling a tent caterpillar right through its silken shelter. The convex white blob you can see on top of this bug (just behind the head) is the egg of a parasitic fly, the larva of which will hatch through the thin bottom of the egg and burrow into the doomed stink bug. The parasitic fly belongs to a specialized group of bug-killing flies (subfamily Phasiinae, family Tachinidae). ❼ Until recently known as *Podisus modestus*, ***P. brevispinus*** is the most common of the six northeastern ***Podisus*** species. It usually has less pointy shoulders than the Spined Soldier Bug, and can be reliably recognized as our only species with both a darkly marked wing membrane and unmarked legs. ❽ This ***Podisus*** is molting from one nymphal stage to the next.

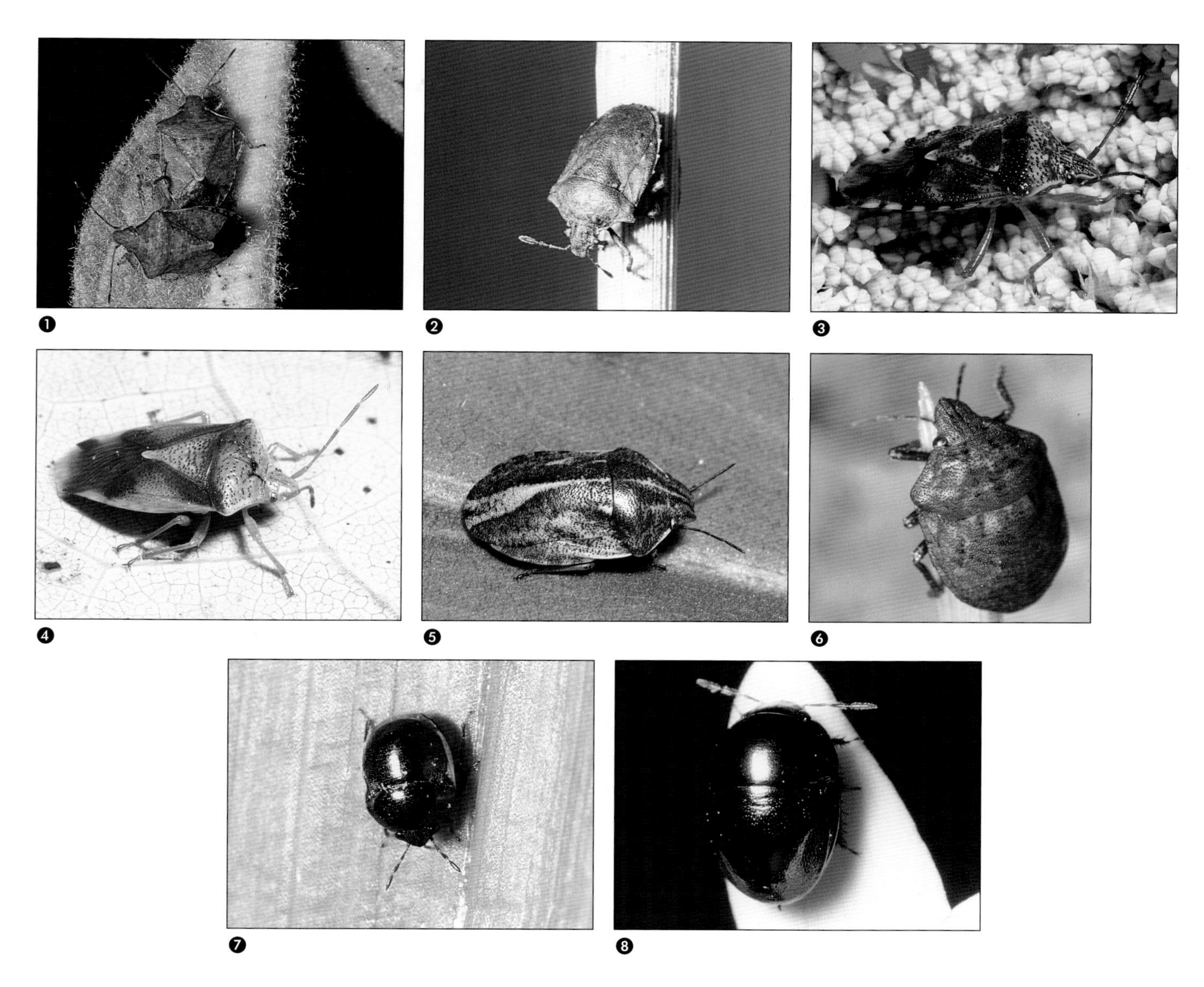

FAMILY PENTATOMIDAE Subfamily Asopinae **(continued)** ❶ ***Podisus serieventris*** can be distinguished from the more common *P. brevispinus* by its heavily speckled hind legs. **Subfamily Podopinae** (TURTLE BUGS) ❷ ***Amaurochrous*** species are called turtle bugs. These phytophagous bugs are rarely encountered. **FAMILY ACANTHOSOMATIDAE** (PARENT BUGS) ❸ The only North American ***Elasmucha, E. lateralis*** (about 8 mm), is a common bug found on birch catkins. Females lay their eggs away from the catkins, later leading the nymphs to the catkin food source. *Elasmucha* females continue to care for their nymphs throughout their development; thus the common name "Parent Bug" given to a European species very similar to our *E. lateralis*. Members of the small family Acanthosomatidae (only two North American genera) are often included with the stink bugs, but differ in having only two tarsal segments (stink bugs have three). ❹ The **Red-crossed Stink Bug** (***Elasmostethus cruciatus***) has the specific name "cruciatus" because of the reddish "X" on its back. Another *Elasmostethus, E. atricornis*, is similar, but darker, with black antennae unlike the pale antennae of the Red-crossed Stink Bug. About 11 mm. **FAMILY SCUTELLERIDAE** (SHIELD BUGS) ❺ Shield bugs look like stink bugs, but have a big shield concealing their wings from the top. This attractive species (***Eurygaster alternata***) is often abundant in wet meadows, especially on sedges. About 9 mm. ❻ The large (about 15 mm) **Shieldbacked Pine Seed Bug** (***Tetyra bipunctata***) often shows up in dry, sandy pinewoods, where it eats pine seeds. **FAMILY THYREOCORIDAE** (NEGRO BUGS) ❼ At first glance, negro bugs look like little black beetles, but similar beetles have a straight line down the back where the wing covers meet (this is never the case in bugs). On closer inspection, negro bugs look like small shield bugs, but they are much smaller, black in color and have spiny tibiae. ***Corimelaena pulicaria***, the most common northeastern negro bug, is a small (about 3 mm) bug that can often be shaken out of bunches of berries or seeds. ❽ ***Galgupha*** species, of which ***G. atra*** is our most common species, lack the white edge that marks most *Corimelaena* species. This one is sitting on a daisy petal. About 5 or 6 mm.

FAMILY CYDNIDAE (BURROWER BUGS) ❶ Burrower bugs look like small, rounded, black or brown stink bugs with heavy spines on the middle and hind legs. Some species are obviously modified for burrowing, with flattened and somewhat shovel-like front legs. ***Pangaeus bilineatus***, our largest burrower bug at 6–8 mm, is an uncommon species sometimes found under logs on sandy beaches, where they probably feed on the roots of dune plants. ❷ The only North American species in the genus ***Sehirus*** is ***S. cinctus***, a shining blue-black species with a contrasting white margin. These bugs are often found on stems and foliage in lush stands of herbaceous plants such as nettles and mint. *S. cinctus* is smaller than *Pangaeus*, but much larger than other burrower bugs in our area. ❸ The burrower bug genus ***Melanaethus*** is mostly southwestern, but this species (***M. robustus***) ranges as far north as Ontario. ❹ ***Amnestus*** species, such as this ***A. pallidus***, are the smallest North American burrower bugs, at only 2.5 mm. **FAMILY REDUVIIDAE** (ASSASSIN BUGS AND THREAD-LEGGED BUGS) Subfamily Reduviinae ❺ ❻ This regular resident of even the cleanest of households is the **Masked Bed Bug Hunter** (***Reduvius personatus***). Nymphs exude a sticky substance that soon accumulates a body-masking layer of junk ranging from the dead bodies of victims to dismembered dust bunnies. Next time you spot some wandering lint in the corner of your living room, take a closer look, but don't touch – this species can give you a painful bite. Subfamily Peiratinae ❼ ***Melanolestes picipes*** is a common, robust (15–20 mm), black assassin bug found throughout much of the eastern United States. ❽ ***Sirthenea carinata*** is a large (18–20 mm) assassin bug found from Ohio and New Jersey south.

FAMILY REDUVIIDAE (continued) Subfamily Harpactorinae ❶ ❷ The bright green, wingless assassin bugs commonly seen in late fall and early spring are usually the overwintering nymphs of ***Zelus luridus***, one of our most common assassin bugs. Look on foliage for these attractive bugs, as often as not with fly or wasp victims impaled on their beaks. Adults (about 16 mm) retain some of the nymph's distinctive green color. Some *Zelus* have paternal care, with the male guarding his eggs and feeding newly hatched nymphs with prey proffered on his beak like invertebrate shish-kebabs. ❸ This ***Zelus longipes*** has impaled a small soldier beetle. ❹ ***Sinea spinipes*** differs from the more common *Sinea diadema* in having blunt tubercles on its pronotal lobes. ❺ The **Spined Assassin Bug** (***Sinea diadema***) is the common brown assassin bug of open areas, such as fields, roadsides and railway embankments, where it can be abundant on flowers. The second long segment of the front leg (the front tibia) has stout spines on its inner surface and the femur has a long spine on the upper surface, which distinguish it from another common brown assassin bug (*Acholla multispinosa*). About 13 mm. ❻ ***Acholla multispinosa***, the common brown assassin bug of wooded areas in the northeast, is the only northeastern species in this genus. The inner surface of the front tibia lacks the spines that characterize the similar *Sinea* species. About 14 mm. ❼ The **Wheel Bug** (***Arilus cristatus***), a common southern insect that occasionally shows up as far north as Canada, is a large (28–36 mm) and distinctly ornamented assassin bug. Handle Wheel Bugs with care, as they can inflict a painful bite. ❽ ***Repipta taurus*** is a brightly and distinctively colored assassin bug found from Pennsylvania south.

FAMILY REDUVIIDAE (continued) ❶ This distinctively colored assassin bug (***Pselliopus cinctus***; 12–13 mm) ranges from Florida north to Massachusetts. ❷ This distinctive little (about 10 mm) red-and-black assassin bug (***Rhynocoris ventralis***) is rarely seen, but shows up regularly in insect traps in Ontario oak savannas. ❸ ***Fitchia aptera*** normally has such tiny wings that its whole, distinctively striped abdomen is exposed. This rarely encountered assassin bug lives on the ground among grasses. From 12–14 mm. **Subfamily Microtominae** ❹ This large (almost 30 mm), strikingly colored bug (***Microtomus purcis***) was coaxed out from under the bark of a dead tree. *Microtomus* has a relatively southern distribution, and only occurs as far north as Indiana. **Subfamily Apiomerinae** ❺ Although *Rhynocoris ventralis* is the only red-and-black assassin bug known from as far north as Ontario, reduviid variety increases as one heads south. This stout-bodied ***Apiomerus crassipes*** (14–19 mm) occurs from Illinois south to Florida. **Subfamily Ectrichodiinae** ❻ The **Scarlet-bordered Assassin Bug** (***Rhiginia cruciata***), which ranges from Illinois south, is the only eastern North American member of the subfamily Ectrichodiinae. **Subfamily Saicinae** ❼ ***Oncerotrachelus acuminatus*** is a small (4–7 mm) but distinctive assassin bug that occurs throughout our area. **Subfamily Triatominae** ❽ "Kissing bugs" in the bloodsucking genus ***Triatoma*** range as far north as Pennsylvania. This is a ***T. dispar***, photographed in Costa Rica. ❾ Although ***Rhodnius prolixus*** only ranges as far north as Mexico in nature, this bloodsucking triatomine is widely maintained in laboratory colonies, and figures prominently in the history of insect physiology research.

FAMILY REDUVIIDAE (continued) Subfamily Stenopodainae ❶ ***Pnirontis*** is a mostly southeastern genus, but ***P. modesta*** ranges as far north as Ontario. ❷ The distinct neck and large size of this assassin bug identifies it as ***Stenopoda cinerea***, a species (and genus) that ranges from eastern Canada south to Florida. Subfamily Emesinae (THREAD-LEGGED BUGS) ❸ Thread-legged bugs often occur in clumps of native grasses, where they prey on small insects. Members of our most common genus (***Barce***), are around 10 mm in length. ❹ ***Barce fraterna*** is the more common of the two northeastern *Barce* species, and the only one that ranges north to Canada. ❺ ❻ The largest species of thread-legged bug found in eastern North America, ***Emesaya brevipennis***, reaches almost 40 mm. Despite this size, these extraordinarily slender insects are hard to see in nature, where their dull colors and slow movements combine with their disproportionately long, but almost invisibly threadlike legs to make an inconspicuous package. You may even have difficulty picking out the whole insect in the above photograph (look for the long midlegs and hind legs), although it is easy enough to see when pinned against a blue background. *Emesaya* species sometimes hang out in spider webs, feeding as kleptoparasites (food thieves) on captured insects. ❼ ***Empicoris*** species are small (about 5 mm) and inconspicuous thread-legged bugs. ❽ ***Ploiaria carolina*** is a small (about 5 mm), slow-moving thread-legged bug.

FAMILY PHYMATIDAE (AMBUSH BUGS) ❶ ❷ A close look at a flower apparently sprouting an insect corpse will often reveal a motionless, cryptically colored ambush bug, with its beak deeply imbedded in the more conspicuous corpse of its prey. ***Phymata pennsylvanica*** is our most common species; a second northeastern species (*P. americana*) has the sides of the abdomen less angulate than *P. pennsylvanica*. ❸ This dark male ambush bug riding piggyback on a paler female is protecting his paternal investment by blocking her from mating with other males. Dark forms occur commonly among ***Phymata*** males, but not among females. **FAMILY NABIDAE** (DAMSEL BUGS) ❹ The **Common Damsel Bug** (***Nabis americoferus***) is one of the most common damsel bugs in North America, and an important natural pest control agent in orchards and gardens. *Nabis* differs from our other common genus of damsel bugs in having a relatively broad body, and a first antennal segment shorter than the head. ❺ ***Nabis americoferus*** overwinters as an adult, and often can be seen on the snow surface on warm winter days. ❻ This black, shining, usually short-winged bug common in meadows across the continent is ***Nabicula subcoleoptrata***, a native species that often feeds on the introduced Meadow Plant Bug (family Miridae). ❼ Like the common *Nabicula subcoleoptrata*, other species of ***Nabicula*** are usually short-winged and have the first segment of the antennae longer than the head. This is ***N. americolimbata***. ❽ ***Pagasa fusca*** is a rarely encountered damsel bug that differs from other Nabidae in having five antennal segments rather than four. This one is brachypterous (short-winged), but the species occurs in both short-winged and long-winged forms. ❾ ***Hoplistoscelis sordidus*** is a distinctive damsel bug with characteristically banded legs and usually very short wings. This one was photographed in southern Ontario, as far north as this eastern species ranges.

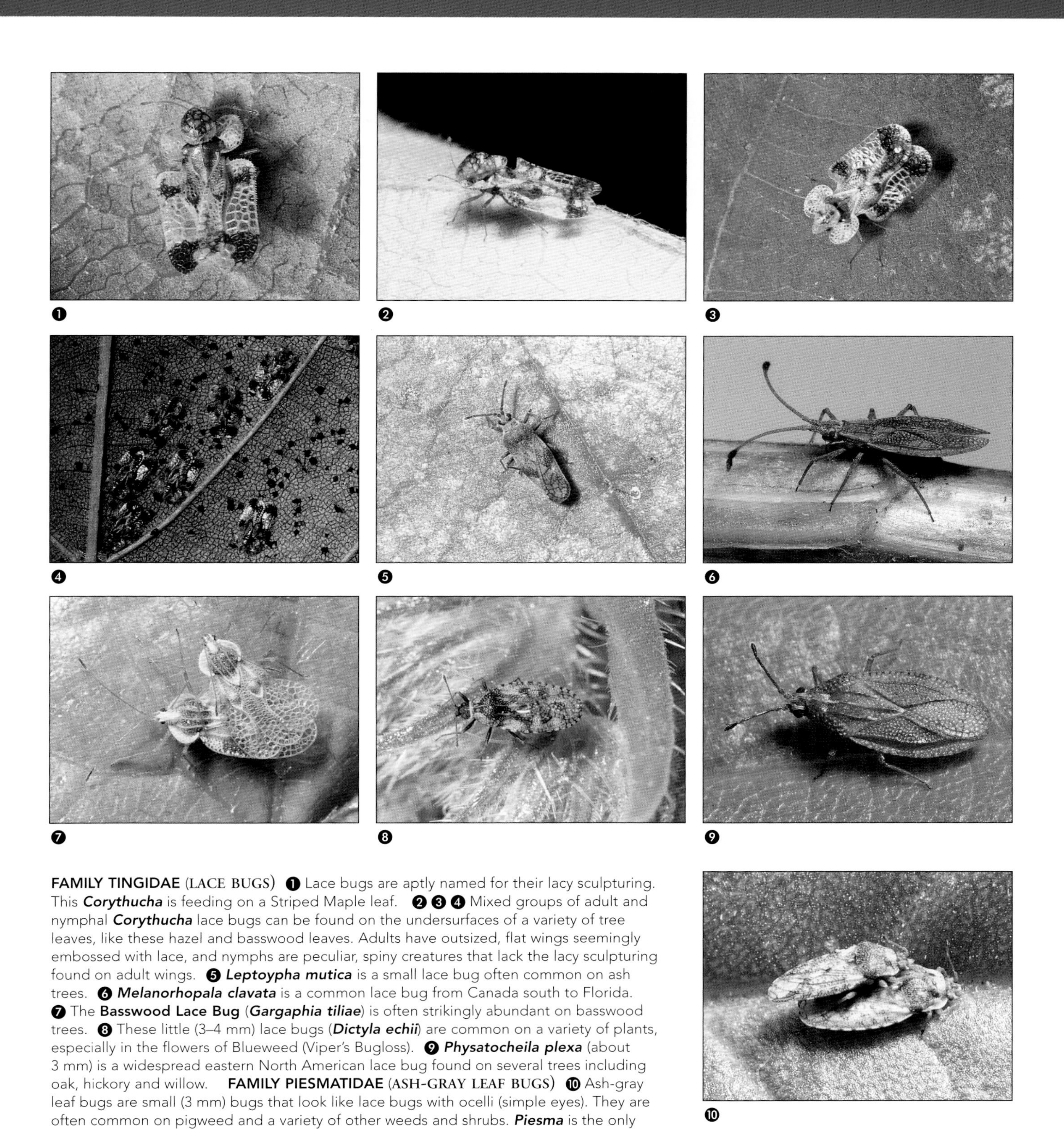

FAMILY TINGIDAE (LACE BUGS) ❶ Lace bugs are aptly named for their lacy sculpturing. This ***Corythucha*** is feeding on a Striped Maple leaf. ❷❸❹ Mixed groups of adult and nymphal ***Corythucha*** lace bugs can be found on the undersurfaces of a variety of tree leaves, like these hazel and basswood leaves. Adults have outsized, flat wings seemingly embossed with lace, and nymphs are peculiar, spiny creatures that lack the lacy sculpturing found on adult wings. ❺ ***Leptoypha mutica*** is a small lace bug often common on ash trees. ❻ ***Melanorhopala clavata*** is a common lace bug from Canada south to Florida. ❼ The **Basswood Lace Bug** (***Gargaphia tiliae***) is often strikingly abundant on basswood trees. ❽ These little (3–4 mm) lace bugs (***Dictyla echii***) are common on a variety of plants, especially in the flowers of Blueweed (Viper's Bugloss). ❾ ***Physatocheila plexa*** (about 3 mm) is a widespread eastern North American lace bug found on several trees including oak, hickory and willow. **FAMILY PIESMATIDAE** (ASH-GRAY LEAF BUGS) ❿ Ash-gray leaf bugs are small (3 mm) bugs that look like lace bugs with ocelli (simple eyes). They are often common on pigweed and a variety of other weeds and shrubs. ***Piesma*** is the only North American genus in this family, and ***P. cinereum*** is the common species.

FAMILY ARADIDAE (FLAT BUGS) ❶ ❷ Perhaps the oddest eating habits among the Hemiptera prevail in a fascinating family of peculiar, flattened bugs found under bark where they feed on fungi using long, thin, threadlike mouthparts. Most of our species are in the genus ***Aradus***, like this scallop-sided ***A. crenatus*** ❶ and these mating ***A. aequalis*** ❷. These are our most common, large (around 10 mm) flat bug species. ❸ ❹ The genus *Aradus* contains over 75 North American species, many of which are quite habitat specific. ***A. robustus*** and ***A. duzeei*** are two relatively small (6–7 mm) flat bugs found in eastern North America. *A. robustus* is the more common species, and is distinctive for its fat and spiny second antennal segment. Look for it under the bark of dead oak trees. ❺ ***Neuroctenus simplex*** is a distinctively blackish flat bug found under the bark of oak and other hardwoods. ❻ ❼ ***Aneurus*** species are wafer-thin, with smoother surfaces and margins than *Aradus* species. *Aneurus*, like these ***A. simplex*** (about 5 mm), are often found in family groups, including eggs, nymphs and adults. **FAMILY LYGAEIDAE** (SEED BUGS) **Subfamily Lygaeinae** ❽ One of the most popular entomological "lab rats" is the **Large Milkweed Bug** (***Oncopeltus fasciatus***), a large (over 17 mm) red-and-black bug easily reared on a diet of milkweed seeds and adequate water. Large milkweed bugs are abundant throughout our area in late summer, but do not normally survive the winter north of the Canadian border. This photograph shows a nymph and an adult on a milkweed leaf. ❾ This easily recognized, bright red seed bug (***Neacoryphus bicrucis***) is common in the southern United States, but scarce as it approaches its northern limits in southern Canada. At 7–9 mm, it is substantially smaller than the familiar and similarly colored milkweed bugs (*N. bicrucis* does not feed on milkweed plants).

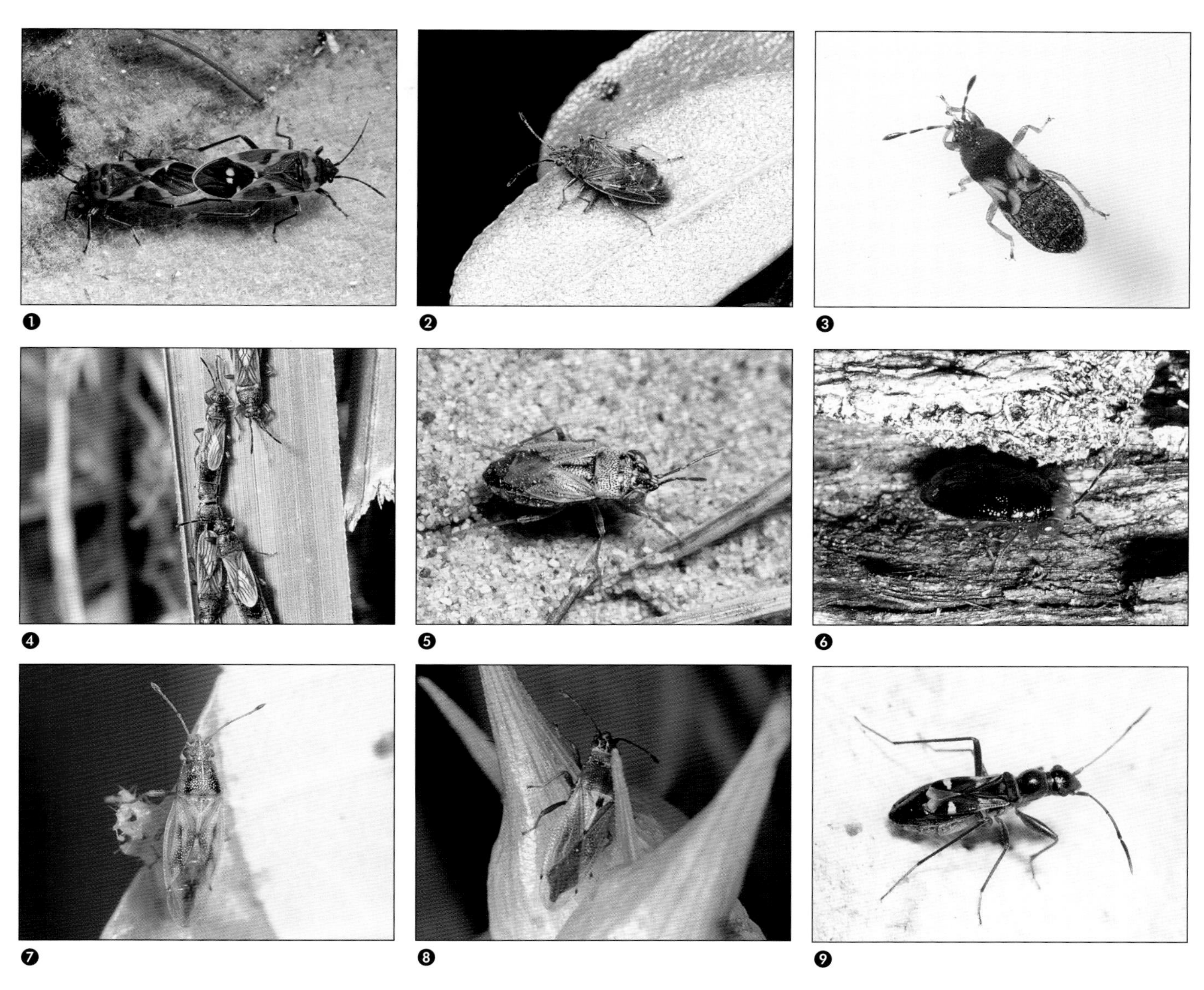

❶ ❷ ❸ ❹ ❺ ❻ ❼ ❽ ❾

FAMILY LYGAEIDAE **Subfamily Lygaeinae (continued)** ❶ Although the Large Milkweed Bug (*Oncopeltus fasciatus*) is a species that doesn't make it through the winter in the northern United States and Canada, the smaller (10–12 mm) **Small Milkweed Bug** (***Lygaeus kalmii***) is a similar red-and-black milkweed bug that breeds continuously throughout our area. **Subfamily Ischnorhynchinae** ❷ The **Birch Catkin Bug** (***Kleidocerys resedae***) makes up in abundance what it lacks in size (these little bugs are only 3–4 mm in length). Although Birch Catkin Bugs feed on the seeds of a number of trees and shrubs, they are often conspicuously abundant on birch trees. They also smell really bad. The adults congregate in the fall as they seek a sheltered place to spend the winter. **FAMILY BLISSIDAE** ❸ The **Hairy Chinch Bug** (***Blissus leucopterus hirtus***) is a common backyard bug that sucks sap from grass crowns and stems. If you have brown patches on your lawn, they could be the result of these little (about 4 mm) bugs sucking the sap from your grass. The Hairy Chinch Bug adult is black with a characteristic white band across its back. ❹ ***Ischnodemus falicus*** is a distinctively shaped bug sometimes abundant in the thatch around the bases of native grasses. About 5 mm. **FAMILY GEOCORIDAE** (BIG-EYED BUGS) ❺ Geocoridae, aptly called big-eyed bugs, are active predators easily spotted as they run around open, dry places like sparse grasslands and sand dunes. ***Geocoris bullatus***, a little (3 mm or so) brownish gray species, is a common big-eyed bug. ❻ ***Isthmocoris piceus*** looks a bit like the more common *Geocoris*, but the bulgy, stalked eyes and striking coloration of this beautiful little bug distinguishes it from its dowdier relatives. About 4 mm. **FAMILY CYMIDAE** ❼ ❽ The small (about 4 mm), elongate species of *Cymus* often sit concealed among the seeds of *Carex* and similar plants. The little dimples all over their backs make these bugs distinctive, despite their small size. The bugs shown here are ***Cymus luridus*** (most common in the north) and ***C. angustatus*** (widespread), respectively. **FAMILY RHYPAROCHROMIDAE** ❾ ***Slaterobius insignis*** (about 6 mm) can be common on sedges and grasses in wet, open areas; it is also sometimes abundant in blueberry bogs.

FAMILY RHYPAROCHROMIDAE (continued) ❶ ***Myodocha serripes*** is perhaps the most distinctively shaped of our common seed bugs. This widespread seed bug often occurs among stems and other debris in open areas or forest margins. About 9 mm. ❷ ***Ligyrocoris diffusus*** is a medium-sized (around 8 mm) seed bug that is often common along roadsides. ❸ ***Emblethis vicarius*** is a distinctively robust, medium-sized (about 7 mm) seed bug common in dry areas like open fields. ❹ ***Heraeus plebejus*** is a common seed bug throughout the northeast, where it is most often found in fields near woodland margins. ❺ ***Cnemodus mavortius*** is a widespread species usually found in dry or gravelly areas. **FAMILY PACHYGRONTHIDAE** ❻ ***Oedancala dorsalis*** is found in wet areas, usually on sedges where it feeds in the seed heads. **FAMILY ARTHENEIDAE** ❼ This **Cattail Bug** (***Chilacis typhae***) was one of a dozen picked out of a single cattail (*Typha*) head. Cattail Bugs were only recently (accidentally) introduced to North America, and occur in cattail heads with a fluffed-out appearance caused by Cattail Caterpillars (*Limnaecia phragmitella* – see Cosmopterygidae, page 193). **FAMILY BERYTIDAE** (STILT BUGS) ❽ Stilt bugs are common little bugs with a "mini-walkingstick" appearance created by the long, stiltlike legs and slender body. We only have a few species, and since they all have spindle-shaped swellings on their antennae, the family is easy to recognize in the field. Some stilt bug species are both predators and plant feeders. ***Neoneides muticus*** (about 10 mm) is by far the most common species. ❾ Although similar in size and shape to the more common *Neoneides muticus*, ***Jalysus*** species like this ***J. wickhami*** have a much shorter head. *Jalysus* also lacks the elephant-like snout found on top of the head of *Neoneides* (although you will need a good hand lens to see this).

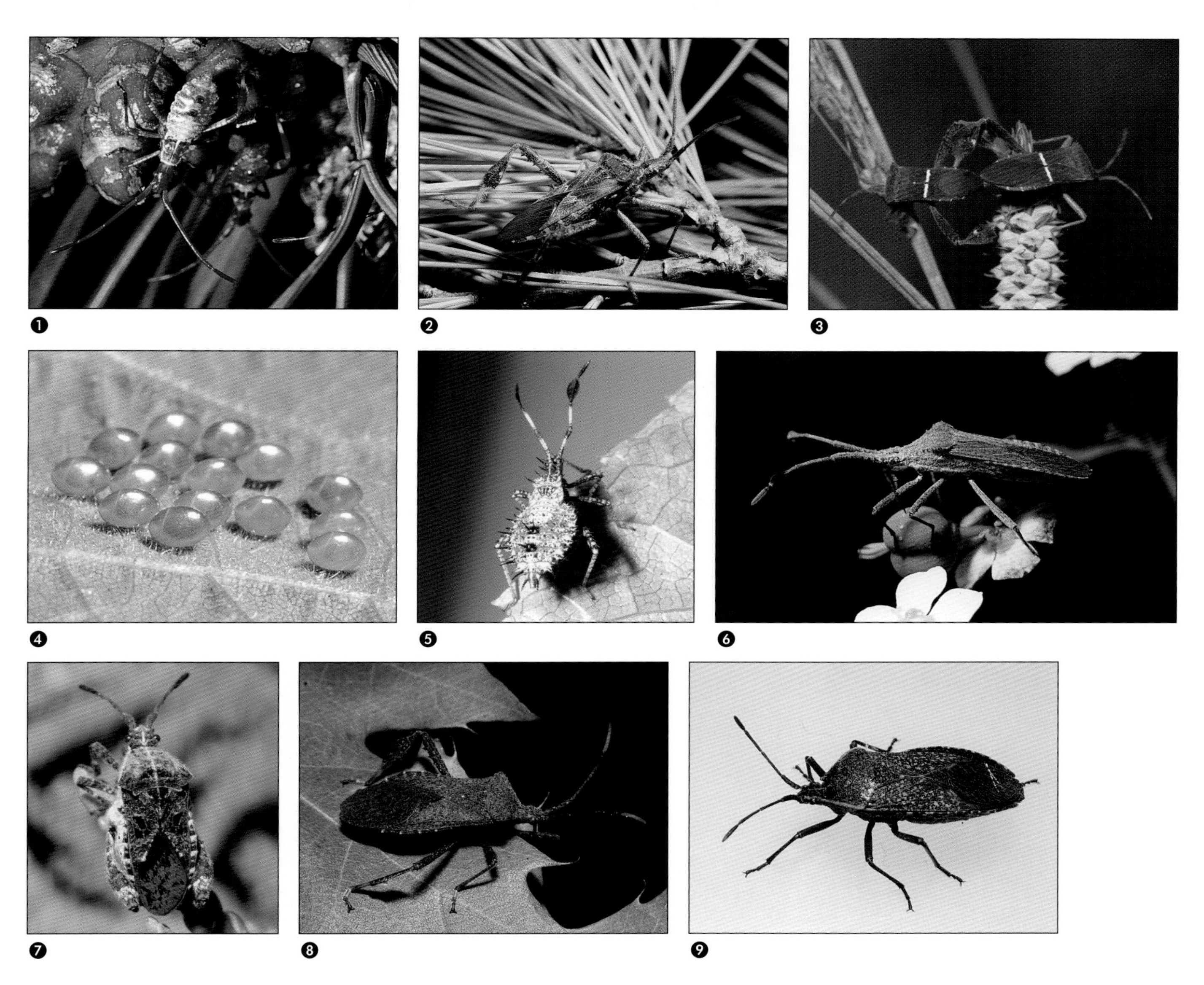

FAMILY COREIDAE (LEAF-FOOTED BUGS) ❶ ❷ Our most conspicuous common coreids are large (20 mm), slow-moving bugs with flattened and expanded hind legs. These big bugs are **Western Conifer Seed Bugs** (***Leptoglossus occidentalis***), a species of leaf-footed bug that didn't show up in eastern North America until about 1980. *L. occidentalis* has long been an abundant pest of conifer seeds in the west. *Leptoglossus* bugs are now among the most common Coreidae in the northeast, and can often be seen on walls and in buildings in late fall as they aggregate for the winter. ❸ The distinctively striped ***Leptoglossus phyllopus*** is a very common coreid in the southern United States, and ranges as far north as New York. ❹ ❺ ❻ The distinctive golden eggs, spiny nymphs and broadened antennae of ***Chariesterus antennator*** make this an unmistakable species. *C. antennator*, our only *Chariesterus*, is uncommon and found in native grasslands where the adults usually feed on Flowering Spurge (*Euphorbia corollata*). These photos were taken in southern Ontario. About 14 mm. ❼ ***Merocoris distinctus*** is an aptly named species because it is distinct for its small size (8–9 mm) and unusually fuzzy appearance. It is common throughout our area, often occurring on flowers. ❽ ***Acanthocephala terminalis*** is one of our largest (over 20 mm) and most spectacular leaf-footed bugs. Its broad hind tibiae explain the common name "leaf-footed," and its orange terminal antennal segment differentiates it from other *Acanthocephala* (more southerly in distribution). These big bugs are distributed as far north as southern Ontario. ❾ The **Squash Bug** (***Anasa tristis***) can be conspicuous because of its size (about 16 mm) and occasional abundance in gardens. It is dull colored and lacks the flattened "leaf feet" which give the Coreidae their common name.

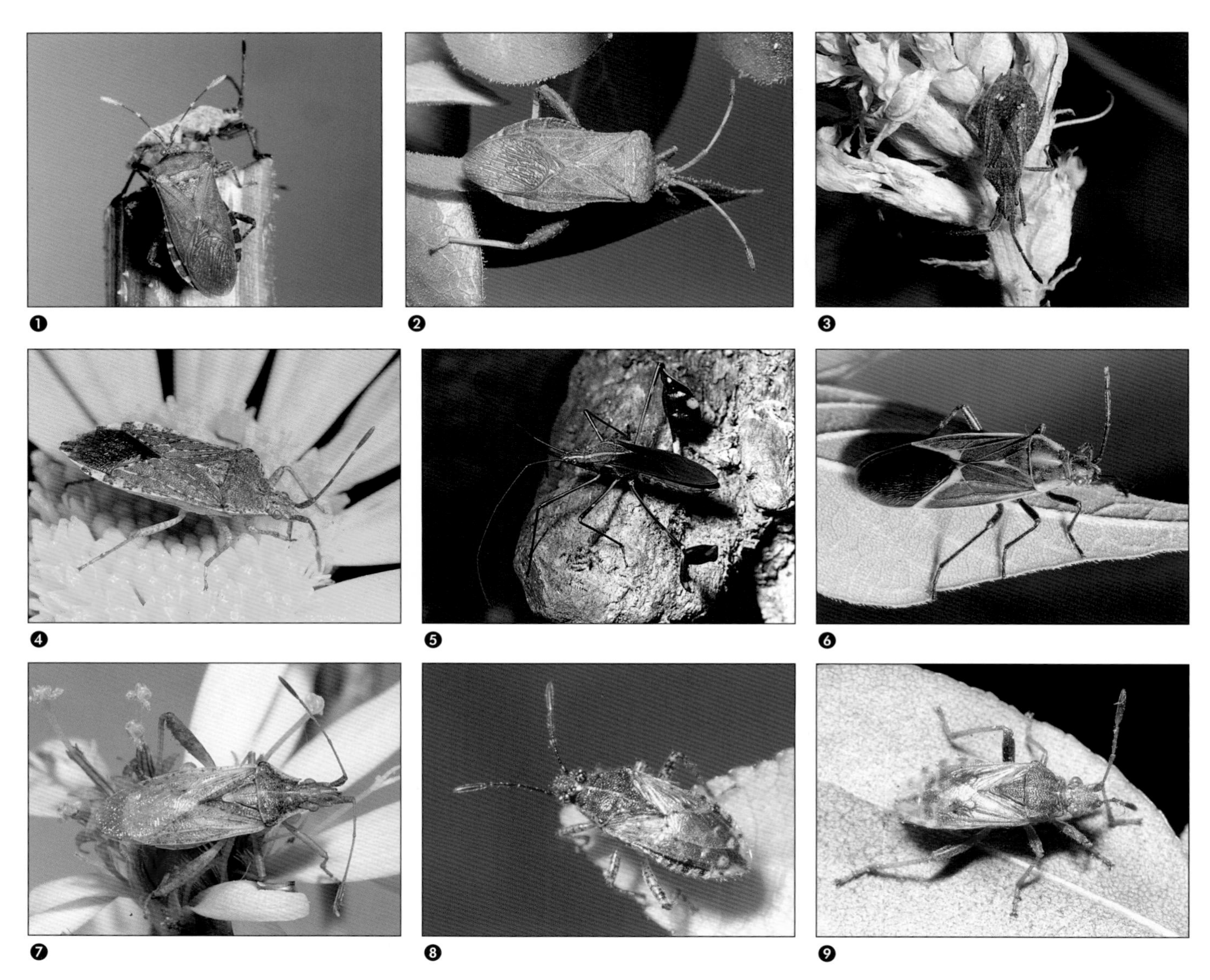

FAMILY COREIDAE (continued) ❶ Squash Bugs (*Anasa tristis*), and the closely related ***A. armigera*** (the **Horned Squash Bug**), are sometimes pests of squash, cucumber and other domestic curcurbits. The Horned Squash Bug (nymph and adult shown here) looks much like the Squash Bug, but unlike *A. tristis*, it has a pair of hornlike processes at the base of the antennae. ❷ Large hind femoral spines distinguish ***Euthochtha galeator*** from the similarly sized (about 16 mm) *Anasa* species, and the unmodified hind tibiae distinguish it from the leaf-footed *Acanthocephala terminalis*. This species is conspicuous in flight because of its brightly colored reddish hind wings. ❸ ***Coriomeris humilis*** is a relatively small (7–9 mm) species that is common in the west, but relatively rare in eastern North America. It is the only *Coriomeris* in the northeast. ❹ ***Althos obscurator*** is a small (10 mm) coreid that is distributed from North Carolina south to Brazil. ❺ Tropical leaf-footed bugs, like this **Passion Vine Bug** (***Diactor bilineatus***) often have spectacularly expanded hind legs. **FAMILY RHOPALIDAE** (SCENTLESS PLANT BUGS) ❻ Scentless plant bugs, so-called because they lack the big scent gland openings found above the hind legs of most other bugs, usually look much like small, dull-colored Coreidae, but the **Box Elder Bug** (***Boisea trivittata***) is a relatively large (about 14 mm), bright red-and-black bug. This common bug, which feeds on seeds of Box Elder or Manitoba Maple, often aggregates on sunny walls while looking for a place to hibernate. Box Elder Bugs do not endanger their host trees, but are sometimes a nuisance when they decide to spend the winter inside your house. ❼ The only common ***Harmostes*** in the northeast (***H. reflexulus***) differs from other genera of scentless plant bugs in having spines on the hind femora, and pointy processes projecting off the thorax alongside the head. About 8 mm. ❽ ***Arhyssus*** species are often found on fall flowers. ***A. lateralis*** seems to be our most common species, but the darker *A. nigristernum* also occurs in our area. ❾ Our most common scentless plant bug, ***Stictopleurus punctiventris***, has a pair of lines or grooves behind the head that end in little loops on each side. These reddish, medium-sized (7–8 mm) bugs are common in wet meadows and other open areas.

FAMILY ALYDIDAE (BROAD-HEADED BUGS) ❶ ❷ ❸ Like leaf-footed bugs, broad-headed bugs are all plant feeders on herbaceous plants. These photos show the most common northeastern ***Alydus*** species (about 15 mm). ***A. conspersus*** is paler and conspicuously "spotted" on the wing membranes; ***A. eurinus*** is the common black broad-headed bug found on goldenrod flowers. Adult *Alydus* are unmistakably bugs, but the nymphs are remarkable ant mimics. ❹ The only northeastern ***Protenor*** (***P. belfragei***) seems to refute the common name "broad-headed," as this distinctive insect has a parallel-sided body and a narrow, somewhat tapered head. ❺ ***Megalotomus quinquespinosus*** looks like an assassin bug, but upon closer inspection it is clearly a plant feeder, exhibiting many wing veins, a big stink gland opening and the broad head that mark it as a broad-headed bug. This handsome bug, our only *Megalotomus*, is common on weedy road margins in forested areas. **FAMILY PYRRHOCORIDAE** (RED BUGS) ❻ Red bugs, sometimes called stainers because of the way some feed on cotton, causing discoloring or staining, are common in Florida, Caribbean and the neotropics. The few species of red bugs that occur in North America belong to the genus ***Dysdercus***. **FAMILY LARGIDAE** (LARGID BUGS) ❼ The only largid bug to reach the northern United States is ***Largus succinctus***, a big (13–17 mm), distinctively colored bug found as far north as southern New York. These plant-sucking bugs are usually found on low vegetation, but this one is on a young pine tree. **FAMILY MIRIDAE** (PLANT BUGS) ❽ Plant bugs, like this ***Taedia scrupea***, have the wing abruptly bent down over the abdomen just before the start of the membranous part of the wing. That wing bend, plus the absence of ocelli and the presence of a couple of closed loops in the wing membrane, characterizes the huge family Miridae. ❾ ❿ The **Meadow Plant Bug** (***Miris dolabratus***, 8–9 mm) is often tremendously abundant in early summer meadows. Meadow Plant Bugs are distinctive in shape and color, although the odd short-winged female may appear quite different from the more common long-winged individuals.

FAMILY MIRIDAE (continued) ❶ **Tarnished Plant Bugs** (***Lygus lineolaris***) are extremely common, nondescript, little (about 5 mm) brown bugs. They are serious pests on a tremendous variety of foliage and fruit, which they disfigure with their feeding punctures. *Lygus* bugs spend the winter as adults and stay active well after the last frost. ❷ Tiny (2 mm or so), flealike jumping bugs are sometimes tremendously abundant in lawns, especially late in the season when one species (***Halticus bractatus***, the **Garden Flea Hopper**) often abounds in lawns with lots of white clover. The rotund individual shown here is a female with evenly thickened wings (without membranes); the more slender males have typical plant bug wings. ❸ The **Four-lined Plant Bug** (***Poecilocapsus lineatus***) is a common backyard plant bug, sometimes occurring in colorful dense populations on garden plants such as currants and daisies. ❹ ***Phytocoris pallidicornis*** is just one common species in this enormous genus of around 200 North American species, almost all of which are predacious. ❺ This bright red species (***Coccobaphes frontifer***) can often be found adorning maple twigs – try shaking some branches onto a white sheet to look for these bright bugs. ❻ This large (8 mm) introduced species (***Megaloceroea recticornis***) is often common on grasses, especially Timothy Grass. ❼ As the species name "binotatus" suggests, ***Stenotus binotatus*** has two distinct markings on the pronotum. These are very common roadside insects, where they feed on grasses. *S. binotatus* was accidentally introduced from Europe about 100 years ago. ❽ The **Alfalfa Plant Bug** (***Adelphocoris lineolatus***) is a very common pest introduced from Europe around 1940. It now occurs across the continent and can be a major pest of alfalfa grown for seed. ❾ ***Capsus ater*** is usually an entirely black species, but strikingly bicolored variants like this one are occasionally encountered. This introduced (European) species is common on roadside grasses.

FAMILY MIRIDAE (continued) ❶ ***Plagiognathus*** is a very diverse plant-feeding genus. ❷ ***Blepharidopterus provancheri*** is predacious on small invertebrates. A closely related species (the Black-kneed Mirid, *B. angulatus*) is a well-studied European import to North America that feeds on several pests including aphids, the Codling Moth, and plant-feeding mites. ❸ ***Hyaliodes vitripennis*** (5 mm) is a predator of aphids and other small arthropods. ❹ ***Dicyphus famelicus*** is a predacious mirid found on host plants with hairy leaves, to which it is adapted by having modified claws to grip the hair shafts and walk on "tiptoes" over the plant's defensive hairs. Some of its close relatives are beneficial predators of greenhouse pests, especially whiteflies. ❺ A thickly haired first antennal segment characterizes members of the genus ***Neurocolpus***, like this ***N. nubilus*** eating a small midge. About a third of the world's 10,000 plant bug species are at least partly predacious. ❻ ❼ ***Prepops***, with over 20 North American species, includes some of our most brightly colored plant bugs, like this ***P. rubellicollis*** drinking from a raspberry, and this flower-sipping ***P. fraternus.*** *P. fraternus* is normally found on sumac. ❽ This ***Pseudoxenetus regalis*** was photographed on a leaf surface as it ran about in a convincingly antlike fashion. ❾ This medium-sized (6–7 mm) plant bug (***Collaria meilleurii***) can be very abundant in wet sedge meadows. ❿ ***Trigonotylus coelestialium*** is a distinctively green plant bug with contrasting pinkish antennae. It is common on a variety of wild grasses, especially in the northern United States and Canada.

FAMILY MIRIDAE (continued) ❶ Jumping tree bugs, like this ***Corticoris signatus***, have been traditionally treated as a separate family (Isometopidae) but are now treated as a subfamily of Miridae (Isometopinae). These rarely seen, little (about 2 mm) bugs have ocelli, unlike other Miridae. ❷ Jumping tree bugs, like this ***Myiomma cixiiforme***, feed on scale insects. **FAMILY ANTHOCORIDAE** (MINUTE PIRATE BUGS) ❸ Although this minute pirate bug (***Anthocoris musculus***) is almost twice the size of the Insidious Flower Bug (*Orius insidiosus*), it is still a small bug (4 mm). Try shaking willow branches to collect these little predators. ❹ The **Insidious Flower Bug** (***Orius insidiosus***) is a little (2 mm) black-and-white bug, commonly found on a variety of flowers and commercially available for use as a biological agent for the control of thrips and other pests in greenhouses. **FAMILY LYCTOCORIDAE** ❺ ***Lyctocoris campestris*** (3–4 mm) often occurs in sheltered places including mammal nests, barns, and houses. Although a predator like the related anthocorids, it sometimes bites people. **FAMILY CERATOCOMBIDAE** ❻ ***Ceratocombus vagans*** (until recently included in the Dipsocoridae) is a tiny (about 1 mm), rarely noticed bug, and the only northeastern member of this small family. This was one of many found in a snow-covered log. **FAMILY ENICOCEPHALIDAE** (GNAT BUGS OR UNIQUE-HEADED BUGS) ❼ Gnat bugs, like this ***Systelloderes biceps***, are bizarre-looking insects with a long head divided into bulbous regions like a miniature balloon animal. The common name "gnat bug" refers to the propensity of these little (3–4 mm) predacious bugs to form large swarms like some small flies (gnats). **FAMILY BELOSTOMATIDAE** (GIANT WATER BUGS) ❽ **Giant Water Bugs** or **Giant Electric Light Bugs** (***Lethocerus americanus***, 50–60 mm) lay masses of eggs on stones, plants, planks or logs in wet places near the ponds and slow streams in which the nymphs and adults live. Another entirely unjustified name for this species is "toe biter"; like all aquatic Hemiptera (except Corixidae), giant water bugs can inflict a painful bite if handled carelessly but, unless you handle a captured bug using your toes, your feet are safe from belostomatid beaks. This *Lethocerus americanus* is just a nymph, but it is still big enough to take a frog as prey.

FAMILY BELOSTOMATIDAE (continued) ❶ Adult giant water bugs have a formidable appearance because of their huge, grasping forelegs and a stout beak. ❷ This pair of giant water bugs (***Lethocerus uhleri***, 45 mm) has climbed out of the water and up a stem to mate and lay eggs. The male will guard the female and insist on repeated matings to ensure paternity, and he will stay on to tend and guard the eggs after the female's departure. Each male marks his nursery stem with his individual chemical marker, produced in large thoracic scent glands. ❸ ***Belostoma flumineum***, the most common *Belostoma*, is about half the size of the Giant Electric Light Bug (*Lethocerus americanus*). Instead of sticking her eggs on inanimate objects like *Lethocerus*, the female *Belostoma* glues them to the back of a male, where his movements provide ventilation and protection. The eggs hatch in 10 days, with the nymphs emerging from a flaplike lid at the top of each egg. Some of the eggs on this male have already hatched, and you can see the small nymphs nearby. ❹ Although most eastern giant water bugs belong to the genera *Belostoma* and *Lethocerus*, one small ***Abedus*** species (***A. immaculatus***, 12–15 mm) occurs in the southeast. This individual was found in a dried out ditch in South Carolina. **FAMILY NEPIDAE** (WATER SCORPIONS) ❺ Water scorpions are slow-moving aquatic bugs that normally sit on aquatic vegetation with a long, tail-like air tube running to the surface, waiting to ambush any small organism that moves within reach. Our most common water scorpions are 20–40 mm long sticklike insects in the genus ***Ranatra***, like this ***R. fusca***. This water scorpion is sitting motionless and almost invisible below a pond weed leaf, seemingly watched over by a relatively small fisher spider sitting on the leaf. **FAMILY CORIXIDAE** (WATER BOATMEN) ❻ Water boatmen, like this nymph and this adult ***Sigara*** resting on the bottom of a small stream, feed by ingesting small particles. ❼ The short, scooplike front legs of water boatmen assist both with gathering food and in stridulation, making sounds that are both species- and sex-specific by rubbing against the head. The red blobs on this ***Sigara*** are parasitic mites. Most of our small (5–7 mm) water boatmen are in the genus *Sigara*.

FAMILY CORIXIDAE (continued) ❶ In contrast to almost all other true bugs, Corixidae have uniformly textured wings, without veins in the membrane. This small species (***Sigara lineata***) has distinctively pigmented wings, but most water boatmen are difficult to identify to species. ❷ Most of our relatively large (8–9 mm) water boatmen belong to the common genus ***Hesperocorixa***. **FAMILY NOTONECTIDAE** (BACKSWIMMERS) ❸ Our most common backswimmers, relatively large (8–16 mm) bugs in the genus ***Notonecta***, literally hang out at the water surface, suspended at about a 45-degree angle. This is ***N. insulata***, a large backswimmer with distinctive black pigmentation against a pale, yellowish background. ❹ This handsome species (***Notonecta irrorata***) is our largest (about 14 mm) backswimmer, and is easily recognized by its distinctive pigmentation. ❺ ***Notonecta undulata***, probably our most common backswimmer, is a medium-sized (about 12 mm) species variably colored in black and white. The scutellum (the big triangular part of the back just behind the head) is black, and the area behind that is usually mostly white. One of the two backswimmers in this photograph has just impaled a midge larva. ❻ ***Notonecta lunata*** is a small, pale backswimmer distinguished from *N. undulata* by its pale scutellum. ❼ This is a nymph of ***Notonecta borealis***, a large (adults about 12 mm) backswimmer with a relatively northern distribution. ❽ We have only two genera of backswimmers in eastern North America, the common *Notonecta* and the smaller and more slender ***Buenoa*** shown here. Unlike *Notonecta*, which rests at the water surface, *Buenoa* species rest submerged in hydrostatic balance. The rows of hairs under the abdomen serve to hold the bug's air supply. **FAMILY PLEIDAE** (PYGMY BACKSWIMMERS) ❾ This tiny (about 3 mm) ***Neoplea striola*** shows the silvery bubble of air held underneath the bug by a special coating of hairs. Pygmy backswimmers sometimes leave the water to clean those vital hairs and coat them with chemicals that keep them from becoming fouled.

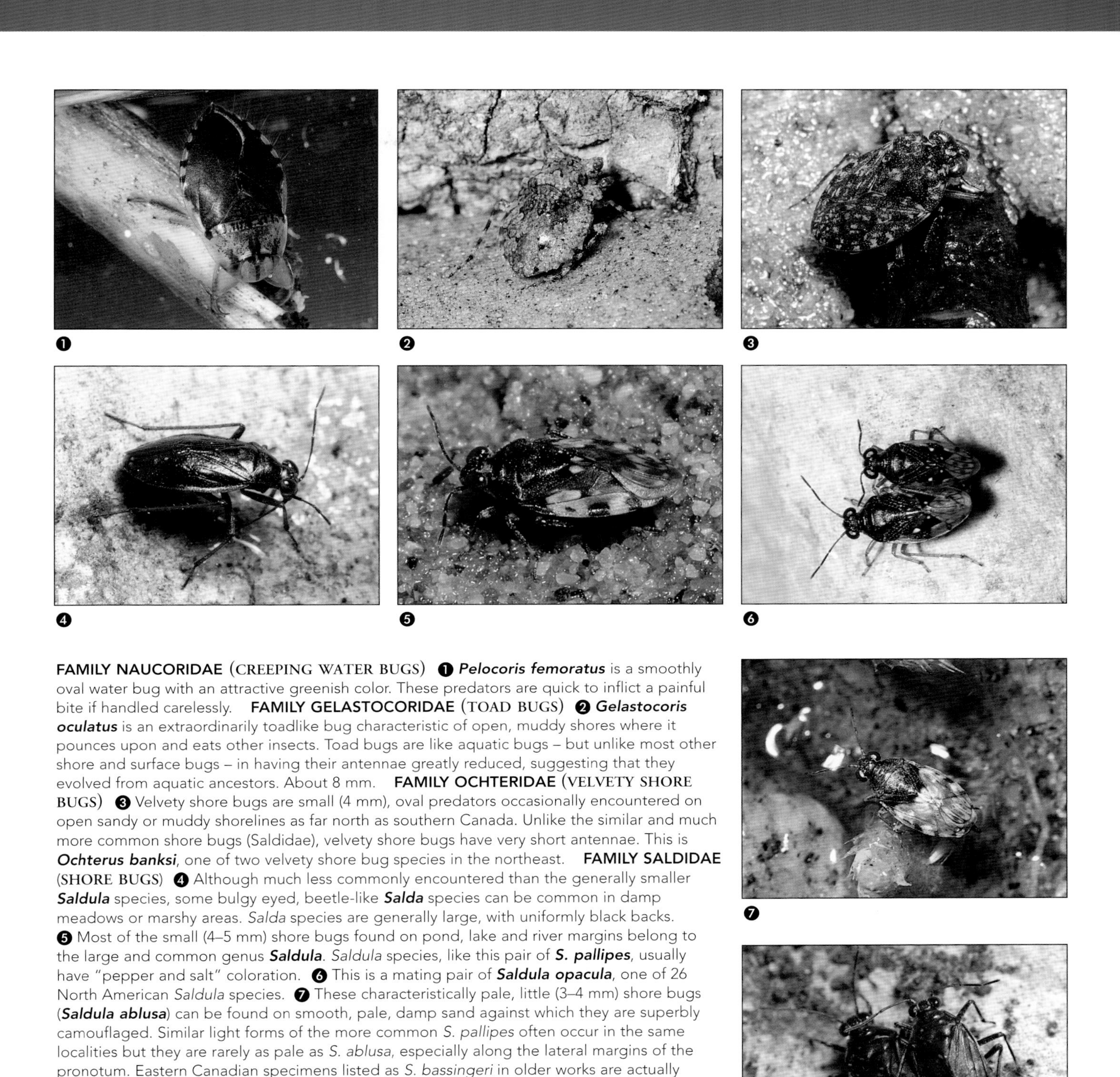

FAMILY NAUCORIDAE (CREEPING WATER BUGS) ❶ ***Pelocoris femoratus*** is a smoothly oval water bug with an attractive greenish color. These predators are quick to inflict a painful bite if handled carelessly. **FAMILY GELASTOCORIDAE** (TOAD BUGS) ❷ ***Gelastocoris oculatus*** is an extraordinarily toadlike bug characteristic of open, muddy shores where it pounces upon and eats other insects. Toad bugs are like aquatic bugs – but unlike most other shore and surface bugs – in having their antennae greatly reduced, suggesting that they evolved from aquatic ancestors. About 8 mm. **FAMILY OCHTERIDAE** (VELVETY SHORE BUGS) ❸ Velvety shore bugs are small (4 mm), oval predators occasionally encountered on open sandy or muddy shorelines as far north as southern Canada. Unlike the similar and much more common shore bugs (Saldidae), velvety shore bugs have very short antennae. This is ***Ochterus banksi***, one of two velvety shore bug species in the northeast. **FAMILY SALDIDAE** (SHORE BUGS) ❹ Although much less commonly encountered than the generally smaller ***Saldula*** species, some bulgy eyed, beetle-like ***Salda*** species can be common in damp meadows or marshy areas. *Salda* species are generally large, with uniformly black backs. ❺ Most of the small (4–5 mm) shore bugs found on pond, lake and river margins belong to the large and common genus ***Saldula***. *Saldula* species, like this pair of ***S. pallipes***, usually have "pepper and salt" coloration. ❻ This is a mating pair of ***Saldula opacula***, one of 26 North American *Saldula* species. ❼ These characteristically pale, little (3–4 mm) shore bugs (***Saldula ablusa***) can be found on smooth, pale, damp sand against which they are superbly camouflaged. Similar light forms of the more common *S. pallipes* often occur in the same localities but they are rarely as pale as *S. ablusa*, especially along the lateral margins of the pronotum. Eastern Canadian specimens listed as *S. bassingeri* in older works are actually *S. ablusa*. ❽ The "penta" in the name of this genus (***Pentacora***) refers to the five closed loops in the wing membrane of these relatively large (about 6 mm), handsome bugs. Look for them on exposed rocky shorelines or on rocks emerging from rivers and streams, and on hard-packed sand of Great Lakes beaches. This is ***P. ligata***, a common species in the Great Lakes region.

❶

❷

❸

❹

FAMILY SALDIDAE (continued) ❶ ***Lampracanthia crassicornis*** is rarely encountered, but can be abundant in Sphagnum bogs. The beetle-like body and swollen black antennal segments make this species easy to recognize. ❷ ***Micracanthia humilis*** is a small (about 3 mm) shore bug found on pond margins and in marshy meadows across the continent. **FAMILY HEBRIDAE** (VELVET WATER BUGS) ❸ Velvet water bugs (Hebridae) are tiny (around 2 mm) bugs, with a distinctive black-and-white velvety appearance. Our two genera of velvet water bugs, *Hebrus* and *Merragata*, require a microscope to separate (*Merragata* has five-segmented antennae; *Hebrus* has four-segmented antennae). *Hebrus* species are normally found along shorelines. ***Merragata hebroides***, the species shown here to the left of the relatively huge water strider, is a common little surface bug almost invariably found on duckweed. **FAMILY VELIIDAE** (SMALLER WATER STRIDERS OR RIFFLE BUGS) ❹ ***Microvelia*** species are only a couple of millimeters long, and most tiny, blackish pond surface bugs are wingless individuals of this genus. Like some common rove beetles, *Microvelia* species can scoot across the water surface by releasing water repellent glandular secretions onto the water surface. These ***Microvelia americana*** (four females and one male) are dining on a dead mayfly. ❺ ❻ The robust, broad-shouldered water striders which appear like constantly moving black dots on the water surface of eddies in swift streams are called riffle bugs. Riffle bugs have the middle tarsus (foot) deeply cleft, with flat tarsal claws and a bunch of basal, feather-like tarsal hairs that help them maintain their position on a constantly moving surface film. All our riffle bugs are 3–4 mm long, black and usually wingless. ***Rhagovelia*** is our only genus and ***R. obesa***, shown here, is the only species that gets as far north as Canada.

❺

❻

FAMILY MESOVELIIDAE (WATER TREADERS) ❶ The small green bugs invariably seen running rapidly across the surface of still waters, or scurrying across floating plants, are water treaders (Mesoveliidae). One very common species (***Mesovelia mulsanti***) normally occurs as entirely wingless populations, easily recognized as the only greenish, wingless, slender, 3–4 mm long bugs common on floating vegetation. The water treaders in this photo are probably eating aphids found on the same floating leaves. ❷ Winged water treaders are rare, but are distinctive for the complete lack of veins in the wing membrane. Wing polymorphism is a common occurrence in surface bugs, with wingless forms often predominating and winged forms just turning up now and then. **FAMILY HYDROMETRIDAE** (WATER MEASURERS OR MARSH TREADERS) ❸ Look very closely at the surface near the edge of a calm pond for water measurers, which look much like minute (about 7 mm) walking sticks, slowly and methodically walking along the water surface, perhaps stalking the mosquito larvae which serve as their major menu items. There is only one genus, with five eastern species, of which one (***Hydrometra martini***) is common in the northeast. **FAMILY GERRIDAE** (WATER STRIDERS) ❹ ❺ Gerridae are assisted in their water walking by fine, water-repellent hairs on their feet. Like most other surface bugs, they also have their claws inserted up above the foot, instead of at the end of the foot as in most other insects. Water striders use the entire water surface the way orb-web spiders use their webs. Any terrestrial insect unlucky enough to fall onto a pond surface, and to struggle and vibrate the surface the way a trapped insect vibrates a spider's web, is likely to attract hungry water striders like the large (over 10 mm), brownish ***Limnoporus dissortis*** seen here impaling a drowning lady beetle and a terrestrial bug. ❻ Most water striders, like this common species (***Gerris comatus***) eating a water lily leaf beetle, are relatively small (7–9 mm) bugs in the genus *Gerris*. Males of *G. comatus* are easily identified by the two beardlike tufts of long hairs under the abdomen (just before the tip). ❼ One of our largest (about 11 mm) water strider species, ***Aquarius remigis***, is abundant in the backwaters of rivers and streams. Most other water striders occur on lakes, ponds and the calmer parts of rivers and streams. As is true for other surface bugs, adult water striders sometimes have fully developed wings and sometimes lack wings altogether. *A. remigis* adults, like this copulating pair, are almost always wingless. The female of this pair is eating a shore fly (family Ephydridae). ❽ A close-up view of a water strider's head.

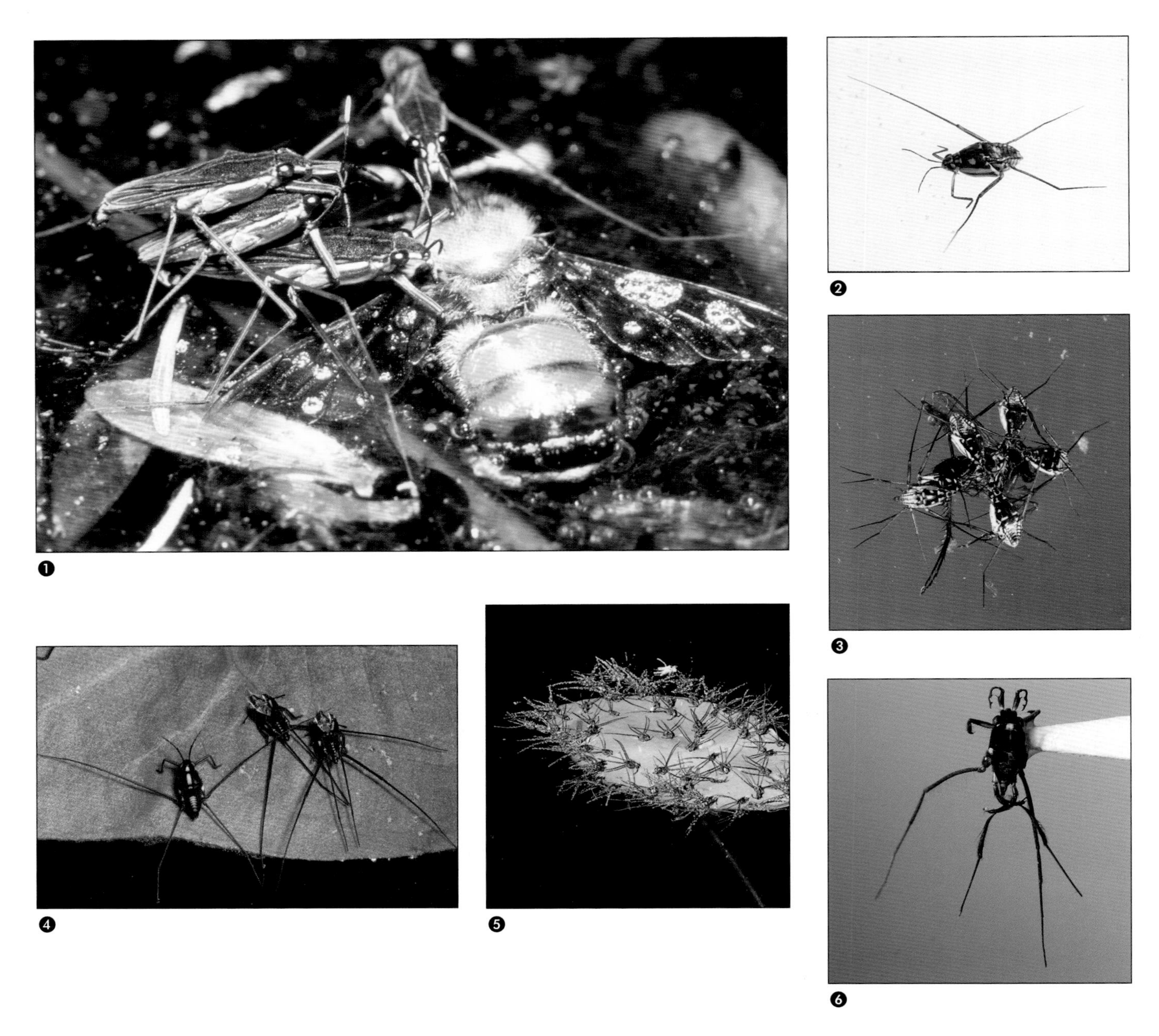

FAMILY GERRIDAE (WATER STRIDERS) **(continued)** ❶ These fully winged ***Gerris buenoi*** include a female feeding on a drowned honeybee while two males attempt to mate with her. Water striders are generally predators, but they will opportunistically feed on dead insects as well. *G. buenoi* usually has a fairly distinctive pattern of three stripes on the pronotum (right behind the head). ❷ ***Trepobates*** species, like this ***T. subnitidus***, are small (about 4 mm) rounded water striders. They are rarely seen and when they are encountered, they are easily mistaken for immature *Gerris*. *Gerris* species are thinner and have the inner margin of the eye indented. ❸ These ***Trepobates subnitidus*** are feeding on a long-legged fly (Dolichopodidae). ❹ ***Metrobates hesperius***, the only northeastern *Metrobates*, is a squat little (3.5–5 mm) bug with a tiny pronotum and distinctive coloration (top of head orange, mesonotum with three grayish stripes). This individual has just moulted while clinging to a floating leaf in a slow-moving river. ❺ Floating vegetation, like this pondweed leaf, are often festooned with the exuviae of water striders that molted while clinging to the leaf. ❻ ***Rheumatobates*** water striders are found throughout our area, but are infrequently encountered. These little (2.5–3 mm) striders are strongly sexually dimorphic, with males (shown here) adorned by thick, curved antennae and legs that are strikingly modified into thickened, twisted, armed appendages that differ markedly from species to species. This is ***R. rileyi***, the only species in the genus known as far north as Canada.

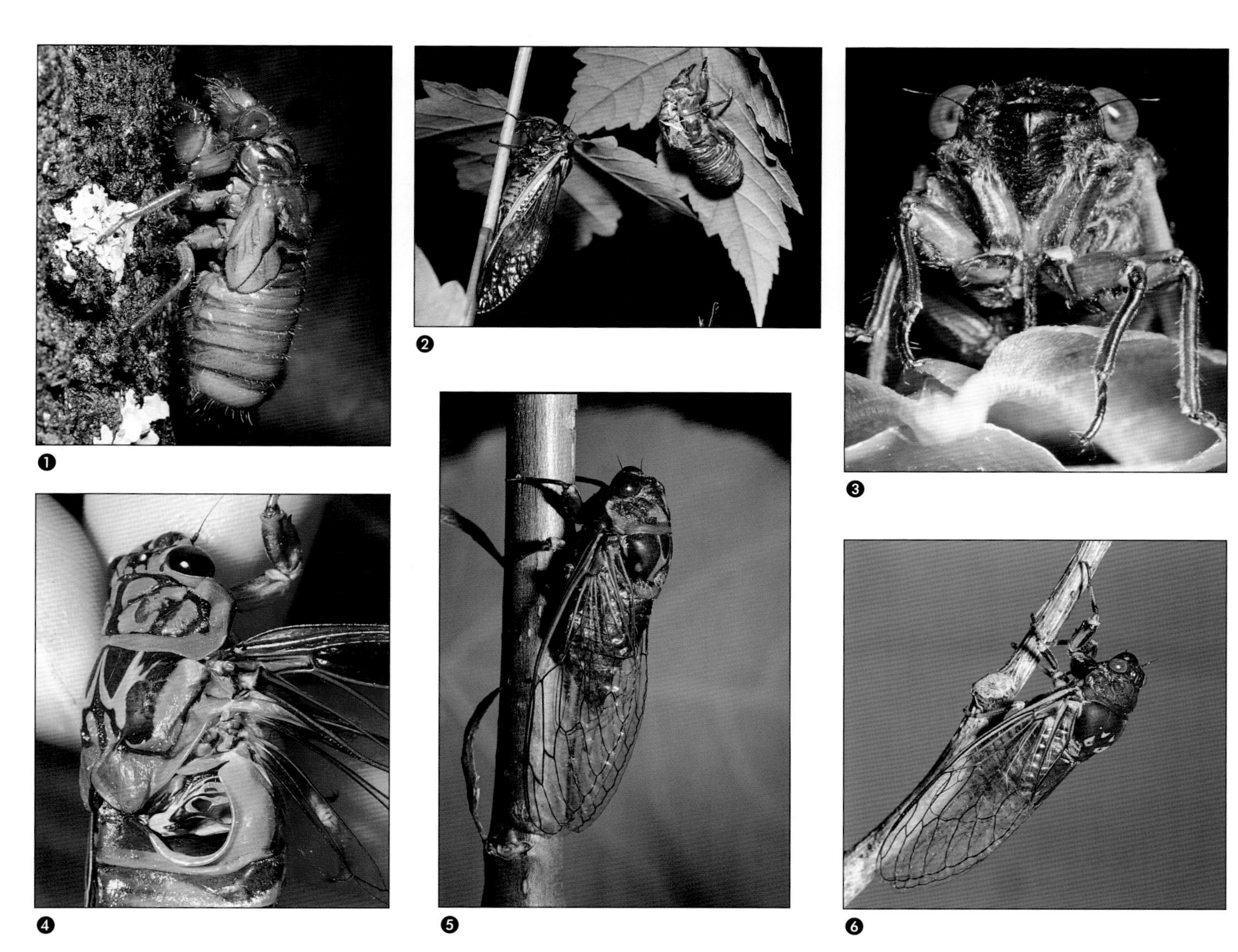

SUBORDER AUCHENORRHYNCHA

SUPERFAMILY CICADOIDEA

FAMILY CICADIDAE (CICADAS) ❶ ❷ Cicadas, by far the largest (usually 30–40 mm) North American homopterans, spend their short adulthood sucking sap from tree branches. Nymphal cicadas suck sap from roots for several years before climbing onto tree trunks or foliage to molt into clear-winged, brightly colored adults. The distinctively robust cast nymphal skins, split lengthways up the back, often persist long after the adult cicadas have emerged. These **Linnaeus' Seventeen-year Cicadas** or **Seventeen-year Locusts** (***Magicicada septendecim***) were among millions emerging in a small area where this species appears only once each 17 years. Periodical cicadas (*Magicicada* species) are restricted in space and time, with some emerging every 13 years and some every 17 years. *Magicicada* occurs only in the eastern United States and perhaps extreme southern Canada. ❸ The bulging front of this ***Magicicada's*** head hides the muscles that work to pump sap up through the beak. ❹ Cicada males have flat, drumskin-like plates, called tymbals, just inside the base of the abdomen. The tymbal is "wobbled" by muscular action, resulting in a sound that is modified by adjacent structures, and amplified by a hollow air chamber that takes up most of the abdomen. The sound-producing organs are hidden by flaps in our common Dog-day Cicadas (*Tibicen canicularis*) and periodical cicadas, but the right tymbal is exposed for view in this tropical species. ❺ Eastern North America's most common cicadas, well known to most people because of their noisy abundance during late summer, are in the genus *Tibicen*. ***Tibicen canicularis*** (the **Dog-day Cicada**) is a common species throughout much of our area. Dog-day Cicadas take several years to mature, but broods overlap and adults appear every year. ❻ This brightly colored cicada (***Okanagana canadensis***), called a "small-headed cicada" because the head is narrower than the thorax, is characteristic of mixed forests and pine forests. *Okanagana* species emerge earlier in the season and are more northern in their distribution than the more familiar *Tibicen* species. ❼ This **Translucent Cicada** (***Neocicada hieroglyphica***), a southeastern cicada, has just transformed from nymph to adult, leaving the empty nymphal skin on a North Carolina fence post.

❼

FAMILY CICADELLIDAE (LEAFHOPPERS) **Subfamily Cicadellinae** ❶ The brilliantly colored ***Graphocephala coccinea***, one of the most common of North America's roughly 2,500 leafhopper species, often occurs on the leaves of raspberries and various ornamental shrubs. Leafhoppers have rows of spines running down their hind legs, unlike similar spittlebugs, which have a cluster of spines at the end of the hind tibia. ❷ The brightly striped ***Cuerna striata*** is one of our most easily identified leafhoppers. This common species overwinters in the adult stage, and can often be found in dense aggregations among grass thatch or leaves early in spring. ❸ ***Cuerna fenestella*** is a grassland species that occurs in widely separated eastern and western tallgrass habitats. ❹ This recently discovered species of ***Thatuna***, photographed on a tree trunk in Virginia, represents the first eastern North American record in this genus. ❺ This sharpshooter (***Paraulacizes irrorata***) is one of our largest leafhoppers (about 12 mm). Sharpshooters range as far north as southern Ontario. ❻ The powdery white patches of the side of this female **Blue Sharpshooter** (***Oncometopia orbona***) are made up of brochosomes, structurally complex particles produced in the bug's malphigian tubules. She will later use her hind legs to scrape the powdery brochosomes off her forewings and on to her newly deposited eggs. ❼ The common ***Draeculacephala zeae*** is a relatively large, distinctively pointy-headed leafhopper. This species is often extremely abundant in wet meadows. ❽ ***Neokolla hieroglyphica*** is a common leafhopper in patches of native grassland.

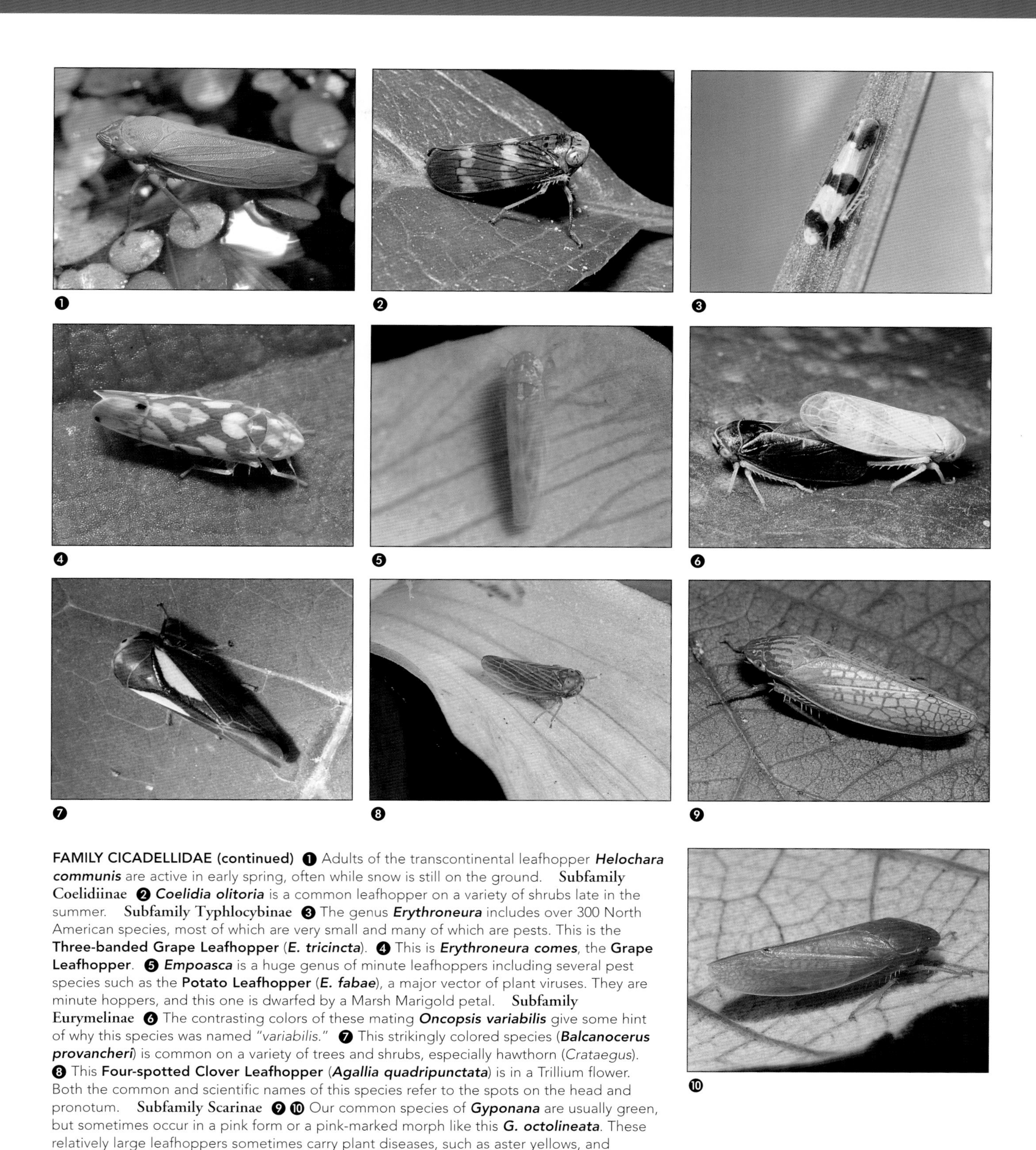

FAMILY CICADELLIDAE (continued) ❶ Adults of the transcontinental leafhopper ***Helochara communis*** are active in early spring, often while snow is still on the ground. Subfamily Coelidiinae ❷ ***Coelidia olitoria*** is a common leafhopper on a variety of shrubs late in the summer. Subfamily Typhlocybinae ❸ The genus ***Erythroneura*** includes over 300 North American species, most of which are very small and many of which are pests. This is the **Three-banded Grape Leafhopper** (***E. tricincta***). ❹ This is ***Erythroneura comes***, the **Grape Leafhopper**. ❺ ***Empoasca*** is a huge genus of minute leafhoppers including several pest species such as the **Potato Leafhopper** (***E. fabae***), a major vector of plant viruses. They are minute hoppers, and this one is dwarfed by a Marsh Marigold petal. Subfamily Eurymelinae ❻ The contrasting colors of these mating ***Oncopsis variabilis*** give some hint of why this species was named "*variabilis.*" ❼ This strikingly colored species (***Balcanocerus provancheri***) is common on a variety of trees and shrubs, especially hawthorn (*Crataegus*). ❽ This **Four-spotted Clover Leafhopper** (***Agallia quadripunctata***) is in a Trillium flower. Both the common and scientific names of this species refer to the spots on the head and pronotum. Subfamily Scarinae ❾ ❿ Our common species of ***Gyponana*** are usually green, but sometimes occur in a pink form or a pink-marked morph like this ***G. octolineata***. These relatively large leafhoppers sometimes carry plant diseases, such as aster yellows, and sometimes damage twigs as they lay eggs. *Gyponana* species often come to lights on late summer nights.

FAMILY CICADELLIDAE (continued) Subfamily Aphrodinae ❶ The distinctively striped ***Amblysellus curtisii*** is the only member of this mostly western genus to reach eastern Canada. ❷ ***Balclutha*** includes several similar species in the northeast. ❸ ***Athysanus argentarius***, originally from Europe, is now common in North American lawns and meadows. ❹ ***Texananus marmor*** is a striking species common in western North America, but it is not widespread in the east. It is locally abundant on the alvars of Ontario's Bruce Peninsula. ❺ ***Penthimia americana*** is a distinctively squat leafhopper found on broadleaf trees and shrubs in early summer. This odd hopper is sometimes placed in its own subfamily. ❻❼❽ The subfamily Aphrodinae is a huge group with far more than a thousand North American species. The species shown here are ***Norvellina seminuda***, ***Scaphoideus*** sp., and ***Flexamia delongi***. ❾ The genus ***Limotettix*** is a very large and diverse genus with many species, such as this colorful ***L. nigrax***, found in western and eastern North America.

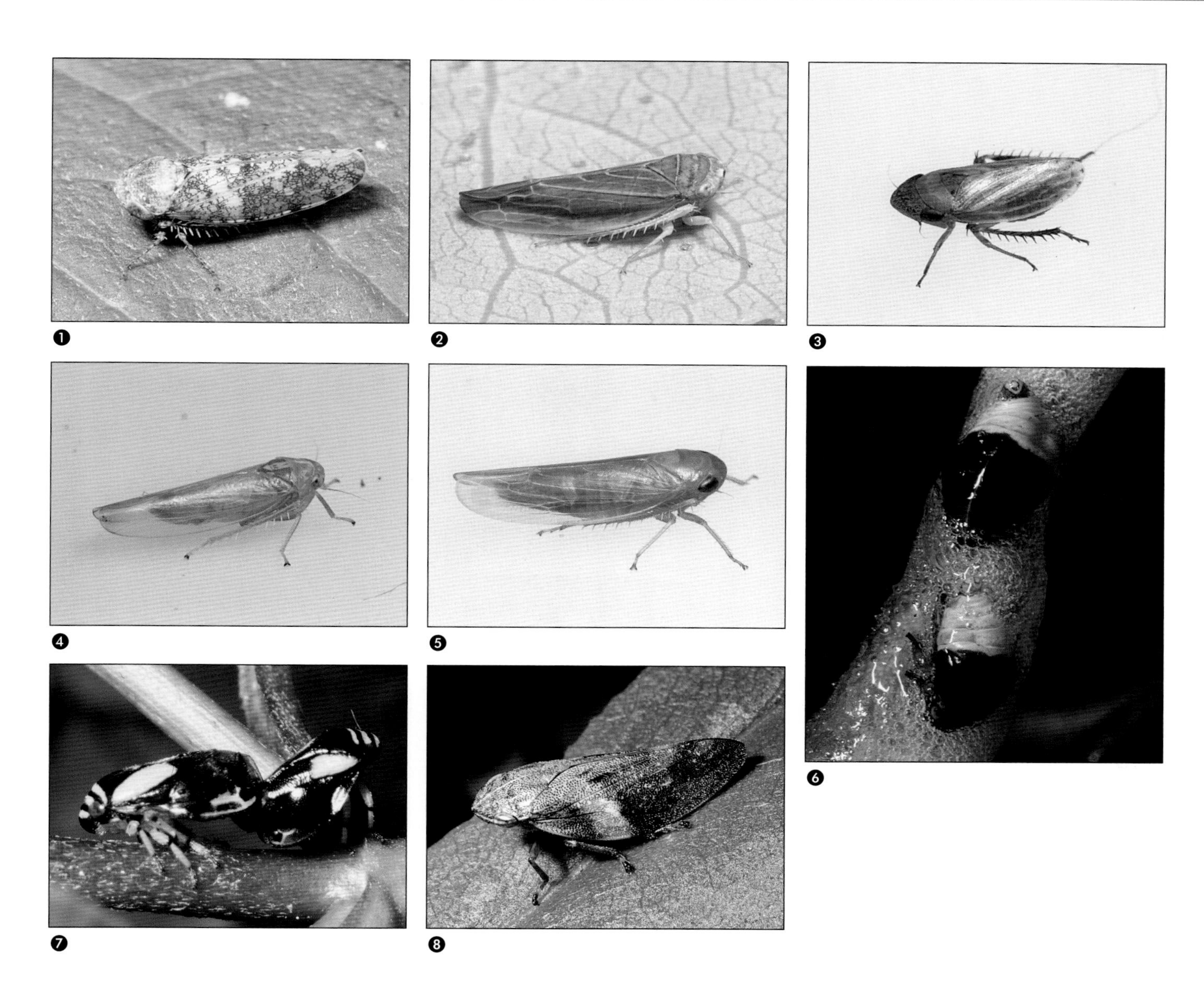

FAMILY CICADELLIDAE (continued) These species illustrate the diversity of the huge subfamily Aphrodinae. ❶ This is ***Paraphlepsius collitus***. ❷ ***Idiodonus morsei*** is a boreal species, found across Canada. ❸ The several species of ***Anoscopus*** now found in the northeast, were all accidentally introduced from Europe. ❹ ❺ Although leafhoppers occur in a rainbow of colors, many, such as this ***Opsius stactogalus*** and this ***Chlorotettix*** sp., are green insects that blend with the surrounding vegetation. **FAMILY CERCOPIDAE** (SPITTLEBUGS) ❻ These **Heath Spittlebug** (***Clastoptera saintcyri***) nymphs are sucking xylem sap from a blueberry plant while rotating their abdominal tips to whip up a frothy spittle shelter from a mixture of digested plant sap, air and sticky glandular secretions. *Clastoptera* is sometimes treated as part of the family Clastopteridae, separate from the Cercopidae. ❼ The **Dogwood Spittlebug** (***Clastoptera proteus***) is a common insect on its host plant. ❽ Adults of the **European Alder Spittlebug** (***Aphrophora alni***) can be found in numbers on and around alder and willow; nymphs occur on herbaceous plants at ground level. *Aphrophora* spittlebugs are common, relatively large (8–11 mm) spittlebugs with distinctive pits on the front wings. The North American spittlebugs are sometimes split into three families, with *Aphrophora* in the Aphrophoridae and *Clastoptera* in the Clastopteridae.

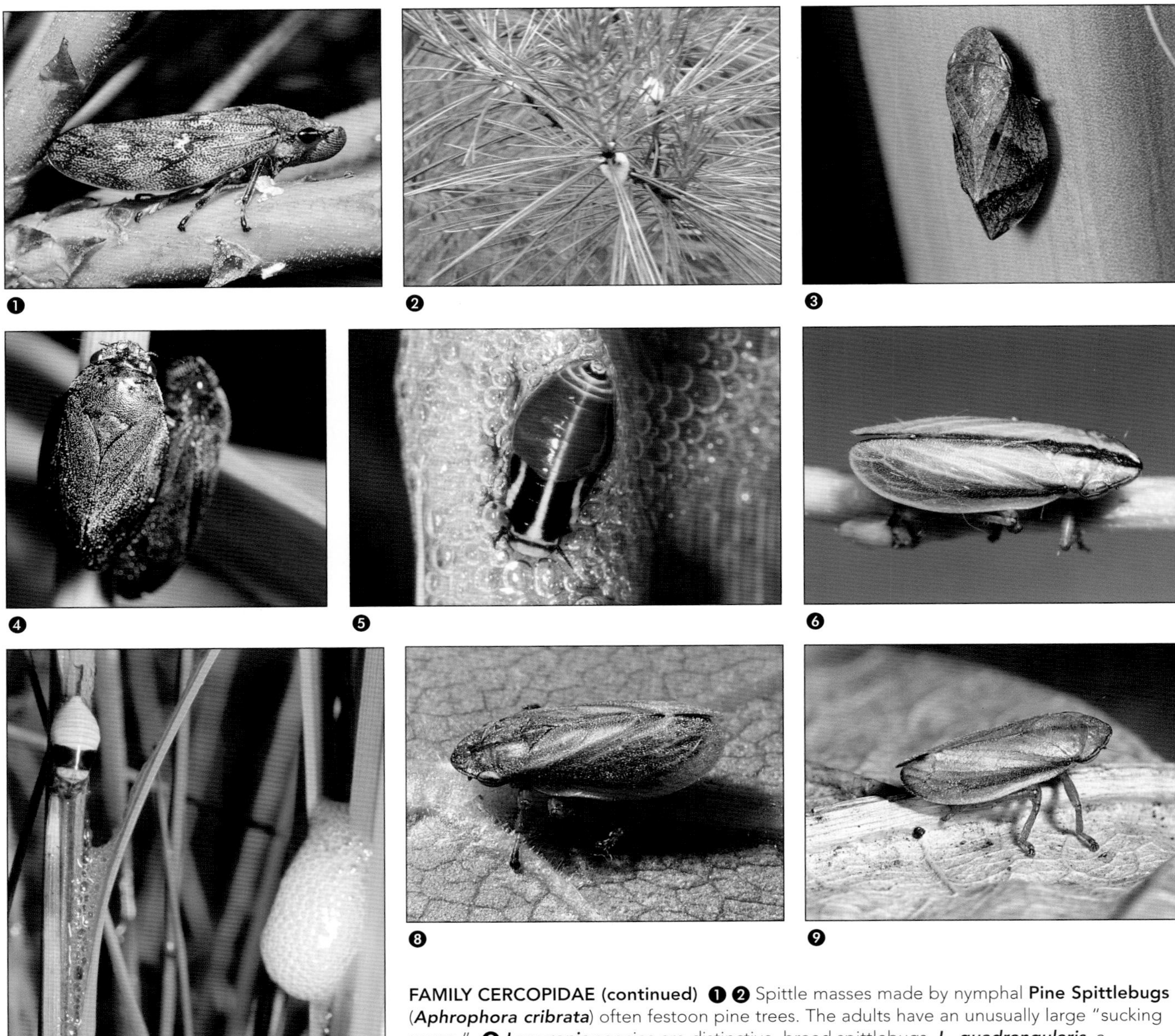

FAMILY CERCOPIDAE (continued) ❶ ❷ Spittle masses made by nymphal **Pine Spittlebugs** (***Aphrophora cribrata***) often festoon pine trees. The adults have an unusually large "sucking pump." ❸ ***Lepyronia*** species are distinctive, broad spittlebugs. ***L. quadrangularis***, a common species with a characteristic flattened appearance, is known as the **Diamond-backed Spittlebug** because of the triangular to diamond-shaped pattern on the front wings. Around 7 mm. ❹ ❺ The **Black Spittlebug** (***Prosapia ignipectus***) is a stout, dark species about 8 mm in length, with bright red markings underneath. This uncommon spittlebug feeds on Little Bluestem, a native grass; nymphs are usually hidden under the soil surface. ❻ ❼ ***Paraphilaenus parallelus*** is known only from a few peatlands in Ontario and Wisconsin, but where it occurs it is often abundant. These photos, showing the black-striped adult, the nymph and the nymph's spittle shelter, are from a large fen in Ontario. About 8 mm. ❽ The **Prairie Spittlebug** (***Philaenarcys bilineata***) is a relatively stout, medium-sized (5–8 mm) spittlebug distributed across the continent in grassy habitats, but in the east it occurs mostly in the boreal region. ❾ The **Lined Spittlebug** (***Neophilaenus lineatus***) is a small (4–6 mm) species introduced from Europe, and now common through much of the northeast.

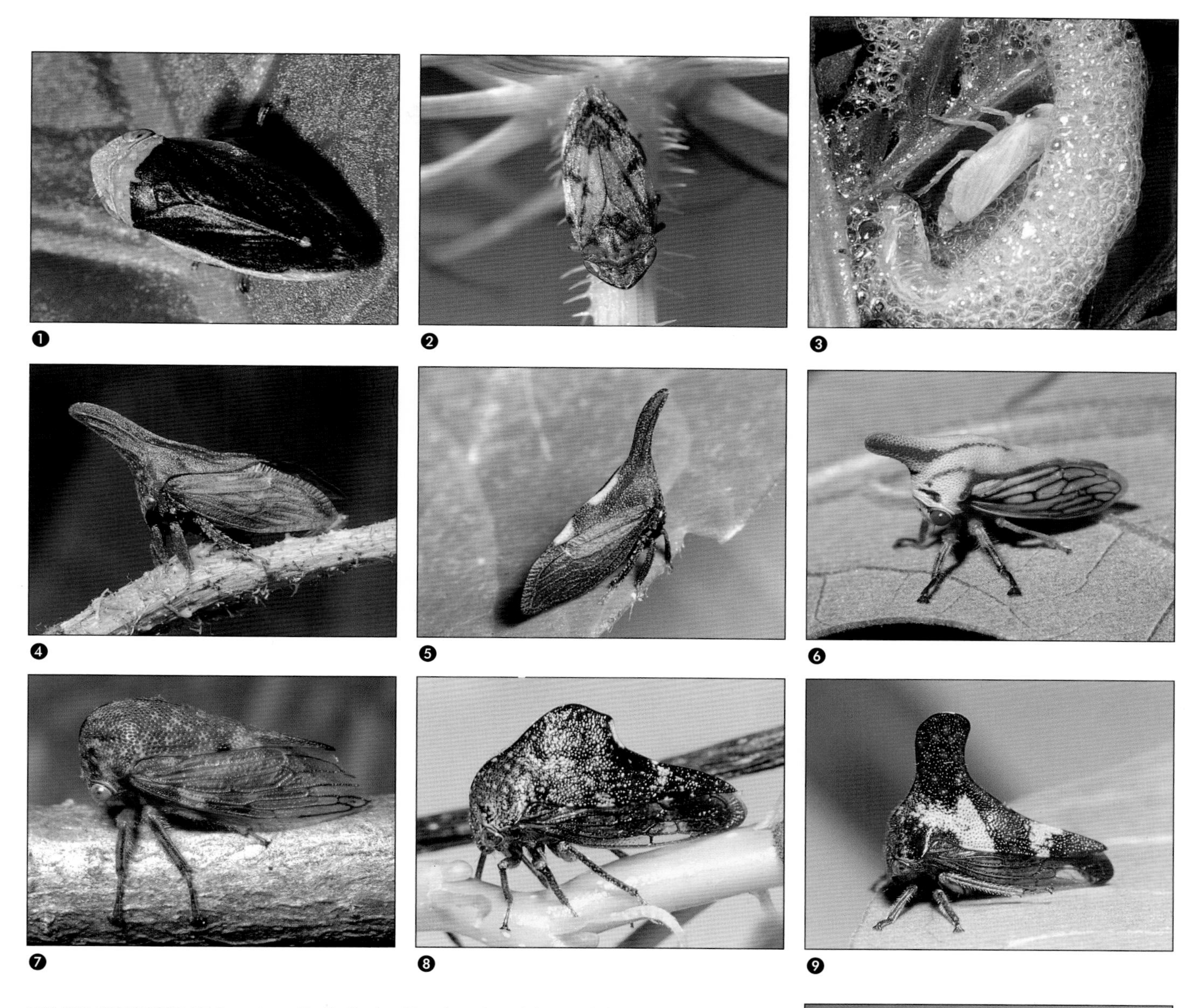

FAMILY CERCOPIDAE (continued) ❶ ❷ The **Meadow Spittlebug** (***Philaenus spumarius***) was accidentally introduced from Europe, and is now responsible for most of the spittle masses you see on herbaceous plants. Occurring in almost every color from black to brown to mottled yellow, the Meadow Spittlebug (about 6 mm in length) is a polymorphic species best recognized by its smooth wings and some other structural characters rather than color. Other spittlebugs are less variable in color. ❸ **Meadow Spittlebugs** (***Philaenus spumarius***) form a special spittle chamber in which they molt from nymph to adult. **FAMILY MEMBRACIDAE** (TREEHOPPERS) Subfamily Membracinae ❹ Treehoppers usually have a big shieldlike structure, derived from the pronotum, which covers much of the body and often has spectacular processes or spines. ***Campylenchia latipes***, a species commonly found on goldenrod and other plants in disturbed areas, has a striking, ridged pronotal horn projecting over its head. ❺ ***Enchenopa binotata*** is a common treehopper that forms dense aggregations along twigs of a variety of shrubs and vines. ❻ ❼ **Oak Treehoppers** (***Platycotis vittata***, 10–12 mm) often occur in large groups on oak twigs, usually including both long-horned and short-horned individuals and often including both striped and spotted individuals. Subfamily Smiliinae ❽ Some of our largest and most boldly crested treehoppers are in the large genus ***Telamona***, including this birch-feeding species, ***T. gemma***. ❾ This big (about 10 mm) treehopper, ***Glossonotus crataegi***, occurs on hawthorn (*Crataegus*), as suggested by its specific name. ❿ ***Thelia bimaculata*** is found on only one species of locust tree (*Robinia pseudoacacia*), but it is often abundant on its host tree.

FAMILY MEMBRACIDAE (continued) ❶ ❷ These distinctively colored ***Acutalis brunnea*** treehoppers are unusually small for the Membracidae at only 4 or 5 mm. The spittle above this courting pair holds a nymphal Meadow Spittlebug (*Philaenus spumarius*; family Cercopidae). ❸ This minute (3–4 mm) ***Micrutalis calva*** looks much like an *Acutalis*, but it has pale wing veins. ❹ ❺ The distinctively shaped ***Smilia camelus*** can be found on oak and locust trees. ❻ ***Atymna querci***, as the name suggests, feed on oaks (*Quercus*). ❼ Treehoppers in the genus ***Carynota***, like this ***C. stupida***, feed on a variety of trees. ❽ This can-opener-shaped species (***Entylia carinata***) is one of the most common treehoppers in our area, and can be found on a variety of plants including goldenrod and thistle. ❾ ***Publilia concava*** is a common treehopper, often occurring on weeds in dense aggregations including both adults and nymphs. Many adult treehoppers stay with their offspring and provide some degree of maternal care, and ants often protect aggregations of treehoppers from which they obtain sugary honeydew.

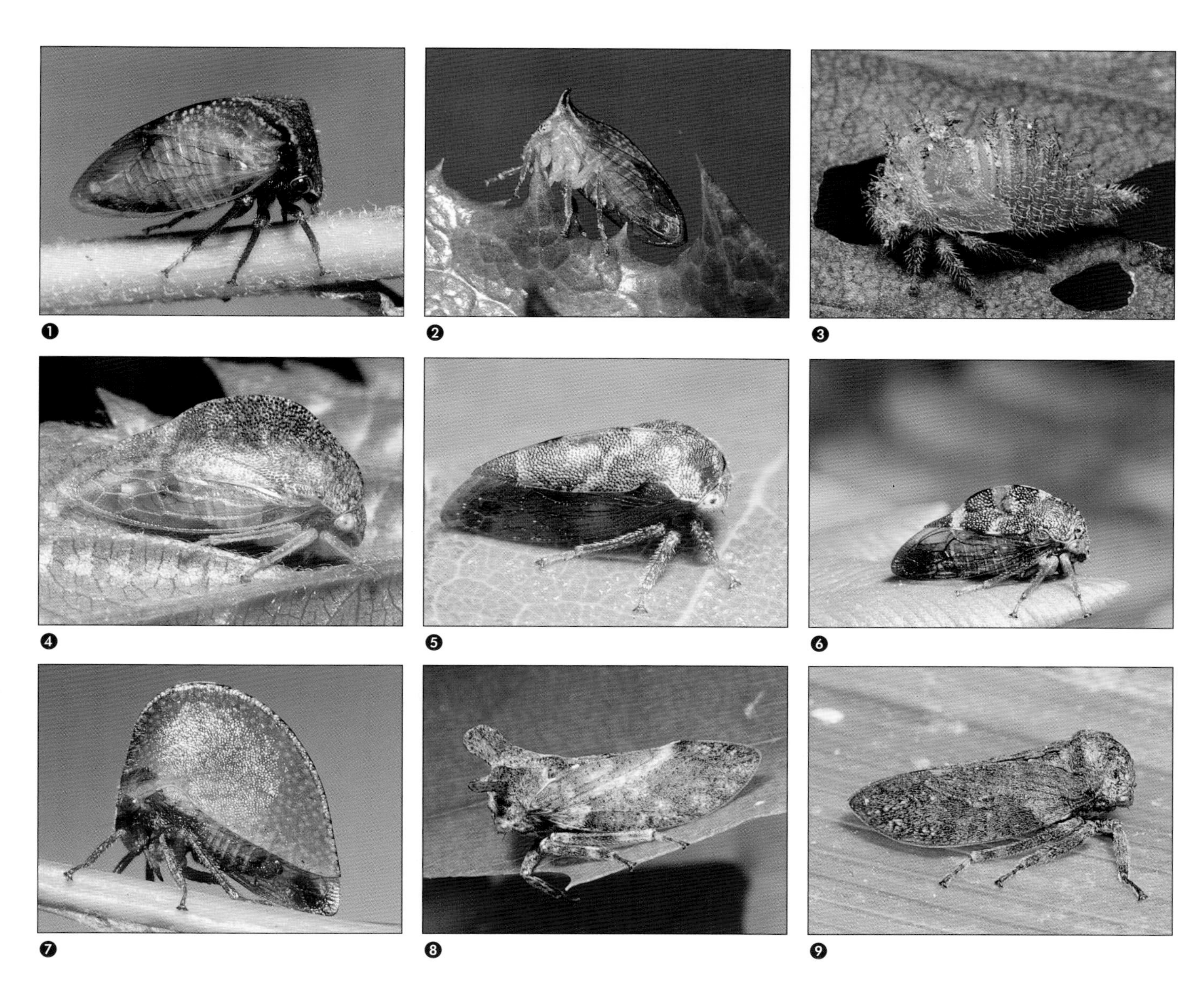

FAMILY MEMBRACIDAE (continued) ❶ Some of the most common treehoppers in our area are ***Ceresa*** species, such as the adult ***C. basalis*** shown here, characterized by a broad, triangular pronotum covering most of the adult body. ❷ ❸ The green nymphs of **Buffalo Treehoppers** (***Ceresa alta***), which lack the conspicuous spined pronotum of the adults and are instead covered with multispiked processes, are often common in weeds near fruit trees. They move up to the fruit trees when adults, laying eggs in the twigs with sharp ovipositors, often killing the outer part of the twig in the process. ❹ ❺ ❻ ***Cyrtolobus*** is one of our largest treehopper genera. These species (***C. fulginosus***, ***C. vau*** and ***C. tuberosus***, respectively) all feed on oak. ❼ The spectacular helmet-shaped ***Archasia auriculata*** is one of three similar *Archasia* species found on oak trees in the northeast. Subfamily Stegaspidinae ❽ ***Microcentrus***, like this distinctively ornamented ***M. perditus***, are oddballs among the Membracidae because their pronotal shields are so much smaller than those of other treehoppers. *M. perditus* normally occurs on Burr Oak; the more common *M. caryae* is found on hickory as well as oak and other trees. ❾ ***Microcentrus caryae*** occurs on a variety of trees but it is most common on hickory. Some entomologists recognize a separate family, Aetalionidae, for *Microcentrus* and the related southern genus *Aetalion*.

SUPERFAMILY FULGOROIDEA (PLANTHOPPERS)

FAMILY ISSIDAE ❶ ***Thionia bullata*** is a relatively large (about 6 mm) and attractive member of the small family Issidae. ❷ ❸ ***Acanalonia*** species are relatively large (7–10 mm) planthoppers that often abound on shrubs and vines. This distinctively colored species (***A. bivittata***) is usually green with black stripes on top, but sometimes occurs as a pink form. These bright planthoppers are sometimes treated as a subfamily of Issidae and sometimes as the separate family Acanaloniidae. ❹ This brightly colored planthopper (***Acanalonia conica***) lacks the distinct stripes of the more common *A. bivittata*. *Acanalonia conica* used to have a much more southerly distribution, and has only recently become common as far north as southern Ontario. **FAMILY FLATIDAE** ❺ Flatidae are exotic-looking planthoppers somewhat similar in shape and size to Acanaloniidae, but with a more pronounced wedge-shape when at rest and with distinctive rows of parallel wing veins near the wing margin. This is ***Anormenis septentrionalis***, a species that just barely gets as far north as southern Ontario. ❻ Although most flatids are broadly wedge-shaped, ***Cyarda melichari*** has distinctively narrow wings. This species occurs along the east coast of the United States. ❼ Planthopper nymphs, like this ***Metcalfa pruinosa***, often produce abundant white, waxy filaments. ❽ These two pale-colored ***Ormenoides venusta*** are sharing a southern Ontario twig with a purplish ***Metcalfa pruinosa***, our most common flatid.

FAMILY DERBIDAE ❶ ❷ ❸ Derbidae is a mostly tropical family of delicate planthoppers that feed on woody fungi; most northeastern species are distinctively colored *Otiocerus* and *Cedusa* species. ***Otiocerus*** species are long-winged, generally pastel-colored planthoppers such as the pinkish ***O. degeerii***, the yellowish ***O. wolfei*** and the red-margined ***O. stollii***. Look for them in hardwood forests. ❹ The strikingly colored ***Amalopota uhleri*** is a widespread but rarely encountered derbid. ❺ ***Cedusa*** species are easily overlooked because of their small size (about 2 mm), but are common insects. Several species of *Cedusa* are uniformly blue-black with a waxy bloom, with only a few species (like ***C. maculata*** ❻) easily recognized on the basis of color pattern. **FAMILY DICTYOPHARIDAE** ❼ ***Scolops*** species (like this ***S. sulcipes***), with their long, skinny "noses," are certainly our most distinctive planthoppers. They occur on grasses throughout southern Canada and the eastern United States. Some entomologists include *Scolops* and *Phylloscelis* in the Fulgoridae. ❽ The distinctive, long-snouted *Scolops* are by far the most common Dictyopharidae, but you might also encounter strikingly stout, black-winged planthoppers in the genus ***Phylloscelis***. ***P. atra*** occurs among native grasses and on mountain mint (*Pycnanthemum*). **FAMILY CALISCELIDAE** ❾ ***Bruchomorpha*** species, our most common Caliscelidae, are small (about 3 mm), shining bugs that can easily be mistaken for little beetles. Caliscelids are sometimes included in the family Issidae.

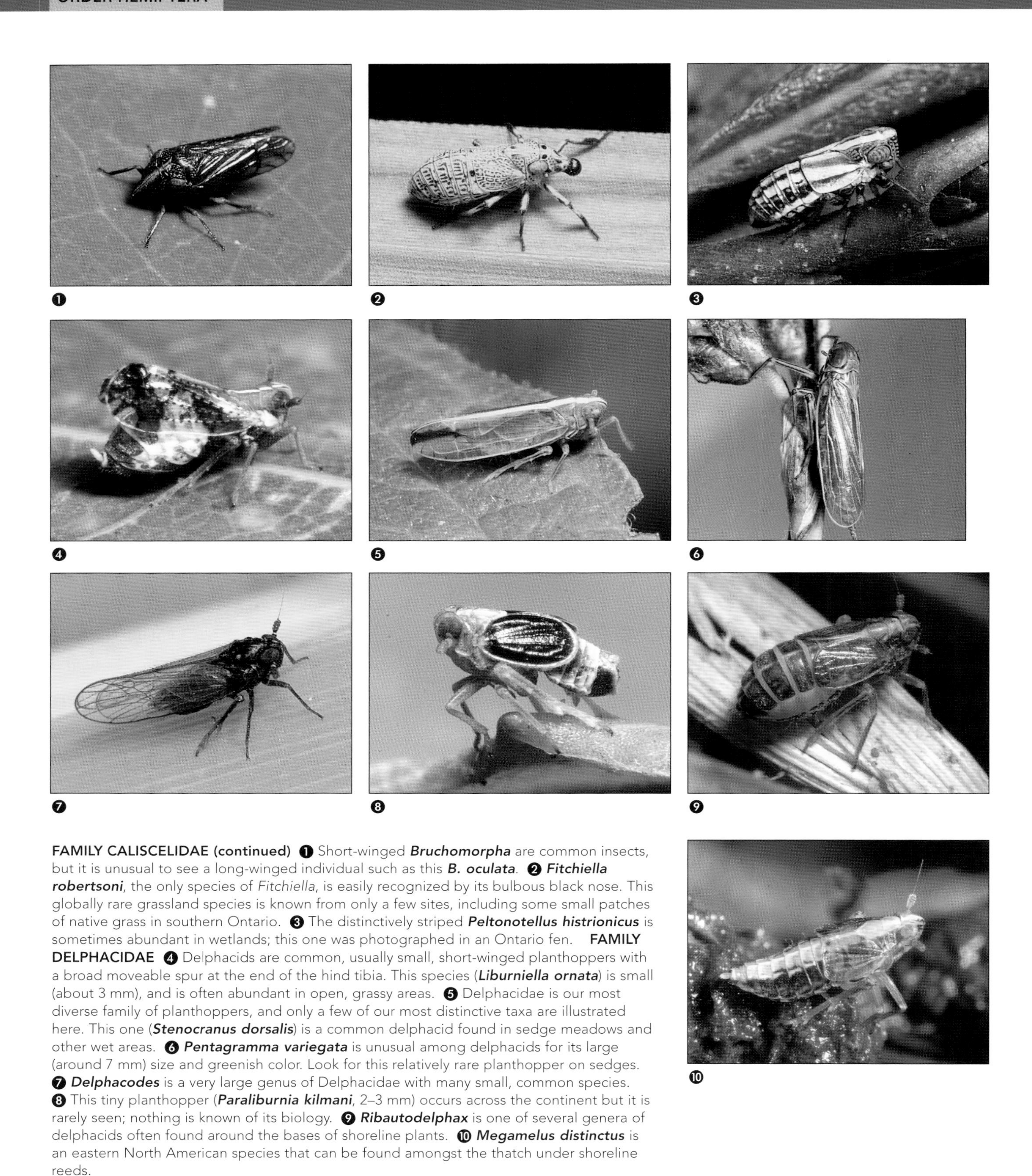

FAMILY CALISCELIDAE (continued) ❶ Short-winged ***Bruchomorpha*** are common insects, but it is unusual to see a long-winged individual such as this ***B. oculata***. ❷ ***Fitchiella robertsoni***, the only species of *Fitchiella*, is easily recognized by its bulbous black nose. This globally rare grassland species is known from only a few sites, including some small patches of native grass in southern Ontario. ❸ The distinctively striped ***Peltonotellus histrionicus*** is sometimes abundant in wetlands; this one was photographed in an Ontario fen. **FAMILY DELPHACIDAE** ❹ Delphacids are common, usually small, short-winged planthoppers with a broad moveable spur at the end of the hind tibia. This species (***Liburniella ornata***) is small (about 3 mm), and is often abundant in open, grassy areas. ❺ Delphacidae is our most diverse family of planthoppers, and only a few of our most distinctive taxa are illustrated here. This one (***Stenocranus dorsalis***) is a common delphacid found in sedge meadows and other wet areas. ❻ ***Pentagramma variegata*** is unusual among delphacids for its large (around 7 mm) size and greenish color. Look for this relatively rare planthopper on sedges. ❼ ***Delphacodes*** is a very large genus of Delphacidae with many small, common species. ❽ This tiny planthopper (***Paraliburnia kilmani***, 2–3 mm) occurs across the continent but it is rarely seen; nothing is known of its biology. ❾ ***Ribautodelphax*** is one of several genera of delphacids often found around the bases of shoreline plants. ❿ ***Megamelus distinctus*** is an eastern North American species that can be found amongst the thatch under shoreline reeds.

FAMILY CIXIIDAE These broad, somewhat flattened planthoppers are diverse out of proportion to their relative rarity. Look for adult cixiids, such as this paler brown this ***Cixius pini*** ❶ and this ***C. basalis*** ❷ on leaves and tree trunks. Nymphs are usually concealed beneath the ground, feeding on plant roots. ❸ Several planthoppers, like these nymphs under the bark of a dead pine tree, produce conspicuous, long wax filaments or "tails." ❹ ***Oliarus*** species often occur on tree trunks. **FAMILY ACHILIDAE** ❺ Achilids, like this ***Catonia***, are distinctive among planthoppers for their overlapping wings, but otherwise they look much like Cixiidae. They are relatively rare but are most likely to be found among loose bark, where some have the very unusual (for homopterans) habit of feeding on fungi.

❼

SUBORDER STERNORRHYNCHA

SUPERFAMILY APHIDOIDEA

FAMILY ANOECIIDAE ❻ Clusters of winged adults and wingless nymphs of **dogwood aphids** (***Anoecia*** sp.) are common under fall foliage of Red Osier Dogwood (*Cornus sericea*). *Anoecia* species alternate between dogwood and the roots of herbaceous plants. **FAMILY APHIDIDAE** (COMMON APHIDS OR PLANTLICE) ❼ ***Uroleucon cirsii*** was introduced from Europe, and is now a very common species in eastern North America where it is found only on thistles (*Cirsium* spp.). ❽ Winged and wingless aphids often occur together, as in this group of ***Uroleucon***. Note the drop of defensive fluid at the tip of one of the long black cornicles (tubes sticking out of the back of the abdomen) on the pale individual.

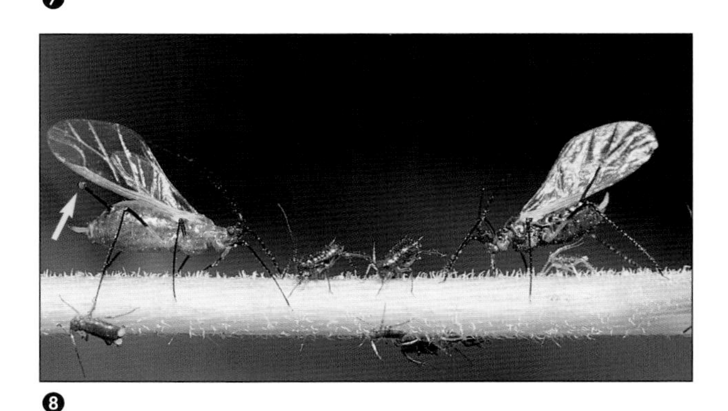

❽

FAMILY APHIDIDAE (continued) ❶ These wingless ***Aphis nerii*** (an introduced aphid species) give a bright splash of color to this milkweed stem. Their long, thin beaks are sunk deeply into the stem as they pump up sugar-rich phloem sap. ❷ Aphids, like these **Spotted Poplar Aphids** (***Aphis maculatae***), are often attended by ants that protect the aphids from parasitoids and predators "in exchange" for their sugar-rich waste honeydew. ❸ This soybean pod is covered with **Soybean Aphids** (***Aphis glycines***), an Asian species first noticed in northeastern North America in the summer of 2000. Soybean aphids are easily recognized as the only yellow aphids found on soybean, where they can occur in such numbers that whole soybean fields seem to glisten with aphid honeydew. ❹ The tiny (1–2 mm) aphids tapping a vein on this melon leaf are ***Aphis gossypii***, the **Melon Aphid** or **Cotton Aphid**. This worldwide pest attacks a variety of plants, causing distorted leaves and other problems. ❺ ***Pterocomma bicolor*** is a distinctive and appropriately named species of aphid, with a bluish body and contrasting pinkish cornicles and legs. **FAMILY LACHNIDAE** ❻ **Giant Willow Aphids** (***Tuberolachnus salignus***), here tended by *Crematogaster* ants, are among our largest aphids. They kick and flick their hind legs when disturbed, probably an innate response meant to deter parasitic wasps. ❼ The large genus ***Cinara*** includes most of the conifer-feeding aphids such as these pine aphids. Pine aphid colonies can often be spotted from a distance because of the black, sooty mold growing on the honeydew-spattered foliage below aphid-encrusted branches. **FAMILY DREPANOSIPHIDAE** ❽ The relatively short beak is clearly visible on this winged *Euceraphis* individual. ***Euceraphis*** species, including the introduced **European Birch Aphid** (***E. punctipennis***), often abound on birch trees. ❾ ***Myzocallis punctata*** is a transcontinental species in the large genus *Myzocallis*. Look for it on oak leaves.

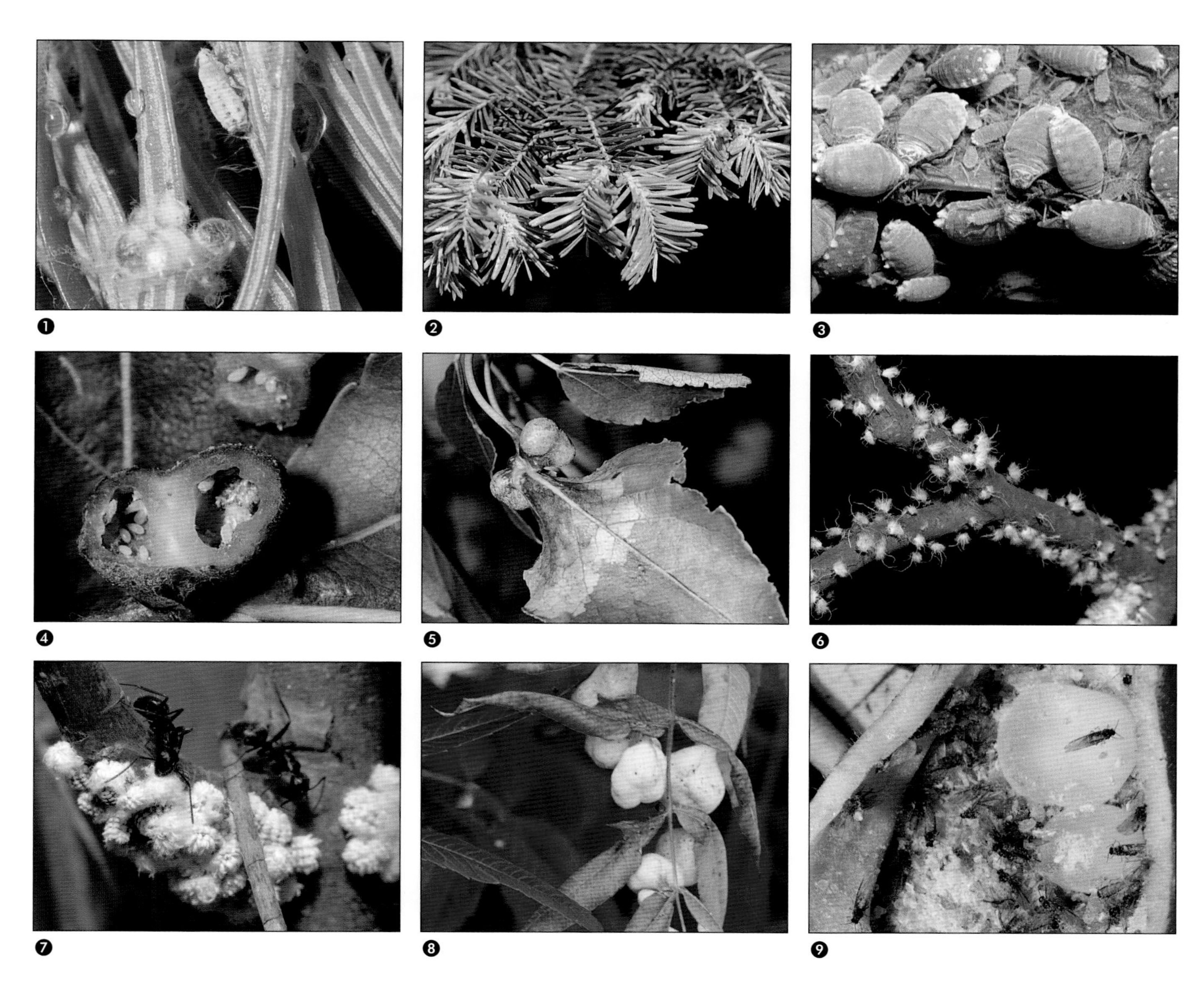

FAMILY MINDARIDAE ❶ ❷ The **Balsam Twig Aphid** (***Mindarus abietinus***) is the most common species in this small aphid family. These odd aphids can be found on tender young balsam needles, often in a sticky waxy mess. **FAMILY PEMPHIGIDAE** (WOOLLY APHIDS, GALL-MAKING APHIDS AND OTHERS) ❸ Although males and sexual females of Pemphigidae are very small, the asexual females are often relatively large aphids that lack the conspicuous, tubelike cornicles seen on common aphids. Most members of this family alternate between host plants, and this species probably has a stage that lives on roots of coniferous trees. ❹ ❺ The eggs of Poplar Petiolegall Aphids (*Pemphigus populitransversus*) and closely related ***Pemphigus*** species, such as this ***P. populiglobuli***, hatch in spring, and the tiny nymphs induce the formation of hollow, ball-shaped galls by feeding on poplar and aspen leaf petioles. ❻ **Beech Blight Aphids** (***Grylloprociphillus imbricator***) sometimes build up incredible populations on American Beech (*Fagus grandifolia*). This tree, in southern Ontario, was so loaded with aphids the surrounding ground was entirely gray-black with honeydew and sooty mold. ❼ **Woolly Alder Aphids** (***Paraprociphilus tessellatus***) often form cottony masses on alder trees. Some Woolly Alder Aphids migrate from alder to maple in fall, where sexual forms produce eggs. ❽ ❾ **Sumac Gall Aphids** (***Melaphis rhois***) induce conspicuous galls that not only hold hundreds of aphids, but also contain remarkable-looking spheres of wax-coated honeydew.

FAMILY HORMAPHIDIDAE ❶ The **Witch Hazel Gall Aphid** (***Hormaphis hamamelidis***) makes galls on the leaves of Witch Hazel, and is usually abundant wherever Witch Hazel is found.

SUPERFAMILY PHYLLOXEROIDEA

FAMILY ADELGIDAE (PINE AND SPRUCE APHIDS) ❷ The **Balsam Woolly Adelgid** (***Adelges piceae***) is an accidentally introduced European adelgid that has devastated eastern fir forests over the past decade. The waxy covering has been removed from this female, exposing her clutch of eggs. ❸ ❹ ❺ Pine and spruce aphids make common and familiar galls on conifers, and often have complex life cycles involving different kinds of host trees. Their tiny, wax-encrusted nymphs often occur near branch tips early in spring, and can be found later in the year inside galls that have been induced by nymphal feeding. Several species of ***Adelges*** attack conifers. These photos show wax-covered nymphs of the **Pale Spruce Gall Adelgid** (***A. laricis***) early in the spring, a young gall and an old gall after the aphids have left. *A. laricis* has a two-year life cycle, alternating between spruce and larch. **FAMILY PHYLLOXERIDAE** (PHYLLOXERANS) ❻ ❼ **Grape Phylloxera** (***Daktulosphaira vitifoliae***) forms galls on leaves and roots of grapevines, and can cause premature defoliation, reduced shoot growth and reduced yield and quality of French–American hybrid grapevines. ❽ **Hickory Gall Phylloxerans** (***Phylloxera caryaecaulis***) induce growing hickory leaves to produce tough, globular galls in which they feed until late summer.

SUPERFAMILY PSYLLOIDEA

FAMILY PSYLLIDAE ❾ Hackberry psyllids (***Pachypsylla*** spp.) make distinctive galls on the leaves of hackberry. One of these galls has been cut open to expose a psyllid nymph.

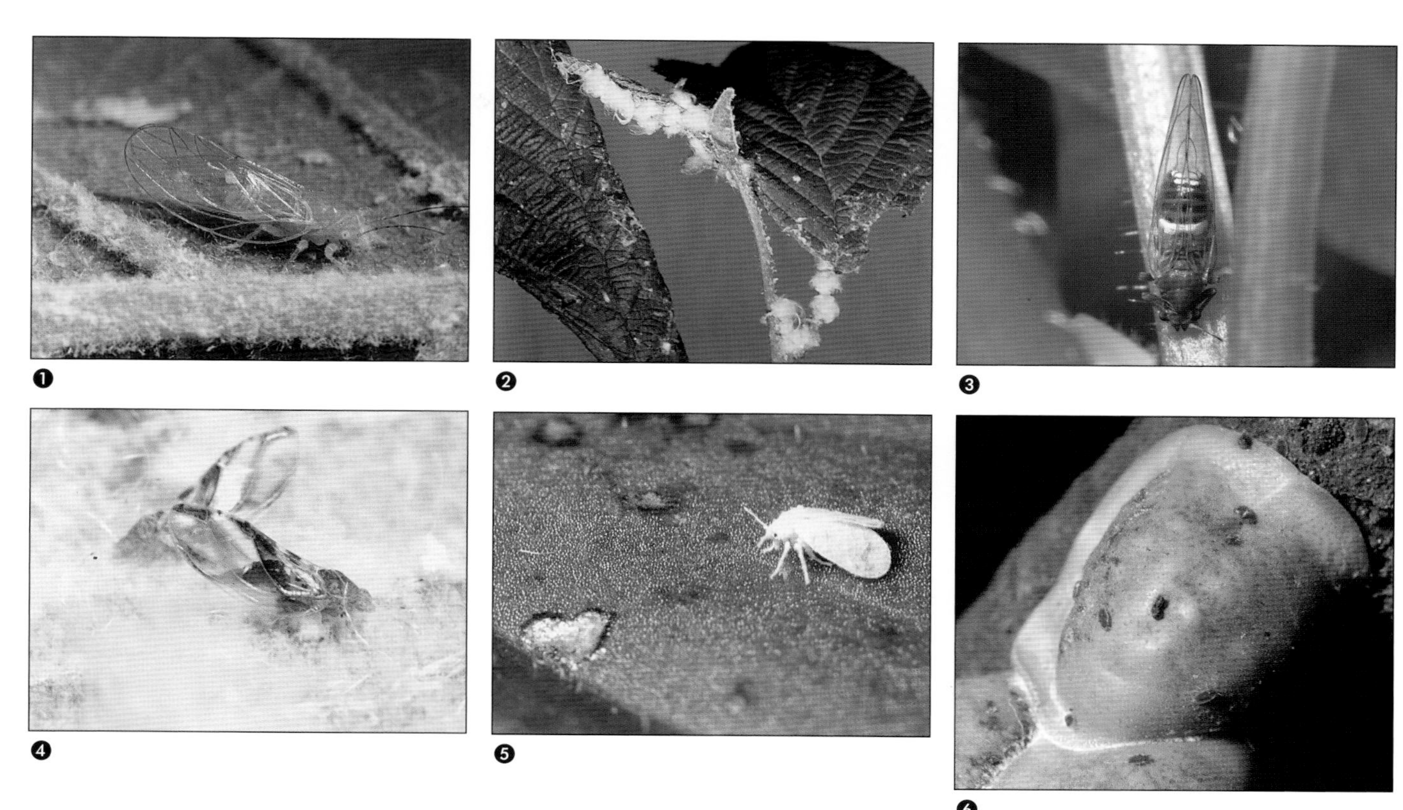

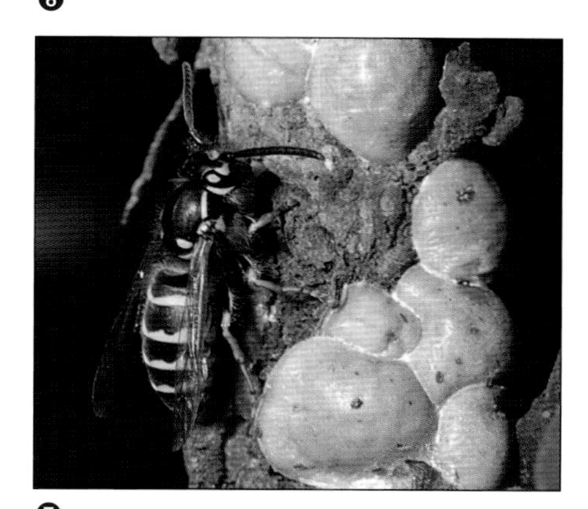

FAMILY PSYLLIDAE (continued) ❶ ❷ The **Cottony Alder Psyllid** (***Psylla floccosa***) is an aptly named psyllid that aggregates in woolly masses on alder branches during early summer. Don't confuse it with the Woolly Alder Aphid (*Paraprociphilus tessellatus*, family Pemphigidae), which appears later in the season. **FAMILY TRIOZIDAE** ❸ Although most of our common jumping plant lice are in the family Psyllidae, ***Trioza*** and a couple of related genera are treated as a different family (Triozidae). ❹ Nymphs of the **Blackberry Psyllid**, ***T. tripunctata***, develop in gregarious groups enveloped by cottony white fluff on the undersides of blackberry leaves. Adults, which have strikingly patterned wings, spend the winter in nearby conifer trees.

SUPERFAMILY ALEYRODOIDEA

FAMILY ALEYRODIDAE (WHITEFLIES) ❺ Whiteflies are minute (around 1 mm) mothlike homopterans covered with a fine, white, waxy powder. Whiteflies usually occur in large numbers, often on curled or discolored leaves.

SUPERFAMILY COCCOIDEA (SCALE INSECTS AND MEALYBUGS)

FAMILY COCCIDAE (SOFT SCALE INSECTS) ❻ **Magnolia Scales** (***Neolecanium cornuparvum***) are our largest scale insects (over 10 mm). The convex, shell-like blobs that sometimes encrust the branches of Magnolia trees are waxy secretions that conceal typical female scale insects, or, during the winter, scale insect eggs. Those eggs will hatch in spring into tiny active young that still have legs and antennae, and are called crawlers (look closely for the crawlers in this photograph). Female crawlers find themselves a good piece of branch where they settle down and degenerate, quickly losing appendages and secreting a wax house. Male crawlers develop into a small two-winged insect, a bit like a small fly but without functional mouthparts and with a thin tail-like structure. ❼ Many other insects, such as yellowjacket wasps, are attracted to honeydew like that egested by these **Magnolia Scales** (***Neolecanium cornuparvum***). ❽ **Wax Scales** (***Ceroplastes*** sp.), which are mostly southern in distribution, produce conspicuously thick waxy shelters. Look for these convex scales on holly and citrus plants.

FAMILY COCCIDAE (continued) ❶ Like other homopterans, **Cottony Maple Scales** (***Pulvinaria innumerabilis***) produce copious amounts of honeydew. This sugary waste product attracts ants but sometimes also encourages sooty fungus to grow all over the sticky leaves of heavily infested trees. The cotton-like white stuff in this picture is the waxy egg sac of this female scale. **FAMILY DIASPIDIDAE** (ARMORED SCALE INSECTS) ❷ The **Pine Needle Scale** (***Chionaspis pinifoliae***), is a very common pest of pines, but can also occur on needles of spruces and firs. The oyster-shell-shaped wax covers on these overwintering female scales conceal eggs that will hatch in spring. Newly hatched nymphs (called crawlers) insert hairlike mouthparts at their new feeding sites, and then produce their own armor-like waxy covering. Male nymphs develop into winged adults but females stay in place. These scales can completely cover needles, causing plant discoloration due to needle and branch death. ❸ The **Euonymus Scale** (***Unaspis euonymi***) is a common pest of evergreen euonymus and sometimes other ornamentals (Bittersweet, Pachysandra). Both sexes secrete protective waxy shelters (armor) but males produce elongate, white shells with three longitudinal ridges. The female's armor is oyster shell-shaped and dark brown. ❹ The **Oystershell Scale** (***Lepidosaphes ulmi***) is a cosmopolitan pest of trees and shrubs. If you were to look under one of these seashell-like waxy shelters during the summer months you would find a legless, eyeless adult female sucking sap with her long, thin beak; if you look during the winter you would find only eggs and the shriveled remains of the adult female. Eggs hatch in spring. **FAMILY PSEUDOCOCCIDAE** (MEALY BUGS) ❺ Mealy bugs, such as the **Longtailed Mealy Bug** (***Pseudococcus longispinus***) are common pests of greenhouse and household plants, where the waxy, wingless females can festoon a wide variety of foliage. They build up numbers quickly by skipping eggs and giving birth to live young. Mealy bugs are much like scales, but are a bit more mobile and lack the hard shell. **FAMILY DACTYLOPIIDAE** ❻ The white waxy blobs on this prickly pear cactus (*Opuntia* sp.) are **Cochineal Bugs** (***Dactylopius confusus***), and the brilliant purple fluid oozing out of the bug being punctured with my forceps is loaded with carminic acid, which protects the bugs from all but a few specialized predators. Cochineal bugs were harvested and cultivated as the major source of red pigments for almost 300 years. Prickly pear cacti and their associated bugs are common in the southeast, but are increasingly rare northward. The Eastern Prickly Pear Cactus is an endangered species in Canada.

❶

❷

❸

FAMILY MARGARODIDAE (GIANT COCCIDS AND GROUND PEARLS). ❶ The **Cottony Cushion Scale** (***Icerya purchasi***) is a serious citrus pest accidentally introduced from Australia to California in the late 1800s; it now also occurs in the eastern U.S. and the Caribbean. The cottony, fluted parts are the egg sacs, each containing about 1,000 eggs. **FAMILY KERMESIDAE** (GALL-LIKE SCALES) ❷ ***Allokermes galliformis***, the only member of the family Kermesidae that ranges north to Canada, forms large clusters on oak twigs, often attended by wasps, ants and flies (like this tachinid) feeding on the copious amounts of honeydew egested by the female scales. **FAMILY ORTHEZIIDAE** (ENSIGN SCALES) ❸ Female ensign scales are unusually long-legged and active for scale insects, and often carry a distinctive armature of white, overlapping waxy plates. Some are found in litter and soil during at least part of their life cycles, and a few have the unusual (for Sternorrhyncha) habit of feeding on fungal hyphae.

THRIPS
ORDER THYSANOPTERA

❶

❷

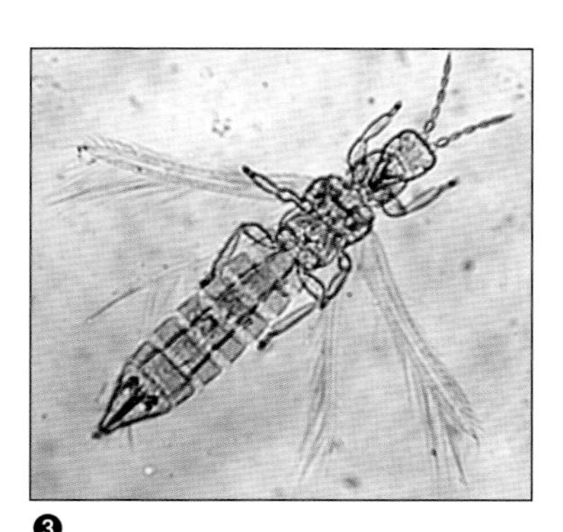

❸

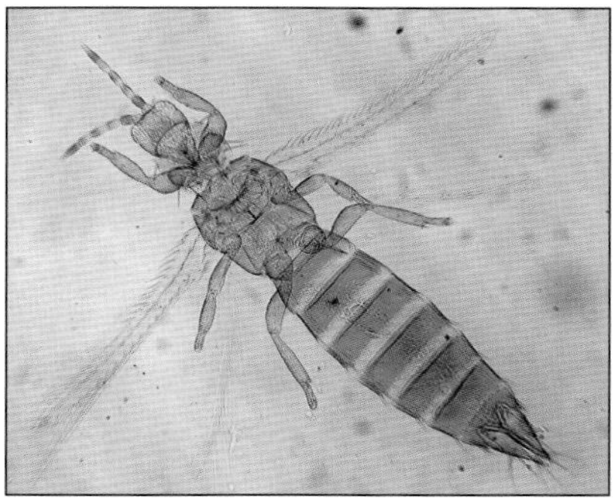

❺

FAMILY PHLAEOTHRIPIDAE (TUBE-TAILED THRIPS) ❶ Thrips are odd little hemipteroids with asymmetrical mouthparts – the right mandible is absent, but the left mandible is like a pickaxe that rips a hole through which the contents of the target (another insect, a pollen grain, a plant) is imbibed. They live under bark, on fungi and in many other habitats. Daisy flowers are almost invariably speckled with a common species called the **Clover Thrips** (***Haplothrips leucanthemi***). ❷ Thrips are tiny, somewhat torpedo-shaped insects with bladder-like feet (blobs at the tips rather than the usual insect claws) and thin wings fringed with long hairs. You can see the wings folded over the backs of these three winged adult ***Haplothrips***. Nymphal thrips are often red, orange or yellow in contrast with the dark adults. **FAMILY THRIPIDAE** (COMMON THRIPS) ❸ **Onion Thrips** (***Thrips tabaci***) is a tiny (about 1 mm) pest of many crops and a significant vector of plant viruses. ❹ **Flower Thrips** (***Frankliniella tritici***) is a tiny (about 1 mm) slender, pale pest of a wide variety of plants. **FAMILY AEOLOTHRIPIDAE** ❺ The **Banded Thrips** (***Aeolothrips fasciatus***) is a widespread predaceous thrips about 1.6 mm in length.

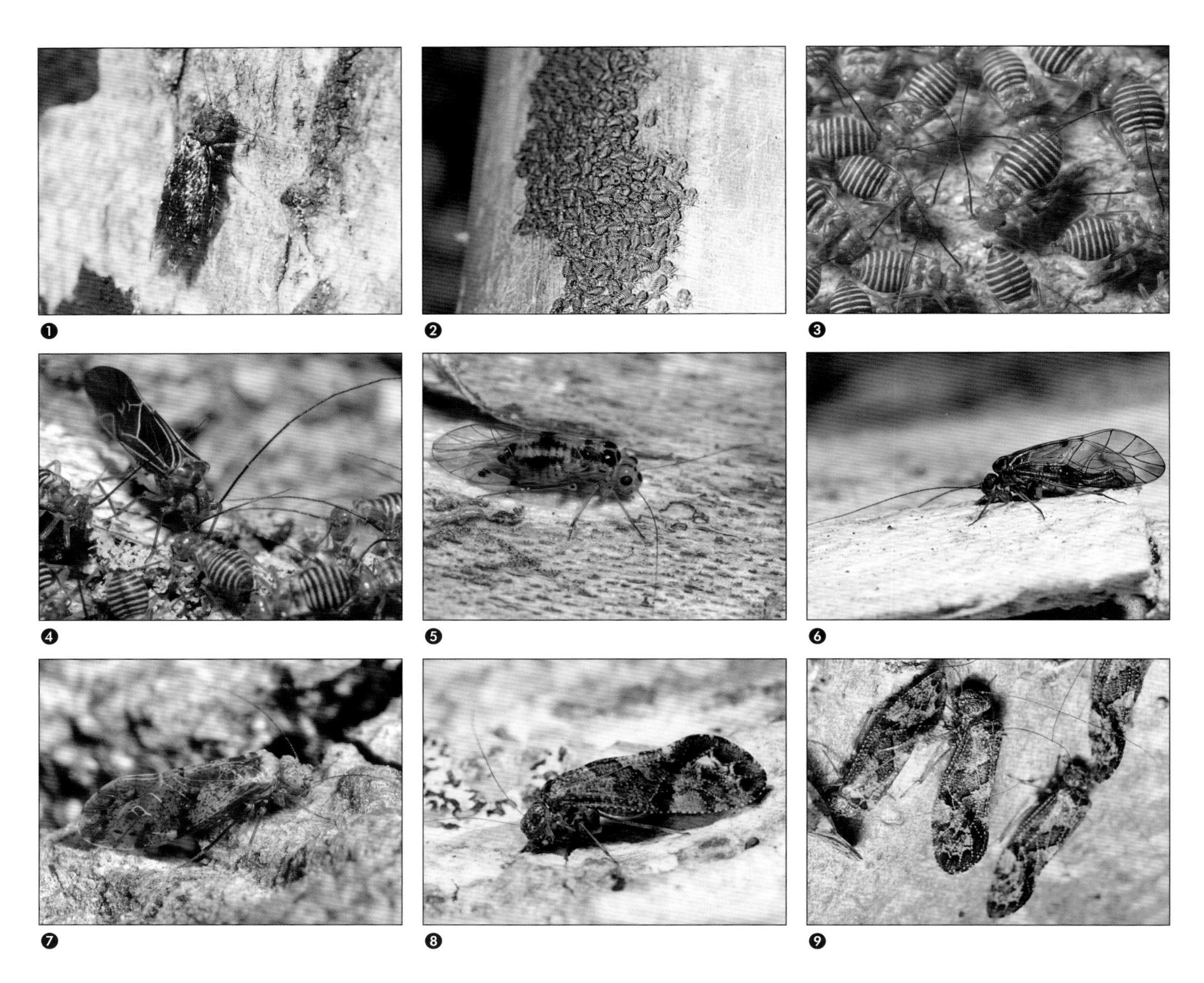

FAMILY LEPIDOPSOCIDAE ❶ The aptly named Lepidopsocidae, like this tiny ***Echmepteryx hageni***, look like tiny moths because of their narrow, scale-covered wings. This species is common under bark. **FAMILY PSOCIDAE** ❷ ❸ These nymphal ***Cerastipsocus venosus*** have formed tightly packed "herd" that looks, from a distance, like a patch of loose bark on an otherwise bare tree trunk. When approached, the herd moves as one. One of these nymphs is burdened with a bright red mite. ❹ This close up of part of a ***Cerastipsocus venosus*** "herd" shows an adult and several nymphs feeding on lichen on a maple tree trunk. ❺ ***Psocus leidyi*** is one of many psocids common on standing dead trees. ❻ ***Metylophorus novaescotiae***, like related species, can be found on trees, often among dead leaves. ❼ This ***Trichadenotecnum alexanderae*** is grazing lichens from the bark of a dead beech tree. **FAMILY MYOPSOCIDAE** ❽ Barklice are easily recognized as the order Psocoptera because of their distinctive bulging faces and long antennae. Look for them on tree trunks, where several species such as this ***Lichenomima*** feed on algae and lichens. ❾ ***Lichenomima*** species often occur in large numbers in late summer on a variety of trees.

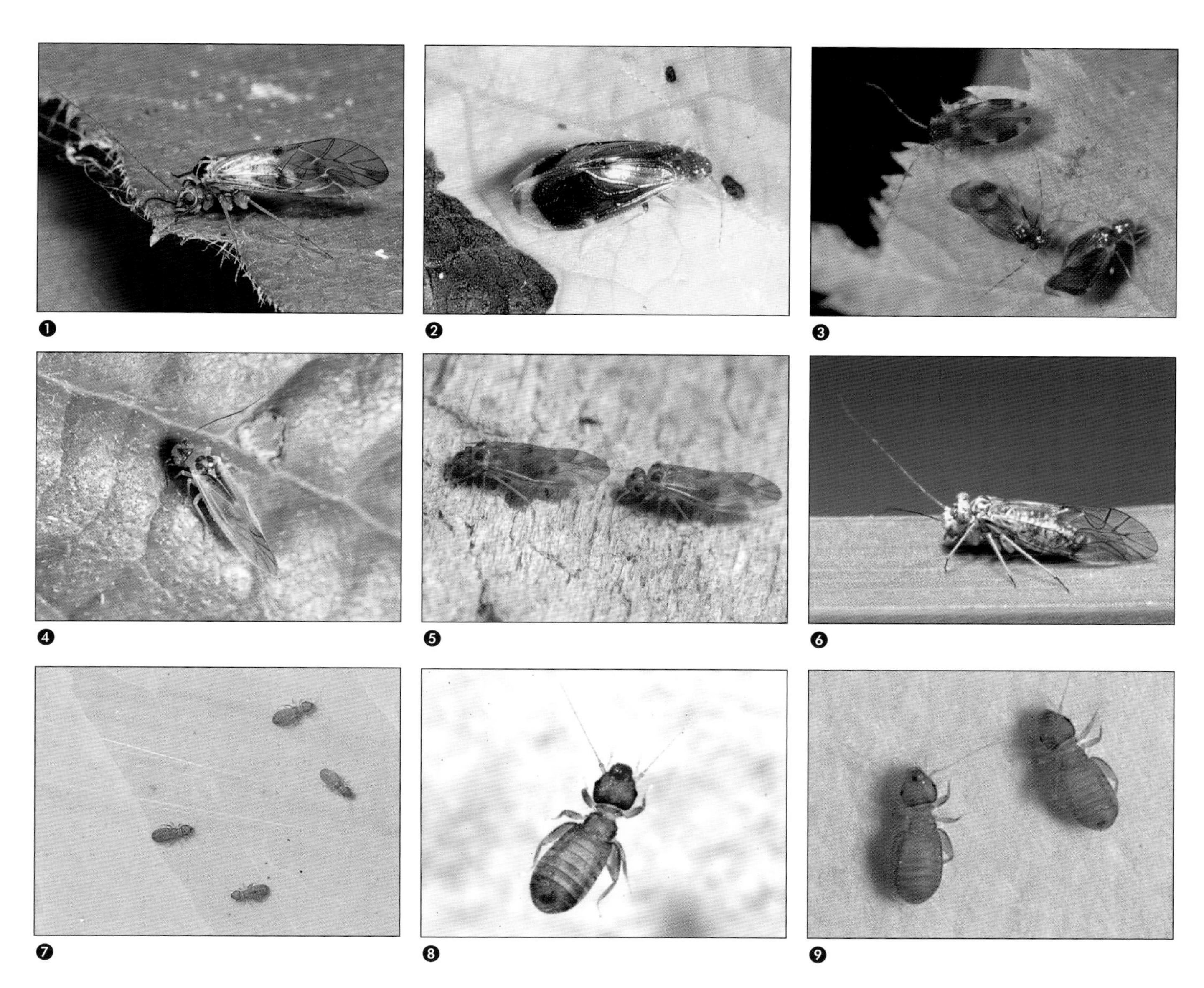

FAMILY DASYDEMELLIDAE ❶ ***Teliapsocus conterminus*** is a common barklouse on birch trees. Like most barklice, they hold their wings like a steep-sided roof over the body. **FAMILY AMPHIPSOCIDAE** ❷ ❸ ***Polypsocus corruptus*** is a distinctive barklouse that seems to mimic a small beetle. Look for them on foliage where they graze pollen and other material off leaf surfaces. **FAMILY CAECILIIDAE** ❹ Members of the family Caeciliidae (like this ***Caecilius***) are usually found on leaves, sometimes along with the Jumping Plant Lice (Psyllidae) they superficially resemble. **FAMILY PERIPSOCIDAE** ❺ These ***Peripsocus subfasciatus*** are gnawing on the surface of an old willow log. **FAMILY MESOPSOCIDAE** ❻ This apparently two-headed bark louse is a mating pair of ***Mesopsocus***, with the male's antennae and legs tucked out of sight. As is true for most *Mesopsocus* species, the female has no wings. **FAMILY TROGIIDAE** ❼ Some Psocoptera, called booklice, are wingless or almost wingless and often occur inside buildings, where they feed on stored products or starchy materials like book bindings. Several species, such as the **Larger Pale Booklouse** (***Trogium pulsatorium***, about 1 mm), are common in houses. **FAMILY LIPOSCELIDAE** ❽ ❾ ***Liposcelis*** species are common indoor pests, especially where temperature and humidity are high. These minute (about 1 mm), entirely wingless booklice have distinctively swollen hind femora. Some entomologists think that Liposcelidae are closely related to true lice (Phthiraptera).

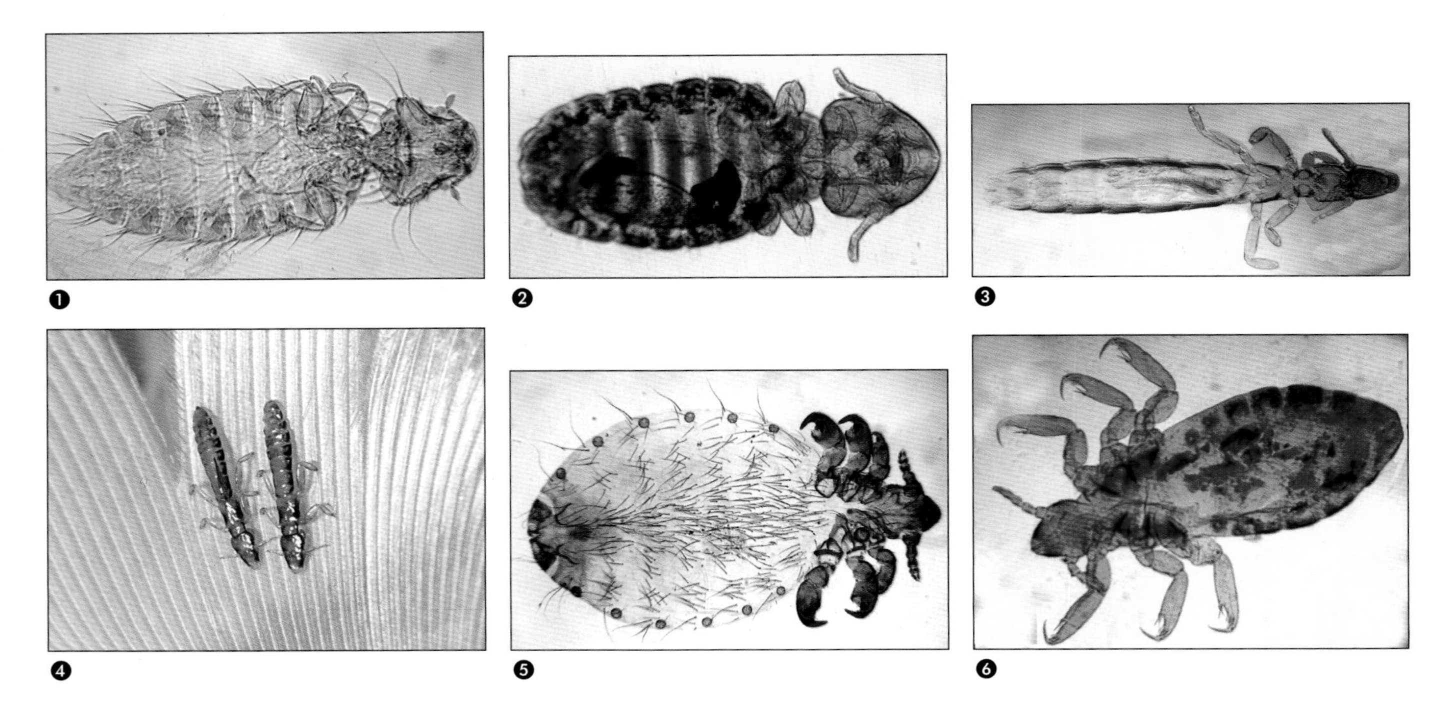

❶ ❷ ❸ ❹ ❺ ❻

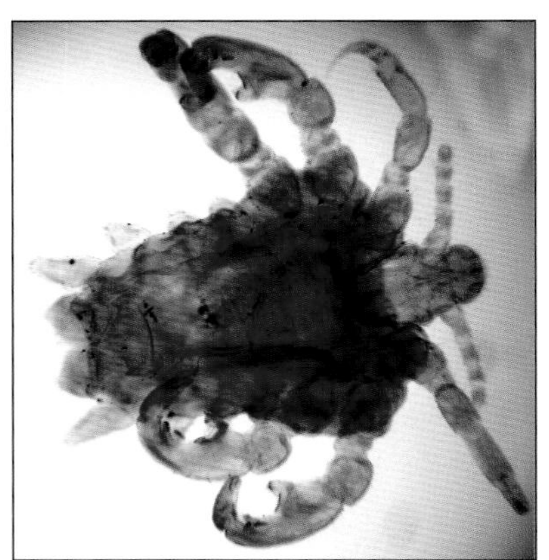

❼

Lice are parasitic relatives of the barklice that have lost their wings, shortened their antennae, and become flattened and otherwise adapted to their parasitic existence. You might get lucky and find some lice on your own body, but the average naturalist is more likely to encounter one of the many species of lice that feed on birds.

SUBORDER AMBLYCERA

FAMILY MENOPONIDAE ❶ The **Shaft Louse** or **Chicken Louse** (***Menopon gallinae***) is a small (about 1.5 mm) chewing louse often found on chicken, ducks and pigeons throughout the world.

SUBORDER ISCHNOCERA

FAMILY TRICHODECTIDAE ❷ The **Cattle-biting Louse** (***Bovicola bovis***) is one of the common pests in this family. A similar species (*B. equi*) is found on horses. **FAMILY PHILOPTERIDAE** ❸ The **Slender Pigeon Louse** (***Columbicola columbae***) is a common ectoparasite of pigeons. ❹ Bird Lice, such as these ***Anaticola crassicornis*** (**Slender Duck Louse**), usually feed on the feathers and skin of their host.

SUBORDER ANOPLURA (SUCKING LICE)

FAMILY LINOGNATHIDAE ❺ The four North American species of ***Linognathus*** occur on domestic animals such as dogs and cattle. This species (**Dog-sucking Louse**, ***L. setosus***) occurs only on dogs. **FAMILY PEDICULIDAE** ❻ ***Pediculus humanus*** is larger (around 3 mm) and longer than the Crab Louse (the only other louse commonly found on people). Those that lay eggs (nits) in hair are called **Head Lice**; those that lay eggs on clothing are called **Body Lice**. Some consider head and body lice to be different species. **FAMILY PHTHIRIDAE** ❼ The **Crab Louse** or **Pubic Louse** (***Phthirus pubis***) is a small (about 1.5 mm), squat louse with crab-like claws used to cling to coarse body hairs.

CHAPTER 7

Butterflies and Moths

Those strange and mystical transmigrations that I have observed in Silkworms, turned my philosophy into Divinity ... Ruder heads stand amazed at those prodigious pieces of Nature, Whales, Elephants, Dromedaries and Camels; ... but in these narrow Engines there is more curious mathematics.

– Sir Thomas Brown, 1635, Religio Medici

Metamorphosis

Everyone knows that caterpillars have no wings, that they transform into pupae and that the pupae transform into butterflies and moths. This fantastic type of transformation – known to entomologists as complete metamorphosis – is so commonplace that it is impossible to ignore. You may never actually watch the development of an insect from larva to adult, but in order to be completely free from exposure to the concepts of metamorphosis you would have to avoid reading, watching TV and even extensive conversation. You would have to live in a cocoon (so to speak) to miss the television ads, movies, children's books, novels and educational media that regularly portray the familiar phenomenon of a caterpillar changing into a butterfly or moth.

Metamorphosis is thus a well-known fact of life, but it is less widely appreciated that the majority of species in the animal kingdom belong to one big related group of insects, all of which have a similar and special type of metamorphosis known as complete metamorphosis. Flies, wasps, beetles, moths, caddisflies and a few smaller orders all belong to this one big group (Endopterygota or Holometabola), and all have wingless larvae that transform first into pupae and then into winged adults.

Metamorphosis is an abrupt, drastic change in form that probably originated to expedite the transition from a young insect that can't fly to an adult that can. If that transition took place in a uniform fashion, with each succeeding stage having slightly larger wings, it would not be metamorphosis. Look at some bugs or grasshoppers and note how the young ones (nymphs) have small, unobtrusive wing buds, and the adults have fully developed wings. There are no intermediate stages with awkward-sized wings, not quite big enough to fly but big enough to get in the way. To find immature insects with wings like that, you would have to look in the fossil record, which shows that some ancient lineages of winged insects did not have metamorphosis, and thus didn't dispense with the intermediate wing stages. Those lineages are now extinct. All surviving winged insects use a metamorphic molt to bypass the hazards of adolescence.

Like all living winged insects, grasshoppers have an abrupt, drastic change in form from a stage with wing buds to a form with fully developed wings. This change in form takes place in one "metamorphic" molt from a nymph (with wing buds) to an adult (with wings). Grasshoppers, dragonflies, bugs and other insects that have nymphs with externally developing wings have only one metamorphic molt – from nymph to adult.

Insects that start out as larvae without external evidence of wings, like caterpillars, have to undergo two drastic changes, or metamorphic molts. In these groups, wings develop internally until the mature larva (without external wing buds) undergoes a metamorphic molt to the pupa (with external wing buds). The pupa later transforms into a fully winged adult in another metamorphic molt. Insects with two metamorphic molts have complete metamorphosis; those with only a single metamorphic molt are said to have gradual metamorphosis. Groups with complete metamorphosis have certainly been successful (they dominate the planet), perhaps because the development of complete metamorphosis freed the larvae and adult to respond to very different selection pressures, and allowed them to become specialized for very different lives.

Metamorphosis, like other aspects of growth and development, is regulated by hormones – the internal chemical messengers that mediate communication between an individual's nervous system and its other body functions. When an insect grows to the point where it will no longer fit in its inflexible external skeleton, stretch receptors tell the nervous system to release a number of different hormones, called neurohormones, which effectively lead the insect through the complexities of molting to the next stage. One neurohormone triggers glands to release another kind of hormone (molting hormones or ecdysteroids), which in turn initiates the formation of a new external skeleton. Metamorphosis takes place when molting is associated with the activation of genes that code for previously suppressed adult characters, so it follows that there must be some kind of additional chemical messengers telling the insect whether to just molt to another larval stage or to undergo a metamorphic molt.

One of the many internal chemical messengers involved with metamorphosis is a hormone secreted by tiny, glistening white glands (the corpora allata) that sit right behind the insect's brain. In a famous experiment from the 1950s, entomologist Carroll M. Williams removed these glands from a small caterpillar that had not yet reached the stage where it would normally molt to a pupa. The caterpillar molted anyhow, but instead of simply casting its skin to become a larger larva it molted prematurely to a pupa. In another experiment, Williams dissected these little glands out of a small caterpillar and inserted them into a mature caterpillar that would normally molt to a pupal stage. Instead of molting to a pupa, this caterpillar molted to a giant caterpillar. The explanation for these remarkable results was that the tiny white corpora allata produce a hormone, aptly called juvenile hormone, which keeps adult characters from developing. Normally, juvenile hormone secretion drops when the insect has developed to the stage where it is ready to develop into an adult, but addition of juvenile hormone from a younger individual will cause a molt to another, larger larva instead of a molt to a pupa or adult.

Although juvenile hormone actually plays many roles in insect development, it is most famous as a kind of entomological elixir of youth that allows growth without maturation to adulthood. An understanding of this "Peter Pan" effect of juvenile hormone has potential value far beyond the creation of giant caterpillars in entomological laboratories. For example, chemicals that mimic juvenile hormone can be used as pesticides, preventing normal development of pest insects such as mosquitoes.

Plants have protected themselves against insects for millions of years using similar fake hormones. The existence of insect juvenile hormones in plants was discovered quite by accident, when some entomologists were trying to rear bugs in cages lined with paper towels. To their consternation, the bugs molted into giant nymphs and never developed into adults. They managed to figure out that something in the paper towels, which were made from balsam fir, was keeping the insect juvenile. This "paper factor," as they called it, was a juvenile hormone mimic. We now know that similar compounds play a defensive role in many kinds of plants.

Butterflies and Moths (Lepidoptera)

In ancient Rome, the people of the world were classified into two groups: "Romans" and "barbarians." The Romans made up a familiar, identifiable group and the barbarians included everyone else. I'm sure you can think of a current classification of people or things into one definable group and one group that includes "all the rest." The popular division of the Lepidoptera into the familiar butterflies on the one hand and the moths on the other is exactly this kind of artificial classification, with butterflies (and skippers) making up a small, easily recognized group of brightly colored, day-flying insects with clubbed antennae. Moths, then, are simply all Lepidoptera other than butterflies and skippers.

If you have ever grasped a moth or butterfly by the wing, you likely saw its colors transferred to your fingers as a smudge of dustlike scales. Those scales, which are just elaborately flattened hairs, give the butterflies and moths their formal name Lepidoptera (*lepido* = scale, *ptera* = wings). With the exception of the rare and primitive suborder Zeugloptera (mandibulate moths), adult Lepidoptera lack chewing mandibles, and most butterflies and moths have a long proboscis coiled under the head. The proboscis (made from long parts of the maxillae) resembles one of those coiled party favors popular at New Year's Eve gatherings, and can be extended like a straw to suck up nectar. Lepidoptera larvae (caterpillars) have chewing mandibles and usually have two to five pairs of fleshy prolegs that are almost always tipped with circlets of hooks called crochets.

Butterflies and a huge majority of moths belong to a large suborder (Ditrysia) in which the female has two genital openings, one to receive sperm and the other through which the eggs are passed. The former opening leads to a storage pouch which is connected to the usual reproductive system. This contrasts with the "primitive moths" outside the huge suborder Ditrysia that retain the usual arrangement of only a single opening through which the male comes in and the eggs come out. Butterflies and skippers are just two of the many superfamilies within the Ditrysia.

Butterflies and most moths feed with a long proboscis, normally held coiled under the head.

The butterflies (superfamily Papilionoidea) are such conspicuous, brightly colored insects that they have received attention far out of proportion to their diversity. So many amateur and professional entomologists study this little superfamily of around 750 North American species that there is probably more than one butterfly student per butterfly species. Contrast this with a ratio of about one student to one thousand species in the Diptera (true flies). This attention is by no means wasted, for butterflies have served as important study animals for research in ecology, conservation biology and evolutionary biology. Most of the insects on endangered species lists are butterflies because they are among the few invertebrates well-enough known to allow correlation of habitat loss with well-documented declines.

Gossamer-winged butterflies and metalmarks (Lycaenidae and Riodinidae)

Perhaps the most publicized invertebrate on North American endangered species lists is the beautiful Karner Blue (*Lycaeides melissa samuelis*), first described by the well-known writer Vladimir Nabokov. Blues, along with harvesters, hairstreaks, metalmarks and coppers, belong to the large and diverse family **Lycaenidae** (gossamer-winged butterflies, page 180). Blues are often among the first (and most beautiful) insects seen in early spring, and one attractive little species (the Spring Azure, *Celastrina ladon*) is a common harbinger of spring across North America. Spring Azure larvae feed on the hidden petals of developing flowers on a variety of shrubs and trees, especially wild cherry, then transform into pupae that do not hatch until early the following spring. Blue butterflies seen after early spring are not second and third generation Spring Azures, as previously thought, but are other species of blue butterflies, including the similar Summer Azure (*C. neglecta*).

In contrast to the abundant Spring Azure, the famous Karner Blue is highly specific in its choice of host plants. The green, somewhat sluglike larvae feed only on lupine, a plant that thrives in the kinds of open areas made by regular fires. In Ontario, the beautiful Karner Blue once lived in open oak savanna, mostly along the shores of Lake Huron, but it is now extirpated from the province due to some combination of fire control, pine plantations, development, recreational vehicle use and burgeoning deer populations. The Karner Blue was last sighted in Ontario in 1992. Another lupine-feeding lycaenid, the Frosted Elfin (*Callophrys irus*), has not been seen in Ontario since 1986. There is talk now about reintroducing these species to Ontario, but the reintroduction of a conspicuous species like the Karner Blue seems a bit like window dressing. Every beautiful, well-known species like the Karner Blue shares its habitat with thousands of obscure and understudied species, which together make up the ecosystems we should be protecting. Endangered butterflies are valuable warning flags, pointing to greater problems when they disappear.

Among the other species intricately involved with the threatened Karner Blue are ants that tend the butterfly larvae, much the way other ants tend aphids. Aphids maintain their mutualistic relationships with ants by providing a sugary waste product called honeydew. Karner Blue caterpillars and many other Lycaenidae accomplish the same end using special glands that produce tasty secretions of various sorts. Ants are attracted to Karner Blue caterpillars because of sugary secretions produced by abdominal glands, and the caterpillars in turn obtain protection from predators and parasites. Some other lycaenid butterflies cheat on their deals with ants, and use their glandular chemical factories to trick ants into defending them without providing anything nutritious to the ants in return. Many lycaenids carry this trickery to amazing extremes, actually fooling the ants into feeding the caterpillars rather than the other way around. Perhaps the ultimate turnaround in ant-caterpillar relationships is perpetrated by those species that use their glandular gadget bag to become specialized predators in ant nests. The Large Blue (*Maculinea arion*) is an Old World species that produces glandular secretions to mimic chemicals used by a type of ant to recognize its own larvae. The Large Blue caterpillar starts out life feeding on thyme flowers, then drops to the ground and waits for an ant to mistake it for one of its own. The ant carries the caterpillar back to its nest and places it with its brood. The Large Blue caterpillar then feasts, unmolested, upon the ant brood before it spends the winter and pupates in the nest. Despite, or perhaps because of this elaborate life history, the Large Blue went "extinct" (a local extinction is more properly called an extirpation) in Britain, and has since been the subject of elaborate reintroduction schemes.

Spring Azure caterpillar (*Celastrina ladon*), page 180, caption 3.

A taste for eating other insects rather than plants is not restricted to foreign lycaenid species… North America is home to some odd predaceous caterpillars too. One little orangeish butterfly with a widespread eastern North American distribution develops from a caterpillar that eats Woolly Alder Aphids (Pemphigidae). The adult butterfly, or Harvester (*Feniseca tarquinius*), feeds on honeydew and other fluids using its unusually short proboscis. Harvesters are more closely related to African and Southeast Asian butterflies than they are to other North American species.

In addition to many species of blues and our one harvester, North America has a rich lycaenid fauna including hairstreaks, elfins and coppers – all relatively small, delicate butterflies with distinctively banded antennae. Although some are widespread and common, many Lycaenidae are quite habitat-restricted and form sensitive localized populations.

Metalmarks (**Riodinidae**, page 181) are similar to Lycaenidae and are sometimes included as a subfamily of Lycaenidae. Like lycaenid caterpillars, riodinid caterpillars often have sugar glands used in a mutualistic relationship with ants, and some have additional scent-producing or sound-making organs used to call in ants when under threat. Metalmarks are extremely diverse in the tropics, but only a couple of uncommon species occur in northeastern North America.

Swallowtails (Papilionidae)

Swallowtails are among our largest and most attractive insects, and even if you rarely venture into habitats more exotic than backyards or gardens you have probably been awed by one of these distinctively shaped butterflies (family **Papilionidae**, page 181). Perhaps the most commonly encountered swallowtail is the Black Swallowtail (*Papilio polyxenes*), a regular breeder on wild carrot as well as parsley and cultivated carrots.

Like other swallowtails, Black Swallowtails spend the winter as resting pupae, emerging fairly early to lay eggs near the tips of young host plant leaves. The eggs hatch to black-and-white caterpillars that initially avoid predation by resembling bird droppings, and which later develop into striking black-and-greenish yellow caterpillars with rows of orange dots. If you handle or disturb mature swallowtail larvae, be prepared for a spectacular defensive display, as these caterpillars are equipped with a bright orange, forked, foul-smelling eversible organ, called an osmeterium, on the top of the first thoracic segment. The osmeterium presumably deters predators by its smell, startles them with its sudden appearance and bright colors and possibly scares them because of its similarity to a snake's forked tongue.

Black Swallowtail larvae are easily found on carrots, dill or common related weeds like Queen Anne's Lace (*Daucus carota*), and provide a good opportunity to watch the formation of a chrysalis, or butterfly pupa. Like other swallowtails, the Black Swallowtail caterpillar spins a silk girdle in which the pupa will be suspended at about a 30-degree angle from the supporting stem or bark (butterflies in the family Pieridae have the same habit). Like most other butterflies, the pupa (or chrysalis) keeps itself in place by hooking on to a pad of silk, previously put in place by the larva, with a Velcro-like tail called a cremaster.

Male and female Black Swallowtails differ in color because females are mimics of another swallowtail (the Pipevine Swallowtail, *Papilio philenor*) that picks up protective toxic chemicals from its host plant (Dutchman's Pipe). Pretending to be a Pipevine Swallowtail must offer some significant protection to mimics, since several other butterflies, and even one moth, have forms that closely mimic Pipevine Swallowtails.

The Black Swallowtail is one of the most common swallowtails throughout much of northeastern North America. Until very recently it was thought that a big yellow swallowtail called the "Tiger Swallowtail" was equally widespread, but this papilionid turned out to be a mixture of two species now known as the Canadian Tiger Swallowtail (*Papilio canadensis*, abundant across Canada, and common in Ontario far north of the range of other swallowtails) and the Eastern Tiger Swallowtail (*P. glaucus*, common in the eastern United States and southern Ontario). The two species, which do occasionally hybridize in southern Ontario, are almost identical as adults. Larval Eastern Tiger Swallowtails, however, cannot survive on the birch and aspen trees preferred by Canadian Tiger Swallowtails, and Canadian Tiger Swallowtails do not thrive on the Carolinian trees preferred by Eastern Tiger Swallowtails. Each species can feed on some of the same plants (like cherry); they occasionally hybridize and mature caterpillars of both species have a striking appearance due to the possession of two brilliant eyespots on the swollen front part of the body. Very young caterpillars resemble bird droppings. Although common, tiger swallowtail caterpillars are rarely noticed because they feed at night and hide in curled leaves during the day.

Zebra Swallowtail (*Papilio marcellus*), page 183, caption 8.

Whites and sulfurs (Pieridae)

Most North American butterflies are either brush-footed butterflies (Nymphalidae), in which the front legs are usually reduced in both sexes, or gossamer-winged butterflies (Lycaenidae), in which the front legs are normal in the female but reduced in the male. Only two families of butterflies, each a relatively small and easily recognized group, have six fully developed legs in both sexes. Those families are the Papilionidae (swallowtails) and **Pieridae** (whites and sulfurs, page 184).

The best-known butterfly in the family Pieridae is the Cabbage White (*Pieris rapae*), a widely recognized garden butterfly with one or two black spots on its white front wings (males have one spot and females have two). These common introduced butterflies develop from velvety green caterpillars on cabbage and related crops. Another common pestiferous pierid, the native Orange Sulfur (*Colias eurytheme*), is particularly abundant where alfalfa is grown. The widespread distribution and conspicuous abundance of Cabbage Whites and Orange Sulfurs is not typical of the Pieridae on the whole, as many whites and sulfurs have restricted ranges and occur on only a few kinds of plants (usually crucifers or legumes). The West Virginia White (*Pieris virginiensis*), for example, is a small, white butterfly associated with mature hardwood forests, where it feeds on toothworts (*Dentaria* spp.) and flies only in very early spring. This species occurs in scattered populations, and was on the Ontario endangered species list for some years because it was known from only a single site in the province. It was de-listed when enough stable populations were discovered to satisfy authorities that it was not a threatened species.

Brush-footed Butterflies (Nymphalidae)

Most of our common butterflies, including those previously placed in the separate families Danaeidae (milkweed butterflies), Satyridae (nymphs, satyrs and arctics) and Libytheidae (snouts), are now included in the single family **Nymphalidae** (page 184). With the exception of female snouts, the front legs of brush-footed butterflies are reduced, giving these insects a four-legged appearance. The brushlike forelegs are used primarily for "tasting" potential food sources, usually liquid substances imbibed in typical lepidopteran fashion using a long, coiled proboscis made up of two long mouthparts (parts of the maxillae) pressed together and interlocked to form a sort of flexible straw. The butterfly proboscis contains muscles used in coiling, but the spectacular uncoiling that occurs when a butterfly lands on a nectar-filled flower is driven by "blood pressure" as blood (hemolymph) is pumped into the proboscis. The proboscis is normally used to feed on nectar or rotting fruit, but some butterflies are attracted to dung or carrion.

Snouts (subfamily Libytheinae), named for the distinctively long maxillary palpi that give them a long-nosed appearance, are usually common where their host plant, hackberry, occurs. You might have to look twice to see them, though, because they look remarkably like dead leaves when at rest. Snouts do not survive the winter in the northern part of their range but instead disperse northward

Clouded Sulfur (*Colias philodice*), page 184, caption 4.

each summer. The only species of snout in eastern North America ranges as far north as southern Ontario.

Heliconians (subfamily Heliconiinae) are also unable to survive freezing winter weather, but unlike the snouts the heliconians are only rarely encountered in the northeast. Two species, the Gulf Fritillary (*Agraulis vanillae*) and the Zebra (*Heliconius charitonius*), sometimes establish temporary breeding populations as far north as the host plant (passionflowers) occur, and occasional vagrants get collected as far north as Minnesota.

Nymphs, satyrs and arctics (subfamily Satyrinae) are generally dull-colored, slow-flying butterflies that develop as inconspicuous grass or sedge-feeding caterpillars. Most have brownish wings marked with distinctive eyespots. Nymphs and satyrs are common insects, but arctics are often restricted to cool habitats such as mountaintops or bogs. Arctic caterpillars have the remarkable ability to survive extreme cold by dehydrating their tissues and apparently freezing solid, in contrast with the usual insect strategy of lowering the freezing point with natural antifreeze compounds.

Monarchs (subfamily Danainae) are familiar orange-and-black butterflies, famous for their spectacular annual migrations and their fascinating roles as unpalatable models imitated by other butterflies. Perhaps because of this familiarity, and perhaps because of its beauty, the Monarch (*Danaus plexippus*) is frequently chosen as an emblem or state insect. The Entomological Society of Ontario, a society that has been thriving since 1863, was the first group to use the Monarch as its emblem. Vermont now lists the Monarch as its state butterfly, and it is also the state insect of Illinois. The United States Congress recently considered a bill requesting that the Monarch be named the national insect, and the Mexican government has suggested using it as the symbol of free trade between Mexico, the United States and Canada.

The Monarch was first described by a Swedish scientist on the basis of a specimen from the United States, but the species ranges from Canada through South America (it is also found in Australia and New Zealand). Monarchs breed continuously in warmer climates but in the northern United States and Canada they do not survive the winter. In the north, the late summer generation (there are usually two generations a year in southern Ontario) responds to shortening day lengths by storing body fat, and then flying south as part of a unique mass migration. Canadian entomologist Fred Urquhart started studying Monarch migrations in 1937, using wing tags to trace the destination of these beautiful insects. It took him until 1976 to make the dramatic discovery that the vast majority of Monarchs that depart northeastern North America every fall end up spending the winter in a few aggregation sites in the highlands of Mexico, where they festoon fir trees in spectacular millions (Monarchs west of the Rocky Mountains spend the winter in small groves of trees along the California coast). The unique nature of these sites, and their importance, is now reflected in their designation by the International Union for the Conservation of Nature and Natural Resources as "threatened phenomena," and the designation of these areas as "ecological preserves" by the Mexican government. Perhaps these designations, along with local understanding of what a precious tourist attraction the Monarch aggregations can be, will save these sites from the sad fate of much of Mexico's highland forest.

In spring, when the masses of Mexican Monarch migrants have used up their stored fat, those that have survived the winter mate and start their journey back up north. It is a matter of dispute as to whether any migratory individuals actually make it the whole 2,000 miles (3,200 km) or so back to Canada, but most females lay eggs and die along the way, leaving their progeny to finish the return journey. Whether by generational leapfrogging or marathon two-way travel, these beautiful butterflies start to appear in Canada in late May, and have deposited their eggs singly on milkweed leaves by early June. The conspicuous yellow-black-and-white-banded caterpillars appear to adorn most milkweed plants by midsummer. When mature, the caterpillar attaches itself to the milkweed plant with silk threads, then transforms into a spectacular green-and-gold chrysalis. The adult butterfly emerges from this chrysalis about six weeks after the eggs are laid.

The conspicuous colors of all stages in the Monarch's life serve as a warning to birds and other predators that Monarchs are not good to eat. The milky white sap of milkweed plants is loaded with chemicals called cardenolides, which are there to protect the plant from herbivores. Cardenolides muck up the physiology of many insects (they inhibit the membrane-bound sodium pump, causing paralysis), and also have nasty effects on vertebrates, but do not seem to harm the Monarch. The Monarch picks up these chemicals while feeding on milkweed, and advertises the fact with bright colors (aposematic coloring).

Other butterflies without protective chemicals sometimes cheat by imitating the Monarch's red-and-black coloration, thereby becoming unattractive to birds that have already had the unpleasant experience of trying to eat a Monarch. Many insects have aposematic coloring or imitate the aposematic coloring of others, but the Monarch and its common mimic, the normally edible Viceroy (*Limenitis archippus*), have probably been the subject of more study than any other model-mimic relationship. Viceroy adults are orange-and-black butterflies that are slightly smaller than Monarchs and have a black band across the hind wing but are otherwise similar enough to be convincing Monarch mimics. Blue jays unfamiliar with Monarchs will readily eat Viceroys, but once a bird has been made violently ill in an encounter with a poisonous Monarch it will avoid both Monarchs and the similar Viceroys – better safe than sorry! When an edible insect mimics a distasteful insect it is called Batesian mimicry, after the great naturalist explorer Henry Walter Bates who first documented the phenomenon. When one distasteful insect mimics the warning color of another distasteful species it is called Müllerian mimicry. Since some populations of Viceroy are distasteful and others are edible, Viceroy mimicry of Monarchs can be Batesian or Müllerian.

Monarchs are in the small subfamily Danainae (milkweed butterflies), while Viceroys belong to a separate small subfamily of brush-footed butterflies called the admirals (Limenitidinae). The other well-known admiral in our area occurs as two distinct subspecies, the White Admiral (*Limenitis arthemis arthemis*) and the Red-spotted Purple (*L. arthemis astyanax*). The latter form mimics a distasteful swallowtail (the Pipevine Swallowtail, *Papilio philenor*) much as the related Viceroy mimics the Monarch.

The brush-footed butterfly family includes several other subfamilies in addition to the distinctive heliconian, snout, satyr, milkweed butterfly and admiral subfamilies. The fritillary subfamily (Argynninae), for example, is a group of similar orange butterflies, most of which feed on violets. The crescent and checkerspot subfamily (Melitaeinae) is a group of relatively small butterflies that usually develop gregariously, sharing silken webs on asters and related plants. The emperor subfamily (Apaturinae) includes only a couple of species in our area, both swift-flying butterflies that rarely stray far from their host plants (hackberry trees). The most diverse subfamily of brush-footed butterflies is the nominate subfamily (Nymphalinae), which includes the anglewings, tortoiseshells, peacocks, thistle butterflies and others. Some of these, including peacocks and thistle butterflies such as the very widespread Painted Lady (*Vanessa cardui*), are among those insects that cannot tolerate cold winters and must migrate north to repopulate southern Canada and the northern United States anew every summer.

Skippers (Hesperiidae)

Swallowtails, gossamer-wings, whites and sulfurs are all true butterflies in the butterfly superfamily Papilionoidea. Members of another superfamily of day-flying Lepidoptera, called the skippers (superfamily Hesperoidea), are distinctive for their rapid, skipping kind of flight, and differ from true butterflies in having hook-tipped antennae and stout, muscular, mothlike bodies. A skipper's body is so muscular that if you try and prepare a skipper specimen for an insect collection you may have to lance the side of the thorax to cut the muscles before spreading the wings. **Hesperiidae** (page 188), or common skippers, are the only skippers in most of North America, and are generally small, with reddish or brownish wings and strange-looking larvae. The huge larval heads, usually separated from the bare abdomen by a strongly constricted neck, make skipper larvae easy to recognize although they tend to hide inside folded or rolled leaves, and can be hard to find. The easiest one to find in the northern United States and southern Canada is the very common European Skipper (*Thymelicus lineola*).

The European Skipper is a small, bright orange skipper that was first found in North America in 1910 near London, Ontario, and has since spread throughout the northeast (and western Canada) where it now abounds in meadows, fields and roadsides for much of June and July. European skippers spend the winter as eggs on grass stems, the distinctive green larvae occur in partially rolled blades of grass (especially Timothy, *Phleum pratense*) in spring, and the mothlike cocoons that shelter green-and-yellow pupae appear in early June. There are about 300 other skipper species in North America, many of which are similar in color and more challenging to identify than true butterflies.

Moths

Butterflies form a small, familiar group with around 750 mostly colorful, day-flying, well-known North American species. Moths, on the other hand, do not form a single group but make up an artificial assemblage of many distantly related groups with over 10,000 mostly nocturnal species in North America alone. In terms of evolution we can view the whole order Lepidoptera as a tree, and we can view the moths as the same tree with one small branch (the butterflies and skippers) removed from somewhere in the middle of the tree. A "moth" is thus just a lepidopteran that is not a butterfly or skipper.

It is convenient to divide the moths into the relatively large moths, or macrolepidoptera, and the relatively small moths, or microlepidoptera (micromoths). A few micromoths are well known because they impose themselves on our business of growing things or storing things, but most of the smaller micromoths are rarely noticed and many are difficult to identify, even to the family level.

One easy way to see micromoths on the wing is to shake an Eastern White Cedar branch in early summer, and watch for a cloud of tiny, white cedar leaf miner moths (*Argyresthia*, family **Argyresthiidae**, page 190), so called because their larvae mine inside cedar leaves. Closely related moths in the family **Yponomeutidae** (page 190) found on *Euonymus* and *Ailanthus* foliage are called ermine moths, perhaps because one common species is colored like the elegant white ermine trim used on robes worn by European judges. Other micromoths inhabit less sylvan locales. One species of clothes moth in the family **Tineidae** (page 191), for example, has the irksome habit of incorporating fibers from my favorite woolen sweaters into its own little silk and wool case, and other species of minute, dull-colored clothes moths can rapidly develop disturbingly large populations in carpets and coats made from animal fibres. Casemaking Clothes Moths (*Tinea pellionella*) are by no means the only moths with the caddisfly-like habit of making tubular houses. Some Bagworms (**Psychidae**, page 191), for example, make conspicuous baglike cases that hang from the trees on which they feed, such as ornamental red cedar. They pupate right in the bag, and the wingless female spends her whole life in the bag, emitting a pheromone that attracts fully winged males when she is reproductively mature. The baglike female cases can be common and conspicuous on cedar and other trees during the winter months, when they contain the developing eggs and the body of the caterpillar-like mother. Other widespread and common bagworms make inconspicuous tublar cases. Larvae of the small, rarely encountered macrolepidopteran family **Mimallonidae** (page 235) also make cases, and are called sack-bearers because their cases are open ended and sacklike.

Perhaps the most diverse casemaking moths are the microlepidopterans called casebearers (**Coleophoridae**, page 191). These small, often bright and metallic moths include some aptly named orchard pests like Apple Pistol Casebearers (*Coleophora malivorella*) and Cigar Casebearers (*C. spinella*). Casebearer larvae usually start out feeding as miners between the upper and lower leaf surfaces, and only later make a case from which they continue to feed within leaves, but only as far as possible without leaving the case.

Leaf miners occur in several orders, especially Lepidoptera and Diptera, and their characteristic trails can be found on almost any plant. Leaf mines always tell a story, usually starting small and enlarging to the point where the maker has pupated and emerged. Some miners make blotchlike mines; others follow characteristic patterns and even deposit their feces in places and shapes that give clues as to the identity of the mine architect. Many families of small, otherwise difficult to identify, micromoths make characteristic leaf mines. Tiny moths in the family **Gracillariidae** (leaf blotch miners, page 192) are responsible for some of the most familiar and frequently encountered leaf mines, such as the serpentine mines that adorn many poplar leaves and the blotch mines common on oak trees. As is true for leaf miners in many orders, leaf mining caterpillars are usually flattened and have greatly reduced appendages.

A few microlepidopteran families are common and easily recognized because of the remarkable appearance of the adult moths. A common group of unmistakably shaped moths is the plume moth family (**Pterophoridae**, page 196), in which the hind wings are divided into three fringed lobes or plumes. At rest, these moths resemble a "T" as the narrow, plumelike wings are held at right angles to the body. Plume moths are common, and feed on or in many of our common weeds such as thistle, bramble and wild grape. Another unmistakable group of moths is the clear-winged moth family (**Sesiidae**, page 204). The wasplike clear-winged moths develop as borers inside plants, including squash vines and peach trees. Many of them are active during the day, presumably protected by their similarity to stinging before emerging as wasps. A few other small families have similar boring habits, most notably the carpenterworms or **Cossidae** (page 205). Carpenterworms spend two or three years boring in a tree trunk before emerging as stout-bodied, often spotted moths.

Some moth families are characterized by conspicuous caterpillars easily recognized for their eye-catching shapes and colors that warn potential predators (or curious naturalists) of potent defenses. Slug caterpillars (**Limacodidae**, page 203) are consistently bizarre, with a variety of shapes and colors that usually advertise the presence of venom-tipped spines that can cause excruciating pain. My most painful insect encounter, ranking far above the stings of wasps, bees and tropical ants, came when I accidentally grabbed a tropical leaf with a slug caterpillar on its underside. The ensuing pain from hand to shoulder, swollen glands and shortness of breath gave me 20 minutes

The venom-tipped spines of slug caterpillars like this Central American *Parasa macrodonta* can give you a significant jolt of pain.

of real concern. The densely hairy caterpillars in the flannel moth family (**Megalopygidae**, page 235) have even more potent spines hidden away in their furlike coat, and some giant silkworm caterpillars are also quite well defended.

One of the biggest families of little moths includes a variety of species with long labial palps extending in front of the head, giving the appearance of a snout. These snout moths or pyralids (**Pyralidae**, page 197) comprise some 1,400 North American species, including the elongate, cream-colored moths that are so common in meadow areas. Snout moth larvae tend to hide when feeding, sometimes spinning a shelter from silk webbing, often rolling leaves and occasionally boring into plant tissue or other things ranging from stored products to other insects (a few are parasitoids, developing inside scale insects). Some of the common stored product pests in this group infest grains and a remarkable variety of dry foodstuffs. I have even found larvae of one species, the Indian Meal Moth (*Plodia interpunctella*), inside wrapped chocolate bars. Larvae of yet another group of pyralids, called wax moths, infest stored beekeeping equipment, eating the wax and even chewing into the wooden frames before pupating.

The best-known member of the snout moth family, at least in corn-growing areas, is probably the European Corn Borer (*Ostrinia nubilalis*), a major pest of corn and several other crops. One of our most destructive insects, this pyralid spends the winter as a mature larva in a corn stalk or other plant stem. It pupates inside a thin cocoon early in spring, emerging as a rather dull-looking, yellowish adult that soon wings its way to peppers, beans, potatoes and corn where it lays clusters of eggs under lower leaves. Young larvae feed on leaves, often breaking the midrib before boring into the stem. They sometimes move up into the cobs of corn, so every time you husk an ear during sweet corn season you have a good chance of exposing a fat, brown or pinkish caterpillar of the European Corn Borer.

The tiny but gorgeous members of the snout moth subfamily Acentropinae (aquatic snout moths; until recently known as Nymphulinae) are among the most remarkable micromoths. Anyone who has paddled a canoe through a patch of water lilies has seen these tiny, often metallic-spotted moths flying from pad to pad. Their caterpillars, some of which have long gills all over their bodies, can often be found feeding on lily pads under the shelter of cases made from cut out pieces of the lily pads. Other members of this subfamily spend their larval lives in an even less caterpillar-like fashion, using silken webs to cling to rocks in flowing water while competing with caddisfly neighbors for diatoms and other algae.

Two of our most famous moths belong in a family with square-tipped wings, giving them a bell-shape when at rest. At least in Canada, the most familiar member of this family (**Tortricidae**, page 201) is the famous Spruce Budworm (*Choristoneura fumiferana*), even though most of us recognize it only from the severe damage it causes to large areas of our coniferous forests. This native insect periodically occurs in outbreak numbers and appears to play an important role in spruce/fir forest ecology by feeding preferentially on balsam fir during outbreaks, thus allowing regeneration of the more valuable spruce trees. Nonetheless, the Spruce Budworm has been the target of staggering amounts of insecticides dumped over our forested lands. If you live in Canada or the northern United States, take a walk in a spruce/balsam forest and shake a few branches. Even in a non-outbreak year, you should be able to find the small, bell-shaped adults, which are distinctive for their mottled, gray-brown wings. They are usually abundant in late summer, when they lay masses of overlapping eggs on the tree needles. The eggs hatch and spend the winter as tiny larvae in a silken shelter, ready and waiting to mine a couple of needles in early spring before moving on to the developing shoots. By May they will be feeding conspicuously on foliage and you should be able to shake a tree and dislodge numerous brown, pale-spotted larvae, each with a broad yellowish stripe down the side. Budworms usually pupate by the end of June in loose webs among the twigs.

Famous or infamous as the Spruce Budworm may be in Canada, its fame hardly compares with the Codling Moth (*Cydia pomonella*), a member of the Tortricidae you probably know as the "worm in the apple." The bell-shaped, dark gray adult Codling Moths are inconspicuous and rarely noticed, but most of us are familiar with wormy apples. Codling Moths spend the winter as larvae sheltered in silken cocoons on tree trunks. Pupation takes place in early spring and adults emerge in May or June to lay eggs on the fruit or leaves. Larvae burrow into the fruit, feeding briefly before going to the core and eating the seeds. If you bite into a caterpillar in the flesh of an apple rather than the core, it might be another kind of tortricid moth – the Oriental Fruit Moth (*Grapholita molesta*).

Many other members of the family Tortricidae are fruit pests, including one that feeds in grapes (Grape Berry Moth, *Endopiza viteana*) and another that feeds on strawberry foliage (Strawberry Leaf Roller, *Ancylis comptana*), rolling the leaves into characteristic tubular shelters. Another tortricid moth feeds inside the thin-walled seeds of *Sebastiana*, transforming them into the "Mexican jumping beans" that periodically bounce into the air as if someone had carefully implanted them with tiny springs. The bean jumps when the larva, which has eaten out the contents of the seed, throws itself forcibly against the wall of the seed. In nature this is probably a mechanism to move the host seed out of the hot sun and into a sheltered place favorable to the developing moth inside. Similar behavior is found in a small gall wasp, called the Jumping Oak Gall (Cynipidae), except the jumping "bean" is a tiny gall that has dropped from the foliage to the litter. Neither Jumping Oak Galls nor Mexican jumping beans occur naturally in northeastern North America, but jumping beans are occasionally imported and sold in novelty stores in our area.

Although few micromoths are as conspicuously charismatic as the Mexican jumping bean, many local moths live inside seeds, fruits, stems and other plant parts. The producers of the narrow swellings on goldenrod stems (not the ball-shaped ones, which are caused by a fly), the larvae that cause fluffy swelling of cattail heads and the caterpillars often found inside burdock burrs are all common moths. These, however, are small moths, not conspicuous or striking additions to insect collections. For this, we will have to turn to the so-called macrolepidoptera, or macromoths, starting with the most macro of them all, the giant silkworm moths.

Silkworm moths (Saturniidae and Bombycidae)

It is always a special treat to come across an insect displaying the spectacular size and stunning beauty of an ethereal green Luna (*Actias luna*), a huge reddish Cecropia (*Hyalophora cecropia*) or perhaps a magnificent Polyphemus (*Antheraea polyphemus*). These giant silkworm moths (**Saturniidae**, page 205) show up at porch lights or street lamps just infrequently enough so they seem exotic despite their widespread distribution in North America. Lots of insect collections get started by people who can't resist keeping their first giant silkworm moth, and the presence of hand-sized moths on the bulletin boards or displays of classrooms at all levels attests to their attraction. Students and teachers also seem to find the large, brightly ornamented giant silkworm caterpillars and the distinctive silken cocoons they form to be the kind of noteworthy nature discovery that just has to be shared with the class. I still remember my own elementary school teacher bringing in a tree twig bearing the dense brown silken cocoon of a Cecropia Moth. Most of our other silkworm moths form cocoons on the ground, usually wrapped in a leaf, so students or naturalists are more likely to see the exposed Cecropia cocoons than the more cryptic cocoons of other common silkworm moths. My whole classroom waited in anticipation for the emergence of a beautiful moth from that Cecropia cocoon, and we were all surprised when the cocoon yielded some small parasitic flies instead! Had that cocoon yielded a moth, we could have looked at its antennae to find out whether it was a male with broad feathery antennae, or a female with narrower antennae. Had it been a female, we could have put it in a screened cage and allowed it to release a remarkably powerful "perfume" that would have attracted males from miles around, dramatically demonstrating the importance of volatile chemicals in insect interaction. Chemicals that mediate aggregation, alarm and other forms of interaction between members of the same species are called pheromones. One microgram (a microgram is a millionth of a gram, which is in turn only one thousandth of a kilogram) of the powerful pheromone released by female moths is, in theory, enough to "turn on" one billion male moths.

Giant silkworm moths do pupate in silken cocoons, but the silk in your silk tie or shirt was probably unraveled from cocoons made by another family of moths, the true silkworms or **Bombycidae** (page 235). One long-domesticated species is now a helpless, virtually flightless captive of an industry worth hundreds of millions of dollars

Cecropia Moth (*Hyalophora cecropia*), page 206, caption 8.

annually. Each silkworm cocoon is made up of a double silk strand about half a mile (900 m) long, thousands of which are required to make a pound of silk. Silk production, or sericulture, is no longer the huge industry it was before much cheaper synthetic fibers appeared on the scene, but natural silk is a superior product and several countries, including Spain, Italy, Japan and China, maintain a viable sericulture industry. Silk was a valued Chinese monopoly for centuries, with a threat of stiff penalties (death!) for anyone trying to smuggle silkworms out of the country. Silkworm eggs smuggled out of China in the 6th century AD broke the monopoly by providing stock for a European silkworm industry.

Although the domestic silkworm or Mulberry Silk Moth (*Bombyx mori*) now overwhelmingly dominates the global silk industry, several species of wild silkworms (in other moth families) still play a minor role in commercial silk production. Muga silk and tasar (tussah) silk, for example, are spun from the cocoons of two different species of giant silkworm moths (Saturniidae), and form the basis of regional industries worth hundreds of thousands of dollars. It is always a pleasure to find the products of wild silk moths, usually from India or China, for sale in North American shops. There is nothing like a fine, durable, attractive textile made from exotic caterpillar spit! Most members of the true silkworm family are Asian, but North America is home to five species of Bombycidae in addition to the introduced domestic silkworm. All five North American true silkworms have traditionally been treated as another family, Apatelodidae, but are now usually recognized as Bombycidae.

Some attempts to establish a silkworm industry here in North America have had interesting and far-reaching effects. The most famous of these was an 1866 attempt to import a moth that produced lots of silk but was not as fussy about food as the mulberry-feeding true silkworm. Leopold Trouvelot, a French astronomer working at the Harvard Observatory, supposedly imported eggs of the Gypsy Moth (*Lymantria dispar*, family Lymantriidae) in order to cross it with the true silkworm (family Bombycidae) and in doing so produce a "super silkworm," resistant to disease and easier to feed than the true silkworm. It seems hard to believe that any scientist of that era, even an astronomer, really expected to successfully cross such distantly related species. The Gypsy Moth was a failure when it came to improving the silk industry, but a few of the moths that escaped into Trouvelot's backyard proved so well adapted to life in North America that the Gypsy Moth is now a major pest of shade trees.

Tent caterpillars (Lasiocampidae)

Eastern Tent Caterpillars (*Malacosoma americanum*) are easily found 12 months of the year. During the late summer, fall and winter they can be found on twigs of chokecherries and various other trees, inside shining brown rings of eggs that look a bit like slightly flattened, varnished brown spittlebug masses, each egg containing a tiny larva bent into a U-shape to fit in its minute enclosure. Although the eggs are deposited in late spring, and have developed to the point of holding a recognizable caterpillar within a few weeks, they don't hatch until early spring of the following year. The larvae leave the eggs, ready to tackle spring foliage, as soon as foliage starts to appear on cherries and other favored trees and shrubs. Soon thereafter the larvae work together to build communal tents. Each tent starts out as a simple, silk-walled structure used for shelter between forays to the foliage-rich branch tips. As it develops, however, it becomes less of a tent than a multiple-story apartment building, new floors being built over old as the old "apartments" become too small and fouled with feces and cast skins. The net effect is a glistening white mass of silk often surrounding the bases of three or four branches.

Tent caterpillars not only build silk apartment buildings, they also build silk highways. If the light is right, you can see glistening paths resulting from the silk trails deposited by hundreds of caterpillars traveling from nest to foliage. In addition to the silken strands, caterpillars returning from good foliage add chemicals that tell their nestmates that their trails are worth following. These silken highways, along with their chemical road signs, have a downside in the form of predators and parasitoids that find them a convenient route to caterpillar food. At least one kind of predacious stink bug makes a specialty of following silk trails with intent to impale the trail-making caterpillars.

After several weeks of apartment life, commuting to work along silken highways along with numerous roommates and enemies, then returning home to participate in a daily routine of building up an increasingly dirty apartment building, tent caterpillar communal life starts to deteriorate. The nests, by now massive structures packed with feces and cast skins, fall into disrepair and the nest-makers strike out on their own, often with dramatic leaps off the ends of the now defoliated branches of their home tree.

Dispersal of the restless mature larvae takes place in my area in June, about the time public schools are due to close up for the summer holidays, and ever since my public school days I have associated an abundance of wandering tent caterpillar larvae with the last few days of the school year. Individually, these are beautiful red, blue and black caterpillars with a white stripe down the back and white spots on the side, but they are so commonplace that their individual beauty seems easily overlooked. I grew up in an insensitive era, and all the boys in my school knew that if you stepped on these conspicuous caterpillars just right, their insides would shoot out several inches. The schoolyard caterpillars that escaped our innocent sadism hunkered down under bits of debris or other shelter and spun silk cocoons. Although the silk is white, tent caterpillar cocoons have a characteristic yellow, powdery appearance, as though the loose weave of the silk is packed with dried urine. In a sense, that is exactly the case, since the yellow stuff is produced in the caterpillar's malpighian tubules (their excretory system), excreted into the digestive tract and then pumped out into the cocoon.

The **Lasiocampidae** (page 207) is a relatively small family of somewhat fat, fuzzy moths with bodies that look too big for their wings. They lack the coiled proboscis characteristic of other moths, and the big bristle that usually comes off the base of the hind wing is absent. Instead of using this bristle (the frenulum) to link the wings, lasiocampids appear to use an expanded part of the hind wing itself.

Northern Tent Caterpillar (*Malacosoma californicum pluviale*), page 207, caption 9.

The best-known member of the family Lasiocampidae, other than the Eastern Tent Caterpillar itself, is the Forest Tent Caterpillar, *Malacosoma disstria*. Despite being in the same genus as our familiar backyard tent-makers, the Forest Tent Caterpillar lacks the tent-making skills of its urban relatives, and makes only simple silk pads on which to rest and molt. Forest Tent Caterpillars look a bit like Eastern Tent Caterpillars, but have a row of diamond- or keyhole-shaped spots down the back instead of a stripe. Their common name gives you a good clue as to where to find these foliage-feeding insects, which in some years are abundant enough to cause severe defoliation in aspen, maple and other forests.

Forest Tent Caterpillars do not build elaborate silken tents like Eastern Tent Caterpillars, but they do build silk trails from trunk to foliage. Mature caterpillars spin loose, silken cocoons in which to pupate, and those pupae not killed by parasitoids, disease or predators develop into fuzzy brown moths about two weeks later (early July).

If you see large silk tents in late summer, they are not tent caterpillars, and are probably the loose nests of the Fall Webworm (*Hyphantria cunea*), a very common member of the family Arctiidae. Fall Webworms and most other tent-making caterpillars including the smaller Uglynest Caterpillar (*Archips cerasivorena*, family Tortricidae) feed within their tents and therefore make large tents that envelop lots of foliage. Eastern Tent Caterpillars make compact tents used only for shelter. Some other Lasiocampidae (Northern Tent Caterpillars, *Malacosoma californicum pluviale*) make tents similar to those of Eastern Tent Caterpillars.

Sphinx moths (Sphingidae)

Sphinx moths, or hawk moths (**Sphingidae**, page 208), are familiar to almost everyone because of their robust bodies, distinctively long wings and attractive colors. Most sphinx moths are night-flying moths that hover over pale-colored, deep flowers as they use a long proboscis to reach the nectar within, but the day-flying hummingbird moths are among our most familiar sphingids. Aptly named for their size, rapid flight and hummingbird-like habit of hovering over flowers while feeding on nectar, these clear-winged moths are often seen nectaring at vetch flowers.

Many orchids have a particular dependence on the nocturnal visits of pollinating sphinx moths, and the associated sphinx moths have in turn developed a particular dependence on the orchids with which they have co-evolved. When Charles Darwin wrote his famous book on orchid pollination in 1862 he had just received an orchid specimen from Madagascar in which the nectary was around 12 inches

(30 cm) long. He was astounded to see such a long nectary on a flower dependent on moth pollination, and he penned the following prediction: "In Madagascar there must be moths with proboscises capable of extension to a length of between 10 and 11 inches." Darwin further speculated that this still undiscovered giant moth would probably be a sphinx moth. After four decades of dispute about this outrageous prediction, a huge Madagascar sphinx with a proboscis about 12 inches (30 cm) long was discovered and described in 1903 as *Xanthopan morgani praedicta*.

If you grow tomatoes, you are probably familiar with a large green caterpillar called a hornworm because of a conspicuous "horn" sticking up off the end of the abdomen. The Tomato Hornworm and the closely related Tobacco Hornworm are typical sphinx larvae. If you dig around your tomato plants at the end of the season you might find the large, bare, mahogany-brown pupae of these sphinx moths, easily recognized as sphinx pupae because of the way the large, developing proboscis looks like a pitcher handle. If you do find some of these big pupae, keep them cold for a few months to simulate winter, then bring them into a warm room. You might get the rather elegant gray-banded adult moth, or you might rear out some interesting parasitoids that have consumed the hornworm from the inside.

Any time you want to rear out an overwintering insect pupae remember that in nature it would be fatal for an adult to emerge from its pupa on a nice warm day in late November or December. Because of this, insects have various mechanisms that prevent emergence until some cue, such as an adequate accumulation of cold days, tell them winter is over. If your pupa yields parasitoids and not what you expected, you shouldn't be disappointed. Associations between host species and parasitoid species are always interesting, and the parasitoids themselves are often fascinating and poorly known species in the Diptera and Hymenoptera.

Owlet moths (Noctuidae)

Many night-flying insects are routinely waylaid when artificial lights interfere with their navigation systems, and the riot of moths and other insects fluttering around artificial lights is a familiar part of any warm summer night. The most abundant of the stout bodied, medium-sized, relatively dull-colored moths found attracted to lights or fluttering around inside your porch at night are in the speciose moth family **Noctuidae** (page 216). These generally nocturnal moths are sometimes called owlet moths (*noctua* means owl in Latin) because of the way their eyes pick up and reflect the smallest amount of light, shining brightly in contrast with the usually inconspicuous body and forewings. This is a huge group including some of our many common species known as cutworms, armyworms, dagger moths, miller moths and underwings.

Although most noctuids are somber in color, some are strikingly beautiful. Underwings (genus *Catocala*), for example, have drab forewings and brilliantly colored hind wings. In flight, the bright hind wings are highly visible, but upon landing, the hind wings are concealed by the camouflaged forewings, effectively allowing the conspicuous flying moth to "disappear." Another group of unusually bright noctuids is the forester moth group (Agaristinae), one of which is a common day-flying moth that feeds on Virginia Creeper (*Parthenocissus quinquefolia*) and grape vines. Foresters are black with white or yellow spots, and when at rest they look a bit like bird droppings.

The noctuid moths familiar to most people are those that attack our garden plants or crops. Every gardener is familiar with cutworms and armyworms, noctuid moth larvae which bite off tender plant shoots during the night. Every spring, when I till the soil for my garden, I find some bare, greasy-looking caterpillars that have obviously spent the winter as

Eight-spotted Forester (*Alypia octomaculata*), page 220, caption 4.

mature larvae in the soil. They stay in the soil during the upcoming spring days, coming out at night to indulge in their irritatingly wasteful habit of feeding at the base of young shoots. Bare, shining brown cutworm pupae most often show up in June and July, and the dull-looking cutworm moth adults are common in summer. Several different noctuid species are called cutworms or armyworms, the latter group sometimes forming invading hordes that can completely destroy the vegetation in a field or garden before moving to a new feeding site like an advancing army.

Owlet moths can be found in a wide variety of habitats, such as the stems of cattails, water lilies, potatoes and other plants; the foliage of a wide variety of plants; and in many kinds of roots and fruits. A couple of species will probably come to your attention whether you are looking for them or not, especially if you grow corn or cabbage in your garden. The Corn Earworm, known in different areas as the Cotton Bollworm or the Tomato Fruitworm, but known everywhere under the scientific name *Helicoverpa zea*, gets into the fruits of cotton, corn and tomato. Those fat earworm larvae seen amongst the kernels of freshly husked corn would normally move into the soil to spend the winter as pupae in the soil, but not in the northeast. Despite the summertime abundance of this species in my home area of Ontario, northern populations of the Corn Earworm are the result of dispersal from warmer climes south of the border. The caterpillars I flick off corncobs are doomed anyhow, the end of an annual, temporary spread of the species. Some of our other Noctuidae are less regular visitors, including some extraordinarily large noctuids that occasionally stray all the way to Canada from a normally tropical range. The Black Witch (*Ascalapha odorata*), a distinctive 100–150 mm giant, is the best known of these spectacular straying owlet moths.

One whole subfamily of owlet moths, including a garden pest called the Cabbage Looper (*Trichoplusia ni*), is characterized by larvae that crawl along in a looping fashion, much like the inchworms in the family Geometridae. Caterpillars of both groups cling to leaves or twigs with the segmented legs of the thorax, then arch up the back to form a loop as the hind end is slid forward. When this loop is formed, the hind legs are used to hold on to the substrate, the front legs are released and the body is straightened out by moving the front part forward.

As you may recall, caterpillars normally have five pairs of unsegmented prolegs on the abdomen. The loopers in the family Noctuidae have lost the first two pairs to accommodate their loopy locomotion, while the Geometridae have quite independently lost the first two prolegs and lost or reduced the next pair to accommodate the same kind of action. The Cabbage Looper is yet another insect, like the Corn Earworm, that spreads north every summer only to be pushed back south of the border by the Canadian winter. You have to look closely to find the distinctive white-striped, green Cabbage Looper larvae because they feed under leaves, and they don't normally expose themselves by chewing all the way through.

Although most adult owlet moths are not particularly colorful, some are extraordinary in other ways, and some southeast Asian species bear the distinction of being the only moths that can pierce your skin and suck your blood. The bloodsucking moths belong to a genus (*Calyptra*, page 219) that also occurs in eastern North America, but our species have less spectacular habits.

Owlet moths, and related moths such as tiger moths, have thoracic "ears," called tympanic organs. Tympanic organs are simple eardrum-like structures that can be seen by lifting an owlet moth's hind wing and looking for a bare, shining, drumlike plate on the side of the thorax, just below the wing. Owlet moths are able to use those drumlike tympanic organs to pick up on the sonar system that some bats use to detect their major prey, which just happens to be owlet moths. Not only can the moths detect the bat squeaks long before the bats can detect the moths, they can also determine how far away the bat is and what direction it is coming from, enabling them to take appropriate evasive action.

If, for example, a bat is coming in from a long distance to the left, the moth will turn right and fly away. If the bat is very close, the moth might plummet to the ground or take complex evasive action. Each of the moth's ears is like a drumskin with only two wires going to it, one that sends out an impulse when you hit the drum hard and another that is triggered by a lighter wave. That explains how owlet moths know if a bat is far or near, with the nearby bat hitting the drum hard with its strong sonar and the more distant bat hitting it softly with its weaker signal. Repeated blocking and unblocking of the tympanic organs as the moth's wings beat up and down, combined with the left-right asymmetry of stimulation depending on where the bat is, tell the moth where the bat is coming from. Owlet moths, and some related groups such as tiger moths, are thus able take sophisticated evasive action based on amazingly simple structures.

The degree to which noctuid moths and their close relatives (other members of the superfamily Noctuoidea) are dependent on their ears can be surmised by the behavior of specialized parasitic mites that live in moth ears, invading their hosts as they visit flowers. The mites, which destroy the invaded ear as they feed, rarely infest both ears of a moth. New additions to a moth's mite fauna will almost invariably take up residence in a previously infested ear rather than attacking the remaining functional ear. Apparently a one-eared moth can avoid predation, but a deaf moth is an unsafe vehicle for its resident mites.

Virgins, tigers and woolly bears (Arctiidae)

Arctiidae (page 212) can be thought of as Noctuidae dressed up to party, for if you took away the bright stripes, spots and colors of the tiger moths (most Arctiidae are tiger moths) you would have something similar to an owlet moth, right down to the tympana under the hind wings. Tiger moths use their tympanic ears to gain the same military advantages in their war with bats as the related owlet moths do, but tiger moths escalate the arms race a bit. These little moths appear to go on the offensive, making a clicking sound that stops the attacking bats in their tracks. Whether or not the moth is really going on the offensive, it is advertising the fact that it is offensive. We all know some animals which are well protected by poisons or defensive chemicals advertise this fact to predators with bright warning coloration, and the tiger moths are among the most brightly colored and most distasteful of moths. Bright warning colors, however, count for little when faced with a predator that does not hunt by vision. The clicking sound these moths make in response to bat sonar is like a warning "color" bats can hear. There is also some evidence that the moth clicking can interfere with bat sonar.

If bright colors and noisy behavior are evidence of bad taste (another parallel between people and insects) then we should expect the brightest and boldest of tiger moths to be a paragon of bad taste. A good candidate for that status is the aptly named Bella Moth (*Utetheisa bella*), a stunning pink, white and black moth with audacious day-flying habits. It is not surprising to find that birds leave the Bella Moth and its close southern relative the Ornate Moth (*Utetheisa ornatrix*) pretty much alone, but the story goes deeper. A group of scientists at Cornell University found that the larvae of Ornate Moths extract a bitter, toxic alkaloid from the plant material they eat, much the way the more familiar Monarchs sequester toxic cardiac glycosides from milkweed plants. Male moths use the bitter alkaloid to manufacture the pheromones with which they attract a female. The more alkaloid the male has, the more pheromone he can make, and the more he gets to mate.

Bella Moth (*Utetheisa bella*), page 213, caption 9.

There are good reasons why female moths choose males that really stink of alkaloid perfume. For one thing, sex takes a long time in this species, often over eight hours, and if the male has a good supply of defensive alkaloids, predators are less likely to interrupt the act. More importantly, the male actually pumps quantities of the alkaloid poison into the female along with the sperm. This toxic gift is then transferred to the eggs, giving them a chemical defense to tide them over until they can start munching on alkaloid-rich seeds and building up their own arsenal.

Whatever the reasons for their colors, tiger moths are among the most beautiful of insects and are popular among collectors for that reason. Even the larvae tend to be strikingly colored, often densely hairy and conspicuous, such as the fat, fuzzy Woolly Bears often seen hurrying across late fall roadways and paths in search of an overwintering site. Some people think that the relative width of the red band encircling the middle of these otherwise black caterpillars is an indication of the severity of the upcoming winter, but those caterpillars with lots of red are probably just relatively old. In the spring, Woolly Bears pupate to develop into orange-brown moths, spotted with black. Woolly Bears (*Pyrrharctia isabella*), like most Arctiidae, are seldom seen more than one at a time, but some tiger moths are conspicuous for both bright colors and communal feeding habits. The orange, black and white tufted Milkweed Tussock Moth caterpillar (*Euchaetes egle*), for example, forms impressively dense aggregations on milkweed leaves. Other commonly seen arctiid larvae include the long-haired Saltmarsh Caterpillar (*Estigmene acrea*) and the similarly hirsute Yellow Bear or Virginia Tiger Moth (*Spilosoma virginica*). Despite their common names, these are widespread caterpillars that eat a wide variety of grasses and garden crops. The adults are among the many similar white-winged tiger moth species.

Although this may be a small comfort if your garden is overrun with Saltmarsh Caterpillars, tiger moths are not generally pests. One notable exception to this rule is a sort of ugly duckling among tiger moths, a species called the Fall Webworm (*Hyphantria cunea*). The larvae of these whitish moths build large, messy silk tents enclosing a whole limb of foliage. In contrast with the tent caterpillars (Lasiocampidae), Fall Webworm caterpillars feed entirely within their silken shelters and are common on a variety of trees and shrubs in late summer and fall. Fall Webworms have been known to feed on over 600 species of plants, giving them the dubious honor of having the widest known host range of any plant-eating insect; not surprisingly, they are extremely common insects.

All the Arctiidae discussed so far have been tiger moths, which make up one of four subfamilies of Arctiidae. Two other subfamilies are common throughout North America, including urban areas. The wasp moths, formerly placed in their own family, Ctenuchidae, are striking black-and-blue moths, frequently encountered because of their day-flying habits. The Virginia Ctenucha (*Ctenucha virginica*), a black-winged species with a brilliant blue body and an orange head, is one of the half a dozen conspicuous insects you can count on finding on goldenrod flowers; their hairy, black-and-yellow early instar larvae are often conspicuously abundant on grasses. Footman moths or lichen moths (subfamily Lithosiinae) are less likely to show up in your backyard than tiger moths and wasp moths, and are usually understated little arctiids found on rocks or tree trunks, where they feed on lichens. Lichen moths are often stunningly beautiful moths, colored in muted shades of pink, yellow, black and white.

Geometer moths (Geometridae)

Geometer moths (**Geometridae**, page 230, the second largest family of moths after Noctuidae) differ from most other familiar moths in a number of ways. Unlike owlet moths, tiger moths and tent caterpillars, geometers are usually slender-bodied moths, with disproportionately broad, delicate-looking wings often marked by a pattern of fine lines extending from the front wings to the similar hind wings. Adult Geometridae do have big "ears" (tympanic organs) like noctuids and their relatives, but the tympanic organs are on the first abdominal segment rather than the thorax (as in noctuids).

Geometer larvae are the familiar inchworms, slender caterpillars that loop along by drawing the hind end up to the front, anchoring the hind legs, then extending the front end, anchoring the front legs, then repeating the process. Most inchworms have lost the first three pairs of prolegs (out of the normal caterpillar complement of five pairs) in order to expedite this type of motion, but in some species the third pair of prolegs is present but reduced. Many inchworms have a characteristic habit of striking a twiglike pose when disturbed, clinging to the substrate with the hind prolegs and holding the rest of the body out like a tiny branch.

North America's 1,400 species of Geometridae include many common and striking species as well as a number of forest pests such as the Fall Cankerworm (*Alsophila pometaria*). Cankerworms are among those odd moth species with entirely wingless females.

The idea of a wingless moth is an apparent oxymoron, and even to someone familiar with the frequency of wing loss within most of the winged orders, the sight of a fuzzy moth body without any apparent wings seems somehow unnatural. Perhaps that is why students so often bring these funny-looking, fuzzy, female Fall Cankerworms into my entomology labs. (Of course, it could also be that my entomology class runs from fall into early winter, and toward the end of the course adult Fall Cankerworms are among the few conspicuous, active insects still around.) Females are even seen wandering about the snow surface in early December, presumably in search of a tree on which to deposit eggs, and the brownish gray, winged males are

Filament Bearer (*Nematocampa resistaria*), page 230, caption 7.

sometimes seen flying during late November snowfalls.

The grayish, flowerpot-shaped eggs deposited by those fat, fuzzy, wingless females hatch in spring just as the new foliage appears. The larvae are brighter than most inchworms, with a green bottom, a brown top and white stripes in between. Fall Cankerworm caterpillars are also odd in having three pairs of prolegs, rather than two pairs like most Geometridae. The larvae develop on a variety of hardwood trees (sometimes severely defoliating fruit and shade trees), and then pupate underground in silken cocoons. Adults don't start to emerge until there has been a hard frost, and they continue to emerge till early winter. Similar moths emerging in early spring probably belong to another inchworm species, the similar and equally common Spring Cankerworm. Both species are cyclic in their abundance and can be serious pests during peak years.

The most famous of all Geometridae are the Peppered Moths, so named because their wings normally have a salt-and-pepper coloration that makes them almost invisible against the background of lichen-covered tree trunks. Almost every textbook on evolution or ecology tells the story of how British populations of the Peppered Moth (*Biston betularia*) occurred as a common whitish form and a scarce black form before the industrial revolution. The widespread increase of sooty air pollution in the mid-19th century killed off the lichen and blackened the trees in some areas, after which the black form apparently blended in better with the dirty tree trunks than did the white form. Under these conditions, the standard, whitish, salt-and-pepper colored moths were supposedly easy targets for birds, and were selectively eaten instead of the now less conspicuous black forms. In Manchester of 1848 the black form probably made up less than 1 percent of the population, but by 1898 it made up over 90 percent of all pepper moths in the area. This classic textbook story, often cited as one of the best-documented examples of natural selection in action, was only possible because of the legacy left by amateur insect collectors in the form of well-labeled moth collections from which the first record (1811) and subsequent spread of the black form can be documented. Entomology remains one of the few areas where carefully executed amateur work, such as the compilation of well-labeled regional collections, can make important contributions to the advancement of science.

The phenomenon of dark, or melanic, forms of moths increasing in frequency as a result of industrial pollution is called industrial melanism. Hundreds of moth species have melanic forms, and industrial melanism is a widespread phenomenon. The Peppered Moths of North America (also known as Pepper and Salt Geometers) are considered to be a different subspecies than the Peppered Moth of Britain, but show the same dimorphism. The balance of light and dark forms in your area might just offer some insight into local air pollution, as well as provide a simple model of natural selection in action. Peppered moth adults are out all summer and can be found on a wide variety of trees and shrubs. Adults are hard to find during the day, as they remain motionless and almost invisible under the bases of tree branches, but they are often very common and conspicuous around artificial lights in well-wooded areas. The classic pictures of Peppered Moths sitting on tree trunks, by the way, were probably staged by capturing the moths at lights, then placing them on bark of an appropriate color and texture to support the story of industrial melanism. The story remains a good one, but peppered moths of all colors are normally well-concealed under branch bases during the day.

Gypsy Moths and other Lymantriidae

My family recently took an early spring drive from Ontario to Virginia, camping along the way in a couple of state parks infested with Gypsy Moths (*Lymantria dispar*; **Lymantriidae**, page 235). My strongest memory of that trip is of sitting at picnic tables and listening to the constant pitter-patter of frass (bug poop) falling from the overhanging oak branches onto the table and into my coffee cup. I didn't have to look at too many tree trunks to find yellow, fuzzy egg masses, some of which were still surrounded by hundreds of hairy caterpillars. Actually, it was not necessary to go looking for the caterpillars at all, because these lightweight little larvae tend to get blown out of the trees and right into your face. With the aid of a strand of silk, Gypsy Moth larvae can balloon along for miles. Larval dispersal of this sort, along with a tendency to hitch rides on hubcaps and trailers, explains the gradual spread of these pestiferous European lymantriid moths since they escaped captivity in a Boston suburb in the late 1860s. Eastern North American Gypsy Moth populations appear to have depended on larvae for their spread because the fat, white, female adults are unable to fly. The Gypsy Moth took over 100 years to make its way from Boston to Canada, where it now feeds on a wide variety of trees, but retains a preference for oak.

Despite its fame and the millions of dollars spent on Gypsy Moth control programs every year, it usually takes repeated defoliation to kill a tree, and the long-term effect on forests is usually no worse than a shift of species composition away from oak and towards beeches and maples. Foliage-feeding caterpillars might even benefit a forest, killing unhealthy trees, opening up the forest canopy, and returning nutrients to the soil in the form of easily decomposed droppings. Nonetheless, Gypsy Moths are considered a serious problem. Even if defoliation is not fatal, naked trees are considered unacceptable in suburbia and cottage country. Furthermore, the abundant, hairy larvae are a nuisance and sometimes even cause skin rashes and allergic reactions.

The Gypsy Moth is a notoriously variable species, and Siberian moths differ from European Gypsy Moths (the ones we have had in eastern North America for the last century) in some sinister ways. For one thing, the Asian females can fly, and furthermore they have a taste for the kinds of conifer trees on which our western forest industry depends. Not surprisingly, the discovery of some of these particular Russian immigrants in British Columbia led to a flurry of spraying activity. European Gypsy Moths have also been detected in pheromone traps on the west coast, so Gypsy Moths, whether Asian or European in origin, are now very much a North American problem.

With the famous Gypsy Moth hogging all the limelight, it is easy to forget that the family Lymantriidae holds some other common and interesting species. The common name for the family as a whole is tussock moths, a name that refers to a wide range of species including the White-marked Tussock Moth (*Orgyia leucostigma*). The larvae of this common moth are odd-looking caterpillars routinely encountered on a variety of trees ranging from alder to balsam fir. With four tufts of white bristles near one end, the top of this caterpillar looks like a toothbrush. The head looks like a maraschino cherry plunked on the end of the mostly yellow, hairy body, and is flanked by long, black hair tufts that look like forward-reaching arms. The hind end has a couple of bright red tufts and another long, black hair tuft. The whole effect is so striking that naturalists and students often bring these fuzzy caterpillars home to admire, keeping them around till they spin cocoons in late summer or early fall.

Most moths that pupate late in the season enter a pupal diapause, which means that the adults will not emerge until the next spring. Tussock moth cocoons, on the other hand, develop promptly and transform into adults before the onset of winter. Males are normal brown-winged moths, while females are completely wingless insects that lay their overwintering eggs close to where they have emerged from their cocoons, sometimes right on top of their empty cocoons.

White-marked Tussock Moth (*Orgyia leucostigma*) caterpillar, page 236, caption 1.

FAMILY LYCAENIDAE Subfamily Polyommatinae (BLUES) ❶ ❷ Blues, like this **Eastern Tailed Blue** (***Everes comyntas***) are small, delicate, weak-flying butterflies usually found near their host plants. Males are almost invariably bright blue, while females tend to be duller. Eastern Tailed Blue caterpillars feed on the flowers of clover and related Fabaceae. ❸ ❹ This **Spring Azure** (***Celastrina ladon***) caterpillar is feeding on the flowers of an American Bittersweet (*Celastrus scandens*) while an ant feeds on sugary secretions from the caterpillar's abdominal glands; the adult was photographed in early May in Ontario. The very similar Summer Azure (*Celastrina neglecta*) appears somewhat later in the season and, unlike the Spring Azure, can be seen on the wing all season (it has multiple generations). Subfamily Theclinae (HAIRSTREAKS AND ELFINS) ❺ ❻ The hind wing tails of most hairstreaks look a bit like antennae, and probably serve to deflect predator attention away from the butterfly's real head. The eyelike spot on the hind wing near the tail aids in this deception, as does the way perching hairstreaks rub their hind wings together to give the tail an antenna-like motion. These two hairstreaks (**Striped Hairstreak**, ***Satyrium liparops*** and **Banded Hairstreak**, ***Satyrium calanus***) develop on a variety of forest trees and shrubs. ❼ This is a **Banded Hairstreak** larva. ❽ The **Acadian Hairstreak** (***Satyrium acadicum***) is a common hairstreak that develops on willow. Adults are commonly seen on Butterflyweed (*Asclepias tuberosa*). ❾ Eastern populations of the **Juniper Hairstreak** (***Callophrys grynea***) used to be known as the Olive Hairstreak. Caterpillars are found on Eastern Red Cedar (*Juniperus virginiana*). ❿ The **Coral Hairstreak** (***Satyrium titus***) lacks a hind wing tail, an unusual feature amongst hairstreaks.

FAMILY LYCAENIDAE (continued) ❶ The **Brown Elfin** (***Callophrys augustinus***) is widespread in Canada, but is restricted to cooler mountainous areas in the United States, where it develops on a variety of heath plants such as blueberries, Bearberry and Leatherleaf. **Subfamily Lycaeninae** ❷ The **American Copper** (***Lycaena phlaeas***) is somewhat misnamed since this colorful little butterfly also occurs naturally in Europe and Asia. Larvae feed on sorrel and dock (*Rumex* spp.). ❸ The tiny **Bog Copper** (***Lycaena epixanthe***) occurs as isolated populations in bogs and fens where their caterpillars feed on cranberry plants. ❹ The **Dorcas Copper** (***Lycaena dorcas***) can be common where its food plant, Shrubby Cinquefoil (*Potentilla fruticosa*), occurs. **Subfamily Miletinae** (HARVESTERS) ❺ ❻ One of these **Harvesters** (***Feniseca tarquinius***) is sampling a scat, an unexpected snack choice for a butterfly, but not as unusual as the preferred food of Harvester caterpillars. Harvester caterpillars, our only insectivorous caterpillars, are predators in colonies of Woolly Alder Aphids (*Paraprociphilus tessellatus*). *F. tarquinius* is the only North American species in the subfamily Miletinae. **FAMILY RIODINIDAE** (METALMARKS) ❼ The **Northern Metalmark** (***Calephelis borealis***), the only member of this mostly tropical family to reach the northeast, occurs only in northeastern United States and in a few spots in Missouri and Oklahoma. Larvae of this very rare butterfly develop on Roundleaf Ragwort (*Senecio obovatus*). **FAMILY PAPILIONIDAE** (SWALLOWTAILS) ❽ The two yellow butterflies in the foreground of this group of puddling swallowtails are Eastern Tiger Swallowtails (the black swallowtails in the background are Spicebush Swallowtails, *Papilio troilus*; and the butterfly between the tigers is a Giant Swallowtail, *P. cresphontes*). **Eastern Tiger Swallowtails** (***P. glaucus***) are common butterflies throughout eastern United States and southern Ontario. Elsewhere in Canada this species is replaced by the very similar Canadian Tiger Swallowtail (*P. canadensis*, which was treated as a subspecies of *P. glaucus* until 1991). ❾ The **Palamedes Swallowtail** (***Papilio palamedes***) ranges from New Jersey south to Florida.

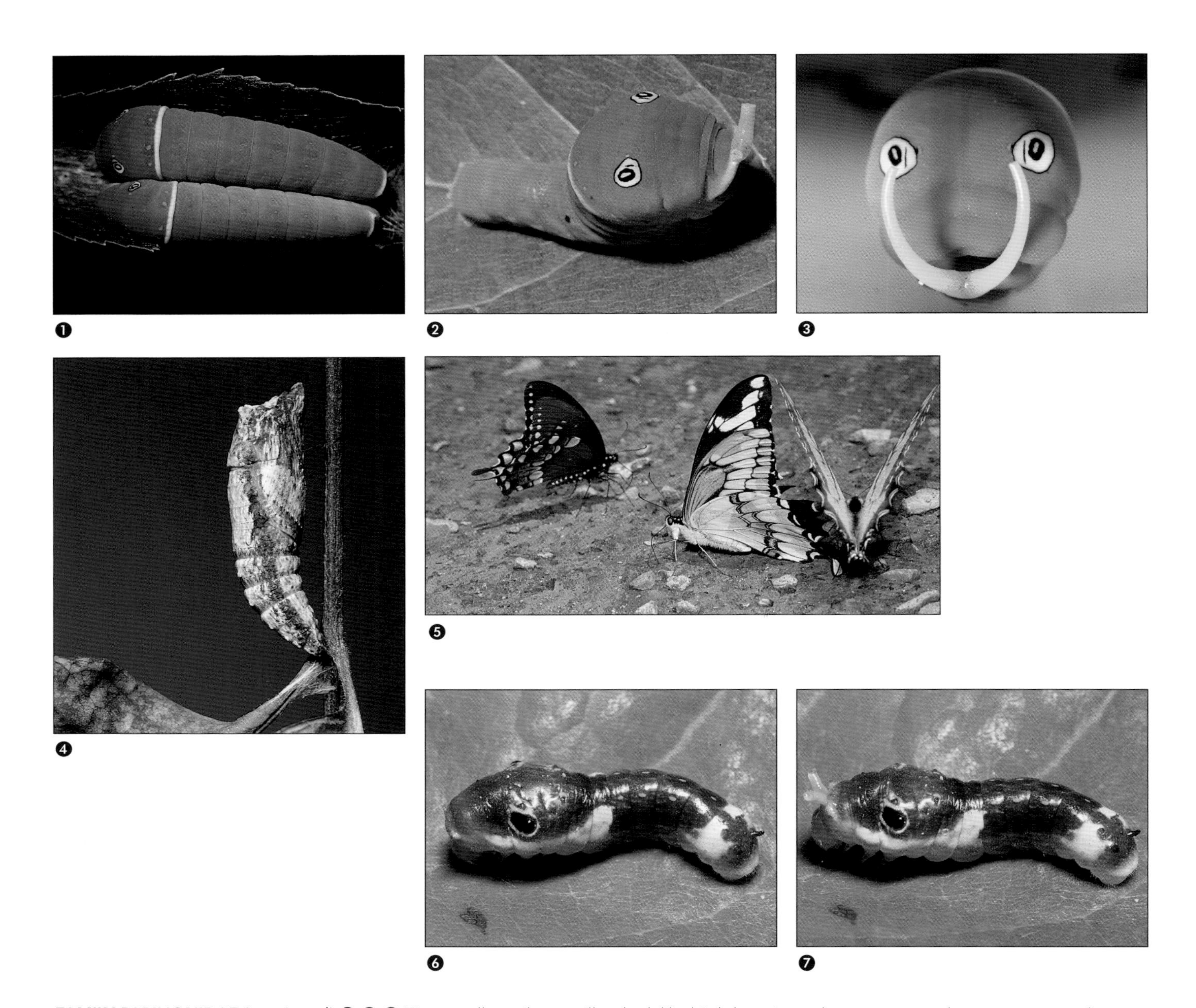

FAMILY PAPILIONIDAE (continued) ❶ ❷ ❸ Tiger swallowtail caterpillars look like bird droppings when very young, but turn green and resemble green snakes when older. The "snake eyes," complete with convincing pupils, are just pigment spots on the caterpillar's thorax. The bright tongue that appears from behind an alarmed caterpillar's head (a forked and odorous process called an osmeterium) provides both an olfactory and visual shock for would-be predators. ❹ Swallowtail pupae are slung from twigs or stems with a distinctive silken "girdle."
❺ These swallowtails are "puddling," probably in search of sodium salts on some damp soil. The mostly black individual is a ***Papilio troilus***, the **Spicebush Swallowtail**. The two larger butterflies in front of it are a **Giant Swallowtail**, ***P. cresphontes***, and an **Eastern Tiger Swallowtail, *P. glaucus.*** Spicebush Swallowtails occur in Carolinian forests where their host trees (spicebush, sassafras and tulip tree) are found.
❻ ❼ Swallowtail larvae, like these **Spicebush Swallowtail** (***Papilio troilus***) larvae, often have a fleshy, forked process (an osmeterium) normally kept tucked away in a pocket just behind the head. When disturbed, they stick out this odorous structure like a stinky, bright pair of tongues. This combines with the convincing fake eyes a little farther back on the body to give an alarmed Spicebush caterpillar an alarming appearance indeed.

FAMILY PAPILIONIDAE (continued) ❶❷❸❹ Caterpillars of the **Black Swallowtail** (***Papilio polyxenes***) are called **Parsleyworms** because of their preference for cultivated carrots, dill and parsley, but they also feed on other members of the parsley family (Umbelliferae) such as Queen Anne's Lace. Male adults have more extensive yellow markings on their wings than the females, which are mimics of the distasteful Pipevine Swallowtail (*P. philenor*). ❺ **Pipevine Swallowtails** (***Papilio philenor***) develop on poisonous pipevine plants (*Aristolochia* spp.); in doing so, they sequester toxic chemicals that protect all stages of this species from predators. Several other butterflies, including female Black Swallowtails (*P. polyxenes*) and Spicebush Swallowtails (*P. troilus*), mimic Pipevine Swallowtails and are avoided by predators that have had a bad experience with a distasteful Pipevine. ❻❼ Caterpillars of the **Giant Swallowtail** (***Papilio cresphontes***), sometimes called Orange Dogs, look like big, unappetizing, bird droppings. **Orange Dogs** feed on citrus in the south and on Hop Tree (*Ptelea trifoliata*) and Northern Prickly-ash (*Xanthoxylum americanum*) in the north; the northern edge of this species' range is extreme southern Ontario. ❽❾ The beautiful **Zebra Swallowtail** (***Papilio marcellus***) occurs throughout the United States, but occurs in Canada only in extreme southern Ontario. Caterpillars feed on Pawpaw.

FAMILY PIERIDAE (WHITES AND SULFURS) ❶ ❷ The **Cabbage White** (***Pieris rapae***) is a common insect both in North America and in its native Europe. This is an adult female (males have only one dot on the forewing). The green caterpillars are frequent pests of cabbage and broccoli. ❸ The **Mustard White** (***Pieris oleracea***) is a common native white that develops on rock cress (*Arabis* spp.) and related plants in the mustard family, including toothwort (*Dentaria* spp). Toothwort is also the food plant for caterpillars of the very similar, but much less common, West Virginia White (*P. virginiensis*). Mustard Whites can be found all season, but West Virginia Whites occur only in early spring. ❹ Several butterfly species indulge in "puddle parties" involving dozens of adult males imbibing liquids from wet soils, usually in search of sodium salts. Females are less likely to puddle since they get most of their sodium salts from males at the time of mating. The **Clouded Sulfur** (***Colias philodice***) is found almost everywhere clover, alfalfa and related legumes occur. ❺ The **Orange Sulfur** (***Colias eurytheme***) is similar to the Clouded Sulfur (*C. philodice*) and often occurs at the same time and place. Both species sometimes reach pest levels in clover and alfalfa fields. ❻ This **Pink-edged Sulfur** (***Colias interior***) was photographed in an Ontario bog, rich in the caterpillar's food plants (blueberry, *Vaccinium* spp.). **FAMILY NYMPHALIDAE** (BRUSH-FOOTED BUTTERFLIES) Subfamily Libytheinae (SNOUT BUTTERFLIES) ❼ The distinctively long-nosed **American Snout** (***Libytheana carinenta***) is the only snout butterfly in North America. Although this butterfly can be found as far north as southern Ontario during the summer, it does not survive the northern winters and instead migrates from the southern United States each spring. Caterpillars feed on hackberry. Subfamily Heliconiinae (HELICONIANS) ❽ The **Zebra** (***Heliconius charitonius***) is one of only two heliconian species occasionally encountered in northeastern North America. Zebras are common from Peru north to Florida and only appear north of South Carolina as rare vagrants. Heliconians feed on passionflowers (*Passiflora* spp.), from which they sequester poisonous and protective compounds. Subfamily Satyrinae (SATYRS AND WOOD NYMPHS) ❾ **Little Wood Satyrs** (***Megisto cymela***), like this one resting on a maple leaf, are common, slow-flying butterflies along forest edges. Like most members of the subfamily Satyrinae, the larvae live on grasses.

FAMILY NYMPHALIDAE (continued) ❶ The **Northern Pearly Eye** (***Enodia anthedon***) can be distinguished from the similar browns by the small second spot on the forewing. This common species occurs deep in the forest, far from the sunny haunts of most other butterflies. Larvae feed on woodland grasses. ❷ **Northern Pearly Eye** butterflies spend the winter as first instar larvae. ❸ Despite its common name, the **Appalachian Brown** (***Satyrodes appalachia***) occurs all the way from Florida to southern Ontario. Caterpillars of this species and the more common Eyed Brown (*S. eurydice*) develop on sedges. The Appalachian and Eyed Browns are very similar, and were only recognized as distinct species in 1970. ❹ The **Common Ringlet** (***Coenonympha tullia***) abounds right across the northern United States, Canada and Eurasia. Adult males can often be seen "patrolling" in search of mates on warm summer days. The caterpillars eat a variety of grasses. **Subfamily Danainae** (MONARCHS) ❺ The distinctively striped milkweed-feeding caterpillars of **Monarchs** (***Danaus plexippus***) are easily recognized. ❻ The wrinkled black mass below this newly formed **Monarch** (***Danaus plexippus***) chrysalis is the cast skin of the Monarch caterpillar. The small silk pad just above the cast skin was spun by the caterpillar before its metamorphosis, and was grabbed onto by the end of the chrysalis before it had entirely left the caterpillar skin. ❼ The bright colors of these mating **Monarchs** (***Danaus plexippus***) warn potential predators that Monarchs contain bitter poisons (cardiac glycosides) acquired from their milkweed food plants. This is an abundant and well-known migratory insect, and eastern North American Monarch populations spend the winter in a few patches of pine forest in one small area of Mexico. **Subfamily Limenitidinae** (ADMIRALS) ❽ Caterpillars in the genus ***Limenitis***, like the **Viceroy** (***L. archippus***), are often common on willow and poplar. They can look remarkably like glistening fresh bird-droppings. ❾ Adult **Viceroys** (***Limenitis archippus***), which may or may not be distasteful themselves, closely mimic the warning coloration of poisonous Monarch butterflies. Viceroys differ from Monarchs in having a black line across the hind wing.

FAMILY NYMPHALIDAE (continued) ❶ The two butterflies in this photograph, the **White Admiral** and the **Red-spotted Purple** (a bit out of focus!), are different forms of the same species (***Limenitis arthemis***). The White Admiral (the one with the white bands) is the subspecies ***L. arthemis arthemis***. Although the Red-spotted Purple (***L. arthemis astyanax***) is generally found farther south than the White Admiral, this photograph was taken in southern Ontario where both forms occur and regularly interbreed. ❷ ❸ The **Red-spotted Purple** (***Limenitis arthemis astyanax***) is a mimic of the chemically defended Pipevine Swallowtail (*Papilio philenor*). **Subfamily Nymphalinae** (ANGLEWINGS AND RELATIVES) ❹ ❺ ❻ ❼ Although anglewings (genus ***Polygonia***) are generally difficult to identify, adults of the common **Question Mark** (***P. interrogationis***) are easily recognized by the long, pinkish hind wing tails and the eponymous mark on the underside of the hind wing. The spiny larvae feed on the underside of host plant leaves including elm and stinging nettle, and the pupae can be found on the same plants. ❽ ❾ The **Red Admiral** (***Vanessa atalanta***) is a common butterfly in urban areas, where the pugnacious males can often be seen defending territories (such as stumps or clearings) on into the evening. Red Admiral caterpillars make rolled leaf nests on stinging nettle (*Urtica* spp.) and related plants.

❶ ❷ ❸ ❹ ❺ ❻

FAMILY NYMPHALIDAE (continued) ❶ The **Painted Lady** (***Vanessa cardui)*** is sometimes called the "Cosmopolitan," a common name that reflects this species' wide range. Painted Ladies do not normally survive the northeastern winter, but instead migrate into our area from the south each spring. ❷ ❸ The distinctively eye-spotted **Buckeye** (***Junonia coenia***) is a common and easily recognized butterfly found from southern Canada to Florida during the spring and summer months. It cannot, however, survive the winter north of the Carolinas, and spreads north anew each spring. ❹ ❺ The **Mourning Cloak** (***Nymphalis antiopa***) is a familiar butterfly in North America as well as Europe and Asia, where it is known as the Camberwell Beauty. Larvae feed gregariously, often defoliating whole branches of willow or poplar trees. This adult is feeding on fermenting sap on a wounded aspen late in the summer and, like other late-season (second generation) Mourning Cloaks, it will spend the winter hibernating under bark or in a similar retreat. This is one of the first butterfly species to fly in spring, often when snow is still on the ground. Subfamily Apaturinae (EMPERORS) ❻ Caterpillars of the **Tawny Emperor** (***Asterocampa clyton***) feed on hackberry along with the similar Hackberry Emperor (*A. celtis*). The fast-flying adults are usually found near the host plants. Subfamily Melitaeinae (CHECKERSPOTS AND CRESCENTS) ❼ ❽ The **Baltimore Checkerspot** (***Euphydryas phaeton***) is an easily recognized eastern North American butterfly that develops on Turtlehead (*Chelone glabra*) and ash (*Fraxinus* spp.).

❼ ❽

FAMILY NYMPHALIDAE (continued) ❶ The **Northern Crescent** (***Phyciodes cocyta***) is a very common butterfly across the northern United States and Canada, where it occurs almost anywhere its host plants (Asters) can be found. The Northern Crescent was only recently recognized as distinct from the Pearl Crescent (*P. tharos*), which only extends as far north as southern Ontario. **Subfamily Argynninae (FRITILLARIES)** ❷ Caterpillars of our larger fritillaries, like this **Aphrodite Fritillary** (***Speyeria aphrodite***, wingspan up to 73 mm) feed only on violets, but are unlikely to be noticed since they visit the host plants only at night. Aphrodite Fritillaries lay eggs on or near old violet plants that are unsuitable as food by the time eggs hatch in late summer. The small larvae hibernate, and feed on new foliage the next spring. ❸ Our smaller fritillaries belong to the mostly northern genus ***Boloria***. The **Silver Bordered Fritillary** (***B. selene***) develops on violet, but other *Boloria* have different host plants. Mating butterflies usually stay coupled for some time, while the male fills the female's mating chamber with fluids and waits for them to harden into a spermatophore. The spermatophore houses the male's sperm and also serves as a sort of butterfly "chastity belt," preventing the female from mating with another male for several days. **FAMILY HESPERIIDAE (SKIPPERS) Subfamily Hesperiinae (BRANDED SKIPPERS)** ❹ ❺ Skippers have conspicuously thick and muscular bodies, relatively short wings and a hooked club at the end of the antennae. Skippers in the subfamily Hesperiinae, like this flower-visiting **Leonard's Skipper** (***Hesperia leonardus***) and dull-colored **Dun Skipper** (***Euphyes vestris***), often rest with their front and hind wings at different angles, probably to aid in regulating body temperature. ❻ These **European Skippers** (***Thymelicus lineola***) are nectaring on an endangered, dune-loving thistle (Pitcher's Thistle, *Cirsium pitcheri*). Eggs of European Skippers were accidentally introduced from Europe with grass seeds, and this species is now very common in all kinds of grassy areas. ❼ **Long Dash Skippers** (***Polites mystic***) are common flower visitors. Larvae live in shelters made from grass leaves. ❽ This **Common Roadside Skipper** (***Amblyscirtes vialis***) has extended its long proboscis along this wet rock, probably in search of salts. This species is usually seen on the ground, typically along roadsides. ❾ Males of the **Hobomok Skipper** (***Poanes hobomok***) are commonly seen perched on raspberry leaves, often along forest trails or edges, as they watch for passing females.

FAMILY HESPERIIDAE (continued) Subfamily Heteropterinae (INTERMEDIATE SKIPPERS) ❶ The **Arctic Skipper** (***Carterocephalus palaemon***) is somewhat misnamed, since it is common across the northern United States and Canada but does not occur in the arctic. Subfamily Pyrginae (PYRGINE SKIPPERS) ❷ ❸ **Silver Spotted Skipper** (***Epargyreus clarus***) larvae feed inside shelters made from the leaves of Black Locust (*Robinia pseudoacacia*), Hog Peanut (*Amphicarpaea bracteata*) and related plants. These large skippers have a wingspan up to 45 mm. ❹ ❺ This **Sleepy Duskywing** (***Erynnis brizo***) nectaring on a yellow flower is colored much like this mating pair of **Dreamy Duskywings** (***E. icelus***), although they differ in fine details such as male genitalia. The Dreamy Duskywing is common in the north, where it feeds on willow, while the oak-feeding Sleepy Duskywing barely gets as far north as the Canadian border. FAMILY HEPIALIDAE (GHOST MOTHS) ❻ Although the photographs in this book are organized according to the traditional and convenient division of the order Lepidoptera into "butterflies" and "moths," the moths do not form a natural group. Ghost moths, like this ***Sthenopis auratus***, are distantly related to most other butterflies and moths and are sometimes placed in their own small suborder (Exoporia). Larvae bore in roots. ❼ This recently introduced ghost moth (***Korscheltellus lupulinus***) is now common in southern Ontario, where it develops in grasses. FAMILY ERIOCRANIIDAE ❽ Eriocraniids are primitive micromoths. This species, ***Dyseriocrania griseocapitella***, is a miner in oak leaves. FAMILY ADELIDAE (FAIRY MOTHS) ❾ Fairy moths, like this male ***Adela caeruleella***, are eye-catching because of their impressively long antennae. Some common species are leaf miners when very young, and case-makers when older. ❿ The day-flying habits, tremendously long antennae, and bright colors of ***Adela ridingsella*** render it distinctive despite its small size.

FAMILY HELIOZELIDAE (SHIELD-BEARER MOTHS) ❶ The small (less than 10 mm) leaf-mining moths in the family Heliozelidae (such as this ***Antispila***) are known as shield-bearer moths because of the shieldlike piece of food-plant leaf used in forming the cocoon, which drops from the host plant to the ground. **FAMILY INCURVARIIDAE** (LEAFCUTTER MOTHS) ❷ **Maple Leaf Cutter** (***Paraclemensia acerifoliella***) caterpillars are leaf miners at first, then they cut two circular pieces out of the maple leaf and make a turtleshell-like portable shelter. They pupate in the case, and the distinctive adults are found in maple forests in spring. **FAMILY PRODOXIDAE** ❸ **Yucca moths** (*Tegeticula* spp.) are highly specialized pollinators of Yucca that seem to occur wherever Yucca grows, even where it is planted in Canadian gardens. This ***Tegeticula yuccasella*** female has a ball of Yucca pollen held in place with her specialized maxillary palpi ("maxillary tentacles"). She will carry this pollen ball to another Yucca, put her eggs in the plant's ovary, and then place the pollen directly on the stigma to guarantee a developing seed for her larva to feed on. ❹ ***Prodoxus decipiens*** is sometimes called the **Bogus Yucca Moth** because it occurs in Yucca but does not carry balls of pollen like true Yucca moths. Larvae develop in stems of Yucca flower heads. **FAMILY ARGYRESTHIIDAE** ❺ ❻ Cedar leaf miners (***Argyresthia*** spp.) are minute moths that often occur in huge numbers on Eastern White Cedar trees. The larvae mine in the leaves, and this species (***A. aureoargentella***) forms white cocoons near the mined tissue. The more common *A. thuiella* pupates in the mine. ❼ ***Argyresthia goedartella*** is one of over 50 North American species in the genus *Argyresthia*, a large genus of very small moths. **FAMILY YPONOMEUTIDAE** (ERMINE MOTHS) ❽ The **Spindle Ermine Moth** (***Yponomeuta cagnagella***) develops in communal tentlike webs on cultivated *Euonymus* foliage. The similar *Y. multipunctella* is found on wild *Euonymus*. ❾ The **Ailanthus Webworm** (***Atteva punctella***) develops in communal tentlike webs on *Ailanthus* foliage.

FAMILY PLUTELLIDAE (DIAMONDBACK MOTHS) ❶ This ***Ypsolopha dentella*** adult is striking a characteristic resting pose. Larvae occur on honeysuckle (*Lonicera* species). ❷ The delicate, meshwork cocoon seen in the background behind this ***Plutella porrectella*** is a typical diamondback moth cocoon. A similar species, *P. xylostella*, is a pest of cabbage. **FAMILY TINEIDAE** (CLOTHES MOTHS) ❸ Larvae of clothes moths can probably be blamed for those holes that just appeared in your best woollen sweater. These little caterpillars and the associated dull moths also appear in wool carpets, stored furs or hides and other dry products of animal origin. Some common species make tubular cases reminiscent of caddisfly cases. **FAMILY PSYCHIDAE** (BAGWORM MOTHS) ❹ This male ***Psyche casta*** has just emerged from its bag-like larval case; the remains of its pupal shell can be seen sticking out the end of the case. Females are wingless and do not leave the cases. ❺ ***Dahlica triquetrella*** is an introduced, and very common, European bagworm that occurs only as a female in North America (both males and females occur in Europe). The female adult (which looks so un-mothlike it was first formally described as a fly!) comes out of the case but lays its eggs in the case. Larvae, like this one, are lichen feeders. ❻ This **Evergreen Bagworm Moth** (***Thyridopteryx ephemeraeformis***) case (about 40 mm in length) might contain a larva, a wingless, legless, eyeless adult female or a mass of eggs. Males, which are winged, emerge from their cases in early spring and track the female's pheromones to find their degenerate mates in their cases. Females never leave their cases; eggs, which hatch in late spring, are laid in the case. **FAMILY COLEOPHORIDAE** (CASEBEARER MOTHS AND THEIR RELATIVES) **Subfamily Coleophorinae** ❼ ❽ Casebearer moth caterpillars start out life as leaf miners, and then build cases from silk and debris. They then feed between the upper and lower leaf surfaces without fully leaving their cases. Most casebearers belong to the large genus ***Coleophora***, some of which are metallic golden moths commonly seen on flowers. ❾ The **Cherry Casebearer** (***Coleophora pruniella***) is an introduced species that makes portable cigar-shaped cases from which to feed on birch buds and foliage. They spend the winter as small caterpillars in cases; in spring they make new cases that are attached to the leaf surface while the larvae mine the inner tissues of the leaf as far as they can reach without leaving their cases.

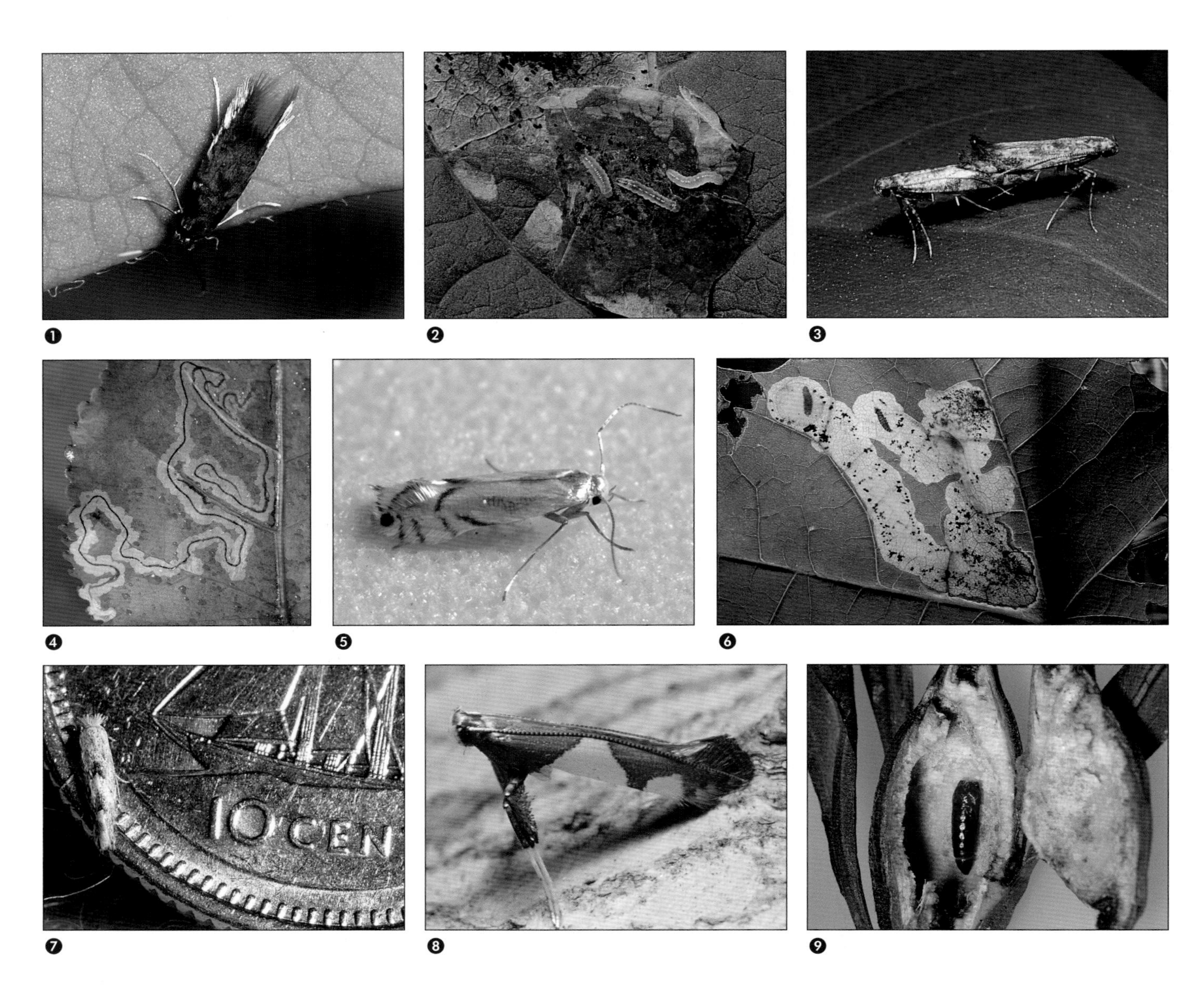

FAMILY COLEOPHORIDAE (continued) Subfamily Momphinae ❶ ***Mompha terminella*** is a minute moth that develops as a leafminer in Enchanter's Nightshade (*Circaea alpina*). **FAMILY GRACILLARIIDAE** (LEAF BLOTCH MINER MOTHS) ❷ A mine formed by **Lilac Leaf Miners** (***Gracillaria syringella***) has been opened up to show some of these tiny caterpillars and their frass (caterpillar poop). ❸ A pair of adult **Lilac Leaf Miners** (***Gracillaria syringella***). ❹ ❺ The black line up the middle of this elegant mine made by a **Poplar Serpentine Leaf Miner** (***Phyllocnistis populiella***) is a neatly deposited line of frass, and you can follow it from where the larva started its mine to where it is currently feeding near the leaf edge. Although most Gracillariidae make blotchlike leaf mines, members of the genus *Phyllocnistis* make these snake-shaped mines. ❻ **Maple Leafblotch Miners** (***Cameraria aceriella***) are small, flattened caterpillars that feed beneath the upper surfaces of maple leaves, making characteristic broad, frass-strewn mines. ❼ Although the mines made by leafblotch miners and some other micromoth families can be distinctive, most micromoth adults are minute and difficult to identify (this is probably a ***Phyllonorycter***). ❽ Despite its small size, the distinctive posture of this ***Caloptilia*** (standing on a tree trunk) makes it easy to recognize. **FAMILY GELECHIIDAE** (TWIRLER MOTHS) ❾ The elongate, spindle-shaped galls found on goldenrod stems are made by moths in two different families, Gelechiidae and Tortricidae. The **Goldenrod Gall Gelechiid** (***Gnorimoschema gallaesolidaginis***) lays eggs on goldenrod in the fall, and the larvae burrow into the stem and induce these slender galls. The larva cuts a beveled hole in the gall and packs it with debris before pupating in the gall. The gall shown here, cut open in late summer, has only an empty pupal case; you can see the hole through which the moth emerged on the upper left corner.

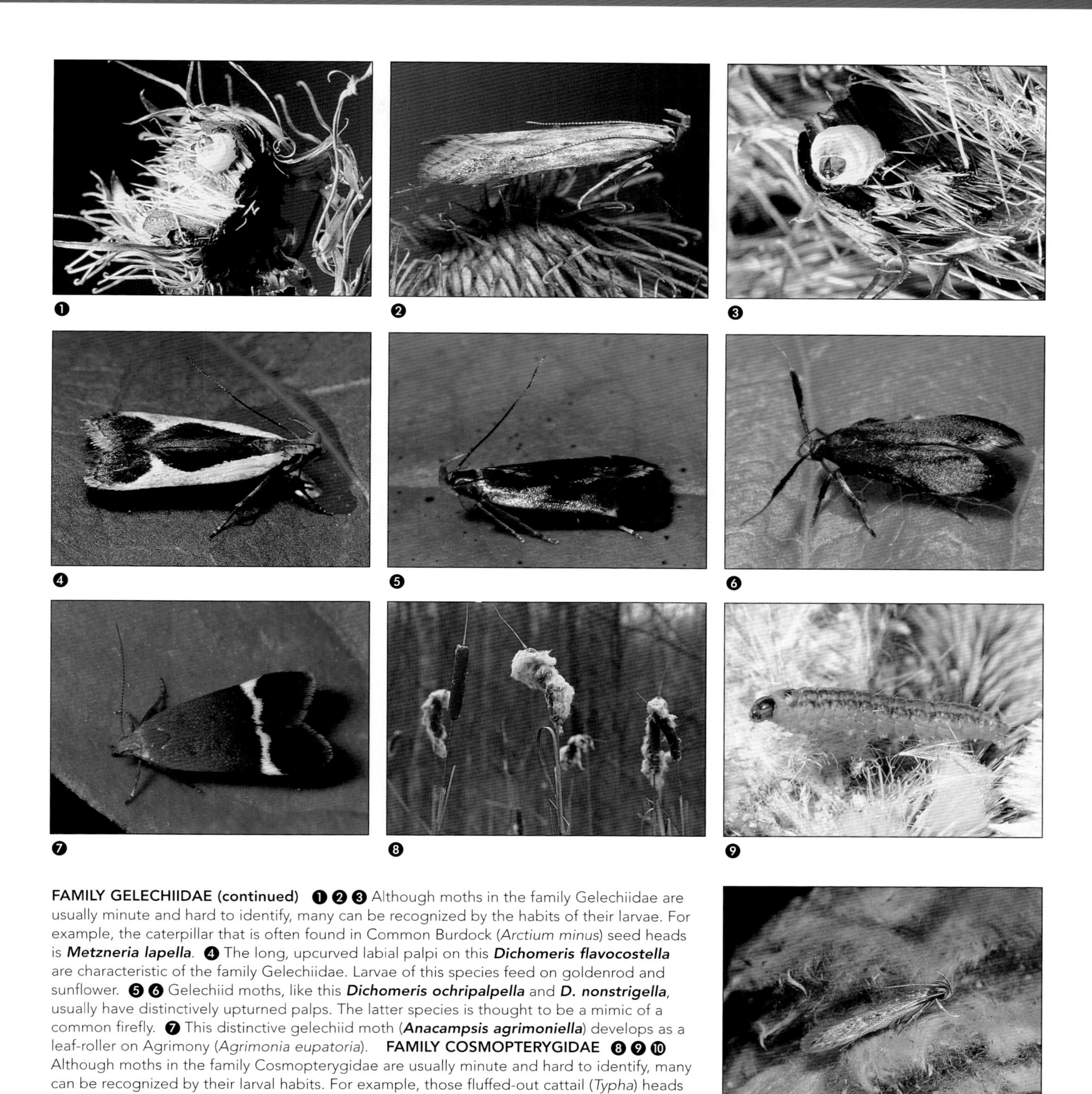

FAMILY GELECHIIDAE (continued) ❶ ❷ ❸ Although moths in the family Gelechiidae are usually minute and hard to identify, many can be recognized by the habits of their larvae. For example, the caterpillar that is often found in Common Burdock (*Arctium minus*) seed heads is ***Metzneria lapella***. ❹ The long, upcurved labial palpi on this ***Dichomeris flavocostella*** are characteristic of the family Gelechiidae. Larvae of this species feed on goldenrod and sunflower. ❺ ❻ Gelechiid moths, like this ***Dichomeris ochripalpella*** and ***D. nonstrigella***, usually have distinctively upturned palps. The latter species is thought to be a mimic of a common firefly. ❼ This distinctive gelechiid moth (***Anacampsis agrimoniella***) develops as a leaf-roller on Agrimony (*Agrimonia eupatoria*). **FAMILY COSMOPTERYGIDAE** ❽ ❾ ❿ Although moths in the family Cosmopterygidae are usually minute and hard to identify, many can be recognized by their larval habits. For example, those fluffed-out cattail (*Typha*) heads that can be seen in any winter wetland are caused by **Cattail Caterpillars**, which are larvae of the **Shy Cosmet** (***Limnaecia phragmitella***). Another member of the Cosmopterygidae has the odd (for a moth) habit developing as a parasitoid of scale insects.

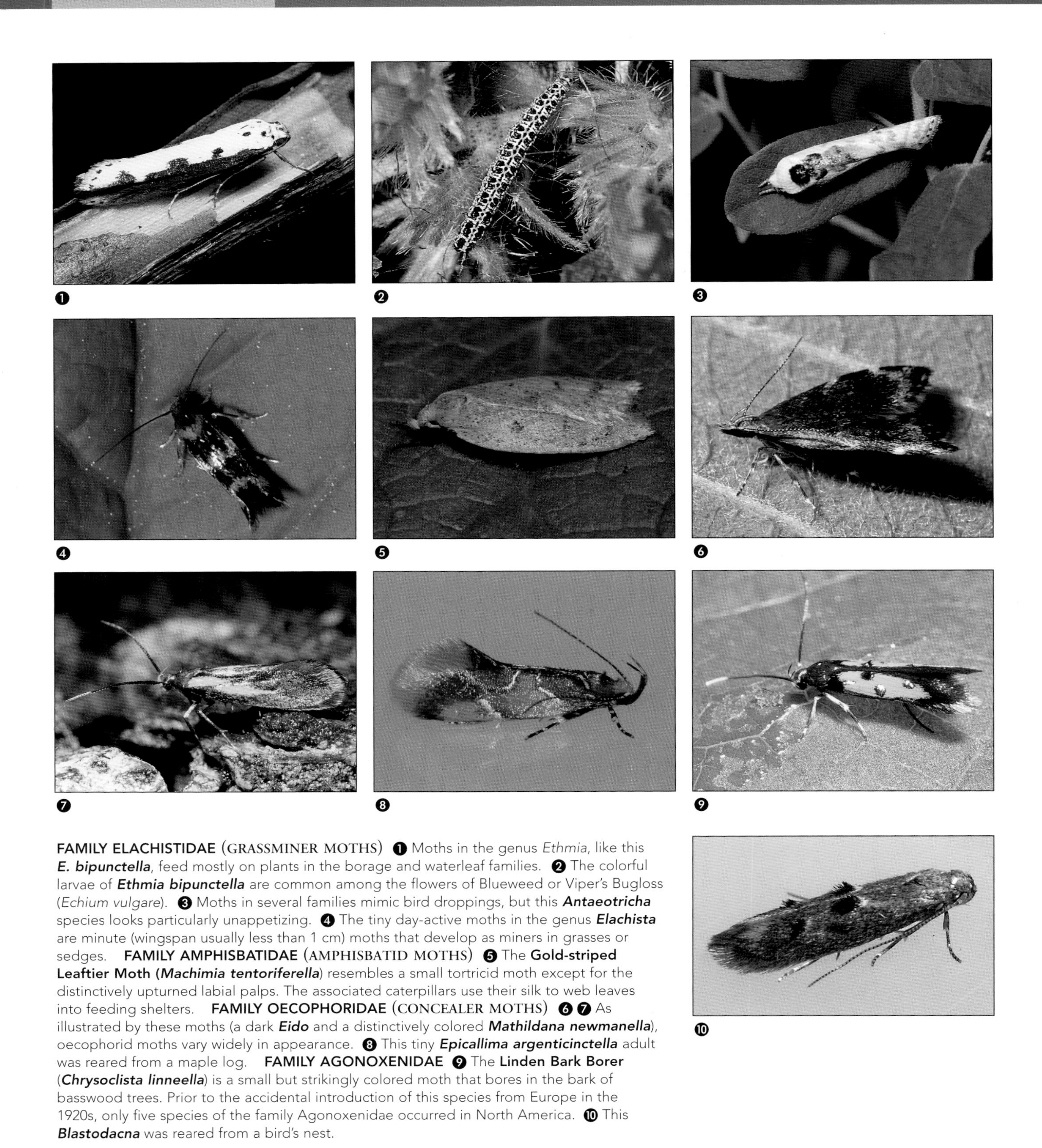

FAMILY ELACHISTIDAE (GRASSMINER MOTHS) ❶ Moths in the genus *Ethmia*, like this ***E. bipunctella***, feed mostly on plants in the borage and waterleaf families. ❷ The colorful larvae of ***Ethmia bipunctella*** are common among the flowers of Blueweed or Viper's Bugloss (*Echium vulgare*). ❸ Moths in several families mimic bird droppings, but this ***Antaeotricha*** species looks particularly unappetizing. ❹ The tiny day-active moths in the genus ***Elachista*** are minute (wingspan usually less than 1 cm) moths that develop as miners in grasses or sedges. **FAMILY AMPHISBATIDAE** (AMPHISBATID MOTHS) ❺ The **Gold-striped Leaftier Moth (*Machimia tentoriferella*)** resembles a small tortricid moth except for the distinctively upturned labial palps. The associated caterpillars use their silk to web leaves into feeding shelters. **FAMILY OECOPHORIDAE** (CONCEALER MOTHS) ❻ ❼ As illustrated by these moths (a dark ***Eido*** and a distinctively colored ***Mathildana newmanella***), oecophorid moths vary widely in appearance. ❽ This tiny ***Epicallima argenticinctella*** adult was reared from a maple log. **FAMILY AGONOXENIDAE** ❾ The **Linden Bark Borer** (***Chrysoclista linneella***) is a small but strikingly colored moth that bores in the bark of basswood trees. Prior to the accidental introduction of this species from Europe in the 1920s, only five species of the family Agonoxenidae occurred in North America. ❿ This ***Blastodacna*** was reared from a bird's nest.

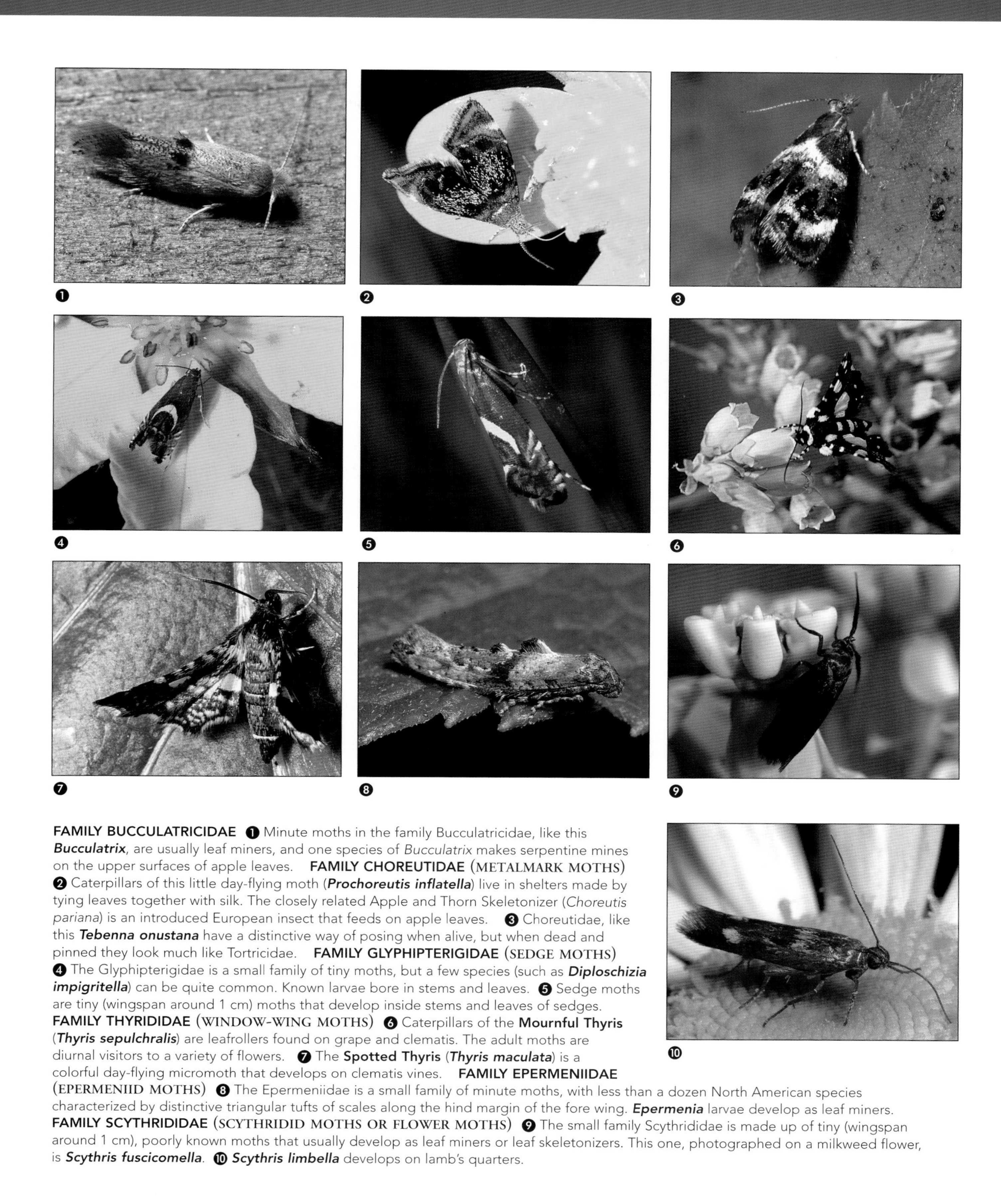

FAMILY BUCCULATRICIDAE ❶ Minute moths in the family Bucculatricidae, like this ***Bucculatrix***, are usually leaf miners, and one species of *Bucculatrix* makes serpentine mines on the upper surfaces of apple leaves. **FAMILY CHOREUTIDAE** (METALMARK MOTHS) ❷ Caterpillars of this little day-flying moth (***Prochoreutis inflatella***) live in shelters made by tying leaves together with silk. The closely related Apple and Thorn Skeletonizer (*Choreutis pariana*) is an introduced European insect that feeds on apple leaves. ❸ Choreutidae, like this ***Tebenna onustana*** have a distinctive way of posing when alive, but when dead and pinned they look much like Tortricidae. **FAMILY GLYPHIPTERIGIDAE** (SEDGE MOTHS) ❹ The Glyphipterigidae is a small family of tiny moths, but a few species (such as ***Diploschizia impigritella***) can be quite common. Known larvae bore in stems and leaves. ❺ Sedge moths are tiny (wingspan around 1 cm) moths that develop inside stems and leaves of sedges. **FAMILY THYRIDIDAE** (WINDOW-WING MOTHS) ❻ Caterpillars of the **Mournful Thyris** (***Thyris sepulchralis***) are leafrollers found on grape and clematis. The adult moths are diurnal visitors to a variety of flowers. ❼ The **Spotted Thyris** (***Thyris maculata***) is a colorful day-flying micromoth that develops on clematis vines. **FAMILY EPERMENIIDAE** (EPERMENIID MOTHS) ❽ The Epermeniidae is a small family of minute moths, with less than a dozen North American species characterized by distinctive triangular tufts of scales along the hind margin of the fore wing. ***Epermenia*** larvae develop as leaf miners. **FAMILY SCYTHRIDIDAE** (SCYTHRIDID MOTHS OR FLOWER MOTHS) ❾ The small family Scythrididae is made up of tiny (wingspan around 1 cm), poorly known moths that usually develop as leaf miners or leaf skeletonizers. This one, photographed on a milkweed flower, is ***Scythris fuscicomella***. ❿ ***Scythris limbella*** develops on lamb's quarters.

FAMILY PTEROPHORIDAE (PLUME MOTHS) ❶ Plume moths, so called because the hind wing is divided into three "plumes" and the front wing is deeply notched, are easily recognized in the field by their T-shaped stance. Most are leafrollers or borers. This milkweed blossom has attracted numerous plume moths, including several ***Geina tenuidactyla*** and one ***Cnaemidophorus rhododactylus***. ❷❸❹❺ The characteristic T-shaped resting posture of these plume moths (***Cnaemidophorus rhododactylus***, ***Amblyptilia pica***, ***Emmelina monodactyla*** and ***Hellinsia homodactylus***, respectively) is typical of the family Pterophoridae. **FAMILY URANIIDAE** ❻ The **Brown Scoopwing** (***Calledapteryx dryopterata***) is one of only two uraniid moths in the northeast, and one of only seven in North America. Larvae feed on *Viburnum*. ❼ Young larvae of the **Gray Scoopwing** (***Callizzia amorata***) feed gregariously inside webs on honeysuckle. ❽ Although North American Uraniidae (formerly placed in Epiplemidae) are relatively inconspicuous insects, some tropical members of the family look like brilliantly metallic swallowtail butterflies. This ***Urania*** was photographed in Bolivia. **FAMILY CARPOSINIDAE** ❾ Members of the small family Carposinidae, like this tiny ***Bondia crescentella***, are mostly borers in fruits and shoots.

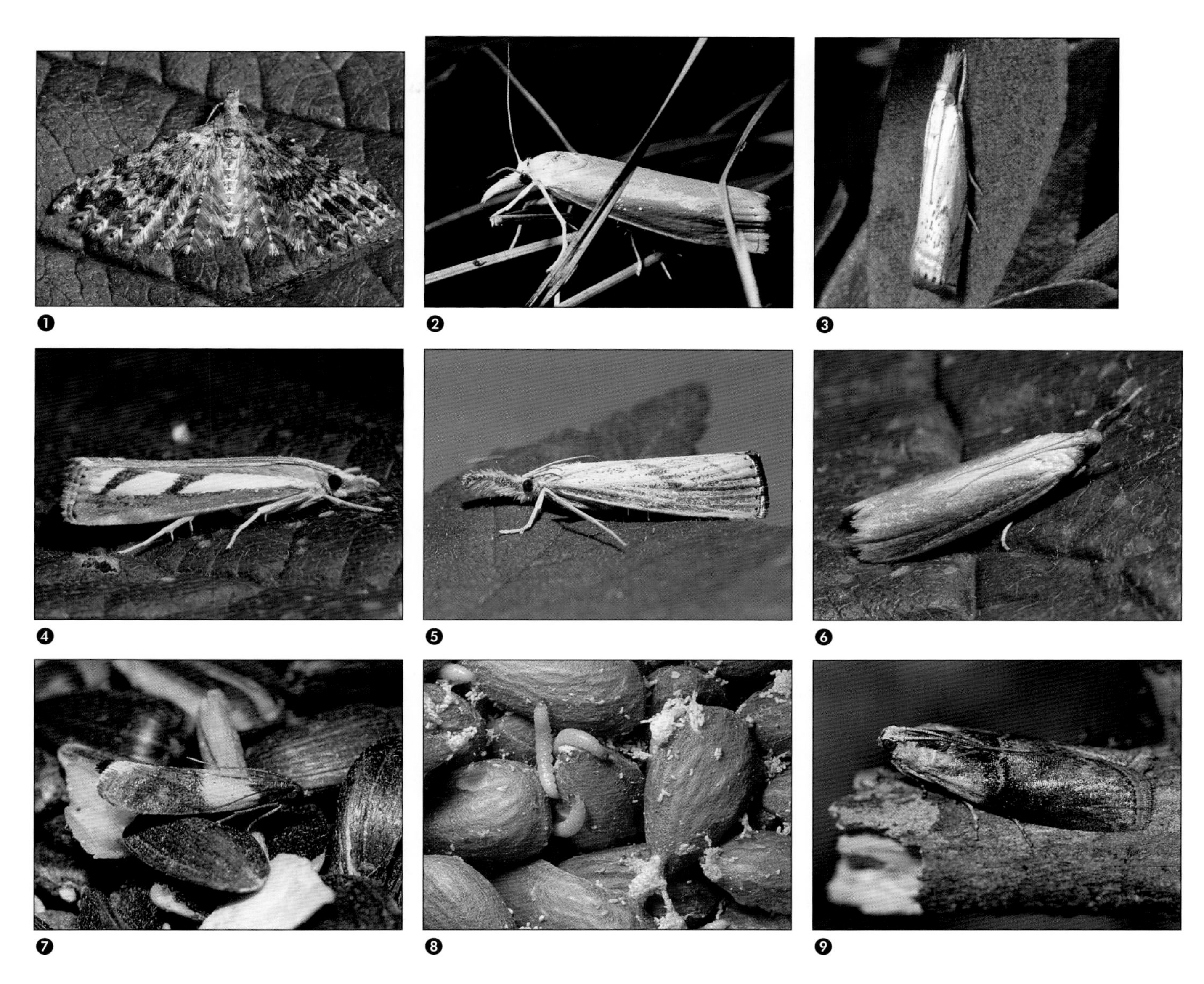

FAMILY ALUCITIDAE (SIX-PLUME MOTHS OR MANY-PLUME MOTHS) ❶ The **Six-plume Moth** (***Alucita adriendenisi***) was one of two North American *Alucita* species described as new to science in 2004, bringing the northeastern alucitid fauna to three species. Larvae of this small (wingspan about 13 mm) moth bore into the flowers or buds of honeysuckle. **FAMILY PYRALIDAE** (SNOUT MOTHS) **Subfamily Crambinae** (CLOSE-WINGS OR GRASS MOTHS) ❷ Sod webworms (larvae of several species of ***Crambus***; this is ***C. perlellus***) live in silken tubes on the soil near grass stems, often doing considerable damage to plant stems and roots. Some *Crambus* species feed on mosses. ❸ This ***Crambus, C. albellus***, occurs in bogs where it conspicuously webs up clusters of leaves on ericaceous shrubs. ❹ ❺ Moths in the subfamily Crambinae, like this brightly banded ***Catoptria latiradiella*** and golden-yellow ***Agriphila ruricolella***, are sometimes called "close-wings" because of the way they hold their wings close to the body. **Subfamily Phyctinae** ❻ The distinctively colored ***Peoria approximella*** is a common moth, but its larval habits remain unknown. ❼ ❽ The **Indian Meal Moth** (***Plodia interpunctella***) is a cosmopolitan and common pest of stored products. This adult was one of many in a bag of sunflower seeds; these larvae were infesting a jar of almonds. ❾ ***Sciota basilaris*** develops on willow and aspen.

FAMILY PYRALIDAE (continued) Subfamily Epipaschiinae ❶ ❷ This adult moth is a ***Pococera*** species; the caterpillars are **Maple Webworms** (***Pococera asperatella***), that sometimes abound on maple, sumac and other trees. Subfamily Pyraustinae ❸ This brightly colored moth (***Anania funebris***) is a day-flying species that develops on goldenrod. ❹ The **Grape Leaf-folder Moth** (***Desmia funeralis***) is a day-flying moth that develops on both wild and domestic grape vines. ❺ Larvae of the **Grape Leaf-folder Moth** (***Desmia funeralis***) spin strands of silk which contract and pull the edges of grape leaves into conspicuous folds or rolls in which the larvae feed on the leaf surface. ❻ Larvae of the glistening, window-winged **Melonworm Moth** (***Diaphania hyalinata***) develop on the foliage of squash and related plants, sometimes also burrowing into the stems. ❼ ❽ Members of the large genus ***Pyrausta***, with almost 60 North American species, can be common on garden flowers. Both of these species (***P. orphisalis*** and the bright pink ***P. signatalis***) feed on Wild Bergamot, Bee Balm and related plants (*Monarda* spp.). ❾ Larvae of the **Basswood Leafroller** (***Pantographa limata***) are solitary leafrollers that make silk and leaf shelters on basswood, rock elms and oak trees.

FAMILY PYRALIDAE (continued) ❶ ❷ ❸ The **European Corn Borer** (***Ostrinia nubilalis***) is a major introduced pest with larvae that burrow in the stalks and cobs of corn and in the stalks of several other plants, often cutting leaves off at their bases and causing significant crop losses. ❹ Larvae of the **Celery Leaftier Moth** (***Udea rubigalis***) web up and eat leaves of celery, beans, spinach and many other plants. ❺ The **Small Magpie Moth** (***Eurrhypara hortulata***) is a European species first introduced into Nova Scotia in 1907 and now widespread in the northeast. Larvae make shelters by rolling leaves of nettle, mint and other plants. ❻ The **Celery Webworm** (***Nomophila nearctica***) is a common backyard snout moth that feeds on a variety of herbaceous plants including grasses, celery and clover. ❼ ***Perispasta caeculalis***, our only *Perispasta*, is a fairly common moth throughout much of North America, but the larvae remain unknown. Subfamily Galleriinae ❽ The **Greater Wax Moth** (***Galleria mellonella***) is one of two related wax moth species that make a mess of stored beeswax, beekeeping equipment and neglected beehives. The caterpillars eat the wax, covering the feeding area with silken webbing. Subfamily Pyralinae ❾ **Clover Hayworm Moth** (***Hypsopygia costalis***) caterpillars sometimes make a nuisance of themselves by feeding on stored hay.

FAMILY PYRALIDAE (continued) **Subfamily Acentropinae (formerly Nymphulinae–AQUATIC SNOUT MOTHS)** ❶ Snout moths in the subfamily Acentropinae are often found near water, where their larvae are aquatic or semiaquatic. Larvae of ***Nymphula ekthlipsis***, for example, are aquatic caterpillars that make cases from sedges. ❷ ❸ Caterpillars of ***Petrophila canadensis*** seem more like caddisflies than moths because they live in moving water. These gilled caterpillars web themselves down to avoid being swept away as they graze diatoms from rock surfaces. ❹ ❺ ❻ Most aquatic snout moths use silk and leaf tissue to make a case. One of these caterpillars is wrapped in a duckweed case as it feeds on duckweed, and the other has made a case from part of a water lily leaf (opened to expose the larva). ❼ ❽ ❾ Aquatic snout moths in the genus ***Parapoynx*** are often abundant on floating vascular plants. The moths shown here are ***P. maculalis*** (on a lily pad), ***P. curviferalis*** (on a sedge leaf) and ***P. allionealis***.

FAMILY PYRALIDAE (continued) ❶ ***Munroessa*** caterpillars feed on water lily leaves when small and then bore into the leaf petioles. This adult *Munroessa* moth has been captured by a water strider. Subfamily Ancylolomiinae ❷ ***Prionapteryx nebulifera*** is a rarely encountered snout moth associated with the Great Lake shorelines. **FAMILY TORTRICIDAE** Subfamily Tortricinae ❸ ❹ **Spruce Budworm** (***Choristoneura fumiferana***) moths lay rows of blue-green eggs on spruce or balsam foliage, where the first stage larvae remain during the winter months, ready to feed and grow rapidly when the new growth appears in spring. This is one of our most important forest pests. ❺ ❻ The genus ***Choristoneura*** includes the famous Spruce Budworm (*C. fumiferana*), as well as several other common moths such as the **Jack Pine Budworm** (***C. pinus***) and the **Oblique-banded Leafroller** (***C. rosaceana***), a common pest of apple. ❼ The **White Triangle Tortrix** (***Clepsis persicana***) develops on a huge variety of deciduous and coniferous trees. ❽ The **Black-patched Clepsis** (***Clepsis melaleucana***) is a common moth that develops on a variety of plants ranging from apple to trillium. ❾ ❿ ***Archips striana*** and ***A. dissitana*** are among about 24 North American species in the large genus *Archips*. Larvae are leafrollers, and some species are economically important pests of crops, forest trees, and ornamentals.

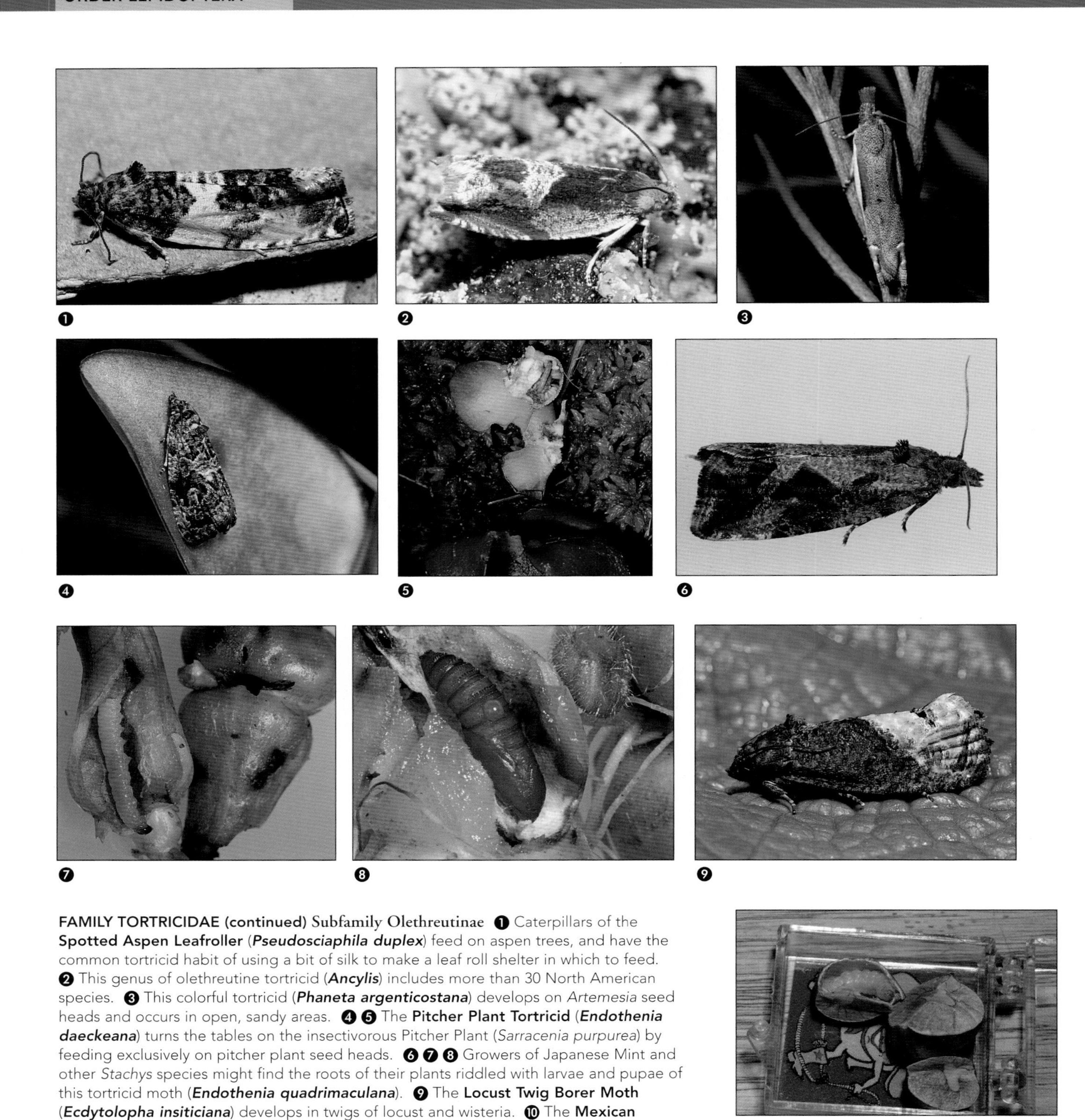

FAMILY TORTRICIDAE (continued) Subfamily Olethreutinae ❶ Caterpillars of the **Spotted Aspen Leafroller** (***Pseudosciaphila duplex***) feed on aspen trees, and have the common tortricid habit of using a bit of silk to make a leaf roll shelter in which to feed. ❷ This genus of olethreutine tortricid (***Ancylis***) includes more than 30 North American species. ❸ This colorful tortricid (***Phaneta argenticostana***) develops on *Artemesia* seed heads and occurs in open, sandy areas. ❹ ❺ The **Pitcher Plant Tortricid** (***Endothenia daeckeana***) turns the tables on the insectivorous Pitcher Plant (*Sarracenia purpurea*) by feeding exclusively on pitcher plant seed heads. ❻ ❼ ❽ Growers of Japanese Mint and other *Stachys* species might find the roots of their plants riddled with larvae and pupae of this tortricid moth (***Endothenia quadrimaculana***). ❾ The **Locust Twig Borer Moth** (***Ecdytolopha insiticiana***) develops in twigs of locust and wisteria. ❿ The **Mexican Jumping Bean Moth** (***Cydia deshaisiana***) belongs to the same widespread genus as the Codling Moth (*C. pomonella*), a major pest of apples. These "Mexican Jumping Beans" were purchased at an Ontario novelty store, and the side of one seed was removed to expose the leaping larva within.

FAMILY TORTRICIDAE (continued) ❶ Caterpillars of the **Grape Berry Moth** (***Endopiza viteana***), as the name suggests, are found in grapes. ❷ ❸ The **Goldenrod Gall Tortricid** (***Epiblema scudderiana***) is one of two types of moths that induce elliptical galls on goldenrod. Unlike the Goldenrod Gall Gelechiid (*Gnorimoschema gallaesolidaginis*), which overwinters as an egg, the Goldenrod Gall Tortricid overwinters in the goldenrod stem just below the main gall cavity. These caterpillars contain "antifreeze" compounds that protect them from freezing at temperatures down to almost –40°. ❹ **Maple Trumpet Skeletonizer** (***Epinotia aceriella***) caterpillars spin silken webs across the undersides of maple leaves, crumpling the leaves into shelters in which the caterpillars "skeletonize" the leaves by feeding on the parenchyma cells. These common caterpillars are normally concealed inside trumpet-shaped tubes, made from silk and frass, within their leaf shelters, but they leave the tubes to feed and to drop to the ground and pupate in fall. **FAMILY LIMACODIDAE** (SLUG CATERPILLAR MOTHS) ❺ ❻ Slug caterpillars, like this **Spiny Oak Slug** (***Euclea delphinii***), glide along like slugs because their thoracic legs are short and their prolegs are reduced to suckers. Most slug caterpillars have stinging hairs or spines, and careless contact with a Spiny Oak Slug is intensely painful. Despite the common name, Spiny Oak Slugs feed on a wide variety of hardwood trees. ❼ The **Crowned Slug** (***Isa textula***), one of our most attractive slug moth caterpillars, feeds on a variety of host trees. This individual was parasitized, and was ruptured by the emergence of a tachinid maggot shortly after being photographed. ❽ The oval, green, sluglike caterpillar of ***Lithacodes fasciola*** can be seen gliding along the foliage of a wide variety of trees and shrubs. ❾ The short, humped, rocker-shaped caterpillar of the **Skiff Moth** (***Prolimacodes badia***) is one of the most bizarre caterpillars found in our region. They feed on a wide variety of trees and shrubs.

FAMILY LIMACODIDAE (continued) ❶ **Saddleback caterpillars** (***Sibine*** spp.) are well-armed insects that should be recognized and avoided. Although this is a photograph of a tropical species, the common northeastern Saddleback Caterpillar (*S. stimulea*) has the same color pattern and general appearance. ❷ The pose struck by this adult ***Tortricidia testacea*** is typical for this species. ❸ Several slug caterpillar moths have patches of green on the wings and thorax, but the green pattern on the **Smaller Parasa** (***Parasa chloris***) is unusually extensive. ❹ ***Apoda biguttata*** is distributed from southern Canada to Florida. Larvae feed on beech, hickory and oak. **FAMILY SESIIDAE** (CLEARWING MOTHS) ❺ Most clearwing moths are day-flying moths that look and act like wasps. The **Raspberry Crown Borer Moth** (***Pennisetia marginata***) is a striking yellowjacket (Vespidae) mimic not only in color, shape and size, but also in behavior. Larvae feed on raspberry and blackberry foliage. ❻ This clearwing moth (***Paranthrene tabaniformis***) is a striking mimic of a blackjacket wasp (Vespidae). Larvae bore in willow and poplar branches. ❼ This female **Peachtree Borer** (***Synanthedon exitiosa***) looks impressively like a spider wasp (Pompilidae). The black and yellow males resemble potter wasps (Vespidae, Eumeninae). Larvae are serious pests that bore into trunks of peach trees and other *Prunus* species. ❽ Clearwing moths in the large genus *Synanthedon* (about 40 North American species) bore into a wide variety of trees. This is ***S. proxima***. ❾ This female **Virginia Creeper Borer** (***Albuna fraxini***), photographed in southern Ontario, developed as a borer in roots of Virginia Creeper.

FAMILY SESIIDAE (continued) ❶ This male clearwing moth (***Synanthedon fatifera***) has its proboscis unfurled into a silverweed flower. **FAMILY ZYGAENIDAE** (SMOKY MOTHS AND BURNETS) ❷ The Zygaenidae is a small family in North America, with only 22 species. The most common species, the **Grape Leaf Skeletonizer** (***Harrisina americana***), is usually found on grapevines, where its larvae feed in groups that skeletonize leaves. **FAMILY COSSIDAE** (CARPENTERWORMS AND LEOPARD MOTHS) ❸ Caterpillars in the small family Cossidae are wood borers that sometimes seriously damage trees. Larvae of the **Poplar Carpenterworm** (***Acossus centerensis***) favor poplar trees, sometimes riddling the wood with tunnels. ❹ The **Carpenterworm** (***Prionoxystus robiniae***) is a pest of a wide range of hardwood trees throughout the U.S. and southern Canada. ❺ Carpenterworms (larvae of moths in the family **Cossidae**) excavate galleries in branches, trunks and roots of host trees. **FAMILY SATURNIIDAE** (GIANT SILKWORM MOTHS) ❻❼❽ The **Luna** (***Actias luna***) is one of our largest and most widely recognized moths. Caterpillars feed on a variety of trees before pupating in oval silken cocoons disguised by leaves on the forest floor. As is the case for other cocoon-making moths, when the adult emerges from the pupa it pushes its way to freedom through a weakening in the cocoon created by its saliva. ❾❿ **Io Moth** (***Automeris io***) caterpillars are things of beauty, but tips of the fine spines that cover the body contain poisons that can cause extreme pain on contact. Look for these dangerous caterpillars on a variety of trees. Adult Io Moths have impressive eyespots on the hind wing, normally concealed under the front wings but undoubtedly startling to potential predators when unexpectedly exposed.

FAMILY SATURNIIDAE (continued) ❶ ❷ ❸ Caterpillars of the **Rosy Maple Moth** (***Dryocampa rubicunda***) are sometimes called **Green-striped Mapleworms**, although they can be significant defoliators of oaks as well as maple. Rosy Maple Moths have distinct color forms in different areas, with a pale form common in southern Ontario and the brighter, distinctly banded form common throughout most of the northeastern United States and Canada. ❹ This **Pine-devil Moth** (***Citheronia sepulcralis***) was photographed on a white sheet near a moth-attracting ultraviolet light. Pine-devil Moths occur in coastal pine forests in the eastern United States from Maine to Florida. ❺ The **Imperial Moth** (***Eacles imperialis***) has been recorded from a broad range of coniferous and deciduous trees, but is usually found on White Pine. ❻ ❼ ❽ This enormous caterpillar, almost 100 mm in length, is a mature larva of the common **Cecropia Moth** (***Hyalophora cecropia***). Cecropias pupate inside spindle-shaped, brown silken cocoons on low twigs, often clearly visible after the leaves have fallen in late autumn. These reddish-bodied adults, sometimes called **Robin Moths**, are often common near new housing developments with many young hardwood trees. Males, like this one, have broad and feathery antennae with which to detect the female's powerful pheromones. ❾ Half a dozen similar species of oakworm moths (***Anisota*** species) occur in eastern North America.

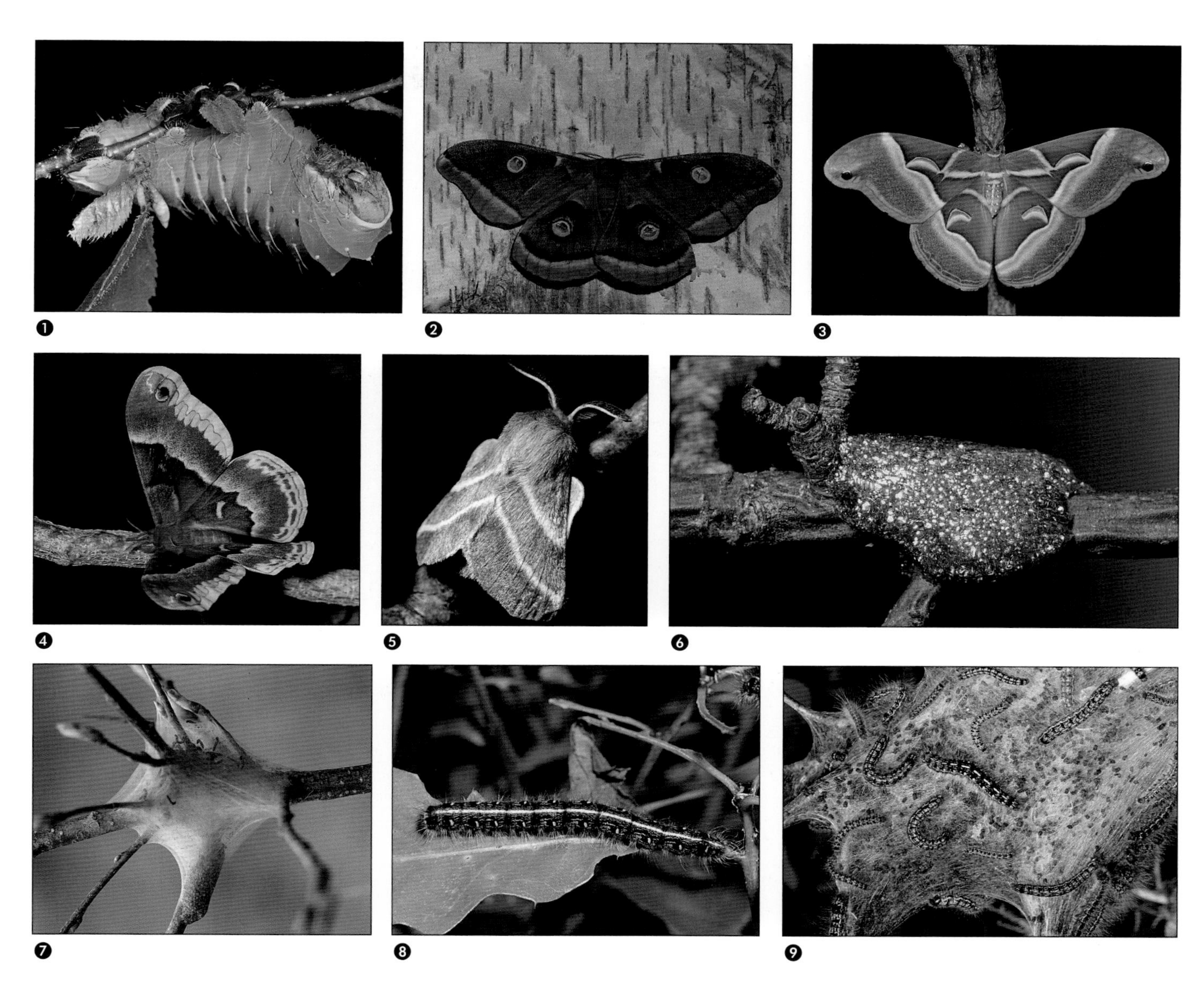

FAMILY SATURNIIDAE (continued) ❶ ❷ Caterpillars of the **Polyphemus Moth** (***Antheraea polyphemus***) feed on a variety of trees and shrubs. Adults are common at porch lights in early summer, often along with Luna (*Actias luna*) and Cecropia (*Hyalophora cecropia*) moths. ❸ The **Ailanthus Silkmoth** (***Samia cynthia***) was deliberately introduced to the eastern United States from China over a hundred years ago in the hopes of using it as a source of silk. It is now widespread and feeds, as the name suggests, on Tree-of-Heaven (*Ailanthus altissima*). ❹ The **Promethea** or **Spicebush Silkmoth** (***Callosamia promethea***) occurs throughout most of the eastern United States, reaching southern Canada, although it is not common in the northern part of its range. Caterpillars are found on a wide range of hardwood trees. This is a female; males are much darker and resemble the unpalatable Pipevine Swallowtail (*Papilio philenor*). Males fly in search of pheromone-steeped females during the day, when they are apparently avoided by predators because of their similarity to a toxic butterfly. **FAMILY LASIOCAMPIDAE** (TENT CATERPILLARS AND LAPPET MOTHS) ❺ ❻ ❼ ❽ **Eastern Tent Caterpillars** (***Malacosoma americanum***) are common pests on a variety of trees and shrubs. Eggs are laid in distinctive foamy rings around twigs, and larvae emerge from these rings as soon as foliage is available in spring. The distinctive tentlike larval retreats and the single-striped caterpillars are common sights in both urban and rural settings. ❾ The **Northern Tent Caterpillar** (***Malacosoma californicum pluviale***) makes tents much like the Eastern Tent Caterpillar (*M. americanum*), but as the name suggests, this species has a more northern distribution, including northern Ontario and Quebec. Adults are almost identical to the Eastern Tent Caterpillar but the larvae are distinctively colored.

FAMILY LASIOCAMPIDAE (continued) ❶ ❷ Forest Tent Caterpillars (*Malacosoma disstria*) are significant defoliators of maples, aspen and many other trees, and can be overwhelmingly abundant during outbreak years. The larvae, distinctively colored with keyhole-shaped markings, do not make tents like the related Eastern Tent Caterpillar (*M. americanum*), but they are gregarious and they do make silk trails. **❸ ❹** Males of the **Larch Lappet Moth (*Tolype laricis*)** have almost entirely gray wings, in contrast to the gray and white wings and thorax of the larger female. Despite the common name, this species develops on spruce, balsam fir and other conifers as well as larch. **❺ ❻ ❼ ❽** When alarmed, **Lappet Moth (*Phyllodesma americanum*)** caterpillars expose bright orange strips on top of the second and third segments of the thorax. The "lappets" are the fleshy lobes along the sides of the caterpillar's body. Adult moths are distinctive at rest because of the way the side of the hind wing protrudes beyond the front wing. These common insects occur on poplar, aspen and other trees. **FAMILY SPHINGIDAE (SPHINX MOTHS OR HORNWORMS) Subfamily Sphinginae ❾ ❿** Each hind wing of the **Twin-spotted Sphinx (*Smerinthus jamaicensis*)** has a blue "eye" with a central black band dividing it into two blue spots. Larvae are found on several kinds of broadleaf trees.

FAMILY SPHINGIDAE (continued) ❶ ❷ The **One-eyed Sphinx** (***Smerinthus cerisyi***) has a blue "eye" with a central black "pupil" on each hind wing. Larvae are found on pear, plum, poplar and willow trees. ❸ ❹ The **Blinded Sphinx** (***Paonias excaecatus***) has a single blue "eye" on each hind wing, but the eyes lack a black "pupil" like that found in the One-eyed Sphinx (*Smerinthus cerisyi*). Larvae are common on a wide variety of broadleaf trees. ❺ ❻ The **Small-eyed Sphinx** (***Paonias myops***) is smaller and much darker than the Blinded Sphinx (*P. excaecatus*). Larvae are common on birch, poplar, willow and other trees. ❼ ❽ The **Laurel Sphinx** (***Sphinx kalmiae***) feeds on poplar, laurel, ash, lilac and privet. ❾ Sphinx caterpillars, like this **Great Ash Sphinx** (***Sphinx chersis***) larva, often strike a characteristic pose when disturbed. The green bug in the background is *Acanalonia conica*.

FAMILY SPHINGIDAE (continued) ❶ ❷ The **Big Poplar Sphinx** (***Pachysphinx modesta***) is often one of the most common large moths in areas with extensive willow and poplar stands. This large (100–120 mm wingspan), robust moth is sometimes called the **Modest Sphinx**. ❸ ❹ The **Elm Sphinx** (***Ceratomia amyntor***) is sometimes called the **Four-horned Sphinx** because of its strikingly spinulose, four-horned caterpillar. This enormous caterpillar was feeding on a basswood tree, but Elm Sphinx caterpillars also eat elm, birch and cherry. ❺ ❻ Sphinx moths, like the **Waved Sphinx** (***Ceratomia undulosa***) on this tree trunk, are usually well camouflaged while at rest. This specimen was moved to a twig for a close-up photograph. ❼ Caterpillars of the **Walnut Sphinx** (***Laothoe juglandis***), which feed on other nut trees as well as walnut, can make a whistle-like hiss when disturbed.

FAMILY SPHINGIDAE (continued) Subfamily Macroglossinae ❶ **Abbot's Sphinx Moths** (***Sphecodina abbottii***) adults fly in early spring. ❷ ❸ ❹ Some of our most common macroglossine sphinx moths have highly distinctive caterpillars that feed on both Virginia Creeper (*Parthenocissus quinquefolia*) and grape. This bright green and brown **Abbot's Sphinx** (***Sphecodina abbottii***), this brightly yellow-spotted **Pandorus Sphinx** (***Eumorpha pandorus***) and this **Hog Sphinx** (***Darapsa myron***) were all photographed on grape vines. The Abbot's and Pandorus caterpillars have their horns reduced and replaced with strikingly eyelike bulges that would surely startle potential predators. ❺ The **Leafy Spurge Hawkmoth** (***Hyles euphorbiae***) was deliberately introduced to North America for the biological control of Leafy Spurge (*Euphorbia esula*). ❻ The **White-lined Sphinx** (***Hyles lineata***) is found throughout temperate North America, although it is irregularly encountered in the Canadian part of its range. Caterpillars feed on several plants, including fireweed, evening primrose and apple trees. ❼ This ***Darapsa versicolor*** was one of thousands of insects on a white sheet hung behind a mercury vapor lamp during a southern Ontario summer. Larvae of this uncommon sphinx occur on Buttonbush (*Cephalanthus occidentalis*). ❽ ❾ This **Hummingbird Clearwing** (***Hemaris thysbe***) is nectaring at Crown Vetch (*Coronilla varia*). Its rapid flight and habit of hovering over flowers gives this day-active sphinx a remarkable similarity to a hummingbird. The larva feeds on honeysuckle, rose, hawthorn and related plants.

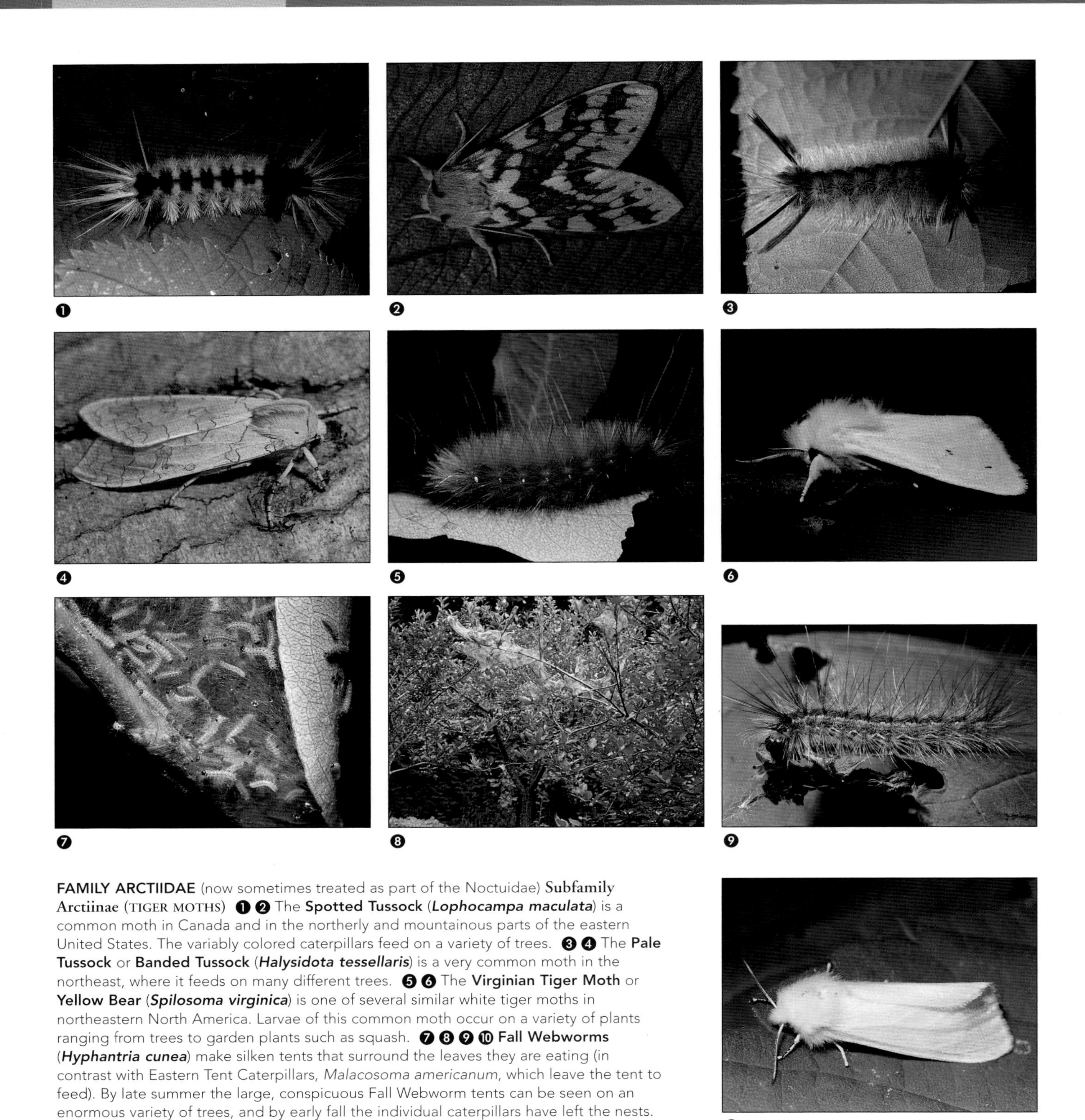

FAMILY ARCTIIDAE (now sometimes treated as part of the Noctuidae) **Subfamily Arctiinae** (TIGER MOTHS) ❶ ❷ The **Spotted Tussock** (***Lophocampa maculata***) is a common moth in Canada and in the northerly and mountainous parts of the eastern United States. The variably colored caterpillars feed on a variety of trees. ❸ ❹ The **Pale Tussock** or **Banded Tussock** (***Halysidota tessellaris***) is a very common moth in the northeast, where it feeds on many different trees. ❺ ❻ The **Virginian Tiger Moth** or **Yellow Bear** (***Spilosoma virginica***) is one of several similar white tiger moths in northeastern North America. Larvae of this common moth occur on a variety of plants ranging from trees to garden plants such as squash. ❼ ❽ ❾ ❿ **Fall Webworms** (***Hyphantria cunea***) make silken tents that surround the leaves they are eating (in contrast with Eastern Tent Caterpillars, *Malacosoma americanum*, which leave the tent to feed). By late summer the large, conspicuous Fall Webworm tents can be seen on an enormous variety of trees, and by early fall the individual caterpillars have left the nests.

FAMILY ARCTIIDAE (continued) ❶ ❷ The **Woolly Bear** or **Isabella Tiger Moth** (***Pyrrharctia isabella***) is familiar to most because of the wandering habits of the conspicuous caterpillars as they seek sheltered sites to spend the winter. ❸ Several tiger moths, including the **Delicate Cycnia** (***Cycnia tenera***) are able to develop on milkweed plants. ❹ ❺ ❻ **Milkweed Tussocks** (***Euchaetes egle***) are the most conspicuous of the milkweed-eating tiger moths. These gregarious caterpillars feed in dense masses, often entirely defoliating milkweed plants. ❼ ❽ The **Cinnabar Moth** (***Tyria jacobaeae***) was introduced from Europe to control the Tansy Ragwort (*Senecio jacobaea*), a weed poisonous to livestock. The bright striped coloration of these Cinnabar Moth caterpillars is clearly a warning to potential predators that the caterpillars, like their host plant, are poisonous. ❾ **Bella Moths** (***Utetheisa bella***) are conspicuous day-fliers that often abound among the legumes on which their larvae usually feed. The bright colors advertise a body full of toxic alkaloids that render adults, larvae and eggs of these moths unpalatable to most predators.

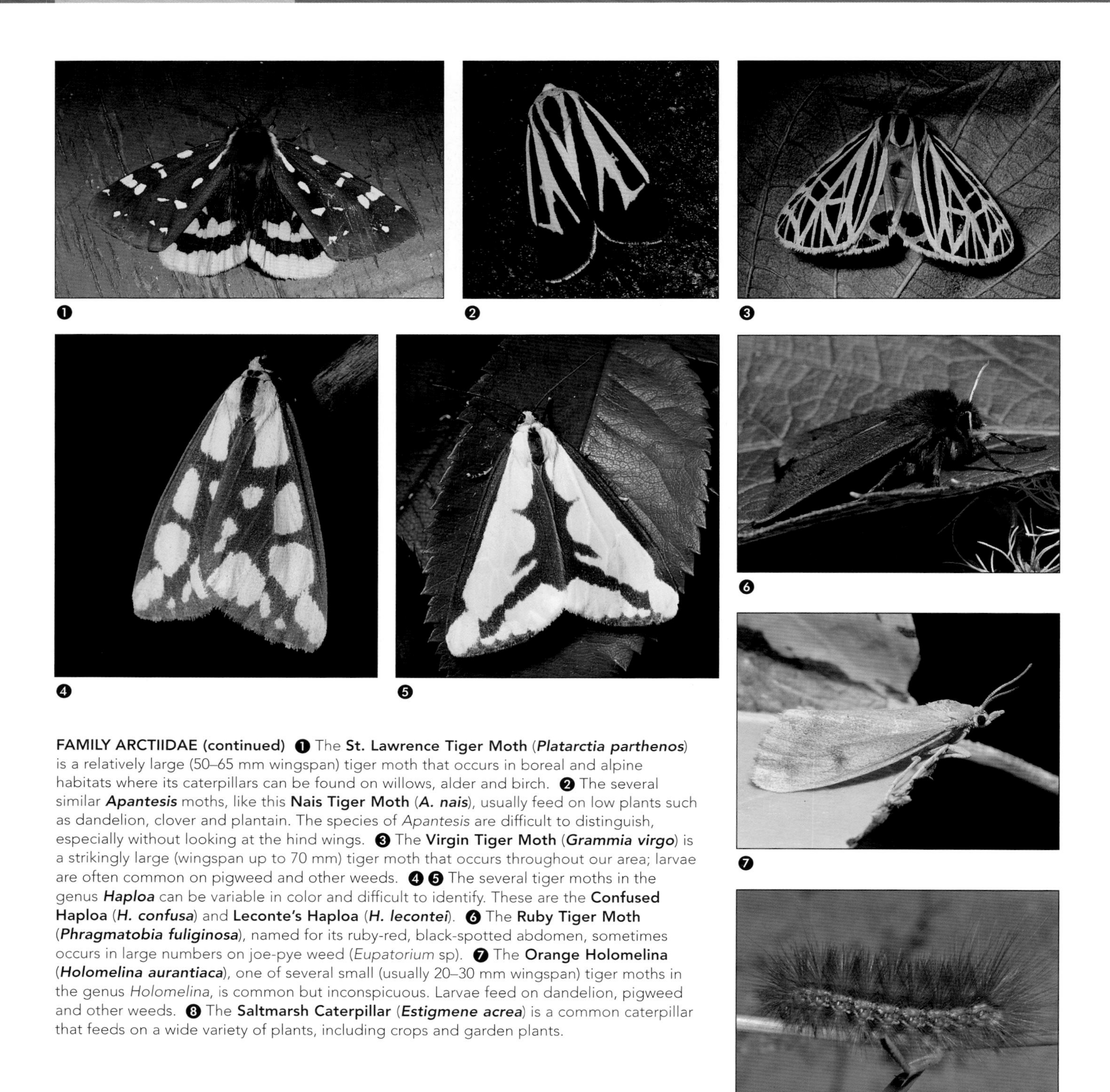

FAMILY ARCTIIDAE (continued) ❶ The **St. Lawrence Tiger Moth** (***Platarctia parthenos***) is a relatively large (50–65 mm wingspan) tiger moth that occurs in boreal and alpine habitats where its caterpillars can be found on willows, alder and birch. ❷ The several similar ***Apantesis*** moths, like this **Nais Tiger Moth** (***A. nais***), usually feed on low plants such as dandelion, clover and plantain. The species of *Apantesis* are difficult to distinguish, especially without looking at the hind wings. ❸ The **Virgin Tiger Moth** (***Grammia virgo***) is a strikingly large (wingspan up to 70 mm) tiger moth that occurs throughout our area; larvae are often common on pigweed and other weeds. ❹ ❺ The several tiger moths in the genus ***Haploa*** can be variable in color and difficult to identify. These are the **Confused Haploa** (***H. confusa***) and **Leconte's Haploa** (***H. lecontei***). ❻ The **Ruby Tiger Moth** (***Phragmatobia fuliginosa***), named for its ruby-red, black-spotted abdomen, sometimes occurs in large numbers on joe-pye weed (*Eupatorium* sp). ❼ The **Orange Holomelina** (***Holomelina aurantiaca***), one of several small (usually 20–30 mm wingspan) tiger moths in the genus *Holomelina*, is common but inconspicuous. Larvae feed on dandelion, pigweed and other weeds. ❽ The **Saltmarsh Caterpillar** (***Estigmene acrea***) is a common caterpillar that feeds on a wide variety of plants, including crops and garden plants.

FAMILY ARCTIIDAE (continued) Subfamily Lithosiinae (LICHEN MOTHS AND FOOTMAN MOTHS) ❶ ❷ ❸ **Scarlet-winged Lichen Moth** caterpillars (***Hypoprepia miniata***) often live on open rocky cliffs and alvars, where they feed on lichen. This striking species is also known as the **Striped Footman Moth**. One of these photos shows a pupa surrounded by a loose silken cocoon and a caterpillar just starting a cocoon; both were exposed by lifting a rock. ❹ The **Painted Lichen Moth** (***Hypoprepia fucosa***) is not as brightly colored as the very closely related Scarlet-winged Lichen Moth (*H. miniata*). ❺ The **Black and Yellow Lichen Moth** (***Lycomorpha pholus***) lays eggs on open rock faces, where its caterpillars feed on lichen. ❻ Larvae of the **Bicolored Moth** (***Eilema bicolor***) eat lichens on spruce, fir and other conifers. Subfamily Ctenuchinae (WASP MOTHS) ❼ ❽ ❾ Our largest and most common wasp moth is the **Virginia Ctenucha** (***Ctenucha virginica***), a brilliantly metallic day-flying moth. Larvae are abundant on grasses, but also feed on irises and sedges. The dark, early instar larvae are abundant in early spring; later stage larvae look quite different.

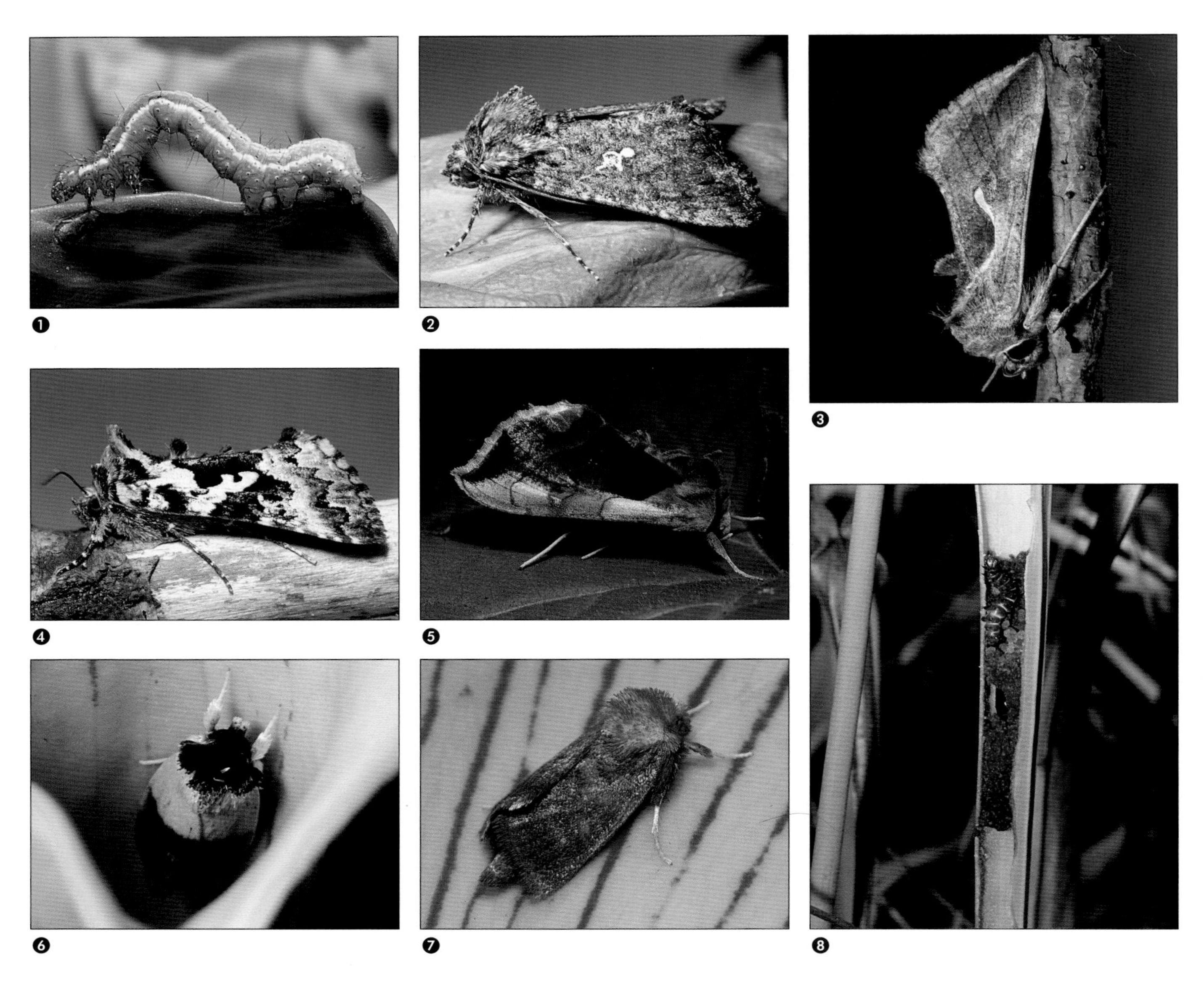

FAMILY NOCTUIDAE (OWLET MOTHS OR CUTWORM MOTHS) **Subfamily Plusiinae** (LOOPERS OR SEMILOOPERS) ❶ ❷ Caterpillars that are missing the middle prolegs, and therefore inch along with a looping motion, are found both in the Geometridae and in some owlet moths. Those in the Noctuidae are sometimes called "semiloopers," and differ from inchworms (Geometridae) in having three equally developed prolegs at the hind end (in the Geometridae only two pairs are fully developed). **Cabbage Looper** or **Cabbage Semilooper** (***Trichoplusia ni***) caterpillars attack a variety of crops, and are serious pests of cabbage and related crucifers. ❸ The **Celery Looper** or **Celery Semilooper** (***Anagrapha falcifera***) is a pest that develops on beets, celery, lettuce and many other low plants. ❹ The **Salt and Pepper Looper** (***Syngrapha rectangula***) is found on spruce and other conifers. ❺ The brilliant metallic highlights of ***Diachrysia balluca*** make it one of our most striking owlet moths. Larvae occur on Trembling Aspen (*Populus tremuloides*) and other plants. ❻ ❼ ❽ Pitcher plant moths (three species of ***Exyra***) live only in insectivorous pitcher plants (*Sarracenia* spp.), spending almost their entire lives in the vessel-like leaves that serve as death traps for most other insects. Adults, like this ***E. semicrocea*** inside a Northern Pitcher Plant (*S. purpurea*) leaf, and the reddish ***E. fax*** photographed against the inner surface of a Yellow Pitcher Plant (*S. flava*), have a unique ability to walk on the inner wall of the leaf, although they always maintain a "heads up" position. The caterpillars (surrounded by frass in this cut-open Yellow Pitcher Plant) feed on the inner surface of the pitcher plant leaf. Larval feeding causes the top of the leaf to wilt. The round hole just below where the caterpillar has been feeding is a drainage hole, covered with a silken screening, and may serve as an exit for the newly emerged adult moth.

FAMILY NOCTUIDAE (continued) Subfamily Acronictinae (DAGGER MOTHS AND OTHERS) ❶ ❷ ❸ Like many caterpillars, larvae of the **Cherry Dagger Moth** (***Acronicta hasta***) change in appearance as they age. Young larvae are bright green with red markings, and older caterpillars are blue-black with red markings. ❹ ❺ **American Dagger Moth** (***Acronicta americana***) caterpillars are common on maple and a variety of other trees including alders and poplars. The adult moth, with a wingspan of up to 65 mm, is our largest dagger moth. Dagger moths (the genus *Acronicta*, with 75 North American species) are usually rather dull-colored moths with characteristic black marks, or "daggers," on the front wings. ❻ ❼ ❽ ❾ Many dagger moth (***Acronicta***) species are more distinctive as caterpillars than as moths. The caterpillars shown here are the **Alder Dagger** (***A. dactylina***), the **Smeared Dagger** (***A. oblinita***), the **Yellow-haired Dagger** (***A. impleta***) and the **Impressed Dagger** (***A. impressa***), respectively. ❿ The **Hebrew** (***Polygrammate hebraeicum***) is a common moth where its host (Black Gum, *Nyssa sylvatica*) occurs.

FAMILY NOCTUIDAE (continued) ❶ ❷ **Harris's Three-spot** (***Harrisimemna trisignata***) caterpillars look a bit like large, animated bird droppings. The black baubles on the hairs projecting forward over the head are head capsules from the cast skins of previous larval stages. Look for these bizarre caterpillars and the distinctively spotted adult moths on willows, apple and other trees and shrubs. **Subfamily Acontiinae** ❸ This bird dropping moth (***Tarachidia erastrioides***) develops on Common Ragweed (*Ambrosia artemisiifolia*). **Subfamily Catocalinae** (UNDERWINGS AND OTHERS; part of a group of subfamilies now sometimes treated as a distinct family, the Erebidae, because they are more closely related to the Arctiidae than to other Noctuidae.) ❹ ❺ Moths in the large genus ***Catocala*** (over 100 North American species) are called underwings because of the contrast between the bright hind wings and the dull front wings. Underwings, like this **Ilia Underwing** or **Beloved Underwing** (***C. ilia***), are well camouflaged with the wings closed, but present a startlingly bright appearance when the hind wings are exposed. ❻ This **Sleepy Underwing** or **Pink Underwing** (***Catocala concumbens***) is conspicuous with its pink hind wings exposed, but effectively "disappears" when it closes its wings. This species develops on willow and poplar. ❼ ❽ The **White Underwing** (***Catocala relicta***) is a common moth across Canada and in the northern United States, where the caterpillars usually feed on Trembling Aspen (*Populus tremuloides*). These large (wingspan 70–80 mm) moths are often attracted to lights on cold fall nights, long after most other moths have stopped flying.

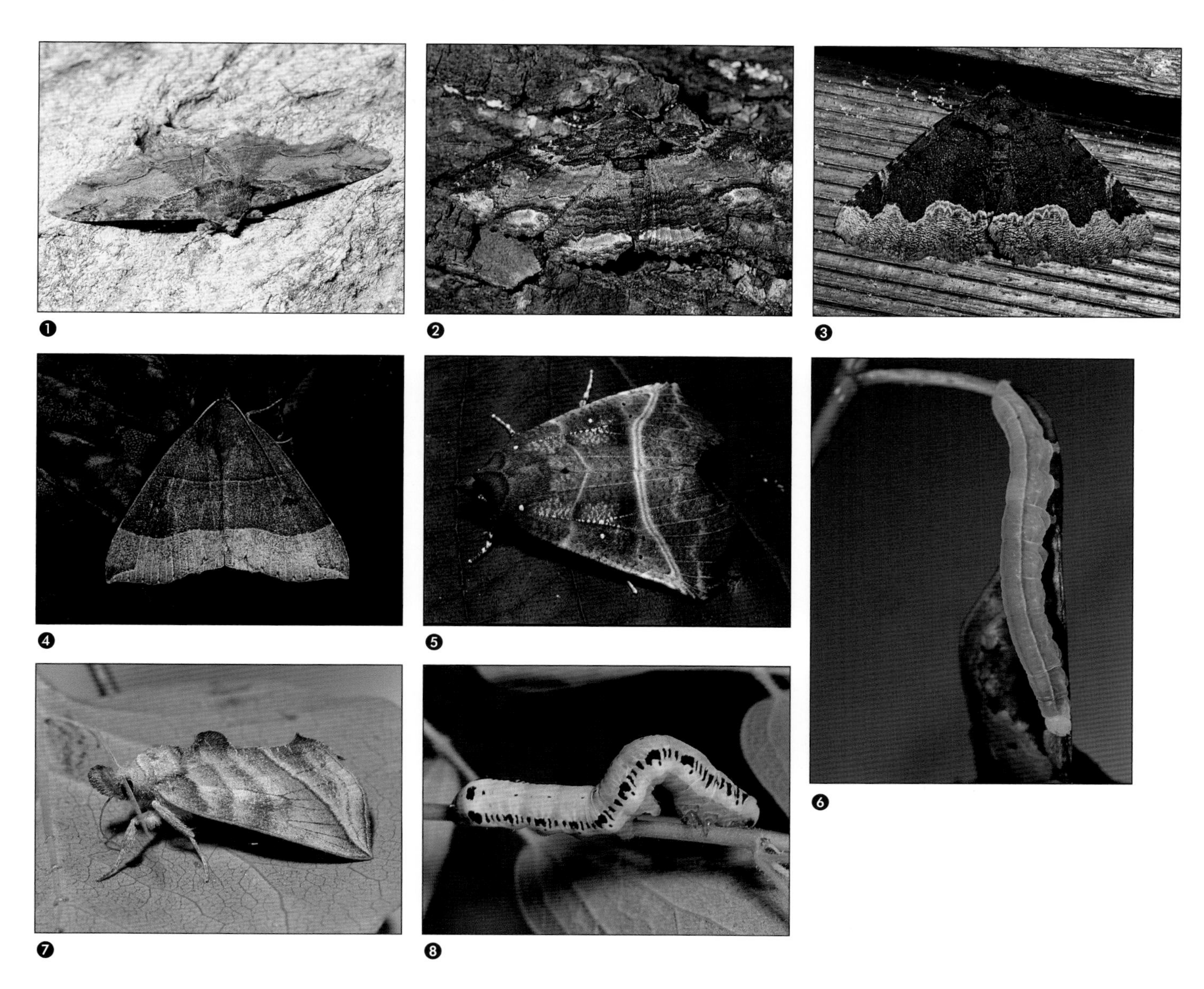

FAMILY NOCTUIDAE (continued) ❶ ❷ Adult moths in the common genus ***Zale*** (20 eastern species) often rest on tree trunks during the day. This pale brown **Colorful Zale** (***Z. minerea***) is easily seen against the contrasting background of a poplar tree, but this darker **Lunate Zale** (***Z. lunata***) is well camouflaged on a lichen-flecked maple trunk. ❸ The **Horrid Zale** (***Zale horrida***) is an easily recognized species in an otherwise difficult genus. Caterpillars develop on Nannyberry (*Viburnum lentago*). ❹ Caterpillars of the **Maple Looper Moth** (***Parallelia bistriaris***) move with a looping motion because the first two pairs of prolegs are reduced, much like the "semiloopers" in the subfamily Plusiinae. ❺ ❻ The **Herald** (***Scoliopteryx libatrix***) is a striking moth that spends the winter in sheltered places like caves, often in impressive numbers. The relatively plain caterpillars feed on poplar and willow. The Herald has a Holarctic distribution (it occurs in Europe and Asia as well as in North America). ❼ ❽ Larvae of the **Canadian Owlet** (***Calyptra canadensis***) feed on Tall Meadowrue (*Thalictrum dasycarpum*).

FAMILY NOCTUIDAE (continued) Subfamily Amphipyrinae ❶ ❷ This **Copper Underwing** (***Amphipyra pyramidoides***) caterpillar has just molted, and is still standing on its shed skin. Peeling bark off a dead tree exposed these adults, with their forewings concealing the copper-colored hind wings. Subfamily Agaristinae (FORESTER MOTHS) ❸ ❹ The **Eight-spotted Forester** (***Alypia octomaculata***) is a familiar day-flying moth. The bright, long-haired caterpillars are common on grape and Virginia Creeper (*Parthenocissus quinquefolia*). ❺ ❻ The **Pearly Wood Nymph** (***Eudryas unio***) and the similar **Beautiful Wood Nymph** (***E. grata***) both feed on grape leaves and related plants. Subfamily Euteliinae ❼ The **Beautiful Eutelia** (***Eutelia pulcherrima***) can be recognized by its uniquely hooked forewing. Caterpillars feed on Poison Sumac (*Rhus vernix*). ❽ The **Dark Marathyssa** (***Marathyssa inficita***) looks remarkably like a branched twig at rest. Larvae feed on Staghorn Sumac (*Rhus typhina*).

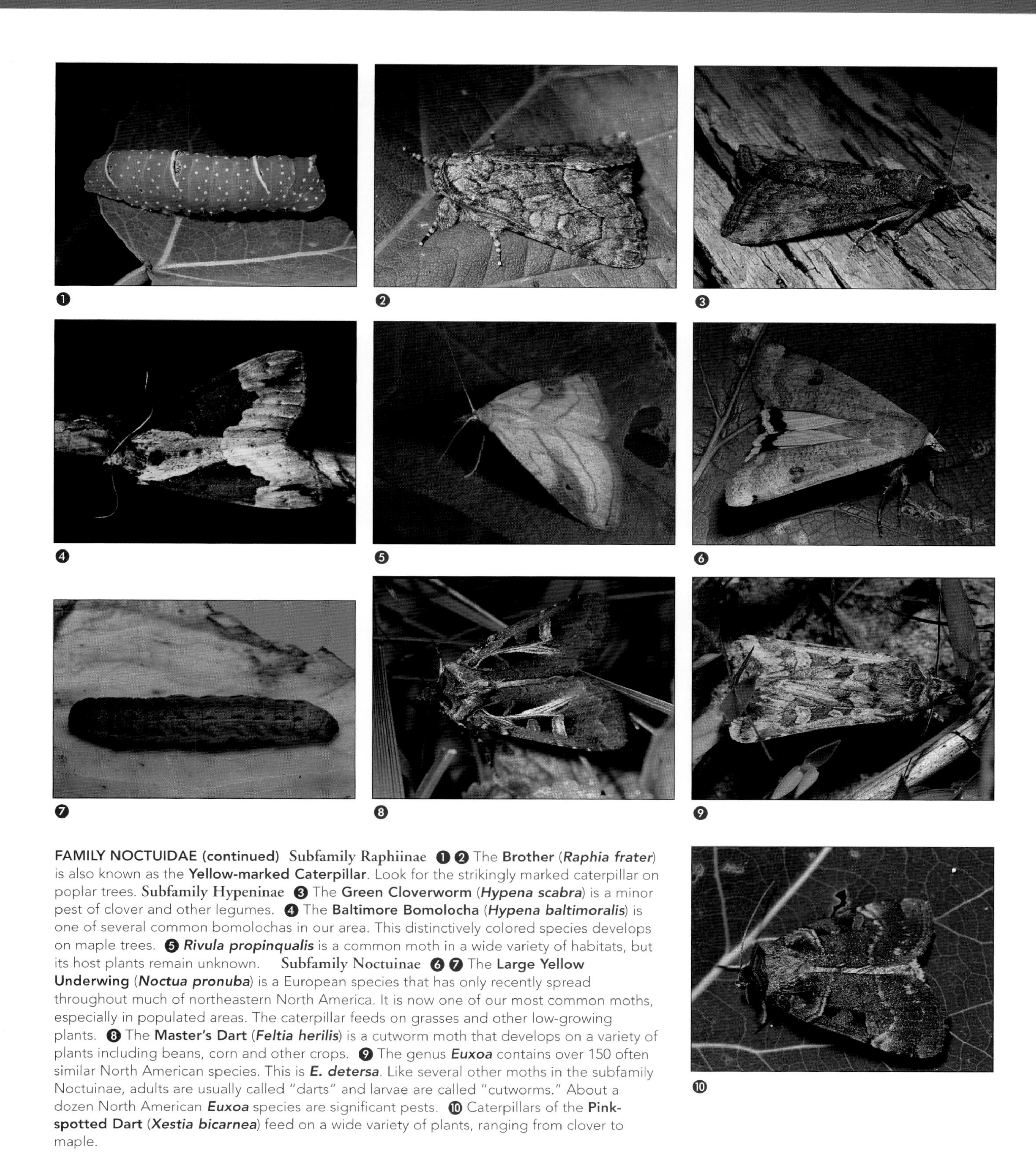

FAMILY NOCTUIDAE (continued) Subfamily Raphiinae ❶ ❷ The **Brother** (***Raphia frater***) is also known as the **Yellow-marked Caterpillar**. Look for the strikingly marked caterpillar on poplar trees. Subfamily Hypeninae ❸ The **Green Cloverworm** (***Hypena scabra***) is a minor pest of clover and other legumes. ❹ The **Baltimore Bomolocha** (***Hypena baltimoralis***) is one of several common bomolochas in our area. This distinctively colored species develops on maple trees. ❺ ***Rivula propinqualis*** is a common moth in a wide variety of habitats, but its host plants remain unknown. Subfamily Noctuinae ❻ ❼ The **Large Yellow Underwing** (***Noctua pronuba***) is a European species that has only recently spread throughout much of northeastern North America. It is now one of our most common moths, especially in populated areas. The caterpillar feeds on grasses and other low-growing plants. ❽ The **Master's Dart** (***Feltia herilis***) is a cutworm moth that develops on a variety of plants including beans, corn and other crops. ❾ The genus ***Euxoa*** contains over 150 often similar North American species. This is ***E. detersa***. Like several other moths in the subfamily Noctuinae, adults are usually called "darts" and larvae are called "cutworms." About a dozen North American ***Euxoa*** species are significant pests. ❿ Caterpillars of the **Pink-spotted Dart** (***Xestia bicarnea***) feed on a wide variety of plants, ranging from clover to maple.

FAMILY NOCTUIDAE (continued) ❶ The **Black Cutworm** (***Agrotis ipsilon***) is a virtually worldwide pest that attacks a variety of crops. Young larvae feed on leaves, and older larvae feed on stems and roots. Like other cutworms, Black Cutworms do most of their aboveground feeding at night, often cutting off seedlings at the base. Several other *Agrotis* species are significant cutworm pests. Subfamily Heliothinae ❷ **Corn Earworm Moths** (***Helicoverpa zea***) are often seen at flowers during the day. Also known as the American Bollworm, Cotton Bollworm and the Tomato Fruitworm, the caterpillars of this species bore into a variety of fruits. Although the Corn Earworm cannot survive the winter in Canada and the adjacent states, it migrates north each year and establishes itself as a serious pest of corn throughout our area (including Canada). The fat caterpillars are often encountered in corncobs, under the husk and at the outer end. ❸ **Primrose Moth** (***Schinia florida***) adults can usually be found in the flowers of their host plant, Common Evening Primrose (*Oenothera biennis*). Subfamily Cuculliinae ❹ ❺ ❻ Several brightly colored ***Cucullia*** species occur on goldenrod, asters and other meadow plants. The white dots on the darker of these two *Cucullia* caterpillars (the **Brown-bordered Cucullia**, ***C. convexipennis***) are its marks of doom, since each will hatch into a parasitic fly (Tachinidae) larva that will burrow into the caterpillar and ultimately kill it. Caterpillars of the **Asteroid** (***C. asteroides***) are distinctly striped in yellow and green, but the adults are very similar to the Brown-bordered Cucullia. Subfamily Psaphidinae ❼ The lichen-colored **Comstock's Sallow** (***Feralia comstocki***) is among our most elegant moths. Caterpillars feed on conifers. Subfamily Hadeninae ❽ Caterpillars of the **Hemina Pinion** (***Lithophane hemina***) occur on birch.

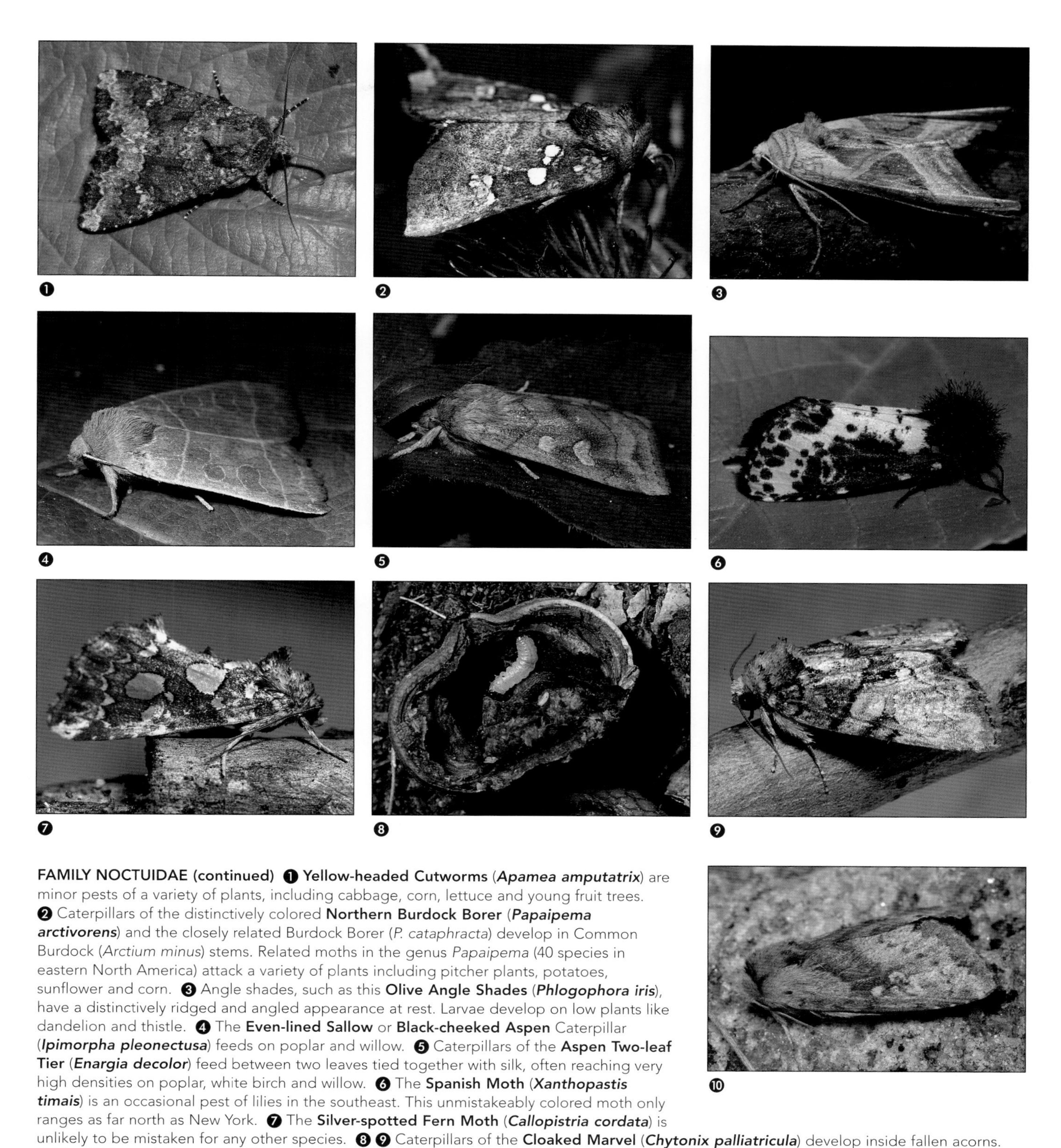

FAMILY NOCTUIDAE (continued) ❶ **Yellow-headed Cutworms** (***Apamea amputatrix***) are minor pests of a variety of plants, including cabbage, corn, lettuce and young fruit trees. ❷ Caterpillars of the distinctively colored **Northern Burdock Borer** (***Papaipema arctivorens***) and the closely related Burdock Borer (*P. cataphracta*) develop in Common Burdock (*Arctium minus*) stems. Related moths in the genus *Papaipema* (40 species in eastern North America) attack a variety of plants including pitcher plants, potatoes, sunflower and corn. ❸ Angle shades, such as this **Olive Angle Shades** (***Phlogophora iris***), have a distinctively ridged and angled appearance at rest. Larvae develop on low plants like dandelion and thistle. ❹ The **Even-lined Sallow** or **Black-cheeked Aspen** Caterpillar (***Ipimorpha pleonectusa***) feeds on poplar and willow. ❺ Caterpillars of the **Aspen Two-leaf Tier** (***Enargia decolor***) feed between two leaves tied together with silk, often reaching very high densities on poplar, white birch and willow. ❻ The **Spanish Moth** (***Xanthopastis timais***) is an occasional pest of lilies in the southeast. This unmistakeably colored moth only ranges as far north as New York. ❼ The **Silver-spotted Fern Moth** (***Callopistria cordata***) is unlikely to be mistaken for any other species. ❽ ❾ Caterpillars of the **Cloaked Marvel** (***Chytonix palliatricula***) develop inside fallen acorns. ❿ The **White-tailed Diver** (***Bellura gortynoides***) develops in the leaves (young larvae) and stems (older larvae) of aquatic plants including cattails and water lilies. Larvae have the unusual habit of diving into the water and swimming to shore before pupating.

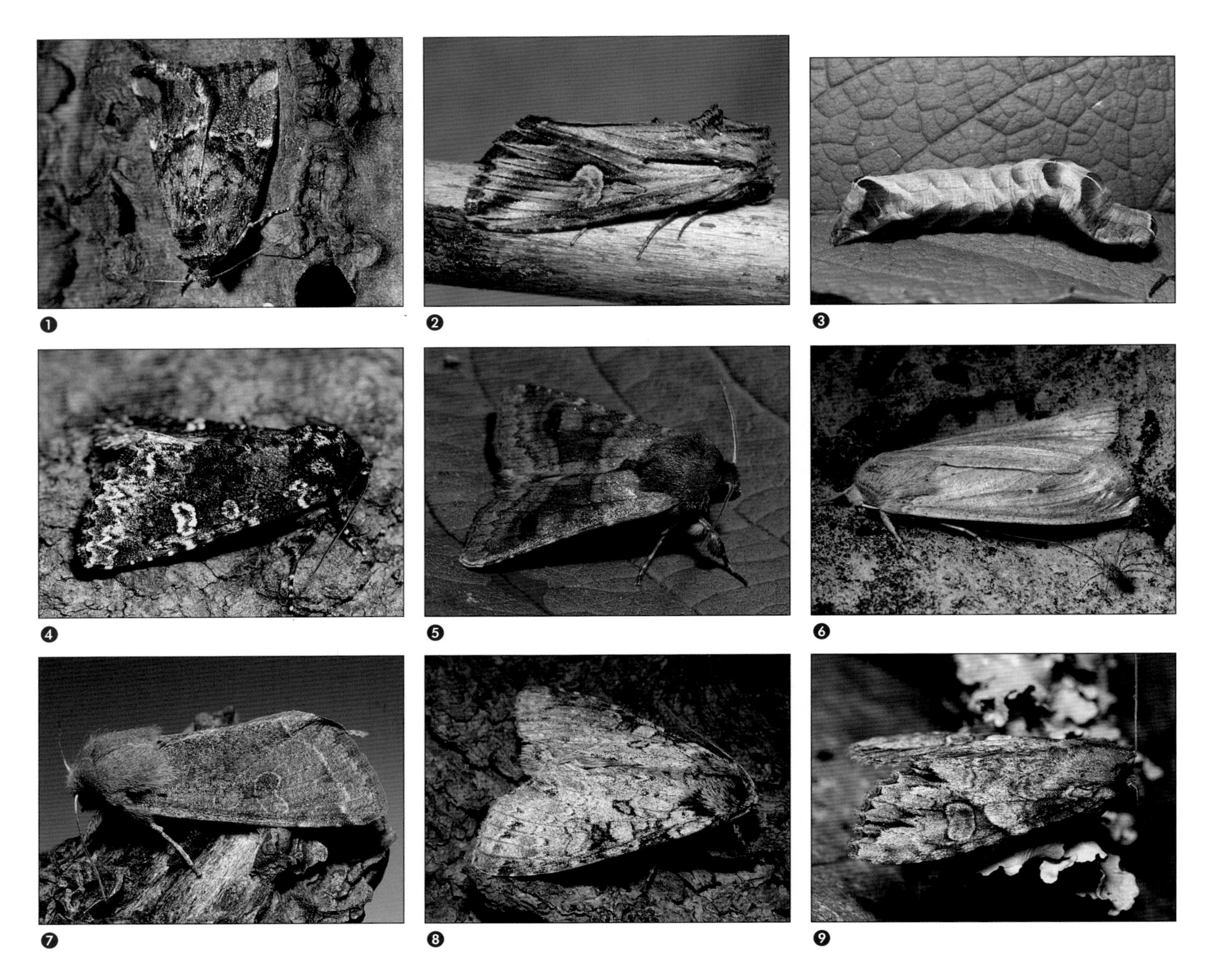

FAMILY NOCTUIDAE (continued) ❶ The **Variegated Midget** (***Elaphria versicolor***) develops on conifer foliage. ❷ The **Grey Half-spot** (***Nedra ramosula***) is one of the few insects that develop on the acrid foliage of Saint-John's-wort (*Hypericum perforatum*) ❸ ❹ **Hitched Arches** (***Melanchra adjuncta***) caterpillars feed on a wide variety of plants ranging from dandelions to elm. ❺ The **Bronzed Cutworm** (***Nephelodes minians***) is common throughout our area, where the caterpillar is a cutworm pest of corn and grasses. ❻ The **Armyworm** (***Mythimna unipuncta***) is an infamous pest also known as the Cereal Armyworm, White-speck and American Wainscot. The caterpillars sometimes build up enormous populations that decimate their food plants (often cereal crops). They then march by the thousands to greener pastures. Armyworms are major pests throughout our area, although they do not survive the winter as far north as Canada, and migrate in from the south each spring. ❼ The **Speckled Green Fruitworm Moth** (***Orthosia hibisci***) develops on a wide variety of fruit trees including apple, cherries and plums. ❽ ❾ Although most insects are known only by their scientific names, moths usually have one or more widely used common names. The common names of these two species – the **Stormy Arches** (***Polia nimbosa***) and the **Fluid Arches** (***Morrisonia latex***) – seem to have been concocted by someone particularly imaginative.

FAMILY NOCTUIDAE (continued) ❶ This **Thinker** (***Lacinipolia meditata***) was photographed during the day on a milkweed plant. Thinker caterpillars occur on a variety of plants ranging from apple to clover. **Subfamily Eustrotiinae** ❷ The **Red-spotted Lithacodia** (***Maliattha concinnimacula***) is one of our most distinctively colored Noctuidae, but the larval food remains unknown. ❸ Adult **Tufted Bird-dropping Moths** (***Cerma cerintha***) look unappetizingly like bird feces. Larvae feed on apple, plum, cherry and other Rosaceae. **Subfamily Pantheinae** ❹ ❺ The **Eastern Panthea** (***Panthea pallescens***) or **Tufted White Pine Caterpillar**, shown here feeding on white pine, is very similar to the Tufted Spruce Caterpillar (*P. acronyctoides*). ❻ The **Laugher** (***Charadra deridens***), a common moth on a variety of hardwoods, seems to blend into tree trunks leaving only the subtle eyespots visible. **FAMILY NOLIDAE (NOLID MOTHS)** (sometimes treated as part of the Noctuidae) ❼ ***Nola cilicoides*** develops on Fringed Yellow Loosestrife (*Lysimachia cilata*). **FAMILY NOTODONTIDAE (PROMINENTS)** ❽ ❾ Notodontid caterpillars, like this larval **Fanned Willow Prominent** (***Notodonta scitipennis***), are distinctive insects with prominent body tubercles. The corresponding adult moths often have prominent tufts along the inner edge of the front wings; you can see them projecting above the wings in this adult Fanned Willow Prominent.

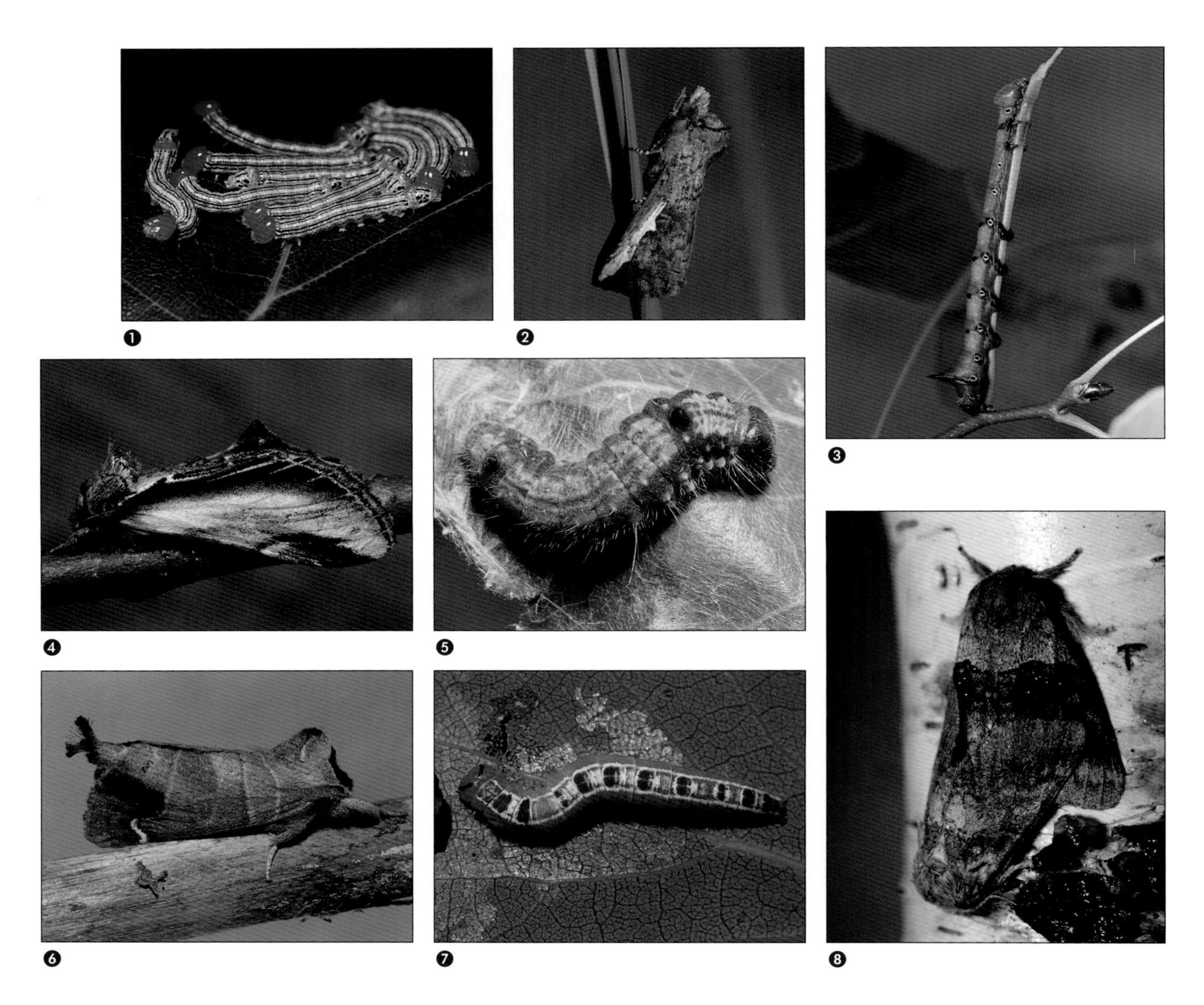

FAMILY NOTODONTIDAE (continued) ❶ ❷ Caterpillars of the **Red-humped Oakworm** (***Symmerista canicosta***) feed gregariously on forest hardwood trees when young, then become solitary in later instars. Adult *Symmerista* species, which are difficult to identify, often rest in characteristic poses that help them blend in with tree twigs. ❸ ❹ Caterpillars of the **Black-rimmed Prominent** (***Pheosia rimosa***) are often called **False Hornworms** because of the sphinx-like "horntail." Look for these attractive caterpillars on willows and poplars. ❺ ❻ The **Sigmoid Prominent** (***Clostera albosigma***) gets both its scientific and common names from the S-shaped white mark on the adult wing. This larva is feeding on a poplar leaf, still partly webbed up into the leaf shelter opened to expose it. ❼ ❽ The brightly marked caterpillars of the **Common Gluphisia** (***Gluphisia septentrionis***) feed on poplars, although these adults are mating on a birch trunk.

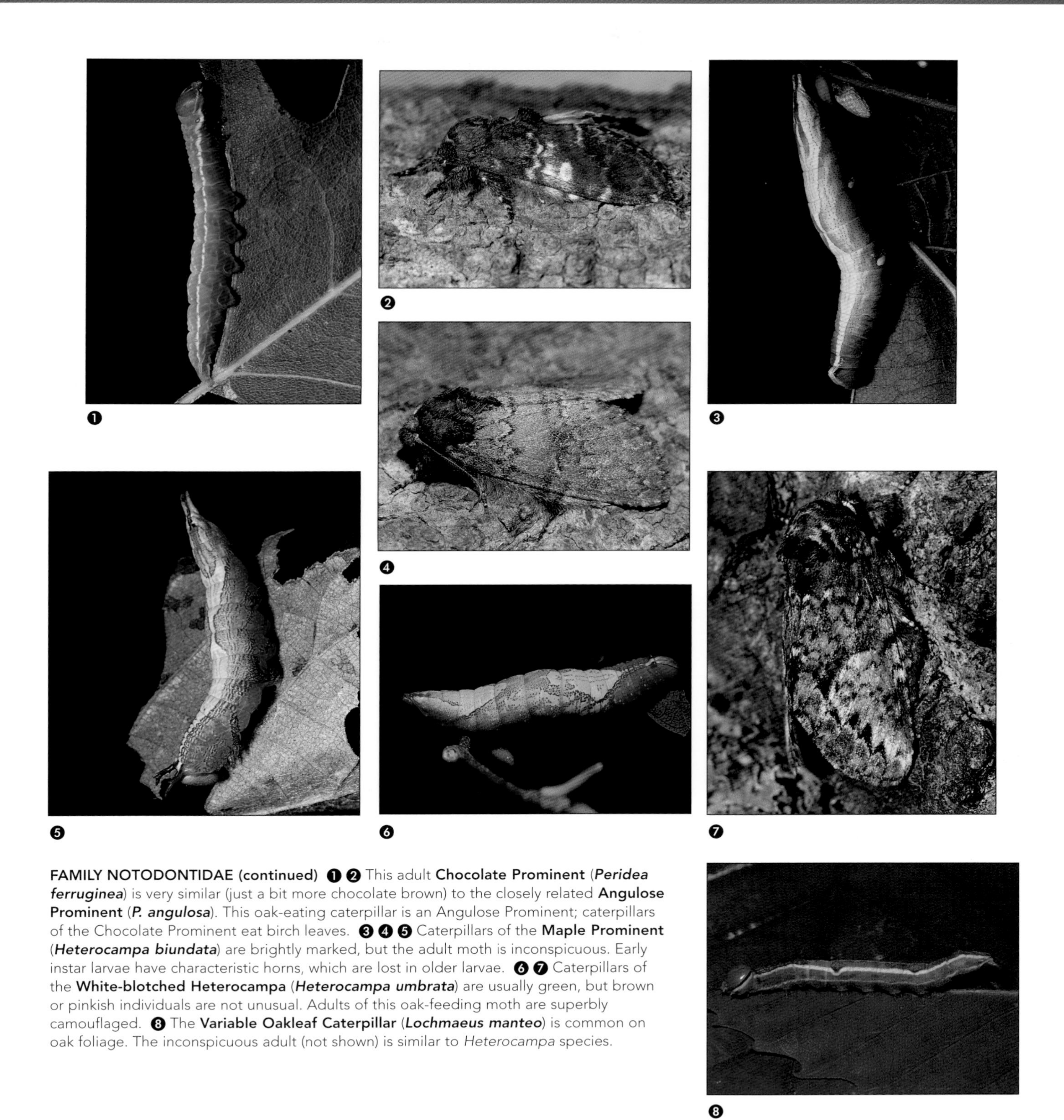

FAMILY NOTODONTIDAE (continued) ❶ ❷ This adult **Chocolate Prominent** (***Peridea ferruginea***) is very similar (just a bit more chocolate brown) to the closely related **Angulose Prominent** (***P. angulosa***). This oak-eating caterpillar is an Angulose Prominent; caterpillars of the Chocolate Prominent eat birch leaves. ❸ ❹ ❺ Caterpillars of the **Maple Prominent** (***Heterocampa biundata***) are brightly marked, but the adult moth is inconspicuous. Early instar larvae have characteristic horns, which are lost in older larvae. ❻ ❼ Caterpillars of the **White-blotched Heterocampa** (***Heterocampa umbrata***) are usually green, but brown or pinkish individuals are not unusual. Adults of this oak-feeding moth are superbly camouflaged. ❽ The **Variable Oakleaf Caterpillar** (***Lochmaeus manteo***) is common on oak foliage. The inconspicuous adult (not shown) is similar to *Heterocampa* species.

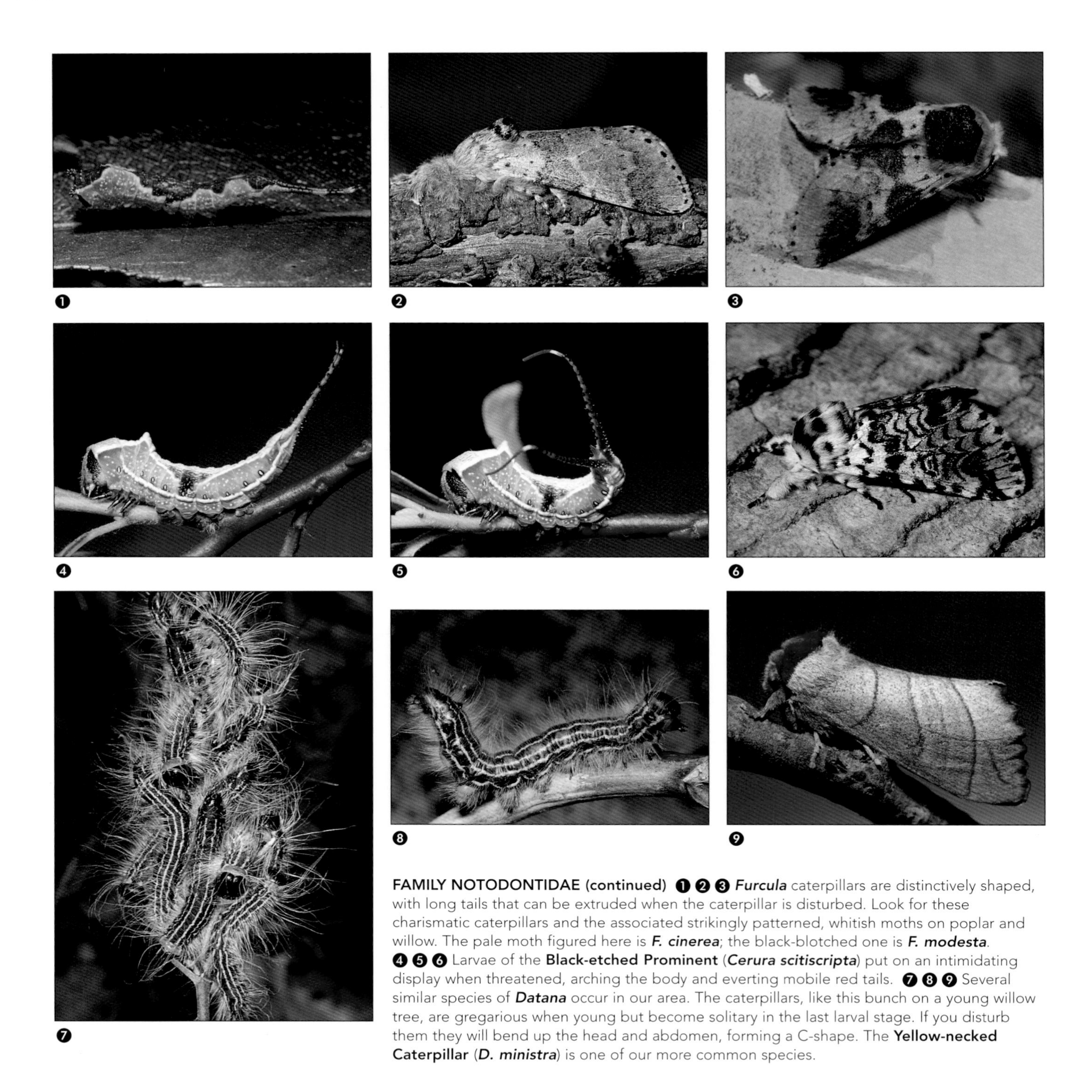

FAMILY NOTODONTIDAE (continued) ❶ ❷ ❸ ***Furcula*** caterpillars are distinctively shaped, with long tails that can be extruded when the caterpillar is disturbed. Look for these charismatic caterpillars and the associated strikingly patterned, whitish moths on poplar and willow. The pale moth figured here is ***F. cinerea***; the black-blotched one is ***F. modesta***. ❹ ❺ ❻ Larvae of the **Black-etched Prominent** (***Cerura scitiscripta***) put on an intimidating display when threatened, arching the body and everting mobile red tails. ❼ ❽ ❾ Several similar species of ***Datana*** occur in our area. The caterpillars, like this bunch on a young willow tree, are gregarious when young but become solitary in the last larval stage. If you disturb them they will bend up the head and abdomen, forming a C-shape. The **Yellow-necked Caterpillar** (***D. ministra***) is one of our more common species.

FAMILY NOTODONTIDAE (continued) ❶ ❷ Some caterpillars in the genus ***Schizura*** are called "unicorn" caterpillars because of the single "horn" that projects forward just behind the green thorax. They are well camouflaged when feeding on leaf edges (first photograph), but they are still heavily parasitized. The greenish objects on the side of this caterpillar are larvae of a parasitic wasp (Braconidae). ❸ ❹ ❺ Adult moths in the genus *Schizura*, including ***S. semirufescens***, the **Morning Glory Prominent** (***S. ipomoeae***) and the **Black-blotched Prominent** (***S. leptinoides***), often enhance their camouflage with cryptic postures. ❻ **Red Humped Caterpillars** (***Schizura concinna***) are bright, gregarious feeders found on a variety of hosts, including ericaceous shrubs. ❼ This **White Streaked Prominent** or **Lacecapped Caterpillar** (***Oligocentria lignicolor***) has been attacked by a parasitic fly (Tachinidae). The three white eggs on its thoracic hump will hatch to parasitic larvae that will consume the caterpillar's internal organs. ❽ The **Elegant Prominent** (***Odontosia elegans***) is indeed an elegant moth, easily recognized as a prominent by the conspicuous brown tuft on the inner margin of the front wing. Elegant Prominents develop on poplars. ❾ Although this mating pair of **White Dotted Prominents** (***Nadata gibbosa***) is hiding under a beaked hazel leaf, the female will later lay eggs on the oak leaves on which the larvae feed.

FAMILY GEOMETRIDAE (INCHWORMS OR GEOMETER MOTHS) Subfamily Ennominae ❶ ❷ Larvae of Geometridae like this **Common Lytrosis** (***Lytrosis unitaria***, arched backwards almost 180 degrees) and this **Pepper and Salt Geometer** (***Biston betularia***) become rigid twig-mimics when alarmed. ❸ ❹ Many inchworms have a characteristic looping gait necessitated by the lack of legs between the true legs at the front end of the body and the two prolegs at the hind end (a third, much smaller pair of prolegs is sometimes present). Adults of the **Pepper and Salt Geometer** (***Biston betularia***) occur in dark and light forms, with the darker forms predominating in some areas where the trees have been darkened by industrial pollution. This phenomenon, called industrial melanism, was first discovered through studies of British populations of Pepper and Salt Geometers. ❺ Like the similar Pepper and Salt Geometer (*Biston betularia*), the **Oak Beauty** (***Nacophora quernaria***) sometimes occurs as blackish, or melanic, individuals (not shown). ❻ **Linden Looper** (***Erannis tiliaria***) caterpillars are attractive, but occasionally reach great enough densities to seriously defoliate a wide variety of hardwoods. Adult females are wingless and appear, along with the winged males, in late fall. ❼ ❽ The **Filament Bearer** or **Horned Spanworm** (***Nematocampa resistaria***) is one of the most distinctive caterpillars in North America. Look for this widespread species on a variety of hardwood and softwood trees.

FAMILY GEOMETRIDAE (continued) ❶ ❷ The **Chain-dotted Geometer** (***Cingilia catenaria***) feeds on a wide variety of trees and shrubs, and is often conspicuously abundant in peatlands. This female is laying eggs on a blueberry leaf. ❸ ❹ ❺ Some geometer moths, like this ❸ **Bluish Spring Moth** (***Lomographa semiclarata***), this ❹ **White Spring Moth** (***Lomographa vestaliata***), ❺ and this brown-banded ***Epelis truncataria***, are butterfly-like day-fliers. ❻ **Porcelain Grays** (***Protoboarmia porcelaria***) are superbly camouflaged as they pose, wings outstretched, on tree trunks. ❼ ❽ The **Gray Spruce Looper** (***Caripeta divisata***) and the similar **Brown Pine Looper** (***C. angustiorata***) occur on variety of conifers across Canada and the northernmost United States. As the common names suggest, spruce loopers prefer spruce, tamarack and fir, while the pine loopers are more likely to be found on pine. ❾ The **Hemlock Looper** (***Lambdina fiscellaria***) is a common moth throughout our area, and can be a pest of various softwood trees.

FAMILY GEOMETRIDAE (continued) ❶ ***Semiothisa*** is a large genus, with almost 90 North American species, most of which are known as "angles." This is ***S. orillata***. ❷ This **Sharp-lined Yellow** (***Sicya macularia***) is striking a characteristic pose. Larvae feed on a variety of trees and shrubs. ❸ The **Fervid Plagodis** (***Plagodis fervidaria***) develops on various trees. This pose, with the abdomen poking up between the wings, is characteristic of the genus. ❹ The **Scallop Moth** (***Cepphis armataria***) is named for the distinctively scalloped hind wing margin. Larvae occur on trees and shrubs including apples and gooseberries. ❺ The **Pale Metanema** (***Metanema inatomaria***) develops on poplar and aspen. ❻ The **Alien Probole** (***Probole alienaria***) develops on various trees including birch, basswood, hawthorn and dogwood. ❼ The **Yellow-dusted Cream Moth** (***Cabera erythemaria***) is a small (less than 30 mm wingspan) geometer that usually occurs on willow, but also feeds on poplar, birch and blueberry. ❽ The **Curved-toothed Geometer** or **Purplish Brown Looper** (***Eutrapela clemataria***) is a relatively large (40–60 mm) geometer found on a variety of trees including birch and poplar. ❾ The **Tulip Tree Beauty** (***Epimecis hortaria***) is a common moth in Carolinian forest. ❿ The **Saddleback Looper** (***Ectropis crepuscularia***) feeds mostly on softwood trees, but the larvae (like many inchworms) can be found on a wide variety of tree species.

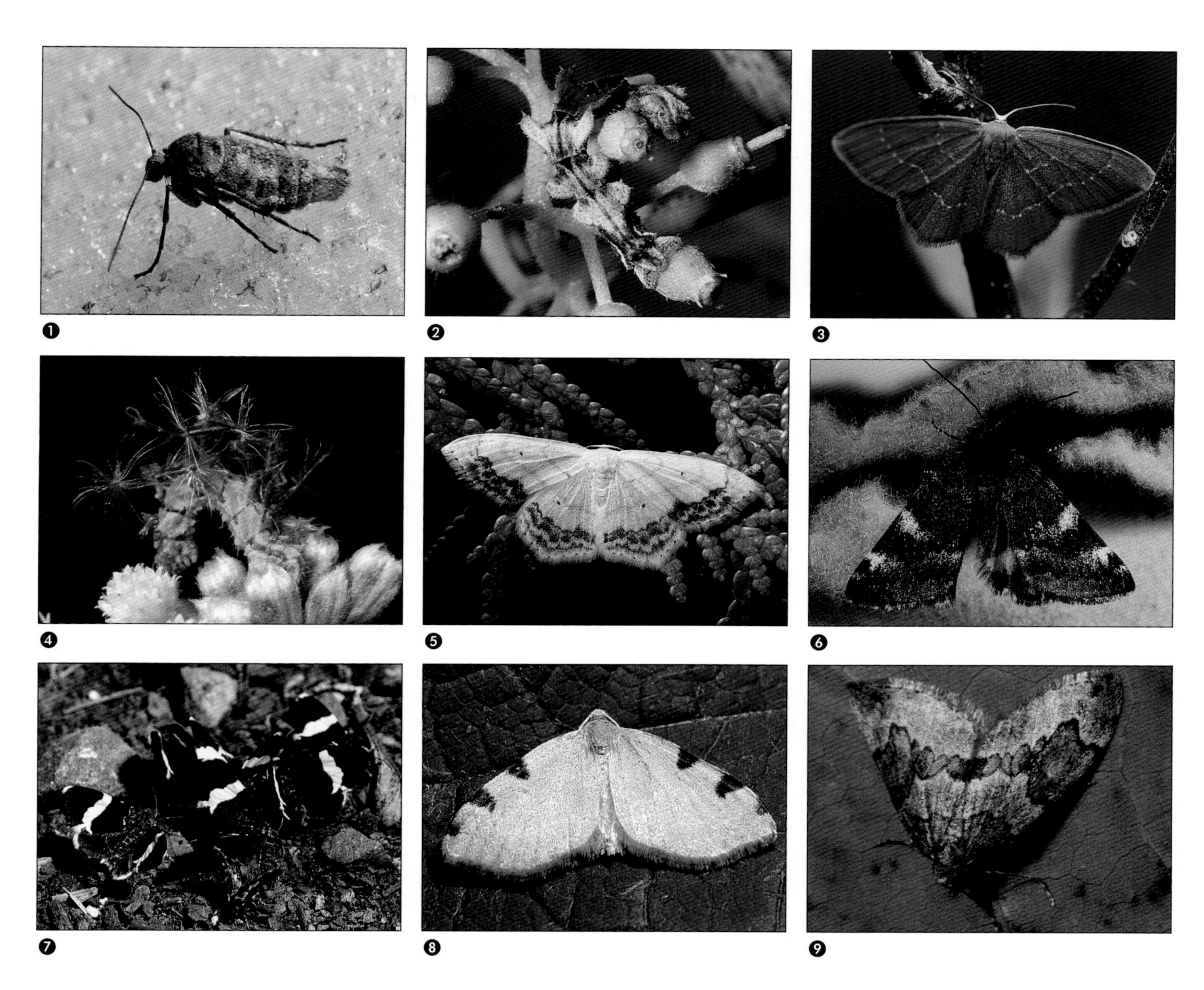

FAMILY GEOMETRIDAE (continued) Subfamily Alsophilinae ❶ This wingless female **Fall Cankerworm Moth** (***Alsophila pometaria***) is walking on the snow on a December day. She has probably already mated with a fully winged male, and might have already laid a ring of eggs around a twig. Caterpillars appear very early in spring, later pupating in underground cocoons where they will remain until they emerge as adult moths in the cold weather of late fall. Subfamily Geometrinae ❷ Geometer moths in the genus *Nemoria* (over 40 North American species) often have distinctive larvae with expanded lateral lobes (one common species, ***N. mimosaria***, is called the **Flanged Looper**). ❸ Many geometrines, like this ***Hethemia pistasciaria***, are delicate green moths. ❹ Larvae of ***Synchlora aerata***, aptly called **Camouflaged Loopers**, usually conceal themselves with stamens and fresh petal fragments from their host flowers. This one, however, has adorned itself with winged seeds. Subfamily Sterrhinae ❺ The **Large Lace Border** (***Scopula limboundata***) is one of many common *Scopula* species in our area. Although they are beautiful moths when examined in detail, from a distance they look like bird droppings. Subfamily Archiearinae ❻ The **Infant** (***Archiearis infans***) appears on warm days in early spring, often flying along forest margins in a butterfly-like fashion while the snow is still on the ground. This birch-feeding species is the only North American *Archiearis*, and one of only two North American species in the subfamily. Subfamily Larentiinae ❼ These **White-striped Blacks** (***Trichodezia albovittata***) are drinking from the same patch of wet (probably urine-soaked) soil as another common day-flying moth, a Spotted Thyris (*Thyris maculata*). White-striped Blacks are common woodland moths that develop on impatiens. ❽ The **Three-spotted Fillip** (***Heterophleps triguttaria***) occurs on maples throughout our area. ❾ The **Autumn Juniper Moth** (***Thera juniperata***) develops on juniper bushes, pupating in fall and emerging as adults late in the season. This adult was photographed in Ontario in November.

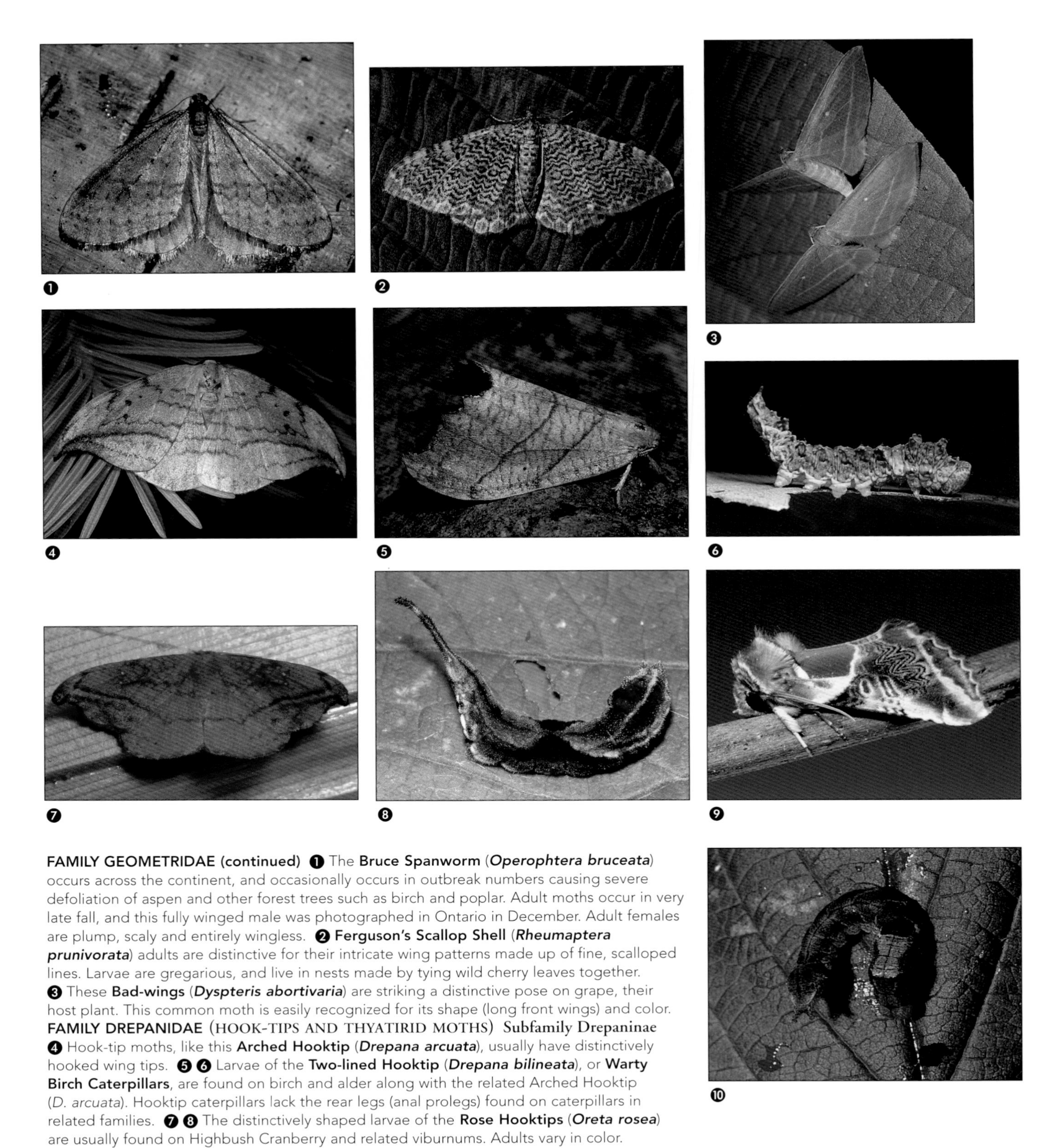

FAMILY GEOMETRIDAE (continued) ❶ The **Bruce Spanworm** (***Operophtera bruceata***) occurs across the continent, and occasionally occurs in outbreak numbers causing severe defoliation of aspen and other forest trees such as birch and poplar. Adult moths occur in very late fall, and this fully winged male was photographed in Ontario in December. Adult females are plump, scaly and entirely wingless. ❷ **Ferguson's Scallop Shell** (***Rheumaptera prunivorata***) adults are distinctive for their intricate wing patterns made up of fine, scalloped lines. Larvae are gregarious, and live in nests made by tying wild cherry leaves together. ❸ These **Bad-wings** (***Dyspteris abortivaria***) are striking a distinctive pose on grape, their host plant. This common moth is easily recognized for its shape (long front wings) and color. **FAMILY DREPANIDAE** (HOOK-TIPS AND THYATIRID MOTHS) Subfamily Drepaninae ❹ Hook-tip moths, like this **Arched Hooktip** (***Drepana arcuata***), usually have distinctively hooked wing tips. ❺ ❻ Larvae of the **Two-lined Hooktip** (***Drepana bilineata***), or **Warty Birch Caterpillars**, are found on birch and alder along with the related Arched Hooktip (*D. arcuata*). Hooktip caterpillars lack the rear legs (anal prolegs) found on caterpillars in related families. ❼ ❽ The distinctively shaped larvae of the **Rose Hooktips** (***Oreta rosea***) are usually found on Highbush Cranberry and related viburnums. Adults vary in color. Subfamily Thyatirinae ❾ ❿ The **Lettered Habrosyne** (***Habrosyne scripta***) is aptly named, since it looks like someone has tried to write on its striking front wing. The caterpillars are usually hidden in rolled leaves of birch or blackberry.

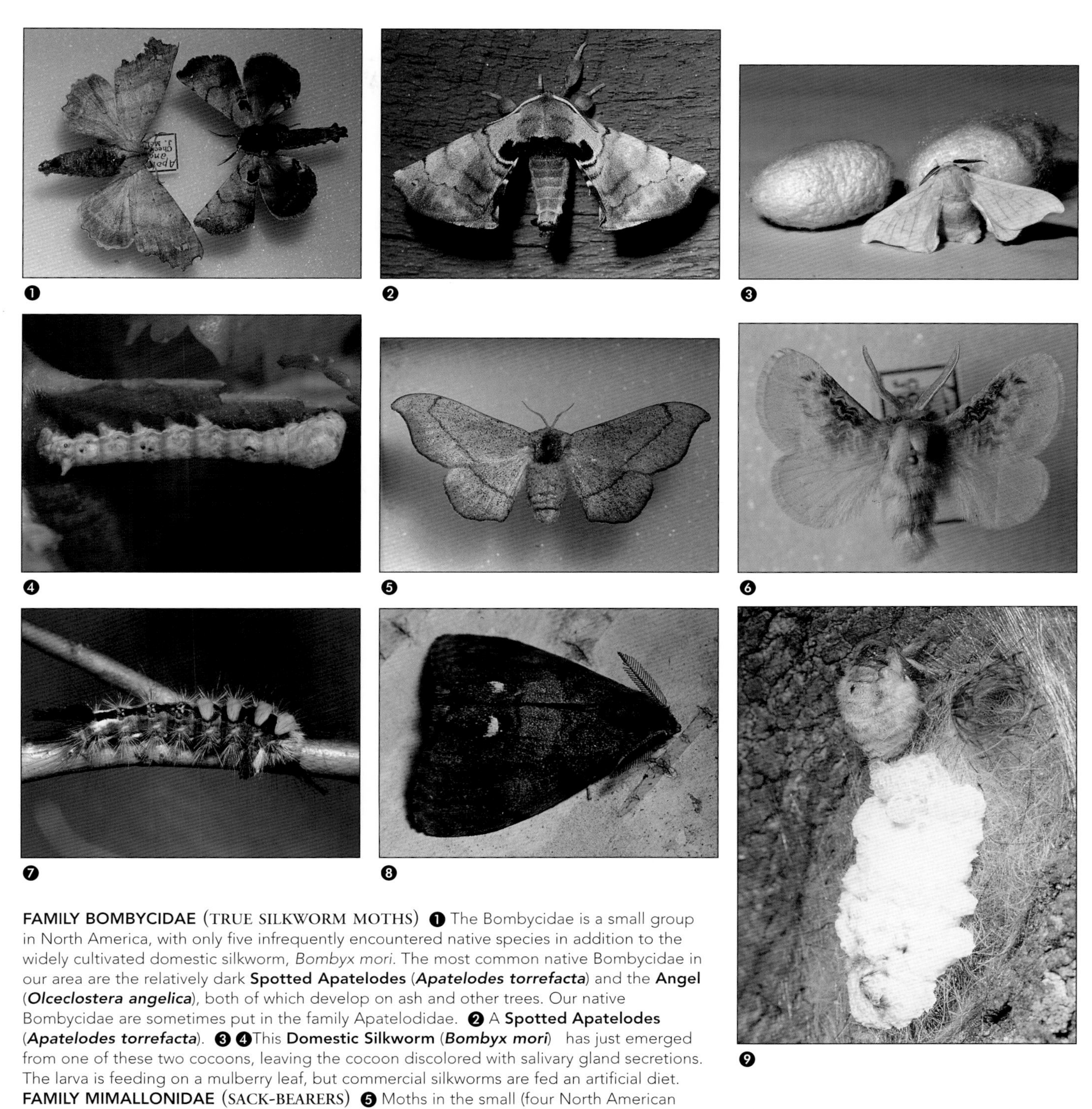

❶ ❷ ❸ ❹ ❺ ❻ ❼ ❽ ❾

FAMILY BOMBYCIDAE (TRUE SILKWORM MOTHS) ❶ The Bombycidae is a small group in North America, with only five infrequently encountered native species in addition to the widely cultivated domestic silkworm, *Bombyx mori*. The most common native Bombycidae in our area are the relatively dark **Spotted Apatelodes** (***Apatelodes torrefacta***) and the **Angel** (***Olceclostera angelica***), both of which develop on ash and other trees. Our native Bombycidae are sometimes put in the family Apatelodidae. ❷ A **Spotted Apatelodes** (***Apatelodes torrefacta***). ❸ ❹This **Domestic Silkworm** (***Bombyx mori***) has just emerged from one of these two cocoons, leaving the cocoon discolored with salivary gland secretions. The larva is feeding on a mulberry leaf, but commercial silkworms are fed an artificial diet. **FAMILY MIMALLONIDAE** (SACK-BEARERS) ❺ Moths in the small (four North American species), rarely encountered family Mimallonidae are called sack-bearers because mature larvae build open-ended "sacks" made of silk and leaves. They overwinter in the sacks, pupating in spring. This is **Melsheimer's Sack-bearer** (***Cicinnus melsheimeri***). **FAMILY MEGALOPYGIDAE** (FLANNEL MOTHS) ❻ Larvae of the small family Megalopygidae (three eastern North American species) are hairy caterpillars that, like the related Limacodidae, are armed with painfully venomous stinging bristles. This is the **Black-waved Flannel Moth** (***Lagoa crispata***), the most common flannel moth in our area. The caterpillar, found on a wide variety of trees and shrubs, is oval and densely hairy, but dull in color. **FAMILY LYMANTRIIDAE** (TUSSOCK MOTHS) (now sometimes treated as part of the Noctuidae) ❼ ❽ ❾ The **Rusty Tussock Moth** (***Orgyia antiqua***) is a Holarctic moth (found in Europe and Asia as well as North America) that feeds on a variety of trees. Females are wingless, and this one is laying a large, white mass of eggs on a tree trunk in early fall. She has not moved far since emerging, and the silk pad on which she is laying eggs is what is left of the cocoon from which she emerged. The eggs will hatch in spring.

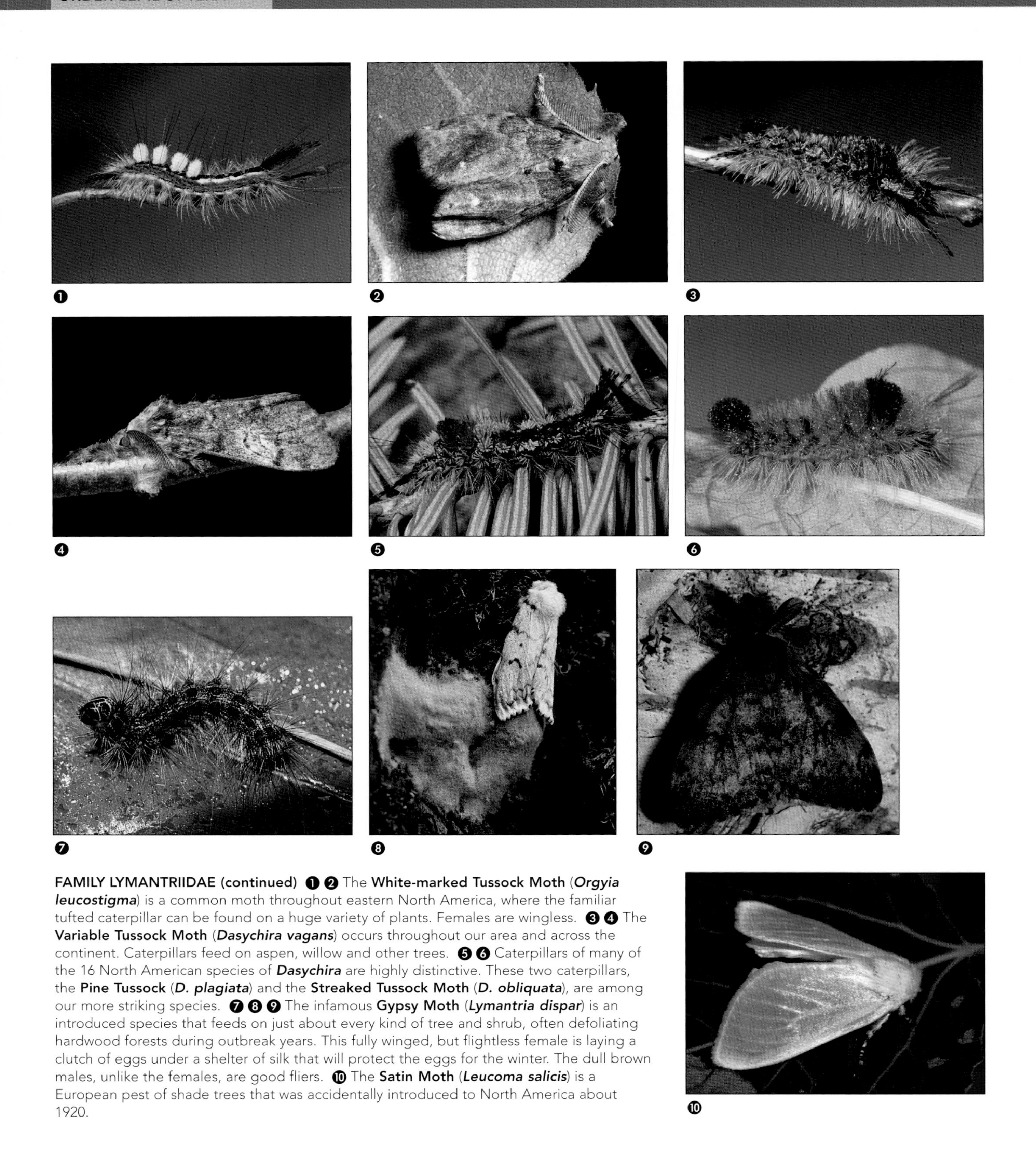

FAMILY LYMANTRIIDAE (continued) ❶ ❷ The **White-marked Tussock Moth** (***Orgyia leucostigma***) is a common moth throughout eastern North America, where the familiar tufted caterpillar can be found on a huge variety of plants. Females are wingless. ❸ ❹ The **Variable Tussock Moth** (***Dasychira vagans***) occurs throughout our area and across the continent. Caterpillars feed on aspen, willow and other trees. ❺ ❻ Caterpillars of many of the 16 North American species of ***Dasychira*** are highly distinctive. These two caterpillars, the **Pine Tussock** (***D. plagiata***) and the **Streaked Tussock Moth** (***D. obliquata***), are among our more striking species. ❼ ❽ ❾ The infamous **Gypsy Moth** (***Lymantria dispar***) is an introduced species that feeds on just about every kind of tree and shrub, often defoliating hardwood forests during outbreak years. This fully winged, but flightless female is laying a clutch of eggs under a shelter of silk that will protect the eggs for the winter. The dull brown males, unlike the females, are good fliers. ❿ The **Satin Moth** (***Leucoma salicis***) is a European pest of shade trees that was accidentally introduced to North America about 1920.

CHAPTER 8

Caddisflies

Caddisfly adults are distinctly moth-like, and even though caddisfly larvae are aquatic they are also caterpillar-like – even to the point of using silk spun from their salivary glands to make cocoons and shelters. Not surprisingly, the order Trichoptera (caddisflies) and the order Lepidoptera (butterflies and moths) are very closely related, and these groups share many special features, including a sex determination system that differs from that of most other insects.

Although caddisflies have much in common with moths, the Trichoptera and Lepidoptera do differ in some obvious ways. For one thing, the name Trichoptera literally means "hairy wings." Adult caddisflies do not have their wings covered by flattened scales, as do similar moths, although caddisfly wings sometimes have a few scattered patches of scales. Furthermore, caddisflies never have a coiled proboscis or sucking tube of the sort you can see below the head of all remotely similar moths (some moths lack this tube, but they are conspicuously scaly and not shaped like caddisflies).

Trichoptera is a relatively small order (only 11,500 or so described species worldwide), but adult caddisflies abound near lights on summer nights, and can usually be found fluttering about or resting on foliage near ponds, rivers, lakes and streams. Females lay gelatinous strings or masses of eggs, usually on underwater objects or objects near the water, although females of some groups go far beneath the surface to deposit eggs, and females in other groups place eggs on foliage far above the water into which the hatching larvae fall. A few species that develop in temporary pools lay their eggs in desiccation-resistant and freeze-resistant gelatinous masses into which the larvae hatch. The gelatinous matrix protects the tiny larvae until their pool is flooded with water early the following spring. Larvae of different trichopteran groups are conspicuously varied in appearance, habitat and behavior, in contrast with the secretive and superficially uniform adults.

Caddisfly larvae look a bit like caterpillars without the several fleshy abdominal prolegs that characterize Lepidoptera larvae. Instead, there is a single pair of hooks on a single pair of prolegs at the hind end of the body. Like caterpillars, caddisfly larvae have huge salivary glands used to produce silk, and it is the variety of ways they use this silk that makes caddisflies such a fascinating group.

Free-living caddisflies and related families

Caddisflies are ubiquitous in all kinds of streams and rivers, so one shouldn't have to look further than the first handy submerged rock or other object to find a few larvae. Caddisfly larvae occur on a variety of substrates, but a rock in a fast, shallow part of a clean stream is likely to yield some particularly interesting specimens. The most conspicuous caddisfly larvae found in that type of habitat are the large (a couple of centimeters), broad-bodied *Rhyacophila* (family **Rhyacophilidae**, page 241) larvae that live unencumbered by a case or other shelter. *Rhyacophila* larvae move around rock surfaces in running water, holding on to the substrate with huge hooks on their anal prolegs as they feed on black fly larvae and other organisms sharing their lotic home. Like black fly larvae, *Rhyacophila* larvae use silk against the possibility of being swept away, usually laying a "thread trail" much like the trails used by some caterpillars.

When it comes time to pupate, free-living caddisfly larvae cease to be free-living and encase themselves in a parchment-like silken cocoon concealed under a shelter made of silk and pebbles attached to the rock surface. The silk lining of a caddisfly cocoon protects a pupa somewhat similar to the pupa found inside a moth cocoon, but the front of the caddisfly pupa has conspicuous bladelike mandibles used to cut free from its silken shelter, and the legs of the caddisfly pupa are free from the body and have long fringes to help it swim to the surface. By this time the pupa is little more than a puppet-like shell occupied by the fully developed adult within, and the pupal mandibles are operated by the adult's muscles and the swimming legs of the pupa are operated by the slender legs of the enclosed adult. All caddisflies pupate in an underwater silken cocoon, usually a loosely spun cocoon integrated with the case or shelter rather than a separate, tightly woven cocoon like that of the free-living caddisflies and their close relatives the saddlecase-makers (Glossosomatidae) and purse-case makers (Hydroptilidae). The latter families spin their silken cocoons precociously, and use them as larval shelters.

It is not that great a leap from making a silk and stone shelter for pupation to making a silk and stone shelter to protect the larva from predation, so it is not too surprising to find some close relatives of the free-living *Rhyacophila* which make saddle-like protective larval cases. In saddle-case makers (*Glossosoma*, family **Glossosomatidae**, page 241), both head and tail project from the saddle-like, or turtleshell-like, flat-bottomed, oval-topped case made of irregular pebbles. These heavily burdened caddisfly larvae slowly scrape food from exposed upper rock surfaces, protected by their pebble shelters. Related caddisflies in the **Hydroptilidae** (page 243) are called purse-case makers. These tiny caddisflies start out life as free living larvae and only later spin purse-like silken shelters, open at each end, that protect the larvae from predation.

Net-spinning caddisflies and other fixed-retreat makers

The most common of the net-spinning caddisfly families, a group so abundant that you can safely predict their occurrence on every rock in any of our local riffles, is the family **Hydropsychidae** (page 241). Hydropsychid larvae, which have characteristic and conspicuous tufts of gills

Adult Net-spinning Caddisfly (Hydropsychidae), page 242, caption 1.

under the abdomen, spin elegant silken capture nets used to filter out different foodstuffs according to the size of the weave. These nets are normally held outstretched by the current and seem to collapse into amorphous muck when removed from the water, so try and find a shallow, rapid stretch of water where you can peer through the surface onto a submerged rock. Here you will see the loosely woven silken seines of hydropsychids attached to pebbles or other debris and held distended by the current. The caddisfly larva itself stays in a pebble and silk shelter alongside the net, from which it advances to eat whatever the current has washed in, much as spiders dash from their web-side shelters in pursuit of web-ensnared prey.

There are other net-spinning caddisflies, but Hydropsychidae are by far the most common. Other common types of nets include the long, narrow, fine mesh nets aptly called "finger-nets," and the funnel-mouthed nets of the "trumpet-net" makers. The elongate finger-nets made by members of the family **Philopotamidae** (page 242) are inconspicuously distended on the undersides of rocks, and although the nets look like strings of silt-covered sludge when out of water, their bright orange (white when preserved in alcohol) makers are usually conspicuously common. Finger-nets are incredibly fine, with millions of mesh openings (100 million per net, according to one published report), and the larvae are uniquely equipped with a spatula-like "lip" (labrum) used to scrape captured particulate matter from the mesh of their nets. Most of the small, uniformly dark brown to black adult caddisflies that abound along rivers and streams throughout much of the season are finger-net caddisflies, as are the wingless female caddisflies that emerge from late fall into early winter. The wingless females belong to a species (*Dolophilodes distinctus*) that emerges throughout the year, but has fully winged females during the summer months and entirely wingless females during the winter months.

Although much less abundant than the finger-net makers, the trumpet-net makers (some species in the family **Polycentropodidae**, page 242) make the most conspicuous nets in the order. *Neureclipsis* species spin enormous (over 12 cm) nets with the upstream end flared like a trumpet and the tapered downstream end recurved to form a retreat for the relatively small (less than 21 mm) larva. The flared end of the net, which may be several centimeters across, is guyed out with silken strands; even so, these huge nets are only found in relatively slow water. Broad, clean, relatively shallow rivers can be so full of these conspicuous nets it is hard to imagine how any small organism drifting down the river could avoid being captured and eaten by waiting net-makers.

Neureclipsis larvae eat both algae and other arthropods, but some related Polycentropodidae are strictly predacious, using silken tubes as retreats and silken strands to detect the movement of potential prey, much the way some spiders use similar silken constructions to detect terrestrial prey. Some of these species also use their silken tubes a bit like pumps, undulating their bodies to bring water through the tubes. This has freed some Polycentropodidae from a dependence on running water and allowed them to live in ponds and lakes – habitats not normally inhabited by net-spinning caddisflies. Some members of a much less commonly encountered, closely related family (Psychomyiidae) have similar habits, although most northeastern psychomyiids are stream-dwellers that consume fine organic particles from within their meandering silken tubes. A small related family, Dipseudopsidae, includes only a few species that make fixed silk and sand tubes within the bottom sediments of lakes and sandy streams.

Most net-making caddisflies depend on running water to distend their nets and continually replenish their food supplies. Net-makers and other caddisflies also depend on flowing water to continually supply fresh, oxygen-rich water in which they can breathe through their gills or skin. Despite this limitation, if you peer into your local pond (even a temporary woodland pool), the chances are good that you will soon see caddisfly larvae. Most, if not all, of these will be tube-case caddisfly larvae crawling around with portable cases dragging behind, or even swimming freely using long, fringed legs stuck out the fronts of tiny tubular cases.

Portable-case making caddisflies

As any trout fisherman will tell you, you can remove a caddisfly from its case by gently squeezing the case to force the larva to show its head, then pulling the larva head first out of the case. Try doing that, not to impale the exposed larva on a fishhook, but to watch its behavior when it is put back in an aquarium. If there is another caddisfly in its case in the aquarium, the exposed larva might attack it from the rear. The attacked larva will probably be driven out the front of its case, but will then likely come around to the rear and attack the intruder to reclaim its case. It might be more interesting, from our point of view, to provide a piece of glass tubing of about the right size to serve as a case. The exposed individual will be quick to take over this unoccupied and undefended transparent case. If it fits well we can see how the larva is held in the case by fleshy swellings just behind the thorax, and how the abdomen behind those humps is wiggled or undulated. The net effect of this positioning and undulation is not unlike a pump, drawing water in the front of the tube and out the back. So you see, caddisflies really are creatures of running water.

It is just that the tube-case makers are able to use their remarkable silk-based houses to create their own flow, even in a pond.

One needn't look far for evidence that the variety of caddisfly larval dwellings rivals the variety of human abodes. The most common caddisfly family (the huge, common and diverse family **Limnephilidae** or northern caddisflies, page 243) is found in a variety of habitats and includes some pond species that make portable tube-cases from pieces of debris, crisscrossed log-cabin–style. Anglers have long used these common insects as bait, and the name "caddisflies" probably came from an old fisherman's term for limnephilid larvae that make tubes covered with bits of debris, similar to the way that sellers of cloth called "caddis men" once peddled their wares with bits and pieces conspicuously pinned to their clothing. Pond-living species in another common Trichoptera family (**Phryganeidae** or large caddisflies, page 245) usually make smooth cylinders walled with carefully cut pieces of plant debris, sometimes arranged in an elegant spiral.

The greatest variety of caddisfly houses is found among the tube-case makers in running water. Here, the challenges of dealing with the current, and the many predators therein, leads some to burden their dwellings with extra ballast pieces, others to build solid stone homes and yet others to build flattened cases with broad flanges pressed against the substrate. One of our commonest caddisfly larvae builds a case of small sand grains that looks almost identical to a snail shell; yet another group makes flat, flanged, somewhat shield-shaped sand cases that sit camouflaged flat against the surface of the sand. Snail-case caddisflies (**Helicopsychidae**, page 246) and flat caddisflies (**Molannidae**, page 246) have particularly distinctive cases, but there is considerable overlap in case morphology between most tube-making caddisfly families. Superficially similar cylindrical stone cases, for example, are found in rarely encountered species of the small families **Odontoceridae** (page 246), **Sericostomatidae** and **Brachycentridae** (page 246), as well as most common **Lepidostomatidae** (page 246), some **Leptoceridae** (page 245) and many extremely common **Limnephilidae** and similar **Apataniidae** (page 245) and **Uenoidae** (page 244). The large family **Limnephilidae** certainly has the widest variety of case morphologies, but it is followed closely by the Leptoceridae, or long-horned caddisflies. Long-horned caddisflies are so named because the adult antennae are conspicuously long and slender. Larval antennae are also long relative to those of other caddisflies, although they are still inconspicuous at only around 10 times as long as wide. Some long-horned caddisflies construct flanged cases like those of molannids, some make stone cases like small limnephilids, and a few even make spiral cases like tiny phryganeids. Some of the species that make spiral cases swim using fringed hind legs that stick out the front of the case. These small (usually 20–25 mm) caddisflies feed on aquatic vegetation and are able to swim from plant to plant.

A *Molanna* larva in its case, seen from underneath, page 246, caption 2.

Making a case in running water must be a tricky thing, especially making a neat cylindrical case. Caddisflies are obliging architects, and will often make a case of any available materials right under your watching eyes. Put a naked tube-case maker in an aquarium full of purple gravel, and the odds are it will expose its architectural secrets as it makes a lurid purple case. You might see it first secrete a neat circle of silk on the substrate in front of its head then, when the silk has hardened, you might see it stand on the hoop of silk to pop it up over its head. The tube case is built backwards on that initial hoop, using the purple gravel, diamonds, emeralds, ground glass or whatever you have provided. Tube-dwelling caddisfly groups normally make their cases from bits and pieces of wood, stone, or leaves but one very odd genus in the rarely encountered (in the northeast) family Calamoceratidae takes the prefab approach. Members of the genus *Heteroplectron* simply excavate the center of an appropriately sized twig and line the resultant tube with silk.

FAMILY RHYACOPHILIDAE (FREE-LIVING CADDISFLIES) ❶ Larvae of free-living caddisflies (***Rhyacophila*** spp.) have long, claw-bearing prolegs at the end of the body. As suggested by the business-like mandibles on this individual, most are predators. ❷ This free-living caddisfly (***Rhyacophila fuscula***) is munching its way through a group of net-spinning caddisflies (Hydropsychidae) in a shallow stream. ❸ ***Rhyacophila*** larvae are common predators in clean streams throughout our area, with several distinctively colored species often living together in the same stream. ❹ This pre-pupal free-living caddisfly (***Rhyacophila fuscula***) has encased itself in a tight, translucent, silken cocoon prior to pupation. ❺ This adult caddisfly (***Rhyacophila fuscula***) has her abdomen arched down as she lays eggs on a rock sticking out of a swift, shallow stretch of Ontario's Grand River. A couple of black flies (Simuliidae) and several midges (Chironomidae) are laying eggs on the same wet rock. **FAMILY GLOSSOSOMATIDAE** (SADDLE-CASE MAKERS) ❻ Saddle case-making caddisflies are closely related to free living caddisflies (and purse-case makers), and pupate within similar tightly woven, semi-permeable silken cocoons. The saddle case has been removed from one of these individuals, exposing the cocoon. **FAMILY HYDROPSYCHIDAE** (NET-SPINNING CADDISFLIES) ❼ This photograph, taken through the surface of a swiftly flowing stream, shows a variety of insects on the upper surface of a rock, including several small black fly larvae, several midges hidden by silken tubes, and a number of cuplike silken nets held open by the current. These are capture nets made by net-spinning caddisflies. ❽ Net-spinning caddisfly larvae usually stay hidden in pebble and silk shelters alongside their nets, but here we can see one individual larva partially out if its shelter. ❾ The tufts of gills under this larva's body are characteristic of the family Hydropsychidae, and the flat, scalelike bristles scattered over its body mark it as a member of the rarely encountered genus ***Parapsyche***. Our common net-spinning caddisflies belong to the genera *Hydropsyche* and *Cheumatopsyche*.

FAMILY HYDROPSYCHIDAE (continued) ❶ Most running waters abound in net-spinning caddisflies, and adult ***Hydropsyche*** species are often very common insects near rivers and streams. ❷ Mating caddisflies are common sights near stream margins. ❸ Net-spinning caddisflies in the genus ***Cheumatopsyche*** often develop in warmer, more polluted waters than related genera. This one was photographed near the Detroit River, in the city of Windsor. ❹ The three North American species of ***Macrostemum*** (formerly *Macronema*; this is the most common, ***M. zebratum***) develop in large, swiftly flowing rivers. The bright colors and long antenna distinguish them from other hydropsychids. **FAMILY PHILOPOTAMIDAE** (FINGER-NET CADDISFLIES) ❺ Larval finger-net caddisflies spin fine, silken, tubelike capture nets on the undersides of rocks in clean streams. The nets collapse into virtual invisibility when the rocks are disturbed, and this larva has moved away from its collapsed net. The characteristic bright orange-yellow color of living ***Chimarra*** larvae, like this one, quickly fades to white in preserved specimens. ❻ The small (6–8 mm), dark brown to black adults of common finger-net caddisflies, especially species in the very common genus ***Chimarra***, are often abundant on vegetation close to running water. ❼ This pair of finger-net caddisflies includes a wingless female and a fully winged male. ***Dolophilodes distinctus*** is an unusual species because adults emerge well on into the winter months, and adult females emerging in the winter months are always entirely wingless. Winged females occur during warmer months. **FAMILY POLYCENTROPODIDAE** (TRUMPET-NET AND TUBE-MAKING CADDISFLIES) ❽ This aquarium photograph shows a bright orange finger-net caddisfly larva (Philopotamidae) beside a pale, speckled-headed trumpet-net caddisfly larva (***Polycentropus*** sp.). *Polycentropus* species occur over a much wider range of habitats than finger-net caddisflies, including both running and still waters. *Polycentropus* larvae are predators that make silken shelters (trumpet-like or tube-like) from which they dart out to capture other organisms. ❾ Larvae of ***Polycentropus*** are the most common benthic caddisflies in large lakes, and the associated adults are often seen along the shores of the Great Lakes.

FAMILY POLYCENTROPODIDAE (continued) ❶ ❷ The first of these photographs shows a stretch of river bottom peppered with the spectacular trumpet-shaped silken nets of ***Neureclipsis*** larvae, showing how each net is turned back upon itself to taper to a tiny larval shelter beside the relatively huge net opening. The net is held distended by the current, but the opening is also guyed out with silken strands. The second photograph shows a small (about 10 mm) larva removed from its retreat. **FAMILY HYDROPTILIDAE** (PURSE-CASE MAKERS) ❸ ❹ These tiny (about 3 mm total length) purse-case caddisfly larvae were photographed on the limestone of Georgian Bay; the minute adult was photographed on a nearby leaf. **FAMILY LIMNEPHILIDAE** (NORTHERN CADDISFLIES) ❺ ❻ ❼ ❽ ❾ Tube-case making caddisflies in the large family Limnephilidae (almost 300 North American species) make a variety of cases. Those in running water often have heavy cases incorporating rocks, while those living in ponds are more likely to have light cases made of pieces of leaves or roots.

FAMILY LIMNEPHILIDAE (NORTHERN CADDISFLIES) **(continued)** ❶ Members of the large (100 or so North American species) genus ***Limnephilus*** are most likely to be found around ponds or marshes. ❷ ***Pycnopsyche*** species are medium-sized (around 20 mm), orange-brown caddisflies that abound around rivers and streams. ❸ ***Pycnopsyche*** larvae, which have cases made of leaves or sticks, can often be seen feeding on dead leaves that accumulate in slow-flowing parts of waterways. ❹ This pair of ***Frenesia*** was photographed near a small spring in Ontario on a mild December day. Northeastern *Frenesia* emerge unusually late in the season, normally November. ❺ The distinctively scalloped wings mark this northern caddisfly as ***Nemotaulius hostilis***, the only North American *Nemotaulius*. Larvae live in ponds and make tube-cases from leaves and twigs. ❻ The distinctive, silvery wing-pattern marks this northern caddisfly as the only northeastern ***Hesperophylax***, ***H. designatus*** (20 mm). Larvae make a cylindrical case of rock fragments and live in small streams throughout the northeast. ❼ Although many caddisflies are difficult to identify even to the family level in the field, a few species, such as this ***Platycentropus radiatus***, are unmistakably patterned. Larvae, which occur in a variety of still and flowing waters, have log-cabin style cases made of crisscrossed pieces of sedge or grass. **FAMILY UENOIDAE** ❽ ❾ The larvae of caddisflies in the small genus ***Neophylax*** (such as ***N. concinnus***) make heavy stone cases with large rock fragments on the side, and are often found in small, clear streams where they graze diatoms and other fine material from rock surfaces. Larvae of this species develop during the winter, and then seal off their cases in spring and wait till late summer or fall before developing into adults. The sealed cases can withstand extreme conditions, so this species sometimes abounds in intermittent streams.

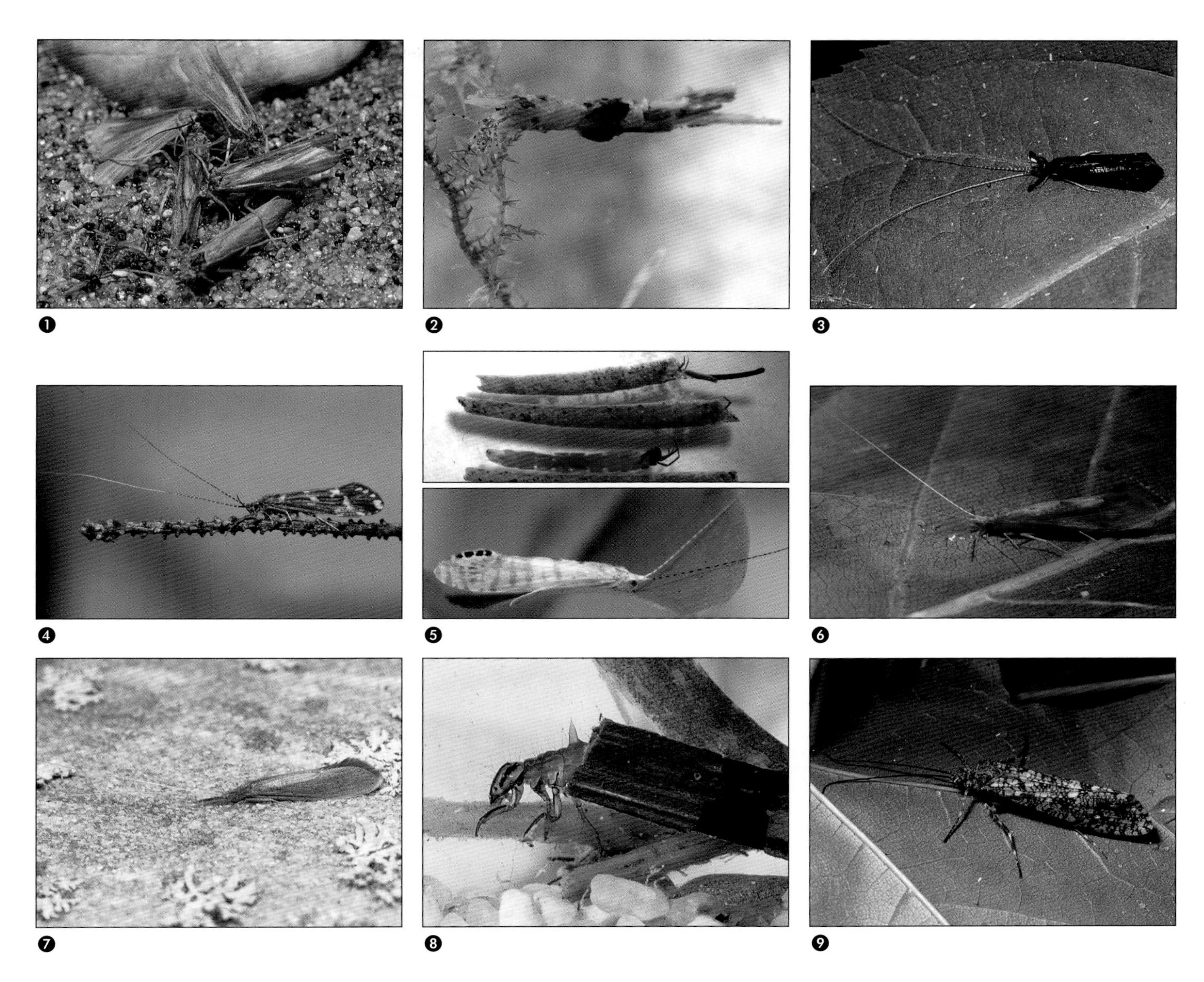

FAMILY APATANIIDAE ❶ These ***Apatania*** adults were part of a frenetic cluster of 40 courting caddisflies on the shore of Lake Huron. Larvae of *Apatania* live in cold lakes where they make tapered rock fragment cases. Apataniid caddisflies used to be treated as part of the Limnephilidae. **FAMILY LEPTOCERIDAE** (LONG-HORNED CADDISFLIES) ❷ ❸ Larval long-horned caddisflies are usually quite small, and make a variety of cases. ***Mystacides sepulchralis*** larvae, for example, are less than 10 mm long and make tubular cases from a mixture of materials. The associated adults are distinctively shaped, shining blue-black caddisflies. ❹ Members of this genus (***Ceraclea***) develop from small (less than 12 mm), stout-bodied larvae that usually make wide, flanged cases from small sand grains. ❺ Long-horned caddisflies in the genus *Nectopsyche* (formerly included in *Leptocella*) are beautiful insects, usually white with characteristic markings. ***Nectopsyche exquisita*** adults can sometimes be seen flying by the thousands just above the surface of shallow lakes. Larvae make long and narrow cases, often with a twig or conifer needle extending beyond one end. ❻ Adult long-horned caddisflies in the genus ***Triaenodes*** (about 12 mm) often occur in great numbers around lights at night. Their larvae, which make distinctive tapered cases made of spirally arranged leaf fragments, can often be seen swimming, case and all, amongst vascular plants along the shallow edges of ponds, lakes and rivers. ❼ The small (5–9 mm) long-horned caddisflies in the genus ***Oecetis*** develop as predators in a variety of waters. **FAMILY PHRYGANEIDAE** (LARGE CADDISFLIES) ❽ Larvae of Phryganeidae are large, often over 40 mm long, and usually have yellow heads with black stripes. Most genera have cases made of leaf fragments, either stacked in rings like the one in this photograph or assembled in spiral fashion. This larva is partly out of its case, and shows the large "hump" on the first abdominal segment. ❾ Adult Phryganeidae, like this ***Agrypnia improba***, are usually around 20 mm in length.

FAMILY PHRYGANEIDAE (continued) ❶ Large caddisflies in the genus ***Phryganea*** develop in lakes, marshes and slow-flowing rivers where their large (up to 43 mm) larvae make spiral cases from cut-to-fit leaf and bark fragments. **FAMILY MOLANNIDAE** ❷ This photo, taken through the glass bottom of an aquarium, shows the larva of a flat-case caddisfly (***Molanna***) and its distinctive sand-grain case in ventral view. The case, about 20 mm long, completely shields the larva from the top. ❸ Flat-case caddisfly larvae are difficult to see on the sandy lake and creek bottoms where they normally live. **FAMILY HELICOPSYCHIDAE** (SNAIL-CASE CADDISFLIES) ❹ The distinctively snail-shell-shaped cases of Helicopsychidae (one genus, ***Helicopsyche***) are common in most running waters, but their small size (less than 7 mm in diameter) and sand-grain composition render them inconspicuous. When this species (***H. borealis***) was first discovered in 1866, it was not recognized as an insect larva and was formally described as a snail. **FAMILY ODONTOCERIDAE** ❺ ***Psilotreta***, and other members of the small family Odontoceridae, spend their larval lives hidden away in the gravel or sand in the bottoms of streams, but come together to pupate (in their cases) in masses on rocks. **FAMILY LEPIDOSTOMATIDAE** ❻ ❼ ***Lepidostoma*** species (Lepidostomatidae) are small caddisflies that develop from larvae that look much like small Limnephilidae. They are often very common in seeps and springs. **FAMILY BRACHYCENTRIDAE** ❽ One of these two small (about 6 mm) round cases of the widely distributed, but uncommon, genus ***Micrasema*** is still occupied; the larva has been pulled out of the other one. Brachycentrid larvae have a distinctive crease across the pronotum; the other eastern genus (*Brachycentrus*) has square-sided cases. ❾ This characteristically square-cased ***Brachycentrus*** was one of many in a shallow woodland spring.

CHAPTER 9

Lacewings, Antlions, Fishflies and Related Insects

Nerve-winged insects

Green lacewings are easily recognized by their jewel-like eyes, delicate diaphanous green wings and slow, fluttering flight. These common predators are familiar to most naturalists and gardeners – you might even have seen them for sale as natural alternatives to insecticides for control of aphids and other insect pests in home gardens. Lacewings don't belong to any of the big, familiar orders of insects such as the flies (order Diptera), wasps (order Hymenoptera) or butterflies and moths (order Lepidoptera), but are instead part of the small order Neuroptera. The name Neuroptera means "nerve-winged insects," referring to the rich patterns of wing veins or "nerves" that characterize most species in this order.

The green lacewing family, **Chrysopidae** (page 252), North America's largest and most familiar group of nerve-winged insects, includes several bright-eyed species with large, somewhat iridescent, green or greenish yellow wings. The wings are richly veined, and some nocturnal green lacewings have "ears" in the larger veins that allow them to detect the ultrasonic sounds made by hunting bats. Look under the leaves of trees for the peculiar eggs deposited by green lacewings, each at the end of a long, threadlike stalk affixed to the underside of a leaf. The long stalks probably serve to protect the eggs from being devoured by predators and precocious siblings who, deterred by the stalk, will presumably move on to more appropriate prey. Lacewing larvae and most lacewing adults are voracious predators of soft-bodied insects such as aphids. Larval lacewings (and other Neuroptera) have long mandibles with hollow ventral grooves closed by long maxillae to make sickle-like tubes. These syringe-like mouthparts are used to puncture prey and suck out the liquefied contents.

Some lacewings are commonly found in colonies of woolly aphids, where their larvae imbibe the contents of individual aphids before hoisting the husks of their aphid prey onto "trash packets" on their Velcro-like hairy backs. Using corpses as camouflage makes this voracious predator hard to spot among its woolly flock, where it is very much the wolf in sheep's clothing. Other green lacewing larvae use different materials, including protective trichomes (plant hairs) clipped from sycamore leaves, to make trash packets. The trash packets presumably protect lacewing larvae from predators, and deter ants that would otherwise attack the larvae or chase them away from ant-tended aphid colonies. When mature, the green lacewing larva uses silk spun out of its anus to weave a tight silken cocoon, which holds the pupa until it pops a cap off the cocoon and crawls out to transform into an adult. Neuropteran silk, although spun out of the anus, is produced in the excretory ducts (malphigian tubules) that discharge into the digestive tract.

Brown lacewings (family **Hemerobiidae**, page 252), unlike lacewings in the larger and more familiar green lacewing family, don't put their eggs on stalks, and their aphid-eating larvae never cover themselves with the bodies of their prey. Brown lacewing cocoons are made of silk spun from the anus, but they are loosely woven, in contrast to the tight silken cocoons of green lacewings.

Green and brown lacewings aren't the only nerve-winged insects out there fighting the homopteran hordes. If you look closely at a whitefly infestation you might spot some tiny neuropterans, not much bigger than the whiteflies themselves and covered with a fine whitish powder, a bit like whiteflies. These little predators are called dusty-wings (**Coniopterygidae**, page 253), and are often overlooked because of their small size. It is often possible to coax a few dozen dusty-wings into fluttering flight by giving a tree branch a good whack with a stick.

The order Neuroptera includes a few other familiar, fascinating families every naturalist should recognize. Some are known for their conspicuous or distinctive adult forms, and some are familiar only because of their larvae. Antlions, which have a family name (**Myrmeleontidae**, page 253) derived from the Greek for ant (*myrmex*) and lion (*leon*), are well known because of the conical pits made by the strange, squat larvae of some species. If it were not for their charismatic larvae, however, antlions would make up a little known group characterized by mostly nocturnal, weak-flying adults with long, slender bodies like damselflies. Antlion adults are more or less the same size as damselflies, but they have clubbed antennae (unlike the inconspicuous little antennae of damselflies) and the wings are usually partly wrapped around the long, soft body. Most people have never seen an antlion adult, but the conical pits made by their larvae are familiar sights in sand or dry, friable soils. Look for circular pits about an inch (2.5 cm) deep and 1–2 inches (2.5–5 cm) in diameter.

Ants or other insects that stumble into one of those conical depressions are often aided in a one-way descent by sand-flicking motions made by the bizarre, fat, hairy antlion larva (doodlebug) concealed at the bottom of the pit. The fate of an insect landing in such a pit is to have its body contents sucked out through the doodlebug's disproportionately long, wicked-looking mandibles. Try sifting the sand at the apex of an antlion pit through your fingers then, when you have had a look at the exposed and ungainly fat doodlebug exposed on your hand, release it on the sand and watch it back around in circles to make a new pit. It will later pupate in a spherical, sand-covered, silken cocoon at the bottom of its pit

The "nerves" of nerve-winged insects are of course veins full of haemolymph, as attested to by the biting midge (Ceratopogonidae) sucking "blood" from the veins of this Goldeneye Lacewing.

before transforming into the delicate, nocturnal adult. Adult antlions live at most for a few summer weeks before scattering eggs on the sand and dying soon thereafter. The larvae burrow into the sand to spend the winter, resurfacing in spring to build their pits and await prey. Larvae can survive long periods of famine by reducing their metabolic rates, and antlions may spend anywhere from one to three years in the larval stage.

Although most species of antlions await prey while concealed under plant material or soil without making pits, pit-making antlions are found throughout the warmer parts of the world. The common pit-making antlion of northeastern American beaches is a widespread species simply known as the Common Antlion, or *Myrmeleon immaculatus*. Antlion larvae are sometimes attacked by wasps that allow themselves to be grasped between the doodlebug's long mandibles. The wasps, members of the family Chalcididae with enormously swollen hind legs, use this apparently dangerous ploy to deposit an egg between the doodlebug's jaws where it will develop into an internal parasitoid. At least in the Great Lakes area, antlion larvae are parasitized by a bee fly species that pops its eggs into the antlion pits.

Spongillafly (*Climacia areolaris*), page 254, caption 7.

Like antlions, which are usually found in dry, sandy areas, almost all Neuroptera are strictly terrestrial insects. The one exception to this generalization is the spongillafly family (**Sisyridae**, page 254), the larvae of which develop in freshwater sponges where they pierce the sponge cells and imbibe the contents through their elongated mouthparts. Adult spongillaflies look a bit like small brown lacewings. One of our two genera is dull brown like a brown lacewing, while the other has beautiful cream and brown wings. These pretty little nerve-winged insects are commonly seen around lights at night, along with other Neuroptera such as green lacewings and antlions.

Nerve-winged insects other than the green and brown lacewings, dusty-wings, antlions and spongillaflies tend to be rather rare and obscure, especially in the north. If you are an observant naturalist, you might spot one of our few species of mantisflies (**Mantispidae**, page 255), so called because of their uncanny resemblance to small praying mantises with clear wings. North American mantisfly larvae are parasitoids of spider eggs, and usually develop within a single egg sac. Larvae of many species board adult spiders for a while as they wait for the opportunity to penetrate newly formed egg cases. The odds of any one newly hatched mantisfly larva finding an appropriate host must be quite small since, although most mantisflies lay an enormous number of eggs, they are relatively rare insects.

Beaded lacewing (**Berothidae**, page 255) larvae live as guests in termite nests, where for some reason they remain unmolested despite the fact that they dine on termites. The obnoxious table manners of beaded lacewings extend beyond their habit of eating their hosts, as berothid larvae apparently use blasts of toxic gas from the anal area to knock out their termite prey. Beaded lacewings are primarily southern in distribution, and are rare in the northeast. They can be common at lights in the southeastern states. The same is true for the owlfly family (**Ascalaphidae**, page 255), a mostly southern family that is rare in our area, although one species does reach the Canadian border and a bit beyond. Owlfly adults look like robust damselflies with long, clubbed antennae. Owlfly larvae, which look much like larvae of the closely related antlions, sit concealed on the surface of the ground while they wait for passing insects to ambush. You are most likely to spot adult owlflies resting on thin twigs, with their antennae stretched forward flat against the twig, and their wings partially wrapped around the twig. Another neuropteran family occasionally encountered in our region is the **Dilaridae** (page 255), or pleasing lacewings. It would be pleasing indeed to encounter one of these rare, mothlike neuropterans. Females have a distinctively long ovipositor; males have plumose antennae. Pleasing lacewings have been seen as far north as Michigan.

The neuropteran family – in fact the insect family – that would currently be the biggest thrill to see in eastern North America is the giant lacewing family, **Polystoechotidae** (page 255). These spectacular (over 50 mm) nerve-winged insects are somewhat of an enigma because, despite their abundance in old (pre-1950) collections from eastern North America, nobody has seen a live one here in the last 40 or 50 years. The whole giant lacewing family has only four species, one of which, the Giant Lacewing (*Polystoechotes punctatus*), used to be found across the continent, north to the Arctic Circle and south to Central America. This pretty species now seems to have completely disappeared from eastern North America, although they still occur in remote mountainous areas in the west.

What happened? One guess is that giant lacewings are victims of light pollution. If these large, wide-ranging insects are fatally attracted to lights, they might need large areas free of artificial lights. It may be that the introduction of non-native predators, such as some foreign ground beetle species, has resulted in decimation of *Polystoechotes punctatus* larvae, pushing their populations below sustainable levels. Another possible explanation for the extirpation of giant lacewings is that these insects are somehow tied into fire ecology, and our unnatural suppression of natural forest fires has helped eliminate them. Like several other insects (especially some flies and beetles), giant lacewings are attracted to smoke, and in past years it was not unusual to find them flopping around campfires. Nobody knows why they are attracted to smoke. It could be to mate, or it could be to lay eggs in a newly burned forest where the larvae will find some special food. We don't know what giant lacewing larvae eat, but an educated guess is that they feed on soil arthropods.

The orders Neuroptera, Megaloptera and Raphidioptera (snakeflies; not present in eastern North America) are all closely related to beetles (Coleoptera), and have at some point or another been combined under the name Neuroptera. True Neuroptera larvae are mostly terrestrial, and have long, hollow mandibles in contrast with the massive-jawed aquatic larvae of Megaloptera.

Snakeflies (Order Raphidioptera) comprise one of the two North American insect orders absent in eastern North America (the other one is the Grylloblattodea, rock crawlers). Snakefly adults and larvae are predators; the characteristically long-necked adults strike at exposed prey, and larvae feed on invertebrates under loose bark. The female

A snakefly from western Canada.

snakefly's long thin ovipositor serves to deposit eggs under bark and in crevasses. This odd order, sometimes placed in the Neuroptera and sometimes placed in the Megaloptera, has a peculiar distribution that includes western North America, Asia, and Europe but skips eastern North America.

Dobsonflies and alderflies: The order Megaloptera

If you enjoy fishing, you are probably familiar with the big, awkward-looking adult fishflies and dobsonflies (family **Corydalidae**, page 256) one often sees fluttering along over the water, a bit like big-jawed stoneflies without cerci. If you have ever fished for trout, you are probably most familiar with the bright, black-and-white winged fishflies (*Nigronia*) characteristic of trout streams, and if you fish for bass, you probably know the huge, sickle-jawed gray dobsonflies (*Corydalus*) usually associated with larger rivers, or the smaller mottled gray fishflies (*Chauliodes*) common in ponds and lakes.

Fishfly and dobsonfly larvae, also called hellgrammites, are obviously predacious larvae with conspicuously stout mandibles and many thin gills arising along the length of the body. Next time you are about to put a hellgrammite on a hook, look at the top of the tip of its abdomen. If there is a long pair of flexible breathing tubes, you probably got it from a pond or lake, and if you caught it in well-oxygenated running water it probably belongs to a genus that lacks elongate breathing tubes. Larvae pupate under rocks and logs near the shoreline, and one can often find larvae or pupae of the large dobsonflies (*Corydalus*) in great numbers under riverside rocks in early spring.

Pupae of Megaloptera are remarkably mobile insects that resemble adults with incompletely developed wings but functional mandibles. Dobsonfly females lay their eggs in large masses on objects over the water, and rivers with good dobsonfly populations are often identifiable by the whitish scars left on overhanging branches by generations of egg masses.

Although of less interest as bait, the smaller black adult alderflies (**Sialidae**, page 256) are frequently so abundant along trout streams that most anglers recognize them. They look like robust, black lacewings. The larvae, which are found in streams, look a bit like small hellgrammites, but with a long, tapered tail. Eggs are usually laid on emergent vegetation, and adult females can often be seen clinging to stems over newly laid batches of eggs, perhaps offering some protection from the tiny parasitic wasps that can almost invariably be seen exploring the surfaces of unprotected egg masses.

An alderfly female with an egg mass, page 256, caption 10

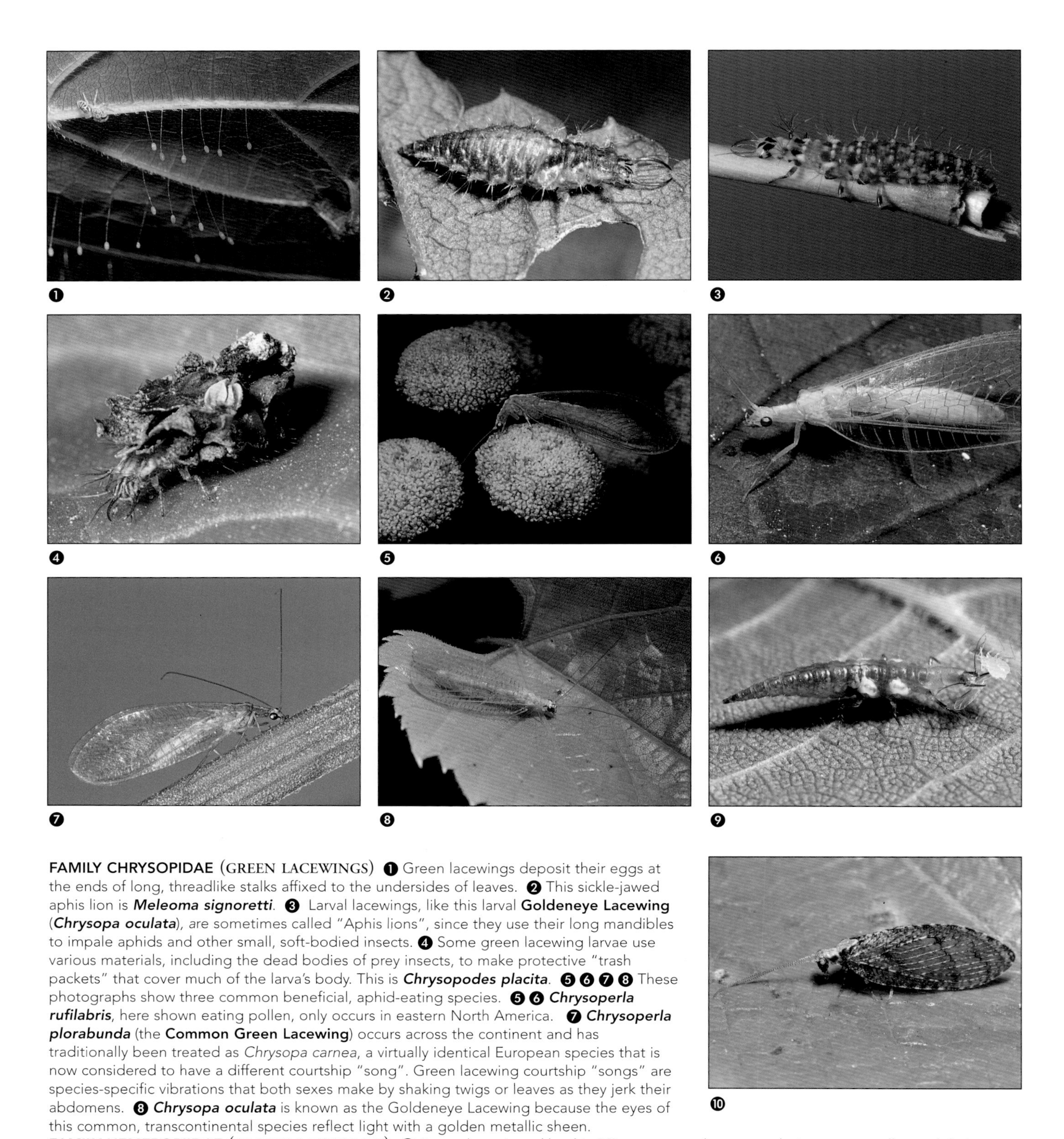

FAMILY CHRYSOPIDAE (GREEN LACEWINGS) ❶ Green lacewings deposit their eggs at the ends of long, threadlike stalks affixed to the undersides of leaves. ❷ This sickle-jawed aphis lion is ***Meleoma signoretti***. ❸ Larval lacewings, like this larval **Goldeneye Lacewing** (***Chrysopa oculata***), are sometimes called "Aphis lions", since they use their long mandibles to impale aphids and other small, soft-bodied insects. ❹ Some green lacewing larvae use various materials, including the dead bodies of prey insects, to make protective "trash packets" that cover much of the larva's body. This is ***Chrysopodes placita***. ❺❻❼❽ These photographs show three common beneficial, aphid-eating species. ❺❻ ***Chrysoperla rufilabris***, here shown eating pollen, only occurs in eastern North America. ❼ ***Chrysoperla plorabunda*** (the **Common Green Lacewing**) occurs across the continent and has traditionally been treated as *Chrysopa carnea*, a virtually identical European species that is now considered to have a different courtship "song". Green lacewing courtship "songs" are species-specific vibrations that both sexes make by shaking twigs or leaves as they jerk their abdomens. ❽ ***Chrysopa oculata*** is known as the Goldeneye Lacewing because the eyes of this common, transcontinental species reflect light with a golden metallic sheen.

FAMILY HEMEROBIIDAE (BROWN LACEWINGS) ❾ Brown lacewings, like this ***Micromus*** sp., do not put their eggs on stalks, and their predacious larvae do not cover themselves with the bodies of their prey, as do some hairy green lacewing larvae. ❿ Brown lacewings, like this ***Wesmaelius*** (***Kimminsia***) ***disjuncta***, are all predators of soft-bodied insects, especially aphids, mealybugs and their relatives.

FAMILY HEMEROBIIDAE (BROWN LACEWINGS) **(continued)** ❶ Our most common brown lacewings belong to ***Hemerobius***, a genus that includes native species as well as species that have been introduced for the control of aphids on forest trees. Although most lacewings spend the winter in cocoons, some *Hemerobius* species, like this ***H. humulinus***, spend the winter as adults and are therefore among the last insects to be seen on the wing in fall, and the first to be seen in spring. ❷ ***Micromus posticus*** is a common beneficial insect. Like most brown lacewings, it is active in relatively cool weather, and can be seen on the wing on crisp autumn days. ❸ ***Sympherobius amiculus***, a small brown lacewing that goes through several generations in a season, has been used for the biological control of scale insects. **FAMILY CONIOPTERYGIDAE** (DUSTY-WINGS) ❹ Dusty-wings are common predators of small insects and insect eggs, but they are so small (less than 3 mm) that they are often overlooked. **FAMILY MYRMELEONTIDAE** (ANTLIONS) ❺ The conical pits made by the larvae of the **Common Antlion** (***Myrmeleon immaculatus***) are familiar sights in sand or dry, friable soils. ❻ **Common Antlion** (***Myrmeleon immaculatus***) larvae, or doodlebugs, are normally concealed at the bottom of conical pits, with only the long, wicked-looking mandibles exposed. ❼ The sand-covered sphere in this photograph is the silken cocoon of a **Common Antlion** (***Myrmeleon immaculatus***). The cocoon and larvae were sifted from antlion pits. ❽ Adult **Common Antlions** (***Myrmeleon immaculatus***) are weak fliers, and are rarely noticed.

FAMILY MYRMELEONTIDAE (continued) Although the familiar pit-forming Common Antlion (*Myrmeleon immaculatus*) is by far our most abundant antlion, we have several other species that do not form pits. ❶ ❷ ***Dendroleon obsoletum*** adults are distinctive for their boldly spotted wings. Larvae can sometimes be found in dry tree holes. ❸ ❹ ❺ The northeastern ***Brachynemurus*** species are uncommon antlions that differ from one another by wing patterns and thorax color. Shown here are ***B. nebulosum*** ❸, ***B. signatus*** ❹ and ***B. abdominalis*** ❺. ❻ The strikingly long abdomen on this antlion is characteristic of ***Brachynemurus abdominalis*** males. **FAMILY SISYRIDAE (SPONGILLAFLIES)** Spongillaflies develop as parasites of freshwater sponges. There are two small genera of spongillaflies in the northeast, the attractively pigmented ***Climacia*** (***C. areolaris***) ❼ and the more uniformly brown ***Sisyra*** ❽. ❾ Spongillafly larvae (***Sisyra*** sp. shown here) have straight, slender, stylet-like mandibles that stick out in front of the head. These odd neuropteran larvae live inside masses of freshwater sponge.

FAMILY MANTISPIDAE (MANTISFLIES) ❶ ***Climaciella brunnea*** is the most commonly encountered mantisfly in the northeast and across the continent. This striking mimic of *Polistes* wasps is variable in color, and seems to mimic different paper wasp species in different parts of its wide range. Newly hatched larvae attach themselves to wolf spiders, and then transfer to the spider's eggs in which they develop as parasitoids. ❷ **Leptomantispa pulchella** is normally a western species, but this one was found in an Ontario grassland. ❸ ***Dicromantispa sayi*** is a rare mantisfly that varies in color from yellowish to almost black. Like the more common *Climaciella brunnea*, *D. sayi* larvae can find and attack spider egg cases without first hitching a ride on an adult spider, but most board spiders and feed from within the spider's book lungs before moving on to the egg sac. ❹ The **Green Mantisfly** (***Zeugomantispa minuta***) occurs from Venezuela north to Wisconsin, but it is very rare in the northern part of its range. ❺ ***Dicromantispa interrupta*** is larger than the other three or four northeastern mantispid species, and has spotted wings. Like other mantispids, it lays enormous numbers of eggs – one captive individual was observed to lay 4,728 eggs. Larvae develop in the egg sacs of a variety of hunting spiders. **FAMILY BEROTHIDAE** (BEADED LACEWINGS) ❻ Beaded lacewings (***Lomamyia*** sp.), which develop as predators in termite nests, are relatively southern in distribution. **FAMILY ASCALAPHIDAE** (OWLFLIES) ❼ Owlflies are common in the southern United States, Mexico and Central America, but they are rare in the northeast (although they occur as far north as Ontario). This photograph was taken in Central America, but it shows the general features common to all owlflies (long, clubbed antennae, robust thorax and dragonfly-like shape). **FAMILY DILARIDAE** (PLEASING LACEWINGS) ❽ Male **Pleasing Lacewings** (***Nallachius americanus***) have distinctively feather-like antennae. This small, mothlike insect is the only member of the Dilaridae in eastern North America. **FAMILY POLYSTOECHOTIDAE** (GIANT LACEWINGS) ❾ **Giant Lacewings** (***Polystoechotes punctatus***) used to be widespread throughout our area, including localities north to 56 degrees, but this species has apparently disappeared from eastern North America. Labeled specimens in the University of Guelph insect collection show that Giant Lacewings were commonly attracted to lights in the southern Ontario city of Guelph up to about 1940, but they seem to have disappeared not only from Guelph but from all of eastern North America by about 1950.

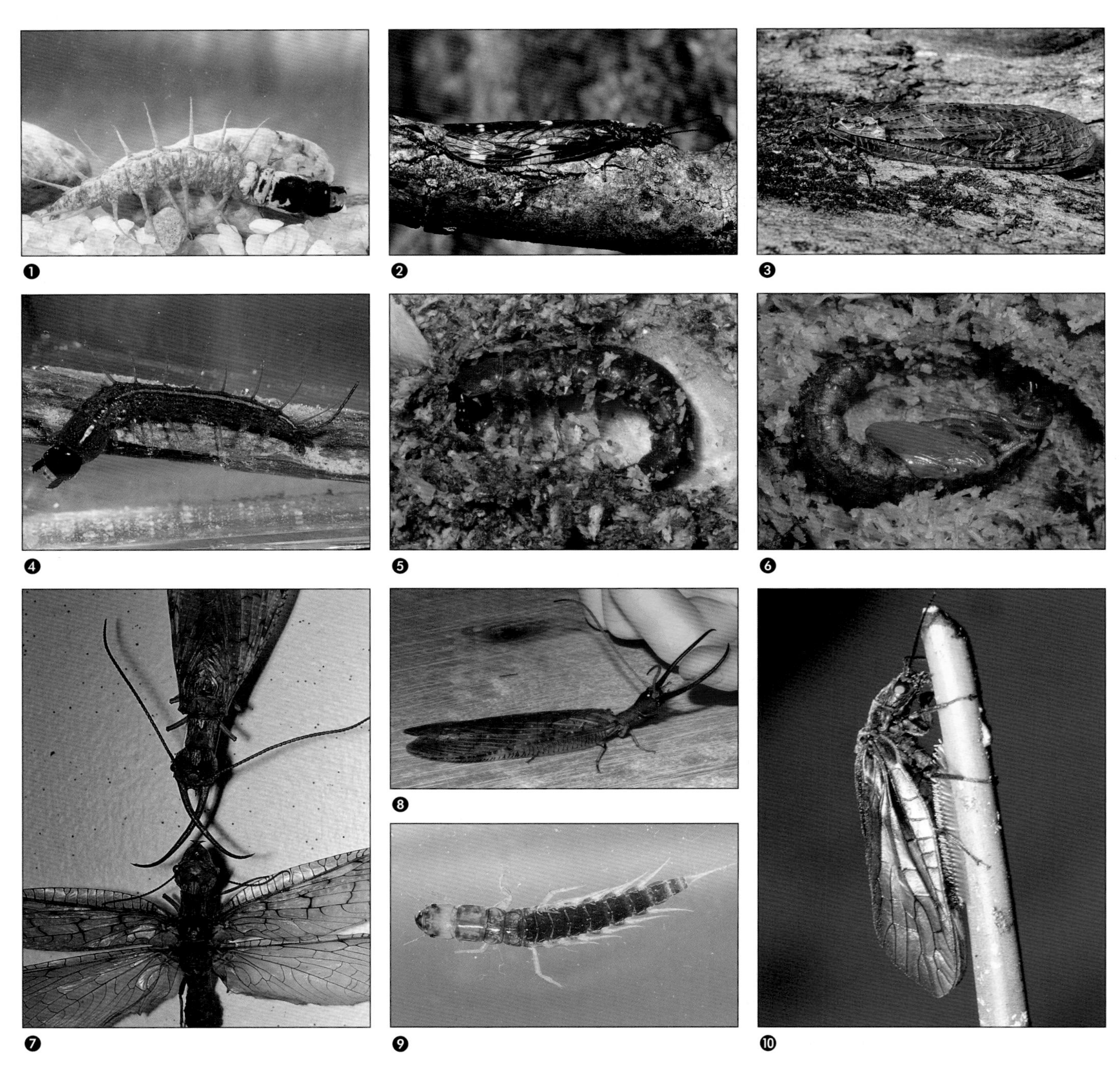

FAMILY CORYDALIDAE (DOBSONFLIES AND FISHFLIES) ❶ The voracious-looking, predacious larvae of fishflies and dobsonflies, called hellgrammites by fishermen, are recognized by their stout mandibles and the many thin gills that extend down the body. This is the common lotic (running water) species ***Nigronia serricornis***. ❷ Adults of the distinctively pigmented fishfly ***Nigronia serricornis*** are common sights along trout streams. ❸ ❹ Fishflies in the genus ***Chauliodes*** develop in lentic (still) waters, where their larvae use long, hoselike "tails" like flexible and retractable snorkels. Corydalid genera (*Corydalus, Nigronia, Neohermes*) that live in well oxygenated running water lack such long breathing tubes. ❺ ❻ Mature fishfly larvae leave the water and make a pupal chamber on shore, often under bark. ❼ Dobsonflies (***Corydalus cornutus***), with a wingspan of almost 14 cm, are the largest northeastern Megaloptera. Only the males have spectacularly long (but weakly muscled) mandibles. ❽ The long mandibles of male ***Corydalis*** species are weakly muscled but can be sharp, and this one (a Central American species) painfully punctured the author's finger. FAMILY SIALIDAE (ALDERFLIES) ❾ Alderfly (***Sialis*** spp.) larvae, which are found in streams, look a bit like small hellgrammites, but with long tapered tails. ❿ Alderfly (***Sialis*** spp.) eggs are usually laid on emergent vegetation, and adult females can often be seen clinging to stems over newly laid batches of eggs, perhaps offering some protection from the tiny parasitic wasps that can almost invariably be seen exploring the surfaces of the egg masses.

CHAPTER 10

Beetles

Many great biologists past and present, including Charles Darwin, spent formative years as avid fans of the beetles, often commenting later how much their development as scientists was influenced by their experiences as beetle collectors. This inordinate impact of beetles on generations of young biologists certainly reflects the ubiquity of these conspicuous insects, but it also results from the ease with which nascent biologists can preserve their observations of nature as attractive and diverse pinned collections of hard, shiny beetles.

Even for naturalists with no inclination to save a bunch of pinned exoskeletons, beetles are attractive subjects for study. They are found almost everywhere in the world and seem to do almost everything, often colorfully illustrating basic principles of ecology and evolution. There are more named species of beetles than there are named species of any other group, with the described beetles making up about a quarter of all described animal species. Beetles occur from pole to pole, utilize almost every terrestrial and freshwater habitat, and are generally easy to recognize and observe. The size range of beetles, from 0.25 mm up to 20 cm, is equivalent to that between the tiniest shrew and the largest whale, and the shape and color range is astounding.

Everyone recognizes beetles as beetles, probably because of their familiar armor-sheathed bodies. "Coleoptera" is from the Greek for "sheath wings," referring to the hard forewings that, unlike the wings of other insects, normally meet in a straight line down the beetle's back; they cover the intricately folded hind wings and encase much of the beetle's body. A beetle with its forewings (elytra) snapped shut is like an insectan tank, able to resist attack while penetrating all sorts of semisolid media. When a beetle opens its forewings and unfolds its hind wings in an intricacy of associated joints and struts, it is like a transformation from a tank – or a submarine, in the case of aquatic beetles – into an airplane.

Although the Coleoptera is divided into four suborders, members of two suborders are exceedingly rare. The beetles encountered by the average naturalist belong to two main suborders: Adephaga and Polyphaga. Adephaga (over 40,000 species worldwide) is made up of a few mostly predacious families in which the basal parts of the hind legs (the hind coxae) are fused to the body and completely divide the underside of the first abdominal segment. The other suborder, Polyphaga (over 300,000 species worldwide), is made up of a great diversity of families in which the coxae are mobile and the first abdominal sternite is not divided. These two big subgroups are easily separated by their behavior in the field, so there is usually no need to peer between beetle hind legs unless you need to identify specimens that have been removed from their natural habitat. Most of the dark, shiny beetles that go scrambling for cover when you turn over a rock or log are in the largest family of Adephaga, the ground beetles (Carabidae), while most beetles found on foliage and flowers are in various families of Polyphaga. Both the Adephaga and Polyphaga include abundant aquatic beetles.

The other two beetle suborders are small, rare and unlikely to be encountered by most naturalists. *Micromalthus debilis*, the only species in the family **Micromalthidae** (one of two families in the suborder Archostemata) is among the few insects that apparently reproduce as larvae. Populations are usually made up entirely of virgin larviform females that give birth to larval offspring. "Normal" adult beetles, which look a bit like tiny (2 mm) checkered beetles, appear only occasionally and have been rarely collected in eastern North America. *Micromalthus* develops in decaying wood, and although these little beetles are not often noticed they are widespread, possibly due to the transportation of oak timbers and telephone poles.

The smallest beetle suborder, Myxophaga, does not occur in the northeast and is represented in North America only by three species of minute bog beetles (**Sphaeriidae**, well under 1 mm) and one western North American aquatic species called a skiff beetle. Skiff beetles (**Hydroscaphidae**) look much like small (1–1.5 mm) rove beetles, but live in running water among filamentous algae.

SUBORDER ADEPHAGA

One of the peculiarities of beetle morphology is the way the basal segment (coxa) of the hind leg extends backward over all or part of the first abdominal segment. In the suborder Adephaga, the hind coxae extend all the way across the first abdominal segment, seemingly dividing it into two halves. This odd bit of underside morphology combines with a generally predacious appearance to make the Adephaga an easily recognized suborder. Although it contains only about 10 percent of the Coleoptera, the Adephaga includes some of the most popular beetle families. As suggested by the subordinal name (*adephagos* means "voracious" in Greek), Adephaga are mostly predators both as larvae and as adults. Many Adephaga, including both aquatic and terrestrial groups, are accomplished stinkers that avoid predation by using abdominal (and sometimes thoracic) glands to produce defensive compounds.

Ground beetles (Carabidae)

The great majority of terrestrial Adephaga belong to the ground beetle family (**Carabidae**, page 285), the most diverse family of Adephaga and a ubiquitous group in almost every terrestrial habitat. You can easily expose dark, shiny ground beetles by turning over rocks, logs or other shelters where these normally nocturnal beetles spend the daylight hours. The majority of North America's 1,700 species of ground beetles are predators, so the Carabidae is one of the most beneficial groups of insects despite the few species that eat seeds or other plant material. The predacious species eat a variety of invertebrates. Some use strikingly elongated mouthparts to dine on the inside of snail shells, and some are elegantly adapted to a diet of springtails. Adults of one species (*Loricera pilicornis*) have traplike jaws that snap shut when small prey set off a hair-trigger between the beetle's jaws. Carabid larvae are usually big-jawed, active predators found in the same habitats as the adults, but some are parasitoids that can develop on a single pupa of another beetle.

Most of the big, shiny blue-green or black beetles common in urban gardens are typical ground beetles known as

In all members of the Adephaga, like this *Calosoma* (Carabidae), the hind coxa extend across the first abdominal sternite.

"searchers," some of which were introduced to North America from Europe because of their propensity to search out and destroy pest caterpillars and slugs. Searchers (*Carabus* and *Calosoma*) usually have distinctive dimples along their wing covers and, like most other ground beetles, use abdominal glands to impart a pungent and persistent odor when handled. Some other ground beetles protect themselves with even more potent chemical defenses that burn as well as stink.

Bombardier beetles (*Brachinus* spp.) are attractive little ground beetles with bluish or black wing covers and reddish yellow legs, prothorax and head. When threatened, these beetles point the tip of the abdomen at the threat and release hot defensive chemicals that blast out in a smokelike burst accompanied by an audible pop. This chemical explosion is too small in scale to pose a serious threat to a vertebrate, but even a little burst of this beetle's caustic chemicals, released at incredible temperatures of around 212°F (100°C), can burn and stain your fingers.

The secret to the bombardier beetle's pyrotechnic performance is a sort of two-chambered gun at the end of its abdomen. Just as bullets in a gun are normally stored in a magazine separate from the firing chamber, the defensive chemicals in the bombardier beetle are secreted into an inner storage chamber. Just as a soldier might cock his gun, moving a bullet into the firing chamber in preparation to fire the weapon, a threatened beetle moves the stored chemicals through a one-way valve into an outer chamber in preparation for blasting its attacker. Once in the thick-walled outer chamber of the beetle, the chemicals are "fired" much the way a bullet in the firing chamber of a gun is fired when the firing pin causes heat and energy-releasing reactions in the gunpowder. The beetle does not have a firing pin, but it does have a set of enzymes in that outer chamber that "fire" (oxidize) the chemicals (hydroquinones plus hydrogen peroxide), releasing heat and gaseous oxygen that propels hot material (p-benzoquinone) out of the beetle's anus with an audible pop. Bombardier beetles are often common under shoreline debris, especially near bodies of water frequented by whirligig beetles (**Gyrinidae**). The larvae of bombardier beetles are parasitoids of whirligig pupae, which are found under domelike mud shelters along shorelines.

While most carabids warrant the name "ground" beetles, there are some common and conspicuous exceptions. Some brightly colored little carabids, for example, are day-active beetles that hunt in foliage. The most common of these (*Lebia* and other members of the tribe Lebiini) have a striking resemblance to common leaf beetles (**Chrysomelidae**), and some species develop as parasitoids on the pupae of leaf beetles. Several kinds of ground beetle are associated with pond and stream margins. The most attractive of these are resplendent round beetles that look so unlike other Carabidae they were previously placed in their own family (Omophronidae, the round sand beetles). Splashing water over the sandy margins of waterways will often expose a great variety of seldom-seen, subterranean shoreline inhabitants, including round sand beetles.

Large numbers of individuals and reasonable numbers of species of ground beetles can be easily collected using pitfall traps (cans sunk into the soil, flush with the soil surface), and most ground beetles are easy to recognize and sort out of trap samples. For these reasons, and because there are good guides to the identification of ground beetle species in most regions, ground beetles have been the focus of a large proportion of "arthropod biodiversity" studies over the past decade. At the same time, many leading taxonomists, ecologists and evolutionary biologists have been ground beetle specialists, resulting in an exceptional body of literature on this fascinating group of beetles. The ground beetles and the closely related tiger beetles are now among the most thoroughly studied of all insects.

Tiger beetles and wrinkled bark beetles (Cicindelidae and Rhysodidae)

Adult tiger beetles (**Cicindelidae**, page 289) are often common on open ground such as sun-dappled woodland paths, sandy dunes and beaches. Huge eyes and formidably large three-toothed mandibles conspicuously crisscrossed at their tips give these colorful hunters a characteristic bulgy-headed and toothy appearance.

The most familiar northeastern tiger beetle is an iridescent insect that regularly provides a flash of color to springtime walks in the woods. *Cicindela sexguttata* is a metallic green species called the Six-spotted Tiger Beetle despite the fact that some individuals have only five, two or even no white spots. This common cicindelid usually appears early in spring after spending the winter as an adult hidden in the burrow or shelter where it pupated the previous fall. Like other adult tiger beetles, Six-spotted Tiger Beetles are voracious hunters that hang out on sunny vantage points where they can spot prey and potential predators. *Homo sapiens* evidently falls in the latter category, so most of us first notice tiger beetles as they make short, rapid escape flights in response to our movements. With a careful approach, however, Six-spotted Tiger Beetles can be observed on exposed rocks, soil, logs and even tree trunks in open forests. Most other tiger beetles occur on open sand or other types of bare ground, usually near the specific substrates required by their bizarre burrowing larvae.

Tiger beetle larvae are ambush predators, striking out at prey from their burrows with an enormous, big-jawed head that is also used to dig a burrow by scooping the diggings out and tossing them away from the burrow entrance. Upon backing into the burrow, the larva blocks the burrow entrance with its massive head and prothorax, its body bent at right angles and extending down into the burrow. A pair of stout hooks on a big back-hump just behind the middle of the larva's abdomen serves much the same purpose as the stop in your kitchen drawer, which prevents the drawer from pulling all the way out and falling on your foot. If a small organism comes too close to the burrow entrance, the massive front end of the larva can whip out of the burrow with lightening speed – at least to the limit of the "stop." The heavy, bucket-like head is flung, often backwards, out of the burrow in a couple of hundredths of a second. With luck, it pulls back into the burrow with a captured insect, usually an ant or a fly, which it mashes in its mandibles before liquefying it with regurgitated digestive fluid. Then, like a heavily mustachioed man straining soup through the hairs on his upper lip, the tiger beetle larva consumes its liquefied victim while filtering out the remaining chunks through setae (hairs) on its labrum (upper lip). Adult tiger beetles also liquefy their food, using their formidable mandibles to masticate prey much the way you might eat a pulpy orange, squeezing and swallowing the juice rather than chewing and swallowing the solid bits.

Most northeastern North American tiger beetles are day-active beetles in the genus *Cicindela*, but another genus, *Megacephala*, includes two large, nocturnal species that occur in the eastern United States (but not Canada). Some entomologists include the tiger beetles in the Carabidae because of some evidence that they are closely related to particular subgroups of ground beetles. Other kinds of evidence (different sets of molecular characters, for example) suggest conflicting patterns of relationship, and entomologists are not in agreement about whether the most closely related lineage is within the Carabidae (making tiger beetles Carabidae) or not. In the lack of consistent evidence to the contrary, it seems best to assume that the Cicindelidae is a good family and to continue to treat it as such.

A tiger beetle, (*Cicindela formosa*), page 291, caption 5.

Although almost all northeastern terrestrial Adephaga are ground beetles or tiger beetles, a few members of the suborder Adephaga are placed in the very small family **Rhysodidae** (wrinkled bark beetles, page 292). Wrinkled bark beetles are elongated beetles with antennae that look like strings of beads (unlike the threadlike antennae of ground beetles), and with wrinkle-like grooves on the thorax and elytra. These rarely encountered beetles feed on the amoeba-like plasmodial stage of slime molds, using a bladelike lower lip to cut food from moist, dead wood. Some authors include the Rhysodidae in the Carabidae.

Water tigers and predacious diving beetles (Dytiscidae)

Just as virtually all the terrestrial Adephaga are ground beetles, the great majority of aquatic Adephaga fall in one large family, the **Dytiscidae** (predacious diving beetles, page 292), with just short of 500 North American species. The **Dytiscidae** are aquatic beetles with a streamlined, convex body and flattened, hair-fringed swimming legs. While different members of this family live in all sorts of aquatic habitats ranging from spring streams to huge lakes, the best place to look for predacious diving beetles is a shallow, weedy pond. Here you should be able to find a variety of dytiscids, ranging from the tiny (less than 2 mm) *Liodessus* species up to the giant (over 30 mm) *Dytiscus* and *Cybister* species.

Dytiscidae are easily seen as they swim to and from the surface with speedy, smooth, synchronous strokes of their hairy hind legs. A predacious diving beetle must make periodic visits to the water surface, where it pauses, hind end up, to replenish the air bubble kept under its wing covers. That bubble is more than just the air store it appears to be, and it lasts much longer than you might expect because it acts like a temporary gill (a "physical gill"). As oxygen is used up and the carbon dioxide level in the bubble increases, more oxygen diffuses in from the water. The bubble eventually collapses due to nitrogen loss, forcing the beetle back to the surface.

One of the beetles you are most likely to notice surfacing for air in ponds is a medium-sized (about 13 mm) predacious diving beetle (*Acilius semisulcatus*) with a black-margined square emblazoned on its yellow prothorax, and curved bands crossing the back of its attractively mottled elytra. Female *Acilius* have prominent furrows on the elytra, which help the male hang on while mating. Males, on the other hand, are slick and smooth with nonridged elytra.

Predacious diving beetle larvae, called "water tigers," are somewhat spindle-shaped larvae with long sickle-like mandibles for injecting digestive juices and withdrawing digested body contents of other aquatic animals. Water tigers are not difficult to keep in aquariums if you separate them from one another and if you have the patience to keep up with their voracious appetites. Most breathe at the surface using posterior spiracles, but a few breathe through their skin or long abdominal gills. Water tigers, like most aquatic beetle larvae, leave the water altogether to pupate in cells dug into damp soil under pond-side logs or rocks. Pale, soft-bodied adults (teneral adults) are often encountered while still in their pupal cells, where they spend some time hardening up after the larval-pupal molt.

Crawling and burrowing water beetles (Haliplidae and Noteridae)

All beetles in the suborder Adephaga have the bases of their hind legs (the hind coxae) enlarged and extended right across the first abdominal segment, but members of one small aquatic family carry this coxal enlargement to a conspicuous extreme. The hind coxae of crawling water beetles (**Haliplidae**, page 294) are expanded to form huge plates extending over much of the underside of the abdomen, where they cover part of the air store that these small (2–3 mm), oval, convex water beetles periodically take from the surface. The air cavity under the coxal plates is continuous with another air store under the elytra, where it is in contact with the abdominal spiracles. Haliplidae are common in mats of filamentous algae, usually in still waters, where both adults and larvae have the unusual (for Adephaga) habit of eating algae. A few crawling water beetles, including one very rare species, occur in swift waters. The endangered Hungerford's Crawling Water Beetle (*Brychius hungerfordi*) is known from only a few streams in Michigan and Ontario.

Noteridae, the burrowing water beetle family (page 295), is another small and entirely aquatic family of Adephaga. Burrowing water beetles with known life cycles occur around the roots of aquatic plants, where the larvae tap roots to obtain oxygen. They pupate in underwater cocoons filled with air from punctured plants.

Whirligigs (Gyrinidae)

Even by the standards of the weird water-surface world, adult whirligig beetles (**Gyrinidae**, page 295) are special. Water striders and the other bugs, springtails, spiders and flies that walk on top of water are essentially aerial creatures, dependent on their water-repellent properties to avoid puncturing the surface film. Whirligig beetles, on the other hand, literally straddle the line between the water below and the air above, retaining the ability to enter either environment as circumstances dictate. You may have noticed shiny black whirligigs waltzing on the surface of a calm pond, gyrating all over the surface of an evening-calmed lake or perhaps massed in a whirling aggregation of hundreds of beetles in a calm spot along the side of a windy bay. If you have tried to catch one, you will realize that they can dive instantaneously and are as adept at swimming below the

Crawling water beetles in the genus *Peltodytes*, page 295, caption 2.

water as moving on the surface. If you do catch one, be prepared for a bit of a stink. Whirligigs produce strong-smelling chemical secretions (aromatic terpenes originating in abdominal glands), which probably explains why they can whirl around on the open surface of the water unmolested by fish.

There are only two common genera of whirligig beetles in the northeast, each with its own smell. The big (over 9 mm) whirligigs in the genus *Dineutus* smell like rotting apples, while the many species of small (less than 8 mm) whirligigs in the genus *Gyrinus* simply stink. The nasty smells produced by whirligigs and related members of the suborder Adephaga result from an impressive repertoire of chemicals, usually produced by glands at the abdominal apex (pygidial glands). Many pygidial products are familiar steroids (like testosterone) and others are complicated compounds unique to the adephagan arsenal.

If you do get a whirligig beetle in hand (and you can stand the smell) you will see that the hind and middle legs are reduced to comical-looking short paddles, while the front legs are normal and relatively long. The eyes are divided into an upper pair for dealing with whatever is above the water, and a lower pair for dealing with aquatic vision. A close look between the upper and lower eyes will reveal short, clubbed antennae on which the third segments are expanded out to form ear- or dishlike structures. These structures help explain why it is possible to watch a dense mass of gyrinids gyrating for hours without ever seeing them bump into one another. Just as a dish antennae picks up radio waves, the whirligig antenna "hears" the vibrating waves made by other whirligigs; the same device picks up on the motion of potential prey such as drowning terrestrial insects.

Whirligig larvae are elongate, whitish, aquatic insects flanked with rows of feathery gills that allow them to respire without leaving the mud at the bottom of the pond, lake or stream. Their long, syringe-like mandibles serve both to inject poisons into and suck the contents out of bloodworms and other prey. They leave the water to pupate, usually in convex, oval pupal shelters on plant stems or the shoreline itself.

SUBORDER POLYPHAGA

The great majority of beetles belong in the suborder Polyphaga. The name "Polyphaga" refers to the great variety of feeding habits in this diverse group, but the Polyphaga also vary enormously in almost every other attribute, including size, shape, habits and habitats.

Riffle beetles, water pennies, long-toed water beetles, pill beetles and their relatives (Superfamily Byrrhoidea)

Riffles – swiftly flowing, relatively shallow stretches of stream – are astounding for their productivity and diversity. No matter what the season, if you carefully examine a rock or piece of wood out of a local riffle you will find a variety of insects, usually including both inconspicuous larvae and long-legged, slow-moving adults of the riffle beetle family (**Elmidae**, page 296).

Adult riffle beetles and other adult beetles that live in running water face different challenges than those faced by pond beetles. All adult beetles breathe through open spiracles that must be brought into contact with air, an apparent problem in rivers and streams where the water moves too fast for most beetles to be able to swim up to the surface. Riffle beetles deal with this problem by absorbing

Riffle beetles (*Macronychus glabratus*), page 296, caption 2.

oxygen directly from the water using a permanent air bubble as a sort of physical gill through which carbon dioxide diffuses out and oxygen from oxygen-rich running water diffuses in. As long as the bubble can be held together and prevented from collapsing, it serves the adult beetle much the way normal gills serve many immature aquatic insects.

The spiracles of riffle beetles open into air held in place by dense patches of pile-like hairs, so tightly packed that the air trapped by the hairs forms a thin, permanent bubble. Dense mats of hair or other structures that hold a permanent air layer, called a plastron, have the advantage of allowing respiration both in and out of water, but they also seem to limit riffle beetles to well-oxygenated streams, rivers and wave-washed shore areas of large lakes.

Long-toed water beetles (**Dryopidae**, page 296) are closely related to riffle beetles but are larger and have short, clubbed antennae. Most adult dryopids are found clinging to woody debris in the same riffles as the related Elmidae, but some occur on streamside vegetation. Larvae, unlike riffle beetle larvae, are terrestrial and burrow in wet organic soil.

Even if you don't have the patience to seek out the elusive Elmidae and related Dryopidae, your riffle searching will certainly yield some ubiquitous disklike organisms called water pennies (**Psephenidae**, page 296). Although they are normally held tight to rock surfaces like little suction cups, if you flip a water penny over you will see that it is a flattened beetle larva with tufts of abdominal gills and six well-developed legs normally covered by the broad, disklike top of the body. Many naturalists are familiar with the fascinating larvae of water penny beetles, but few have seen the terrestrial adults. The most common species is an inconspicuous black beetle that stays close to the water, usually clinging to wet surfaces of riverside rocks.

The same rocks might be home to minute marsh-loving beetles (**Limnichidae**, page 297) that develop in shoreline organic matter. These attractive little (usually 1–2 mm) semi-aquatic beetles have a turtle-like ability to retract the head and legs "into the shell." Travertine beetles (**Lutrochidae**, page 297), larger but somewhat similar in shape to Limnichidae, form a small group with both larvae and adults often found on woody debris in alkaline streams.

Related beetles in the family **Heteroceridae** (variegated mud-loving beetles, page 297) occur in the mud or sand along the edges of both still and running water where they burrow into wet soil. Splashing water up along the shore to flood their burrows can flush them out in large numbers, but adult beetles so exposed usually take flight within seconds of emerging. Another small family, the ptilodactylid beetles (**Ptilodactylidae**, page 297) can be found along the same bodies of water. Some ptilodactylids have aquatic larvae and terrestrial adults, but most larvae occur in damp wood or decaying vegetation, and adults are usually found on streamside vegetation.

The superfamily Byrrhoidea is made up mostly of aquatic or semi-aquatic families, but the family **Byrrhidae** (page 298) itself is a group of terrestrial beetles known as pill beetles because of their very convex, pill-like appearance, enhanced when they pull in their legs and "play dead." Both adults and larvae of most pill beetles feed in moss or leafy liverworts, with many species restricted to particular habitat types such as sandy shorelines or patches of moss in coniferous forests.

Cicada parasites, plate-thighs and marsh beetles (Superfamilies Dascilloidea and Scirtoidea)

A large proportion of all insects are parasitoids that develop on – and kill – a single host, and many families of flies and wasps are diverse groups made up entirely of parasitoids. Beetles, on the other hand, seem to resort to parasitism relatively infrequently, and parasitic beetle families are generally small, with few, rare species. The **Rhipiceridae** (page 298), for example, includes only five species in North America, and the biology of only one species is known. *Sandalus niger* lays huge numbers of eggs, yielding active larvae that apparently seek out cicada nymphs to parasitize. Larvae are found on tree roots with their hosts; adults are found on tree trunks. Rhipiceridae is the only eastern family in the small superfamily Dascilloidea.

Marsh beetles (**Scirtidae**, page 298) are small (2–4 mm) beetles that often abound on foliage near ponds and streams. The larvae, which have exceptionally long antennae, are aquatic or semi-aquatic, often living among submerged rotting leaves in tree holes and similar habitats. Marsh beetles are grouped in the small superfamily Scirtoidea with two rather obscure families, the plate-thigh beetles (**Eucinetidae**, page 344) and the aptly named minute beetles (**Clambidae**). Plate-thigh beetles are rarely encountered fungus beetles with platelike bases on their hind legs. Minute beetles are common in compost and litter, but they are so small (usually a millimeter or so) that they are easily overlooked. Like the similar Leiodidae, many minute beetles are convex and can roll into a ball.

Water scavengers and histers (Superfamily Hydrophiloidea)

Ponds and other still waters always abound in beetles, especially predacious diving beetles (Dytiscidae) and water scavenger beetles (**Hydrophilidae**, page 299). Although our fauna of 258 North American (150 or so northeastern) species of water scavenger beetles is a bit less than half the size of our dytiscid fauna, northeastern hydrophilids range from 1–40 mm and are more diverse in shape and habits than Dytiscidae. Some species of water scavenger beetles have a striking resemblance to predacious diving beetles, but these two common families differ in some consistent ways. Hydrophilidae lack the elegant swimming style of the Dytiscidae and instead crawl or do a sort of coleopteran "dog paddle" with alternating strokes of the middle and hind legs. Furthermore, they break the water surface to replenish their air supply, not with the tip of the abdomen as in the Dytiscidae, but with short, expanded antennae that are generally exceeded in length by the long maxillary palpi. Instead of tucking an air store under the forewings like a diving beetle, a water scavenger beetle keeps its air bubble in place using a pile of hairs on its underside, creating its characteristic silver-bellied appearance.

The common name "water scavenger beetle" is somewhat inappropriate for this diverse family, because some adult water scavengers feed on plants or insects, and the larvae are almost always predators. A few species are found in seemingly unlikely places like cow dung, and several occur in rivers and streams, but larvae and adults of most water scavenger beetles occur in ponds. Pupation usually takes place in mud cells on the shoreline, and newly emerged adults sometimes stay in the mud cells for a few days before heading back to the water.

Larval aquatic water scavenger beetles hatch from eggs protected by silken shelters, either placed over single eggs or woven into silken cases that keep groups of eggs surrounded by air even when submerged. Although the silken egg case is usually attached to aquatic plants, some hydrophilid females carry their egg cases attached to the body, and the female of one very large, common water scavenger beetle produces a floating egg cocoon with a sail-like structure that projects above the water surface, where it apparently functions in gas exchange between the cocoon and the air. The architect of those remarkable sailing egg cocoons is the same big black hydrophilid that often ends up attracted to lights when flying from pond to pond. Perhaps because of this habit, and perhaps because of their large size (around 40 mm), these shiny black beetles (*Hydrophilus*) are common showpieces in insect collections. Inhabitants of ephemeral, isolated habitats like ponds are often fully winged and good at dispersal, so it shouldn't come as a surprise to find not only Hydrophilidae, but also Dytiscidae and other pond insects commonly flying into lights.

Hister beetles (**Histeridae**, page 301) are terrestrial beetles closely related to Hydrophilidae. Histers are usually shiny black, maggot-eating beetles, with elbowed antennae and elytra that seem too small for the body. These terrestrial beetles are found almost anywhere maggot hunting is good, and can be especially common in carrion, dung and under bark. Some common species found under bark are strikingly flattened beetles with conspicuous, pincer-like mandibles used to prey on mites, insect larvae and other subcortical neighbors. Several hister beetles live in ant nests, where some feed on detritus, others feed on ant brood or prey brought back by their ant hosts, and some solicit worker ants to regurgitate food. Histerid adults are able to retract the head deep into the prothorax and pull the appendages tightly against their hard, smooth bodies when attacked; ants sometimes just pick up the resultant seedlike package and drop it unharmed.

Carrion beetles, rove beetles and their relatives (Superfamily Staphylinoidea)

Every living thing has to die sometime. When an animal (or person) is transformed from a living being to a corpse, the odds are that the resultant ephemeral, protein-rich body will become home to a few generations of specialized carrion

A water scavenger beetle, (*Hydrobius melaenus*), page 299, caption 3.

insects. Cadavers may indeed be inevitable, but they are unpredictable in their occurrence and are exploited mostly by two orders of insects with very different preadaptations to temporary and irregular food sources. Flies, with their rapid development times and ephemeral but highly mobile adults, are the masters of short-lived resources like cadavers or dung. Beetles, characteristically with longer development times and long-lived adults, come a close second in the scavenger sweepstakes. Flies cover a lot of territory for little time; beetles usually cover a little territory for a long time. Beyond these different means to essentially the same ends, beetles and flies directly compete for their shares of desirable dead bodies.

One way to open a window on the weird assemblage of insects that specialize on dead bodies is to take a small carcass, such as a mouse, and set it over the mouth of a can buried in the soil up to its lip. The can will serve as a pitfall trap for insects attracted to the carrion, and if you add water (with some salt as a preservative and some dishwashing soap to break the surface tension) to your pitfall trap, you can leave it for days and see what accumulates. The most colorful – and smelliest – of the diverse pile of beetles invariably trapped will be the bright red and black burying beetles and other carrion beetles (**Silphidae**, page 302) – nature's undertakers.

Burying beetles or sexton beetles (*Nicrophorus* spp.), usually work in pairs to quickly dig the soil out from under small cadavers, sinking the bodies in the resulting holes and completely entombing the deceased in cryptlike chambers. Beetles bury bodies not out of altruistic concern over the aesthetic undesirability of having little carcasses lying around, but because it is to their advantage to conceal their food from flies and other competitors. Despite the industrious undertakings of *Nicrophorus* beetles, the average dead animal has accumulated a few maggots or fly eggs before its burial. Burying beetles avoid competition with flies on their buried food by killing off the fly eggs and maggots with the aid of the small reddish mites that can almost invariably be found clinging to *Nicrophorus* adults. These acarine hitchhikers "pay" for their ride by eating fly eggs and pre-empting competition between rapidly developing maggots and burying beetle larvae. The mites eat burying beetle eggs as well as fly eggs, and under some circumstances might lower the density of *Nicrophorus* as well as their competitors.

Young burying beetles have all sorts of other advantages that give them a good start on life. Not only do they have a private carcass protected by maggot-killing mites, they also enjoy a degree of maternal care unusual in the insect world. The mother burying beetle hangs around in the crypt, chews up the carcass to make a tasty, hairless, spherical carrion casserole, then feeds her larvae after they move into a pocket she makes on the outside of the prepared body.

Despite this coddled upbringing, some burying beetles have a hard time making a go of it. *Nicrophorus vespilloides*, for example, can only bury cadavers in Sphagnum moss, and this species disappears when its required bog habitats disappear. Another species, the impressively large (25–35 mm) American Burying Beetle (*Nicrophorus americanus*), has disappeared from much of its formerly extensive range in eastern North America. The American Burying Beetle has been thoroughly studied and widely reintroduced in the States since its addition to the U.S. endangered species list in 1989, but there is still disagreement about what almost drove it to extinction. Light pollution, disease, competition and habitat loss have all been suggested, but the most likely explanation is that the disappearance of large predators has led to an explosion of vertebrate scavengers that compete with American Burying Beetles for the passenger pigeon–sized carrion they require.

A burying beetle, (*Nicrophorus carolinus*), page 302, caption 5.

The Silphidae includes a variety of patrons of putrefaction other than burying beetles, the most familiar of which are large, flattened beetles common under older carrion, like *Necrophila americana*. These nonburying carrion beetles avoid competition with flies by specializing on older carcasses that have become too dry for most maggots.

Most Silphidae are associated with dead bodies, but some eat plant material, and eastern North America was home to at least one predacious species for a little while. A predacious silphid (*Dendroxena quadrimaculata*) was introduced from Europe to northeastern United States as a potential biological control agent for the larvae of another exotic insect, the Gypsy Moth (Lymantriidae), but it does not seem to be established.

Dead bodies are attractive resources, and carrion specialists can be found in a wide variety of major beetle families such as Cleridae, Dermestidae and Scarabaeidae, as well as the small carrion beetles (**Leiodidae**, page 303, subfamily Cholevinae – sometimes treated as the family Leptodiridae). These distinctively shaped, small (usually 3–4 mm) beetles sometimes live in fungi, ant nests or mammal nests, but the ones you are most likely to encounter are attracted to dead bodies.

The most abundant beetles on corpses of all sorts, however, are not the descriptively named small carrion beetles, sexton beetles or carrion beetles, but the ubiquitous and distinctively shaped rove beetles (**Staphylinidae**, page 304).

Beetles, overall, are well-known insects, and generations of beetle enthusiasts have left relatively few undiscovered species of North American Coleoptera. Staphylinidae provide a challenging exception to that generalization, and constitute one of the last great frontiers in North American beetle taxonomy. More than 40,000 species of Staphylinidae have been described worldwide so far, and there is a good chance that the family will ultimately ring in at way over 100,000 species. At last count, the Staphylinidae was the biggest family in North America with well over 3,000 species, and the second biggest family in the world (after the weevils, but just barely).

Turn over a rock or log, or examine some decaying material, and you will frequently find numbers of these fast, elongated, short-winged beetles. They somehow manage to look fierce because of the scorpion-like way they hold the tip of the abdomen up in the air, but most are dangerous only to the insects on which they feed. Rove beetles are usually predacious, but some feed on fungal spores or tissue, some eat pollen, some live as specialized parasitoids inside pupae or puparia and some are specialized guests in ant nests and termite nests. Larvae of one distinctively broad-headed genus (*Stenus*) are invariably common along the edges of ponds and streams where they prey on springtails with the aid of a sticky lower lip. When knocked into the water, some *Stenus* adults release chemicals from the end of the abdomen that lower the surface tension behind the body and cause the beetle to zip forward across the water surface.

Some chemicals produced by other rove beetles are less charming than the propellant produced by *Stenus*. Striking red and blue rove beetles in the genus *Paederus*, for example, produce a unique defensive chemical (pederin) that can be irritating to insect watchers. One of my graduate students recently returned from a collecting trip with his face covered with lesions and swollen beyond recognition after he swatted some *Paederus* that landed on his face while he was collecting moths at a nightlight. Unlike blister beetles, which pump their toxic blood out at the leg joints, *Paederus* beetles must be crushed or roughly brushed to release their toxic chemical.

Both the brightly colored *Paederus* species and the distinctive shoreline *Stenus* species are day-active, and are often seen in exposed situations. Most other Staphylinidae shun the light, either staying in concealed areas or straying out only under the cover of darkness. Look for predacious rove beetles in fleshy fungi, soil or leaf litter, under bark and around dung, carrion and in other decaying material rich in maggots and other potential prey. Several rove beetles are myrmecophiles that live in ant nests, where they are often tightly integrated with the colony and produce glandular secretions that influence ant behavior. Myrmecophilous rove beetles include predators, scavengers and species that induce worker ants to regurgitate food.

Some groups of beetles now treated as subfamilies of Staphylinidae have such distinctive shapes they have been traditionally treated as separate families. The shining fungus beetles (Scaphidiinae), for example, are distinctively convex, tapered beetles that often occur in large numbers in fungi and slime molds. Short-winged mold beetles (Pselaphinae) also have short elytra, but these small beetles (usually only a couple of millimeters long) otherwise look unlike other rove beetles because of their abruptly clubbed antennae and narrow pronotum. Although the Pselaphinae is a big subfamily (650 species in North America), short-winged mold beetles are rarely noticed because of their small size and cryptic habits. Most are predators found in dark, damp places including caves, ant nests, leaf mold and peatlands. The species found in ant nests, where they eat ant larvae, are like some ant-associated rove beetles in producing "appeasement substances" that induce ants to treat them as members of the colony. Other small, short-winged beetles are found in other cryptic habitats, including mammal nests. The **Leiodidae**, for example, includes a minute, flattened, short-winged ectoparasitic beetle (*Platypsyllus castoris*; sometimes put in Platypsyllidae or Leptinidae) that lives only in beaver fur. The family Leiodidae also includes other beetles that live in mammal nests, carrion and fungi. The small carrion beetles (sometimes treated as Leptodiridae) and the round fungus beetles are part of the Leiodidae.

The superfamily Staphylinoidea includes a couple of other families that are rarely noticed because of their minute size. The antlike stone beetles (**Scydmaenidae**, page 307), for example, are often less than a millimeter in length and top out at about 2.7 mm. They occur in moist forest habitats such as rotting logs and leaf litter, where they are specialized predators of oribatid mites. Some antlike stone beetles deal with their well-armored prey with specialized devices such as suckers used to separate prey from the substrate, and mandibles used to open mites in "can opener" fashion so that body fluids can be guzzled out of the opening.

Feather-winged beetles (**Ptiliidae**) are common in compost heaps and a wide variety of other moist, organically rich substrates, but with a size range of 0.3–1.2 mm, they are easily overlooked. Most feed on fungi, and most have feathery hind wings reminiscent of the fringed wings used by some other really minute insects (such as fairyflies and small thrips) to swim through the air.

"Lamellicorn" beetles (Superfamily Scarabaeoidea)

Several families of beetles, including some of our largest and best-known species, are characterized by one-sided antennal clubs made up of several lobes or "lamellae." These beetles are sometimes called "lamellicorn" beetles.

Scarabaeidae

If there was a "most familiar beetle award" it might well go to the common June beetle or May beetle, known in its larval stage as the juicy white grub of lawn and garden, and known in its adult stage as that familiar large, shiny brown beetle that bashes against lights during warm May and June nights. Several similar species, all commonly called June beetles, feed on vegetation at night and develop from fat, white, C-shaped larvae (grubs) that feed on roots of grasses and other plants. Carefully examine the next adult June beetle that careers into your screen door. The end of its antennae will seem to have a distinct, but asymmetrical, club made up of three lobes. When you put the beetle down and it prepares to spread its elytra and unfold its hind wings for take-off, that distinctive antennal club will open out to expose sensory surfaces on three separate leaflike lobes (lamellae). Lamellate antennae that can be closed are unique to a group of beetles called "scarabs" (**Scarabaeidae**, page 285, and related families).

Many familiar scarabs, like June beetles (*Phyllophaga* and *Serica* spp.) Rose Chafers (*Macrodactylus subspinosus*) and Japanese Beetles (*Popillia japonica*), have root-feeding larvae and leaf-eating adults but others abound in a spectrum of less savory habitats such as foods previously processed by other animals and egested as odorous waste material. Egested waste material, also known as dung, feces, poo and more additional synonyms than anything else in the English language, is potentially one of the most ubiquitous substances on the planet, so we should appreciate the armies of insects that prevent us from being knee-deep in the stuff. This was strikingly illustrated not long after cattle were first brought to Australia. Kangaroos and other native Australian animals produce relatively small, dry droppings, so native Australian dung beetles were not adapted to deal with the big juicy cow flops that were soon covering about a fifth of an acre per cow with droppings every year. Plenty of flies were living in the dung, but it was taking five years for a dropping to decompose, in contrast to the few summer weeks it takes for a cow flop to disappear in North America. The upshot of all this was the loss of millions of dollars worth of pasture under cow dung. That problem has now been partly solved by the careful introduction of a scarabaeid cleanup crew from Africa, including beetles adapted to deal with the big droppings of big mammals.

Next time you are out in a pasture, stake out a fresh cow flop and watch for some scarab beetles. It probably won't be long before you hear the low buzz of scarabs winging their way to your odoriferous location, closing their wings and their antennal clubs as they drop their heavy bodies into the excremental arena. "Arena" is an appropriate word, because the dung-eating game is a competitive one, full of pitfalls and interesting strategy. Dung beetles must protect their food from flies and other beetles; they must also protect themselves from the predators and parasites attracted to concentrations of such coprophagous (dung-eating) life. Coprophagous scarabs have developed a variety of ways to lay claim to their slice of fresh cow pie, often by partitioning pieces of dung off the mother lode and storing them in private, cool, moist places where the scarab larvae can feed unmolested. This may simply involve digging a hole under the dung, or it may involve a variety of dung rolling and nesting behaviors.

Dung-rolling scarabs, or tumblebugs (subfamily Scarabaeinae), move bits of dung some distance from the original messy mass, facilitating the trip by forming bits of dung into balls that they can roll or tumble along using their long hind legs. The dung balls are put in underground nests, one egg to the dung ball, so the larvae are ensured a protected place to feed and develop in relative freedom from enemies or competitors. Of course, whenever something (or somebody) saves or stores some sort of resource, there are brigands on hand to steal the savings. So it is with the nesting scarab beetles. Some of the ubiquitous little flies attracted to dung are specialized thieves that ride around on scarabs, depositing their eggs in the food stores those scarabs so solicitously sequester. Scarab-riding flies in the family Sphaeroceridae are quite common in North America, but were only formally discovered and brought to the attention of science a few years ago. The many kinds of wasps and flies, and the few beetles, that develop on the food stored by other insects for their young are called kleptoparasites.

Some of our common dung scarabs are large and brightly colored beetles adorned with spectacular horns and processes on the head and pronotum. Tropical scarabs are often especially enormous and conspicuously adorned, but even in North America the Scarabaeidae includes the heaviest and most heavily horned of beetles. Charles Darwin commented at length on horned beetles, concluding that the horns are only strongly developed in males because they are used to impress females. He was a little bit off the mark on this one, as the horns are now known to serve in a variety of male-male interactions. Beetle horns usually serve either to pry an opponent up as part of a fight, or to actually pick up an opponent and throw it to the ground. Some of most impressively armed scarabs are massive species in the subfamily Dynastinae, a subfamily normally associated with rotting wood. The largest dynastines are tropical, but an eastern North American dynastine (the Eastern Hercules Beetle, *Dynastes tityus*), is the heaviest North American beetle.

Most of the 1,500 or so North American scarab species are either root and foliage feeders or dung feeders, but there are many exceptions. Some scarabs, such as the beelike *Trichiotinus* and *Euphoria* species, are pollen feeders as adults. *Euphoria* beetles fly with a loud buzz, holding their elytra closed (other beetles hold their elytra extended), extending their hind wings through special notches at the wing bases. Species of a closely related genus (*Cremastocheilus*) live in ant nests where they keep their ant hosts appeased with glandular secretions. Many other scarabs, including some of the largest species, develop in rotting wood. *Osmoderma*, for example, includes two northeastern species of shiny black, flat-topped beetles commonly found as huge white grubs in rotting wood at the heart of standing trees. *Osmoderma* beetles look and smell like patent leather, so much so that one species is known as the Odor-of-leather Beetle (*O. eremicola*).

Ceratocanthidae, Geotrupidae, Trogidae, Glaphyridae, Hybosoridae and Ochodaeidae

Beetles with lamellate antennae have been traditionally divided into three families (scarabs, stag beetles and bess beetles), but newer classifications divide the scarabs into several families. Most species remain in the Scarabaeidae, but three "splinter groups" now recognized as separate families are frequently encountered throughout northeastern North America. Pill scarabs, or contractile scarabs (**Ceratocanthidae**, page 311), have the remarkable ability to roll into a seedlike sphere. Not much is known about their biology, but some have been collected in tree holes and in bess beetle (Passalidae) burrows and tropical species appear to be associated with termite nests. Earth-boring scarabs (**Geotrupidae**, page 312) are more common than pill scarabs, but are rarely seen because of their subterranean habits. Adult earth-boring scarabs dig deep burrows; some burrows are provisioned with various kinds of organic matter for their larvae. Skin beetles or hide beetles (**Trogidae**, page 312) can be found on old, dry carcasses. They look a bit like warty June beetles that withdraw their antennae and play dead when disturbed.

Three additional small scarab families occur in the northeast. The day-flying **Glaphyridae** (bumble bee scarab beetles) are sometimes found on flowers along the Atlantic coast; the **Hybosoridae** and **Ochodaeidae** (sand-loving scarab beetles, page 312) are nocturnal and rarely encountered.

Lucanidae and Passalidae

Scarabs are not the only large beetles to be found in rotting wood, nor are they the only beetles with a lamellate, asymmetrical antennal club, although the ability to fold up the lobes of the antennae into a compact ball seems to be special to the scarabs. Careful searching of rotting wood will often turn up two little families of large beetles with permanently open lamellate clubs, the Lucanidae and Passalidae.

Lucanidae (stag beetles, page 312) are notable mainly for the male's large mandibles, which are sometimes half as long as the beetle's body and branched like the antlers of a stag. Males battle for territory using their huge mandibles, with the biggest male generally ending up the victor. Although only 10 of the 30 North American stag beetle species occur in the northeast, they range in size from 8–40 mm and the larger ones are spectacular beetles with long life cycles. Large lucanids sometimes spend up to six years as wood-eating larvae, in striking contrast to some beetles

(Leiodidae) that live in ephemeral slime mold sporocarps and have a larval stage lasting only a couple of days.

The family **Passalidae** (bess beetles, page 313) includes only a single species (*Odontotaenius disjunctus*) in eastern North America, a large, parallel-sided, shiny black beetle with a remarkable repertoire of sounds. Adults stridulate by rubbing the top of the abdomen against special areas under the wings, and larvae stridulate using modified hind legs. Bess beetles live in groups in rotting wood where adult males and females both masticate the wood and feed it to their squeaking larvae. Their stridulatory "language" of 14 calls includes signals for mating and for distress, and even characteristic post-mating songs. Bess beetles need to stay in touch with their family groups since both adults and larvae feed on adult feces processed by microorganisms that digest cellulose. Although cellulose is broken down by microbes outside the beetle's body, this relationship parallels the way many termites depend on internal microflora to process their cellulose-rich diets.

Fireflies, click beetles and their relatives (Superfamily Elateroidea)

This large superfamily brings together two dissimilar groups. The soldier beetle group (formerly treated as the superfamily Cantharoidea) is made up mostly of predacious beetles with leathery, flexible elytra and soft bodies. The click beetle group (Elateroidea in the narrow sense) is made up mostly of very hard-bodied, elongate, generally plant-eating beetles. Although both of these groups are included in the Elateroidea, they are discussed separately. One uncommon family, the **Artematopodidae** (page 313), does not fit easily into either group. The two eastern species of Artematopodidae, which look like click beetles that cannot "click," apparently develop in moss mats.

Cantharidae, Lampyridae, Phengodidae and Lycidae

Many of the beetle families now included in the big superfamily Elateroidea are somewhat flattened beetles, with visibly soft, flexible forewings. Members of the most familiar family with this general appearance (Lampyridae, or fireflies) have their heads distinctively shielded by a broad pronotum; members of the other common family of soft-winged beetles, the **Cantharidae** (soldier beetles, page 313) have their heads clearly visible from above. Soldier beetles are commonly found standing guard on goldenrods or other pollen-rich flowers, and some species regularly patrol foliage in search of aphids. Despite their diversity and occasional bright colors, this relatively ordinary behavior puts soldier beetles in the shadow of their brilliantly interesting cousins, the **Lampyridae** (fireflies or lightningbugs, page 315).

Most of us appreciate the association between flashing fireflies and warm summer nights, although perhaps without appreciating that those familiar mobile flashes originate in light organs at the abdominal tips of adult beetles. All firefly larvae and some firefly eggs also produce a bit of light, albeit rarely noticed except by those who take nocturnal strolls through dark swamps far from urban lights. You may not have had the pleasure of picking a glowing larva out of some damp moss, but your childhood might well have included a jar of blinking firefly adults, the closest thing to stars you were ever able to catch.

Fireflies produce their efficient, almost heat-free light on demand by combining two substances (luciferin and luciferase) with a high-energy compound (ATP) in their abdomen. They do it, generally, for sex. The usual case

A soldier beetle, (*Chauliognathus marginatus*), page 313, caption 8.

among eastern North American species is a sort of luminescent Morse code arrangement, with the males of each species flying around flashing at a rate, number and duration that transmits his identity to any watching females of the same species. When the females, usually on the ground, flash back a matching code, a liaison usually follows. Sometimes, however, female fireflies turn their luminescent charms into deadly seduction by faking the code of another species. When the male responds and homes in on his hot prospect, he ends up a cold meal, truly the result of a fatal attraction. The *femme fatale* gets more than just a simple snack by luring in one particular kind of male victim, since the relatively small male victims (*Photinus* species) are rich in defensive chemicals that are appropriated by the larger killer females (*Photuris* species) and used to protect their eggs. Other predators avoid fireflies, and the clear drops of blood that sometimes appear on firefly elytra when you handle them are loaded with bitter or poisonous defensive chemicals.

Some firefly females have carried the sit-and-wait strategy to such an extreme they have lost their wings and look like larvae with compound eyes (larvae never have compound eyes). A similar extreme sexual dimorphism is found in the small, related family **Phengodidae** (glowworms, page 317). Phengodidae, which are rare in the northeast, are called glowworms because the larvae and the larviform females – but not the adult males – are usually luminescent. (The term glowworm is also used for bioluminescent fly larvae in the family Mycetophilidae.) Larval Phengodidae are specialized predators of millipedes, and an attacking glowworm deals with a millipede's defensive compounds by coiling around the millipede's head and dispatching its victim with a strategically placed bite below the head. The glowworm then drags the millipede underground, sucks out the insides of the millipede's head and eats the millipede's internal organs with the careful exception of the defensive glands.

Glowworms and fireflies are the only bioluminescent beetles in northeastern North America, but light organs are found in some exotic Coleoptera, including some incredible neotropical click beetles (Elateridae) equipped with "headlights" (two bright lights on the front of the thorax).

Most fireflies are nocturnal, as befits an insect equipped with a natural flashlight, but some common species are lightless and diurnal. *Ellychnia* species, for example, are lightless fireflies commonly seen on tree trunks during the day (even during the winter). Lightless fireflies are easily confused with beetles in a smaller family (76 North American species), the net-winged beetles or **Lycidae** (page 317). Net-winged beetles are usually similar in shape to fireflies, but the tip of the abdomen is never modified into a conspicuous yellow light organ, and the wings are crisscrossed by a netlike grid of ridges. Furthermore, they are not predators like fireflies, but instead eat mostly nectar or honeydew as adults, and myxomycetes or fungi as larvae.

Net-winged beetle adults are protected from predation by distasteful chemicals, and they advertise this fact with bright colors and conspicuous, often communal behavior. Several other insects mimic these bright warning (aposematic) colors, either as Müllerian mimics (other distasteful insects) or Batesian mimics (perfectly edible insects pretending to be distasteful). The most convincing lycid mimics in the northeast are lichen moths (Arctiidae, also distasteful), but several beetle families also include remarkable mimics of lycid models. One group of southwestern long-horned beetles (Cerambycidae) includes several palatable lycid mimics that mount the much more common, distasteful net-winged beetle models as if to initiate mating. Instead of mating, however, the cerambycid nips the lycid at the base of the elytra and drinks its blood. Before sequestering protective chemicals from the lycid, the cerambycid is a Batesian mimic, but after drinking the lycid's blood it presumably becomes a Müllerian mimic.

Elateridae, Throscidae, Eucnemidae and Cerophytidae

Click beetles (**Elateridae**, page 317) are distinctively shaped, parallel-sided beetles, with characteristic sharp points at the hind corners of the large, mobile prothorax. To see what these common beetles can do with that flexible thorax, look on foliage or dead wood for one of the thousand or so North American click beetle species. When you find one, give it a poke and watch as it quickly folds its legs against its body and drops to the ground. If it lands on its back, watch it flex its thorax so that only the head and back (elytra) are touching the ground. When the click beetle bends like this it is pushing a long peg that projects backward from the underside of the prothorax against a peg-hold or catch between the middle legs. When that peg slips off the peg-hold and pops into a groove just behind the peg-hold with a distinct clicking sound (snap your fingers if you have trouble envisaging how this works), the front and back parts of the beetle snap up into a jackknife shape, and the beetle is flipped into the air, spinning end over end. The jump takes place in a fraction (1/2000) of a second, putting the beetle in a virtually instantaneous takeoff speed of around 8 feet (2.5 m) per second and an initial acceleration that exceeds the "force of gravity" by several hundredfold. This makes it difficult for potential predators to hold on to clicking click beetles, which is probably the main value of the "jumping" mechanism.

Click beetle jumping is especially impressive when you consider that jumping without the use of any appendages is a talent almost unique to this family of beetles (the legless leaping larvae of some flies, such as Cheese Skippers and other Piophilidae, are also prodigious jumpers). Despite the remarkable acrobatic attributes of adult click beetles, the Elateridae have a generally bad reputation because of their tough, wiry larvae (wireworms) that sometimes attack roots and newly planted seeds. Some click beetles are found in rotten wood, and some are predators.

The click beetles most often noticed by naturalists are large (about 40 mm) widespread species with huge conspicuous eyespots on the top of the prothorax. These spectacular, harmless beetles, aptly called eyed click beetles, are likely to be found near dead wood, where their large, wiry larvae feed on other insects. Dead trees in old apple orchards are great places to look for eyed click beetles.

Although click beetles are generally distinctive insects, a couple of small families have similar shapes and similar click mechanisms. The **Eucnemidae**, tellingly called false click beetles (page 320), look remarkably like unusually robust click beetles from above, but the head is deflexed under the helmet-like prothorax. False click beetles are wood borers, sometimes called "cross borers" because they bore across the grain of dead, usually white-rotten hardwoods.

The small family **Throscidae** (throscid beetles, page 320), sometimes called false metallic wood-boring beetles, includes three small northeastern genera that look like very small click beetles, often with slightly clubbed antennae. Not much is known about throscid biology, but larvae are usually found in rotting wood, and one European species has been described as a liquid feeder on the outer portions of ectotrophic mycorrhizal roots.

The very rare family Cerophytidae, aptly called the rare click beetles, includes only one eastern species which looks like a broad click beetle with saw-toothed antennae and seems to be associated with rotting trees.

Checkered, soft-winged and bark-gnawing beetles (Superfamily Cleroidea)

Checkered beetles (**Cleridae**, page 321) are distinctively shaped, hairy beetles often found on pollen-rich flowers or on fresh-cut wood. Some of our most colorful clerids are common sights on freshly cut logs occupied by other wood-boring insects, since the larvae of predacious species pursue wood-boring larvae in their burrows, and adult clerids often consume adults of wood-boring insects. Several necrophagous clerids are common on dried-out animal carcasses, and one small species has earned the name Red-legged Ham Beetle (*Necrobia rufipes*) because of its failure to discriminate between dead animals in nature and bits of dead animals stored for human consumption.

Soft-winged flower beetles (**Melyridae**, page 322) are related to the Cleridae, but are softer, more convex and broadened toward the hind end of the body. Both adults and larvae of this small family appear to be omnivores, feeding on both plant and animal material. One introduced species (*Malachius aeneus*, the Scarlet Malachite Beetle) has been treated in the literature as a pest of wheat, even though the larvae are ground-dwelling predators that feed on other arthropods. Adults feed on pollen of flowers such as buttercups as well as grass pollen, and are unlikely to significantly damage crops. Scarlet Malachite Beetles are common in northeastern North America, but they have recently become very rare in Europe and have been attracting considerable attention as a possible endangered species in Great Britain.

Some male soft-winged flower beetles are assisted in mating by an enormous swelling near the base of the antennae, and males of several species have deformities near the tips of the elytra that presumably direct the female to glandular products secreted by the male. Females appear to consume these products as a necessary prelude to mating.

The Cleridae, Melyridae and the related family **Trogossitidae** (bark-gnawing beetles, page 323) make up the superfamily Cleroidea. The distinctively shaped bark-gnawing beetles are mostly, as the name suggests, found under bark, but one species (the Cadelle, *Tenebroides mauritanicus*) is a pest in granaries worldwide.

Jewel beetles and flat-headed borers (Superfamily Buprestoidea)

Europeans call beetles in the family **Buprestidae** (page 323) "jewel beetles," a common name that brings to mind glass-topped cabinets full of gleaming metallic green, blue and red beetles displayed like jewels on pins, or perhaps shiny pieces of beetle elytra incorporated into brooches or other jewelry. North Americans use the more descriptive common name "metallic wood-boring beetles," calling up less romantic images of metallic and efficient timber pests.

Adult Buprestidae are sometimes brilliantly jewel-like, and most species have at least a bit of a metallic glint, but members of this family are most easily recognized by the distinctive torpedo-like shape, with the tail tapered and the head sunk into the prothorax as though pulled down into the neck of a turtleneck sweater. A buprestid beetle's "turtleneck" is part of the tough integument that makes these beetles among the most heavily armored insects, as befits beetles that often stand exposed on sunny parts of dead or dying trees.

Most larval Buprestidae burrow in sapwood, where they make distinctive, flattened burrows. The large, white larvae of these beetles are called "flat-headed borers" because the prothorax of the somewhat club-shaped body is flattened top to bottom. This contrasts with the nonflattened larvae of the so-called "round-headed wood borers" (Cerambycidae, or long-horned beetles).

Because of their boring habits, some Buprestidae are important pests in orchards and forests. Some forest species bore in burned trees, and adults of these species fly in from miles around in response to the infrared radiation created by forest fires. A few buprestids, such as the Flat-headed Apple Tree Borer (*Chrysobothris femorata*) and the Bronze Birch Borer (*Agrilus anxius*), are common backyard pests. Bronze Birch Borers are often responsible for the death of ornamental birch trees, and Flat-headed Apple Tree Borers can deal the deathblow to already weakened hardwoods of various species. Look for their flat-headed larvae in your recently demised trees as you saw them up for firewood.

If you have ash trees, watch for the recently introduced Emerald Ash Borer (*Agrilus planipennis*), a serious pest of ash trees currently spreading rapidly through parts of the northeastern United States and southern Canada. Look for D-shaped emergence holes in the bark of recently killed ash trees, and look at the wood surface under the bark for a dense network of snake-shaped burrows packed with fine sawdust. The sawdust is the frass of the buprestid beetles that killed the tree. Most buprestids bore in dead or dying wood, but Emerald Ash Borers attack healthy trees.

Most of North America's 762 buprestid species are wood borers, but some develop as flattened miners between the upper and lower leaf surfaces, and others live in stems or galls. Adults of the small leaf-mining species usually feed on the foliage mined by their larvae. Some metallic wood borers live in wood as larvae, but feed on pollen as adults. Bright yellow and black *Acmaeodera* species and inconspicuous *Anthaxia* species are among the adult buprestids frequently found on flowers.

Blister beetles to darkling beetles (Superfamily Tenebrionoidea)

Some beetle superfamilies are convenient groupings of obviously similar beetles, like the scarab superfamily or the weevil superfamily, but others are mere putative branches of the beetle phylogenetic tree and have little obvious use for nonspecialists. The Tenebrionoidea is one such group, with over 20 northeastern families ranging widely in size, shape and biology.

Blister beetles (**Meloidae**, page 326), probably the best known and most interesting of the tenebrionoid beetles, are usually elongate, somewhat soft and plump beetles that always seem to look like their wing covers are not quite shut. This appearance is enhanced in the big metallic blue *Meloe* species, which have short wings that overlap at the base in a distinctly unbeetle-like fashion. *Meloe* beetles, sluggish and obese insects commonly found in late fall or early spring, are sometimes called oil beetles because of the smelly, oil-like fluid that oozes from their joints when they are disturbed. *Meloe* males are especially odd-looking because of a bottle opener–shaped swelling in the middle of each antenna, used to grip the female's antenna during copulation.

The yellowish fluid that appears at blister beetle leg joints when a beetle is disturbed ("reflex bleeding") is laced with a potent irritant called cantharidin, which can cause blisters or even oozing lesions if it comes in contact with skin. If you must handle a blister beetle try to choose a virgin female, since males produce cantharidin and transfer it to females during copulation. The cantharidin offered to the female

A blister beetle, (*Epicauta cinerea*), page 327, caption 4.

enhances the male's mating success, and ultimately ends up protecting the eggs resulting from the successful mating.

Cantharidin has also played a role, although a shady one, in human mating strategies. One famous species of Spanish blister beetle, called the Spanish Fly (*Lytta vesicatoria*), is sometimes commercially harvested, ground and sold as a source of irritating cantharidin for medicinal and other uses. Spanish Fly has been infamous since the days of Roman orgies as an aphrodisiac with legendary powers to excite and incite. More recently, animal breeders have capitalized on the irritating properties of cantharidin to excite reluctant studs. Don't try this at home, cantharidin is *very* dangerous and can be fatal even in tiny doses – less than one thousandth of an ounce is enough to cause severe kidney damage or even death.

A variety of insects are attracted to dead blister beetles, and some steal the cantharidin-laced fluid directly from the living beetle. Blister beetles are almost invariably molested by tiny biting midges (see page 428) that nip away at the beetle's leg joints and lap up its blood. The biting midges that do this are quite specialized, and are in their own subgenus, appropriately named *Meloehelea*. It's assumed the flies gain some advantage by imbibing this chemical cocktail, just as Monarch butterflies (*Danaus plexippus*) gain a defensive advantage by eating the toxin-laced milkweed plant.

Even aside from the horrible haemolymph exuded by these insects, the Meloidae are fascinating beetles, with really peculiar life cycles. The first blister beetle larvae discovered were found on the hairs of bees, and were formally described as *Pediculus apis*, or "bee lice." Years later "bee lice" were found to be newly hatched blister beetle larvae that had waited on a flower till they managed to hitch a ride on a passing bee. Had the bee returned to its nest with a "bee louse" (a first instar meloid larva, properly called a triungulin) attached, the triungulin would have dropped from the bee into its nest cell containing an egg sitting on top of a mass of pollen and nectar. The triungulin would then eat the egg (merely an appetizer), before proceeding to eat the pollen and nectar, all while developing into a very different, fat, immobile larval form. This kind of development, involving two very different kinds of larval form, is called hypermetamorphosis.

The Meloidae is a big family with over 300 (mostly western) North American species, not all of which are associated with bees. The big oil beetles attack particular species of solitary bees, but some of our other common meloids have triungula that dig into the soil to find grasshopper egg pods. Adult blister beetles in the large genus *Epicauta* eat leaves, but their larvae eat grasshopper eggs. Since some species of *Epicauta* eat the foliage of vegetables and other crops, these are among the few insects that can be pests as adults, but beneficial as larvae.

Active, host-seeking triungulin larvae, like those found in the blister beetles, also occur in a strange little family known as the wedge-shaped beetles (**Ripiphoridae**, page 328). The short-lived adult wedge-shaped beetles have a bizarre appearance, sometimes truly wedge-shaped, with the wings projecting off the broad end of the wedge. Many North American wedge-shaped beetles have active first stage larvae that, like the similar blister beetle larvae, hang out on flowers waiting for the opportunity to hitch rides with bees or wasps back to their nests. Once in the nest, instead of eating both the bee's egg and the pollen stored in the nest like *Meloe* larvae do, the wedge-shaped beetle larva waits for the host's egg to hatch, then burrows into the bee or wasp larva. It consumes the larva from the inside for a while, then rips its way out of its still-living host, wraps itself around the poor host's neck like a spiny collar, and proceeds to eat away at its victim from the outside, eventually killing it.

Blister beetles and wedge-shaped beetles are generally "unmistakably shaped" – a phrase that applies without exception to the closely related family **Mordellidae** (page 328), called tumbling flower beetles for the distinctive habit of launching their apparently unbalanced bodies into comical tumbles when disturbed. Common on flowers and occasionally on dead wood, tumbling flower beetles usually develop in plant stems. One common species, the Gall Beetle (*Mordellistena unicolor*) develops in goldenrod galls, killing the larva of the gall-making insect – usually a fly.

The concept of going to a store to buy insects may seem a bit strange, but many insects are available to the studious shopper. Various predators and parasites are available from specialized garden supply houses, and pet stores sell a number of easily reared insects as food for fish and other animals. If you go into a pet store and request some mealworms, it will be assumed that your interest is in feeding the insects to vertebrate pets. Many purchasers may not realize that those mealworm larvae thriving in a jar of dry oatmeal will soon transform into fascinating pupae, with developing adult appendages held immobile against their bodies like the appendages of a long dead pharaoh resting in his sarcophagus. Mealworm pupae develop into attractive brown to black beetles in the family **Tenebrionidae** (page 329), or darkling beetles (usually *Tenebrio molitor*, the Yellow Mealworm). Most of North America's almost 1,200 darkling beetle species are dull black or brown beetles associated with desert environments, and most of the relatively few northeastern species live in other arid habitats such as hard fungi, dead trees or dry foodstuffs.

Many darkling beetles look like dull ground beetles with notched eyes and slightly clubbed antennae, but some are distinctively shaped. Forked Fungus Beetles (*Bolitotherus cornutus*), for example, are stout, warty darkling beetles that stay close to hard shelf fungi or bracket fungi, rarely even moving between fungi on different trees in the same forest. Males of these bizarre, nocturnal beetles have large or small prominent, forward-pointing horns, but females have only blunt tubercles. If you scan fungi at night you will often see both sexes together and possibly even hear them as the male approaches the female head to head, and clambers over her so that his underside rests on the warty tubercles on top of her thorax as he stridulates by rubbing against her tubercles.

If you handle these slow-moving, hard-bodied insects you might notice some brown stains on your hands. Like many darkling beetles, Forked Fungus Beetles can produce defensive chemicals that burn or discolor skin, but their first defensive response is to pull in their appendages and drop to the ground.

Other, much smaller darkling beetles with horned males (*Neomida bicornis*) can be found in softer shelf fungi than those preferred by Forked Fungus Beetles. Tiny beetles in the tenebrionoid family **Ciidae** (minute tree-fungus beetles, page 332), which occur on the same fungi as Forked Fungus Beetles, sometimes have similarly horned males.

A few groups of darkling beetles are superficially dissimilar from other Tenebrionidae, and have only been recently transferred into this large and diverse family. The subfamilies Alleculinae (comb-clawed beetles) and Lagriinae (long-jointed bark beetles) have been traditionally treated as separate families, but beetle specialists are now in general agreement that comb-clawed beetles are just Tenebrionidae with comblike claws, and long-jointed bark beetles are just Tenebrionidae with a conspicuously long last antennal segment.

A few small families are superficially similar to darkling beetles, so much so that one small family (**Melandryidae**, page 333) is called the false darkling beetles, although they do not have notched eyes like real darkling beetles. False darkling beetles, which often occur under bark, are so variable in shape and size the family seems to have been treated as a convenient place to put beetles that don't fit anywhere else.

Some of the beetles traditionally placed in the Melandryidae are now in the small families **Tetratomidae** (page 332) and **Synchroidae** (page 333), and another little group of beetles, now treated as **Scraptiidae** (page 333), has at different times been put in the Melandryidae, Mordellidae and Anthicidae. This sort of confusing change in classification sometimes reflects mere differences of opinion among specialists, but more often reflects our growing knowledge of the true relationships between various groups. Good classifications are accurate reflections of relationships, with each named taxon (such as a family) a natural (monophyletic) group including all of the descendents of a single ancestor.

Many of the small families in the superfamily Tenebrionoidea are made up of distinctively shaped beetles associated with dead wood or fungi. Ironclad beetles (**Zopheridae**, page 332), for example, are rock-hard beetles often found on fungi growing on birch logs. The same fungi often house the much smaller hairy fungus beetles (**Mycetophagidae**, page 333). Adults and larvae of the families **Colydiidae** (colydiid beetles, page 334), **Pyrochroidae** (fire-colored beetles, page 334), **Pythidae** (dead log beetles, page 335), **Boridae** (conifer bark beetles, page 335) and **Salpingidae** (narrow-waisted bark beetles, page 335) occur under the bark of dead trees. The latter three families (traditionally all included in the Salpingidae) are rarely encountered, and colydiids are so small they usually escape notice, but the flattened larvae of some fire-colored beetles are very common and are probably our most conspicuous subcortical beetle larva. The relatively large flattened larvae found under bark throughout our area are usually either whitish and soft larva of common fire-colored beetles (*Dendroides* spp.) that feed on a mixture of wood and fungal tissue, or superficially similar reddish brown larvae of a large flat bark beetle (*Cucujus clavipes*, Cucujidae) that eats other subcortical insects. These flattened larvae can be found under bark even during the winter months, when antifreeze compounds in their blood helps them avoid freezing.

Although larvae of typical fire-colored beetles (subfamily Pyrochroinae) are found under bark, larvae in the subfamily Pedilinae (sometimes treated as the family Pedilidae or as part of the Anthicidae) occur in less easily observed habitats such as soil. Adult pedilines, on the other hand, are easily found on flowers and are sometimes attracted to dead insects. Some species are strongly attracted to dead blister beetles which serve as a source of the valuable defensive chemical cantharidin. Males of these cantharidin-seeking beetles often score extra points in the mating game after acquiring a good store of cantharidin, which they are able to transfer to the female (and thereby to her eggs) during copulation. Females will sometimes mate only with males that are appropriately perfumed with cantharidin and thus have a proven potential to transfer protective chemicals to her and her eggs. Similar habits are found in the antlike flower beetles (**Anthicidae**, page 335), a group of small, omnivorous beetles with a distinctly antlike shape.

The pediline beetles that steal cantharidin from blister beetles are often superficially similar to another family of tenebrionoid beetles, the **Oedemeridae** (page 336). Oedemerids are called false blister beetles because, like blister beetles, they make their own protective cantharidin and can blister your skin if handled carelessly. Most false blister beetle larvae live in wet rotting wood and one species (the Wharf Borer, *Nacerdes melanura*) sometimes causes damage to wharves and pilings as it burrows into very wet wood.

The other families of Tenebrionoidea in northeastern North America are rarely encountered. Adult **Aderidae** (page 336), or antlike leaf beetles, are small beetles usually found on the undersides of leaves; larvae occur in dead wood or leaf litter. **Stenotrachelidae** (page 337), or false long-horned beetles, are elongate, narrow-headed beetles with some resemblance to flower longhorns (Cerambycidae, Lepturinae). Not much is known about their habits.

Carpet beetles, museum beetles, death watch beetles and their relatives (Superfamily Bostrichoidea)

The curse of every insect collector's life, and the nemesis of professional collection curators, is a rather pretty little beetle covered with multicolored scales as an adult and long, dense, spiky hairs as a larva. The attractive adult beetles can be found on flowers in spring, but if you have a collection of dried insects, your first clue as to their presence will probably be telltale piles of powder under your disintegrating prize specimens. That powder is a distinctive calling card left by larval carpet beetles (*Anthrenus* spp.), in the family **Dermestidae** (page 337), that will destroy your collection of pinned insects if it is not stored in tight cabinets and protected with repellents.

Carpet beetle adults are most likely to be found outdoors feeding on pollen during the spring and summer months, but typically come indoors in the late autumn, when they lay eggs in the dusty corners left by even the most fastidious housekeepers. The resultant hairy larvae feed on a variety of high-protein foods, ranging from dead insects to wool carpets. Sometimes known as "woolly bears" (a name officially applied to tiger moths), these little larvae resemble small porcupines as they bristle with spearheaded and easily detached hairs. Other dermestid pests have similarly hirsute larvae.

Most of us are not too fond of dermestid beetles, but curators of vertebrate collections sometimes deliberately encourage their activities to turn dry vertebrate carcasses into clean vertebrate skeletons. The dermestids put to good use to clean up skeletons in vertebrate museums are normally *Dermestes maculatus*, maintained in cultures for that purpose, and not the same species as the dermestids that terrorize insect collections (mostly *Anthrenus* species, but sometimes other genera).

The habit of eating dried animal matter is the rule among these generally dull-colored, hairy beetles, although several species are pests that eat cereals, and one dermestid species (the Khapra Beetle, *Trogoderma granarium*) is the number one pest of stored grains in the tropics and subtropics. Khapra Beetles do not occur in our area, but the six or seven species of *Trogoderma* that do occur in northeastern North America are all significant pests of stored grains, dried foods and other stored products. Most other dermestid beetles found in North American homes feed on wool carpets or other high protein household items, such as dry pet food.

Just as different families of aquatic beetles tend to converge on the same streamlined body forms, beetle families in which the adults are wood borers have repeatedly converged on a parallel-sided, cylindrical form. The horned powder post beetles or branch and twig borers (**Bostrichidae**, page 337) usually exemplify this form, with the head bent down like that of a ram in charging position and the leading edge of the thorax bulging and bristling with rasplike teeth.

Most of the 70 or so North American bostrichid species bore into dead trees, but the ones that are most likely to come to your attention are those that deviate from a "boring" lifestyle. One normally branch-boring species, for example, occasionally bores into telephone cables, allowing water in and necessitating expensive repairs. Some large tropical species are common in bamboo or other woods used to make products for export, and cause considerable consternation when the adult beetles appear in North American houses after emerging from imported furniture, carvings or bamboo souvenirs. Other bostrichids are serious causes for concern when they infest stored products, and one small species (the Lesser Grain Borer, *Rhyzopertha dominica*) is among the worst pests of stored grain. Most bostrichids are easily recognized by their hooded appearance and their straight, short antennae with a loose club of three or four segments, but powder post beetles, a small group now treated as a subfamily (Lyctinae) of Bostrichidae, lack the hooded appearance. Powder post beetles are narrow, distinctly parallel-sided beetles that bore in seasoned wood, including untreated lumber. Adults in this small subfamily are rarely encountered, but you might have come across small piles of dry powdery beetle poop near the perfectly circular holes left by adults emerging from wood where the larvae developed.

Another beetle family made up mostly of small, somewhat cylindrical species with occasionally strange eating habits is the **Anobiidae** (page 338), a family including some aptly named pests such as the Cigarette Beetle (*Lasioderma serricorne*), Furniture Beetle (*Anobium punctatum*) and Drugstore Beetle (*Stegobium paniceum*). These and most other species in the family Anobiidae are collectively known as death-watch beetles, a name that is most accurately applied to an introduced species (*Xestobium rufovillosum*) often found boring in old oak woodwork. Death-watch beetles make a characteristic clicking sound by bashing their heads against the tops of their thoraces, resonating against the wood floors of their tunnels, to attract mates. The characteristic *click-click* of a death-watch infestation was long ago thought to be a harbinger of death, a superstition that might well reflect the possibility of the roof caving in due to beetle damage, but likely developed because the only people to notice the sound would have been those sitting up all night with a dying relative in an oak-timbered home.

The 464 (mostly southwestern) North American anobiid species usually have a head hung low under the prothorax and often have antennal lobes or an asymmetrical antennal club. One subgroup of Anobiidae is made up of long-legged, strikingly convex beetles called spider beetles (Ptininae, traditionally treated as the family Ptinidae). Spider beetles are frequently encountered because several members of this small family are common indoors, where they are minor pests on a variety of stored products. Some species look remarkably like small spiders or shiny mites due to their small heads, globose bodies and long legs.

Laricobius nigrinus, page 339, caption 6.

The superfamily Bostrichoidea includes a rare family, the **Nosodendridae** (wounded tree beetles, page 339). Nosodendrids are black, oval beetles with distinctive tufts of hairs on the elytra. They are associated with slime fluxes or fermenting sap flows and occur throughout our area, but they are very rarely encountered.

Superfamilies Derodontoidea and Lymexyloidea

The small family **Derodontidae** (tooth-necked fungus beetles, page 339) is now placed in its own superfamily (Derodontoidea) even though some scientists think it is closely related to the Nosodendridae. Tooth-necked fungus beetles are generally obscure beetles in the 2–3 mm range, including some fungus-eating species, with the edge of the pronotum distinctively jagged or "toothed." The derodontids that have been attracting a lot of attention lately, however, are predacious species that are being used as biological control agents in the fight against homopteran pests. One species was introduced from Europe to eastern North America to help control Balsam Woolly Adelgids, and another was recently brought from the west coast to help in the battle against Hemlock Woolly Adelgids (Adelgidae).

The **Lymexylidae** (ship-timber beetles, page 339) is another small family currently put in its own superfamily (Lymexyloidea). Only two of the 50 world species of ship-timber beetles occur in North America, where their larvae tunnel in heartwood and sapwood of fallen trees and feed on ambrosia fungi growing on the walls of their tunnels. As is true for other beetles associated with ambrosia fungi (such as the ambrosia beetles in the Scolytinae), this is a symbiotic association and female beetles inoculate wood with fungal spores carried in specialized structures on their ovipositors. The Sapwood Timberworm (*Elateroides lugubris*) is very common in *Populus* logs although adults are rarely seen. The Chestnut Timberworm (*Melittomma sericeum*) occurs most commonly in oak (despite the common name!); adults often appear at lights. The family name refers to a European species that was once a pest of ship timbers, and does not refer to the biology of North American species.

Beer beetles, lady beetles and their relatives (Superfamily Cucujoidea)

Dining or drinking outside on a warm August afternoon is all you need to do to get on familiar terms with picnic beetles or beer beetles (*Glischrochilus* spp.). These shiny, reddish yellow or orange-spotted, black beetles have a fondness for fermentation that leads them to cavort in your ripe fruit and belly flop in your beer. Other members of the small family **Nitidulidae** (sap beetles, page 340) are less conspicuous than beer beetles, and are denizens of a variety of decomposing digs. They are frequently found under bark, in damaged ears of corn, compost and even on cadavers. One recently introduced member of the family (the Small Hive Beetle, *Aethina tumida*) even lives in beehives, often causing considerable damage. Small Hive Beetles were first found in Florida in 1998, but have now spread through most of the eastern United States.

Sap beetles and about 16 other northeastern families are grouped together in the superfamily Cucujoidea. Most cucujoid families are small, and are made up of small beetles that rarely impose themselves on our daily lives, although they often have intriguing names, like lizard beetles, pleasing fungus beetles, handsome fungus beetles, cryptic slime mold beetles, minute bark beetles and flat bark beetles. The most important and most familiar of these families is the lady beetle family (**Coccinellidae**, page 345).

Many of North America's almost 500 lady beetle species are familiar round, red, convex consumers of aphids, mealybugs and other soft-bodied insects. Convergent Lady Beetles (*Hippodamia convergens*), for example, eat almost 60 aphids per day and consume up to 500 aphids before laying hundreds of eggs destined to hatch into voracious aphid-eating larvae. That sounds like good news for the gardener, especially since you can easily purchase bags of thousands of these pretty beetles to release in your garden.

The Convergent Lady Beetle, so named for the convergent white marks on its pronotum (just behind the head), is a native beetle widespread in North America but commercially harvested only in the western U.S. where well-fed, fat-laden, adult beetles migrate from the valleys to the mountains in spring. For the next nine months Convergent Lady Beetles occur in huge aggregations in the mountains, quietly existing on their stored fat until dispersing over the surrounding valleys the following February or March. These enormous aggregations are an easy mark for commercial operators who harvest millions of beetles for sale to gardeners all over North America for "natural" control of pest aphids, even though there is no evidence that the released beetles serve a useful purpose or even stay in the area where they are released. Before rushing out to buy captured Californian Convergent Lady Beetles for release in your garden you should get to know the several species of introduced and native lady beetles that already live there.

Although we still have many common native lady beetles, such as the pink Spotted Lady Beetles (*Coleomegilla maculata*) that seem so abundant in early spring, most of the lady beetles now encountered east of the Rocky Mountains are introduced species such as the Seven-spotted Lady Beetle (*Coccinella septempunctata*). The Seven-spotted Lady Beetle ("C-7") is characterized by seven spots – three on each wing cover and one right behind the middle of the prothorax. This is the original "Ladybird of Europe," the object of lore and legend for the last five centuries. A number of efforts were made to introduce C-7 from Europe to the United States, ultimately resulting in its establishment in several eastern states and its subsequent spread to become the dominant member of the northeastern lady beetle fauna by the early 1980s. The Seven-spotted Lady Beetle has now, unfortunately, completely replaced the formerly common nine-spotted native species of *Coccinella* in much of our area.

C-7 is by no means the only exotic lady beetle unleashed on North America in the name of biological control. One of the first insects successfully introduced into any country for the biological control of a pest was the much-celebrated Vedalia Beetle (*Rodolia cardinalis*), a little lady beetle from Australia. The Vedalia Beetle is often credited with saving the

Several of the most common lady beetle species in North America, like this *Hippodamia variegata*, are introduced species.

California citrus industry in the 1880s. A scale insect, the Cottony Cushion Scale (*Icerya purchasi*) apparently got into California from Australia in the 1860s, and was threatening to wipe out the whole industry. An entomologist who was sent to Australia to look for natural enemies of this devastating pest sent back two very successful candidates, an obscure-looking predacious fly and the beautiful Vedalia Beetle. The fly played an important role in the spectacular control of the pest scale (especially in coastal areas), but the more conspicuous lady beetle got most of the credit.

There is something very appealing about a visible, charismatic beetle that conspicuously crunches its way through numerous pests on a daily basis. Perhaps that is why the Vedalia Beetle was only the first of 179 species of Coccinellidae to be intentionally imported to North America. Some 17 of those species are now established, along with another nine or ten exotic lady beetles that were apparently introduced accidentally. One of these, a beautiful Asian beetle called *Harmonia axyridis*, seems to have established itself in Louisiana and Mississippi in 1991, and has been spreading like crazy ever since, reaching Canada in 1994. Commonly called the Multicolored Asian Lady Beetle, this large lady beetle is highly variable in color and spotting, but it can usually be recognized by a white M-shaped mark on the pronotum.

By 1999, Multicolored Asian Lady Beetles were abundant enough in northeastern North America to be considered pests, threatening native species and irritating householders with massive home invasions in the fall. Like many lady beetles, this species overwinters in masses of adults in sheltered places. In 2000 (U.S.) and 2001 (U.S. and Canada) this species reached incredible levels of abundance, and population densities reached the point where swarms of lady beetles were observed doing "un-lady-beetle-like" things such as nipping people and damaging fruit. It was probably no coincidence that these lady beetles prey on another Asian invader, the Soybean Aphid (*Aphis glycines*), which invaded the northeastern United States in 2000 and reached awesome densities in both Canada and the United States in 2001.

Invasive exotic insects, even "beneficial" predators like the Multicolored Asian Lady Beetle, often turn out to be serious threats to native biodiversity and sometimes also cause unexpected problems. Asian Lady Beetles, for example, have emerged as a surprising new problem for our wine producers. High densities of Asian Lady Beetles can not only damage grapes before harvest, but also contaminate harvested grapes. Like other lady beetles, these brightly colored insects produce defensive chemicals to deter predators. Those same chemicals, when accidentally crushed with the grapes, cause off-tastes and are currently a serious problem for some wine producers.

Lady beetles are best known as aphid eaters, but many northeastern lady beetles attack scale insects and mealybugs. Among the most common of these is a tiny black beetle with red spots, the pretty *Brachiacantha ursina*. Larvae of the large genus *Brachiacantha*, unlike the familiar, conspicuous larvae of the common red lady beetles, stay underground and are rarely seen. We don't know much about what these little beetles eat, but at least some larvae in the genus are found in ant nests where they eat scale insects. Many small, black lady beetles eat scale insects and mealybugs.

Only half a dozen northeastern lady beetles deviate from the predacious habits found in most of the family. A few small species are fungus feeders, and one little group of lady beetles, the subfamily Epilachninae, is made up of leaf-eating species including important pests that attack members of the melon, potato and bean families in much of the world (except South America and Australia). Leaf-eating lady beetles are less convex than most family members, and they tend to be dirty yellow in color rather than bright black and red like most of their more reputable relatives.

The most common North American leaf-eating lady beetle, the Mexican Bean Beetle, (*Epilachna varivestis*) has spiny larvae that skeletonize leaves by feeding on the tissue between the veins. The adults, which chew holes in leaves, have a strong "reflex bleeding" response, and will release droplets of noxious fluids from glands at their leg joints if you poke them. Other adult lady beetles produce similar repellent fluids from their "knees" (the joint between the tibia and femur), while lady beetle larvae release defensive compounds from abdominal glands. Additional defensive alkaloids, aptly called cocinellines, occur in lady beetle haemolymph.

Boring long-horned beetles and beautiful leaf beetles (Superfamily Chrysomeloidea)

The shiny flecks of coleopteran color that decorate so many leaves and flowers can usually be attributed to two closely related families, the leaf beetle family (Chrysomelidae) and the long-horned beetle family (Cerambycidae). These two huge families make up most of the superfamily Chrysomeloidea, which also includes the small families **Bruchidae** (bean weevils or seed beetles, page 351; sometimes treated as a subfamily of Chrysomelidae), **Megalopodidae** (megalopodid leaf beetles), and **Orsodacnidae** (orsodacnid leaf beetles, page 351). **Megalopodidae** and Orsodacnidae are rarely noticed, but the Bruchidae often draw attention to themselves by destroying dried seeds or seeds of agricultural crops. Bean weevils are distinctively compact little beetles with a small head and a "bum" (pygidium) that sticks out prominently behind the ends of the elytra. All of the roughly 150 North American species develop inside seeds or seed envelopes, usually in legume seeds. Some bruchids, including the Pea Weevil (*Bruchus pisorum*) and the Bean Weevil (*Acanthoscelides obtectus*) are major pests with worldwide distributions.

The 1,500-plus species of leaf beetles (**Chrysomelidae**, page 351) in North America seem to come in almost every shape and color imaginable, but those most familiar to vegetable gardeners around the world are round, convex, yellow beetles with black stripes. If you grow potatoes in your vegetable garden you have probably spent a few hours picking Colorado Potato Beetles (*Leptinotarsa decemlineata*) and their rotund larvae from potato leaves. Next time you pick one of these conspicuous beetles off a potato plant, look closely at its legs. Even without a hand lens or a microscope, you should be able to see that the third tarsal segment (or tarsomere) of each leg – apparently the next segment up from the one with claws – is conspicuously expanded and bilobed, or heart-shaped. (If you looked at that expanded third tarsomere with a microscope, you would see a small fourth tarsomere hidden between its lobes, between the bilobed third tarsomere and the clawed fifth tarsomere.) That big, spongy-looking bilobed tarsomere is a characteristic of the leaf beetles and other Chrysomeloidea, but weevils have similar legs. The two big tarsal lobes bear specially modified hairs which help these plant-eating beetles walk on plant stems and smooth leaves.

A Colorado Potato Beetle (*Leptinotarsa decemlineata*), page 351, caption 8.

Most North American pest insects came from other continents and achieved pest status here when they arrived without their natural predators, parasitoids and diseases. A few pest leaf beetles, including the Northern Corn Rootworm (*Diabrotica barberi*) and Colorado Potato Beetle, are exceptions to this rule. The Colorado Potato Beetle originated in the New World (probably in Mexico), where it fed on a common weed called Buffalo Bur. Buffalo Bur is in the same genus as the potato (*Solanum*), and when potatoes were introduced to the southwest in the early 1800s, Colorado Potato Beetles found potatoes to be a palatable food source. The rest is history. By the late 1800s, these beetles were major pests from coast to coast, and a few years later, they became established in Europe – an accidental exchange for all the pests spreading in the other direction. Colorado Potato Beetles have now made it all the way to Russia.

The seemingly inexorable spread and increasing economic impact of these potato foliage-munching beetles has attracted a lot of scientific attention, making the Colorado Potato Beetle an exceptionally well-studied organism. Considerable research has gone into its control using pesticides, predators, parasitoids and, more recently, the development of genetically modified potatoes that contain a gene for the production of a bacterial toxin that kills both adult and larval beetles. Still, unless you happen to grow genetically modified potatoes, you should be able to find a few potato beetles in your home garden, along with a variety of other interesting members of the leaf beetle family.

Most yards and gardens are regular leaf beetle zoos, harboring dozens of species such as Spotted Cucumber Beetles (*Diabrotica undecimpunctata*), Striped Cucumber Beetles (*Acalymma vittata*), Spotted Asparagus Beetles (*Crioceris duodecimpunctata*), Imported Willow Leaf Beetles (*Plagiodera versicolora*) and Northern Corn Rootworm Beetles (*Diabrotica barberi*). If you live in a corn-growing area, the pale green adults of the Northern Corn Rootworm are probably abundant on many of your plants late in the season, just before they head off into cornfields to lay eggs on the ground, where they will remain until hatching the next spring. If the newly hatched larvae find newly planted corn they burrow into roots and cause serious damage, but if there are no nearby corn roots the rootworm larvae die.

Another corn rootworm species, the Western Corn Rootworm (*Diabrotica virgifera*), has a similar life cycle.

Both of these native pests have been traditionally controlled by simply not planting corn in the same field in consecutive years, but the situation has become more complex in recent years as populations of Northern Corn Rootworms have started to lay eggs that don't hatch for two or more years, thus "outsmarting" growers rotating corn with soybeans or other crops. Western Corn Rootworms, which do occur in northeastern North America, despite the name and a southwestern origin, have another way to beat the system. They lay eggs in soybean fields (and elsewhere), giving rise to larvae that will die if soybeans are planted in the same field the next year but will be there waiting if corn is planted in last year's soybean field following traditional rotation practice.

The Chrysomelidae can be usefully divided into a number of subfamilies, some of which are distinct and easily recognized groups. Flea beetles (subfamily Alticinae), for example, have the first long part of the hind leg (the femur) conspicuously swollen much like a flea's hind leg, enabling the sort of energetic jump that makes both fleas and flea beetles difficult to catch. Flea beetles abound on many plants, but your grapes, potatoes, tomatoes or eggplants would be good places to look for these evasive little beetles, especially early in the season. You might first be tipped off to their presence by leaf damage that looks like the result of a shotgun blast – the leaves will be peppered with lots of little holes.

Tortoise beetles, so-called because of the tortoiseshell-like covering that extends over the head and most of the legs, make up another easily recognized leaf beetle subfamily (Cassidinae). One common garden species, the Golden Tortoise Beetle (*Charidotella sexpunctata*) has a chameleon-like ability to change color from gold to reddish brown by merely altering the water content of its cuticle. Look for Golden Tortoise Beetles on Bindweed (*Convolvulus arvensis*) or Morning Glory (*Ipomoea violacea*) foliage, where their larvae hold a tail-like appendage covered with masses of feces and shed skins over the body like a disgusting (but protective) parasol. If you spot these larvae under their fecal camouflage and disturb them as they chew away at the surfaces of leaves, they will wave their tacky tail at you. To a small predator or parasitoid, such a sticky excremental threat must be quite intimidating; nevertheless, tortoise beetle larvae are heavily parasitized by tachinid flies.

Tortoise beetles are closely related to the distinctive leaf beetle subfamily Hispinae, or leaf-mining leaf beetles, unmistakably flattened, longitudinally ridged beetles that often occur in large numbers on the leaves of some trees. Hispine larvae live between the upper and lower surfaces of leaves, usually on trees, although some common species occur on goldenrod and other herbaceous plants.

Leaves of waterlilies, water shields or other floating plants usually support a couple of other easily recognized groups of leaf beetles, including slow-moving adults and larvae of Waterlily Leaf Beetles (*Galerucella nymphaeae*, subfamily Galerucinae) as well as the more conspicuous, fast-flying long-horned leaf beetles in the subfamily Donaciinae that are almost invariably present in the flowers of waterlilies. The larvae of these elongated, long-horned leaf beetles feed on the submerged parts of waterlily plants where they breathe by tapping into plant roots with stout spines that hold the larval spiracles, in essence using their host plants as living snorkels. Adult long-horned leaf beetles look a bit like the true long-horned beetles in the family Cerambycidae, but similar Cerambycidae have longer antennae and the base of the antenna is partially surrounded by the eye.

Long-horned beetles (**Cerambycidae**, page 360) are popular with collectors and naturalists, in part because of their bright colors and relatively large size, but more importantly because most species in this diverse family can be easily identified in the field based only on shape and color. Adults are called long-horned beetles because most have conspicuously long antennae with bases partially surrounded by the beetle's eyes. If you pick up a cerambycid to have a close look, it will make a characteristic "squeak" by rubbing the pronotum along a series of grooves on the mesonotum. A few other beetles squeak, but none that resemble cerambycids.

Larvae are called round-headed borers because of the swollen and rounded thorax that gives the body a clublike shape, and because they are borers in either dead wood or living woody or herbaceous plants. The larger species that live in dead wood (not the most nutritious of foods) have very long development times – sometimes decades – and some species have symbiotic microorganisms that help them digest woody meals.

The 344 northeastern long-horned beetle species make up less than 50 percent of the North American fauna, which is in turn about 5 percent of the world's 20,000 species. World long-horned beetles have an awesome size range from 2–150 mm, and northeastern North American cerambycids top out at a respectable 60 mm or so. Our most common species are smaller but some, such as the mottled gray and black sawyer beetles (*Monochamus* spp.), are still conspicuously large at 25 mm – especially the males with their strikingly long antennae, which can add another 50 mm. Sawyers are often abundant around freshly cut conifer logs, and are familiar to most cottagers and campers for their habit of showing up at artificial lights at night. The large, club-shaped larvae of sawyers make a "sawing" sound as they feed inside piles of pine logs.

Adult long-horned beetles only live a few weeks, feeding on leaves, fungi, bark, sap, fruit or flowers. Most long-horned beetles found on flowers belong to a subfamily (Lepturinae, the flower longhorns) of broad-shouldered beetles with distinctively long heads and constricted necks to facilitate their pollen-eating habits. Umbelliferous flowers, like Queen Anne's Lace (*Daucus carota*), seem to support a particularly impressive diversity of flower longhorns.

Some of our most common long-horned beetle species (in a number of subfamilies) are borers in sumac, raspberry, apple, poplar, elm and even milkweed. The common, robust red and black milkweed longhorns are remarkable for having their eyes divided into separate upper and lower halves, a peculiar morphology reflected in the genus name *Tetraopes* (*tetra* = four, *opa* = eyes). Larvae of *Tetraopes* bore into milkweed roots, and the bright red and black color of adult milkweed beetles warn potential predators that *Tetraopes*, like other similarly colored insects that feed on milkweed, are loaded with protective chemicals. Adults of another familiar cerambycid, the brilliant black and yellow Locust Borer (*Megacyllene robiniae*), are frequently seen on goldenrod flowers where their bright colors provide a protective similarity to the stinging wasps that frequent the same yellow blooms. Many other long-horned beetles are cryptically colored and almost invisible against the bark of their host trees.

Long-horned beetles have been much in the news lately because of a couple of newly arrived immigrants, especially a big, highly invasive Asian species accidentally imported to North America in pallets, packing crates or other wood products. The Asian Longhorn, or Starry Sky Beetle (*Anoplophora glabripennis*), so named because of the starlike white spots on its black body, spends its larval life in a variety of hardwoods, and has the potential to devastate maple forests. This is such a serious pest that no expense is spared to remove and destroy infested trees whenever it establishes local populations in North America. So far, monitoring for these beetles and eradication of occasional local populations seems to have prevented the Asian Longhorn from becoming widely established in the United States or Canada, but persistent populations in places like New York's Central Park suggest that this dangerous beetle is here to stay.

A less impressive-looking recently introduced long-horned beetle, the Brown Spruce Longhorn Beetle (*Tetropium fuscum*), is now apparently attacking healthy trees in a small area of Nova Scotia. In Europe, where this beetle is native, it attacks dead or weakened trees. Like the Asian Longhorn, the Brown Spruce Longhorn has the potential to spread throughout our area with possibly dire effects on the forestry industry.

At the same time as unwanted exotic long-horned beetles are appearing in North America, some native species are getting harder to find. A recent study of Cerambycidae of Ontario, Canada, showed that 30 of the 214 species known from the province had not been collected there since 1950, at least in some cases because their forest habitats have disappeared.

Weevils, engravers and ambrosia beetles (Superfamily Curculionoidea)

"I fear no weevil" ... or so says one of my favorite bumper stickers. While weevil bumper stickers might not be commonplace, weevils do show up regularly on cartoons, product logos and in popular literature because most people are familiar with these charismatically beaked beetles. Incorporation of weevils into our popular culture probably does not reflect a general awareness that the biggest weevil family (**Curculionidae**, page 370) is the biggest family of living things (there are more species of weevils than of vertebrates), or that weevils include many serious pests of food and fiber. Instead, it is the typical weevil's easily recognizable shape, with an oftentimes comically elongate snout, that has made "weevil" a household word.

North America is home to around 3,100 of the world's 60,000 weevil species, most of which belong in the family Curculionidae. These are very hard-shelled beetles, with distinctively elbowed and clubbed antennae usually arising about halfway along a distinct beak (some groups found in wood or under bark have a very short beak).

Despite their beaks, weevils are chewing insects, like other beetles, and have functional (although sometimes small) mandibles mounted at the very tip of the drill-like snout. The mandibles are the penetrating bits on the end of the drill, used to bore through the protective shells of nuts, fruits, bark and other plant parts so weevils can feed on the softer tissues within. Female weevils insert their eggs deep into plant tissues using the same drill holes. Acorn weevils, for example, have especially long snouts they use to chew into acorns, making neat little holes for the weevil to later lay her eggs. Weevil larvae in general are legless, C-shaped grubs, usually living inside plant material such as fruits, shoots and roots, although some develop inside small seeds and grains, and some feed externally on roots, leaves and reproductive structures.

A granary weevil can complete its larval development hidden within a kernel of grain, but many weevils are more noticeable. You might, for example, have noticed fruit scarred by the beak of a common weevil called the Plum Curculio (*Conotrachelus nenuphar*). Adult females chew small feeding incisions in fruit and then lay their eggs in the incisions. Curculios then compound this injury by chewing crescent-shaped incisions around the egg holes. Either this damage or the feeding activity of the larvae causes the fruit to drop off without developing. The larvae develop in the dropped fruit and then pupate in the soil if the fruit have not been gathered and disposed of before they mature. The warty-looking adults spend the winter in the soil.

Northeastern North America is graced with a wide variety of "evil weevils." White Pine Weevil (*Pissodes strobi*) larvae, for example, will tunnel in the terminal leader of a young White Pine (*Pinus strobus*) tree, killing the leader. One of the lateral branches then becomes the terminal leader, and the result is a kinked tree of limited value. White Pine Weevils also attack spruce trees and (occasionally) other kinds of pine. Similar damage on other species of pines may can be caused by a small moth, the European Pine Shoot Moth, *Rhyacionia buoliana.*

Other weevils are serious pests of alfalfa, and even the humble strawberry plant has its share of weevil pests. The Strawberry Root Weevil (*Otiorhynchus ovatus*) is an interesting beetle that has its elytra fused together, forming an almost indestructible capsule. This weevil feeds on the roots of strawberry, yew, juniper and a variety of other plants. Another weevil, called the Strawberry Clipper Weevil or Strawberry Bud Weevil (*Anthonomus signatus*), punctures strawberry buds with its long beak, sticks eggs in the feeding hole, then does a bit of chewing damage around the stem of the flower bud causing the flower bud to wilt or drop off (thus the name clipper weevil). The weevil larva develops and pupates inside the fallen flower. When the pupa transforms into an adult, the adult makes a small circular exit hole in the side of the bud, then feeds on foliage for a while before looking for a sheltered place to spend the winter.

The Strawberry Bud Weevil is just one of the 100 North American species in *Anthonomus*, an important genus that includes the famous Boll Weevil (*A. grandis*). Boll Weevils feed on the flower buds of cotton much the way Strawberry Bud Weevils feed on strawberry buds, thus nipping cotton development "in the bud." This serious pest has probably been responsible for more pesticide use than any other species, and still causes great reductions in cotton yield. The name *Anthonomus* is from the Greek words for "flower ulcer," aptly descriptive since the feeding holes made on buds by these weevils turn into very characteristic blister-like wounds.

Pests get a lot of press, but the majority of the Curculionidae are innocuous insects, and a few members of this enormous family are even beneficial species that attack dandelions, thistles, purple loosestrife and other plants generally considered to be weeds in northeastern North

A weevil, (*Sphenophorus costipennis*), page 371, caption 5.

America. Many weevil species have been moved around the world for use in the biological control of weeds.

The enormous family Curculionidae is usually divided into several subfamilies, some of which are so distinct that they have been traditionally treated as separate families. The stout, cylindrical wood-boring weevils with reduced beaks are often treated as the families Scolytidae and Platypodidae, although new classifications treat them as subfamilies Platypodinae and Scolytinae of the Curculionidae. If you have ever knocked the bark off a piece of firewood and noticed striking, centipede-like engravings on the surface of the wood, then you have at least seen evidence of these common beetles. Bark or engraver beetles in the subfamily Scolytinae make those familiar markings by tunneling under bark. Both males and females bore into the wood, with the male usually starting things off by making a mating chamber. The "body" of the centipede-shaped engraving is the main gallery made by the female beetle as she tunnels away from the mating chamber, laying eggs along the side of her tunnel. When those eggs hatch, the legless larvae burrow out at right angles to the mother's gallery and keep going until they pupate at the ends of their tunnels. Those larval tunnels are the "legs" of the centipede-like pattern, and the new adults emerge from round holes they chew through the bark at the tips of those "legs."

Different species of engraver beetles leave different patterns in different species of trees. One of our notorious engraver beetles is the Smaller Elm Bark Beetle (*Scolytus multistriatus*), a species that was accidentally introduced to North America along with an associated fungus that causes Dutch elm disease. Dutch elm disease had decimated the stately elms around my southern Ontario home by the time I became an avid insect collector, leaving lots of standing dead elm trees just at the right stage to support a tremendous diversity of beetles and other insects feeding on wood, fungi and each other. One needed only to peel the loose bark off one of those elm trees to see that, at least from an insect collector's peculiar perspective, Dutch elm disease was a cloud with a little silver lining.

Despite our long history of accidentally introducing European pest insects to North America, newly introduced insects are continually appearing in the North American fauna. One of the more recently introduced engraver beetles is a now-common insect called the Pine Shoot Beetle (*Tomicus piniperda*). These small, brown beetles appear in very early spring, when females make galleries and lay eggs under the bark of pine stumps, logs or trunks of severely weakened pine trees. Larvae construct horizontal feeding galleries under the bark, pupate at the end of their galleries and transform into adults that fly to the crowns of living pine trees where they burrow into healthy shoots. Adult feeding in pine shoots causes most of the damage associated with these beetles, but they also carry fungi (blue stain fungi) that stain wood, lowering its value.

The transmission of fungal diseases by elm bark beetles and other Scolytinae makes them important beetles, but the association between these beetles and fungi goes far beyond passive transport. Some of our important forest pests are bark beetles that use symbiotic fungi to help them overcome the tree's defenses. Some bark beetles attack trees in great groups, carrying with them a fungus that inhibits the flow of sticky pitch, thus thwarting the tree's natural defense. Bark beetles are responsible for the great majority of insect-caused tree deaths in North American timber-producing forests.

Some Scolytinae, as well as closely related beetles in the subfamily Platypodinae (pinhole borers), go beyond enlisting the aid of fungi in attacking tree hosts, and actually cultivate particular fungi as food for adult and larval beetles. Adults of these beetles, collectively called "ambrosia beetles" make narrow tunnels where they introduce and cultivate a fungus called ambrosia (*Ambrosiella* sp.). Ambrosia was the food of the gods in classical mythology, and was supposed to impart immortality with a few delicious mouthfuls. Other fungi, such as truffles, have been described as the food of the gods, but this beetle-cultivated pabulum now lays claim to the term "ambrosia." The beetles' ambrosia grows on the walls of their tunnels, sometimes creating characteristic black stains.

Although most weevils are in the huge family Curculionidae, another six or seven families of more or less long-snouted beetles are normally referred to as "weevils" or snout beetles. The most remarkable of these are the leaf-rolling weevils (**Attelabidae**, page 377) that carefully cut leaves before shaping them into cigar-like rolls in which the larvae develop. Some beetles (genus *Pterocolus*) in the attelabid subfamily Rhynchitinae (leaf and bud weevils) also develop in leaf rolls, but they don't roll their own – they steal leaf rolls made by leaf-rolling weevils in the subfamily Attelabinae. (Some authors treat the leaf and bud weevils as a separate family Rhynchitidae; others treat the leaf-rolling weevils as well as the leaf and bud weevils as subfamilies of the Curculionidae.)

The **Ithyceridae** (page 378) is as small as a family can get – it includes only one species (the New York Weevil, *Ithycerus noveboracensis*, 12–18 mm) that damages various hardwoods by feeding on twigs and buds as an adult and on roots as a larva. The straight-snouted weevils and pear-shaped weevils (**Brentidae**, page 379) make up a rather heterogeneous group of weevils traditionally placed in different families, including the small pear-shaped weevils formerly placed in the Apionidae, as well as the larger straight-snouted weevils. Straight-snouted weevils (sometimes called primitive weevils) are often found under the bark of trees, where their larvae develop as wood-borers. The many species of pear-shaped weevils live in a variety of plants and are often very common on their particular hosts. Hollyhock seed heads, for example, are almost always home to several individuals of a tiny introduced pear-shaped weevil, the Hollyhock Weevil (*Rhopalapion longirostre*).

Fungus weevils (**Anthribidae**, page 378) have a much broader beak than most other weevils, and have straight antennae, unlike the elbowed antennae of similar true weevils. The hundred or so North American species of fungus weevils includes some species that warrant the common name "fungus weevils," but also include many species that bore in dead or dying trees, and some species that develop in weeds like common ragweed. Fungus weevils are rarely noticed because of their cryptic habits, but they are not rare insects. One other odd little weevil family, the **Nemonychidae**, or pine flower snout beetles, is truly a rare group, occurring only for brief periods on male pine flowers in early spring where the female beetles lay eggs and the beetle larvae feed on pollen.

ORDER STREPSIPTERA

Twisted-winged Parasites

Parasites, by definition, are organisms that feed on other organisms but don't kill them as a matter of course. Lice, fleas, biting flies and a few other insects that suck blood from people and other animals are typical insect parasites. Most insects that attack other insects, on the other hand, either kill multiple victims outright (predators) or use single individuals as hosts on or in which they develop, consuming and killing the host slowly (parasitoids). Relatively few insects, mostly blood-feeders like the tiny flies that feed on blister beetle blood, are truly parasitic in or on other insects, and of these almost none develop inside insect hosts without killing them. Members of the bizarre order Strepsiptera, or twisted-winged parasites, provide the main exceptions to this generalization.

Strepsiptera (page 380) start out life as active, first stage larvae somewhat similar to those of blister beetles or wedge-shaped beetles. They attack a great variety of hosts, but those species associated with bees and wasps hitch rides to the nests, usually on the surface of the bee but sometimes by going right into the bee along with the nectar the bee is feeding on. Once in the bee's nest, the twisted-winged parasite larva burrows into a bee larva, where it becomes a legless internal parasite. Unlike wedge-shaped beetles, twisted-winged parasites stay inside their hosts at least until adulthood. Females of North American species are eyeless, wingless and legless and never leave the host, but males leave the host upon maturity. The strange twisted-winged adults referred to by the common name of the order are males, which have big eyes, antler-like antennae, large, twisted hind wings, and tiny front wings that look and function like the halters of flies. Twisted-winged parasites are widespread, but you are not too likely to see any winged males unless you rear them from their distinctively distorted parasitized hosts, because they live just long enough to find a host parasitized by a female.

Mating in the parasitic Strepsiptera is strange, even for insects, as only the female's head region projects from the host. Males do manage to reach the necessary parts, and the resultant young emerge by the thousands from pores on the side of the female's abdomen, leaving the host's body through the opening at their mother's head to seek new hosts and start the cycle afresh.

Everything about these insects seems as twisted as their wings, so it is not surprising that entomologists are not in agreement as to what kind of insects twisted-winged parasites really are. Some point to the similarities between wedge-shaped beetles and twisted-winged parasites as evidence that both are beetles, while others argue that the Strepsiptera is an ancient, diverse and distinctive order of insects quite separate from the beetles or any other order of holometabolous insects. Some recent evidence suggests that Strepsiptera might be closely related to flies, and that the knoblike front wings of Strepsiptera function in much the same way as the similar knoblike hind wings (halters) that characterize the Diptera.

A male twisted-winged parasite (*Pseudoxenos* sp.) on its host, page 380, caption 4.

SUBORDER ARCHOSTEMATA

FAMILY CUPEDIDAE (RETICULATED BEETLES) ❶ Fossil evidence shows that the ancient suborder Archostemata has been around for over 200 million years, but it currently includes only two families and less than half a dozen North American species. This is ***Tenomerga concolor*** (7–11 mm), the most common of the 2 northeastern Cupedidae. Adults can be found under bark, but larvae are borers in oak and pine. Another eastern member of the Archostemata, the tiny (2 mm) and rare *Micromalthus debilis* (Micromalthidae), lives in rotting wood; larval females sometimes reproduce without ever developing into adults.

SUBORDER ADEPHAGA

FAMILY CARABIDAE (GROUND BEETLES AND ROUND SAND BEETLES) Subfamily Carabinae ❷ **Fiery Searchers** (***Calosoma scrutator***) hunt caterpillars, often in tree foliage. Look for them on tree trunks at night or under bark during the day. These large (about 30 mm) ground beetles are often common along the shores of the Great Lakes. ❸ ***Carabus maeander*** is a distinctively welted searcher commonly found under bark. ❹ ***Carabus serratus*** is a widespread searcher most often found near streams or rivers. ❺ This snail-eating ground beetle (***Scaphinotus*** sp.) was partially exposed by removing the bark from a dead tree. The narrow front end and long mouthparts allow it to feed through the opening in the snail's shell. Subfamily Nebriinae ❻ Members of the genus ***Notiophilus*** are small (4–6 mm), distinctively bulgy-eyed beetles that use superb vision to hunt springtails and other swift prey. ❼ ***Loricera pilicornis*** adults use long, stout bristles near the base of the antennae to trap springtails and other prey. This individual was spotted on the snow surface on a late winter day.

FAMILY CARABIDAE (continued) Subfamily Omophroninae ❶ ❷ Round sand beetles (***Omophron***, 5–8 mm) are so distinctive they used to be put in their own family (Omophronidae). These photos show a golden ***O. tessellatus*** and a darker ***O. americanus*** photographed on the same Lake Huron beach. Subfamily Scaritinae ❸ Pedunculate ground beetles such as this large ***Scarites subterraneus*** (25 mm) are distinctively shaped, with a marked gap between the elytra and prothorax. This big-jawed predator can usually be found under debris along the edges of fields or beaches. ❹ The large (20–33 mm) ground beetles in the genus ***Pasimachus*** resemble stag beetles with threadlike (rather than clubbed) antennae. Their big jaws are usually used to consume insect larvae. ❺ ***Dyschirius pallipennis*** (4 mm) is usually found along sandy shorelines. Subfamily Elaphrinae ❻ Marsh and bog ground beetles (***Elaphrus*** spp.) are shaped much like tiger beetles (Cicindelidae), but are smaller (6–8 mm) and very different in their habits. *Elaphrus* species are common on muddy margins of ponds and lakes. Subfamily Brachininae ❼ ***Brachinus***, or bombardier beetles, like this pair of ***B. janthinipennis*** (6–9 mm), are capable of blasting you with a burst of hot chemicals that can stain your fingers yellow – handle them with care. These bright beetles are common along shorelines where their larval hosts (whirligig beetle pupae) can be found. Subfamily Trechinae ❽ ❾ Many of the small ground beetles that abound along shorelines belong to the large genus ***Bembidion***. These photos show ***B. levettei*** and ***B. confusum*** on the sandy shores of Lake Huron. The common name for *Bembidion* and related genera is "minute ground beetles" because they are often only a few millimeters long.

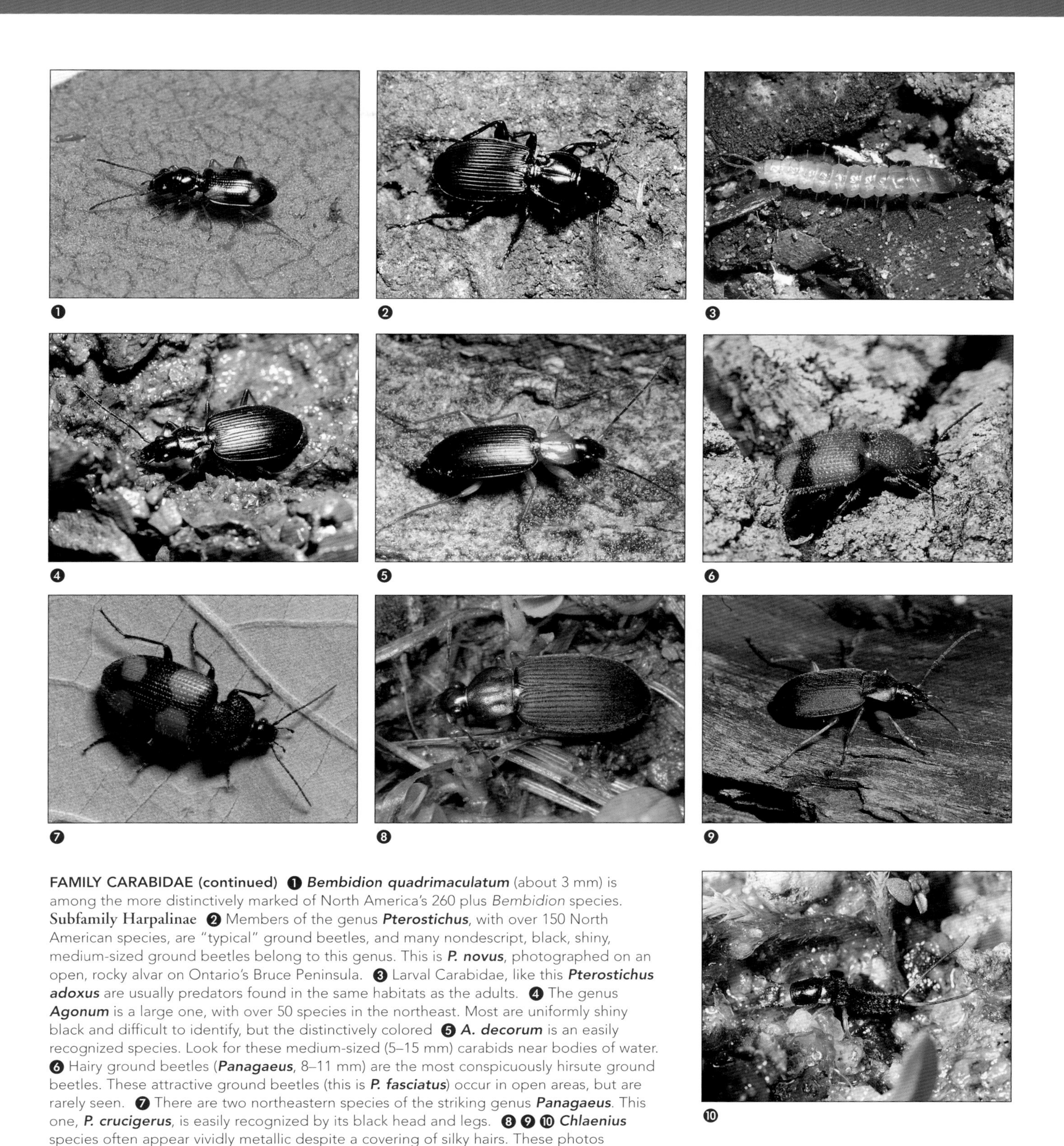

FAMILY CARABIDAE (continued) ❶ ***Bembidion quadrimaculatum*** (about 3 mm) is among the more distinctively marked of North America's 260 plus *Bembidion* species. Subfamily Harpalinae ❷ Members of the genus ***Pterostichus***, with over 150 North American species, are "typical" ground beetles, and many nondescript, black, shiny, medium-sized ground beetles belong to this genus. This is ***P. novus***, photographed on an open, rocky alvar on Ontario's Bruce Peninsula. ❸ Larval Carabidae, like this ***Pterostichus adoxus*** are usually predators found in the same habitats as the adults. ❹ The genus ***Agonum*** is a large one, with over 50 species in the northeast. Most are uniformly shiny black and difficult to identify, but the distinctively colored ❺ ***A. decorum*** is an easily recognized species. Look for these medium-sized (5–15 mm) carabids near bodies of water. ❻ Hairy ground beetles (***Panagaeus***, 8–11 mm) are the most conspicuously hirsute ground beetles. These attractive ground beetles (this is ***P. fasciatus***) occur in open areas, but are rarely seen. ❼ There are two northeastern species of the striking genus ***Panagaeus***. This one, ***P. crucigerus***, is easily recognized by its black head and legs. ❽ ❾ ❿ ***Chlaenius*** species often appear vividly metallic despite a covering of silky hairs. These photos show two of our most common *Chlaenius*, ***C. tricolor*** (11–13 mm) and the bright green ***C. sericeus*** (12–16 mm). Both are common under objects along shorelines. Larval *Chlaenius*, like the associated adults, often occur amongst shorline debris.

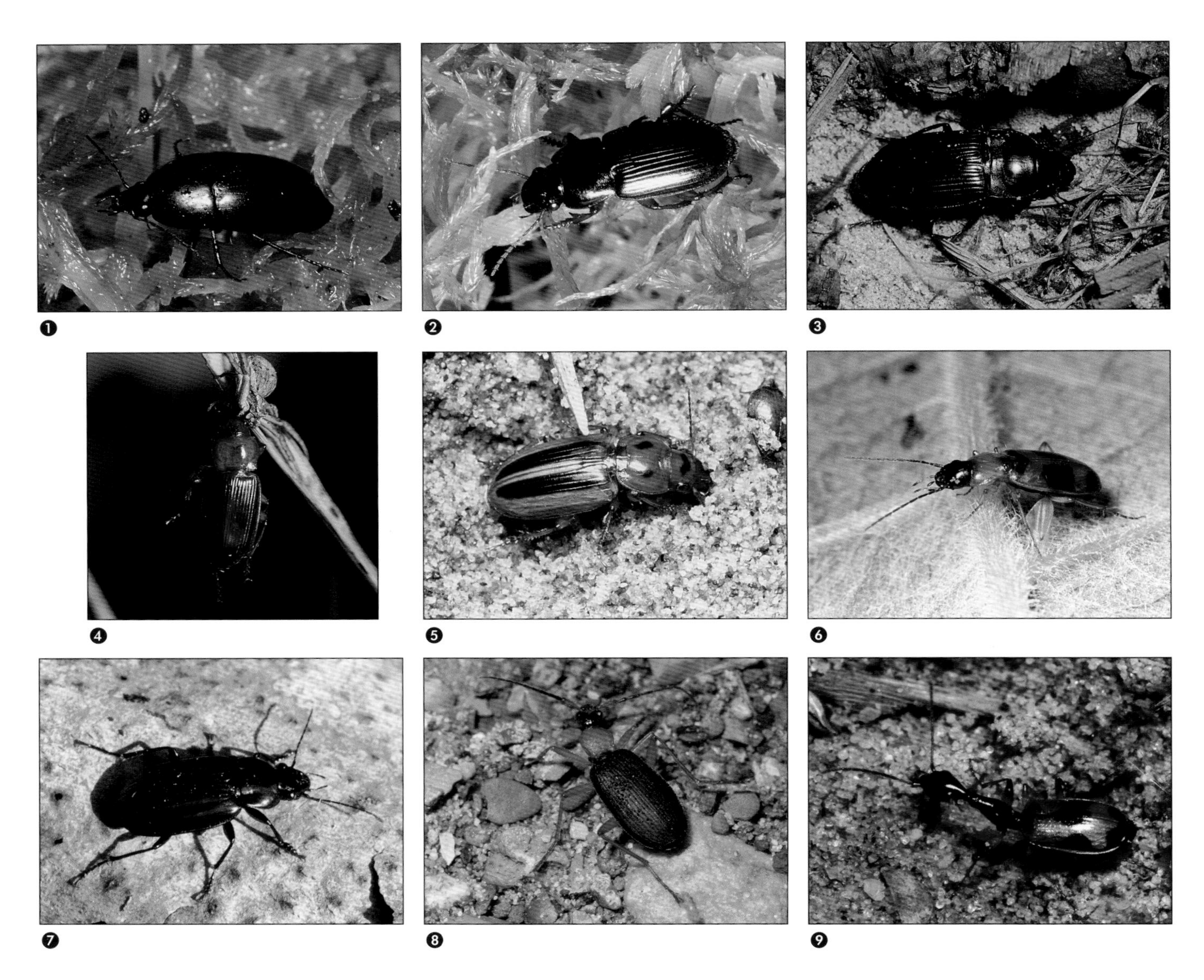

FAMILY CARABIDAE (continued) ❶ ❷ Sphagnum bogs are great places to look for ground beetles, like this ***Oodes amaroides*** ❶ and ***Harpalus affinis*** ❷ . One recent study of an Ontario bog yielded 65 species of Carabidae. ❸ ❹ ***Harpalus*** (5–25 mm) is one of the largest genera of ground beetles, with over 35 northeastern species. Most *Harpalus*, like this ***H. caliginosus***, are uniformly colored, stout black beetles, but ***H. erraticus***, like this one being consumed by a crab spider (*Xysticus*), is pale brown. ❺ The large genus ***Stenolophus*** includes 16 northeastern species, some of which are common at lights at night, and some of which are considered pests because they eat seeds rather than the invertebrates that make up the diet of most Carabidae. This photo shows ***S. lineola*** (6–7 mm), an easily recognized beetle found in open, wet areas. ❻ Some ground beetles, including ***Badister*** species like the ***B. neopulchellus*** (5–6 mm) shown here, are commonly attracted to lights at night. ❼ Individuals of ***Diplocheila striatopunctata*** (13-18 mm) sometimes have distinct purple stripes on the elytra, and are sometimes entirely black. This individual was found in a half-submerged log in a maple swamp. ❽ The **False Bombardier Beetle** (***Galerita janus***) often occurs in the same shoreline habitats as bombardier beetles, but is much larger (over 15 mm; bombardier beetles are less than 10 mm). ❾ ***Colliuris pensylvanica*** (6–7 mm), a distinctively shaped beetle with a long, almost cylindrical prothorax, is most common in open grasslands, but also occurs under rocks along shorelines. This species shows up at lights on a regular basis.

FAMILY CARABIDAE (continued) ❶ ***Tetragonoderus fasciatus*** is a small (4–5 mm) ground beetle often found in sandy places. ❷ ❸ Although most ground beetles are dark beetles that stay close to the soil surface, the small (3–10 mm) but distinctively shaped species of ***Lebia*** can often be found searching for prey on flowers and foliage. These photographs show ***L. ornata*** on a leaf and ***L. viridis*** feeding on dandelion pollen. Larval *Lebia* are parasitoids that attach themselves to the outsides of leaf beetle pupae. ❹ The brightly striped ***Lebia solea*** is often seen on foliage along woodland trails. ❺ This brightly colored ***Calleida punctata*** (7–9 mm) is eating Fall Webworm (*Hyphantria cunea*) caterpillars right through their nest. **FAMILY CICINDELIDAE** (TIGER BEETLES) ❻ The most familiar of eastern North America's tiger beetles is ***Cicindela sexguttata***, called the **Six-spotted Tiger Beetle** despite the fact that some individuals have four, two or even no white spots. This bright green species is often common along woodland trails in early spring. ❼ Male tiger beetles, like these ***C. sexguttata***, often use their huge mandibles to grip the female long after mating, guarding her from competing males. ❽ This larval **Six-spotted Tiger Beetle** (***Cicindela sexguttata***), was pulled from its burrow to show its massive head and the hooked hump on its fifth abdominal segment. ❾ ***Cicindela patruela*** resembles the closely related and much more common Six-spotted Tiger Beetle, but has more extensive white markings. This generally rare species occurs throughout eastern North America as far north as Ontario, where it occurs on sandy soils in mixed pine-oak forests.

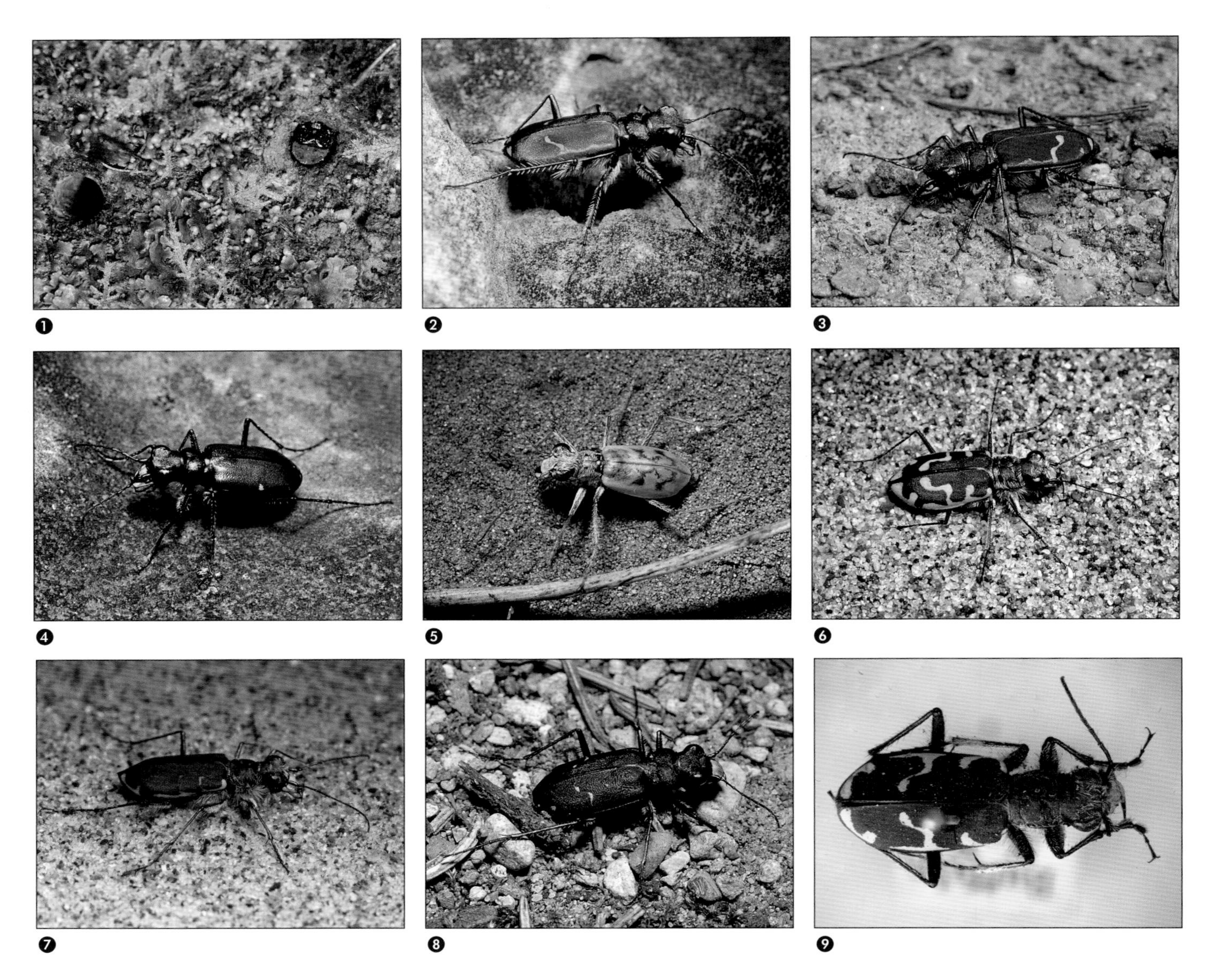

FAMILY CICINDELIDAE (continued) ❶ One of these tiger beetle burrows is open and one is blocked with the massive head and prothorax of a tiger beetle larva. The rest of the body is bent at right angles and extends down into the burrow. ❷ The **Purple Tiger Beetle** (***Cicindela purpurea***) resembles *C. limbalis* (the Green-margined Tiger Beetle), a similarly colored species that has shoulder patches and much more extensive markings. ❸ The **Green-margined Tiger Beetle** (***Cicindela limbalis***) lives on clay soils across Canada and the northeastern states. ❹ ***Cicindela denikei*** occurs on open, rocky alvars in northwestern Ontario and adjacent Manitoba and Minnesota. This scarce species, sometimes treated as a subspecies of *C. sexguttata*, usually lacks spots but sometimes has a few marginal spots like *C. sexguttata*. ❺ The **White Tiger Beetle** (***Cicindela lepida***) occurs in inland areas of open, loose and deep undisturbed sand during late summer or early fall. Pale and superbly camouflaged, they are virtually invisible against a sandy background, and are easiest to spot by their shadows. ❻ The increasingly rare **Beach Dune Tiger Beetle** (***Cicindela hirticollis hirticollis***), which has bare cheeks and a G-shaped shoulder mark unlike the otherwise similar *C. repanda*, is associated with white sand shorelines. ❼ *Cicindela hirticollis*, like several other tiger beetle species, occurs as differently colored forms sometimes recognized as subspecies in different areas. The relatively dark Beach Dune Tiger Beetles (*C. hirticollis*) found along the Atlantic coast from Rhode Island north and on the beaches of the northern Great Lakes, like this one from Manitoulin Island, are ***C. hirticollis rhodensis***. The more extensively marked beetles found along much of the Atlantic coast and southern Great Lakes belong to the typical subspecies *C. hirticollis hirticollis*. Several other subspecies of this widespread beetle occur in western North America and Mexico. ❽ ***Cicindela longilabris*** is the big, black tiger beetle typical of boreal regions of Canada, but it also occurs in the northern United States. This species prefers sunny spots in open coniferous forests. ❾ This specimen of ***Cicindela ancocisconensis*** was collected at Buffalo, New York, but this uncommon species has yet to be recorded across the border in Ontario. It occurs from Georgia to Quebec and is usually found on open sand or gravel along streams or rivers. This species resembles the common, smaller *C. repanda*, but has three teeth on its labrum while *C. repanda* has at most one.

FAMILY CICINDELIDAE (continued) ❶ ***Cicindela scutellaris*** is one of our most abundant tiger beetles on inland dunes, blowouts and open sand roads. It varies in background color from purple to green, but northeastern populations can usually be recognized by the white markings restricted to the edge of the elytra. ❷ Sandy shores throughout the northeast usually support huge numbers of our most common tiger beetle, ***Cicindela repanda***, especially in spring and late summer. ❸ ***Cicindela duodecimguttata***, the **Twelve-spotted Tiger Beetle**, likes a mixture of moist sand and organic soil. *C. duodecimguttata* seems to prefer sheltered areas, as opposed to the open beaches frequented by *C. repanda*. ❹ On the widespread species ***Cicindela tranquebarica*** the hind part of the shoulder marking is elongated, like a finger pointing onto the middle of the wing cover. *Cicindela tranquebarica* can be found in all sorts of open sandy and gravelly habitats early in the spring, but eggs are laid early and this species is rarely seen from late spring until late summer. ❺ ***Cicindela formosa*** is distinctive for its large size (16–18 mm) and the cream-colored markings that form a broad band around the edge of the elytra and extend onto the elytra as thick, finger-like extensions. This is a conspicuous tiger beetle of inland dunes and sand blowouts. ❻ ***Cicindela punctulata*** can be found throughout the late summer and early fall on almost any dry, sunny area with a mixture of open sand and scattered grasses. Gravel pits and open farm tracks are good bets for these small (10–14 mm), dark tiger beetles with inconspicuous punctures and variable tiny white markings. ❼ ***Cicindela dorsalis*** occurs on open, sandy beaches of the east coast of the United States. ❽ ***Megacephala*** is a night-active genus including two large (20 mm or more) species found in the northeastern United States, but not Canada. All other northeastern tiger beetles are in the genus *Cicindela*. ***Megacephala virginica*** is uniformly dark green; ***M. carolina*** has a pale patch at the end of each elytron. ❾ ***Megacephala carolina*** is a night-active beetle, often found around street lights where it feeds on other insects attracted to the lights. Although most abundant in the south, it occurs as far north as Illinois and Maryland.

FAMILY RHYSODIDAE (WRINKLED BARK BEETLES) ❶ The small family Rhysodidae is very closely related to the Carabidae (ground beetles). Wrinkled bark beetles, like this ***Clinidium*** (5–8 mm), live under the bark of dead trees and feed on slime molds in their inconspicuous plasmodial stage. ❷ ***Omoglymmius americanus*** (6–8 mm) is the only species of wrinkled bark beetle that ranges as far north as eastern Canada. **FAMILY DYTISCIDAE** (PREDACIOUS DIVING BEETLES) ❸ Predacious diving beetles range in size from giants like this ***Dytiscus verticalis*** (almost 40 mm long) down to minute forms only a millimeter or so in length. ❹ Our biggest diving beetles are in the genera *Cybister* and *Dytiscus*; males are among those diving beetle species with parts of their front legs modified into scaly pads. These suction cup-like pads provide one solution to a gripping problem faced by such slick and streamlined aquatic insects. The male uses his tarsal discs to cling to the back of the convex, smooth female. The tarsal disks of male ***Dytiscus*** (like this one) are round; those of male *Cybister* are oval. ❺ In some female ***Dytiscus***, like ***D. fasciventris*** (about 25 mm), the elytra is conspicuously ridged, in contrast with the male's smooth surface. ❻ ❼ The long mandibles of this ***Dytiscus*** larva are poised to impale small fish or other aquatic organisms. Diving beetle larvae usually have open spiracles at the tip of the abdomen and come to the surface to breathe. ❽ Members of the widespread genus ***Hydaticus*** often occur in shallow ponds. This is ***H. bimarginatus***, a species which only occurs as far north as New York. ❾ Adults of ***Hydaticus aruspex*** (12–15 mm), a Holarctic species with a northeastern range extending from Pennsylvania to Hudson Bay, seem to spend the winter on land. They appear in ponds in very early spring; this one was found in a partially ice-covered ditch.

FAMILY DYTISCIDAE (continued) ❶ This pair of diving beetles includes a male with his smooth elytra covered with tiny dimples, and a female with prominently furrowed elytra. Her furrows are there to give him a good purchase on his mate's otherwise slick and streamlined body. These medium-sized (12–15 mm) diving beetles are ***Acilius semisulcatus***, a distinctive and common species with a black-margined rectangle emblazoned on its yellow prothorax, and curved bands crossing the back of its attractive black and yellow mottled elytra. ❷ Larval predacious diving beetles, called water tigers, have sickle-shaped mandibles used to impale prey. The mandibles are like two-way hypodermic needles, injecting digestive juices then withdrawing the digested body contents of the prey (in this case, a mosquito pupa). ❸ This ***Agabus*** larva has its mandibles sunk deeply into a mosquito larva. ❹ ❺ Diving beetle adults, like this black ***Agabus*** (about 8 mm long) and this beautifully sculptured ***Colymbetes sculptilis*** (about 16 mm long), are air-breathing insects that periodically hang out at the surface, hind end up, while replenishing the air bubbles kept under their elytra. ❻ Diving beetles have spiracles that open into a periodically replenished air store concealed under the elytra. Submerged beetles, like this ***Rhantus binotatus*** (5–6 mm long), often have a bubble of air visible at the tip of the elytra.

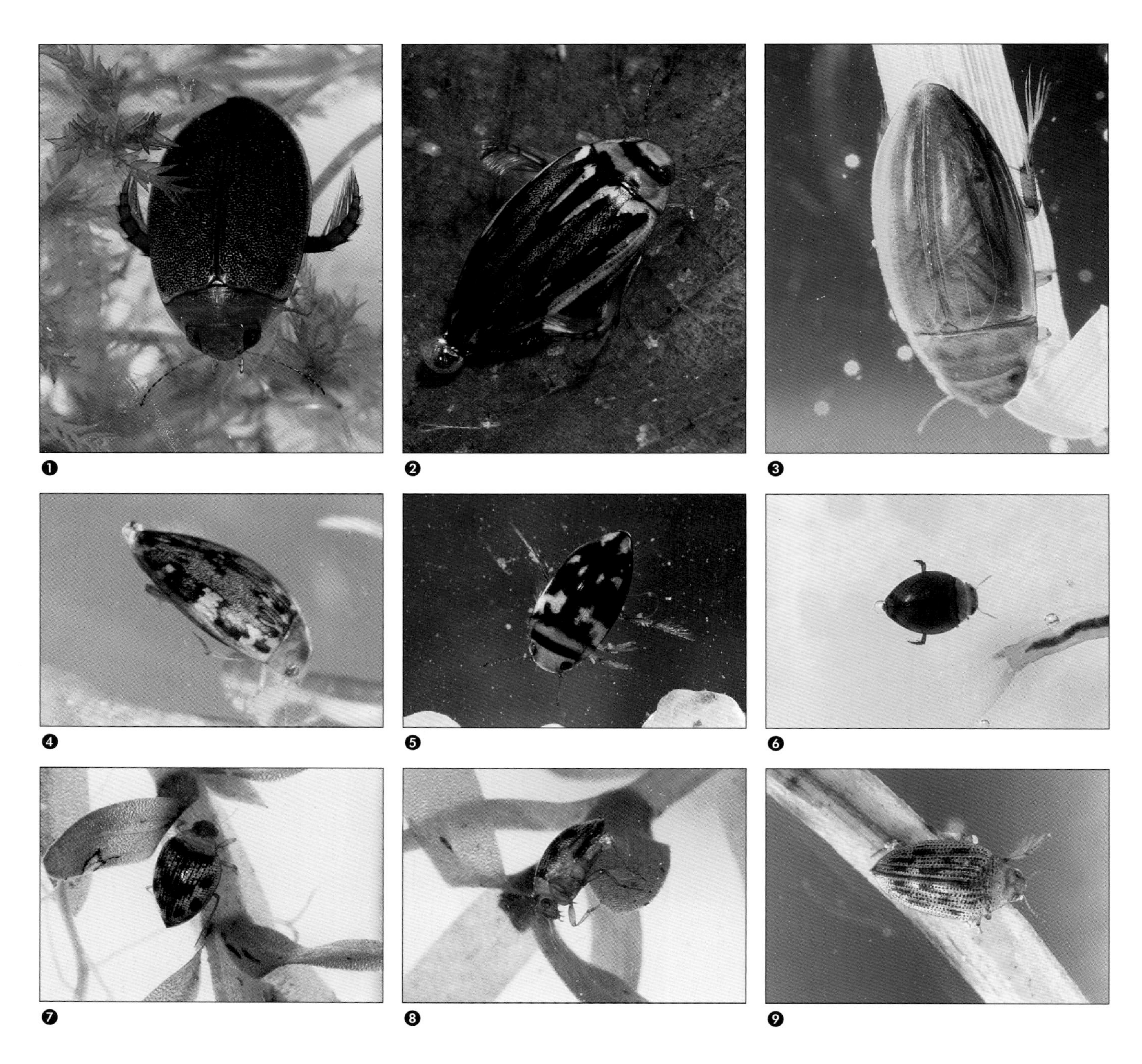

FAMILY DYTISCIDAE (continued) ❶ ***Graphoderus liberus*** (about 11 mm) is found in boggy ponds and lakes in forested areas across the continent. ❷ ***Coptotomus longulus*** (about 4 mm) is a common resident of permanent ponds throughout northeastern North America and across the continent. ❸ ***Matus ovatus*** (about 8 mm) is normally found among aquatic plants or decaying leaves in muddy ponds. ❹ ***Laccophilus maculosus*** (about 6 mm) is one of our most common diving beetles. ❺ ***Heterosternuta*** is a large genus (over 130 North American species) made up mostly of small (less than 5 mm) species like this little ***H. wickhami***. ❻ Despite its small size (only a couple of millimeters long), ***Hydrovatus pustulatus*** is distinctive because of its roly-poly, almost hemispherical shape. **FAMILY HALIPLIDAE (CRAWLING WATER BEETLES)** ❼ ❽ Crawling water beetles, so-called because of a disinclination to swim like the more dexterous predacious diving beetles, are small (2–6 mm) distinctively shaped aquatic beetles with a unique way of storing the air supply that they periodically replenish at the surface. The base of each hind leg is expanded out to form a huge plate that covers much of the abdomen. As you can see, the air supply of this ***Haliplus immaculicollis*** is tucked in between the leg plate and the abdomen. ❾ The largest of our three genera of crawling water beetles is the genus *Haliplus*, with 20 species in the northeast. This is ***H. subguttatus***.

FAMILY HALIPLIDAE (continued) ❶ Haliplid larvae occur amongst algal mats, and this ***Peltodytes*** has filaments of algae trapped between its porcupine-like body spines. Other haliplid larvae are similar in shape, but lack the body spines. ❷ The two black pronotal spots seen on these beetles are characteristic of the common genus ***Peltodytes***, like this ***P. edentulus*** (left) and ***P. tortulosus***. ❸ **Hungerford's Crawling Water Beetle** (***Brychius hungerfordi***) lives in clean streams, unlike the pond-loving species of *Haliplus* and *Peltodytes*. This endangered species, known only from a couple of streams in Michigan and one in Ontario, is different in shape from the common *Haliplus* and *Peltodytes*. **FAMILY NOTERIDAE** (BURROWING WATER BEETLES) ❹ Burrowing water beetles look much like small predacious diving beetles, but these relatively rare beetles have the odd habit of burrowing around the roots of aquatic plants where the larvae tap the roots for their air stores. They even pupate in underwater cocoons filled with air from punctured plant roots. The most common species in this little family (only five species in northeastern North America) is ***Hydrocanthus iricolor*** (4–5 mm). **FAMILY GYRINIDAE** (WHIRLIGIG BEETLES) ❺ ❻ Whirligigs have the unique ability to straddle the line between the water below and the air above, retaining the option of entering either environment as circumstances dictate. The many species of small (less than 8 mm) whirligigs in the genus ***Gyrinus*** produce malodorous defensive chemicals. ❼ Most whirligig beetles occur on lakes and ponds, but this species (***Dineutus discolor***) often occurs in huge aggregations in rivers. Whirligigs spread out evenly over the water surface while hunting (usually in the evening), but aggregate in dense masses while at rest. Dense aggregations are probably an antipredator strategy, allowing the pooling of defensive chemicals. ❽ Whirligig larvae, such as this common lake and pond species (***Dineutus assimilis***), are flanked with rows of feathery gills that allow them to respire without leaving the mud at the bottom of the pond, lake or stream, where their long, syringe-like mandibles serve to inject poisons into, and suck the contents out of bloodworms and other prey. Each larva will molt to a pupa in a convex, oval mud cell along the shoreline.

SUBORDER POLYPHAGA

SUPERFAMILY BYRRHOIDEA

FAMILY ELMIDAE (RIFFLE BEETLES) ❶ Riffle beetles breathe through spiracles opening into thin, silvery bubbles held permanently in place by incredibly dense body hairs. These little beetles are characteristic of moving, oxygen-rich water. The five northeastern ***Dubiraphia*** species tolerate a relatively wide range of conditions, including aquatic habitats ranging from cold springs to lake margins. *Dubiraphia* species are smaller than most riffle beetles, with only one species (the two-striped ***D. bivittata***) exceeding 2 mm in length. ❷ ***Macronychus glabratus*** (about 3 mm), the only North American species in this genus, is easy to recognize by its shape and strikingly long legs, and is usually easy to find on sticks and logs submerged in clean riffles. ❸ ❹ ***Stenelmis*** is our largest genus of riffle beetles, with 17 northeastern species including the very common ***S. crenata*** (about 3 mm). This larva has its tail-end gill chamber open, with the characteristic gill tufts exposed and clearly visible. ❺ This tiny (1–2 mm) three-striped riffle beetle (***Optioservus trivittatus***) occurs in clean, swift streams. **FAMILY DRYOPIDAE** (LONG-TOED WATER BEETLES) ❻ Long-toed water beetles look like relatively large (5 mm) riffle beetles (Elmidae) with short, clubbed antennae cupped into earlike processes. Although not as common as riffle beetles, long-toed water beetles occur regularly on submerged debris in small streams. Three genera occur in northeastern North America, but only ***Helichus*** is regularly encountered. **FAMILY PSEPHENIDAE** (WATER PENNY BEETLES) ❼ ❽ The head, legs and gills of this water penny beetle larva (***Ectopria leechi***) are concealed and protected by the armored upper body surface. *Ectopria* larvae are a little smaller and more elongate than larvae in the more common water penny genus *Psephenus*, and they don't form the unbroken penny shape of *Psephenus*. Adults of *E. leechi* often rest on vegetation near streams. ❾ ***Ectopria thoracica*** is the only one of the three northeastern *Ectopria* species with a bicolored pronotum.

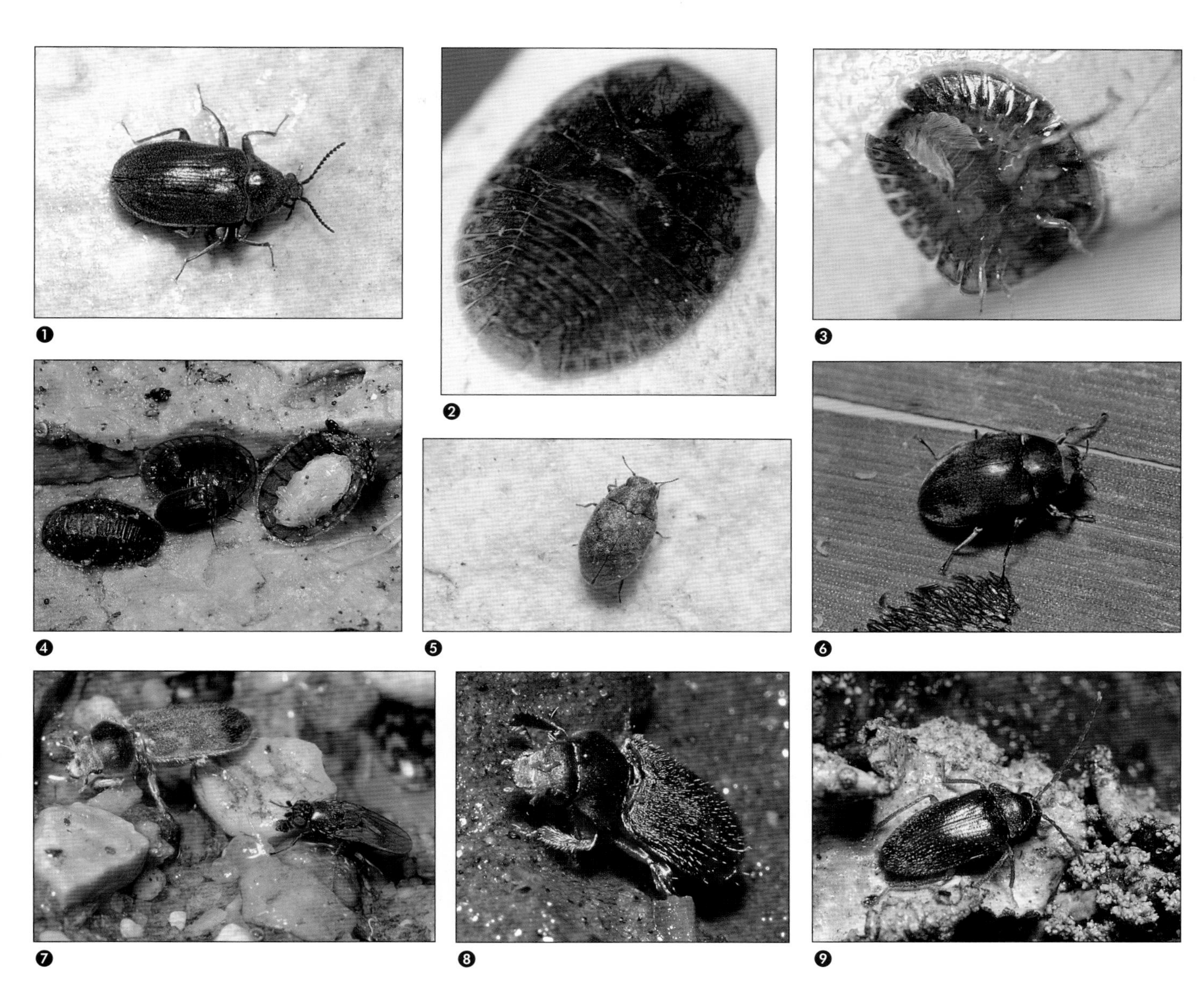

FAMILY PSEPHENIDAE (continued) ❶ ❷ ❸ ❹ These photographs show the adult, larva and pupa of the most common of the five northeastern water penny species, ***Psephenus herricki***. The inconspicuous adults (3–6 mm) are usually found on wet rocks surrounded by moving water. The name of the family refers to the familiar disklike larvae, which appear to stick tightly to the substrate like little suction cups, although they actually grip the substrate with a fringe of little hooks all around the edge of the broad, flat body. Flip a water penny over to see the abdomen with tufts of gills, and the thorax with six well-developed legs. Water penny larvae migrate to the water's edge to pupate, and the last of these four photos shows how they pupate under the same disklike shelters that protect the larvae. **FAMILY LIMNICHIDAE** (MINUTE MARSH-LOVING BEETLES) ❺ This minute marsh-loving beetle (***Limnichites huronicus***, 2 mm) was photographed on wet rocks along Lake Huron's Georgian Bay. Similar *Limnichites* are common along rivers. **FAMILY LUTROCHIDAE** (TRAVERTINE BEETLES) ❻ Travertine beetles are rarely encountered, but adults of the only northeastern species (***Lutrochus laticeps***, 3–3.5 mm) occur in or near running waters. Both adults and larvae normally feed on algae and waterlogged wood, but this one appears to be snacking on newly hatched deer fly larvae on a leaf overhanging a small stream. **FAMILY HETEROCERIDAE** (VARIEGATED MUD-LOVING BEETLES) ❼ ❽ Variegated mud-loving beetles live in burrows along the water's edge, and can be exposed by splashing water onto their burrows. Once you flush them out of their burrows you will have little time to observe their characteristic digging tools and variegated elytra, as they usually take flight quickly. Variegated mud-loving beetles often show up at artificial lights in huge numbers. Most members of this small family belong to the genus ***Heterocerus*** (3–7 mm) (the other insect in this picture is a sphaerocerid fly). **FAMILY PTILODACTYLIDAE** (PTILODACTYLID BEETLES) ❾ Most species in the small family Ptilodactylidae belong to the genus ***Ptilodactyla*** (4–6 mm), and have a heart-shaped scutellum and distinctive antennae. Larvae of *Ptilodactyla* develop in damp decaying wood or leaves, but one eastern ptilodactylid species is aquatic.

FAMILY BYRRHIDAE (PILL BEETLES) ❶ If you disturb one of these extremely convex beetles it will pull in its appendages and take on a pill-like appearance. Some pill beetles, like this ***Cytilus alternatus*** (5 mm) are common on sandy lakeshores, where both adults and larvae graze on mosses. ❷ The largest pill beetles are in the largest genus of the family Byrrhidae, the genus ***Byrrhus***. This is ***B. americanus*** (9 mm), one of six northeastern *Byrrhus* species.

SUPERFAMILY DASCILLOIDEA

FAMILY RHIPICERIDAE (CEDAR BEETLES OR CICADA PARASITE BEETLES)

❸ ***Sandalus niger*** (about 20 mm) is a parasitoid of cicada nymphs and one of only two northeastern members in the superfamily Dascilloidea. Larvae are found on tree roots with their hosts; adults are found on tree trunks.

SUPERFAMILY SCIRTOIDEA

FAMILY SCIRTIDAE (MARSH BEETLES) ❹ ❺ ❻ Marsh beetles are small (2–4 mm), and are often extremely common on foliage near ponds and streams. The larvae, which have exceptionally long antennae, are aquatic or semiaquatic, often living among submerged rotting leaves. Seven of the 17 northeastern marsh beetle species belong to the genus ***Cyphon***. ❼ Larvae of this attractive marsh beetle, ***Prionocyphon discoideus*** (4 mm), live in tree holes. ❽ ***Prionocyphon limbatus*** differs from our other *Prionocyphon*, *P. discoideus*, in antennal and elytral color. ❾ The jumping marsh beetles in the genus ***Scirtes*** have expanded hind legs. Larvae of our two northeastern species – this is ***S. tibialis*** (3 mm) – live among floating vegetation in ponds and calm lake margins. ❿ The only North American species of ***Microcara*** is an infrequently encountered northeastern species (***M. explanata***, 5mm).

SUPERFAMILY HYDROPHILOIDEA

FAMILY HYDROPHILIDAE (WATER SCAVENGER BEETLES) ❶ Water scavenger beetles have distinctively long maxillary palpi and short, clubbed antennae. The big (almost 40 mm) species shown here (***Hydrophilus triangularis***) is often attracted to lights when flying from pond to pond. The only other conspicuously large (over 25 mm) water scavenger beetle in the northeast is the relatively scarce *Dibolocelus ovatus*; other northeastern water scavenger beetles are less than 20 mm in length. ❷ Water scavenger beetles break the water surface to replenish their air supply, not with the tip of the abdomen as in the Dytiscidae, but with short, expanded antennae. Instead of tucking an air bubble under the forewings the way a diving beetle would, a water scavenger beetle keeps its air supply held in place using a dense mat of hairs underneath its body. This gives water scavenger beetles, like this ***Enochrus***, a characteristic silver-bellied appearance in life. *Enochrus* is a diverse (11 northeastern species) genus of small (3–8 mm) beetles. Stomping around the edge of a pond will usually drive some of these common beetles to the surface. ❸ Those silver bubbles carried by water scavenger beetles and other aquatic insects are more than just simple "tanks" of air. An aquatic beetle's bubble provides a surface through which carbon dioxide diffuses out and oxygen diffuses in, allowing the beetle to continually replenish its air store as long as the bubble lasts. The distinctively convex species shown here is ***Hydrobius melaenus*** (about 8 mm), a common beetle on dead wood submerged in slow or still water, and one of three northeastern *Hydrobius* species. ❹ ❺ ❻ Water scavenger beetles in the widespread and common genus ***Tropisternus*** are medium-sized (about 10 mm), streamlined beetles that look like shiny black or greenish black predacious diving beetles. Unlike predacious diving beetles they have short, clubbed antennae, and unlike most other water scavenger beetles they have a conspicuous keel that runs between the leg bases, projecting backward over the middle of the abdomen. These photos show the distinctively yellow-margined ***T. lateralis*** and one of several uniformly greenish black *Tropisternus* species.

FAMILY HYDROPHILIDAE (continued) ❶ Almost all medium-large (12–17 mm) northeastern water scavenger beetles belong to the genus ***Hydrochara***. As you can see by peering through the air bubble in this photo, *Hydrochara* species lack the extended and conspicuous keel that runs along the underside of the somewhat similar but smaller (less than 12 mm) *Tropisternus* species. Both genera are common in ponds throughout our area. ❷ The common name "water scavenger beetle" is somewhat inappropriate for this highly diverse family, because adult Hydrophilidae often feed on plants, and the larvae, like the ***Hydrochara*** larva in this photograph, are big-jawed, stout-bodied predators. ❸ Although most aquatic water scavenger beetles live in still waters some, including several of North America's 26 species of ***Berosus***, occur in shallow, slow-flowing water as well. *Berosus* larvae are distinctive for their long abdominal gills. ❹ The Hydrophilidae includes several genera of minute beetles which can be difficult to identify in the field. Some, such as the genus ***Laccobius***, are identifiable by their attractive colors despite their small size (this ***L. agilis*** is about 3 mm). ❺ ***Anacaena limbata*** (2 mm) sometimes occurs in huge numbers around the margins of ponds. ❻ Not all water scavenger beetles are aquatic insects. A few live in seemingly unlikely places, and cow dung seems like a particularly unlikely place to look for members of a family we normally associate with fresh water. Nonetheless, if you poke around in a fresh cow patty you have a good chance of coming up with one of three species of ***Sphaeridium***, the largest (5–7 mm) and most colorful of our terrestrial Hydrophilidae. All three species are introduced and are often abundant. ❼ The distinctively shaped and sculptured members of the genus *Helophorus* , like this small ***Helophorus*** (***Rhopalelophorus***) sp., are often placed in their own family, the Helophoridae. ❽ Although *Helophorus* species are aquatic, they commonly fly from pond to pond and are sometimes encountered in flight, at lights, or even walking on the snow surface like this ***Helophorus (Helophorus) grandis*** (introduced from Europe; 6–7 mm).

FAMILY HYDROPHILIDAE (continued) ❶ The little (2–6 mm) beetles in the genus ***Hydrochus*** are often placed in their own family (Hydrochidae) because of their striking dissimilarity to other water scavenger beetles. They look a bit like riffle beetles (Elmidae), but occur in weedy ponds. **FAMILY HYDRAENIDAE** (MINUTE MOSS BEETLES) ❷ Minute moss beetles are found in seeps and shore areas, where they feed on algae. These little (1–2 mm) beetles resemble tiny water scavenger beetles (Hydrophilidae) but have six to seven abdominal segments (five in Hydrophilidae). The more common of the three northeastern genera are shown here. ***Hydraena*** (the larger and paler of these two, with very long palpi) has 11 northeastern species; ***Ochthebius*** (lower image of live beetle) has only five. **FAMILY HISTERIDAE** (HISTER BEETLES OR CLOWN BEETLES) ❸ Hister beetles are usually shiny black maggot-eating beetles, with elbowed antennae and elytra that seem too small for the body. They are found anywhere the maggot hunting is good, and can be especially common in carrion, dung and under bark. Many species of the relatively fat, little (3–7 mm) typical hister beetles in the genera ***Hister*** and *Margarinotus* (previously included in the genus *Hister*) occur on dead animals. ❹ ❺ ❻ Some common hister beetles found under bark are strikingly flattened beetles, and our one large (7–10 mm), shiny ***Hololepta*** species (***H. aequalis***) is among the most spectacular beetles found under bark. The pincer-like mandibles leave little doubt as to how these predacious beetles make their living. The same specimen flipped over shows the heavy mite load this species often carries. ❼ Many of our subcortical (under bark) hister beetles are small, elongated, flattened beetles like the ***Platysoma lecontei*** shown here. We have about 50 genera and 350 species of hister beetles in North America; many are small and difficult to identify. ❽ These beetles were exposed by removing some bark from a recently felled maple. The small one is a minute hister beetle (***Aeletes simplex***, about 1 mm); the larger beetle is a sap beetle (***Carpophilus sayi***, Nitidulidae). ❾ A few hister beetles that live in dead wood are cylindrical, probably to facilitate movement through burrows made by the other bark beetles they eat. This is ***Teretrius latebricola*** (about 2 mm).

FAMILY HISTERIDAE (continued) ❶ Many hister beetles, like this ***Hypocaccus patruelis*** (about 4 mm) photographed under a dead fish washed up on a Great Lakes beach, are associated with carrion, where they feed mostly on fly maggots.

SUPERFAMILY STAPHYLINOIDEA

FAMILY SILPHIDAE (CARRION BEETLES) ❷ Burying beetles (genus ***Nicrophorus***) dig the soil out from under a small, dead animal, quickly sinking the body in the resulting hole and completely entombing it in a cryptlike chamber far from competing, carrion-loving flies. This ***N. tomentosus***, a medium-sized (15–23 mm) species distinctive for the dense yellow fuzz ("tomentosity") on its prothorax, is trying to inter a road-killed frog. ❸ ***Nicrophorus*** adults, like this ***N. tomentosus***, invariably carry numerous small reddish mites (*Poecilochirus* spp). These acarine hitchhikers are regular associates of burying beetles and, although they eat both fly and beetle eggs, they purportedly "pay for their ride" by eating fly eggs that would otherwise hatch as maggots able to compete with burying beetle larvae. ❹ Some burying beetles are fussy about the plots they choose to inter corpses, and this species (***Nicrophorus vespilloides***, 10–28 mm) prefers to bury bodies in soft Sphagnum moss. ❺ ***Nicrophorus carolinus*** *(18–25 mm)* is widespread in the United States but is restricted to the southern coastal states in the eastern part of its range. ❻ This ***Nicrophorus sayi*** has just landed, and has not yet folded its hind wings under its elytra. *N. Sayi* is a common species throughout northeastern North America. ❼ This **American Burying Beetle** or **Giant Burying Beetle** (***Nicrophorus americanus***) was taken in a light trap being used to survey agricultural pests in southern Ontario in 1972. American Burying Beetles have not been seen in Canada since. ❽ Although burying beetles (*Nicrophorus*) are by far the best known Silphidae, carrion beetles in other genera deposit their eggs in the soil near dead bodies rather than on purposefully buried carcasses. ***Necrodes surinamensis*** (13–18 mm), the only North American species in the genus *Necrodes*, prefers larger carcasses and can become conspicuously abundant around road-killed deer. ❾ ***Thanatophilus lapponicus*** (8–14 mm) as the species name seems to suggest, has a relatively northern distribution. It ranges from Alaska and the Yukon through to Ontario, Quebec and the northern United States.

FAMILY SILPHIDAE (continued) ❶ ❷ The family Silphidae includes the well-known burying beetles, as well as a variety of large, flattened beetles common under more dessicated dead bodies. ***Necrophila americana*** (16–20 mm) is a very common and easily recognized species. Both larvae and adults can be found under flattened roadside fauna. ❸ This carrion beetle (***Oiceoptoma inaequale***, 13–15 mm) spends the winter as an adult and is active in early spring, capitalizing in part on the recently thawed bodies of animals that died during the winter. ❹ ❺ This early spring species (***Oiceoptoma noveboracense***, 13–15 mm) is usually seen in pairs – the male on top of the female, with one of her antennae gripped in his mandibles. He maintains this position for some time before copulation occurs, and he hangs onto the female until she has laid eggs. This pair was photographed on a road-killed deer. **FAMILY AGYRTIDAE** (AGYRTID BEETLES) ❻ The only eastern species in this small family (sometimes treated as part of the Silphidae) is ***Necrophilus pettiti***, an uncommon species found on decaying fungi from Ontario to Florida. **FAMILY LEIODIDAE** (SMALL CARRION BEETLES AND ROUND FUNGUS BEETLES) **Subfamily Cholevinae** ❼ ***Prionochaeta opaca*** is a relatively large (5–6 mm) "small carrion beetle" that develops in the nests of small mammals. There is only one *Prionochaeta* species. ❽ Small carrion beetles (subfamily Cholevinae, sometimes treated as the family Leptodiridae) are small (2–5 mm), distinctively shaped beetles found on carrion, fungi and other decomposing material. A few occur in mammal or ant nests, where they feed on decomposing material. Although we have half a dozen genera in the northeast, they are cryptic in their habits and are rarely seen except by those who trap insects using dung and carrion. This bicolored species is ***Catops basilaris***. **Subfamily Leiodinae** ❾ This round fungus beetle (***Anogdus*** sp., 2–3 mm) lives in underground fungi.

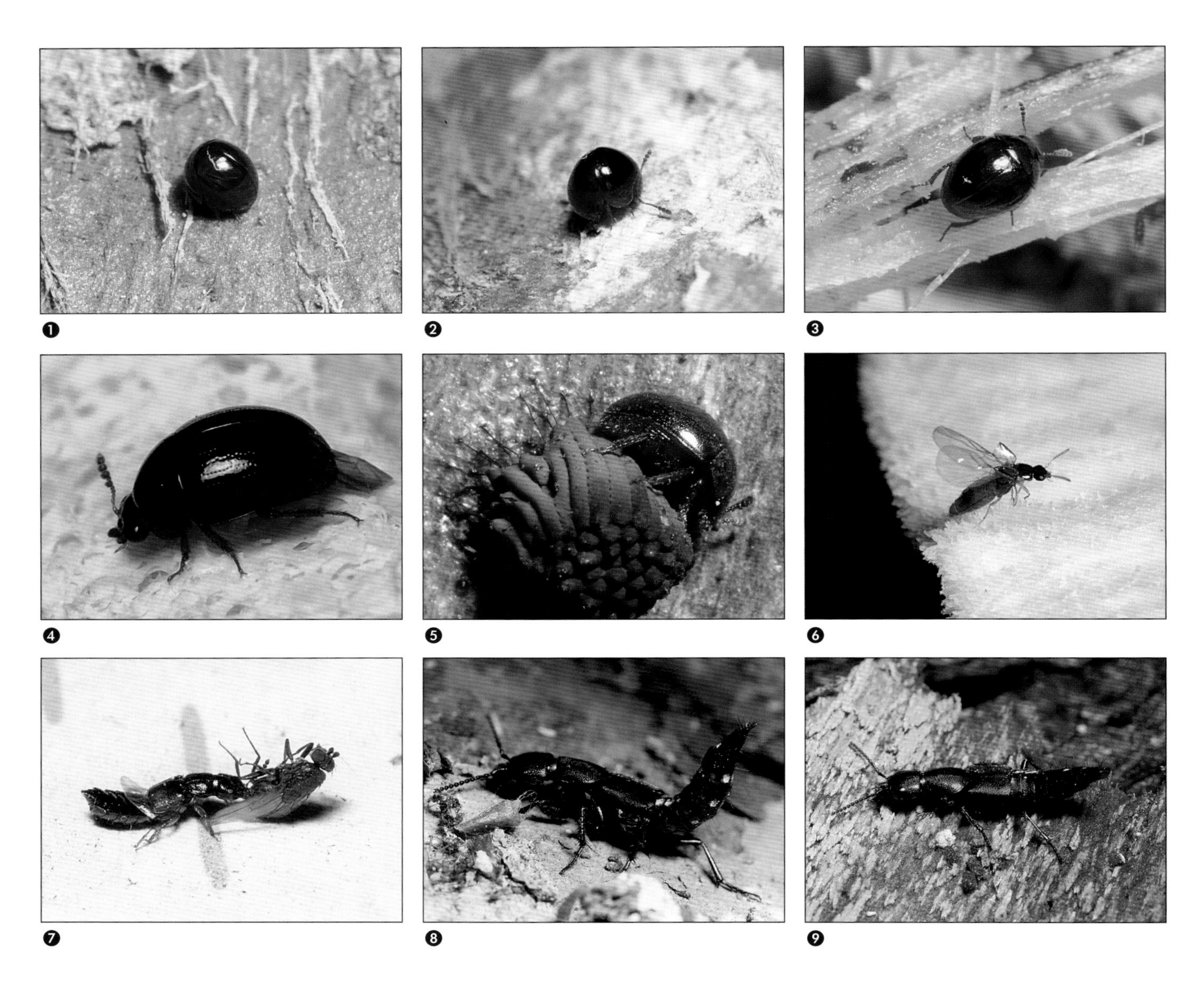

FAMILY LEIODIDAE (continued) ❶ ❷ Many round fungus beetles, like this tiny (2 mm) ***Agathidium***, have the ability to roll up into a ball when disturbed. *Agathidium*, with 40 North American species (14 northeastern), is the largest leiodid genus. ❸ Round fungus beetles are minute, very convex beetles. Most of the northeastern species that cannot roll into a ball are in the genus ***Leiodes*** (2–6 mm). ❹ This leiodid (***Anisotoma*** sp., 2–3 mm) is on a yellow slime mold (*Fuligo septica*). Slime molds (myxomycetes) are protozoan animals that periodically group together and form fungus-like sporocarps. Several beetles, incuding all members of the Sphindidae, many members of the Leiodidae and five other beetle families, develop only in slime mold fruiting bodies (sporocarps). ❺ This ***Anisotoma*** is feeding on sporocarps of the slime mold *Stemonitis splendens*. Larvae develop in the slime mold sporocarps, but must develop very quickly as the sporocarps don't last long. Some *Anisotoma* can complete larval development in only a couple of days. **FAMILY STAPHYLINIDAE (ROVE BEETLES)** ❻ Rove beetles are elongate beetles that usually have their hind wings intricately folded under characteristically short, truncate elytra. The hind wings are unfolded when the beetle takes flight. The Staphylinidae is an enormous family with over 3,000 North American species, many of which are small and difficult to identify. **Subfamily Oxytelinae (Spiny-legged rove beetles)** ❼ This little (3 mm) spiny-legged rove beetle is eating a small fly (Sphaeroceridae). You can see part of the beetle's hind wings sticking out behind the very short elytra. **Subfamily Staphylininae (Large rove beetles)** ❽ ❾ Many rove beetles have a distinctive manner of running around with the end of the abdomen curved up almost at right angles to the rest of the body. This is the aptly named ***Platydracus violaceus*** (13–16 mm).

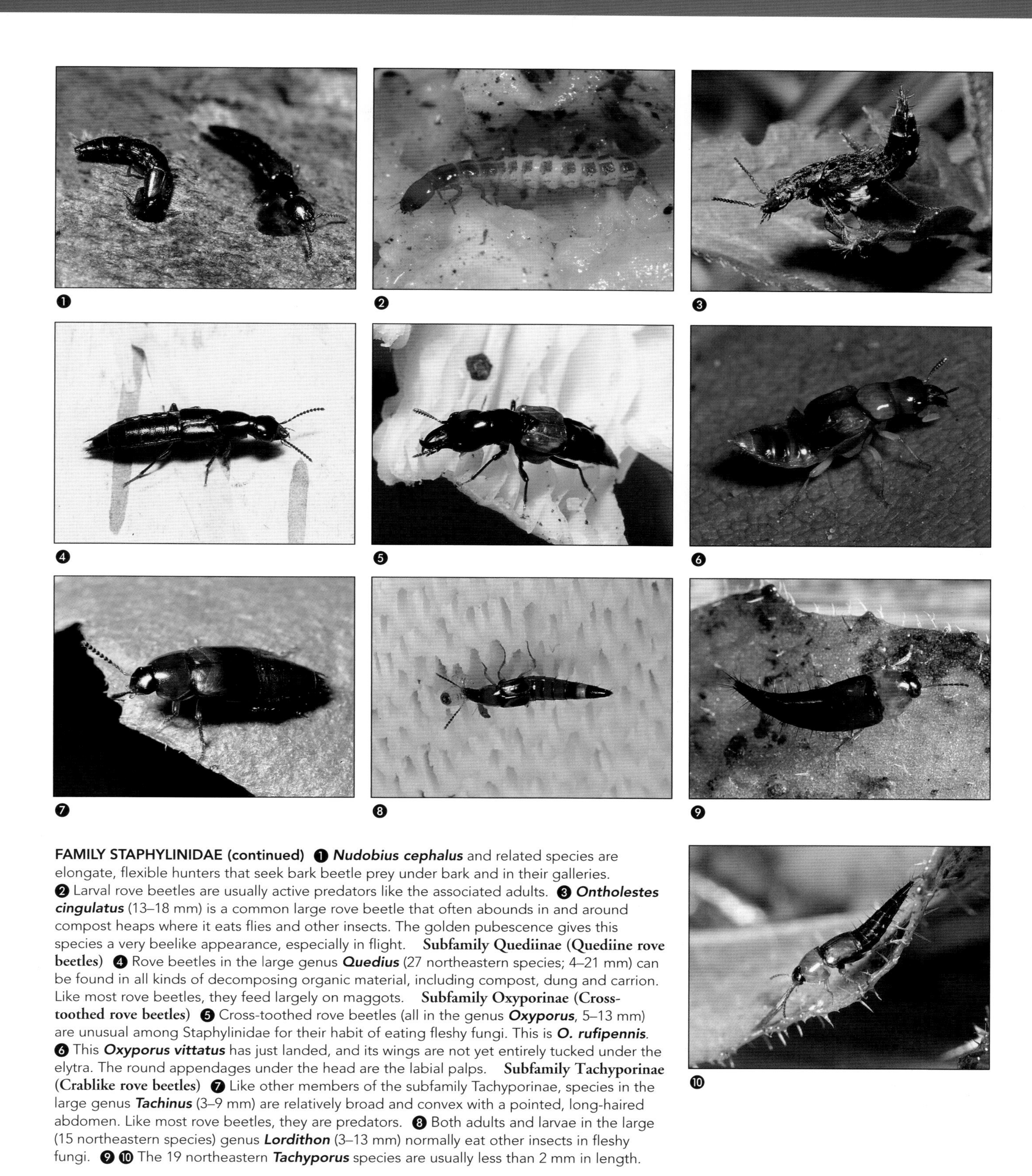

FAMILY STAPHYLINIDAE (continued) ❶ ***Nudobius cephalus*** and related species are elongate, flexible hunters that seek bark beetle prey under bark and in their galleries. ❷ Larval rove beetles are usually active predators like the associated adults. ❸ ***Ontholestes cingulatus*** (13–18 mm) is a common large rove beetle that often abounds in and around compost heaps where it eats flies and other insects. The golden pubescence gives this species a very beelike appearance, especially in flight. **Subfamily Quediinae (Quediine rove beetles)** ❹ Rove beetles in the large genus ***Quedius*** (27 northeastern species; 4–21 mm) can be found in all kinds of decomposing organic material, including compost, dung and carrion. Like most rove beetles, they feed largely on maggots. **Subfamily Oxyporinae (Cross-toothed rove beetles)** ❺ Cross-toothed rove beetles (all in the genus ***Oxyporus***, 5–13 mm) are unusual among Staphylinidae for their habit of eating fleshy fungi. This is ***O. rufipennis***. ❻ This ***Oxyporus vittatus*** has just landed, and its wings are not yet entirely tucked under the elytra. The round appendages under the head are the labial palps. **Subfamily Tachyporinae (Crablike rove beetles)** ❼ Like other members of the subfamily Tachyporinae, species in the large genus ***Tachinus*** (3–9 mm) are relatively broad and convex with a pointed, long-haired abdomen. Like most rove beetles, they are predators. ❽ Both adults and larvae in the large (15 northeastern species) genus ***Lordithon*** (3–13 mm) normally eat other insects in fleshy fungi. ❾ ❿ The 19 northeastern ***Tachyporus*** species are usually less than 2 mm in length.

FAMILY STAPHYLINIDAE (continued) Subfamily Steninae (Water skaters) ❶ ❷ Almost all of the distinctively bulgy-eyed rove beetles invariably found on the wet margins of ponds and streams belong to the huge genus ***Stenus*** (70 northeastern species). Some streamside *Stenus* species act like the natural equivalents of chemically propelled toy boats – if you throw one in the water it will release hydrophobic chemicals from the end of its abdomen in order to zip across the water surface. One of the beetles in these pictures has just captured a chironomid midge, but springtails are the more usual prey. Subfamily Paederinae (Paederine rove beetles) ❸ The beautiful red and blue rove beetles in the genus ***Paederus*** produce a unique defensive chemical (pederin), which is strong enough to repel predators and blister careless insect collectors. ***P. littorarius*** (about 5 mm), the most common of the four northeastern species, often abounds in debris along the edges of ponds and swamps. ❹ ***Homaeotarsus sellatus*** (8–10 mm), one of 15 northeastern *Homaeotarsus* species, is a common rove beetle among wet debris along shorelines. Subfamily Aleocharinae (Obscure rove beetles) ❺ The subfamily Aleocharinae is huge and difficult, with over 1,200 small to minute and relatively uniform North American species. This adult is eating a small fruit fly (*Chymomyza*); larvae of many obscure rove beetles develop as ectoparasites on fly pupae. ❻ This ***Atheta coriaria*** was one of many purchased as biological control agents for use against fungus gnats at a local greenhouse. *Atheta* is a very large and very difficult genus of very small rove beetles, with around 75 northeastern species. Subfamily Omaliinae (Ocellate rove beetles) ❼ Some rove beetles are pollen feeders and can be very abundant on some kinds of flowers, such as this Striped Maple (*Acer pensylvanicum*) blossom. Ocellate rove beetles like this ***Eusphalerum*** (2 mm) differ from other rove beetles in having simple eyes (ocelli) near the rear edge of the compound eyes, and in having relatively long elytra. Subfamily Scaphidiinae (Shining fungus beetles) ❽ Look for this relatively large (4–5 mm) shining fungus beetle (***Scaphidium quadriguttatum***) on fungus-covered wood. ❾ Most of our shining fungus beetles are small to minute beetles, often less than 2 mm, like this ***Scaphisoma*** crawling out of a soft shelf fungus.

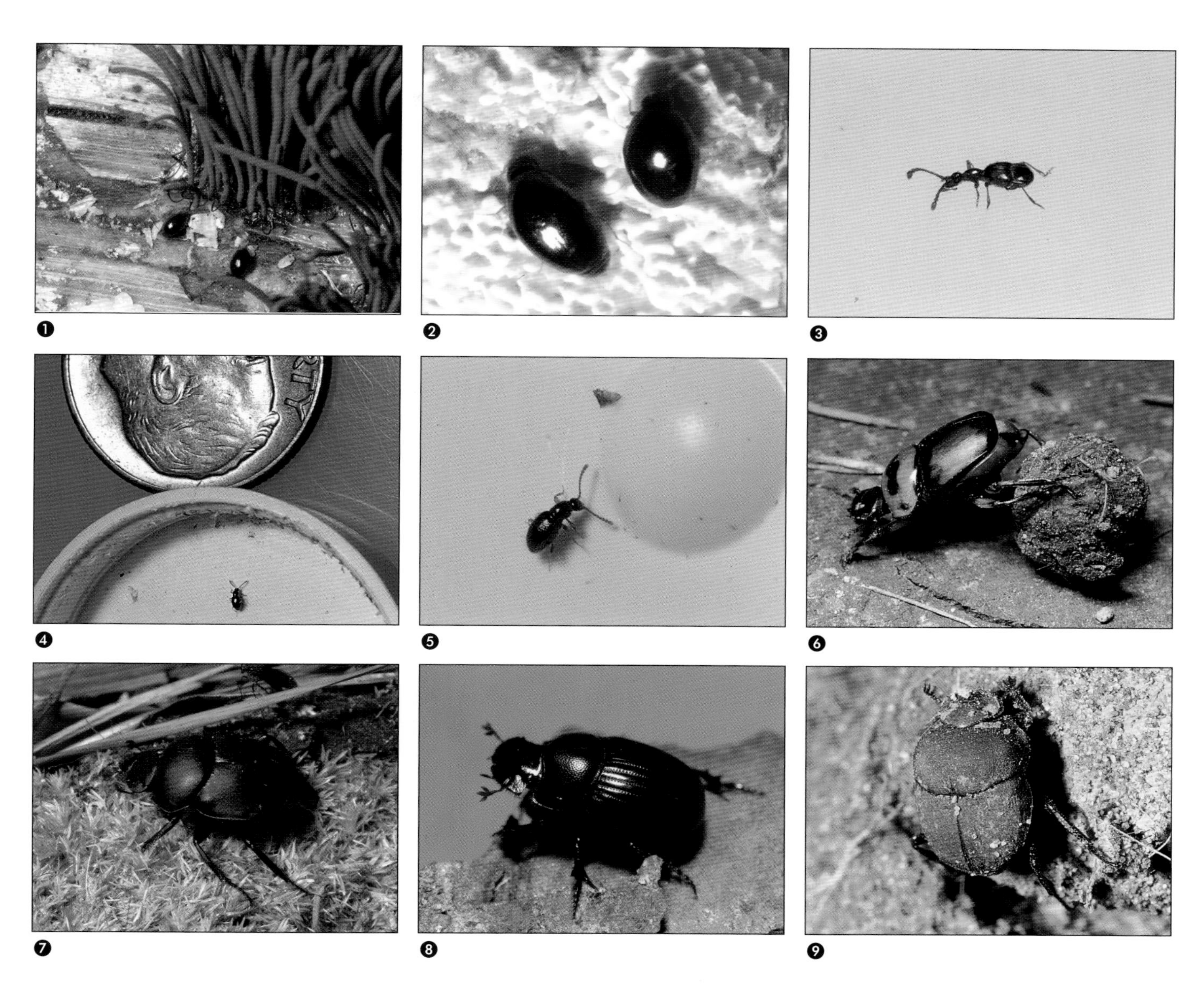

FAMILY STAPHYLINIDAE (continued) ❶ ❷ The minute (about 1 mm) beetles at the base of this slime mold are in the genus ***Eubaeocera***. Subfamily Pselaphinae (Short-winged mold beetles) ❸ Although there are 71 genera and over 500 species of short-winged mold beetles in North America, these tiny (usually 1–2 mm) beetles are not often seen. They are relatively common in peatlands, and this photo shows ***Pselaphus fustifer***, one of several species found in Ontario fens. **FAMILY SCYDMAENIDAE** (ANTLIKE STONE BEETLES) ❹ ❺ This tiny antlike stone beetle (***Euconnus*** sp.) was photographed in the lid of a snap-top pill bottle. The lid is only 2.5 cm across – antlike stone beetles are very small insects! Adults and larvae eat oribatid mites (the tiny beetle-like mites that abound in soil and leaf litter).

SUPERFAMILY SCARABAEOIDEA

FAMILY SCARABAEIDAE Subfamily Scarabaeinae (Dung beetles and tumblebugs) ❻ Members of the genus ***Canthon*** are called "tumblebugs" for their habit of packing a bit of dung into a ball then rolling the ball away from the excremental mother lode to bury it safely away from most predators and parasites. The three northeastern species of *Canthon* are uniformly colored, rather dull beetles (10–20 mm); this colorful neotropical ***C. 5-maculatus*** was photographed in Bolivia. ❼ Three species of dung-rollers in the genus ***Canthon*** occur in the northeastern United States. This species, ***C. laevis***, is southeastern. ❽ ***Copris minutus*** ranges from southern Canada to Florida, but most of the 200-plus species in this genus are tropical. Both northeastern species of *Copris* frequent cow dung. ❾ This ***Melanocanthon bispinatus*** (7–8 mm) was found as it worked on a rather desiccated deer dropping in southern Ontario. This species is normally a dung roller, but it also buries dead insects, and related *Melanocanthon* sometimes make balls out of fungi or carrion.

FAMILY SCARABAEIDAE (continued) ❶ ❷ One of these ***Melanocanthon bispinatus*** is rolling along a dead scarab beetle (*Anomala undulata*), while the other pair of beetles is rolling a bit of dung. *M. bispinatus* occurs from Ontario south to Florida. ❸ This conspicuously horned male dung beetle is an introduced European species (***Onthophagus taurus***) first noticed in North America in the mid-1970s, but now common in pastures through much of the eastern United States. Females, which lack the large forked horn seen on this male, burrow directly below dung and sequester balls of dung in chambers at the end of the burrow. The genus *Onthophagus* is enormous, but only about 13 of the 1,500 world species occur in northeastern North America. ❹ This ***Onthophagus orpheus canadensis*** was one of many in a deer dropping along an Ontario woodland trail. ❺ This dung beetle (***Phanaeus vindex***, 14–22 mm) is carrying a couple of kleptoparasitic flies (*Norrbomia frigipennis*, Sphaeroceridae) waiting for the opportunity to lay their eggs on its dung store. Females of the two northeastern *Phanaeus* do not have horns on their heads, but males usually do. This species ranges from New York south to Florida. ❻ This ***Phanaeus vindex*** male has a strikingly long pronotal horn, but some male *Phanaeus* have no horns. This sort of dimorphism is common amongst scarabs, and may reflect alternative mating strategies. The big horned males are equipped to fight, perhaps driving other males away from their mate's burrow. Small-horned males of some species sneak into female burrows from underground, and mate while the big boys are fighting or guarding the burrow entrance. **Subfamily Aphodiinae (Aphodiine dung beetles)** ❼ Aphodiine dung beetles are usually smaller than other scarabs. These ***Aegialia conferta***, photographed on a Great Lakes beach, are only about 4 mm long. ❽ Aphodiine dung beetles, like this ***Aphodius rubripennis*** (7–8 mm), are often extremely common insects in cattle dung, and flying adults sometimes seem to fill the air in early spring. The lobes of the antennae are spread apart to detect odors while in flight, and closed to cover sensitive antennal surfaces while digging through the dung. ❾ The genus ***Aphodius*** includes over 200 North American species, almost a quarter of which occur in northeastern North America. Most burrow in and under excrement and store their fecal food stores underground. This species (***A. distinctus***, 5 mm) is common in cattle dung.

FAMILY SCARABAEIDAE (continued) Subfamily Trichiinae (Flower scarabs and others) ❶ ❷ If you reach into the rotting wood in the middle of a dead or damaged hardwood tree and find a large (around 30 mm), C-shaped beetle larva, it is probably one of our two species of ***Osmoderma***. This one, ***O. scabra***, has sculptured elytra; *O. eremicola* has smooth elytra. ❸ ***Trichiotinus*** species, like this ***T. affinis*** (7–10 mm), are the most common flower scarabs in the northeast. Their somewhat beelike appearance probably protects these conspicuous flower-frequenting beetles from predation. Larvae live in rotting wood. ❹ The only northeastern species of ***Gnorimella*** (***G. maculosa***, 12–14 mm) is a strikingly spotted species usually seen on flowers or nearby foliage. ❺ Adult flower beetles in the genus ***Valgus*** (two northeastern species, 4–8 mm) are often seen on spring flowers; larvae are found in termite-infested wood. Subfamily Cetoniinae (Bumble flower beetles) ❻ **Bumble Flower Beetles** (***Euphoria inda***, 13–16 mm) look and sound like slow-flying bumblebees while in flight. Look for them flying low to the ground in early spring. Larvae develop on buried decayed plant material or rotting wood. ❼ **Bumble Flower Beetles** (***Euphoria inda***) occasionally occur in a dark form previously known as a separate subspecies (*E. inda nigripennis*). ❽ ***Euphoria sepulchralis*** is more southern in distribution than the common *E. inda*. Look for this species on milkweed flowers. ❾ ***Euphoria fulgida*** (13–18 mm) is one of two brilliant green northeastern beetles in the Cetoniinae. *E. fulgida* occurs throughout northeastern North America, often on lakeshores. The similar Green June Beetle (*Cotinus nitida*) is a larger beetle, over 20 mm in length, that is an occasional turf pest as far north as New York.

FAMILY SCARABAEIDAE (continued) Subfamily Rutelinae (Shining leaf chafers) ❶ The **Japanese Beetle** (***Popillia japonica***, 9–11 mm) was accidentally introduced to North America with nursery stock about 80 years ago, and is now a serious pest of a variety of plants. Larvae feed on roots, and adults feed on foliage and fruit. ❷ **Spotted Pelidnota** (***Pelidnota punctata***, 25 mm) adults can often be found feeding on grape foliage. Larvae live in rotting roots and wood. ❸ The **Goldsmith Beetle** (***Cotalpa lanigera***, 20–26 mm) develops on plant roots, sometimes doing significant damage. ❹ ***Anomala*** is a large genus, with almost 50 North American species. Most are brownish in color, but ***A. marginata*** often has distinct metallic highlights. This species occurs from Ontario to Florida. Subfamily Melolonthinae (June beetles and chafers) ❺ ❻ The white grubs familiar to anyone who has hand-tilled a garden are the typical C-shaped larvae of June beetles or May beetles. Most are in the large genus ***Phyllophaga***, with over 200 eastern North American species. Larvae feed below ground and adults feed above ground. ❼ June beetle adults usually feed on fruit or foliage, often at night. This subtly iridescent beetle feeding on *Potentilla* foliage is one of 19 northeastern species in the genus ***Serica***. ❽ Larval **European Chafers** (***Amphimallon majalis***) feed on roots, often seriously damaging lawns. Adults of this introduced pest can be conspicuous for brief periods in spring when they form noisy mating flights exactly at dusk, often aggregating by the thousands on prominent trees. ❾ This large (25 mm) lined June beetle is ***Polyphylla variolosa***, the northernmost of the of the two northeastern species in the genus. Larvae are sometimes pests of turf.

FAMILY SCARABAEIDAE (continued) ❶ **Rose Chafers** (***Macrodactylus subspinosus***, 8–10 mm) are commonly seen aggregating and mating on various kinds of flowers in sandy areas where the larvae develop on roots of a wide variety of grasses and other plants. ❷ Chafers in the genus ***Dichelonyx*** (nine northeastern species, 7–12 mm) develop on tree roots. Adults can be found on a variety of hardwood and conifer foliage. ❸ The six northeastern species of ***Hoplia*** (6–12 mm) feed on the blossoms and foliage of a variety of plants. This species is often found on *Crataegus* blossoms. **Subfamily Dynastinae (Rhinoceros beetles, hercules beetles and elephant beetles)** ❹ ❺ The subfamily Dynastinae contains some of the largest beetles in the world, such as the fist-sized **Hercules Beetle** (***Dynastes hercules***) male shown here. Hercules Beetles do not occur in our area (this photo was taken in Costa Rica), but a closely related *Dynastes* (***D. tityus***, the **Eastern Hercules Beetle**) does range as far north as Indiana, where you can find their massive larvae in decaying wood. At 40–50 mm, this is the largest scarab, and the heaviest beetle, in North America. This pair of pinned specimens shows the dimorphism between the horned male and the hornless female. ❻ ***Strategus*** males are fairly large (about 30 mm), with impressive pronotal horns. Although mostly distributed in the south, one species (***S. antaeus***, the **Ox Beetle**) occurs as far north as Rhode Island. ❼ Compared to the spectacular *Dynastes* beetles, most northeastern Dynastinae are relatively nondescript beetles. This species (***Tomarus relictus***, 18–24 mm) is usually found on sandy lakeshores.
FAMILY CERATOCANTHIDAE (CONTRACTILE SCARABS) ❽ ***Germarostes globosus*** (5–6 mm) is one of two species of contractile scarabs found in the northeast, both in the genus *Germarostes*. Larvae of this genus have been collected in bess beetle (Passalidae) burrows in rotting logs, perhaps feeding on bess beetle frass.

FAMILY CERATOCANTHIDAE (continued) ❶ These ***Ceratocanthus relucens*** from Costa Rica show clearly why this family is called 'contractile scarabs.' ❷ This ***Germarostes aphodioides*** (about 4 mm) was exposed by peeling the bark off an old fungus-covered log in southern Ontario. **FAMILY GEOTRUPIDAE** (EARTH-BORING SCARABS) ❸ Earth-boring scarabs burrow deeply below dung or carrion, and adults are rarely seen. Most North American species belong to the genus ***Geotrupes*** (10–17 mm). ❹ The seven northeastern species of ***Bolboceras*** are similar to one another and difficult to identify to the species level without dissecting male genitalia. **FAMILY TROGIDAE** (SKIN BEETLES) ❺ Skin beetles are usually found on old, dried-out carcasses that have been abandoned by most other scavengers, although this individual was found under the bark of a dead tree. Adults withdraw their antennae and remain motionless when disturbed so you have to look closely to see them. Out of the 23 northeastern skin beetle species, 22 belong to the genus ***Trox***. **FAMILIES OCHODAEIDAE AND HYBOSORIDAE** ❻ The families Ochodaeidae (***Ochodaeus*** sp., the brown pinned beetle) and Hybosoridae (***Hybosorus illigeri***, the black pinned beetle) were formerly treated as part of the Scarabaeidae. *Ochodaeus* species occur from Canada to Florida. *Hybosorus illigeri*, the only eastern North American hybosorid, was accidentally introduced from Europe to the United States in the 1800s. Both of these poorly known beetle families are rarely encountered except at lights. **FAMILY LUCANIDAE** (STAG BEETLES) ❼ These male **Pinching Bugs** (***Pseudolucanus capreolus***, 30 mm) are using their large mandibles to fight for prime space on a rotting stump. Despite the name and formidable size, these common beetles are harmless and don't have the muscle to give you much of a nip. ❽ ❾ This male **Antelope Beetle** (***Dorcus parallelus***, 20 mm) was seen at night on an exposed tree trunk; the smaller-jawed female was taken from an elm log. Male Antelope Beetles have distinctive processes on the inner surfaces of their large mandibles.

FAMILY LUCANIDAE (continued) ❶ This small (10–13 mm) stag beetle (***Platycerus piceus***) does not show the conspicuous sexual dimorphism found in most other stag beetles, and the mandibles are similar in size in both male and female. ❷ The male of this pair of stag beetles (***Ceruchus piceus***, 10–15 mm) has enlarged and oddly shaped mandibles quite unlike those of the female. Like the other eight or nine northeastern stag beetle species, they live in rotting logs. **FAMILY PASSALIDAE** (BESS BEETLES) ❸ ❹ The only northeastern bess beetle is the **Horned Passalus** (***Odontotaenius disjunctus***), a large (30–40 mm) beetle common in colonies in rotten logs. Both larvae and adults communicate by stridulating. ❺ Larval and adult bess beetles live together in rotting wood, where larvae are dependant on microorganism-rich wood that has been conditioned by adult feeding.

SUPERFAMILY ELATEROIDEA

FAMILY ARTEMATOPODIDAE (ARTEMATOPODID BEETLES) ❻ ***Macropogon piceus*** is one of three eastern species in this small, rarely encountered family. Larvae of some other members of this group develop in moss growing on rocks, but the biology of this species is unknown. **FAMILY CANTHARIDAE** (SOLDIER BEETLES) ❼ The bright yellow and black ***Chauliognathus pensylvanicus***, which usually abounds on goldenrod flowers late in the year, is our most familiar soldier beetle. ❽ ***Chauliognathus marginatus*** appears earlier in the summer than the more common and widely distributed *C. pensylvanicus*. This photograph was taken in an Ontario tallgrass prairie.
❾ ***Cantharis*** is a large genus of soldier beetles common on foliage where they eat aphids and other insects. *Cantharis* species sometimes look a bit like fireflies, but unlike fireflies their heads are visible from above and only partly concealed by the pronotum.

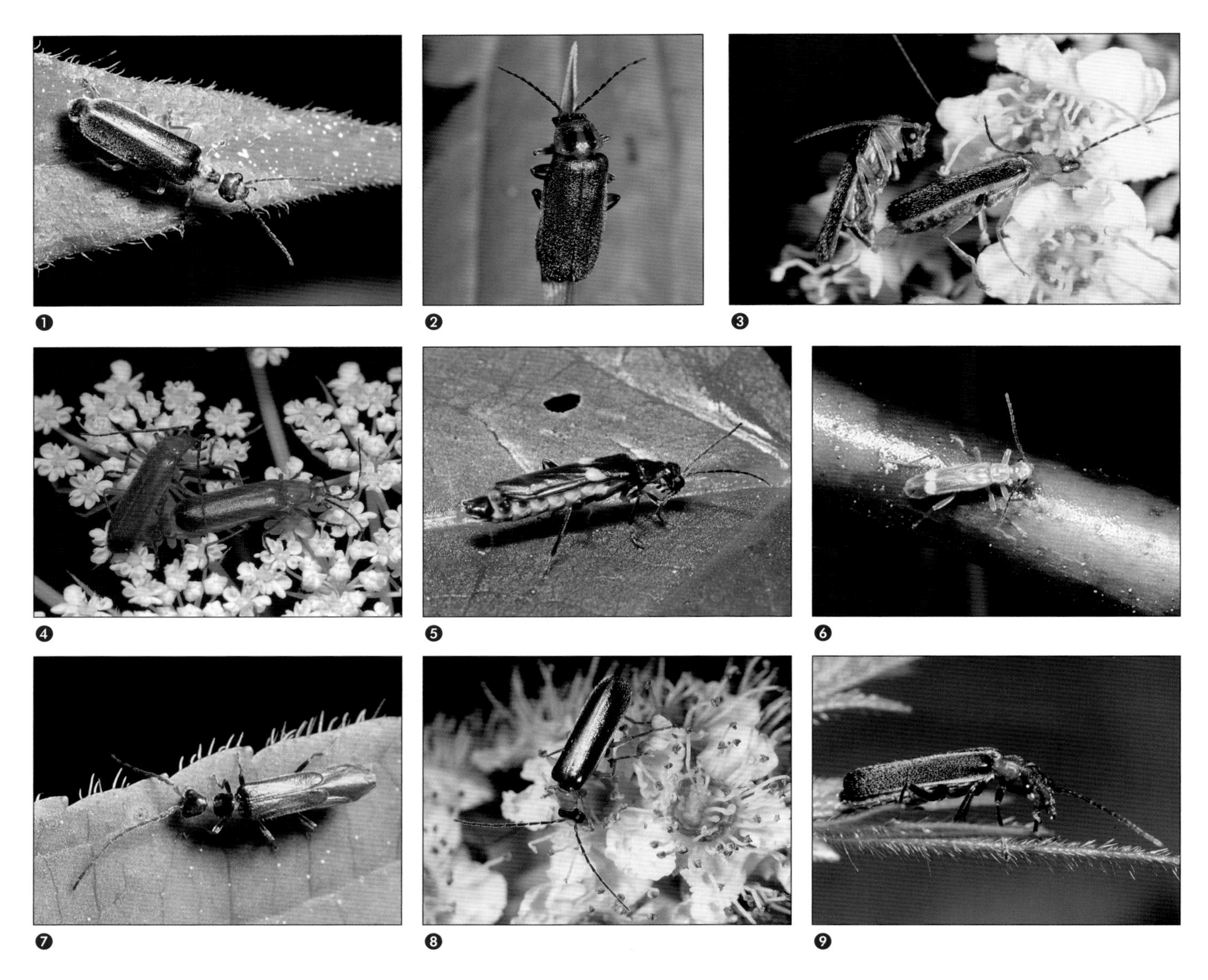

FAMILY CANTHARIDAE (continued) ❶ ❷ These beetles belong to groups that have long been treated as part of the genus *Cantharis*, but are now put in the genera ***Rhagonycha*** and ***Ancistronycha***. The beetle on the leaf ❶ is ***R. oriflava***; the other ❷ is ***A. bilineata***. ❸ Soldier beetles are often found on flowers where, like the female of this pair of ***Rhagonycha imbecillis***, they feed on pollen. ❹ ***Rhagonycha fulva*** is a European beetle recently introduced to North America, and now common across the continent. Copulating pairs of this species are a frequent late summer sight in Ontario meadows. ❺ Some soldier beetles have very short elytra, but do not fold their hind wings away under the elytra like similarly short-winged rove beetles. This is ***Trypherus frisoni***, a relatively small (7 mm) soldier beetle often common on vegetation along woodland streams. ❻ This tiny (3 mm) soldier beetle (***Malthinus occipitalis***) has soft, yellow-tipped elytra that don't quite cover the flying (hind) wings. *Malthinus* is a small genus with only two northeastern species. ❼ This tiny (about 3 mm) soldier beetle belongs to the largest genus of soldier beetles in North America, although only 18 of the more than 120 North American species of ***Malthodes*** occur in the northeast. ❽ ❾ Most of the soldier beetles with a distinct "neck" (the exposed and narrowed back of the head) belong to the large and difficult genus ***Podabrus***, with almost 40 northeastern species.

FAMILY CANTHARIDAE (continued) ❶ ❷ Soldier beetles in the genus ***Silis*** are relatively small (4–6 mm) and broad beetles with an orange pronotum and black elytra. Males have notches at the hind angles of the pronotum. ❸ Larvae of soldier beetles are predacious, like most associated adults, and usually have a distinctively velvety appearance. **FAMILY LAMPYRIDAE** (FIREFLIES OR LIGHTNINGBUGS) ❹ ❺ When you look at this firefly from the top, the head is concealed or largely concealed by the broad pronotum, but when you look from the side you can see the big eyes characteristic of nocturnal organisms that use vision to hunt for mates and prey. Like most firefly adults, this ***Photuris pensylvanica*** (about 10 mm), has light-producing organs in bright yellow terminal abdominal segments. ❻ Larval fireflies, like this heavily armored ***Photuris pensylvanica*** larva, glow as they undertake their nocturnal searches for snails and similar prey. ❼ This brightly colored firefly pupa was exposed in its pupal chamber by lifting bark off a dead log, and the light organ near the tip of the abdomen glowed in response to the disturbance. ❽ Fireflies normally have the head concealed under the pronotum, but the head of this ***Photinus*** is clearly visible as it feeds on pollen. ❾ Most of our small fireflies belong to the large genus ***Photinus*** (17 northeastern species). Most ***Photinus*** adults are accomplished flashers, but this species (***P. indictus***, 6–8 mm) lacks a developed light organ.

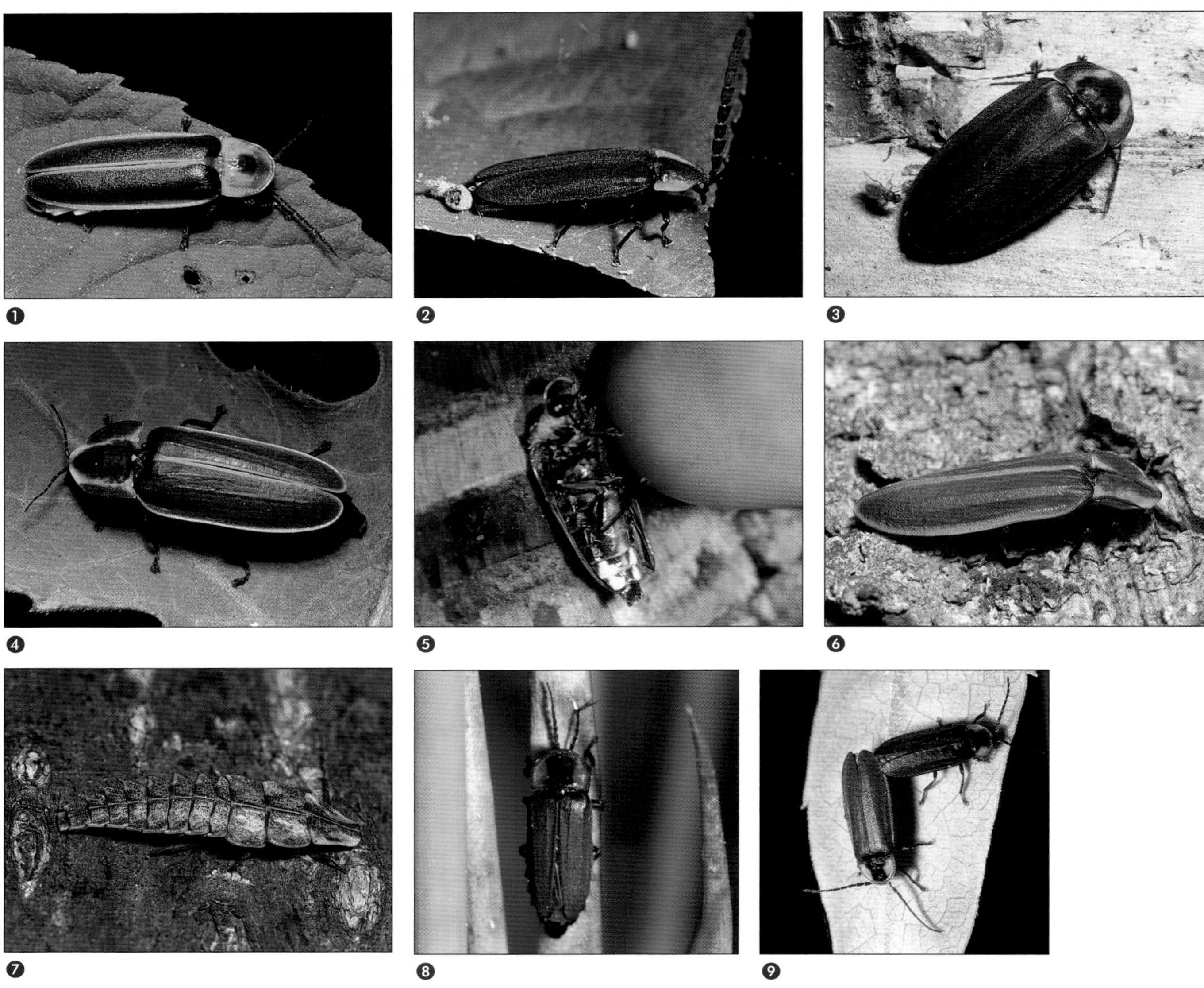

FAMILY LAMPYRIDAE (continued) ❶ The **Big Dipper Firefly** (***Photinus pyralis***, 9–15 mm) is common in lawns, parks and other open areas where it is active earlier in the evening than most other fireflies. Males fly relatively low (making them among the more easily captured fireflies) and appear as characteristic J-shaped streaks of light as they flash in flight. Females respond from low vegetation with brief flashes to guide the males in, and the resulting union leads to eggs that hatch into earthworm-eating larvae. ❷ Not all fireflies are nocturnal. This common day-active species (***Lucidota atra***, 8–11 mm) has greatly reduced light organs, in contrast with the conspicuous light organs of nocturnal fireflies. Males in this genus have long, flattened antennae. ❸ The **Winter Firefly** (***Ellychnia corrusca***, 10–14 mm) is a broad firefly commonly seen on tree trunks from fall to spring. Adults have no light organ but the larvae and pupae, like all firefly juveniles, do emit light. This species is sometimes so abundant on maple trunks in late winter and spring it becomes a pest in sap buckets. The little fly eyeing up this beetle is a phorid fly – some phorid flies are significant parasitoids of Winter Fireflies. ❹ ❺ ❻ ❼ These photos show the top and bottom of ***Pyractomena*** (8–15 mm) fireflies. Females, like this upside-down one, have the light organ divided; in males it extends right across the underside of the abdomen. *Pyractomena* larvae can sometimes be seen glowing beneath the water surface of lakes and ponds as they submerge for brief periods in search of aquatic snails. This is ***P. borealis***, the most common of the 10 northeastern species of *Pyractomena*. ❽ ❾ ***Pyropyga*** fireflies usually occur in damp areas like sedge meadows or river margins. These little (4–8 mm) fireflies have no light organs and, unlike the otherwise similar *Lucidota atra*, the antennae are not flattened and the antennae of both sexes are similar in length. ❿ This distinctively short-winged adult firefly (***Phosphaenus hemipterus***) is known from Europe and a few places in eastern Canada. This one was photographed in Ontario.

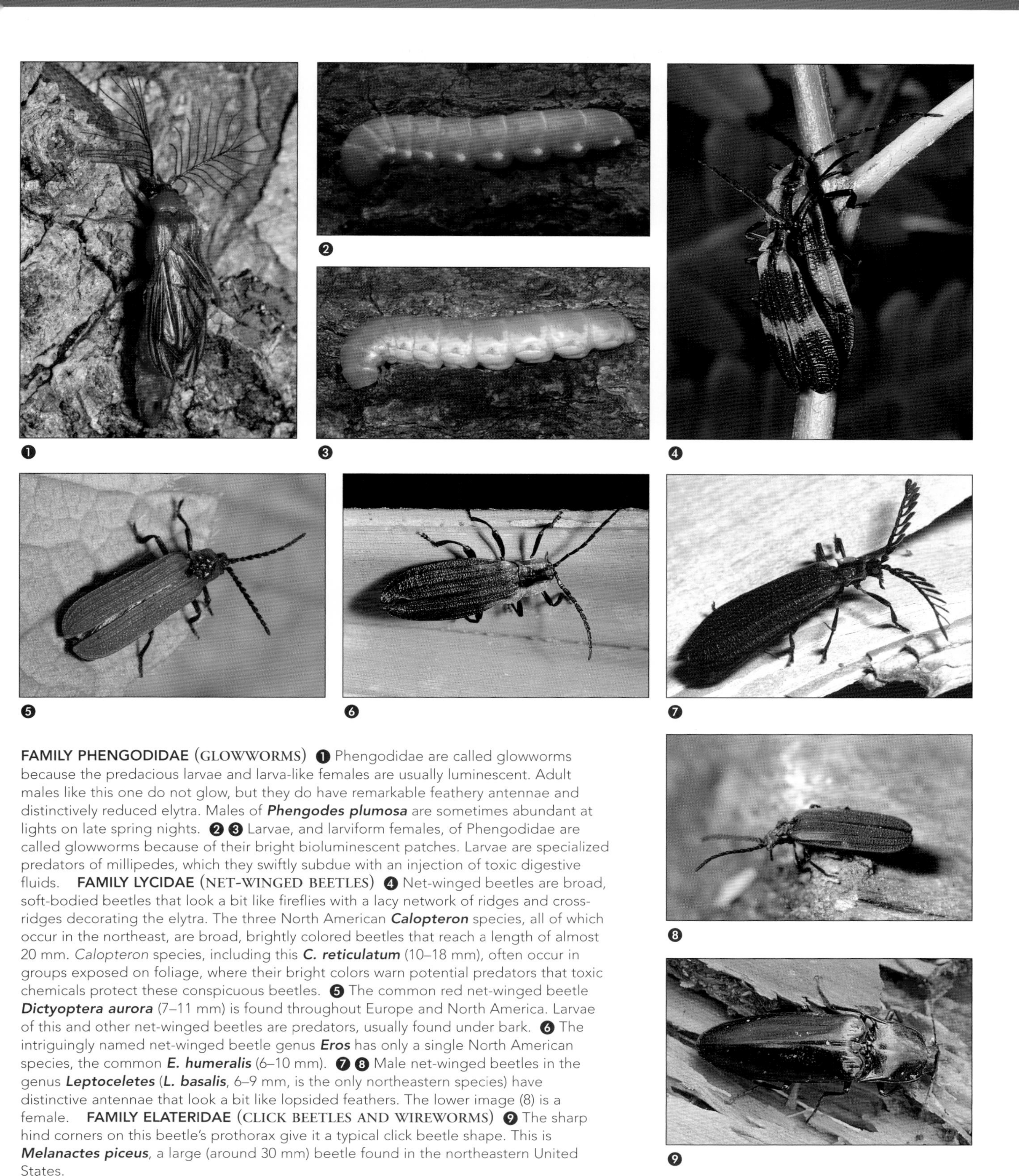

FAMILY PHENGODIDAE (GLOWWORMS) ❶ Phengodidae are called glowworms because the predacious larvae and larva-like females are usually luminescent. Adult males like this one do not glow, but they do have remarkable feathery antennae and distinctively reduced elytra. Males of ***Phengodes plumosa*** are sometimes abundant at lights on late spring nights. ❷ ❸ Larvae, and larviform females, of Phengodidae are called glowworms because of their bright bioluminescent patches. Larvae are specialized predators of millipedes, which they swiftly subdue with an injection of toxic digestive fluids. **FAMILY LYCIDAE** (NET-WINGED BEETLES) ❹ Net-winged beetles are broad, soft-bodied beetles that look a bit like fireflies with a lacy network of ridges and cross-ridges decorating the elytra. The three North American ***Calopteron*** species, all of which occur in the northeast, are broad, brightly colored beetles that reach a length of almost 20 mm. *Calopteron* species, including this ***C. reticulatum*** (10–18 mm), often occur in groups exposed on foliage, where their bright colors warn potential predators that toxic chemicals protect these conspicuous beetles. ❺ The common red net-winged beetle ***Dictyoptera aurora*** (7–11 mm) is found throughout Europe and North America. Larvae of this and other net-winged beetles are predators, usually found under bark. ❻ The intriguingly named net-winged beetle genus ***Eros*** has only a single North American species, the common ***E. humeralis*** (6–10 mm). ❼ ❽ Male net-winged beetles in the genus ***Leptoceletes*** (***L. basalis***, 6–9 mm, is the only northeastern species) have distinctive antennae that look a bit like lopsided feathers. The lower image (8) is a female. **FAMILY ELATERIDAE** (CLICK BEETLES AND WIREWORMS) ❾ The sharp hind corners on this beetle's prothorax give it a typical click beetle shape. This is ***Melanactes piceus***, a large (around 30 mm) beetle found in the northeastern United States.

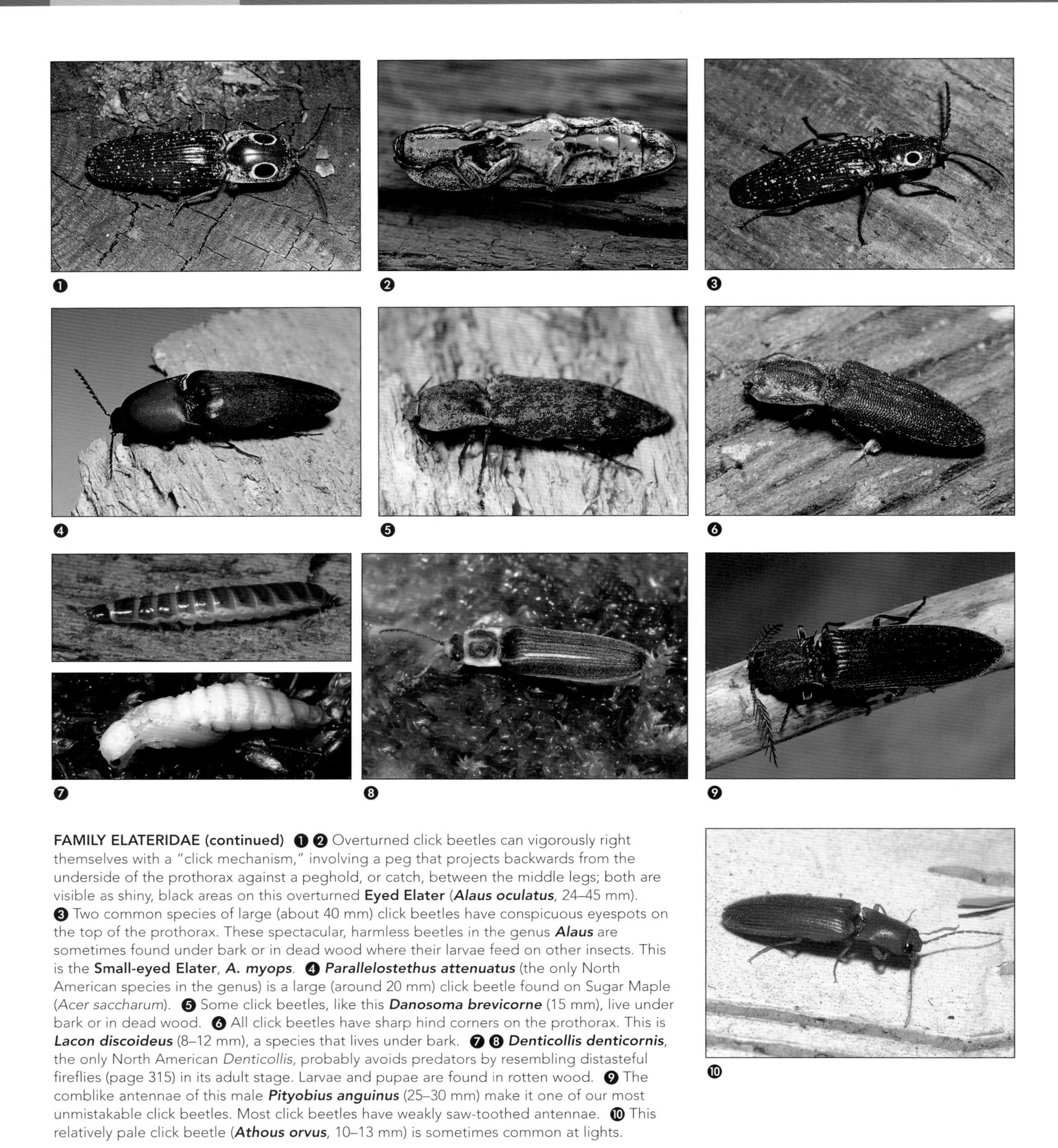

FAMILY ELATERIDAE (continued) ❶ ❷ Overturned click beetles can vigorously right themselves with a "click mechanism," involving a peg that projects backwards from the underside of the prothorax against a peghold, or catch, between the middle legs; both are visible as shiny, black areas on this overturned **Eyed Elater** (***Alaus oculatus***, 24–45 mm). ❸ Two common species of large (about 40 mm) click beetles have conspicuous eyespots on the top of the prothorax. These spectacular, harmless beetles in the genus ***Alaus*** are sometimes found under bark or in dead wood where their larvae feed on other insects. This is the **Small-eyed Elater**, ***A. myops***. ❹ ***Parallelostethus attenuatus*** (the only North American species in the genus) is a large (around 20 mm) click beetle found on Sugar Maple (*Acer saccharum*). ❺ Some click beetles, like this ***Danosoma brevicorne*** (15 mm), live under bark or in dead wood. ❻ All click beetles have sharp hind corners on the prothorax. This is ***Lacon discoideus*** (8–12 mm), a species that lives under bark. ❼ ❽ ***Denticollis denticornis***, the only North American *Denticollis*, probably avoids predators by resembling distasteful fireflies (page 315) in its adult stage. Larvae and pupae are found in rotten wood. ❾ The comblike antennae of this male ***Pityobius anguinus*** (25–30 mm) make it one of our most unmistakable click beetles. Most click beetles have weakly saw-toothed antennae. ❿ This relatively pale click beetle (***Athous orvus***, 10–13 mm) is sometimes common at lights.

FAMILY ELATERIDAE (continued) ❶ ❷ Some adult click beetles eat other insects, as you can see by these two ***Dalopius*** adults eating a dead moth and a live aphid. ❸ Click beetles, like this ***Prosternon medianum***, can sometimes be seen on flowers, where they feed on pollen. ❹ About 150 of North America's 800 or so click beetles, including this common species now known as ***Pseudanostirus triundulata***, were until recently included in the big genus *Ctenicera*. Their larvae live in forest litter where they eat various things, including other insects. ❺ Click beetles, like this ***Ctenicera hornii***, are often seen on foliage. ❻ ❼ Some of our prettiest click beetles belong to the genus ***Ampedus***, including 34 northeastern species. The striking black and yellow ***A. vitiosus*** and the black-tipped ***A. apicatus*** are among many *Ampedus* species frequently found under bark. ❽ ❾ The thin, wiry larvae (wireworms) of some click beetles, including several species of ***Agriotes***, are serious pests that attack roots and seeds. This adult beetle is ***A. fucosus***.

FAMILY ELATERIDAE (continued) ❶ This small click beetle ***Cardiophorus cardisce*** (6 mm) occurs throughout our area, but is particularly common under debris along the sandy shores of the Great Lakes. ❷ Click beetles vary in size from the huge-eyed elaters, that can exceed 40 mm, down to several little species only a few millimeters in length. ***Horistonotus curiatus***, one of the smaller click beetles at about 4 mm, is usually seen on oak foliage. ❸ Like many other insects, click beetles are attracted to fermenting sap runs. The click beetles feeding at this wounded poplar tree (the wound was made by an emerging long-horned beetle) include a ***Melanotus***, a ***Ctenicera*** and an ***Agriotes***. ❹ This ***Melanotus*** is an aberrant specimen, with the left and right elytra differently colored. Bilateral asymmetry like this is uncommon in insects, occurring most often in gynandromorphs (individuals that are half male, half female). **FAMILY EUCNEMIDAE** (FALSE CLICK BEETLES) ❺ Beetles in the small and rarely encountered family Eucnemidae are sometimes called false click beetles because of their similarity to Elateridae, and are sometimes called cross-borers because their larvae burrow across the grain of wood. This ***Isorhipis ruficornis*** is depositing eggs on a recently cut willow log. ❻ False click beetles, like this ***Dromaeolus cylindricollis***, are more robust than similar click beetles. This species bores in hemlock, sycamore and beech trees. ❼ This pair of ***Stethon pectorosus*** (5–10 mm) were among many mating on a dead elm tree near Lake Erie. **FAMILY THROSCIDAE** (THROSCID BEETLES) ❽ Throscid beetles, like this ***Aulonothroscus punctatus*** (3–4 mm) are small beetles with a click-beetle-like jumping ability. They are not common, but can sometimes be seen running around on foliage at dusk. ❾ ***Trixagus carinicollis*** (2–3 mm) is a small throscid with eyes divided into upper and lower halves.

SUPERFAMILY CLEROIDEA

FAMILY CLERIDAE (CHECKERED BEETLES) ❶ ❷ Checkered beetles are not always "checkered," but their shape – broad head, narrow thorax and broad abdomen and their long, sparse hairs make the family easy to recognize. ***Trichodes nutalli*** (8–10 mm), a common pollen-feeding species, can be marked with red, orange or yellow. (The smaller beetle sharing the flower with the red individual is a western species of soft-winged flower beetle, Melyridae). Larvae of some *Trichodes* species are predators in bee's nests, while others develop in grasshopper egg pods. ❸ This decidedly uncheckered checkered beetle (***Zenodosus sanguineus***, 5–6 mm) is an easily recognized member of the subcortical (under bark) community where it feeds on other insects. ❹ Most checkered beetles are predators or pollen-eaters, but some opt for other protein sources. Members of the aptly named genus ***Necrobia*** (3–6 mm) can be found on carrion, bones, and sometimes on stored meat products. All three species are cosmopolitan and probably introduced to North America; this is ***N. violacea***. ❺ Our only ***Chariessa***, ***C. pilosa*** (8–13 mm), is an odd-looking checkered beetle with a relatively narrow head partially hidden by the prothorax. The larvae of these somewhat firefly-like beetles pursue and kill wood-boring insects in their burrows; the adults eat a variety of other insects. ❻ ***Cymatodera*** is the largest genus of checkered beetles in North America, but only three of the 110 or so North American species occur in our area. This is ***C. bicolor*** (6–10 mm), a predator with larvae that search out wood-boring or gall-making insect larvae in their burrows. ❼ This distinctive beetle is a small ***Neorthopleura thoracica***, a species that ranges from 3–12 mm in length and occurs throughout our area. ❽ This ***Phyllobaenus humeralis*** (4–5 mm) is feeding on false nectaries in a Death Camas (*Zigadenus glaucus*) flower. *Phyllobaenus* species are very common beetles that develop as predators of insect larvae in in galls, stems and wood. ❾ A few checkered beetles, like this ***Phyllobaenus pallipennis*** (3–5 mm), have elytra (front wings) that are not quite long enough to cover the flight wings (hind wings).

FAMILY CLERIDAE (continued) ❶ ***Madoniella dislocatus*** is a small (4–6 mm) checkered beetle usually found on dead elm and oak trees. ❷ This checkered beetle (***Thanasimus undatulus***, 7 mm) is dispatching a click beetle. Adults and larvae of our three northeastern species of *Thanasimus* are usually found on tree trunks, and normally feed on bark beetles. ❸ Recently cut pine logs are good places to look for a variety of beetles, including the bark beetle predator ***Thanasimus dubius*** (8 mm). ❹ Several of our prettiest checkered beetles belong to the genus ***Enoclerus***, and ***E. muttkowskii*** (9–10 mm) is probably the most attractive of them all. These typically shaped clerids are usually found around dying trees, where they feed on bark beetles. ❺ ***Enoclerus nigripes*** is a relatively small (5–7 mm) checkered beetle found on tree trunks infested with the small bark beetles on which they feed. ❻ ***Enoclerus rosmarus*** (4–7 mm) is a common clerid beetle found on the ground and low foliage in wet areas. ❼ ***Priocera castanea***, an unusual checkered beetle without an antennal club, occurs under loose bark and in the burrows of the other wood-inhabiting beetles on which they prey. **FAMILY MELYRIDAE** (SOFT-WINGED FLOWER BEETLES) ❽ The **Scarlet Malachite Beetle** (***Malachius aeneus***, 6–8 mm) has become one of our most common soft-winged flower beetles since its accidental introduction from Europe. It has recently become very rare in much of its native range, however, and it is the subject of research and reintroduction efforts in Great Britain. ❾ This little soft-winged flower beetle (***Anthocomus equestris***, 2–3 mm) is often found in houses, and was probably introduced to North America from Europe. Some related native species develop as predators in bark beetle burrows. The pocket-like distortion at the end of this male's elytra holds glands and associated tufts of hairs. Females feed on the contents of the males' elytral pockets before mating.

FAMILY MELYRIDAE (continued) ❶ ❷ Male soft-winged flower beetles in the large genus ***Collops*** usually have some of the basal antennal segments grossly expanded, as you can see on this male ***C. vittatus*** (5 mm). Female *Collops*, like this ***C. vittatus*** feeding on the false nectary of a Grass of Parnassus (*Parnassia palustris*) flower, have relatively simple antennae. ❸ ***Collops quadrimaculatus*** (5 mm), named for the four elytral spots, is one of our most common soft-winged flower beetles and, along with other *Collops* species, is a significant predator of agricultural pests. ❹ This soft-winged flower beetle sharing a Lakeside Daisy (*Hymenoxys herbacea*) with a solitary bee is a tiny ***Attalus terminalis*** (2 mm), one of 14 small northeastern species of *Attalus*. **FAMILY TROGOSSITIDAE** (BARK-GNAWING BEETLES) ❺ The distinctively shaped bark-gnawing beetles are mostly, as the name suggests, found under bark, but one species of ***Tenebroides***, the **Cadelle** (***T. mauritanicus***, 6–10 mm), is a pest in granaries worldwide. ❻ ***Tenebroides bimaculatus*** is the most distinctively colored of the 10 northeastern species of *Tenebroides*. ❼ Although this distinctively metallic beetle (***Temnochila virescens***, 9–18 mm) belongs to the family called bark-gnawing beetles, it is really a "beetle-gnawing" beetle, as it eats other beetles, usually under bark or in galleries in pine trees. ❽ ***Grynocharis quadrilineata*** (5–10 mm), the only northeastern *Grynocharis*, can be found under bark throughout North America's northeastern states and provinces.

SUPERFAMILY BUPRESTOIDEA

FAMILY BUPRESTIDAE (JEWEL BEETLES OR METALLIC WOOD-BORING BEETLES) **Subfamily Buprestinae (Sculptured buprestids)** ❾ Most species in the family Buprestidae warrant the common name "metallic wood-boring beetles" because of their habit of burrowing in sapwood, where they make distinctive, flattened burrows. The large, white larvae of these beetles are called "flat-headed borers" because the front part of their somewhat club-shaped body is flattened top to bottom. This contrasts with the nonflattened larvae of the so-called "round-headed wood borers" (Cerambycidae, or long-horned beetles), which are also wood-boring insects.

FAMILY BUPRESTIDAE (continued) ❶ ❷ Because of their tree-boring habits, some Buprestidae are important pests. The beetles mating and laying eggs in these photos belong to our most common genus, ***Chrysobothris***, with 20 northeastern species including a number of pests attacking a wide variety of trees. *C. femorata*, the Flat-headed Apple Tree Borer, is a common pest in this genus. ❸ Flat-headed borers (also known as jewel beetles and metallic wood-borers), are recognized by their tough, metallic-looking armor and their general shape, especially the characteristic way the head fits into the prothorax like someone's head partly pulled into a turtleneck sweater. Some northeastern buprestids, like this ***Cypriacis striata***, are especially deserving of the common name "jewel beetles." Most species of *Cypriacis* and the very similar genus *Buprestis* are associated with dead or dying conifers. ❹ This ***Buprestis salisburyensis*** is laying eggs in a telephone pole. ❺ This ***Buprestis maculativentris*** was among many emerging from structural wood in an Ontario cottage. ❻ This ***Actenodes acornis*** was captured and paralyzed by a crabronid (*Cerceris fumipennis*) wasp in southern Ontario. The genus *Actenodes* had not been known in Canada prior to this collection of the wasp and its prey. ❼ The **Flat-headed Pine Heartwood Borer** (***Chalcophora virginiensis***), one of three northeastern species of *Chalcophora*, is one of our biggest Buprestidae at 25–30 mm. Look for them on pine trees. ❽ The **Hemlock Borer** (***Phaenops fulvoguttata***) develops in pine, tamarack and balsam, as well as hemlock. Some related species are able to sense and respond to the infrared radiation typical of forest fires. Adults eat other insects killed by fire and lay their eggs on freshly burned trees. ❾ The **Flat-headed Hardwood Borer** (***Dicerca divaricata***), the most common of the 16 northeastern *Dicerca* species, develops in dead or dying hardwood trees. This one is laying eggs on the trunk of a dead basswood tree.

FAMILY BUPRESTIDAE (continued) ❶ This common jewel beetle (***Poecilonota cyanipes***) develops in dead or dying willow or poplar trees. ❷ Larvae of this small (5–8 mm) flat-headed borer (***Anthaxia inornata***) live beneath the bark of coniferous trees, but the adults are commonly seen on flowers such as this Lakeside Daisy (*Hymenoxys herbacea*). ❸ ***Anthaxia viridifrons*** is usually found on elm or hickory trees. **Subfamily Agrilinae (Branch and leaf buprestids)** ❹ ***Agrilus*** is a big genus with almost 60 northeastern species of little (mostly 4–10 mm) beetles that bore in branches, twigs, trunks and sometimes even roots. Adults can often be found feeding on the leaves of the host plants. This is the **Bronze Birch Borer** (***A. anxius***) a widespread and common species on birch and poplar. ❺ ***Agrilus vittaticollis*** is a widespread species that develops on pear, apple, chokecherry and related trees. ❻ The **Two-lined Chestnut Borer** (***Agrilus bilineatus***) is a common native species throughout eastern North America, where it develops on various oak (*Quercus*) species as well as American Chestnut. It is sometimes abundant enough to be considered a pest. ❼ ❽ The recently introduced **Emerald Ash Borer** (***Agrilus planipennis***) is a serious pest that arrived from China in 2002 and has since been killing ash trees in Ontario, Michigan, Ohio, Indiana and Maryland. This recently killed ash tree is covered with frass-filled burrows made by larval Emerald Ash Borers (the bark was removed for this photo). ❾ Most beetles in the large genus ***Agrilus*** are black to bronze in color but a few, like the Emerald Ash Borer (*A. planipennis*) and this ***A. cyanescens***, are strikingly metallic. *A. cyanescens*, which feeds on Honeysuckle (*Lonicera*) and a variety of other trees and shrubs, was accidentally introduced from Europe around 1920. It is usually well under 10 mm in length, in contrast to the Emerald Ash Borer which is an unusually large *Agrilus* at 8–15 mm.

FAMILY BUPRESTIDAE (continued) ❶ ***Agrilus subcinctus***, a small (about 4 mm) species, occurs on Poison Ivy (*Rhus radicans*). ❷ ❸ Not all buprestids are borers, despite the common name for this family. Some of the smallest Buprestidae develop as miners between the upper and lower leaf surfaces, and then feed on the same kinds of leaves as adults. These oak leaves show the typical feeding damage of adult ***Brachys ovatus*** and ***Brachys tessellatus*** (both about 5 mm). ❹ ***Brachys aeruginosus*** (3.5 mm) develops as a miner in oak leaves. ❺ This tiny black beetle eating the upper surface of a legume leaf is ***Pachyschelus laevigatus*** (2 mm). **Subfamily Polycestinae (Yellow-marked buprestids)** ❻ Some metallic wood borers live in wood as larvae, but forego wood in favor of flowers as adults. The two northeastern ***Ptosima*** (our only members of the subfamily Polycestinae) are yellow and black beetles usually found on flowers where they feed on pollen. Larvae of this species, ***P. gibbicollis*** (5–8 mm), develop on Redbud (*Cercis canadensis*). **Subfamily Acmaeoderinae** ❼ ❽ Adult yellow-marked buprestids are usually seen on composite flowers, like the yellow coneflower that has attracted this **Flat-headed Baldcypress Sapwood Borer** (***Acmaeodera pulchella***, 6–10 mm). The more discretely spotted ***A. tubulus*** (5–7 mm) is the most common of the three northeastern *Acmaeodera* species.

SUPERFAMILY TENEBRIONOIDEA

FAMILY MELOIDAE (BLISTER BEETLES) ❾ The brown larva under this solitary bee's wing is a triungulin larva of a blister beetle (*Meloe*) hitching a ride to a bee's nest; once there it will kill the bee's eggs and develop on the pollen in the nest. Early spring bees often accidentally pick up dozens of triungulin larvae as they visit flowers. This is a bee (***Nomada*** sp.) that lays eggs on food stored by other bees, so it is a kleptoparasitic bee carrying a kleptoparasitic beetle.

FAMILY MELOIDAE (continued) ❶ Adult blister beetles in the genus ***Meloe*** (**oil beetles**) are fat, metallic beetles with short, overlapping wings. The male of this pair has strongly modified antennae, used to hold the female's antennae during mating. ❷ Blister beetles, like this ***Meloe*** female, are aptly named because they contain chemicals (cantharidin) that can cause painful blistering. ❸ Blister beetles like this dead oil beetle are strongly protected by cantharidin, which is probably why this individual ended up being regurgitated in the middle of a trail. Dead or injured oil beetles are invariably fed upon by specialized beetles (some Pyrochroidae) and flies (some Ceratopogonidae) that sequester the oil beetle's defensive chemicals for their own use. ❹ Blister beetles in the genus ***Epicauta*** feed on plants while adults, but feed on grasshopper eggs as larvae. This is a big genus with well over 100 North American species; this distinctively striped ***E. cinerea*** is one of 18 northeastern species in the genus. ❺ ***Epicauta*** species, like this ***E. atrata***, are often seen on flowers. Most incidents of skin lesions caused by blister beetle contact are attributable to *Epicauta* species, especially a common eastern North American species called the Old Fashioned Potato Beetle (*E. vittata*). ❻ If you watch open sandy areas in early spring you might see this red-winged blister beetle (***Tricrania sanguinipennis***) scurrying across the sand. Its larvae develop in the nests of certain solitary bees (*Colletes*). ❼ ❽ Coneflowers (*Rudbeckia*) are often visited by blister beetles, especially ***Zonitis*** and ***Nemognatha***, both of which develop as kleptoparasites (food thieves) in the nests of solitary bees. The species shown here are the reddish ***Z. bilineata*** and the black-headed ***N. nemorensis***. *Nemognatha* species have a long, beak used to suck the nectar from flowers (you can see it extending backwards under the head in this photo; no other beetle has a beak like this). ❾ The famous Spanish Fly (*Lytta vesicatoria*) is a Spanish beetle in the genus ***Lytta***, a genus also represented by three large (9–25 mm), colorful eastern North American species (this is ***L. aenea***). Adults eat foliage and fruits; larvae live in bees' nests.

FAMILY MELOIDAE (continued) ❶ Male blister beetles manufacture a defensive blistering chemical (cantharidin) and transfer it to the females during mating. This is the widespread eastern North American species ***Lytta aenea***. **FAMILY RIPIPHORIDAE** (WEDGE-SHAPED BEETLES) ❷ Wedge-shaped beetles in the genus ***Ripiphorus*** lay hundreds of eggs on flower buds. Larvae attach themselves to flower-visiting solitary bees to ride back to the bee's nest where they parasitize bee larvae. This is ***R. luteipennis***, photographed in an Ontario tallgrass prairie. ❸ Elongate wedge-shaped beetles in the genus ***Macrosiagon***, like this ***M. limbatum***, are parasitoids of wasp larvae. Larvae gain access to the nests of solitary wasps by attaching themselves to flower-visiting adult wasps and hitching a ride back to the wasp's nests. ❹ The smallest of our three ripiphorid genera, ***Pelecotoma*** includes only the one species ***P. flavipes*** (4–5 mm). Look for this small, slender beetle on dead beech trees where it parasitizes larval *Ptilinus* (Anobiidae). **FAMILY MORDELLIDAE** (TUMBLING FLOWER BEETLES) ❺ Tumbling flower beetles like this ***Mordella atrata*** (4–6 mm) are distinctively shaped insects with the memorable habit of launching their apparently unbalanced bodies into comical tumbles when disturbed. Common on flowers and occasionally on dead wood, tumbling flower beetles usually develop in plant stems where some are predators and others consume plant tissue. ❻ This tumbling flower beetle is inserting its eggs in a flower base. ❼ A few tumbling flower beetles, like this ***Yakuhananomia bidentata*** (8–13 mm), are usually found on the trunks of dead hardwood trees. ❽ Tumbling flower beetles are often common on Queen Anne's Lace (*Daucus carota*) flowers. This silver-spotted ***Hoshihananomia octopunctata*** (8–9 mm) is one of our prettiest mordellids. ❾ ***Hoshihananomia oculata*** is a relatively large (7–9 mm) tumbling flower beetle that occurs on dead branches, or nearby foliage, of trees and shrubs.

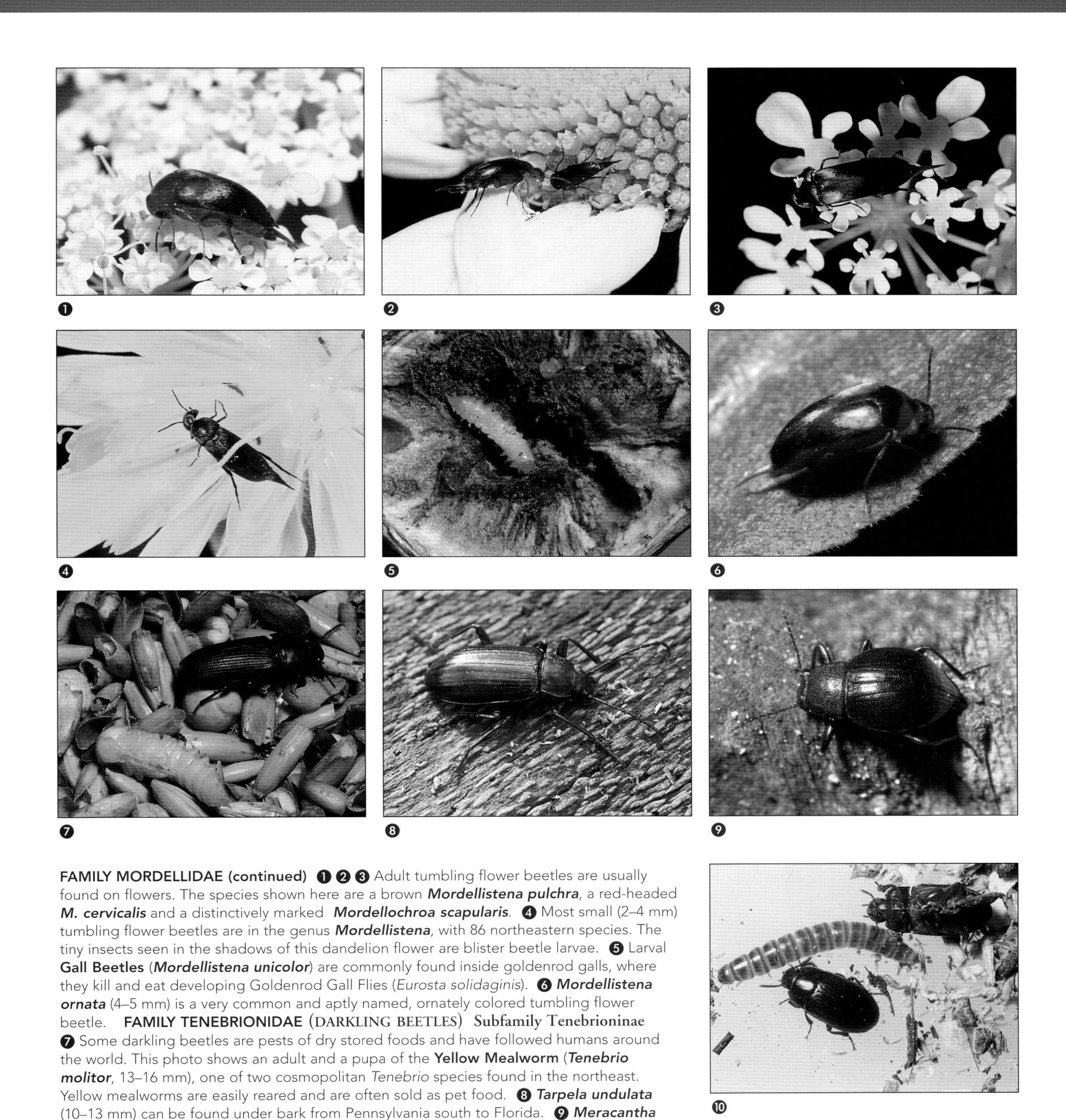

FAMILY MORDELLIDAE (continued) ❶ ❷ ❸ Adult tumbling flower beetles are usually found on flowers. The species shown here are a brown ***Mordellistena pulchra***, a red-headed ***M. cervicalis*** and a distinctively marked ***Mordellochroa scapularis***. ❹ Most small (2–4 mm) tumbling flower beetles are in the genus ***Mordellistena***, with 86 northeastern species. The tiny insects seen in the shadows of this dandelion flower are blister beetle larvae. ❺ Larval **Gall Beetles** (***Mordellistena unicolor***) are commonly found inside goldenrod galls, where they kill and eat developing Goldenrod Gall Flies (*Eurosta solidaginis*). ❻ ***Mordellistena ornata*** (4–5 mm) is a very common and aptly named, ornately colored tumbling flower beetle. **FAMILY TENEBRIONIDAE** (DARKLING BEETLES) Subfamily Tenebrioninae ❼ Some darkling beetles are pests of dry stored foods and have followed humans around the world. This photo shows an adult and a pupa of the **Yellow Mealworm** (***Tenebrio molitor***, 13–16 mm), one of two cosmopolitan *Tenebrio* species found in the northeast. Yellow mealworms are easily reared and are often sold as pet food. ❽ ***Tarpela undulata*** (10–13 mm) can be found under bark from Pennsylvania south to Florida. ❾ ***Meracantha contracta*** (11–14 mm) is a distinctively stout darkling beetle found on dead wood. This is the only *Meracantha* species. ❿ These larval and adult **Lesser Mealworms** (***Alphitobius diaperinus***, 5–7 mm) were found infesting spoiled grains in a pet cage.

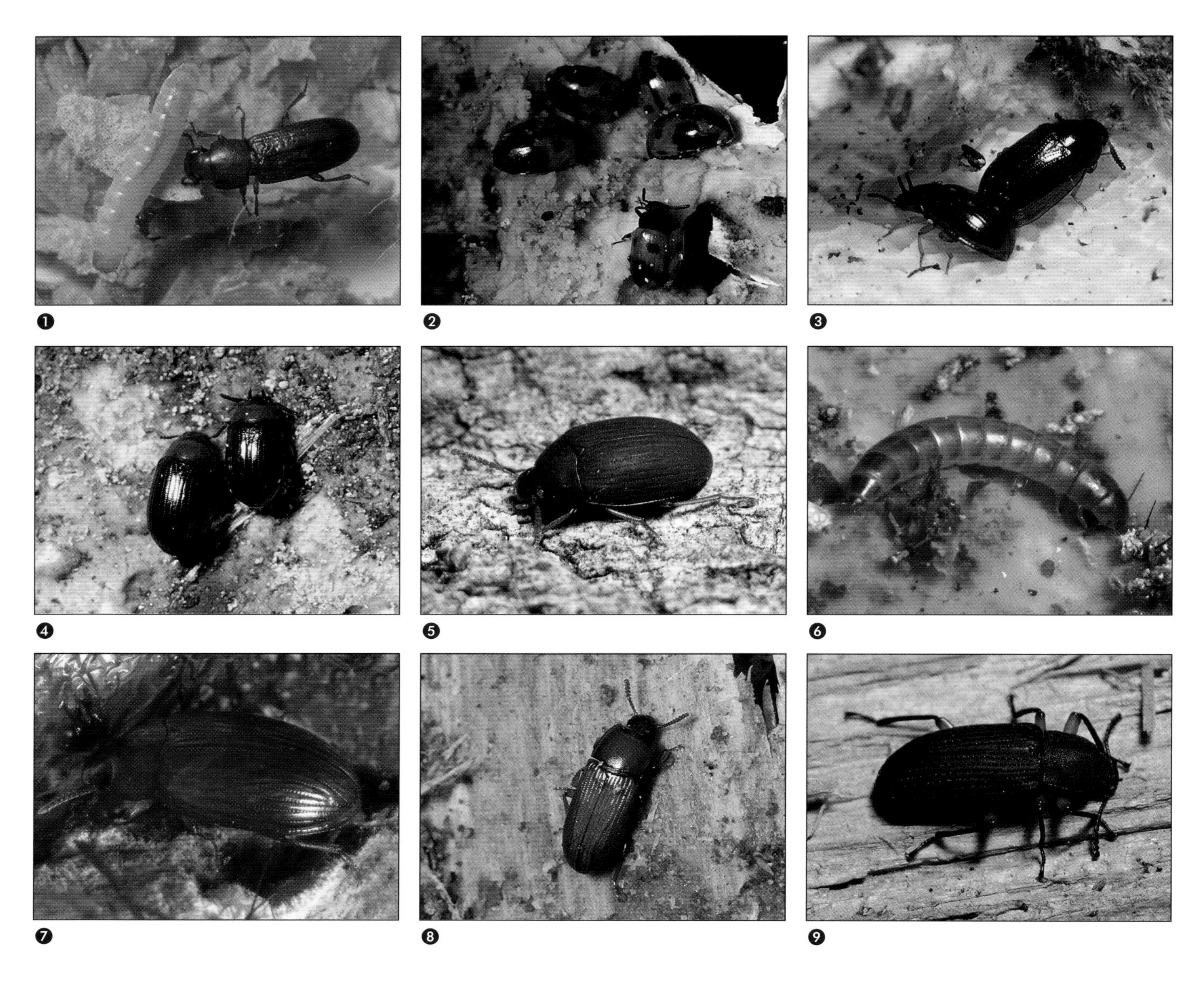

FAMILY TENEBRIONIDAE (continued) ❶ These larval and adult **Confused Flour Beetles** (***Tribolium confusum***, 4–5 mm) were among thousands in a bag of feed. Confused Flour Beetles are the most common pests in North American flour mills. They are remarkably well adapted to this extreme environment, and can obtain all of their water requirements by absorption from the air. Subfamily Diaperinae ❷ This darkling beetle (***Diaperis maculata***, 5–7 mm) has the lady-beetle-like habit of forming aggregations in the hundreds under bark. The two northeastern species of *Diaperis* feed on fungi. ❸ ❹ ***Neomida bicornis*** is common throughout our area, but northern and southern populations differ in color. Southern populations are mostly red and green, but the all-green individuals occur with increasing frequency as you move north; Canadian specimens are generally entirely green. The individuals with horns are males. Look for this species inside old, dry fungi on decayed logs. ❺ Although many of the 1,100 or so western North American darkling beetles are associated with desert areas, the smaller eastern fauna (140 species) is usually associated with bark or fungi. This is ***Platydema ruficorne*** (3–6 mm), one of many small subcortical darkling beetles. ❻ ❼ This teneral adult darkling beetle (***Platydema*** sp.) developed from this active larva found in a dried out fungus. ❽ ***Uloma punctulata*** (7–9 mm) is a common species under the bark of fallen pine trees. Subfamily Coelometopinae ❾ ***Haplandrus fulvipes*** is a medium-sized (8–11 mm) darkling beetle distinctive for its bright orange femora. Like most other eastern tenebrionids, it occurs under bark.

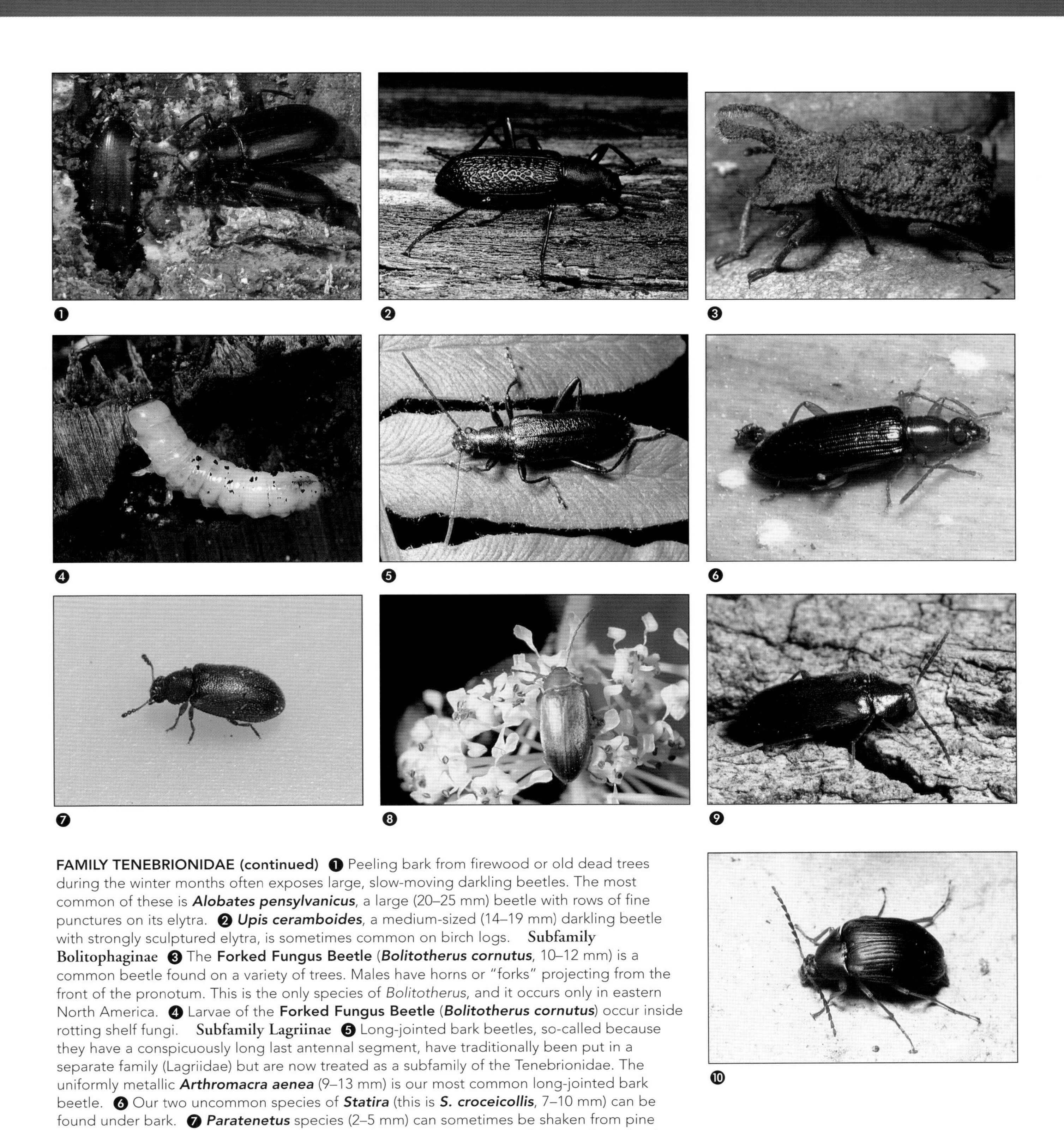

FAMILY TENEBRIONIDAE (continued) ❶ Peeling bark from firewood or old dead trees during the winter months often exposes large, slow-moving darkling beetles. The most common of these is ***Alobates pensylvanicus***, a large (20–25 mm) beetle with rows of fine punctures on its elytra. ❷ ***Upis ceramboides***, a medium-sized (14–19 mm) darkling beetle with strongly sculptured elytra, is sometimes common on birch logs. **Subfamily Bolitophaginae** ❸ The **Forked Fungus Beetle** (***Bolitotherus cornutus***, 10–12 mm) is a common beetle found on a variety of trees. Males have horns or "forks" projecting from the front of the pronotum. This is the only species of *Bolitotherus*, and it occurs only in eastern North America. ❹ Larvae of the **Forked Fungus Beetle** (***Bolitotherus cornutus***) occur inside rotting shelf fungi. **Subfamily Lagriinae** ❺ Long-jointed bark beetles, so-called because they have a conspicuously long last antennal segment, have traditionally been put in a separate family (Lagriidae) but are now treated as a subfamily of the Tenebrionidae. The uniformly metallic ***Arthromacra aenea*** (9–13 mm) is our most common long-jointed bark beetle. ❻ Our two uncommon species of ***Statira*** (this is ***S. croceicollis***, 7–10 mm) can be found under bark. ❼ ***Paratenetus*** species (2–5 mm) can sometimes be shaken from pine branches in large numbers. **Subfamily Alleculinae** ❽ ❾ Members of the small subfamily Alleculinae are called comb-clawed beetles because the lower surface of each tarsal claw has a row of teeth, giving it a comblike appearance. Comb-clawed beetles have traditionally been treated as the family Alleculidae. These photos show common species in the genera ***Isomira*** (***I. sericea*** on a flower, 6 mm) and ***Mycetochara*** (in a rotting tree, 6 mm). ❿ Some of our largest comb-clawed beetles are in the genus ***Pseudocistela***, like this robust, oval ***P. brevis*** (9 mm).

FAMILY TENEBRIONIDAE (continued) ❶ ***Androchirus erythropus*** (10 mm) is one of two North American species of *Androchirus*, both of which occur in the northeast. **FAMILY ZOPHERIDAE** (IRONCLAD BEETLES) ❷ The elongate and flattened species shown here with its legs retracted is the only eastern North American member of the family Zopheridae, the **Eastern Ironclad Beetle** (***Phellopsis obcordata*** 10–14 mm). This rock-hard beetle is normally associated with shelf fungi on birch logs. The short, stout beetle in this picture is the Forked Fungus Beetle (*Bolitotherus cornutus*), a darkling beetle associated with shelf fungi on a variety of trees. **FAMILY MONOMMATIDAE** (MONOMMATID BEETLES) ❸ Only one species in the small family Monommatidae occurs in eastern North America. This ***Hyporhagus punctulatus*** (5 mm) was found in South Carolina, about as far north as the family occurs. Some authors now include this family in the Zopheridae. **FAMILY CIIDAE** (MINUTE TREE-FUNGUS BEETLES) ❹ The tiny beetle in front of this relatively huge male Forked Fungus Beetle (Tenebrionidae) is a minute tree-fungus beetle. Minute tree fungus beetles usually abound inside bracket fungi, and can often be found in the larval burrows of Forked Fungus Beetles. ❺ Males of several species of minute tree-fungus beetles (such as the common ***Ceracis thoracicornis***, 1–2 mm) have prominent forks on the head and pronotum, not unlike those found on some darkling beetles found in the same kinds of fungi. ❻ Minute tree-fungus beetles, like this ***Cis***, are usually found on polypore fungi. **FAMILY TETRATOMIDAE** (POLYPORE FUNGUS BEETLES) ❼ The two eastern species of ***Penthe*** (10–14 mm), including ***P. pimelia*** shown here, are distinctively black, broad beetles found under bark or on shelf fungi. ❽ ***Penthe obliquata***, distinctive for the orange setae on the scutellum, occurs under bark throughout our area. ❾ Adult polypore fungus beetles in the genus ***Synstrophus***, like this ***S. repandus*** (6–7 mm) are often seen feeding on fungi after dark.

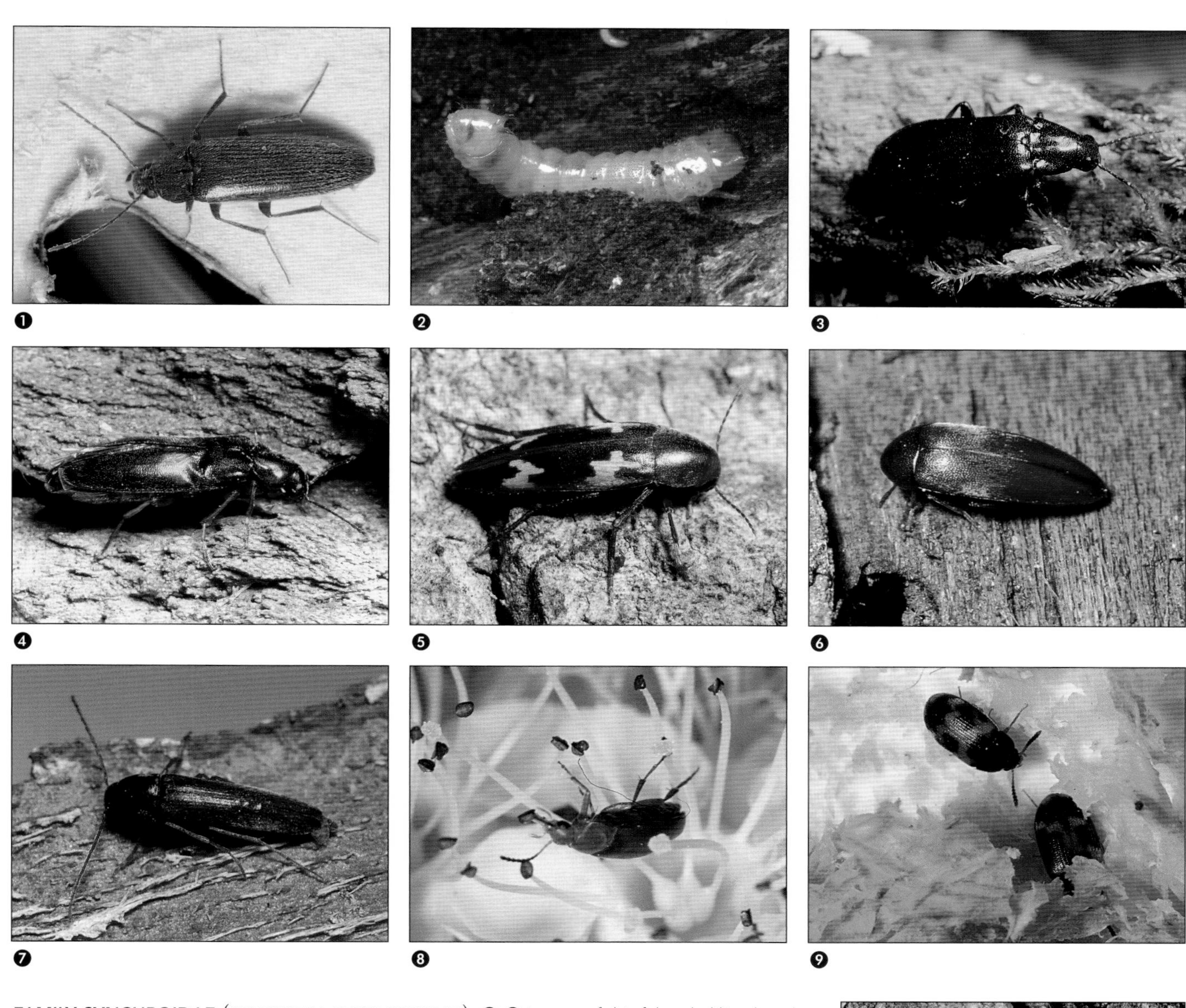

FAMILY SYNCHROIDAE (SYNCHROA BARK BEETLES) ❶ ❷ Larvae of this false darkling beetle (***Synchroa punctata***, 7–13 mm) live in dead tree limbs, but adults are most likely to come to your attention at lights during the night. **FAMILY MELANDRYIDAE** (FALSE DARKLING BEETLES) ❸ ❹ Like many adult beetles, false darkling beetles are usually active at night. Look for the larger (7–11 mm) species like this all-black ***Melandrya striata*** or this spotted ***M. connectens*** on dead wood. ❺ The distinctively marked darkling beetle ***Dircaea liturata*** (6–11 mm) can be found on dead trees or tree branches during the night. ❻ Members of the genus ***Orchesia*** are small (3–5 mm) beetles with spectacular jumping ability. They are usually found on fungus-covered dead wood. ❼ This slender ***Serropalpus*** (three similar North American species; 6–18 mm) was spotted running over a recently debarked pine branch. **FAMILY SCRAPTIIDAE** (SCRAPTIID BEETLES) ❽ The dozen or so small northeastern species now treated as Scraptiidae have at some time or another been put in the families Melandryidae, Mordellidae and Anthicidae. Scraptiids like this ***Anaspis rufa*** (3–4 mm) are common on flowers. **FAMILY MYCETOPHAGIDAE** (HAIRY FUNGUS BEETLES) ❾ Hairy fungus beetles, like these ***Mycetophagus flexuosus***, are often common in soft shelf fungi that have begun to decay. ❿ These two beetles, ***Mycetophagus punctatus*** and a partially hidden *Glischrochilus sanguinolentus* (Nitidulidae) were photographed after dark near the underside of a shelf fungus. Many fungus beetles are active after dark, and can be easily observed with a flashlight.

FAMILY MYCETOPHAGIDAE (continued) ❶ ***Typhaea stercorea*** is a stored product pest that feeds on molds in stored grains and similar foodstuffs. Like most stored product pests, it is found throughout the world. **FAMILY COLYDIIDAE** (COLYDIID BEETLES) ❷ Colydiidae is a diverse group of mostly very small beetles common under bark where most species, like this ***Bitoma crenata*** (2 mm) feed on fungi. ❸ ***Aulonium tuberculatum*** can be found under pine bark throughout the eastern United States. ❹ The long, cylindrical shape of this ***Colydium lineola*** (about 5 mm) is an adaptation for life as a predator in the galleries of wood-boring insects. **FAMILY PYROCHROIDAE** (FIRE-COLORED BEETLES) **Subfamily Pyrochroinae** ❺ ❻ The aptly named ***Dendroides concolor*** (seen here with a pupa of the same species) is probably the most common northeastern fire-colored beetle. Look for these beetles and their flattened larvae under bark of damp, decaying logs. ❼ ❽ ***Dendroides canadensis*** (large eyes, red legs, black head) and the superficially similar ***Neopyrochroa femoralis*** (small eyes, partially black legs, red head) are large (10–16 mm) and distinctive beetles. Both are found under bark. **Subfamily Pedilinae** ❾ Some authors include pediline beetles (like this ***Pedilus***) in the Anthicidae, while others treat them as the family Pedilidae.

FAMILY PYROCHROIDAE (continued) ❶ ❷ The distinctively necked pediline beetles are often found on flowers, but some species (like this bunch of male ***Pedilus terminalis***; only males have the tips of the elytra yellow) are strongly attracted to dead blister beetles. ❸ ***Stereopalpus vestitus*** is a common species in sandy areas along the Great Lakes. ❹ ***Macratia*** species (3 in the northeast) are small (2–5 mm) slender beetles sometimes found on vegetation near water. **FAMILY SALPINGIDAE** (NARROW-WAISTED BARK BEETLES) ❺ Narrow-waisted bark beetles, such as the distinctively long-nosed ***Rhinosimus viridiaeneus***, are infrequently encountered beetles that usually occur under bark. **FAMILY PYTHIDAE** (DEAD LOG BEETLES) ❻ Dead log beetles, like this ***Pytho***, occur under conifer bark, but are infrequently collected. **FAMILY BORIDAE** (CONIFER BARK BEETLES) ❼ ***Boros unicolor***, the more common of the two North American species of Boridae, occurs under the bark of dead pine trees. They withdraw their legs and "play dead" when exposed. This kind of trick is common among insects, and is called thanatosis. **FAMILY ANTHICIDAE** (ANTLIKE FLOWER BEETLES) ❽ Antlike flower beetles are distinctively shaped, small (usually 2–4 mm) beetles often found, as the name suggests, on flowers. The adult beetles feed on pollen, other arthropods, fungal hyphae and spores; larvae feed on a similar range of foods on the ground. ***Anthicus***, with 13 northeastern species, is our largest genus. ❾ ***Malporus formicarius***, as its name suggests, is a particularly ant-shaped beetle common on flowers.

FAMILY ANTHICIDAE (continued) ❶ ❷ Some antlike flower beetles (all species of ***Notoxus*** and the one species of *Mecynotarsus*) have a peculiar pronotal horn projecting over the deflexed head. These photos show ***N. anchora*** from the top and ***N. desertus*** from the side. Like some Pyrochroidae, *Notoxus* males are attracted to dead blister beetles from which they obtain defensive cantharidin.
FAMILY OEDEMERIDAE (FALSE BLISTER BEETLES) ❸ False blister beetle adults like this ***Oxycopis thoracica*** (about 7 mm) feed on pollen, storing it in a special sac off to the side of the esophagus. ❹ Like many false blister beetles, ***Asclera ruficollis*** has distinctive dimples on the top of the prothorax. This species is common on early spring flowers. ❺ The large (13–16 mm) false blister beetles in the genus ***Ditylus*** look so different from other Oedemeridae they are sometimes put in their own family, the Ditylidae. ***D. caeruleus*** is the only eastern species.
❻ The **Wharf Borer** (***Nacerdes melanura***, 8–13 mm) is an introduced species that develops in damp rotting wood, like wharf pilings and driftwood along both fresh and salt water. ❼ ***Xanthochroa erythrocephala***, like other all other false blister beetles, is an obligate pollen feeder. This was one of many individuals on a swamp dogwood bush in South Carolina. Female oedemerids do not mate until they have filled their intestinal sac with pollen, where the pollen is digested and used to produce eggs. **FAMILY ADERIDAE** (ANTLIKE LEAF BEETLES)
❽ Antlike leaf beetles, like this ***Zonantes hubbardi*** (2–3 mm), are small, infrequently encountered insects usually found on the undersides of leaves. ❾ Antlike leaf beetles in the genus ***Elonus*** have the base of the head deeply arched, unlike superficially similar *Zonantes*.

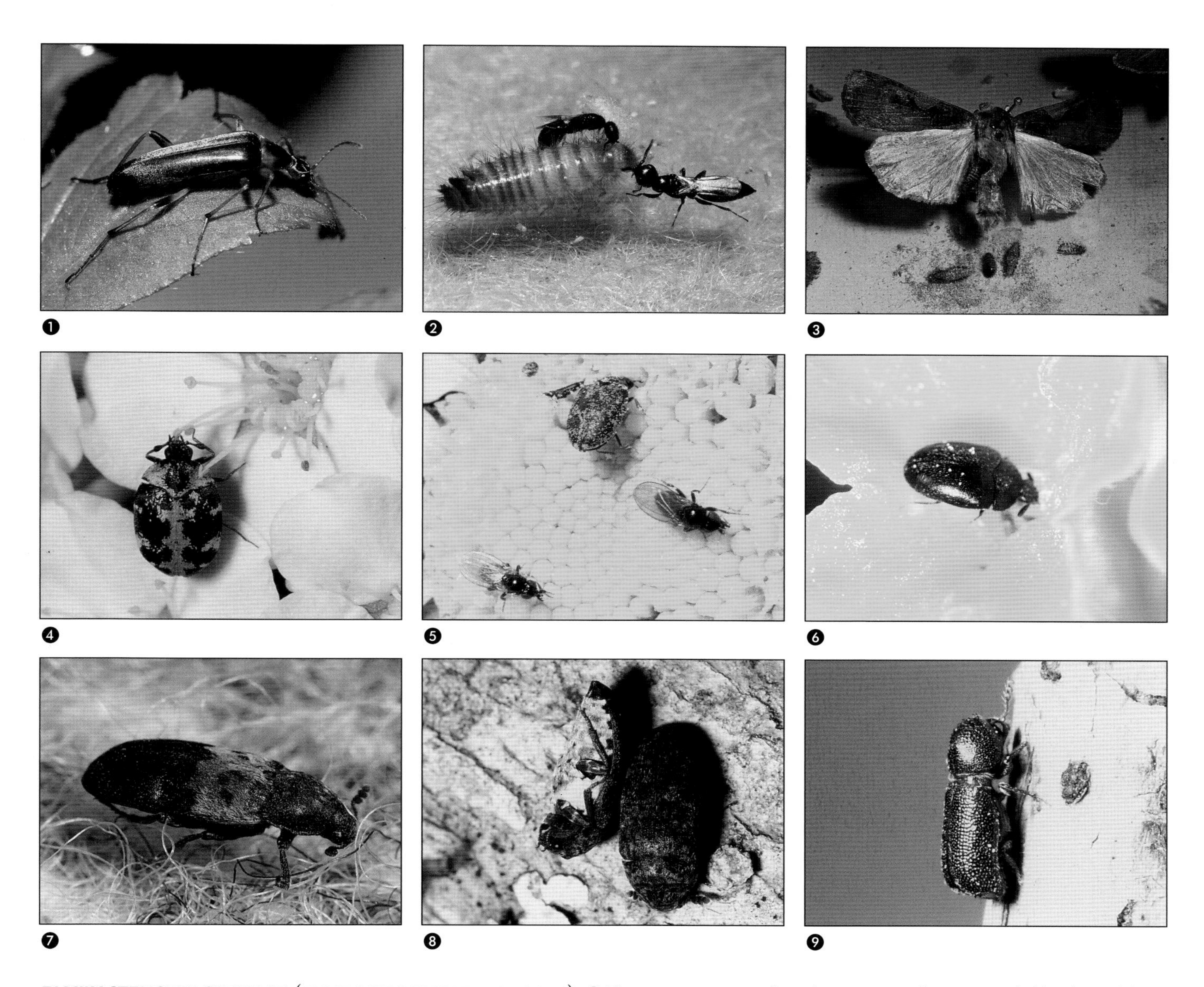

FAMILY STENOTRACHELIDAE (FALSE LONG-HORNED BEETLES) ❶ The most common of northeastern North America's half a dozen false long-horned beetles is ***Cephaloon lepturoides***. These distinctively shaped beetles usually occur on tree foliage.

SUPERFAMILY BOSTRICHOIDEA

FAMILY DERMESTIDAE (DERMESTID BEETLES) ❷ This hairy dermestid beetle larva (***Anthrenus museorum***) is being attacked by parasitic wasps (Bethylidae). Dermestid beetles are the insect collector's nightmare, as these little (around 2 mm) beetles will swiftly destroy a collection of pinned insects unless the specimens are stored in tight cabinets and protected with repellents. ❸ If dermestid beetles invade your collection, the first clue to their presence will usually be telltale piles of powder often mixed with cast larval skins under your disintegrating specimens. ❹ Carpet beetle adults feed on pollen during the spring and summer months, but typically come indoors in the late autumn, and lay eggs in dusty corners. The hairy larvae feed on a variety of high-protein foods, ranging from dead insects to wool carpets. This pollen-eating beetle is a brightly scaled **Buffalo Carpet Beetle** (***Anthrenus scrophulariae***). ❺ This dermestid beetle (***Anthrenus castaneae***) is eating daisy pollen beside a couple of milichiid flies (*Paramyia nitens*). This beetle is similar to the *Anthrenus* most often found ravaging insect collections (*A. museorum*). ❻ Many dermestid beetles, including the little (3–5 mm) black ***Attagenus*** species, feed on both pollen and stored products. ❼ Their quest for high protein food often brings dermestid beetles into conflict with us as they destroy our wool carpets, stored grains and dry, high-protein foodstuffs. The **Larder Beetle** (***Dermestes lardarius***) often occurs in dry dog food. ❽ There are nine species of ***Dermestes*** in the northeast, some of which (like this ***D. caninus***) commonly occur on carrion. Like many other species, *D. caninus* reacts to disturbances by folding up its legs and "playing dead," (thanatosis). **FAMILY BOSTRICHIDAE** (BRANCH AND TWIG BORERS) ❾ The stout, cylindrical beetles in the small, wood-boring family Bostrichidae have a distinctively toothed hood (pronotum) over the oddly down-turned head. This species (***Xylobiops basilaris***, 4–7 mm) breeds in a variety of hardwoods.

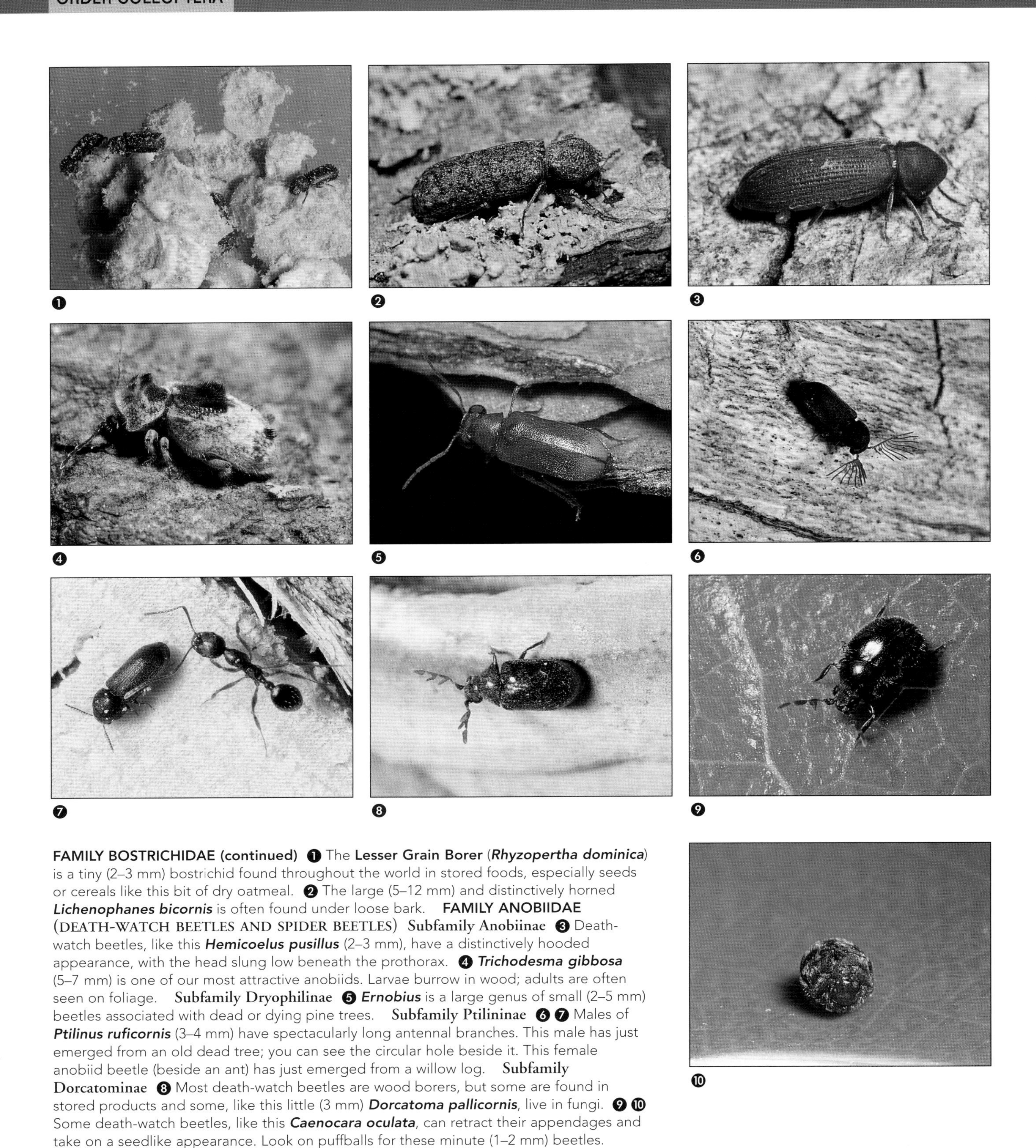

FAMILY BOSTRICHIDAE (continued) ❶ The **Lesser Grain Borer** (***Rhyzopertha dominica***) is a tiny (2–3 mm) bostrichid found throughout the world in stored foods, especially seeds or cereals like this bit of dry oatmeal. ❷ The large (5–12 mm) and distinctively horned ***Lichenophanes bicornis*** is often found under loose bark. **FAMILY ANOBIIDAE (DEATH-WATCH BEETLES AND SPIDER BEETLES)** Subfamily Anobiinae ❸ Death-watch beetles, like this ***Hemicoelus pusillus*** (2–3 mm), have a distinctively hooded appearance, with the head slung low beneath the prothorax. ❹ ***Trichodesma gibbosa*** (5–7 mm) is one of our most attractive anobiids. Larvae burrow in wood; adults are often seen on foliage. Subfamily Dryophilinae ❺ ***Ernobius*** is a large genus of small (2–5 mm) beetles associated with dead or dying pine trees. Subfamily Ptilininae ❻ ❼ Males of ***Ptilinus ruficornis*** (3–4 mm) have spectacularly long antennal branches. This male has just emerged from an old dead tree; you can see the circular hole beside it. This female anobiid beetle (beside an ant) has just emerged from a willow log. Subfamily Dorcatominae ❽ Most death-watch beetles are wood borers, but some are found in stored products and some, like this little (3 mm) ***Dorcatoma pallicornis***, live in fungi. ❾ ❿ Some death-watch beetles, like this ***Caenocara oculata***, can retract their appendages and take on a seedlike appearance. Look on puffballs for these minute (1–2 mm) beetles.

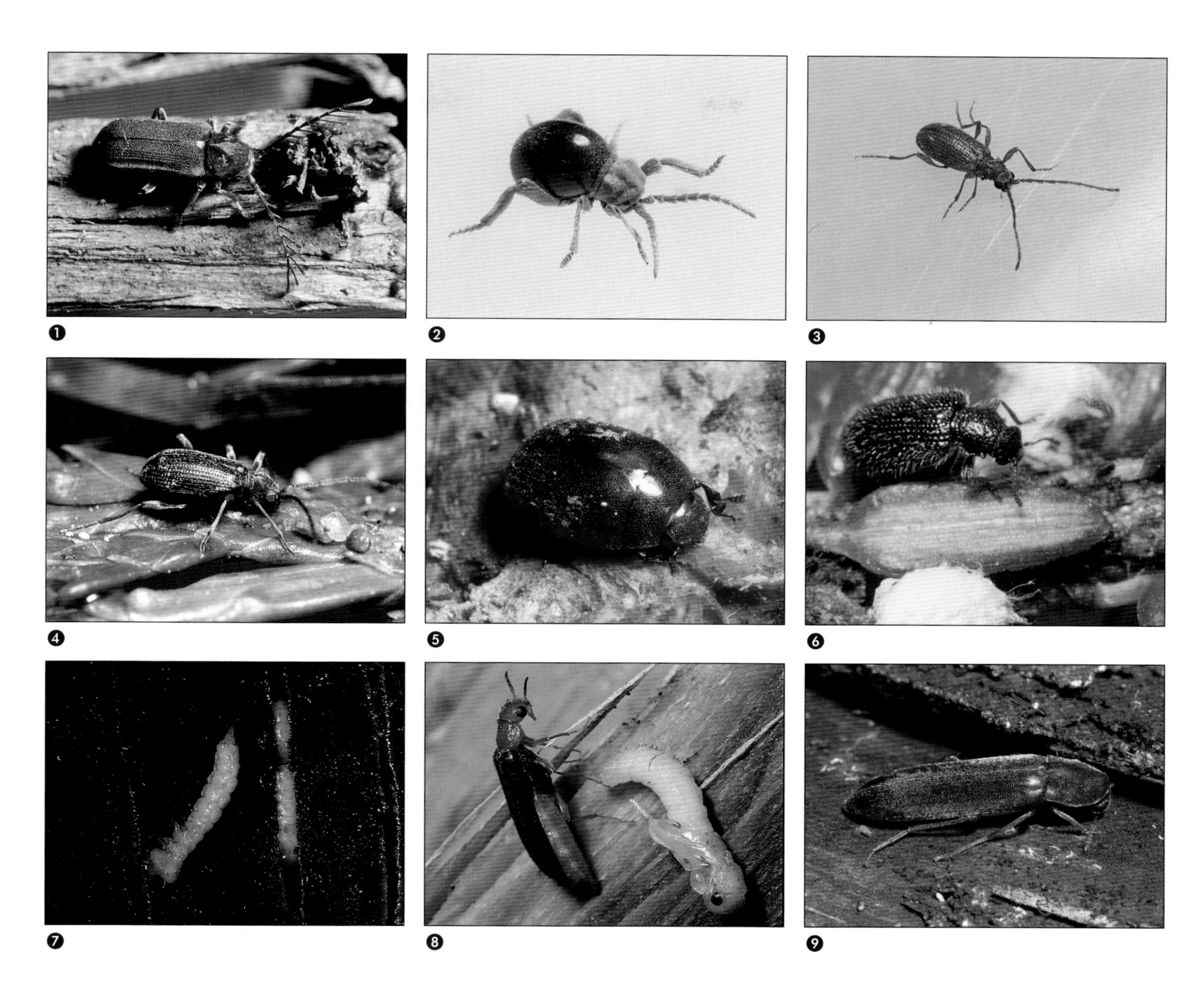

FAMILY ANOBIIDAE (continued) Subfamily Eucradinae ❶ Males of ***Eucrada humeralis*** (4–6 mm) have spectacular antennae with long branches. These attractive anobiids are sometimes common on dead oak and beech trees. **Subfamily Ptininae (Spider beetles)** ❷ The **Northern Spider Beetle** (***Mezium affine***, 3 mm) is a common stored products pest but it is rarely seen outdoors. The similar American Spider Beetle (*M. americanum*) is not as shiny, and has the fuzzy pronotal collar (the narrow ring behind the pronotum) divided into sections. ❸ Spider beetles feed on an enormous variety of foods, and this **Eastern Spider Beetle** (***Ptinus raptor***; 3–4 mm) is often found feeding on stored grains and other dry foodstuffs indoors. Spider beetles are sometimes treated as a separate family, the Ptinidae. ❹ Some of the 11 northeastern species of ***Ptinus*** are found both indoors and outdoors. This is the **White-marked Spider Beetle**, ***P. fur*** (2–4 mm), a pest species that occurs throughout the world. **FAMILY NOSODENDRIDAE** (NOSODENDRID BEETLES) ❺ Nosodendrids use their stout and flattened front legs to dig into wet, fermenting wood, usually in sap flows or fungus infested wood. This is ***Nosodendron unicolor***, the only eastern *Nosodendron*, and one of only two eastern species in the family.

SUPERFAMILY DERODONTOIDEA

FAMILY DERODONTIDAE (TOOTH-NECKED FUNGUS BEETLES) ❻ This ***Laricobius nigrinus*** is eating on a Hemlock Woolly Adelgid (*Adelges tsugae*). *Laricobius nigrinus* was introduced from British Columbia to the eastern United States as a biological control agent for this serious new threat to the eastern hemlock forests. One of the other two *Laricobius* that now occurs in the northeast (*L. erichsoni*) was introduced from Europe to help control, the Balsam Woolly Adelgid (*Adelges piceae*). Other derodontid genera are fungus feeders.

SUPERFAMILY LYMEXYLOIDEA

FAMILY LYMEXYLIDAE (SHIP-TIMBER BEETLES) ❼ ❽ Adult ship-timber beetles are rarely seen, but peeling the bark off damaged poplar trees will often reveal larvae in their characteristic tunnels. These photos show larvae, a pupa and an adult of the **Sapwood Timberworm** (***Elateroides lugubris***). ❾ ***Melittomma sericeum***, one of only two species of northeastern Lymexylidae, occurs under bark.

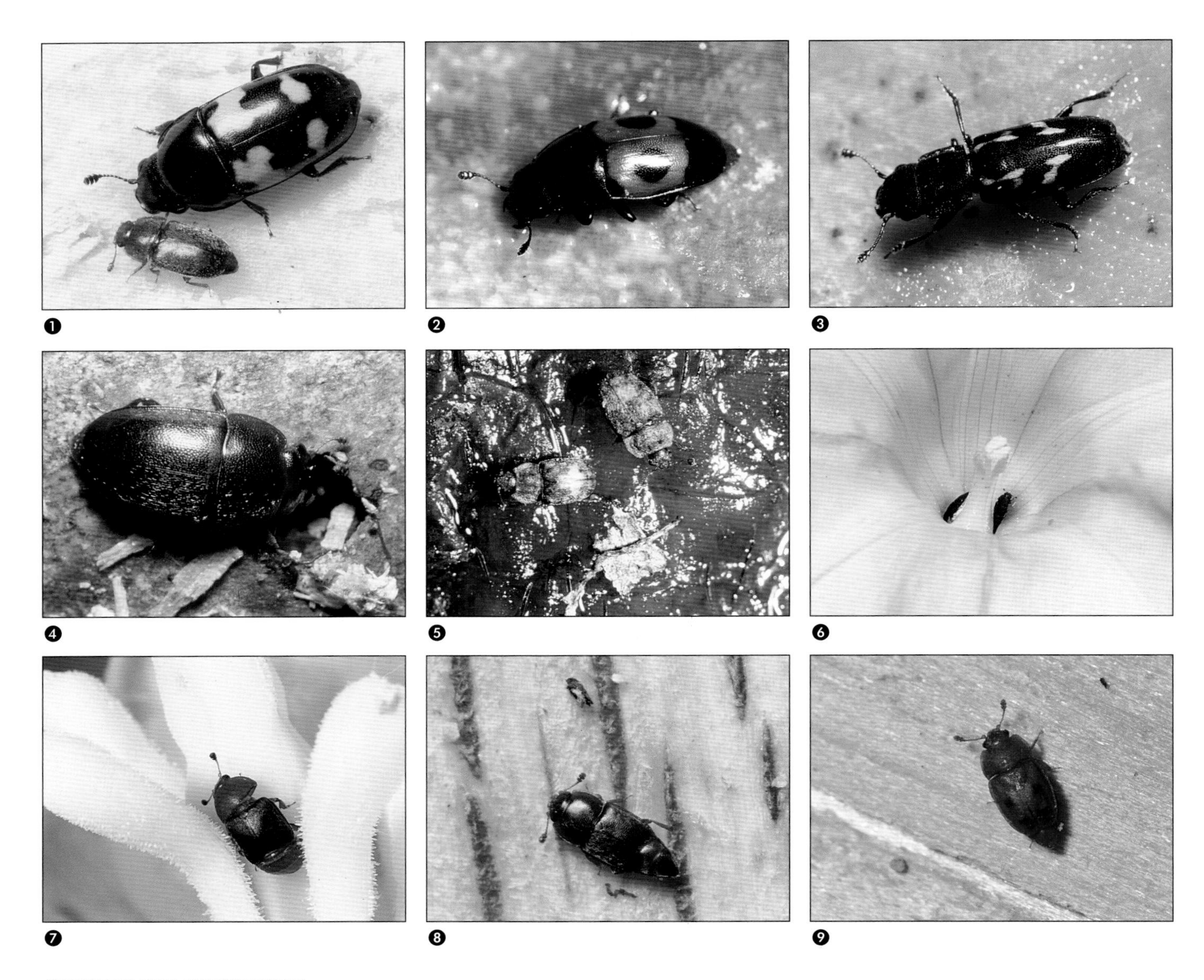

SUPERFAMILY CUCUJOIDEA

FAMILY NITIDULIDAE (SAP BEETLES) ❶ These two sap beetles, a small ***Epuraea*** and a relatively large (6 mm) ***Glischrochilus***, have been attracted to sap flowing from a willow tree. ❷ ❸ The eight northeastern species of ***Glischrochilus*** are called beer beetles or picnic beetles because of their attraction to fermenting items ranging from beer to compost. This boldly spotted ***G. sanguinolentus*** is on a slice of watermelon; the darker ***G. vittatus*** is on a sappy maple stump. ❹ This relatively large sap beetle (***Cryptarcha ampla***, 5–8 mm) is taking advantage of the sap flowing from an exit-hole made by a Poplar Borer (*Saperda calcarata*, Cerambycidae). ❺ Some sap beetles, like this necrophagous species (***Omosita colon***, 2–3 mm) are commonly found on dead animals. This is a close-up of a dead beaver's tail. ❻ Some flower-loving sap beetles are rarely seen except on their preferred flowers. ***Conotelus obscurus***, for example is common in Morning Glory (*Ipomoea violacea*) and related plants (like this Bindweed, *Convolvulus arvensis*). This little (4 mm) beetle has very short elytra, and looks a bit like a rove beetle with clubbed antennae. ❼ This sap beetle (***Carpophilus melanopterus***) can usually be found in flowers of *Yucca* species, like this one on the University of Guelph campus. Other *Carpophilus* species (17 in the northeast) have different habits, and some occur in the flowers of prickly pear cacti (*Opuntia* spp). ❽ ***Carpophilus sayi*** is a common sap beetle under the bark of recently fallen hardwoods. ❾ The 26 northeastern species of ***Epuraea*** are common under bark and at sap runs, but can be difficult to identify.

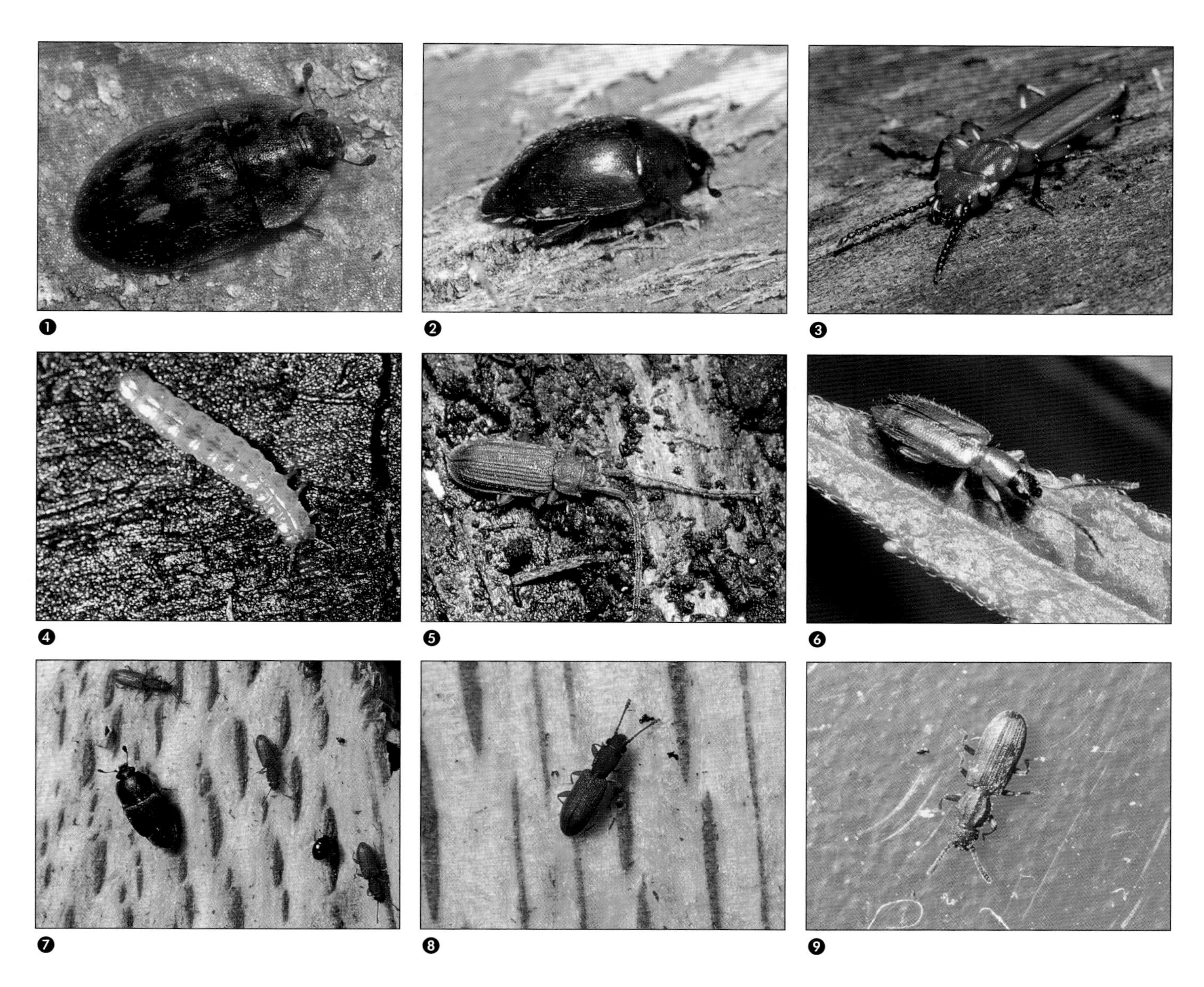

FAMILY NITIDULIDAE (continued) ❶ ***Lobiopa setosa*** (5–7 mm) is one of the many species of sap beetle found under bark of various trees. ❷ Males of ***Amphicrossus ciliatus*** (3–5 mm) have odd hair pencils sticking out of their elytra. This species, like many other Nitidulidae, is found in both sap and in flowers. **FAMILY CUCUJIDAE (FLAT BARK BEETLES)** ❸ ❹ Both adults and larvae of flat bark beetles are modified for life under bark. This common and colorful species is our largest flat bark beetle, ***Cucujus clavipes*** (10–14 mm). **FAMILY SILVANIDAE (SILVANID FLAT BARK BEETLES)** ❺ Flat bark beetles are common under the bark of dead trees. This is ***Uleiota dubia***, the most common large (5 mm) silvanid flat bark beetle in the northeast. ❻ Not all flat bark beetles live under bark, and this species (***Telephanus velox***) is common on the ground and on low vegetation, where it could easily be mistaken for a small (4 mm) ground beetle. ❼ Silvanid flat bark beetles are often common under bark along with other subcortical beetles such as sap beetles and hister beetles. This photo shows four ***Silvanus*** beetles, a couple of species of *Carpophilus* (Nitidulidae) and a small hister (*Aeletes simplex*, Histeridae). ❽ This small (2 mm) ***Silvanus*** is one of five northeastern *Silvanus* found under bark. ❾ The **Sawtoothed Grain Beetle**, ***Oryzaephilus surinamensis*** (2–3 mm), and the similar Merchant Grain Beetle (*Oryzaephilus mercator*) are significant stored products pests.

FAMILY PASSANDRIDAE (PARASITIC FLAT BARK BEETLES) ❶ Larvae of the distinctly flattened ***Catogenus rufus*** are ectoparasites on the pupae of long-horned wood borers (Cerambycidae). **FAMILY LAEMOPHLOEIDAE** (LINED FLAT BARK BEETLES) ❷ ❸ Species of ***Laemophloeus***, like this ***L. biguttatus*** and this somewhat paler ***L. fasciatus*** (both about 3 mm) are wafer-thin beetles found under the bark of dead trees, where they appear to feed on fungi. **FAMILY MONOTOMIDAE** (ROOT-EATING BEETLES) ❹ Despite the name "root-eating beetles," the most commonly encountered members of the small family Monotomidae are tiny (around 3 mm) beetles, like these ***Rhizophagus***, that live under bark or in dead wood where they probably eat other insects. **FAMILY SPHINDIDAE** (CRYPTIC SLIME MOLD BEETLES) ❺ Cryptic slime mold beetles are small (1.5–3.5 mm) beetles that feed only on slime molds, and occur on or in slime mold sporocarps. This ***Odontosphindus clavicornis*** was found on a dead poplar tree. ❻ These cryptic slime mold beetles – a spore-covered adult and larva of ***Sphindus americanus*** (2 mm) – were among dozens dislodged from a small fruiting body of a brown slime mold (*Stemonitis splendens*). **FAMILY LANGURIDAE** (LIZARD BEETLES) ❼ Lizard beetles, like the **Clover Stem Borer** (***Languria mozardi***, 4–9 mm) are distinctively narrow, parallel-sided beetles that develop in the stems of plants. ❽ The narrow larvae of this lizard beetle (***Acropteroxys gracilis***, 6–12 mm) bore in the stems of ragweed, nettle and chicory. **FAMILY ENDOMYCHIDAE** (HANDSOME FUNGUS BEETLES) ❾ Adult handsome fungus beetles, like this ***Endomychus biguttatus*** (4 mm), sometimes spend the winter in groups under bark or under logs, and can be found in colorful masses in spring.

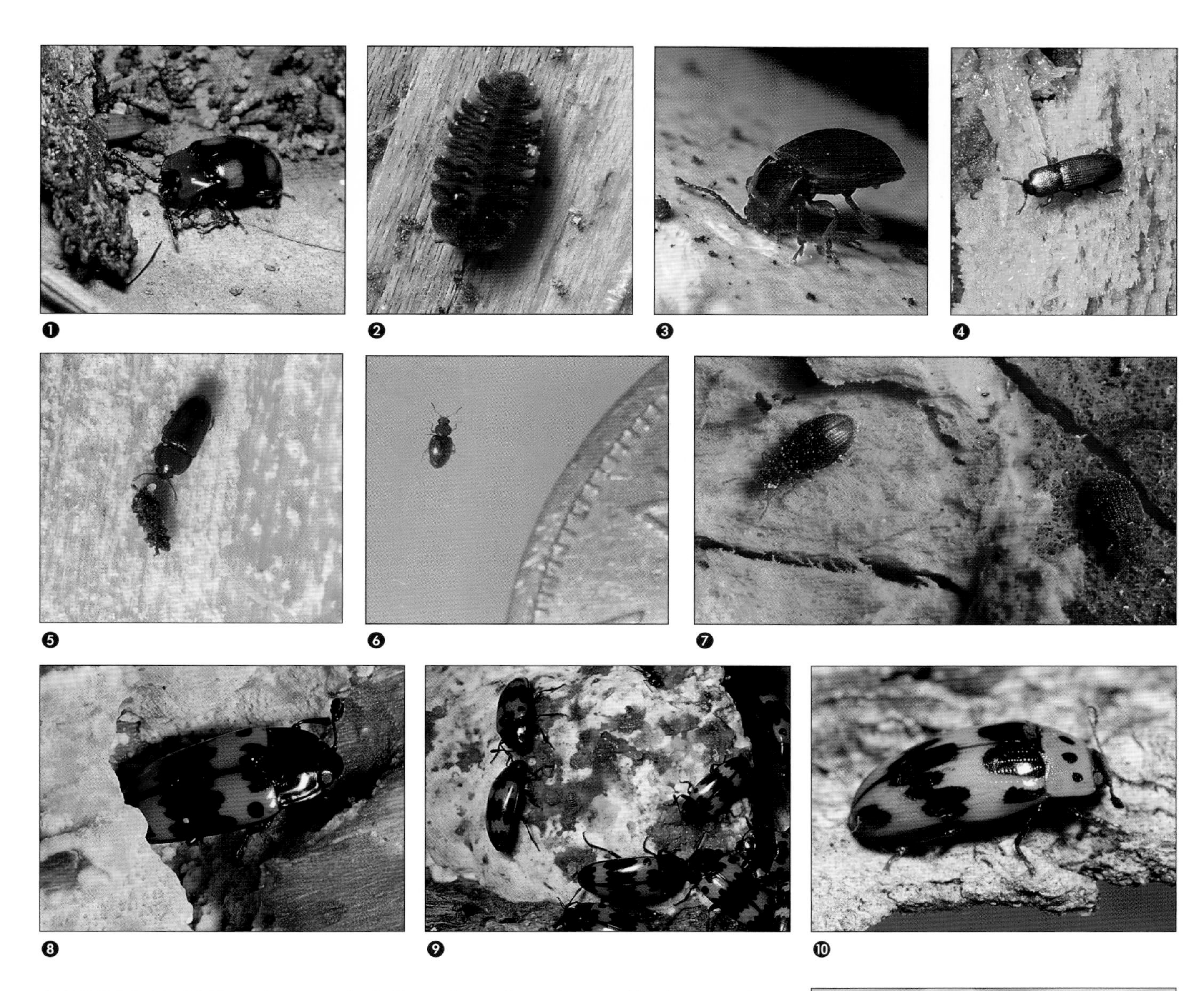

❶ ❷ ❸ ❹ ❺ ❻ ❼ ❽ ❾ ❿ ⓫

FAMILY ENDOMYCHIDAE (continued) ❶ ❷ Handsome fungus beetles, like this ***Mycetina perpulchra*** (4 mm), usually develop in woodland fungi, and can often be found on fungi under bark. ❸ This handsome fungus beetle (***Lycoperdina ferruginea***, 5–6 mm), a species always associated with puffballs, is feeding on the surface of a Giant Puffball (*Calvatia gigantea*). **FAMILY CERYLONIDAE** (MINUTE BARK BEETLES) ❹ Minute bark beetles, like ***Cerylon castaneum***, are small (around 2 mm), rarely noticed beetles that feed on fungi under bark. ❺ ***Philothermus glabriculus***, the only northeastern *Philothermus*, occurs under hardwood bark. **FAMILY LATRIDIIDAE** (MINUTE BROWN SCAVENGER BEETLES) ❻ Minute brown scavenger beetles are common beetles that often abound in leaf litter, compost heaps and some stored products where they feed on conidia of fungi and Myxomycetes, but they are easily overlooked because of their small size. This ***Corticarina***, seen beside the edge a Canadian quarter, is not much more than 1 mm. ❼ These two minute brown scavenger beetles (***Enicmus tenuicornis***, 1.6 mm) are on a rotting fungus. **FAMILY EROTYLIDAE** (PLEASING FUNGUS BEETLES) ❽ ❾ Members of this family are called pleasing fungus beetles, perhaps because encountering one of our two large (10–21 mm) and distinctively colored ***Megalodacne*** species is a "pleasing" experience. This is ***M. fasciata***, a common species that hides near fungi during the day and moves out onto the fungi to feed at night. ❿ ***Ischyrus quadripunctatus*** is a distinctively colored pleasing fungus beetle often found in groups on dead trees, usually associated with soft fungi. ⓫ Soft bracket fungi such as *Pleurotus* are often riddled with brightly colored pleasing fungus beetles, of which the most common are species of ***Triplax*** (2–7 mm).

FAMILY EROTYLIDAE (continued) ❶ ***Tritoma sanguinipennis*** is one of the most common and widespread erotylids. ❷ Pleasing fungus beetles in the genus ***Tritoma***, like this ***T. pulchra*** (3–6 mm), usually develop in mushrooms. **FAMILY PHALACRIDAE** (SHINING FLOWER BEETLES) ❸ Shining flower beetles are little (1–3 mm), brilliantly shiny, black beetles that often abound on flowers, especially composites. Larvae feed in the flower heads or eat fungal spores. **FAMILY EUCINETIDAE** (PLATE-THIGH BEETLES) ❹ Six small (2–4 mm) species of plate-thigh beetles, all in the genus ***Eucinetus***, occur in northeastern North America. This one was photographed as it scurried from beneath the cover of a slime mold sporocarp. **FAMILY CORYLOPHIDAE** (MINUTE FUNGUS BEETLES) ❺ Minute fungus beetles are, as the name suggests, really tiny. This ***Orthoperus***, photographed on the underside of a hard bracket fungus, is less than 0.5 mm long. Several other families of beetles are also extremely small. Ptiliidae (feather wing beetles) range from 0.4–1.5 mm and Clambidae (minute beetles) never exceed 1 mm. These and other minute, rarely noticed families are not illustrated here. ❻ ***Clypastraea lunata*** is larger than most minute fungus beetles at just over 1 mm. It's found under bark. **FAMILY CRYPTOPHAGIDAE** (SILKEN FUNGUS BEETLES) ❼ Some species of silken fungus beetles occur in the nests of bees and wasps, and are sometimes phoretic on adult hosts. Look closely at this bee to see a silken fungus beetle (***Antherophagus***) clamped tightly to the right antenna. ❽ Silken fungus beetles are small (0.8–6 mm, usually about 2 mm) beetles with a loose, three-segmented antennal club. They are common but normally stay out of sight as they feed on mold and decaying vegetation. **FAMILY BYTURIDAE** (FRUITWORM BEETLES) ❾ The tiny family Byturidae (fruitworm beetles) has only one northeastern species, the **Raspberry Fruitworm** (***Byturus unicolor***). Fuzzy golden adult fruitworm beetles are common pollinators of woodland plants such as this trillium; larvae feed on Red Raspberry (*Rubus idaeus*) receptacles.

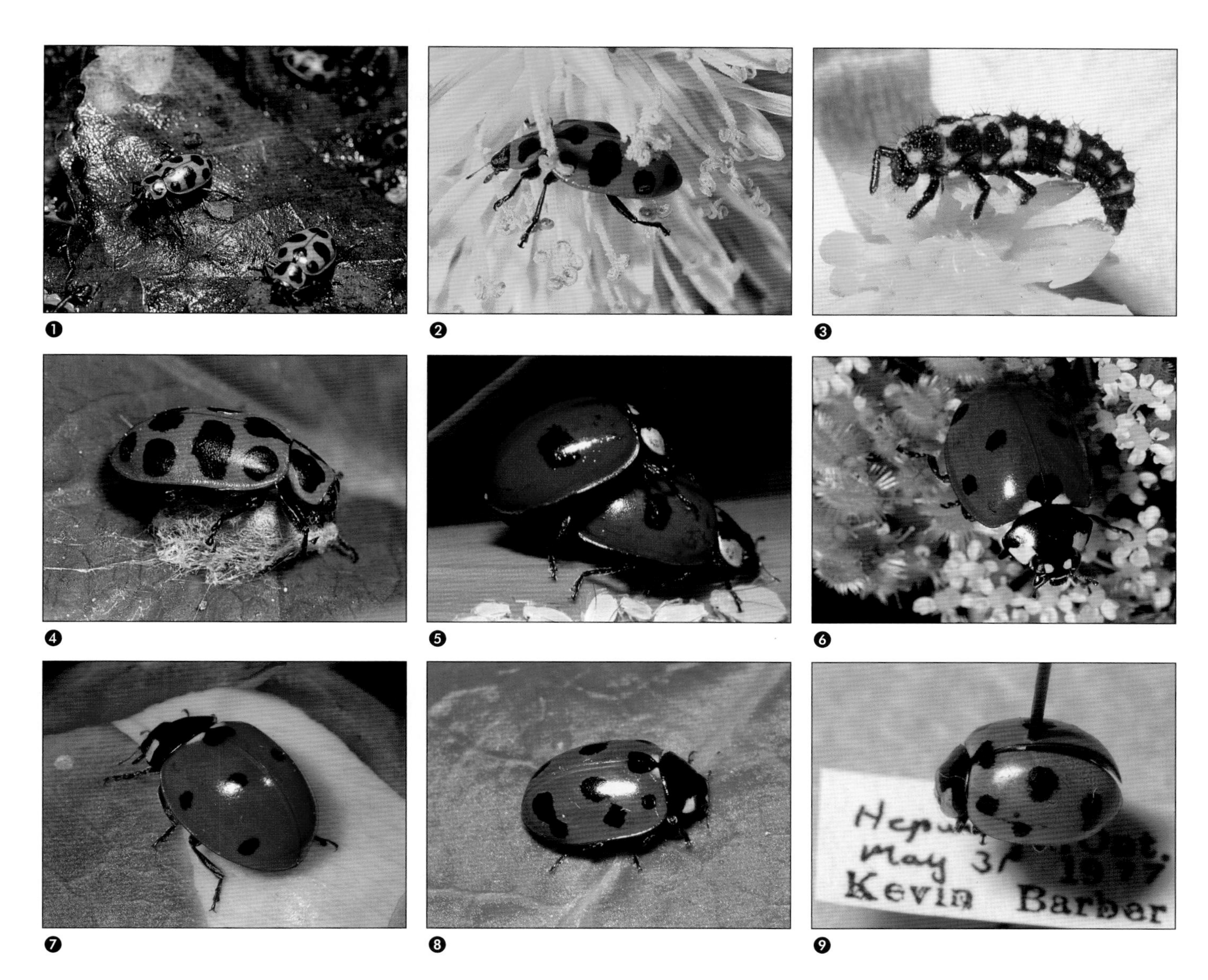

❶ ❷ ❸ ❹ ❺ ❻ ❼ ❽ ❾

FAMILY COCCINELLIDAE (LADY BEETLES) Subfamily Coccinellinae (Black-spotted lady beetles) ❶ ❷ ❸ ❹ The **Spotted Lady Beetle** (***Coleomegilla maculata***) is one of the most common native lady beetles in eastern North America. These 4–7 mm long, pinkish, multispotted beetles form overwintering aggregations among leaves at the bases of trees, and venture out onto the dead leaves as soon as the spring sun melts the snow around the tree trunks. Both adult and larval Spotted Lady Beetles often eat pollen; this adult is on a dandelion, and this larva is feeding on Arrowhead (*Sagittaria*) pollen. The fourth photograph shows a Spotted Lady Beetle sitting on top of a cocoon of a parasitic wasp (*Perilitus coccinellae*, Braconidae) that developed inside the beetle. The lady beetle is not quite dead, and can still twitch enough to scare off other insects that might eat the parasitic wasp's cocoon. ❺ The **Two-spotted Lady Beetle** (***Adalia bipunctata***, 3.5–5.5 mm) is one of our most common native lady beetles. Although both its common and scientific names refer to the common two-spotted form, this species is notoriously variable, and can have four spots, transverse markings or even a black background color. There is only one North American species of *Adalia*. The female of this pair is eating aphids; the male is otherwise occupied. ❻ ❼ ***Coccinella*** species usually fit the popular image of "typical" lady beetles – rounded, convex, reddish orange beetles with black markings. The most common *Coccinella* is the introduced **Seven-spotted Lady Beetle** (***C. septempunctata***) widely known as "C-7". ❽ ***Coccinella undecimpunctata*** (the **Eleven-spotted Lady Beetle**, or **C-11**) is a European species that was common throughout the northeast from the late 1960s through to the early 1980s. it disappeared from much, if not all, of northeastern North America after the establishment of the Seven-spotted Lady Beetle (*C. septempunctata*). This photograph was taken in New Zealand, where C-11 was introduced and still thrives. ❾ ***Coccinella novemnotata*** (the **Nine-spotted Lady Beetle**, or **C-9**) was one of the most common lady beetles in eastern North America prior to the introduction of the European Seven-spotted lady Beetle (C-7, C. *septempunctata*). C-9 had disappeared from much, if not all, of our area by the early 1980s.

FAMILY COCCINELLIDAE (continued) ❶ ❷ ❸ ❹ Nine-spotted Lady Beetles (*Coccinella novemnotata*) have effectively disappeared, but these four native *Coccinella* species still occur in the northeast. The large (around 7 mm) **Transverse Lady Beetle** (***C. transversoguttata***) has become an increasingly rare species over the past 20 years, although it is known from across the continent and throughout the northeast. The small (4–5 mm) **Three-banded Lady Beetle** (***C. trifasciata***, with distinctive bands) is currently the most frequently encountered native ***Coccinella***. It occurs across the continent, including all northeastern states and provinces. The **Hieroglyphic Lady Beetle** (***C. hieroglyphica***, 4–5 mm, with a W-shape across the front of the elytra) and the **Monticola Lady Beetle** (***C. monticola***, 5–7 mm, each elytron with a dash and a dot) have relatively northern distributions, occuring across the continent, including the northern United States and much of Canada. ❺ ***Myzia pullata***, the only eastern North American member of the genus *Myzia*, is a native predator of aphids on woody plants. This one was photographed in the midst of a large colony of White Pine Aphids (*Cinara strobi*). ❻ Although the **Convergent Lady Beetle** (***Hippodamia convergens***), so named for the convergent white marks on its pronotum (just behind the head) occurs naturally in the northeast, huge numbers of Convergent Lady Beetles are imported from California and sold as putative biological control agents for use against aphids and other pests. The imported beetles have been harvested from overwintering aggregations and normally disperse widely before feeding and reproducing.

FAMILY COCCINELLIDAE (continued) ❶ The genus ***Hippodamia*** includes nine species of lady beetles in the northeast, including the widespread **Thirteen-spotted Lady Beetle** (***H. tredecimpunctata***, 5–6 mm). ❷ The introduced ***Hippodamia variegata*** (4–5 mm) has become very common in recent years. ❸ ***Hippodamia glacialis*** is our largest *Hippodamia* (6–8 mm). ❹ The widespread **Parenthesis Lady Beetle** (***Hippodamia parenthesis***) has a distinctive white spot on the middle of the pronotum. ❺ The **Five-spotted Lady Beetle** (***Hippodamia quinquesignata***) (4–7 mm) occurs across the continent but is uncommon in the east. ❻ The northeast is home to two large (7–10 mm), distinctive species in the genus ***Anatis***, both common on coniferous trees. The **Eye-spotted Lady Beetle** (***A. mali***) is aptly named for the adult beetle's yellow-ringed, eyelike spots. ❼ ❽ The **Fifteen-spotted Lady Beetle** (***Anatis labiculata***) normally has seven black spots on each elytron, and one spot in the middle, but in older individuals the 15 spots are hard to differentiate from the dark, purplish background color. ❾ The large, voracious larvae of the genus ***Anatis*** (this is ***A. labiculata*** is eating a plant bug) are black with distinctive yellow highlights.

FAMILY COCCINELLIDAE (continued) ❶ ❷ If you spot a spotless lady beetle under 6 mm, it is likely to be the only northeastern species of ***Cycloneda*, *C. munda***. Similarly immaculate *Harmonia axyridis* specimens occasionally turn up, but they are much larger (8 mm or more), and usually have a white M-shape on the pronotum (right behind the head) rather than the white-edged pronotum of *Cycloneda*. ❸ The distinctively patterned ***Neoharmonia venusta*** (our only *Neoharmonia*, 5–6 mm) has a taste for beetle larvae rather than the usual lady beetle fare of aphids. Look for them on willow trees infested with Willow Leaf Beetles (*Plagiodera versicolora*, Chrysomelidae). ❹ The **Multicolored Asian Lady Beetle** (***Harmonia axyridis***; 5–8 mm) has been one of our most common lady beetle species since the mid-1990s. The cluster of lady beetles shown here is made up mostly of Multicolored Asian Lady Beetles, but also includes three dark red Seven-spotted Lady Beetles (*Coccinella septempunctata*). ❺ **Multicolored Asian Lady Beetles**, like the two color forms mating here, can occur in a wide variety of color forms, but can usually be recognized by a white M-shaped mark on the pronotum (behind the head). ❻ Larval and pupal **Multicolored Asian Lady Beetles** are as brightly colored as the adults. ❼ Like many lady beetles, the **Multicolored Asian Lady Beetle** has a black form as well as the common red forms. ❽ ❾ These **Fourteen-spotted Lady Beetles** (***Propylaea quatuordecimpunctata***) belong to one of two species (in different genera) to go by that common name. This European lady beetle has been spreading throughout our area since the late 1990s, and should probably be called the "Introduced Fourteen-spotted Lady Beetle."

FAMILY COCCINELLIDAE (continued) ❶ ***Calvia quatuordecimguttata*** is one of two species that goes by the common name **Fourteen-spotted Lady Beetle**. This beetle comes in a wide variety of color forms and occurs naturally both in Europe and North America.
❷ ❸ Northeastern North America is home to two widespread species of distinctive, yellow and black ***Mulsantina***, both 3–5 mm in length. The first species shown here (***M. hudsonica***) has a middle black line where the elytra meet; the other species (***M. picta***) lacks this line. ❹ The only northeastern North American species of ***Anisosticta***, ***A. bitriangularis***, is a small (3–4 mm), distinctively colored, yellow and black species that occurs across the continent from Labrador to Alaska and south to Mexico. Look for it in wet places, like fens and bogs. ❺ These two individuals (a larva and an adult) belong to the only northeastern North American species of ***Psyllobora*** (***P. vigintimaculata***, 2–3 mm). Unlike most lady beetles, which are predators, *Psyllobora* species are mildew-feeders and have mandibles armed with rakelike rows of small teeth used in gathering up spores and conidia. **Subfamily Scymninae (Yellow-spotted lady beetles, dusky lady beetles and Australian lady beetles)** ❻ The genus ***Hyperaspis*** is a large genus of tiny (usually 2–3 mm), shiny, mostly black beetles. ***H. bigeminata*** is distinctive for having a bright orange spot right at the hind end of each elytron. ❼ A couple of dozen *Hyperaspis* species occur in the northeast, including several similar species with red dots on the elytra. This male ***Hyperaspis binotata*** differs from the female in having a white margin on the pronotum.
❽ Larvae of the small (usually 3–4 mm) beetles in the genus ***Brachiacantha***, like this ***B. ursina*** on a butterfly weed flower, are rarely seen. Not much is known about their biology, but at least some larvae in the genus eat scale insects on underground parts of plants. ❾ ***Brachiacantha dentipes*** is much larger than most other Scymninae at 5–6 mm.

FAMILY COCCINELLIDAE (continued) ❶ ***Didion punctatum*** is one of three *Didion* species, all of which are less than 2 mm. ❷ ***Stethorus*** species, two of which occur in the northeast, are minute (1.4–1.6 mm) predators of mites. *Stethorus* is one of several genera of tiny and superficially similar "dusky lady beetles." ❸ **Mealybug Destroyers** (***Cryptolaemus montrouzieri***, 3–5 mm; adult and larva shown here), nicknamed "crypts" by insectary workers, were first introduced from Australia to California in 1892 to control scale insects, and are now widely used for control of mealybugs and scale insects in greenhouses. Although crypts are known to occur in the wild in the southern United States, they probably cannot survive northeastern winters outside greenhouses. **Subfamily Chilocorinae (Red-spotted lady beetles)** ❹ The **Twice-stabbed Lady Beetle**, ***Chilocorus stigma*** (4–5 mm), is so aptly named one can almost see the blood dripping from the bright reddish marking on each of the contrasting black elytra. Twice-stabbed Lady Beetles are common on tree trunks and branches, where they feed on scale insects and other soft, slow-moving prey. A very similar species, *C. kuwanae*, was recently imported from Korea to eastern North America as part of a biological control program for the Euonymus Scale (*Unaspis euonymi*). ❺ This **Twice-stabbed Lady Beetle** (***Chilocorus stigma***) was feeding on Pine Needle Scales (*Chionaspis pinifoliae*), visible in the background along these pine needles. ❻ This is a female ***Exochomus marginipennis*** (about 3 mm); on males, the sides of the pronotum are a reddish yellow. **Subfamily Epilachninae (Plant-eating lady beetles)** ❼ This accidentally introduced European plant-eating lady beetle (***Subcoccinella vigintiquatuorpunctata***, 5 mm) is the only plant-eating lady beetle in the northeast other than our two *Epilachna* species. ❽ ❾ Larval and adult **Mexican Bean Beetles** (***Epilachna varivestis***; 6–8 mm) feed on the undersides of leaves of beans and related plants, transforming the leaves into characteristic lacy networks. Our only other *Epilachna*, the Squash Beetle (*E. borealis*, not shown), has four spots on its pronotum. Despite the apparent implications of its species name, *E. borealis* has a more southerly distribution and does not range north of the Canadian border.

SUPERFAMILY CHRYSOMELOIDEA

FAMILY BRUCHIDAE (BEAN WEEVILS) ❶ Seed beetles, like this ***Meibomeus musculus*** (2 mm) have an "exposed butt" beyond the end of the elytra. These small beetles glue their eggs to the outsides of seeds, and larvae burrow inside the seeds. ❷ ***Amblycerus robiniae*** is one of our larger seed weevils at 7–8 mm. As the species name suggests, this beetle develops in seeds of locust trees. **FAMILY ORSODACNIDAE** (ORSODACNID LEAF BEETLES) ❸ ❹ ***Orsodacne atra***, the only northeastern orsodacnid, has several color forms which have been erroneously treated as different species. At least 11 different species names have been given to this species, of which only the oldest (*O. atra*) is correct. The other names are junior synonyms. **FAMILY CHRYSOMELIDAE** (LEAF BEETLES) **Subfamily Eumolpinae (Oval leaf beetles)** ❺ The **Dogbane Beetle** (***Chrysochus auratus***, 8–11 mm) is a gemlike leaf beetle found on Dogbane (*Apocynum androsaemifolium*). ❻ Like many leaf beetles, the **Grape Rootworm** (***Fidia viticida***) is host-specific, feeding only on grapes. **Subfamily Chrysomelinae (Broad-bodied leaf beetles)** ❼ ❽ **Colorado Potato Beetle** (***Leptinotarsa decemlineata***, 6–11 mm) is a native Mexican beetle that shifted from its original host (native *Solanum* species) to cultivated potato plants, allowing it to spread throughout much of North America and Europe to become a common and serious pest. ❾ The **False Potato Beetle** (***Leptinotarsa juncta***) looks like the Colorado Potato Beetle (*L. decemlineata*), but this native eastern *Leptinotarsa* feeds on Horse Nettle (*Solanum carolinense*) and Ground Cherry (*Physalis heterophylla*).

FAMILY CHRYSOMELIDAE (continued) ❶ Swamp Milkweed (*Asclepias incarnata*) and the related Butterfly Weed (*A. tuberosa*) are often home to the large (8–12 mm), convex **Milkweed Leaf Beetle** (***Labidomera clivicollis***). ❷ Marsh Marigolds (*Caltha palustris*) are host to this little (4 mm) leaf beetle (***Hydrothassa boreela***). ❸ Some leaf beetles have been moved around to help suppress populations of their host plants. The **Klamathweed Beetles** (***Chrysolina quadrigemina*** and ***C. hyperici***), for example, were introduced to North America to control a noxious rangeland weed (Klamathweed or St. John's Wort, *Hypericum perforatum*) . Both species are now common throughout our area, and it is rare to find a *Hypericum* that is not chewed ragged by these shining beetles. ❹ The boldly striped ***Zygogramma suturalis*** is a native leaf beetle that has been used for the biological control of Ragweed (*Ambrosia artemisiifolia*). ❺ This elongate leaf beetle (***Prasocuris phellandrii***, 5–6 mm) is found on Marsh Marigold (*Caltha palustris*) and Water Parsnip (*Sium suave*). ❻ ❼ The **Imported Willow Leaf Beetle** (***Plagiodera versicolora***, 4 mm) is a European import that can be extremely abundant on willow trees. Larvae eat away the upper surfaces of the leaves, giving them a skeleton-like appearance. ❽ The **Cottonwood Leaf Beetle** (***Chrysomela scripta***, 6–9 mm) is found on poplar and willow. ❾ The **Alder Leaf Beetle** (***Chrysomela mainensis***, 6–8 mm) feeds on alder.

FAMILY CHRYSOMELIDAE (continued) ❶ ❷ ❸ Larvae, pupae and adults of ***Chrysomela interrupta*** occur on alder. ❹ ❺ ❻ ❼ ❽ ❾ ❿ Leaf beetles in the descriptively named genus ***Calligrapha*** are usually associated with particular plants, and are often difficult to identify unless you know the host. The species illustrated here are ***C. alni*** (on alder), ***C. philadelphica*** (on dogwood – note the all-black pronotum), ***C. multipunctata*** (on willow – note the yellow part between the head and pronotum), the boldly black and yellow marked ***C. pnirsa*** (on basswood), the distinctively colored ***C. rowena*** (on dogwood), the black striped ***C. californica*** (on ragweed and other plants) and the reddish-brown striped ***C. lunata*** (on rose).

FAMILY CHRYSOMELIDAE (continued) Subfamily Cryptocephalinae (Cylindrical leaf beetles, warty leaf beetles and short-horned leaf beetles) ❶ Leaf beetle larvae in several subfamilies use their own excrement to form protective shields or coverings, but larvae in the subfamily Cryptocephalinae carry this habit to an extreme. The eggs hatch under a fecal blanket and proceed to use their own waste to make a case that is added to as they grow. Most occur on the ground where they feed on dead leaves, but some are found on foliage. ❷ This ***Bassareus mammifer*** is carefully manipulating an egg, coating it with successive layers of protective feces. ❸ This compact little beetle (***Cryptocephalus venustus***, 4–5 mm) is a typical cylindrical leaf beetle, and one of 19 northeastern *Cryptocephalus* species. ❹ ***Cryptocephalus notatus*** (3–5 mm) has several color forms, some of which are treated as subspecies. This is the typical subspecies, *C. notatus notatus*. ❺ *Pachybrachis* is a large genus with 33 northeastern species. This is ***Pachybrachis nigricornis*** (4 mm). ❻ ***Anomoea laticlavia*** (7–8 mm) is a distinctively colored leaf beetle with short antennae. Look for this species on members of the pea family. ❼ ❽ ❾ If you disturb a warty leaf beetle it will pull in its appendages (second photo) and drop (third photo), looking so much like a caterpillar dropping it is almost unrecognizable as an insect. This is ***Neochlamisus gibbosus*** (4 mm), common on raspberry bushes.

FAMILY CHRYSOMELIDAE (continued) **Subfamily Donaciinae (Long-horned leaf beetles)** ❶ ❷ The long yellow "tusks" on this long-horned leaf beetle (***Donacia piscatrix***, 3–9 mm) larva are on its hind end, and are used to puncture the submerged parts of aquatic plants to obtain air while the larva feeds on the plant tissue. This species is a minor pest of cultivated waterlilies, and adults are common on lily pads. ❸ Long-horned leaf beetles, like these ***Donacia subtilis***, look a bit like long-horned beetles (Cerambycidae) but they don't have notched eyes like cerambycids. Besides that, you are unlikely to see a cerambycid sitting on a lily pad like these *Donacia*. ❹ Most of our species of long-horned leaf beetles are either in the genus *Donacia* (29 northeastern species) or ***Plateumaris*** (15 northeastern species, including ***P. nitida***, shown here). ❺ ***Plateumaris rufa*** adults usually have a coppery reddish metallic luster, but some individuals are blue, green or purple. ❻ ***Plateumaris shoemakeri*** adults can often be found in large numbers in Marsh Marigold flowers. ❼ ***Neohaemonia*** adults are more aquatic than most other donaciine beetles, and often occur underwater on submerged plants. **Subfamily Synetinae (Punctate leaf beetles)** ❽ Punctate leaf beetles are small (4–8 mm), elongate leaf beetles that feed on roots as larvae, and foliage and fruits as adults. ***Syneta ferruginea*** feeds on a variety of hardwoods. *Syneta*, with three northeastern species, is the only northeastern genus in the subfamily Synetinae. **Subfamily Criocerinae (Shining leaf beetles)** ❾ The **Threelined Potato Beetle** (***Lema daturaphila***, 6–8 mm) feeds on many species in the potato family (Solanaceae).

FAMILY CHRYSOMELIDAE (continued) ❶ The **Cereal Leaf Beetle** (***Oulema melanopus***, 5 mm) is a pest species introduced from Europe in the 1960s. ❷ Asparagus is almost always decorated with two accidentally introduced European leaf beetles, both about 6–7 mm in length. The **Asparagus Beetle** (***Crioceris asparagi***, orange with black spots) has larvae that feed within the asparagus seeds. Larvae of the **Spotted Asparagus Beetle** (***Crioceris duodecimpunctata***), a pretty blue-black beetle with orange borders and yellow spots, feed on the foliage. Both species can emit a high-pitched squeak when disturbed, presumably to startle predators. ❸ The **Lily Leaf Beetle** (***Lilioceris lilii***) is an introduced European species disliked by lily growers, although some might argue that the pest beetle is as pretty as the host. **Subfamily Galerucinae (Skeletonizing leaf beetles)** ❹ Members of the subfamily Galerucinae are called skeletonizing leaf beetles because of the way many larvae in this group feed on the upper surfaces of leaves, resulting in a "skeletonized" leaf. ❺ The **Striped Cucumber Beetle** (***Acalymma vittata***, 5–6 mm) is a pest of cucumber, squash, melon and related plants. Adults chew up foliage and flowers, and are often abundant in flowers. ❻ Larvae of the **Spotted Cucumber Beetle** or **Southern Corn Rootworm** (***Diabrotica undecimpunctata***, 6–7 mm) feed on the roots of corn and grasses; adults feed on foliage and flowers. Despite the "southern" moniker, this is an abundant pest as far north as Ontario. ❼ The male **Western Corn Rootworm** (***Diabrotica virgifera***) is usually smaller and darker than the prominently striped female. ❽ The **Northern Corn Rootworm** (***Diabrotica barberi***, 4–5 mm) is a serious pest of corn. Larvae develop only on corn roots; adults feed on corn foliage and silk. These beetles are usually so abundant in corn growing areas they are among the most common beetles on goldenrod and other flowers in the late summer and fall. This is a native midwestern insect that switched from feeding on native grasses to corn when corn (originally domesticated in Mexico) became widely planted in the late 1800s; it now occurs throughout northeastern North America. ❾ The distinctively striped ***Ophraella conferta*** feeds on goldenrod, but this individual was found overwintering in leaf litter.

FAMILY CHRYSOMELIDAE (continued) ❶ ***Diabrotica cristata*** (4–6 mm), can be conspicuously abundant on flowers in natural grasslands. ❷ ❸ The **Bean Leaf Beetle** (***Cerotoma trifurcata***, 4–5 mm), a common leaf beetle on beans and related legumes, varies in background color from yellow to orange. ❹ Several similar species of the leaf beetle genus *Trirhabda* (5–12 mm), including the common ***T. canadensis***, feed on goldenrod. ❺ ❻ The **Viburnum Leaf Beetle** (***Pyrrhalta viburni***, 5–6 mm) is a recently introduced pest now common on garden *Viburnum* bushes. Winter twigs of *Viburnum* are often conspicuously pocked with the bulging excrescences left by Viburnum Leaf Beetles, which lay eggs in excavations in twigs before covering the excavations with blobs of feces. ❼ ❽ **Loosestrife Leaf Beetles** (***Galerucella calmariensis***, shown here, and the similar *G. pusilla*, 3–5 mm) were deliberately introduced from Europe to North America to control Purple Loosestrife (*Lythrum salicaria*). These abundant beetles now conspicuously damage most stands of loosestrife. ❾ This ***Galerucella calmariensis*** female is depositing feces on each of her recently laid eggs, presumably offering them some protection against egg parasitoids.

FAMILY CHRYSOMELIDAE (continued) ❶ The **Waterlily Leaf Beetle** (***Galerucella nymphaeae***, 5–6 mm) is an abundant beetle wherever white or yellow water lilies occur, and the tar-black larvae of these soft, slow-moving beetles conspicuously skeletonize a high proportion of lily pads. **Subfamily Alticinae** ❷ These flea beetles (***Alticus*** sp.) are mating while feeding on Purple Loosestrife (*Lythrum salicaria*). ❸ Flea beetles are often flealike both in small size and the ability to jump using swollen hind femora. Adult feeding damage often takes the form of circular holes, and damaged plants look like they have been blasted with buckshot. ❹ This **Striped Willow Leaf Beetle** (***Disonycha alternata***, 7–8 mm) is feeding on a willow bud. ❺ ❻ Larvae of the **Sumac Leaf Beetle** (***Blepharida rhois***, 5–7 mm), like many other leaf beetle larvae, use their excrement as a protective armor. ❼ ❽ Several of the 11 northeastern flea beetles in the genus ***Phyllotreta*** are significant pests, especially of cabbage and related plants. These mating **Horseradish Flea Beetles** (***P. armoraciae***, 2 mm) feed on horseradish; the similar ***P. zimmermanni*** feeds on radish, cabbage and other crucifers.

FAMILY CHRYSOMELIDAE (continued) ❶ ***Systena frontalis***, the **Red-headed Flea Beetle**, is a polyphagous pest distinguishable from similar pest flea beetles by its large size (about 5 mm) and reddish head. ❷ This relatively large flea beetle (***Kuschelina vians***, 4–7 mm), like other members of the genus *Kuschelina* (11 northeastern species), has a strikingly inflated and ball-like hind foot. **Subfamily Hispinae (Leaf-mining leaf beetles)** ❸ ❹ ❺ ❻ Larval leaf-mining beetles are flattened to live between the upper and lower surfaces of leaves, and this flattening is carried over to the adult form. The four species shown here are mating pairs of ***Baliosus nervosus*** (5–7 mm) and the darker ***Sumitrosis inaequalis*** (3–5 mm), and single individuals of ***Odontota dorsalis*** (the **Locust Leafminer**) and ***Microrhopala vittata***. Most of these species develop in the leaves of a variety of trees, but *M. vittata* usually occurs on goldenrod. **Subfamily Cassidinae (Tortoise beetles**, sometimes now included in the Hispinae**)** ❼ ❽ This common garden tortoise beetle (***Plagiometriona clavata***, 7–8 mm) has a variety of hosts, and often abounds on garden Solanaceae such as Chinese lantern (*Physalis* sp.). The flattened and fringed larvae, like other tortoise beetle larvae, have a recurved tail used to hold a protective shield of excrement over the body. ❾ ❿ Two species of ***Physonota*** (7–12 mm) occur in the northeast, one on sunflowers and the other (***P. unipunctata***, shown here) on bee balm (*Monarda* spp.). The excrementally defended young larvae feed in groups.

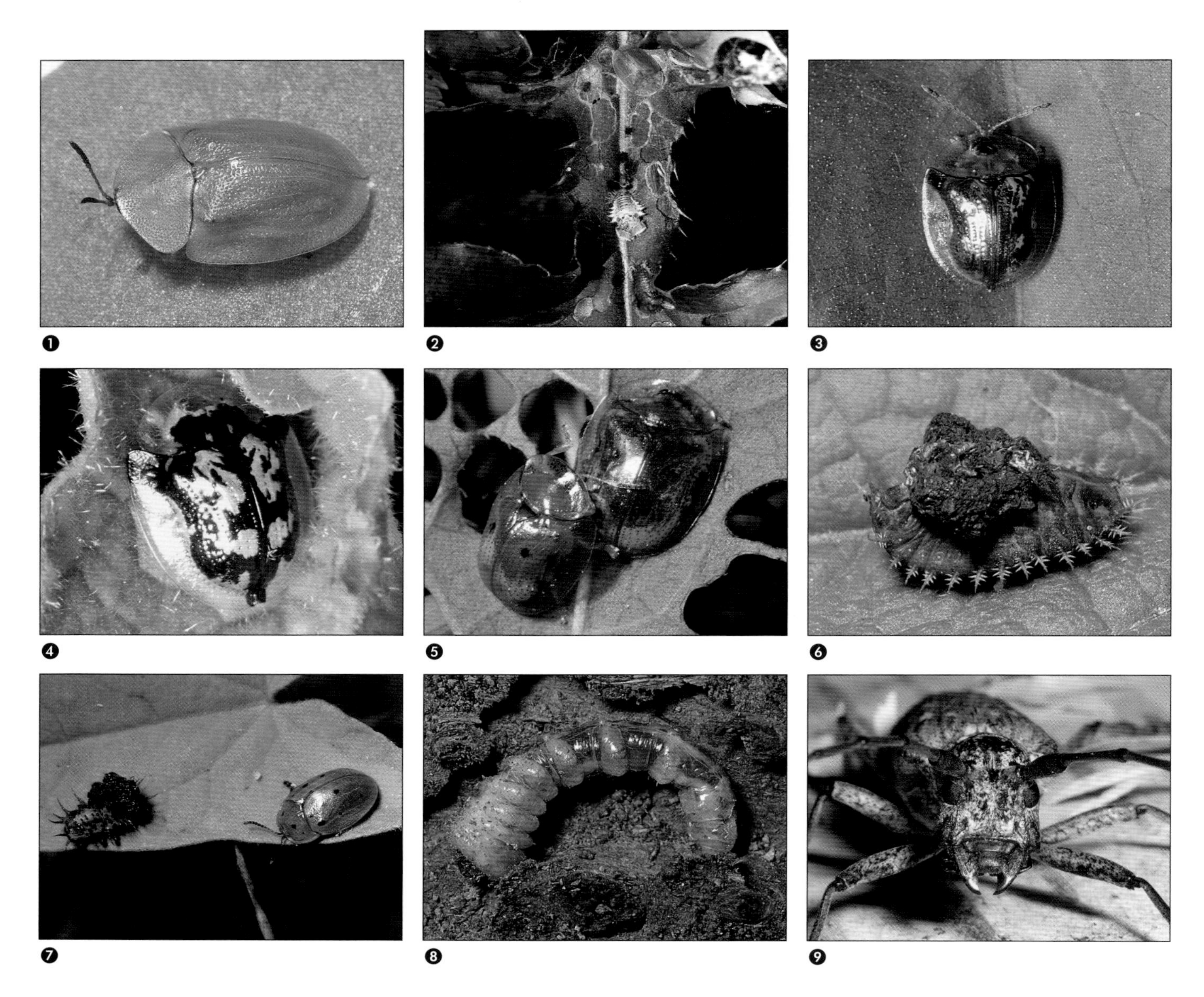

FAMILY CHRYSOMELIDAE (continued) ❶ ❷ The green tortoise beetles found on thistle belong to the genus ***Cassida*** (this is ***C. rubiginosa***, 7–8 mm). All three northeastern *Cassida* species are introduced from Europe. ❸ ❹ When disturbed, tortoise beetles like these **Mottled Tortoise Beetles** (***Deloyala guttata***, 5–7 mm) withdraw their legs and antennae under their expanded "shell." This species is common on Bindweed (*Convolvulus arvensis*), Morning Glory (*Ipomoea violacea*) and related plants. ❺ ❻ The **Golden Tortoise Beetle** (***Charidotella sexpunctata bicolor***, often called *Metriona bicolor*, 5–6 mm) changes color from gold to red when excited. Larvae, with their protective fecal parasols, are common on Morning Glory (*Ipomoea violacea*) and related plants. Pinned specimens of these beautiful beetles undergo a one-way transformation from gemlike specimens to drab pinned beetles as they dry out in insect collections. ❼ The **Argus Tortoise Beetle** (***Chelymorpha cassidea***) occurs on a variety of plants, but is most common on Morning Glory (*Ipomoea violacea*), Bindweed (*Convolvulus arvensis*) and related species. Like other tortoise beetles, the larva carries a protective fecal shield on its tail. **FAMILY CERAMBYCIDAE (LONG-HORNED BEETLES) Subfamily Lamiinae (Flat-faced longhorns)** ❽ Long-horned beetle larvae are called round-headed wood borers because of the round, clublike front end. Peeling the bark off a pine log exposed this larva. ❾ Long-horned beetles like this **Northeastern Sawyer** (***Monochamus notatus***, 23–35 mm) have notched eyes, with the antennae arising in the notch. Sawyers belong to a subfamily called flat-faced longhorns because of the way the downward directed mandibles give a flat-faced appearance.

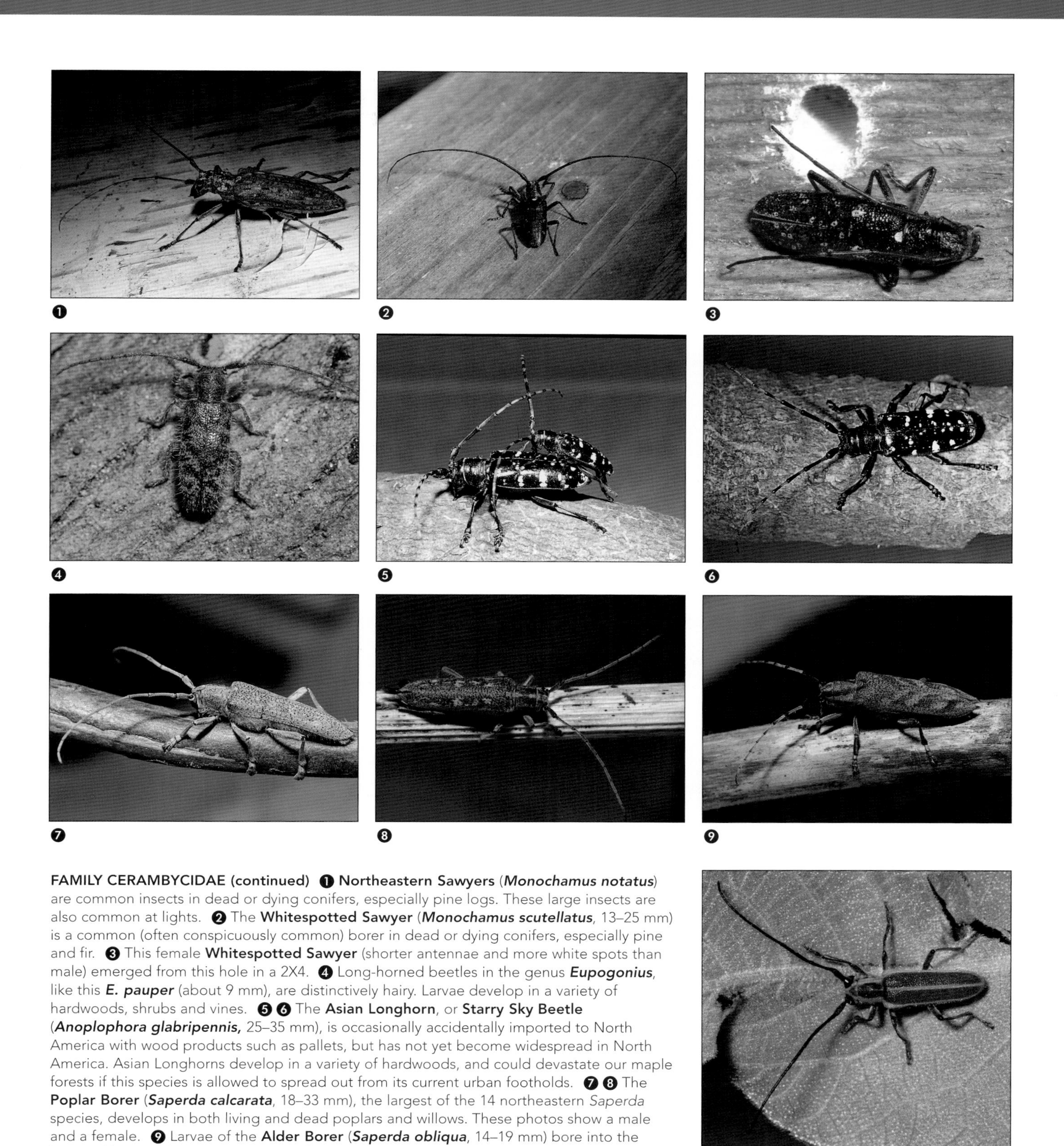

FAMILY CERAMBYCIDAE (continued) ❶ **Northeastern Sawyers** (***Monochamus notatus***) are common insects in dead or dying conifers, especially pine logs. These large insects are also common at lights. ❷ The **Whitespotted Sawyer** (***Monochamus scutellatus***, 13–25 mm) is a common (often conspicuously common) borer in dead or dying conifers, especially pine and fir. ❸ This female **Whitespotted Sawyer** (shorter antennae and more white spots than male) emerged from this hole in a 2X4. ❹ Long-horned beetles in the genus ***Eupogonius***, like this ***E. pauper*** (about 9 mm), are distinctively hairy. Larvae develop in a variety of hardwoods, shrubs and vines. ❺ ❻ The **Asian Longhorn**, or **Starry Sky Beetle** (***Anoplophora glabripennis,*** 25–35 mm), is occasionally accidentally imported to North America with wood products such as pallets, but has not yet become widespread in North America. Asian Longhorns develop in a variety of hardwoods, and could devastate our maple forests if this species is allowed to spread out from its current urban footholds. ❼ ❽ The **Poplar Borer** (***Saperda calcarata***, 18–33 mm), the largest of the 14 northeastern *Saperda* species, develops in both living and dead poplars and willows. These photos show a male and a female. ❾ Larvae of the **Alder Borer** (***Saperda obliqua***, 14–19 mm) bore into the bases of living alder and birch trees. ❿ The distinctively striped ***Saperda lateralis*** develops on a variety of hardwoods from Canada to South Carolina.

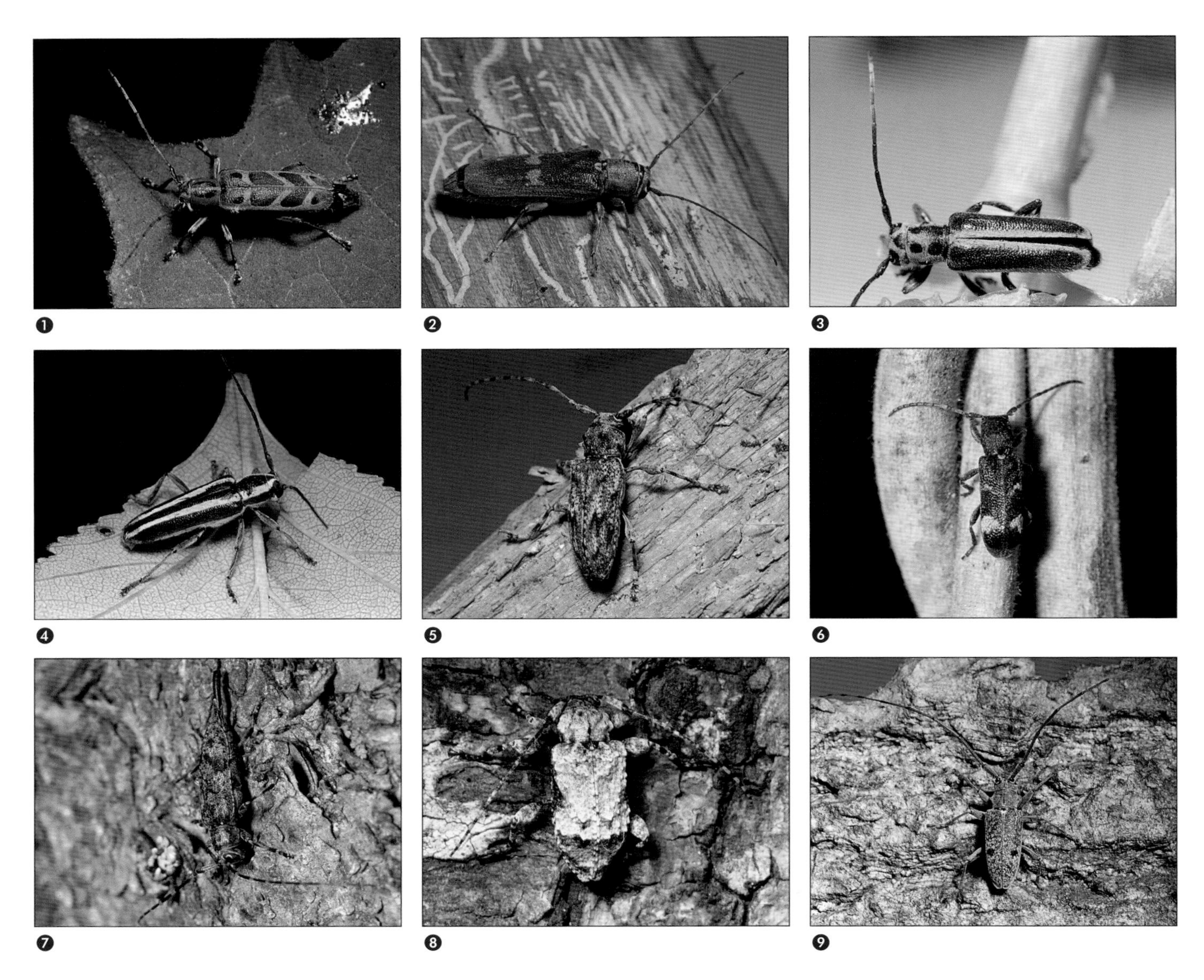

FAMILY CERAMBYCIDAE (continued) ❶ The **Elm Borer** (***Saperda tridentata***, 8–17 mm) is common in diseased elms. ❷ This female **Hickory Saperda** (***Saperda discoidea***) is large and strikingly colored; males are small with grayish elytra. ❸ The brightly colored **Woodbine Borer** (***Saperda puncticollis***, 8–12 mm) is one of the few insects to develop on Poison Ivy (*Rhus radicans*). Larvae of this attractive species also develop in dead or dying grape and Virginia Creeper (*Parthenocissus quinquefolia*). ❹ The **Round-headed Apple Tree Borer** (***Saperda candida***, 10–21 mm) is a common and colorful pest that can damage living apple, quince, hawthorn and related trees. ❺ The M-shaped markings on the elytra make ***Aegomorphus modestus*** (10–16 mm) an easily recognized species. Larvae occur in older dead wood. ❻ The **Currant Tip Borer**, ***Psenocerus supernotatus*** (4–8 mm), sometimes occurs in great numbers on Dutchman's Pipe (*Aristolochia* sp.) vines. Larvae develop in the dead branches of a variety of shrubs and vines. ❼ The long egg-laying tube on this female ***Acanthocinus pusillus*** (7–13 mm) is used to insert eggs in the bark of dead or dying conifers. ❽ This superbly camouflaged ***Leptostylus asperatus*** (9–13 mm) only occurs as far north as Maryland. ❾ The distinctively spotted ***Microgoes oculatus*** occurs under the bark of various trees and shrubs.

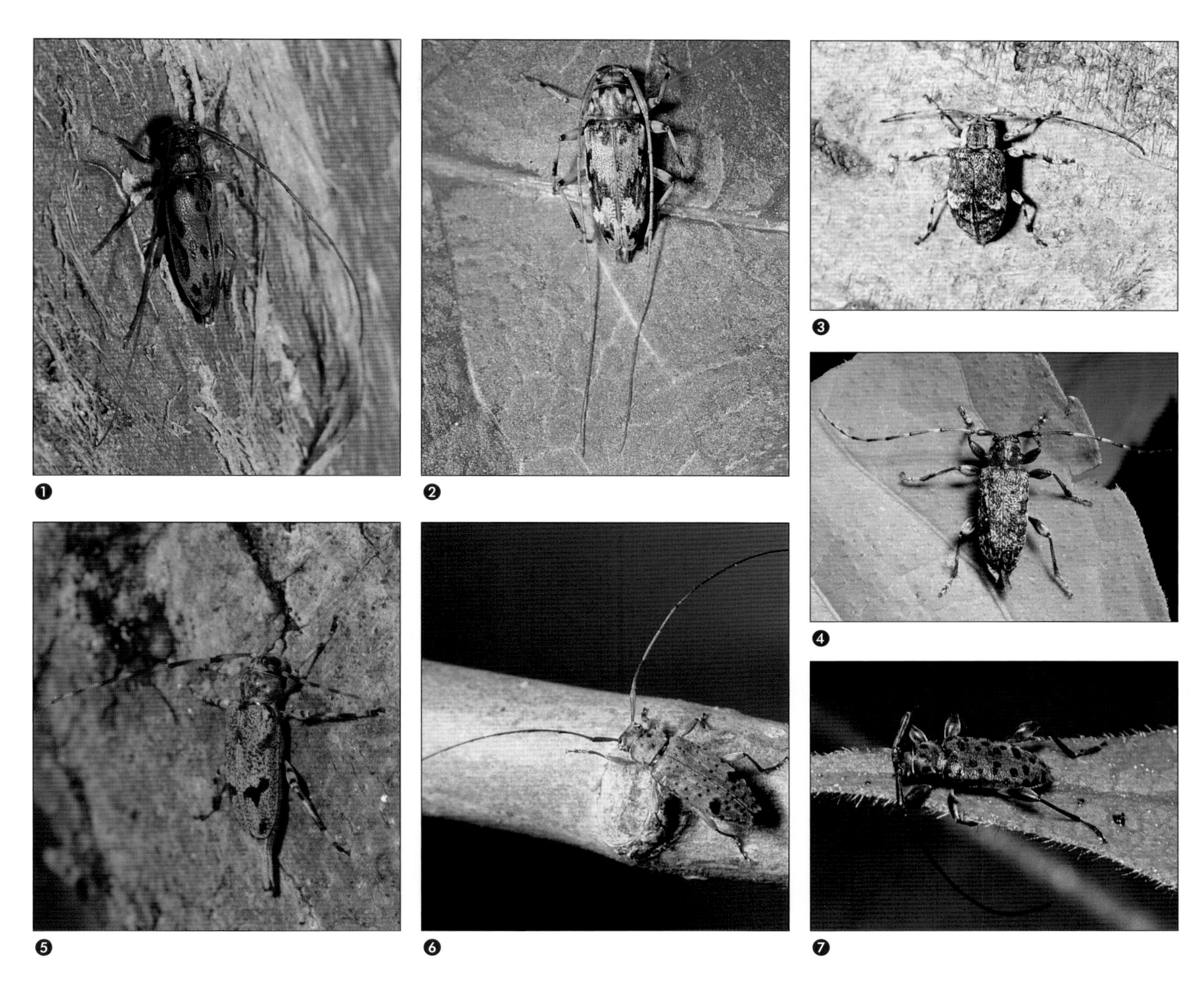

FAMILY CERAMBYCIDAE (continued) ❶ ❷ The genus ***Lepturges*** includes several species with similar color patterns. This distinctively spotted ***L. pictus*** (8–11 mm) was found on its host tree (Hackberry *Celtis occidentalis*), and this paler ***L. confluens*** (5–10 mm) was attracted to an ultraviolet light. Larvae of *L. confluens* develop in hickory and other hardwoods. ❸ This ***Astylopsis macula*** (6–9 mm) was one of many camouflaged against the smooth bark of a healthy ironwood tree. ❹ ***Oplosia nubila*** (9–14 mm), the only *Oplosia* species, is restricted to northeastern North America, where it develops under the bark of decaying trees. ❺ This female ***Urographis despectus*** has a conspicuous ovipositor used to lay eggs on hickories. ❻ ***Hyperplatys aspersa*** is a small (4–10 mm) species that develops under the bark of various kinds of dead trees. ❼ The distinctively spotted ***Hyperplatys maculata*** (4–8 mm) develops in a variety of hardwoods.

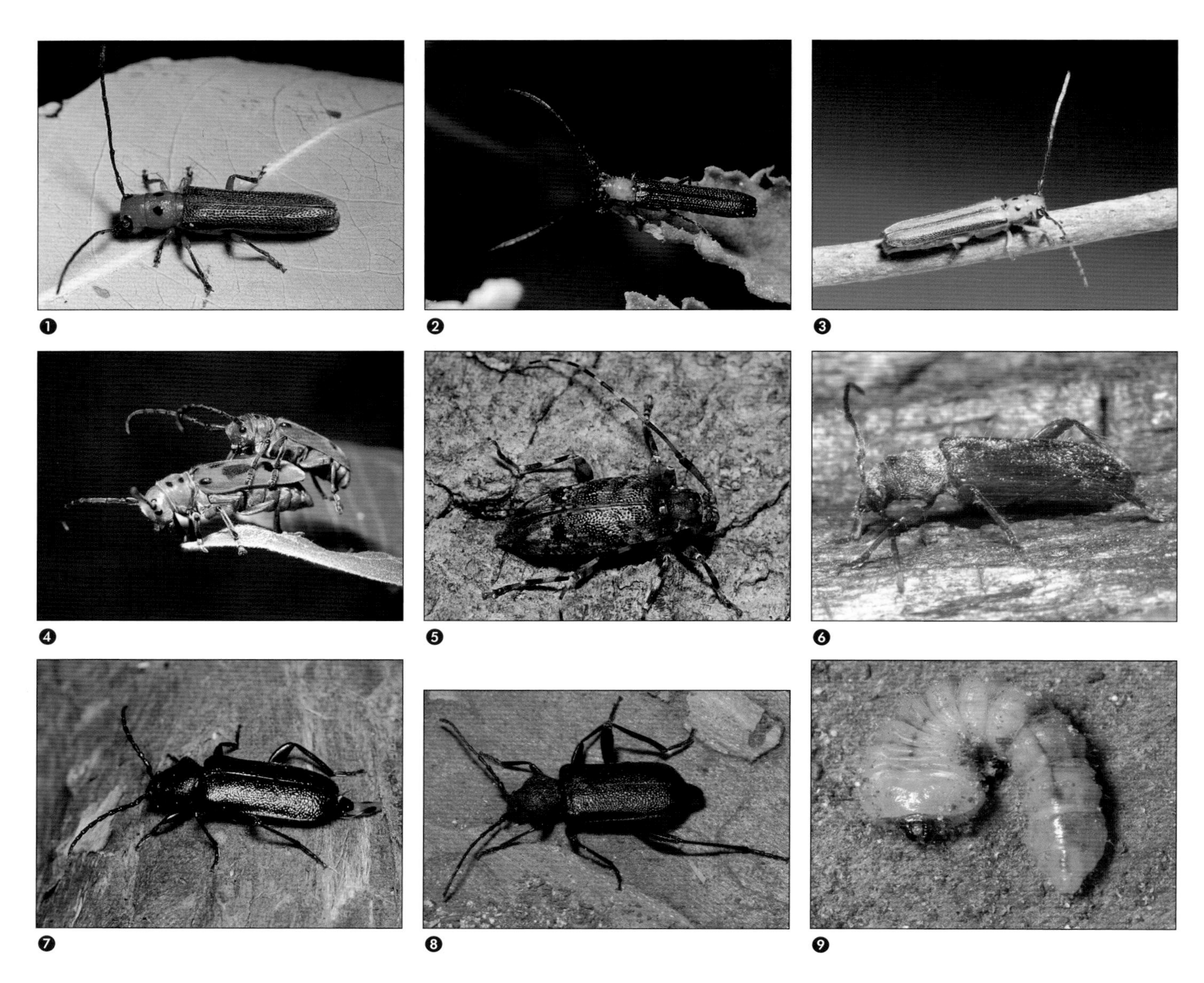

FAMILY CERAMBYCIDAE (continued) ❶ Larvae of the long, cylindrical beetles in the genus ***Oberea*** live in twigs and stems of various plants. This is the **Sumac Stem Borer** (***O. ocellata***, 9–17 mm). ❷ ***Oberea affinis*** is a common beetle that develops in raspberry and blackberry stems. ❸ Although ***Oberea tripunctata*** looks almost identical to the Dogwood Twig Borer (*O. praelonga*) it does not develop in dogwood and its host plants remain unknown. ❹ The **Red Milkweed Beetle** (***Tetraopes tetrophthalmus***, 8–15 mm), a common borer in milkweed roots, has its eyes completely divided by the antennae. Eggs are laid in grass stems near milkweed plants; larvae drop to the ground and burrow into the milkweed. ❺ Larvae of ***Sternidius variegatus*** (7–11 mm) bore into the branches of a variety of trees, shrubs and vines in the eastern United States and southern Ontario. Subfamily Cerambycinae (Round-necked longhorns) ❻ This ***Obrium rubidum***, a relatively scarce and poorly known species, is ovipositing on a bare section of an old log. ❼ This ***Callidium frigidum*** is using her telescoping ovipositor to stick eggs under the bark of a recently cut cedar post. ❽ ❾ This adult **Shining Blue Longhorn** (***Callidium violaceum***) recently emerged from a bark-covered beam; this larva was in the same beam. This introduced species can reinfest rough-hewn timber year after year.

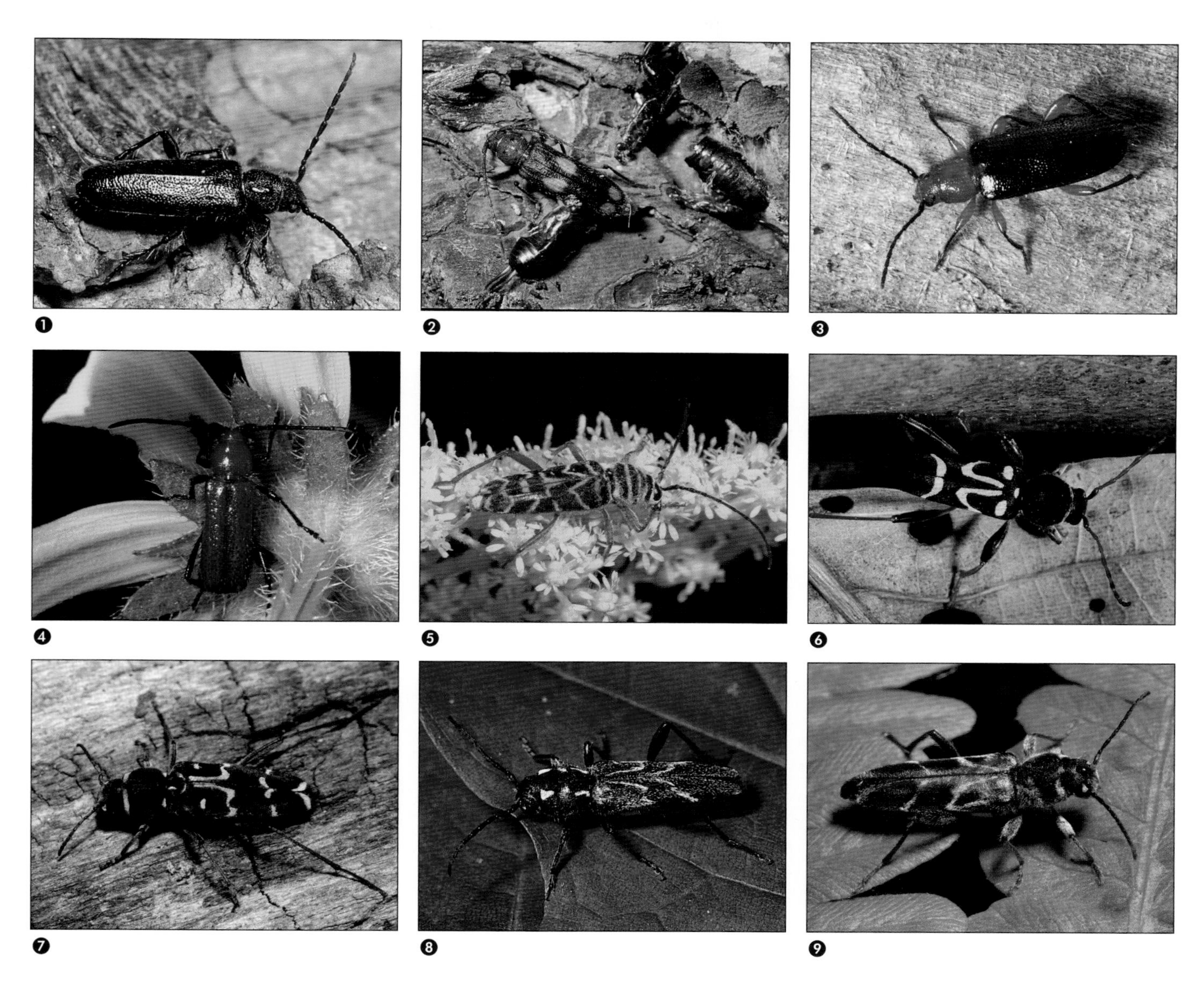

FAMILY CERAMBYCIDAE (continued) ❶ Long-horned beetles are usually easily identified, but a few variably or dull-colored species can be tricky. This **Fir Tree Borer** (***Semanotus litigiosus***, 8–12 mm) compounded its cryptic appearance by emerging from a pine log, although the normal hosts for this species are fir and spruce. ❷ This distinctively marked **Ash and Privet Borer** (***Tylonotus bimaculatus***, 9–18 mm) is exposed, along with some European Earwigs, by peeling some loose bark off a damaged portion of a living ash tree. ❸ Larvae of this long-horned beetle (***Phymatodes amoenus***, 5–7 mm) develop in dead grape vines. ❹ Larvae of the distinctively colored ***Batyle suturalis*** (7–9 mm) bore in small twigs of oak and other hardwoods; adults are sometimes abundant on composite flowers. ❺ ❻ ❼ Many long-horned beetles are bright black and yellow, presumably to deter wasp-shy predators. The best known of these is the **Locust Borer** (***Megacyllene robiniae***, 11–28 mm), a common pest of Black Locust (*Robinia pseudoacacia*) trees. The other photos show the similarly colored ***Clytus ruricola*** (10–15 mm) and ***Xylotrechus undulatus*** (10–16 mm). ❽ The **Birch and Beech Girdler** (***Xylotrechus quadrimaculatus***, 8–15 mm) kills branches of beech and birch by girdling them (chewing rings around the living tissue) before laying eggs in the outer parts of the girdled branches. ❾ The **Gall-making Maple Borer** (***Xylotrechus aceris***) bores in living maple twigs, causing distinctive swellings around the feeding larvae.

FAMILY CERAMBYCIDAE (continued) ❶ ❷ ❸ Several round-necked longhorns are marked and shaped to take on a somewhat antlike appearance. These photos show dark and light forms of ***Cyrtophorus verrucosus*** (7–11 mm), and a ***Euderces picipes*** (4–8 mm) eating pollen on a Queen Anne's Lace flower. ❹ ❺ ❻ A few round-necked longhorns have the distinctively shortened elytra, leaving the membranous hind wings exposed. These photos show a mating pair of ***Molorchus bimaculatus*** (5–8 mm), and a male and female (with a red pronotum) of ***Callimoxys sanguinicollis*** (7–12 mm). ❼ The **Spined Bark Borer** (***Elaphidion mucronatum***, 13–20 mm) is a common species, developing under the bark of a variety of trees and shrubs. ❽ The **Oak Twig Pruner** (***Anelaphus parallelus***) damages live twigs of many shrubs and hardwood trees in eastern Canada and United States. ❾ The **Flat Powder Post Beetle** (***Smodicum cucujiforme***, 7–10 mm) is a pale longhorn beetle with a scientific name that reflects its superficial similarity to flat bark beetles (Cucujidae and related families), and a common name that reflects its similarity to the rarer powder post beetles (Lyctinae, Bostrichidae). Larvae bore in the dry heartwoods of hardwood trees.

FAMILY CERAMBYCIDAE (continued) Subfamily Lepturinae (Flower longhorns) ❶ ❷ Most Cerambycidae seen on flowers are flower longhorns (Lepturinae). These photos show a female and male (darker) of ***Leptura subhamata*** (11–17 mm), a species that develops in dead hemlock and pine. ❸ ***Lepturopsis biforis*** larvae develop in many kinds of dead trees. ❹ Long-horned beetles are important pollinators, and this ***Judolia montivagans*** (7–12 mm) is loaded with pollen on its relatively narrow head and prothorax. This broad shouldered shape, with a narrow prothorax, is typical for flower longhorns. ❺ ***Brachyleptura champlaini*** (8–12 mm) has distinctly shortened elytra that diverge at their ends. Larvae develop in pine. ❻ ***Bellamira scalaris*** is a large (usually over 20 mm) longhorn beetle often found at lights at night. Larvae are found in decaying wood. ❼ ❽ ❾ These flower longhorns (***Trigonarthris*** sp., 11–17 mm; ***Analeptura lineola***, 6–12 mm; ***Typocerus velutinus***, 9–16 mm) are munching on pollen of Queen Anne's Lace (*Daucus carota*) flowers.

FAMILY CERAMBYCIDAE (continued) ❶ ❷ These small (5–8 mm), mostly black flower longhorns (***Grammoptera subargentata*** and ***Pidonia ruficollis***) could be easily mistaken for pedilid beetles. ❸ Most long-horned beetles are distinctive in color and shape but some, like ***Pidonia ruficollis***, are variable in color (compare this mating pair to the individual in the previous picture). ❹ ❺ The flower longhorn ***Pseudogaurotina abdominalis*** (9–16 mm, mating in a wild rose) and the more robust ***Gaurotes cyanipennis*** (9–13 mm) are remarkably similar, brilliantly metallic beetles. ❻ ❼ Long-horned beetles are occasionally variable in color, as you can see from these two **Red-shouldered Pine Borers** (***Stictoleptura canadensis***, 10–15 mm). The one with the red shoulders is a female, the other is a male. ❽ The **Elderberry Borer** (***Desmocerus palliatus***, 18–26 mm), one of our most easily recognizable longhorns, develops in the living roots of elderberry. ❾ This female **Mottled Long-horned Beetle** (***Anthophylax attenuatus***, 12–17 mm) is laying eggs in a dead maple tree. ❿ ***Anthophylax cyaneus***, a borer in various hardwoods, is one of three similar metallic *Anthophylax* in eastern North America.

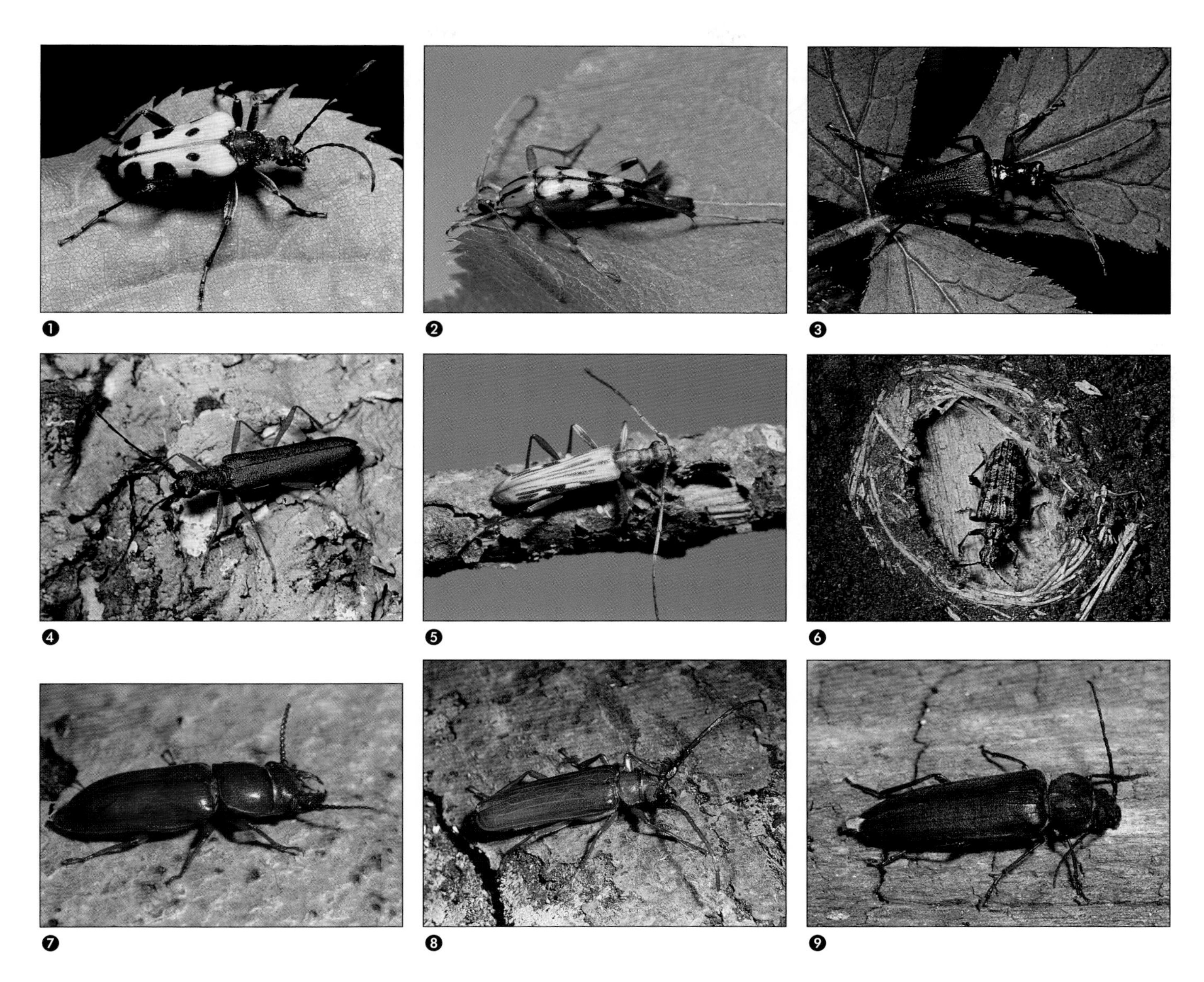

FAMILY CERAMBYCIDAE (continued) ❶ Larvae of ***Evodinus monticola*** (8–13 mm) live in conifer bark. ❷ ***Strangalia luteicornis*** (9–14 mm) is a common and distinctive species that develops in hardwood trees and shrubs. ❸ ***Stenocorus schaumii*** is one of our larger common flower longhorns (up to 30 mm). Larvae live in hardwoods. ❹ The **Oak Bark Scaler** (***Encyclops caerulea***, 7–11 mm) is a delicate and distinctively colored flower longhorn. As the common name suggests, it lives in the outer bark of oak and other hardwoods. ❺ The slender ***Leptorhabdium pictum*** (9–17 mm) lives in various hardwoods. ❻ This **Ribbed Pine Borer** (***Rhagium inquisitor***, 9–21 mm) was exposed in its characteristic overwintering shelter under the bark of a dead pine tree. Larvae make the ringlike shelters from shredded bark, and pupate within the rings in fall. Subfamily Parandrinae (Aberrant longhorns) ❼ Beetles in the small subfamily Parandrinae, with only three North American members, are referred to as aberrant longhorns because they look more like darkling beetles or stag beetles than longhorns. This is the **Pole Borer** (***Parandra brunnea***, 9–18 mm), a pest of shade trees. Subfamily Prioninae (Tooth-necked longhorns) ❽ Called tooth-necked longhorns because of the distinctive spines along the side of the pronotum, prionines are our largest Cerambycidae and can reach lengths up to 60 mm. This is the **Brown Prionid** (***Orthosoma brunneum***, 25–50 mm), which develops in well-decayed wood. Some tooth-necked longhorns develop in the roots of living trees. Subfamily Aseminae (Asemine longhorns) ❾ The **Opaque Sawyer** (***Asemum striatum***, 10–16 mm) breeds in a variety of conifers and is often attracted in numbers to freshly cut spruce lumber. Members of the small subfamily Aseminae (seven or eight northeastern species) are relatively dull-colored beetles.

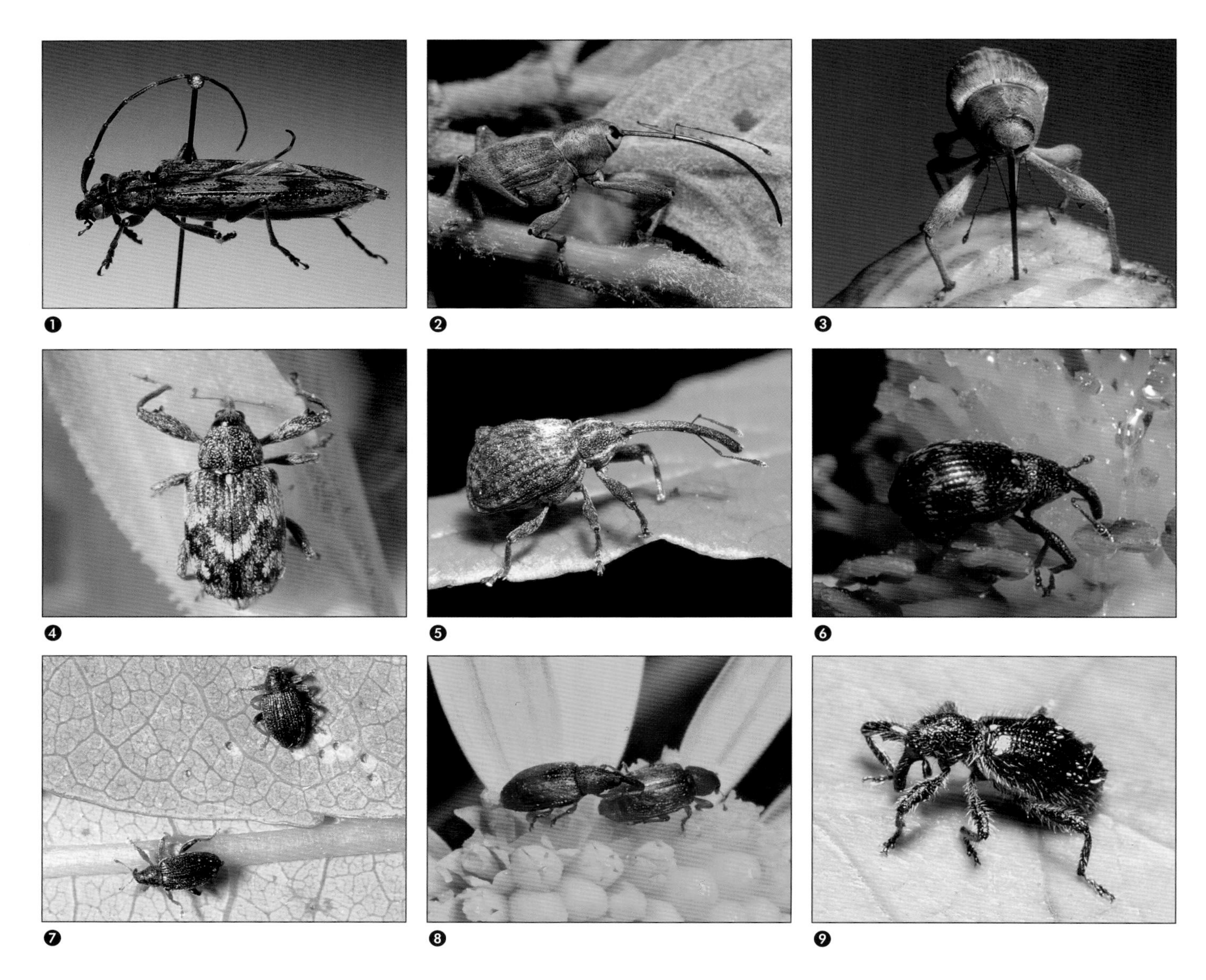

FAMILY CERAMBYCIDAE (continued) Subfamily Disteniinae (Disteniine longhorns) ❶ The Disteniinae are distinctive from all other long-horned beetles, so much so they are sometimes treated as a separate family (Disteniidae). The only northeastern species in the subfamily (***Distenia undata***) ranges as far north as Pelee Island, Ontario. The chisel-shaped mandibles of this group are unique in the family.

SUPERFAMILY CURCULIONOIDEA

FAMILY CURCULIONIDAE Subfamily Curculioninae (Acorn and nut weevils, fruit and seed weevils, flea weevils, antlike weevils and others) ❷ ❸ Weevils have their chewing mouthparts (mandibles) at the end of a long beak, and this acorn weevil (***Curculio***) uses its vertically moving mandibles to drill holes in nuts. Eggs are laid in the holes made with the long beak. There are 14 northeastern acorn and nut weevils in the genus *Curculio*. ❹ Many fruit and seed weevils are in the large and important genus ***Anthonomus***, with 39 northeastern species including several pests that lay eggs in buds or developing fruit. The infamous Boll Weevil (*A. grandis*) is in this genus, as are several northeastern pests such as the Strawberry Bud Weevil (*A. signatus*) and Cranberry Weevil (*Anthonomus musculus*). The species illustrated here, ***A. nebulosus*** (4 mm), occurs on hawthorn (*Crataegus* sp.). ❺ This is the **Apple Curculio** (***Anthonomus quadrigibbus***), which develops in early-season dropped apples, or in the fruits of hawthorn and crabapple. ❻ The **Strawberry Bud Weevil** (***Anthonomus signatus***) is a small (2–3 mm) weevil that develops in strawberry buds, which drop from the plant when adult weevils chew through the stems just below the buds. ❼ These **Willow Flea Weevils** (***Isochnus rufipes***, 2 mm) can jump like flea beetles (Chrysomelidae) using enlarged hind legs, and also cause "shot hole" feeding damage similar to flea beetle damage. Larvae mine in the leaves. ❽ ***Phyllotrox nubifer*** is a small (2–3 mm) widespread weevil that sometimes abounds on composite flowers. ❾ The aptly named antlike weevils are small (3–5 mm) beetles that usually bore in twigs of dead or dying trees, although one northeastern species (*Oopterinus perforatus*) lives only in the galls made by gall wasps (Cynipidae) on the roots of oaks. Four of the five northeastern antlike weevil species (like this ***Myrmex scrobicollis***) are in the genus *Myrmex*.

FAMILY CURCULIONIDAE (continued) Subfamily Cryptorhynchinae (Hidden snout weevils) ❶ Larvae of the **Poplar-and-willow Borer** (***Cryptorhynchus lapathi***, 8–10 mm) usually occur in willow bark; adults are often common on twigs. ❷ Although the white, sticky sap of milkweed is toxic to many insects, it does not seem to bother this ***Rhyssomatus lineaticollis*** (5–6 mm). Larvae live in milkweed stems. Subfamily Zygopinae (Twig and stem weevils) ❸ Zygopine weevils are big-eyed, fast-flying weevils that abound in conspicuous diversity in the neotropics. They are not common in the northeast, but this ***Acoptus suturalis*** (with mites) was photographed on a small beech tree in southern Ontario. Subfamily Dryophthorinae (Billbugs and grain weevils) ❹ The **Timothy Billbug** (***Sphenophorus zeae***, 7–9 mm) is one of 31 northeastern species in the large genus *Sphenophorus*, which includes most of our billbugs. ❺ This conspicuous billbug (***Sphenophorus costipennis***, 8–13 mm) is common along shorelines near the sedge plants in which its larvae develop. ❻ The two northeastern ***Rhodobaenus*** species, (this is ***R. quinquepunctatus***, 5–8 mm) are among our most distinctive weevils. ❼ The **Cocklebur Weevil** (***Rhodobaenus tredecimpunctatus***, 7–11 mm) is a common species on a variety of weeds. Subfamily Erirhininae (Marsh weevils) ❽ The tiny **Duckweed Weevil** (***Tanysphyrus lemnae***, 1 mm) develops within duckweed (*Lemna* sp.) plants. ❾ **Granary Weevils** (***Sitophilus granarius***) and the closely related Rice Weevils (*S. oryzae*) are major pests of stored grains. These small (2–4 mm) and inconspicuous weevils are able to develop within single grains.

FAMILY CURCULIONIDAE (continued) ❶ The **Horsetail Weevil** (***Grypus equiseti***, 5–7 mm) develops within horsetail (*Equisetum* sp.) plants. **Subfamily Ceutorhynchinae (Minute seed weevils)** ❷ Minute seed weevils like this ***Rhinoncus*** are small (usually 2–3 mm) but distinctively shaped, robust beetles. ❸ ❹ Some minute seed weevils are crop pests, but ***Glocianus punctiger*** (3 mm) is a beneficial insect that attacks dandelion seed heads. ❺ The **Cabbage Seedpod Weevil** (***Ceutorhynchus obstrictus***), like this one on a rapeseed (canola) flower, is a European species that has been in western North America for decades, but first appeared in eastern North America just a few years ago. These little (3–4 mm) weevils are now serious pests of canola in much of the continent. ❻ Larvae of this rotund weevil (***Mononychus vulpeculus***) are common in the seed heads of the Blue Flag (*Iris versicolor*). At 5 mm, this is the largest member of the minute seed weevil subfamily. **Subfamily Lixinae (Cylindrical weevils)** ❼ The biggest genus of cylindrical weevils is ***Lixus***, with 18 species of mostly large, attractive weevils. This is ***L. rubellus*** (8–11 mm), which is found on Smartweed (*Polygonuum amphibium*). ❽ ***Larinus planus*** was introduced to North America for the biological control of Canada Thistle (*Cirsium arvense*). **Subfamily Mesoptiliinae (Wedge-shaped weevils)** ❾ The **Red Elm Bark Weevil** (***Magdalis armicollis***, 3–7 mm) is a minor pest of elm. The small teeth on the front corners of the pronotum (just behind the head) are characteristic of *Magdalis*, which includes most of the 16 northeastern species in this small subfamily. **Subfamily Molytinae (Molytine weevils)** ❿ The **White Pine Weevil** (***Pissodes strobi***, 5–9 mm) lays eggs in the terminal buds of pine and other conifers in early spring, causing stunted or drooping terminal leaders by midsummer.

FAMILY CURCULIONIDAE (continued) ❶ **White Pine Weevil** (***Pissodes strobi***) adults mate and feed in early spring, leaving characteristic feeding punctures that often lead to conspicuous resin flow. ❷ ❸ The **Plum Curculio** (***Conotrachelus nenuphar***, 5–7 mm) is somewhat misnamed, since it belongs to the large genus *Conotrachelus* rather than the genus *Curculio*. Larvae, which are legless and somewhat C-shaped like other weevil larvae, develop in plums and related fruits, causing premature fruit drop. ❹ This ***Conotrachelus anaglypticus*** (3–5 mm) does a fair imitation of a bird dropping when it pulls in its appendages and takes on this typical pose. ❺ Several species of pine weevils are forestry pests, including the Pales Weevil (*Hylobius pales*, 7–10 mm), the similar Pine Root Collar Weevil (*Hylobius radicis*, 10–12 mm) and the **Seedling Debarking Weevil** (***Hylobius congener***, 6–9 mm) shown here. These beetles can kill seedlings by feeding all around the base. ❻ Adults of this attractive cylindrical weevil (***Lepyrus palustris***, 9–10 mm) spend the winter under bark, and can often be found in numbers under willow bark. **Subfamily Entiminae (Broad-nosed weevils)** ❼ Three species of the genus ***Polydrusus***, including the **Pale Green Weevil** (***P. impressifrons***, 5 mm) have been accidentally introduced from Europe to North America, and they are now among the most common weevils on poplar, willow and birch trees in southeastern Canada and the northeastern United States. ❽ The **European Snout Beetle** (***Phyllobius oblongus***, 5 mm) is now common in northeastern North America, feeding on a variety of trees. All three North American species in this genus were accidentally introduced from Europe.

FAMILY CURCULIONIDAE (continued) ❶ The 15 northeastern species of *Otiorhynchus*, including the **Black Vine Weevil** (***Otiorhynchus sulcatus***, 8–11 mm) Strawberry Root Weevil (*Otiorhynchus ovatus*) and several similar stout, flightless species, were all accidentally introduced to North America, and all are known only from females. Larvae eat roots, and some species are significant pests. ❷ The **Rough Strawberry Root Weevil** (***Otiorhynchus rugostriatus***, 6–8 mm) feeds on foliage as adults, but does more significant damage while feeding on roots in the larval stage. Hosts include strawberry, rhododendron and raspberry. ❸ The **Imported Long-horned Weevil** (***Calomycterus setarius***, 4 mm), was accidentally imported from Japan and is now common throughout the eastern United States. ❹ The **Asiatic Oak Weevil**, ***Cyrtepistomus castaneus*** (5–6 mm), is a common introduced pest of oak. ❺ This introduced broad-nosed weevil, the **Diaprepes Root Weevil** (***Diaprepes abbreviatus***, 10–19 mm) has become a serious pest of a variety of plants, especially citrus, in the southern U.S. (its host range exceeds 270 species), but it has not yet spread into the northeast. ❻ The **Clover Root Weevil** (***Sitona hispidulus***, 3–5 mm) is one of several introduced (European) *Sitona* species that are now common and widespread pests of Fabaceae. Adults eat foliage; larvae feed on roots. **Subfamily Baridinae (Flower weevils)** ❼ Weevils in the subfamily Baridinae are called flower weevils because adults of many genera, such as this ***Odontocorynus*** (3–5 mm), can be found in flowers. This is the largest subfamily of weevils, with over 500 North American species. ❽ ***Glyptobaris lecontei*** (4 mm), the only North American *Glyptobaris*, occurs on hawthorn (*Crataegus* spp.) throughout the eastern United States. **Subfamily Cyclominae (Underwater weevils)** ❾ This underwater weevil (***Listronotus lutulentus***, 4 mm) was found on arrowhead plants (*Sagittaria* sp.) in the middle of a shallow lake. Most underwater weevils develop in aquatic plants, but one species is a pest of carrot and dill.

FAMILY CURCULIONIDAE (continued) ❶ ***Listronotus***, the only eastern North American genus in the Cyclominae, is usually found in wetlands. This ***L. delumbis*** (4 mm) was photographed on a Broad-leaved Arrowhead (*Sagittaria latifolia*) in a pond. **Subfamily Cossoninae (Bark weevils)** ❷ Bark weevils, like this ***Cossonus platalea*** (5–7 mm), often occur under bark of trees recently killed by bark beetles (Scolytinae). These distinctively shaped beetles are frequently found in large numbers in sawdust-packed galleries running across the wood grain. **Subfamily Hyperinae (Clover and alfalfa weevils)** ❸ All 11 northeastern species of clover and alfalfa weevils belong to the genus ***Hypera***, including several introduced species that are considered serious crop pests. This is the **Lesser Clover Leaf Weevil** (***H. nigrirostris***, 3–4 mm), a common species on red clover. ❹ The **Clover Leaf Weevil** (***Hypera punctata***, 5–10 mm) damages clover and alfalfa as it feeds on the growing tips and skeletonizes leaves. **Subamily Scolytinae (Bark beetles)** ❺ Bark beetles normally feed in dead or dying trees, which are sometimes dead or dying because of mass attack by adult bark beetles and associated fungi. Some of the most important bark beetle pests of conifer trees are in the genus ***Dendroctonus*** (3–9 mm), such as this ***D. valens*** (**Red Turpentine Beetle**). ❻ Some bark beetles are called engraver beetles because they make engravings under bark, often with a "centipede" shape. The body of the "centipede" is the adult beetle's gallery, and the "legs" are the narrower larval galleries. This is a dead willow tree with the bark removed. ❼ Some bark beetles feed on fungi called ambrosia fungi that they introduce into their galleries. This ambrosia-feeding bark beetle ***Trypodendron retusum*** (4 mm) occurs on poplar. ❽ The **Chestnut Brown Bark Beetle** (***Pityogenes hopkinsi***) is a small (about 2 mm) beetle often found in large numbers on dead or dying White Pine (*Pinus strobus*). Each male bores through the bark to make a nuptial chamber where he will be joined by several females. ❾ Peeling bark from a recently damaged spruce tree exposed this bark beetle, ***Dryocoetes autographus*** (3–5 mm).

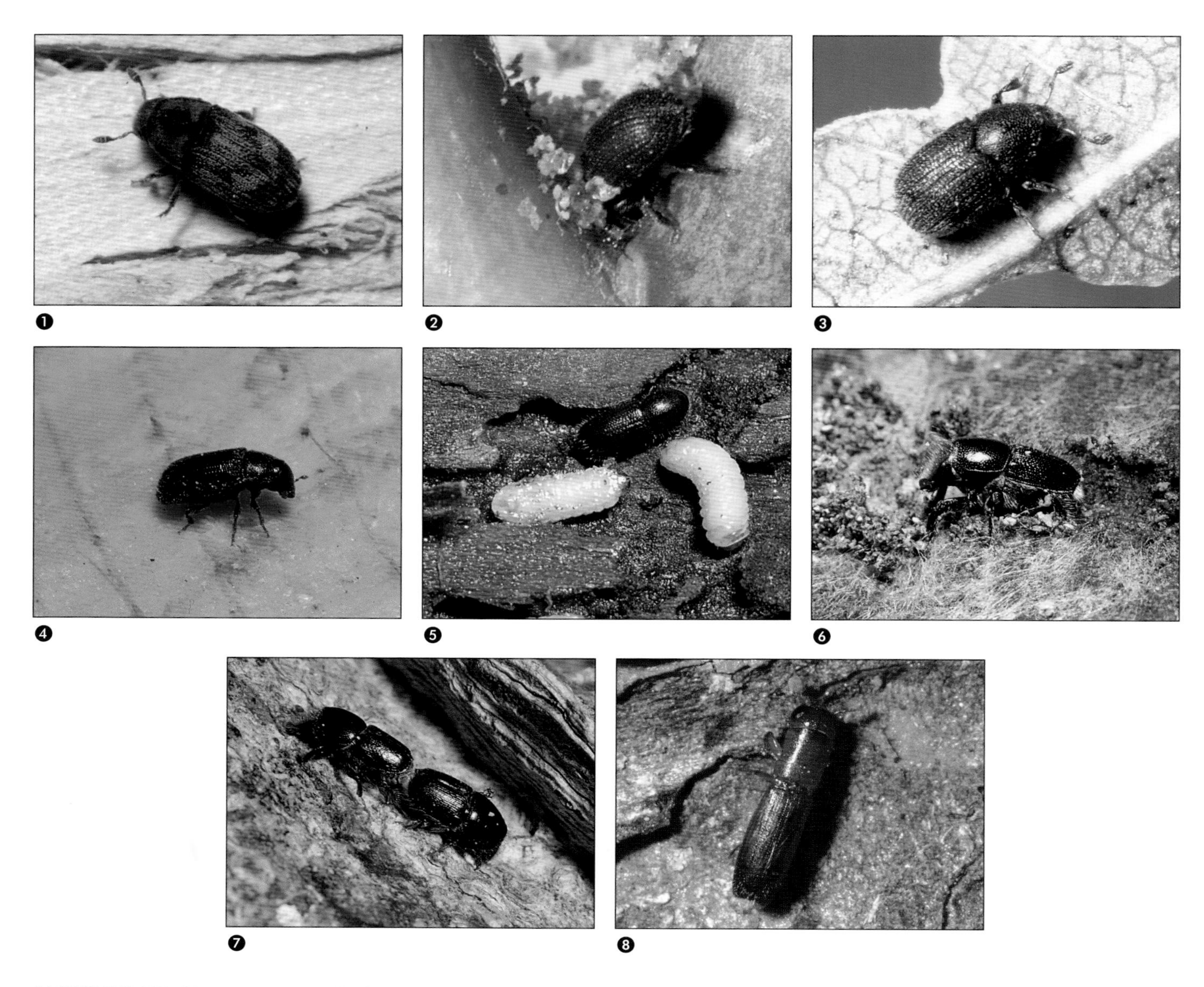

FAMILY CURCULIONIDAE (continued) ❶ This bark beetle (***Hylesinus aculeatus***, 2–3 mm) has a distinctive pattern created by a dense covering of flattened hairs or scales. Larvae develop on ash trees. ❷ ❸ This little (2 mm) scolytid (***Phloeotribus scabricollis***) is burrowing into a Hop Tree (*Ptelea trifoliata*) on Ontario's Pelee Island. *Phloeotribus* species have an asymmetrical, long-lobed antennal club quite unlike the round and compact club of most bark beetles. ❹ ❺ The **Pine Shoot Beetle** (***Tomicus piniperda***) is common in early spring on recently cut pine stumps and logs where they lay eggs in galleries under bark. Larvae tunnel out at right angles to the adult gallery, ultimately emerging through the bark as new adults; these adults fly to the crowns of living pine trees where they feed in the shoots and cause serious damage. Pine Shoot Beetles are native to Europe and Asia, and were accidentally introduced to North America in the early 1990s. ❻ Males of some ***Scolytus*** species, like this **Hackberry Bark Beetle** (***S. muticus***) burrowing under the bark of a fallen Hackberry (*Celtis occidentalis*) tree, have dense hairs on the front of the head. ❼ The female of this pair of **Hackberry Bark Beetles** will make a short egg gallery under the bark. Larvae will mine across the grain, ultimately pupating in a pupal cell deep in the wood. **Subfamily Platypodinae (Pinhole borers)** ❽ Larvae of pinhole borers make distinctive round holes in living trees, where the larvae live on ambrosia fungi in their galleries. All North American species are in the genus ***Platypus*** (2–8 mm). This group is relatively southern, and eastern North American *Platypus* species are not known north of Indiana.

FAMILY ATTELABIDAE Subfamily Attelabinae (Leaf-rolling weevils) ❶ ❷ ❸ Leaf-rolling weevils, like this ***Attelabus bipustulatus*** (3–4 mm), cut out pieces of leaves and roll them up into compact cigar-like cylinders. One egg is placed in the moist center of each cylinder, and larvae feed on the leaf tissue. These three photographs show the elaborate cutting and rolling of an oak leaf, and a mating pair on the completed roll. This is the most common of the four northeastern species of leaf-rolling weevils, and is found from Ontario to Florida. ❹ ***Himatolabus pubescens*** (4–6 mm) is a characteristically fuzzy species that usually makes its leaf rolls from hazelnut or alder leaves. ❺ ***Homoeolabus analis*** females make cigar-like leaf rolls from pieces of oak leaves; larvae develop in the rolls. This is our only *Homoeolabus*; it is found throughout the eastern U.S. and southeastern Canada. Subfamily Pterocolinae (Thief weevils) ❻ Where there is thrift there is theft, and whenever an insect sequesters a resource it seems there is some other insect waiting around to steal it. This little green beetle (***Pterocolus ovatus***, 3 mm) waits until a leaf-rolling weevil (one is out of focus in the background of this picture) has made a leaf-roll and put her egg safely inside. The *Pterocolius* then lays her eggs in the same leaf roll; her larva will kill the original inhabitant and develop in the roll. Subfamily Rhynchitinae (Tooth-nosed snout weevils) ❼ **Sunflower Head-clipping Weevils** (***Haplorhynchites aeneus***, 4–7 mm) develop in the flower heads of sunflowers and related plants, causing the flowers to droop and fall to the ground. The weevil larvae develop in the decapitated flower head. ❽ Larvae of ***Eugnamptus angustatus*** are miners in dead leaves; adults are common on oak leaves. ❾ The **Rose Curculio** (***Merhynchites bicolor***) develops in the reproductive parts of rose flowers. Adults feed on flower buds and the tips of shoots.

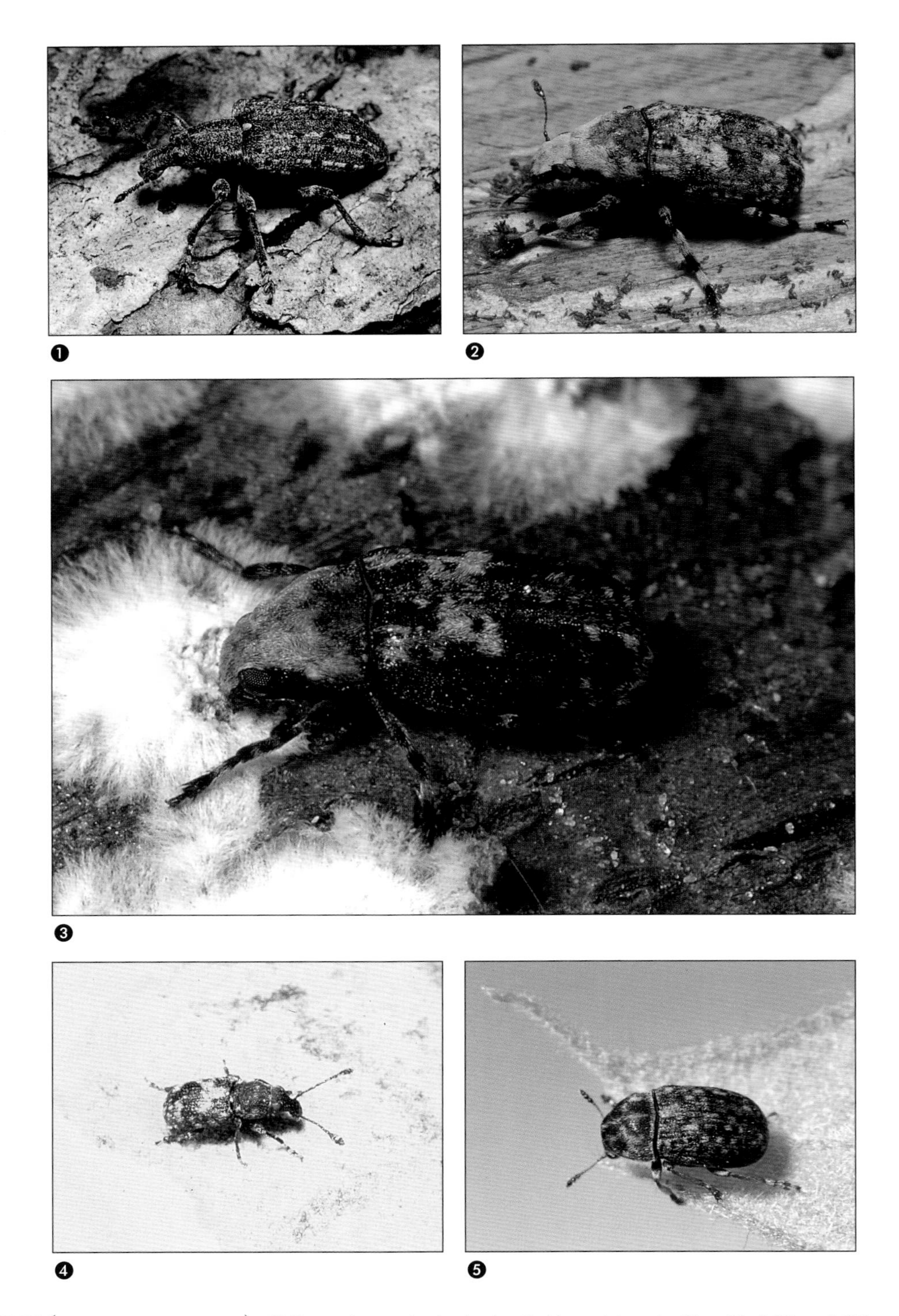

FAMILY ITHYCERIDAE (NEW YORK WEEVILS) ❶ The only species in the family Ithyceridae, the **New York Weevil** (***Ithycerus noveboracensis***, 12–18 mm) damages various hardwoods by feeding on twigs and buds. **FAMILY ANTHRIBIDAE** (FUNGUS WEEVILS) ❷ ❸ Fungus weevils have a broad, flat beak and (unlike similar true weevils) the antennae are not elbowed. This is ***Euparius marmoreus*** (5–9 mm), the most commonly encountered fungus weevil. ❹ Many fungus weevils are small, wood-boring beetles, like this ***Allandrus populi*** (3–4 mm) that develops in aspen trees. Adults of this species are sometimes abundant on damaged aspen branches. ❺ Fungus weevils in the genus ***Trigonorhinus*** (2–3 mm) are sometimes found on corn smut.

FAMILY BRENTIDAE (PRIMITIVE WEEVILS OR STRAIGHT-SNOUT WEEVILS AND PEAR-SHAPED WEEVILS) ❶ ❷ Primitive weevils are distinctively shaped wood boring beetles often found under bark. Females and some males have a long, weevil-like beak, but some males have a short beak with conspicuous mandibles. The only eastern species, the **Oak Timberworm** (***Arrhenodes minutus***, 7–22 mm) lays eggs in damaged, but living, trees. ❸ This big-jawed male **Oak Timberworm** (***Arrhenodes minutus***) is guarding a smaller female. ❹ The **Hollyhock Weevil** (***Rhopalapion longirostre***), and related small (1–3 mm) pear-shaped weevils are sometimes treated as a separate family (Apionidae). Hollyhock Weevils are found only on hollyhocks, where the adults feed on foliage and larvae feed on seeds.

❶

❷

❸

FAMILY STYLOPIDAE ❶ This *Polistes* wasp (Vespidae) has been parasitized by ***Xenos pecki***, a twisted-winged parasite. Females (which never leave the host) and pupae can be seen projecting between the tergites of parasitized hosts. Although probably more closely related to flies than beetles, the life cycle of twisted-winged parasites is similar to some beetles, and both beetles and Strepsiptera fly with their hind wings. For these reasons, Strepsiptera have been traditionally grouped with or near the beetles. ❷ This wasp (*Ancistrocerus adiabatus*, Vespidae) has been parasitized by a number of twisted-winged parasites (***Pseudoxenos*** sp.). The narrow body sticking out of the wasp's abdominal plates is a female who will never leave the host. The fatter object causing a conspicuous bulge between the tergites farther back on the abdomen is a male pupa due to hatch into a winged adult like the one crawling beneath the left side of this wasp's abdomen in search of a female. ❸ This male strepsipteran (***Pseudoxenos*** sp.). has popped the cap off its puparium and is trying to emerge from its eumenid wasp host. The much flatter structure projecting from between the preceding abdominal segments is a female strepsipteran. ❹ This adult male twisted-winged parasite (***Pseudoxenos*** sp.) is seeking a sessile female in the abdomen of this potter wasp. The large, white wings are the parasite's hind wings; the front wings are reduced to solid knobs (halters).

❹

CHAPTER 11

Flies, Scorpionflies and Fleas

TRUE FLIES (ORDER DIPTERA)

Some insects are recognized by almost everyone as "flies" – a general impression reflected in the common names of familiar insects such as flower flies, black flies, robber flies and horse flies. Common concepts of what a fly should look like are also reflected in some misleading common names for flylike insects in other orders, such as scorpionflies and whiteflies; these differ from true flies in having two pairs of wings. Aristotle, one of the great thinkers of ancient Greece, gave true flies the Greek name "Diptera," meaning "two winged." He recognized that this group of insects is different from other winged insects in having only a single pair of wings, and noted "no insect with only two wings has a sting in the rear." However similar a fly might appear to a bee or wasp, flies can't sting (although some can inflict a painful bite). For Aristotle's time, over 2,000 years ago, this was a remarkably sound example of the predictive value of correct classification.

With the exception of the rarely seen male twisted-winged parasites (which have only the wings of the third thoracic segment developed), orders of winged insects other than Diptera have wings on both the second and third thoracic segments. Loss of one or both pairs of wings has occurred within several orders, so some Hemiptera (male scales), a few barklice, a few mayflies and even some grasshoppers have only a single pair of wings. Diptera, however, is the sole order in which the only the second pair of wings is uniformly lost or vestigial. Flies only have wings on the second thoracic segment, which is essentially a big box of flight muscles dwarfing the rest of the thorax. The third thoracic segment, which supports the second pair of wings in most insects, is small because in flies the second pair of wings is reduced to a pair of little knobs, called halters. Halters were once thought to function as balancers, like a tightrope walker's pole. If you damage a fly's halter, it will indeed affect its balance, but this is because the halters sit in a nerve-filled socket; from here the halters send complex messages to the fly about its flight position. So the halter is really more like the gyroscopic flight indicator on an airplane than the simple balance pole of a tightrope walker.

Adult flies frequently fuel up for flight at sugar stations, including flowers or, more often, honeydew deposits left by aphids or related phloem-sucking bugs. Most higher flies have spongelike lower lips (made up of the fused labial palps) that serve perfectly to pick up honeydew and other semiliquid foods, such as feces, but adult flies never have chewing mouthparts like beetles or grasshoppers. Fly larvae are legless feeding machines, usually socking away enough protein to allow the resulting adults to spend their brief but active lives in search of sugar, a mate and a good place to deposit eggs or maggots. Ephemeral though they may be, adult flies are the most conspicuous and ubiquitous of all the animals sharing our terrestrial environment.

Although the 25,000 or so described species of flies in North America is a few thousand fewer than our described species of beetles, flies are less well-known than beetles and may yet emerge as our largest order. In the northern half of North America there are more species of flies than there are species of any other order of living things, and many species of flies are overwhelmingly abundant. This is a tremendously important group of organisms, and it behooves every naturalist to know a bit about the major families of Diptera.

The order Diptera includes two huge suborders: Nematocera (literally "long horned") and Brachycera ("short horned"). If you can picture a mosquito on one hand and a horse fly or house fly on the other, then you are already familiar with these two suborders. Members of the Nematocera are like mosquitoes in being generally long-legged, fragile-looking flies with long, multisegmented antennae. Brachycera, on the other hand, are generally more robust flies with short, few-segmented antennae like those of a house fly or horse fly. Although these two suborders are easily recognized, only the Brachycera is a "real" group, and the Nematocera is just all the flies left over when the short-horned flies are taken out. The suborder Nematocera will undoubtedly be eventually broken into several more natural suborders, one or more of which will prove to be more closely related to the Brachycera than to other "Nematocera." In the meantime, the old division of the Diptera into the long-horned and short-horned groups is convenient.

SUBORDER NEMATOCERA
Midges, mosquitoes and other long-horned flies

Crane flies (Tipulidae and similar families)

Perhaps the most conspicuous of the very common Nematocera are the gangly-legged flies commonly called crane flies in North America, but usually called daddy long legs in Britain (**Tipulidae**, page 425; sometimes divided into the Tipulidae, Limoniidae, Pediciidae and Cylindrotomidae). Many people think crane flies are just big, old mosquitoes – quite a misconception since no mosquito or other winged insect grows in size once it has molted to the adult stage. Crane flies don't suck blood, although some crane flies do have long mosquito-like beaks used for taking nectar. If in doubt, have a close look at the top of a crane fly or mosquito thorax, just above the wing base. Crane flies have a characteristic V-shaped suture, while mosquitoes are covered with scales that densely clothe both the proboscis and the wing veins.

Crane flies are most common in cool, shady places, but they occur in a wide variety of environments, including

Crane flies, *Limonia*, page 425, caption 3.

seashores, mountaintops and deserts. The tipulids that most often attract the attention of non-entomologists are probably the familiar leatherjackets, very large and conspicuous crane flies named for the thick-skinned larvae that feed on roots and other underground plant material. Introduced European leatherjackets (*Tipula paludosa* and related species) have recently become common pests of turf in northeastern North America. Like all Diptera larvae, leatherjacket larvae are legless, and like other crane fly larvae, they have a reduced head that is at least partially retracted into the body.

It is hard to point to the leatherjackets or any other particular kind of crane fly as being "typical" because there are at least 15,000 species of crane flies (1,500 in North America) ranging in size from a few millimeters to several centimeters, and ranging in color from brown to bright orange and black. Biologically, the family shows a fascinating diversity, perhaps best illustrated by briefly summarizing the life histories of a few of the most common, or most interesting, crane flies.

Almost every kind of aquatic or semiaquatic habitat is home to some kinds of larval crane flies, including many that eat decaying plant material and a few that eat other insects. Aquatic species usually have a single pair of spiracles at the tip of the abdomen, each spiracle surrounded by a number of lobes fringed with water-repellent hairs. Crane fly larvae often hang from the water surface by these glistening rosettes of hydrophobic hairs, their spiracles high and dry and their heads down in the water. The lobes surrounding the spiracles can close up, capturing a bubble of air over the spiracles before the larva submerges.

Some unusual aquatic crane flies (in the genus *Antocha*) lay eggs along the waterline of rocks surrounded by clean, running water. *Antocha* eggs hatch into aquatic larvae that forgo the use of open spiracles, and instead respire by using the entire body as a gill, absorbing oxygen through the body wall. These ubiquitous riffle inhabitants use silk and debris to build tubular shelters from which they feed on the fine particulate organic matter that accumulates on submerged rock surfaces.

"Unusual" is a term we might use for many crane flies, and the observant naturalist can find unusual crane flies in a wide variety of habitats. For example, if you dig around some streamside sand and gravel you will probably find some golden-haired larvae (*Hexatoma*) that morph readily from club-shaped to cylindrical, then back again. These predacious larvae move through the substrate by pumping out one end of the body to form a swollen anchor, then pulling the rest of the body along to the anchor before contracting, poking ahead, and repeating the process. If you dig around in Sphagnum moss you might find another kind of crane fly larva with long, finger-like processes all over its body (*Phalacrocera*, in the small subfamily Cylindrotominae), and if you rip apart some moist logs you will find yet another unusual type of crane fly larva (*Ctenophora*), modified for life in that relatively hard substrate. Each of these various forms of larvae has a soft, legless body with a head capsule excavated at the back and retractable into the body.

Perhaps the most unusual of the unusual crane flies are those with wingless adults. Next time you are out on a relatively warm winter day, watch for stout little (5 mm or so) wingless crane flies strolling on the surface of the snow. Snow-walking crane flies (several species of *Chionea*) have lost their wings but retain their distinctive halters, making it easy to separate them from other wingless winter insects such as the similarly sized snow scorpionflies and the much smaller snowfleas. Wingless flies occur in many groups, especially on mountaintops and islands, but the wingless snow-walking crane flies are the ones most likely to be noticed in eastern North America.

Although the vast majority of insects that look like crane flies are indeed Tipulidae, a few other families have the same leggy, primitive appearance. The ones that fly on nice days right into the winter, for example, usually belong to a family called the winter crane flies (**Trichoceridae**, page 428). Look for them dancing over compost heaps – where the larvae live – on mild days throughout the winter months.

The phantom crane fly family (**Ptychopteridae**, page 428) has only a few species, some of which are common near nutrient-rich springs, fens or stream margins where their long, rat-tailed larvae breathe using retractile, snorkel-like breathing tubes. The peculiar pupae, which have one long breathing tube arising from one side of the thorax, are found in the same habitats. The most common ptychopterids are leggy, crane fly–shaped flies banded in black and white that seem to float through the air with their legs outstretched, challenging the observer to maintain focus on their disruptively colored body for more than a few seconds. These large flies are strikingly beautiful insects, truly phantom-like as they drift in and out of your vision.

Crane flies, winter crane flies and phantom crane flies are relatively common flies, in contrast to a fourth "crane fly" family called the primitive crane flies (**Tanyderidae**). Primitive crane flies look much like some species of common crane flies with attractively pigmented wings, but differ in wing venation (crane flies have two complete anal veins, primitive crane flies have one). Our one eastern species, *Protoplasa fitchii*, is very rare (or at least rarely seen by entomologists), but has been found developing in sand and gravel along the edges of a few rivers from Quebec and Nova Scotia south to Florida.

Bloodthirsty females: The biting Nematocera

Although both the long-horned and short-horned flies include blood-feeding groups, it is among the former that this habit seems most general. Among biting Nematocera, as well as the horse flies, deer flies and snipe flies among the Brachycera, only the female exhibits bloodthirsty tendencies. Blood serves to feed the protein-hungry developing eggs of the female, while the males feed on carbohydrate-rich foods such as nectar and honeydew. Most of us have been bitten by at least three common families of Nematocera: the black flies, the mosquitoes and the punkies.

Punkies or biting midges (Ceratopogonidae)

When a North American complains about sand flies, he or she is almost certainly talking about members of the family **Ceratopogonidae** (page 428), tiny flies also known as biting midges, punkies, no-see-ums and several unprintable local names. The term "sand fly" can be a bit misleading, since it is also used to refer to a primarily tropical group of biting flies in the moth fly family (Psychodidae). "Punky" is a unique name, although "no-see-um" is certainly descriptive of some of these vicious little biters, which are often so small they pass right through window screens.

Ceratopogonid midges, like these *Atrichopogon* on a white water lily, are significant pollinators.

Punky larvae look much like common nonbiting midge larvae and are often found in the same moist or wet places. Most, including all the aquatic ones, differ from midge larvae in lacking prolegs; others have spines and projections all over their bodies. Aquatic punkies are long and slender, and move with a rapid whiplike motion that makes them easy to recognize in the field. Some ceratopogonid larvae are carnivores, while others eat detritus. Adult female punkies usually need a protein meal, such as invertebrate or vertebrate blood, for egg development.

Only a few genera of punkies attack humans, but these can cause a lot of grief with their tiny mouthparts. Each of the seemingly countless punkies which tormented you that whole evening at the beach last summer homed in on your hot breath (your carbon dioxide), landed on your exposed parts, then thrust a pair of serrated blades (laciniae of the maxillae) into your skin to serve as anchors. That done, she snipped a hole in your skin and lacerated the underlying capillaries with her tiny, scissor-like mandibles and proceeded to "spit" into the open wound using a strawlike "tongue," the hypopharynx. Her "spit," or salivary fluid, prevented your blood from clotting and also caused the allergic reaction that makes fly bites so itchy. Your blood was imbibed through a channel behind the fly's upper lip (labrum), possibly until the little sucker was swollen to multiples of her original size.

As might be expected from an organism with the bad habit of spitting into your blood, punkies can transmit disease, and tropical punkies are vectors for disease-causing viruses and nematodes. Fortunately, unlike mosquitoes, punkies are not significant carriers of human diseases in North America. They do carry some mammal diseases, such as equine onchocerciasis, equine encephalitis and the bluetongue virus of sheep and cattle; they also transmit several bird diseases, and at least one common North American frog-sucking punky (*Forcipomyia fairfaxensis*) has been implicated as a carrier of a frog virus.

Many punkies feed only on insect blood. *Meloehelea*, for example, is an aptly named subgenus of Ceratopogonidae made up of species that bite *Meloe* and other blister beetles (Meloidae). These otherwise well-protected, generally toxic beetles are often plagued by swarms of the minute beetle-biting female flies, which presumably benefit from imbibing the toxin-laced meloid blood. Blister beetles seem to be the most common insect victims of punkies in northeastern North America, but in warmer areas dragonfly and damselfly wing veins are often festooned with feeding ceratopogonids, and I have even seen punkies biting crane flies.

Punkies that attack vertebrates or large insects such as dragonflies or walkingsticks are bloodsucking parasites, but punkies that bite small insects are often bloodsucking predators that kill their hosts. Many predacious punkies preferentially attack swarming males of other flies such as chironomid midges, but a few feed on males of their own species. Females of one genus (*Probezzia*) are a bit like gory vampires, as the female flies into a male swarm, mates, then pierces her mate through the head and sucks out his brains (along with his other body contents).

Some biting midge females use their biting mouthparts to puncture pollen rather than other animals, and both sexes of many species feed on nectar. Flower-visiting punkies are often important pollinators. Water lily blossoms, for example, are often crowded with tiny *Atrichopogon* species, and cacao (the tropical tree that chocolate is obtained from) is quite dependent on pollination by a related ceratopogonid fly. Next time you are scratching your punky bites and cursing ceratopogonids in general, take a bite of a chocolate bar and reflect on the fact that the Ceratopogonidae made that chocolate possible.

Mosquitoes (Culicidae)

One group of flies you have probably exchanged some body fluids with is the **Culicidae** (mosquitoes; page 429). Adult female mosquitoes penetrate people (and other victims) with essentially the same biting equipment as their tiny punky cousins, but the bits and pieces are elongated into the mosquitoes' trademark long "beak" or proboscis. In contrast to the punkies, which crudely slice your skin with their scissor-like mandibles, mosquitoes have long needle-like mandibles and long, serrated maxillae that, together with a central two-barreled tube made from the upper lip plus the tongue (labrum-epipharynx plus the hypopharynx), are inserted through your skin and delicately snaked through to a venule or arteriole.

Like punkies, mosquitoes stop your blood from clotting by secreting anticoagulant saliva that may cause an itchy welt and, worse yet, might carry some unwelcome microscopic guests. Different mosquitoes harbor different microbes, so correct identification of the mosquito biting you can tell you a lot about the potential impact of the bite. Mosquitoes capable of transmitting yellow fever look quite different from the mosquitoes that carry malaria, which in turn differ in appearance from the vectors of encephalitis and other diseases. Mosquito-borne diseases are major problems throughout the tropics and a growing concern in North America, but you can still practice mosquito identifications in your backyard with minimal risk of an infectious injection. Just expose an arm outside on a pleasant spring evening and look closely at the two-winged visitors to your warm and carbon-dioxide-rich aura. But to be on the safe side, swat them before they bite. "Minimal risk" is a relative idea and, as we shall see, it changes from year to year and place to place as new mosquito-borne diseases appear and old ones reappear.

Most of the mosquitoes that inevitably appear as you relax in your backyard on spring and summer evenings are likely to be bloodthirsty females of the large genus *Aedes*, relatively large mosquitoes with a pointed abdomen. Similar to other mosquitoes, they are covered by scales much like those found on butterflies and moths, and the scales of some species even impart color just the way that scales give color to butterfly wings. Odds are that your visitor will be clothed in the alternating white and black colors characteristic of many *Aedes* species.

All mosquitoes develop in standing water where the larvae can breathe at the surface, and most feed on microorganisms and other finely particulate organic material, either by filtering surface film or by wriggling down into the water in search of a snack. *Aedes* larvae fall in the latter

Yellow Fever Mosquito (*Aedes aegypti*) larva and pupa, page 430, caption 3.

group, and often abound in temporary, organically rich waters like woodland pools. *Aedes* mosquitoes lay their eggs one by one, often in patches of low ground subject to periodic inundation, where the eggs hatch in response to flooding. In the north, huge masses of *Aedes* eggs can accumulate at perfect patches of pond margin at just the right level and just the right aspect to warm up quickly in spring. The eggs sit waiting until it rains or meltwater fills the depression the following spring and, when conditions are right, they hatch and the larvae develop quickly in this temporary habitat.

The eggs of *Aedes* have some remarkable ways of determining when conditions are right to hatch. Some respond to water temperature, but others hatch in response to the low oxygen tension created when microorganisms become active in the water, indicating that there will be food for the hatching larvae. Some *Aedes* lay their eggs and then die on the spot. When the dead bodies of adult mosquitoes start to decay in the floodwaters of the following spring, their eggs hatch into larvae that feed in the microbially enriched water, closing a cycle both macabre and elegant.

The mosquitoes that develop from these sit-and-wait type eggs can be incredibly abundant near flood plains, salt marshes or snow pools. Some *Aedes* species are especially adept at rapidly developing in ephemeral meltwater pools during the short northern summers, and every northerner seems to have stories of mosquito masses so thick that exposed body parts get blackened with hungry hordes within seconds. Saltmarsh mosquitoes (*Aedes sollicitans* – sometimes called *Ochlerotatus sollicitans*) are common in brackish water along the entire east coast of North America and sometimes occur in brine dumps near inland salt mines; they are also famous for occurring in vicious hordes that rival the density and ferocity of their northern relatives.

Aedes species are the major carriers of several serious viral diseases such as dengue and yellow fever. Yellow fever, a virus now normally found only in monkeys, is a particularly horrible disease that repeatedly devastated eastern North American cities for centuries after it arrived from Africa along with its mosquito vector. Yellow fever is still present in tropical America and Africa, but if you are traveling to yellow fever areas, you are probably vaccinated against this disease. We have had no major urban outbreaks of yellow fever in the Americas for over 50 years, but many areas now have high densities of *Aedes aegypti*, the Yellow Fever Mosquito, waiting to spread the virus when conditions are right. In many cases *A. aegypti* has reinvaded areas in Latin America where the species was eradicated years ago in the fight against yellow fever. Ironically, the source of some of those reinvading Yellow Fever Mosquitoes seems to have been eastern North America, where *A. aegypti* remains common in the south and along the coast north to New England, although yellow fever has long been eradicated in the United States.

The spread of certain *Aedes* species, especially *A. aegypti*, has been accompanied by millions of cases of a worrisome mosquito-borne disease called dengue. People often pick up dengue viruses ("breakbone fever") from mosquitoes in holiday hot-spots like the Caribbean or Rio de Janeiro. Classical dengue fever, which causes fevers, sore joints and other problems, but is not usually fatal, is widespread in warm parts of the world, and past outbreaks have occurred in the southern United States and Puerto Rico. A closely related life-threatening disease called dengue hemorrhagic fever was first recognized in 1954, and has been spreading in the Americas since the mid-1970s.

Although we do not currently have a dengue problem in North America, we do have mosquitoes perfectly able to transmit the virus. *Aedes albopictus*, the Asian Tiger Mosquito, was probably accidentally introduced to the southeastern United States with shipments of used tires from Asia (water standing in discarded tires makes a great habitat for the larvae of this species). These strikingly striped mosquitoes

Culex territans, page 431, caption 3.

have now spread all the way to southern Canada, where they are vicious biters of people and animals, and have the ability to transmit not only dengue but also a number of other viral diseases. A similar species with similar habits, the Rockpool Mosquito (*A. japonicus*), was recently accidentally introduced from Japan to the United States, and has also made its way northward to Canada (it first appeared in Ontario in 2004).

The Yellow Fever Mosquito, Rockpool Mosquito and Asian Tiger Mosquito are beautifully pigmented mosquitoes that lay eggs in containers such as rain barrels, old tires and flower vases in cemeteries. *Aedes* larvae, like most mosquito larvae, rest suspended from the water surface by the water-repellent tip of a conspicuous, snorkel-like breathing tube. They periodically wriggle down into the water where they filter feed on aquatic microorganisms using a complex assemblage of mouth brushes. Particles are grabbed by a sort of muscular mustache made of two large brushes inserted on the upper lip (labrum). The brushes are opened by hydraulic pressure, and then pulled shut by muscles, pulling food particles on to combs on top of the mandibles. The mandibles then open, passing the food to other bristles at the middle of the labrum, then close and pick up the food using another bundle of bristles near the base of the mandibles. All this flicking and filtering takes place hundreds of times per minute as they wriggle through microbe-rich water.

Meanwhile, back in your safe suburban yard, you might also enjoy the attentions of a delicate little *Culex* mosquito with a blunt tail end, unlike the pointed abdomen of the more common *Aedes*. The little mosquitoes frequently found inside houses in late fall or early spring are usually *Culex pipiens*, sometimes aptly called the Northern House Mosquito. They are also sometimes called Rainbarrel Mosquitoes, a name that conjures up images of the big wooden barrel my grandfather had under his eavestrough downspout, a natural aquarium full of "wrigglers" (mosquito larvae), "tumblers" (mosquito pupae) and the characteristic raftlike clusters of *Culex* eggs.

Culex species can carry various diseases, including St. Louis encephalitis and the recently arrived West Nile virus. Both of these viruses are normally found in birds and are occasionally transmitted to us by *Culex* mosquitoes.

Although St. Louis encephalitis doesn't usually cause serious problems, it can cause symptoms ranging from mild fever and headaches right through to paralysis and even death. An outbreak of St. Louis encephalitis in the mid-1970s led to well over 2,000 North American cases, with 171 deaths in the United States and half a dozen deaths in eastern Canada.

Recently, some *Culex* species have been in the news because of their role in moving the West Nile virus from birds to people. West Nile virus, unknown in North America before an outbreak in New York in 1999, is usually a relatively mild, flulike disease. Like the flu, however, it can be serious. West Nile virus has caused thousands of illnesses and several deaths in our area, and it serves as a chilling reminder that North America is by no means immune to outbreaks of mosquito-borne diseases. The appearance of this disease should serve as a wake-up call, reminding us of the potential threat posed by accidentally introduced organisms, including exotic mosquitoes and the 100 or so known mosquito-borne viruses that can infect people.

Culex species also carry another well-known parasite found here in northeastern North America, a cosmopolitan nematode called dog heartworm that wise dog owners will certainly have discussed with their vets. These nematodes can kill your pets painfully by clogging up their hearts and pulmonary arteries. The first diseases to be conclusively associated with mosquitoes were human diseases caused by filarial worms (nematodes) that clog up lymph nodes, often causing elephantiasis, a debilitating swelling of the lower limbs or genitals. Fortunately, we have no problems with mosquito-borne human filarial diseases in North America … at the moment.

The world's most important genus of mosquito – in fact the most important genus of insects – includes all the species capable of transmitting human malaria. Mosquitoes in the genus *Anopheles* are easy to recognize by their attractively pigmented wings and exceptionally long maxillary palpi (giving them the appearance of having three long beaks).

When *Anopheles* mosquitoes settle down on your arm and start to bite, they sometimes seem to stand on their heads, quite unlike other common mosquitoes. Watching an *Anopheles* at work, you can be thankful that you are doing this particular type of insect observation at home in a temperate country because, although *Anopheles* species are common in North America, malaria is now very rare here. Malaria used to occur all the way north to southern Canada, and as recently as the 1930s malaria was a major problem in North America. We still have local *Anopheles* species that could serve as carriers (vectors), and people regularly get malaria abroad and bring their infections back to the United States or Canada. The odds of one of those imported cases of malaria being picked up by our local *Anopheles* and starting a North American malaria epidemic are low enough so we can say that our area is effectively free of malaria, even though half the world's population lives in the 100 or so countries that are not.

More than 300 million people suffer from malaria in any given year, every case resulting from *Anopheles* mosquito injections of protist parasites (four different species of the genus *Plasmodium*) that reproduce only inside *Anopheles* mosquitoes. Millions of people die from malaria annually, so check with your doctor before you go off on a tropical nature tour. The drugs he or she will prescribe carry their own risks, but none as serious as malaria. *Anopheles* mosquitoes generally bite between dusk and dawn, so sleeping under a mosquito net (especially one treated with insecticides) can greatly lower your risk.

If you are planning to travel in the tropics, don't be lulled into thinking that malaria eradication programs have made the world safe for tourists. Check the current statistics on the area you plan to visit. Lots of factors, including population pressures, cessation of expensive mosquito control programs, mosquito resistance to pesticides, dam construction, irrigation and deforestation can lead to unexpected problems. Sri Lanka provides a spectacular illustration of the kind of malaria resurgence recently documented in several countries. Malaria was virtually eradicated from Sri Lanka by 1963, but in 1987 there were almost 700,000 cases, almost 200,000 of which were the particularly dangerous kind caused by *Plasmodium falciparum.*

Malaria in much of South America has been on the increase in recent decades as well, with cases in Brazil rising to over 600,000 per year by 1990 and then going up another 35 percent between 1993 and 1995, with lots of drug-resistance and an increasing proportion of *P. falciparum* malaria. Similar statistics are available for most other tropical South American countries, although Ecuador reported a steep drop in malaria cases coincident with increased use of DDT in the 1990s. New and promising malaria vaccines are now being tested, and malaria is being tackled in various places through programs like the "Roll Back Malaria" program launched in 1998 by the World Health Organization, UNICEF, UNDP and the World Bank. Still, *Anopheles* is a genus to be aware, and wary, of.

Malaria is not just a human health problem; it is also an issue in conservation biology. Bird malaria is common in continental North America where birds are adapted to it, but the accidental introduction of mosquitoes carrying bird malaria to Hawaii helped to drive many native bird species to extinction.

Malaria mosquitoes differ from other mosquitoes in more than just their long palpi and other adult peculiarities. *Anopheles* eggs are laid on the surface of the water as in *Culex*, but they are laid singly rather than in rafts. Each egg looks a bit like one of those canoes with big float pads on either side, and ultimately hatches into a larva that lacks the long siphon characteristic of most of our other mosquitoes. *Anopheles* larvae lie stretched out parallel with the surface, suspended from the surface tension. They take both food and air at the surface, so they require access to the surface film with both their spiracles and mouthparts. This presents *Anopheles* larvae with a bit of a physical problem, since the spiracles are on the back (dorsal), and the mouthparts normally face the opposite direction (ventral). Their solution to this problem is to rotate the head a remarkable 180 degrees when feeding. This allows the mouthparts to break the surface film and filter out floating particles at the same time as the dorsal spiracles are at the surface.

Most routinely encountered eastern North American mosquitoes belong to the three common biting genera (*Culex*, *Anopheles* and *Aedes*), but our mosquito fauna includes several other genera, some of which do not even bite. The delicate little *Wyeomyia smithii*, for example, develops inside pitcher plants, where its larvae feed in the microbe-rich soup made up of the plant's trap fluids and the decomposing bodies of its victims. Adults do not bite, at least not in northeastern populations. Southeastern populations of *W. smithii* do bite, and other members of this mostly Neotropical genus are major biters in Central and South America.

Larvae of this mosquito genus (*Toxorhynchites*) are predators; adults do not bite, page 432, caption 2.

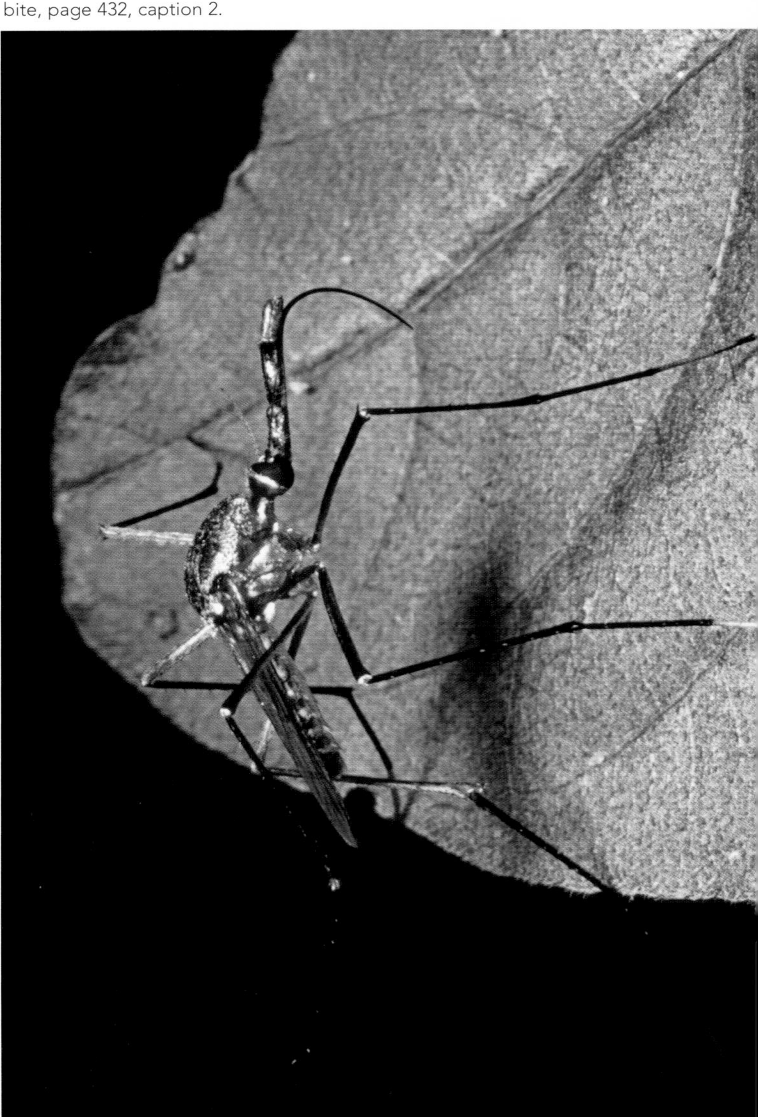

Black fly larvae, page 433, caption 2.

Our largest mosquitoes, with predacious larvae as long as 15 millimeters, belong to the nonbiting genus *Toxorhynchites*. *Toxorhynchites* larvae eat other mosquito larvae in tree holes; in doing so they acquire enough protein to free the adults from a biting lifestyle. Another common local mosquito with an odd lifestyle is a biting species called *Coquillettidia perturbans*; the larvae poke their tail-like breathing tubes into the roots of cattail plants. Both larvae and pupae breathe air from the cattails, in effect using them like giant snorkels. Adults of *C. perturbans* are vicious biters and can be major pests near large marshes.

Not quite mosquitoes: Frog midges and phantom midges (Corethrellidae and Chaoboridae)

Mosquitoes are by no means restricted to biting birds and mammals, and several species seem to prefer cold-blooded victims. For example, *Culex tarsalis* rarely bites people, but these common mosquitoes often throng green frogs. The tiny females of a closely related family, **Corethrellidae** (page 432), are specialized feeders on male tree frogs, which they locate by their loud spring singing. Although infrequently encountered, corethrellids are not rare flies, and a protist disease carried by *Corethrella* (our only corethrellid genus) infects most males of some species of eastern North American hylid tree frogs. The response of *Corethrella* to sound is remarkable – they even respond to tape-recordings of frog song – since no other insect cues in on the calls of its vertebrate hosts (although some parasitic flies are attracted to the singing of cicadas, katydids and crickets). Larval *Corethrella* are aquatic predators of mosquito larvae and similar aquatic prey.

Corethrellids and mosquitoes are very closely related to a family of nonbiting flies called the phantom midges. The term phantom does not refer to ghostlike adults, as is the case for the black-and-white banded phantom crane flies, but instead refers to the transparent larvae of the genus *Chaoborus* (family **Chaoboridae**, page 432) which drift ghostlike and all but invisible through the water column in ponds, lakes and sometimes even in containers such as water barrels. You can often see spectacular numbers of phantom midge larvae by shining a powerful flashlight through the water at night, but during the day you will probably have to dip some water into a jar in order to see them. It's not that they are that small – at 5–6 millimeters they are larger than most midges – it is just that they really are almost transparent.

Once you focus in on a phantom midge lazily drifting up and down with the aid of its silvery anterior and posterior air bubbles, look at the way its head is bent down and dominated by huge hairy, prehensile antennae that are used to grasp planktonic prey. Given that antennae are usually the primary sense organs in insects, an insect that grasps prey with its antennae seems as strange as a vertebrate that grasps prey with its eye.

The planktonic habits of our most common phantom midges, in the genus *Chaoborus*, are unusual since most aquatic insects occur on substrates or the water surface. One atypical member of the Chaoboridae (*Eucorethra underwoodi*) occurs at the surface of temporary pools in coniferous forests, where it feeds mostly on terrestrial insects trapped in the water surface.

Black flies (Simuliidae)

Compact, humpbacked female black flies (**Simuliidae**, page 432) have a propensity to occur in enormous numbers near the flowing waters in which the larvae develop, often descending in hungry hordes on any tiny bit of your exposed flesh where they can collectively cut into your skin in a manner akin to that described earlier for punkies. The bites do not hurt (black fly saliva includes an anesthetic), but nonetheless can result in pools of blood and itchy welts all around your belt line, cuffs and neck. Aside from all that, the constant activity of hundreds of black flies crawling over your skin, repeatedly landing and taking off, and noisily buzzing around your ears is enough to make you wax nostalgic about past experiences with swarms of relatively delicate mosquitoes.

Fortunately, both the seasonal and daily activities of the most annoying black flies tend to be very limited. Seasonal activity seems to be timed to try the dedication of early season trout fishers, and daily activity is just that – daily. Unlike mosquitoes, many of which are most active in the evening or after dark and will usually bite indoors or out, black flies are daytime insects and rarely bite indoors. The males, which don't bite, often form conspicuous swarms. As is often the case among insects that form male swarms, the males have huge eyes with big upper portions made up of conspicuously enlarged facets. Swarming males of several fly families have similarly enlarged eyes that usually take up much of the head, often meeting above the antennae.

After a female black fly has flown into a bunch of big-eyed males with predictable results, she proceeds to lay strings or masses of eggs in running water, often on objects such as branches or leaves dangling in the current. Black flies are associated with moving water, such as your favorite trout stream, because of the unique habits of their larvae. Most black fly larvae look a bit like old-style punching bags, fat at the bottom and narrow at the top. Just as a punching bag stays in place while it is knocked backward, thanks to a heavy bottom fixed to a stand, a black fly larva stays in place while it is buffeted and bent backward by flowing water thanks to a fat bottom firmly hooked to a little pad of silk previously placed on the substrate by the larva. The head end of the larva leans with the current, holding out two fans that arise from the labrum or upper lip. The labral fans are very large in species that live in slow waters, small in inhabitants of torrential waters, and entirely reduced in some species that live in headwater springs. As the current flows around the larva's head, tiny particles of food are strained out by the fans, which are constantly being folded through the other mouthparts to ingest the food.

Black fly larvae pupate in slipper-like cocoons on the same submerged surfaces on which they spend their larval lives. Each pupa ultimately releases an adult fly in a glistening bubble of air. When you see hordes of black flies rising off the surface of a creek or river, each has just ridden to the surface in its own private air-bell, which exposes a dry and ready-to-fly adult as it bursts at the surface.

Most black fly species prefer to feed on bird blood rather than that of humans. Nonetheless, these stout little flies are major pests in some areas, and they do serve as vectors for some disease-causing organisms. In eastern North America, the diseases carried by black flies are mostly diseases of birds and animals, including filarial nematodes that infect cattle (bovine onchocerciasis) or protozoans that affect birds (leucocytozoonosis).

In tropical countries, however, black flies carry some nasty human diseases. The worst of these is river blindness or onchocerciasis, caused by a parasitic nematode and transmitted to millions of people by river-inhabiting black flies. Larval nematodes, delivered to human hosts by biting black flies, grow into big adult worms that collect in nodules within their host's body. The nodules themselves are not a big deal, but the millions of baby nematodes (microfilariae) they produce can create serious problems. Some microfilaria get picked up by biting black flies, where they develop into larval worms that can be transmitted to other hosts, but most concentrate just under the skin where they eventually die. Some 18 million people – mostly in Africa, but also in tropical America and the Near East – are in a position to tell you that concentrations of microfilaria under the skin cause a terrible itch. Many of those people have impaired vision or cannot see at all, because microfilaria often migrate to the eyes and cause "river blindness."

Moth flies, sand flies, wood gnats and minute black scavenger flies (Psychodidae, Anisopodidae and Scatopsidae)

Moth flies (family **Psychodidae**, page 433) often abound where there is some kind of wet, microorganism-rich film such as sink drainpipes, wet rocks along streams, beaver dams and even the bacterial beds of sewage plants, where enormous numbers of these little flies often serve to keep the beds from getting clogged. One author recently calculated that sewage plants in the San Diego area of California are so full of these little flies they discharge 20 million psychodids a day into the ocean.

Closer to home, hairy, mothlike, pointy-winged moth flies are common sights in many kitchens and bathrooms, where they breed on slimy surfaces deep inside sink drains. Most North American moth flies don't bite, but some primarily tropical members of the psychodid subfamily Phlebotominae are blood feeders notorious for carrying leishmaniasis, an increasingly common protozoal disease found in warmer parts of the world. Phlebotomine larvae develop in relatively dry soils and are often called sand flies, the same common name North Americans sometimes use for unrelated biting midges (Ceratopogonidae), and New Zealanders apply to yet another family (Simuliidae). Phlebotomine sand flies occur as far north as southern Canada, but North American species feed mostly on reptiles and small burrowing mammals, and do not currently pose a health threat.

A wood gnat (*Sylvicola*, Anisopodidae), page 434, caption 3.

Tropical blood-feeding sand flies can give you a number of diseases, including both visceral and cutaneous leishmaniasis. With the latter, as a friend of mine experienced while on an insect collecting expedition to Peru, your flesh deteriorates in a rather gruesome fashion. The protists (species of *Leishmania*) that cause leishmaniasis have a form that hides within the cells of the victim, and my friend was only cured with a dangerous and expensive course of antimony compounds. He was lucky, as most leishmaniasis sufferers live in areas such as lowland tropical America, Central Asia or Africa, where the cost of treatment is out of reach for many.

Some species of North American nonbiting moth flies occur in huge numbers in home compost heaps, a habitat they share with common members of the small families **Anisopodidae** (wood gnats, page 434) and **Scatopsidae** (minute black scavenger flies, page 434). Minute black scavenger flies are usually small and inconspicuous, but wood gnats are often attractive flies with patterned wings. Wood gnats are of special interest to fly specialists because they are probably more closely related to Brachycera than to other Nematocera.

Nonbiting midges (Chironomidae, Blephariceridae, Deuterophlebidae, Thaumaleidae and Dixidae)

Common midges, net-winged midges, mountain midges, solitary midges and gall midges make up a heterogeneous assemblage of delicate little long-horned flies, most of which swarm and none of which bite. "Midge" is kind of a handy term to refer to the little flies that a happy motorcyclist gets to scrape off his teeth after a summer night's ride, those that hover in a swarm over your head as you walk along a country road, or most of the little flies attracted to your porch light.

The vast majority of routinely encountered midges belong to the common midge family, **Chironomidae** (page 439), a family that includes over half the total species of animals (and the vast majority of the individual animals) that live in fresh water. North America's 2,000 or so species of chironomid midges, which look a bit like mosquitoes with neither scales nor biting mouthparts, are mostly aquatic but occupy a variety of habitats ranging from streams to soils. Their numbers are so high (sometimes tens of thousands per square meter) that they are of incredible importance in both aquatic and terrestrial food chains.

Aquatic chironomids abound in every imaginable wet situation except the open ocean, and even here chironomids have been more successful than most insects. Chironomids are among the most abundant of seashore insects, and a few have become specialized for a marine or intertidal existence. Some male marine midges have reduced wings, useless for flight but great for propelling them along the sea surface. Females of the same species completely lack wings, and adult females of some marine species are legless and look like larvae.

Larval midges have a hard head capsule, one pair of fleshy prolegs right behind the head, another pair of prolegs at the tail end, and no spiracles. Huge numbers of "bloodworms," aquatic midge larvae with a red pigment resulting from an oxygen storage mechanism not unlike our own hemoglobin, often occur in polluted water or anoxic mud. Bloodworms feed by netting microorganisms from the rich organic material in which they live. Other chironomids utilize a variety of foodstuffs in a myriad of different aquatic habitats. Some feed on dead leaves, some mine in living vegetation, some eat microorganisms, some are predators and some live as parasites of other animals such as stoneflies, mayflies and caddisflies. One kind of midge larva (*Metriocnemus knabi*) is invariably found living in the debris in the bottom of pitcher plants. Another has a symbiotic relationship with a kind of colonial alga, the larva obtaining food and shelter and the alga gaining advantages in reproduction, dispersal and substrate placement.

A careful look into almost any aquatic habitat should expose some chironomid larvae swimming or clinging to substrates, but common chironomids are normally concealed in silken tubes, nets or cases reminiscent of some caddisfly cases. Common pond and lake midges that make silken cases start out life as free-living young larvae, only later becoming sedentary homebodies in much the way small caddisflies in the family Hydroptilidae (purse-case makers) only inhabit cases during the later larval instars. Caddisflies and midges both produce silk using labial glands.

Midges in the family Chironomidae are especially abundant at high latitudes and cool temperatures, and they are among the most common adult insects during the winter months throughout North America. Although members of this genus have been recorded out and about at temperatures as low as 3°F (–16°C), you are most likely to find adult midges such as the common winter-active *Diamesa* species walking on the snow near running waters during mild March days. A few midges are adapted to high temperatures as well as low temperatures, and at least one species can dry up in the larval stage and survive an extraordinary range of conditions in the dehydrated state, resuming activity when wetted.

Among the most improbable-looking midges are three poorly known families that live in torrential running water or thin films of water. Two of these families, the **Blephariceridae** (net-winged midges, page 434; widespread) and **Deuterophlebidae** (mountain midges; not in eastern North America), are called torrent midges because their larvae are modified into a series of flat, sucker-like discs that allow them to cling to rocks in torrential mountain streams.

Some of the most habitat-restricted insects are found only in thin films of water, like those frequently found along the limestone faces of the Niagara Escarpment. Such seepage faces are well worth looking closely at, here and elsewhere, for the fascinating insect communities they support. If you look amongst the crane fly larvae, moth fly larvae, dance fly larvae, common midge larvae and soldier fly larvae that usually live a fairly sedentary existence on this type of wet surface, you might find attractively pigmented, midgelike larvae responding to your approach by zooming away with a very fast, sidewinder type of motion. These are seepage midge larvae, in the small, highly specialized family **Thaumaleidae** (page 435). Unlike common midges, which breathe through their skin, Thaumaleidae have two pairs of open spiracles on the dorsal surface of the body, which is always exposed to the air above the thin film of water in which they live. They feed on algae and diatoms at the bottom of the film while breathing at the top of the film, quite a remarkable specialization for a whole family.

Another small family of "midges," the **Dixidae** (dixid midges, page 435) has larvae that sometimes occur along with seepage midges at the edges of seeps, although they feed mostly in the surface of the water like some mosquito (*Anopheles*) larvae, rather than on the substrate like seepage midges. Dixid midge larvae are somewhat similar to *Anopheles*, but the thorax is not swollen and the body is normally bent into a deep U-shape; adults look a bit like mosquitoes, too, but they don't bite and there are no scales on the wings.

Gall midges (Cecidomyiidae)

A lot of insects that feed on plants are able to induce their vegetative victims to undergo some unusual tissue development, often producing a tumor-like growth that the insect can live in and feed on. There are flies, wasps, moths, beetles, aphids, thrips and other insects that can induce these odd growths, called galls, but by far the largest and least known of the gall-making families is the **Cecidomyiidae** (page 435), or gall midges.

Recent estimates put the Canadian gall midge fauna at 1,600 species, of which only a hundred are actually recorded from Canada, including only 40 known as larvae. One well known species is responsible for the fat, gray, pine-cone-like structures that abound on Pussy Willow (*Salix discolor*) and various other willow species across the continent. These familiar structures are not normal willow growths, but are instead unique plant structures induced by larval salivary secretions of a particular species of gall midge (*Rhabdophaga strobiloides*). Willow cone galls are particularly conspicuous on winter days. If you collect one, cut into it to see the characteristic, legless and apparently headless pink larva of the gall midge that induced the gall. Other familiar galls induced by gall midges include leaf cluster galls on goldenrod, grape leaf petiole galls and an enormous variety of other galls appearing on the stems, flowers and leaves of many trees, shrubs and herbs.

Gall midge larvae normally feed within the galls by piercing plant cells with their mandibles and consuming the fluid, but some species feed only on symbiotic fungi within their galls. Galls can be inhabited by a single larva, as with the willow cone gall, or by multiple larvae. If you collect galls in late winter, keep them until spring and you might get some typical gall midge adults – tiny, delicate, long-legged flies with only a few wing veins; you might also end up with an assortment of inquilines, predators and parasitoids. Inquilines, some of which are gall midges that don't make their own galls, are "guests" in galls induced by the primary gall-makers.

As you might expect from such a large family, Cecidomyiidae exhibit a wide range of habits, and many non-galling gall midges can be found in fungi, under bark and in other habitats. Some odd subcortical, fungus-eating cecidomyiids exhibit a peculiar life cycle, in which daughter larvae are sometimes produced inside mother larvae, eventually consuming the mother. This mode of larval reproduction, called paedogenesis, allows rapid population buildups in short-lived food sources like fungi.

Some of the plant-feeding members of the gall midge family are serious pests. Hessian Flies (*Mayetiola destructor*), for example, are introduced cecidomyiids that cause considerable economic damage to wheat as they feed between leaf sheaths and stems. Other species, including a recently introduced pest of broccoli called the Swede Midge (*Contarinia nasturtii*), lay eggs in flowers or other plant tissue. Some gall midges are beneficial insects, and several species are specialized predators and parasitoids of mites, aphids, scales and other small pest insects. Aphid-eating "gall midges" are often conspicuous as little orange larvae grazing among bunches of backyard aphids, and one such non-galling gall midge (in the genus *Aphidoletes*) is sold for biological control of aphids in greenhouses.

Lovebugs, glowworms and fungus gnats (Bibionidae, Mycetophilidae and Sciaridae)

March flies (**Bibionidae**, page 437) have big-eyed males that form conspicuous swarms in early spring or late fall. Females flying into these masculine masses are promptly mounted, after which they leave the swarm attached to a male and often remain attached to the male for some time. Male insects often have "anticuckolding" devices to protect their mating investment and to ensure that their sperm is not displaced when the female takes a second mate. This probably explains why many March flies stay attached to their mates so long, sometimes for days, and why one abundant species (*Plecia nearctica*) is called the Lovebug.

Like most other Bibionidae, Lovebugs live in habitats rich in decaying plant matter, such as pastures full of cow dung. However, unlike other Bibionidae, Lovebugs have recently become abundant enough in the southeastern United States

A dark-winged fungus gnat (Sciaridae, *Eugnoriste occidentalis*), page 439, caption 4.

to become a serious nuisance. If you drive to Disney World in fall or spring, you may well drive your car through clouds of paired flies and have to spend some effort scraping copulating corpses off your windshield.

Although adult-only activities are most likely to draw your attention to March flies, larvae of this family often occur in conspicuous numbers as well. Sometimes larval Bibionidae can be seen gliding along the ground as a glistening group of larvae, piled on top of one another and moving along as an aggregate mass that rolls along the ground like a living conveyer belt, the larvae at the front of the migrating mass continually being overridden by those behind. Similar mass larval movement is found in some Sciaridae.

One of the few flies to attain economic importance as a tourist attraction is a member of the fungus gnat family (**Mycetophilidae**, page 437). The New Zealand Glowworm, a fungus gnat with the wonderfully descriptive Latin name *Arachnocampa luminosa*, has luminescent larvae with light organs that can be shut off in response to disturbance, so you must be very quiet if you join a tour to see the spectacular numbers of luminescent larvae in certain New Zealand caves. The ceilings of the best caves are festooned with glowworms dangling glistening, sticky silken threads below their bodies, enhancing the spectacular light show they provide. Small insects (usually midges) attracted to the luminescence are caught in the threads, reeled in and eaten. Other *Arachnocampa* decorate caves and moist mossy banks in eastern Australia, and more distantly related predacious and bioluminescent fungus gnats occur in North America.

Bioluminescent eastern North American species of the widespread genus *Orfelia* cannot switch their lights on and off like *Arachnocampa*, but they spin silken webs and capture prey, as do many related species in the subfamily Keroplatinae (now often treated as a separate family). You can often find their mucous-like silken webs or tunnels on fungi or under bark. Look for larvae sliding along in their silken tubes. Silk is also used in other groups of fungus gnats, including a recently discovered northeastern species in which the pupae dangle from cave roofs by glistening silken threads; related species pupate in silk webs or cocoons.

Adult fungus gnats, which look a bit like angular mosquitoes with big coxae, conspicuous leg spurs and relatively thick antennae, abound in cool, moist, forest floor habitats. Larvae live in a variety of habitats but most species are, as the common name suggests, fungus feeders. As any would-be consumer of wild mushrooms will tell you, distinctive black-headed fungus gnat larvae are particularly abundant in fleshy fungi.

Members of a related family, the **Sciaridae** (page 439), or dark-winged fungus gnats, also attack fleshy fungi, and one of the major pests of commercial mushroom production (*Lycoriella mali*) is in this group. Unless you grow mushrooms, the sciarids you are most likely to encounter are delicate little black-winged insects (some of the 65 North American species of *Bradysia*) that thrive in the moist soil of potted plants, sometimes damaging roots. Some sciarids have a bizarre life cycle in which several larvae form one cocoon, and then develop into immobile adults that mate and lay eggs right in the cocoon. Others have unusual sex determination mechanisms in which the sex of the progeny is governed by the genetics of the mother, with some mothers having all daughters and others producing only males. Although sciarids are extremely abundant in leaf litter, rotting wood and soils, where they are major decomposers, most species are inconspicuous and rarely noticed. This is a poorly known family, and much remains to be discovered about the North American dark-winged fungus gnats.

The rare long-horned flies (Axymyiidae and Nymphomyiidae)

Some long-horned flies are so rarely encountered even most professional entomologists have never seen one. So it is with the extraordinarily habitat-restricted families **Axymyiidae** (page 439) and **Nymphomyiidae**. Our one eastern North American species of Axymyiidae (*Axymyia furcata*) develops in rotting wood, but not just any rotting wood. It has to be a pale, bare, waterlogged but not submerged log that is soft enough to penetrate with a jackknife, but too hard to easily pry open. When you do find the right kind of log, often partially submerged in a small, permanent woodland stream, the presence of little piles of wet sawdust on the log surface might indicate that an axymyiid larva resides within. The larva uses its massive mandibles to excavate a flask-shaped chamber in the wood where it presumably feeds on microorganisms or fungi while breathing through a long, flexible "tail." Adults emerge in early spring and are rarely seen.

Nymphomyiidae are perhaps the strangest of all flies. Larvae, which look like small midge larvae with eight long prolegs, live in moss in small streams from Maine north to James Bay. Adults are bizarre, slender, tiny (around 2 mm) flies with clubbed antennae and skinny, long-fringed, almost feather-like wings. Adults shed their wings, much the way winged termites and a few ectoparasitic flies shed their wings, along a special weakening. The secondarily wingless adult flies occur under water, where the female lays eggs that remain attached to her until she dies. The intriguing morphology and life cycle of nymphomyiids is matched by their unusual distribution, since the closest relatives to our eastern North American species are found in Japan, the Himalayas and the Russian Far East.

SUBORDER BRACHYCERA Short-horned flies

The "lower Brachycera"

Horse flies, deer flies and relatives (Tabanidae, Athericidae, Rhagionidae and Pelecorhynchidae)

Larval horse flies (family **Tabanidae**, page 441) are like spineless subterranean serpents that use fanglike mandibles to impale other animals above and below wet ground, paralyzing and liquefying their victims with venomous saliva. Horse fly larvae usually eat other invertebrates, but sometimes strike small toads from underfoot, and they can even give the curious naturalist a bit of a bite if carelessly handled.

Only the most curious of naturalists is likely to have dug up larval tabanids, but blood-feeding adult female horse flies and deer flies are familiar to nearly everyone. Females of most (but not all) members of the family Tabanidae can slice your skin with their stabbing maxillae and scissor-like mandibles before imbibing the blood so liberated from your ruptured capillaries. Tabanidae and other Brachycera are called short-horned flies because, unlike the more primitive long-horned flies (Nematocera), they have compact antennae with relatively few segments. Larval horse flies, and larvae of all other short-horned flies, have mandibles that work parallel to one another in the vertical plane like fangs, unlike larvae of long-horned flies, which have mandibles that work against one another in a horizontal plane.

Anyone who has taken a midsummer walk near a forest or bog is probably familiar with the persistent circling and painful bites of deer flies (*Chrysops*), which tend to batter you around the head and shoulders despite liberal applications of insect repellent. Repellents that work well against mosquitoes are of little use against most horse flies and deer flies, because these day-active, big-eyed flies make extensive use of visual cues to locate hosts, and to quickly home in on their favorite body parts. Different Tabanidae usually attack their hosts on different parts of the body, with horse flies often preferring legs or lower regions and most deer fly species attacking from above. An entomologically astute way to deal with deer flies is to periodically swing an insect net around your head, nabbing the flies before they bite. This not only saves a bit of blood loss, but it also yields great collections of beautiful deer flies, usually with iridescent eyes patterned in red and green, colorful bodies and distinctively patterned or banded wings.

Deer fly larvae live in wet or saturated soil (often under water) where their presumably omnivorous feeding habits allow them to attain much greater densities than is normal for the entirely predacious, often cannibalistic, horse fly larvae. Deer flies pupate above the water line, where the pupae twist their auger-like front ends partly up through the soil surface before emerging. Patches of shoreline mud are often briefly studded with the delicate empty pupal cases of recently emerged tabanids.

The term "horse fly" usually refers to biting members of the Tabanidae other than the conspicuous deer flies, but common names such as cleg, greenhead and moose fly are also used to refer to some tabanids. As some of the names imply, these flies normally feed on animals other than humans, and they can be serious pests of livestock.

Although they are not important carriers of human disease in North America, eastern North American deer fly species have been shown experimentally to support the development of African eyeworms, disease-causing filarial nematodes that require deer flies as intermediate hosts (a disturbing fact, even though the nematodes, which cause a disease called loiasis, are currently only found in equatorial Africa). Horse flies are also potential mechanical vectors of disease organisms that do not require the flies as intermediate hosts, since they tend to be intermittent biters that move quickly from host to host with blood from the last bite still clinging to their mouthparts. Anthrax, a disease familiar to every North American since the bioterrorism scares of 2001, can be transmitted by tabanids, and the same is true for several livestock and wildlife diseases.

Male tabanids, which do not bite, are rarely encountered, especially considering the irritating abundance of the biting females. If you do come across male tabanids – innocuous creatures with huge eyes that meet along the top of the head – they will probably be freshly emerged from soil-bound pupae. Mature male tabanids sometimes swoop down to take water from exposed surfaces and are sometimes seen enjoying a sugary snack of nectar or honeydew, but otherwise spend most of their time away from the gaze of curious human observers. Male tabanids (and many other insects) often aggregate and hover around high, isolated landmarks such as exposed treetops as they await arriving females. Similar habits are found in other groups of flies, and a good isolated hilltop can be a great place to collect an astonishing variety of rarely encountered insects, including mate-seeking male Tabanidae.

Once mated, female Tabanidae have the peculiar habit of depositing their eggs in masses on objects overhanging potential larval habitats, into which the larvae drop upon hatching. Deer flies tend to be very fussy about what kinds of plants they lay eggs on, but cattail leaves are generally good places to look for clumps of eggs. Look closely for tiny wasps (Scelionidae and Trichogrammatidae) wandering over the egg masses. These little wasps, which develop as parasitoids inside the eggs, are almost invariably present. Females of one rare eastern North American tabanid (*Goniops chrysocoma*) lay their eggs under leaves, and apparently protect their eggs from parasitoids by straddling the egg mass and firmly grabbing the leaf surface with the tarsal claws. The female *Goniops* remain over the eggs, sometimes vibrating their wings against the leaf, till the eggs hatch.

A similar habit is found in the closely related aquatic family **Athericidae** (page 445), in which some species (in the genus *Atherix*) lay their eggs on top of egg masses previously deposited by other females, or even on top of the other females that have died on the mass after laying their eggs. Leaves hanging over running water are sometimes weighted down with masses of eggs and dead or immobile female flies. The eggs drop into running water where they develop into strange-looking predacious larvae with eight pairs of large prolegs. Adults of the only northeastern North American genus of Athericidae (*Atherix*) have innocuous nectar-feeding habits, but a mostly tropical group of blood-feeding athericids (*Suragina*) occurs in the southwest.

The **Rhagionidae** (snipe flies, page 445) is another widespread horse fly–like group that includes a few bloodthirsty species. Rhagionids, which look a bit like horse flies with slender, pointed abdomens, spend their larval lives in moist soil where they probably eat other soil invertebrates.

The most commonly noticed eastern North American snipe flies belong to the large and widespread genus *Chrysopilus*, some of which are large, brightly colored, slow-moving species that stand conspicuously on low foliage. Most other common snipe fly adults stand on tree trunks in a characteristic head-down stance that has earned them the nickname "downlooker flies." Little is known about the diets of most adult rhagionids, but some western species of the widespread genus *Symphoromyia* are vicious biting flies known as Rocky Mountain Black Flies.

Horse flies, snipe flies, athericids and the small family **Pelecorhynchidae** (page 446) form a group of closely related families usually called the "Tabanomorpha." The one rare eastern North American genus of Pelecorhynchidae (*Glutops*) looks like a small, dull, nonbiting horse fly without the conspicuously large calypters (flaps under the base of the wing) found in the Tabanidae. Some dipterists now include the Pelecorhynchidae within the small family Rhagionidae, at the same time taking *Chrysopilus* and *Symphoromyia* out of Rhagionidae and putting them into a family called Spaniidae.

The family classification of Tabanomorpha was recently enriched by the discovery of a new family (Oreoleptidae), described in 2005 from larvae and associated reared adults from torrential waters in the Rocky Mountains.

Soldier flies, subcortical flies and small-headed flies (Stratiomyidae, Xylophagidae, Xylomyidae and Acroceridae)

Fly feet usually end in five short segments (called tarsomeres), with a pair of claws at the end of the last tarsomere. Each claw usually sits over a broad adhesive leaflike structure or pad (called a pulvillus) densely covered with hollow hairs that ooze sticky fluids. Most flies have only two pulvilli, with a thin, almost hairlike structure (empodium) in between, but the empodium is fat and padlike in a few fly families. The apparent possession of three pulvilli (two pulvilli plus a padlike empodium in between) is a distinctive characteristic of the horse flies and their relatives plus a handful of small but distinctive fly families including the Stratiomyidae, Xylophagidae, Xylomyidae and Acroceridae.

Soldier flies (**Stratiomyidae**, page 446) are often bright and conspicuous, and presumably got their common name because their neat, colorful, angular appearance reminded someone of a soldier in uniform. Some soldier flies are striking mimics of bees and wasps, including common, bee-sized species (*Stratiomys meigenii* and similar species)

Soldier fly larvae (*Berkshiria albistylum*) and a springtail under bark.

that have bright yellow patches on each side of an otherwise dark abdomen. The bright yellow patches look just like loaded pollen baskets carried by similarly shaped and colored bees.

Most of our conspicuous soldier flies, including the large black and yellow ones common on flowers, breed in water where the distinctively flattened, rough-surfaced ("shagreened") larvae eat algae or small aquatic animals. Soldier flies pupate inside the shelter of the tough, calcium carbonate-studded larval skin, and the immobile stratiomyid larvae found along pond margins are usually cocoon-like exoskeletons of the mature larvae, with pupae safely ensconced inside (a similar pupation strategy is found in the higher Brachycera).

Some soldier flies, including some small metallic green species, are common in compost heaps and other deposits of decay; other species live under bark where they feed on bark beetle larvae. The dung-inhabiting species *Hermetia illucens* (the Black Soldier Fly), sometimes used for biological control of house fly larvae, has window-like transparent patches at the base of the abdomen that create an illusion of a wasp-waist when the fly is in flight. This, combined with their buzz and wasplike flight, gives flying Black Soldier Flies an uncanny similarity to stinging wasps.

A few soldier flies prey on other insects under bark, as do the related small families **Xylophagidae** (page 449) and **Xylomyidae** (page 449). Although both the Xylomyidae and Xylophagidae are rarely noticed in the adult form, their larvae are common under the bark of recently fallen trees (the prefix "xylo" is from the Greek word for wood), where most species feed on the larvae of other insects. Some Xylomyidae adults are striking wasp mimics; the more common xylophagid adults are usually slender, mostly black, flies.

Small-headed flies (**Acroceridae**, page 449) have three tarsal pads like Stratiomyidae and related families, but otherwise these families could hardly be more different. Adult acrocerids appear at first to be more like the product of someone's twisted imagination than the end product of natural selection, and these absurdly microcephalic insects obviously don't have much use for well-developed eyes or antennae. Small-headed flies deposit large numbers of eggs, usually over a wide area of potential spider habitat, but sometimes right on or in spider webs. Acrocerid eggs hatch into tiny, active larvae, called planidia, able to actively jump or crawl along in search of hosts, sometimes making their way along the silken strands of spider's webs.

All small-headed flies, at least all of those we know about, are parasitoids of spiders. Once the parasitoid larva has burrowed into the host, usually penetrating the host through a leg joint, it attaches itself to the spider's book lung and just sits there in a state of arrested development for months or sometimes even years as the host continues to develop. After that long period of inactivity, the fly larva enters a period of explosive development and voracious feeding, in some cases resulting in death of the host in as little as 24 hours. The small-headed fly larva then pupates, often in the dead spider's webbing, and emerges as an adult a week or so later. Small-headed flies are not usually common, but occasionally occur in good numbers on prominent, exposed rocks or other irregular objects such as discarded moose antlers.

Robber flies, stiletto flies, mydas flies and window flies (Asilidae, Therevidae, Mydidae and Scenopinidae)

North America is home to almost 1,000 species of **Asilidae** (robber flies, page 450), including some of our largest and most spectacular flies. Most robber flies are day-active ambush predators that perch on leaf tips, logs or other sunny vantage points where they use their large, well-separated eyes to scan for appropriate prey. They generally dart out from their perches to capture other insects, which they impale with stout, forward-facing mouthparts, usually penetrating the victims through the neck or other chinks in the prey's chitinous armor. Robber flies use their mouthparts not only to impale prey, but also to inject a mixture of powerful nerve poisons and enzymes that liquefy the tissues of their victims. Some of the more robust robber flies will inflict a painful bite if you confine them in your hand but, fortunately, none of these well-armed flies normally attacks vertebrates.

Robber flies are common enough that you have probably seen one with a bee or other victim impaled on its robust beak. A stout, forward-facing, beardlike tuft of bristles separates the fly's face from its struggling prey, and protects the face and eyes of the predator as it dines. This tuft, called a mystax, is a characteristic of the family Asilidae and helps you recognize these variably sized and colored flies as robber flies.

Many robber flies eat almost any insects (or sometimes other arthropods) appropriate to the size of the flies, which themselves range from a couple of millimeters up to 5 or 6 cm in length. Some robber fly species do show definite prey preferences. A few are well known as "bee catchers" because of their taste for honey bees; others consume beetles, wasps or even dragonflies. Smaller species eat springtails or tiny midges. The larvae of robber flies are not well known, but most species seem to be subterranean predators in dry or sandy soils. Most female robber flies work the entire abdomen down into the soil, sand or wood in order to deposit eggs, but a few lay their eggs on plants, and some of our most common woodland species develop in decaying wood or under bark. One introduced species common in backyards and gardens lays its eggs one at a time while flying.

Many flies are remarkable bee and wasp mimics, presumably utilizing their intimidating appearance to deter sting-shy predators, and some robber flies are great mimics of spider wasps, bumble bees and leafcutter bees. Bumble-bee-like robber flies (mostly *Laphria* species) are common sights along woodland paths, where they sit on exposed leaves or twig tips and scan a discrete search area for potential prey. After darting out to grab a victim, usually a small beetle, they regularly return to the same perch.

One group of robber flies, called the slender robber flies (subfamily Leptogastrinae), is so distinct from other robber flies that some consider it a separate family (the Leptogastridae). Slender robber flies have a slow, almost helicopter-like, flight that contrasts with the darting flight of other robber flies. They usually snatch their prey off leaf surfaces or tree trunks (much the way many damselflies feed), in contrast with other robber flies that usually take their prey on the wing (much the way dragonflies feed).

A robber fly (*Laphria sacrator*) eating a long-horned beetle, page 453, caption 6.

Three other eastern North American families, the Mydidae, Therevidae and Scenopinidae, are closely related to the robber flies. **Mydidae** (page 456), or mydas flies, are large flies with distinctively clubbed antennae. This group is mostly southern and western, and the only common mydid in northeastern North America is a huge black species (*Mydas clavatus*) with a red band across the base of the abdomen and a general resemblance to a big spider wasp. *M. clavatus* is sometimes common on Great Lakes beaches, especially around stumps or logs partly buried in sand. They lay eggs in the sand, where their larvae are apparently predators on beetle larvae. Adult habits are not well known, but they seem to be nectar feeders. The Mydidae has the dubious honor of including the only fly on the U.S. endangered species list. The Delhi Sands Flower-loving Fly (*Rhaphiomidas terminatus abdominalis*) is a protected species known only from a small and threatened patch of land in California.

Therevidae (page 456), or stiletto flies, are like mydas flies in having predacious larvae and nonpredacious adults. Neither family has a deep depression on the top of the head, but stiletto flies otherwise look like small to medium-sized robber flies. Forest stiletto flies (a few species) are characteristically sluggish, but the many species that frequent open, sandy areas tend to make short, fast flights when disturbed.

Scenopinidae (page 456), or window flies, are small, blackish flies similar to the closely related Therevidae. The only scenopinid regularly encountered in the east is *Scenopinus fenestralis*, which is often seen on windows in houses, where its larvae supposedly develop as predators of carpet beetle larvae.

Bee flies (Bombyliidae)

Adults and larvae of robber flies and horse flies feed on other animals, with adult robber flies attracting our attention for their conspicuous assassinations of insects, and adult horse flies infamous for their unwelcome attacks on vertebrates. Bee flies (**Bombyliidae**, page 457), in contrast, do all their protein feeding as specialized larvae that attack other insects. Most are ectoparasites that attach to their hosts and consume them from the outside, but a few are endoparasites (developing inside the host) and some are predators (consuming many hosts) or kleptoparasites (consuming the host's food).

Bee flies usually attack hosts found in burrows or nests on or in the ground, especially solitary wasps and bees. Adult bee flies, usually conspicuous for their robust, fuzzy (or scaly) bodies and their swept-back and often patterned wings, are nectar feeders, but their larvae consume a variety of other insects including larval bees, larval beetles, caterpillars, ground-living fly larvae, antlion larvae and grasshopper eggs. Perhaps the most familiar of bee flies are the very common early spring *Bombylius* species. *Bombylius major* and *B. pygmaeus* are both long-beaked, rotund, furry bee flies with the front halves of their outstretched wings conspicuously black. These pretty harbingers of spring, along with many other bee flies, spend their larval lives as deadly guests in the underground nests of solitary bees.

Adult bee flies are delicate, velvety insects with no special tools for inserting eggs into well-protected solitary bee nests, so their attack strategies cannot involve inserting eggs directly into bee larvae surrounded by layers of earth or mud. Instead, bee fly females usually hover over their host's nest and repeatedly dip down to the substrate to drop an egg. *Bombylius* species, and many related bee flies, have sticky eggs that get coated with soil or sand in a special chamber in the female's abdomen before they are deposited in the exposed, often dry areas frequented by potential host bees. The eggs hatch into active first instar larvae that penetrate the host's nest before settling down to become sedentary parasitoids. This kind of drastic transformation from an active young larva into a conspicuously different, sedentary older larva is found in many groups of parasitic insects, and is called hypermetamorphosis.

Once inside the host's nest, the sedentary bee fly larva feeds by affixing its mouthparts to the skin of the bee larva, sucking out the body contents without making a visible wound. The great naturalist Fabre once described this mode of feeding as a "perfidious kiss," and compared it to suckling on a "teatless breast ... which, instead of milk, yields fat and blood." No horror tale of spine-tingling vampirism can compare with this true story of victims sucked entirely dry by the perfidious kiss of a bee fly. What could be more horrific than a killer able to penetrate the most fortified of nurseries, only to eat the baby alive without even leaving a wound? After a period of feeding the host looks like a deflated balloon, and the fat, soft bee fly larva appears to be imprisoned in the host's nest. Escape is dependent on the pupal stage, which is equipped with an impressive array of spines and spikes used to drill out of the nest.

Despite the fascinating lives of bee flies, we know remarkably little about the biology of most species. Most known species are ectoparasitoids of bees and wasps, as described above, but others are predators of grasshopper eggs, and some attack moths or other insects. Bee flies in the genus *Lepidophora* are kleptoparasites, or food thieves, developing on the paralyzed prey stored away by some kinds of wasps. The immature stages of most North American Bombyliidae remain unknown, so bee fly study is yet another area where amateur naturalists can make significant contributions.

Dance flies and long-legged flies (Empididae and Dolichopodidae)

Most dance flies (**Empididae**, or empids, page 460) are little predators that look a bit like small robber flies, although they lack the robber fly's stout beard and bulgy eyes, and usually have piercing mouthparts that face downward or backward unlike the stout forward-facing robber fly beak. A few species feed on pollen but, in general, both adults and larvae prefer living protein. Some common species are aquatic predators as larvae and some prey on small soil organisms, but most dance fly larvae are unknown. Dance fly adults prey on a variety of arthropods, with different species eating emerging aquatic insect adults, tree trunk insects, leaf-surface insects, insects stolen from spider webs and flying insects. Some empids are specialized raiders of swarms of flies, and help control mosquitoes and black flies by decimating their swarms.

Swarm-raiding dance flies sometimes raid the mating swarms of long-horned flies, and then use their captive's carcasses in their own mating rituals. In many dance fly genera, females take no food other than prey offered to them by the males as nuptial gifts. Sometimes this gift giving is highly ritualized, with the proffered prey elaborately wrapped in silk produced in the male's front legs. Many male empids fly up and down in their mating swarm carrying silk-wrapped prey that attracts, and perhaps distracts, the females. Males of some species carry this gift wrapping procedure to a ludicrous extreme, dancing their way up to females with nicely wrapped but empty packages in the form of glistening silken balloons, sometimes with bits of prey inside or stuck to the silken surfaces.

Most empid swarms, like most swarms in other insect groups, are male swarms. Females fly into the swarms and choose a mate with a good-looking nuptial gift. Sex roles are reversed in some empids, including a very common, conspicuous eastern North American species in which the females do the swarming, and the gift-bearing males do the choosing. Females of *Rhamphomyia longicauda* are spectacular insects with broad wings and big, black, feather-like scales all over their long legs. When they form their evening swarms, they hold their legs outstretched and add to their apparent size by inflating balloon-like sacs out of the sides of their abdomens. Males, with their precious protein-rich prey, presumably select the fittest females based on their apparent size.

The Empididae is a diverse group, so diverse that some entomologists divide it up into a number of families, some of which are apparently more closely related to other flies than to other "dance flies." Some Empididae, for example, are closely related to another distinctive family of flies called the long-legged flies (**Dolichopodidae**, or dolichopodids, page 462).

Long-legged flies are ubiquitous on foliage in temperate regions. One can be confident that a scan of a hedge or shrub will take in one of these usually metallic green flies. Male Dolichopodidae often have a huge genital capsule that curves forward under the abdomen. Dance flies, in contrast, usually hold their male parts above or behind the rest of the abdomen.

Long-legged flies differ from dance flies in many other ways. For one thing, dolichopodids do not impale their prey in the way that robber flies and dance flies do. Instead, they usually have a lower lip or labellum (a spongelike structure derived from the labial palps) that can spread out to expose two lobes covered with hard, crushing or ripping structures. One species that abounds on lily pads grabs leaf-mining midge larvae between the lobes of its labellum, and slurps them up the way kids eat spaghetti.

Most long-legged flies specialize on soft-bodied prey, such as small flies, springtails or scales, and some of the common adult dolichopodids on pond surfaces are significant predators of mosquito larvae. Although most dolichopodids are associated with wet places, one genus can almost invariably be found on tree trunks, usually standing head-up in a characteristic stance that has led one entomologist to christen this genus (*Medetera*) the "woodpecker flies." Larvae of these common little long-legged flies are important predators of bark beetle larvae; most other dolichopodid larvae are aquatic or semiaquatic predators, although the family does include one plant-feeding genus (*Thrypticus*).

Mating in the long-legged flies often involves elaborate courtship, and males of some of our common, beautifully metallic species (in the huge genus *Dolichopus*) have enlarged, flaglike structures on their front legs. They wave these modified legs in conspicuous mating dances, usually exposed to view on leaves or on the ground.

Despite the fascinating behavior, beauty and abundance of long-legged flies, Dolichopodidae are not well known, and remarkably few of North America's 1288 described species are known as larvae.

The "higher Brachycera"

All the short-horned flies discussed above belong to the so-called "lower Brachycera." Many of the lower brachyceran families are distinctive-looking insects like horse flies and robber flies, and many are common, attractive flies familiar to most naturalists. The rest of the Brachycera, the so-called "higher" Brachycera, or infraorder Muscomorpha, comprise a diverse group of flies in which the main part of the antenna (the flagellum) is usually made up of a fat basal portion and a hairlike apical piece (the arista). The higher Brachycera can be divided into two easily recognized groups: one enormous group characterized by a unique feature called a frontal suture (a semicircular "crack" arcing up over the bases of the antennae), and one smaller group of distinctive families that lack the frontal suture. Before looking at these interesting groups, let's examine why they are thought of as "higher" Brachycera.

When biologists refer to some group as "higher" or more "advanced," they are usually referring to one or two conspicuous features present in the "advanced" group, but absent in the "lower" group. Thus the wingless insects are often referred to as lower or primitive relative to the winged insects; insects with complete metamorphosis are sometimes called the "higher" orders compared with those which lack complete metamorphosis; and the Nematocera are considered primitive relative to the Brachycera because brachyceran short antennae have evolved from more primitive long antennae.

It follows, then, that to refer to some flies as the "higher" Brachycera it must be because they have features which are "advanced" relative to "primitive" features in the "lower" Brachycera. A quick look at familiar higher brachycerans like house flies, fruit flies and hover flies reveals that the adult flies have short antennae with fat basal segments and a long, hairlike arista, unlike most lower Brachycera. This is indeed an advanced character, but similar aristate antennae have developed independently in a few small groups of lower Brachycera. The clearest evidence that the higher Brachycera form a natural group comes from derived characters of the larvae. Larval characters of the higher Brachycera are unique and distinctive, and the higher Brachycera is best characterized as an "advanced" group on the basis of the distinctive and specialized larvae, called maggots.

Maggots are legless, pointy-headed, usually cream-colored fly larvae, familiar to most people as writhing masses of blow flies or house flies in pockets of putrefication such as fly-blown carcasses or the soupy goop in the bottoms of dirty garbage cans. The tapered front end of a maggot ends in two fanglike mandibles used to rip material up and move it into the mouth, usually along with a lot of fluid.

Like most other maggots, house fly and blow fly maggots are not simply eating the sewage and other disgusting stuff they live in. Instead, they are a lot like miniature baleen whales swimming in an organically rich ocean. Most people know how baleen whales feed by filtering out small organisms from the ocean, and everybody is familiar with the way whales have to come to the surface to breathe. Few people, however, know that maggots have a special filter every bit as intricate as a whale's baleen, which they use to filter microorganisms out of the media in which they live. Furthermore, most maggots are like whales in that they must come to the surface to breathe, although they breathe through a pair of siphon-like processes at the back end of the body. The tips of these processes are usually surrounded by water-repellent hairs that close up over the spiracles as the maggot dives into its little ocean of decay. When fully mature, maggots tend to abandon their unsanitary but food-rich larval soup, and migrate to drier territory to transform into pupae.

One of the things that makes maggots maggots, and defines the higher Brachycera as a real group, is a neat trick these insects have for protecting their pupae. Instead of casting the larval skin before pupation, the way butterflies and most other insects with complete metamorphosis do, maggots simply transform to a pupa right inside the skin of the last larval stage. This skin hardens to a brown, seedlike shell called a puparium, which protects the pupa but also creates a potential problem. How does the delicate adult fly get out of this hard shell? In most cases, it simply pops a cap off the puparium by pumping a big, spiny, balloon-like swelling, called a ptilinum, out the front of its head. After it has served to pop the lid off the puparium, and perhaps served as an earth-moving tool in the case of species that pupate underground, the ptilinum is withdrawn into the head. The retraction of the ptilinum leaves a distinct semicircular scar or suture over the tops of the antennae, called a frontal suture or ptilinal suture.

A few members of the higher Brachycera lack this special escape mechanism and thus have adults that lack the ptilinal suture. The subgroup that lacks this specialized device is called the Aschiza, and the subgroup characterized by the distinctive ptilinal suture is descriptively referred to as the Schizophora. Once again, we have divided a group into the lower or primitive ones that lack some important character (Aschiza), and the higher or advanced ones that have it (Schizophora). As is often the case, the group with the important new character is infinitely more diverse. There are vastly more winged insects than wingless insects, more insects with complete metamorphosis than simple metamorphosis and more Brachycera than Nematocera; the higher Brachycera are more diverse than "lower" Brachycera and, of course, there are vastly more Schizophora than Aschiza.

The Aschiza lacks striking unifying features like the ptilinal suture of the Schizophora, but is nonetheless an easily recognized group simply because of its distinctive components. Each of the families of Aschiza – the big-headed flies, coffin flies, flat-footed flies, spear-winged flies and flower flies – stands out as singularly different from other flies.

The Aschiza: Coffin flies to flower flies

Two of the families of Aschiza are overwhelmingly abundant and diverse, each in their own contrasting environment. Flower flies or hover flies (family Syrphidae) are the supreme symbols of pleasant, sunny days in the country. These colorful flies seem to be an inevitable result of combining sun and flowers. In marked contrast to this, scuttle flies (family Phoridae) are found almost everywhere, but seem most often to be associated with dark, dank places like forest floors, caves and even coffins.

Scuttle flies (Phoridae)

Phoridae (page 474) are odd little flies, distinctive in the field for their unique manner of darting or scuttling about on the surfaces of leaves or debris. If they have wings (and some do not) almost all the heavy veins seem squeezed onto the basal front half of the wing. The small head is usually covered with large, backward-pointing bristles, giving the appearance of a "slicked back" hairstyle, a punk impression often enhanced by the presence of large, spiny palpi sticking out from the mouthparts like armed clubs.

Although phorids are sometimes called "scuttle flies," another widely used common name is "coffin flies." That common name really refers specifically to the habits of one ghoulish species (*Conicera tibialis*), females of which are able to track you down when you are "six feet under," and lay eggs to start a subterranean population on your nutritious cadaver. You might also share your body with scuttle flies before your demise, since many phorids breed in meat, damaged fruit and other things you might just ingest along with fly eggs or larvae. Don't let that worry you, though. Despite one person's report of all life stages of a phorid appearing in his feces over a year-long period after he accidentally ate some maggoty fruit, phorids are unlikely to routinely survive ingestion. Some phorids can infest people through wounds or their eyes, but the vast majority of phorids attack other invertebrates, not vertebrates.

Many phorids are scavengers or guests in the nests of social insects, and the females of some of the ant- or termite-associated species are wingless and strongly modified for their specialized habitat. The legless, wingless females of one recently discovered phorid species mimic ant larvae, and are fed and tended by the workers of their army ant hosts. Many other phorids are parasitoids of adult ants, some species parasitizing healthy ants and others laying their eggs in injured individuals.

Species of flies that attack healthy ants are often major enemies of ants, which respond to the presence of these little flies with what appears to be a panic response. Given the range of possible outcomes of attention from parasitic phorids, the ants have every reason to allow visits from these little flies to disrupt their normal activities. One possible outcome, as the common name "ant decapitating fly" suggests, is that the fly's visit will result in the ant losing its head. If you happen to see a carpenter ant's head pop off for no apparent reason, the odds are that a phorid fly female previously inserted a single egg into the hapless ant. The egg hatches into a tiny maggot that burrows in and feeds on the internal contents of the ant, ultimately consuming the contents of the head and pupating therein. Before pupating, however, the maggot produces enzymes that dissolve the connective tissue between the ant's head and thorax, decapitating the ant. Pupation takes place in the shelter of the detached head capsule and, about three weeks later, the adult fly emerges. Look for adult female ant-decapitating flies hovering around ant trails or nests, waiting for an opportunity to dart in and nail another hapless victim with a fatal egg.

Several aspects of ant behavior provide conspicuous evidence of the importance of parasitic phorids. Foraging leafcutter ants, for example, are protected from ever-present parasitic phorids by specialized smaller worker ants that "ride shotgun" on top of the leaf pieces these ubiquitous Neotropical ants conspicuously carry. Many other ants simply stop foraging when parasitic phorids appear on the scene. This is the case among fire ants, including the aggressive imported fire ant *Solenopsis molesta*. Imported fire ants originated in Brazil, but have now spread into the southern United States where they out-compete native ants, sting viciously and represent a serious pest problem. Several species of ant-decapitating fly (in the genus *Pseudacteon*) attack fire ants in Brazil, but no native phorids attack these foreign ants in North America. Entomologists are hoping that the importation of ant-decapitating flies from Brazil to the United States will disrupt imported fire ant foraging and give native ants (not attacked by these particular ant-decapitating flies) a competitive upper hand. Imported ant-decapitating flies were released for the first time in Florida in 1997.

If at this point you were asking, "What *don't* phorids get into?" you would be posing a good question. There are phorids that live as thieves in ant and termite nests, thrive in snail slime (on the snail), eat slug eggs, eat spider eggs, develop in dung, inhabit ant nests, infest beehives, parasitize beetles and damage mushrooms. Phorid specialist Dr. Brian Brown, at the Natural History Museum of Los Angeles County, has recently discovered a group of ant-associated phorids in which the female has a long proboscis used to penetrate an ant's anus, and to feed inside the ant's digestive tract! Larvae of these long-beaked phorids develop as parasitoids inside ants. Brown suggests that phorids are the ambulance chasers of the insect world, since so many parasitize injured or dying ants. Next time you see ant colonies in combat, watch along the sidelines for the waiting phorids.

Perhaps the most interesting thing about the family Phoridae is what we don't know. If you take a closer look at the forest floor or any other place characterized by damp, dark organic material, the chances are pretty close to 100 percent that you will see at least one distinctive little coffin

Some phorids, like this *Myriophora*, develop as parasitoids or scavengers in millipedes. This one was attracted to a spirobolid millipede found injured on a trail.

fly, and probably over 90 percent that nobody knows what that particular species does for a living, and nobody has ever seen its larval stage. Chances are very good that the species you are looking at is unknown to science, a species never formally described and named. When Dr. Brown was a graduate student at University of Guelph in the 1980s, he made a collection of phorids right on the university campus. He found around 250 species (23 genera), many of which were new to science.

Flower flies (Syrphidae)

Naturalists often treat butterflies and dragonflies as "honorary birds" because of the appeal these conspicuous, small groups have for those with otherwise ornithological interests. Following the same logic, we should consider flower flies or hover flies (family **Syrphidae**, page 464) honorary butterflies. Syrphidae are conspicuous day-flying insects, well known for their skilled hovering, frequency on flowers, bright colors and abundance in familiar environments. Flower flies are mostly bare or fuzzy (as opposed to bristly) flies, easily recognized by their behavior and general appearance. They are easily recognized in the field by their familiar flight patterns and bright appearance, and all members of the Syrphidae have a sort of family insignia in the form of a distinctive squiggle or false vein (spurious vein) running through the middle of the wing. Generations of naturalists have been attracted to these diverse, colorful flies, and as a result, syrphids are among our best known flies. Nonetheless, the immature stages, and thus the biology, of most North American species remain unknown.

One of the most thoroughly known of all flower flies is the widespread and familiar Drone Fly (*Eristalis tenax*), a common introduced syrphid that looks remarkably like a bee as an adult and lives out its maggot days in disgustingly bacteria-rich liquids. A Drone Fly larva is called a rat-tailed maggot because of its long, retractile breathing tube, which serves as a snorkel, keeping the maggot's two posterior spiracles in contact with the surface while its mouth is down in the murky, microorganism-rich muck. From a distance, puddles in feedlot manure piles and similarly contaminated pools of water often glisten with thin, silvery breathing tubes of rat-tailed maggots, which usually disappear into the murky depths as your approach disturbs the myriad of maggots.

Drone Flies are extremely common insects, especially in early spring and late fall, when flowers seem to attract many "bees" but few other insects. The remarkable resemblance between Drone Flies and bees (especially drones, or male bees), exacerbated by the loud hovering of Drone Fly males as they stake out a territory, has led to confusion for centuries, and is the probable explanation for an ancient Greek recipe for making bees out of the carcass of an ox. This recipe, which provided good instructions for the creation of a bacteria-rich soup suitable for rat-tailed maggots, was taken seriously until the idea of spontaneous generation of life was debunked in the 17th century.

Although the "bees" that emanate from putrefying puddles neither sting nor produce honey like their hymenopteran models, flower flies and their bee models do have some similarities. They look alike, sound alike and, furthermore, both bees and flower flies are significant pollinators. Like bees, Drone Flies eat nectar for energy and pollen for protein. The hairy bodies of Drone Flies serve as portable pollen feasts for the fly as well as important brushes for the pollination of the yellow-flowered crops frequented by these flies. Many other syrphid larvae have snorkel-like breathing tubes, sometimes short, sometimes long and retractile like *Eristalis*, and sometimes even modified into a spikelike tap used to breathe through the roots of aquatic plants as in the genus *Chrysogaster*.

Most adult flower flies feed on nectar, but some seem to fuel up on alternative sugar sources. Honeydew, the ubiquitous sticky-sweet excrement of aphids and other homopterans (order Hemiptera), is a major food source for many adult flies. Some of the most remarkable wasp mimics among the Syrphidae rarely occur on flowers, but instead feed on honeydew and pollen on leaf surfaces. Look closely at the wasps you see twitching along leaves splashed by sunlight in a hardwood forest. They might turn out to be flower flies, perhaps one of our big-legged *Xylota* species. As the prefix "xylo" suggests, *Xylota* larvae are xylophagous, which means they live in wood. Many of our most spectacular and rarest syrphids are xylophagous species dependent on large, old trees with rot holes and other particular types of decay.

Many Syrphidae, including some of our most spectacular xylophagous species, mimic stinging Hymenoptera, such as bald-faced hornets and yellowjacket wasps, in appearance, sound and behavior. Several common flower flies are convincing bumble bee mimics, including some members of the rat-tailed maggot genus (*Eristalis*), a species (*Merodon equestris*) that lives inside the damaged bulbs of narcissus and daffodils, and a common, big, fuzzy species (*Volucella bombylans*) with plumose antennae. Larvae of the big, beelike *Volucella* live by scavenging inside the nests of yellowjacket wasps and bumble bees.

Other flower flies, especially those of the strange genus *Microdon*, have an even closer association with Hymenoptera. *Microdon* adults are squat syrphids, often resembling solitary bees even to the extent of having fake pollen baskets on the hind legs. The apparently unsegmented larvae of *Microdon*, which are so sluglike they were misidentified as mollusks by the first to discover them, live as guests in ant nests. Although *Microdon* larvae eat ant brood, ants tolerate them and sometimes even "groom" the adult flies as they emerge from their puparia in the ant nests.

Most flower flies, including nearly all the small to medium-sized, frail-looking black and yellow species ubiquitous in gardens, are predators as larvae. Look for these greenish, leechlike maggots among groups of aphids on fruit trees or flowers. They are inactive during the day, but at night they move blindly among the aphids, grasping victims using typical maggot mouth hooks, then holding the doomed aphids up off the surface to consume the body contents.

Predacious flower flies are not only important predators, they also make up one of the most important groups of pollinators. This group also includes some of our most common flower fly species, such as the medium-sized, brightly colored species *Syrphus ribesii.* This familiar fly, which could be described as the "typical" flower fly, often draws attention when groups of noisily whining males hover in a single spot, usually in a sun splash under a tree. Swarming males apparently do not feed, since their abdomens are translucent and their guts are conspicuously empty. Short-snouted *Syrphus* females frequent flowers, but only to feed on pollen. Honeydew, rather than nectar, seems to be the favored sugar source in *Syrphus*. Some other flower flies obtain all their energy and nutrients from pollen.

Many syrphids use their hovering skills to deposit eggs amongst the aphids that serve as prey for their blind, legless larvae.

The most easily recognized genus of flower flies is probably the widespread genus *Rhingia*. *Rhingia* adults sport conspicuous and distinctive snouts that conceal long, bent beaks they can extend into long-tubed flowers in search of nectar. Although the biology of the one common North American *Rhingia* species (*R. nasica*) is unknown, European species deposit masses of eggs on vegetation overhanging piles of cow dung. Upon hatching, the larvae drop into the dung to develop.

Other syrphids have similar habitat requirements. One of the most abundant flies in really rank compost piles, for example, is a fascinating little flower fly (*Syritta pipiens*). Despite their humble origins, the pretty little striped adults of *Syritta* are fun to watch as they hover, fly sideways, fly backward and dart straight after competitors or potential mates with astounding speed. Look for *Syritta* on small flowers, like forget-me-nots.

Most flower fly larvae feed either on aphids or on decaying plant material, but a few species feed either opportunistically or exclusively on plants. Some exclusively plant-feeding species (totally black flies in the genus *Cheilosia*) are quite host-specific, and a few of these have been shuffled around the world by entomologists because of their desirable habits of feeding on noxious weeds.

Big-headed flies (Pipunculidae)

Although the Syrphidae are often referred to as hover flies, the real masters of hovering are in the family **Pipunculidae** (page 474), commonly called big-headed flies. Big-headed flies resemble small flower flies, but with huge heads enlarged out of all proportion to the rest of the body. These superb fliers can hover in small spaces (even the small spaces within a folded insect net) and can commonly be seen hovering among dense vegetation where they use their enormous eyes to scan for leafhoppers. If you watch one long enough, you might see it swoop down to grab an unsuspecting hopper and haul it up into the air to insert an egg between the hopper's armored segments.

Big-headed fly larvae consume the entire contents of their victims, then emerge to pupate nearby. Despite the fascinating biology of big-headed flies, and the obvious value of parasitoids that attack pestiferous hoppers, host data are available for less than 10 percent of the North American species, and many of the species have never even been described in the adult form.

When pipunculid expert Dr. Jeff Skevington was a graduate student at University of Guelph in the early 1990s, he looked at all available New World (North, Central and South American) specimens of the common genus *Pipunculus*. Almost half of the 29 species he found turned out to be new to science. The one (and only) North American *Pipunculus* species for which hosts are known attacks leafhopper nymphs (Cicadellidae), but some other genera of big-headed flies attack the nymphs of planthoppers (Fulgoroidea) or spittlebugs (Cercopidae). Only adult spittlebugs are vulnerable to attack, since nymphs are protected by spittle masses.

Flat-footed flies and spear-winged flies (Platypezidae and Lonchopteridae)

In comparison with the Phoridae, Pipunculidae and Syrphidae, other Aschiza are relatively rare and insignificant. The **Lonchopteridae** (page 476), or spear-winged flies, so named for their pointed wings, includes only one genus and 35 species worldwide (seven in North America), contrasting with the 2,500 species of Phoridae so far described. One cosmopolitan species of *Lonchoptera* (*L. bifurcata*) is very common in damp, grassy areas, where its larvae develop in decomposing vegetation. North American populations of *L. bifurcata* appear to be almost all female.

Enlarged, sometimes spectacularly flattened and modified hind tarsi characterize most of the 200 odd species of **Platypezidae** (page 475), or flat-footed flies. Platypezidae can be quite common on leaves, often frenetically running around in a fashion similar to some Phoridae. Look for velvety black flies on sun-splashed leaves in otherwise shady, moist forests. As might be expected for a fly family generally associated with decaying fungi, they are particularly abundant in fall. One group of flat-footed flies (genus *Microsania*) is attracted to smoke, and males often swarm above campfires. Nobody knows why these flies are attracted to campfires, but perhaps they develop in fungi associated with recently burned woods.

The Schizophora: Fruit flies to house flies

Most adult flies, including house flies, fruit flies, blow flies and thousands of other kinds of flies, have the remarkable ability to open up a fissure at the front of the head and pump out a relatively large, spiny, balloon-like structure, called a ptilinum. The ptilinum serves to pop the cap off the hard pupal shelter (puparium), which is simply the inflated, hardened skin of the last stage larva. All flies that use a ptilinum to emerge from the puparium belong to Schizophora, a group named for the conspicuous suture or "schism" that arches over the antennae where the deflated ptilinum has been withdrawn into the head of the adult fly.

The Schizophora is usually divided into two different-looking subgroups, the calyptrates and the acalyptrates. Calyptrates are generally big, bristly flies characterized by a large flap of membrane (a calypter) under the base of the wing and a distinctive button (a small round swelling called the greater ampulla) on the side of the thorax. House flies and blow flies are typical calyptrates. The acalyptrates, on the other hand, make up a phenomenally diverse group of generally smaller flies with small, inconspicuous calypters and almost never with a greater ampulla. The more familiar of these two groups, the calyptrates, will be discussed first.

House flies and their relatives (Muscidae, Fanniidae, Scathophagidae and Anthomyiidae)

One species of calyptrate fly is familiar to the extent that this is the only insect envisaged by most people when they refer to "flies." House flies (*Musca domestica*) are ubiquitous pests known to all.

House Flies seem to be invariable associates of human-kind, breeding in dung and other kinds of microorganism-rich materials, where the maggots rasp at the putrefying matter with their mouthhooks as they filter-feed on masses of nutritious bacteria. Adult House Flies live everywhere humans do, and deserve a closer look than one normally takes to aim a flyswatter. Let's home in on that fly that just landed on your donut. Don't be put off by the thought that it probably dragged its hairy, bacteria-covered feet through something disgusting not long before it decided to take a stroll across the donut, and don't let the way it is vomiting on your sugary snack affect your interest – that is just the way adult House Flies feed.

The fly's spongelike mouthparts, made up mostly of two fleshy, grooved lobes (labella) at the end of the lower lip (labium), would have been pushed on to the surface of your donut in response to contact between the fly's feet and your snack's sugary surface (yes, flies taste with their feet). Liquids and dissolved food particles are imbibed through the tiny grooves (pseudotracheae) on the spongy mouthparts, or drawn directly up the channel between the upper and lower lips. Since the surface of your donut is probably a bit crusty, the fly will probably have facilitated its feeding by depositing saliva and stomach contents (spitting and vomiting) on the sugary surface.

Now that you are likely no longer interested in eating your snack, you can devote full attention to *Musca domestica* and have a look at some of its finer points. It is hard to believe that this pretty little fly, with its feathery antennae, subtly gray-and-black-striped thorax and orange-sided abdomen, probably spent its maggothood in some bacteria-rich medium such as the grunge in the bottom of a dirty garbage can, feces or other decaying material.

Lesser House Flies (*Fannia canicularis*) and Latrine Flies (*F. scalaris*) are somewhat smaller household flies with similar disgusting habits. *Fannia* species are sometimes treated as part of the House Fly family (**Muscidae**, page 476) but really belong in a separate family **Fanniidae** (page 478). The little flies seen endlessly circling in the middle of a room or barn, usually under a "swarm marker" like a hanging light, are likely to be Lesser House Flies.

House Flies are objectionable and dangerous for their unsanitary habits, but we can at least be thankful that they do not bite like some other members of the large (around 700 North American species) family Muscidae. If you think a House Fly has bitten you, the culprit was probably a Stable Fly (*Stomoxys calcitrans*), a muscid fly that looks remarkably like a House Fly with a forwardly directed lance held under its chin. That characteristic "lance" is the stiffened lower lip (labium). Unlike the grooved, spongy lower lip of House Flies, the Stable Fly labium ends in thorny labella used by both sexes to rasp painfully through your skin, often biting through sock-clad ankles. Females need a blood meal in order to develop eggs, which are laid in moist decaying material such as urine-soaked straw or rotting vegetation along lakeshores. A couple of other North American muscids have similar biting habits, feeding preferentially on livestock (the Horn Fly, *Haematobia irritans*) or moose (the Moose Fly, *Haematobosca alcis*).

Although the House Fly family (Muscidae) includes a few pest species, most of the adult muscids you see frequenting leaves as you take a stroll through the forest are innocuous insects that developed as predacious larvae in some kind of decaying material, and feed as adults on pollen, decaying material or other insects. Larval muscids often start out feeding on dung or other organic matter, but switch to predation as they grow older.

Gardens are home to some very common predacious muscids, especially a small species (*Coenosia tigrina*) with a gray, distinctively black-spotted abdomen. These abundant flies often sit conspicuously on foliage while consuming recently captured flies, leafhoppers or other insects. The genus *Lispe* is another group of deadly predators in the family Muscidae. These house-fly-like flies, distinctive for their huge palpi, occur near the water's edge where they often pounce upon immature aquatic insects that are leaving the water to transform into terrestrial adults.

Although muscids are probably the most common predators in urban gardens and along shorelines, flies in the closely related family **Scathophagidae** (page 478) often outnumber the Muscidae in rural areas. The most familiar scathophagid is the common, fuzzy, yellowish orange Pilose Yellow Dung Fly (*Scathophaga stercoraria*), which abounds in cow pastures where its larvae feed opportunistically on other insects in fresh cow pies. Despite the abundance of dung-associated species like the Pilose Yellow Dung Fly, and despite the Greek roots of the family name (*Scato* = dung, *phage* = to eat), most of North America's 150 or so scathophagid species are not associated with excrement. Many are specialized stem or leaf miners, some develop only in piles of seaweed and others live in rivers and streams. Larvae of one common aquatic species (*Hydromyza confluens*) tunnel in submerged parts of water lilies. Other aquatic species are generally predators, as are the adults of most or all scathophagids.

Anthomyiidae (page 479) are generally described as plant-eaters, rather than microbe-eaters like close relatives in the family Muscidae, but the common anthomyiid flies known as root maggots demonstrate that there is a fuzzy line between eating bacteria and eating plants. Gray, slender, dull-looking adult root maggot flies are ubiquitous in gardens, and the larvae are common pests of a variety of root crops. The Onion Maggot fly (*Delia antiqua*), for example, lays its eggs on onions, which are destroyed by the maggots and associated microorganisms. Root maggots feed partially on the plant tissue, but mostly on decay that sets in when your rutabagas or onions are damaged and inoculated with bacteria by the fly larvae. Onion Maggot larvae have remarkable structures called pharyngeal filters that allow them to filter out masses of microbes from the gooey messes they feed in. Many other maggots, including those of House Flies and Stable Flies, have similar, baleen-like structures used to concentrate bacteria for ingestion.

Although many anthomyiid maggots are filter feeders in dung, rotting fungi, rotting seaweed and other bacteria-rich material, others feed on living plant tissues as leaf miners, stem miners and gall-makers. A few exceptional species develop as parasitoids of grasshoppers, and members of two genera (*Eustalomyia* and *Leucophora*) develop as kleptoparasites in the nests of solitary wasps and bees. North America's 500 or so anthomyiid species generally look like slender muscids; they differ in that the last ("sixth") wing vein (Cu2 + 2A) goes all the way to the wing margin. Anthomyiids other than the odd subfamily Fucelliinae also differ from Muscidae in having distinctive erect hairs on the under surface of the scutellum.

Fucellia is eastern North America's only genus in the anthomyiid subfamily Fucelliinae, a group normally associated with marine wrack, but also found in freshwater algal deposits along the Great Lakes. The anthomyiid *Fucellia tergina* joins *Scathophaga intermedia* (Scathophagidae), *Coelopa frigida* (Coelopidae), several *Thoracochaeta* species (Sphaeroceridae), *Orygma luctuosum* (Sepsidae) and some *Chersodromia* species (Empididae) to make up the highly specialized community of flies found in huge numbers in windrows of seaweed, or wrack, found along rocky Atlantic shorelines.

Cluster flies, bluebottles, screwworms and blow flies (Calliphoridae)

Just as Onion Maggots (*Delia antiqua*) feed both on living onion tissue and the microorganisms that they filter out of rotting onions, blow flies (**Calliphoridae**, page 481) often straddle a fine line between consuming animal tissue and eating microbes. Leave a bit of meat on the ground next time you are outside for a summer barbecue, and watch how quickly it attracts large, shiny blow flies, which waste no time in dumping their cargoes of whitish eggs on the meat. Blow fly maggots start out by feeding on fluid serum and bacteria between the muscle fibers, all the time secreting alkaline waste containing digestive enzymes. The combined effort of masses of maggots breaks the tissue down to a gooey mass that allows the maggots to rasp and filter away in the usual microbe-munching way.

Disgusting though it may sound, not all blow fly maggots restrict their activity to dead meat, and they sometimes consume the flesh of living individuals. This is not always a bad thing. During World War I, wounds of injured soldiers left on the battlefield for hours did not develop certain infections that plagued those whose wounds were more promptly treated. This was because wounds that were exposed for some time often ended up full of blow fly maggots that consumed the necrotic tissue and secreted something that promoted healing. Observations of this fortuitous blow fly activity led to the surgical use of sterile maggots for cleaning out deep wounds, and ultimately to the modern use of sterile maggots in the treatment of some kinds of difficult infections. The "active ingredient" of blow fly secretions, called allantoin, also has a medicinal use.

If you are stranded on a tropical island and find you have to use maggots to clean out a festering wound, watch that your pet maggots don't start into your living flesh after eating up your tasty necrotic bits. Some blow fly larvae start off feeding on microbially rich body surfaces, such as festering wounds or contaminated hair pressed against an animal's body, then burrow into the flesh of the animal.

Other species have become truly parasitic and feed only in or on living vertebrates. The most famous of these is the Primary Screwworm (*Cochliomyia hominivorax*), a devastating pest of livestock that attacks animals through minor wounds. Primary Screwworm flies are multimillion-dollar pests from Illinois through to South America, but more northern localities have only the Secondary Screwworm (*Cochliomyia macellaria*), a much less serious pest usually associated with dead tissues. When screwworms, or other flies, invade a living vertebrate it is called myiasis.

Northeastern North America is home to some other flesh-eating blow flies, such as the aptly named *Bufolucilia silvarum*, which attacks the flesh of frogs, toads and salamanders; and *Protocalliphora* species with blood-sucking maggots that attack nestling birds. A similar blow fly, called the Congo Floor Maggot (*Auchmeromyia senegalensis*), does have a people-sucking maggot, but fortunately this species is restricted to Africa. Yet another African blow fly, called the Tumbu Fly (*Cordylobia anthropophaga*), attacks people much the same way that rodent bot flies attack rodents. Tumbu Flies lay their eggs on contaminated sites, such as dirty diapers hanging on a clothesline, or perhaps a patch of sand where someone has urinated. The larvae hatch out and await an opportunity to latch on to a vertebrate (like you) where they burrow in and develop under the skin, each one creating a boil-like swelling.

If you are unfortunate enough to meet your demise under suspicious circumstances, perhaps due to a stab wound or an implausible tumble off a high cliff, the different blow flies that immediately set to work to liquefy your flesh might also leave clues that help police unravel when, where and how you died. Combined with our knowledge of development time, habitat preference and distribution of each species, the presence of different kinds and sizes of blow fly maggots and puparia can tell a story that might put your murderer in jail. For example, if there are puparia of a typically urban species of blow fly in your corpse at the bottom of a rural cliff, perhaps accompanied by three-day-old larvae of a more rural species, a forensic entomologist might suggest that you were murdered in town and dumped off the cliff about three days earlier.

Although blow flies are generally nasty creatures associated with stench and unpleasantness, not even sterile suburbanites should think that they can remain unaffected by the family Calliphoridae. Cluster flies, atypically dull calliphorids colored in shades of gray and densely clothed in fine yellow hair, are earthworm-eaters as maggots and household pests as adults. If you have a large lawn loaded with worms, look for thousands of adult cluster flies clustering on your south-facing walls in fall. These fuzzy flies will later move up through vents into attic spaces to spend the winter. Warm spring days will bring them out of the attic and into the house, where they accumulate on windows and generally make a nuisance of themselves. Accumulations of dead fly bodies in windows can lead to infestations of dermestid beetles that breed in the fly bodies, but the adult flies themselves are generally harmless. Cluster flies (*Pollenia* species) are by far the most common calliphorids that consume other invertebrates, but one other introduced species (the localized and rarely encountered *Bellardia agilis*) is an earthworm parasitoid, and a handful of rarely encountered native blow fly species are internal parasitoids of land snails. The Calliphoridae is a large family with over 1,000 species worldwide, but less than 75 species are known from North America.

Flesh flies (Sarcophagidae)

Flesh flies (**Sarcophagidae**, page 482) are ubiquitous, easily recognized and easily observed flies, yet the Sarcophagidae remains a group unfamiliar to most naturalists. Perhaps people are put off by the common name "flesh flies" given to these common gray-and-black-striped flies, or by the association of some of our common species with unattractive materials like carrion or dung. Female flesh flies also have the sometimes disconcerting habit of depositing living larvae (rather than eggs), which is one reason the flesh-eating Sarcophagidae are able to utilize the ephemeral bodies of insects and other small invertebrates as well as more long-lasting vertebrate carrion.

It is a small step from eating dead insects to eating live but immobile insects, and one major group of Sarcophagidae specializes on paralyzed prey stolen from hunting wasps. Many species in the subfamily Miltogramminae develop on paralyzed arthropods stored in the nests of solitary wasps, and a large proportion of the arthropods stung, paralyzed and stored away as food for wasp larvae end up being consumed by these kleptoparasitic maggots. Adult Miltogramminae usually deposit larvae in wasp's burrows or on recently captured prey, and these little gray and black flies can almost invariably be seen near hunting wasps, often circling them like little satellites (thus the common name "satellite flies").

The rest of the Sarcophagidae is quite diverse. Some flesh flies have moved on to eating insects before they are dead or paralyzed, and can be important parasitoids of insects such as tent caterpillars and pest grasshoppers. One sarcophagid species found in our area (*Emblemasoma auditrix*) homes in on cicada songs, parasitizing the songsters; others have been reared from land snails, millipedes and even spiders. Many flesh fly species are associated with disgusting substrates such as decaying meat or dog feces. Some develop only inside turtle eggs, and one species feeds only on thin-skinned young vertebrates, producing a boil-like sore where the maggot feeds just under the skin. Unlike most Calliphoridae, these parasitic Sarcophagidae (*Wohlfahrtia vigil)* do not require dirty or wounded tissue to initiate feeding, and *Wohlfahrtia* females occasionally deposit larvae on clean human babies.

Some sarcophagids (*Fletcherimyia* species and *Sarcophaga sarraceniae*) are regular inhabitants of pitcher plants, where you can usually find a single large larva feeding on the drowned insects in the pitcher. Sarcophagid larvae differ from other maggots in having a deeply recessed, pitlike hind end, with the spiracles opening at the bottom of the pit, so they are easily recognized.

Killer calyptrates (Tachinidae and Rhinophoridae)

We are surrounded by killer flies. Most of these killers are not conspicuous predators like the fierce robber flies frequently seen clutching recently dead victims, but are instead parasitoids that develop at the expense of their doomed host's lives. Despite the fact that parasitoids are extremely common insects (close to half of all insect species are parasitoids), the bizarre variety of attack strategies and life cycles of these insects seems more the realm of science fiction than natural history. Fiction writers have not been unaware of the possibility of such a macabre killer-victim relationship. For example, the popular movie *Alien* involved a creature from another planet that inserted its embryos in human hosts, where the young developed until they emerged almost fully grown at the expense of the life of the human host. For us, *Alien* is frightening fiction. For most animals, *Alien* is merely a pale model of the way things truly are.

Although the hosts of parasitoid flies cover a wide range of invertebrates, the flies themselves belong to relatively few families. Most parasitoid flies belong to the huge family **Tachinidae** (page 484), with 1,200 or so species in North America. Almost all Tachinidae are parasitoids of other insects, but a few species attack other arthropods, and a very few are parasites (which do not kill their hosts) rather than parasitoids (which do). Adult tachinid flies are incredibly diverse in size, shape and color, often with a beauty belying their categorization as killer flies.

The typical tachinid is house-fly-like in general size and shape, but with long, stout bristles all over its hind end, and with the upper plates of its abdomen wrapped right around the body to meet or overlap the lower plates. This well-defended abdomen, entirely encased in spiked armor, serves to deliver eggs to a potentially resistant recipient. The most basic mode of egg delivery, a mode now abandoned by most species of tachinid, is the simple deposit of an undeveloped egg on the surface of the host. When the fly larva emerges from the egg it burrows into the host, sometimes directly through the bottom of the egg. This system has some obvious disadvantages for the tachinid. The eggs are vulnerable outside the host's body, where they could be bitten or scraped off. Furthermore, the adult female tachinid has to find a host to lay an egg on, something the host could avoid by hiding, by only coming out at night or by having a protective coat of hairs or scales. The strategies tachinid flies have evolved for circumventing these problems are partly responsible for the great morphological and behavioral diversity of this group of "killer flies."

Some tachinids have solved the problem of exposed, vulnerable eggs by inserting eggs right into the body of their host, a strategy that demands modification of the female abdomen into some sort of piercing structure. You can often guess what a female tachinid does by looking closely at her business end. Some common species, for example, have structures that look – and function – much like can openers. These species inject their eggs into hard-bodied insects such as adult beetles. Other species almost completely bypass the vulnerable egg stage by depositing eggs that are almost ready to hatch, thus allowing the larvae to burrow into the host immediately after the egg has been deposited.

Of course, if a larva can burrow into its host, why not also give the larva the job of finding the host? Many species that attack hosts inaccessible to adult flies have adopted this strategy. For example, some tachinids that attack stem-boring caterpillars deposit their eggs outside the host's burrow, perhaps in response to the odor of caterpillar excrement (frass) in the burrow. The eggs hatch as active tachinid larvae able to burrow through the frass to attack the concealed host. Other species with active larvae use an ambush technique, depositing eggs where the appropriate hosts are likely to occur. Newly hatched larvae wait for a potential host to pass by, then they attack and burrow into their doomed victim.

By this time, you must be counting yourself fortunate in not belonging to a species subject to attack by these killer flies. Otherwise, you would have to worry not only about having eggs laid in or on your body, but also about being chased by attack maggots or being ambushed by specialized maggots hiding in your favorite haunts. This, however, would not be the worst of it. You would also have to worry about attack by killer flies with every bite of food you ate. Many tachinids produce thousands of tiny eggs that they deposit on the foodstuff of the host. These tiny eggs survive consumption, and then hatch into larvae that proceed to eat the host from the inside out.

Different attack strategies lead to specializations of larvae as well as adults. Those tachinid larvae that spend time as free-living larvae before entering hosts are specially modified maggots with hard plates that protect the body from damage and desiccation, often with false legs to assist in locomotion. Those that attack hard-bodied insects, such as beetles, often have sawlike structures to aid in penetrating the host's armor.

The need for extraordinary maggot modifications, however, doesn't end upon penetration into a host. These flies are successful killers partly due to their marvelous mechanisms for dealing with the hostile environment inside

Epalpus signifer, a tachinid parasitoid of caterpillars, page 485, caption 1.

another animal's body. If there were parasitoids inserting their eggs directly in you and me, our bodies would react to this effrontery by healing over the surface wound and attacking the inserted egg the same way any foreign object is attacked by our body's defensive system. Insect hosts react in much the same way but, surprisingly, the attacking tachinid turns these responses to its own advantage. In species that adopt this strategy, the wound-healing process of the host is distorted so that the "scab" does not close the entry hole, but instead fuses with the capsule formed around the invader by the host's internal defenses. The ultimate effect is that the host produces a tube running from the tachinid larva to the surface of the host. Since growing tachinid larvae have to breathe air through their posterior spiracles, the tube serves as a sort of snorkel, allowing the parasitoid to breathe at the surface of the host while dining on the host's entrails. Other tachinids have different breathing strategies. Many tap the internal air supply of their host, analogous to a parasitoid sticking a breathing tube into your lung while feeding on your other internal organs.

Most parasitic flies are tachinids. Other calyptrate families, such as the Sarcophagidae and Calliphoridae include a very few parasitic species, but the only entirely parasitic calyptrate family other than Tachinidae is the small family **Rhinophoridae** (page 492). Rhinophorids are parasitoids of woodlice or sowbugs (terrestrial isopods), which are penetrated by the fly's first instar larva. Both North American species of Rhinophoridae are introduced, and one is a distinctively black-winged fly often common in urban areas and greenhouses. Females of the common species *Melanophora roralis* have a white spot at the end of the wing; males do not.

Bots and warbles (Oestridae)

An entomologist friend of mine once returned from Costa Rica with a boil-like swelling on his upper arm. He knew that this was no ordinary wound because he could look down into a crater at the top of the swelling and see the hind end of a massive bot fly maggot. He knew the bot fly maggot needed to breathe through that hole, and if he covered it with a slab of meat the bot would probably burrow up through the meat and out of his arm, but he preferred to leave it in place to develop. About a month later the mature larva popped out of the disgusting, oozing, swelling; it soon formed a puparium, and a large beelike adult emerged from it about three weeks later.

Human Bot Flies (*Dermatobia hominis*) do not normally occur in our region (except when they arrive here in the bodies of returning travelers), but they are common in much of South America and Central America, where they have the remarkable habit of attaching their eggs to mosquitoes and other flies. When an egg-encumbered mosquito lands on the skin of almost any warm-blooded animal, body heat causes the bot fly egg to instantly hatch into a little larva that promptly burrows under the skin, where it feeds on the spot until maturity.

Although there are no reproducing populations of Human Bot Flies in eastern North America, big, fuzzy, beelike Emasculating Bot Flies (*Cuterebra emasculator*), which cause similar boil-like swellings on chipmunks, are very common in the northeast. Mice, squirrels and rabbits are heavily parasitized by other *Cuterebra* species.

Emasculating Bot Flies lay eggs in areas frequented by chipmunks. Here they await the warmth of a passing chipmunk, then hatch into active maggots that latch onto the rodent host and make their way inside its body through a convenient orifice. Once inside the chipmunk, the bot fly larvae develop as excruciatingly large parasites before popping out through the chipmunk's belly (through a breathing hole already maintained by the larva) and pupating in the soil. Despite the common name, Emasculating Bot Flies attack both sexes and rarely emasculate their victims.

Although several bot fly species occasionally attack humans, in North America bot and warble flies are primarily wildlife or livestock problems, and human attacks are accidental and rare. Although adult **Oestridae** (page 492) have neither functional mouthparts nor stingers, one common group in this family seems to strike fear into the hearts of livestock. When adult warble flies or cattle grubs (two introduced species of *Hypoderma*) arrive to painlessly attach their eggs to cattle hairs, the cattle gallop wildly and often injure themselves in an attempt to escape. This behavior is well enough known to have its own special term – "gadding." It is almost as if the cattle know those eggs will hatch into spiny larvae, which will burrow under their skin and torment them with subcutaneous migrations culminating in boil like swellings (warbles), each containing a massive maggot breathing through a hole in the cow's hide. Like the human bot, the warble ultimately pops out, perforating the cow's hide in the process, in order to pupariate in the ground. The exit holes take a while to heal, sometimes providing an avenue for attacks by screwworm flies (Calliphoridae).

Horses, sheep and deer all have their share of bot fly woes. Horse bot flies are honeybee-like flies (four accidentally introduced species of *Gasterophilus*) that develop as large, spiny larvae in horse stomachs. Some horse bot species lay their eggs in or near the horse's mouth, but the most common species in North America (*G. intestinalis*) places its eggs on the horse's foreleg hairs or other places where the horse will likely lick them off. The eggs hatch in response to the warmth of the tongue, and the larvae ultimately end up in the horse's stomach where they attach themselves to the stomach lining to feed. When ready to pupariate, the mature larvae pass out with the feces.

Deer, caribou, sheep and the occasional unlucky shepherd have to deal with a couple of genera of particularly nasty Oestridae that have the horrible habit of squirting streams of live larvae up the noses of their hosts in a swift aerial attack. The larvae work their way into the throat or sinuses, where they develop until they are sneezed out at maturity. Our most common nasal bot fly maggots (*Cephenemyia phobifer*) develop – probably painfully – as parasites in the throat and sinus cavities of deer, ultimately leaving the nose with a snort or sneeze when they are ready to pupate. All our native nasal bot flies attack cervids (caribou, deer, moose), but an introduced species (*Oestrus ovis*) attacks sheep.

Pupiparous flies (Nycteribiidae, Streblidae, Glossinidae and Hippoboscidae)

In marked contrast to the Oestridae, in which the larvae do all of the eating and adults often do not even have functional mouthparts, some groups of blood-feeding flies do all of their feeding as adults. A couple of oddball families that suck bat blood (**Nycteribiidae**, page 493 and **Streblidae**, page 493), the tsetse fly family (**Glossinidae**) and the common, widespread louse fly family (**Hippoboscidae**, page 493) essentially bypass the vulnerable larval stage by giving birth to single, completely developed larvae, which they sustain almost to maturity using internal "milk glands." Flies that deposit mature, ready-to-pupate larvae are called "pupiparous."

Tsetse flies, the famous vectors of African sleeping sickness and related livestock disease, no longer occur in North America, although fossil evidence suggests that they once did. The only pupiparous flies most North American naturalists are likely to see are louse flies, different species of which suck blood from a variety of birds and mammals. Louse flies are flattened flies with a somewhat crablike appearance reinforced by the way they rapidly scuttle through fur and feathers. Some species are wingless, or they lose their wings after finding a host. Although they occasionally alight on humans, you are most likely to see hippoboscids sliding smoothly along the feathered surface of a recently killed bird, or perhaps scuttling about the wool of a sheep. The large, entirely wingless louse fly sometimes found on sheep is an introduced species called the Sheep Ked (*Melophagus ovinus*). This species, sometimes inappropriately called the "sheep tick," lays mature larvae that immediately form puparia stuck to the wool. Sheep Keds were once common pests throughout the northeast but, at least in Ontario, they have become relatively rare in recent years.

Even as Sheep Keds seem to be disappearing (probably due to widespread use of systemic parasiticides such as Ivermectin), another Old World Louse Fly periodically threatens to establish itself in North America. *Hippobosca longipennis*, the Dog Fly, is a louse fly that bites dogs and a wide variety of other hairy animals and which is occasionally accidentally introduced to North America with zoo animals. Every known introduction so far has been swiftly eradicated.

Icosta americana, page 493, caption 2.

Acalyptrate flies: The ultimate in ubiquity and diversity

"Time flies like an arrow, fruit flies like a banana."

When Groucho Marx coined that quip he may have been thinking about the way rotting bananas, or any other fruit, will soon attract a cloud of tiny, pale-colored fruit flies or vinegar flies in the family Drosophilidae, a group famous for a much-manipulated member called *Drosophila melanogaster. D. melanogaster,* better known as the Laboratory Fruit Fly, is about as close as you can get to a "typical" acalyptrate fly, although a huge group with no better definition than "all Schizophora that are not calyptrates" cannot really have a single typical member.

Acalyptrate flies usually contrast in size and appearance with calyptrate flies just the way the tiny *D. melanogaster* contrasts with big, bristly bluebottle flies, but acalyptrates range from under a millimeter up to a couple of centimeters in length, and include parasitoids, scavengers, predators, plant eaters, aquatics, desert dwellers, wingless species and phoretic species. The North American acalyptrates are divided into around 50 families, although the great majority of the routinely encountered acalyptrate flies belong to a handful of large, abundant or conspicuous families including the Sciomyzidae, Sphaeroceridae, Ephydridae, Drosophilidae, Chloropidae and Tephritidae.

Apple maggots, gall flies and their relatives (Tephritidae, Platystomatidae, Richardiidae, Ulidiidae, Lonchaeidae and Pallopteridae)

Fruit flies in the family **Tephritidae** (page 493) are usually attractive flies, about the size of a house fly, but often brightly colored with spotted or banded wings. One of our most common fruit flies, the Apple Maggot fly (*Rhagoletis pomonella*), even has the bands on each wing forming a broad F-pattern as if to announce that it is a **F**ruit **F**ly. A more likely explanation for the wing-banding pattern is that, seen from behind, it makes the adult fly look remarkably like a jumping spider (the bands look like spider legs).

Wing patterns in fruit flies might sometimes scare away predators, but their more normal role is a sexual one. Tephritids go in for a variety of sexual displays including dancing (ritual movement), perfume (pheromones), kissing (exchange of oral fluids) and music (wing fanning sounds), as well as bright colors and patterns.

Fruit fly females have a hard egg-laying tube, called an ovipositor although it is just the hardened, pointed tip of the abdomen and not a true ovipositor like that of grasshoppers or sawflies. Whatever its origin, the Apple Maggot ovipositor is sharp enough to puncture an apple's skin and insert one egg at a time into the developing fruit. You can't easily tell by looking at the outside that an apple has been attacked by one of these flies, but the inside of the apple will become marked by rotten-brown "tracks" left by the developing maggots (Apple Maggots are often called railroad worms because of these tracks). Maggoty apples drop prematurely, and the mature maggot leaves the rotting apple to burrow down into the ground where it pupates inside the inflated skin of the last larval stage (the puparium). Winter is spent as an underground puparium, and some puparia stay underground for two, three or even four years before emerging. Similar, closely related species attack other fruits such as cherries and rose hips.

Bearing in mind that maggots usually feed by filtering out microorganisms from their mucky environments, we might

Urophora cardui, a tephritid introduced for the biological control of thistles, page 494, caption 5.

well ask what happened to this bacteria-feeding habit in the fruit flies. This is a good question, because many fruit fly maggots have a perfectly functional microbial filter, along with other modifications that suggest that bacteria are important to them. Some adult fruit flies have special mechanisms by which the females smear their eggs with microorganisms, and some larvae have special pouches that house bacteria. Those rotten-brown "tracks" left by railroad worms are no accident, but instead are the result of inoculation of the fruit with bacteria, which are fed upon and spread by the maggot as it moves through the fruit.

Not all fruit flies are fruit eaters, and many tephritids feed in other kinds of plant tissue. Species in the subfamily Tephritinae, for example, stick their characteristic fruit-fly-type ovipositor into flower buds, often greatly reducing the number of seeds a plant can develop. A fruit fly that develops in buds of a rangeland weed called Spotted Knapweed (*Centaurea maculosa*) was recently introduced from Europe to North America to control the weed. The fly, a pretty little species called *Urophora quadrifasciata*, is now common throughout eastern Canada and United States, and knapweed flower heads are frequently infested with *Urophora* larvae. Ironically, the introduced *Urophora* flies might be helping the spread of knapweed by attracting mice that eat the tasty fly larvae along with knapweed seeds. The seeds (which are still viable even after digestion) are later deposited with the mouse's feces or in pellets regurgitated by owls that eat the mice.

Other fruit flies deviate even farther from the frugivory suggested by the common name for the family, feeding as leaf miners or gall formers. By far the most familiar of these is the very common species that forms ball-shaped galls on goldenrod. Every meadow seems to have both goldenrod and the ubiquitous spherical galls induced when these flies lay their eggs in the stems of developing plants. Look for a gall without any holes created by emerging flies or hungry birds, and cut it open with a penknife. Winter galls usually contain both the fat maggot of a Goldenrod Gall Fly (*Eurosta solidaginis*) and a side-tunnel almost to the surface, although they sometimes instead contain a larva of a predacious beetle or parasitic wasp with no side tunnel. Goldenrod Gall Fly maggots are in a state of arrested development (diapause) in the gall throughout the winter, but resume development in spring, when they pupate and develop into boldly patterned adult fruit flies. Gall fly maggots make the side tunnel to provide an exit route for the soft, adult fly that, unlike beetle or wasp adults, has no mandibles to chew a route to the surface. One kind of parasitic wasp larvae (*Eurytoma obtusiventris*) spends the winter inside fly puparia formed prematurely in the heart of the gall before the maggot has made its exit tunnel; a related larger, wasp (*Eurytoma gigantea*) kills the fly at an earlier stage and sits unencumbered in the middle of the gall. The beetle most commonly found in goldenrod galls is a predacious tumbling flower beetle (*Mordellistena unicolor*).

Tephritids stick their eggs into plants using a lancelike ovipositor made up of the last few abdominal segments (mostly a stiff segment 7, called the oviscape). Similar modifications are found in several closely related acalyptrate families, including the superficially similar Platystomatidae, Otitidae and Richardiidae.

The family **Platystomatidae** (page 495) includes three genera in the northeast, but only the genus *Rivellia* is common. *Rivellia* species have the remarkable habit of developing inside root nodules of legumes, a habit also found in some Micropezidae.

The mostly tropical family **Richardiidae** (page 496) is rare in northeastern North America, but one strikingly antlike genus (*Sepsisoma*) occurs as far north as Ontario. Ant mimicry is found in a wide variety of flies, and presumably offers some protection from predation.

Unlike the Richardiidae and Platystomatidae, which are primarily southern groups, the **Ulidiidae** or picture-winged flies (also widely known as **Otitidae**, page 496) is a diverse group of tephritid-like flies common throughout our area. Some have the tephritid habit of using their sharp, stout ovipositors to lay eggs in living plant tissue such as the roots of onions or sugar beets, but most otitids develop under bark or in decaying plant material. **Lonchaeidae** (page 497) is another common family with a tephritid-like ovipositor, and the small, shiny blue-black flies commonly seen with their long ovipositors slid under the bark of recently fallen trees are lonchaeids. Other members of this family feed in conifer cones, and some are pests in fruits and nuts.

The small family **Pallopteridae** (flutter flies, page 498), often seen displaying outstretched, attractive, brown-patterned wings on woodland foliage, also belongs to the tephritoid group. Most pallopterids use a prominent ovipositor to deposit eggs in bark beetle galleries where the larvae develop as predators, but some species have been reared from stems and flowers, and a Japanese pallopterid species mines in the leaves and stems of ferns.

Lab rats, spittle-suckers, fungus flies and mystery flies (Drosophilidae, Diastatidae, Camillidae and Chyromyidae)

The most famous flies in the world are arguably the Laboratory Fruit Flies (*Drosophila melanogaster*), those easily reared, rapidly reproducing entomological "lab rats" that have changed our world by facilitating fundamental discoveries in genetics and molecular biology. *Drosophila* species are also familiar to many as the little yellowish orange flies that hover in clouds over fruit that has been left on the table too long.

D. melanogaster is arguably the best-known organism on the planet, but it belongs to a family (**Drosophilidae**, or vinegar flies, page 498) that is full of surprises. Even the genus *Drosophila*, a huge genus with over 100 species in North America and perhaps five times that many in Hawaii, exhibits many variations on the common drosophilid theme of feeding on microorganisms in decaying fruits.

Many drosophilids are found in fungi, a few occur only in tree wounds, some species are leaf miners, a few are predators on mealybugs, and some exotic species develop in unexpected places like crustacean gills or frog eggs. Perhaps the strangest common North American drosophilid (*Cladochaeta inversa*) develops inside the spittle masses of some spittlebugs, especially alder spittlebugs, where the fly maggots seem to cling to the nymphal spittlebug. This association is unusual because the drosophilid maggot apparently does no damage to the spittlebug, and both the fly and bug emerge from the spittle mass to develop normally. The drosophilid is either living on the spittle or it is a true parasite of the spittlebug. Insects that are true parasites of other insects and do not kill their hosts – as opposed to parasitoids that do kill their hosts – are very rare.

A few infrequently encountered acalyptrate flies are often described as "drosophilid-like flies," but we know so little about them I think of them as "mystery flies." The **Camillidae**, for example, are drosophilid-like flies that have been collected only a few times in North America, almost invariably on windows. Larvae might develop in mouse nests, but immature stages remain unknown. **Diastatidae** (page 498), another drosophilid-like family, has about half a dozen described North American species, but at least that many remain to be described, and we have yet to discover what diastatid larvae do. **Chyromyidae** (page 498) is another rare family of drosophilid-like flies, with less than 10 tiny, usually yellowish North American species. Adults are sometimes common on oak foliage and some species are associated with bird nests, but the larvae of chyromyids remain unknown.

Shore flies (Ephydridae)

The **Ephydridae** (page 499), properly called shore flies because of their dominance in shoreline habitats, are generally small (only a few millimeters in length), dark-colored flies superficially similar to the closely related Drosophilidae. Shore flies, however, usually have pectinate aristae (the thin terminal part of the antenna has long hairs on the upper surface only) and a prominently bulging face that gives most species of this large family a distinctive appearance. Even without a close look at the face or antennae, ephydrids are usually recognizable as the hordes of little flies commonly found on mud or wet shorelines. Those hordes of shore flies reach astronomical proportions in apparently hostile habitats like salt lakes or even pools of crude petroleum, where they feed on other insects blundering into their sticky turf.

Most shore fly larvae behave like typical maggots and feed by filtering microorganisms from their surroundings. A few compete with other maggots in really rich substrates like dung, but most are associated with water. Adult females of some shore flies crawl into the water, a bubble of air stored under the wings, and walk around in search of a good place to lay eggs. When they finish their business and release their grip on the bottom, the bubble carries them to the surface and back to a normal aerial existence.

Shore fly larvae often have fleshy prolegs for aquatic locomotion, and a long posterior tube for taking air from the surface. Even those shore fly larvae that have abandoned their ancestral microbe-munching habits in favor of mining in plants or attacking other arthropods tend to remain associated with water. For example, one tiny shore fly (*Lemnaphila scotlandae*) spends its larval life mining in the tiny floating leaves of duckweed, and the widespread *Hydrellia griseola* makes a pest of itself by mining in irrigated cereal crops. Larvae of the genus *Notiphila* live under water and breathe by puncturing roots of aquatic plants to obtain oxygen. Yet another specialized shore fly (*Trimerina madizans*) develops inside egg masses of wetland spiders.

Although most adult and many larval shore flies feed on algae and other shoreline microorganisms, both adults and larvae of our most distinctive shore fly genus (*Ochthera*) are predators; adults have huge, mantis-like front legs used to grab small insects. Globular springtails seem to be a favorite food for adult *Ochthera*, but these robust shore flies also dig out midge larvae from shoreline mud.

Eye flies and grass flies (Chloropidae, Anthomyzidae and Opomyzidae)

If you have ever found yourself virtually blinded by the unwelcome activity of dozens of little shiny black flies flitting in and out of your eyes, the odds are good that the nuisance was caused by eye flies (**Chloropidae**, page 499), nonbiting flies that feed on mucous membranes and animal secretions (some Drosophilidae have similar habits). It may not increase your affection for these irritating acalyptrates to know that some entomologists call them "pecker gnats," a name that probably has something to do with the fact that these little flies have a habit of taking refreshment in large numbers on moist parts of male dogs. Not surprisingly, eye flies have been implicated in transmitting pink eye and other infections of humans and livestock.

Eye fly maggots breed in decaying material, but most larval Chloropidae are associated with grass stems. A few chloropid larvae are parasitic or predacious, but so many species feed in grass stems or similar habitats the family is generally referred to as the grass flies. Some, like the pestiferous Frit Fly (*Oscinella frit*) and the Wheat Stem Maggot (*Meromyza americana*), feed on living plant tissue; many others are secondary invaders, feeding on decomposing plant material or even the frass of other grass flies.

One interesting group of tiny chloropids always seems to appear when large invertebrate predators, like robber flies, spiders or mantids, are dining on chemically protected prey. These little chloropids in the genus *Olcella* use specific kinds of punctured prey (stink bugs, stinging wasps, rove beetles) like fly singles' bars, where they imbibe leaked juices and link up with the opposite sex.

The Chloropidae is a huge group with almost 300 North American species and a corresponding diversity of habits. This is in marked contrast to the other grass-associated families **Opomyzidae** (page 500) and **Anthomyzidae** (page 501), each of which has less than 20 described North American species. One recently introduced opomyzid, *Geomyza tripunctata*, a distinctively colored little fly with spots at the wing tips, is now common in eastern North American lawns and gardens. Anthomyzids are poorly known, strikingly slender acalyptrates that occur mostly in open, damp, grassy areas, although a few, including some of the larger and more colorful species, live in forests. All three acalyptrate families associated with grasses include species with wings that are greatly narrowed or shortened.

Kleptoparasites and ectoparasites (Milichiidae, Braulidae and Carnidae)

A number of acalyptrate species visit relatively large predators long enough to consume excess fluids as the predator feeds, sort of like visiting a giant at dinner time to scavenge scraps under his table (but only if his menu includes attractive items). This kind of specialized scavenging, a type of kleptoparasitism, is found in a few Chloropidae and in several members of the family **Milichiidae** (page 501). Some milichiid adults frequent spider webs where they snack on half-eaten stink bugs, some are associated with big robber flies and their prey, and many visit flowers. Larval Milichiidae are scavengers in various kinds of decomposing materials.

Some small kleptoparasitic acalyptrate flies are not mere visitors at their hosts' tables, but are instead permanent guests and constant food thieves. Bee lice (**Braulidae**, page 502) are wingless acalyptrate flies that spend their adult lives clinging to adult honey bees. The tiny, mitelike adult bee lice feed on liquids stolen from their host's mouths. Larval bee lice develop in beehives, where they eat wax and pollen.

Adults of a species (*Carnus hemapterus*) in the closely related family **Carnidae** (page 501) are also wingless inhabitants on larger organisms, but there the similarity to Braulidae stops. Adult *C. hemapterus* start out fully winged, but they shed their wings upon arrival in an appropriately sheltered bird nest where the secondarily wingless adults feed on the skin or skin secretions of nestling birds, and possibly their blood. Larvae of *Carnus* are also found in bird nests, where they feed on decomposing material; other carnid larvae feed on a variety of organic materials including dung, carrion and debris in nests. Carnids are small (1–3 mm) flies, and are rarely noticed.

Leaf, stem and root miners (Agromyzidae, Psilidae, Tanypezidae and Strongylophthalmyiidae)

Leaf-mining insect larvae are conspicuous for the various blotches, trails and other patterns they create while feeding and defecating between upper and lower leaf surfaces, but the associated adult moths, flies and other insects are often small and nondescript. Although mining maggots are found across the whole fly spectrum, from midges to muscoids, most distinctive Diptera mines are the work of the large acalyptrate family **Agromyzidae** (page 502). Agromyzid females use their ovipositors to poke plants for two reasons, one of which is to puncture plant tissue to release the sap they feed on. The other reason, of course, is to insert an egg, usually between the upper and lower surface of the leaf. A few species close off the puncture with a bit of feces once the egg is in place. Larvae lie sideways as they feed within the leaf, leaving a characteristic translucent trail or "mine" as they feed.

If you want to get to know your backyard mines, note the host, shape of the mine, position of the mine and deposition of the frass (feces). You might, for example, have an attractive serpentine mine starting out small where the egg was deposited and gradually enlarging to the point where the larva has left the leaf to pupate. Your mine might have frass deposited in regular piles along the trail, or you might have a blotchlike mine over the midrib with the feces in a single mass and the puparium glued to it. To find out what made the mine, hold the leaf up to the light and see if the larva is still there. If it is, you should be able to tell by shape if it is a maggot or some other leaf miner, such as a caterpillar or beetle larva. To associate an adult insect with the mine, keep the leaf in a closed container until the adult emerges. Keep the reared adult associated with a photograph of the mine and you could have some useful scientific information, since many North American agromyzids are unknown or not yet associated with their mines.

Although all agromyzid larvae occur in living plant tissue, not all are leaf miners. A relative few agromyzids develop in roots, stems, cambium of trees or seed heads. Similar habits are found in the small family **Psilidae** (page 502), with only about 30 North American species, compared with the 500 or so North American agromyzid species. The root mining activity of one psilid, the Carrot Rust Fly (*Psila rosae*) renders it a pest of carrot, celery and related plants.

The families **Tanypezidae** (page 503) and **Strongylophthalmyiidae** (page 503) – sometimes treated as a single family Tanypezidae – are closely related to the Psilidae, but in both cases the natural habitats of the larvae remain unknown. North America is home to only one named and one unnamed species of the mostly Old World family Strongylophthalmyiidae, and two species in the mostly Neotropical family Tanypezidae. Both families can be locally abundant in wooded areas where they often conspicuously strut around on exposed leaves.

Silver flies, aphid killers and dead leaf miners (Chamaemyiidae, Cremifaniidae and Lauxaniidae)

Chamaemyiidae (page 503), called silver flies or aphid flies, are usually considered beneficial insects and have been utilized for the biological control of both aphids and scale insects. Aphid fly maggots can be found munching their way through aphid colonies, dining on scales in concealed places like the bases of grasses, and consuming gall-making aphids inside their galls. Adult aphid flies are usually silvery gray with black spots, but some are tiny, shiny black flies. Members of the closely related family **Cremifaniidae** (sometimes included in the Chamaemyiidae) are parasitoids of aphids, and one species was introduced to North America to control Balsam Woolly Adelgids (*Adelges piceae*).

The **Lauxaniidae** (page 503) is a diverse family, especially in the tropics where they superficially resemble numerous other fly families. North American lauxaniids are usually small to medium-sized gray, black or orange flies, often with the strange habit of mining in dead or decaying leaves. Larvae of many species develop between the upper and lower surfaces of fallen leaves, much the way leaf mining agromyzids mine in living leaves. Not just any leaf will do – different species feed on dead leaves (and the associated microbial communities) under different tree species.

Snail killers and similar flies (Sciomyzidae and Dryomyzidae)

Arguably our most attractive and intriguing acalyptrate is a bright, velvety yellow fly with a strikingly polka-dot body and contrastingly patterned wings. *Poecilographa decora* (the Spotted Marsh Fly) is the only species in the genus *Poecilographa* and, remarkably, nobody knows what it does for a living.

An insect with unknown biology is normally no big deal, since the majority of our insects remain unknown as larvae. *Poecilographa*, however, is a marsh fly (**Sciomyzidae**, page 504), and marsh flies are now among the best known of all flies. We know the larvae of almost all North American genera, in part because the potential of snail-killing Sciomyzidae for biological control of snail-borne diseases like schistosomiasis has stimulated the development of important research programs on these flies, resulting in a prodigious literature on their biology. Still, we don't know what the larvae of *Poecilographa* do, and that represents an intriguing challenge since other local genera are common, conspicuous products of amazing life cycles.

The most easily located, and easily recognized, of all marsh flies are elongate brownish orange flies in the genus *Sepedon*. Every marsh, swamp and pond margin I know of is frequented throughout the season by these distinctive flies, each taking a characteristically attentive, head-down position on a leaf or stem overlooking the water. Numbers of these common flies are often seen feeding on deer fly eggs or dead insects on the same emergent stems where female *Sepedon* lay their eggs. *Sepedon* eggs hatch to remarkable bumpy-looking maggots that drop into the water, where they use a tapered and retractile tube at the hind end for breathing, and a sickle-like pair of mouth hooks at the front end for ripping into aquatic snails.

Sepedon larvae float freely just below the surface, suspended from the surface film by water-repellent hairs at the end of the breathing tube, and aided in maintaining their position by a bubble of air in the gut. Upon encountering a snail, the larva swiftly attacks the mollusk's exposed foot, which is in turn retracted into the shell dragging the fly larva with it. *Sepedon* larvae quickly kill their mollusk victims, consuming a dozen or more before transforming into a floating puparium.

Mating Spotted Marsh Flies (*Poecilographa decora*, Sciomyzidae).

Sepedon's successful strategy of snagging aquatic snails is found in several other marsh flies, and it is a good bet that most of the sciomyzids hanging around your local pond spent their youth in similar pursuits. Most of the stout, orange-bodied flies that abound around still waters belong to the large genus *Tetanocera*. Some *Tetanocera* feed on aquatic snails as do *Sepedon*, but a progression of *Tetanocera* species has moved on to aquatic snails stranded on the shoreline, terrestrial snails and even slugs. Young larvae of the latter species rear up the front parts of their bodies in search of the right genus of slug, sometimes attacking them through the mouth (*T. clara*) and sometimes attacking them through the invaginated eye stalks (*T. valida*); other species slip beneath the slug's mantle.

Some species, including those that attack slugs, develop as parasitoids (with only a single host) rather than predators (many hosts). Another member of the family Sciomyzidae specializes on snail eggs, others eat fingernail clams and many develop as parasitoids of single snails. This range of hosts, habitats and feeding strategies has one common denominator – all known Sciomyzidae (except for one Afrotropical genus that eats aquatic worms) develop in mollusks. It is a good bet that *Poecilographa decora*, the mysterious spotted sciomyzid, will turn out to have a mollusk host. The Spotted Marsh Fly is still easily found along the edges of lakes and ponds in southern Canada, but it has reputedly disappeared or become rare in some parts of United States where it was formerly abundant.

The small family **Dryomyzidae** (page 507) includes some common, medium-sized orange flies, very similar to, and often confused with, the related Sciomyzidae. Although one west coast dryomyzid has the interesting habit of laying eggs in barnacles where the larvae develop as predators, eastern North American dryomyzids are strictly microbial grazers in decomposing material such as rotting fungus and dead invertebrates. Look for dryomyzids, especially the very common *Dryomyza anilis*, in moist woods.

Killer acalyptrates (Conopidae, Stylogastridae and Pyrgotidae and Cryptochetidae)

If you watch bumble bees visiting flowers on a sunny day, there is a good chance that you will see the adult bees attacked by attractive, somewhat wasplike, dipterous dive-bombers in the family **Conopidae** (page 507). Conopids, or thick-headed flies, develop as internal parasitoids of adult insects, mostly bees and wasps that the female flies swiftly parasitize as they visit flowers or return to their nests. Parasitized host bees and wasps go about their business until the conopid larva is almost finished feeding (usually within the host's abdomen), but by the time the conopid larva is ready to pupate the host has died. The host's skeleton becomes a shelter for the parasitoid's puparium.

The oddest flies to be traditionally grouped with the thick-headed flies belong to a genus of gangly-legged species characterized by an enormously long, bent proboscis and a long, skinny female abdomen that looks and acts like a spear gun. The "spear" delivered by this remarkable ovipositor is a long, barbed, harpoon-like egg that is jabbed into another insect, such as a cockroach or cricket.

These conspicuously armed parasitoids, all in the genus *Stylogaster*, occur as far north as Canada but they are most abundant in the Neotropics, where they are associated with army ant raids. If you are ever lucky enough to witness an army ant raid, look along the front edge of the swarm for dozens of *Stylogaster* hovering almost motionless as they wait for the ants to flush out potential cockroach targets. The word "target" rather than "host" is appropriate here, since no one has actually reared out a *Stylogaster* from a speared cockroach. One northeastern North American species of *Stylogaster* has been reared from crickets, but the life cycles of the very abundant Neotropical *Stylogaster* remain a bit of a mystery. Some authors put *Stylogaster* in its own family (**Stylogastridae**, page 508), while others treat it as a subfamily (Stylogastrinae).

Pyrgotidae (page 508), sometimes called "light flies" because they are among the very few higher flies attracted to lights at night, are rarely seen parasitoids that attack adult June beetles and chafers. These large, attractive flies swoop down on their beetle hosts, usually at dusk or during the night, and insert eggs between the beetle's wing covers. Larvae develop inside the adult beetle, ultimately killing it.

The Pyrgotidae, Stylogastridae and Conopidae are the only entirely parasitoid families of acalyptrate flies in eastern North America, although many other acalyptrate families, including Sciomyzidae, Chloropidae, Ephydridae and Drosophilidae, have a few parasitoid species. One species (*Cryptochetum iceryae*) in another parasitoid family, the family **Cryptochetidae**, was introduced from Australia to California in a successful bid to control the Cottony Cushion Scale (*Icerya purchasi*) but it does not appear to have spread eastward despite the presence of appropriate hosts.

Micropezid flies, page 509, caption 4.

Stilt-legged flies (Micropezidae)

Stilt-legged flies (**Micropezidae**, page 509) are extraordinarily long-legged acalyptrates that use their midlegs and hind legs to stand conspicuously elevated above leaves or other surfaces. They often resemble ants or slender wasps, the latter similarity magnified by many micropezids that hold their bright forelegs in front of the body and twitch them like the antennae of sympatric, similarly colored, ichneumonid wasps. Micropezid wings often have bands and pigment patterns that, when the wings are held over the body, seem to create the silhouette of an ant or wasp.

Stilt-legged flies are easy to watch, since they frequent exposed leaves and other surfaces where curious naturalists can get an eyeful of their strange sex lives. Many species have elaborate courtships starting with the inflation of sacs along the side of the male's abdomen, and culminating in a long copulation involving ritual eye stroking, periodic "kissing" (exchange of oral fluids) and peculiar posturing.

Although stilt-legged flies are among the most ubiquitous of flies in warm parts of the world, the few species that range as far north as Canada are not common, and we know very little about larval habitats. Some live in dung, decaying plants or dead wood, and one important genus (*Micropeza*) seems to be associated with legumes. The only species of *Micropeza* for which the larva is known is a European species recently found in northeastern North America, where we presume it was accidentally introduced. *Micropeza corrigiolata* develops inside the nitrogen-fixing root nodules of legumes, and the larva hollows out the nodules as it feeds.

Head butting, head measuring and leaping larvae (Clusiidae, Piophilidae and Diopsidae)

The **Clusiidae** (page 509) are small and rarely noticed, but these pretty little flies are easy to find once you know where to look, and their interesting behavior makes them worth looking for. Males of many clusiid species stake out a territory on large, pale fallen trees, and use their territories as mating arenas, or leks. Mated females usually lay their eggs in fallen trees as well, but usually in softer, wetter places than the lek sites chosen by, and defended by, males.

You should be able to find male clusiids patrolling and defending their little lengths of log on any spring day. The genus *Clusiodes*, for example, has about a dozen described North American species that regularly face off head-to-head as they hold their preferred bits of log surface. Some Australian species of *Clusiodes* do the same thing, but make the interactions more interesting by developing disproportionately wide heads with vibrissae (bristles that stick off the lower front corners of the head) that spiral like bedsprings. When two males meet, they flip their unusually long antennae out to the sides of their expanded heads and face each other. If one of the males cannot make the same width as his opponent, he abruptly leaves. If the flies are similar in width, they lock spiral vibrissae and fight until the vanquished male is forced to leave. In the meantime, small males without enlarged heads or spiral vibrissae lurk around the edges of the lek, presumably awaiting the chance to mate while the macho broad-heads are fighting it out.

This sort of head widening and interaction is found in a variety of flies at home and abroad, and some exotic members of the Diopsidae, Richardiidae, Platystomatidae and Ulidiidae have incredibly wide heads that may even be wider than the length of the body. The **Diopsidae** (page 510), formally called the stalk-eyed flies, are mostly Old World, but two moderately broad-headed species (*Sphyracephala brevicornis* and *S. subbifasciata*) occur in swampy areas in northeastern North America.

Some male flies duke it out with lengthened, rather than widened, heads. Males of one common North American skipper fly (**Piophilidae**, page 511) stake out their territories on old carcasses, and can be especially common in early spring after the melting snow has exposed small winter-killed mammals.

Males of *Prochyliza xanthostoma* have a combination of long antennae and a long snoutlike head that they apparently use in head-measuring competitions much like those of some Clusiidae. Males approach each other face-on, then rear up to seem as tall as possible as they push heads together. Females, with unmodified heads, sit on the sideline and presumably copulate with the fittest males. Less spectacular territorial battles take place on hotly contested surfaces of discarded moose antlers, home to specialized skipper flies that complete their development within the antlers. The common name "skipper fly" refers to the way a piophilid maggot can arc its head around to grab its butt-end, pull tight, then release its hold to spring the straightened body up into the air. One pest piophilid species, called the Cheese Skipper (*Piophila casei*), is named for its habit of larval skipping on the surface of infested high-protein food products. All North American skipper flies feed on dead materials, ranging from fungi to carrion, but larvae of some European species (sometimes placed in a separate family, Neottiophilidae, nest skipper flies) feed on the blood of living nestling birds.

Acalyptrate rarities (Aulacigastridae, Odiniidae, Periscelididae, Asteiidae and Acartophthalmidae)

Fly specialists, or dipterists, are noted for their peculiar conversations, with spikes of excitement surrounding discussions of esoteric topics such as fly genitalia (used in species identification), maggot husbandry or the latest discovery of a slime flux. Slime fluxes, which develop when bleeding tree wounds ferment, always interest dipterists because they attract a wide variety of insects, including some enticingly rare acalyptrate families.

If you are lucky enough to find a good flux, perhaps on a small tree recently damaged by the emergence of long-horned wood borers or other large beetles, look closely at the attending flies, especially their heads. Along with the ubiquitous drosophilids, you are likely to see some small flies with a rainbow-like pattern of stripes across the front of the head. These are **Aulacigastridae** (page 511), specifically *Aulacigaster neoleucopeza*, the larvae of which develop in tree wounds.

You are unlikely to see a North American aulacigastrid anywhere but a tree wound, but elsewhere members of this family have a variety of odd habitats. Some live in bromeliads, and one tropical species found on *Heliconia* leaves is a great rove beetle mimic, even to the point of running along with its elongate abdomen held up like that of a rove beetle.

A small family of flies usually associated with bleeding trees is the family **Odiniidae** (page 511); its larvae sometimes live in the same beetle borings that cause the tree wounds. Odiniids are strikingly bristly little flies, often with spotted wings and bodies. The heads of some (the genus *Traginops*) are memorably elongated, in contrast with other slime flux specialists in the family **Periscelididae** (page 511), which sometimes have distinctly broad heads. Males seem to use their enlarged heads in interactions with other males, probably in defense of their bit of sappy bark, where mating takes place and their larvae develop.

Two other rarely collected families, the **Asteiidae** (page 512) and **Acartophthalmidae** (page 512), are sometimes associated with tree wounds. Not much is known about these tiny (1–3 mm) flies, but asteiids apparently breed in fungi or in frass of other insects.

Caves, carrion, compost and grasshopper eggs (Heleomyzidae and Curtonotidae)

Although many flies seem to abound almost everywhere and at any time, others seem to have been pushed to the periphery and specialize in odd places and odd times. The **Heleomyzidae** (page 512), for example, is a diverse family of medium-sized acalyptrates usually armed with a conspicuous row of spines along the leading edge (costa) of the wing. Heleomyzids are most abundant in very early spring and late fall, when they can be easily found on compost heaps, dung, rotting fungi and other decomposing material. The easiest place to find them in the summertime is deep in a cave, where small-eyed heleomyzids are often the most conspicuous and common insect inhabitants. Other species occur in bird's nests, mammal burrows and older carrion. One small, picture-winged genus (*Trixoscelis*) seems to be associated with open, grassy areas on sandy soils. *Trixoscelis* is common in western North America but rare in the east, where it has been most often collected in cemeteries and similar habitats.

Curtonotidae (page 513) are distinctive, somewhat humpbacked flies that resemble heleomyzids right down to the stout costal spines, but have conspicuously plumose aristae. North America's only species of curtonotid (*Curtonotum helvum*) is a pale, slow-moving fly found on open sandy areas, where the larvae probably develop underground on grasshopper egg pods.

On and off the wrack (Coelopidae, Sepsidae, Canacidae and Tethinidae)

If you have ever taken a stroll along a stretch of rocky shoreline, you have probably noticed windrows of decomposing kelp, or "wrack," along the high tide line. Far from being mere trash piles of rotting algae, these rapidly decomposing deposits play an important role in cycling the awesome primary productivity of offshore kelp beds back into the marine environment. Tons of kelp is washed up along the high tide line at each monthly high tide, giving a highly specialized insect community one month to reduce the resultant wrack piles into a soluble goop that is reclaimed by the next high tide. The rapid decomposition of wrack is thanks to the activities of the specialized maggots of seaweed flies, along with the associated bacteria that they spread through the piles of kelp.

If you have ever stepped in a wrack pile, you probably noticed clouds of thousands of tiny flies flying around your feet, and hundreds of larger, strikingly flattened flies scurrying into the interstices between the rotting fronds. The tiny flies belong to a wrack-restricted genus (*Thoracochaeta*) in the large family Sphaeroceridae, and the larger, flat flies belong mostly to a family of flies found only in wrack piles. The family **Coelopidae** (page 507), or seaweed flies, is found throughout the world, but only in wrack piles. *Coelopa frigida*, the common seaweed fly of Atlantic coasts, is a greasy looking fly that seems completely water repellent and unperturbed by occasional inundation.

Some of the stout, flattened flies scuttling around on the surface of the seaweed piles are not Coelopidae, but a similar specialized subfamily of seaweed flies in the family **Sepsidae** (page 513). Sepsids, or antlike scavenger flies, are normally (as the common name suggests) ant-shaped flies that are conspicuously common on dung and other ubiquitous decomposing materials, although some frequent flowers, and the odd subfamily Orygmatinae is absolutely restricted to wrack piles. The Orygmatinae is an isolated lineage containing only the remarkably specialized shoreline species *Orygma luctuosum*. Most other antlike scavenger flies abound on all sorts of decomposing material, where they conspicuously display by flicking their outstretched wings as they strut about.

Another seashore acaylptrate family, the **Tethinidae** (beach flies, page 513) frequents smaller deposits of seaweed and other organic material on sandy shores. The small, usually silvery gray, adult tethinids are common inhabitants of saltwater shores, but larval habits are not known. A very closely related acalyptrate family, the **Canacidae**, is found (at least in North America) on marine shores. These tiny, ephydrid-like flies occur in the intertidal zone where they presumably feed on living algae exposed at low tide.

The most inconspicuously successful of flies (Sphaeroceridae)

The great British fly specialist Harold Oldroyd once described the **Sphaeroceridae** (page 514) as the "most inconspicuously successful of all flies." Few, if any, other insects are as common – or as rarely noticed – as the Sphaeroceridae. To test that claim, you could put a bit of soapy water in a bowl, then leave that bowl in any nondesert terrestrial environment; after a day or two, check out the insects drowned in the bowl (this is a standard sampling technique, called pan-trapping). Try it in your lawn, a bush, a bog, Siberia, Patagonia, Africa or wherever, and you will find sphaerocerids. These are *very* common insects, but you have to look closely to notice them.

Sphaeroceridae are the black or brown *Drosophila*-sized flies running and jumping along the surface of your compost heap, dead mushrooms, dog droppings, dead bodies, leaf litter and almost anywhere else their maggots can find sufficient moisture and microbes to keep them grazing. Sphaerocerids differ from other acalyptrates in having short, swollen first tarsomeres on the hind legs. This is a handy characteristic for those wanting to recognize the family, though no one knows what function it serves.

The prediction of drowned sphaerocerids in every bowl of soapy water is a safe one, one that I have checked out through the use of pan traps in all sorts of habitats from swamps to mountaintops, and localities ranging from the shores of the Arctic Ocean to those of the Beagle Channel. Pan traps take very different species in different places and habitats, and often take as many as 50 sphaerocerid species in a single site.

What are all these little flies doing? For the most part, nobody knows. We do know that some specialize in odd little pockets of decay, like millipede dung or the green glands of crabs. Many breed only in fungi, in carrion or in special kinds of decaying vegetation such as seaweed. Others are not so fussy, and seem to get into everything from dung to drainpipes to home composters. Two distantly related groups of North American sphaerocerids with the remarkable habit of hitching rides on dung beetles were only discovered a few years ago. These hitchhiking flies are specialized kleptoparasites that lay their eggs in the dung balls gathered and stored by the dung beetles.

Flies, especially acalyptrate flies, remain a fascinating frontier for both the professional entomologist and the observant naturalist, and are likely to remain so for many years to come. Many North American species still await discovery, and the biology of the majority of flies remains unknown, posing a great challenge to all.

SCORPIONFLIES (ORDER MECOPTERA)

Recognition of an insect as a member of the scorpionfly order (Mecoptera) is easy, as members of this order have peculiar long faces with mouthparts at the end of a usually elongate beak. The term "scorpionfly" is applied to the whole order even though the strikingly scorpion-like tail that the name refers to is only seen in one common family of Mecoptera, the **Panorpidae** (common scorpionflies, page 515). The scorpionfly "sting" is just the bulbous male genitalia threateningly flexed back up over the insect's body, and it is harmless (except perhaps to a female scorpionfly).

Common scorpionflies (all in the genus *Panorpa*) frequent low vegetation in deciduous forests, where they exhibit the apparently mundane habit of feeding only on dead insects. Some spice up their dining experiences by stealing from spider webs, and the dead scorpionfly males regularly seen in webs suggest that males benefit in some way that makes the risk worthwhile. As we all know, the need to impress females regularly leads to male-specific high-risk behavior, so it is not surprising to find that males in possession of insect carcasses are often successful in acquiring receptive females. Males without the "real thing" to offer prospective mates instead spit up on leaves and use the presumably odorous spittle to lure females, which feed on the salivary mass during a prolonged copulation. Males unable to either obtain dead insects or to generate a good pile of saliva have little mating success, and are restricted to trying forced copulations.

Species in another group of scorpionflies, called the hanging scorpionflies (**Bittacidae**, page 516), don't take the risk of invading spider webs for food, and instead hang from twigs or leaves by their long front legs and serve almost as living spider webs themselves, intercepting flying insects with their hind legs. Hanging scorpionflies also take prey on the wing by flying up stems and grabbing insects with their hind legs, but you are most likely to come across them while they are just hanging around or making frenetic short flights from hang-out to hang-out.

These odd insects, which look a bit like huge (about 20 mm) mosquitoes or four-winged crane flies, often dangle in patches of stinging nettle. Like common scorpionflies, they exhibit complex courtship behavior involving the presentation of a nuptial gift in the form of a dead insect. Successful mating leads to the deposition of rather large eggs in the soil. The caterpillar-like larvae of hanging scorpionflies and common scorpionflies feed at least partly on dead insects.

A snow scorpionfly, *Boreus brumalis*, page 516, caption 4.

Other than the hanging and common scorpionflies, the only Mecoptera most people are likely to encounter are snow scorpionflies (**Boreidae**, page 516). These black, flightless scorpionflies can be seen walking along the snow surface on relatively mild winter days, although it takes a sharp eye to spot these small (about 4 mm) insects. Snow scorpionflies often jump straight up when you approach them for a close look, landing back in the snow with their appendages folded against the body so that they resemble inanimate specks of dirt on the snow. Look for snow scorpionflies on late winter days when the temperature hovers around the freezing mark. Their black, sun-absorbing color, along with an antifreeze substance in their blood, allows them to remain active to 21°F (–6°C), but in this case "active" is a relative term. They don't raid spider webs or perform as acrobatic predators like other scorpionflies. In fact, both larvae and adult snow scorpionflies just eat moss. Larvae pupate in silken cocoons (much like those of fleas). Snow scorpionfly females have their wings entirely reduced, but males have stout, bristle-like wings used to grab and manipulate the female.

Although most entomologists will only encounter the three scorpionfly families discussed above, two "rare" families also occur in eastern North America. "Rare," of course, is one of those words one should always be suspicious of, and which usually means "rarely encountered." The earwig scorpionfly family, **Meropeidae** (page 516), provides an excellent case in point. These odd-looking insects, which look like large-winged earwigs with a typical scorpionfly head attached, used to be considered extremely rare. In recent years, however, the widespread use of the Malaise trap (a tentlike trap that snares flying insects) has yielded so many earwig scorpionflies they are now known to be common inhabitants of eastern hardwood forests. One Canadian entomologist even took some live adults out of his trap and noted that they make a sound by stridulation. Beyond that, nothing is known about them, and no one has ever seen the larva. There are only two species of Meropeidae, one in eastern North America (*Merope tuber*) and one in Western Australia (*Austromerope poultoni*). The other rarely encountered family of scorpionflies is the **Panorpodidae** (short-faced scorpionflies, page 516), which occurs in the southern Appalachians and is characterized by a markedly shorter beak than other scorpionflies.

FLEAS (ORDER SIPHONAPTERA)

Almost anyone who has ever had a pet has at some time encountered, and probably failed to appreciate, the fantastic fleas routinely associated with cats, dogs and other mammals. In the event that you have been able to get your hands on one of these tiny, heavily armored, compressed, jumping insects, you probably instinctively crushed it between thumb and forefinger, albeit with great difficulty because of its hard body plates and disklike form. A flea's disk-shape not only renders it almost squish-proof, it also helps it move between hairs that must appear to it as a dense forest of trees would appear to us.

Another conspicuous feature of most fleas is the presence of backward-projecting comb-like structures (ctenidia) behind the head, and usually on the cheek as well. The gaps between the teeth of these combs usually correspond more or less with the width of the host's hair, and probably help the flea stay on board despite scratching and other grooming activities aimed at dislodging it. The flea's head, a small structure with small eyes and short antennae normally hidden away in grooves, is barely visible to the naked eye. If you have been bitten, however, you won't need to actually see the two bladelike structures (maxillary blades) designed to rip open a hole and shoot in a bit of saliva to keep the blood from clotting, nor will you have to see the knifelike labrum that projects down from the roof of the mouth, joining the maxillary blades to form a tube used to suck your blood. You will be too well aware of how effective these structures are, especially if your allergic reaction to flea saliva leaves you with itching red welts as fleabite souvenirs. Most household fleas, on both dogs and cats, are Cat Fleas (*Ctenocephalides felis*) although there is a related species called the Dog Flea (*C. canis*).

A cat flea, *Ctenocephalides felis*, page 517, caption 2.

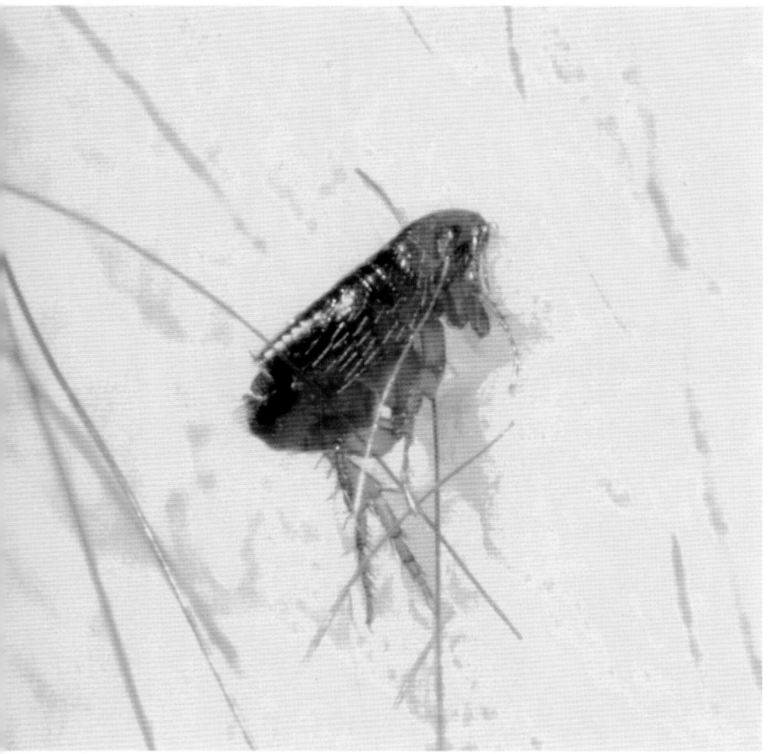

Fleas have a complete life cycle involving an egg, a legless larva, a pupa and an adult. Most female fleas lay only a few, relatively large eggs at a given time, but produce hundreds of eggs over their lifetime. A Cat Flea can lay a couple of dozen eggs per day over a period of about a month. The eggs usually fall to the floor below your pet's sleeping mat, where they hatch in a few days. The larvae feed on detritus, particularly the bloody excrement of adult fleas. Obviously, a good way to keep fleas out of your house is to vacuum regularly to pick up both the larvae and the dirt on which they feed.

A couple of weeks after hatching from the egg, the flea larva will spin a silken cocoon (using silk produced in mandibular glands) and pupate inside the cocoon. Flea cocoons usually look dirty because the newly spun silken cocoon remains sticky long enough to become covered with a camouflaging layer of dust. Depending on temperature, that pupa might transform into an adult flea in as short a time as a week, or as long a time as a year.

Although adult Cat Fleas prefer to bite dogs and cats, they quite happily feed on you and me as well. If they happen to emerge into a world free of vertebrate hosts, for example the carpet of a house that used to have pets but has been empty for a while, they can survive for long periods (several months) waiting for your ankles to appear on the horizon. Flea adults sometimes emerge from their pupae inside their silken cocoons, but remain for long periods in the cocoons as pre-emergent adults until conditions are right to head out in search of a blood meal.

You are not too likely to suffer anything more serious than an itching welt due to the unwelcome attentions of a household flea, but fleas do sometimes carry disease. Tens of thousands of North Americans get cat scratch disease, carried by Cat Fleas (but more often transmitted in a cat scratch). Other kinds of fleabites can have more serious repercussions up to and including the contraction of Bubonic Plague. Plague, also known as Black Death, is the infamous disease that decimated much of the world's population during the Middle Ages. In the past, the bacteria which causes plague was transmitted from rats to humans by rat fleas and spread like wildfire due to the ubiquity of both rats and fleas (plague also has a pneumonic form, rapidly spread by coughing). Nowadays, you are not too likely to be bitten by a rat flea, but the disease is still with us, waiting in the wings as it were. Other rodents, such as ground squirrels, can and do carry the plague bacteria. There can be rather a lot of fleas in a single ground squirrel nest (one scientist recorded 48,996 fleas from 143 ground squirrels), and you can get plague if one of them bites you.

All sorts of fascinating specialized insects are found only in mammal burrows, so the investigation of this habitat is quite a legitimate entomological pursuit. If you indulge in it, do try not to get bitten by a flea, as I know of at least one prominent entomologist who did contract plague while searching for beetles in mammal burrows in the southwest. Two hundred and ninety six cases of plague were reported in the United States between 1970 and 1991, and ground squirrels were the most frequently implicated sources of infection. Outbreaks of rat-flea-vectored plague in urban areas still occur – the most highly publicized ones in recent years being the 1994 outbreaks in the Indian states of Mharashtra and Gujarat – and there are recent reports of a multidrug-resistant strain in Madagascar (the disease is normally treatable with modern antibiotics). It is probable that a few hundred people die of plague each year – not much compared to the millions annually killed by malaria, but still a chilling reminder of the continuing presence of a disease that once killed tens of millions of people.

Although some fleas attack birds, most are associated with mammals. The comfortable environments of dog dens, gopher burrows and other kinds of nests are great places for fleas, and some species do especially well in nests containing newborn animals. This has led to an interesting relationship between the European Rabbit Flea (*Spilopsyllus cuniculi*) and its host. European Rabbit Fleas don't mate or lay eggs except after dining on the blood of a pregnant female rabbit or her newborn babies. Female fleas respond to hormones in the blood of a pregnant female rabbit by developing their own ovaries before they move to the new litter to mate, feed and lay eggs.

Although most fleas, like Rabbit Fleas and Cat Fleas, move freely on and off their hosts, some fleas attach themselves permanently to their hosts. Female Chigoes or sand fleas, (*Tunga penetrans*) for example, start out their adult lives as tiny (1 mm) insects with the notorious habit of fixing themselves to peoples' feet, usually under toenails, where they cause distinctive (and ugly) swellings. The swellings almost entirely enclose the developing female fleas, which can grow to the size of a small pea, leaving only a small pore through which eggs and feces get out. Mating usually takes place before the female is embedded, but for what it is worth, the chigoe male's penis is the longest, relative to body size, in the animal kingdom. Chigoe fleas are very common in the tropics, with infestation rates exceeding 50% in some parts of the Caribbean, Africa and South America. Chigoe infestations are uncommon in North America. There are other fleas that attach themselves permanently to their hosts, including the Sticktight Fleas that attack poultry in warmer parts of the New World including the southeastern United States. Female Sticktight Fleas (*Echidnophaga gallinacea*) have long mouthparts and can sometimes be seen in clusters firmly affixed to the head and neck of domestic poultry.

One of the most remarkable things about most fleas is their incredible strength and ability to jump, which is only partly dependent on the muscles of the enlarged hind legs. The key to the "super bounceability" of the flea is somewhat akin to that of those high-density rubber balls ("superballs") you probably played with as a child. Fleas have a superball-like mass of a rubbery substance called resilin right above their hind legs. They are able to cock their legs in a position that compresses this mass the way a super-ball gets compressed when you hurl it against a wall. When they release the compressed resilin as part of their jump, whammo! A flea's jump is far in excess of what could be accomplished by muscle action alone.

The use of resilin for spring-loaded action is not by any means unique to fleas. In fact, morphologists tell us that the jumping mechanism of the flea is probably a modification of a type of flight mechanism used by most flying insects. If we think about where fleas came from, this shouldn't come as too much of a surprise. Fleas evolved from an ancestor held in common with flies and scorpionflies, and are probably very closely related to the snow scorpionflies (**Boreidae**). Flea larvae are legless, similar to those of flies, many flies feed on mammals as do most fleas, and the flies and scorpionflies include several wingless lineages other than fleas. The fleas are merely one of those wingless lineages, highly specialized for an ectoparasitic existence. Flies, fleas and scorpionflies are sometimes together referred to as the "Antliophora".

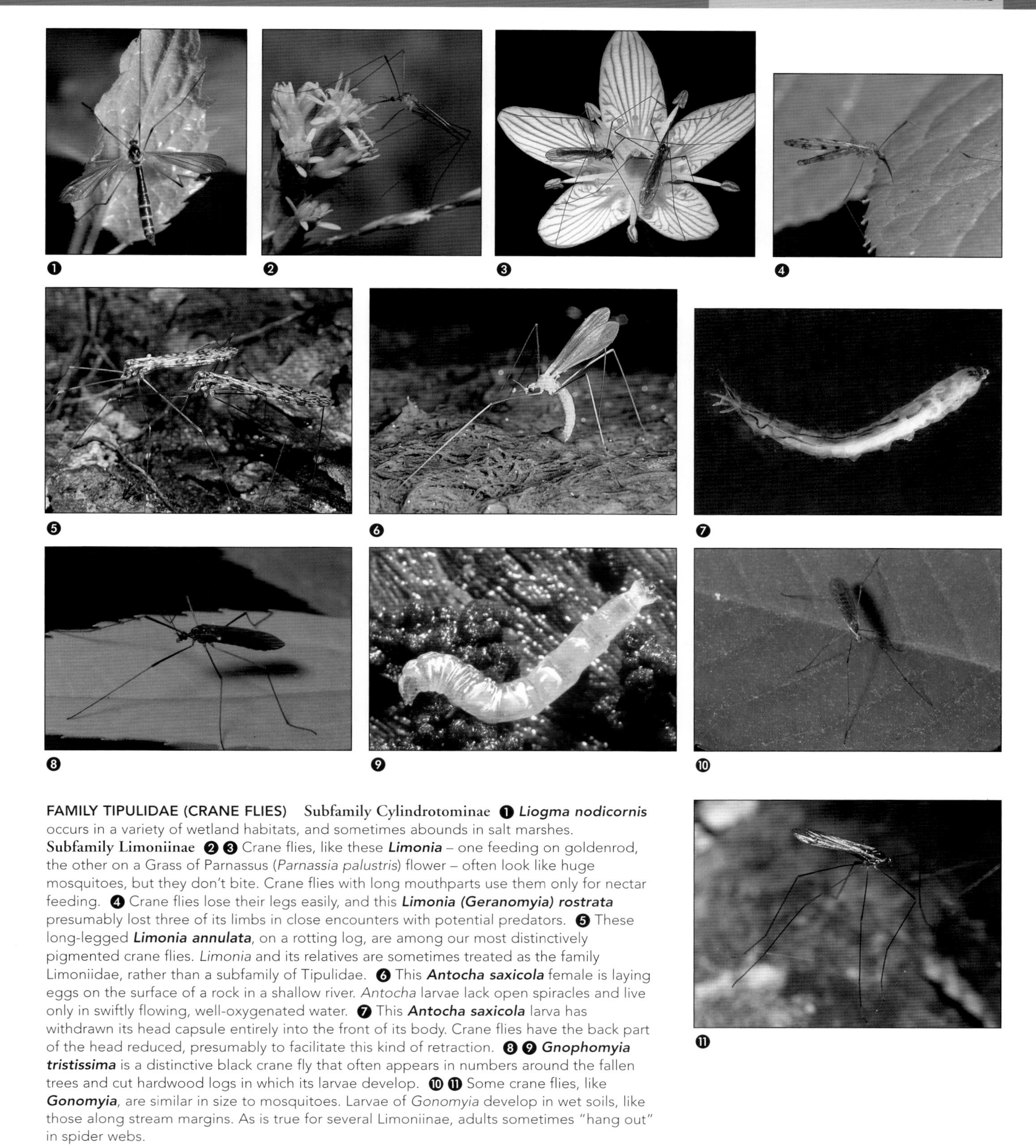

FAMILY TIPULIDAE (CRANE FLIES) Subfamily Cylindrotominae ❶ ***Liogma nodicornis*** occurs in a variety of wetland habitats, and sometimes abounds in salt marshes. Subfamily Limoniinae ❷ ❸ Crane flies, like these ***Limonia*** – one feeding on goldenrod, the other on a Grass of Parnassus (*Parnassia palustris*) flower – often look like huge mosquitoes, but they don't bite. Crane flies with long mouthparts use them only for nectar feeding. ❹ Crane flies lose their legs easily, and this ***Limonia (Geranomyia) rostrata*** presumably lost three of its limbs in close encounters with potential predators. ❺ These long-legged ***Limonia annulata***, on a rotting log, are among our most distinctively pigmented crane flies. *Limonia* and its relatives are sometimes treated as the family Limoniidae, rather than a subfamily of Tipulidae. ❻ This ***Antocha saxicola*** female is laying eggs on the surface of a rock in a shallow river. *Antocha* larvae lack open spiracles and live only in swiftly flowing, well-oxygenated water. ❼ This ***Antocha saxicola*** larva has withdrawn its head capsule entirely into the front of its body. Crane flies have the back part of the head reduced, presumably to facilitate this kind of retraction. ❽ ❾ ***Gnophomyia tristissima*** is a distinctive black crane fly that often appears in numbers around the fallen trees and cut hardwood logs in which its larvae develop. ❿ ⓫ Some crane flies, like ***Gonomyia***, are similar in size to mosquitoes. Larvae of *Gonomyia* develop in wet soils, like those along stream margins. As is true for several Limoniinae, adults sometimes "hang out" in spider webs.

❶ ❷ ❸ ❹ ❺ ❻ ❼ ❽ ❾

❿

FAMILY TIPULIDAE (continued) ❶ The small tipulids in the aptly named ***Erioptera (Erioptera) chlorophylla*** complex are chlorophyll green. ❷ Species in the small (four species) genus ***Epiphragma***, like this ***E. fasciapenne***, have distinctively patterned wings. ❸ ***Limnophila (Prionolabis) rufibasis*** is one of over a hundred North American species in the genus *Limnophila*. ❹ ❺ The first of these two photos of a crane fly larva (***Hexatoma***) shows the typical horizontally opposed mandibles of the suborder Nematocera. The second image shows how part of the larva can expand into a round anchor that facilitates larval locomotion through wet gravel or sand. Subfamily Pedciinae ❻ ❼ The aquatic larva of ***Pedicia***, like that of most aquatic crane flies, has open spiracles on a disc at the hind end of the body. The large adults of *Pedicia* are familiar sights near woodland pools or slow-moving rivers. Subfamily Tipulinae ❽ Although the family is enormously diverse, all crane flies have a V-shaped suture on the top of the thorax. This is ***Tipula (Nippotipula) abdominalis***. ❾ ❿ These photos show a female and male **European Leatherjacket**, ***Tipula (Tipula) paludosa***. Females of this recently introduced turf pest use the pointed tip of the abdomen to lay eggs in the soil, where the larvae feed on grass roots.

FAMILY TIPULIDAE (continued) ❶ ❷ ❸ ❹ The enormous genus *Tipula* is divided into over 30 subgenera. These photographs show the subgenera ***Schummelia***, ***Pterelachisus***, ***Yamatotipula*** and ***Platytipula***, respectively. ❺ This female wood-boring crane fly (***Ctenophora***) is using the hardened tip of her abdomen to insert eggs into dead wood. ❻ ❼ ***Nephrotoma*** are often relatively colorful and shiny for Tipulidae. The 40 North American species usually develop among roots in dry soil. ❽ ❾ Species of ***Chionea*** (a male and female, respectively, of ***C. valga*** are shown here) are entirely wingless crane flies often seen walking on the snow surface.

FAMILY TRICHOCERIDAE (WINTER CRANE FLIES) ❶ ❷ Most of the small, fully winged "crane flies" that appear on early or late winter days, often walking on top of the snow, belong to the family Trichoceridae, or winter crane flies. Unlike crane flies, winter crane flies have ocelli (simple eyes) just above the base of the antennae (you will need a good hand lens to see the ocelli). Most winter crane flies are in the genus ***Trichocera***. These larvae are in a bracket fungus. **FAMILY PTYCHOPTERIDAE** (PHANTOM CRANE FLIES AND THEIR RELATIVES) ❸ **Phantom Crane Flies** (***Bittacomorpha clavipes***) are often common around forest seeps, but their outstretched, banded legs and slow, drifting flight give them a phantom-like appearance. ❹ Larval Ptychopteridae, like these **Phantom Crane Flies** (***Bittacomorpha clavipes***), have a long, snorkel-like breathing tube. **FAMILY CERATOPOGONIDAE** (PUNKIES OR BITING MIDGES) ❺ The biting midge family (punkies, sand flies, no-see-ums) is familiar to most people for the burning bites of the few pest species, like this ***Culicoides***. ❻ Many punkies bite invertebrates, including dragonflies, stick insects and even other flies. Members of the specialized ***Atrichopogon*** subgenus ***Meloehelea*** imbibe the toxic blood of blister beetles such as these *Epicauta*.
❼ Many biting midges feed from the wing veins of other insects, including dragonflies, damselflies and lacewings. This one is sucking haemolymph from the wing of a green lacewing (Neuroptera, Chrysopidae). ❽ ❾ Most Ceratopogonidae, like these ***Palpomyia***, are predators of other insects. ❿ This tiny female ***Forcipomyia fairfaxensis*** is imbibing blood from the head of a green frog, possibly along with the viruses this fly is known to transfer from frog to frog.

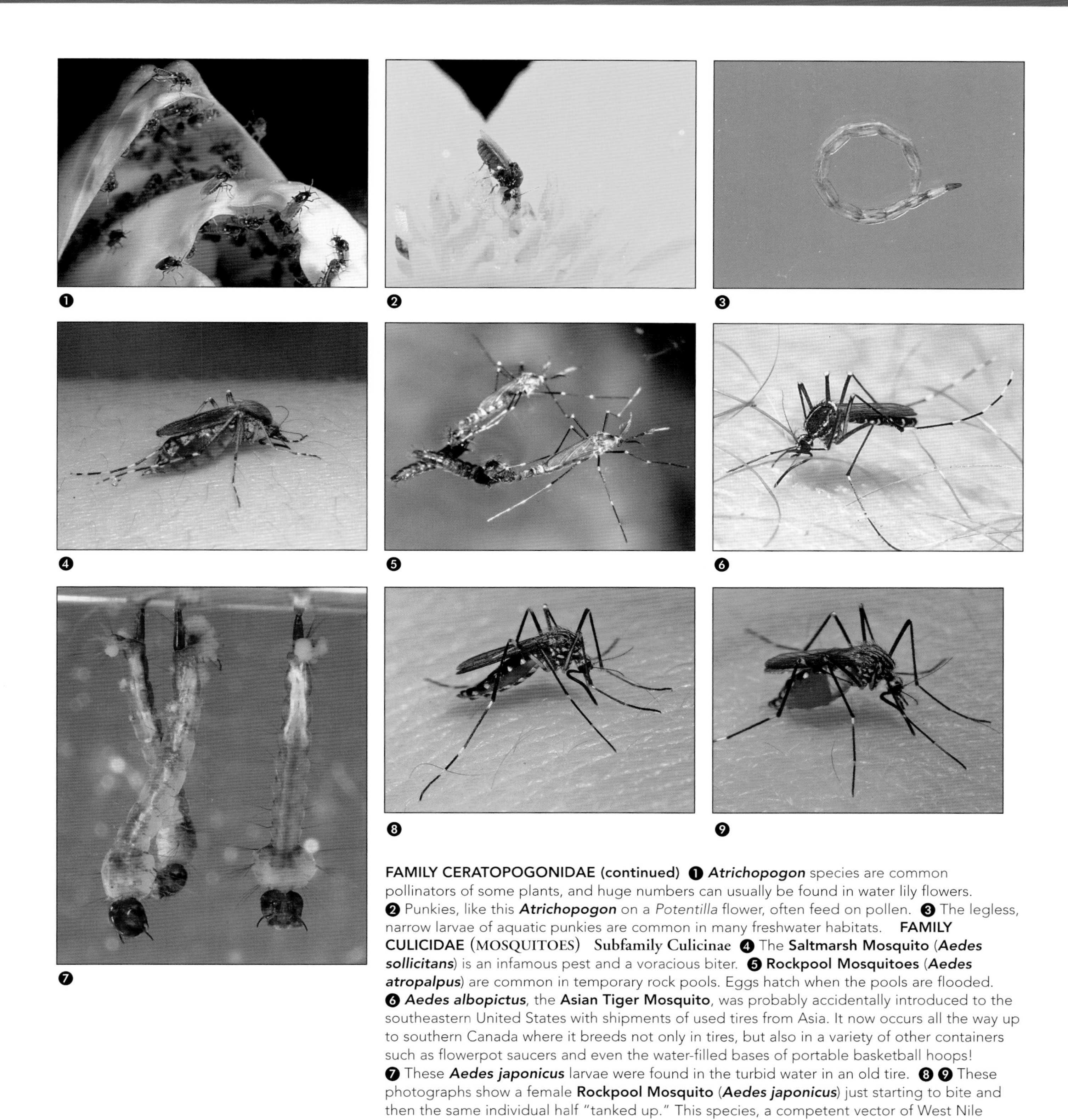

FAMILY CERATOPOGONIDAE (continued) ❶ ***Atrichopogon*** species are common pollinators of some plants, and huge numbers can usually be found in water lily flowers. ❷ Punkies, like this ***Atrichopogon*** on a *Potentilla* flower, often feed on pollen. ❸ The legless, narrow larvae of aquatic punkies are common in many freshwater habitats. **FAMILY CULICIDAE** (MOSQUITOES) Subfamily Culicinae ❹ The **Saltmarsh Mosquito** (***Aedes sollicitans***) is an infamous pest and a voracious biter. ❺ **Rockpool Mosquitoes** (***Aedes atropalpus***) are common in temporary rock pools. Eggs hatch when the pools are flooded. ❻ ***Aedes albopictus***, the **Asian Tiger Mosquito**, was probably accidentally introduced to the southeastern United States with shipments of used tires from Asia. It now occurs all the way up to southern Canada where it breeds not only in tires, but also in a variety of other containers such as flowerpot saucers and even the water-filled bases of portable basketball hoops! ❼ These ***Aedes japonicus*** larvae were found in the turbid water in an old tire. ❽ ❾ These photographs show a female **Rockpool Mosquito** (***Aedes japonicus***) just starting to bite and then the same individual half "tanked up." This species, a competent vector of West Nile virus, was first seen in North America in 1998. It now ranges from Ontario to Virginia.

FAMILY CULICIDAE (continued) ❶ ❷ Females of the **Eastern Treehole Mosquito** (***Aedes triseriatus***), like the one shown here, can carry several dangerous viruses, including eastern equine encephalitis and West Nile virus. Larvae are normally found in tree holes, but can also develop in artificial containers such as tires and gutters. Some authors put this species in the genus *Ochlerotatus*. ❸ These two radically different-looking organisms suspended from the water surface are the larval ("wriggler") and pupal ("tumbler") stages of ***Aedes aegypti***, the **Yellow Fever Mosquito**. The pupa has broad breathing tubes (respiratory horns) on the thorax, while the larva breathes through a siphon at the tip of the abdomen. ❹ This **Yellow Fever Mosquito** (***Aedes aegypti***) is just emerging from its pupa. ❺ ❻ Both male and female mosquitoes can be common on flowers. These photographs show male and female ***Aedes*** on *Potentilla* flowers. ❼ ❽ Although this female mosquito (***Coquillettidia perturbans***) is innocuously feeding on milkweed nectar, this species also bites humans, and can do so in irritatingly large numbers near the marshes where the larvae breed. *Coquillettidia* larvae and pupae do not breathe at the water surface, but instead, in order to obtain air, pierce the underwater roots of emergent aquatic plants with a uniquely stout and tapered posterior siphon. ❾ ***Culex*** mosquitoes lay their eggs in rafts, in contrast to the single eggs of *Anopheles* or *Aedes*.

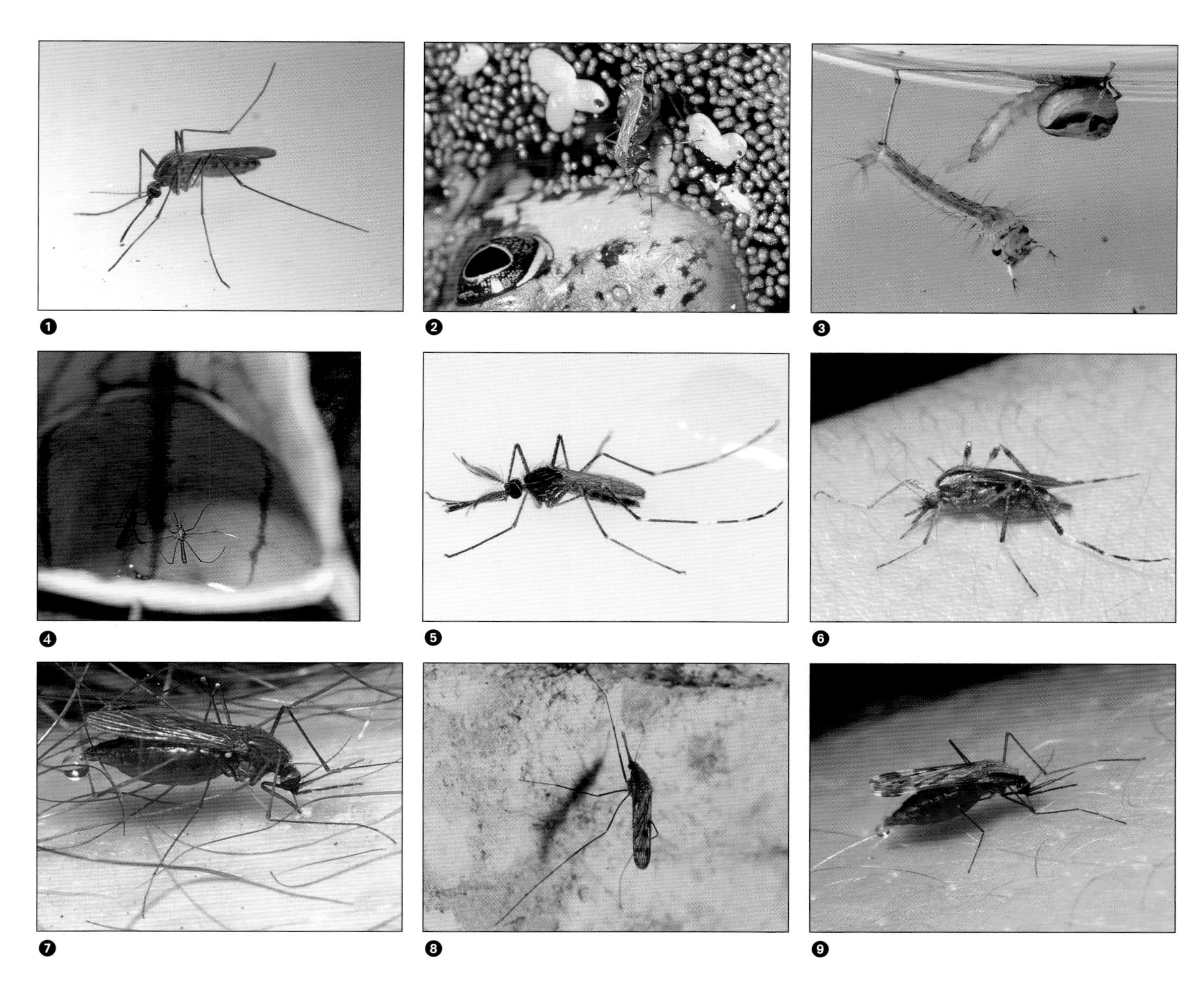

FAMILY CULICIDAE (continued) ❶ The **Northern House Mosquito** (***Culex pipiens***) feeds mostly on birds, but can be a significant "people biter" around dusk – a problem, as this mosquito is a significant carrier of West Nile virus. Adults of this species often spend the winter in houses. ❷ ❸ Many mosquitoes do not bite, and many of those that do bite prefer victims other than humans. ***Culex territans***, for example, prefers to bite frogs. ❹ These ***Wyeomyia smithii*** females are laying eggs in a pitcher plant. Larvae develop inside pitcher plants, where they feed in the microbe-rich soup made up of the plant's trap fluids and the decomposing bodies of its victims. Adults of this species do not bite in northeastern North America, but do bite in the southeast. ❺ ***Orthopodomyia*** species, like this male ***O. signifera***, are beautifully marked mosquitoes that breed in tree holes. This is a mostly neotropical genus, with only a couple of species ranging north to Canada; they are not pests. ❻ ***Psorophora ciliata***, or the **"Gallinipper,"** is the largest eastern mosquito likely to bite you, and it is equipped to bite right through a fairly heavy shirt. Larvae start out as filter feeders, later becoming active predators of other mosquito larvae and other aquatic invertebrates. Subfamily Anophelinae ❼ ***Anopheles***, the infamous genus responsible for malaria transmission, can be recognized by the long palpi, which look like an extra pair of antennae between the antennae and the mouthparts. This female ***A. barberi*** is concentrating her blood meal by pumping excess fluid out of her anus. ❽ Several mosquito species can be found in sheltered places during the fall and winter months. This ***Anopheles earlei*** is resting on the ceiling of a cave in Ontario. ❾ ***Anopheles punctipennis*** is easily recognized by the distinctive pattern of dark and light scales on its wings.

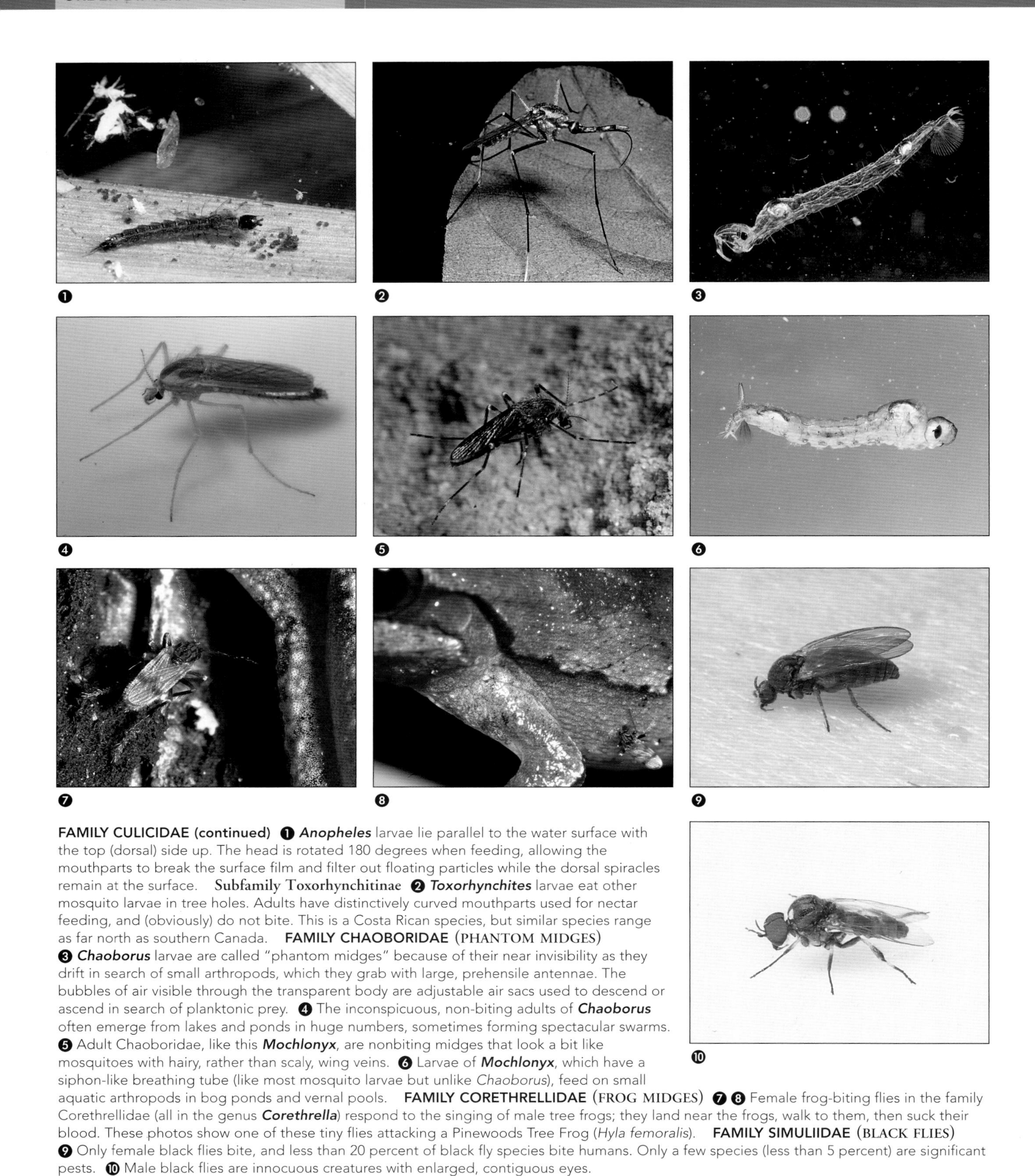

FAMILY CULICIDAE (continued) ❶ ***Anopheles*** larvae lie parallel to the water surface with the top (dorsal) side up. The head is rotated 180 degrees when feeding, allowing the mouthparts to break the surface film and filter out floating particles while the dorsal spiracles remain at the surface. Subfamily Toxorhynchitinae ❷ ***Toxorhynchites*** larvae eat other mosquito larvae in tree holes. Adults have distinctively curved mouthparts used for nectar feeding, and (obviously) do not bite. This is a Costa Rican species, but similar species range as far north as southern Canada. **FAMILY CHAOBORIDAE** (PHANTOM MIDGES) ❸ ***Chaoborus*** larvae are called "phantom midges" because of their near invisibility as they drift in search of small arthropods, which they grab with large, prehensile antennae. The bubbles of air visible through the transparent body are adjustable air sacs used to descend or ascend in search of planktonic prey. ❹ The inconspicuous, non-biting adults of ***Chaoborus*** often emerge from lakes and ponds in huge numbers, sometimes forming spectacular swarms. ❺ Adult Chaoboridae, like this ***Mochlonyx***, are nonbiting midges that look a bit like mosquitoes with hairy, rather than scaly, wing veins. ❻ Larvae of ***Mochlonyx***, which have a siphon-like breathing tube (like most mosquito larvae but unlike *Chaoborus*), feed on small aquatic arthropods in bog ponds and vernal pools. **FAMILY CORETHRELLIDAE** (FROG MIDGES) ❼ ❽ Female frog-biting flies in the family Corethrellidae (all in the genus ***Corethrella***) respond to the singing of male tree frogs; they land near the frogs, walk to them, then suck their blood. These photos show one of these tiny flies attacking a Pinewoods Tree Frog (*Hyla femoralis*). **FAMILY SIMULIIDAE** (BLACK FLIES) ❾ Only female black flies bite, and less than 20 percent of black fly species bite humans. Only a few species (less than 5 percent) are significant pests. ❿ Male black flies are innocuous creatures with enlarged, contiguous eyes.

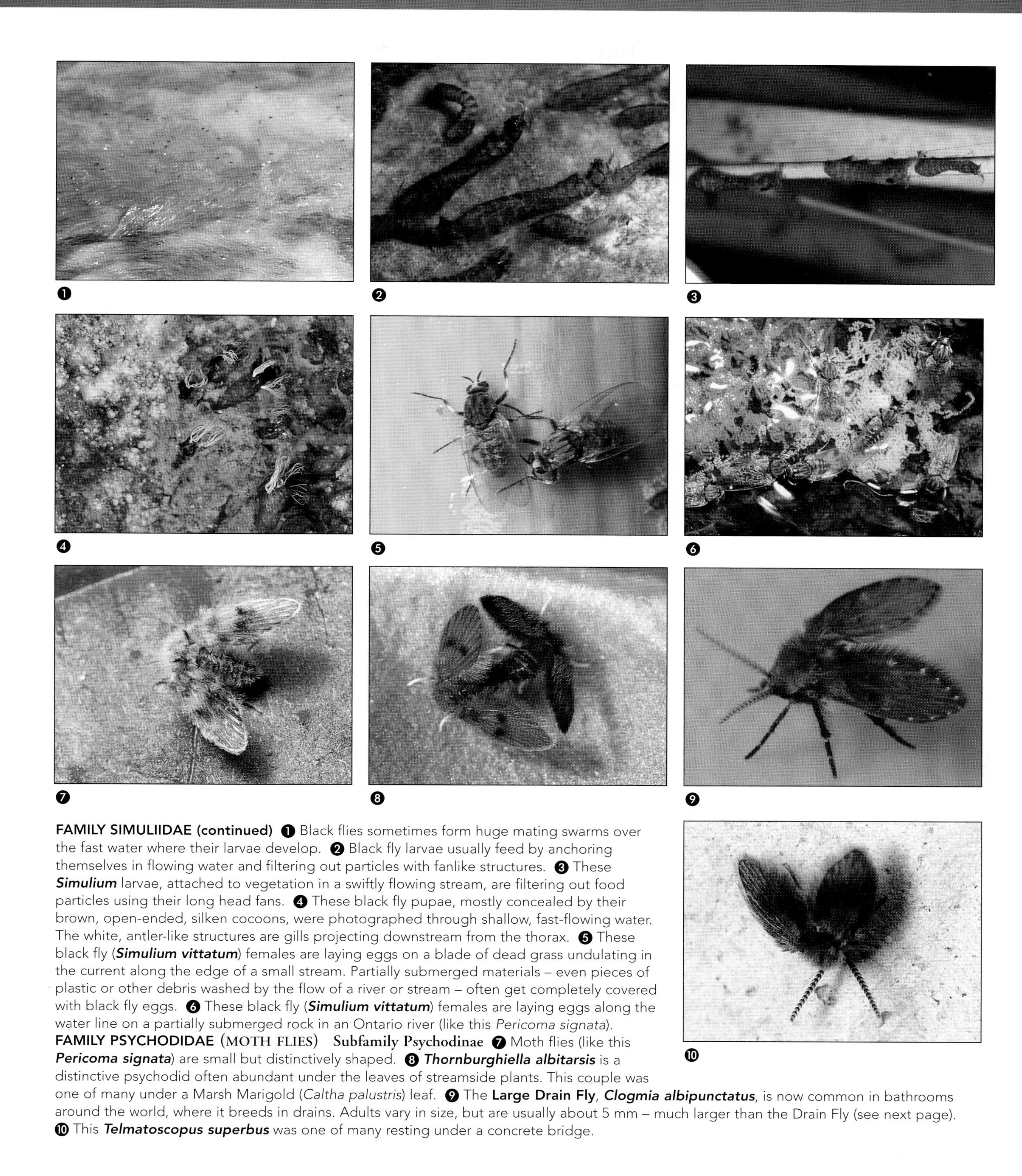

FAMILY SIMULIIDAE (continued) ❶ Black flies sometimes form huge mating swarms over the fast water where their larvae develop. ❷ Black fly larvae usually feed by anchoring themselves in flowing water and filtering out particles with fanlike structures. ❸ These ***Simulium*** larvae, attached to vegetation in a swiftly flowing stream, are filtering out food particles using their long head fans. ❹ These black fly pupae, mostly concealed by their brown, open-ended, silken cocoons, were photographed through shallow, fast-flowing water. The white, antler-like structures are gills projecting downstream from the thorax. ❺ These black fly (***Simulium vittatum***) females are laying eggs on a blade of dead grass undulating in the current along the edge of a small stream. Partially submerged materials – even pieces of plastic or other debris washed by the flow of a river or stream – often get completely covered with black fly eggs. ❻ These black fly (***Simulium vittatum***) females are laying eggs along the water line on a partially submerged rock in an Ontario river (like this *Pericoma signata*).
FAMILY PSYCHODIDAE (MOTH FLIES) **Subfamily Psychodinae** ❼ Moth flies (like this ***Pericoma signata***) are small but distinctively shaped. ❽ ***Thornburghiella albitarsis*** is a distinctive psychodid often abundant under the leaves of streamside plants. This couple was one of many under a Marsh Marigold (*Caltha palustris*) leaf. ❾ The **Large Drain Fly**, ***Clogmia albipunctatus***, is now common in bathrooms around the world, where it breeds in drains. Adults vary in size, but are usually about 5 mm – much larger than the Drain Fly (see next page). ❿ This ***Telmatoscopus superbus*** was one of many resting under a concrete bridge.

FAMILY PSYCHODIDAE (continued) ❶ ***Psychoda alternata***, one of the most widespread and common North American moth flies, can be found along the edges of streams and ponds as well as on artificial biofilms such as those formed on sewage plant trickling filters. This species is often called the **Drain Fly** because of its habit of breeding in bathtub and sink drains. **Subfamily Phlebotominae** ❷ Sand flies (blood-feeding psychodids in the subfamily Phlebotominae) normally do not bite people in northeastern North America, but tropical sand flies are significant disease vectors. This is a ***Lutzomyia***, photographed in Cuba. **FAMILY ANISOPODIDAE (WOOD GNATS)** ❸ Common wood gnats in the genus ***Sylvicola*** develop in decomposing vegetation, including backyard compost. **FAMILY SCATOPSIDAE (MINUTE BLACK SCAVENGER FLIES)** **Subfamily Scatopsinae** ❹ Minute black scavenger flies in the family Scatopsidae often occur in decomposing plant material, although they are easily overlooked because of their small size (usually 1–3 mm). ❺ ***Scatopse notata*** is a common species that often develops large populations in back-yard compost. **Subfamily Aspistinae** ❻ ❼ Scatopsids in the genus ***Aspistes*** can be common on large sand dunes, and the male of this mating couple is disappearing under the sand with his burrowing mate. Larvae of ***Aspistes*** (and all scatopsids in the small subfamily Aspistinae) remain unknown. **FAMILY BLEPHARICERIDAE (NET-WINGED MIDGES)** ❽ Net-winged midge larvae are superbly adapted for life in fast water; each section is made up of two sucker-like disks that cling to the rock. ❾ The associated adults are inconspicuous, long-legged insects with oddly creased wings; look for them hanging under leaf tips. Females are predators, with grasping claws and sharp mandibles (absent in male). ***Blepharicera*** is the only genus in eastern North America.

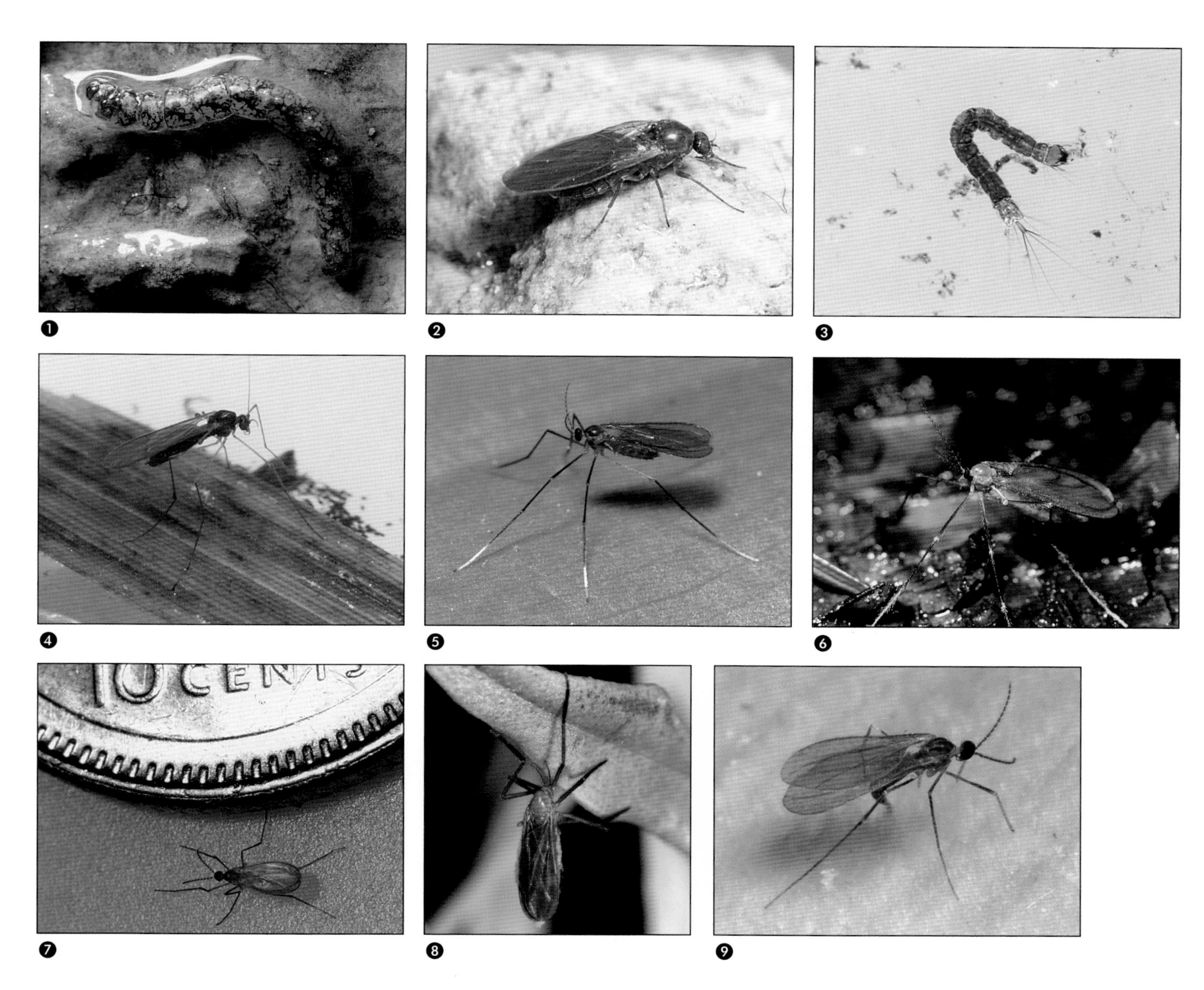

FAMILY THAUMALEIDAE (SEEPAGE FLIES) ❶ ❷ Thaumaleidae larvae are characteristic of thin films of water, in which they zip along like little sidewinders. The associated adults are secretive and rarely seen. **FAMILY DIXIDAE** (DIXID MIDGES) ❸ ❹ Dixid midge larvae like this ***Dixella*** often take a characteristic "U" position in the surface film. Associated adults are inconspicuous and much less often noticed than the larvae. **FAMILY CECIDOMYIIDAE** (GALL MIDGES) ❺ ❻ ❼ Gall midges are usually tiny flies, with few wing veins. ❽ This brightly colored gall midge can be recognized as a ***Planetella*** by the hoodlike thorax projecting over the head. Larvae develop in sedges. ❾ Despite the family name "gall midge" some cecidomyiids live in wood, fungi or other plant tissue. This female is laying eggs in a recently cut willow log.

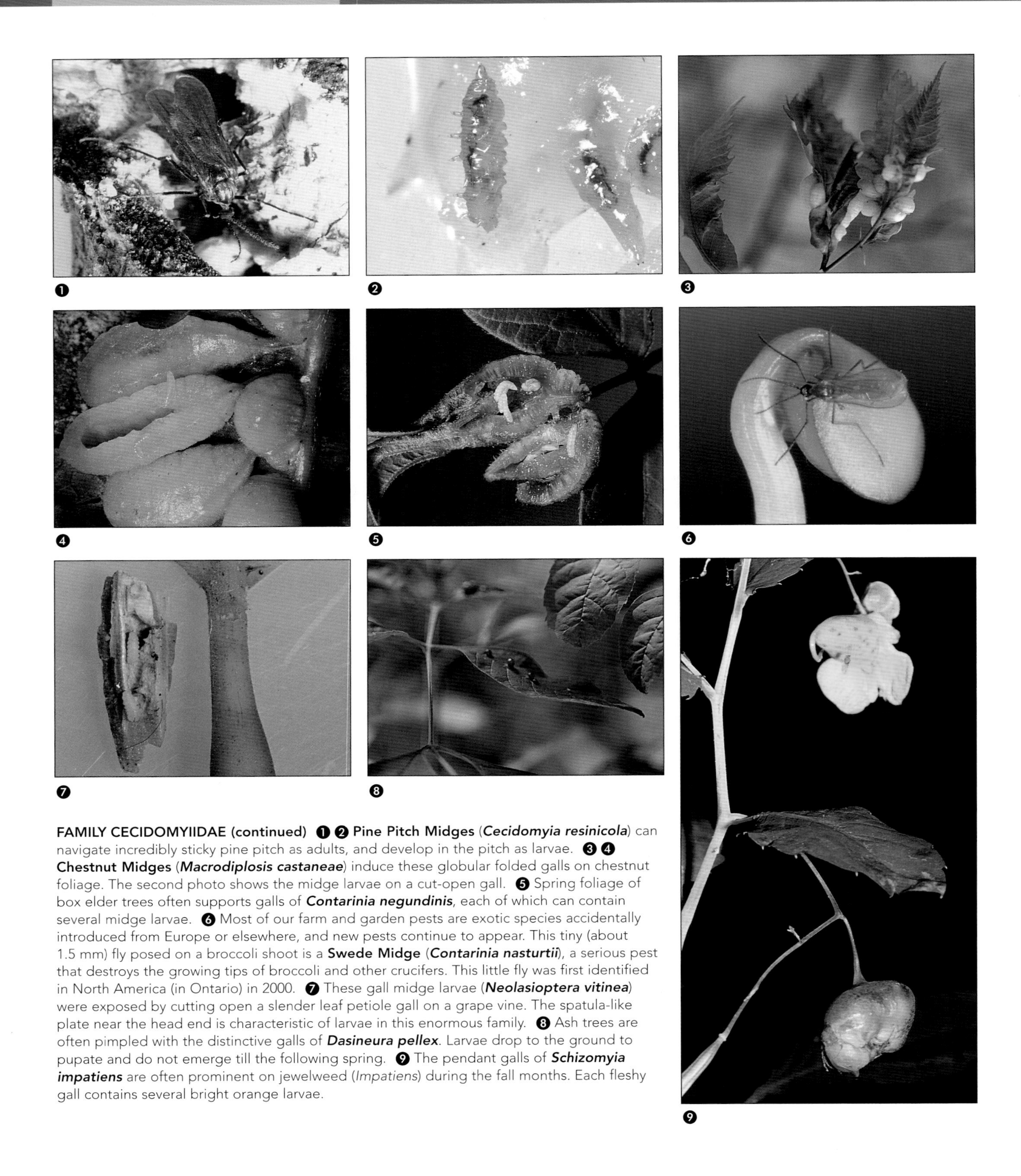

FAMILY CECIDOMYIIDAE (continued) ❶ ❷ **Pine Pitch Midges** (***Cecidomyia resinicola***) can navigate incredibly sticky pine pitch as adults, and develop in the pitch as larvae. ❸ ❹ **Chestnut Midges** (***Macrodiplosis castaneae***) induce these globular folded galls on chestnut foliage. The second photo shows the midge larvae on a cut-open gall. ❺ Spring foliage of box elder trees often supports galls of ***Contarinia negundinis***, each of which can contain several midge larvae. ❻ Most of our farm and garden pests are exotic species accidentally introduced from Europe or elsewhere, and new pests continue to appear. This tiny (about 1.5 mm) fly posed on a broccoli shoot is a **Swede Midge** (***Contarinia nasturtii***), a serious pest that destroys the growing tips of broccoli and other crucifers. This little fly was first identified in North America (in Ontario) in 2000. ❼ These gall midge larvae (***Neolasioptera vitinea***) were exposed by cutting open a slender leaf petiole gall on a grape vine. The spatula-like plate near the head end is characteristic of larvae in this enormous family. ❽ Ash trees are often pimpled with the distinctive galls of ***Dasineura pellex***. Larvae drop to the ground to pupate and do not emerge till the following spring. ❾ The pendant galls of ***Schizomyia impatiens*** are often prominent on jewelweed (*Impatiens*) during the fall months. Each fleshy gall contains several bright orange larvae.

FAMILY CECIDOMYIIDAE (continued) ❶ A tiny midge called ***Rhabdophaga strobiloides*** induces willow "cone" galls, possibly the most familiar midge galls. Many other gall midges cause a variety of other willow galls. ❷ Some Cecidomyiidae are beneficial predators, and these ***Aphidoletes*** were purchased for use as biological control agents in a local greenhouse. **FAMILY BIBIONIDAE** (MARCH FLIES) **Subfamily Bibioninae** ❸ March flies usually develop underground where they feed on roots and other organic material, and ***Dilophus*** females like this one use their large fore tibial spines to dig a deep (several centimeters) burrow in which eggs are laid. *Doliphus* adults are common on flowers. ❹ ❺ Like many insects in which the males form swarms, March fly males are holoptic (they have enlarged eyes that meet at the top). These photographs show a male and a copulating pair of the common (over 50 North American species) genus ***Bibio***. Swarms of male *Bibio* are common sights in early spring and late fall. **Subfamily Pleciinae** ❻ This is a male of the only North American ***Penthetria***, the widespread ***P. heteroptera***. ❼ This is the famous "**Lovebug**" (***Plecia nearctica***), in a characteristic pose. Lovebugs are tremendously abundant in the southern United States although *Plecia* is a mostly tropical genus, with only two species in North America. **FAMILY MYCETOPHILIDAE** (FUNGUS GNATS) **Subfamily Mycetophilinae** ❽ Our most commonly encountered fungus gnats belong to the typical genus ***Mycetophila***, with almost 100 described North American species. Fungus gnats can often be seen in frenetic mating aggregations on the lower surfaces of shelf fungi. ❾ Many fungus gnats are active on warm winter days, and adults can often be located in sheltered areas even in the coldest months. This group of ***Mycetophila*** adults was exposed by peeling loose bark off a dead tree in January.

FAMILY MYCETOPHILIDAE (continued) Subfamily Sciophilinae ❶ ***Boletina*** species are often seen walking or mating on the snow surface. ❷ ❸ The largest, and among the most colorful, fungus gnats in northeastern North America belong to the genus ***Leptomorphus***. This dark species is ***L. subcaeruleus*** and the orange species is ***L. bifasciatus***. ❹ This ***Docosia*** is a particularly colorful fungus gnat. ❺ Many fungus gnats are found in caves. ***Gnoriste macroides***, distinctive for its long mouthparts, occurs in caves but is often seen on flowers near cave entrances. ❻ This fungus gnat (an undescribed ***Speolepta*** sp.) lives only deep in caves, where its pupae are suspended from ceilings by silken threads. Subfamily Keroplatinae ❼ Keroplatinae, like this large flower-visiting fungus gnat (***Asindulum montanum***) are sometimes treated as a separate family, Keroplatidae. ❽ ***Macrocera*** species, some of which live in caves, have exceptionally long legs and antennae. Like other Keroplatinae, *Macrocera* larvae are predators. ❾ Larvae of Keroplatinae often make slimy tube-like shelters from which they feed on fungi or other animals, sometimes snaring prey in mucous webs or snares. Some species, like this New Zealand glowworm (***Arachnocampa luminosa***) are bioluminescent, their light attracting small flies into sticky strands suspended below the predaceous glowworm larva.

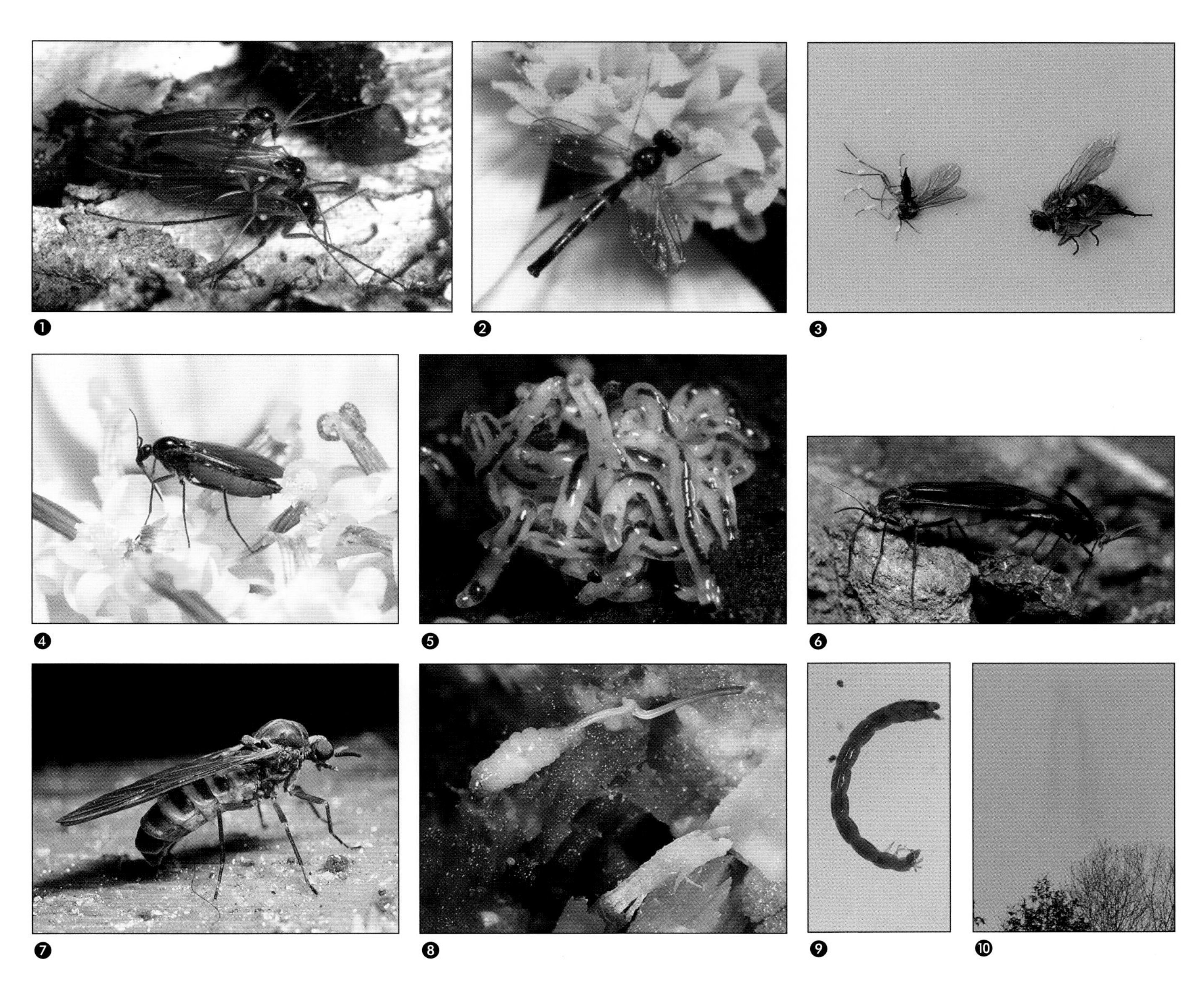

FAMILY MYCETOPHILIDAE (continued) Subfamily Ditomyiinae ❶ These ***Ditomyia*** are mating and laying eggs in the bark of an oak tree. Ditomyiinae is now often treated as a separate family, Ditomyiidae. Subfamily Lygistorrhininae ❷ The only North American species in the subfamily Lygistorrhininae (now widely treated as a separate family, Lygistorrhinidae) is the southeastern species ***Lygistorrhina sanctaecatharinae***. **FAMILY SCIARIDAE** (DARK-WINGED FUNGUS GNATS) ❸ Most people who have houseplants have seen dark winged fungus gnats in their potting soil, and yellow sticky cards used to monitor greenhouse insects usually trap a few of these common little flies. This card has trapped a dark-winged fungus gnat (***Bradysia*** sp., left) and a predacious fly in the family Muscidae. ❹ Dark-winged fungus gnats are very common, although usually overlooked because of their small size. Some of the genera with long mouthparts, like ***Eugnoriste occidentalis***, are frequent on flowers. ❺ Larvae of some Sciaridae occur in masses under bark or in fungi. Some sciarid larvae pupate in a communal cocoon before developing into immobile adults that mate and lay eggs right in the cocoon. ❻ Some dark-winged fungus gnats, such as this mating pair of ***Odontosciara nigra***, can be very abundant on rotting wood. Females of this species (with a red abdomen) are hotly contested by competing males; once in copula, males try and stay attached for a long period. **FAMILY AXYMYIIDAE** (AXYMYIIDS) ❼ Axymyiidae, found only in permanently wet wood, is one of the most habitat-restricted families of insects. This female ***Axymyia furcata*** (eastern North America's only species of axymyiid) is laying eggs in a half-submerged log. ❽ Larvae of ***Axymyia*** have a long, retractile breathing tube with a hard, drill-like apex. **FAMILY CHIRONOMIDAE** (MIDGES) Subfamily Chironominae ❾ The red color of this ***Chironomus*** larva comes from hemoglobin that helps them respire in low-oxygen environments. Note the complete head capsule (typical of Nematocera), the prolegs right behind the head and at the tip of the abdomen (typical of chironomid midges) and the long ventral tubules near the end of the body. ❿ Many midges form enormous male swarms above prominent markers. The smokelike columns above these treetops are made up of thousands of male chironomids.

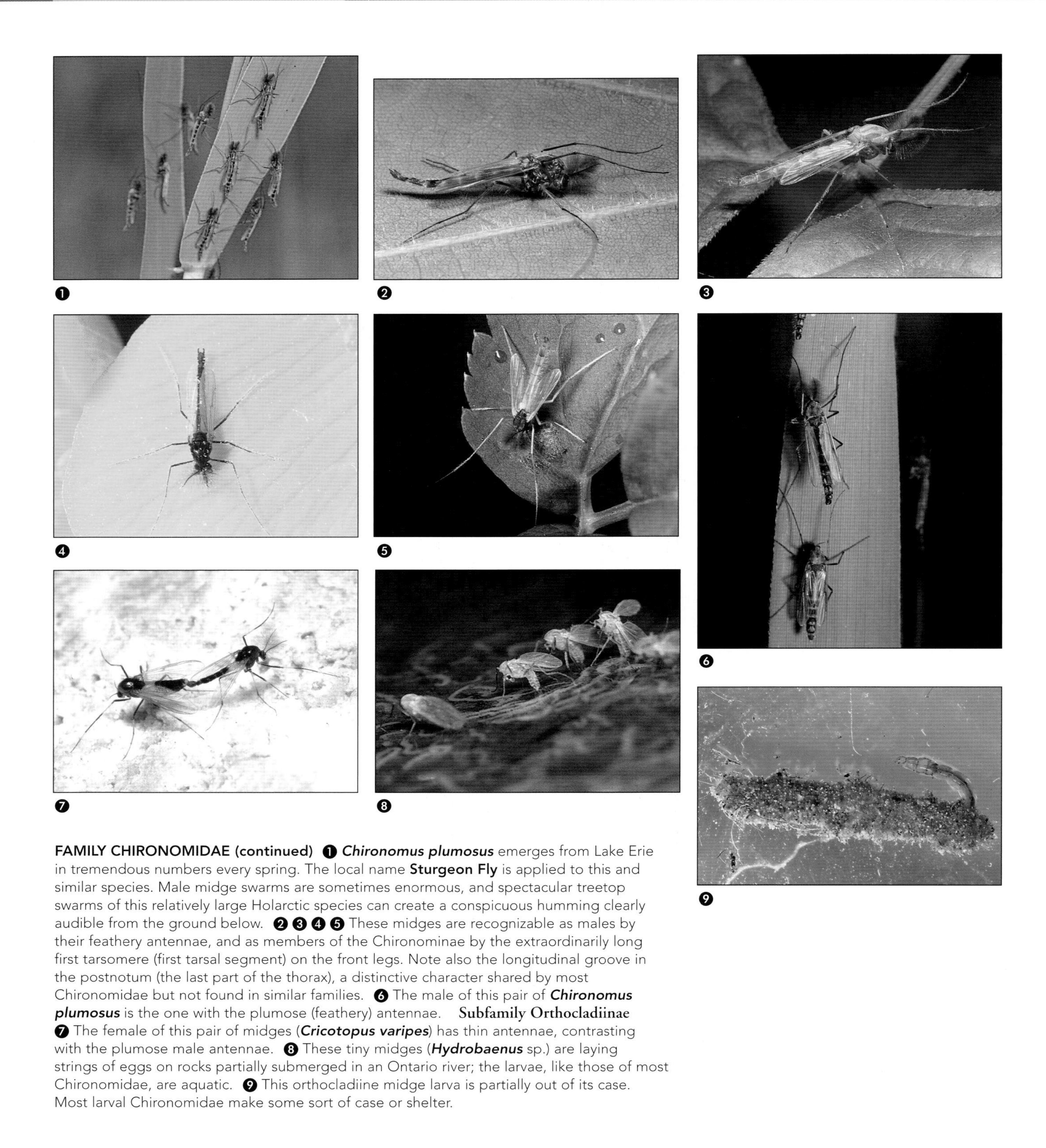

FAMILY CHIRONOMIDAE (continued) ❶ ***Chironomus plumosus*** emerges from Lake Erie in tremendous numbers every spring. The local name **Sturgeon Fly** is applied to this and similar species. Male midge swarms are sometimes enormous, and spectacular treetop swarms of this relatively large Holarctic species can create a conspicuous humming clearly audible from the ground below. ❷❸❹❺ These midges are recognizable as males by their feathery antennae, and as members of the Chironominae by the extraordinarily long first tarsomere (first tarsal segment) on the front legs. Note also the longitudinal groove in the postnotum (the last part of the thorax), a distinctive character shared by most Chironomidae but not found in similar families. ❻ The male of this pair of ***Chironomus plumosus*** is the one with the plumose (feathery) antennae. Subfamily Orthocladiinae ❼ The female of this pair of midges (***Cricotopus varipes***) has thin antennae, contrasting with the plumose male antennae. ❽ These tiny midges (***Hydrobaenus*** sp.) are laying strings of eggs on rocks partially submerged in an Ontario river; the larvae, like those of most Chironomidae, are aquatic. ❾ This orthocladiine midge larva is partially out of its case. Most larval Chironomidae make some sort of case or shelter.

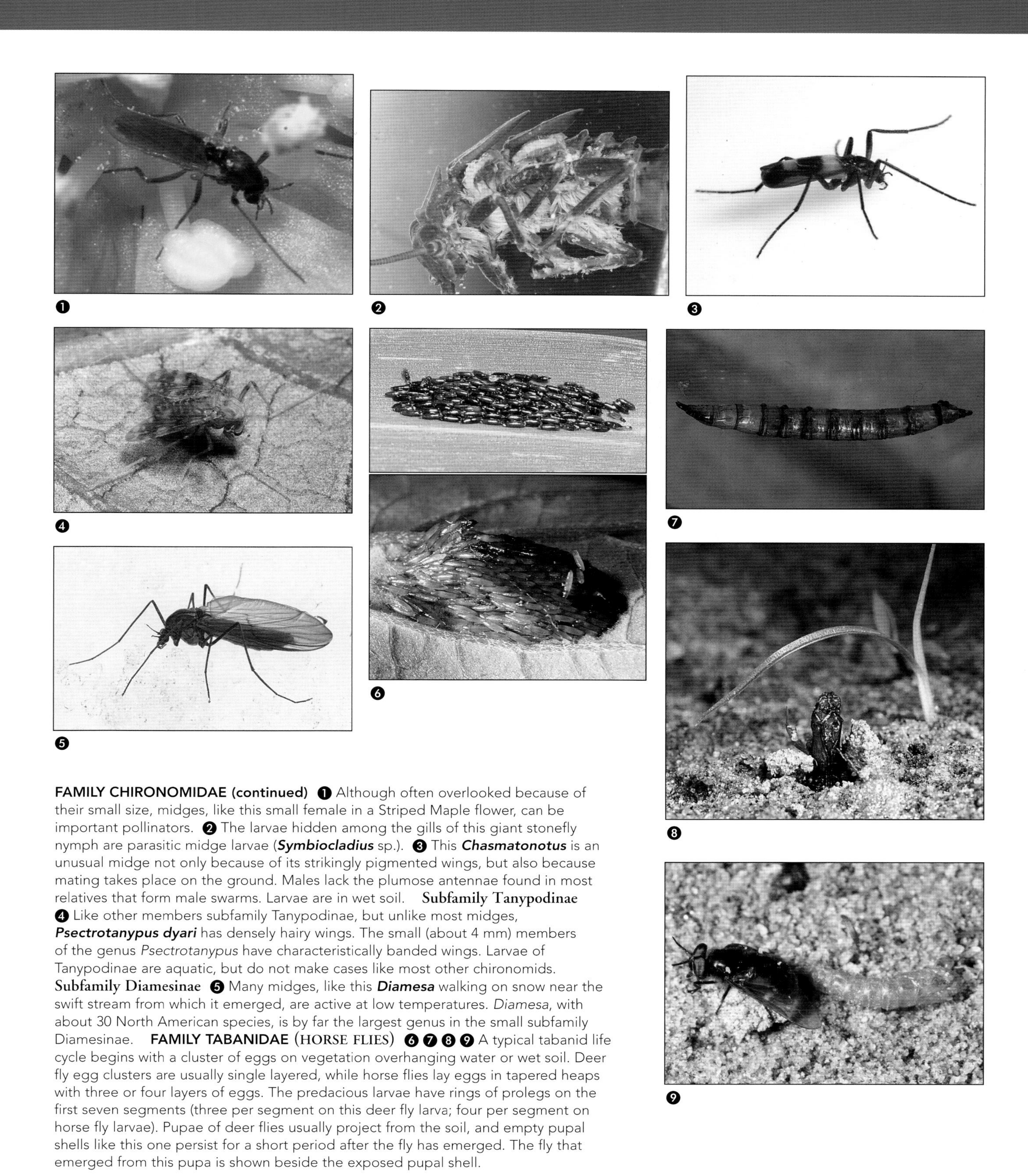

FAMILY CHIRONOMIDAE (continued) ❶ Although often overlooked because of their small size, midges, like this small female in a Striped Maple flower, can be important pollinators. ❷ The larvae hidden among the gills of this giant stonefly nymph are parasitic midge larvae (***Symbiocladius*** sp.). ❸ This ***Chasmatonotus*** is an unusual midge not only because of its strikingly pigmented wings, but also because mating takes place on the ground. Males lack the plumose antennae found in most relatives that form male swarms. Larvae are in wet soil. **Subfamily Tanypodinae** ❹ Like other members subfamily Tanypodinae, but unlike most midges, ***Psectrotanypus dyari*** has densely hairy wings. The small (about 4 mm) members of the genus *Psectrotanypus* have characteristically banded wings. Larvae of Tanypodinae are aquatic, but do not make cases like most other chironomids. **Subfamily Diamesinae** ❺ Many midges, like this ***Diamesa*** walking on snow near the swift stream from which it emerged, are active at low temperatures. *Diamesa*, with about 30 North American species, is by far the largest genus in the small subfamily Diamesinae. **FAMILY TABANIDAE (HORSE FLIES)** ❻ ❼ ❽ ❾ A typical tabanid life cycle begins with a cluster of eggs on vegetation overhanging water or wet soil. Deer fly egg clusters are usually single layered, while horse flies lay eggs in tapered heaps with three or four layers of eggs. The predacious larvae have rings of prolegs on the first seven segments (three per segment on this deer fly larva; four per segment on horse fly larvae). Pupae of deer flies usually project from the soil, and empty pupal shells like this one persist for a short period after the fly has emerged. The fly that emerged from this pupa is shown beside the exposed pupal shell.

FAMILY TABANIDAE (continued) ❶ The business end of a female deer fly is like a diabolical pair of scissors attached to a sponge. ❷ ❸ ❹ ❺ ❻ ❼ ❽ ❾ ***Chrysops***, the deer fly genus, includes over 80 North American species. These photos show females of some common northeastern species: ❷ ***C. aestuans***, ❸ ***C. excitans***, ❹ ***C. frigidus***, ❺ ***C. aberrans***, ❻ ***C. vittatus***, ❼ ***C. indus***, ❽ ***C. montanus*** and ❾ ***C. niger***.

FAMILY TABANIDAE (continued) ❶ ❷ ❸ ❹ These photos are of the female deer flies (***Chrysops***) ***C. univittatus***, ***C. brunneus***, ***C. geminatus*** and ***C. moechus***, respectively. ❺ Only female Tabanidae bite, and the rarely encountered males, like this male ***Chrysops aberrans***, are distinctive for their large eyes. This one is drinking from an early morning dewdrop. ❻ ❼ ***Tabanus atratus*** is a huge (25–30 mm) black horse fly that makes quite an impression when it bites, but these massive flies are rarely abundant and in any case are more likely to bite livestock than people. Larvae live along the margins of ponds and ditches. These photos show a male and a female. ❽ ❾ As is true for other Tabanidae, horse fly males are non-biting insects with large eyes meeting on the top of the head, while female eyes are smaller and separated. These photos show a male and female of ***Tabanus reinwardtii***. This species is common across the continent, but rarely bites people. ❿ The enlarged eyes on this male ***Tabanus superjumentarius*** give it a strikingly big-headed appearance.

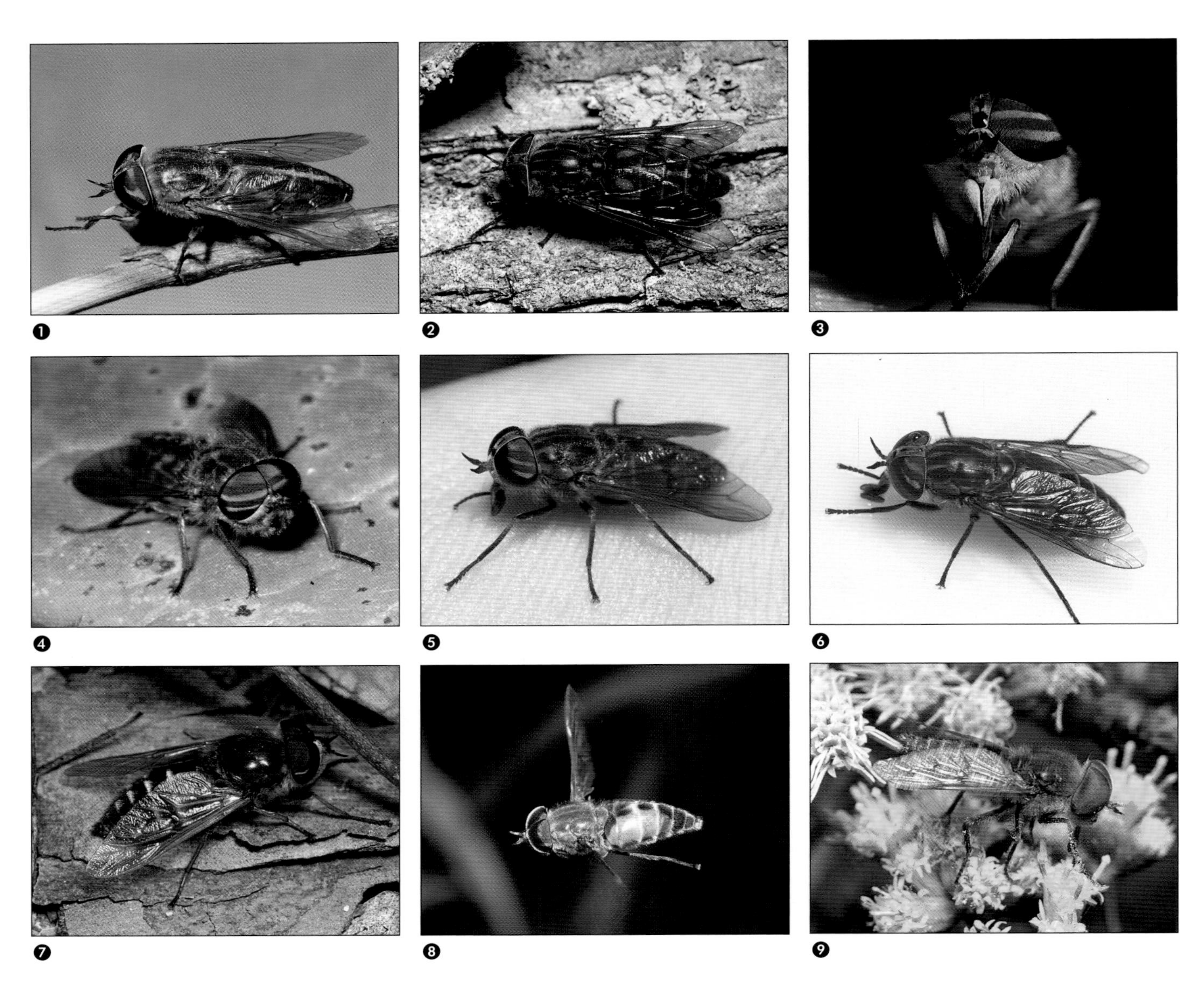

FAMILY TABANIDAE (continued) ❶ Most of the relatively large and noisy pest Tabanidae, the flies most of us know as horse flies and greenheads, are in the genera *Hybomitra* and *Tabanus*. This is the **Striped Horse Fly** (*T. **lineola***). ❷ ❸ Many of our most common nuisance horse flies are in the genus *Hybomitra*; ***H. lasiophthalma*** is among our worst horse fly pests. ❹ Male horse flies, like this ***Hybomitra microcephala***, have enlarged and contiguous eyes. ❺ ❻ ❼ These photos show females of ***Hybomitra trepida*** and ***H. pechumani***, respectively, and a big-eyed male ***H. sodalis***. ❽ Some horse flies form groups of large-eyed males that alternate periods of hovering and of darting at rivals or prospective mates. This ***Hybomitra lasiophthalma*** was hovering at eye level along a peatland pool. ❾ ***Atylotus*** species differ from similar Tabanidae in having relatively pale eyes covered with microscopic hairs. This genus includes several relatively rare species, such as this fen-restricted ***A. palus***.

❶ ❷ ❸ ❹ ❺ ❻ ❼ ❽ ❾ ❿

FAMILY TABANIDAE (continued) ❶ ❷ ***Atylotus bicolor*** is one of the most common eastern *Atylotus* species. These photos show a male (eyes meeting on the top of the head) and female (eyes separated). ❸ ***Stonemyia*** is an entirely North American genus of unknown biology. This species (***S. rasa***) can be common in open forests, but its larvae remain unknown. **FAMILY ATHERICIDAE** (ATHERICID FLIES) ❹ ❺ ❻ ***Atherix lantha*** adults, like this female and male, are rarely noticed, but the odd-looking predacious larvae are common in a variety of running waters. **FAMILY RHAGIONIDAE** (SNIPE FLIES) ❼ ❽ ***Chrysopilus ornatus*** (with clear wings) and ***C. thoracicus*** (with black wings) are probably our most striking snipe flies. ❾ ❿ ***Chrysopilus***, our largest rhagionid genus, also includes many smaller species such as ***C. quadratus*** and the inconspicuous ***C. modestus***.

FAMILY RHAGIONIDAE (continued) ❶ Biting members of the snipe fly genus ***Symphoromyia*** are known as Rocky Mountain black flies. ❷ Male ***Symphoromyia*** do not bite. Biting male flies are found only in Muscidae and some closely related "higher Diptera". ❸ ❹ ***Rhagio tringarius*** (male and female shown here) is an introduced European species that seems to have replaced the similar native *R. hirtus*, and is now the most common large *Rhagio* in the northeast. ❺ This is ***Rhagio hirtus***, a native *Rhagio* that was common until recently, but seems to have been largely replaced by the introduced *R. tringarius*. ❻ ***Rhagio mystaceus*** is one of eastern North America's most common snipe flies. Males seem to stake out territories on tree trunks, although mating usually occurs on foliage or other surfaces. **FAMILY PELECORHYNCHIDAE** (PELECORHYNCHIDS) ❼ ***Glutops***, the only eastern genus in the family Pelecorhynchidae, is rarely collected; this specimen was photographed in the Canadian National Collection of Insects. **FAMILY STRATIOMYIDAE** (SOLDIER FLIES) **Subfamily Stratiomyinae** ❽ Soldier fly adults are often seen sitting motionless on flowers or vegetation near water. The ***Odontomyia cincta*** adult shown here on a daisy is a typical aquatic soldier fly. ❾ Larvae of this soldier fly (***Hedriodiscus dorsalis***) usually occur on the submerged parts of aquatic vascular plants, from which they scrape organic material. ❿ *Hedriodiscus* males, like this ***H. varipes***, are often common on flowers near ponds and lake margins.

FAMILY STRATIOMYIDAE (continued) ❶ The yellow patch on this ***Stratiomys adelpha*** looks like a bee's pollen basket, enhancing this harmless fly's similarity to a stinging bee. ❷ Males of *Stratiomys*, like this ***S. meigenii***, have much larger eyes than females. ❸ ***Stratiomys obesa*** is one of the largest and most common eastern *Stratiomys*. ❹ This ***Stratiomys*** larva is feeding on organic material in a shallow, slow-moving river. The dirty-looking streaks on the rock are silken cases inhabited by midge larvae (Chironomidae). ❺ These ***Stratiomys*** larvae are breathing at the water surface, using the snorkel-like last abdominal segments that end in open spiracles surrounded by water-repellent hairs. ❻ This small soldier fly larva (***Odontomyia***) is floating among some duckweed. The tuft at its hind end is made up of water-repellent hairs that surround a breathing slit (a transverse cleft that conceals the spiracles). ❼ This beautiful species (***Caloparyphus tetraspilus***) develops in thin films of water, like those created when a spring runs down a cliff face. ❽ Thin films of water, like the spring seeping over this clay, support diverse insect communities. The large, flat larva pictured here is a ***Euparyphus*** soldier fly, while the small black larva is a moth fly (*Pericoma*, Psychodidae) and the fat larva partially out of its tube is a crane fly (Tipulidae). ❾ ***Myxosargus nigricornis***, the only eastern species in the genus *Myxosargus*, occurs from southern Ontario to Florida.

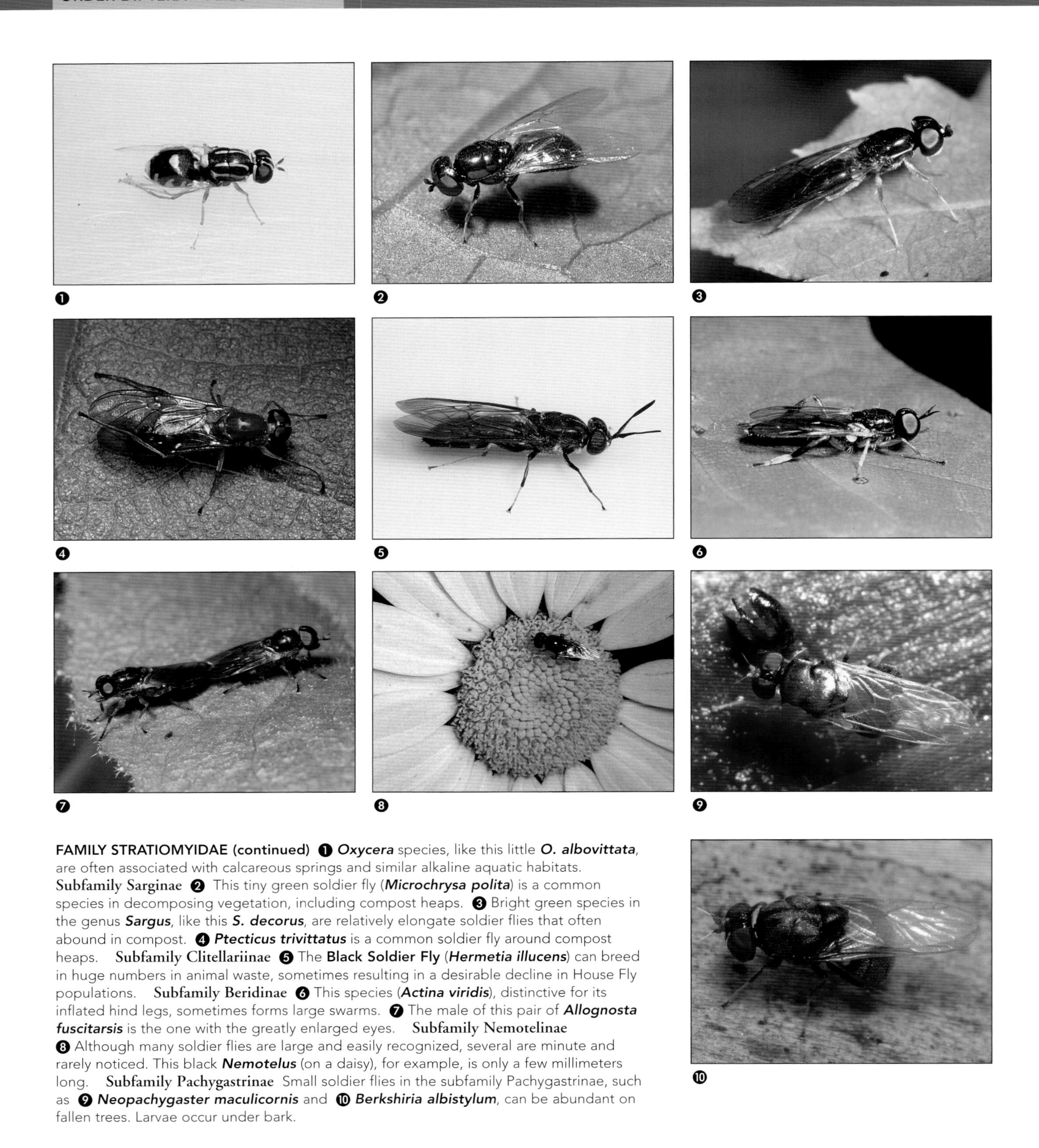

FAMILY STRATIOMYIDAE (continued) ❶ ***Oxycera*** species, like this little ***O. albovittata***, are often associated with calcareous springs and similar alkaline aquatic habitats. **Subfamily Sarginae** ❷ This tiny green soldier fly (***Microchrysa polita***) is a common species in decomposing vegetation, including compost heaps. ❸ Bright green species in the genus ***Sargus***, like this ***S. decorus***, are relatively elongate soldier flies that often abound in compost. ❹ ***Ptecticus trivittatus*** is a common soldier fly around compost heaps. **Subfamily Clitellariinae** ❺ The **Black Soldier Fly** (***Hermetia illucens***) can breed in huge numbers in animal waste, sometimes resulting in a desirable decline in House Fly populations. **Subfamily Beridinae** ❻ This species (***Actina viridis***), distinctive for its inflated hind legs, sometimes forms large swarms. ❼ The male of this pair of ***Allognosta fuscitarsis*** is the one with the greatly enlarged eyes. **Subfamily Nemotelinae** ❽ Although many soldier flies are large and easily recognized, several are minute and rarely noticed. This black ***Nemotelus*** (on a daisy), for example, is only a few millimeters long. **Subfamily Pachygastrinae** Small soldier flies in the subfamily Pachygastrinae, such as ❾ ***Neopachygaster maculicornis*** and ❿ ***Berkshiria albistylum***, can be abundant on fallen trees. Larvae occur under bark.

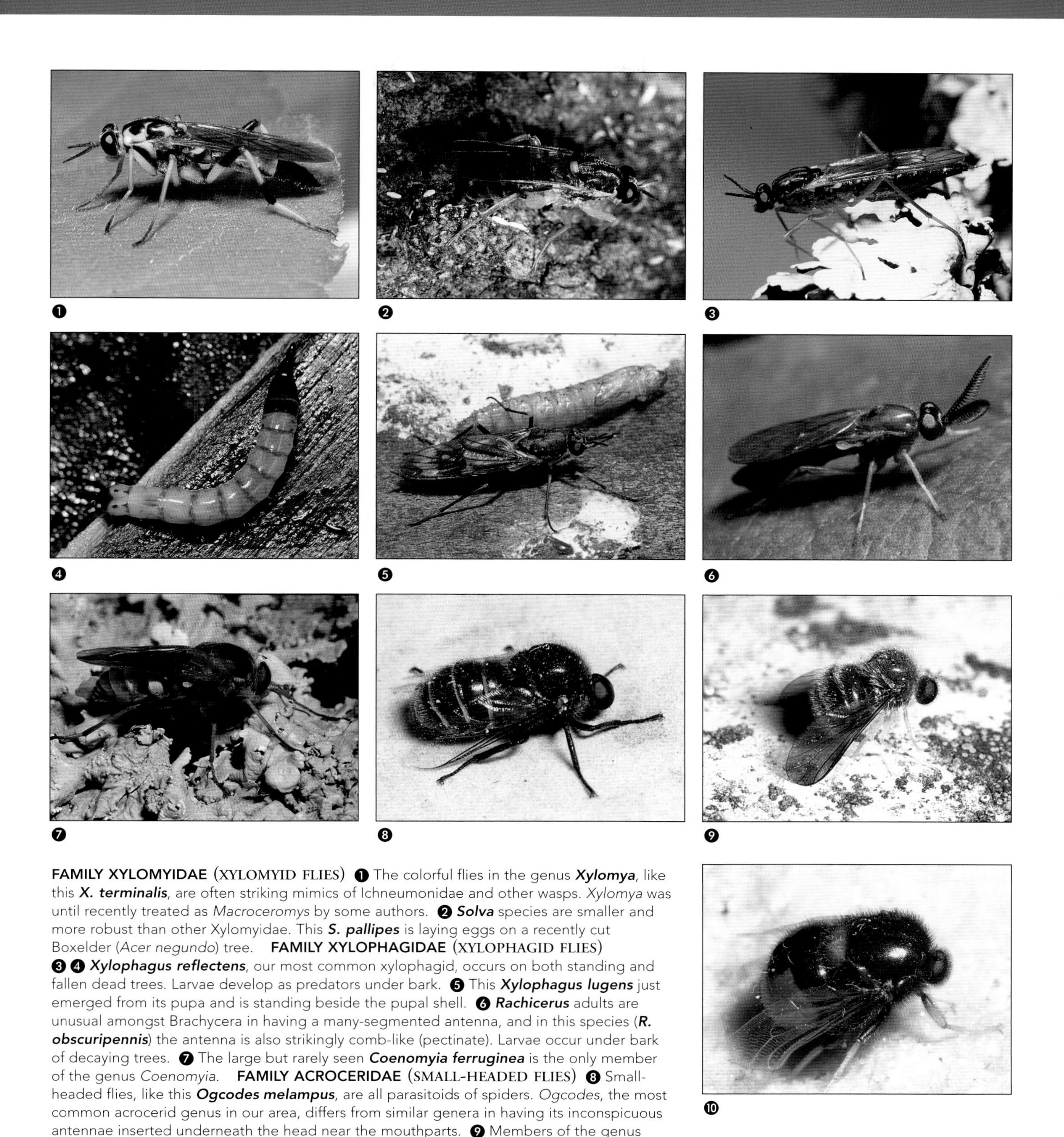

FAMILY XYLOMYIDAE (XYLOMYID FLIES) ❶ The colorful flies in the genus ***Xylomya***, like this ***X. terminalis***, are often striking mimics of Ichneumonidae and other wasps. *Xylomya* was until recently treated as *Macroceromys* by some authors. ❷ ***Solva*** species are smaller and more robust than other Xylomyidae. This ***S. pallipes*** is laying eggs on a recently cut Boxelder (*Acer negundo*) tree. **FAMILY XYLOPHAGIDAE** (XYLOPHAGID FLIES) ❸ ❹ ***Xylophagus reflectens***, our most common xylophagid, occurs on both standing and fallen dead trees. Larvae develop as predators under bark. ❺ This ***Xylophagus lugens*** just emerged from its pupa and is standing beside the pupal shell. ❻ ***Rachicerus*** adults are unusual amongst Brachycera in having a many-segmented antenna, and in this species (***R. obscuripennis***) the antenna is also strikingly comb-like (pectinate). Larvae occur under bark of decaying trees. ❼ The large but rarely seen ***Coenomyia ferruginea*** is the only member of the genus *Coenomyia*. **FAMILY ACROCERIDAE** (SMALL-HEADED FLIES) ❽ Small-headed flies, like this ***Ogcodes melampus***, are all parasitoids of spiders. *Ogcodes*, the most common acrocerid genus in our area, differs from similar genera in having its inconspicuous antennae inserted underneath the head near the mouthparts. ❾ Members of the genus ***Turbopsebius*** (this is ***T. sulphuripes***) lay their eggs right in the funnel-webs of their preferred host spiders (Agelenidae). ❿ This fat ***Pterodontia flavipes***, a rarely encountered small-headed fly, was photographed on a rocky Lake Huron shore.

FAMILY ASILIDAE (ROBBER FLIES) **Subfamily Apocleinae** ❶ ❷ These two photographs show the front and side of a robber fly (***Promachus***) head. Note the beardlike tuft (the mystax) projecting forward above the beak, the dip between the eyes and the stout beak dripping with neurotoxic, protein-dissolving saliva. ❸ ❹ ***Promachus bastardii*** is a large and aggressive robber fly found on sand dunes. The female has a tapered abdomen and the male has a conspicuously expanded silver abdominal tip. ❺ ***Promachus vertebratus*** is a large robber fly with a specific name that describes the vertebrae-like pattern on the abdomen. ❻ ❼ This male ***Efferia albibarbis*** has a silvery band just in front of his bulbous genitalia; the female (here eating a small muscid fly) has a very long and tapered abdominal tip. *Efferia* is now split up by some, putting this species in the genus *Albibarbefferia*. ❽ ❾ Some of our largest robber flies belong to the genus ***Proctacanthus***. This distinctively reddish species (***P. hinei***) is associated with sand dunes, and one of the females shown here has her abdomen buried deep in the sand as she lays her eggs. The other one has just captured a bumble bee.

FAMILY ASILIDAE (continued) ❶ ❷ Our most abundant large robber fly, ***Proctacanthus milbertii***, occurs in a variety of open, dry habitats. The one eating a dragonfly is a male; the one eating a yellowjacket wasp is a female. The tip of the female's abdomen is equipped with spiny plates (called acanthophorites) used for digging in the sand while laying eggs. ❸ This ***Proctacanthus milbertii*** has captured a rove beetle, and in doing so it has attracted some kleptoparasitic acalyptrate flies (*Olcella*, family Chloropidae). You can see a little female kleptoparasite, engorged with beetle blood, just below the fly's beardlike mystax. ❹ This ***Proctacanthella cacopiloga*** is consuming a green plant bug. *P. cacopiloga* is mostly western, but disjunct populations occur around the Great Lakes. **Subfamily Dasypogoninae** ❺ Large robber flies in the genus ***Diogmites basalis*** have a characteristic shape and posture. They are often seen hanging from the vegetation by one or both forelegs as they consume recently captured insects. **Subfamily Asilinae** ❻ The 18 eastern North American species of *Machimus* include many of our most abundant woodland robber flies. This is a mating pair of ***M. paropus***. ❼ ❽ ❾ Our most common genus of robber fly is the genus ***Machimus***, until recently treated as part of the genus *Asilus*. These medium-sized asilids take a variety of prey, including moths, spittlebugs and mosquitoes. ❿ ***Neoitamus*** females, like this ***N. flavofemoratus*** consuming a weevil, have an elongate abdominal tip used for putting eggs in flower heads or leaf sheaths.

FAMILY ASILIDAE (continued) ❶ ***Asilus sericeus*** is the only North American species currently included in the otherwise Old World genus *Asilus. A. sericeus* ranges through the Atlantic states and occurs as far northwest as southern Ontario. **Subfamily Laphystiinae** ❷ Small, robust, silvery flies in the genus ***Laphystia*** are often common on sand dunes. This ***L. flavipes*** has impaled a stiletto fly (*Cyclotelus rufiventris*) through the abdomen. **Subfamily Stichopogoninae** ❸ ❹ Small species in the genus ***Stichopogon*** are among the more common robber flies on beaches. These photos show ***S. argenteus*** and ***S. trifasciatus*** (eating a jumping spider). Both were photographed on Great Lakes beaches. ❺ ***Lasiopogon*** species usually have a short (early) flight period, and different species are associated with particular habitats such as Great Lakes beaches. The species shown here, ***L. marshalli***, is so far known only from flat rocks along the New River, Virginia. **Subfamily Dioctriinae** ❻ ***Eudioctria*** species are superficially similar to the more common *Dioctria*, but lack the club-shaped hind femur and hind tibia of *Dioctria*. ❼ ***Dioctria baumhaueri*** adults often impale their prey, in this case a small bee, through the oral cavity. This very common species has the unusual habit of laying eggs in flight. **Subfamily Stenopogoninae** ❽ This ***Cyrtopogon falto*** female has worked her abdomen down into the leaf litter to lay eggs that will develop into predacious larvae. ❾ This ***Cyrtopogon falto*** is eating a dance fly (*Platypalpus*).

FAMILY ASILIDAE (continued) ❶ This small ***Cyrtopogon*** robber fly has impaled a drosophilid fruit fly. ❷ Tiny robber flies in the genus ***Holopogon*** can often be seen perched on the ends of twigs or other small vantage points. Subfamily Trigonomiminae ❸ Like most predators, robber flies usually have relatively low population densities. Tiny robber flies in the genus ***Holcocephala***, however, are often extremely abundant in sedge meadows, where they use their disproportionately large eyes to scan for small insects such as springtails and midges. Subfamily Laphriinae ❹ Most of the robust robber flies that look like bumble bees belong to the genus ***Laphria***. This pair of ***L. sacrator*** is copulating in the tail-to-tail position typical for the Asilidae. They are able to fly remarkably well while linked this way. ❺ ❻ ❼ ***Laphria*** and a few closely related genera differ from other robber flies in having a flattened, knifelike beak. These photos show how ***L. sacrator*** can slip its beak between the armored plates of a weevil, a long-horned beetle and a click beetle, respectively. ❽ ***Laphria thoracica***, another large bumble bee mimic, lacks the dense yellow leg hairs that characterize *L. sacrator*. ❾ ***Laphria saffrana***, a striking mimic of the Southern Yellowjacket (page 576), is a southeastern species characteristic of the coastal plains. This individual has landed on the ground, a typical habit of the species.

FAMILY ASILIDAE (continued) ❶ This ***Laphria flavicollis***, a species with an all-black abdomen, has just impaled a small scarab beetle. ❷ This ***L. flavicollis*** is eating a muscid fly. ❸ This distinctively red-legged ***Laphria sadales*** has slipped its knifelike beak between the wing covers of a click beetle. ❹ This **Laphria** has impaled a minute beetle between the wing covers. ❺ This **Laphria** (one of a group of similar black species in the *L. cana* group) has impaled an otitid fly. ❻ This ***Laphria sicula*** is eating a small syrphid (*Toxomerus*). This species is difficult to distinguish from other small black *Laphria* without examination of the male genitalia. ❼ This ***Laphria sicula*** has captured a long-legged fly (Dolichopodidae). ❽ Some authors divide **Laphria** into several genera. This species (***L. gilva***), for example, is sometimes placed (along with *L. sicula* and related species) in the genus *Choerades*. ❾ Although **Laphria** most frequently feed by darting out from their perches to impale flying beetles, other insects are taken as well. This ***L. index*** is eating a crane fly.

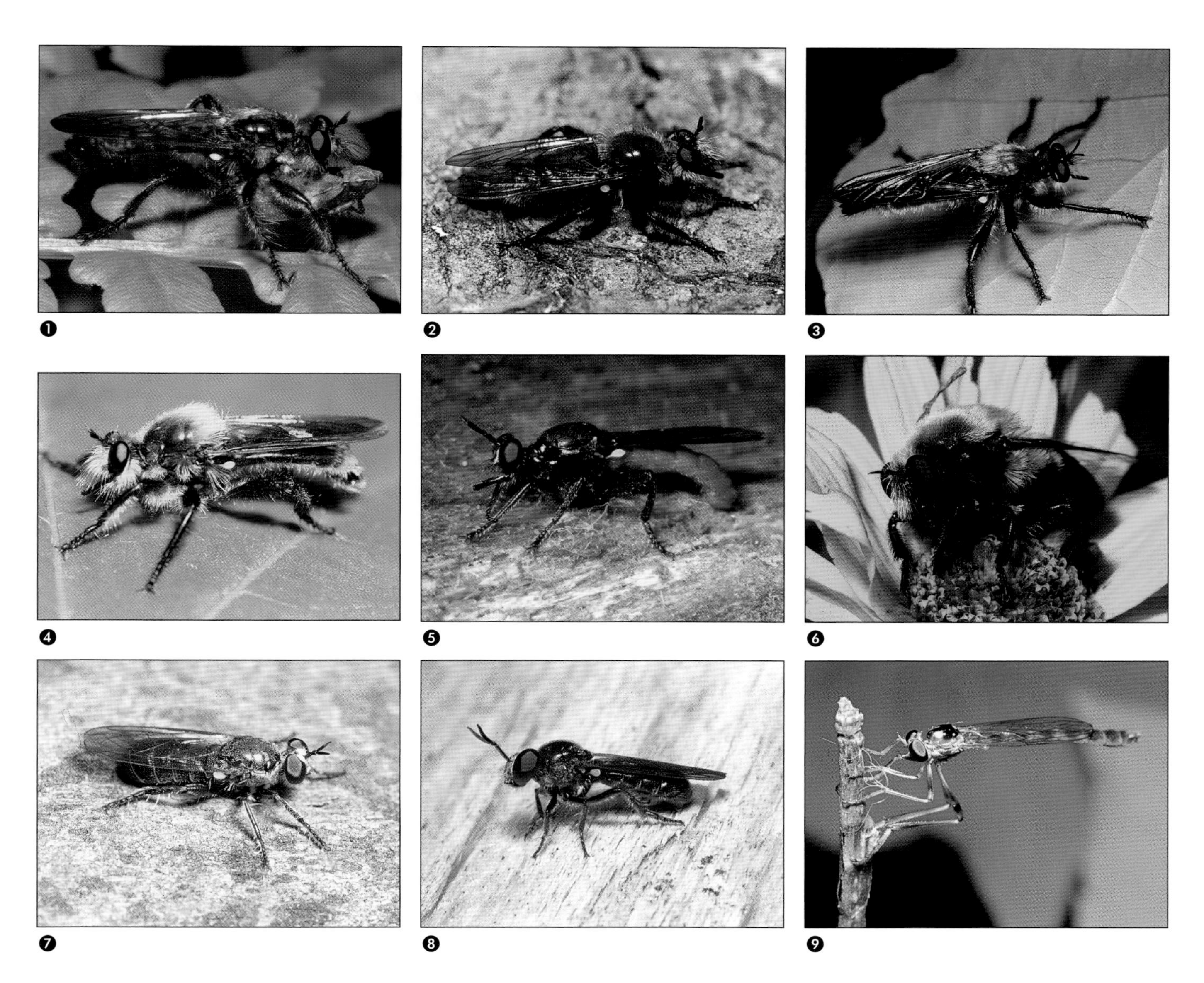

FAMILY ASILIDAE (continued) ❶ This ***Laphria index*** has impaled a spittlebug between its protective forewings. ❷ ***Laphria*** larvae often develop as predators in dead wood, and this species (***L. posticata***) has been reared from pine stumps. ❸ ***Laphria sericea*** is sometimes placed in the genus *Hirtochona*, but most entomologists still put this and most other big fuzzy robber flies in the genus *Laphria*. ❹ ***Laphria janus*** is another bumble-bee-like robber fly sometimes put in the genus *Hirtochona*. ❺ The eastern North American genus ***Lampria***, with three small (10–15 mm) eastern North American species, includes some of our most attractive robber flies. The aptly named ***L. bicolor*** usually occurs in open areas along forest edges. ❻ *Dasylechia* is our only big, fuzzy bumble-bee-like genus other than *Laphria*. ***Dasylechia atrox***, the only species in the genus, is an extremely rare fly with a short, rounded beak unlike the bladelike beak of *Laphria*. (This is a staged photograph… robber flies rarely land on flowers.) ❼ ***Atomosia puella*** is a small robber fly usually found on bare fallen logs. ❽ This small robber fly in the genus ***Cerotainia*** is watching for small prey on a tree trunk. **Subfamily Leptogastrinae** ❾ Slender robber flies (subfamily Leptogastrinae) like this ***Tipulogaster glabrata*** can often be seen flying slowly up and down tree trunks as they search for prey to snatch off the surface. Most other robber flies normally attack flying prey.

FAMILY ASILIDAE Subfamily Leptogastrinae (continued) ❶ ***Beameromyia vulgaris***, one of our smallest robber flies, measures only about 6 mm from head to tail. **FAMILY MYDIDAE** (MYDAS FLIES) ❷ ***Mydas clavatus***, one of our largest flies at 25–30 mm, is distinctive for its long antennae and black and orange color. Look for this species around dead wood in sandy areas. **FAMILY THEREVIDAE** (STILETTO FLIES) ❸ This stiletto fly looks a bit like a robber fly with neither a beard (mystax) nor a depression between the eyes. Species in the genus *Thereva*, like the ***T. strigipes*** shown here, usually occur in wooded areas. ❹ Fuzzy white *Spiriverpa* species, like this ***S. albiceps***, are characteristic of open areas of white sand. This pair was photographed on a beach along the north shore of Lake Superior. ❺ ***Cyclotelus rufiventris*** is a common therevid on inland dunes and similar sandy areas where the predacious larvae lurk underground. ❻ ***Ozodiceromyia notata*** is normally seen on pristine white sand or nearby low vegetation. ❼ Some stiletto flies are quite abundant, although only in the right habitat and the right season. During spring and early summer, for example, this species (***Pallicephala variegata***) is common on the dunes and beaches of Manitoulin Island and other northern Great Lakes shores. ❽ The genus *Penniverpa*, with only the single species ***P. festina***, has a more southerly distribution than the other species illustrated here. **FAMILY SCENOPINIDAE** (WINDOW FLIES) ❾ The **Window Fly**, ***Scenopinus fenestralis***, is often found on the inside surfaces of house windows, but is rarely encountered out of doors.

FAMILY BOMBYLIIDAE (BEE FLIES) Subfamily Bombyliinae ❶ ❷ Bee flies can often be seen hovering over the ground, occasionally dipping down to deposit an egg. This one (***Lordotus gibbus***, a western North American species) attacks the larvae of solitary ground nesting wasps. ❸ Although the mouthparts of this bee fly look threatening, bee flies never bite and the mouthparts are used for nectar feeding. Larvae of this species (***Systoechus vulgaris***) are predators of grasshopper eggs. ❹ ***Anastoechus*** species develop on grasshopper egg pods, and can be abundant in open areas. Only one species, ***A. barbatus,*** ranges north to Canada. ❺ ❻ *Bombylius* species, especially ***B. pygmaeus*** (shown here on leaf litter) and ***B. major*** (shown here taking nectar from a Coltsfoot [*Tussilago farfara*] flower), are welcome harbingers of spring. The larvae of these common and widespread species develop as parasitoids of solitary bees. ❼ Unlike the other three *Bombylius* species illustrated here, ***B. comanche*** is a rarely encountered species in the northeast. This specimen was photographed in southern Ontario, near the northern limit for this mostly midwestern fly. ❽ Known larvae of ***Sparnopolius*** are underground ectoparasites on larvae of scarab beetles. ❾ ***Aldrichia*** species have not yet been associated with host insects.

FAMILY BOMBYLIIDAE (continued) Subfamily Anthracinae ❶ ***Anthrax georgicus*** (widely known as *A. analis*) is an internal parasitoid of tiger beetle larvae, often occurring at high enough densities to have a significant impact on tiger beetle populations. ❷ Most members of the genus ***Anthrax***, including the ***A. irroratus***, shown here resting on wood, attack larvae and pupae of bees and wasps. ❸ This well-armed ***Anthrax irroratus*** pupa has developed at the expense of the dead and deflated bee larva seen beside it. ❹ This ***Anthrax albofasciatus*** probably developed as a parasitoid of a ground-nesting bee or wasp. ❺ This large bee fly (***Xenox tigrinus***) is resting on an old stone wall, where it was probably seeking bee hosts nesting in the stone or structural beams. Subfamily Phthiriinae ❻ Small bee flies in the genus ***Geron*** are parasitic on moth caterpillars. These long-beaked little flies are often common on yellow composite flowers. ❼ Tiny bee flies in the genus ***Tmemophlebia*** (formerly part of *Phthiria*) develop as parasitoids of correspondingly tiny micromoth larvae. This is ***T. coquilletti***. ❽ Some bee flies, such as this pollen-spattered ***Apolysis sigma*** are most often seen at flowers. Nobody knows where the larvae of this genus live. Subfamily Systropodinae ❾ Members of the genus ***Systropus***, shown here in copula, do not look much like typical bee flies. These elongate flies, which look remarkably wasplike in flight, are parasitoids of slug caterpillars (Limacodidae).

FAMILY BOMBYLIIDAE (continued) Subfamily Toxophorinae ❶ Both North American species of ***Lepidophora*** develop either as parasitoids or as kleptoparasites (food thieves) in nests of solitary wasps. This is ***L. lutea***. ❷ Members of this small bee fly genus (***Toxophora***) develop inside potter wasp (Vespidae) nests, either as parasitoids of the wasps or as parasitoids of other parasitoids of potter wasps (a parasitoid that attacks a parasitoid is called a hyperparasitoid). This is ***T. amphitea***, the only member of this mostly southwestern genus with a range extending north to southern Ontario. Subfamily Exoprosopinae ❸ Some of our largest, most colorful and fastest bee flies are in the genera *Exoprosopa* and *Poecilanthrax*. This ***Exoprosopa*** probably developed as a parasitoid of one of the solitary wasps it shares its typical sandy habitat with. ❹ This bee fly (***Exoprosopa fascipennis***) is shooting her eggs into the burrow of a solitary hunting wasp; her larvae will develop as ectoparasitoids on the wasp larvae. ❺ This large ***Poecilanthrax bicellata*** probably developed as an internal parasitoid in a cutworm moth pupa. ❻ Members of the large genus ***Villa*** are probably the most common bee flies in eastern North America. *Villa* species develop as internal parasitoids of a variety of insects, including horse flies. ❼ This ***Villa*** is laying eggs in an old gravel pit peppered with the nests of hunting wasps and ground-nesting bees. ❽ ***Hemipenthes sinuosa*** develops as a hyperparasite on the larvae of parasitic wasps and parasitic flies inside moth caterpillars. ❾ ***Chrysanthrax*** is a mostly southwestern genus, with only a couple of species reaching the northeast and only one (***C. dispar***) reaching eastern Canada. Known larvae are external parasitoids on subterranean larvae of scarab beetles ("white grubs").

FAMILY BOMBYLIIDAE (continued) ❶ ***Dipalta banksi*** is a parasitoid of antlion pupae, and occurs with its host in sandy areas around the Great Lakes. Subfamily Lomatiinae ❷ ***Ogcodocera leucoprocta*** is another distinctive bee fly that remains unknown in the larval stage. Subfamily Ecliminae ❸ ***Thevenemyia*** species, like this ***T. funesta***, are usually found hovering around the bark of dead trees, probably in search of wood-boring beetle larvae to parasitize. **FAMILY EMPIDIDAE** (DANCE FLIES) Subfamily Empidinae ❹ This dance fly female (***Rhamphomyia longicauda***) is preparing for her crepuscular dance by inflating abdominal sacs that will combine with her outstretched and feathered legs to give her as large a presence as possible in the all-female swarm. Males arrive in the swarm with a precious prey item, and presumably choose the biggest, most fecund-looking female. ❺ Males of ***R. longicauda*** lack the leg scales and abdominal sacs seen on females. ❻ Empidinae females cannot hunt for themselves. The female (below) in this pair is dining on a "nuptial gift," a prey item presented to her by the male (above). ❼ The genus ***Rhamphomyia*** is enormous, with over 150 named North American species and many more that have not yet been named. This male, with strongly modified hind legs, is probably ***R. basalis***. ❽ There are more than 80 described North American species in the genus ***Empis***, and many more await discovery. The fly seen here on a hawthorn blossom is either ***E. leptogastra*** or a similar undescribed species. ❾ ***Empis*** species like this pollen-spattered fly are common visitors to buttercup (*Ranunculus*) and other flowers.

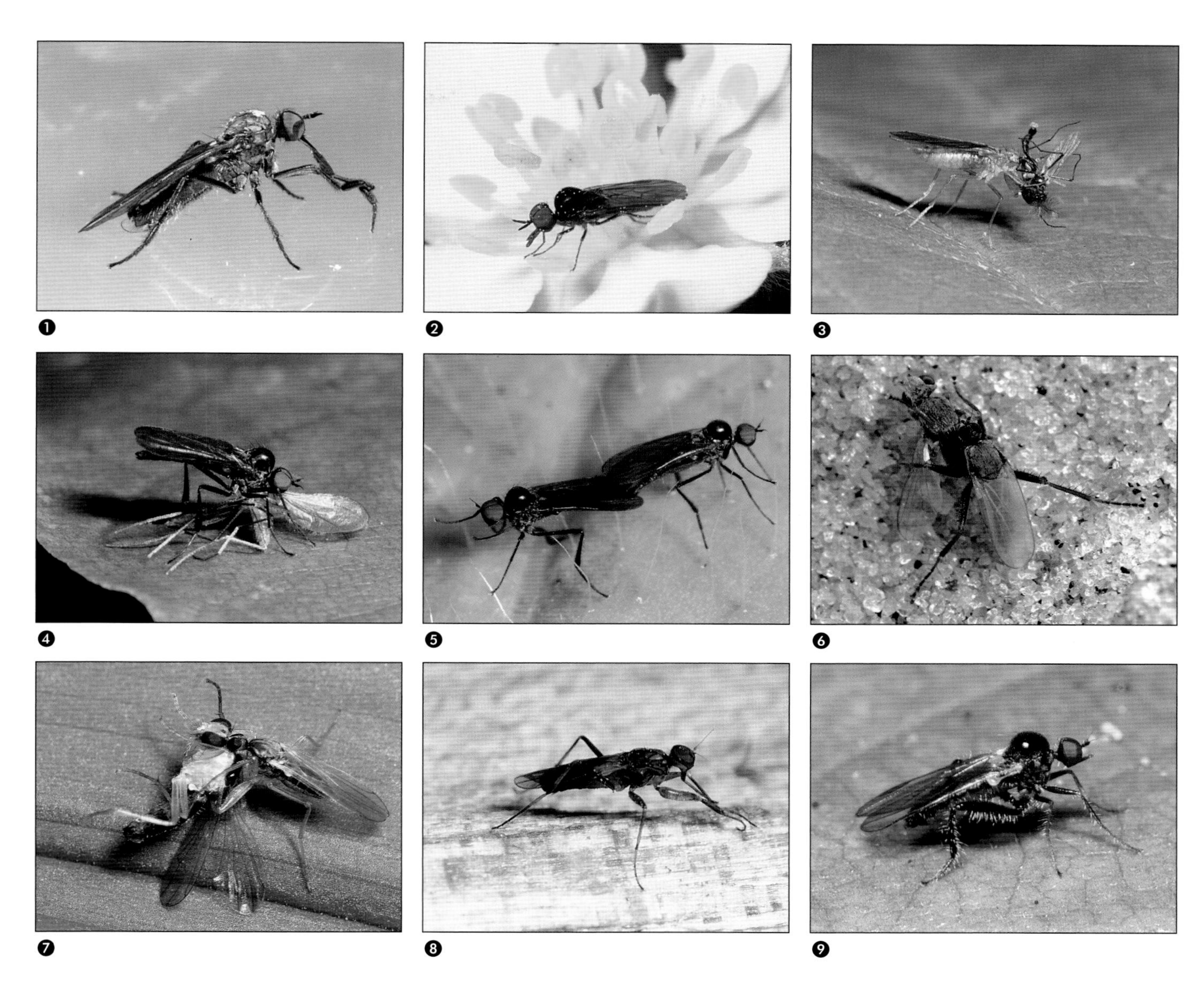

FAMILY EMPIDIDAE (continued) ❶ Males in this genus (***Hilara***) wrap captured insects in silk produced in their swollen front legs before presenting the prey item to the female as a nuptial gift. Prey-carrying males can be seen swarming millimeters above the surface of most lakes and rivers on any spring evening, but they are hard to catch and harder to photograph in the field. This male was photographed indoors. ❷ Dance flies are common, if inconspicuous flower visitors, and this ***Iteaphila*** is feeding on the pollen of a Virginia Strawberry (*Fragaria virginiana*) blossom. **Subfamily Hemerodrominae** ❸ This ***Chelipoda*** is using its swollen raptorial forelegs to eat a minute midge. **Subfamily Ocydrominae** ❹ ❺ One of these humpbacked ***Bicellaria*** dance flies is eating a midge; the others are mating tail-to-tail, a position that demands that the male twist his tail 180 degrees. **Subfamily Tachydrominae** ❻ Fast-running empids in the genus ***Chersodromia*** are common seashore insects, where they actively hunt other insects. This ***C. inusitata*** has captured a *Tethina* (Tethinidae), another characteristic seashore fly. ❼ Dance flies are often seen with small insects impaled on their beaks. These photos show ***Platypalpus holosericus*** eating a scathophagid fly. *Platypalpus* species, especially *P. holosericus*, are among the most common empids. ❽ Some empids, like this ***Tachydromia***, are antlike predators always found on tree trunks and similar surfaces. Although fully winged, they are reluctant to fly. **Subfamily Hybotinae** (now often treated as family Hybotidae rather than a subfamily of Empididae) ❾ Most empids have downward-directed beaks but some, like this ***Hybos reversus***, have a forward-facing beak like a robber fly.

FAMILY EMPIDIDAE (continued) ❶ ❷ Ninebark (*Physocarpus*) flowers attract a variety of empids, including the aptly named genus ***Anthalia***. The dark fly with the short and broad antenna is an *Anthalia* male; the orange flies feeding on pollen are *Anthalia* females. **Subfamily Clinocerinae** ❸ Adults of the dance fly genus ***Clinocera*** can often be seen sitting in groups on wet rocks, waiting to prey on other aquatic insects as they emerge from the water. Larvae of this species (***C. fuscipennis***) develop in thin films of water (madicolous habitats).
❹ These ***Clinocera binotata*** are mating in the middle of a small stream. **FAMILY DOLICHOPODIDAE (LONG-LEGGED FLIES) Subfamily Microphorinae** ❺ The subfamily Microphorinae has been traditionally placed in the Empididae but some authors have treated it as a family (Microphoridae) and it is now treated as a subfamily of Dolichopodidae. Although these ***Microphor*** are on a False Solomon's Seal flower, *Microphor* adults are often kleptoparasitic, stealing prey from spider webs. **Subfamily Dolichopodinae** ❻ Dolichopodid larvae are characteristically shaped, with a truncate hind end surrounded by four or more lobes. ❼ ❽ ❾ Dolichopodids have unusual mouthparts used to grind and ingest small prey including midge larvae and worms.

FAMILY DOLICHOPODIDAE (continued) ❶ This distinctively spot-winged dolichopodid (***Tachytrechus vorax***) often abounds on large expanses of wet sand. ❷ ❸ Males of the enormous genus ***Dolichopus*** (over 300 North American species) often have spectacularly ornamented forelegs used in sexual display. Ornamented hind legs, like those of the ***D. remipes*** male shown in the second photo, are more unusual. ❹ The flaglike black and white foreleg tarsomeres on this male ***Dolichopus dakotensis*** are clearly visible as it takes up honeydew from a leaf. **Subfamily Plagioneurinae** ❺ The only North American member of the Plagioneurinae, ***Plagioneurus univittatus***, ranges from Michigan and Ontario south to the neotropics. **Subfamily Rhaphiinae** ❻ Long-legged flies in the genus ***Rhaphium*** have a distinctive apical arista. **Subfamily Neurigoninae** ❼ This tree-trunk frequenting dolichopodid (***Neurigona***) is eating a globular springtail. **Subfamily Hydrophorinae** ❽ Some dolichopodids, especially ***Hydrophorus*** species, are accomplished water walkers and can often be seen skating along the surfaces of still waters in search of living or drowned prey. ❾ ***Liancalus genualis*** is found on or near thin films of water where springs seep along cliff faces.

FAMILY DOLICHOPODIDAE (continued) Subfamily Sciapodinae ❶ Long-legged flies usually occur near water, but members of this group (the Sciapodinae) are common predators in yards, gardens and even greenhouses. Most of the pretty little garden dolichopodids, like this one, are in the large genus ***Condylostylus***. Subfamily Medeterinae ❷ ***Medetera*** adults are commonly seen striking a "woodpecker pose" on tree trunks. This female is laying eggs on a dead spruce; her larvae will feed on beetle larvae in the wood. Subfamily Diaphorinae ❸ ***Argyra*** is a widespread genus of long-legged flies, including almost 50 North American species. ❹ This tiny dolichopodid (probably the southeastern species ***Chrysotus crosbyi***) is only a few millimeters long, but the silvery reflective male palpi can be seen catching the light from some distance away. This specimen was photographed in an Ontario butterfly house, where it was feeding voraciously on whiteflies.

MUSCOMORPHA – ASCHIZA

FAMILY SYRPHIDAE (FLOWER FLIES OR HOVER FLIES) Subfamily Syrphinae ❺ Flower flies like this ***Syrphus*** belong to a larger group known as Aschiza because there is no suture or "crack" arching over the antennae. ❻ This ***Syrphus*** female was attracted to the sap flow created by weevils feeding and laying eggs in a small willow. ❼ ❽ Syrphidae are often called hover flies, for obvious reasons. These photos show a hovering ***Eupeodes americanus*** and ***Toxomerus geminatus***, respectively. ❾ Small flower flies in the genus ***Toxomerus*** make up the majority of backyard and garden syrphids. The female of this mating pair will soon deposit eggs that will hatch into aphid-eating larvae; so, like most syrphids, these colorful little flies (***T. geminatus***) should be welcome in the garden.

FAMILY SYRPHIDAE (continued) ❶ ***Toxomerus geminatus*** *is* probably the most common northeastern flower fly, especially in gardens. ❷ ***Toxomerus marginatus*** differs from our other very common *Toxomerus*, *T. geminatus,* in having a yellow margin on the sides of the abdomen. ❸❹❺❻ The common name "flower fly" is an apt one for many (but not all) members of this family, and syrphids are important pollinators. These photos of syrphine syrphids on yellow flowers are: ❸ a ***Platycheirus*** on Marsh Marigold (*Caltha palustris*), ❹ a ***Sphaerophoria*** on Lakeside Daisy (*Hymenoxys acaulis*), ❺ a ***Melanostoma*** on Coltsfoot (*Tussilago farfara*) and ❻ a ***Melangyna*** on goldenrod (*Solidago* sp). ❼ Species of ***Chrysotoxum***, like this ***C. flavifrons***, are among the most convincing wasp mimics in the subfamily Syrphinae. ❽ All ***Chrysotoxum*** species have strikingly long antennae, but in ***C. pubescens*** the first two segments (the scape and pedicel) are especially long relative to the third (the first flagellomere). Larvae probably feed on aphids like other Syrphinae, but details of *Chrysotoxum* biology remain unknown. ❾❿ Males and females of ***Xanthogramma flavipes*** have slightly different color patterns; the male is the one with eyes meeting on top of the head. The biology of *X. flavipes*, the only North American member of the genus, remains unknown but European *Xanthogramma* larvae have been found in ant nests.

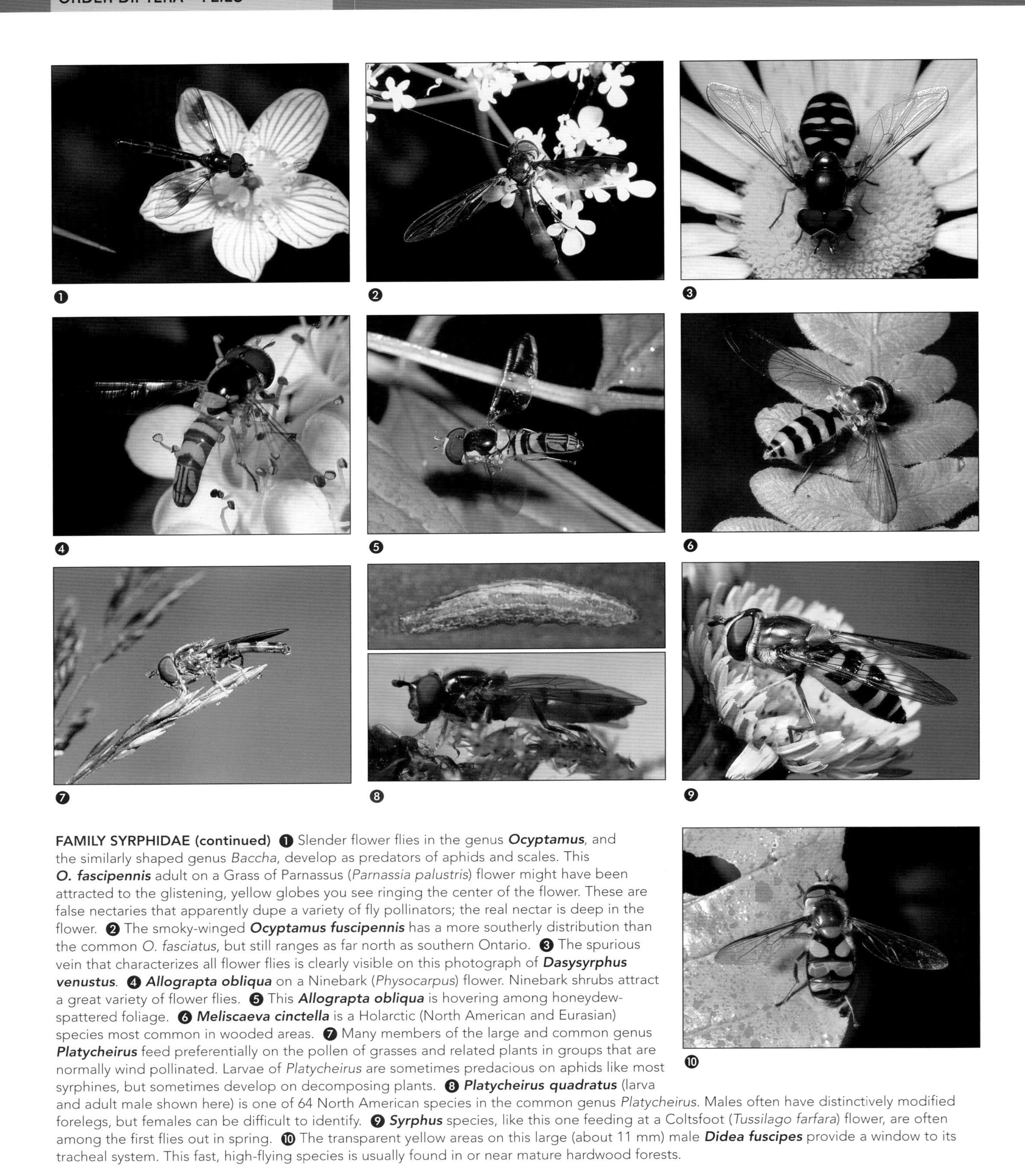

FAMILY SYRPHIDAE (continued) ❶ Slender flower flies in the genus ***Ocyptamus***, and the similarly shaped genus *Baccha*, develop as predators of aphids and scales. This ***O. fascipennis*** adult on a Grass of Parnassus (*Parnassia palustris*) flower might have been attracted to the glistening, yellow globes you see ringing the center of the flower. These are false nectaries that apparently dupe a variety of fly pollinators; the real nectar is deep in the flower. ❷ The smoky-winged ***Ocyptamus fuscipennis*** has a more southerly distribution than the common *O. fasciatus*, but still ranges as far north as southern Ontario. ❸ The spurious vein that characterizes all flower flies is clearly visible on this photograph of ***Dasysyrphus venustus***. ❹ ***Allograpta obliqua*** on a Ninebark (*Physocarpus*) flower. Ninebark shrubs attract a great variety of flower flies. ❺ This ***Allograpta obliqua*** is hovering among honeydew-spattered foliage. ❻ ***Meliscaeva cinctella*** is a Holarctic (North American and Eurasian) species most common in wooded areas. ❼ Many members of the large and common genus ***Platycheirus*** feed preferentially on the pollen of grasses and related plants in groups that are normally wind pollinated. Larvae of *Platycheirus* are sometimes predacious on aphids like most syrphines, but sometimes develop on decomposing plants. ❽ ***Platycheirus quadratus*** (larva and adult male shown here) is one of 64 North American species in the common genus *Platycheirus*. Males often have distinctively modified forelegs, but females can be difficult to identify. ❾ ***Syrphus*** species, like this one feeding at a Coltsfoot (*Tussilago farfara*) flower, are often among the first flies out in spring. ❿ The transparent yellow areas on this large (about 11 mm) male ***Didea fuscipes*** provide a window to its tracheal system. This fast, high-flying species is usually found in or near mature hardwood forests.

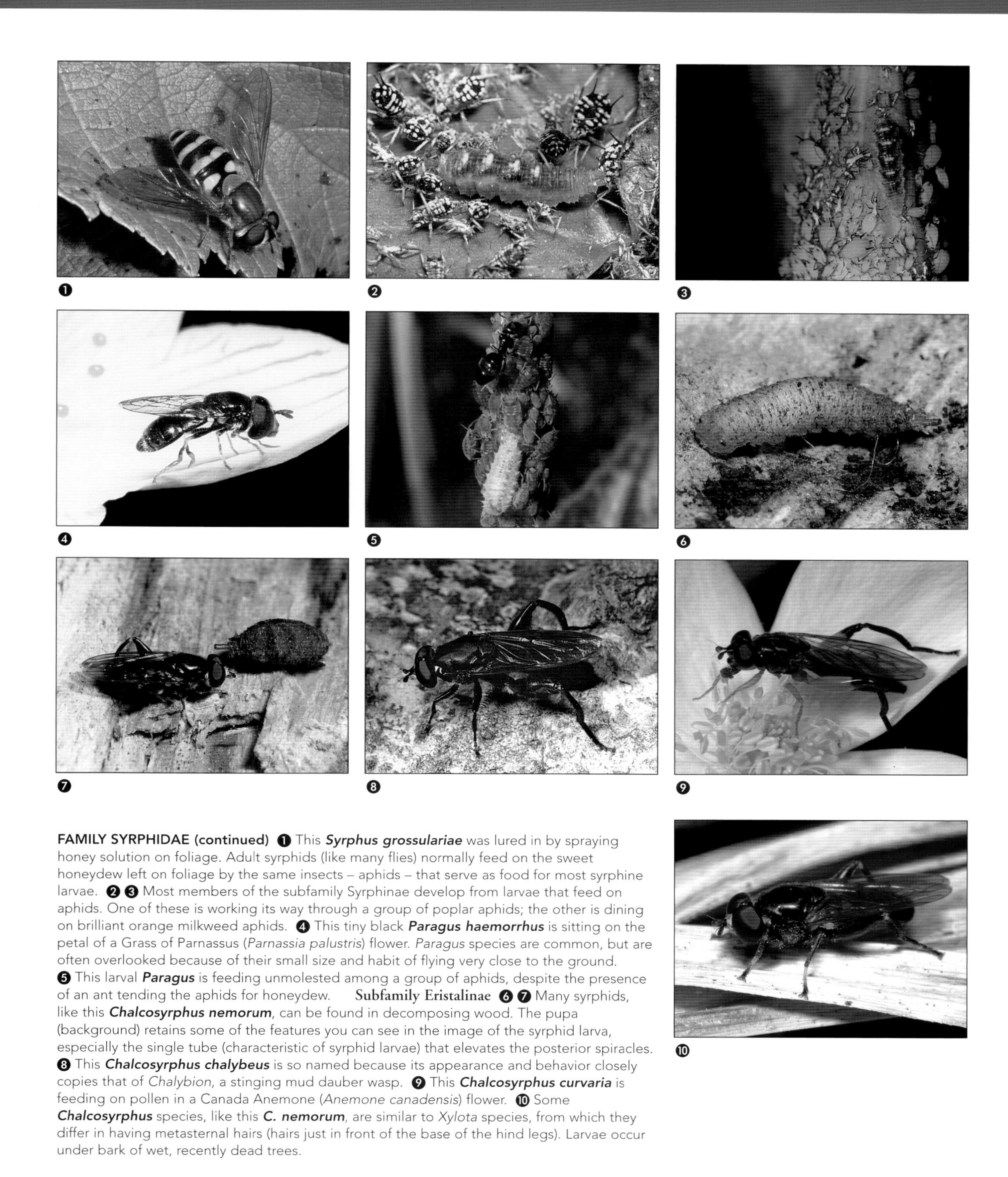

FAMILY SYRPHIDAE (continued) ❶ This ***Syrphus grossulariae*** was lured in by spraying honey solution on foliage. Adult syrphids (like many flies) normally feed on the sweet honeydew left on foliage by the same insects – aphids – that serve as food for most syrphine larvae. ❷ ❸ Most members of the subfamily Syrphinae develop from larvae that feed on aphids. One of these is working its way through a group of poplar aphids; the other is dining on brilliant orange milkweed aphids. ❹ This tiny black ***Paragus haemorrhus*** is sitting on the petal of a Grass of Parnassus (*Parnassia palustris*) flower. *Paragus* species are common, but are often overlooked because of their small size and habit of flying very close to the ground. ❺ This larval ***Paragus*** is feeding unmolested among a group of aphids, despite the presence of an ant tending the aphids for honeydew. Subfamily Eristalinae ❻ ❼ Many syrphids, like this ***Chalcosyrphus nemorum***, can be found in decomposing wood. The pupa (background) retains some of the features you can see in the image of the syrphid larva, especially the single tube (characteristic of syrphid larvae) that elevates the posterior spiracles. ❽ This ***Chalcosyrphus chalybeus*** is so named because its appearance and behavior closely copies that of *Chalybion*, a stinging mud dauber wasp. ❾ This ***Chalcosyrphus curvaria*** is feeding on pollen in a Canada Anemone (*Anemone canadensis*) flower. ❿ Some ***Chalcosyrphus*** species, like this ***C. nemorum***, are similar to *Xylota* species, from which they differ in having metasternal hairs (hairs just in front of the base of the hind legs). Larvae occur under bark of wet, recently dead trees.

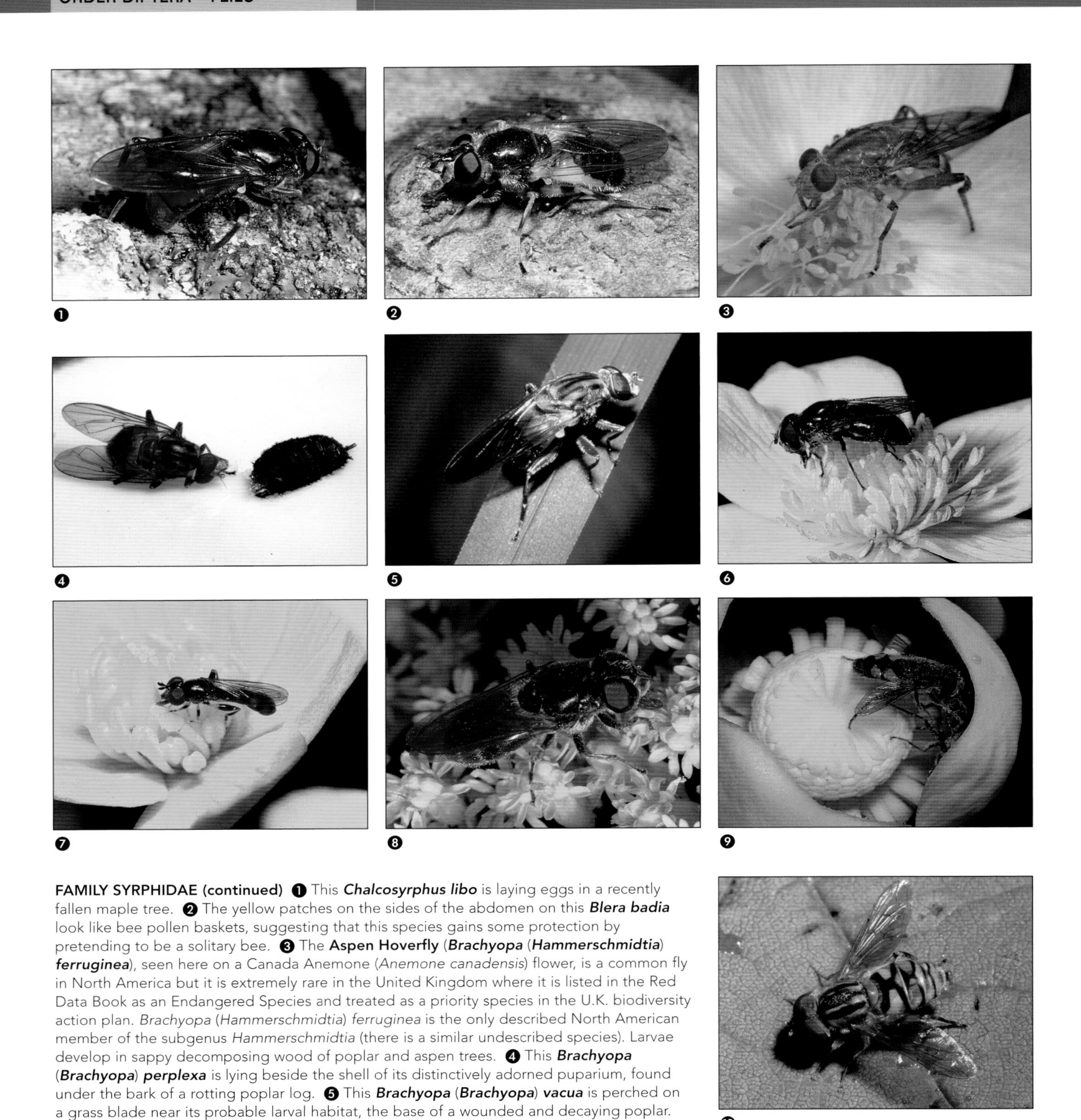

FAMILY SYRPHIDAE (continued) ❶ This ***Chalcosyrphus libo*** is laying eggs in a recently fallen maple tree. ❷ The yellow patches on the sides of the abdomen on this ***Blera badia*** look like bee pollen baskets, suggesting that this species gains some protection by pretending to be a solitary bee. ❸ The **Aspen Hoverfly** (***Brachyopa*** (***Hammerschmidtia***) ***ferruginea***), seen here on a Canada Anemone (*Anemone canadensis*) flower, is a common fly in North America but it is extremely rare in the United Kingdom where it is listed in the Red Data Book as an Endangered Species and treated as a priority species in the U.K. biodiversity action plan. *Brachyopa* (*Hammerschmidtia*) *ferruginea* is the only described North American member of the subgenus *Hammerschmidtia* (there is a similar undescribed species). Larvae develop in sappy decomposing wood of poplar and aspen trees. ❹ This ***Brachyopa*** (***Brachyopa***) ***perplexa*** is lying beside the shell of its distinctively adorned puparium, found under the bark of a rotting poplar log. ❺ This ***Brachyopa*** (***Brachyopa***) ***vacua*** is perched on a grass blade near its probable larval habitat, the base of a wounded and decaying poplar. ❻ ❼ Marsh Marigold (*Caltha palustris*) flowers attract large numbers of a few species of flower flies. Black syrphids in the genus ***Cheilosia*** (***C. rita*** shown here) and the tiny, somewhat petiolate-waisted ***Neoascia metallica*** are common visitors. ❽ The genus ***Cheilosia*** includes over 75 North American species, some of which have the unusual (for a hover fly) habit of developing on living plants, and some of which develop on fungi. This is ***C. nigroapicata***. ❾ Yellow water lilies are often visited by large numbers of ***Parhelophilus rex***, which probably play a major role in pollination. ❿ ***Parhelophilus laetus*** is one of nine North American species in the genus, which usually develop in aquatic or semiaquatic habitats such as the leaf sheaths of cattails (*Typha*).

FAMILY SYRPHIDAE (continued) ❶ The large genus ***Eristalis*** is best known for the familiar **Drone Fly** (***E. tenax***), an originally European species that is now one of our most common syrphids. Larvae (rat-tailed maggots) develop in pockets of bacteria-rich water. ❷ The darkly pigmented ***Eristalis dimidiata*** is often among the first flies to become active in spring. ❸ ❹ ***Eristalis flavipes*** and ***E. anthophorina*** are obviously bumble bee mimics. *Eristalis flavipes* is named for its distinctively orange tarsi; *E. anthophorina* has entirely black legs. The 17 North American *Eristalis* species develop from "rat tailed" larvae, so-called because of the aquatic or semiaquatic larva's long telescopic breathing tube. ❺ ***Eristalis transversa*** is a common species differing from otherwise similar *Eristalis* species by its yellow scutellum. ❻ As with many male flies, male syrphids (like the upper of these two ***Eristalis arbustorum***) have much larger eyes than the female. ❼ ***Hiatomyia cyanescens*** is the only eastern North American species of the mostly western genus *Hiatomyia*. ❽ ***Mallota posticata*** is a common bumble bee mimic that develops in rot-holes in old trees. ❾ ***Pterallastes thoracicus***, a widespread eastern species, is the only North American species in the genus.

❶ ❷ ❸ ❹ ❺ ❻ ❼ ❽ ❾ ❿

FAMILY SYRPHIDAE (continued) ❶ Members of the genus ***Criorhina*** (this is ***C. nigriventris***) are among the best bumble bee mimics in the Syrphidae. *Criorhina* are active in early spring; this one was photographed as it worked the flowers on a gooseberry bush. ❷ Although ***Volucella bombylans*** is a great bumble bee mimic, even to the point of having alternate color forms that mimic different bumble bees, it has been reared from yellowjacket wasp nests. Female flies secrete a chemical that prevents the wasps from attacking when they enter the nests to lay eggs. Larvae usually develop as scavengers, but sometimes attack the wasp's brood. ❸ This ***Sphecomyia vittata*** is feeding on pollen in a Canada Anemone (*Anemone canadensis*) flower. ❹ ❺ Some Syrphidae, like this ***Ceriana abbreviata*** and this ***Sphiximorpha willistoni***, respectively, are strikingly similar to potter wasps (Vespidae). Many flower flies are impressive mimics of stinging yellowjackets and related wasps. The beauties shown here are: ❻ ***Temnostoma aequale*** (until recently treated as *T. vespiforme*, which is now considered to be an Old World species), ❼ ***Temnostoma alternans***, ❽ ***Spilomyia longicornis***, ❾ ***Spilomyia sayi*** (previously known as *S. quadrifasciata*), and ❿ ***Somula decora***.

❶ ❷ ❸ ❹ ❺ ❻ ❼ ❽ ❾

FAMILY SYRPHIDAE (continued) ❶ Many of the wasplike syrphids develop in dead wood. This ***Temnostoma balyras*** female is laying eggs on the moist surface of a decomposing log. ❷ ***Temnostoma barberi*** is one of several *Temnostoma* species with a close similarity to wasps in the family Vespidae. ❸ ***Temnostoma*** species, like this pair of ***T. alternans***, are commonly seen on wild rose. ❹ ❺ ❻ ***Sericomyia*** species have aquatic larvae with long breathing tubes (like *Eristalis*), and are often found near bogs and other wetlands. Three of the most common species are shown here, ***S. chrysotoxoides*** (on foliage), ***S. militaris*** (on a Grass of Parnassus [*Parnassia palustris*] flower) and ***S. lata*** (on white blossoms). ❼ The distinctively striped adults of ***Helophilus*** are among the first flies to take to the wing in spring. Larvae are aquatic. ❽ ***Lejops chrysostomus*** is a small flower fly found in marshy areas. ❾ ❿ Some flower flies, including the fuzzy and beelike **Narcissus Bulb Fly** (***Merodon equestris***) and the much smaller **Lesser Bulb Fly** (***Eumerus funeralis***; until recently treated as *E. tuberculatus*), are introduced pests that develop in bulbs of various plants in the lily family.

❿

FAMILY SYRPHIDAE (continued) ❶ Adults of ***Chrysogaster*** are often seen on flowers like this *Physocarpus*; larvae of some *Chrysogaster* species have their hind spiracles modified into a spikelike tap used to breathe through the roots of aquatic plants. ❷ Although the only North American species of ***Rhingia*** (***R. nasica***) is unknown as a larva, a similar European species develops in cow dung. The distinctive *Rhingia* snout normally conceals the long proboscis, here directed down into a flower. ❸ Delicate, narrow-waisted species of ***Sphegina***, like this ***S. petiolata***, are often seen on Marsh Marigold (*Caltha palustris*) flowers. ❹ The wasp-waisted syrphids in the genus ***Sphegina***, like this ***S. lobulifera***, are unusual among flower flies in preferring moist, shady places. ❺ ***Sphegina flavomaculata*** adults are sometimes common on woodland flowers in early spring. ❻ *Sphegina* species, like this ***S. rufiventris***, develop in rotting wood and are most likely to be found in dense, shaded woodland. ❼ This ***Sphegina lobata*** is eating pollen on a Sweet Cicily (*Osmorhiza claytoni*) flower. ❽ ***Lejota aerea*** is one of the many eristaline syrphids associated with fallen trees. *Lejota* is a small genus, with only two North American species. ❾ ❿ ***Syritta pipiens***, an introduced species which breeds in compost and decaying vegetation, is one of the most common backyard and garden flower flies.

FAMILY SYRPHIDAE (continued) ❶ Males of ***Teuchocnemis*** (2 eastern species; this is ***T. lituratus***) have a prominent spur under the hind femur. ❷ Like the more common compost-inhabiting *Syritta pipiens*, members of the genus ***Tropidia*** develop in dung and other decomposing material. The flange on the hind femur (the thick part of the hind leg) on this male ***T. quadrata*** is a characteristic of the genus. ❸ This ***Pelecocera pergandei***, distinctive for its hatchet-shaped antennae, was photographed along the Ontario shores of Lake Huron. ❹ ***Copestylum*** is a diverse and common genus in the tropics, but a relatively scarce group in the northeast. This ***C. florida*** was photographed in South Carolina. Similar species range as far north as southern Ontario. ❺ ***Eristalinus aeneus*** is a widespread introduced species, found from Ontario to Florida. ❻ Larvae of the scarce species ***Ferdinandea buccata*** occur in sappy tree wounds, such as those found around exit holes left by carpenter moth larvae (Cossidae). ❼ ***Orthonevra*** adults are often common near the small bodies of water where their larvae develop. ❽ As suggested by the "xylo" in the generic name, ***Xylota bicolor*** develops in dead wood. ❾ Unlike other pollen-eating flower flies, ***Xylota*** species are rarely seen on flowers but instead graze pollen off surfaces like this sumac leaf. This is ***X. ejuncida***. ❿ ***Brachypalpus oarus*** is our only species in the subgenus *Brachypalpus*. Like the closely related genus *Xylota*, it is associated with dead wood.

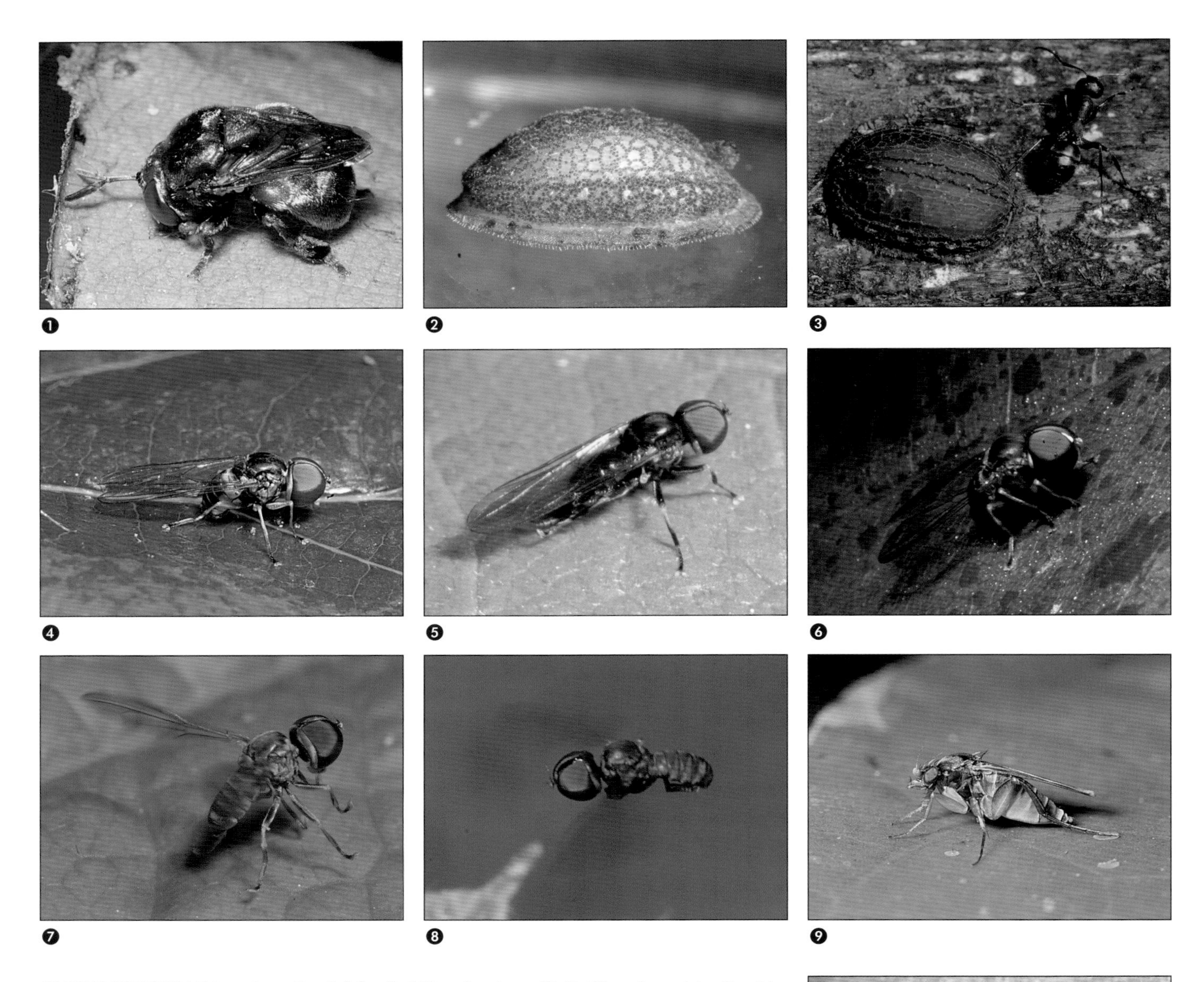

FAMILY SYRPHIDAE (continued) Subfamily Microdontinae ❶ ❷ ***Microdon*** adults like this ***M. globosus*** are squat syrphids, often resembling solitary bees even to the extent of having a fake pollen basket on the hind legs. The sluglike larvae of *Microdon* live in ant nests. ❸ This mollusk-like *Microdon* larva was found in a carpenter ant (*Camponotus noveboracensis*) colony, where it developed on the ant's brood. **FAMILY PIPUNCULIDAE** (BIG-HEADED FLIES) ❹ Big-headed flies, like this ***Elmohardyia atlantica***, are parasitoids of adult leafhoppers and related hoppers. ❺ Although the widespread ***Pipunculus hertzogi*** is the most common species in the large genus *Pipunculus*, it has never been associated with a host. Other *Pipunculus* species are parasitoids of leafhoppers in the family Cicadellidae. ❻ Many flies, including big-headed flies like this ***Eudorylas***, can be attracted to sunny leaf surfaces sprayed with a honey and water solution. Known members of the huge genus *Eudorylas* are parasitoids of leafhoppers. ❼ This ***Eudorylas*** is just taking off from a honeydew-spattered leaf. ❽ Big-headed flies, like this ***Eudorylas*** can hover in small spaces as they scan for potential leafhopper hosts or mates. **FAMILY PHORIDAE** (SCUTTLE FLIES) ❾ The shape and head bristling of this scuttle fly are typical of this large and biologically diverse family. ❿ Scuttle flies, like this ***Dohrniphora gigantea*** feeding on a dead moth, have distinctive wing venation with a few strong veins squeezed up against the base. This is a tropical species, but the genus does occur throughout much of North America.

FAMILY PHORIDAE (continued) ❶ This ***Anevrina*** was one of many emerging from a rabbit burrow in early spring. ❷ This scuttle fly (***Phora*** sp.) is feeding on a bird dropping. Phorid adults are common on all sorts of decomposing materials. ❸ ***Gymnophora*** species are relatively large, relatively bare phorids that develop on dead or dying snails. This is ***G. luteiventris***. ❹ Although the ubiquitous genus ***Megaselia*** has slightly fewer than 200 described North American species, this is only a small fraction of the real diversity in this enormous and poorly known genus. ❺ These larval ***Megaselia*** are feeding on grasshopper eggs. **FAMILY PLATYPEZIDAE (FLAT-FOOTED FLIES)** ❻ This male flat-footed fly (***Bertamyia notata***) is only a few millimeters long, but still stands out because of the silvery reflective patches on its body. This extraordinarily widespread species ranges from Canada to South America. ❼ ***Polyporivora polypori*** is a distinctive fly named for its unusual habit of developing in hard fungi (polypore fungi). Larvae feed on fungal tissue, then construct a silk cocoon and pupariate partially exposed on the undersurface of the bracket fungus host. Most flat-footed flies feed on the gill tissue of soft fungi and pupariate in the ground. ❽ Flat-footed flies, like this female ***Platypeza***, often occur on fungi in late fall. ❾ ❿ ***Callomyia*** and ***Agathomyia***, two of the more common genera of Platypezidae, often abound on fungi during the autumn months.

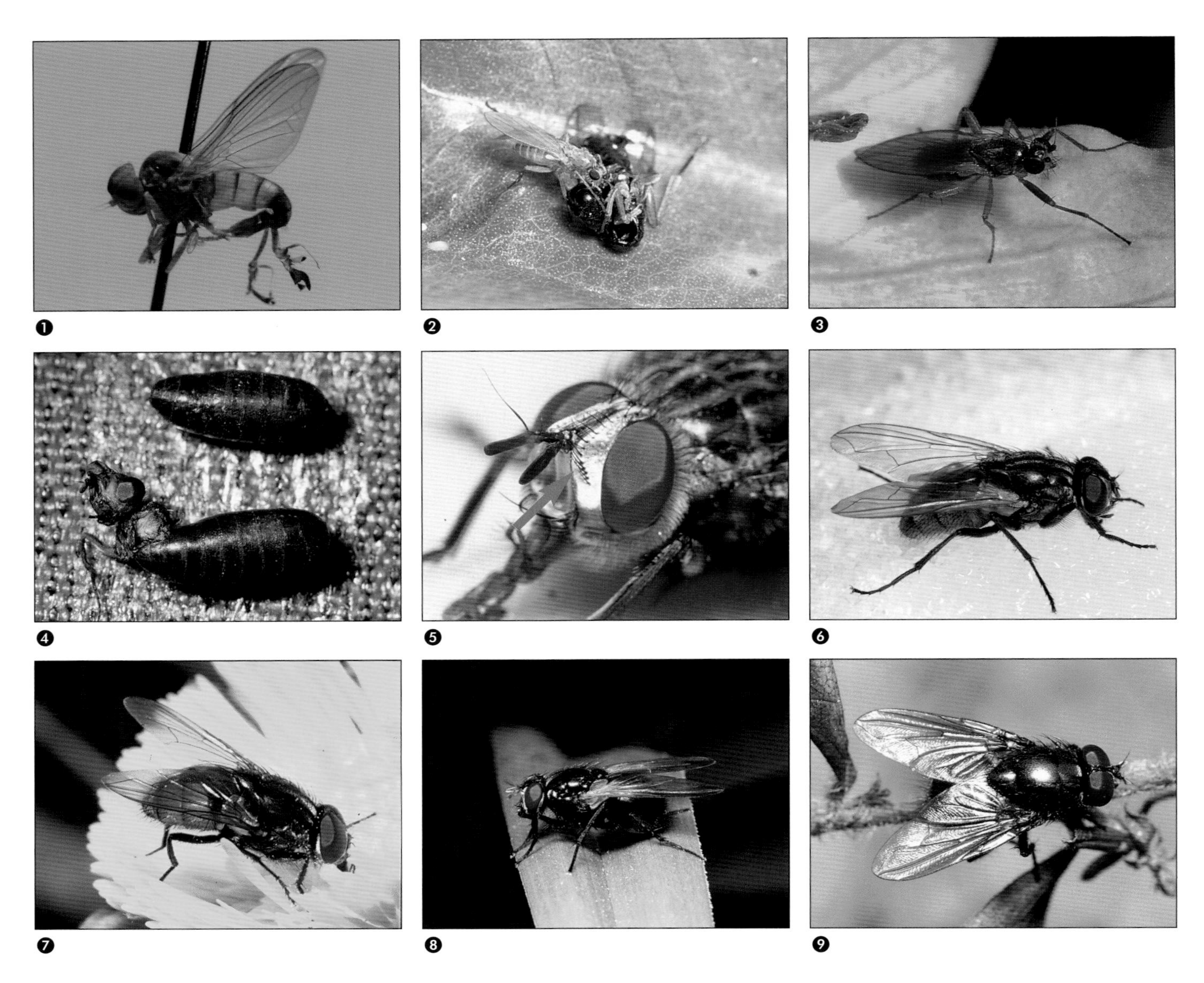

FAMILY PLATYPEZIDAE (continued) ❶ All platypezids have the hind tarsus expanded to some extent, but few can match the elaborate tarsal ornamentation of male ***Calotarsa pallipes***, a species that occurs from Ontario and Quebec south to the Carolinas.
FAMILY LONCHOPTERIDAE (SPEAR-WINGED FLIES) ❷ The distinctively spear-winged ***Lonchoptera bifurcata*** is a common species in damp, grassy areas. This one is scavenging on a dead sawfly. New World populations of this cosmopolitan species occur in female-only (parthenogenic) populations. ❸ ***Lonchoptera uniseta*** is a North American species that, unlike North American populations of the much more common *L. bifurcata*, includes both males and females.

SCHIZOPHORA

FAMILY MUSCIDAE (MUSCID FLIES) ❹ The lower of these two house fly puparia has just "hatched" and the emerging adult still has a balloon-like sac (the ptilinum) inflated in front of its head. That sac was used to push the cap off the puparium. ❺ When the balloon-like ptilinum is drawn back into a newly emerged adult fly's head it leaves a scar arching over the antenna, called a frontal suture or ptilinal suture, as shown on this close-up of a tachinid head. All flies with that distinctive suture are called "Schizophora." **Subfamily Muscinae** ❻ This **House Fly** (***Musca domestica***) was photographed on a bowl of sugar. ❼ This **Face Fly** (***Musca autumnalis***) is feeding on an early spring flower – Coltsfoot (*Tussilago farfara*). Face Flies, so called for their irritating habit of massing on the faces of livestock, are common earlier and later in the year than the similar House Fly. ❽ ***Mesembrina latreillii***, distinctive for its shiny body and contrasting yellow wing bases, is an especially conspicuous fly in northern hardwood forests. Larvae develop in dung. ❾ This metallic muscid (***Eudasyphora cyanicolor***) looks superficially like a blow fly. This species, which flies later in the season than most other flies, lays eggs on dung.

FAMILY MUSCIDAE (continued) Subfamily Stomoxyinae ❶ ❷ The conspicuous biting mouthparts of the **Stable Fly** (***Stomoxys calcitrans***) usually stick out in front of the fly's head, but can be swiftly swung down to painfully puncture your skin. ❸ **Horn Flies** (***Haematobia irritans***) are serious livestock pests with mouthparts much like those of Stable Flies (*Stomoxys calcitrans*). Adult horn flies leave their hosts only for brief egg-laying forays on their host's still steaming dung. Larvae develop quickly in the fresh and moist "meadow muffins." Subfamily Coenosiinae ❹ Many muscid flies hunt and kill other insects, and ***Coenosia tigrina*** is one of the most common predators in yards and gardens. ❺ ***Macrorchis ausoba*** is a common fly that develops in soil or detritus. The large tarsal pads, or pulvilli, are distinct in this photograph. Flies use their pulvilli, which are covered with tiny hooklike bristles, to walk on smooth surfaces. ❻ ***Lispe tentaculata***, like other members of the genus, wait along the edge of the water to pounce upon newly emerged aquatic insects like this prong-gill mayfly. ❼ ***Lispoides aequifrons*** can usually be found on rocks surrounded by running water. ❽ The distinctive gray-and-black-striped species of ***Graphomya*** are among the most common muscid flies on flowers. Subfamily Azeliinae ❾ This ***Hydrotaea*** was photographed as it snacked on my sandwich. Adults of this large genus are often attracted to trampled vegetation; larvae develop in a variety of materials where they usually feed on other insect larvae.

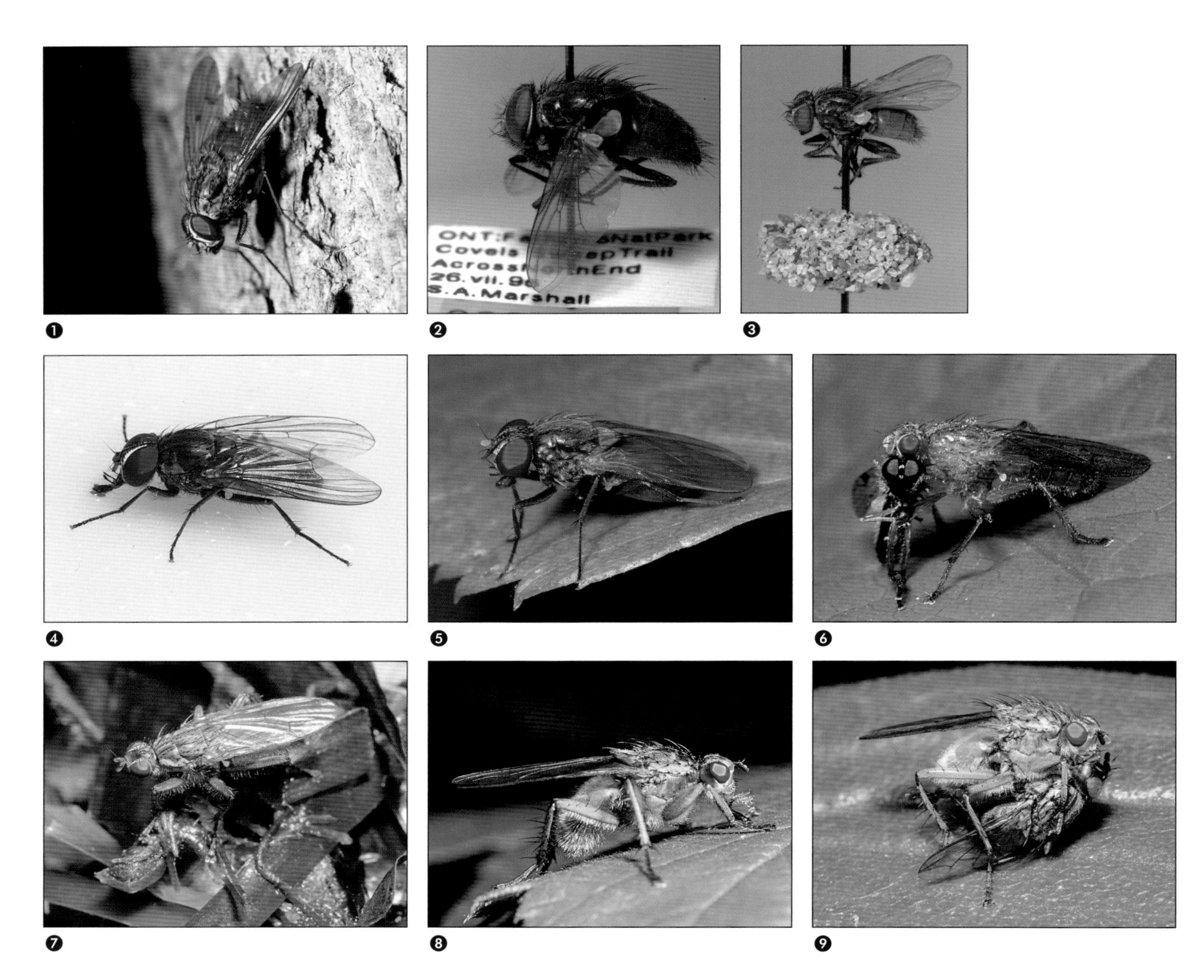

FAMILY MUSCIDAE (continued) Subfamily Phaoniinae ❶ Some muscids, like this ***Helina troene***, are often seen on tree trunks. Known *Helina* hatch from their eggs as almost mature (third instar) larvae, ready to capture and eat other organisms under bark, in tree wounds and similar habitats. Subfamily Reinwardtiinae ❷ ***Muscina pascuorum***, one of the more common muscids in the subfamily Reinwardtiinae, develops in a diversity of decaying materials. Subfamily Mydaeinae ❸ Many muscid flies, like this ***Myospila meditabunda***, pupariate within a cocoon made from sand or soil stuck together with sticky secretions. *Myospila meditabunda* is a common species in both Europe and North America that, like most Mydaeinae, develops in dung. **FAMILY FANNIIDAE** (FANNIID FLIES) ❹ The little flies that seem to circle endlessly in the middle of a room, usually under a marker like a light, are likely to be **Lesser House Flies** (***Fannia canicularis***), like this one. ❺ Although the Fanniidae is best known for a few nuisance species such as the Lesser House Fly (*Fannia canicularis*) and Latrine Fly (*F. scalaris*), the 100 or so North American species of ***Fannia*** develop in a wide range of substrates including fungi, carrion, bird nests, wasp nests and decaying wood. This cosmopolitan species, ***F. incisurata***, has been reared from dung, dead bodies, bird and wasp nests, and even human intestinal tracts. **FAMILY SCATHOPHAGIDAE** (SCATHOPHAGID FLIES) ❻ Fuzzy, often yellowish ***Scathophaga*** species are common and conspicuous predators. This ***S. nigrolimbata*** has captured a small soldier fly. ❼ ***Scathophaga intermedia*** is a member of the "wrack fauna," and is found only in strands of decaying algae washed up on the seashore. ❽ **Pilose Yellow Dung Flies** (***Scathophaga stercoraria***) are common predacious flies that usually breed in fresh ungulate dung. Males of this species guard their mates while they lay eggs in cattle droppings, each dung pat serving as a territory controlled by a dominant male and contested by other males lurking nearby. ❾ This **Pilose Yellow Dung Fly** is giving a deadly kiss to a flesh fly, consuming its body contents through its oral opening.

FAMILY SCATHOPHAGIDAE (continued) ❶ This ***Cordilura (Cordilura) angustifrons*** male is feeding on honeydew; a small aphid is visible just in front of his head. ❷ ❸ ***Cordilura*** species are elongate, relatively bare scathophagids that develop in the stems of sedges and related monocots. ❹ The male and female of this mating pair of ***Cordilura (Achaetella) varipes*** show sexual dimorphism in color and other characters. Most of the 50 North American species in genus *Cordilura* develop as borers in sedge (*Carex* and *Scirpus*). ❺ Some ***Cordilura*** species, like this ***C. gracilipes***, are conspicuously colored. ❻ ***Spaziphora cincta*** adults are found along the shorelines of large lakes where their aquatic, predacious larvae live. ❼ This ***Acanthocnema*** species (an undescribed species from Ontario) can be found only on algae-covered rocks surrounded by running water. The larvae are aquatic. ❽ ***Hydromyza confluens*** is a common fly anywhere water lilies occur. Adults are predators, but larvae tunnel in the underwater stems. FAMILY ANTHOMYIIDAE (ANTHOMYIID FLIES OR ROOT MAGGOT FLIES) Subfamily Anthomyiinae ❾ Anthomyiidae differ from the otherwise similar Muscidae in details of the wing venation. The last wing vein reaches the wing margin in anthomyiids but not muscids. ❿ This anthomyiid fly (***Chirosia betuleti***, a species recently introduced to North America), is laying eggs in a young fern at the fiddlehead stage. The maggots develop inside conspicuous swellings on the fern fronds.

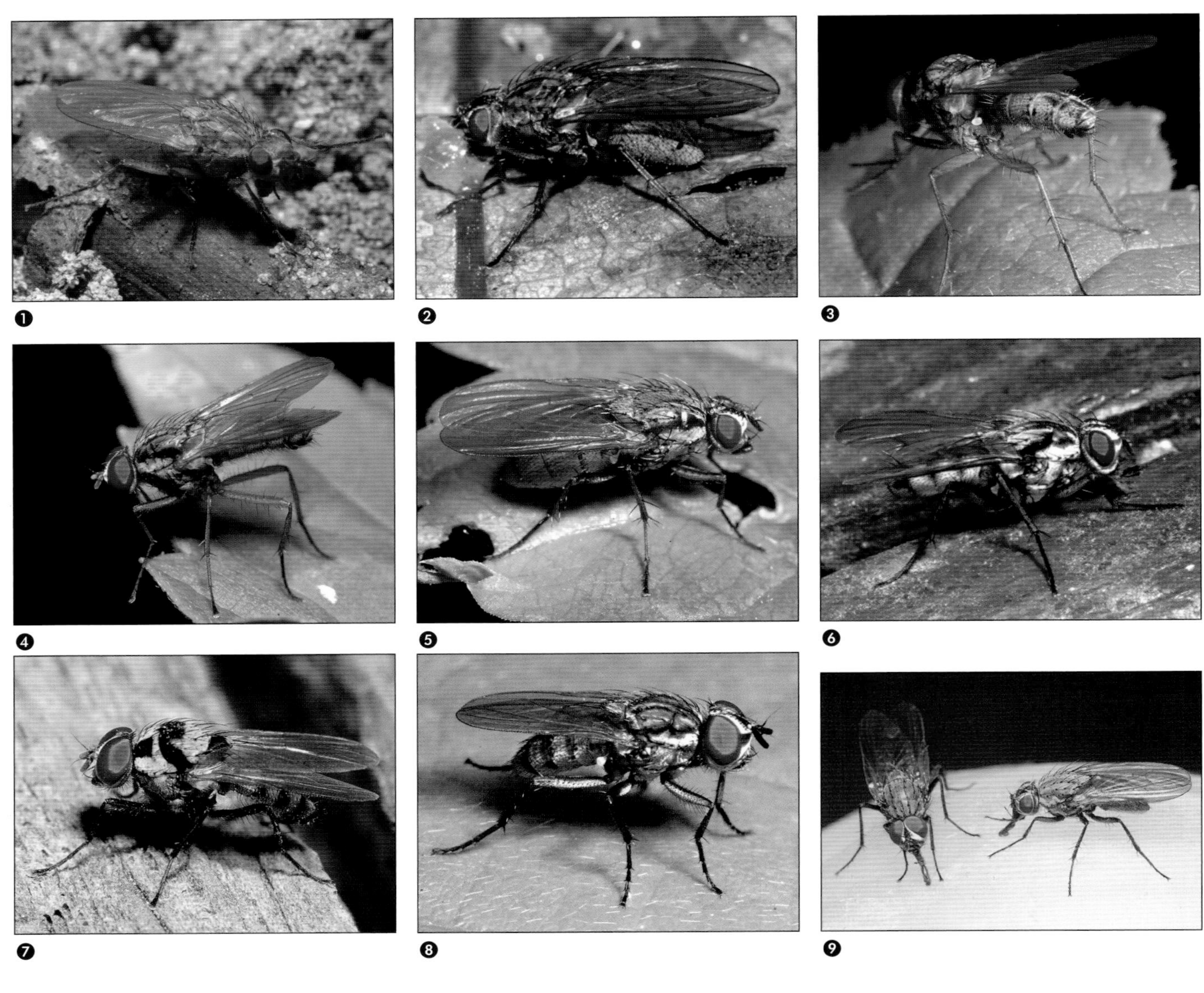

FAMILY ANTHOMYIIDAE (continued) ❶ ❷ This ***Eutrichota*** was found crawling out of a rabbit burrow as a teneral specimen, newly emerged from its puparium. The second photo is the same fly an hour later. ❸ Almost all non-coastal Anthomyiidae differ from similar fly families in having fine erect bristles pointing down from the underside of the scutellum, visible as pale hairs under the scutellum of this ***Eutrichota***. ❹ ***Eutrichota*** are common flies that probably develop as scavengers in dung and other substrates. Some have been found in animal burrows. ❺ ***Hydrophoria*** species are common flies that develop in a range of substrates including dung and sawfly cocoons. This large genus is sometimes divided, putting this species in the genus ***Zaphne***. ❻ Members of the genus ***Eustalomyia***, like this ***E. festiva***, are kleptoparasites that develop in the nests of twig- or dead wood-nesting wasps (Crabronidae). *Eustalomyia* larvae consume paralyzed prey stored by wasps for consumption by wasp larvae. ❼ ***Anthomyia*** species, like this brightly colored ***A. oculifera***, develop in a variety of habitats including bird's nests, decaying plants and fungi. ❽ ***Leucophora*** species develop as kleptoparasites in the nests of solitary bees (Andrenidae), where they consume the pollen collected by the host bee. ❾ As is the case for many calyptrate flies, male **Onion Maggot** (***Delia antiqua***; often treated as *Hylemya antiqua*) flies have much larger eyes than females. Both of these individuals are using their spongy labellae to feed on a rotting onion; the male is facing forward. ❿ ***Pegomya*** is a large genus including species that develop in fungi and others that develop as leaf or stem miners.

❿

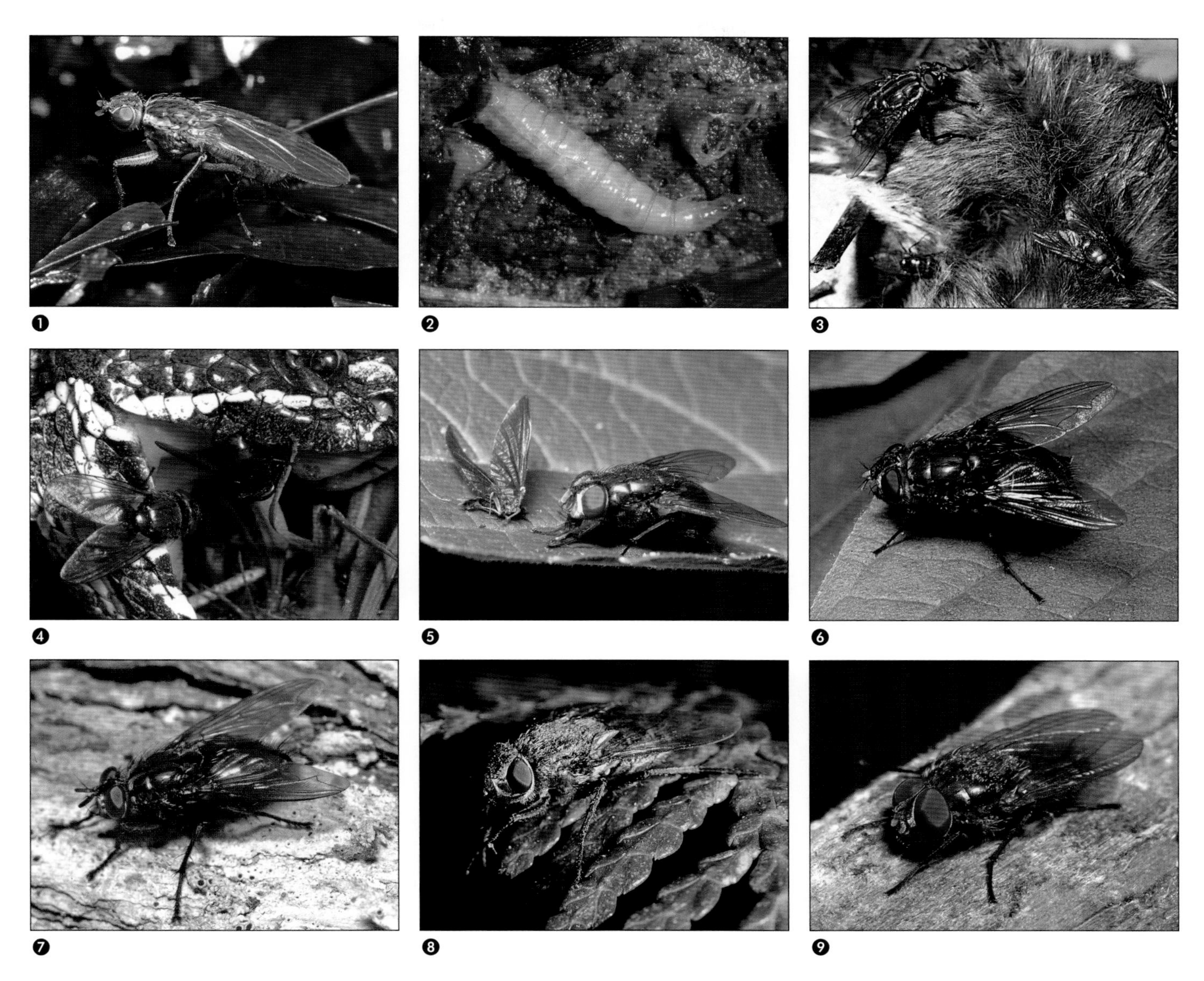

FAMILY ANTHOMYIIDAE (continued) Subfamily Fucelliinae ❶ ***Fucellia tergina*** is one of the most abundant flies in east coast wrack piles. **FAMILY CALLIPHORIDAE** (BLOW FLIES) ❷ Larval blow flies are typical maggots, tapered to a pointed front end with an internal head skeleton. The two round disks on the broad hind end of this maggot each have three spiracular slits for breathing. The rings of little black hooks (creeping welts) that circle the body assist with crawling. Subfamily Luciliinae ❸ Blow flies (Calliphoridae like this ***Lucilia***) and flesh flies (Sarcophagidae, like this *Sarcophaga* (*Robineauella*) are the most conspicuous flies on dead animals or their parts. Metallic green blow flies in the genus *Lucilia* are often the most common flies on fresh carrion. ❹ This **Greenbottle Fly** (***Lucilia sericata***, widely known as *Phaenicia sericata*) has settled on a rattlesnake head. Sterile cultures of Greenbottle Flies are now widely used in "maggot therapy," in which live maggots are used to clean up difficult wounds and infections of living flesh and bone. The same species is a pest in areas where it routinely parasitizes sheep ("sheep strike"), but it normally develops in carrion in northeastern North America. ❺ This ***Lucilia illustris*** is snacking on a dead mayfly subimago, but this species is normally found in association with more substantial carrion and sometimes even occurs in wounds on living vertebrates. A superficially similar blow fly (*Bufolucilia silvarum*) sometimes lays eggs on healthy amphibians, which usually die following an infestation of blow fly maggots. Subfamily Calliphorinae ❻ Blow flies in the genus ***Calliphora*** have a variety of habits, but most North American species normally breed in carrion. This species, the charmingly named ***C. vomitoria***, is most common in wooded areas in both Europe and North America. ❼ The species name of this blow fly (***Cynomya cadaverina***) gives a good clue to its necrophagous habits. Subfamily Polleniinae ❽ **Cluster Flies** (***Pollenia rudis***) are atypical blow flies covered with yellowish, silky hairs. These common parasitoids of earthworms spend the winter as adults, often sheltering in large numbers in houses. ❾ Although most of the cluster flies that create problems in homes are *Pollenia rudis*, eastern North America is now home to several other ***Pollenia*** species. This is ***P. pediculata***.

FAMILY CALLIPHORIDAE (continued) ❶ Flies like this cluster fly (***Pollenia*** sp.) have less than a week to live after getting infected by the fungus *Entomophthora muscae*. They typically climb to the apex of a stem or other prominent place before dying, thus assisting the dispersal of the fungus. **Subfamily Chrysomyiinae** ❷ This **Black Blow Fly** (***Phormia regina***) was attracted to a road-killed snake. Like many other blow flies, this species normally develops in carrion but occasionally attacks the flesh of living vertebrates (myiasis). ❸ This bird blow fly (***Protocalliphora*** sp.) emerged from one of many puparia found in a barn swallow nest. *Protocalliphora* maggots develop in bird nests, where they suck blood from nestlings. ❹ The **Secondary Screwworm** (***Cochliomyia macellaria***) is a minor pest that occasionally attacks living animals but is usually associated with dead tissues. This one is on a dead fish on the shore of Lake Erie. ❺ Four species of the Old World genus ***Chrysomya*** have been accidentally introduced to the New World in the past few years, and two species that first appeared in the neotropics in the 1970s have since spread to eastern North America. The **Oriental Latrine Fly** (*Chrysomya megacephala*) had spread at least as far north as Georgia by 2001. Another *Chrysomya*, the Hairy Maggot (*C. rufifacies*) has now spread all the way north to Ontario. This ***C. megacephala*** is laying eggs on a Carrion Plant (*Stapelia flavopurpurea*), which uses the odor of rotting flesh to dupe calliphorids into serving as pollinators. **FAMILY SARCOPHAGIDAE (FLESH FLIES)** **Subfamily Sarcophaginae** ❻ Sarcophagidae deposit newly hatched larvae rather than eggs, as you can see from a close look at the tip of the abdomen of this dung-feeding species. ❼ Larvae of most Sarcophagidae feed on flesh (living or dead), but this genus (***Ravinia***) is very common in dung, where they feed on both dung and the larvae of other flies. ❽ ***Oxysarcodexia***, like the closely related *Ravinia*, is a widespread genus with several dung-associated species. ❾ Turtle eggs are often attacked by larvae of specialized flesh flies such as this ***Tripanurga importuna***, which make their way down through the soil to feed on the eggs.

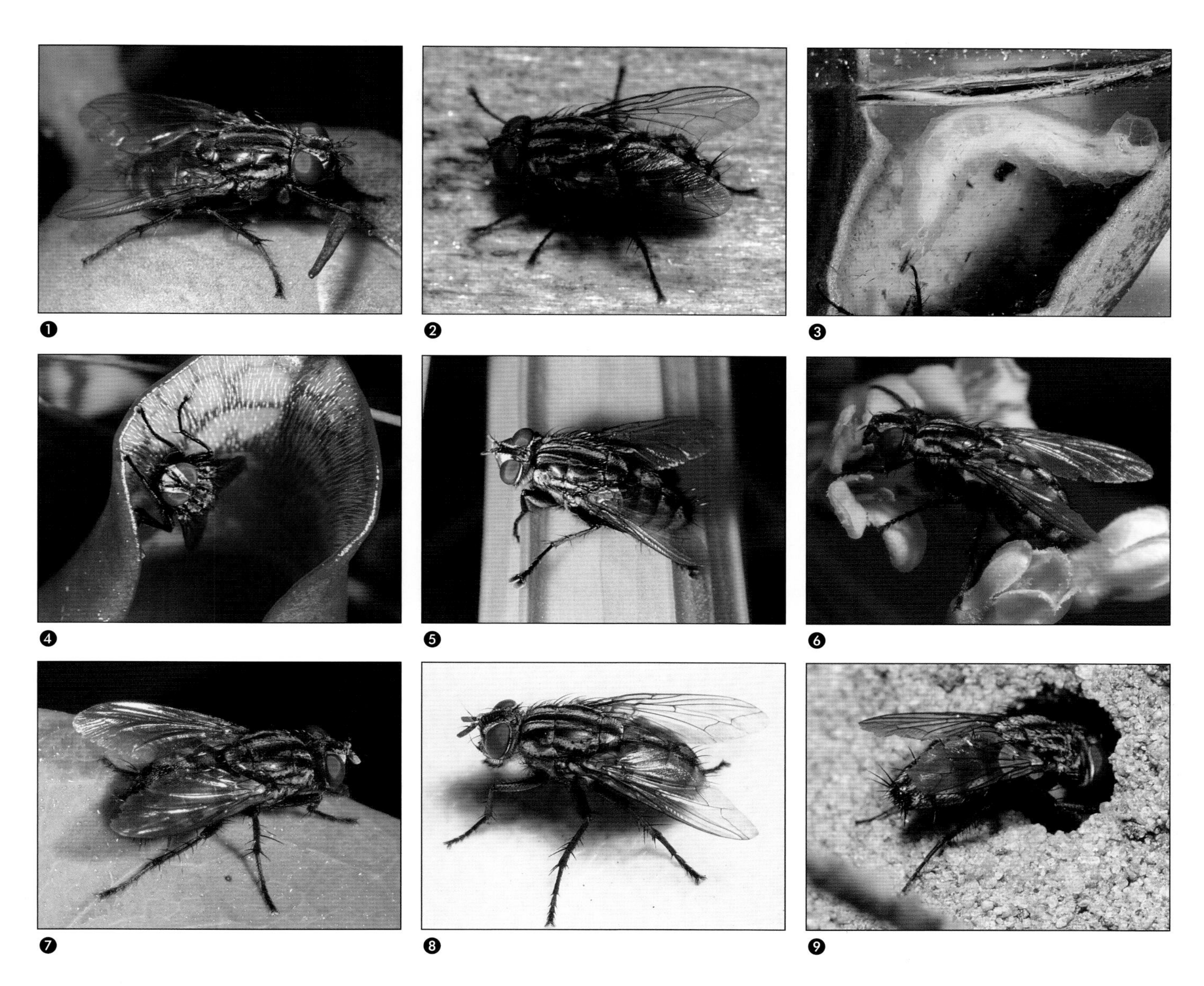

FAMILY SARCOPHAGIDAE (continued) ❶ This ***Boettcheria*** adult has landed on a peony head. Larvae of this common genus usually develop as parasitoids inside other insects. ❷ ***Helicobia rapax*** is a common small flesh fly that develops both as a scavenger and an opportunistic parasitoid of insects and snails. ❸ ❹ Some flesh flies develop only inside pitcher plants, where they feed on the other insects trapped and drowned in the pitcher plant fluids. One of these photos shows an adult resting in a Northern Pitcher Plant (*Sarracenia purpurea*); the other photo shows the larva of the most common pitcher plant sarcophagid, ***Fletcherimyia fletcheri***, in a cutaway view of a pitcher plant. Like all sarcophagid maggots, it has its hind spiracles deep in a cuplike pit. ❺ This is one of several ***Fletcherimyia*** species that develops in Yellow Pitcher Plants (*Sarracenia flava*). ❻ Flesh flies are common insects at flowers, and this ***Sarcophaga*** (***Robineauella***) is taking nectar at an American Bittersweet (*Celastrus scandens*) blossom. ❼ ***Sarcophaga melanura*** is an introduced European species now common in eastern North American yards and gardens. ❽ The **Friendly Fly** (***Sarcophaga aldrichi***) is a parasitic fly often conspicuously abundant during peak years for its host, the Forest Tent Caterpillar (*Malacosoma disstria*). **Subfamily Miltogramminae** ❾ Members of the flesh fly subfamily Miltogramminae, like this ***Senotainia*** entering a wasp's burrow, are mostly kleptoparasites that steal food from solitary wasps and bees. Some *Senotainia* deposit larvae on paralyzed prey before it is concealed by the host wasp, but most go into open host burrows to deposit larvae.

FAMILY SARCOPHAGIDAE (continued) ❶ ***Senotainia trilineata*** is a widespread kleptoparasite of a huge range of ground-nesting wasps in the families Pompilidae, Sphecidae, Crabronidae and Vespidae, feeding on paralyzed prey including aphids, bees, beetles, caterpillars, flies, katydids and spiders. A related European species is a parasitoid of adult honey bees, and can be a very serious pest in southern European apiaries. ❷ This wasp (*Astata bicolor*) captured and paralyzed a stinkbug nymph. When it flew slowly towards its nesting site carrying its prey, it was spotted and closely followed by the kleptoparasitic ***Senotainia***, seen here depositing larvae on the prey shortly after the wasp landed. ❸ The distinctively silver-headed ***Metopia argyrocephala*** is unusual among kleptoparasitic flies because it deposits larvae in the nests of both bees and wasps, and can thus presumably develop on either paralyzed prey or stored pollen. ❹ This ***Phrosinella*** is searching the ground for a wasp nest stocked with paralyzed prey. Unlike most other common kleptoparasitic Sarcophagidae, *Phrosinella* females dig their way into closed nests before releasing their larvae (larvipositing). ❺ Kleptoparasitic miltogrammine sarcophagids, like this tiny (about 4 mm) ***Sphenometopa tergata***, can almost invariably be seen perched on convenient vantage points near the nests of their prospective host wasps. ❻ ***Oebalia minuta*** is a kleptoparasite like other miltogrammines, but instead of putting larvae on the host's prey or in the host's nest, it lays eggs directly on the adult female wasps. ❼ Adult Sarcophagidae, like this ***Amobia*** on a *Euphorbia* flower, are often attracted to flowers. *Amobia* is a kleptoparasite in solitary wasp nests. **FAMILY TACHINIDAE** (PARASITIC FLIES) Subfamily Tachininae ❽ ***Hystricia abrupta***, one of the most spectacular tachinids of northeastern North America, is a parasitoid of tiger moth caterpillars. ❾ Members of the genus ***Ormia*** are nocturnal parasitoids of singing Orthoptera (katydids and crickets). Hosts are located by their songs, using the inflated prosternum (the breastlike bulge underneath the front of the thorax) which functions as an "ear." Most *Ormia* are southern, but this species (***O. reinhardi***) occurs as far north as Ontario.

FAMILY TACHINIDAE (continued) ❶ ***Epalpus signifer***, a conspicuous and common early spring tachinid, is a parasitoid of caterpillars (Noctuidae). ❷ ***Jurinia pompalis*** is a conspicuously attractive tachinid frequently found on meadow flowers, but the host remains unknown. ❸ ***Copecrypta ruficauda*** develops as a parasitoid of cutworm larvae and pupae. ❹ ***Aphria ocypterata*** is often seen perched on low vegetation on dunes and undisturbed beaches, presumably in search of caterpillar hosts. ❺ ***Strongygaster triangulifera*** is a small tachinid that parasitizes an unusually wide range of beetle hosts, ranging from weevils to lady beetles. ❻ The bulky, strikingly small-eyed ***Microphthalma michiganensis*** is a common fly on open sand dunes, where the sluggish adults are often seen on the sand, and the larvae parasitize white grubs (scarab larvae). This adult has recently emerged from its puparium, and the ptilinum is still partially protruding from the front of the head. ❼ ***Genea*** species, like this ***G. texensis*** are parasitoids of snout moths and other small moths. ❽ ***Phytomyptera*** species are small flies that develop as parasitoids of micromoth caterpillars and pupae. ❾ ***Peleteria*** species are parasitoids of various caterpillars. This is ***P. anaxias***.

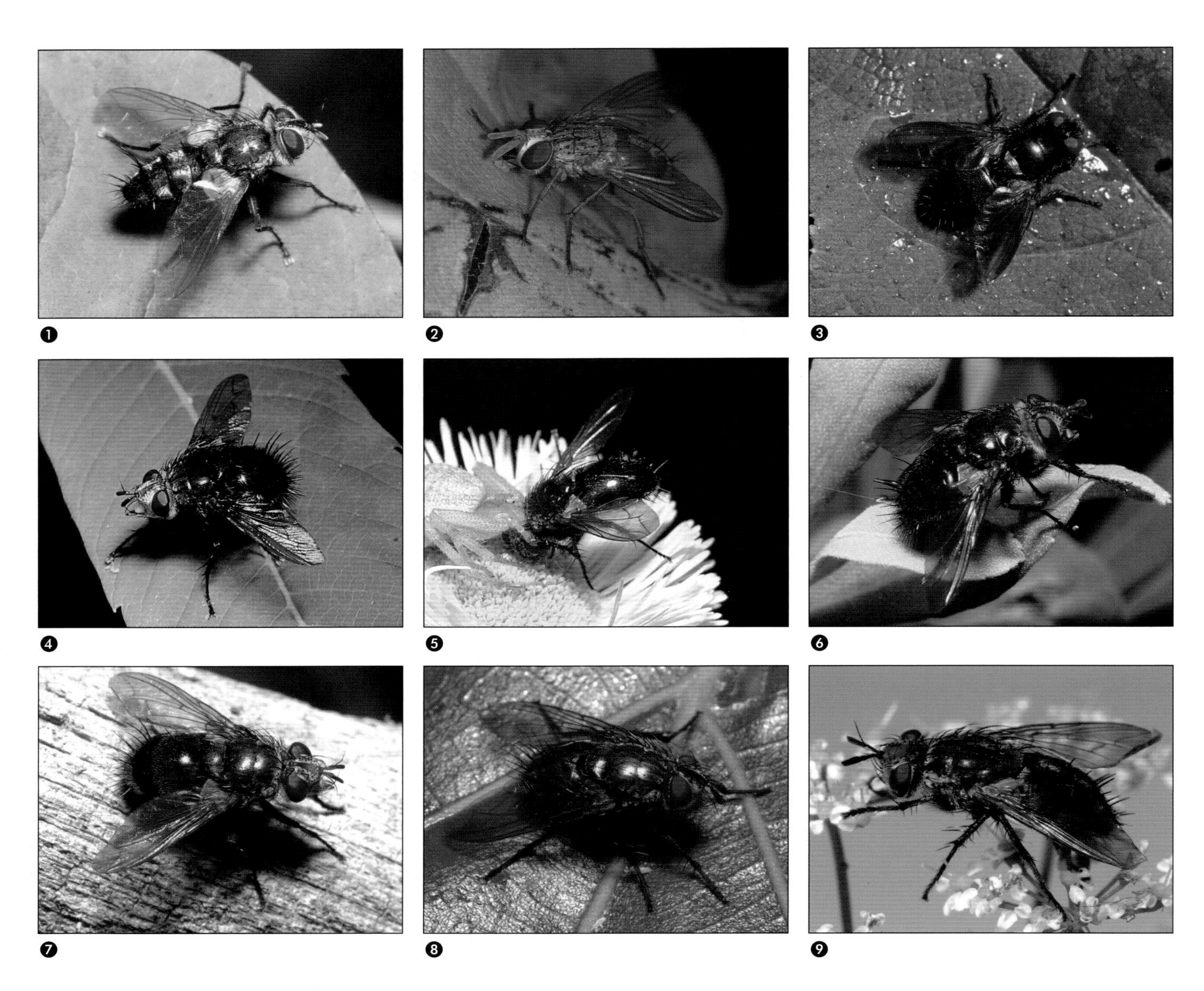

FAMILY TACHINIDAE (continued) ❶ This ***Neomintho celeris*** is perched on the edge of a leaf as it searches for a host, probably a small caterpillar. ❷ ***Leskia depilis*** is a parasitoid of small moths such as melonworms (Pyralidae). ❸ Almost all species in the huge family Tachinidae are parasitoids of other insects, but this distinctive black species (***Loewia foeda***) is a parasitoid of centipedes. *L. foeda* is a European species first found in North America in 1972. ❹ ***Pararchytas decisus*** is a large (around 10 mm) tachinid frequently seen on flowers throughout northeastern North America. This is a common insect but, remarkably, almost nothing is known of its biology. Larvae of most Tachininae wait in ambush for passing hosts such as caterpillars, but larval hosts have not been recorded for any of the three North American *Pararchytas* species. ❺ ***Archytas*** species, like this ***A. aterrimus*** captured by a crab spider, parasitize caterpillars in several families. This species is an important parasitoid of the Fall Armyworm (*Spodoptera frugiperda*); a closely related species (*A. marmoratus*) helps keep the Corn Earworm (*Helicoverpa zea*) under control. ❻ ❼ ***Archytas aterrimus*** is a large (8–17 mm), common tachinid that develops in a variety of caterpillar hosts in the families Arctiidae, Lasiocampidae, Noctuidae, Notodontidae and Psychidae. These photographs show a female (on tree trunk) and a male (on leaves). As is true for most Tachininae, female *Archytas* lay eggs away from their hosts. Larvae remain on the spot until they ambush a passing host. ❽ ***Chrysotachina longipennis*** is one of a few metallic green tachinids that resemble unusually bristly blow flies. This species occurs from Pennsylvania to Florida but was only recently (2002) discovered and named. Hosts remain unknown. ❾ ***Linnaemya comta*** is an important parasitoid of pest cutworms and armyworms (Noctuidae).

FAMILY TACHINIDAE (continued) ❶ This ***Paradidyma*** probably developed as a parasitoid of a small caterpillar. ❷ ***Siphona*** species are small (3–5 mm) parasitic flies with a strikingly long proboscis. Most attack caterpillars, but some have the unusual habit of parasitizing crane fly larvae. Tachinidae rarely parasitize other flies. ❸ This distinctively long-legged tachinid fly (***Cholomyia inaequipes***) parasitizes weevils. **Subfamily Exoristinae** ❹ ***Exorista*** species attack a wide range of caterpillars and similar sawfly larvae, including Gypsy Moths (*Lymantria dispar*), Monarch Butterflies (*Danaus plexippus*), Forest Tent Caterpillars (*Malacosoma disstria*) and many others. ❺ ***Hemisturmia parva*** is a small parasitoid that attacks small caterpillars, including Spruce Budworms (*Choristoneura fumiferana*) and other Tortricidae. ❻ ***Austrophorocera*** species are parasitoids of medium-sized caterpillars, including slug moths (Limacodidae) and prominents (Notodontidae). ❼ ***Anisia flaveola*** is rarely collected, but it can be a common fly in deciduous forests populated by its host camel crickets (*Ceuthophilus* spp.). ❽ Although this species is called ***Smidtia fumiferanae***, it attacks many hosts in addition to *Choristoneura fumiferana* (Spruce Budworm). This species was included in *Winthemia* until recently. ❾ The somewhat metallic sheen of this ***Winthemia abdominalis*** is unusual for the genus *Winthemia*. Members of the large southern genus *Chrysoexorista* are similarly reflective in life, but fade soon after death. ❿ The genus ***Winthemia*** is a speciose and common genus of tachinids that deposit large eggs on the outside of caterpillar hosts. This is ***W. occidentis***.

FAMILY TACHINIDAE (continued) ❶ Flies in the genus ***Panzeria*** usually parasitize relatively large caterpillars, such as tiger moths (Arctiidae) and owlet moths (Noctuidae). One North American *Panzeria* species has been used in Europe as a biological control agent against Fall Webworms (*Hyphantria cunea*). This is ***P. ampelus***. ❷ ***Hineomyia setigera*** is one of several northeastern tachinids in which the apex of the abdomen is bright red. Hosts remain unknown. ❸ This tachinid (***Phebellia cerurae***) has just put an egg on this *Furcula* caterpillar (Notodontidae), and looks as if it is about to add another one. ❹ ***Phebellia*** species (eight in North America, this is ***P. helvina***) lay large, planoconvex eggs that are fully embryonated and ready to hatch once attached to a caterpillar host. ❺ Some tachinids, like this ***Gonia distincta***, deposit huge numbers of tiny eggs that must be consumed by the right host before they hatch. ❻ This ***Nilea*** was attracted to the honeydew produced by these Oak Kermes scales. *Nilea* species are parasitoids of caterpillars. ❼ ***Lespesia*** is a large tachinid genus with over 30 North American species, including some of the most common parasitoids of butterfly and moth caterpillars. ❽ ***Belvosia unifasciata*** is a parasitoid of cutworm caterpillars and their pupae. ❾ ***Oswaldia albifacies*** is a relatively small tachinid that has been reared from caterpillars of snout moths (Pyralidae).

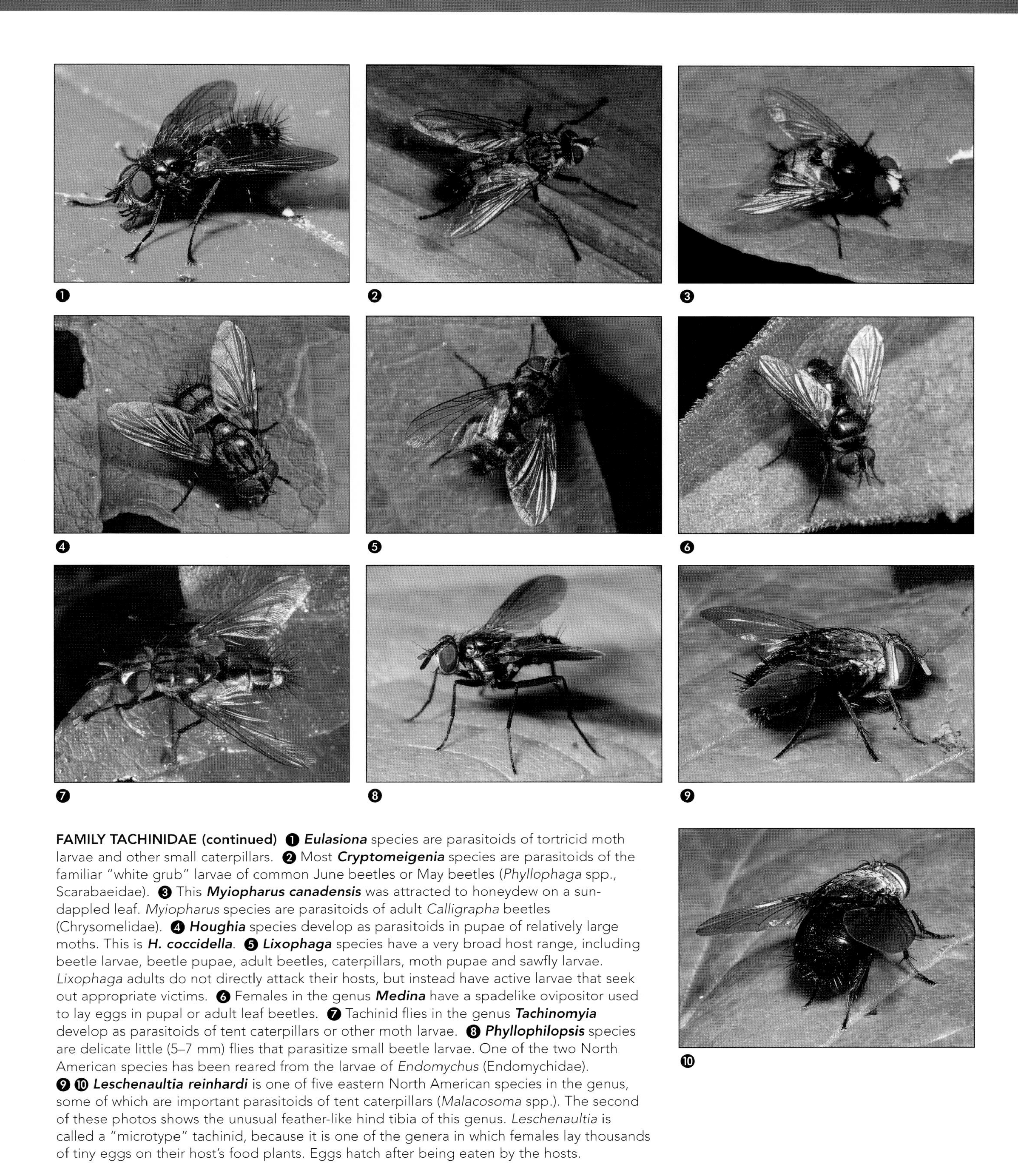

FAMILY TACHINIDAE (continued) ❶ ***Eulasiona*** species are parasitoids of tortricid moth larvae and other small caterpillars. ❷ Most ***Cryptomeigenia*** species are parasitoids of the familiar "white grub" larvae of common June beetles or May beetles (*Phyllophaga* spp., Scarabaeidae). ❸ This ***Myiopharus canadensis*** was attracted to honeydew on a sun-dappled leaf. *Myiopharus* species are parasitoids of adult *Calligrapha* beetles (Chrysomelidae). ❹ ***Houghia*** species develop as parasitoids in pupae of relatively large moths. This is ***H. coccidella***. ❺ ***Lixophaga*** species have a very broad host range, including beetle larvae, beetle pupae, adult beetles, caterpillars, moth pupae and sawfly larvae. *Lixophaga* adults do not directly attack their hosts, but instead have active larvae that seek out appropriate victims. ❻ Females in the genus ***Medina*** have a spadelike ovipositor used to lay eggs in pupal or adult leaf beetles. ❼ Tachinid flies in the genus ***Tachinomyia*** develop as parasitoids of tent caterpillars or other moth larvae. ❽ ***Phyllophilopsis*** species are delicate little (5–7 mm) flies that parasitize small beetle larvae. One of the two North American species has been reared from the larvae of *Endomychus* (Endomychidae). ❾ ❿ ***Leschenaultia reinhardi*** is one of five eastern North American species in the genus, some of which are important parasitoids of tent caterpillars (*Malacosoma* spp.). The second of these photos shows the unusual feather-like hind tibia of this genus. *Leschenaultia* is called a "microtype" tachinid, because it is one of the genera in which females lay thousands of tiny eggs on their host's food plants. Eggs hatch after being eaten by the hosts.

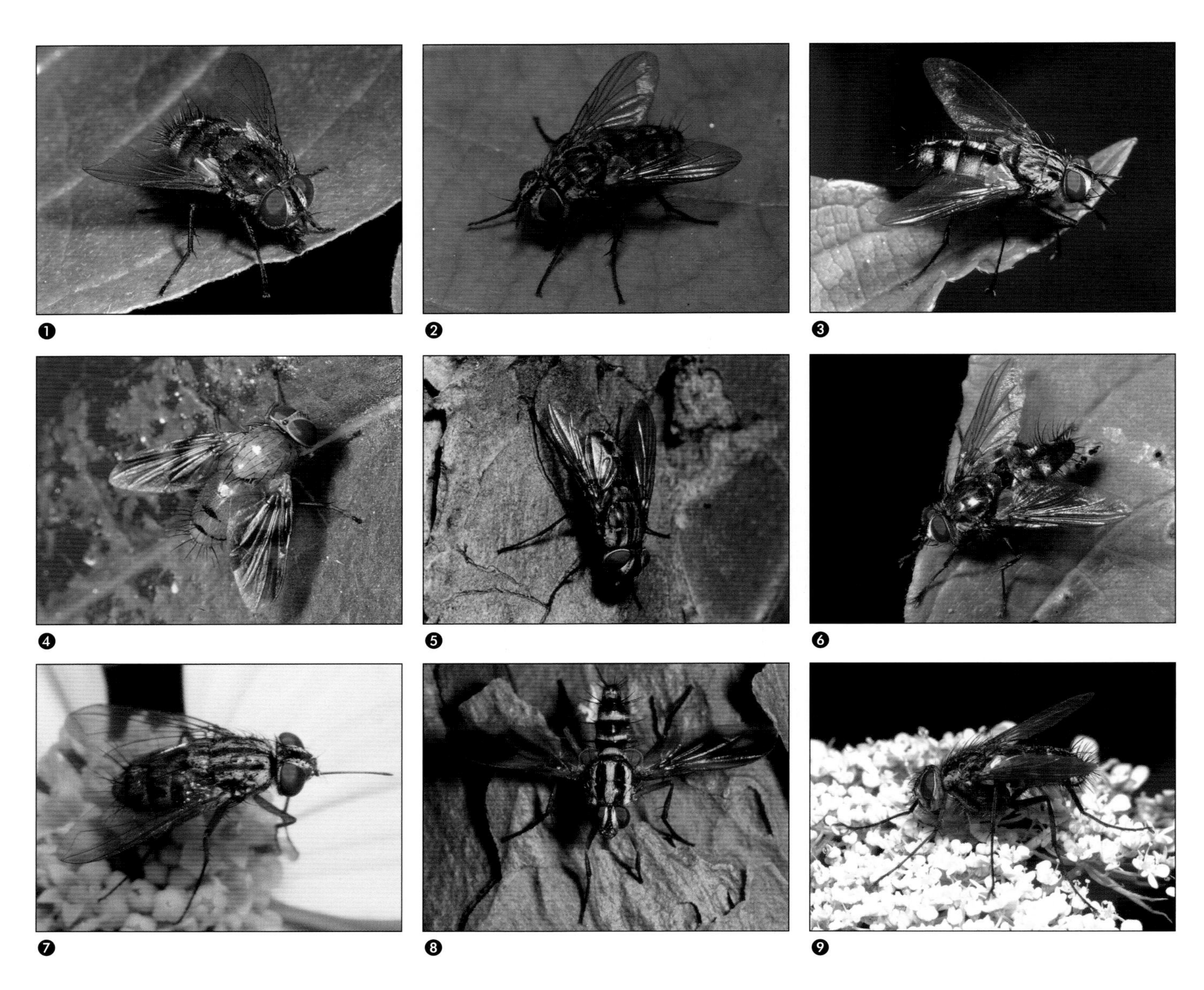

FAMILY TACHINIDAE (continued) ❶ ***Carcelia*** species are parasitoids of a wide variety of caterpillars, including several pests such as Satin Moths (*Leucoma salicis*), Gypsy Moths (*Lymantria dispar*), Fall Webworms (*Hyphantria cunea*), Forest Tent Caterpillars (*Malacosoma disstria*) and Eastern Tent Caterpillars (*M. americanum*). Female flies attach their eggs to host hairs by stalks; larvae burrow into the host upon hatching. ❷ ***Compsilura concinnata*** was deliberately introduced from Europe to North America in 1906 for the control of Gypsy Moths (*Lymantria dispar*), but it is now blamed for declining populations of giant silkworm moths and other non-target hosts. With over 200 known hosts in three insect orders, *C. concinnata* is a remarkably polyphagous species. It attacks its host by gripping and ripping with a can-opener-like modification of its abdomen before injecting maggots into the host. Larvae leave the spent host to pupate less than two weeks later. ❸ ***Calolydella lathami***, the only North American species in this otherwise Neotropical genus, occurs from Ontario and Quebec south to the Carolinas. It probably parasitizes beetles. Subfamily Dexiinae ❹ This little tachinid (***Oestrophasia signifera***) probably developed as a parasitoid inside a cricket. ❺ Tachinids in the genus ***Billaea*** are often seen on tree trunks, like this one perching head-down on a pine tree. *Billaea* species are parasitoids of wood-boring beetles, and an Asian species is a likely candidate for biological control of the Asian Longhorn Beetle (*Anoplophora glabripennis*) in North America. ❻ ***Uramya*** species (there are two in the northeast; this is ***U. limacodis***) are common, slender tachinids usually seen as they strike a characteristic pose along the edge of a broad leaf. They parasitize seemingly well-protected spiny or bristly caterpillars such as slug caterpillars, puss moths and tiger moth caterpillars. ❼ The strikingly long-beaked species of ***Prosenoides*** occur as far north as Ohio, but this ***P. flavipes*** was photographed in Florida. ❽ ***Euantha litturata*** is a striking tachinid seen on tree trunks from Pennsylvania on south. ❾ ***Ptilodexia*** is a large genus of long-legged beetle parasitoids. Although this one is nectaring on a Queen Anne's Lace (*Daucus carota*) flower, members of this genus are most often seen on tree trunks. *Ptilodexia* species, like other Dexiinae, have active larvae that do the job of hunting down a host, often by burrowing through soil or rotting wood.

FAMILY TACHINIDAE (continued) ❶ Tachinid flies in the genus ***Zelia*** are parasitoids of scarabs and other beetles. Note the pseudoscorpion hitching a ride by clamping itself onto the hind leg of this ***Z. vertebrata***. ❷ ***Campylocheta*** species are parasitoids of cutworms and inchworms. ❸ ***Thelaira americana*** is a widespread species found across Canada and throughout eastern North America. ❹ ***Euthera tentatrix*** is a parasitoid of stink bugs. Subfamily Phasiinae ❺ ***Hemyda aurata*** is a parasitoid of predacious stink bugs such as spined soldier bugs (*Podisus* spp.). ❻ ***Cylindromyia*** species, like this ***C. interrupta***, are usually parasitoids of stink bugs. ❼ This pair of ***Xanthomelanodes flavipes*** probably developed as parasitoids of assassin bugs. ❽ ❾ ***Trichopoda*** species, like this ***T. pennipes***, parasitize a wide range of true bugs, and are sometimes used in the biological control of plant-feeding stink bugs and leaf-footed bugs. The generic name *Trichopoda* is from the Greek for "hairy" and "foot," and the species name "*pennipes*" is from the Latin for "feather." The second of these photos shows the feathery hind legs alluded to by the scientific names.

FAMILY TACHINIDAE (continued) ❶ ***Gymnosoma*** are common, bare tachinids that develop as parasitoids of stink bugs. ❷ ❸ ***Gymnoclytia occidua*** females are less colorful than the males. ❹ ***Phasia fenestrata*** is a large, slow-flying tachinid that appears in early spring when it can sometimes be seen on willow catkins. There are no host records for this species, but it probably parasitizes large stink bugs. ❺ The genus ***Euclytia*** includes a single described North American species, ***E. flava***, which is a parasitoid of stink bugs. **FAMILY RHINOPHORIDAE** (WOODLOUSE FLIES) ❻ Our only common rhinophorid fly, ***Melanophora roralis***, is an internal parasitoid of terrestrial isopods (woodlice or sowbugs). The female of this mating pair of ***Melanophora roralis*** has distinguishing white spots at the tips of her wings. **FAMILY OESTRIDAE** (BOT FLIES) ❼ The 26 North American ***Cuterebra*** species, such as this ***C. fontinella***, are known as rodent bots although these bumble bee-sized flies parasitize rabbits as well as rodents. Eggs are laid near nests or runs, where the warmth of a passing host triggers hatching into a maggot that adheres to a host and invades it through a mouth or nose. After a week or so in the host's head, the bot maggot migrates through the host's tissue to develop under the skin, leaving a breathing hole that is ultimately enlarged to allow the maggot to pop out and pupariate. ❽ ***Gasterophilus intestinalis*** is a horse bot that sticks its eggs on a horse's foreleg hairs where the horse will likely lick them off. The eggs hatch in response to the warmth of the tongue, and the larvae ultimately end up in the stomach attached to the stomach lining. ❾ This is the common eastern North American species of deer nose bot fly (***Cephenemyia phobifer***, a.k.a. the "snot bot") that develops in the throat and sinus cavities of deer, ultimately leaving the nose with a snort or sneeze to pupate in the ground.

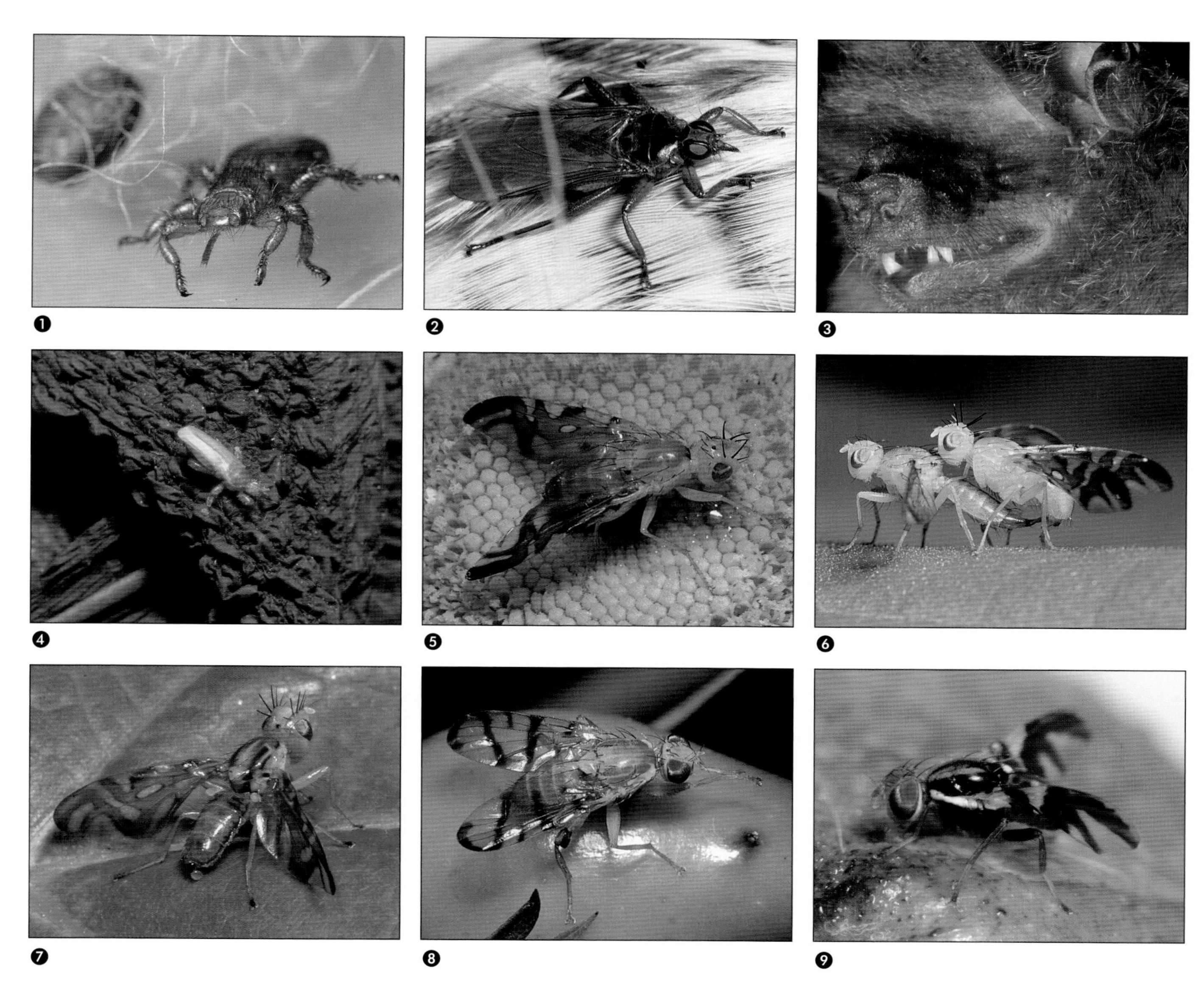

FAMILY HIPPOBOSCIDAE (LOUSE FLIES) ❶ **Sheep Keds** (***Melophagus ovinus***) were once common bloodsucking ectoparasites of sheep in much of the world, but these wingless flies are now rare in eastern North America. (This photo was taken in Chile.) Keds spend their whole lives on their hosts, where they deposit mature larvae that immediately form seed-like puparia among the wool. (One puparium can be seen in the background of this photo.) ❷ This louse fly (***Icosta americana***) was photographed on a road-killed grouse. Louse fly females do not lay eggs, but instead deposit mature larvae that pupate immediately. This species attacks a wide variety of birds, and can spread West Nile virus from bird to bird. **FAMILY NYCTERIBIIDAE** (WINGLESS BAT FLIES) ❸ Unless you work with bats, you are unlikely to see the tiny, entirely wingless, spider-like bat parasites in the mostly Old World tropical family Nycteribiidae. Although wingless bat flies are recorded from eastern North America, this one (just below the bat's ear) is Australian. Adult bat flies rarely leave their bat hosts except to deposit mature larvae on the bat roost. **FAMILY STREBLIDAE** (BAT FLIES) ❹ Bat flies are rarely collected, but members of this mostly tropical group have been recorded as far north as Virginia in eastern North America. These very small, blood-feeding bat parasites are pupiparous (they lay mature larvae rather than eggs) like louse flies and wingless bat flies. **FAMILY TEPHRITIDAE** (FRUIT FLIES) ❺ The **Sunflower Maggot** (***Strauzia longipennis***) is a common fruit fly that develops in the stems of sunflowers and related plants. This is a male, with large head bristles used in jousting with other males for position on sunflower stems. ❻ Like all fruit flies, the tip of the abdomen of the female **Sunflower Maggot** (***Strauzia longipennis***) is modified into a hard egg-laying tube. ❼ ***Strauzia intermedia*** is closely related to the Sunflower Maggot, but develops in Tall Coneflower (*Rudbeckia laciniata*). ❽ The genus ***Rhagoletis*** includes a number of important species, including the Apple Maggot (*R. pomonella*) and Cherry Fruit Fly (*R. cingulata*). This is the **Rose Hip Maggot** (***R. basiola***) laying eggs on a rose hip, in which its larvae will develop. ❾ The **Apple Maggot** (***Rhagoletis pomonella***) is a native species originally associated with wild hawthorn, but now divided into hawthorn- and apple-host races, with the latter a major pest of apples.

FAMILY TEPHRITIDAE (continued) ❶ ❷ ❸ Three species in the attractive genus ***Euaresta*** occur in eastern North America. ***E. bella*** is a beneficial species that develops only on Common Ragweed (*Ambrosia artemisiifolia*). ***E. festiva*** develops on Giant Ragweed (*A. trifida*) and ***E. aequalis*** develops on Cocklebur (*Xanthium strumarium*). ❹ This little fruit fly (***Urophora quadrifasciata***) was introduced from Europe to North America to control pest knapweeds and it is now widespread in both United States and Canada. ❺ ❻ ***Urophora cardui*** was introduced to North America from Europe for the control of Canada Thistle (*Cirsium arvense*). Canada Thistle is as widespread as ever, but now it is often decorated with the large, ball-like galls that house developing fruit fly larvae. ❼ ❽ ❾ ❿ Goldenrod ball galls are the most familiar fruit fly galls, and can be found in almost any patch of goldenrod. The photograph of the cut open gall was taken in late winter, after the larva had tunneled almost to the gall surface and formed a puparium. In late spring the adult **Goldenrod Gall Fly** (***Eurosta solidaginis***) will push its way out of the puparium and the gall by inflating a sac out of the front of its head; the close-up photograph of a gall's surface shows a fruit fly head with the sac (ptilinum) fully inflated. The ptilinum will later be withdrawn, leaving only a suture above the antennae.

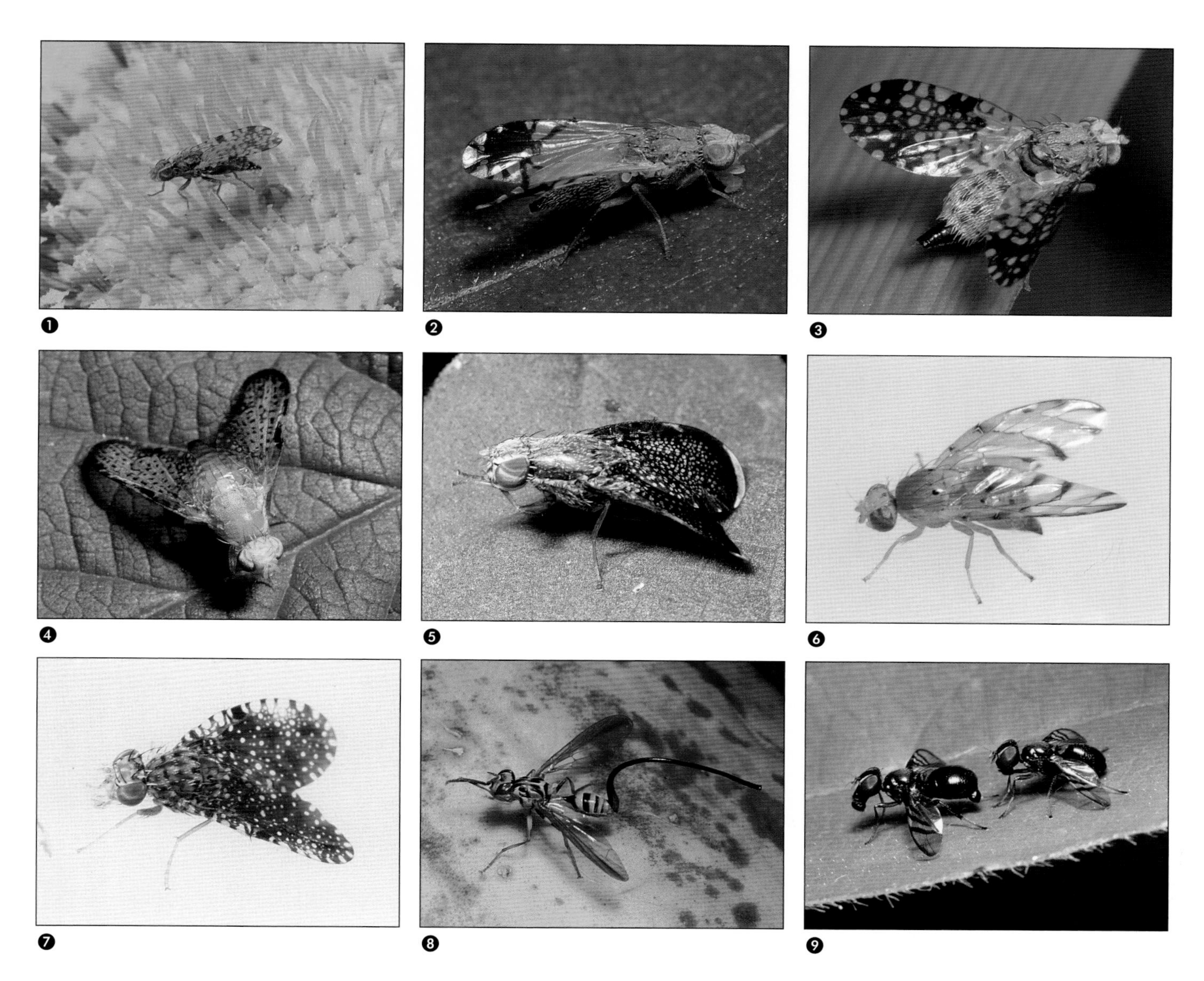

FAMILY TEPHRITIDAE (continued) ❶ ***Dioxyna picciola*** is a common tephritid on flowers of Asteraceae, in which its larvae develop. ❷ The widespread genus ***Trupanea*** includes 21 North American species. The widespread species ***T. actinobola*** has an unusually large host range, developing on asters, goldenrod and other plants. ❸ ***Campiglossa albiceps*** is a widespread and common species that develops on asters. ❹ ***Icterica seriata*** develops in *Bidens* (beggarticks). ❺ ***Eutreta noveboracensis*** is the most common northeastern species in this large genus. Larvae live in goldenrod (*Solidago* spp.) rhizomes. ❻ ***Tomoplagia obliqua*** is a small tephritid that breeds in flower heads of ironweed (*Veronia* spp.). ❼ Hosts for the two North American ***Xanthomyia*** species remain unknown. This ***X. platyptera*** is from a tallgrass prairie reserve in Ontario. ❽ Although this fruit fly (the **Papaya Fruit Fly**, ***Toxotrypana curvicauda***) does not occur in northeastern North America, it does occur in Florida where it uses its spectacularly long ovipositor to insert eggs into papaya fruits. **FAMILY PLATYSTOMATIDAE** (PLATYSTOMATID FLIES) ❾ These ***Rivellia*** are engaged in courtship behavior in which their outstretched, banded wings are conspicuously displayed. Larvae of *Rivellia* develop in the root nodules of legumes, a single larva inside each nodule.

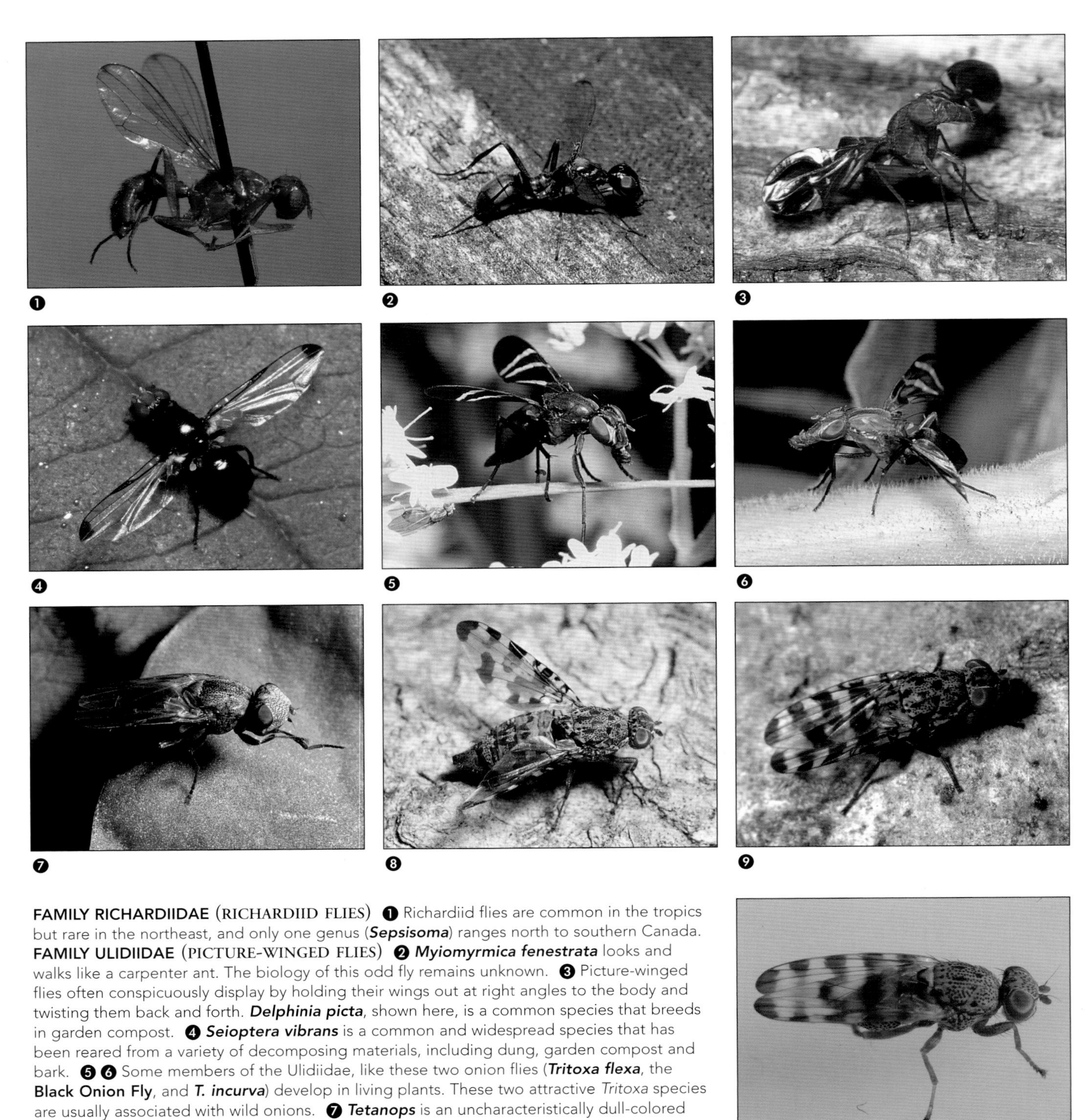

FAMILY RICHARDIIDAE (RICHARDIID FLIES) ❶ Richardiid flies are common in the tropics but rare in the northeast, and only one genus (***Sepsisoma***) ranges north to southern Canada. **FAMILY ULIDIIDAE** (PICTURE-WINGED FLIES) ❷ ***Myiomyrmica fenestrata*** looks and walks like a carpenter ant. The biology of this odd fly remains unknown. ❸ Picture-winged flies often conspicuously display by holding their wings out at right angles to the body and twisting them back and forth. ***Delphinia picta***, shown here, is a common species that breeds in garden compost. ❹ ***Seioptera vibrans*** is a common and widespread species that has been reared from a variety of decomposing materials, including dung, garden compost and bark. ❺ ❻ Some members of the Ulidiidae, like these two onion flies (***Tritoxa flexa***, the **Black Onion Fly**, and ***T. incurva***) develop in living plants. These two attractive *Tritoxa* species are usually associated with wild onions. ❼ ***Tetanops*** is an uncharacteristically dull-colored member of the Ulidiidae, but its bulbous and spotted head makes it easy to recognize. Look for it in gardens, where some *Tetanops* species attack the roots of living plants. ❽ ❾ ***Pseudotephritis*** species like this ***P. vau*** (with its wings raised) and this ***P. approximata***, are commonly seen on exposed, dead wood. ❿ ***Pseudotephritina*** species can be distinguished from the similar *Pseudotephritis* by the shining black patches along the side of the scutellum.

FAMILY ULIDIIDAE (continued) ❶ ❷ Some Ulidiidae, like these two ***Callopistromyia*** species (***C. strigula*** and ***C. annulipes***), develop under bark and can often be seen conspicuously displaying with outstretched wings on newly fallen trees. ❸ ❹ The Ulidiidae includes several genera of relatively small, banded-winged flies such as this ***Herina*** and this ***Chaetopsis***, respectively. Although they are common flies, little is known of their biology. ❺ ***Idana marginata***, the largest ulidiid in eastern North America (over 10 mm), develops in compost. ❻ The genus ***Otites*** is mostly western, but this species (***O. michiganus***) occurs in Michigan and Ontario as well as the western states and provinces. ❼ ***Melieria*** species are very common in marshes, and can often be seen perched on cattail (*Typha*) leaves. ❽ This ***Euxesta*** species occurs in salt marshes along the east coast. Other members of this genus (over 30 North American species) are more widespread and one species is a significant pest of corn in the southeast. **FAMILY LONCHAEIDAE** (LONCHAEID FLIES) ❾ Although this ***Lonchaea*** (Lonchaeidae) is laying eggs in a bracket fungus, most of these common, shiny blue-black flies develop under bark or in dead wood.

FAMILY LONCHAEIDAE (continued) ❶ This ***Lonchaea*** has just emerged from its puparium, found under the bark of a dead tree. **FAMILY PALLOPTERIDAE** (FLUTTER FLIES) ❷ ***Toxoneura superba***, our most common species in this small family, can often be seen conspicuously displaying on fall foliage. **FAMILY DROSOPHILIDAE** (SMALL FRUIT FLIES OR VINEGAR FLIES) ❸ The most familiar member of the Drosophilidae is the **Laboratory Fruit Fly, *Drosophila melanogaster***. This is a "brown eye" mutant from laboratory stock. ❹ ***Drosophila*** species are common on fungi, including poisonous mushrooms like this Fly Agaric (*Amanita muscaria*). ❺ Many drosophilids, such as this ***Mycodrosophila***, are associated with fungi. Most *Mycodrosophila* develop in bracket fungi. ❻ This ***Chymomyza amoena*** is strutting around a cut maple stump, conspicuously flicking its wings. Larvae of this species are found under loose, decaying bark. ❼ ***Scaptomyza***, with 22 species, is a common genus of Drosophilidae. They often occur in flowers, such as this trillium. **FAMILY DIASTATIDAE** (DIASTATID FLIES) ❽ The Diastatidae is a small, poorly known family. This ***Diastata*** species was found in a forested wetland. **FAMILY CHYROMYIDAE** (CHYROMYID FLIES) ❾ The Chyromyidae is a small, rarely encountered family, but this species of ***Gymnochiromyia*** is sometimes common amongst oak foliage in southern Ontario. Larvae probably develop in decaying organic material.

FAMILY EPHYDRIDAE (SHORE FLIES) ❶ Shore flies in the genus ***Hydrellia*** commonly feed on floating dead insects, like this mayfly. Larvae of *Hydrellia* usually develop as miners in the leaves of aquatic plants. ❷ Larvae of ***Notiphila*** usually develop in microbe-rich but oxygen-poor mud around the roots of aquatic plants. They breathe by using plant roots like living snorkels, piercing them with taplike posterior spiracles. ❸ Shore flies in the genus ***Parydra*** are usually the most common flies on muddy shores, where their larvae feed on diatoms. ❹ The mantis-like forelegs of predacious ***Ochthera*** adults make these common shore flies easy to identify. Their larvae are predators too, feeding mostly on midge larvae. ❺ One of our smallest shore flies, ***Lemnaphila scotlandae***, develops as a leaf miner within the tiny floating leaves of duckweed. ❻ ***Setacera atrovirens*** is a common shore fly on mats of filamentous algae. ❼ This distinctively mustached ***Cirrula gigantea*** adult is on a pile of eel grass (*Zostera*) on an Atlantic coast beach. Larvae of both North American species of *Cirrula* are filter feeders in salt marsh algal mats. **FAMILY CHLOROPIDAE (GRASS FLIES AND EYE FLIES)** ❽ Chloropidae are usually small and nondescript, but some are brightly colored like this green ***Meromyza***, photographed on a blade of grass. ❾ Most grass flies, like this ***Epichlorops***, are bare and shiny, often with a distinct shiny triangle on the front of the head.

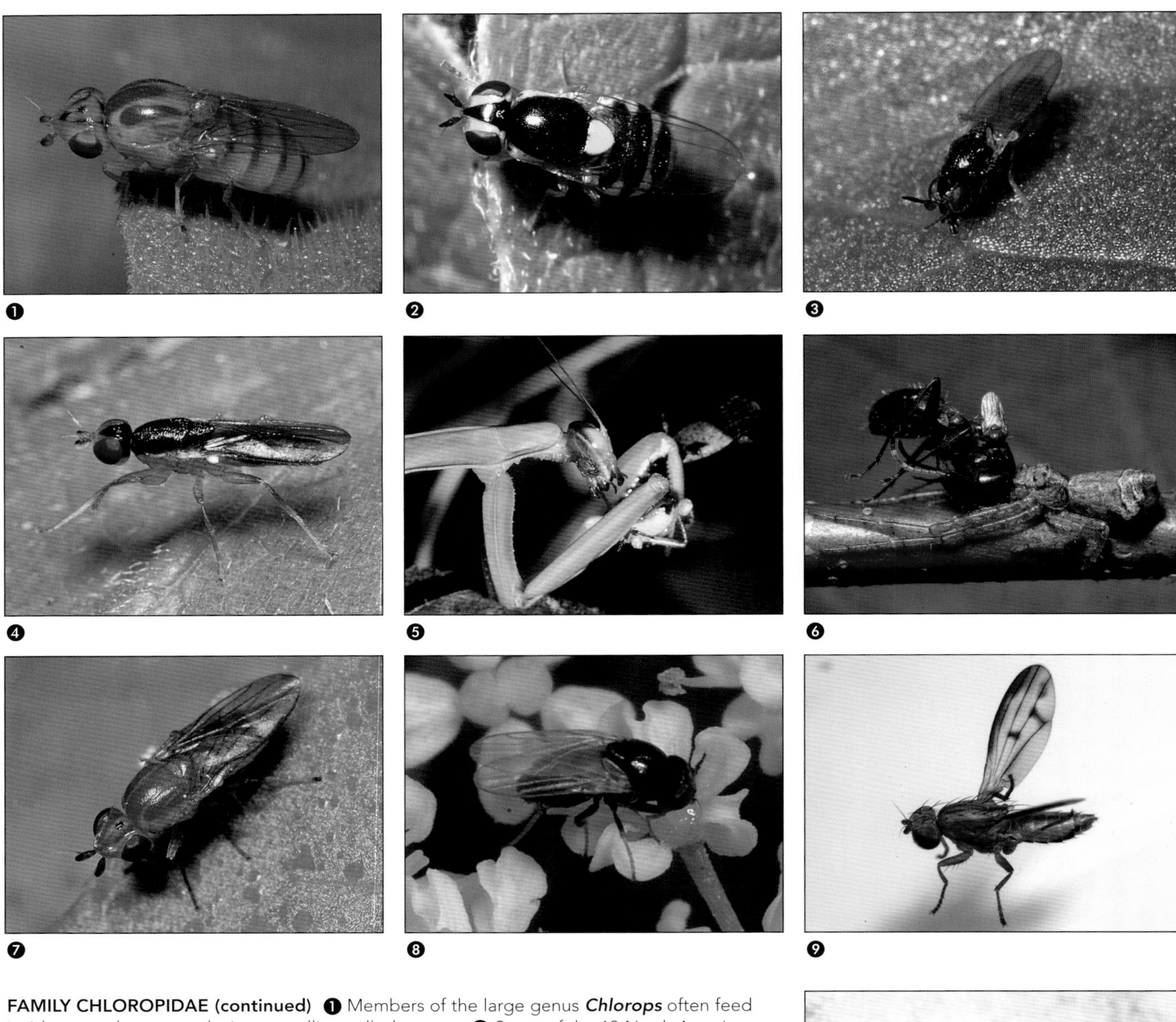

FAMILY CHLOROPIDAE (continued) ❶ Members of the large genus ***Chlorops*** often feed inside grass shoots, producing a swelling called a gout. ❷ Some of the 13 North American species of ***Thaumatomyia*** are common, variably colored flies. The genus is poorly known, but at least one species develops as a predator of gall-forming aphids. ❸ ***Elachiptera*** species usually develop as secondary invaders of damaged or decayed shoots and stems. ❹ ***Cetema*** species, like this ***C. subvittata***, are locally abundant among woodland grasses. ❺ The little black flies sharing a stink bug snack with this mantis are kleptoparasitic ***Olcella quadrivittata***. ❻ This kleptoparasitic chloropid fly (***Olcella cinerea***) has been attracted to a crab spider (*Tmarus angulatus*) eating an ant. ❼ ***Ectecephala*** is a widespread chloropid genus characterized by a white arista with a yellow base. ❽ The small, white-winged adults of ***Siphonella oscinina*** are often seen on flowers; larvae are predators and have been reared from spider egg sacs. **FAMILY OPOMYZIDAE** (OPOMYZID FLIES) ❾ Both species of ***Opomyza*** found in North America, including this ***O. petrei***, are accidental introductions from Europe. ❿ This recently introduced opomyzid (***Geomyza tripunctata***) is now a common insect in eastern North American yards and gardens.

FAMILY OPOMYZIDAE (continued) ❶ ***Geomyza apicalis*** is now less common than the more recently introduced species *G. tripunctata*. Both species develop in the shoots and stems of grasses. **FAMILY ANTHOMYZIDAE** (ANTHOMYZID FLIES) ❷ ❸ Anthomyzidae, like the pale ***Anthomyza*** shown here, are usually associated with grassy areas but some, like this dark ***Ischnomyia spinosa*** occur in woodlands. ❹ Most acalyptrate flies, like this anthomyzid, are relatively poorly known, and this genus (called ***Fungomyza*** because of its association with fungus) was only recently discovered. The individual in this photo became one of the type specimens used to describe the eastern North American species ***F. buccata*** in 2004. The only *Fungomyza* in our area, *F. buccata*, occurs from New England to Florida. **FAMILY MILICHIIDAE** (MILICHIID FLIES) ❺ Small flies in the family Milichiidae, like this ***Pholeomyia***, are often found on flowers. ❻ ❼ ***Paramyia nitens*** is a common milichiid, often seen on flowers like this daisy. These small flies are also regular visitors to spider webs, where they feed on fluids leaking from recently impaled prey like this green stink bug. ❽ Some Milichiidae are known kleptoparasites that occur on or around predators like this *Laphria* robber fly. The small flies under and near the much larger robber fly in this picture are milichiids in the genus ***Neophyllomyza***, and they are probably waiting for the robber fly to catch something so they can steal a snack. **FAMILY CARNIDAE** (CARNID FLIES) ❾ These adult ***Carnus hemapterus*** (four fat females, about 2 mm; four males, about 1 mm), started their adult lives as fully winged flies but they shed their wings upon arrival in an appropriately sheltered bird nest where they fed on blood or skin exudates of nestling birds. *Carnus* larvae consume decomposing material in bird nests. Larvae of the other two North American carnid genera feed on a variety of organic materials including dung, carrion and debris in nests.

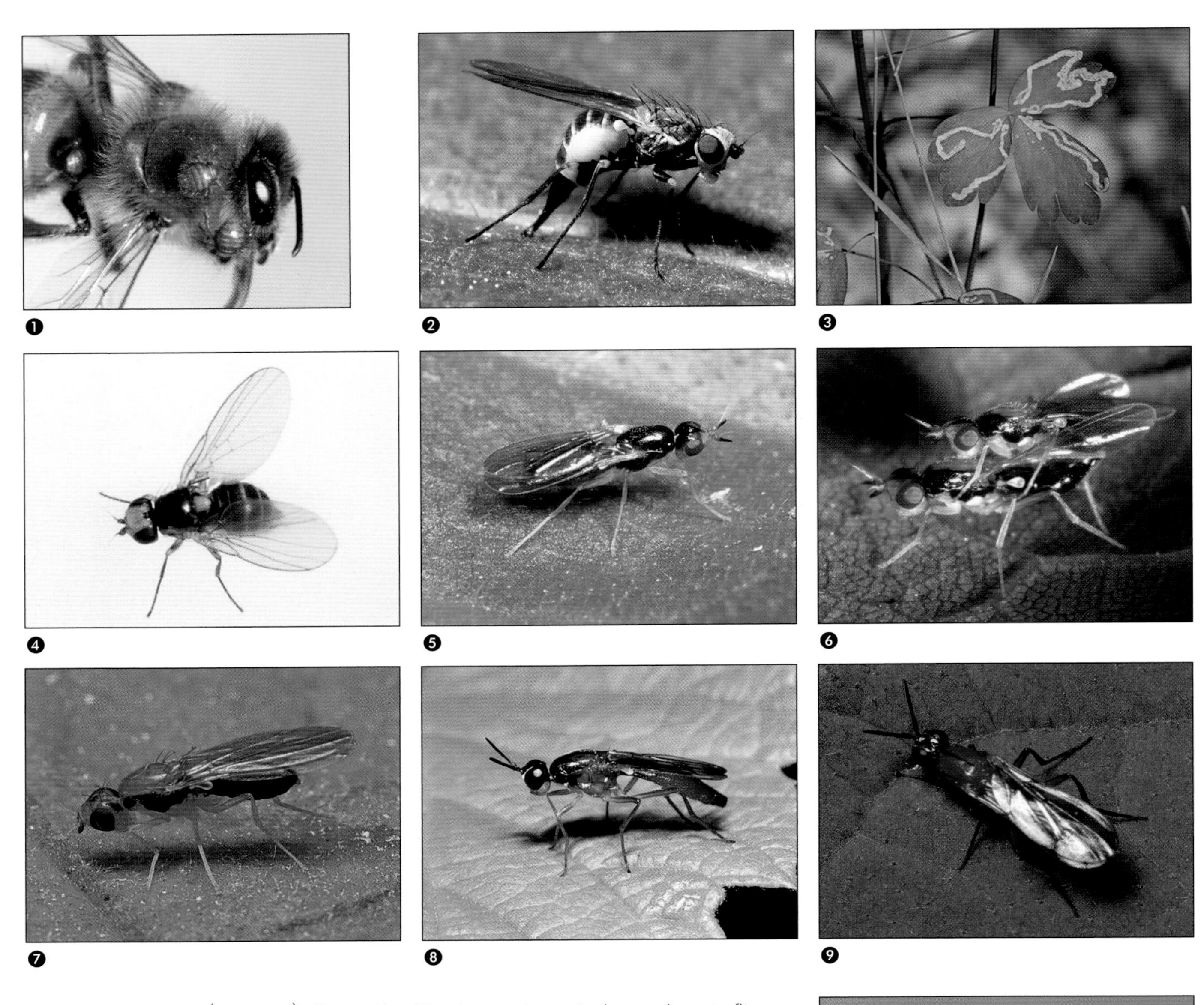

❿

FAMILY BRAULIDAE (BEE LICE) ❶ **Bee Lice** (***Braula coeca***) are wingless acalyptrate flies that spend their adult lives clinging to adult honey bees. The tiny, mitelike adult bee lice feed on liquids stolen from their host's mouth. **FAMILY AGROMYZIDAE** (LEAF-MINING FLIES) ❷ Female Agromyzidae use the stiffened tip of the abdomen to insert eggs into host plants, in which the larvae usually develop as leaf or stem miners. Adult females also feed by puncturing leaves with the ovipositor before feeding on the plant juices. This image shows a black and yellow ***Napomyza*** (on an *Aster* leaf). ❸ Most species of Agromyzidae are easiest to identify on the basis of their distinctive mines, like these columbine leaf-miner mines (***Phytomyza*** sp). ❹ The **South American Pea Leafminer**, ***Liriomyza huidobrensis***, is an introduced pest of chrysanthemum, pea, spinach, celery, pepper and other crops. It has only recently been recognized in the northeast, and it is now a serious pest in Ontario. A very similar native species (*L. langei*) occurs in the west. **FAMILY PSILIDAE** (RUST FLIES) ❺ ❻ ❼ ***Psila*** species, like these ***P. fallax***, ***P. collaris*** and ***P. lateralis*** respectively, develop in the stems and roots of various plants. ❽ Distinctive for their remarkably long antenna (long first flagellomere), *Loxocera* species are known to develop in sedge stems. This is ***L. cylindrica***. ❾ ***Loxocera cylindrica***, the most common species of *Loxocera* in eastern North America, varies widely in color. ❿ Wings of ***Chyliza***, like this ***C. erudita***, are abruptly bent so they conform to the body, giving these flies a beetle-like appearance.

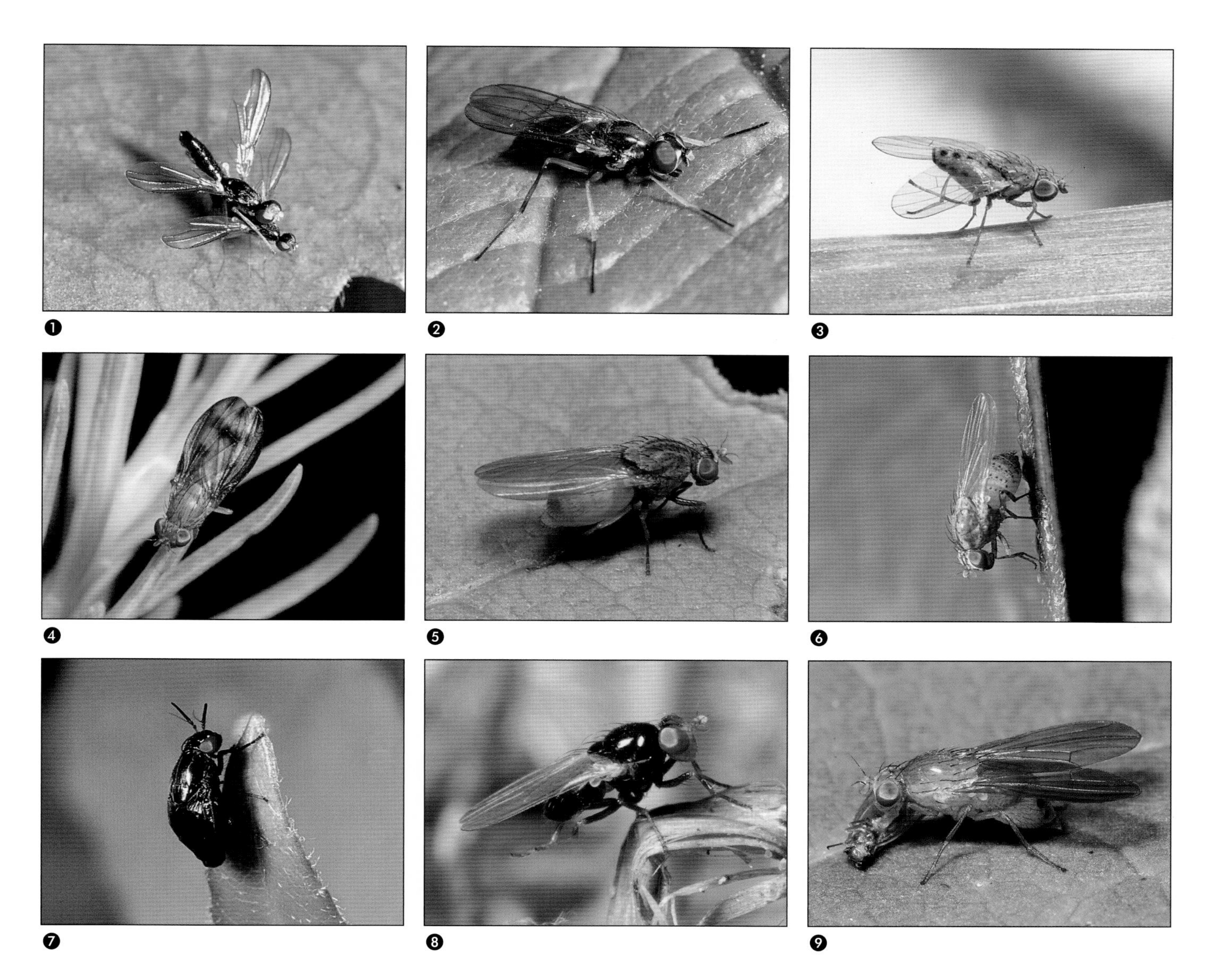

FAMILY STRONGYLOPHTHALMYIIDAE (STRONGYLOPHTHALMYIID FLIES) ❶ ***Strongylophthalmyia angustipennis*** is the only named North American member of the small, mostly Oriental, family Strongylophthalmyiidae. Little is known of its biology, but a closely related undescribed species is associated with decaying poplar trees in northern Ontario. **FAMILY TANYPEZIDAE** (TANYPEZID FLIES) ❷ Tanypezids are common flies in South and Central America, but only two species of ***Tanypeza*** occur in eastern North America. These long-legged flies are often seen on exposed foliage in forested areas. This species, ***T. longimana***, is common right up to northern Ontario. **FAMILY CHAMAEMYIIDAE** (APHID FLIES) ❸ Chamaemyiidae, like this ***Chamaemyia***, develop as predators of aphids and related insects. **FAMILY LAUXANIIDAE** (LAUXANIID FLIES) ❹ ❺ ❻ These are three of our common genera of Lauxaniidae – ***Homoneura***, ***Minettia*** and ***Poecilolycia***. Lauxaniids usually develop as miners in dead leaves. ❼ Several flies in different families, including this ***Steganolauxania***, seem to deliberately mimic small beetles by bending their wings against their abdomens. ❽ Some lauxaniid flies, like this ***Pseudocalliope***, develop in bird's nests. ❾ *Sapromyza* species develop in witches' broom, abnormally dense growths of accessory shoots caused by different fungi on different plants. This female ***Sapromyza*** sp. (an undescribed species near *S. rotundicornis*) seems to be feeding on a dead jumping plantlouse (Psyllidae).

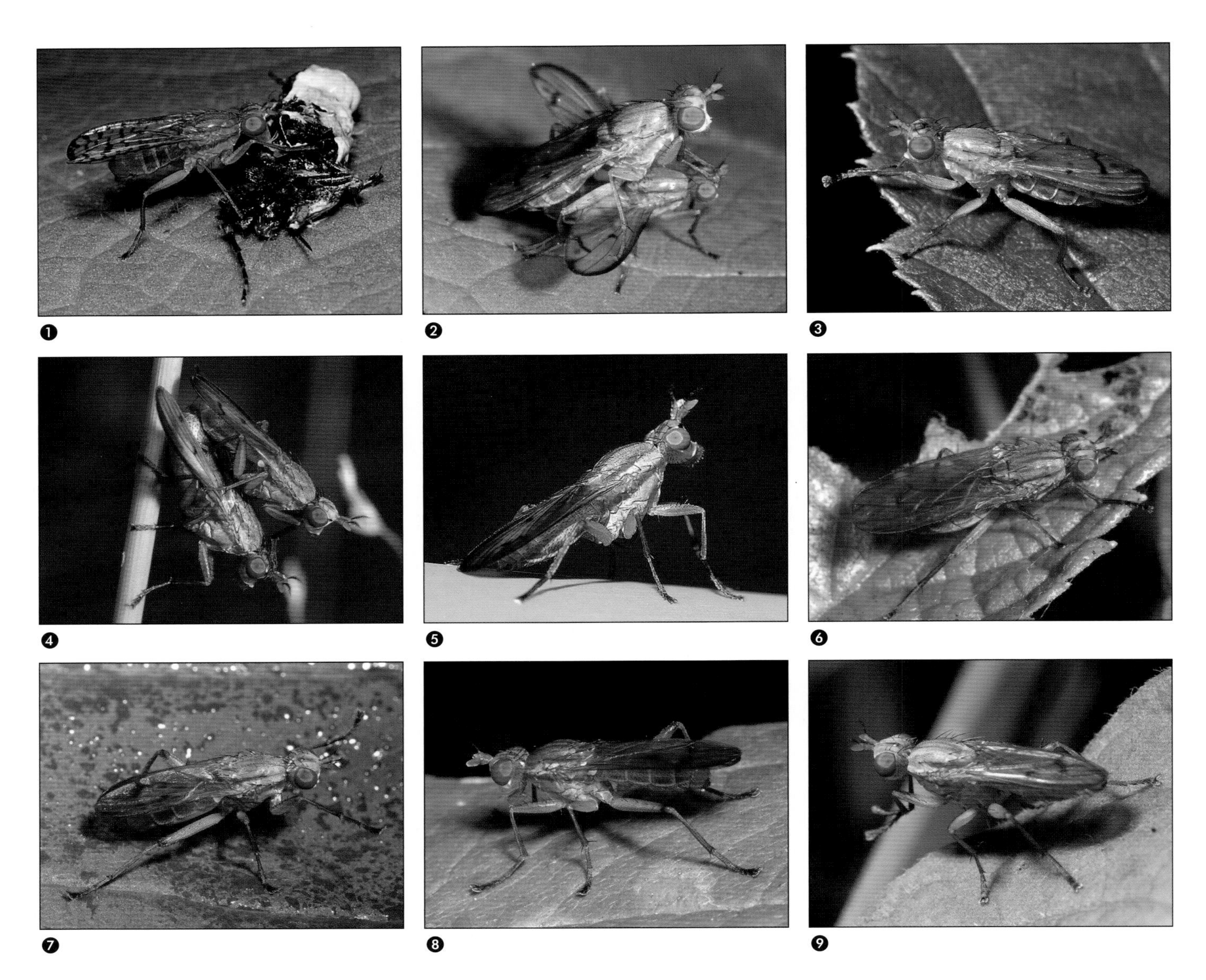

FAMILY SCIOMYZIDAE (MARSH FLIES) ❶ Members of the large marsh fly genus ***Tetanocera*** are parasitoids or predators of snails and slugs. ***T. valida*** larvae live in slugs, entering them through their invaginated eye stalks. This adult is feeding on a bird dropping loaded with insect parts. ❷ ❸ ***Tetanocera plebeja*** is a common fly in a variety of forest and wetland habitats, where it develops as a predator or parasite on slugs. The red blotch on the front of the single fly's head is a mite. ❹ This pair of ***Tetanocera plumosa*** is mating in a typical head-down position on a meadow stem. Unlike most other *Tetanocera*, which attack snails in the water, larvae of *T. plumosa* also attack snails stranded along marshy shores. ❺ ❻ ❼ ❽ ❾ ***Tetanocera*** adults are commonly seen perched, usually head-down, on cattails and other emergent aquatic vegetation amongst which their larvae attack aquatic snails. The species shown here are ***T. loewi***, ***T. ferruginea***, ***T. annae***, ***T. mesopora*** and ***T. montana*** respectively.

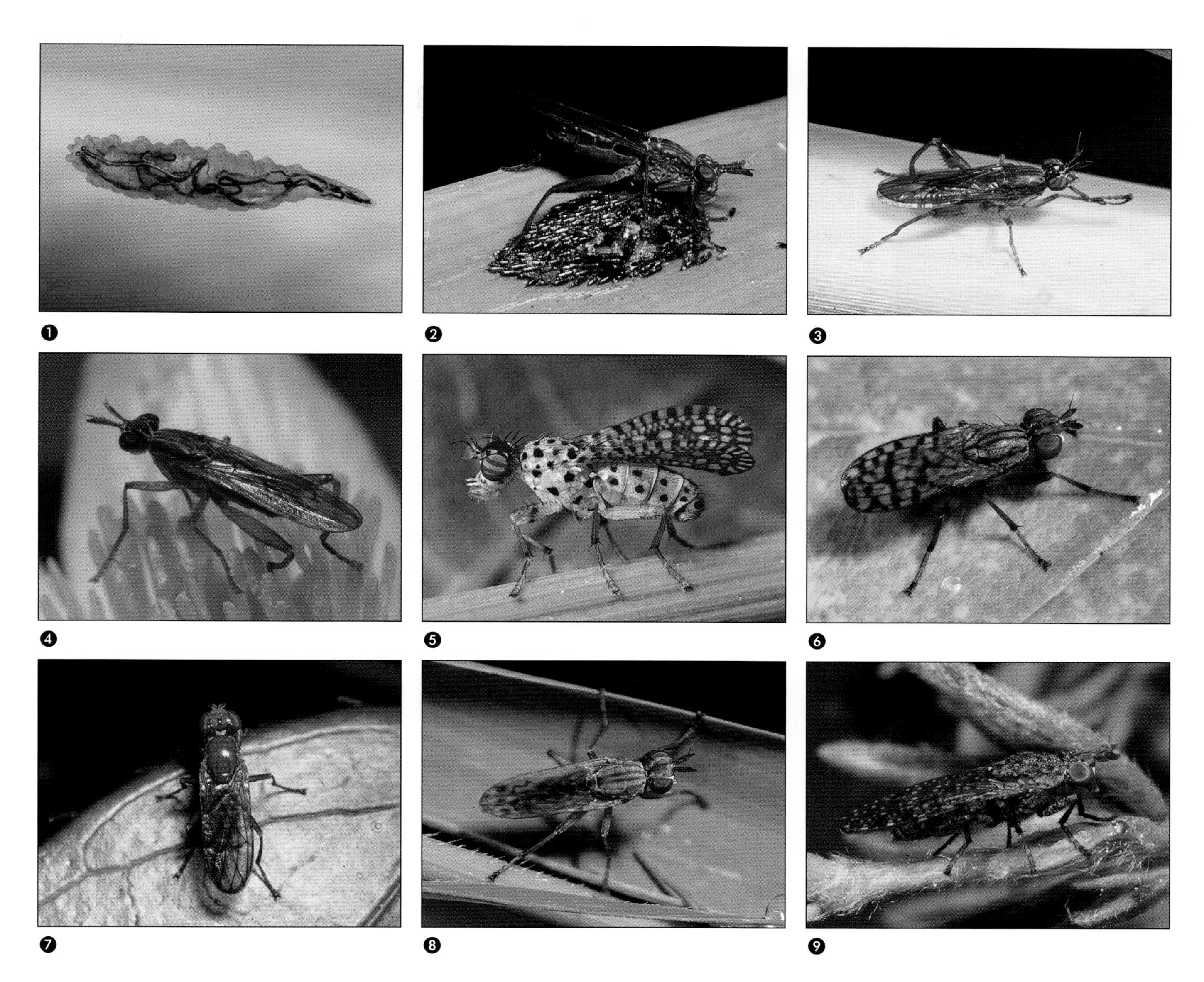

FAMILY SCIOMYZIDAE (continued) ❶ This translucent marsh fly larva is a predator of aquatic snails. You can see the mouth hooks at the narrow end of the body and the trachea opening to spiracles at the broad hind end. ❷ Adult marsh flies often scavenge on dead insects or insect eggs. This common species, ***Sepedon fuscipennis***, is eating horse fly eggs on a cattail leaf. North American *Sepedon* larvae are predators of aquatic snails. ❸ Members of the genus ***Sepedon***, like this ***S. armipes***, have distinctively long antennal bases (the pedicel is elongate). ❹ This ***Sepedon spinipes*** is perched on a Marsh Marigold (*Caltha palustris*) flower. ❺ The beautiful **Spotted Marsh Fly** (***Poecilographa decora***) is a formerly widespread species that has become rare in some areas. The biology of this species remains a mystery. ❻ ***Pherbecta limenitis***, the only member of the genus *Pherbecta*, is a rare fly associated with pristine fens. ❼ Members of this marsh fly genus (***Renocera***) feed on fingernail clams (Sphaeriidae). ❽ ***Elgiva*** species, like this ***E. solicita***, are common near the fresh water where their aquatic larvae pursue snails. ❾ ***Dictya***, the most common genus of marsh flies with spotted wings, differs from most similar sciomyzids in having a black arista (the hairlike end of the antenna), and in having a black dot on the middle of the white face. Larvae feed on freshwater snails.

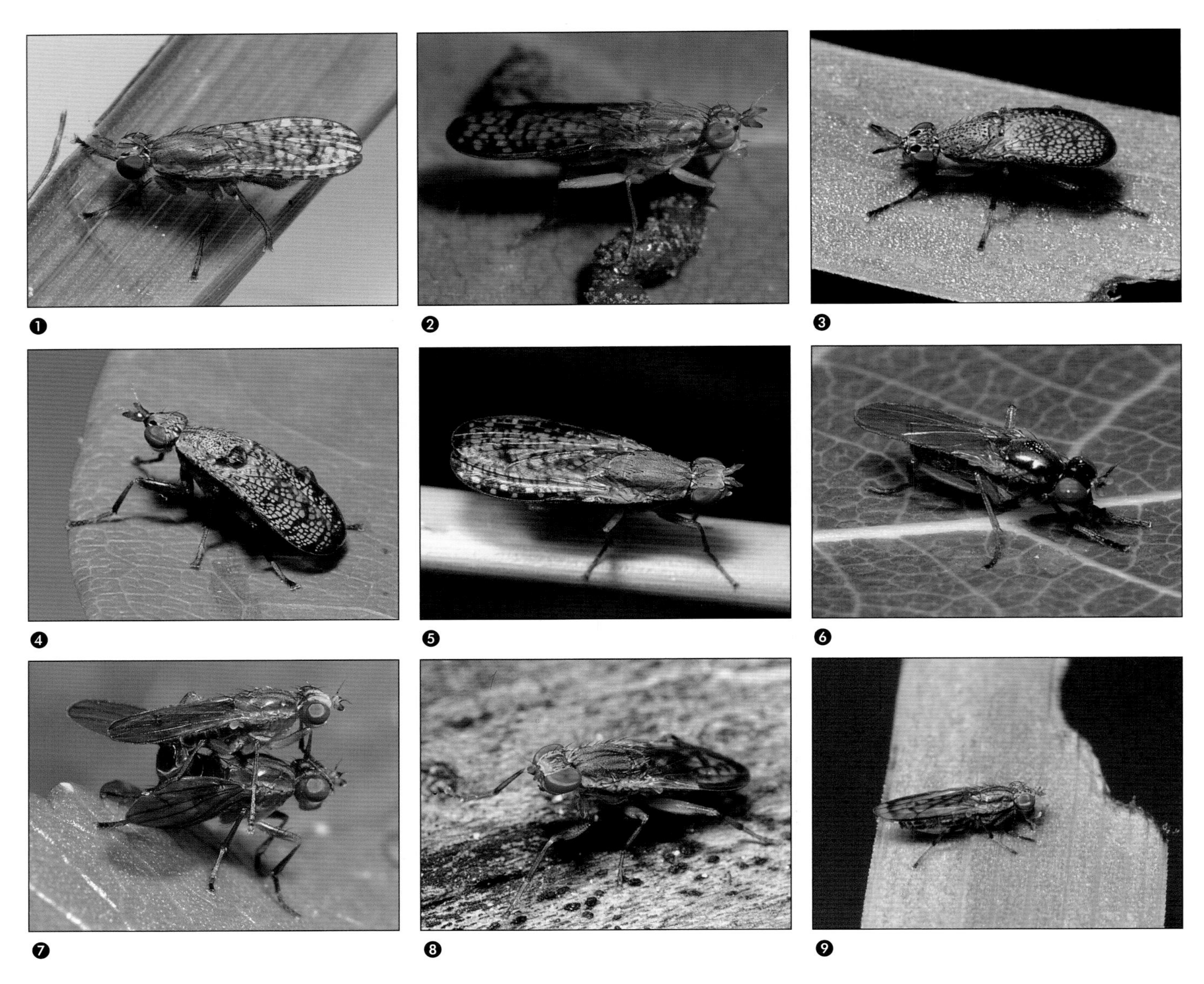

FAMILY SCIOMYZIDAE (continued) ❶ ***Limnia boscii*** is a common fly in marshes where its hosts (aquatic snails) occur. Most of North America's 17 species of *Limnia* remain unknown as larvae although some species of the genus are common. ❷ This ***Limnia loewi*** is snacking on a dead worm. The larval habits of this species are unknown, but adults occur in forests and swamps where the probable larval hosts (terrestrial or semiaquatic snails) occur. ❸ This rarely seen marsh fly genus (***Dictyacium firmum***) can be very abundant in fens, but nothing further is known of its biology. ❹ The only currently described North American species in this genus, ***Euthycera arcuata***, feeds on terrestrial snails. ❺ Our only species in this genus, ***Trypetoptera canadensis*** feeds on terrestrial snails. ❻ The small sciomyzids in the genus ***Anticheta*** lay their eggs on the eggs of aquatic and terrestrial mollusks. This is ***A. melanosoma***. ❼ ***Anticheta borealis***, here mating on a Marsh Marigold (*Caltha palustris*) in Ontario, often occur in wet areas where their larvae feed on egg masses of *Oxyloma* snails. ❽ ***Pherbellia***, the largest genus of marsh flies, includes some of the smallest and least often seen species of Sciomyzidae. This species (***P. albovaria***) is found in woodlands and usually parasitizes terrestrial snails. ❾ The small (about 3 mm) widespread sciomyzid ***Pherbellia nana*** usually develops as a parasitoid or predator of aquatic snails, and can be common in wet areas.

FAMILY SCIOMYZIDAE (continued) ❶ ***Atrichomelina pubera*** larvae develop in living or dead snails stranded along shorelines. ❷ Larvae of ***Sciomyza*** develop as parasitoids of snails, each larva developing in one specific host and ultimately killing it. Known hosts include semi-terrestrial Succineidae and aestivating aquatic *Stagnicola*. **FAMILY DRYOMYZIDAE** (DRYOMYZID FLIES) ❸ ***Dryomyza anilis***, by far the most common eastern dryomyzid, occurs in forested areas where it develops in decaying material. **FAMILY COELOPIDAE** (SEAWEED FLIES) ❹ ***Coelopa frigida*** is the only northeastern species of the Coelopidae, a family of flies found on seaweed-strewn seacoasts around the world. Larvae develop in piles of rotting seaweed, or wrack. **FAMILY CONOPIDAE** (THICK-HEADED FLIES) **Subfamily Conopinae** ❺ ***Physocephala*** is the most common genus of Conopidae. Look for these thick-headed flies on or near flowers, where they attack bees and wasps in flight. Some *Physocephala* species are parasitoids of honey bees and bumble bees. ❻ ❼ Small Conopidae in the genera ***Thecophora*** ❻ and the more common ***Zodion*** ❼ are often seen on flowers along with the Hymenoptera they parasitize. ❽ Thick-headed flies in the genus ***Myopa*** are often seen striking this peculiar pose on foliage, presumably as they search for suitable host bees. ❾ This ***Physoconops*** has been captured by a crab spider. One member of this relatively rare genus is known to parasitize leafcutter bees.

FAMILY CONOPIDAE (continued) Subfamily Dalmanniinae ❶ Some Conopidae, like this ***Dalmannia***, are convincing mimics of the small solitary bees that they parasitize. **FAMILY STYLOGASTRIDAE** ❷ Mating is quite a feat for this southern Ontario pair of ***Stylogaster neglecta***, thanks to the extremely elongate female abdomen. ❸ ***Stylogaster*** females hover in search for prey insects into which they jab barbed, harpoon-like eggs. This one was photographed as it hovered over an army ant swarm front in Costa Rica, but similar species occur in northeastern North America. *Stylogaster* is often treated as the subfamily Stylogastrinae of the family Conopidae. **FAMILY PYRGOTIDAE** (LIGHT FLIES) ❹ ***Pyrgota undata*** is a parasitoid of adult June beetles (*Phyllophaga*). Look for these spectacular nocturnal flies at your porch lights in early spring. ❺ ***Pyrgotella chagnoni*** flies during the daytime and early evening, when it attacks adult chafer beetles (*Dichelonyx*) in flight. ❻ Pyrgotids in the mostly southern genus ***Boreothrinax***, like other pyrgotids, attack adult scarab beetles and are unlike most other acalyptrate Diptera in their normally nocturnal habits. **FAMILY NERIIDAE** (NERIID FLIES) ❼ The Neriidae are sometimes inappropriately called "cactus flies" because one species breeds in decaying cacti in the southwestern United States. Most species breed in sappy wood or other decomposing material. This ***Nerius plurivittatus*** male is guarding his mate from competing males as she lays eggs in a wounded tree in Costa Rica. Both North American species of this mostly tropical family are southwestern. **FAMILY PSEUDOPOMYZIDAE** ❽ ***Latheticomyia***, the only pseudopomyzid genus known from North America, ranges from southwestern United States to South America (this photo is from Bolivia). Pseudopomyzids have not yet been recorded from eastern North America. **FAMILY ROPALOMERIDAE** (ROPALOMERID FLIES) ❾ Ropalomerid flies are mostly neotropical, but one species (*Rhytidops floridensis*) occurs around sappy stumps and tree wounds in the southeastern United States. This is a similar species (***Willistoniella pleuropunctata***) from Costa Rica.

FAMILY MICROPEZIDAE (STILT-LEGGED FLIES) Subfamily Taeniapterinae ❶ ***Rainieria antennaepes*** is a common long-legged, wasplike micropezid fly. Like many other members of the mostly tropical subfamily Taeniapterinae, this species waves its white-tipped front legs much the way wasps wave their antennae. The swellings on the sides of this male's abdomen probably play a role in releasing chemicals that attract females. ❷❸❹ ***Taeniaptera trivittata*** is eastern North America's only member of a large, otherwise Neotropical genus. This species only gets as far north as southern Ontario. Micropezid flies often have elaborate mating behavior, including the transfer of oral fluids between flies. ❺ This female ***Taeniaptera trivittata*** is laying eggs in a rotting stem; the white eggs and a puparium of the same species can also be seen in the photo. Subfamily Calobatinae ❻ ***Compsobata***, a mostly northern genus, includes several species commonly found in wet meadows. This ***C. univitta*** is feeding on a small dead beetle. ❼ The structure beside this ***Compsobata univitta*** adult is the shell of its puparium, found in wet, rotting wood. The adult fly used its ptilinum to force the cap off the puparium, leaving the open end seen near the fly's left middle leg. Subfamily Micropezinae ❽ Those ***Micropeza*** species with known larvae develop in the root nodules of legumes. **FAMILY CLUSIIDAE** (CLUSIIDS) ❾ The two eastern North American species of ***Clusia*** (***C. czernyi*** shown here) are the largest and most commonly encountered clusiids in our area.

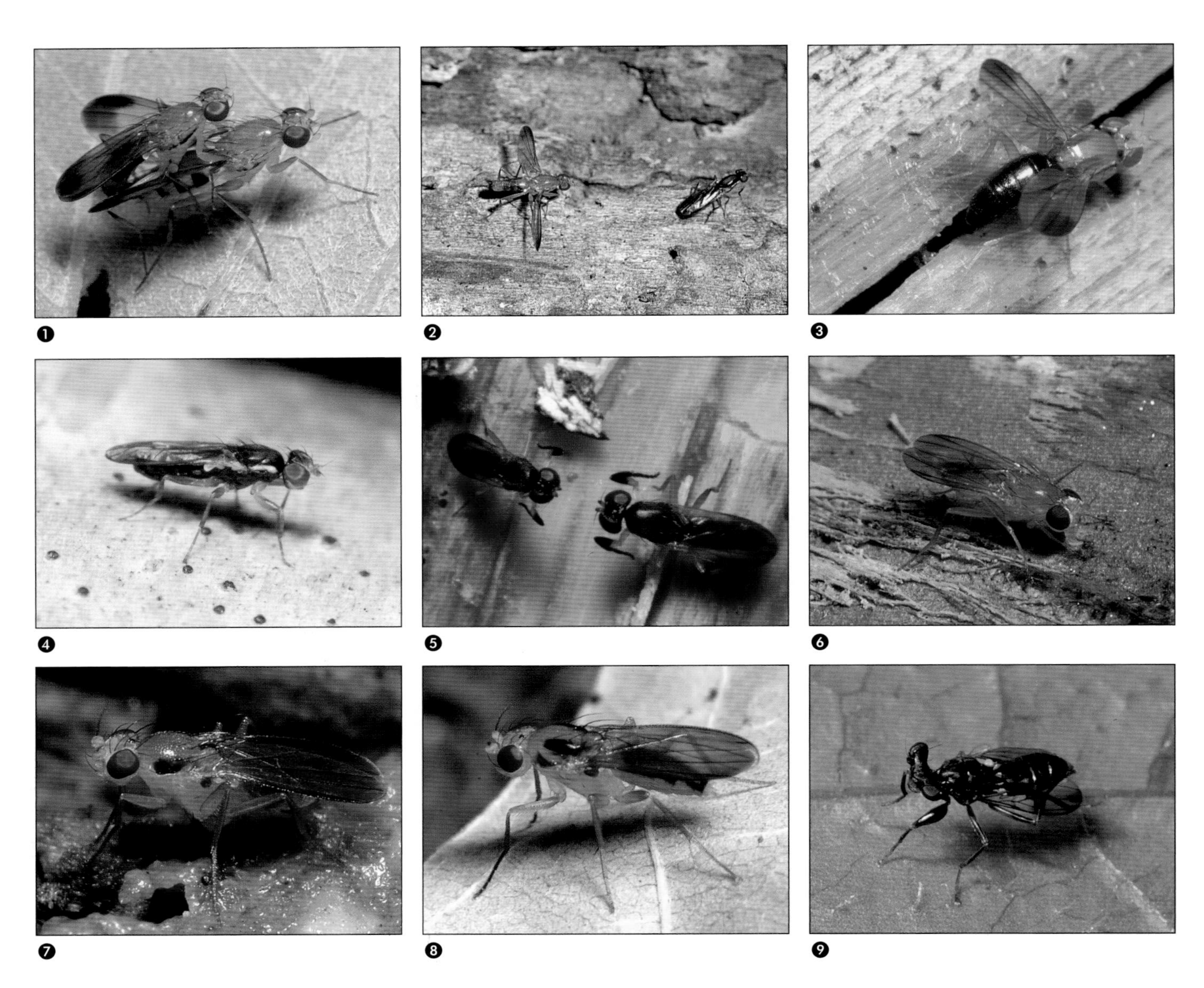

FAMILY CLUSIIDAE (continued) ❶ Although normally seen on exposed wood, these ***Clusia lateralis*** are mating on the undersurface of a leaf. Clusiid larvae develop in dead wood, where they probably prey on other insects. ❷ Clusiidae can often be found strutting and wing-flicking on the surfaces of fallen logs. Males normally challenge other males of the same species, but this ***Clusia lateralis*** is displaying to a ***Clusiodes americanus***. When males of the same species meet, they sometimes size each other up closely before the smaller one gives way. ❸ This ***Clusiodes clandestinus*** female is laying eggs in crack of a dead log, where her larvae will probably prey upon beetle larvae. ❹ Members of the large genus ***Clusiodes***, most of which are at least partly orange, are common insects on fallen trees where males vigorously defend their territories. This unusually dark species is ***C. ater***. ❺ Male clusiids often fight over mating territories, usually by butting heads. Males of the eastern North American species ***Heteromeringia nitida*** go into battle with one another by folding their front legs back at the tibial-femoral joint before "boxing" using their black "elbows." ❻ ❼ ❽ ***Sobarocephala flaviseta***, ***S. latifrons*** and ***S. setipes*** are among the 15 North American members of the huge genus *Sobarocephala*, which includes almost 150 named and hundreds of unnamed species in Central and South America. **FAMILY DIOPSIDAE** (STALK-EYED FLIES) ❾ Diopsids, or stalk-eyed flies, are mostly an Old World group, with only two species in North America. ***Sphyracephala brevicornis***, which is found throughout the northeast, overwinters in the adult stage and can often be found around skunk cabbage in early spring or late fall.

FAMILY PIOPHILIDAE (SKIPPER FLIES) ❶ Male ***Prochyliza xanthostoma*** have long antennae and an elongated head, which they use in territorial battles. ❷ These two ***Prochyliza xanthostoma*** males are sizing each other up over some territory on a desiccated carcass. Females of this species, like most other piophilids, are small, shiny black flies without head modifications. **FAMILY AULACIGASTRIDAE** (AULACIGASTRID FLIES) ❸ ❹ Despite its small size, the distinctively striped eyes of ***Aulacigaster neoleucopeza*** make this tree wound fly easily recognizable. The second photo shows this fly's tree wound home. **FAMILY PERISCELIDIDAE** (PERISCELIDID FLIES) ❺ This ***Periscelis wheeleri*** was attracted to a small poplar tree damaged by poplar borers. Periscelidids are rare flies, normally seen only at tree wounds. ❻ ***Cyamops*** is an odd, rarely encountered little genus currently placed in the Periscelididae. The three eastern North American species are superficially very similar to the Australian species illustrated here. ❼ ***Stenomicra*** is an odd genus of enigmatic little flies that are sometimes treated as their own family (Stenomicridae). Tropical species (like this one) are common in furled *Heliconia* leaves, but little is known of the one described eastern North American species (*S. angustata*). **FAMILY ODINIIDAE** (ODINIID FLIES) ❽ Flies in the family Odiniidae are rarely seen except at tree wounds. This ***Odinia*** was found on a poplar tree damaged by long-horned beetles. ❾ The two species of ***Traginops*** have a distinctively pointy-headed appearance because of an expanded ocellar plate. Like other Odiniidae, this ***T. irrorata*** is associated with bleeding tree wounds.

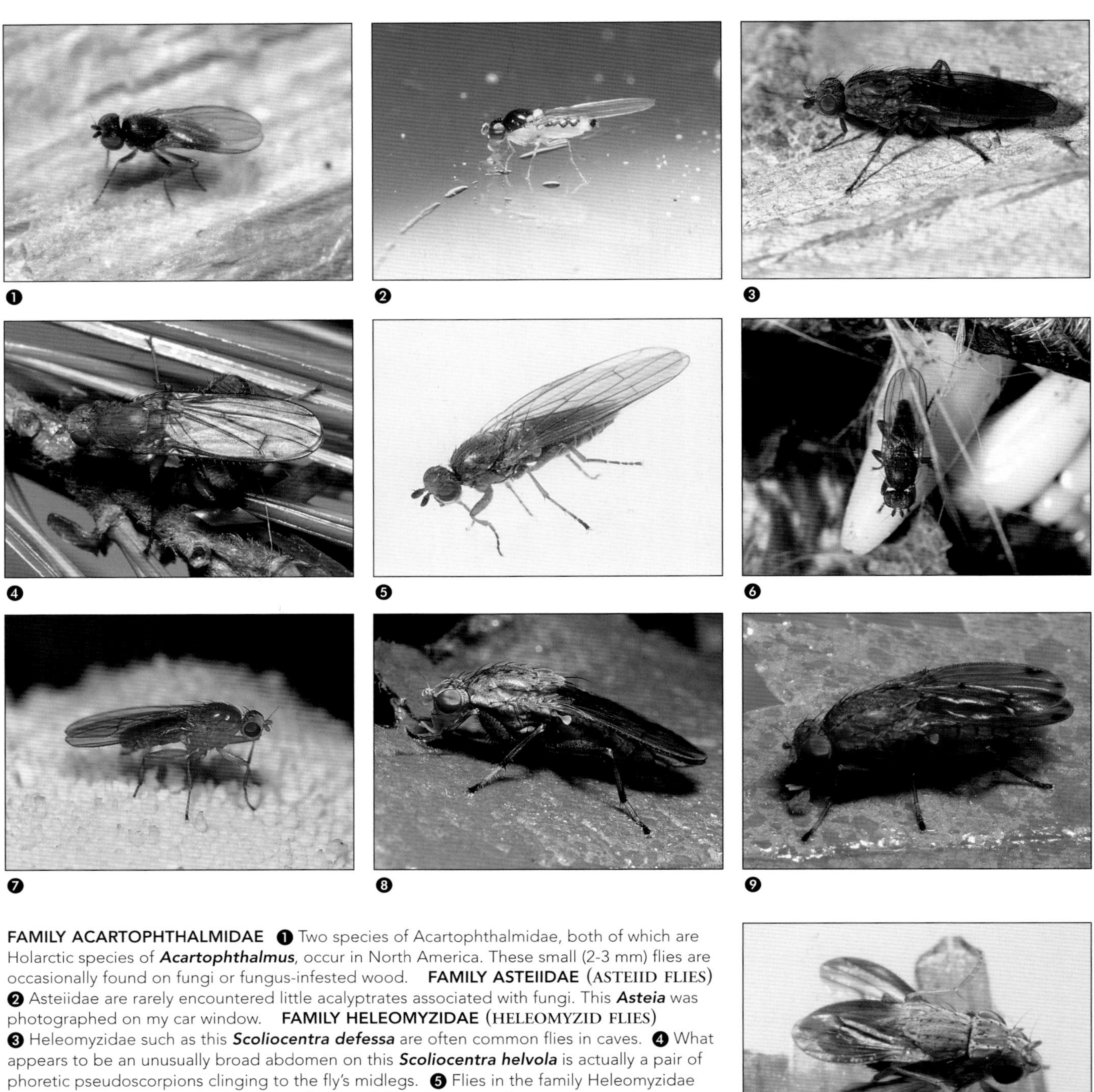

FAMILY ACARTOPHTHALMIDAE ❶ Two species of Acartophthalmidae, both of which are Holarctic species of ***Acartophthalmus***, occur in North America. These small (2-3 mm) flies are occasionally found on fungi or fungus-infested wood. FAMILY ASTEIIDAE (ASTEIID FLIES) ❷ Asteiidae are rarely encountered little acalyptrates associated with fungi. This ***Asteia*** was photographed on my car window. FAMILY HELEOMYZIDAE (HELEOMYZID FLIES) ❸ Heleomyzidae such as this ***Scoliocentra defessa*** are often common flies in caves. ❹ What appears to be an unusually broad abdomen on this ***Scoliocentra helvola*** is actually a pair of phoretic pseudoscorpions clinging to the fly's midlegs. ❺ Flies in the family Heleomyzidae are most common in the cool months of spring, fall and occasionally even winter. This ***Orbellia*** was photographed as it walked on the snow surface. ❻ Some heleomyzids, such as this ***Neoleria*** on the tooth of a dead cat, develop in carrion. ❼ ❽ Some heleomyzids, including the closely related genera ***Allophyla*** and ***Suillia***, develop in fungi. These photos show an ***A. atricornis*** sitting on a mushroom, and an ***S. barberi*** on a leaf. ❾ Most Heleomyzidae, like this ***Suillia quinquepunctata***, have a row of conspicuous spines along the costa (the leading edge of the wing). ❿ ***Trixoscelis***, often treated in its own family Trixoscelididae, has several common species in the west, but includes only a single, rare eastern North American species (***T. fumipennis***), which occurs in sparsely grassed areas.

FAMILY CURTONOTIDAE (CURTONOTID FLIES) ❶ ***Curtonotum helvum*** (Curtonotidae) occurs on sand dunes, where it probably develops in the eggs of short-horned grasshoppers. The spiny costa and humpbacked posture of this fly make it easy to recognize. FAMILY SEPSIDAE (ANTLIKE SCAVENGER FLIES) ❷ ❸ Antlike scavenger flies (Sepsidae) are ant-shaped flies common on dung and other decomposing material. Many species, including all members of this genus (***Sepsis***) have a dark spot at the end of the wing, conspicuously displayed with a wing-flicking motion as they walk around. ❹ Although most common Sepsidae have a distinctly antlike shape, there are exceptions, such as this small (about 3 mm) flower-frequenting species ***Saltella sphondylii*** (on a daisy). As suggested by the pollen grains all over the female of this pair of *Saltella*, acalyptrate flies can play a significant role in pollination. ❺ ***Themira annulipes***, like most Sepsidae, breeds in dung. Antlike scavenger flies in the genus *Themira*, our largest genus of Sepsidae, are common on a variety of decaying materials. ❻ ❼ ***Nemopoda nitidula*** is a Holarctic species found both in Europe and North America. ❽ ***Orygma luctuosum***, the only North American species in the subfamily Orygmatinae, is a distinctively small-eyed, somewhat flattened fly that abounds in wrack piles along the Atlantic Coast. It also occurs in Europe. FAMILY TETHINIDAE (BEACH FLIES) ❾ Tethinidae are small (1.6–3 mm) flies usually associated with ocean beaches. The silvery gray ***Tethina*** species are often common, although inconspicuous, along Atlantic beaches.

FAMILY SPHAEROCERIDAE (SPHAEROCERID FLIES) **Subfamily Sphaerocerinae** ❶ ***Sphaerocera curvipes*** is a common fly on horse manure, a habit that probably led to the name "small dung flies" sometimes inappropriately applied to the whole family. **Subfamily Limosininae** ❷ ❸ The clouds of tiny flies that invariably occur on the seaweed washed up on marine shores are made up mostly of the seaweed-specializing sphaerocerid genus ***Thoracochaeta***. ❹ This ***Coproica ferruginata*** was one of a number of sphaerocerid species visiting a dead fish on a Lake Erie beach. ❺ This sphaerocerid being eaten by a rove beetle is one of the most widespread and abundant of all insects. Sweep a net around almost anywhere and you are likely to pick up a few dozen ***Leptocera erythrocera***, along with many other common sphaerocerids that develop in a variety of decaying organic material. ❻ ***Spelobia*** is an enormous and ubiquitous genus of generally similar flies, but this common introduced species (***S. ochripes***, 2 mm) photographed on snow is easily recognized by its orange head. ❼ Both North American species of ***Ceroptera*** are kleptoparasitic flies that hang out on adult scarab beetles awaiting the opportunity to lay their eggs on the scarab's larval food store. *Ceroptera sivinskii* ranges from Florida north to southern Canada, and ***C. longicauda*** (shown here) is southeastern. **Subfamily Copromyzinae** ❽ The white-winged ***Norrbomia frigipennis*** riding on this dung-rolling scarab beetle are awaiting an opportunity to scoot down into the scarab's burrow and lay eggs on the scarab's dung ball. ❾ ***Lotophila atra*** is a very common sphaerocerid fly on dog feces and in similar urban habitats.

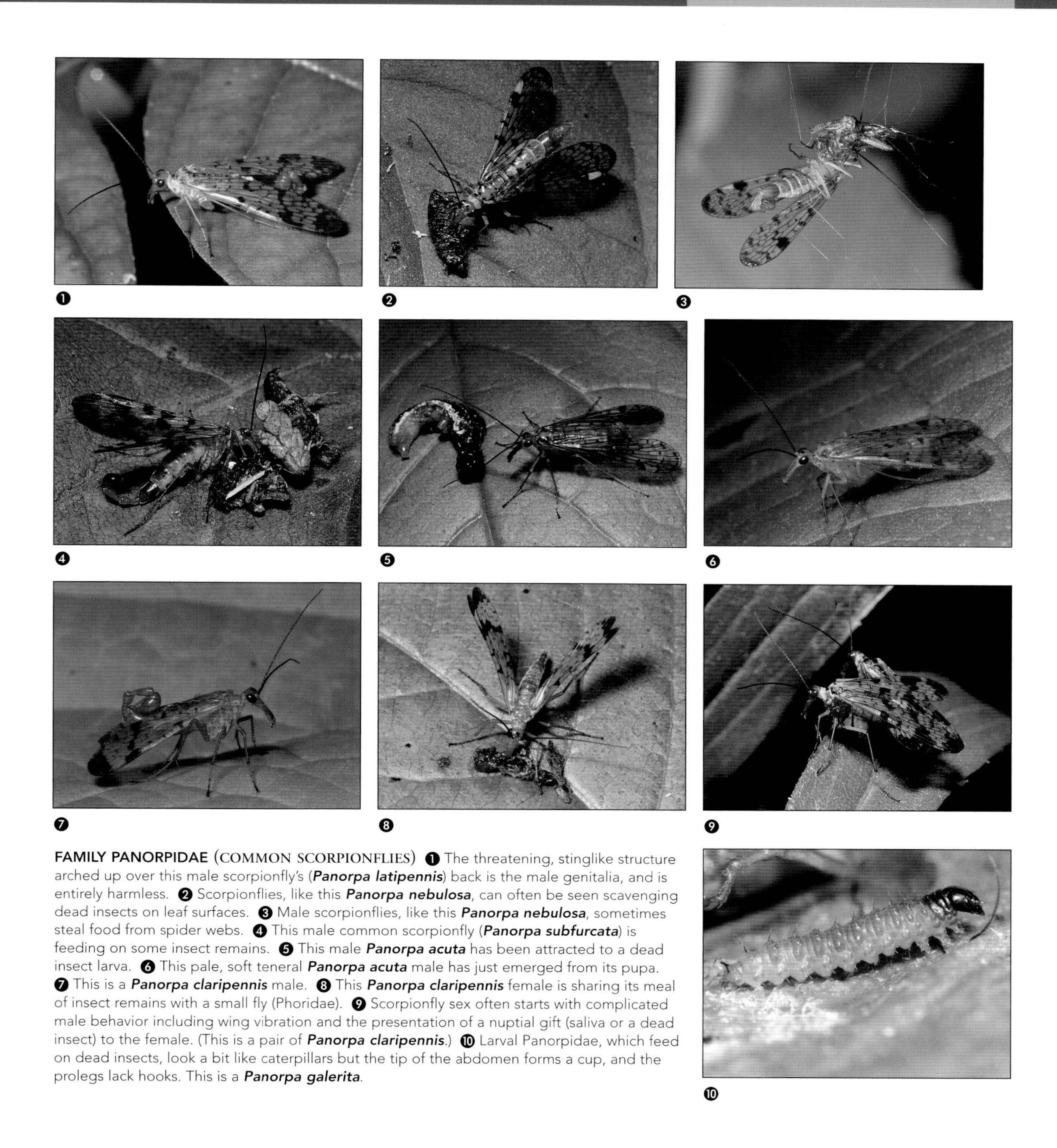

FAMILY PANORPIDAE (COMMON SCORPIONFLIES) ❶ The threatening, stinglike structure arched up over this male scorpionfly's (***Panorpa latipennis***) back is the male genitalia, and is entirely harmless. ❷ Scorpionflies, like this ***Panorpa nebulosa***, can often be seen scavenging dead insects on leaf surfaces. ❸ Male scorpionflies, like this ***Panorpa nebulosa***, sometimes steal food from spider webs. ❹ This male common scorpionfly (***Panorpa subfurcata***) is feeding on some insect remains. ❺ This male ***Panorpa acuta*** has been attracted to a dead insect larva. ❻ This pale, soft teneral ***Panorpa acuta*** male has just emerged from its pupa. ❼ This is a ***Panorpa claripennis*** male. ❽ This ***Panorpa claripennis*** female is sharing its meal of insect remains with a small fly (Phoridae). ❾ Scorpionfly sex often starts with complicated male behavior including wing vibration and the presentation of a nuptial gift (saliva or a dead insect) to the female. (This is a pair of ***Panorpa claripennis***.) ❿ Larval Panorpidae, which feed on dead insects, look a bit like caterpillars but the tip of the abdomen forms a cup, and the prolegs lack hooks. This is a ***Panorpa galerita***.

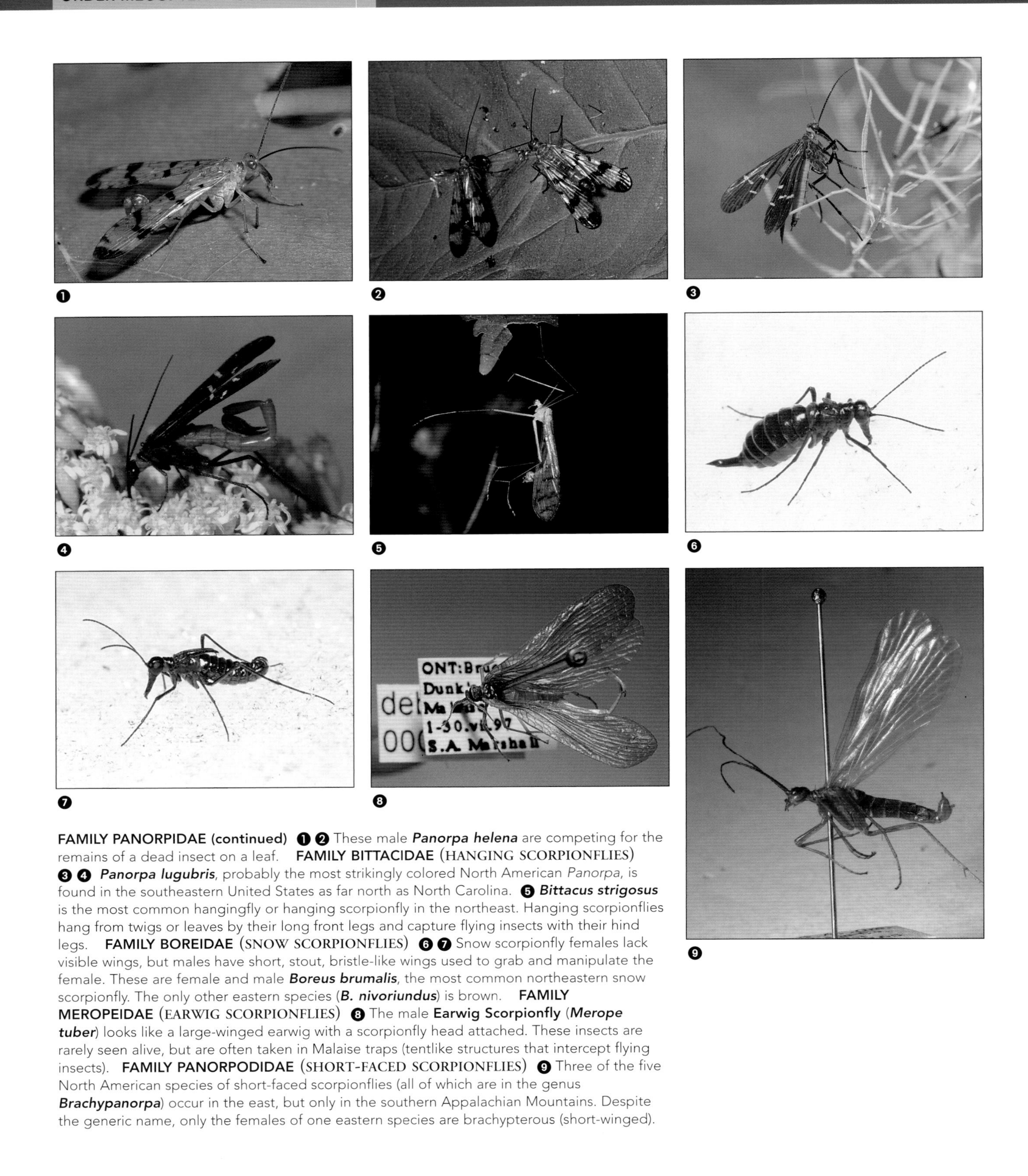

FAMILY PANORPIDAE (continued) ❶ ❷ These male ***Panorpa helena*** are competing for the remains of a dead insect on a leaf. **FAMILY BITTACIDAE** (HANGING SCORPIONFLIES) ❸ ❹ ***Panorpa lugubris***, probably the most strikingly colored North American *Panorpa*, is found in the southeastern United States as far north as North Carolina. ❺ ***Bittacus strigosus*** is the most common hangingfly or hanging scorpionfly in the northeast. Hanging scorpionflies hang from twigs or leaves by their long front legs and capture flying insects with their hind legs. **FAMILY BOREIDAE** (SNOW SCORPIONFLIES) ❻ ❼ Snow scorpionfly females lack visible wings, but males have short, stout, bristle-like wings used to grab and manipulate the female. These are female and male ***Boreus brumalis***, the most common northeastern snow scorpionfly. The only other eastern species (***B. nivoriundus***) is brown. **FAMILY MEROPEIDAE** (EARWIG SCORPIONFLIES) ❽ The male **Earwig Scorpionfly** (***Merope tuber***) looks like a large-winged earwig with a scorpionfly head attached. These insects are rarely seen alive, but are often taken in Malaise traps (tentlike structures that intercept flying insects). **FAMILY PANORPODIDAE** (SHORT-FACED SCORPIONFLIES) ❾ Three of the five North American species of short-faced scorpionflies (all of which are in the genus ***Brachypanorpa***) occur in the east, but only in the southern Appalachian Mountains. Despite the generic name, only the females of one eastern species are brachypterous (short-winged).

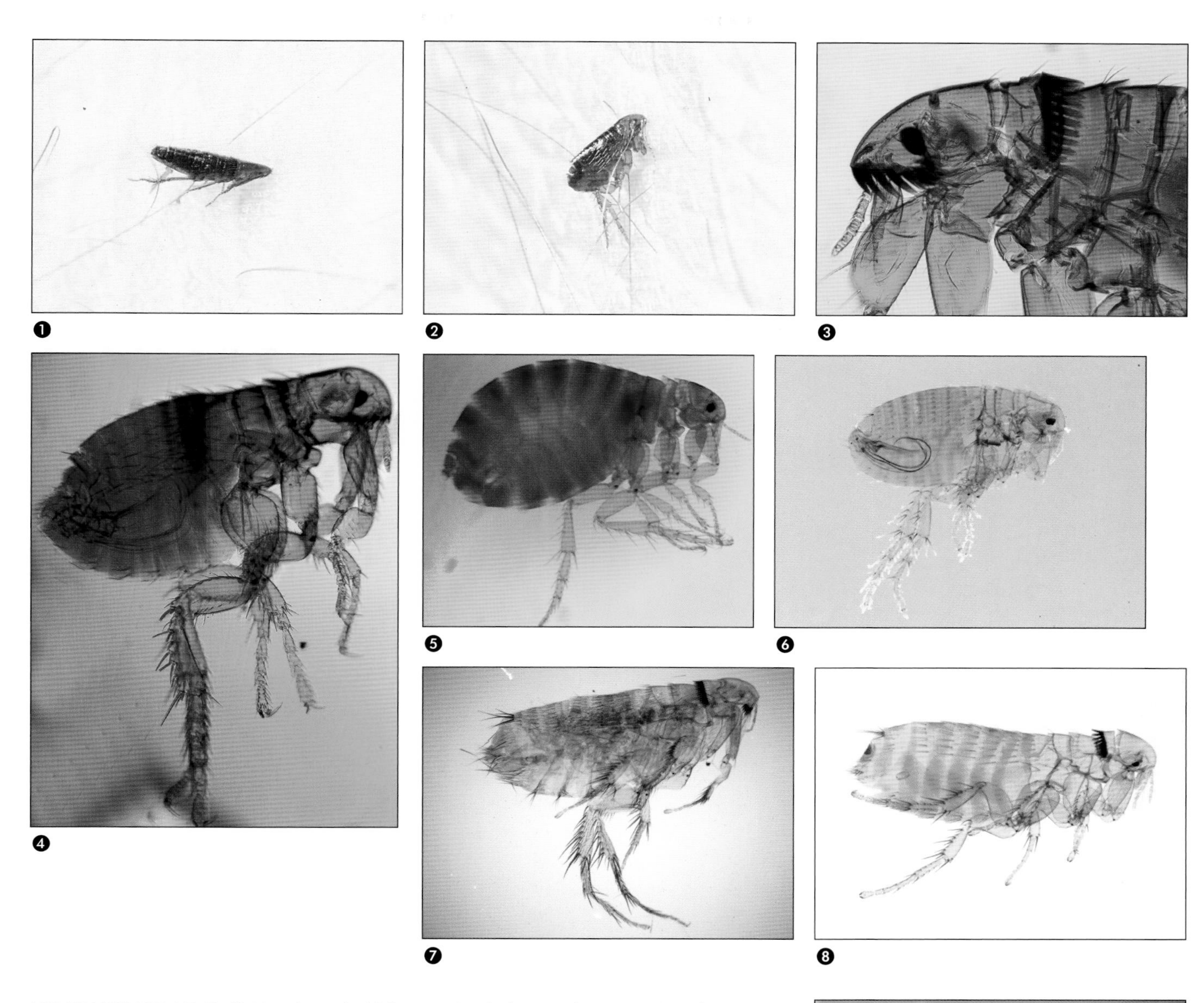

❶ ❷ ❸ ❹ ❺ ❻ ❼ ❽ ❾

FAMILY PULICIDAE ❶ ❷ Most household fleas, on both dogs and cats, are **Cat Fleas** (***Ctenocephalides felis***, shown here) although there is a related species called the Dog Flea (*C. canis*). ❸ **Cat Fleas** (***Ctenocephalides felis***) are cosmopolitan pests that bite a wide range of animals. Cat Fleas are the most common fleas of dogs, cats and people in North American homes. This close-up shows the combs, made up of rows of flat spines along the hind margin of the prothorax and the lower margin of the head, which help the flea stay on its hairy hosts. ❹ **Dog Fleas** (***Ctenocephalides canis***) were once common pests of domestic dogs but, at least in North America, they have been largely replaced by the very common Cat Flea (*C. felis*). Dog Fleas remain common on foxes, coyotes and wolves. ❺ **Human Fleas** (***Pulex irritans***) bite humans and their domestic animals (pigs, cats, dogs) throughout the world, but are relatively rare in North America at present. If you are bitten by a household flea in North America it is much more likely to be a Cat Flea (*Ctenocephalides felis*). ❻ The **Oriental Rat Flea** (***Xenopsylla cheopsis***), found in tropical and subtropical areas throughout the world, is the most important carrier of plague and also a vector for murine typhus. This is a male, and the coiled structures visible inside the end of the abdomen are his penis rods and other genitalic components. Like the human flea, this species lacks the combs found on the pronotum and/or gena of most other fleas **FAMILY HYSTRICHOPSYLLIDAE** ❼ This ***Hystrichopsylla tahavuana*** female was found on a red squirrel in Ontario's Algonquin Park. Note the two sets of backwards-facing combs on the check (gena) and pronotum. **FAMILY CERATOPHYLIDAE** ❽ This is a female ***Orchopeas leucopus***, a common flea on *Peromyscus* mice throughout North America. The seedlike structure inside her abdomen is a spermatheca (sperm storage organ), and the dark patch at the end of her abdomen is a patch of sensillae used to detect a potential host by body warmth. **FAMILY CTENOPHTHALMIDAE** ❾ This ***Tamiophila grandis*** female was found on a chipmunk.

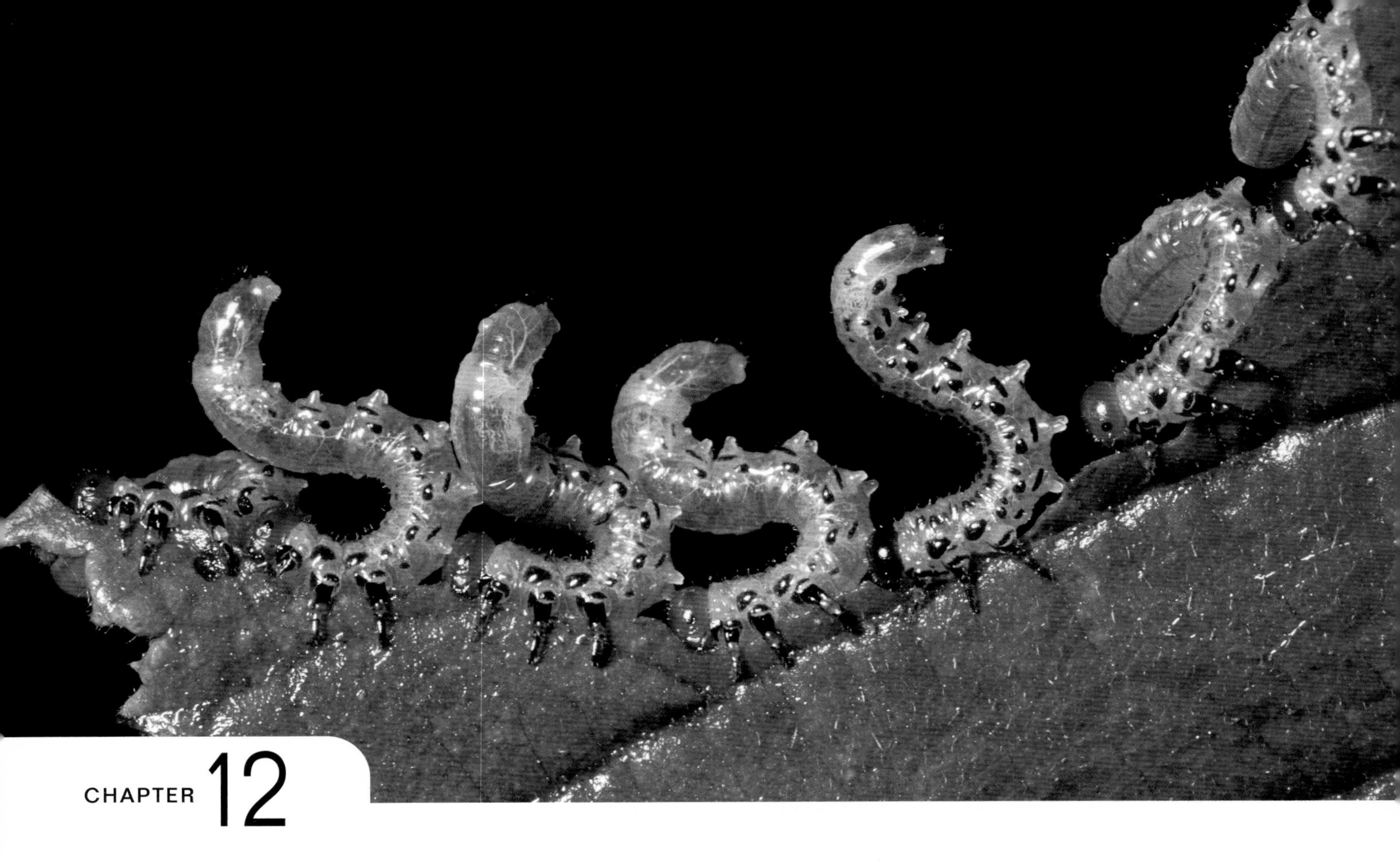

CHAPTER 12

Sawflies, Wasps, Bees and Ants

Ants, wasps, bees and sawflies form the order Hymenoptera, a group that vies with the huge insect orders Diptera, Coleoptera and Lepidoptera for the distinction of being the largest order of living things. Although a few hymenopterans are significant pests, the Hymenoptera is for the most part a beneficial group of insects that pollinate many of the plants we depend on, regulate pest populations, provide honey and serve as important components of all terrestrial ecosystems. Despite all this, there is no single common name that applies to all Hymenoptera, and most people would be hard-pressed to name any distinctive attributes that characterize the whole order.

Most other insect orders have familiar diagnostic characteristics, often reflected in their scientific names. "Coleoptera," for example, means "sheath wings," "Lepidoptera" means "scaled wings" and "Diptera" means "two wings." The order name "Hymenoptera" can be interpreted as meaning "membranous wings" (*hymen* = membrane; *ptera* = wings), or as meaning "married wings" *Hymeno* = god of marriage), referring to the way the front and hind wings of bees and wasps are linked by little hooks, called hamuli, on the leading edge of the hind wing.

Many insects, however, have membranous wings, and nobody routinely recognizes insects as belonging to the Hymenoptera by looking for hamuli. Hymenoptera are usually recognizable as Hymenoptera because they are membranous-winged, hard-bodied insects lacking the mouthpart and wing specializations of other orders. The two distinct suborders of Hymenoptera, the primitive, paraphyletic Symphyta and the distinctively wasp-waisted Apocrita, are individually easier to characterize than the order taken as a whole.

Sawflies, horntails and wood wasps: Suborder Symphyta

Most gardeners have had the experience of watching a cherished plant, perhaps a currant bush, columbine, mountain ash or rose, virtually denuded almost overnight by groups of bare, caterpillar-like larvae that seem to work in concert to consume leaf after leaf from the edges on in. These common, often gregarious foliage feeders are larvae of the primitive Hymenoptera suborder Symphyta (sawflies, horntails and wood wasps). Sawfly larvae often look like hairless caterpillars, but they have more prolegs than caterpillars (over five pairs) and these prolegs lack the little hooks (crochets) found on caterpillar prolegs. Sawfly larvae also have what appear to be single eyes rather than the cluster of tiny ocelli that characterize caterpillars.

Most sawflies use silk produced in their salivary glands to spin a cocoon to protect the pupa just as many moth caterpillars do, but adult sawflies are not at all mothlike. Adult sawflies look like hard, shiny wasps without a constricted waist, but with a sawlike ovipositor that flips down like a jackknife to saw into, and lay eggs in, leaf edges or stems. This "saw" is similar to the ovipositor of Orthoptera and more primitive insects, but entirely unlike the egg-laying structures found in other common insect orders with complete metamorphosis.

Most conspicuous sawfly pests of home and garden, including Imported Currantworms (*Nematus ribesii*), Mountain Ash Sawflies (*Pristiphora geniculata*), Columbine Sawflies (*Pristiphora aquilegiae*) and Pear Slugs (*Caliroa cerasi*) belong to the large common sawfly family **Tenthredinidae** (page 539), but it is not hard to find distinctive members of other sawfly families. Small pine trees, for example, are often webbed up with the silken shelters made by larvae of a web-spinning sawfly (**Pamphiliidae**, page 538) called the Pine False Webworm (*Acantholyda erythrocephala*). Pine trees are also frequently plagued by brightly colored conifer sawfly (**Diprionidae**, page 537) larvae that form conspicuous aggregations at branch tips. Roses are attacked by many sawflies including some common **Argidae** (page 541), in which the adult antenna consists mostly of one seamless segment, unlike the multisegmented antennae of other sawflies. Oak leaves are sometimes attacked by members of the **Pergidae** (page 542), small sawflies with only six antennal segments. Pergids are the common sawflies of Australia, where their larvae are known as "spitfires" for their habit of forming conspicuous clusters that react to disturbance by "spitting" and flicking defensive fluids at intruders, but the single North American genus (*Acordulecera*) is infrequently encountered.

Look on honeysuckle, willow and elm for the largest northeastern North American sawflies, the **Cimbicidae** (page 537). The solitary, fat cimbicid larvae that feed on willow leaves respond to disturbance by dropping from the foliage and curling up into a spiral position. If you pick one up it will squirt out a whitish defensive fluid from glands just above its spiracles. Adults of our most common Cimbicidae are robust, handsome insects that would look a bit like bald bumble bees but for their clubbed antennae. Cimbicids, like other sawflies, cannot sting, but their bee-like buzz and muscular, sharp-spined legs discourage handling.

From these examples we can see that most Symphyta are simple foliage-feeding insects, but a few groups of sawflies eschew such exposed habits in favor of cryptic lives inside plant tissue. These relatively hidden groups include the few gall-making Tenthredinidae, those members of the odd family **Xyelidae** (page 542) that develop within the buds or male cones of pine or fir trees, and the slender stem sawflies (**Cephidae**, page 542) that develop inside grass stems or small twigs. Wood-boring Symphyta, the wood wasps (**Xiphydriidae**, page 542) and the large yellow and black horntails (**Siricidae**, page 542) are often seen on tree trunks, where females use their ovipositors to insert eggs and more into the wood. Wood wasp and horntail larvae feed on fungi within tree trunks, and the adults use their ovipositors to inject a fungal inoculum and associated nutritious mucus into the tree, thus conditioning the wood for larval feeding. This is an unusual habit for the order, and the only other North American Hymenoptera that feed on fungi are the fungus gardening ants.

It is only a small step from feeding on growing tips or fungi in stems or trunks to feeding on the concentrated pockets of protein provided by other inhabitants of pith or wood, and some Symphyta take that step by deliberately placing their eggs in the burrows of other insects. Their larvae attack other larvae in those burrows, thus taking a first step into the parasitoid life strategy that characterizes most of the rest of the Hymenoptera. Such Symphyta are called parasitic wood wasps (family **Orussidae**, page 542) because their larvae develop as parasitoids of wood-boring beetles and horntail larvae. Parasitic wood wasps are of special interest because the common ancestor of all "higher Hymenoptera" (Hymenoptera other than the Symphyta) was probably something like a modern-day parasitic wood wasp that developed a novel abdominal constriction, giving its abdomen unprecedented mobility. The higher Hymenoptera comprise the suborder Apocrita, and differ from Symphyta in having the unique "wasp waist" that most people envisage when they think of an ant, wasp or bee.

Wasp-waisted Hymenoptera: Suborder Apocrita

Most Hymenoptera, including ants, wasps, parasitic wasps and bees, have a narrow-waisted appearance created by a characteristic constriction between the first and second abdominal segments. These wasp-waisted insects (the suborder Apocrita) have attracted the attention of many great scientists, not because of their attractive hourglass figures, but because of their abundance, complex behavior and great importance as pollinators, parasitoids and predators.

Interesting Apocrita are unavoidable elements in everyone's environment, and they certainly warrant a closer look.

Larval wasp-waisted Hymenoptera usually specialize on high protein diets involving minimal indigestible matter, so they are able to get by without regular defecation. The larval intestine is closed and waste material is voided in one neat package (a meconium) just before pupation, clearly an adaptive feature for larvae that develop in enclosed spaces such as the inside of a host or nest. The great majority of wasps in the suborder Apocrita, including all the primitive lineages within this suborder, live as parasitoids on or in other insects.

Killer wasps: The parasitoid Hymenoptera

Insects with larvae that complete their development at the expense of a single prey or host organism that is killed in the process are called parasitoids. They are not parasites, because parasites do not normally kill their hosts, and they are not predators because predators develop at the expense of more than one prey item. The first few lineages to arise in the suborder Apocrita are parasitoid lineages collectively referred to by the loose term "Parasitica." Two of these parasitoid lineages, the superfamilies Ichneumonoidea and Chalcidoidea, are enormous groups that together include over 7,500 described North American species.

Superfamily Ichneumonoidea

The superfamily Ichneumonoidea, made up of the families Braconidae and Ichneumonidae, includes most of the commonly noticed parasitic wasps, such as the large, brightly colored, long-legged species often seen erratically flying along hedges or shady paths in search of hosts.

Ichneumonidae (page 543) are generally larger and longer-bodied than Braconidae, and also differ in host range, habitats and habits. This huge family includes a wide variety of ectoparasitoids and endoparasitoids of several arthropod orders, but most ichneumonids parasitize caterpillars and sawflies, often pupating inside the host's skin, sometimes inflating the larval skin of the caterpillar host to form a protective tent. Unlike braconids, many ichneumonids parasitize the pupal stage, and several common wingless ichneumonid species search on foot for their pupal hosts. Several ichneumonids (but very few braconids) have the peculiar habit of entering hosts not to parasitize them, but to parasitize other parasitoids within the hosts (hyperparasitism). Another common interaction between parasitoid species (found in many parasitoid families) is multiparasitism, where more than one species of parasitoid compete on or in the same host. Multiparasitism can result in the death of one of the parasitoids, and many parasitoid larvae are equipped to kill off competitors within their hosts.

Some of our most conspicuous parasitoids are long, slender wasps with a body length of 4 cm or more, excluding the 8–10 cm long trailing ovipositor. Look for these impressive ichneumonid wasps (family Ichneumonidae, genus *Megarhyssa*) as they position themselves like drilling rigs on trees inhabited by horntail larvae (family Siricidae). An ovipositing female wasp will spread her legs out to form a solid platform and hold her abdomen tower-like into the air while her long, thin ovipositor is directed by two sheathlike filaments down between her middle legs as it drills deep into the wood. Each egg is deformed into a long, thin strand to pass through the narrow ovipositor and onto an immobilized horntail larva, where the legless wasp larva will later feed externally upon the doomed host. One of the cues used by the female *Megarhyssa* to choose a drilling site is a fungus that horntail females inject into the wood when they lay eggs.

Horntail-infested trees are often busy places, with females of *Megarhyssa* and closely related but smaller *Rhyssa* wasps

Megarhyssa atrata (Ichneumonidae) laying eggs on a tree trunk, page 544, caption 1.

sometimes taking half an hour or more to drill the tree and insert their eggs. Other ichneumonid wasps sometimes use the *Rhyssa* drill holes to insert their nondrilling ovipositors and lay eggs on the sawfly hosts next to the *Rhyssa* eggs. The eggs of the invading wasps hatch quickly into larvae that kill the *Rhyssa* before moving on to develop on the sawfly hosts.

The basic parasitoid strategy of feeding on hosts that are immobilized, usually by an injection of venom through the ovipositor, works well as long as the hosts live in concealed, protected places (like horntail larvae), but presents problems when potential hosts live on foliage and other exposed places. An injured or immobilized foliage-feeding sawfly, for example, might drop to the ground and be eaten by ants or ground beetles, which presents a problem for would-be parasitoids of foliage-feeding sawflies and other unconcealed potential hosts.

There seem to be two possible solutions to this problem, both of which are used by a great many wasps. One solution is to paralyze the host, then move it to a protected place where the wasp larva can feed in safety. This is the strategy that has led to the diversification of the familiar stinging wasps that will be discussed later. The second solution, one which seems to have been hit upon many times and now characterizes an incredible number and variety of parasitoids, is for the parasitoid larva to develop in a nonparalyzed host while allowing the host to go about its normal business of feeding and protecting itself. Such parasitoids develop a remarkably close relationship with their hosts, sometimes even regulating the development and behavior of the particular species they attack.

The host specificity demanded by this kind of close parasitoid-host relationship is one reason there are so many parasitic wasp species, most attacking only a particular host or a narrow range of hosts. The family Ichneumonidae alone includes over 3,300 described species in North America, and the world fauna is estimated at 60,000 species – more than the world total of vertebrate species, but probably only slightly more than the total for the closely related family Braconidae.

Although most ichneumonids are larger than braconids, these two enormous groups of slender, wasplike parasitoids overlap enough in appearance to make field identification difficult. **Braconidae** (page 546) usually have a short abdomen with the first two tergites (upper plates) behind the wasp waist fused into a single long plate. Furthermore, the wings of braconids usually have two similarly sized, closed cells (they look like windows) right under the stigma (the dark thickening along the leading edge of the wing), and have at most a single cross-vein connecting the two main long veins in the middle of the wing (m-cu cross-vein; ichneumonids almost always have two).

Braconid wasps are common insects, and one need only try and rear out a few common caterpillars to see living proof of the effect these ubiquitous parasitoids can have on other insect populations. Cabbage Butterflies or Cabbage Whites (*Pieris rapae*), for example, are heavily parasitized by a braconid species (*Cotesia glomerata*) that was deliberately imported from Europe to help control this introduced pest butterfly. Eggs of the braconid are laid into very young, relatively defenseless caterpillars. They hatch into internal parasitoids that collectively consume the caterpillar's contents, slowly at first but eventually developing in the mature caterpillar before leaving the spent host to spin silk cocoons surrounding the caterpillar's corpse. Braconids, like most Hymenoptera, produce silk using larval labial glands.

Many Braconidae have similar habits, and it is not unusual to see the deflated husks of parasitized caterpillars surrounded by braconid cocoons, or perhaps a dead (or almost dead) lady beetle sitting on top of its killer's cocoon. A few groups of Braconidae forego exposed cocoons in favor of other pupal shelters. One group, aptly called the Aphidiinae, parasitizes aphids and pupates inside the hollow, somewhat inflated skeleton of its host (called an aphid "mummy"); yet another group (Alysiinae) parasitizes maggots and pupates within the host's puparium (the inflated skin of the last larval stage of a maggot). Adults of the latter group have conspicuous out-turned mandibles that they use to pry open an exit from the host puparium.

Superfamily Chalcidoidea

Next time you have the good fortune to be outdoors on a warm summer day with an insect net in hand, try sweeping the net through some foliage a few times. In addition to dozens of flies, a few beetles, a couple of bugs and the odd large wasp, your net will almost certainly contain numerous minute metallic members of the huge, almost entirely parasitic, wasp superfamily Chalcidoidea. Most chalcidoids are too small to easily identify with the naked eye, and some species that develop inside single insect eggs are almost incredibly tiny – about one-third of a millimeter in length (smaller than some single-celled organisms!). Species in the family **Trichogrammatidae** (page 548), for example, attack the eggs of moths and other insects, and are sometimes used like living insecticides. Huge numbers are reared on the eggs of easily cultured moths and then released to control pests without incurring the environmental costs often associated with chemical insecticides. Trichogrammatids attack the eggs of most terrestrial arthropod orders, usually selecting hosts by habitat and egg size, rather than by host species.

A few species of Trichogrammatidae parasitize the eggs

of aquatic insects, seeking out their hosts by swimming under water. Swimming is an unusual habit for the almost entirely terrestrial order Hymenoptera, but is also seen in couple of other families of egg parasitoids. Fairyflies (**Mymaridae,** page 548), for example, are delicate chalcidoid wasps that normally use their narrow, fringed, short-veined wings to "swim" through the air, but sometimes use them like paddles as they swim through the water in search of host eggs. Similar habits are found in some Scelionidae, a group of egg parasitoids in another superfamily (Platygastroidea).

Egg parasitoids are usually too tiny for field identification, but many other chalcidoids can be identified either by appearance or habit. The family **Chalcididae** (page 548), for example, is a common group of relatively large (often 5–6 mm long), strikingly sculptured wasps with enormously swollen hind femora. One European species uses those massive hind legs to hold apart the mandibles of antlion larvae while inserting eggs between the doomed doodlebug's head and thorax, but most North American chalcidids attack larvae or pupae of flies or moths.

Members of the genus *Chalcis* oviposit in the eggs of aquatic soldier fly (Stratiomyidae) species that lay eggs near water but develop as larvae in the water. The parasitic wasp larva remains dormant inside the newly hatched fly larva as it leaves the egg and drops into the water, and remains dormant throughout the aquatic larval stages of the fly. Only when the soldier fly larva has left the water to pupate along the shore does the wasp larva start to feed on its host, developing quickly and pupating within the calcified skin of the soldier fly's last larval stage. The adult wasp later escapes by cutting round exit holes with its mandibles. Wasps in the related chalcidoid family **Leucospidae** (page 549) also have swollen and toothed hind legs, but they are even larger than chalcidids and usually look like small yellowjacket wasps with long ovipositors curved up over their bodies. These uncommon wasps are most likely to be found on flowers frequented by their hosts (bees and wasps).

Another chalcidoid family with relatively large members is the **Perilampidae** (page 549), which includes several stout, metallic species similar to small cuckoo wasps (Chrysididae). Perilampids lay their eggs on foliage or other surfaces, and leave the job of finding hosts to their tough, flattened first instar larvae (called planidia). A few find their hosts (wood-boring beetles) inside dead wood, but most find their hosts *inside* other insects. Perilampid planidia usually locate and penetrate caterpillars in search of pupae of other parasitoids (tachinid flies or ichneumonoid wasps), which they attack and parasitize (as hyperparasitoids). The **Eupelmidae** (page 550) is another chalcidoid family rich in hyperparasitic species, many of which are facultative hyperparasitoids able to develop either on other parasitoids or their hosts. Eupelmid adults, which are sometimes wingless, have a remarkable jumping mechanism involving rapid arching of the thorax with a concomitant explosive kick of the midlegs.

Perilampid wasp, page 549, caption 2.

within the nest-making, food-storing Hymenoptera, and which has independently appeared several times within the bees alone. Even so, it is a mistake to think of bees just as social insects. The great majority of bees are solitary, behaving much like digger wasps with a preference for pollen rather than prey.

Most female bees accumulate pollen on their finely branched or feathered body hairs, and then use their midlegs to rake out the pollen and transfer it to special structures (usually on the hind legs) for storage and transport. Check out a busy honey bee next time you are out in your garden. If she has been at work for a while, she will have large yellow masses of pollen held in special basket-like structures on the hind legs.

If honey bees are "busy as a bee" in packing their pollen baskets with high-protein pollen and their crops with nectar (to be evaporated and stored as honey), what can we conclude of bees with hardly any pollen-holding hairs? With the exception of some bare-bodied little yellow-faced bees (Colletidae) that transport pollen in their crop (part of the digestive tract used for food storage), these ill-equipped bees are the bare-faced thieves of the bee world. These are species that steal from other bees rather than work for themselves – it seems that wherever there are nests there are cuckoos of a sort.

A leafcutter bee arriving at its nest.

The attractive, orange or brownish, wasplike Apidae (genus *Nomada*) that are so common in early spring provide a case in point. A *Nomada* female will slip into a host's nest and stick her eggs into the wall of the host's larval cell before the host has finished stocking it with pollen. When the host (usually an *Andrena* bee) finishes stocking her cell and lays *her* egg in the nest, the *Nomada* larva will kill the host's egg and take over the nest, like a cuckoo chick in a bird's nest. *Nomada* bees are commonly called cuckoo bees, but there are many other genera that exhibit this kind of kleptoparasitic behavior. In fact, there are more kleptoparasitic species of bees than social ones.

Pollen is a food source that hardly needs to be subdued with the use of a stinger. It follows, then, that the bee stinger is entirely a defensive weapon, and the only kinds of bees likely to sting you are the social Apidae (bumble bees and Honey Bees) that use stingers for colony defense. Even among the social bees, most species are reluctant to sting. Bumble bees, for example, pack an impressive weapon but rarely sting people.

If a bee stings you, it is probably the introduced domestic species everyone knows as the Honey Bee. The Honey Bee stinger is barbed, and it remains in your skin along with a poison sac (and much of the bee's insides) after your doomed attacker has departed. If you are stung, it is helpful to calmly scrape the stinger out, both to reduce the amount of venom injected and to get rid of a volatile chemical that excites other bees to sting you.

Most people suffer only a local swelling as a result of a sting, but if you get stung by a bee (or anything else) and suffer nonlocal effects such as difficulty breathing, seek medical attention immediately. Some people are dangerously allergic to bee and wasp venom, so much so that about 50 people die annually due to insect stings in the United States. On the other hand, it has long been appreciated that bee stings can also be beneficial. Hippocrates, the Father of Medicine, wrote about the therapeutic value of bee stings, and bee venom is currently touted by some as a treatment for inflammatory diseases such as arthritis and bursitis.

Honey Bees feed on a mixture of nectar and pollen, usually evaporating the nectar into concentrated, decomposition-resistant honey. A foraging Honey Bee worker carries nectar to the nest in her honey stomach (the enlarged crop) before transferring it to nest workers who add enzymes to break the nectar down to simpler sugars. The nectar is then concentrated to form honey by evaporation on the bee's tongue or in the nest.

A typical Honey Bee colony includes a fertilized queen, tens of thousands of her daughters (worker bees and a few virgin queens) and a small number of males (drones). Young workers stay in the hive and handle house cleaning, ventilation and colony defense, while older workers forage for nectar and pollen. Drones live for a single mating flight during which, if they get lucky, they will enjoy an explosive, and fatal, copulation with a new queen. The successful drone ejaculates with such force he leaves the end of his phallus inside his royal conquest before dropping to the ground, spent and doomed. The queen acquires a lifetime supply of sperm (several million) by mating with several males during her mating flight, after which she never leaves the colony except as part of a swarm. The queen uses the sperm to fertilize eggs (around 1,500 a day during the summer) destined to become females (workers and potential new queens), but drones hatch from unfertilized eggs. Colonies reproduce when the old queen departs with a swarm of workers, often clustering conspicuously on tree branches as some of the workers scout for new quarters. The workers that remain in the old colony raise a new queen by feeding developing larvae a special diet.

Unlike the familiar, introduced Honey Bees, which form elaborate perennial colonies that persist for years, North America's native social bees (bumble bees in the genus *Bombus* and a few species of primitively social Halictidae) form colonies that last only one season. Mated female bumble bees spend the winter in sheltered places, emerging in spring to start colonies in old mouse nests or similar cozy cavities. After cleaning up her new home and building an initial wax honey pot, a foundress female *Bombus* forages for nectar and pollen, ultimately filling the wax pot with honey and building up a supply of pollen on the surrounding floor. She then lays a batch of eggs on the pollen, covers them with a wax sheet and proceeds to incubate the eggs by lying over them and raising her body temperature well above the surrounding temperature. Her solicitous maternity continues through the development of the larvae, as she feeds them a mixture of pollen and honey until they have pupated within silken cocoons.

Once the new *Bombus* queen's first brood has hatched and taken over the tasks of foraging and nest maintenance, she is free to focus on laying eggs, quickly building up her force of workers. She continues to lay fertilized eggs until the end of summer, when she lays some unfertilized eggs that give rise to short-lived males, and she allows some of her daughters to develop into the new queens that will start the following year's colonies. The foundress female dies shortly thereafter.

Not all our familiar fuzzy bumble bees are industrious gatherers of pollen. Species of cuckoo bumble bees in the subgenus *Psithyrus* are social parasites that invade other *Bombus* nests, leaving their progeny to be reared by host workers.

There are differing opinions about the subfamily classification of the Apidae, with some authors using only three subfamilies (with honey bees, bumble bees, digger bees, stingless bees and orchid bees all in the Apinae). The more traditional approach is to divide the family into seven subfamilies, including the familiar honey bees (subfamily Apinae), cuckoo bees (subfamily Nomadinae) and bumble bees (subfamily Bombinae) discussed above. The other four subfamilies are the digger bees (Anthophorinae), carpenter bees (Xylocopinae), orchid bees (Euglossinae) and the stingless bees (Meliponinae). Of these, only the digger bees and carpenter bees occur in northeastern North America.

Digger bees are solitary bees that stock underground nests with pollen, much like the Andrenidae. Carpenter bees nest either in solid wood (large carpenter bees, *Xylocopa*) or in hollowed out stems (small carpenter bees, *Ceratina*). The Common Carpenter Bee (*Xylocopa virginica*) is a large, bumble-bee-like insect that has been extending its range north over the past few years, and which now ranges north to southern Ontario. The nonstinging males are often seen hovering near nesting sites, and females are usually seen going in and out of perfectly round holes on unpainted wood surfaces. Carpenter bees frequently form nesting aggregations at the same sites year after year. Newly mated young females either make new tunnels or clean up old ones in spring before dividing each tunnel into six to eight cells, each with a supply of pollen and a single egg.

Two important subfamilies of Apidae, the stingless bees and orchid bees, are tropical groups that do not reach northeastern North America. Orchid bees (subfamily Euglossinae) are stunningly beautiful, often metallic, bees that occur from Texas south to Argentina. Males (but not females) are normally attracted to orchids, which they pollinate as they gather fragrances apparently used in courtship. Male orchid bees are sometimes encountered as they aggressively patrol small territories, but these massive metallic bees are also easily attracted to bits of paper soaked with aromatic compounds like methyl salicylate, eugenol or cineol. The males are elaborately equipped to mop up attractive compounds with their front legs before transferring them to hind tibial slits leading to glandular pouches. Orchid bees are normally solitary, but may nest in aggregations or small family groups. Stingless bees (subfamily Meliponinae), on the other hand, are the native social bees of Central and South America, with colonies often including tens of thousands of workers. They are ubiquitous throughout the tropics, not only on the flowers where they gather pollen but also on dung, carrion, sap flows and other ephemeral resources. Most gather pollen and make honey much like the other social bees.

Social bees, like honey bees, bumble bees and stingless bees, are usually so conspicuous it is easy to forget that most bees are solitary. With the exception of kleptoparasitic species, each female solitary bee nests in her own burrow or other cavity, placing one egg in each of several cells stocked with pollen and nectar. The nest cells are usually lined with some sort of impervious waxy or cellophane-like secretion, and most bees transport pollen using pollen baskets on specially modified hind legs. Leafcutter bees (**Megachilidae**, page 564), however, are a bit different. Not only do they transport their pollen in conspicuous rows of modified hairs under the abdomen rather than on the legs, they use a variety of materials to line their nests.

The most familiar leafcutter bees are the common *Megachile* species that cut conspicuous circular chunks out of leaves for use as nest liners. Related genera have other strategies. Two recently introduced leafcutter bees belong to a genus known as wool carder bees because of the way they gather woolly plant hairs (especially from the common garden plant Woolly Lamb's Ears, *Stachys lanata*) to line their nest cells. Yet another recently introduced megachilid bee, the Giant Resin Bee (*Megachile sculpturalis*), lines its nest with resin gathered from trees. Resin bees and wool carder bees are accidental additions to our fauna, but other leaf cutter bee species have been deliberately introduced to North America because of their value as pollinators.

Every family seems to have its black sheep, and some members of the leafcutter bee family (genera *Stelis* and *Coelioxys*) are kleptoparasites. *Coelioxys* species are common kleptoparasites that use their strikingly pointed abdomens to break into *Megachile* nests, where their larvae will kill off the original occupants and steal the stored food. *Stelis* is a kleptoparasite in the nests of other genera of leafcutter bees.

The short-tongued bees (Halictidae, Colletidae, Andrenidae and Melittidae)

The bees that do not have a specialized long "tongue" with a prominent and elongated glossa are sometimes treated as the "short-tongued bees." The short-tongued bees do not form a natural group, and are instead simply those groups of bees that do not have a long tongue.

The sweat bee family (**Halictidae**, page 562) includes one wasplike kleptoparasitic genus (*Sphecodes*), but the family is otherwise made up of mostly small bees that gather pollen in the usual way and store it in ground nests (usually) or nests in dead wood. Members of this large (500 North American species) family often nest in aggregations, and a few species exhibit overlap of generations and cooperative care of their young. Halictids vary in color from black to green, but most of the common bright green bees belong to this family. The front wings of halictids have a distinctively arched "basal vein" (first section of medial vein).

Like the Halictidae and Apidae, the small family **Colletidae** (page 564) includes a genus of relatively bare wasplike bees, the yellow-faced bee genus *Hylaeus. Hylaeus* species, however, are not kleptoparasites despite their suspiciously kleptoparasite-like lack of pollen-gripping branched hairs. Yellow-faced bees carry pollen mixed with nectar in their crops, depositing their liquid cargoes in cells plastered with impermeable, silklike or cellophane-like walls. *Colletes*, the only other northeastern North American genus in the family Colletidae, has earned the name "plasterer bee" for its similar food storage habits. *Colletes* species are hairy bees that carry pollen outside the body.

The largest family of bees in North America is the family **Andrenidae** (page 563), with around 1,200 North American species of generally dark-colored bees, the majority of which are in the genus *Andrena. Andrena* bees are very common first thing in the spring, when they stock their underground nest burrows with pollen of willow and other early spring flowers.

The only other family of bees in eastern North America, the **Melittidae** (page 566), is made up of a few rarely encountered, small, andrenid-like species. The best place to look for melittid bees is on flowers of yellow loosestrife (or other members of the genus *Lysimachia*) on which our most common genus (*Macropis*) is entirely dependent for pollen and for resins to line its nest cells. Bees that are tied to a single kind of flower are called "monolectic," and *Macropis* is one of our few monolectic bees. Other bees are either oligolectic (normally take pollen from a few closely related flowers) or polylectic (gather pollen from whatever happens to be in bloom at the time). Most bees are polylectic.

SUPERFAMILY VESPOIDEA

Velvet ants, cow killers and scarab hunters (Mutillidae, Tiphiidae, Scoliidae and minor vespoid families)

Velvet ants (**Mutillidae**, page 579) are extraordinarily hard-bodied, antlike insects that usually parasitize pupae of ground-nesting insects in open, sandy areas. Only female velvet ants are wingless, and only the females search out the immature wasps and bees that usually serve as hosts for their ectoparasitic larvae. Most mutillids in the northeast are brightly colored insects that hunt in broad daylight, but nocturnal species are common in the south. The loud colors of many mutillids are often augmented by an ability to stridulate loudly, and some species are said to "scream" when picked up. The colors and the noise both advertise potent defenses, although it is a false warning in the case of the fully winged but stingless males.

In addition to dispensing with the encumbrance of wings, female velvet ants are very heavily armored and almost impervious to the stings of hosts trying to defend their nests. They also carry some impressive weapons of their own. In some parts of the United States, these insects are called "cow killers" because of the legendary potency of their long stings. Such painfully efficient defensive weapons are well known among many of the large, familiar social wasps, but are rarely encountered among solitary Hymenoptera.

Most velvet ants attack ground-nesting Hymenoptera, but some similar antlike parasitoids in the family **Tiphiidae** (page 578) attack tiger beetle larvae. Wingless females in the genus *Methocha* seemingly throw themselves into the jaws of death by walking into the "strike range" of a tiger beetle larva waiting to ambush prey from its burrow. Some wasps then slip into the beetle burrow undetected, but even if the beetle larva's jaws clamp down on the wingless wasp, the wasp's thorax is strategically narrowed so that the tiger beetle cannot crush it. The beetle's enormous mandibles merely hold the wasp in place while she delivers a paralytic sting to her host's thorax. The wasp then drags the paralyzed tiger beetle larva deeper into the burrow and adorns it with a single egg destined to hatch into a larval wasp that will rapidly develop at the expense of the doomed tiger beetle. Having laid her egg, the wasp closes up the burrow to protect her own progeny from other kinds of parasites, parasitoids and predators. *Methocha* parasitism can kill as many as a quarter of the tiger beetles in a given population.

Other members of the family Tiphiidae, as well as the closely related family **Scoliidae** (page 579), are mostly ectoparasitoids of other soil-inhabiting beetle larvae such as white grubs (Scarabaeidae). The small and rarely encountered vespoid families **Rhopalosomatidae** (page 578) and **Sapygidae** (page 579) also hunt prey on or in the ground. Rhopalosomatids develop as ectoparasitoids on nymphal crickets, and sapygids are parasitoids or kleptoparasites in bee and wasp nests. The small (9 world species) and extremely rare vespoid family **Sierolomorphidae** (page 578), also ranges as far north as eastern Canada. Nothing is known of sierolomorphid biology but it is a good bet that they are parasitoids of beetle larvae, as are similar small tiphiid wasps.

Spider wasps (Pompilidae)

Spider wasps (**Pompilidae**, page 571) are the mostly black, long-legged, somewhat frenetic wasps commonly seen flicking their wings in a characteristic fashion as they run back and forth over open ground. Although some spider wasps are simple parasitoids of spiders, most North American species thoroughly paralyze their arachnid victims and then drag them to some kind of nest, usually a burrow in the ground but occasionally a mud nest or a cavity in dead wood. Unlike some spider-hunting Sphecidae and Crabronidae that stock their cells with multiple spiders, spider wasp nests always include a single spider and a single egg per nest cell.

If you scan sand dunes or similar places on a sunny summer day you are bound to see common spider wasps as they dig burrows and drag paralyzed spiders across the ground and down the burrows. A few species are associated with other habitats. Most northeastern species have a wide host range, but some specialize on web-spinning spiders, others chase down running spiders and one interesting northeastern species (*Anoplius depressipes*) hunts fisher spiders (*Dolomedes* species). Some routinely trim the legs off spiders for transport, and others attack burrow-inhabiting spiders that don't have to be transported at all. Some spider wasps are entirely kleptoparasitic, depositing eggs only on spiders captured and paralyzed by other spider wasps.

Hornets, yellowjackets and potter wasps (Vespidae)

The most commonly encountered and most dangerous wasps in the family **Vespidae** (page 573) are social species that make paper nests of various shapes and sizes, either hung in exposed locations, concealed underground or hidden in cavities. The familiar social wasps called yellowjackets can often be seen scraping away at weathered trees or boards as they gather up wood fibers to masticate into pulp used in making paper nests. The nests are started by the queen, who lays her first eggs in a comb of egg-carton-like cells suspended from a short stalk and surrounded by an envelope of paper. Those first eggs hatch into worker females that take over building up the nest, gathering insect prey or other high-protein food (such as masticated carrion) and feeding the larvae with a high-protein mush of food fragments. In our temperate climate, the onset of fall induces wasp queens to start laying unfertilized eggs destined to develop into males (as is the case throughout the Hymenoptera, unfertilized eggs develop into males) and to start putting eggs in large cells where they will develop into new queens. The old queen dies in the fall, but the new queens mate and spend the winter in sheltered places.

Nuisance yellowjackets, such as the introduced *Vespula germanica*, become particularly irksome in late summer and autumn after the old queen is dead. There are no more developing larvae to feed, so the workers stop gathering high protein food for the larvae and start homing in on sugar sources for themselves. Who hasn't had a yellowjacket doing the backstroke in his or her drink at an August picnic, or noticed the hordes of wasps on rotting windfall apples later in the fall? These irksome individuals are destined to die soon because social wasps of temperate regions survive the winter only as mated queens that individually start colonies the next spring.

Wasps in the family Vespidae, all of which have longitudinally folded, apparently narrow wings, are divided into four subfamilies. The social Vespidae – the yellowjackets (Vespinae) that make enclosed paper nests, and the paper wasps that establish small colonies in open paper nests hung from stalks (Polistinae) – are the most commonly noticed vespids, but the great majority of vespid species are solitary wasps in the subfamily Eumeninae, the potter wasps. The name "potter wasp" only accurately describes a few members of the subfamily that use mud to make elegant, clay-pot-like nests that they stock with paralyzed caterpillars and leaf beetle larvae. Most other Eumeninae nest in twigs, burrows or other cavities. Potter wasps lay a single egg in each "clay pot," usually hanging the egg from the roof of the pot by a slender thread. The pot is then stocked with paralyzed host larvae and closed off with a mud wall. Wasps in the other North American subfamily of Vespidae (Masarinae) are called pollen wasps because species of this group are solitary wasps that, like bees, have substituted pollen for arthropod prey. Pollen wasps and bees are the only major groups of pollen-feeding Hymenoptera (a few ants also eat pollen). Pollen wasps occur in western North America and in Europe, but they are rare in North America east of Kansas. This odd distribution pattern is found in only a few major groups, including the Masarinae and the snakeflies (order Raphidioptera).

A potter wasp (*Eumenes crucifer*), page 573, caption 1.

Ants (Formicidae)

Ants (**Formicidae**, page 579) are the ultimate social insects, and comprise the only family of Hymenoptera that is entirely social. With the exception of those species that have abandoned their social skills in favor of permanent nest parasitism, ants form colonies that have a division of labor, with some individuals specialized for reproduction and others devoted to care of the colony. There is an overlap of generations, and immatures are taken care of by adults.

Ants constitute almost a third of the animal biomass in Amazonian rain forest, and similarly tip the scales in other major terrestrial habitats throughout the world. The world's 9,500 or so ant species arguably include the most important predators, scavengers and soil turners globally, even as they are among the most familiar household and backyard insects right here at home.

Ants, like other social Hymenoptera, have legless larvae confined to nests where they are supplied with food by the worker adults. This is in marked contrast to the other major social taxon, the termites, which have mobile nymphs as we would expect in a group with incomplete metamorphosis. Termites also have a sort of sexual equality, in that males and females play equal roles as reproductives (king and queen), workers and soldiers. Hymenoptera have an odd sexual determination system (haplodiploidy) wherein males typically hatch from unfertilized eggs (they therefore have only one set of chromosomes, from the mother, and are referred to as haploid), and females hatch from fertilized eggs (they therefore have two sets of chromosomes, one set from the mother and one set from the father, and are referred to as diploid). Ants, like other social Hymenoptera, produce males only occasionally, and males live only long enough to supply the reproductive females with their lifetime supply of sperm. Otherwise, colonies of social wasps, bees and ants are made up of females.

The strange sex determination system in Hymenoptera, in which males are haploid and females are diploid, may help explain why sociality has popped up so often in this order. Because hymenopteran full sisters get exactly the same set of genes from their father (remember, he has only one set to give; this is very different from a pair of human sisters, who on average share only half of their father's genes), full sisters are more closely related to each other than they would be to their own daughters. That suggests that more of a hymenopteran's genes are perpetuated if she contributes to the rearing of a sister than if she expends similar efforts taking care of her own progeny. This, combined with the basic aculeate wasp habit of stinging prey and storing it in a nest, is thought to have set the stage for the independent origin of sociality in several groups of Hymenoptera.

Unlike the familiar yellowjacket wasp colonies, ant colonies don't collapse in autumn, but instead survive the winter in the soil, in wood or in other protected nest localities. Colonies reproduce themselves by periodically producing winged sexual forms (reproductives), sometimes in huge numbers which form spectacular mating swarms.

After the aerial orgy that marks the beginning and end of a male ant's active life, the mated female ant typically finds an appropriate locality, sheds her wings and then breaks her wing muscles down to produce food, which she regurgitates to feed the first batch of larvae that will soon pupate in the new nest, sometimes in silk cocoons. If you have ever disturbed an ant nest, you have likely seen ant workers carrying these immobile pupae to safety. Most pupae will emerge as workers, which will take over the activities of foraging for food and taking care of the nest. Many ant species deviate from this basic life history, adopting a parasitic strategy of invading the nests of other ants rather than founding their own colonies.

Adult ant activity is described nicely (if not accurately) by Ogden Nash in the following lines:

> *The ant has made himself illustrious*
> *Through constant industry industrious*
> *So what?*
> *Would you be calm and placid*
> *If you were full of formic acid?*

Formic acid is just one of many ant-produced chemicals serving a complicated array of functions ranging from communication to defense, and it is produced in just one kind of abdominal gland in just one subfamily of ants (Formicinae). Those medium-sized black ants crawling over your unopened peonies, the tens of thousands of inhabitants of each wood ant mound in nearby forests and most of the common ants you see on sidewalks and roads are typical formicine ants in the genus *Formica*. These black or black-and-red ants are the most commonly noticed ants, sometimes called sidewalk ants because of their conspicuous foraging habits. *Formica* ants can be tremendously abundant, and some have interconnected colonies known as supercolonies, one of which (in Japan) was reported to contain 306 million workers. You need only pick up one or two *Formica* ants to get a good whiff of formic acid. When they start biting you and squirting formic acid into the bite, it should be clear why these stingless ants produce such an irritating chemical.

Ants often produce strong-smelling chemicals, many of which are characteristic of particular groups of ants. One of the most impressive ant odors emanates from common yellowish ants, called foundation ants because they frequently nest in the "fill" that passes for soil around

Carpenter ants and a treehopper, page 580, caption 4.

suburban house foundations. These little ants (*Acanthomyops* species) can be easily recognized by their strong citronella odor. Foundation ants are harmless insects that don't forage in houses or do any damage, but their habit of nesting in house foundations or even basements often brings them into contact with insect-fearing souls. They are also frequently noticed during their impressively large reproductive flights, which typically take place on warm, humid summer or fall days, usually just after a rain.

Another group of ants that frequently clash with homeowners is the genus of large, wood-inhabiting ants that includes the familiar black carpenter ants. Carpenter ants have different sizes of workers within the same colony, with larger ones (called "majors") conspicuously big-headed and over 10 mm in length. People often complain that our most common carpenter ant, *Camponotus pennsylvanicus*, the Black Carpenter Ant, is "eating" their deck or other household woodwork. It always comes as a surprise to them to hear that ants never eat wood, and carpenter ants only chew their galleries as a nesting refuge; from here they launch foraging columns in search of dead insects and other food. The large ants sometimes seen tending aphids are carpenter ants.

Aphid tending is common among ants, which guard their homopteran charges in exchange for honeydew. Some common little brown ants in the genus *Lasius* carry aphids (specifically Corn Root Aphids, *Anuraphis maidiradicis*) to weed roots in spring, transfer them to corn roots in the summertime and store the aphid eggs in the ant nests during the winter. If that can be compared to bringing your cows in from the pasture to winter in an earthen stable, then another kind of formicine ant has a trick analogous to storing the cow's milk in a massive indoor mammary. These ants, the Honeypot Ants (*Myrmecocystus* species) of the American southwest, use some individual ants as rotund, swollen reservoirs for the honeydew collected by other workers.

Several species of ants are slave-makers that depend on workers from other ant colonies to do their bidding. The most spectacular slave-making ants in northeastern North America are red slave-making ants in the genus *Polyergus*, spectacular ants sometimes called Amazons. Amazon queens start new colonies by invading *Formica* nests, killing the queen and then taking over the worker multitude. The adopted workers care for the Amazon queen, rearing out her progeny of *Polyergus* workers. The *Formica* workers, now slaves, handle all the work of foraging, brood rearing and nest building, while the *Polyergus* workers do nothing but raid other *Formica* nests to maintain a steady supply of slave pupae. An Amazon raid is exciting to watch, with a milling, hyperactive mass of bright raiding females flowing like a huge amoeba toward and into a *Formica* ant nest. They soon emerge carrying captured pupae, leaving the raided colony strewn with injured or dead *Formica* workers. Although *Polyergus* is the most conspicuous obligate colony raider, many other North American ants are slave-makers or otherwise dependent on other ant colonies. Some are social parasites that have no worker caste of their own, instead taking over the nests of host species and using host workers to rear their brood.

The ants mentioned so far all belong to the subfamily Formicinae, a group notable for being "full of formic acid," but also a group that has lost its sting. To encounter that notorious type of defensive ovipositor (remember that ants really are just specialized wasps) we need to look at other subfamilies such as the Myrmicinae. Myrmicine ants have a wasp-waist (pedicel) that is two-segmented and thus appears to have two lobes instead of the single lobe that characterizes other ants. This large subfamily of stinging ants includes famous ants such as the leafcutter ants that slice disks out of leaves to carry into huge underground nests, where they are used as a substrate on which to grow edible fungi. Leafcutter ants in the genera *Atta* and *Acromyrmex* are the dominant herbivores and the most destructive plant pests of Central America and South America, where their economic impact runs into the billions of dollars. Although neither *Atta* nor *Acromyrmex* occur in northeastern North America, the less conspicuous Northern Fungus Gardening Ant (*Trachymyrmex septentrionalis*) occurs as far north as New York, where it grows its fungus gardens on plant fragments as well as other debris such as caterpillar dung. One of the remarkable features of leafcutter biology is the use of antibiotics from strains of *Streptomyces* bacteria to combat unwanted fungi in their nests. The antibiotic-producing bacteria are passed from generation to generation within colonies.

The big-jawed harvester ants, the principle granivores of the southwestern United States, are also in the Myrmicinae. The only harvester ant species that occurs in eastern North America has conspicuously dimorphic workers, including spectacularly big-headed workers that excel at milling and crushing seeds as well as colony defense. The eastern species (Florida Harvester Ant, *Pogonomyrmex badius*) feeds as much on other insects as on seeds, but southwestern species of *Pogonomyrmex* are seed specialists. Some *Pogonomyrmex* species ("Pogos") are popular ant farm pets, although they have a potent sting.

Most Myrmicinae are less specialized than harvester or leafcutter ants, usually foraging on other insects (dead or alive) or scavenging a variety of materials including the contents of your kitchen. *Solenopsis* is an infamous group of small myrmicine ants distinctive for their 10-segmented antenna with a two-segmented club, but better known for the obnoxious habits of a few species such as the minute Thief

Ant (*S. molesta*), and the famous Red Imported Fire Ant (*S. invicta*).

Thief Ants are named for their habit of robbing food and brood from neighboring ants, but they are also both crop and household pests. Imported fire ants were accidentally introduced from South America into the southeastern states, where they form mounds in open areas such as pastures, schoolyards and parks. These aggressive stinging ants occur as far north as North Carolina, but they are particularly common in Florida. Red Imported Fire Ants are currently the targets of an innovative biological control program involving the importation of ant-decapitating flies from Brazil to the United States. Ant-decapitating flies are parasitic Phoridae whose larvae decapitate host ants as they leave the ant's body to pupate (some species pupate in the ant's head capsule). The presence of ant-decapitating flies effectively disrupts fire ant foraging, giving native ants (not attacked by these particular ant-decapitating flies) a competitive upper hand.

Near my home in Ontario, the most common myrmicine ant is an attractive little ant with a distinctive heart-shaped abdomen. These little ants, in the genus *Crematogaster*, are often seen out tending aphids, but can usually be exposed at home under the odd rock and in almost every old log or stump in nearby woodlots. When exposed, they run about comically with the shiny, heart-shaped abdomen bent forward and brandished above the head. They don't sting like most myrmicines, but they do have a painful bite.

Most northeastern ants are either in the stinging ant subfamily Myrmicinae or in the carpenter ant subfamily Formicinae, but some of the 700 or so species of North American ants fall into a few smaller subfamilies, including the Dolichoderinae, Ponerinae, Ecitoninae and Pseudomyrmecinae. Odorous House Ants (*Tapinoma sessile*), so-called because they exude a stench like rotting coconuts from their anal glands, are among the relatively few northeastern ants in the subfamily Dolichoderinae. Another northeastern dolichoderine, *Dolichoderus mariae*, nests in spongy hummocks in Sphagnum bogs.

The subfamily Ponerinae includes some intimidating, large and well-armed predators in the tropics, but in North America most ponerines are inconspicuous, slender ants that make small colonies, usually in rotting wood or soil. They do have a sting, but they don't have a two-segmented wasp-waist like our common stinging ants in the Myrmicinae. Members of one group of ponerines, called the trap jaw ants (*Odontomachus* species), have enormous jaws normally held outstretched. Fine hairs midway between the jaws serve as triggers to snap the jaws shut. Trap jaw ants are mostly tropical, but naturally range north to Alabama, and regularly appear as adventives in tropical gardens in the northeast. Bullet Ants (*Paraponera clavata*), huge ants that make small colonies in Neotropical rainforests, are also traditionally included in the Ponerinae (although this placement has recently been questioned). Bullet Ants do not occur in temperate North America, but they are very common ants in primary rainforests and living colonies are on display at North American insectaria including the Cincinnati Zoo. They must be handled with extreme care, since a sting from one of these huge ants delivers a debilitating injection of a potent nerve toxin called poneratoxin. Although Bullet Ants arguably have the worst stings of any insect, the most venomous animal toxin is reputedly produced by another ant, species of harvester ants (*Pogonomyrmex*) with venom over 20 times as potent as honey bee venom.

Like the ponerine ants, ants in the subfamily Pseudomyrmecinae are mostly tropical. A few species occur in southeastern North America, and a few introduced species occasionally show up as accidental introductions in tropical greenhouses (zoos and insectaria) in the northeast. These small ants are sometimes called twig ants because they nest in preformed cavities such as hollow twigs.

The subfamily Ecitoninae includes the famous New World army ants (African army ants are in a different subfamily, the Dorylinae). Although this group is best known for some spectacular Neotropical species, North America is home to many species in the inconspicuous army ant genus *Neivamyrmex*, which ranges north to Illinois and North Carolina. Army ants do not have permanent nests, but instead establish temporary "bivouacs" when the queen is ready to produce a batch of eggs. The colony maintains this temporary nest throughout a "statary" phase, during which the queen lays tens of thousands of eggs, and workers make local raids in search of food (often other ants) for the voracious developing larvae. Once the larvae have pupated, the colony enters a "nomadic" phase that lasts until the queen is once again swollen with mature eggs. Unlike most ants, army ant queens do not take a mating flight, and only the male reproductives are winged. In the Neotropics, where army ants abound, the large, winged male army ants are often conspicuous at lights.

Army ant columns are regular sights in neotropical forests, and it is not unusual to come across a fan-shaped swarm front of the common *Eciton burchelli*. Any small organism encompassed by the swarm front is flushed out and captured, and the swarm front is always alive with scrambling prey along with a fantastic assortment of opportunistic parasitoids, such as Stylogastridae and Tachinidae, that seek hosts along the swarm front. Army ants do sting, but they are not the dangerous animals of movie lore, and if you tuck your pants into a pair of good boots you can stand in the midst of a raid with impunity.

SUBORDER SYMPHYTA (SAWFLIES)

FAMILY DIPRIONIDAE (CONIFER SAWFLIES) ❶ Larval sawflies, like this **Introduced Pine Sawfly** (***Diprion similis***), look like caterpillars with distinct eyespots and with more than five prolegs. ❷ This adult **Introduced Pine Sawfly** (***Diprion similis***) has just emerged from a brown, silken cocoon, which still has its circular cap hanging open. ❸ These **European Pine Sawfly** (***Neodiprion sertifer***) larvae are defoliating a Jack Pine (*Pinus banksiana*) tree. European Pine Sawflies attack a variety of pine species. ❹ ❺ Male conifer sawflies often detect female pheromones using branched antennae, like those on the male of this pair of ***Monoctenus***. *Monoctenus* larvae develop on Eastern White Cedar (*Thuja occidentalis*) and Common Juniper (*Juniperus communis*). **FAMILY CIMBICIDAE** (CIMBICID SAWFLIES) ❻ ❼ The large (18–25 mm) adults of the **Elm Sawfly** (***Cimbex americana***) are among our most distinctive insects. The female abdomen is banded like a wasp and the male abdomen is black or red-brown and black. ❽ ❾ **Elm Sawfly** (***Cimbex americana***) larvae feed on both elm and willow foliage. These distinctive robust larvae can squirt a defensive fluid from glands above the spiracles. Sawflies currently treated as Elm Sawflies are part of a species complex with differently colored larvae.

FAMILY CIMBICIDAE (continued) ❶ **Elm Sawfly** (***Cimbex americana***) larvae are usually green or yellowish, but this individual is quite pink. Note the single simple eye, characteristic of sawfly larvae but different from the multiple ocelli of otherwise similar moth larvae. ❷ The conspicuously hairy adults of ***Trichosoma triangulum*** are the only northeastern sawflies that rival Elm Sawflies (*Cimbex americana*) in size. These common insects develop on alder, birch, poplar and a few other broadleaf trees. ❸ ❹ ❺ This **Honeysuckle Sawfly** (***Zaraea inflata***) is using its sawlike ovipositor to insert eggs within the margin of a honeysuckle (*Lonicera*) leaf. The Cimbicidae is a small family with only seven northeastern species, four of which are in the genus *Zaraea*. **FAMILY PAMPHILIIDAE** (WEB-SPINNING SAWFLIES) ❻ Larvae of web-spinning sawflies, like ***Pamphilius middlekauffi***, feed on leaves of a variety of trees and shrubs, sometimes spinning silk and leaf shelters similar to those made by some caterpillars. The forward oriented and distinctively wide head is characteristic of Pamphilidae. ❼ ***Onycholyda luteicornis*** is a distinctively colored sawfly often found on the leaves of cherry, serviceberry, raspberry and related plants. ❽ ❾ The **Pine False Webworm** (***Acantholyda erythrocephala***) is an introduced pest of pine. Males and females differ widely in color (the female has a red head), and the larvae spin messy, frass-strewn, tubelike silken shelters along pine twigs.

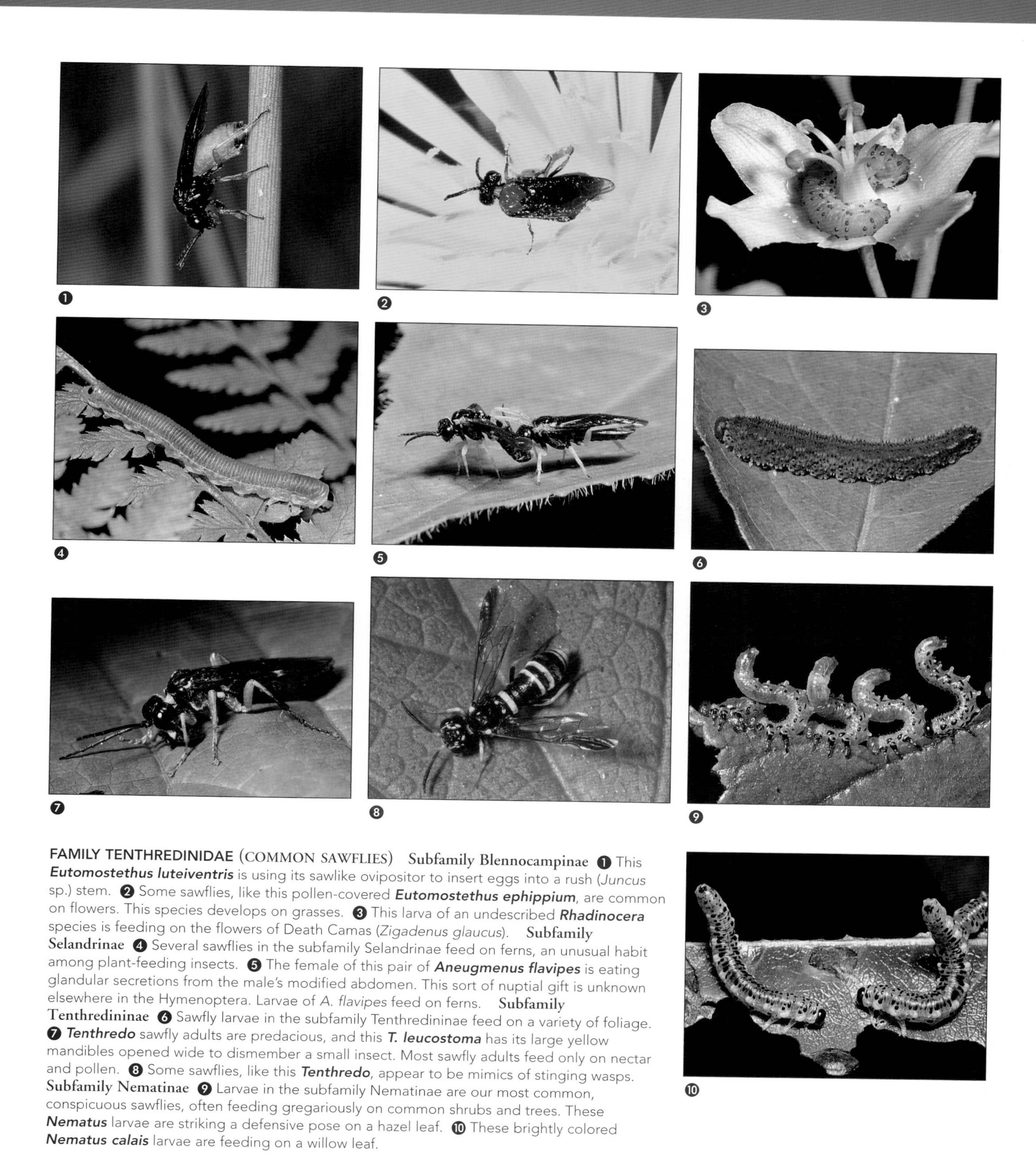

FAMILY TENTHREDINIDAE (COMMON SAWFLIES) Subfamily Blennocampinae ❶ This ***Eutomostethus luteiventris*** is using its sawlike ovipositor to insert eggs into a rush (*Juncus* sp.) stem. ❷ Some sawflies, like this pollen-covered ***Eutomostethus ephippium***, are common on flowers. This species develops on grasses. ❸ This larva of an undescribed ***Rhadinocera*** species is feeding on the flowers of Death Camas (*Zigadenus glaucus*). Subfamily Selandrinae ❹ Several sawflies in the subfamily Selandrinae feed on ferns, an unusual habit among plant-feeding insects. ❺ The female of this pair of ***Aneugmenus flavipes*** is eating glandular secretions from the male's modified abdomen. This sort of nuptial gift is unknown elsewhere in the Hymenoptera. Larvae of *A. flavipes* feed on ferns. Subfamily Tenthredininae ❻ Sawfly larvae in the subfamily Tenthredininae feed on a variety of foliage. ❼ ***Tenthredo*** sawfly adults are predacious, and this ***T. leucostoma*** has its large yellow mandibles opened wide to dismember a small insect. Most sawfly adults feed only on nectar and pollen. ❽ Some sawflies, like this ***Tenthredo***, appear to be mimics of stinging wasps. Subfamily Nematinae ❾ Larvae in the subfamily Nematinae are our most common, conspicuous sawflies, often feeding gregariously on common shrubs and trees. These ***Nematus*** larvae are striking a defensive pose on a hazel leaf. ❿ These brightly colored ***Nematus calais*** larvae are feeding on a willow leaf.

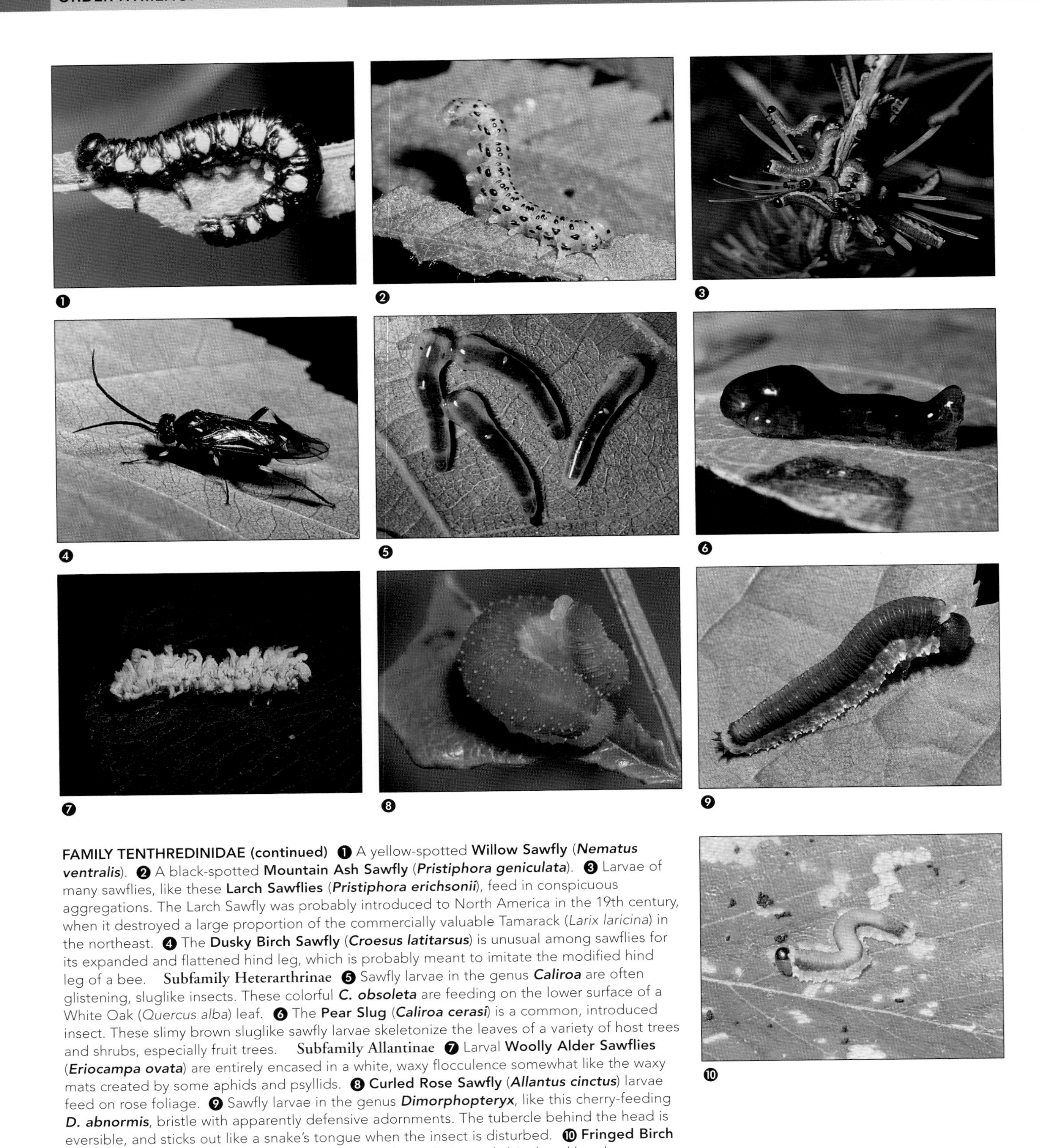

FAMILY TENTHREDINIDAE (continued) ❶ A yellow-spotted **Willow Sawfly** (***Nematus ventralis***). ❷ A black-spotted **Mountain Ash Sawfly** (***Pristiphora geniculata***). ❸ Larvae of many sawflies, like these **Larch Sawflies** (***Pristiphora erichsonii***), feed in conspicuous aggregations. The Larch Sawfly was probably introduced to North America in the 19th century, when it destroyed a large proportion of the commercially valuable Tamarack (*Larix laricina*) in the northeast. ❹ The **Dusky Birch Sawfly** (***Croesus latitarsus***) is unusual among sawflies for its expanded and flattened hind leg, which is probably meant to imitate the modified hind leg of a bee. Subfamily Heterarthrinae ❺ Sawfly larvae in the genus ***Caliroa*** are often glistening, sluglike insects. These colorful ***C. obsoleta*** are feeding on the lower surface of a White Oak (*Quercus alba*) leaf. ❻ The **Pear Slug** (***Caliroa cerasi***) is a common, introduced insect. These slimy brown sluglike sawfly larvae skeletonize the leaves of a variety of host trees and shrubs, especially fruit trees. Subfamily Allantinae ❼ Larval **Woolly Alder Sawflies** (***Eriocampa ovata***) are entirely encased in a white, waxy flocculence somewhat like the waxy mats created by some aphids and psyllids. ❽ **Curled Rose Sawfly** (***Allantus cinctus***) larvae feed on rose foliage. ❾ Sawfly larvae in the genus ***Dimorphopteryx***, like this cherry-feeding ***D. abnormis***, bristle with apparently defensive adornments. The tubercle behind the head is eversible, and sticks out like a snake's tongue when the insect is disturbed. ❿ **Fringed Birch Sawfly** (***Dimorphopteryx melanognathus***) larvae have distinctively bicolored heads.

FAMILY TENTHREDINIDAE (continued) ❶ Birch leaves are often disfigured by blotch mines made by the larvae of the small European sawfly ***Messa nana***, which was first recorded in North America in 1967. ❷ The **Violet Sawfly** (***Amestastegia pallipes***) feeds under the leaves of violets during the night, often skeletonizing the foliage. ❸ Sawfly larvae, like this ***Macremphytus***, have simple eyespots and have more than five pairs of prolegs. ❹ This adult **Dogwood Sawfly** (***Macremphytus testaceus***) is on its host plant (*Cornus* sp.). **FAMILY ARGIDAE (ARGID SAWFLIES)** ❺ This **Poison Ivy Sawfly** (***Arge humeralis***) is one of the few insects that feed on Poison Ivy (*Rhus radicans*). ❻ The long, one-piece antennal flagellum is characteristic of the small family Argidae, like this ***Arge cyra*** laying eggs in the edge of an Ironwood (*Ostrya virginiana*) leaf. ❼ ❽ The **Imported Rose Sawfly** (***Arge ochropa***) inserts rows of eggs on young rose branches. Larvae start off by skeletonizing rose leaves, and then progress to eating the whole leaves. The S-shaped pose struck by this larva is characteristic. ❾ The **Birch Sawfly** (***Arge pectoralis***) inserts eggs along the edges of birch leaves in spring; the young larvae feed gregariously on the foliage.

FAMILY PERGIDAE (PERGID SAWFLIES) ❶ Pergidae is an uncommon family in North America, with only a few species of the genus ***Acordulecera***, usually found on oak foliage. Unlike similar sawfly families they have only six antennal segments. **FAMILY XYELIDAE** (XYELID SAWFLIES) ❷ The distinctive antennae, with a large basal part and a thin end, identify this odd little sawfly (***Xyela*** sp.) as a member of the small family Xyelidae. Larvae develop in staminate pine flowers, like the one supporting this pollen-covered adult. **FAMILY CEPHIDAE** (STEM SAWFLIES) ❸ Members of the small family Cephidae are slender and compressed insects that usually develop inside the stems of grasses, willows and berry bushes. This one (***Hartigia trimaculata***) probably developed in either rose or raspberry stems. **FAMILY XIPHYDRIIDAE** (WOOD WASPS) ❹ This wood wasp is laying an egg inside the trunk of a small, dead maple tree. All ten North American species of wood wasps are in the genus ***Xiphydria***. **FAMILY SIRICIDAE** (HORNTAILS) ❺ ❻ The female of this pair of horntails (**Pigeon Tremex**, ***Tremex columba***; about 40 mm) will use her dagger-like ovipositor to insert an egg, along with symbiotic fungi and a mucus secretion that promotes fungal growth, 5–10 mm into a dying tree. Horntails of both sexes have a hornlike protuberance at the end of the abdomen, but only the females have an ovipositor. ❼ Larval horntails feed on a mixture of fungi and wood deep within damaged or diseased trees. ❽ This black horntail (***Urocerus cressoni***) occurs naturally across Canada and the United States, and has been recorded as an "adventive" in Britain, where it occasionally emerges from imported North American lumber. **FAMILY ORUSSIDAE** (PARASITIC WOOD WASPS) ❾ This parasitic wood wasp (***Orussus terminalis***) has lowered his antennae into a beetle burrow on a dead log, perhaps in search of a female attracted to the log by the presence of potential hosts such as larval jewel beetles (Buprestidae). Females also investigate burrows with their antennae before uncoiling a concealed ovipositor and threading it into the burrow to reach a host larva. Parasitic wood wasps may look like sawflies and horntails, but their parasitic habits betray a closer relationship to the parasitic wasps in the suborder Apocrita. The bumpy head of this individual is characteristic of the family.

SUBORDER APOCRITA (PARASITIC WASPS)

SUPERFAMILY ICHNEUMONOIDEA

FAMILY ICHNEUMONIDAE (ICHNEUMONID WASPS) **Subfamily Ophioninae** ❶ Although most members of the huge family Ichneumonidae are difficult to identify, this large species (***Thyreodon atricolor***) is an easily recognized, day-active, slow-flying parasitoid of sphinx moth caterpillars. ❷ ❸ Ichneumonid wasps use their ovipositors as egg-laying devices and so lack the specialized sting found in aculeates such as bees and yellowjacket wasps. Some Ophioninae nonetheless deliver a startlingly stinglike poke, sometimes along with a venom injection, using the sharp ovipositors that normally function to inject eggs into caterpillar hosts. ***Enicospilus*** and *Ophion* are similar large ophionine genera frequently attracted to lights at night as they search for nocturnally active caterpillars. Note the large ocelli, typical of the few groups of night-active Hymenoptera. **Subfamily Tryphoninae** ❹ ***Netelia*** species differ from similar nocturnal ichneumonids in having a small, window-like cell (areolet) in the forewing. Female *Netelia* are often seen with eggs hung from the ovipositor like bunches of grapes, with only an anchor-like stem inside the track-like ovipositor. Eggs are attached to caterpillars using the anchor (much like price tags are sometimes attached to clothing). Larvae develop as external parasitoids. **Subfamily Acaenitinae** ❺ Some ichneumonid wasps, like this ***Spilopteron vicinum*** (a parasitoid of long-horned beetles), are frequent flower visitors. **Subfamily Rhyssinae** ❻ ❼ Our most spectacular parasitic wasps are the enormous ichneumonids in the genus ***Megarhyssa***, known to foresters as "stumpstabbers," which parasitize horntail (Siricidae) larvae inside hardwood trees or logs. These photographs show females of ***M. macrurus*** inserting their ovipositors into tree trunks. Three species of *Megarhyssa* are found on hardwood trees in eastern North America. ❽ ❾ These male ***Megarhyssa*** wasps, which lack the long ovipositor of the females, have detected an emerging virgin female and are attempting to mate with her before she leaves the tree.

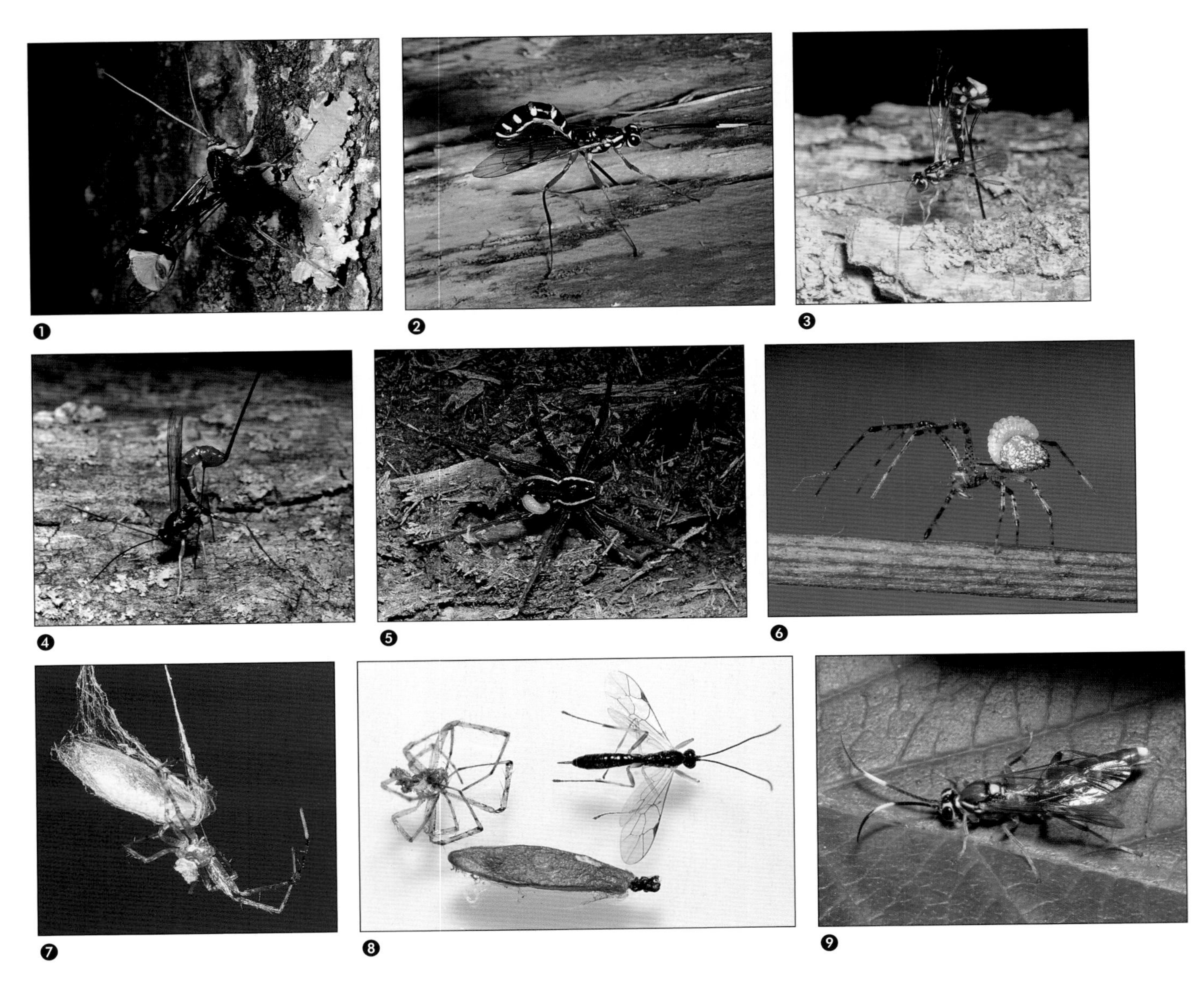

FAMILY ICHNEUMONIDAE (continued) ❶ This impressively long ***Megarhyssa atrata*** female is using her 10 cm ovipositor to parasitize the same host (the horntail *Tremex columba*) in the same dead tree and at the same time as the more common *M. macrurus*. The latter species has a slightly shorter ovipositor (about 8 cm). ❷ ❸ These two female ***Rhyssa*** (***R. lineolata*** and ***R. persuasoria***) are drilling into dead wood in search of horntail hosts. Subfamily Pimplinae ❹ Larvae of ***Dolichomitus irritator*** are parasitoids on a wide variety of wood-boring insect larvae, including long-horned beetles, weevils, metallic wood-boring beetles and wasp moths. ❺ The larva attached to the abdomen of this wolf spider is an ectoparasitic ichneumonid wasp. ❻ ❼ ❽ The larva attached to the abdomen of this sheet-web spider (Linyphiidae) is the ectoparasitic ichneumonid ***Acrodactyla ocellata***, which will kill its spider host before pupating in a silk cocoon in the spider's web. Some wasps in this subfamily induce their host spiders to make special webs to protect the wasp cocoon. The third photograph shows the adult wasp with its drained host and empty cocoon. ❾ ***Theronia hilaris*** is a distinctively colored species that parasitizes budworms, tent caterpillars, tussock moths and many other moths and butterflies. Adult females of this wasp can sometimes be found under bark in midwinter. Ichneumonidae and Braconidae often overcome their host's immune systems with the aid of symbiotic viruses injected into the hosts along with the wasp eggs.

❶ ❷ ❸ ❹ ❺ ❻ ❼ ❽ ❾

FAMILY ICHNEUMONIDAE (continued) Subfamily Cryptinae ❶ ***Endasys*** species attack sawflies in several families (Argidae, Diprionidae, Tenthredinidae). Most members of the enormous subfamily Cryptinae develop as ectoparasitoids attached to the hosts' bodies within the hosts' cocoons or shelters. ❷ This unhealthy-looking forest tent caterpillar is already host to another parasitoid, which is itself about to be parasitized by this hyperparasitic cryptine ichneumonid wasp. ❸ This wingless female ichneumonid wasp (***Gelis*** sp.) has just emerged from a bagworm (*Psyche casta*, Psychidae) case, where it probably developed as a hyperparasite of another (winged) ichneumonid wasp. ❹ ❺ ***Apsilops hirtifrons*** is a cryptine ichneumonid that hunts underwater for its aquatic hosts (aquatic moth caterpillars). One of the individuals shown here is under water, surrounded by a glistening bubble of air. Subfamily Ichneumoninae ❻ Parasitic wasp adults, like this ***Ichneumon***, often feed on honeydew and can be attracted by spraying honey solution on foliage. Like other members of the Ichneumoninae, this species lays eggs in mature caterpillars but does not complete its development until the host has pupated. ❼ This large ichneumonine wasp (***Protichneumon grandis***) is a parasitoid of a common giant silkworm moth, the Rosy Maple Moth (*Dryocampa rubicunda*). ❽ Like other members of the subfamily Ichneumoninae, this wasp developed as an internal parasitoid of a moth or butterfly caterpillar and emerged from the host pupa. The Ichneumoninae is the second largest subfamily of Ichneumonidae, after the Cryptinae. ❾ ***Trogus pennator*** is a large and distinctively colored parasitoid of swallowtail butterflies.

FAMILY ICHNEUMONIDAE (continued) Subfamily Campopleginae ❶ This ***Dusona*** female has a sharp ovipositor for laying eggs directly into exposed caterpillars such as inchworms (Geometridae). Her larvae will develop at the host's expense, but will delay killing the host larva until it is mature and about to pupate. ❷ This campoplegine ichneumonid emerged from a spindle gall made by a goldenrod gall moth. **FAMILY BRACONIDAE (BRACONID WASPS)** Subfamily Euphorinae ❸ Braconid wasps are often used in biological control programs against pest insects. This wasp species (***Peristenus relictus***) was imported to augment native parasitoids as control agents against Tarnished Plant Bugs (*Lygus lineolaris*, Miridae). ❹ This native wasp (***Leiophron lygivorus***) is laying eggs in Tarnished Plant Bug nymphs. ❺ This lady beetle (*Coleomegilla maculata*) was still twitching when it was collected along with this attached silk cocoon containing its braconid parasitoid, ***Dinocampus coccinellae***, which later emerged from the cocoon. Before leaving the beetle in which it developed, the larval wasp used her (this is a parthenogenic species) mandibles to cut the nerves controlling movement in the lady beetle's legs. The lady beetle host was thus immobilized, but remained alive to stand guard over the parasitoid cocoon with its bright warning colors. Subfamily Agathidinae ❻ Common Burdock (*Arctium minus*) seedheads are frequently infested with small moths, which are in turn parasitized by braconid wasps. This braconid female (***Agathis malvacearum***) has penetrated the seedhead with her ovipositor, and is parasitizing the larva of a gelechiid moth (*Metzneria lapella*). Subfamily Doryctinae ❼ Species in the genus ***Spathius*** are parasitoids of larval wood-boring beetles. Subfamily Rogadinae ❽ ❾ The round holes on this dried-out caterpillar formerly housed pupae of parasitic wasps (**Aleoides** sp.) that developed and pupated inside the caterpillar. The second photograph shows a fresher specimen, with some holes visible and one of the parasitic braconid wasp adults still in the act of emerging. Many braconids are gregarious parasitoids, in which numerous progeny of one parent wasp develop in the same host.

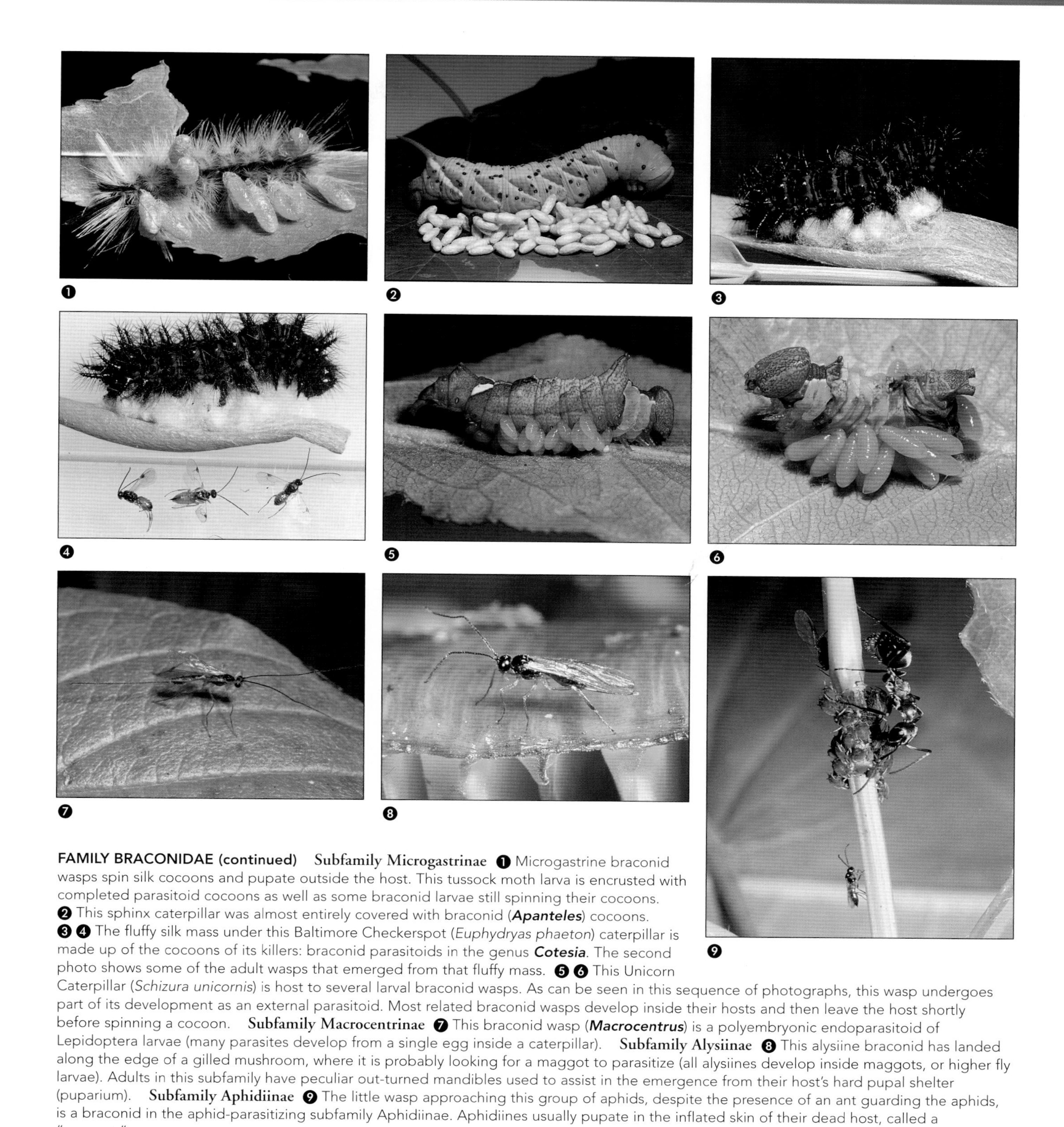

FAMILY BRACONIDAE (continued) Subfamily Microgastrinae ❶ Microgastrine braconid wasps spin silk cocoons and pupate outside the host. This tussock moth larva is encrusted with completed parasitoid cocoons as well as some braconid larvae still spinning their cocoons. ❷ This sphinx caterpillar was almost entirely covered with braconid (***Apanteles***) cocoons. ❸ ❹ The fluffy silk mass under this Baltimore Checkerspot (*Euphydryas phaeton*) caterpillar is made up of the cocoons of its killers: braconid parasitoids in the genus ***Cotesia***. The second photo shows some of the adult wasps that emerged from that fluffy mass. ❺ ❻ This Unicorn Caterpillar (*Schizura unicornis*) is host to several larval braconid wasps. As can be seen in this sequence of photographs, this wasp undergoes part of its development as an external parasitoid. Most related braconid wasps develop inside their hosts and then leave the host shortly before spinning a cocoon. Subfamily Macrocentrinae ❼ This braconid wasp (***Macrocentrus***) is a polyembryonic endoparasitoid of Lepidoptera larvae (many parasites develop from a single egg inside a caterpillar). Subfamily Alysiinae ❽ This alysiine braconid has landed along the edge of a gilled mushroom, where it is probably looking for a maggot to parasitize (all alysiines develop inside maggots, or higher fly larvae). Adults in this subfamily have peculiar out-turned mandibles used to assist in the emergence from their host's hard pupal shelter (puparium). Subfamily Aphidiinae ❾ The little wasp approaching this group of aphids, despite the presence of an ant guarding the aphids, is a braconid in the aphid-parasitizing subfamily Aphidiinae. Aphidiines usually pupate in the inflated skin of their dead host, called a "mummy."

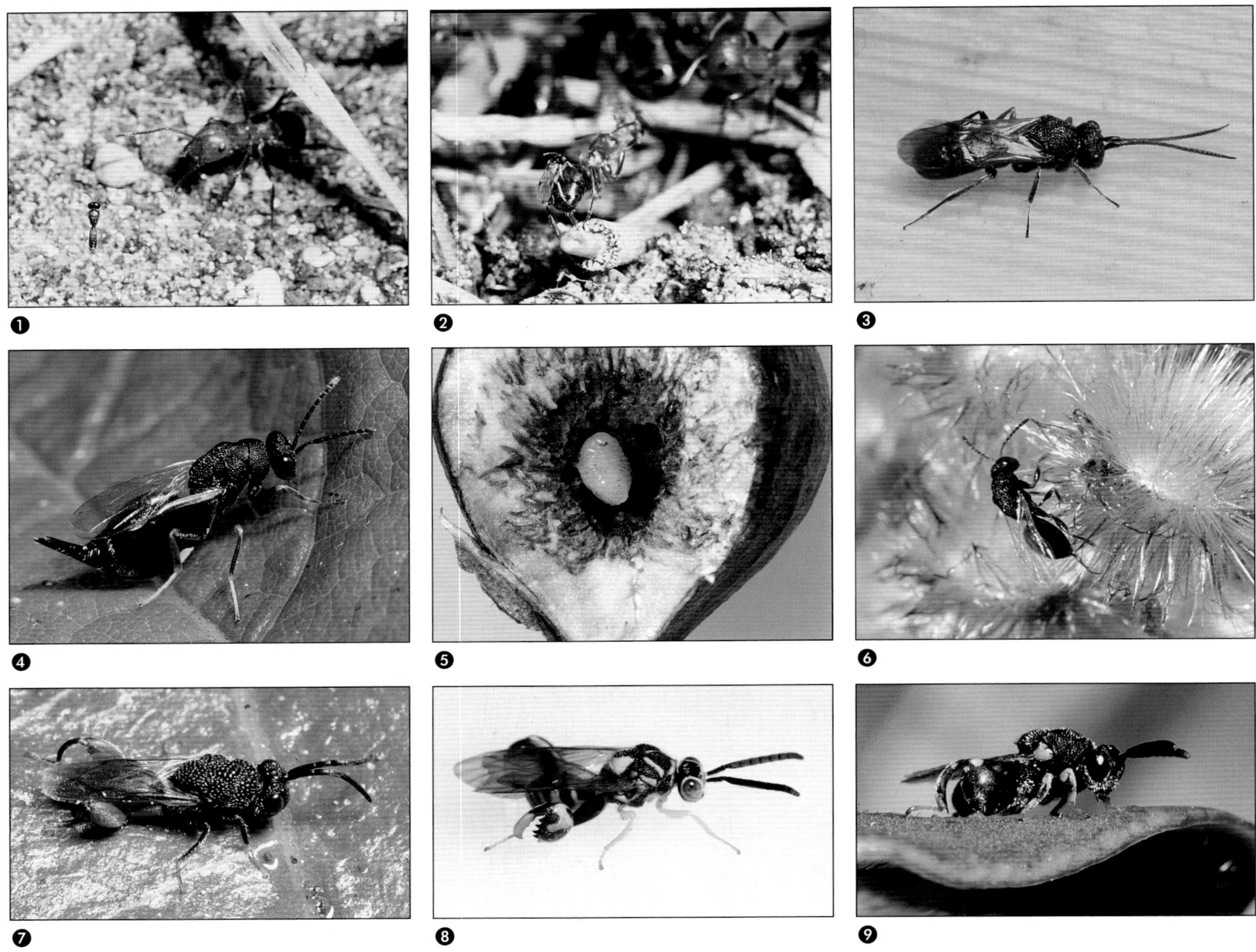

FAMILY BRACONIDAE (continued) Subfamily Neoneurinae ❶ ❷ The first of these photographs shows a small braconid wasp (***Elasmosoma***) hovering just in front of a wood ant (*Formica*). The second photo shows the wasp mounting a worker ant's abdomen and inserting an egg into the ant, now doomed to play host to the wasp's larva. Subfamily Cheloniinae ❸ The robust, punctate wasps in the braconid subfamily Cheloniinae lay their eggs in the eggs of butterflies and moths. The wasp larvae develop as parasitoids within caterpillars, ultimately spinning their own cocoons within the host's pupation retreat.

❿

SUPERFAMILY CHALCIDOIDEA
FAMILY EURYTOMIDAE (SEED CHALCIDS AND THEIR RELATIVES) ❹ ❺ These two images show a larva and an adult of ***Eurytoma gigantea***, a eurytomid parasitoid of Goldenrod Gall Flies (*Eurosta solidaginis*, Tephritidae). The rotund, white larva in the middle of the cut-open goldenrod gall had entirely consumed the gall-maker before the photograph was taken (in midwinter). Other, smaller, *Eurytoma* species complete their development in prematurely formed puparia (inflated and hardened larval skins) of the Goldenrod Gall Fly. ❻ Some Eurytomidae have abandoned parasitism in favor of a vegetarian lifestyle, and this species (***Eurytoma rhois***) develops on sumac (*Rhus*) seeds. **FAMILY CHALCIDIDAE** ❼ Wasps in the family Chalcididae are characterized by conspicuously enlarged hind legs and a short, inconspicuous ovipositor. ❽ This striking chalcidid species (***Conura amoena***) is a parasitoid of hairstreak butterflies. ❾ Chalcidid wasps in the large genus ***Brachymeria*** are parasitoids on a variety of hosts; mostly moths, flies and other wasps. **FAMILY MYMARIDAE** (FAIRYFLIES) ❿ Fairyflies are parasitoids that develop inside the eggs of other insects, and are usually overlooked because of their small size (usually less than 1 mm and as small as 0.2 mm). **FAMILY TRICHOGRAMMATIDAE** ⓫ Minute wasps in the family Trichogrommatidae, some of which are less than 0.2 mm, are egg parasitoids.

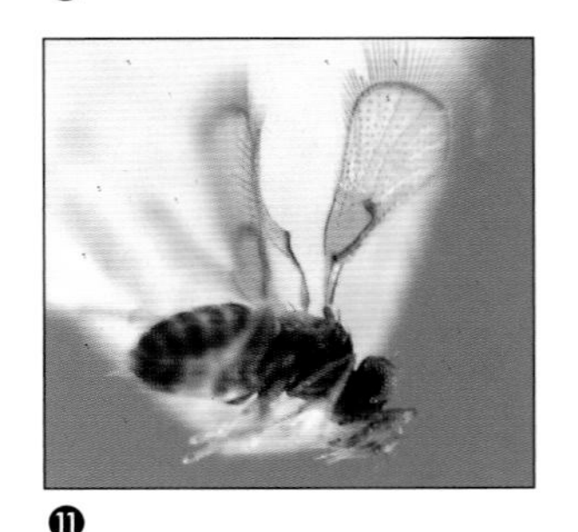

⓫

FAMILY LEUCOSPIDAE ❶ ***Leucospis affinis***, the only northeastern member of the family Leucospidae, looks like a small yellowjacket wasp, but the female has a long ovipositor that is normally curved up over her body. These parasitoids of solitary bees (usually leafcutter bees) use the long ovipositor to drill through the wall of the host's nest cell. **FAMILY PERILAMPIDAE** ❷ This relatively large, brilliantly metallic, perilampid wasp is a hyperparasitoid (a parasitoid of other parasitoids). Perilampids usually lay their eggs on foliage, and leave the job of finding a host to their tough, flattened larvae (called planidia). Planidia usually grab onto caterpillars and penetrate them in search of the other parasitoids. **FAMILY EUCHARITIDAE** ❸ Eucharitid wasps, such as this distinctively sculptured ***Kapala***, lay huge numbers of eggs, often on particular kinds of flowers. The resultant small larvae quest for passing ants, and those lucky enough to successfully hitch a ride back to an ant's nest will develop as parasitoids of ant larvae. **FAMILY APHELINIDAE** ❹ One species of aphelinid wasp (***Encarsia formosa***) is often used in greenhouses for the control of pest whiteflies (Aleyrodidae). You can buy the pupae of these minute parasitoids glued in groups on cards, ready to hang in your greenhouse. **FAMILY PTEROMALIDAE** ❺❻ Many pteromalid wasps are hyperparasitoids (parasitoids that attack other parasitoids). This sphinx caterpillar was parasitized by larvae of a braconid (*Apanteles*), the cocoons of which are being hyperparasitized by this tiny pteromalid. ❼ This pteromalid wasp (***Muscidifurax raptor***) has slipped her ovipositor into the puparium of a House Fly (*Musca domestica*) in order to parasitize the pupa within. This species and another common pteromalid (*Nasonia vitripennis*) that also attacks fly puparia are used in the biological control of House Flies, blow flies and Stable Flies (*Stomoxys calcitrans*). ❽ Many members of the large family Pteromalidae parasitize aphids, and this tiny wasp is near the remains of an aphid host. ❾ Leaves of wild columbine are often mined by tiny flies in the family Agromyzidae, which are in turn attacked by pteromalid wasps (***Seladerma diaeus***). This female is laying an egg through the upper surface of a mine and into the puparium of the fly that made the mine.

FAMILY PTEROMALIDAE (continued) ❶ This ***Schizonotus rotundiventris*** is inserting an egg in the pupa of an Imported Willow Leaf Beetle (*Plagiodera versicolora*, Chrysomelidae). **FAMILY TORYMIDAE** ❷ Torymid wasps often use their long ovipositors to penetrate plant tissue such as galls, developing seeds or nests of host species. This species, ***Monodontomerus obscurus***, is a parasitoid of leafcutter bees. ❸ Members of the large genus ***Torymus*** usually attack insects inside galls, and can often be seen inserting their long ovipositors into rose galls made by cynipid wasps. ❹ Members of the genus ***Podagrion*** parasitize mantid egg masses. This is one of dozens of wasps that emerged from a South American mantid case, but other *Podagrion* species occur in eastern North America where they parasitize both native and introduced mantids. **FAMILY EUPELMIDAE** ❺ Many eupelmids, like this ***Brasema***, develop as ectoparasitoids on the larvae of wood-boring beetles and other insects concealed in plant tissue. ❻ ***Brasema leucothysana***, recently described by Canadian entomologist Gary Gibson, is distributed throughout eastern North America from southern Canada to Florida. Nothing is known about its hosts. ❼ Eupelmid adults, many of which are wingless like this ***Eupelmus vesicularis***, have a remarkable jumping mechanism involving rapid arching of the thorax and a concomitant explosive kick of the midlegs. Members of the large genus *Eupelmus* are parasitoids that usually attack insect larvae or pupae in galls, stems or other plant tissue. **FAMILY EULOPHIDAE** ❽ ❾ Members of the genus ***Sympiesis*** are ectoparasitoids that develop on leaf-mining caterpillars within their mines. ❿ As the scientific name suggests, ***Tetrastichus malacosomae*** is a parasitoid of the tent caterpillar genus *Malacosoma*. This photo shows three female wasps ovipositing in a Forest Tent Caterpillar (*Malacosoma disstria*, Lasiocampidae).

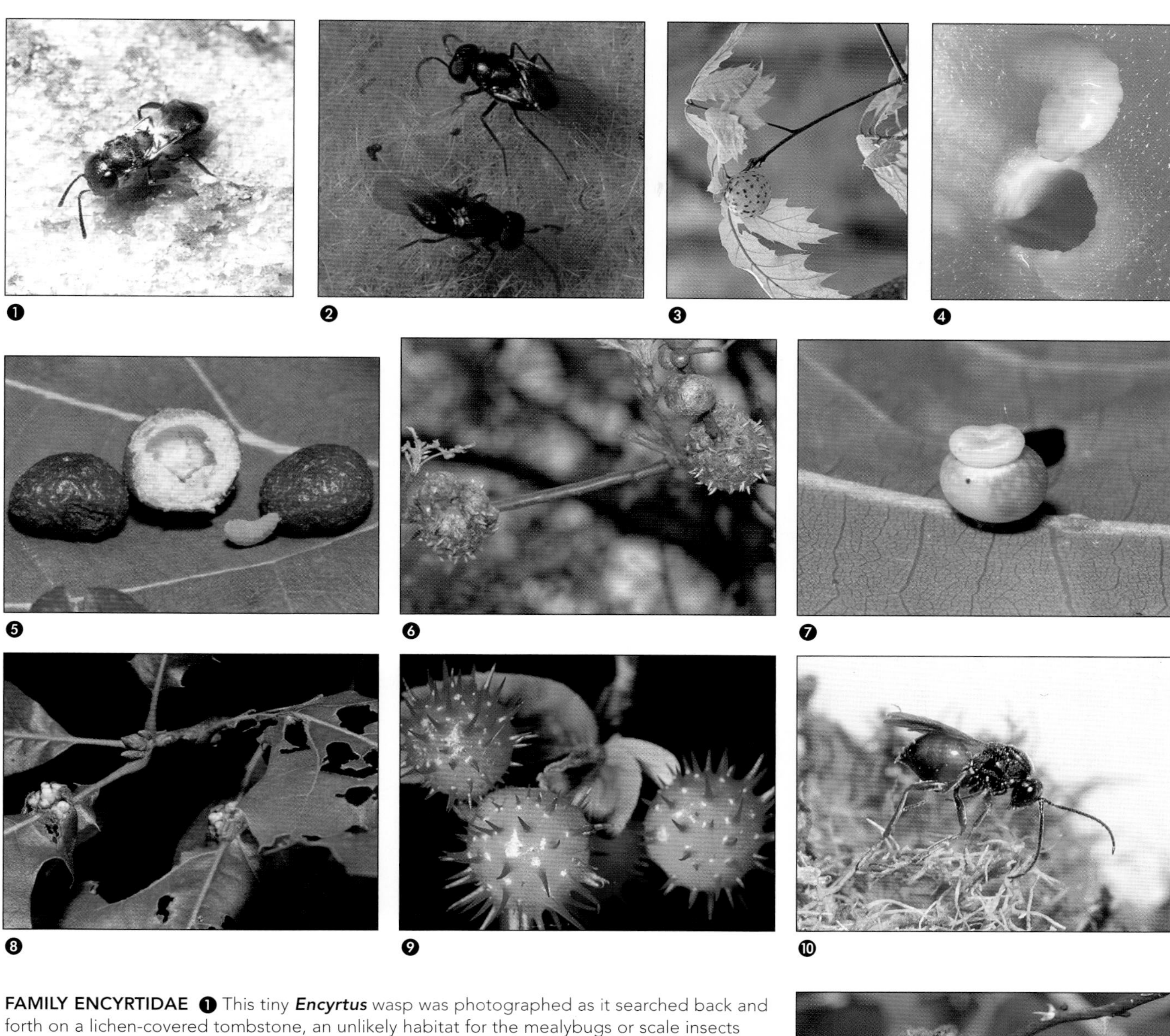

FAMILY ENCYRTIDAE ❶ This tiny ***Encyrtus*** wasp was photographed as it searched back and forth on a lichen-covered tombstone, an unlikely habitat for the mealybugs or scale insects that are the only known hosts for *Encyrtus* species. ❷ ***Ooencyrtus kuvanae*** is a small encyrtid wasp that was first introduced to the eastern United States in 1909 from Japan as an egg parasitoid for control of the Gypsy Moth. It is now a common insect in eastern North America, and an important agent in the biological control of Gypsy Moths.

SUPERFAMILY CYNIPOIDEA

FAMILY CYNIPIDAE (GALL WASPS) ❸ Gall wasps make a wide variety of galls on leaves, stems, buds and roots of oak trees. Several species induce large, apple-like galls collectively called "oak apple galls." ❹ Gall wasp larvae, like other larval Apocrita, lack legs and a discernable head. ❺ These small detachable galls, found on Red Oak leaves, are probably ***Dryocosmus rileyi***. ❻ The **Horned Oak Gall Wasp** (***Callirhytis cornigera***) induces conspicuously spiky twig galls on pin, scrub and black oak. ❼ This spangle gall on a White Oak leaf is probably ***Phylloteras poculum***. ❽ These cynipid galls (probably ***Amphibolips cookii***) are developing on Red Oak leaf petioles. ❾ Several gall wasps in the genus ***Diplolepis*** induce differently shaped galls on different parts of rose bushes. These bright ***D. bicolor*** galls were formed on the leaves of a wild rose. ❿ ⓫ This ***Diplolepis rosae*** has just emerged from a mossy rose gall, or bedeguar gall. The development of these large and distinctive galls follows the deposition of eggs in unopened buds in spring. Each gall has dozens of chambers, and is often occupied by various hangers-on (inquilines, parasites) as well as the gall makers.

❶ ❷ ❸ ❹ ❺ ❻

FAMILY IBALIIDAE ❶ This ***Ibalia anceps*** has inserted its ovipositor into a tree trunk occupied by its host, a sawfly (Siricidae).

SUPERFAMILY TRIGONALYOIDEA

FAMILY TRIGONALYIDAE ❷ Trigonalyid wasps lay thousands of eggs under leaves, often near the edge where they are most likely to be ingested by caterpillars or sawfly larvae. The wasp eggs hatch inside the caterpillar, but only develop if there is another parasitoid inside the caterpillar for the wasp to attack. This is ***Lycogaster pullata***, one of only four North American species in this family.

SUPERFAMILY STEPHANOIDEA

FAMILY STEPHANIDAE ❸ Stephanids are parasitoids of wood-boring insects, including beetles and sawflies. This is ***Megischus bicolor***, the only stephanid in northeastern North America. The long neck and cluster of spines on the head are distinctive for the family.

SUPERFAMILY EVANIOIDEA

FAMILY GASTERUPTIIDAE ❹ Gasteruptiids, like this ***Rhydinofoenus***, appear to have the abdomen inserted so high it seems to arise from the back of the thorax. The apparent abdomen of wasp-waisted Hymenoptera (Apocrita) is made up of abdominal segments 2–10 (the gaster) because the first abdominal segment is fused with the thorax in front of the wasp waist (petiole). ❺ ❻ This female ***Gasteruption*** (long ovipositor) is taking nectar from a *Physocarpus* flower, and the male (no ovipositor) is on an *Aegopodium* flower. ❼ This gasteruptiid female (an Australian species) has her ovipositor deep in a tree trunk. Most gasteruptiids are kleptoparasites that kill the egg or larvae of their bee or wasp hosts, then develop on the stored pollen or prey in the host's nest.

❼

FAMILY EVANIIDAE (ENSIGN WASPS) ❶ Only four of the nine North American species of ensign wasps range as far north as Canada, where they parasitize egg cases of native wood roaches (*Parcoblatta* spp.). Ensign wasps have an unmistakable appearance because of the short, flag-shaped abdomen on a pole-like petiole. This ***Hyptia harpyoides***, photographed on a log in southern Ontario, has a mite on its right mid tibia.

SUPERFAMILY PROCTOTRUPOIDEA

FAMILY PROCTOTRUPIDAE ❷ This ***Proctotrupes*** probably developed as a parasitoid of a large ground beetle larva. Hosts of proctotrupids are not well known, but known hosts are fly and beetle larvae. **FAMILY PELECINIDAE** (PELECINID WASPS) ❸ ❹ The large (20–70 mm) females of our only pelecinid species (***Pelecinus polyturator***) are common parasitoids of June beetle larvae (*Phyllophaga* spp.); the much shorter males are rarely encountered. **FAMILY DIAPRIIDAE** (DIAPRIID WASPS) ❺ This minute ***Basalys*** sp. is probably a parasitoid of fly pupae. ❻ ❼ Members of this very large family of tiny wasps are mostly parasitoids of flies. The minute (2 mm) flightless diapriids (***Platymischus***) shown here are abundant in seaweed tossed up on ocean beaches, where they parasitize the puparia of wrack flies. One of these photos shows two males trying to mate with one female on top of a wrack fly (*Thoracochaeta*, Sphaeroceridae) puparium. **FAMILY ROPRONIIDAE** ❽ ***Ropronia garmani*** is one of only four roproniid wasp species in North America. These rarely seen but distinctively sculptured and shaped wasps have been reared from sawfly cocoons. **FAMILY VANHORNIIDAE** (VANHORNIID WASPS) ❾ ***Vanhornia eucnemidarum***, one of two North American species in the rare family Vanhorniidae, is sometime seen on dying maple trees. The long, forwardly projecting ovipositor is used to parasitize false click beetle (Eucnemidae) larvae in the wood.

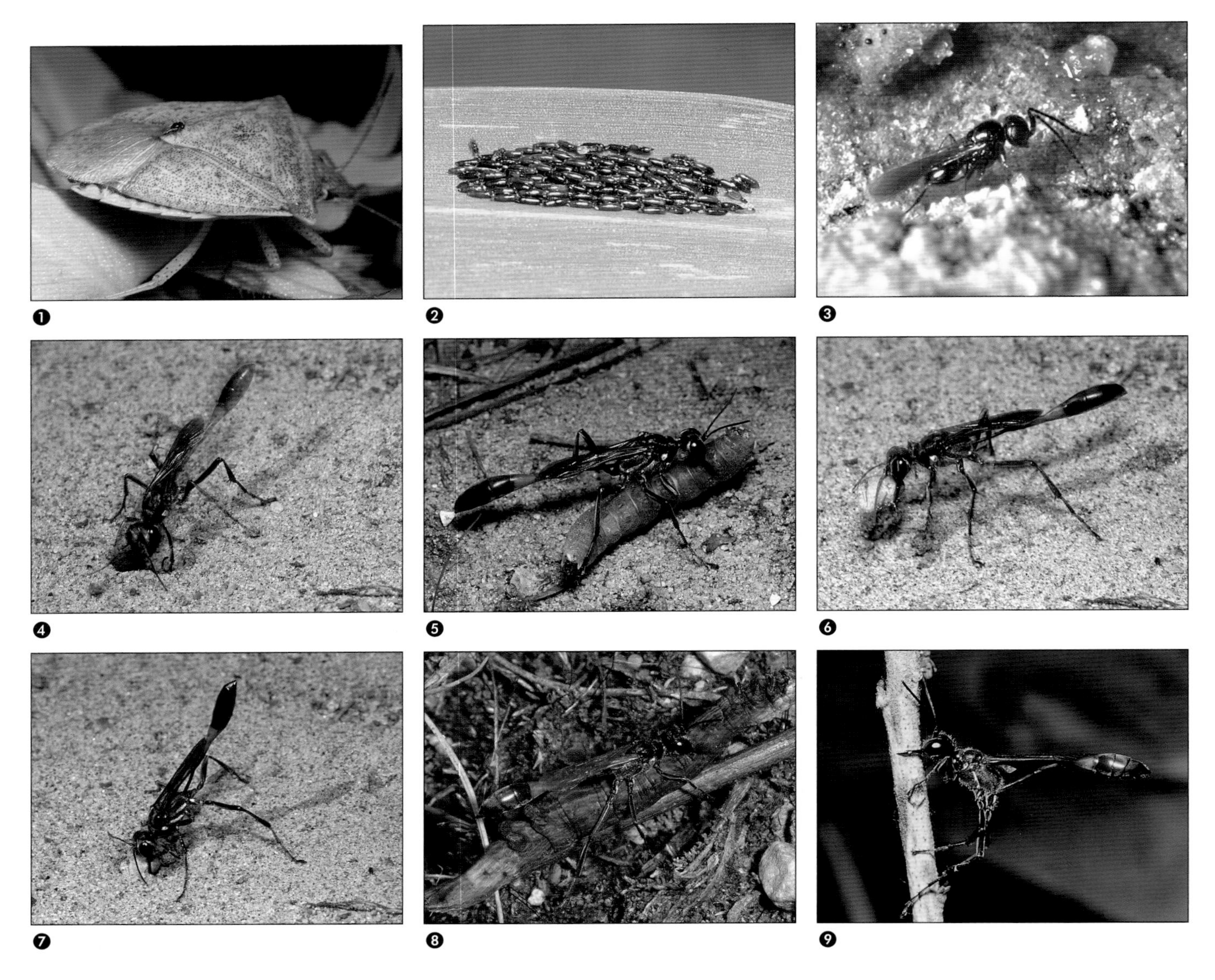

SUPERFAMILY PLATYGASTROIDEA

FAMILY SCELIONIDAE ❶ The Scelionidae is an enormous family of minute egg parasitoids, with an estimated 7,000 species. Some scelionids, like the ***Trissolcus euschisti*** clinging to the back of this stink bug, are phoretic and ride their hosts as they await an opportunity to parasitize their eggs. ❷ Scelionids can usually be found on horse fly and deer fly egg masses, and a really close look at this deer fly egg mass will reveal tiny wasps (***Telenomus emersoni***). ❸ This tiny male scelionid wasp (an undescribed species in the huge genus ***Telenomus***) is near the water line of a mostly submerged rock where females search for eggs of its host, the riffle bug *Rhagovelia obesa*.

SUBORDER ACULEATA (STINGING WASPS, BEES AND ANTS)

SUPERFAMILY APOIDEA (BEES AND RELATED SOLITARY WASPS)

FAMILY SPHECIDAE (THREAD-WAISTED WASPS) ❹ ❺ This ***Ammophila procera*** first dug an underground nest, then stung and immobilized a caterpillar (Notodontidae) before dragging it back to her burrow. ❻ ❼ Most sphecid wasps conceal their prey-filled nests to protect them from kleptoparasites (especially satellite flies, family Sarcophagidae) and other natural enemies. Concealment usually involves kicking sand over the entrance, but some thread-waisted wasps in the genus ***Ammophila*** (like this ***A. procera***) use an appropriate pebble to carefully block the nest entrance. ❽ This ***Ammophila kennedyi*** is struggling to drag a large inchworm (Geometridae) back to her burrow. ❾ Sphecid wasps are sun-loving insects that "sleep" at night or in poor weather. Some wasps sleep in characteristic poses tightly attached to plants; this ***Ammophila urnaria*** is striking a characteristic sleeping pose for the genus.

FAMILY SPHECIDAE (continued) ❶ Like closely related *Ammophila* species, this ***Eremnophila aureonotata*** (the only *Eremnophila* in North America), makes simple ground nests and places a single large caterpillar (usually a prominent moth, Notodontidae) in each nest. Each paralyzed caterpillar will serve as food for a wasp larva that will hatch from an egg attached to the side of the caterpillar. ❷ Sphecid wasps in the genus ***Podalonia***, like this ***P. violaceipennis***, usually hunt for bare caterpillars such as cutworms (Noctuidae), which they dig out of the soil and sting before excavating a nest where they can place an egg and the paralyzed prey. ❸ ***Podium luctuosum*** is a large but rarely collected digger wasp that hunts wood roaches (*Parcoblatta*). ❹❺❻❼ This sequence shows a **Mud Dauber** (***Sceliphron caementarium***) building a cell on its mud nest, loading a paralyzed spider into the completed cell, and capping the cell with freshly gathered mud. The last image is of a cell that has been opened to show some paralyzed spiders and an almost mature Mud Dauber larva. ❽❾ This **Blue Mud Dauber** (***Chalybion californicum***) approached a comb-footed spider in its web, and then paralyzed it with a swift sting to the abdomen. ❿ The **Great Golden Digger** (***Sphex ichneumoneus***) is a common and widespread hunter of large long-horned grasshoppers (Tettigoniidae).

FAMILY SPHECIDAE (continued) ❶ ***Prionyx parkeri*** is a widespread species that stocks each underground nest cell with a single large grasshopper (Acrididae). ❷ Despite its species name, ***Isodontia mexicana*** is widespread in eastern North America and has even been introduced to Hawaii and Europe. It nests in hollow stems and other cavities, which it stocks with paralyzed Orthoptera.
FAMILY CRABRONIDAE (DIGGER WASPS AND THEIR RELATIVES) **Subfamily Pemphredoninae (Aphid wasps)** ❸ This little ***Stigmus*** wasp has provisioned its nest in a dead branch with aphids carried home in her mandibles. Related crabronid wasps in the Pemphredoninae also capture other small prey, including leafhoppers, springtails and thrips. ❹ ***Psen monticola*** often occurs among dense foliage where it searches for homopteran prey, like the paralyzed treehopper (*Entylia carinata*) this female is holding with her middle legs. ❺ ***Diodontus minutus***, one of our smallest crabronids, is not much larger than the aphids on which it preys. Male aphid wasps, like this one, have a silvery face. **Subfamily Crabroninae (Silver mouth wasps, Digger wasps, and organ pipe mud daubers)** ❻ ❼ This female ***Oxybelus*** wasp is using her stinger to carry a paralyzed fly, shish-kebab fashion, back to her nest. Males, like this ***O. emarginatus***, are not so equipped. ❽ This ***Oxybelus bipunctatus*** is about to disappear down her burrow with a paralyzed tachinid fly (*Phasia*) held behind her body. ❾ This brightly colored ***Lestica producticollis*** has stocked her nest – in a rotting log – with small moths. Her nest probably has several cells, each stocked with several moths and one wasp egg.

FAMILY CRABRONIDAE (continued) ❶ Each kind of crabronid wasp has its own preferred prey, which is usually carried in a characteristic fashion. ***Lestica*** species grab small moths from above, seizing the neck with the midlegs. ❷ The male of this pair of ***Lestica confluenta*** is vibrating his wings rapidly, and will continue to do so throughout courtship and copulation. The female will later lay eggs in nests stocked with paralyzed moths in rotting logs or hollow stems. ❸ This female ***Crabro monticola*** is digging out a nest, which she will later stock with numerous paralyzed flies. ❹ Wasps in the large genus ***Crabro*** stock their nests with flies. This female ***C. monticola*** is carrying her prey in typical *Crabro* fashion, gripped with her middle legs. ❺ The front legs of males in the genus ***Crabro***, like this ***C. argusinus***, are often expanded into shieldlike structures that the male places over his partner's eyes during mating. It is thought that females evaluate males by the pattern of light transmitted through the plates. ❻ Crabronid wasps such as this ***Ectemnius cephalotes***, a fly-hunting silver mouth wasp, often frequent Queen Anne's Lace (*Daucus carota*) flowers. ❼ This ***Ectemnius continuus*** has just emerged from a fallen branch, where it developed in a nest packed with paralyzed flies. ❽ The female ***Ectemnius maculosus*** on this goldenrod flower is being mobbed by relatively slender males, seen here hovering over the flower as they await a mating opportunity. ❾ The extra pair of wings appearing to sprout from the abdomen of this ***Ectemnius stirpicola*** female belong to a paralyzed platystomatid fly (*Rivellia*) held beneath her abdomen.

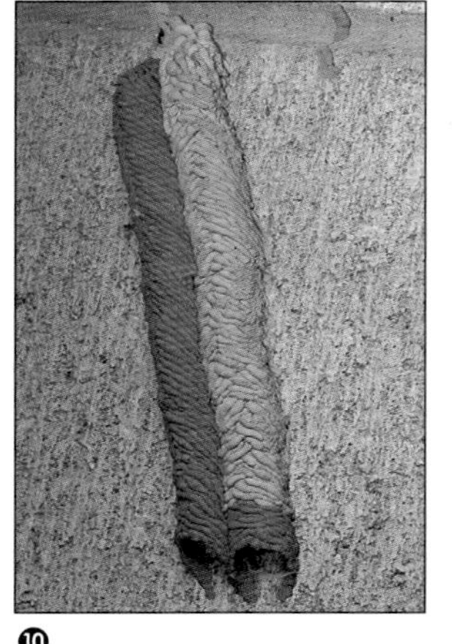

FAMILY CRABRONIDAE (continued) ❶ ❷ The inside of this ***Ectemnius*** nest is packed with paralyzed flies. The wasp larva can be seen feeding on a fly in one of the two cells in one photograph; the other photograph shows flower fly prey and a more mature wasp larva in a single cell. ❸ ❹ **Antenna-Waving Wasps** (***Tachysphex pechumani***), like this female (partly red) and male (all black) are rare wasps, largely restricted to oak savannah habitats where they hunt down grasshopper prey. ❺ ❻ This ***Tachysphex pompiliformis*** is dragging a paralyzed grasshopper into its nest. ❼ This ***Plenoculus davisi*** is carrying a paralyzed mirid bug (*Adelphocoris*). ❽ ***Larra bicolor*** was introduced from South America to help control Southern Mole Crickets (page 79) and other accidentally introduced *Scapteriscus* species (Northern Mole Crickets are attacked by another *Larra* species). Unlike most crabronids, *Larra* wasps are true parasitoids. The host mole crickets are only temporarily paralyzed by the wasp's sting, and resume activity while the wasp larva develops as an ectoparasitoid. ❾ ❿ ⓫ The first of these images shows a ***Trypoxylon politum*** arriving at her nest with a mouthful of mud. The second image shows the completed tubular nest that explains the name "organ pipe mud dauber" and the third shows a nest broken open to expose some paralyzed spiders, wasp cocoons and part of a wasp larva.

FAMILY CRABRONIDAE (continued) ❶ Despite the familiar common name "organ pipe mud daubers," most ***Trypoxylon*** species are cavity nesters, like this ***T. lactitarse*** cleaning up an old beetle burrow in a dead tree. Subfamily Bembicinae (Sand wasps) ❷ Sand wasps are often conspicuously common in open sandy areas, where numerous individuals often nest side by side. This ***Bembix americana*** is heading down a burrow with a large soldier fly (Stratiomyidae) held firmly under her body. She will progressively provision her nest by continuing to add similar flies (including horse flies) as her larva develops, temporarily closing the entrance between visits and sealing it off when the larva is mature and about to spin a cocoon. ❸ ❹ Sand wasps in the genus *Bicyrtes* are similar to those in the much larger genus *Bembix*, but mass-provision their nests (they stock the nest just once, in contrast to progressive provisioning in *Bembix*) with bugs (in contrast to the flies used by *Bembix*). This ***Bicyrtes ventralis*** has brought home a stink bug nymph. ❺ This impressively large **Cicada Killer** (***Sphecius speciosus***) has a conspicuous and intimidating stinger, but it is not an aggressive insect. The sting is used to paralyze the large cicadas this species uses to stock its underground nests. ❻ Sand wasps in the genus ***Microbembex*** commonly nest in large aggregations on open sandy areas from Canada south to Argentina, where they have the unusual habit of progressively provisioning their nests with scavenged dead arthropods. This was one of many ***M. monodonta*** seen carrying dead Stable Flies on a small stretch of Lake Erie beach. ❼ This little sand wasp (***Alysson melleus***) will stock her nest with several leafhoppers (Cicadellidae). Males have a much darker thorax than this female. ❽ Sand wasps in the genus ***Gorytes*** provision their burrows with leafhoppers and related homopterans. This ***G. canaliculatus*** is about to disappear underground with a leafhopper held under her body. ❾ ***Gorytes simillimus*** often nests in sand, stocking each nest with several paralyzed leafhoppers (Cicadellidae). Males, like this one, are more often seen on flowers.

FAMILY CRABRONIDAE (continued) ❶ ***Hoplisoides placidus*** makes small nests in sandy ground, each nest with one to three cells stocked with paralyzed treehoppers (Membracidae). This wasp has a treehopper nymph held beneath her abdomen. ❷ Members of the sand wasp genus ***Nysson*** are kleptoparasites, acting like cuckoos in the nests of *Gorytes* and related sand wasps. This species, ***N. daeckei,*** usually invades the nests of *Gorytes canaliculatus*. ❸ ***Synnevrus plagiatus*** is a rarely encountered wasp that is probably kleptoparasitic on the Golden Digger Wasp (*Sphex ichneumoneus*). ❹ This ***Mellinus abdominalis*** appears to be eating a small fly (Dolichopodidae), which is unusual since most solitary wasps capture arthropod prey only as food for their larvae. *M. abdominalis* is a mostly western species, but this photograph was taken on a sand dune in the Great Lakes (Manitoulin Island). **Subfamily Philanthinae (Bee wolves and weevil wasps)** ❺ Wasps in the genus ***Philanthus*** (bee wolves) provision their nests with bees or wasps. This ***P. sanbornii*** has stung another solitary hunting wasp (a silver mouth wasp) and has set it aside while she reopens a previously constructed burrow. ❻ ❼ This bee wolf (***Philanthus gibbosus***) has captured a halictid bee, which she will place in a nest burrowed into a steep earthen bank. ❽ This bee wolf (***Philanthus lepidus***) is flying toward her burrow with a paralyzed halictid bee held under her body. ❾ This ***Philanthus solivagus*** has set aside her paralyzed prey, a male *Crabro cognatus*, while she cleans up her burrow.

FAMILY CRABRONIDAE (continued) ❶ Although in the "bee wolf" group, this ***Aphilanthops frigidus*** would be better called an "ant wolf," as it provisions its nests strictly with young queen *Formica* ants. ❷ The largest crabronid wasp genus is ***Cerceris***, the weevil wasp genus. About 75 of the 850 or so world species occur in North America. Despite the implications of the common name "weevil wasp," *Cerceris* species stock their nests with a variety of beetles including weevils, leaf beetles, darkling beetles and metallic wood-boring beetles. This is ***C. atramontensis.*** ❸ ***Cerceris nigrescens*** is a true weevil wasp, with a preference for capturing Curculionidae such as alfalfa weevils (*Sitona* spp.). ❹ ***Cerceris fumipennis*** is a jewel beetle (Buprestidae) specialist, and this one has captured a relatively large beetle in the genus *Dicerca*. ❺ This ***Cerceris fumipennis*** is rapidly dropping into her nest burrow with a small jewel beetle (*Agrilus*) held under her body. *Cerceris fumipennis* sometimes nests in aggregations on small patches of appropriately hard-packed, open soil, with each wasp stocking each of several cells with numerous buprestid beetles. **Subfamily Astatinae (Astatine wasps)** ❻ Members of the small solitary wasp subfamily Astatinae provision their nests with Hemiptera (true bugs). This is a male ***Diploplectron peglowi***, a mostly western species that also occurs on dunes around the Great Lakes. Females lack the conspicuous brown spot on the hind wing. ❼ ***Astata*** species provision their nests with stink bugs, and this ***A. unicolor*** is using her mandibles to grip her paralyzed stink bug prey by its beak. ❽ Male ***Astata***, unlike females and unlike other Crabronidae, have enlarged eyes that meet at the top of the head. This is a male ***A. occidentalis***. **FAMILY AMPULICIDAE (COCKROACH WASPS)** ❾ Cockroach wasps have distinctively long mandibles, used by the female to scoop up a cockroach and squeeze it between her mandibles and "nose" (the long clypeus) while stinging it under its thorax. Although they occur as far north as Canada, cockroach wasps are relatively rare in the northeast. ***Ampulex canaliculata*** ranges throughout eastern North America.

FAMILY HALICTIDAE (SWEAT BEES AND THEIR RELATIVES) ❶ Almost all bees develop on pollen, usually carried back to the nest on special structures on the hind legs, as on this halictid bee (***Agapostemon texanus***). ❷ Halictid bees, like this ***Agapostemon texanus***, often nest in the ground, sometimes in large aggregations in which several females will use a single burrow guarded by a single female. ❸ These whitish, legless halictid bee larvae are feeding on yellowish pollen balls in nest cells exposed by peeling bark off a rotting log. Halictid nests usually have a main central tunnel, with branches leading to single cells containing pollen balls and eggs. ❹ This halictid (***Agapostemon***) is using its "tongue" (made up of the maxillae and labium) to suck nectar from a strawberry blossom. ❺ ❻ Brightly striped ***Agapostemon splendens*** males are common in sandy areas such as beaches and sand dunes. This more conservatively colored *A. splendens* female is preparing a burrow on a popular Great Lakes' swimming beach. ❼ Some halictid bees (***Lasioglossum*** spp.) are attracted to sweat, and earn the name "sweat bees" that is sometimes applied to the whole family. ❽ This ***Lasioglossum leucozonium*** is feeding in a chicory flower. ❾ Females of this minute ***Lasioglossum*** species (***L. vierecki***) have an orange abdomen; males are green.

FAMILY HALICTIDAE (continued) ❶ Members of the distinctively colored genus ***Sphecodes*** are kleptoparasites that develop on the pollen supplies salted away in the nests of other halictid bees. Kleptoparasitism is a common strategy, and it is usual for the kleptoparasitic larva to kill off the host's egg or larva before digging into the stored food. *Sphecodes*, however, is unusual because the adult bee kills the host's egg. ❷ This ***Sphecodes*** is entering the burrow of another bee where it will kill the resident egg before laying its own egg on the host's pollen store. ❸ The antennae projecting behind one wing of this kleptoparasitic bee (***Sphecodes***) belong to a male that popped quickly in and out as the female visited a flower. ❹ This ***Sphecodes*** male looks different from the female, which has the usual (for *Sphecodes*) entirely red abdomen. Most of the 80 North American species of *Sphecodes* are similar in color. ❺ The brown larva between the wings of this halictid bee (*Lasioglossum zonulum*) is a triungulin larva of a blister beetle. The bee will inadvertently take the triungulin back to its nest, where the blister beetle larva will kill the bee's egg and eat the stored pollen. **FAMILY ANDRENIDAE** (ANDRENID BEES) ❻ This bee is taking nectar from an early spring gooseberry blossom. Adults in the family Andrenidae, our largest family of bees, hibernate in their burrows and become active earlier than most other bees. ❼ ❽ The genus ***Andrena***, with over 1,300 species worldwide, is our largest bee genus and includes most commonly encountered andrenid bees. Andrenid bees have two sutures running down the face below the base of each antenna (other bees have only one). ❾ The blister beetle larvae on this ***Andrena*** (between the thorax and abdomen) are kleptoparasites that will ride to the bee's nest, where they will kill the *Andrena* eggs and eat its pollen stores.

FAMILY ANDRENIDAE (continued) ❶ Andrenid bees nest in the ground, where they line their cells with waxlike material and stock each cell with a sphere of pollen. ❷ ***Perdita octomaculata*** is a small but colorful andrenid bee common in sandy areas. **FAMILY COLLETIDAE** (PLASTERER BEES AND YELLOW-FACED BEES) ❸ Yellow-faced bees (***Hylaeus*** species; this is a female ***H. modestus***) look a bit like small wasps, a similarity enhanced by the lack of pollen-carrying devices on their almost bare bodies. Female yellow-faced bees swallow pollen when foraging, later regurgitating nectar and pollen into each cell. The bee's egg floats on top of the liquid mixture. Both eastern North American genera of Colletidae (bare *Hylaeus* and hairy *Colletes*) line, or "plaster," the inner walls of their nest cells with cellophane-like shells remarkably similar in composition to sheets of polyester. ❹ Most yellow-faced bees (***Hylaeus*** spp.) have some yellow or white marks on the face, markings that are much more extensive on males (shown here) than on females. ❺ Plasterer bees (***Colletes***) stock each cell with liquid food and then hang one egg from the upper wall of each cell (other bees put their eggs right on top of the gathered pollen). ❻ ***Colletes inaequalis*** are among our earliest bees, and can often be seen forming nesting aggregations on bare soil while the snow is still on the ground. This female has attracted the attentions of two males. **FAMILY MEGACHILIDAE** (LEAFCUTTER BEES, MASON BEES AND THEIR RELATIVES) ❼ Male ***Megachile***, like this colorful ***M. latimanus*** clinging to a Viper's Bugloss (*Echium vulgare*) flower, often have spectacularly enlarged and sculptured front legs. ❽ Leafcutter bees in the genus ***Megachile*** swiftly and neatly cut pieces out of leaves to use as nest lining, sometimes significantly damaging roses and other plants in the process.

FAMILY MEGACHILIDAE (continued) ❶ ❷ ❸ ❹ This leafcutter bee (***Megachile frigida***) has flown to its nest (in a rotting deck board) carrying an almost circular piece of rose leaf, which will be added to other leaf pieces to make the walls of tubular cells. Like other bees, leafcutter bees stock their cells with pollen, but, unlike other bees, they carry the pollen in combs of bristles under the abdomen. ❺ This pointy-tailed megachilid bee (***Coelioxys***) is about to sneak into a *Megachile frigida* nest, where it will lay an egg on the *Megachile* pollen store. All members of the genus *Coelioxys* are food thieves (kleptoparasites) and do not gather their own pollen. ❻ Members of the genus ***Osmia*** are sometimes called mason bees because they make small, earthen cells in various enclosed spaces, like the beetle burrow being used by this one. Some *Osmia* are important pollinators, especially of fruit trees. ❼ ❽ Wool carder bees line their nests with material scraped from the leaves of Woolly Lamb's Ears (*Stachys lanata*) or similar fuzzy plants. This female ***Anthidium manicatum*** has just brought a load of plant fuzz to her nest in the perforated steel leg of a gas barbecue. *A. manicatum* is a recent introduction to North America, probably brought in by accident in wood pallets. ❾ This ***Anthidium oblongatum*** is one of two recently introduced wool carder bees that now occur in northeastern North America. ❿ ***Hoplitis anthocopoides*** is an introduced bee that nests on exposed rock surfaces. This male is standing beside the mortar and pebble nest from which it has just emerged.

FAMILY MEGACHILIDAE (continued) ❶ **Giant Resin Bees** (***Megachile sculpturalis***) were accidentally introduced from Asia to the southeastern United States in the early 1990s and have since spread all the way to southern Ontario. Unlike most other Megachilidae, these large (about 2 cm) solitary bees line their nests with resin, not leaves. **FAMILY MELITTIDAE** (MELITTID BEES) ❷ ❸ Members of the small, rarely encountered family Melittidae nest in the soil like similar andrenid bees. ***Macropis*** species line their nest cells with oil from loosestrife flowers (genus *Lysimachia*), and stock the cells with pollen from the same plants. This female (***M. nuda***) is feeding at a Swamp Candle (*Lysimachia terrestris*) flower, and this male has his head (but not his long antennae) concealed in a partially opened flower on the same plant. **FAMILY APIDAE** Subfamily Xylocopinae (Carpenter bees) ❹ The **Common Carpenter Bee** (***Xylocopa virginica***, 25 mm) looks a bit like a bumble bee with a bare abdomen. ❺ Large carpenter bees (genus *Xylocopa*) nest in solid wood, and females can often be seen going in and out of holes in decking, fences, and similar structures. This **Common Carpenter Bee** is about to enter a circular hole it had previously chewed into the lower surface of a wooden porch beam. ❻ Small carpenter bees (genus ***Ceratina***) are common in early spring, when they gather pollen to stock nest cells in hollowed out stems. Subfamily Nomadinae (Cuckoo bees) ❼ Members of the large genus ***Nomada*** are attractively colored bees with relatively few body hairs and no tools for gathering and transporting pollen. They don't need pollen baskets or dense hairs since they are all pollen thieves, developing as kleptoparasites in the nests of other bees. *Nomada* bees are often conspicuously abundant in early spring, along with the *Andrena* bees that are their main hosts. *Nomada* bees lay two or more eggs in each host cell; the first *Nomada* larva to hatch eats the host's egg and proceeds to develop on the pollen store. ❽ This ***Nomada*** bee is seeking nectar at an early spring willow catkin; you can see the very long tongue that characterizes the Apidae and Megachilidae. ❾ Wasps and bees, like this ***Nomada*** bee, can often be found "sleeping" with their mandibles firmly clasped to a leaf.

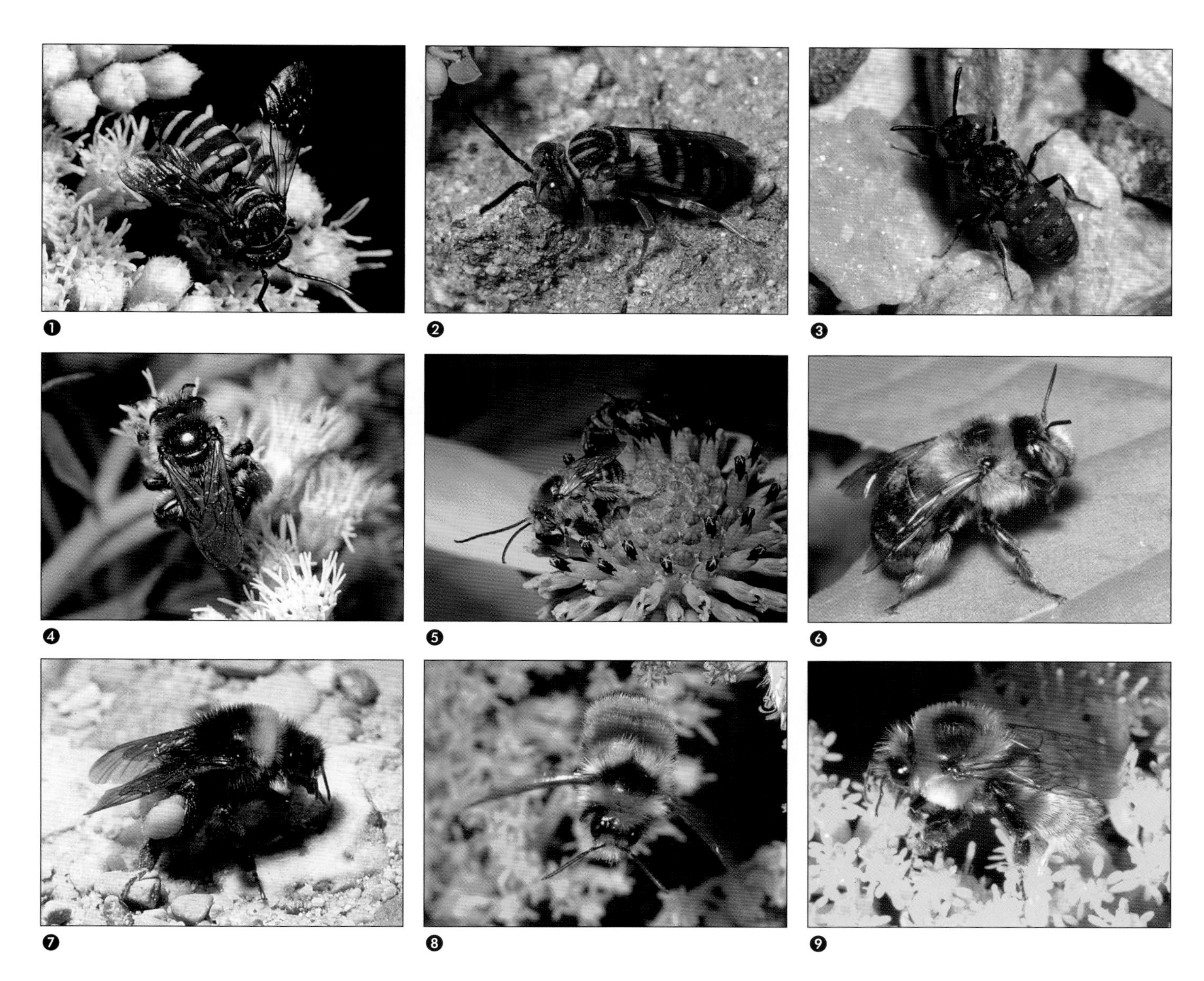

FAMILY APIDAE (continued) ❶ The robust, attractive bees in the genus ***Epeolus*** are common on flowers, although these bees do not gather pollen to feed their larvae. They are kleptoparasites that lay their eggs between the inner and outer linings of plasterer bee (*Colletes*) cells. The larval *Epeolus* will kill the *Colletes* egg and eat the pollen store put in place by its mother. ❷ This kleptoparasitic bee (***Epeolus pusillus***) is searching for the nest of a *Colletes* bee in which to lay her eggs. ❸ ***Holcopasites calliopsidis*** is a kleptoparasite of andrenid bees. This one has landed near an andrenid nest and is hunting with her wings folded out of the way along the sides of her body. **Subfamily Anthophorinae (Digger bees)** ❹ Bees in the genus ***Melissodes***, like this ***M. rusticus***, usually make underground nests with long lateral burrows, each leading to a cell stocked with pollen. ❺ ***Melissodes*** males have unusually long antennae. ❻ **Digger bees** in the large genus ***Anthophora*** (over 450 species worldwide; this is a female of *A. furcata* or *A. terminalis*) dig into soil, where their larvae develop in cells with waterproof linings. They are not social, but sometimes nest in aggregations. **Subfamily Bombinae (Bumble bees)** ❼ This ***Bombus terricola*** has already loaded her pollen baskets and is now taking up a bit of sodium from a wet gravel road. ❽ This ***Bombus ternarius*** is about to land in a goldenrod flower. ❾ Bumble bee males play no role in maintaining the colony. Once they leave the nest they never return, instead spending their nights on flowers.

FAMILY APIDAE (continued) ❶ ❷ ❸ Bumble bees (genus *Bombus*) are important pollinators, and are common sights on a variety of flowers. The species shown here are a ***Bombus ternarius*** on a damaged yellow composite flower, a ***B. affinis*** on a pink *Eupatorium* flower and a ***B. fervidus*** on a Butterflyweed (*Asclepias tuberosa*). ❹ ❺ ❻ ❼ ***Bombus*** species are found on a variety of flowers. These are, respectively, ***B. frigidus*** on Blueweed (*Echium vulgare*), ***B. griseocollis*** on goldenrod (*Solidago*), ***B. vagans*** on coneflower (*Echinacea*) and ***B. borealis*** on clover (*Trifolium*). ❽ Bumble bees are cold-tolerant insects, but this ***Bombus impatiens*** appears to be stupefied as it clings to a flower on a cold fall day. This species, distinctive for the yellow band restricted to its first abdominal segment, is an important pollinator and colonies are often moved around to augment pollination of blueberry and other crops. ❾ Bumble bees are social bees that form small colonies, often in abandoned mouse nests or similar cavities. This is the inside of a ***Bombus impatiens*** colony. The brown spheres under these bees are "honey pots" full of nectar and pollen.

FAMILY APIDAE (continued) ❶ The "black sheep" of the bumble bee subfamily (Bombinae) is the subgenus ***Psithyrus*** (cuckoo bumble bees). Although *Psithyrus* bees look like their relatives in the genus *Bombus*, they lack pollen baskets and do not collect their own pollen. Instead, they lay their eggs in the nests of other bumble bees, leaving them to be raised by the host workers. All *Psithyrus* females are queens, as they have no worker caste of their own. **Subfamily Apinae (Honey bees)** ❷ ❸ The **Honey Bee** (***Apis mellifera***) is one of the most familiar and beneficial of all insects because of its importance as a pollinator and as a source of honey and beeswax. Large feral colonies of this introduced social bee sometimes occur in cavities such as hollow trees. **Subfamily Euglossinae (Orchid bees)** ❹ Orchid bees are common in South and Central America, but only range as far north as Texas. This male has been attracted to a paper soaked with essential oils; he would normally seek similar compounds in orchid flowers. **Subfamily Meliponinae (Stingless bees)** ❺ Stingless bees do not occur in North America, but they abound in South and Central America. These social bees often build up their nest entrances into narrow, easily guarded resinous tubes.

SUPERFAMILY CHRYSIDOIDEA (CUCKOO WASPS AND THEIR RELATIVES)

FAMILY CHRYSIDIDAE (CUCKOO WASPS) ❻ This brilliantly colored and strongly punctate cuckoo wasp (***Chrysis*** sp.) is searching for nests of solitary bees or wasps. Most cuckoo wasps use the telescoping segments at the tip of the abdomen to deposit single eggs in brood cells of solitary bees, where the hatching larvae either attach to the host and develop as ectoparasitoids, or eat the host then move on to eating the host's food supply. ❼ Cuckoo wasps, like this ***Chrysis***, are often seen feeding at flowers or on honeydew. ❽ ***Holopyga ventralis*** is a kleptoparasite in the nests of crabronid wasps (*Bicyrtes*), where it consumes the paralyzed prey after killing the host. ❾ ***Trichrysis tridens*** is a kleptoparasite that develops on spiders stored in the nests of Organ Pipe Mud Daubers (*Trypoxylon politum*, Crabronidae).

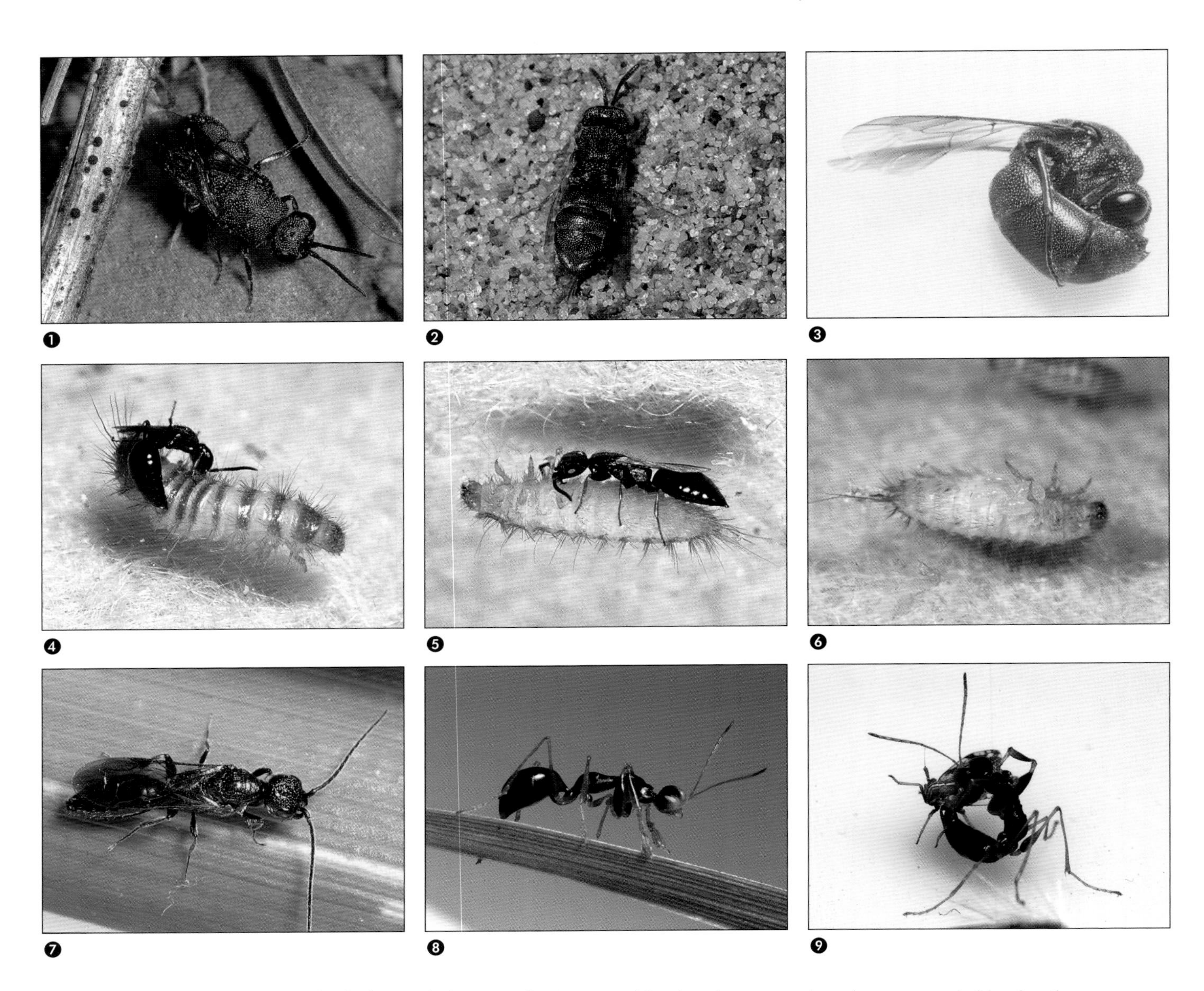

FAMILY CHRYSIDIDAE (continued) ❶ This ***Hedychrum confusum*** is searching for a host ground-nesting wasp, probably a beetle wasp (*Cerceris*, Crabronidae) to parasitize. ❷ ***Parnopes fulvicornis*** is a small cuckoo wasp often seen flying along sandy beaches where its host (*Microbembex monodonta*, Crabronidae) abounds. They land frequently, usually bouncing up and down as they stick their unusually long mouthparts into the sand. *Parnopes* eggs are laid as the host provisions its nest, and the parasitic larvae develop after the host larva has spun its cocoon. ❸ Cuckoo wasps are unique among the aculeate wasps in having the sting greatly reduced, but these spectacularly sculptured wasps are by no means defenseless. The underside of the abdomen is deeply concave so these pretty wasps can roll up like armadillos, entirely encased in impenetrable shining armor. **FAMILY BETHYLIDAE** (BETHYLID WASPS) ❹ ❺ ❻ This bethylid wasp (***Laelius pedatus***) will first sting a carpet beetle larva (*Anthrenus*, Dermestidae) into submission, then use her mandibles to shave part of its hairy underside before laying eggs on the bare patch. Her clutch of larvae will develop as external parasitoids on the carpet beetle larva. ❼ ***Pristocera armifera*** is a parasitoid of larval click beetles (Elateridae). The relatively large (9 mm), fully winged male wasps carry the tiny, wingless, antlike females during their mating flight. **FAMILY DRYINIDAE** (PINCHER WASPS) ❽ Most female dryinids, like this ***Gonatopus***, are equipped with long front legs ending in conspicuous pincers (chelae). ❾ This wingless female dryinid wasp (***Gonatopus***) is attacking a planthopper (*Liburniella ornata*, Delphacidae), using the elaborate pincers on her long front legs to hold her host as she stings it and paralyzes it before injecting an egg between its segments. Her larvae will develop internally, but will protrude cystlike from the host's abdomen for a while before leaving the host to spin a cocoon. Male dryinids have wings and lack the elaborate leg modifications of the female.

FAMILY DRYINIDAE (continued) ❶ ❷ This female ***Pseudogonatopus*** first grabbed a small planthopper nymph with her short, stout mandibles, and then spread the hapless hopper's legs apart using her modified front legs as she bent her abdomen forward to paralyze and parasitize her host. Dryinidae are unique among stinging wasps (Aculeata) in that the stinger (ovipositor) is used to deliver both venom and the egg.

SUPERFAMILY VESPOIDEA

FAMILY POMPILIDAE (SPIDER WASPS) ❸ ❹ ❺ Most spider wasps are dark, smoky-winged wasps, often seen frenetically searching the ground for spider prey. This female ***Arachnospila scelesta*** is making a burrow in the sand, where she will place a paralyzed spider. One larval wasp will develop on this spider. ❻ ❼ Some spider wasps, like this ***Episyron biguttatus***, drag their prey to the general area of the nest, park it on a twig or leaf while the nest is prepared, then return and make a short flight down with the retrieved prey. ❽ This ***Anoplius*** species (subgenus *Pompilinus*) is exhibiting typical pompilid behavior as it walks backward, dragging a relatively huge paralyzed victim gripped in its mandibles. ❾ This ***Anoplius*** has just captured an amaurobiid spider on a rocky lakeshore.

FAMILY POMPILIDAE (continued) ❶ Spider wasps in the genus ***Auplopus*** (and a few close relatives in the small tribe Auplopodini) trim the legs off their victims for ease of transport. These same wasps have the unusual habit of building mud nests, a habit not found in other spider wasps. Mud cells are constructed in concealed places like hollow logs before spiders are captured, and capped with mud once the paralyzed prey and wasp egg are in place. ❷ This ***Auplopus*** has only partially trimmed the appendages from her paralyzed spider prey. ❸ ***Auplopus mellipes*** is unusual among species of *Auplopus* because it usually makes its mud nests high on the trunks of living trees. ❹ This ***Auplopus architectus*** was seen going in and out of a shelter under a small overhanging rock on a Georgian Bay cliff, presumably an appropriate place for a mud nest stocked with a paralyzed spider. ❺ ***Agenioideus humilis*** is found both on natural cliffs, like this escarpment, and on the artificial clifflike habitats provided by rock walls and masonry. ❻ ***Agenioideus humilis***, the only native *Agenioideus* in northeastern North America, preys on spiders in the family Araneidae (orb weavers), in contrast with the introduced *A. cinctellus*, which hunts Salticidae (jumping spiders). ❼ ***Agenioideus cinctellus*** is a European spider wasp recently discovered in northeastern North America. This one is dragging its jumping spider prey up the wall of an old stone building. ❽ This kleptoparasitic spider wasp (***Evagetes crassicornis***) has just emerged from the burrow of another spider wasp, where she has left eggs on paralyzed prey placed there by the other wasp. ❾ Some spider wasp species specialize in stealing spiders from other spider wasps en route to their nests, and some, like this distinctively colored ***Ceropales maculata***, are kleptoparasites on their relative's prey. When a *Ceropales* female spots another spider wasp hauling a paralyzed spider back home, she sneaks up and slips an egg into the spider's book lungs. Her egg hatches first, and her larva kills the egg of the host wasp before developing on the host's prey.

❶ ❷ ❸ ❹ ❺ ❻ ❼ ❽ ❾ ❿

FAMILY VESPIDAE (YELLOWJACKETS, HORNETS AND THEIR RELATIVES) Subfamily Eumeninae (Potter wasps and mason wasps) ❶ **Mason wasps (*Monobia quadridens*)** stock nest cells with paralyzed caterpillars, dividing the cells from one another with mud partitions. This large (2 cm) wasp sometimes takes over abandoned nests of other solitary wasps and bees, but usually nests in wood borings. ❷ ❸ ❹ Potter wasps in the genus *Eumenes*, like this ***E. crucifer*** on goldenrod flowers and this ***E. fraternus*** on *Eupatorium* flowers, are strikingly shaped wasps that usually make elegant potlike nests. ❺ One of the two adjacent cells in this potter wasp (***Eumenes***) nest has been broken open to show the paralyzed caterpillar prey and the developing potter wasp larva. ❻ ***Eumenes verticalis*** is unusual for its inconspicuous mud nests made on exposed rock surfaces. ❼ This mason wasp (***Euodynerus foraminatus***) is gathering a ball of mud that it will use to construct partitions between brood cells in its nest, probably in beetle-bored dead wood. ❽ This ***Ancistrocerus antilope*** is gathering a ball of mud or clay for use in dividing a hollow twig into brood cells stocked with paralyzed caterpillars. ❾ This ***Ancistrocerus waldenii*** is using mud to build up a nest in the crevasse of a large rock. No other *Ancistrocerus* builds this kind of exposed nest. ❿ This ***Ancistrocerus catskill albophaleratus*** is carrying a paralyzed caterpillar back to its nest.

FAMILY VESPIDAE (continued) ❶ Most ***Ancistrocerus*** nest in pre-existing cavities such as hollow stems or abandoned burrows made by other Hymenoptera. This species (***A. unifasciatus***) nests in abandoned mud dauber (Sphecidae) nests. ❷ The male of this pair of mason wasps (***Ancistrocerus antilope***) is the one with the yellow face. ❸ Some mason wasps, like this ***Ancistrocerus antilope***, chew their way into silk and leaf shelters made by caterpillars, snatching caterpillar prey from inside their shelters. ❹ This ***Ancistrocerus adiabatus*** is carrying an unusual load of true parasites, called twisted-winged parasites (Strepsiptera). The relatively flat structure sticking out between the wasp's abdominal tergites is a female who will never leave her host; the fat pupa farther back on the abdomen will emerge to a short-lived and fully winged male parasite. ❺ Although most Eumeninae stock their nests with caterpillars, ***Symmorphus*** species like this one usually provision their nests with leaf beetle (Chrysomelidae) larvae. Some *Symmorphus* tear leaf-mining larvae (beetles and moths) right out of their mines. ❻ Potter and mason wasps are often common on flowers, and this goldenrod flower has attracted at least two genera, ***Euodynerus foraminatus*** and the smaller ***Parancistrocerus pedestris***. ❼ This ***Parancistrocerus pensylvanicus*** is feeding in an American Bittersweet (*Celastrus scandens*) flower. ❽ Potter wasps are frequent flower visitors, and this ***Euodynerus leucomelas*** has become covered with dogwood pollen. ❾ Unlike most other potter wasps, ***Stenodynerus kennicottianus*** nests in the hard-packed soil of vertical bluffs and similar habitats.

FAMILY VESPIDAE (continued) ❶ ***Zethus*** is a mostly tropical genus, but ***Z. spinipes*** ranges north to Massachusetts. **Subfamily Polistinae (Paper wasps)** ❷ The northeastern paper wasps, all of which belong to the genus ***Polistes***, make simple, open, nests hung from single stalks. These ***P. dominula*** are nesting under a wood deck. ❸ ***Polistes dominula*** is a recent immigrant to our area, and has only become common in the last few years. ❹ The common native paper wasp of northeastern North America, ***Polistes fuscatus***, occurs throughout our area. ❺ ***Polistes bellicosus,*** like other paper wasps, are hunters that feed masticated prey, such as this long-horned grasshopper, to developing larva. ❻ This ***Polistes metricus*** is chewing on the end of a cut stem, presumably gathering fiber for her paper nest. *Polistes metricus* is found throughout the eastern United States but only reaches southernmost Canada (Point Pelee). **Subfamily Vespinae (Social wasps or yellowjackets)** ❼ This enormous (about 5 cm) **European Hornet** (***Vespa crabro***) queen differs from yellowjackets in its larger size and orange-brown colors. The European Hornet, North America's only true hornet, was introduced from Europe and is now one of the largest wasps in North America. Nests can be found in hollow trees, sheds and other sheltered sites. ❽ This **European Yellowjacket** (***Vespula germanica***) queen was found in an apple orchard as it sought a sheltered place to overwinter. European Yellowjackets have spread rapidly throughout the northeast over the past 30 years, and are now common nuisance wasps that often establish large colonies in enclosed spaces like attics and walls. ❾ This overwintering **European Yellowjacket** queen was exposed under the bark of a fallen tree.

FAMILY VESPIDAE (continued) ❶ Although yellowjackets are mostly predacious, **Common Yellowjacket** (***Vespula vulgaris***) workers will gather food at almost any protein source, such as this dead bird. Carrion feeding is rare among Hymenoptera other than *V. vulgaris* and related vespids, a few stingless bees and some ants. ❷ This **Common Yellowjacket** (***Vespula vulgaris***) queen was exposed by breaking open a rotting log in the middle of winter. She has already mated, and when spring rolls around she will become active and start a new colony, probably underground. ❸ This **Eastern Yellowjacket** (***Vespula maculifrons***) is feeding on a plum. This ground-nesting or cavity-nesting species is a common nuisance wasp, especially late in the season. ❹ Some yellowjackets, like these **Eastern Yellowjackets** (***Vespula maculifrons***), usually nest underground, although this species will also nest in hollow walls and similar cavities. ❺ This **Hybrid Yellowjacket** (***Vespula flavopilosa***) has been attracted to sugary water sprayed on the foliage. Hybrid Yellowjackets nest underground and are rarely abundant enough to become a serious nuisance. ❻ **Southern Yellowjacket** (***Vespula squamosa***) queens do not normally establish their own nests, but instead usually usurp nests from other yellowjackets and use host workers to rear their first broods (it is a temporary social parasite). This individual is a male. Despite the name, this species has been recorded as far north as southern Ontario (Point Pelee National Park). ❼ ❽ This **Blackjacket** (***Vespula consobrina***) worker is arriving at its nest in a dead tree. Blackjackets hunt only for living prey. ❾ This **Blackjacket** (***Vespula consobrina***) male has been attracted to an introduced orchid (Helleborine) where it has picked up a sticky pollen packet on its head.

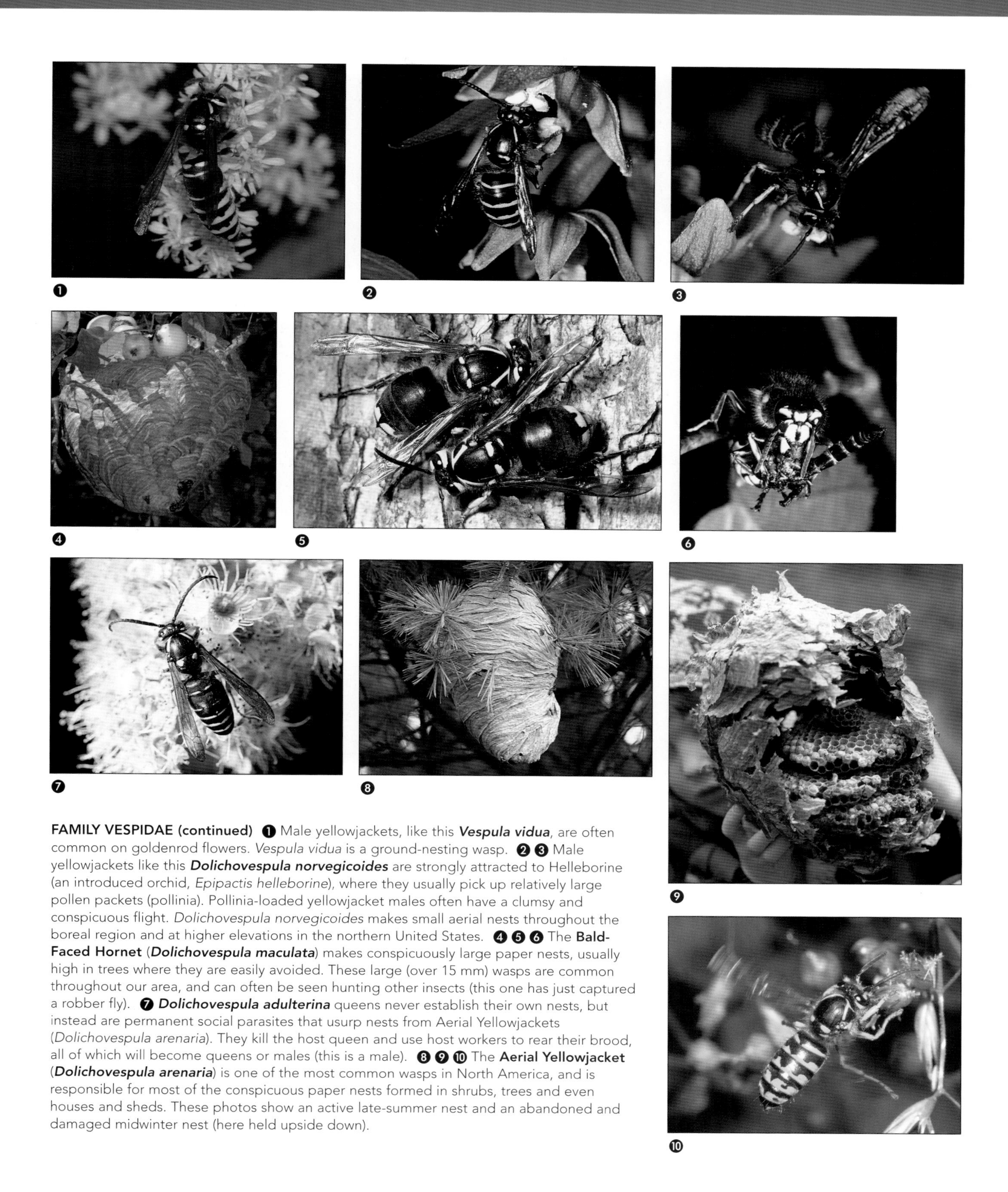

FAMILY VESPIDAE (continued) ❶ Male yellowjackets, like this ***Vespula vidua***, are often common on goldenrod flowers. *Vespula vidua* is a ground-nesting wasp. ❷ ❸ Male yellowjackets like this ***Dolichovespula norvegicoides*** are strongly attracted to Helleborine (an introduced orchid, *Epipactis helleborine*), where they usually pick up relatively large pollen packets (pollinia). Pollinia-loaded yellowjacket males often have a clumsy and conspicuous flight. *Dolichovespula norvegicoides* makes small aerial nests throughout the boreal region and at higher elevations in the northern United States. ❹ ❺ ❻ The **Bald-Faced Hornet** (***Dolichovespula maculata***) makes conspicuously large paper nests, usually high in trees where they are easily avoided. These large (over 15 mm) wasps are common throughout our area, and can often be seen hunting other insects (this one has just captured a robber fly). ❼ ***Dolichovespula adulterina*** queens never establish their own nests, but instead are permanent social parasites that usurp nests from Aerial Yellowjackets (*Dolichovespula arenaria*). They kill the host queen and use host workers to rear their brood, all of which will become queens or males (this is a male). ❽ ❾ ❿ The **Aerial Yellowjacket** (***Dolichovespula arenaria***) is one of the most common wasps in North America, and is responsible for most of the conspicuous paper nests formed in shrubs, trees and even houses and sheds. These photos show an active late-summer nest and an abandoned and damaged midwinter nest (here held upside down).

FAMILY VESPIDAE (continued) ❶ This close-up of some brood cells of an active **Aerial Yellowjacket** (***Dolichovespula arenaria***) nest shows several capped cells and a couple of adult wasps as well as a larva, a pupa, and new adult pulled out of their cells; another pupa has been pulled partly from a cell. **Subfamily Masarinae (Pollen wasps)** ❷ ❸ Pollen wasps are rare in eastern North America, and this pair (***Pseudomasaris zonalis***) was photographed in western Canada (the one with long, yellow antennae is the male). Unlike other North American Vespidae, these unusual solitary wasps stock their burrows or mud nests with pollen and nectar. **FAMILY TIPHIIDAE (TIPHIID WASPS)** ❹ Brightly colored adult tiphiid wasps in the genus ***Myzinum*** are common on flowers; larvae develop as parasitoids on larval scarab beetles. ❺ The antlike females of ***Methocha*** enter burrows occupied by tiger beetle larvae, using their size and shape to slip below the tiger beetle's formidable jaws before stinging it on the neck and thorax. This female ***M. stygia*** will lay a single egg near the base of a tiger beetle larva's hind leg, where her larva will develop as a parasitoid. ❻ This fully winged male ***Methocha stygia*** is much larger than the wingless female held firmly in his mandibles. ❼ Most of our tiphiid wasps belong to the genus ***Tiphia***, species of which attack larval June beetles and other scarab larvae that live underground. One *Tiphia* species was introduced to North America for the biological control of Japanese beetles. **FAMILY SIEROLOMORPHIDAE** ❽ Sierolomorphids, like this ***Sierolomorpha canadensis*** from southern Ontario, look similar to tiphiid wasps, but these rare, poorly known wasps are thought to comprise the most primitive group of vespoids. Nothing is known of their biology. **FAMILY RHOPALOSOMATIDAE (RHOPALOSOMATID WASPS)** ❾ Both males and females of ***Olixon banksii*** (one of only three North American species in the rarely encountered family Rhopalosomatidae) have tiny, paddle-shaped wings. Larval rhopalosomatids develop as parasitoids of nymphal crickets.

FAMILY SCOLIIDAE (SCOLIID WASPS) ❶ The large (20–50 mm) wasp dwarfing the syrphid fly on this honeysuckle flower is ***Campsomeris plumipes***, a scoliid that often abounds in open, sandy areas where it flies low over the ground in search of good places to seek subterranean scarab beetle larvae, on which its larvae will develop as ectoparasites. The fine, parallel wrinkles at the wing tip are distinctive, and allow easy recognition of this family. **FAMILY SAPYGIDAE** (SAPYGID WASPS) ❷ Northeastern species in the small family Sapygidae (this is a ***Sapygina*** sp.) are kleptoparasites that lay their eggs in the nests of leaf cutter bees (Megachilidae). Newly hatched wasp larvae kill the host egg and develop on the stored nectar and pollen. These generally uncommon wasps sometimes kill large numbers of Alfalfa Leafcutter Bees (*Megachile rotundata*), a valuable and commercially propagated pollinator of alfalfa. **FAMILY MUTILLIDAE** (VELVET ANTS) ❸ This day-active little velvet ant female (***Pseudomethoca frigida***) is looking for nests of halictid bees, where its larvae will develop as parasites. ❹ This large velvet ant (***Dasymutilla occidentalis***) develops in the nests of bumblebees (*Bombus fraternus*). Members of the large genus *Dasymutilla* (more than 100 North American species) often attack nests of solitary wasps; females are reputed to have some of the nastiest stings of all insects. ❺ ***Dasymutilla nigripes*** occurs throughout the eastern United States, where it invades the nests of *Philanthus* and other sand wasps. ❻ ❼ ***Dasymutilla vesta*** is a widespread velvet ant that develops in nests of various sand-nesting bees and wasps. Female velvet ants are wingless and antlike but males look more like fuzzy spider wasps. ❽ ***Myrmosa unicolor*** parasitizes larvae of small ground nesting wasps and bees. **FAMILY FORMICIDAE** (ANTS) Subfamily Formicinae ❾ Ants have at least one distinctive node between the thorax and the main part of the abdomen, clearly visible and scale-like in this ***Formica*** worker.

FAMILY FORMICIDAE (continued) ❶ ❷ Ants in the genus ***Formica*** are our most commonly observed ants, especially those species that make large mounds containing tens of thousands of workers. The most common medium-sized (3–6 mm) ants seen in flowers, like this Grass of Parnassus (*Parnassia palustris*) flower, are *Formica* species. ❸ Although a member of the carpenter ant genus, ***Camponotus subbarbatus*** is not a pest. This species only ranges as far north as extreme southern Ontario (Point Pelee). ❹ These **Black Carpenter Ant** (***Camponotus pennsylvanicus***) workers are tending a treehopper for sugary honeydew. Black Carpenter Ants are among our most common ants, and sometimes damage wood structures as they excavate their nests. Carpenter ants are general scavengers and they do not eat wood. ❺ This **Black Carpenter Ant** (***Camponotus pennsylvanicus***) was killed by *Desmidiospora myrmecophila*, a rarely seen *Camponotus*-specific fungus. This individual, found under a rotting log on an alvar on Ontario's Bruce Peninsula, represents the third observation of this fungus since it was discovered in Connecticut in 1891 (the other records are from Idaho and Arizona). ❻ Carpenter ants often tend aphids, treehoppers and other homopterans for honeydew, and they are also associated with larvae and pupae of lycaenid butterflies. This ***Camponotus noveboracensis*** is tending a butterfly (Lycaenidae) pupa. ❼ ***Camponotus noveboracensis*** is a widespread and very common carpenter ant that usually situates its colonies in rotting wood. ❽ Many ants, like these ***Lasius***, have a close relationship with aphids, and tend them in exchange for their sugary excrement (honeydew). ❾ These ants (***Lasius*** sp.) are protecting a blue butterfly (Lycaeinidae) caterpillar from parasites. Many lycaenid larvae secrete an exudate to attract ants.

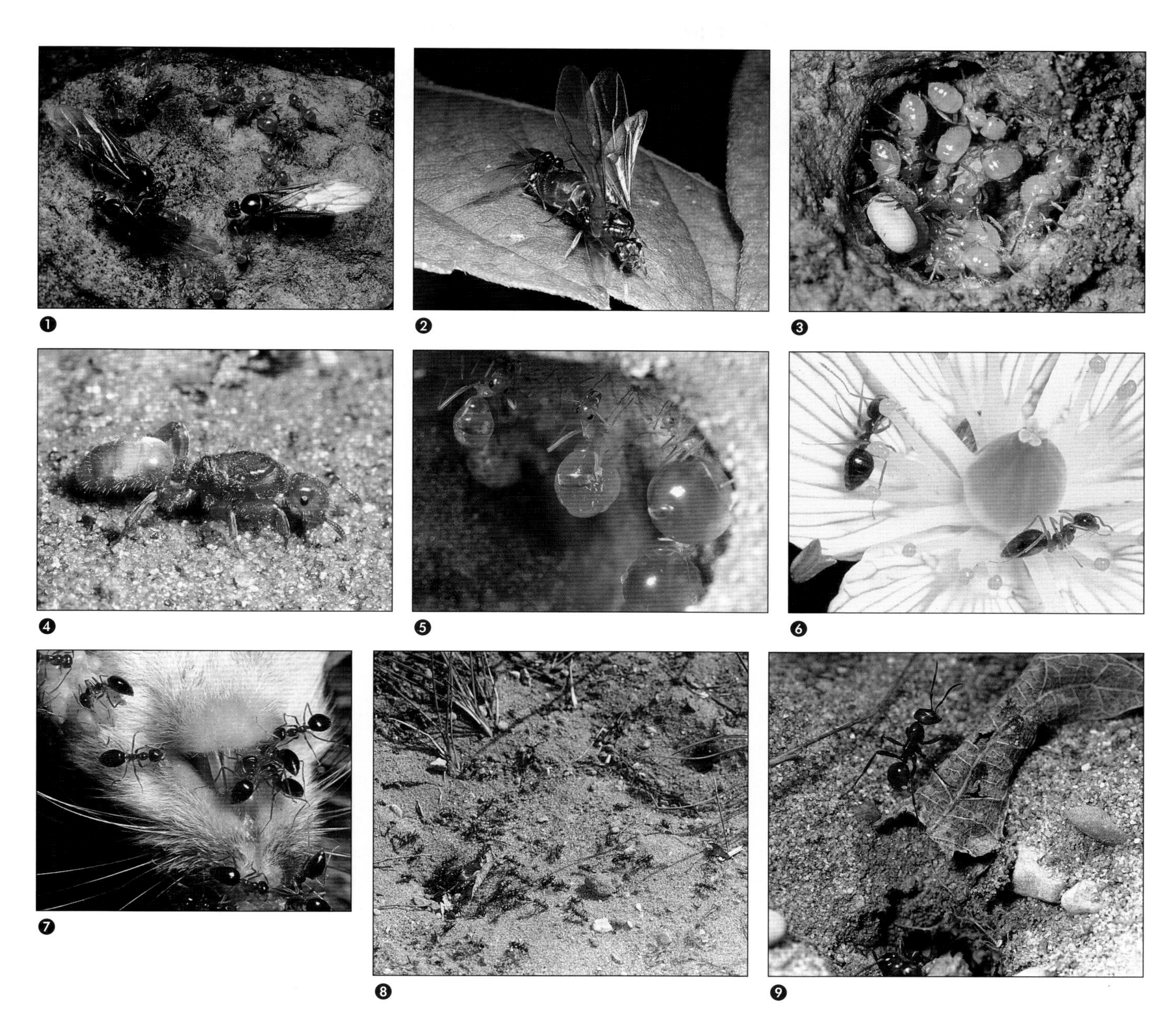

FAMILY FORMICIDAE (continued) ❶ ❷ Most ants are sterile wingless female workers, but colonies reproduce by periodic production of large numbers of winged males and females called reproductives. This enormous ***Lasius*** queen is in copula with a relatively minute male. He will die shortly, and she will shed her wings and start a new colony. ❸ Most of the little yellow ants found under stones or underground are either *Lasius* or the very closely related *Acanthomyops. Acanthomyops*, some of which are known as "foundation ants" for their harmless habit of nesting in house foundations, emit a strong citronella-like odor when disturbed (or crushed). The white object held in the jaws of one of these ants is an ant cocoon. ❹ This ***Acanthomyops latipes*** queen has shed her wings and is seeking a *Lasius* ant colony where she will kill the defending queen and use the colony's worker force to raise her own brood (temporary social parasitism). ❺ Honeypot ants (***Myrmecocystus*** spp.) occur in western North America, and only reach the northeast in zoos and insectaria. Colonies of these remarkable ants include specialized individuals (a storage caste) that hang from the ceiling of the nest and serve as living honey pots, storing quantities of fluids (nectar, honeydew, water) for their sisters. ❻ ❼ ***Prenolepis imparis***, the only native North American *Prenolepis*, is a common, small (2–4 mm) ant that makes small nests in the ground and feeds on all sorts of things including household foods, carrion (like this dead rodent) and flowers (like this Grass of Parnassus, *Parnassia palustris*). ❽ ❾ The handsome reddish ants in the genus ***Polyergus*** are slave-making ants, seen here swarming into a *Formica* nest where they will snatch the *Formica* pupae and bring them home to rear as slaves.

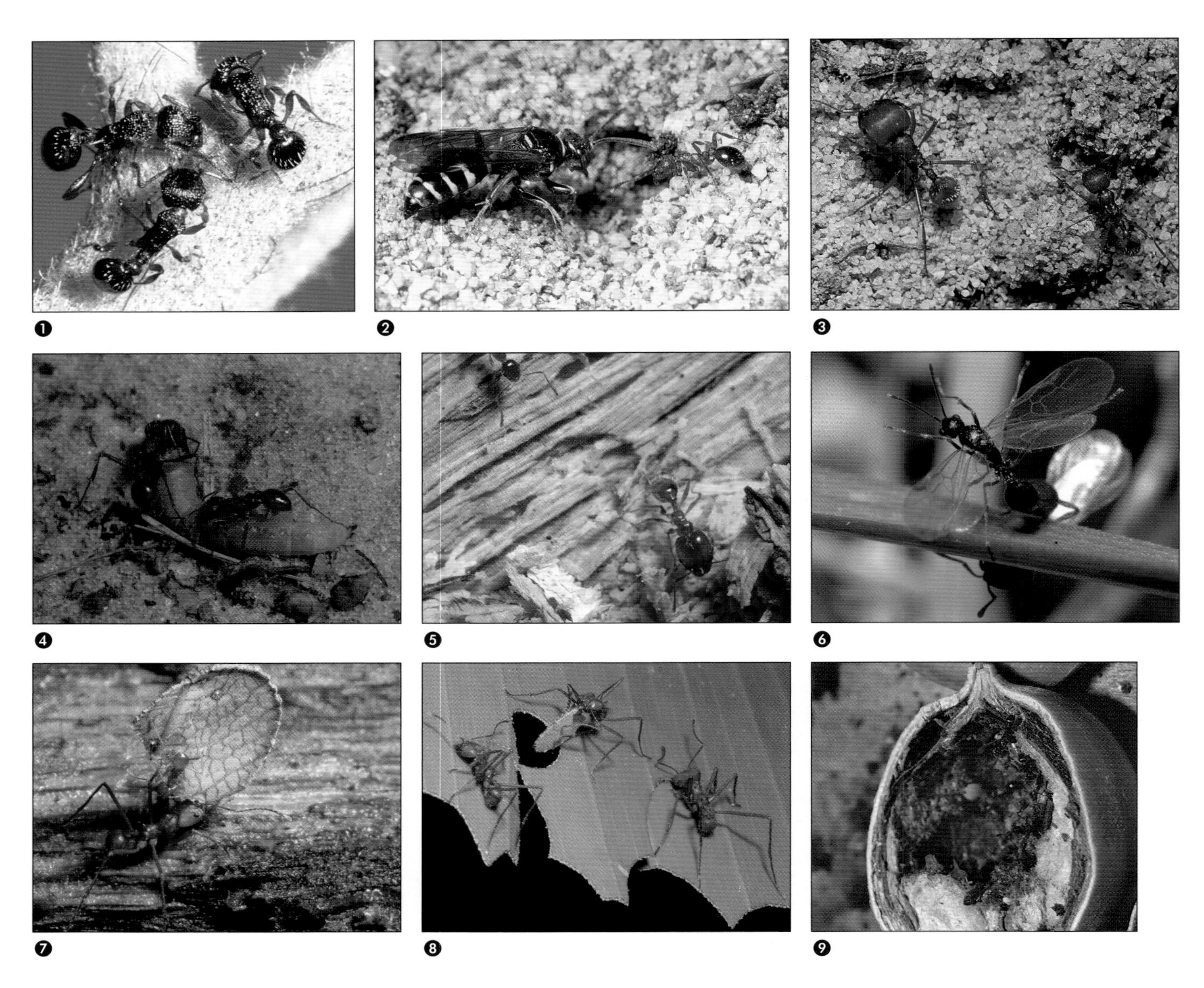

FAMILY FORMICIDAE (continued) Subfamily Myrmicinae ❶ Some plants have close relationships with ants and provide shelter or food for ants that defend the plant from its enemies, assist with dispersal or provide nutrients for the plant. Ant-plant relationships are most elaborate in the tropics, but many common local plants ranging from roses to oaks have inconspicuous extrafloral nectaries that produce sweet exudates to attract ants. The extrafloral nectaries on this Ontario bracken fern are obviously attractive to ***Myrmica*** ants. ❷ This ***Myrmica*** ant seems to have been attracted to a burrow made by a sand wasp (*Gorytes canaliculatus*). ❸ The big-jawed ants in the genus ***Pogonomyrmex*** are called harvester ants because they stock their nests with harvested seeds. "Pogos," which pack a powerful sting, are primarily distributed in the west and south, and only one species (the **Florida Harvester Ant**, ***P. badius***), occurs east of the Mississippi River. These were photographed in North Carolina, near the northeastern limit for this species. ❹ These harvester ants (***Pogonomyrmex badius***) are dragging a small caterpillar back to the nest. This species has both large-headed and normal sized workers, with the former well adapted to mill seeds, although insects make up about 50 percent of this species' diet. The tiny ant in the background is a *Dorymyrmex*. ❺ **Big-headed Ants** (***Pheidole megacephala***), like the larger Florida Harvester Ants, have specialized big-headed ("megacephalic") workers able to crush seeds brought back to the nest by smaller workers. ❻ This pair of ***Stenamma*** ants (only the male is clearly visible in this picture) got together during a mating flight, and will soon complete their copulation and shed their wings. The mated queen will establish a small colony. ❼ ❽ The small ant riding a leaf fragment cut and carried by this leafcutter ant is a specialized worker dedicated to defending her larger sister from phorid fly parasitoids. The leaf fragment will be used to feed a fungus garden tended by even smaller workers deep in an enormous underground nest. This is a Costa Rican ant in a genus (***Atta***) that only ranges as far north as the southern United States, but a less conspicuous relative (the Northern Fungus Gardening Ant, *Trachymyrmex septentrionalis*) ranges north to New York. ❾ Some ants have remarkably tiny colonies, and some species of ***Temnothorax*** routinely house their entire colonies in acorns previously inhabited by acorn weevils. The entrance to this ***T. longispinosus*** colony was a hole left by an acorn weevil (*Curculio*). Another species of *Temnothorax* (*T. duloticus*) is a slave-raider that attacks these colonies.

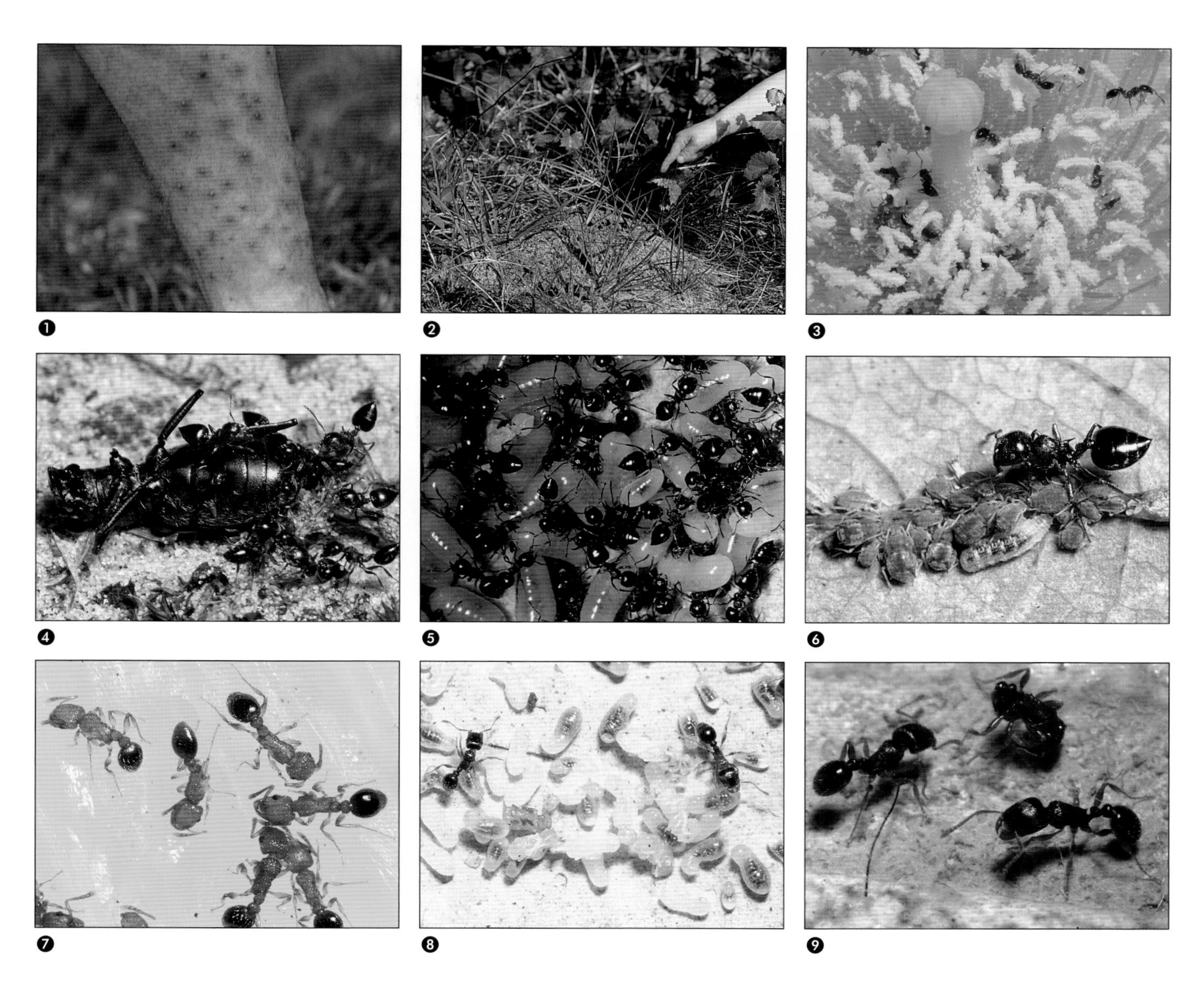

FAMILY FORMICIDAE (continued) ❶ ❷ The **Red Imported Fire Ant** (***Solenopsis invicta***) is a destructive pest accidentally introduced from Brazil around 1945, and now common as far north as North Carolina. These photographs, showing a fire ant mound and my leg after stepping on said mound, were taken in Florida. A native *Solenopsis* species (*S. molesta*, the Thief Ant) occurs throughout the northeast where it is a household pest. ❸ These tiny fire ants are in a prickly pear cactus (*Opuntia*) flower. ❹ ***Crematogaster*** ants, distinctive for their short and heart-shaped abdomens, are commonly seen tending aphids or scavenging on dead insects like this tiger beetle. ❺ These workers and larvae of ***Crematogaster cerasi*** are part of a large colony found under a stone. Despite their small size, these ants can inflict a painful bite. Unlike most Myrmicinae, *Crematogaster* ants have a greatly reduced sting. ❻ Many ants, like this ***Crematogaster***, are attracted to honeydew-producing homopterans. The ants provide some protection to the homopterans in return, but note the predacious flower fly larva in the midst of these aphids, almost below the ant's head. ❼ ***Tetramorium bicarinatum*** is an introduced ant common in greenhouses. ❽ ❾ An introduced *Tetramorium*, the **Pavement Ant** (***T. caespitum***) nests in sidewalks or under paving stones, boards and similar disturbed habitats. This little (about 3 mm) ant can be a household pest as it forages on a variety of foodstuffs.

FAMILY FORMICIDAE (continued) ❶ These ***Aphaenogaster*** (an undescribed species near *A. rudis*) were exposed by peeling bark from a rotting log; some of the workers are rescuing young larvae, and another larva is visible among a stockpile of seeds (and a dead wood louse). *Aphaenogaster* are common ants throughout eastern deciduous forests. Many plants induce ants to transport and store their seeds by producing seeds with attractive edible outgrowths call elaiosomes. Subfamily Dolichoderinae ❷ ***Dolichoderus mariae*** is a common ant in Sphagnum bogs, where it nests in soft Sphagnum hummocks. ❸ The genus ***Tapinoma*** includes two species in the northeast, the common and widespread Odorous House Ant (*T. sessile*) and the tiny **Ghost Ant** (***T. melanocephalum***) shown here swarming over a dead butterfly in an Ontario greenhouse. The Ghost Ant is an introduced tropical species, restricted to heated buildings in temperate climates. Subfamily Ponerinae ❹ The distinctively elongate, little (2–4 mm) ants in the genus ***Ponera***, like this ***P. pennsylvanica***, are predators on other insects, often under bark or in the same rotting wood as the small (usually less than 50 workers) colonies. ❺ ***Amblyopone*** species, like this ***A. pillipes***, form small, simple colonies. Some taxonomists treat *Amblyopone* in a separate subfamily, the Amblyoponinae. ❻ Trap jaw ants (***Odontomachus***) are mostly tropical ants, but they regularly show up as adventives in tropical greenhouses attached to zoos or insectaria. The mustache-like piece across the front of this ant's head is actually an outstretched pair of mandibles able to snap shut when triggered by fine hairs between the mandibles. Subfamily Pseudomyrmecinae ❼ The only North American genus in the Pseudomyrmecinae (***Pseudomyrmex***) occurs in the wild only as far north as North Carolina, but these distinctive ants sometimes show up as accidental introductions to greenhouses (such as butterfly houses) throughout our region. Members of this large, mostly tropical genus nest in preformed cavities such as hollow twigs. Subfamily Ecitoninae ❽ One small New World Army Ant (*Neivamyrmex carolinensis*) ranges north to Illinois; this is a soldier of a more spectacular tropical species, ***Eciton burchelli***, a common army ant from South America north to Mexico.

Non-insect Arthropods

Most terrestrial and freshwater invertebrate species, and all winged invertebrates, are insects. Insects belong to a larger group of invertebrates characterized by external skeletons and the corresponding attribute of jointed legs, as reflected in the formal name "Arthropoda" (*arthro* = joint, *poda* = feet).

The phylum Arthropoda is traditionally divided into two subgroups. One subgroup (Mandibulata) includes insects, crustaceans and other arthropods that have mandibles on the second body segment. The other subgroup (Chelicerata) includes spiders and other arthropods that lack mandibles, but have piercing structures called chelicerae as the first pair of appendages. All terrestrial and freshwater chelicerates belong to the Class Arachnida (spiders, mites, scorpions, pseudoscorpions and their relatives), but the terrestrial and freshwater mandibulates are divided into several groups that different authors recognize at different ranks (so you might see the same group in the literature as a subphylum, a superclass, a class or a subclass). The main non-insect mandibulates that share habitats with insects are the millipedes, centipedes and crustaceans.

This book is primarily about insects, but many non-insect arthropods are routinely confused with insects or encountered in the same environments as insects, so photographs and a few comments are provided here for the most common groups.

Millipedes, centipedes and other myriapods (Classes Diplopoda, Chilopoda, Pauropoda and Symphyla)

Millipedes (class **Diplopoda**, page 590) are familiar, conspicuously multi-legged arthropods usually found in damp places, where they slowly glide their elongate, often cylindrical bodies along on dense rows of relatively short legs. Most of their body segments have two pairs of legs, giving a total of 30 or more pairs. Millipedes are generally scavengers that do not bite (unlike some superficially similar predacious centipedes), but many release noxious or toxic fluids when handled. Some species defend themselves with potent poisons, including hydrogen cyanide. Very few millipedes have any economic impact; although one introduced species (*Oxidus gracilis*) is a common pest in greenhouses.

Centipedes (class **Chilopoda**, page 592) are elongate and multi-legged like millipedes, but each body segment has only a single pair of legs, and some common species have as few as 15 leg pairs. Unlike millipedes, centipedes are swift predators, armed with poison jaws (the modified first pair of legs), which they use to immobilize insects and other small prey. Although some of the larger centipedes (order Scolopendromorpha) that occur in the tropics and the southern United States can inflict a painful bite, northeastern centipedes defend themselves in other ways. The long-legged house centipedes (*Scutigera coleoptrata*, order Scutigeromorpha) rely on impressive speed; the common stone centipedes (order Lithobiomorpha) use their distinctively long hind legs to flick sticky material at attackers; and the long, skinny soil centipedes (order Geophilomorpha) are more inclined to coil up and exude repellent chemicals than to strike out with their large poison jaws.

Although essentially all routinely encountered elongate, multi-legged terrestrial invertebrates are either centipedes or millipedes, two other classes of multi-legged mandibulates are considered as "myriapods." These two classes, the **Pauropoda** (page 591) and **Symphyla** (page 591), are small, rarely noticed groups normally found under logs, in leaf litter and other moist environments. Pauropods are tiny (about 1 mm), pale myriapods with odd antennae split into three branches. Symphylans are a little larger (1–8 mm) and resemble small, pale centipedes. One species of symphylan is known as the garden symphylan (*Scutigerella immaculata*) because it sometimes damages plant roots.

Myriapods have long been considered the insects' closest relatives, but recent work suggests that insects are more closely related to crustaceans than to myriapods, and that myriapods might be more closely related to chelicerates than to insects and crustaceans. Other evidence suggests that the Myriapoda is not even a "natural" group. Even so, "myriapod" remains a convenient term with which to refer to the "myriapodous" (many legged) arthropods.

Crayfish, woodlice and other crustaceans (Class Crustacea)

Most **Crustacea** (page 594) are marine organisms, but some share the freshwater and terrestrial habitats otherwise overwhelmingly dominated by insects. The largest of these are the crayfish, members of the same order (Decapoda) as the crabs, shrimp and lobsters that can be purchased at most supermarkets. Lobster-like crayfish share their freshwater environments with several smaller aquatic crustaceans, most of which are minute, easily overlooked organisms such as the water fleas (order Cladocera; often much less than 1 mm). Fairy shrimps (order Anostraca) are relatively large (around 25 mm) freshwater crustaceans with distinctively stalked eyes and 11 pairs of swimming legs. They appear each spring in temporary woodland pools, where they swim on their backs and feed on microorganisms.

Another shrimp-like freshwater crustacean is the Opossum Shrimp (*Mysis relicta*). This species, in the otherwise almost entirely marine order Mysidacea, is a glacial relict surviving in deep northern lakes on both sides of the Atlantic. It occurs naturally in deep northeastern lakes, and has been accidentally introduced to lakes in the western United States.

The most common macroscopic (over 3 mm) aquatic crustaceans belong to two somewhat similar orders; one made up of depressed (flattened top to bottom, like cockroaches) organisms called isopods (order Isopoda), and the other made up of compressed (flattened side to side, like fleas) organisms called amphipods (order Amphipoda). Both are primarily marine groups with relatively few representatives in terrestrial or aquatic environments.

An isopod, *Caecidotea*, page 594, caption 10.

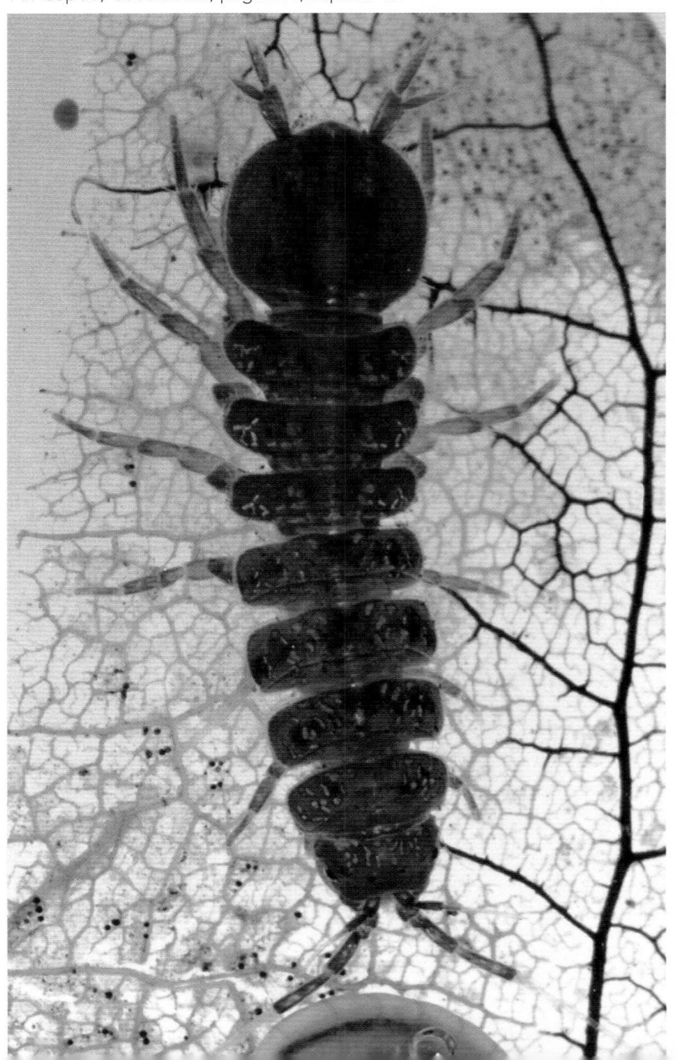

The Isopoda includes the familiar terrestrial woodlice or sowbugs, as well as some species found in fresh water. The northeastern amphipods, on the other hand, are associated either with marine, marine shoreline or freshwater environments. Most amphipods and isopods are omnivores or scavengers, but some feed on living plants.

The class Crustacea is now widely considered to be an artificial construct (not a natural group), and will undoubtedly be split up into two or more classes in the future. The closest relatives to the insects are probably among the crustaceans.

Spiders, mites, scorpions and pseudoscorpions (Class Arachnida)

Spiders, mites, ticks and their relatives (class **Arachnida**, page 596) are easily distinguished from insects by their lack of antennae and by the division of the body into at most two (rather than three) subdivisions. They also usually have more legs than insects.

The most commonly noticed arachnids belong to the order Araneae (spiders), an easily recognized group in which the body has a narrow waist, dividing it into a soft, unsegmented hind part (the abdomen) and a harder, flatter front part (the cephalothorax) that includes the head and legs.

Spiders breathe using trachea much like most insects, but with only one spiracle, which opens on the underside of the abdomen just in front of the spinnerets. Most spiders also have book lungs, comprised of plates or leaves (like the leaves of a book) in a saclike pocket that opens under the base of the abdomen, well in front of the spinnerets. The spinnerets, lobelike appendages under the tip of the abdomen, are the spinning organs from which liquid silk is released. Newly produced silk usually hardens immediately into a tough, insoluble fiber used to capture prey, to make shelters, to float through the air and for a variety of other functions, but it sometimes remains viscous like the sticky globs on the trap lines incorporated into the webs of orb-weaving spiders. Depending on how the silk is drawn out and spun together into compound threads, it can be incredibly strong – much stronger than steel, and highly elastic.

Male spiders, like male dragonflies and damselflies, take an indirect approach to fertilization. Prior to mating, a male spider must transfer sperm from his genitalia to secondary structures on his highly modified pedipalps, which are armlike appendages situated behind the chelicerae and before the first pair of walking legs. The process of getting sperm from the genitalia to the pedipalps (called sperm induction) is incredibly variable among spiders, and sometimes involves stroking of the genitalia using the chelicerae or special-purpose silk structures to incite ejaculation into silken receptacles from which the sperm can be transferred to the pedipalps. Once the pedipalps are loaded, the male spider is faced with the potentially dangerous task of getting close enough to insert them into his predacious mate's copulatory opening (epigynum). Different spiders approach this task in different ways, often involving elaborate courtships and only rarely ending in the postnuptial consumption of the hapless male.

All spiders eat other organisms, and they often interact with insects as predators, prey or hosts. Although almost all have venom glands that open near the tips of their chelicerae, only a few spiders are dangerous to humans. Widows (genus *Latrodectus*, family Theridiidae), the only significantly venomous spiders in northeastern North America, can inflict painful bites followed by non-local effects such as difficulty with breathing and severe abdominal pain (bites are rarely fatal, and there is an antivenin). There are two species of widow in eastern North America. The Southern Black Widow (*L. mactans*) ranges from New England south to Florida, and the Northern Widow (*L. variolus*) occurs through much of eastern North America including southern Canada.

North American species of the small venomous spiders known as violin spiders or recluse spiders (genus *Loxosceles*, family Sicariidae) are mostly midwestern in distribution, but

Crab spider (*Misumena vatia*) and bee.

one species (the Brown Recluse, *L. reclusa*) has been recorded a few times (usually in houses) in the northeastern United States. Bites from this small, retiring spider are rare (occasionally occurring while victims are putting on clothing), but may induce localized tissue damage that can take a long time to heal. The Brown Recluse Spider has never been collected in Canada and probably does not occur that far north.

Most spiders, including all North American spiders, are divided into two groups (infraorders) – the Mygalomorphae and Araneomorphae. Members of the small, relatively primitive infraorder Mygalomorphae (tarantulas and their relatives) are usually large spiders with chelicerae that operate in a vertical plane, like a snake's fangs. A large tarantula with its fangs raised into a striking position is an intimidating sight, but tarantulas are normally neither aggressive nor very venomous. The most northerly orthognath spiders hunt from within silken tubes called purse-webs. Prey are captured through the web, and then dragged right through the silken wall into the shelter. The wall is soon repaired, and the remains of the victims are later discarded through the end of the tube.

Most spiders belong to the other spider infraorder (Araneomorphae or "true" spiders), in which the chelicerae work against each other in a horizontal plane. True spiders are divided into over 100 families, and a review of all true spider families is beyond the scope of this book. The most commonly encountered families are the jumping spiders, wolf spiders, fishing spiders, crab spiders, orb weavers, long-jawed spiders, funnel-web spiders, comb-footed spiders and sheet-web spiders.

Harvestmen (order Opiliones) are often confused with spiders, but these generally long-legged arachnids have a round or oval body that does not have a spider-type waist dividing it into two sections. Unlike all other arachnids, in which males use indirect methods to transfer a sperm package to the female, harvestmen have a specialized elongate organ called a penis, used to directly deliver spermatozoa into the female's gonopore. Most harvestmen are predators, but some are scavengers.

Northeastern North American harvestmen, which belong to a subgroup known as "daddy long-legs," are familiar and conspicuous arthropods, especially in woodlands and particularly in late summer and fall. They cannot puncture your skin, and can be handled with impunity, although they often respond to handling (or threats from a predator) by breaking off a couple of long legs, which continue to twitch and quiver well after their owner has departed. Defensive shedding of appendages ("autotomy") also occurs in many other arthropods, including spiders, but harvestmen provide the most familiar example of the phenomenon.

Mites and ticks form the enormous subclass Acari, with around 40,000 species abounding in a variety of habitats including soils, plants and animal hosts. Early stages ("larvae") have only three pairs of legs like insects, but later stages ("nymphs" and adults) usually have four pairs. Unlike insects or spiders, in ticks and mites the body is not divided into distinct regions.

The mites most likely to bring themselves to your attention are members of the family Trombiculidae. Trombiculids have parasitic immature stages (larvae) called chiggers. Unlike their parasitic larvae, trombiculid adults are predators, feeding on other small arthropods. Chiggers normally attack a variety of wildlife, attaching themselves to a host's skin and secreting saliva that digests a feeding crater in which the chigger feeds on partially digested tissue and lymph.

Some species that abound in shrubby, disturbed areas are notorious for "accidentally" attacking humans rather than their normal wild hosts. Chiggers only survive a day or two on humans, but many an insect collector has woken up the morning after a day of walking in scrubby bush to find numerous red, itchy swollen bites in the armpits, groin, ankles and anywhere clothing fits tightly against the skin.

A nursery web spider, *Pisaurina mira*, page 603, caption 6.

Most chigger bites in the northeast are caused by larvae of *Eutrombicula* species that normally feed on amphibians, reptiles, birds or small mammals. Unlike ticks, chiggers are not implicated in the transmission of any human diseases in North America.

Most of the dozens of families of Acari are made up of minute and inconspicuous mites, but two families of Acari contain relatively large species, known as ticks, that are parasitic during both immature and adult stages (ticks have four stages: egg, larva, nymph, adult). One of these families, the Ixodidae (ixodid ticks or hard ticks), includes most of the ticks that bite people. Ixodid ticks are distinctive for the hard plates that cover the top of the front half of the body, in contrast with the soft ticks (Argasidae) that often infest poultry, other birds or bats.

Adult and immature ticks often seek hosts by hanging out on vegetation and grabbing on with their long forelegs when heat, sight, smell or vibrations tip them off to the presence of a potential host. Ticks tend to be opportunistic, and immature ticks that usually feed on lizards and birds will often latch on to people, as will adults of tick species that normally attack deer. Given the chance, most adult female ticks will feed until they are grossly swollen (sometimes more than 100 times their unfed weight) before dropping off and (much later) laying thousands of eggs.

Ticks usually crawl around your clothing and skin for a while before inserting their mouthparts, so you are likely to spot them either before they have attached or at least before they have had much chance to feed. This is a good thing, since your chance of getting a tick-borne disease is minimized by early removal of ticks. Ticks that have managed to attach should be removed by using fine tweezers to grab the tick's mouthparts as close to your skin as possible; otherwise you could be left with some bits and pieces under your skin that might cause irritation. It is a good idea to wear light-colored clothing and long pants tucked into your boots in tick-infested areas. The same repellents used for protection from mosquitoes will give some protection against ticks.

Ticks are important carriers of disease, including the much publicized Lyme disease, a bacterial infection that has been spreading through the northeast and which has affected over 100,000 North Americans since it was first recognized in the mid-1970s. If you are unlucky enough be bitten by an *Ixodes* tick (usually a deer tick or blacklegged tick) carrying this disease, you might see an expanding bulls-eye-shaped rash around the bite, then experience flu-like symptoms. That is usually as far as it goes, but untreated patients sometimes go on to develop late-stage Lyme disease, with symptoms including cardiac and arthritic problems. It is a good idea to remove ticks as promptly as possible, just in case.

The more common dog ticks or wood ticks (*Dermacentor* species) do not carry Lyme disease, but if they are allowed to feed for a period of time they can transmit another serious bacterial disease called Rocky Mountain spotted fever. Although now more common in eastern North America than western North America, this disease was named for where it was first discovered, and for its characteristic symptoms.

Members of the remaining chelicerate groups are small and infrequently encountered in northeastern North America. Pseudoscorpions (order Pseudoscorpiones) are common, but are typically overlooked because of their small size (usually 3–5 mm). These little predators usually occur under bark, often hitching rides from habitat to habitat on flies and beetles. One pseudoscorpion species (*Chelifer cancroides*) is a common household hunter most likely to come to your attention when accidentally trapped in a bathtub or sink. North American true scorpions (order Scorpiones) are primarily western and southern in distribution.

A predacious mite, page 597, caption 10.

CLASS DIPLOPODA (MILLIPEDES)
ORDER JULIDA

❶ ❷

FAMILY JULIDAE ❶ In millipedes, each trunk segment is "doubled up" so there are two pairs of legs on most segments. Julids, like this ***Ophyiulus pilosus***, have a convex, almost cylindrical shape. As with other Julidae in the northeast, this is an introduced species.
FAMILY PARAJULIDAE ❷ Elongate, convex parajulid millipedes are common under bark. Parajulid segments lack either striations or distinct constrictions like those found in other families in the order Julida.

CLASS DIPLOPODA (MILLIPEDES)
ORDER SPIROBOLIDA

❶ ❷

FAMILY SPIROBOLIDAE ❶ ❷ ***Narceus americanus*** is a common, distinctively colored, large (up to 100 mm) and convex millipede often seen in open areas.

❶

❷

❸

FAMILY PARADOXOSOMATIDAE ❶ ***Oxidus gracilis*** (about 20 mm) is a common introduced pest found in greenhouses. This species does not occur outdoors in Canada or the northern United States. **FAMILY POLYDESMIDAE** ❷ ***Pseudopolydesmus*** species are distinctively colored, flattened millipedes with an ability to roll up into a spiral when disturbed. Polydesmid millipedes are the only millipedes that use cyanide in their defensive secretions. **FAMILY XYSTODESMIDAE** ❸ Xystodesmids are large, flattened millipedes that cannot curl up into a spiral like similar polydesmids. ***Sigmoria trimaculata*** is a distinctive species that reaches almost 40 mm in length.

CLASS SYMPHYLA (SYMPHYLANS)

❶

❶ Members of the small class Symphyla are rarely noticed because of their small size (1–8 mm) but occur under logs, under rocks, or even in garden soils where one species sometimes damages roots of plants.

CLASS PAUROPODA

❶

❶ Myriapods in the small class Pauropoda, distinctive for their peculiar branched antennae, are common arthropods in leaf litter and under bark although they usually escape attention because of their small size (around 1 mm).

CLASS CHILOPODA (CENTIPEDES)
ORDER LITHOBIOMORPHA

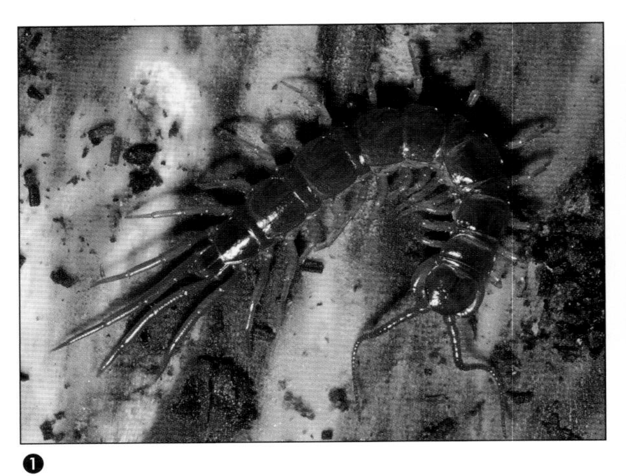

❶

❷

❶ Lithobiomorph centipedes are common brown centipedes with relatively few short legs (15 pairs in adults; hatchlings with 6–8 pairs). They occur under stones and bark, and some defend themselves by launching sticky blobs with their hind legs. ❷ Lithobiomorph centipedes are common predators of subcortical (under bark) larvae. This one is eating a flat bark beetle larva (Cucujidae).

CLASS CHILOPODA (CENTIPEDES)
ORDER GEOPHILOMORPHA

❶

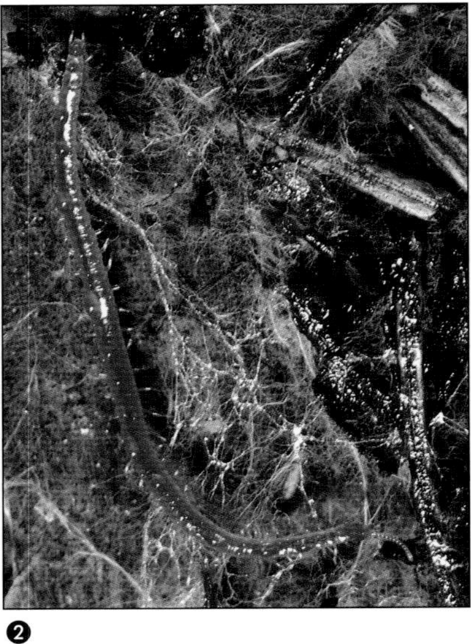

❷

❸

❶ ❷ Geophilomorph centipedes are elongate, eyeless centipedes with lots of legs – at least 29 pairs. They are common in rotting logs and are also found in the soil. ❸ Some centipedes exhibit maternal care, and this female actively herded her young when disturbed.

❶

❶ Scolopendromorph centipedes are mainly southern, and are absent from most of the northeast. They are among the largest centipedes in North America (up to 15 mm) and inflict a painful bite.

❶

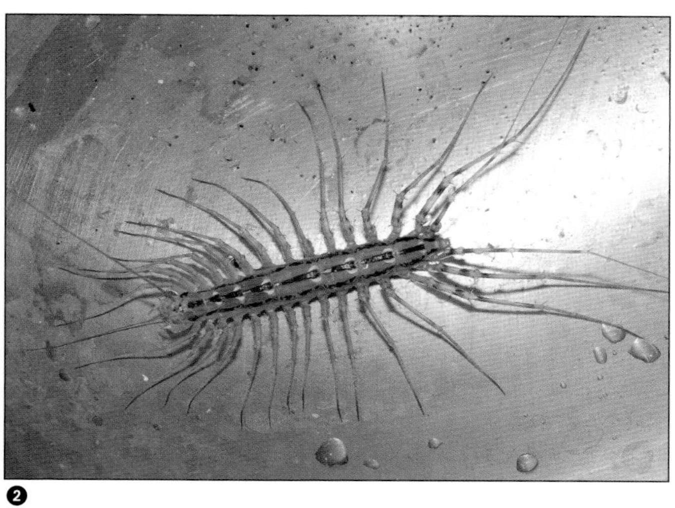

❷

❶ ❷ The **House Centipede** (***Scutigera coleoptrata***) is a common household predator that moves with incredible speed on its very long legs. As is true for most domestic arthropods, this is an introduced species.

❶ ❷ ❸ ❹ ❺ ❻ ❼ ❽ ❾

FAMILY ONISCIDAE ❶ ❷ Isopods are small (up to 20 mm) crustaceans, distinctive for their top-to-bottom flattened appearance. ***Porcellio scaber*** and ***P. spinicornis*** (shown here respectively) are two common isopods, locally known as "woodlice" or "sowbugs." *Porcellio spinicornis* is common on limestone or cement surfaces, including foundation walls and rocky areas. Most northeastern terrestrial isopods are introduced from Europe. ❸ Woodlice have the remarkable habit of molting in two stages, starting with the back half. This one (***Trachelipus rathkii***) has just cast the skin from the front half, which is still pale and soft. ❹ The **Common Striped Woodlouse** (***Philoscia muscorum***) is sometimes seen scavenging for dead insects on leaf surfaces. ❺ ***Oniscus asellus*** is a common woodlouse in disturbed areas. **FAMILY LIGIIDAE** ❻ ***Ligidium elrodii*** occurs throughout eastern North America, and is most common under objects on moist soil near springs. **FAMILY ARMADILLIDIIDAE** ❼ ❽ ❾ ***Armadillidium vulgare***, the common pillbug, is one of two species in our area that can roll up into a ball. Female isopods have a pouch called a "marsupium" (like a kangaroo). Eggs are laid into the marsupium, and when they hatch the newly emerged pillbugs spend from a few hours to several weeks in the marsupium before breaking free. **FAMILY ASELLIDAE** ❿ All of the freshwater isopods in the northeast, like this ***Caecidotea***, belong to the family Asellidae and are restricted to slower parts of springs, streams and underground water, where they feed as general scavengers. There is only one other genus of freshwater isopod in the northeast, and it is much less common than *Caecidotea*.

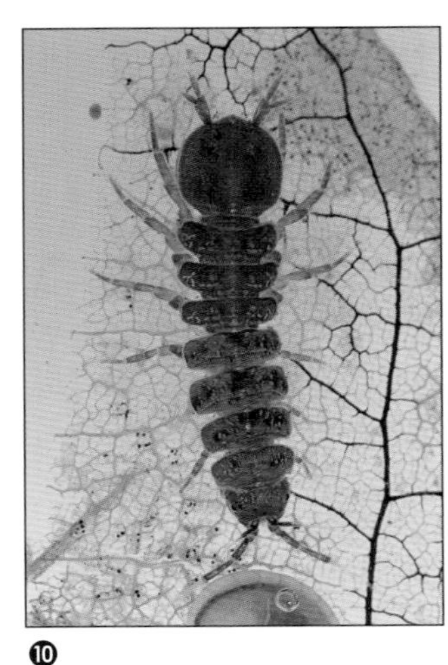

❿

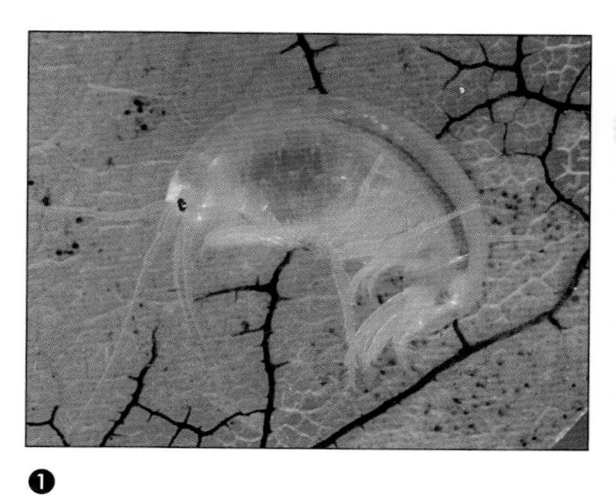
❶

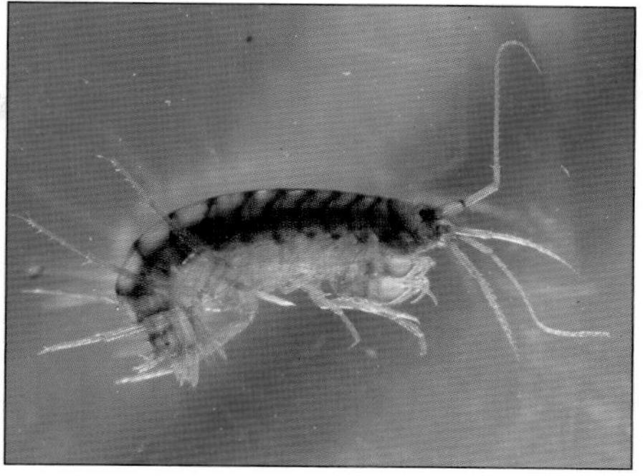
❷

❸

FAMILY CRANGONYCTIDAE ❶ ❷ Amphipods, also known as scuds or side-swimmers, are flattened side-to-side. Scuds are often seen in pairs, with the female riding on the male's back. Like the related isopods, amphipods have a marsupium into which eggs are laid and the new hatchlings remain for a week or so. This ***Crangonyx*** is one of around 50 North American aquatic amphipod species.
FAMILY TALITRIDAE ❸ Sand fleas or sand hoppers (***Orchestia platensis***) are among the most abundant seashore arthropods, especially among seaweed washed up along the Atlantic coast.

CLASS CRUSTACEA (CRAYFISH, SHRIMP AND OTHER CRUSTACEANS)
ORDER DECAPODA

❶

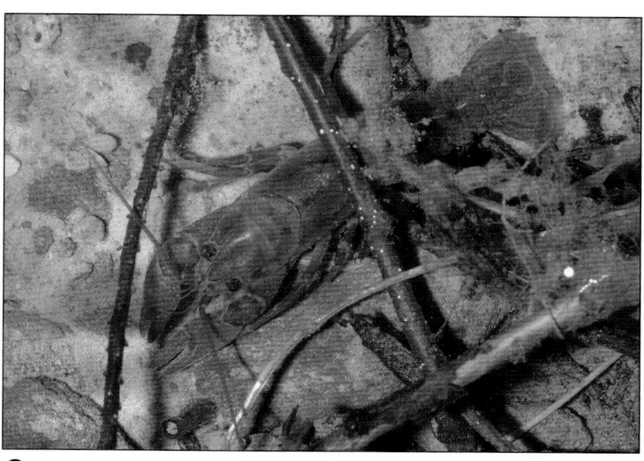
❷

FAMILY CAMBARIDAE (CRAYFISH) ❶ ***Orconectes propinquus*** is one of less than a dozen northeastern crayfish, over half of which are in the genus *Orconectes*. This species is common in open waters, but many other crayfish are usually or typically found in burrows. ❷ The **Eastern Crayfish** (***Cambarus bartoni***) is the common crayfish of shallow stream bottoms of the Appalachian mountains and small streams of the Canadian Shield.

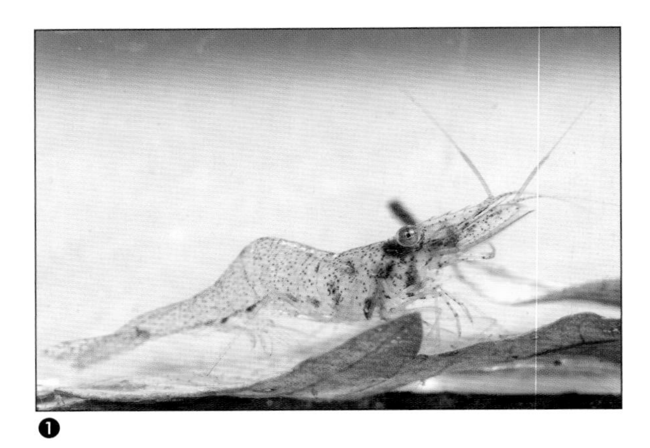

❶

FAMILY PALAEMONIDAE (FRESHWATER PRAWNS AND GLASS SHRIMP) ❶ The **American Freshwater Glass Shrimp** (***Palaemonetes paludosus***) is common in fresh waters near the Atlantic coast from New Jersey south to Florida.

❷

FAMILY BRANCHINECTIDAE (FAIRY SHRIMPS) ❷ Fairy shrimp appear each spring in temporary woodland pools, where they swim on their backs and feed on microorganisms. The female of this pair has a pouch full of eggs (an ovisac) that she will soon drop into the mud; there they will either hatch immediately (summer eggs) or go into a resting stage (winter eggs). Thick-walled winter eggs have been kept in dried pond mud for as long as 15 years and still hatched when added to water.

CLASS ARACHNIDA (SPIDERS, MITES, TICKS AND THEIR RELATIVES) SUBCLASS ACARI (MITES AND TICKS)

❶ ❷ ❸ ❹ ❺ ❻

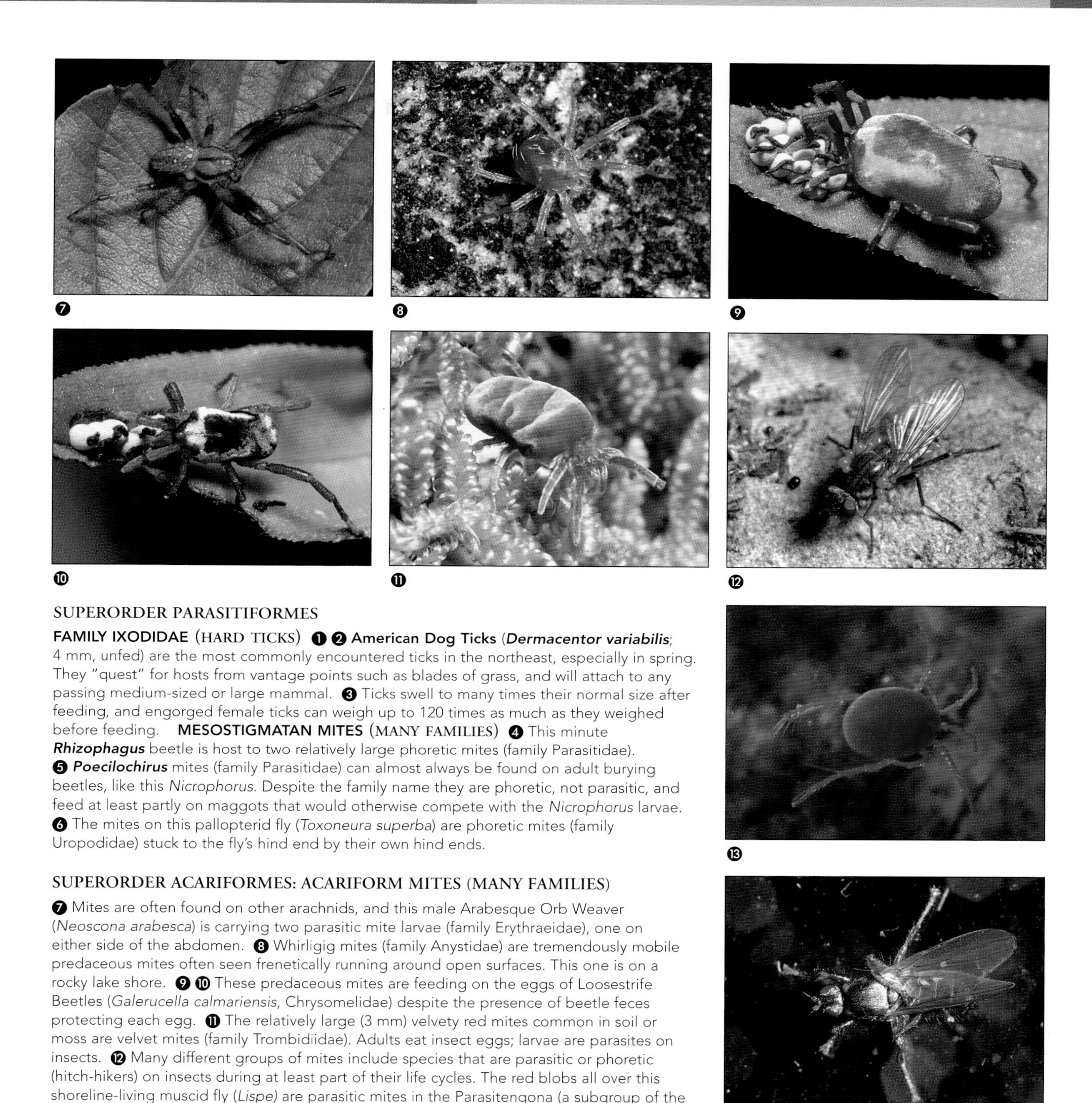

SUPERORDER PARASITIFORMES

FAMILY IXODIDAE (HARD TICKS) ❶ ❷ **American Dog Ticks** (***Dermacentor variabilis***; 4 mm, unfed) are the most commonly encountered ticks in the northeast, especially in spring. They "quest" for hosts from vantage points such as blades of grass, and will attach to any passing medium-sized or large mammal. ❸ Ticks swell to many times their normal size after feeding, and engorged female ticks can weigh up to 120 times as much as they weighed before feeding. **MESOSTIGMATAN MITES** (MANY FAMILIES) ❹ This minute ***Rhizophagus*** beetle is host to two relatively large phoretic mites (family Parasitidae). ❺ ***Poecilochirus*** mites (family Parasitidae) can almost always be found on adult burying beetles, like this *Nicrophorus*. Despite the family name they are phoretic, not parasitic, and feed at least partly on maggots that would otherwise compete with the *Nicrophorus* larvae. ❻ The mites on this pallopterid fly (*Toxoneura superba*) are phoretic mites (family Uropodidae) stuck to the fly's hind end by their own hind ends.

SUPERORDER ACARIFORMES: ACARIFORM MITES (MANY FAMILIES)

❼ Mites are often found on other arachnids, and this male Arabesque Orb Weaver (*Neoscona arabesca*) is carrying two parasitic mite larvae (family Erythraeidae), one on either side of the abdomen. ❽ Whirligig mites (family Anystidae) are tremendously mobile predaceous mites often seen frenetically running around open surfaces. This one is on a rocky lake shore. ❾ ❿ These predaceous mites are feeding on the eggs of Loosestrife Beetles (*Galerucella calmariensis*, Chrysomelidae) despite the presence of beetle feces protecting each egg. ⓫ The relatively large (3 mm) velvety red mites common in soil or moss are velvet mites (family Trombidiidae). Adults eat insect eggs; larvae are parasites on insects. ⓬ Many different groups of mites include species that are parasitic or phoretic (hitch-hikers) on insects during at least part of their life cycles. The red blobs all over this shoreline-living muscid fly (*Lispe)* are parasitic mites in the Parasitengona (a subgroup of the order Acariformes that includes both chiggers and water mites). ⓭ Adult ***Hydryphantes*** (family Hydryphantidae) water mites are common in temporary pools, where they feed on eggs of insects. Hydryphantid mite larvae parasitize adult insects, especially flies. ⓮ Mites are common in piles of seaweed washed up along the Atlantic coast, and they often get there by hitching rides on seaweed flies (phoresy). The mites on this seaweed fly (*Coelopa frigida*) are hypopodes (nymphs modified for phoresy) of the family Histiostomatidae.

❶ Pseudoscorpions are common, but they are usually overlooked because of their small size. These little predators usually occur under bark, but many hitch rides from habitat to habitat on flying insects and are encountered when these insects are collected or closely observed. ❷ ❸ Pseudoscorpions routinely hitch rides from habitat to habitat on winged insects such as this heleomyzid fly. ❹ One pseudoscorpion species (***Chelifer cancroides***) is a common household hunter most likely to come to your attention when accidentally trapped in a bathtub or sink. ❺ Look for the small pseudoscorpion clamped to the left hind leg of this parasitic fly (Tachinidae).

CLASS ARACHNIDA (SPIDERS, MITES, TICKS AND THEIR RELATIVES) ORDER SCORPIONES (TRUE SCORPIONS)

FAMILY BUTHIDAE ❶ ❷ North American scorpions occur in the southern and western parts of the continent. The scorpion on the left (***Centruroides vittatus***) is from Florida; the other scorpion (***Centruroides robertoi***) was photographed in Cuba.

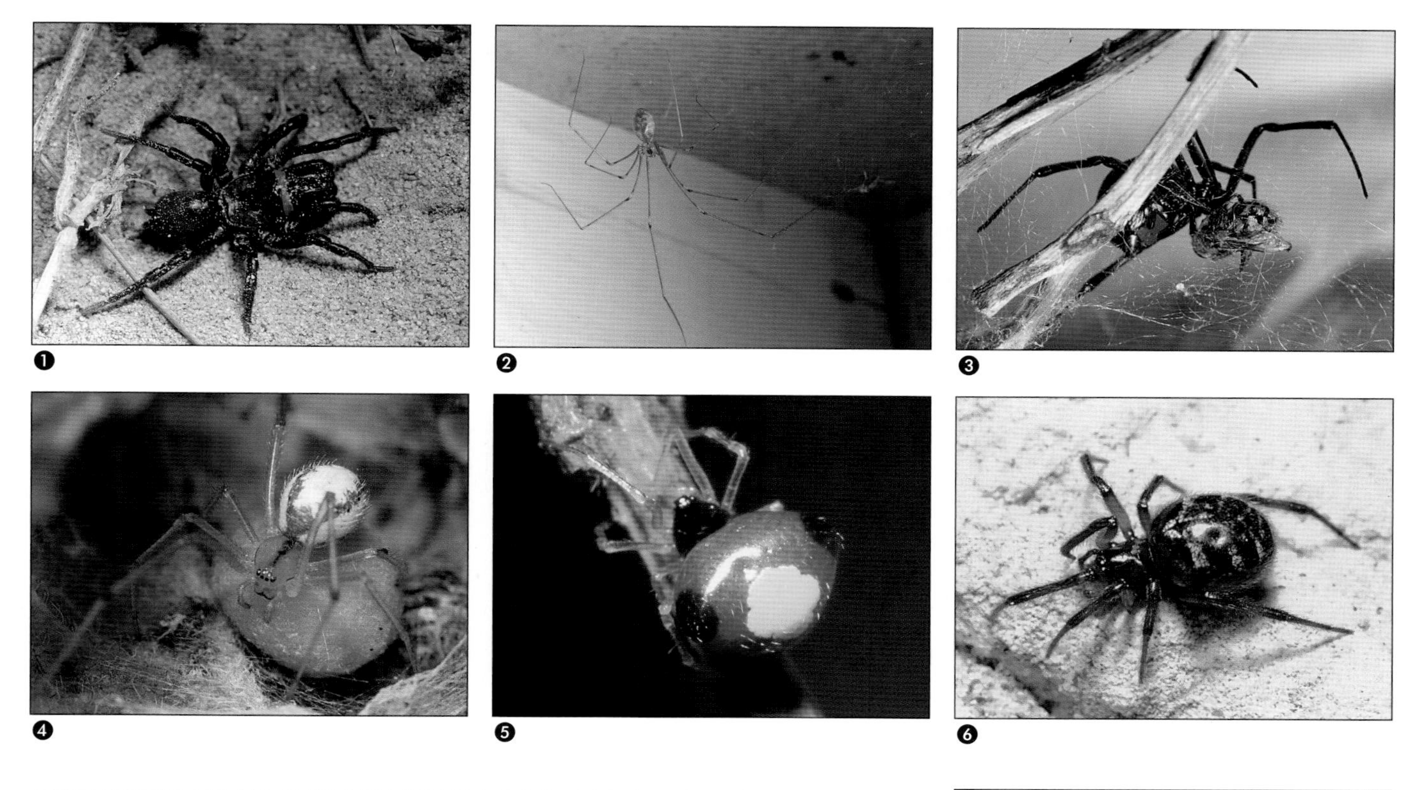

❶ ❷ ❸ ❹ ❺ ❻

❼

INFRAORDER MYGALOMORPHAE (TARANTULAS, TRAPDOOR SPIDERS AND PURSE-WEB SPIDERS)

FAMILY ATYPIDAE ❶ Mygalomorph spiders are mostly southern in distribution, but some purse-web spiders, like this ***Sphodros niger***, range as far north as southern Canada. Purse-web spiders make purse-like silken tubes, often running along a tree trunk to a hole at the base of a tree. If an insect lands on the tube, the spider impales it with its formidable fangs and drags it into the tube.

INFRAORDER ARANEOMORPHAE (TRUE SPIDERS)

FAMILY PHOLCIDAE (CELLAR SPIDERS) ❷ The **Long-bodied Cellar Spider** (***Pholcus phalangioides***, 8 mm) is a prolific nuisance spider found worldwide, often occurring in huge numbers in house basements where they hang in loose webs like this one. When threatened, they vigorously vibrate in their webs. **FAMILY THERIDIIDAE** (COMB-FOOTED SPIDERS OR COBWEB WEAVERS) ❸ The **Black Widow** (***Latrodectus mactans***, 12–16 mm) is one of the most famous of spiders because *Latrodectus* bites can make you feel really awful, with symptoms including severe abdominal pain. Males are small and do not bite, but females often make webs in outhouses and similar places where accidental contact can result in bites. There are several similar *Latrodectus*, including the Western Widow (*L. hesperus*) and Northern Widow (*L. variolus*) with ranges that extend well into Canada. The small males gingerly court females by plucking at their webs, sometimes ending up as meals. ❹ ***Enoplognatha ovata*** is a variably colored species common in gardens and yards. Some individuals are brightly banded like this one; others have a pale abdomen with dark spots. This female (about 6 mm) is guarding her egg sac. Comb-footed spiders have comblike rows of bristles at the ends of the hind legs, used to throw silk over prey tangled in their sticky webs. ❺ ***Theridula emertoni*** is a small (2–3 mm) but distinctively colored comb-footed spider. ❻ Comb-footed spiders in the genus ***Steatoda*** are unusual because males have a stridulatory mechanism reminiscent of that in crickets, with a file on the front of the abdomen and a scraper on the back of the cephalothorax. Their tangled webs are usually woven under rocks. ❼ Members of the enormous genus ***Theridion*** usually hang out in webs among vegetation, but this couple was spotted on the branch of a dead tree. The male, hanging upside down in a characteristic *Theridion* pose, is relatively small; the larger female is devouring a harvestman (Phalangidae).

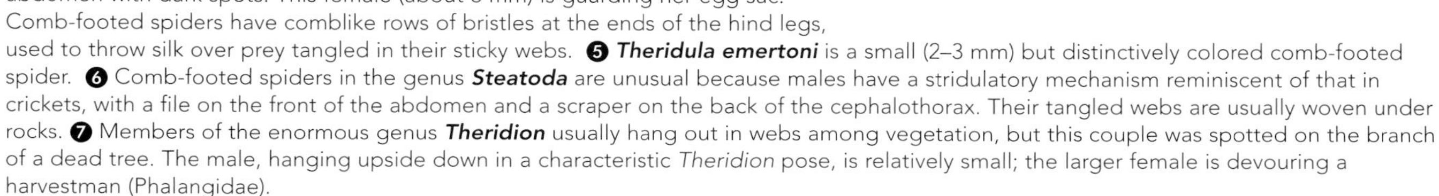

FAMILY LINYPHIIDAE (SHEET-WEB WEAVERS AND DWARF SPIDERS) ❶ This sheet-web weaver is hanging underneath its web where it is safe from predators. Insects landing on the web will be impaled from underneath and dragged through the web ❷ The aptly named "bowl and doily spider" (***Frontinella pyramitela***) makes a web that looks like a bowl suspended over a lacy doily. ❸ This sheet-web weaver (***Neriene*** sp.) has captured a sphaerocerid fly (Sphaeroceridae). ❹ ❺ The male of this pair of sheet-web spiders is the relatively skinny one, with the large palpi held in front of the body. In the second photo, taken from below the pair of spiders, the male has the tip of his left pedipalp inserted in the female's genital opening (epigynum). Part of the pedipalp is inflated like a massive and complex bulb as he inserts a droplet of sperm. ❻ This male ***Pityohyphantes costatus*** is using his large pedipalps to transfer sperm to the female. Members of the genus *Pityohyphantes*, sometimes called hammock spiders, live in trees. ❼ Dwarf spiders (subfamily Erigoninae) are probably our most abundant spiders, but they are so small (usually 1–2 mm) they are usually overlooked. This one is walking on the snow. ❽ ***Hypselistes florens*** is a common and distinctively colored dwarf spider, but it is usually overlooked due to its small size (only 2–3 mm). These little spiders make minute sheet webs close to the ground. **FAMILY ARANEIDAE** (ORB WEAVERS) ❾ The female **Black and Yellow Argiope** (***Argiope aurantia***) is a large, distinctive spider common in gardens and fields, usually seen sitting in a large web with a prominent zigzag ribbon of silk (a stabilimentum) crossing the middle. Males are small and, like other male spiders, have armlike sexual organs (pedipalps) used to transfer sperm to the female. The *A. aurantia* male spontaneously expires immediately after inserting his pedipalps into the female's sexual organs, his dead body protecting his paternity by blocking other males.

❶

❷

❸

❹

❺

❻

❼

❽

FAMILY ARANEIDAE (continued) ❶ The egg sac of the **Black and Yellow Argiope** (***Argiope aurantia***) is a robust and tightly woven sphere, elaborately suspended in a protective silken infrastructure. ❷ These **Black and Yellow Argiope** (***Argiope aurantia***) spiderlings have just hatched from their egg case. ❸ ***Argiope trifasciata***, the **Banded Argiope**, is very common in open grasslands where it makes large orb webs, often with a dense white stabilimentum in the middle. ❹ The **Bridge Orb Weave**r (***Larinioides sclopetarius***) often builds its webs on buildings or bridges, feeding mainly on night-active insects like this midge (Chironomidae). ❺ Most orb-weaving spiders belong to the huge genus ***Araneus*** (over 1,500 species worldwide). The **Garden Cross Spider** (***A. diadematus***), easily recognized by the cross pattern on its abdomen, often makes its web near porch lights where it can feast on insects attracted to the lights. ❻ The **Shamrock Spider** (***A. trifolium***) is another common backyard *Araneus*. ❼ ***Araneus nordmanni*** is a widespread species, found from coast to coast and north to the tree line. Webs are usually spun on coniferous trees. ❽ The **Marbled Orb Weave**r (***Araneus marmoreus***) is common and widespread across the country and throughout northeastern North America, especially in open wetlands and meadows. ❾ The **Arrowshaped Micrathena** (***Micrathena sagittata***) belongs to a group of orb weavers characterized by a hard, shiny and ornamented abdomen.

❾

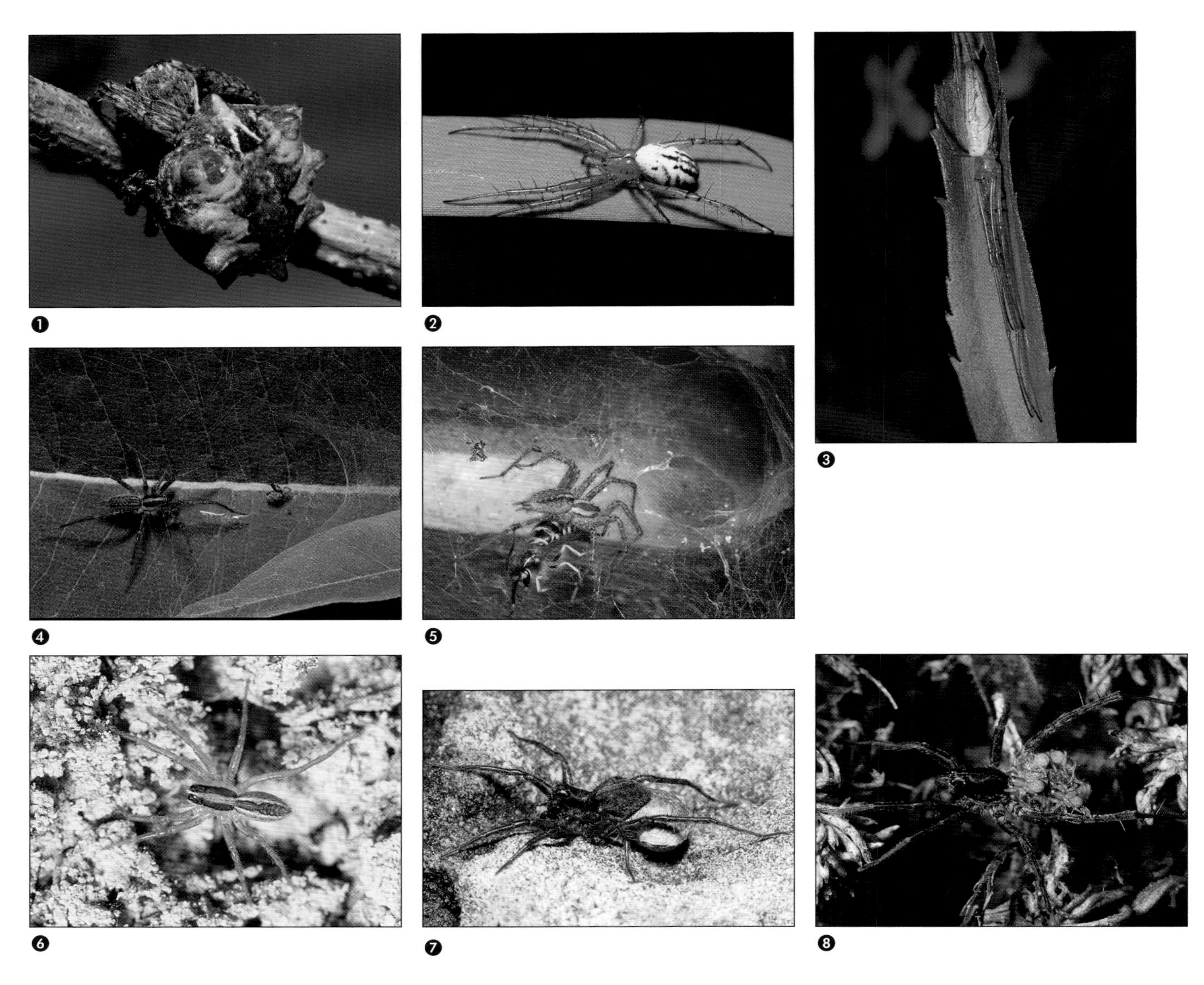

FAMILY ARANEIDAE (continued) ❶ ***Acanthepeira stellata*** is a common orb weaver with a bizarre abdomen. **FAMILY TETRAGNATHIDAE** (LONG-JAWED ORB WEAVERS) ❷ Members of the genus ***Leucauge***, like this ***L. venusta***, are known as orchard spiders. ❸ Long-jawed orb weavers in the genus ***Tetragnatha*** are commonly seen stretched out on grass blades or similar foliage with their long legs extended. **FAMILY AGELENIDAE** (FUNNEL WEAVERS) ❹ Funnel weavers in the genus ***Agelenopsis***, called grass spiders, make funnel-shaped webs in low vegetation. The webs are often conspicuous when they are glistening with dew; the spider hides in the narrow parts of the funnel and dashes out when a potential victim vibrates the web. ❺ This funnel web spider (***Agelenopsis***) is dragging a yellowjacket (*Vespula maculifrons*) male into its funnel-like retreat. **FAMILY LYCOSIDAE** (WOLF SPIDERS) ❻ Most of the larger, conspicuous spiders seen running around on the ground are wolf spiders, like this ***Hogna rabida***. ❼ ❽ Wolf spiders, like these ***Pardosa*** females, are common hunting spiders with the distinctive habit of carrying their large egg sacs behind the body (attached to the spinnerets). The young spiders are carried on the mother's back for some time after emerging. *Pardosa* is a common genus of long-legged wolf spiders. ❾ The **Brush-legged Wolf Spider** (***Schizocosa ocreata***) is a common eastern species that has been extensively studied because of its complex courtship behavior, which involves sound, vibration and chemical communication.

FAMILY LYCOSIDAE (continued) ❶ Pirate wolf spiders (***Pirata*** spp.) are relatively small (less than 10 mm), distinctively marked wolf spiders found in wet areas, including bogs, marshes and pond margins. ❷ ***Arctosa littoralis*** (10–15 mm) is a superbly camouflaged wolf spider common on sand beaches. **FAMILY PISAURIDAE** (NURSERY WEB SPIDERS AND FISHING SPIDERS) ❸ ❹ The huge and intimidating spiders often seen resting on docks with outstretched legs are fishing spiders in the genus ***Dolomedes***, like this ***D. scriptus*** (with eggs) and this ***D. tenebrosus***. Fishing spiders carry their huge egg sacs up front, under the head and cephalothorax, unlike similar wolf spiders (Lycosidae), which carry their egg sacs behind the body. Fishing spider females can have a leg span of over 80 mm. ❺ The **Six-spotted Fishing Spider** (***Dolomedes triton***, 9–20 mm) is a common and distinctively marked fishing spider. Fishing spiders use the whole water surface as a web, picking up on the vibrations made by insects on the surface just as many other spiders detect insects struggling in their webs. ❻ ❼ One of these nursery web spiders (***Pisaurina mira***, about 14 mm) has started to construct her "nursery web" – a shelter made from silk and leaves to hold the mature egg sac and newly hatched young spiders. These attractive spiders are closely related to fishing spiders, but live away from water in low vegetation. Nursery webs are often found in poison ivy. **FAMILY GNAPHOSIDAE** (GNAPHOSIDS) ❽ Gnaphosid spiders like this ***Herpyllus*** are normally nocturnal hunters and remain hidden during the day, often in silken sacs concealed under bark, rock or other shelters. ❾ Many gnaphosids, like this black ***Zelotes***, can be found hiding under stones.

❶ ❷ ❸ ❹ ❺ ❻ ❼ ❽ ❾

FAMILY GNAPHOSIDAE (continued) ❶ ***Sergiolus capulatus*** is a distinctively colored (and fast moving) gnaphosid. **FAMILY AMAUROBIIDAE** (AMAUROBIIDS OR WHITE-EYED SPIDERS) ❷ Amaurobiids live in irregular webs in concealed places. This is ***Amaurobius ferox***. ❸ ***Wadotes***, like other Amaurobiidae, is a nocturnal hunter likely to be found in silken retreats under bark during the day. ❹ This ***Callobius*** has just fought a losing battle with a spider wasp (*Anoplius virginiensis*) and is now destined to serve as food for this wasp's larvae. **FAMILY LIOCRANIDAE** ❺ This ***Agroeca*** is eating a small *Dictyna* (Dictynidae). Until recently, *Agroeca* species were included in the Clubionidae. **FAMILY CLUBIONIDAE** (SAC SPIDERS) ❻ This clubionid was exposed by unrolling the shelter it had made in a dead leaf. ❼ Clubionid spiders like this ***Clubiona*** are often found climbing on plants, where they make silk-lined nests, usually in rolled leaves or under bark. **FAMILY SALTICIDAE** (JUMPING SPIDERS) ❽ Jumping spiders are favorites among naturalists because of their conspicuous daytime activity, bright colors and obvious diversity. These big-eyed spiders leap onto their prey while attached to a silken safety line used to retreat if the mark is missed. This male ***Eris militaris*** has captured a muscid fly (Muscidae). ❾This ***Eris militaris*** female has captured a caddisfly. ❿ The **Zebra Spider** (***Salticus scenicus***) is a common, small (6 mm) jumping spider regularly seen on decks, sidewalks and the outside walls of houses.

❿

FAMILY SALTICIDAE (continued) ❶ Some of the largest and most conspicuous jumping spiders belong to the genus ***Phidippus***. This is a male ***P. clarus***. ❷ This ***Phidippus clarus*** is eating a cluster fly (Calliphoridae). ❸ ***Phidippus whitmani***, one of our most distinctive jumping spiders, occurs throughout most of eastern North America. ❹ This jumping spider (***Phidippus audax***) is being dragged up a stone wall by a spider wasp (*Agenioideus cinctellus*). ❺ This jumping spider (***Phidippus purpuratus***) has captured a small butterfly. ❻ Males of ***Habronattus decorus***, like other males in this large genus of jumping spiders, are conspicuously ornamented and engage in complex courtship displays. ❼ This little jumping spider (***Habronattus viridipes***) has captured a springtail. ❽ Some jumping spiders, like this ***Synemosyna lunata***, are good ant mimics. ❾ ❿ Many male jumping spiders have striking eyebrow-like tufts behind the eyes. This species (***Maevia inclemens***) is known as the **Dimorphic Jumping Spider** because some males are black, but other males and all females are gray with red, black and white markings. These are both males.

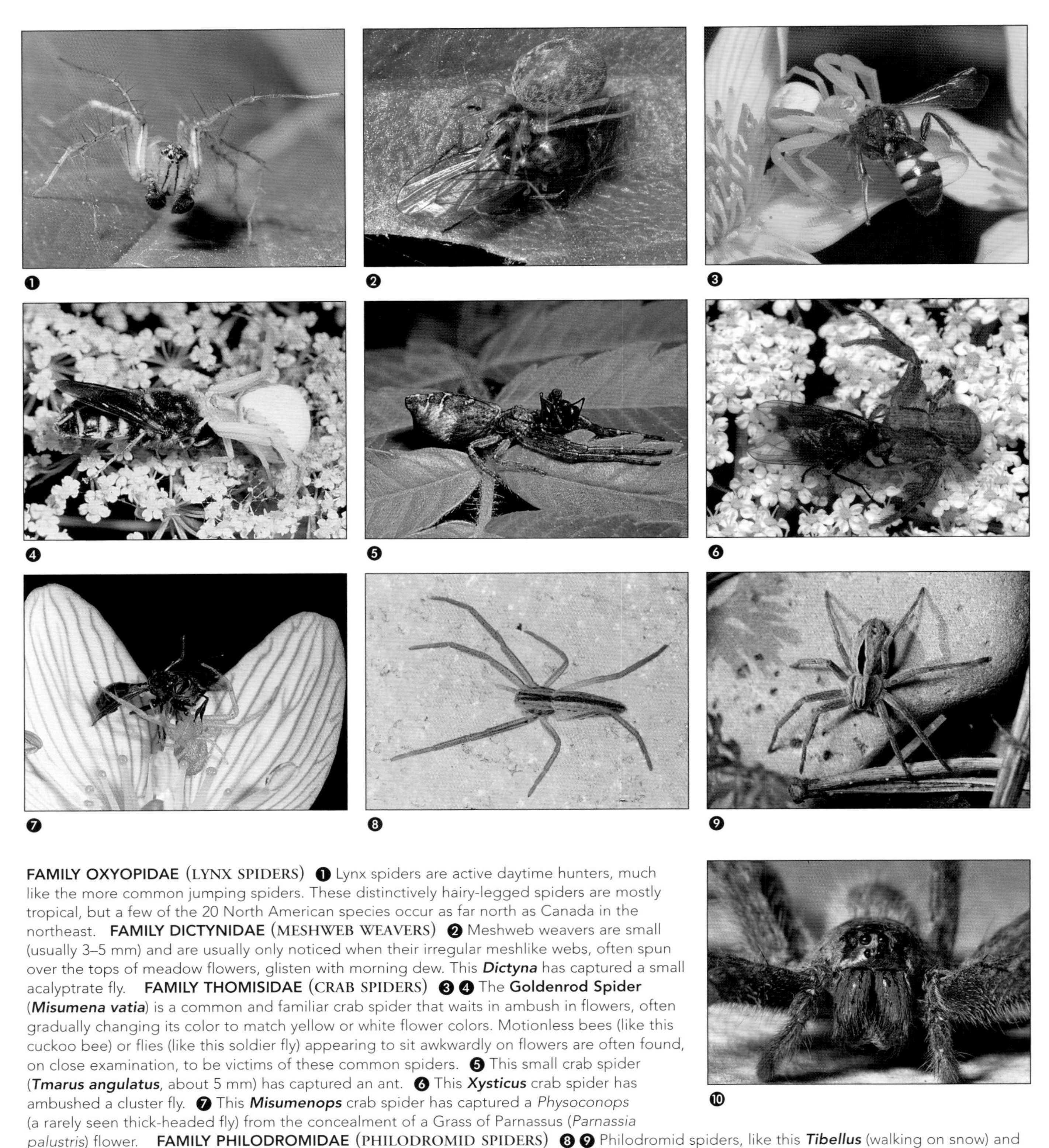

FAMILY OXYOPIDAE (LYNX SPIDERS) ❶ Lynx spiders are active daytime hunters, much like the more common jumping spiders. These distinctively hairy-legged spiders are mostly tropical, but a few of the 20 North American species occur as far north as Canada in the northeast. **FAMILY DICTYNIDAE** (MESHWEB WEAVERS) ❷ Meshweb weavers are small (usually 3–5 mm) and are usually only noticed when their irregular meshlike webs, often spun over the tops of meadow flowers, glisten with morning dew. This ***Dictyna*** has captured a small acalyptrate fly. **FAMILY THOMISIDAE** (CRAB SPIDERS) ❸ ❹ The **Goldenrod Spider** (***Misumena vatia***) is a common and familiar crab spider that waits in ambush in flowers, often gradually changing its color to match yellow or white flower colors. Motionless bees (like this cuckoo bee) or flies (like this soldier fly) appearing to sit awkwardly on flowers are often found, on close examination, to be victims of these common spiders. ❺ This small crab spider (***Tmarus angulatus***, about 5 mm) has captured an ant. ❻ This ***Xysticus*** crab spider has ambushed a cluster fly. ❼ This ***Misumenops*** crab spider has captured a *Physoconops* (a rarely seen thick-headed fly) from the concealment of a Grass of Parnassus (*Parnassia palustris*) flower. **FAMILY PHILODROMIDAE** (PHILODROMID SPIDERS) ❽ ❾ Philodromid spiders, like this ***Tibellus*** (walking on snow) and this ***Thanatus***, were once included as a subfamily of the crab spider family. Philodromids are shaped somewhat like crab spiders, but are more likely to run after prey than to ambush insects (as do crab spiders). **FAMILY CTENIDAE** (WANDERING SPIDERS) ❿ Wandering spiders are tropical or subtropical spiders that occasionally arrive in the northeast as accidental travelers in fruit shipments. This venomous (and aggressive) ***Phoneutria*** was found in a shipment of bananas. Most of the large spiders regularly found in banana shipments are harmless members of another family (Sparassidae, the giant crab spiders).

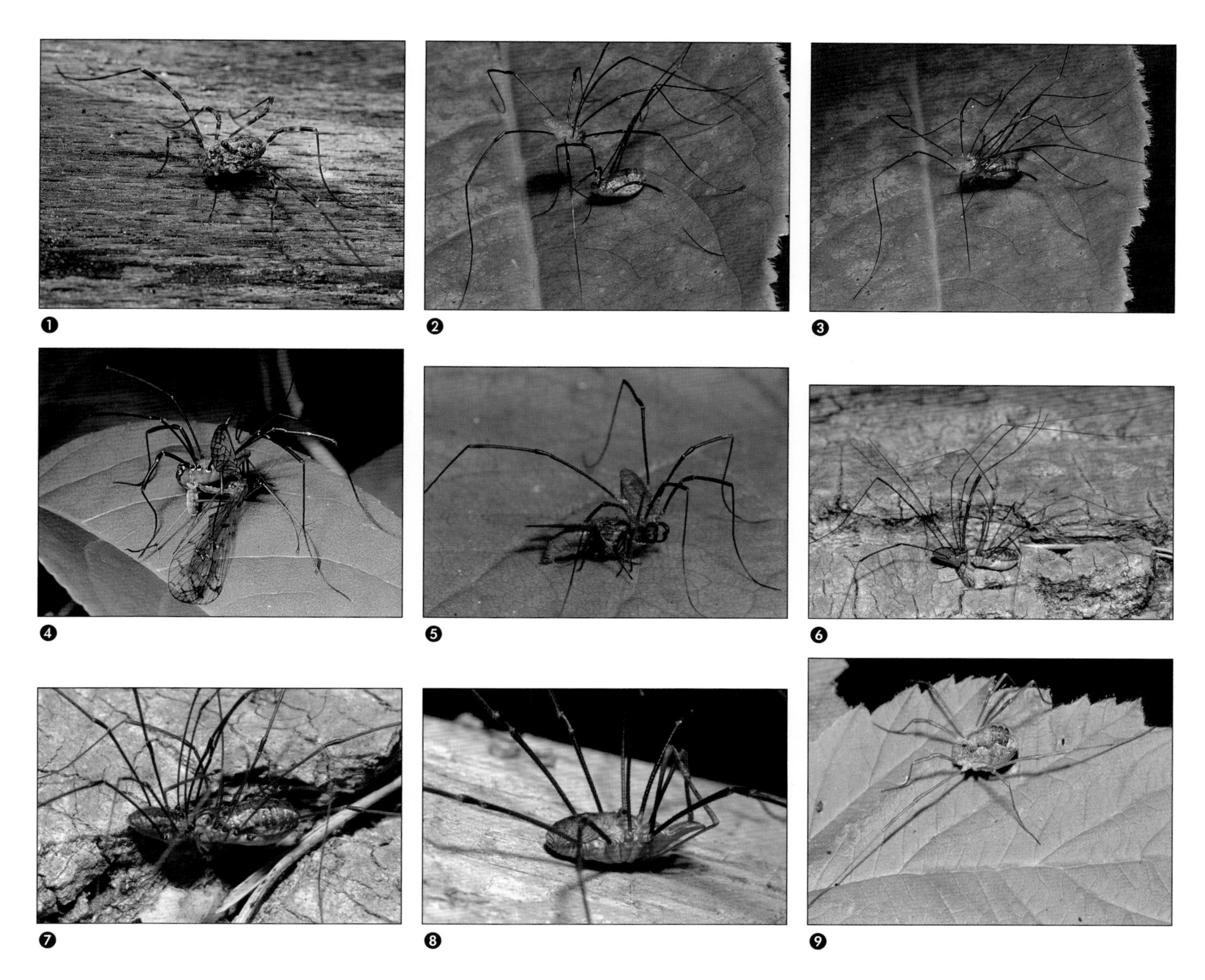

FAMILY PHALANGIDAE ❶ Northeastern North American harvestmen, like this ***Odiellus***, belong to the family Phalangidae, and are known as daddy long-legs. Harvestmen are often confused with spiders, but these generally long-legged arachnids have a round or oval body that does not have a spider-type waist divided into two sections. ❷ ❸ Unlike all other arachnids, in which males use indirect methods to transfer a sperm package to the female, male harvestmen have a specialized elongate organ called a penis, used to directly deliver spermatozoa into the female's gonopore. The female will later use a long, telescoping ovipositor to conceal her eggs in soil or under bark. ❹ ❺ Most harvestmen are predators, but some are scavengers. One of these ***Leiobunum*** has caught a hanging scorpionfly (Bittacidae), and is manipulating its abdomen with his pedipalps while tearing it apart with his chelicerae. The other one (probably ***L. nigropalpi***) is eating an anthomyiid fly. ❻ ❼ Some harvestman males, like this ***Leiobunum***, have enlarged pedipalps that fit into a groove between the female's second and third legs. ❽ ❾ ***Phalangium opilio***, probably introduced from Europe, is a widespread and common harvestman of disturbed areas like roadsides and city parks. Males have distinctively long pedipalps (they look like a fifth pair of legs at the front of the body) and strikingly robust chelicerae.

CHAPTER 14

Observing, Collecting and Photographing Insects

Eastern North America is home to tens of thousands of insect species, providing a rich menu of natural novelties from which you can select subjects for casual observation or detailed study. Recognition of the common families and a few conspicuous species of insects adds another dimension to outdoor activities or nature study, and a more focused familiarity with almost any major group of insects will lead to new discoveries about insect distribution, evolution and biology.

The study of insect diversity has been traditionally centered on the collection, mounting and subsequent identification of insect specimens. Those who need to identify most groups of insects to the species level are still dependent on this traditional approach, but it is increasingly practical to identify larger insects from live specimens, much the way birds are routinely identified in the field. The traditional approach of killing and mounting specimens is no longer necessary, or even appropriate, for many of the more popular groups of North American insects such as dragonflies, butterflies, larger moths, larger true bugs and many groups of beetles.

Every microhabitat is home to different and seasonally variable insect fauna. The bark of this fallen tree, for example, hid a variety of beetles including a couple of spectacular Fiery Searchers (*Calosoma scrutator*).

Pinned or preserved specimens are still extremely important for the documentation of poorly known faunas and for the identification of smaller or more difficult insects. Small flies, parasitic wasps and many immature insects, for example, can often only be identified by careful examination of features not visible in the field. Furthermore, the accumulation of properly labeled, properly preserved pinned insect specimens remains the most important approach to documenting biodiversity and the changes in biodiversity through time. Serious studies of arthropod diversity demand the permanent preservation of specimens as "vouchers" that can be used to confirm identifications, and older identified specimens in collections are routinely used to identify new material by comparison.

The "minimum" equipment you will need for insect observation and collection will depend on your objectives. Many naturalists use close-focusing binoculars or telephoto lenses to take a "birdwatcher's" approach to insect study, but most insects are easily observed up close with no special equipment as long as you avoid rapid movements. Even the most wary insects, such as tiger beetles and dragonflies, can be closely approached if you move very slowly, and most of the photographs in this book were taken close-up in the field with a macro lens.

All you really need to collect insects, either for temporary observation and release, or for a permanent collection, are a few small bottles, such as the plastic film container being used here to house my son's lady beetle collection.

Smaller insects often need to be captured and held for close observation in the field, for later examination with good lights and reference books, for rearing to another stage or for use in a collection of dead insects. The soft plastic containers used for rolls of 35 mm film are excellent vessels for holding captured insects, and are often available from stores that develop film. (I buy smaller snap-top "pill bottles" in bulk, and put one insect in each vial, usually with a small leaf to keep the insect from desiccating.) Captured specimens can be examined and released, or killed in a freezer for subsequent mounting. Remember to keep careful field notes, and to label your bottles (or put them in separate bags with labels). Many naturalists carry a hand lens for field observation of small specimens, but I favor magnifier goggles (available in any electronics shop) since they leave my hands free and can be flipped up and out of the way easily.

Freezing is the safest and cleanest way to kill insects, but professional entomologists often use poison bottles, especially when they need to collect large samples. The most commonly used killing agent is ethyl acetate, a volatile poison that is a component of acetone-free nail polish remover. Good killing bottles can be made using soft plastic canisters (film canisters are fine, centrifuge tubes are better) with some absorbent material soaked with ethyl acetate and stuffed in the bottom of the canisters. Further material should be placed in the canisters to keep specimens dry. If you use a killing bottle, label it clearly and do not use the same bottle for moths as for other insects because moth scales will ruin the other specimens. Periodically empty the contents from your killing bottles into storage containers (I use 35 mm film canisters with some tissue in the bottom) with a label. Soft-bodied insects, such as larvae, are usually collected into 70 percent alcohol or other liquid preservatives.

Insect nets are by no means the essential entomological equipment they are often perceived to be, but they can be useful. Sweeping a net through vegetation can yield hundreds of insects that might otherwise be overlooked, and some large, fast-flying insects are hard to catch without a net. A good insect net can be made by putting a net bag made out of fine-weave polyester cloth on the frame of a fisherman's landing net, or on a homemade frame made from a loop of

Sweeping a net through vegetation can yield hundreds of insects you might otherwise overlook.

Some entomologists use a light aerial net for capturing single insects, and a heavy canvas sweep net for bashing through the vegetation. I prefer an all-purpose net made from tent fly material, sewn with an abruptly tapered tip to make it easy to transfer a sample into a vial or small-diameter bottle as demonstrated here.

heavy wire clamped to a broom handle. I use black mosquito netting since it is tough and relatively transparent, but others prefer white mesh since it is easy to see insects against the white background.

Bulk-sampling is expedited by sewing the net bag so that the tip forms a pocket with the same diameter as your preferred collecting bottle; this net design also makes it simple to remove insects from the net without handling them. Aerial net bags need to be long enough so the net can be flipped over the frame to prevent insects from escaping.

Nets for collecting aquatic samples are stouter and shallower than aerial nets, and most aquatic insect specialists use heavy and expensive "D nets." An inexpensive alternative is to use a food strainer for aquatic insect collecting, clamping or taping it to a stick if a long handle is required. Collecting and observing aquatic insects is easiest if samples (taken with a net or sieve) are dumped into a white or yellow dishpan; individual specimens can be removed using a large eyedropper (like a turkey baster) or a pair of forceps. White pans are also useful for sorting through insect samples from compost, litter or other substrates.

Insects that are too small to pick up between thumb and forefinger, and too awkward to grab with a pair of forceps, can be captured using various devices. The easiest way to get insects out of a net is to swing the net hard to force insects to the apex where they can be easily knocked directly into a small collecting bottle. Transparent collecting bottles also work well to pick up insects directly from baits or other substrates, since most insects will fly up into a transparent tube carefully lowered over the specimen.

A more efficient way to pick up individual small insects is to use a suction device or aspirator, the simplest of which is just a tube running from your mouth to the insect with some sort of screen or collection chamber in between. An efficient aspirator can be made from a clear plastic pen barrel with fine mesh wrapped around one end, which is inserted in a length of tubing. Sucking on the tubing pulls small specimens up through the pen barrel and up against the mesh; blowing on the tubing shoots the capture out of the aspirator into another container.

Small insects can be effectively collected using an aspirator or "pooter," normally a mouth-operated vacuum collector with the collecting chamber separated from the mouth tube by a fine screen. This young collector (my youngest son) is using a standard aspirator to collect beetles in a clean, streamside habitat, but for less sanitary environments I would recommend a battery-operated aspirator, easily made by modifying a rechargeable car vacuum cleaner.

Actively hunting for insects in or on recognizable microhabitats such as fungi, foliage or wood is a good way to make useful observations, but many more species can be taken in traps – such as Malaise traps, pan traps, bait traps and light traps – than can be found by merely searching. Malaise traps are like tents with no sides, but with a barrier in the middle. Insects fly in the open sides and hit the barrier, after which most species fly up toward the light and into a bottle at the apex of the trap (the bottle normally contains a killing agent or liquid preservative). Some kinds of insects drop when they hit a barrier, and can be collected by placing pans of preservative (or just soapy water) under the barrier.

A Malaise trap placed along an edge or on a natural flyway will trap about a quart of insects a week, amounting to millions of specimens over the course of a season. That may sound like a lot, but it is insignificant compared to the number of insects squashed on the radiators of the millions of cars on our roads on any summer evening. Those squashed bugs on your radiator or windshield are lost to science, but it is not too difficult to mount a net in front of your vehicle that diverts potential radiator mush into the apex of the net, where the mass of intact insects can be easily gathered using a modified car vacuum cleaner.

An even more effective sampling device can be made by simply putting disposable plastic bowls on or in the ground and adding some water and a few drops of soap. Called "pan traps," these simple devices attract some insects by their bright colors (yellow is best), while other insects seem to just tumble in and drown. I have used yellow pan traps in several insect surveys, often taking over a thousand species of insects in a few pan traps left in a single habitat for a season. Trapped insects are removed using an aquarium net, daily if the pans have only soapy water in them and less frequently if a preservative (usually salt or propylene glycol) is used. Baits, such as carrion, dung or mushrooms, can be placed near the pan traps to attract large numbers of some specialized groups of insects.

Insects that lead cryptic lives in soil and litter can be coaxed out of their hidden habitats using a number of devices, the most common of which is the Berlese funnel – a large funnel with a screen partition at mid-depth. Litter is placed in the funnel on the screen partition, and a light bulb is hung above the litter. Insects move away from the light and heat, ultimately dropping down into the funnel and into a waiting jar of alcohol. The efficiency of a Berlese funnel can be greatly enhanced by sifting the litter before placing it into the funnel. Sifting separates the coarse debris from the fine debris and small insects; the coarse debris is discarded and the fine debris is put in the Berlese funnel.

Many insects are active at night, and hunting with a headlamp can be very productive. Most night collecting, however, is done by hanging a bright light over a white sheet and watching to see what shows up. Mercury vapor lights and blacklight (ultraviolet) tubes are very effective, and can also be used in traps (hung over funnels) that are very effective but deny you the fun and excitement of watching the sheet to see what is "attracted" (it is generally thought that nocturnal insects are not so much attracted as disoriented by bright lights). Larger moths can be identified on the sheet or photographed for later identification, while smaller insects can be picked off the sheet with an aspirator, forceps or clear tubes.

Collecting insects is only the first, and by far the easiest, step toward making an insect collection. If you opt to make a permanent collection, your specimens must be carefully handled and stored until they are mounted properly on

Malaise traps are a bit like tents with no sides, but with a center barrier that deflects flying insects either up or down according to their habits. Those that move up toward the light when encountering a barrier end up in a trap head, in this case full of alcohol, at the high end of the trap. A well-placed Malaise trap is likely to capture an incredible variety of otherwise rarely seen species.

Pan traps are simply small pans or bowls filled with water and a couple of drops of soap to break the surface tension. Pans set flush with the soil surface can trap thousands of species of insects over the course of a season. Pan traps put in unusual places, like the shelf-brackets on this tree trunk, capture fewer species but often take rare or unusual specimens.

Insects associated with discrete habitats such as tree stumps, fungi, mammal burrows or carrion can be cleanly and effectively sampled using an emergence trap. This one allows insects to enter along the bottom, thus taking both insects emerging from the stump and those attracted to the stump.

Constructing novel collecting devices can be great fun, and most collectors soon find their basements littered with modified vacuum cleaners, light traps, extraction funnels and other devices. This car net, mounted in special brackets on the bumper of my 1967 Rambler, yielded thousands of interesting catches before I traded the car for a newer, bumper-less vehicle.

This photograph shows emergence traps, pan traps, and a Malaise trap in service during an insect survey in Bruce Peninsula National Park.

special insect pins; then they must be labeled properly and stored in pest-proof boxes. Newly collected material can be stored for a few days in a refrigerator before mounting (longer in a freezer), otherwise specimens will quickly dry out and become brittle. Insects should be mounted on special insect pins, which come in sizes from #000 (impossibly flimsy) through to #7 (railway spikes). Most large insects can be pinned with #3 pins, and most small specimens with #1 pins. As a rule of thumb, the space left above the pinned specimen should be exactly 7 mm – approximately the thickness of a thumb.

Specimens too small for a #0 pin should probably be glued to a "point" (an 8 mm long paper triangle pinned at a standard height on a #3 pin) using white glue. Alternatively, some insects (small flies and wasps) can be glued directly to the side of a pin. Try to pin your insects just to the right of the center, so as not to obscure bilaterally symmetrical structures. By convention, beetles are pinned through the right elytron, true bugs through the right half of the scutellum, and flies or wasps through the right side of the main part of the thorax.

The most important part of making an insect collection, and the part most often botched, is labeling. It is very easy to come home after a pleasant day of insect collecting and throw a few bags of specimens in the freezer with a mental note to label them later. If you do not have labels in those bags before they go in the freezer, it is unlikely that you will be able to attach good data to them a week later. Similarly, when you pin specimens you must put at least temporary labels on the groups of specimens from the same dates and localities at the time of mounting. Never leave specimens, mounted or not, without at least a temporary label, because specimens with inaccurate data are worse than useless.

Permanent labels are affixed to the pin under each specimen at a standard height (17 mm), and should follow the long axis of the specimen or point, using a minimum of space while providing a maximum of protection for the specimen. Good labels can be made using the smallest readable font on a laser printer (about 6 point), and should include, at an absolute minimum, the locality, date and collector. Most serious collectors now carry a GPS and record latitude and longitude for their labels. Specimens with further data such as microhabitat or host are much more valuable than those without.

Run the lines together on a label to leave as little white space as possible, and aim for about four or five lines with 17 to 20 characters across. Hand-written labels can be made using 0.01 mm or 0.005 mm indelible drafting pens. Identification or determination labels should always be separate from, and below, the data labels. All labels should be made on heavy paper, preferably acid-free paper with a high linen content. Cover stock, available from any stationery store, is adequate.

Although almost any soft material, ranging from a piece of foam insulation to a sheet of gasket cork, can serve as a temporary surface for arranged insects, longer term storage and organization of insect collections will require boxes lined with high-density foam of the sort sold as camper's sleeping pads. Cardboard or loose-fitting wooden boxes are fine for field boxes or for temporary storage, but museum pests (dermestid beetles, see page 337) can easily penetrate all but the most tight-fitting of boxes unless they are stored in a freezer or other sealed structure.

You can store your mounted catches in temporary boxes in a freezer until the specimens are thoroughly dry, and then move them into a rigid plastic box for permanent storage. Specimens stored this way will mold unless they are absolutely dry. A more common, although more expensive, approach is to use tight-fitting wooden boxes or glass-topped wooden drawers. A good wooden box will prevent ingress of pests while still allowing the pinned specimens to dry out. Boxes, pins, point punches, nets and other entomological supplies ranging from genitalia vials to microscopes are readily available from various suppliers, some of which post catalogues on the web.

If you do make an extensive, properly labeled collection it could be of long-term value to science by serving as an archive of information about distributions and habitat associations the insect species you collected, some of which might be otherwise poorly known or even undescribed. Your collection will outlast you, and you should try to make arrangements for its long-term care or transfer to a permanent institutional collection.

The University of Guelph Insect Collection, for example, is a collection of about two million specimens that started out as the Entomological Society of Ontario collection when the Society was formed in 1863. The initial collections, which remain the core of the modern facility, were personal collections made by keen individuals just like you. Those specimens now tell us a great deal about how the fauna of Ontario has changed over the past 140 years, and they are still used in a wide variety of research programs – they were used to check up on the identifications of many of the photos in this book.

Now that you know how to make an insect collection, here are a few reasons not to make an insect collection, as well as a few suggestions on how to make a collection of insect images instead. The killing of a few hundred (or even a few hundred thousand) adult insects has absolutely no impact on insect populations, so there is no "ecological" argument against insect collecting. There may be an ethical argument against collecting, although it really does not make much sense to be squeamish about killing a few specimens in a killing jar or freezer when you directly kill thousands on your car radiator and indirectly kill millions by buying cotton or almost anything in a grocery store.

Still, I would suggest that the development of a traditional collection of dead insects is inappropriate for most naturalists for several reasons. For one thing, as should be obvious from the above discussion, it is expensive and time-consuming to make a proper collection. Pins cost around 10 cents each, good boxes around $50, and storage space is always at a premium.

Furthermore, it is illegal to collect insects without a permit in most state

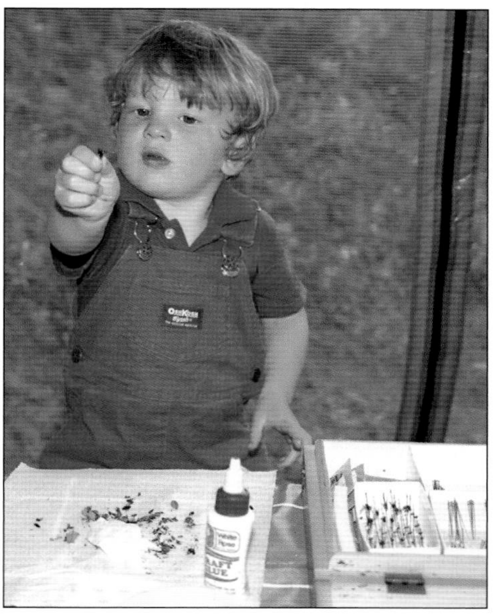

If you collect and kill insects, you must pin them when they are still flexible and not too brittle. Most insect collectors carry a field kit including a box, pins and other equipment for this purpose.

Large insect collections, like the University of Guelph Insect Collection, are usually organized in foam-bottomed unit trays in glass-topped drawers stored in sealed metal cabinets.

parks, provincial parks, national parks and other reserves and conservation areas. Permits are usually difficult or impossible to obtain unless you are a professional with a proven track record, and you can be fined for even possessing or transporting pinned specimens that were taken without proper documentation. These are serious obstacles that should lead you to ask whether you really need to kill and pin specimens. If you want to work on small things, or seriously study biodiversity at the species level, collections are still the only way to go, but if your goal is to become an increasingly well-informed observer of visible six-legged life then I would recommend an image collection rather than a specimen collection. There is much to be said for returning from a day in the field with a digital camera and downloading your day's "collection" from a flash card onto your home computer, and organizing your "specimens" in files and folders rather than on pins in boxes.

Photography, especially macro photography, is changing quickly as new models of digital cameras hit the market, many of which yield remarkably good images right down to the size of leafhoppers and lady beetles. Lightweight, easily used, relatively inexpensive digital cameras are revolutionizing the study of insects, and I hope that they will lead to the same sort of explosion in insect study that we saw in bird study as affordable binoculars (and good field guides) hit the market.

Although digital cameras are likely to be the choice of the new wave of naturalists, there is still much to be said for 35 mm SLR (single lens reflex) cameras. Almost all of the images in this book were taken using slide film in Nikon SLR bodies, usually with either a 60 mm macro lens or a 105 mm macro lens in combination with appropriate close-up rings, teleconverters, extension tubes, flash units, monopods and tripods. Although heavy and expensive relative to digital cameras, SLR setups give superior flexibility and higher resolution. If you plan to shoot insects less than 3 mm in length, or blow up your shots above 8x10, the SLR format remains ideal. The newer digital SLR bodies seem to offer the flexibility of SLR photography with the convenience of digital, and digital SLR cameras are increasingly the choice of professional entomologists. The last few images added to this book were taken with a Nikon digital SLR body fitted with the same lenses and other accessories used with my older film cameras, and any future editions will undoubtedly be enriched by further advances in digital photographic technology in the coming years.

Macro photography is changing fast with the advent of digital cameras, cordless flashes and other advances, but this is the camera I used for most of the pictures in this book. I used either a 60 mm macro or a 105 mm macro and a single flash (connected to the camera with a TTL cord, not shown here), sometimes on a "butterfly bracket," but sometimes hand-held. For higher magnification I used extension rings.

Insect Picture Keys

Tools for Insect Identification

If you have observed, photographed or captured an insect and you want to find out more about it, you need to identify your insect to a useful level. Sometimes just knowing the order is enough, but in most cases the first step towards understanding the habits and importance of an insect is a correct **family** identification. Knowing the family is usually enough to make some useful generalizations, and even an approximate family identification will make it easier to scan photos in search of a generic or specific identification or to search the web for further information.

To identify an insect to family, start with the seven simple questions on this page. They will either send you directly to a set of photographs to which you can compare your insect, or they will direct you to a picture key. The keys are like roads with a pair of signs at each junction; pick the sign that best matches your insect and follow the road to the next junction or to your destination (a family name).

Sometimes a key will lead you to a group of families, or uncertainty about a character will leave you undecided between two families. In those cases, a look at the photographs for each family under consideration should lead you to a correct identification.

Bear in mind that these are simplified keys designed for the northeastern fauna. Some small families that occur in western but not eastern North America are excluded, and some atypical and rarely encountered forms have been ignored to make the keys more user-friendly. Easily seen but occasionally imprecise characters (which work for most, but not all specimens) are often used instead of precise but difficult characters. Despite these caveats, the keys, used in conjunction with the photographs, should guide you quickly and easily to a useful identification for almost all routinely encountered insects.

Look for this icon to find corresponding photographs.

Look for this icon to find further keys.

Which Photos or Picture Keys Should I Start With?

1) Is it one of these five distinctively shaped orders?

With long tails and large, triangular front wings held above the body. ADULT MAYFLIES Pg. 624

Hind legs enlarged for jumping. Mouthparts with stout mandibles for chewing. GRASSHOPPERS AND CRICKETS Pg. 632

Antennae very short, body long, wings outstretched and non-folding. ADULT DRAGONFLIES AND DAMSELFLIES Pg. 628

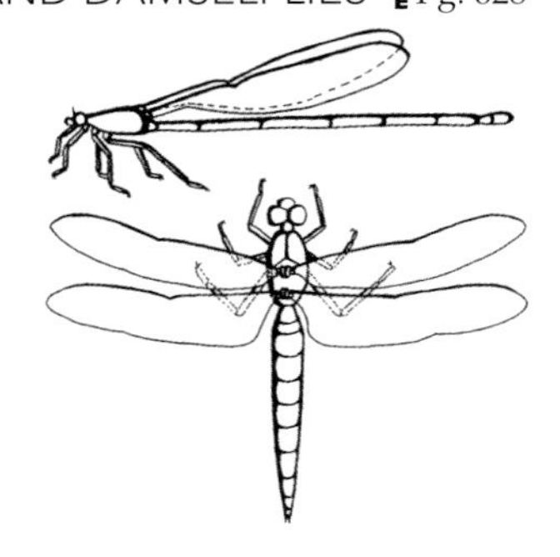

With forceps-like tails; adults with short leathery wings. EARWIGS

Front legs long and grasping, head separated from body by a long, thin neck. MANTIDS Pg. 68

See also photos of similar mantisflies (Neuroptera), page 255.

2) Does it have a hard shell made by the front wings meeting in a straight line down the back? Try the beetle keys. Pgs. 644–650

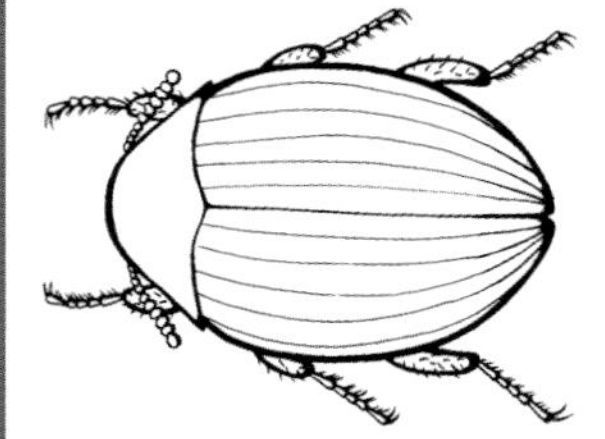

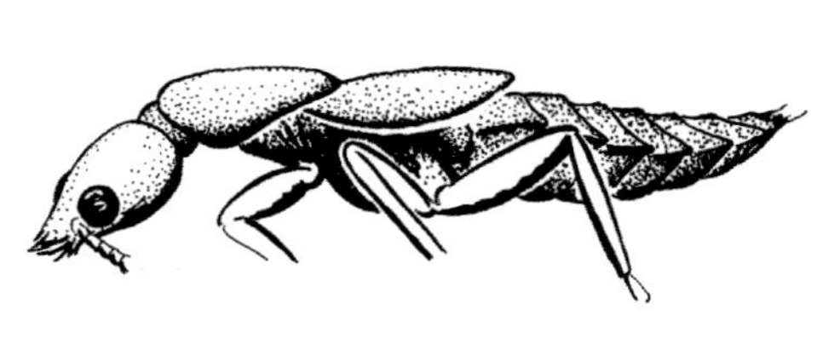

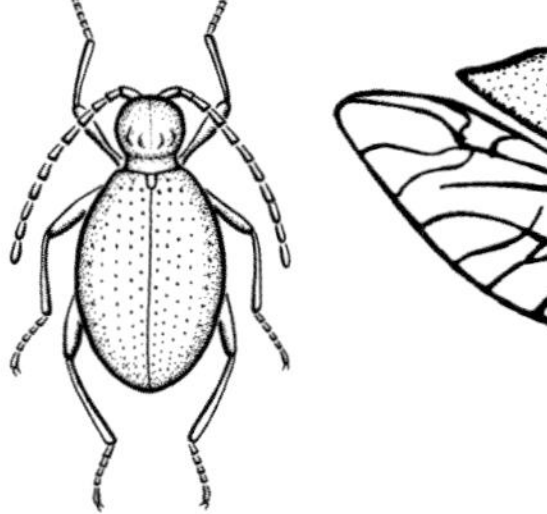

3) Does it have only one pair of wings, with the hind wings reduced to knobs? Try the fly keys. Pgs. 654–660

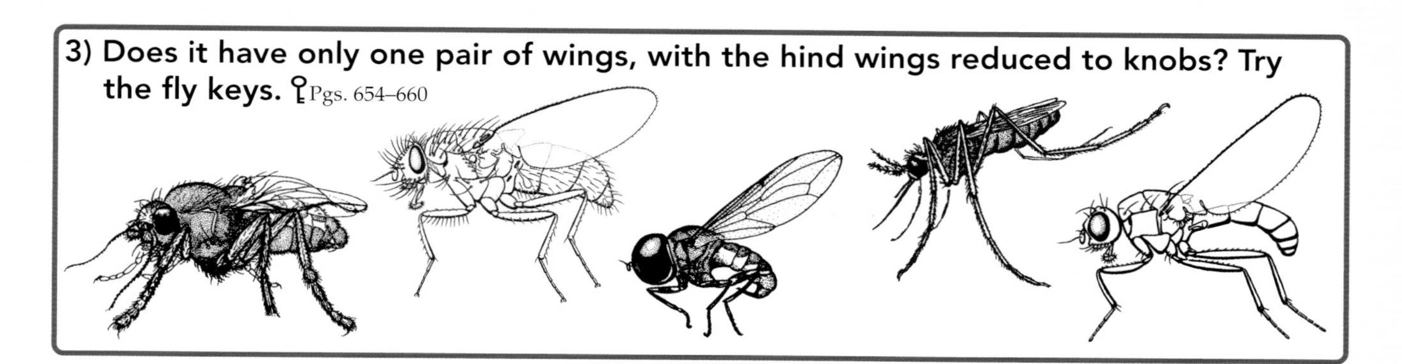

4) Is it a butterfly or moth, with wings and body mostly covered with powdery scales? See the photographs of butterflies and moths. Pgs. 180–236

5) Is it a larva, completely wingless, elongate, no compound eyes and usually worm-like or caterpillar-like? Try the key to larval insects. Pg. 622

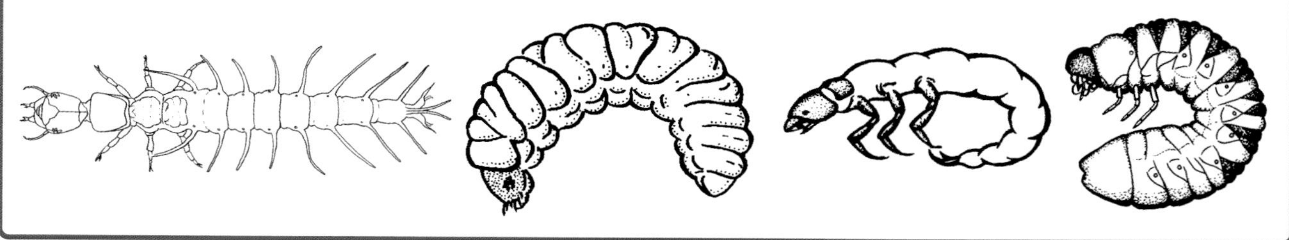

6) Is your insect a flightless adult, or a nymph with small wing buds? Start with the key to common wingless adult insects and nymphs. Pg. 618

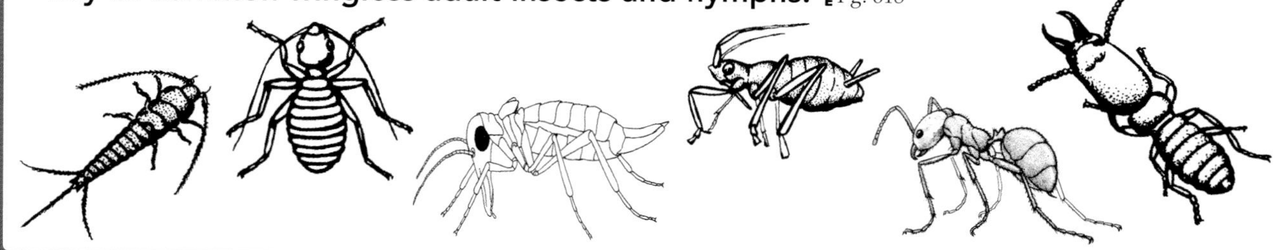

7) Is it a winged insect, but not a mayfly, earwig, dragonfly, damselfly, mantid, grasshopper, beetle, moth, or fly? Try the key to common winged insects. Pg. 620

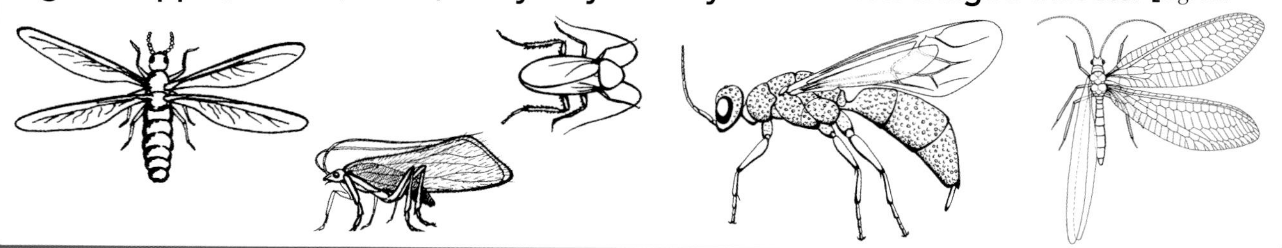

INSECT ORDERS KEY ONE: Wingless or almost wingless adults and common nymphal forms.

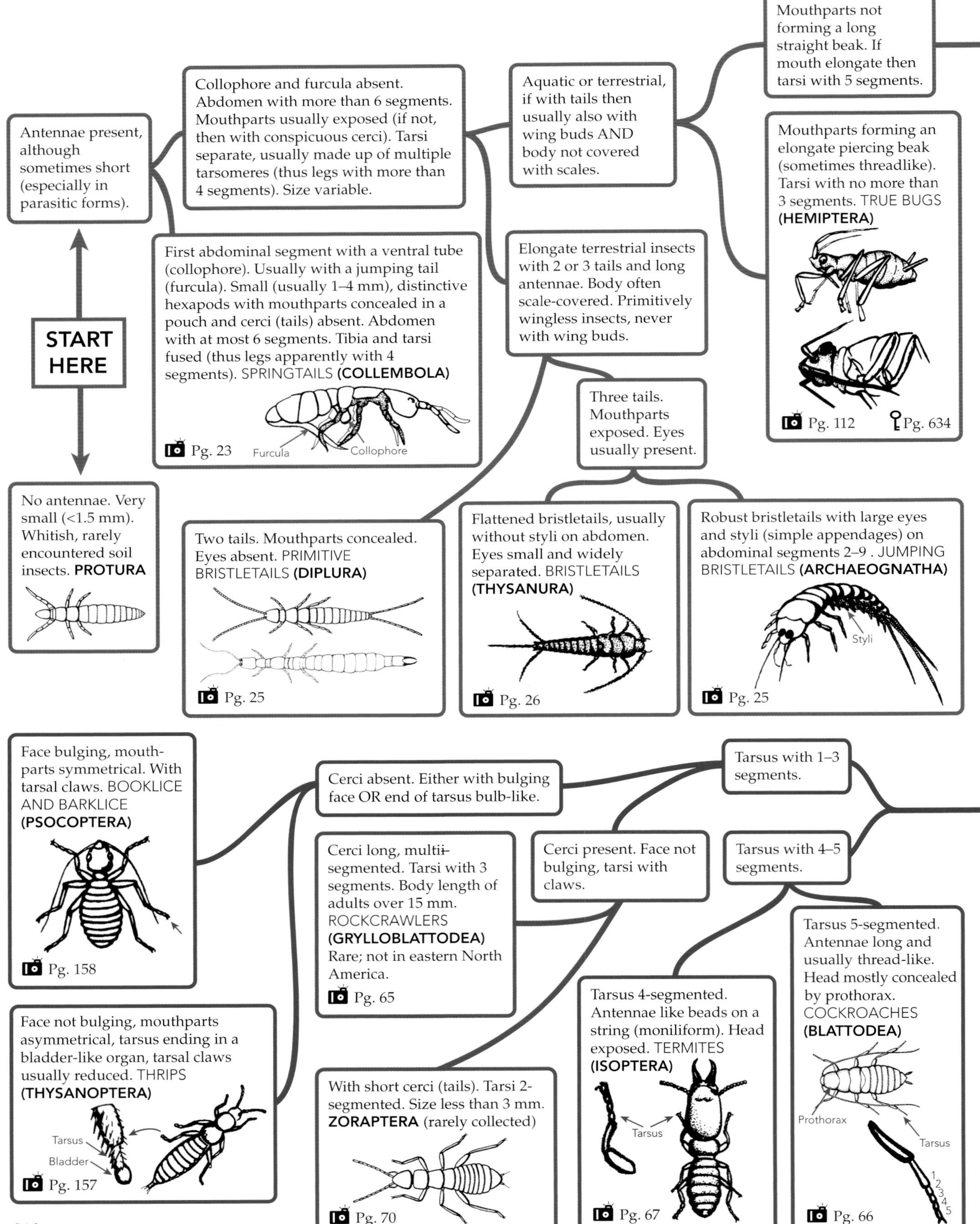

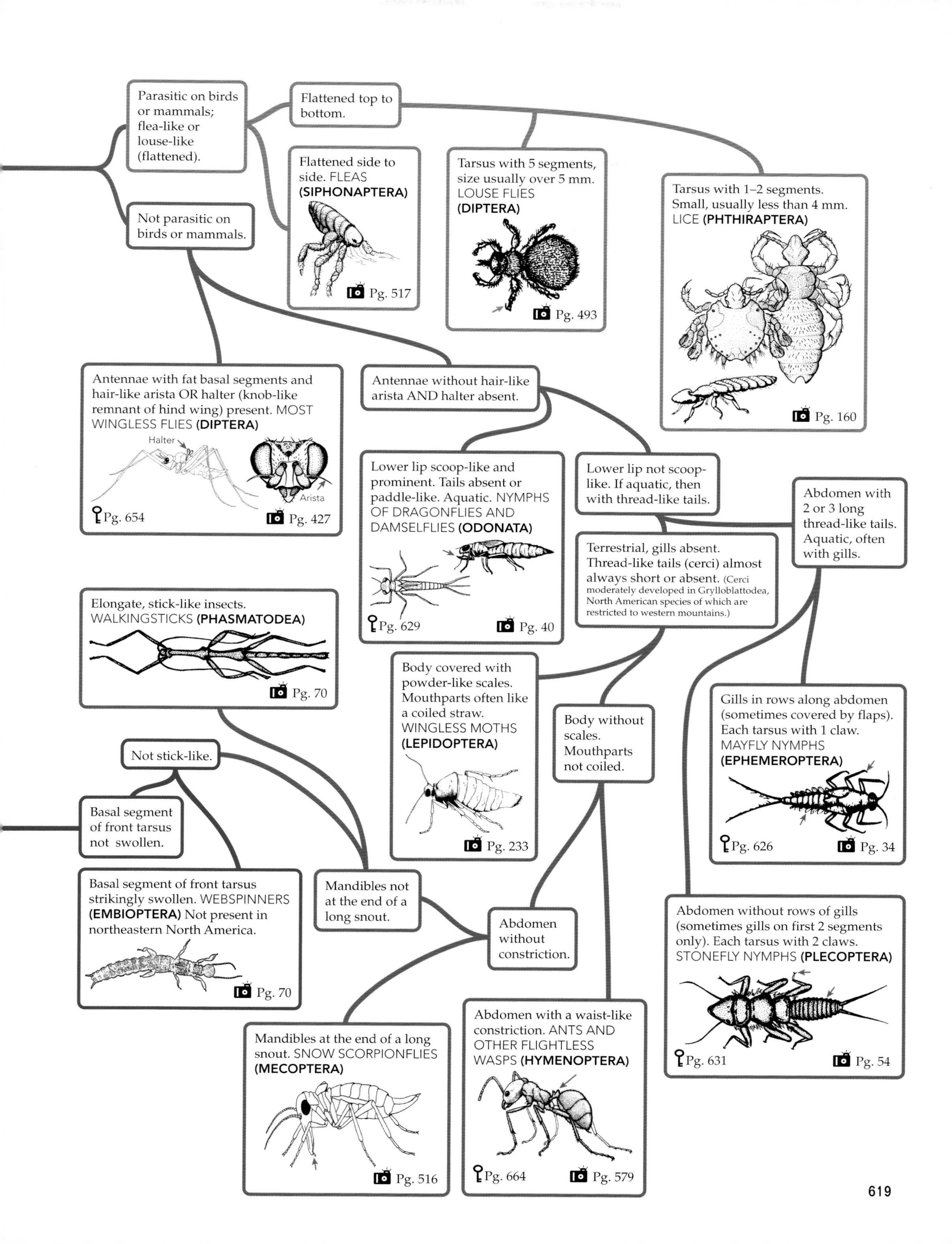
Parasitic on birds or mammals; flea-like or louse-like (flattened).
Flattened top to bottom.
Flattened side to side. FLEAS (SIPHONAPTERA)
Pg. 517
Tarsus with 5 segments, size usually over 5 mm. LOUSE FLIES (DIPTERA)
Pg. 493
Tarsus with 1–2 segments. Small, usually less than 4 mm. LICE (PHTHIRAPTERA)
Pg. 160
Not parasitic on birds or mammals.
Antennae with fat basal segments and hair-like arista OR halter (knob-like remnant of hind wing) present. MOST WINGLESS FLIES (DIPTERA)
Halter
Arista
Pg. 654
Pg. 427
Antennae without hair-like arista AND halter absent.
Lower lip scoop-like and prominent. Tails absent or paddle-like. Aquatic. NYMPHS OF DRAGONFLIES AND DAMSELFLIES (ODONATA)
Pg. 629
Pg. 40
Lower lip not scoop-like. If aquatic, then with thread-like tails.
Abdomen with 2 or 3 long thread-like tails. Aquatic, often with gills.
Terrestrial, gills absent. Thread-like tails (cerci) almost always short or absent. (Cerci moderately developed in Grylloblattodea, North American species of which are restricted to western mountains.)
Elongate, stick-like insects. WALKINGSTICKS (PHASMATODEA)
Pg. 70
Body covered with powder-like scales. Mouthparts often like a coiled straw. WINGLESS MOTHS (LEPIDOPTERA)
Pg. 233
Body without scales. Mouthparts not coiled.
Gills in rows along abdomen (sometimes covered by flaps). Each tarsus with 1 claw. MAYFLY NYMPHS (EPHEMEROPTERA)
Pg. 626
Pg. 34
Not stick-like.
Basal segment of front tarsus not swollen.
Basal segment of front tarsus strikingly swollen. WEBSPINNERS (EMBIOPTERA) Not present in northeastern North America.
Pg. 70
Mandibles not at the end of a long snout.
Abdomen without constriction.
Abdomen without rows of gills (sometimes gills on first 2 segments only). Each tarsus with 2 claws. STONEFLY NYMPHS (PLECOPTERA)
Pg. 631
Pg. 54
Mandibles at the end of a long snout. SNOW SCORPIONFLIES (MECOPTERA)
Pg. 516
Abdomen with a waist-like constriction. ANTS AND OTHER FLIGHTLESS WASPS (HYMENOPTERA)
Pg. 664
Pg. 579

INSECT ORDERS KEY TWO: Winged insects other than flies, moths and butterflies, beetles, grasshoppers and crickets, mayflies, earwigs, mantids, and dragonflies.

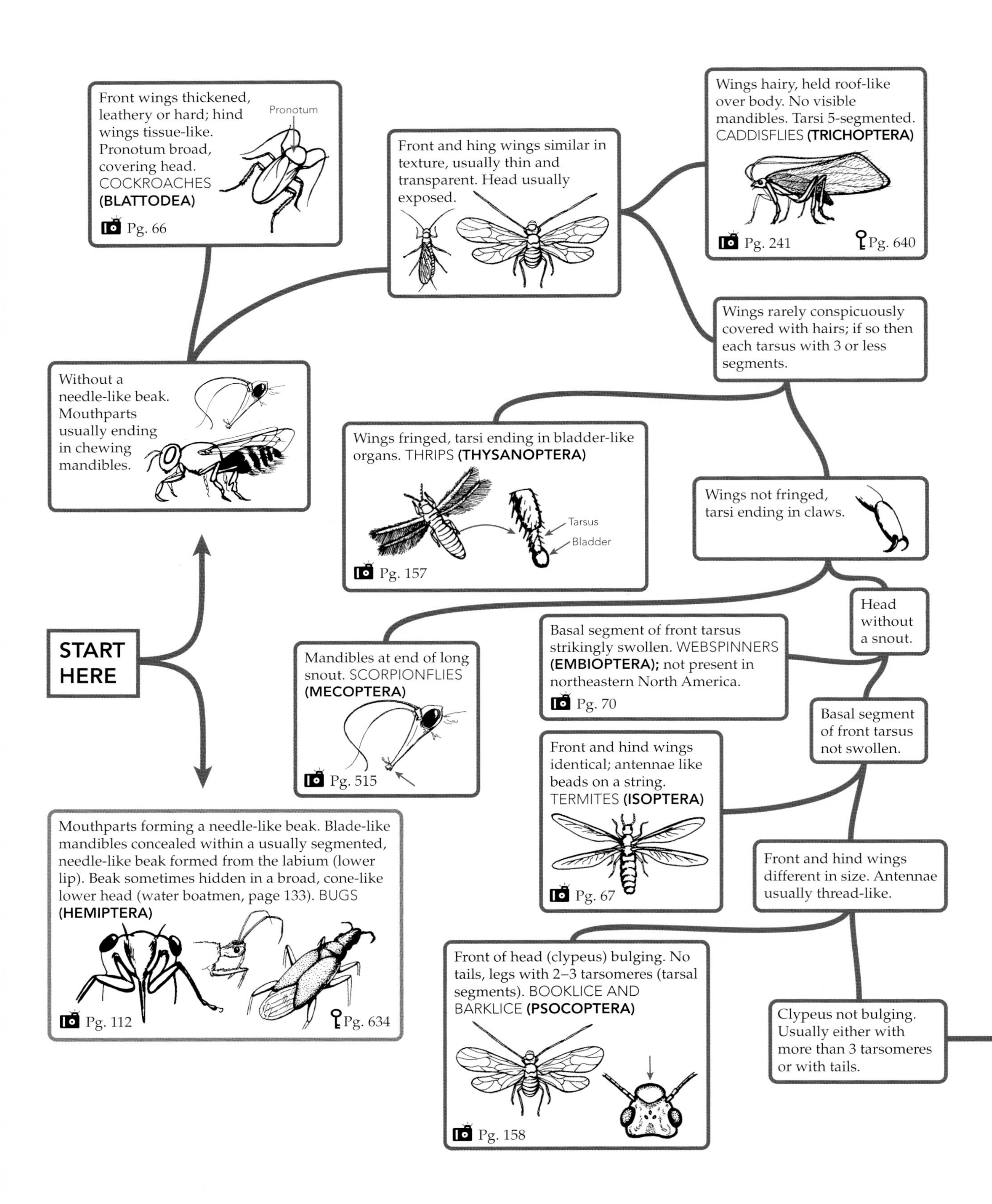

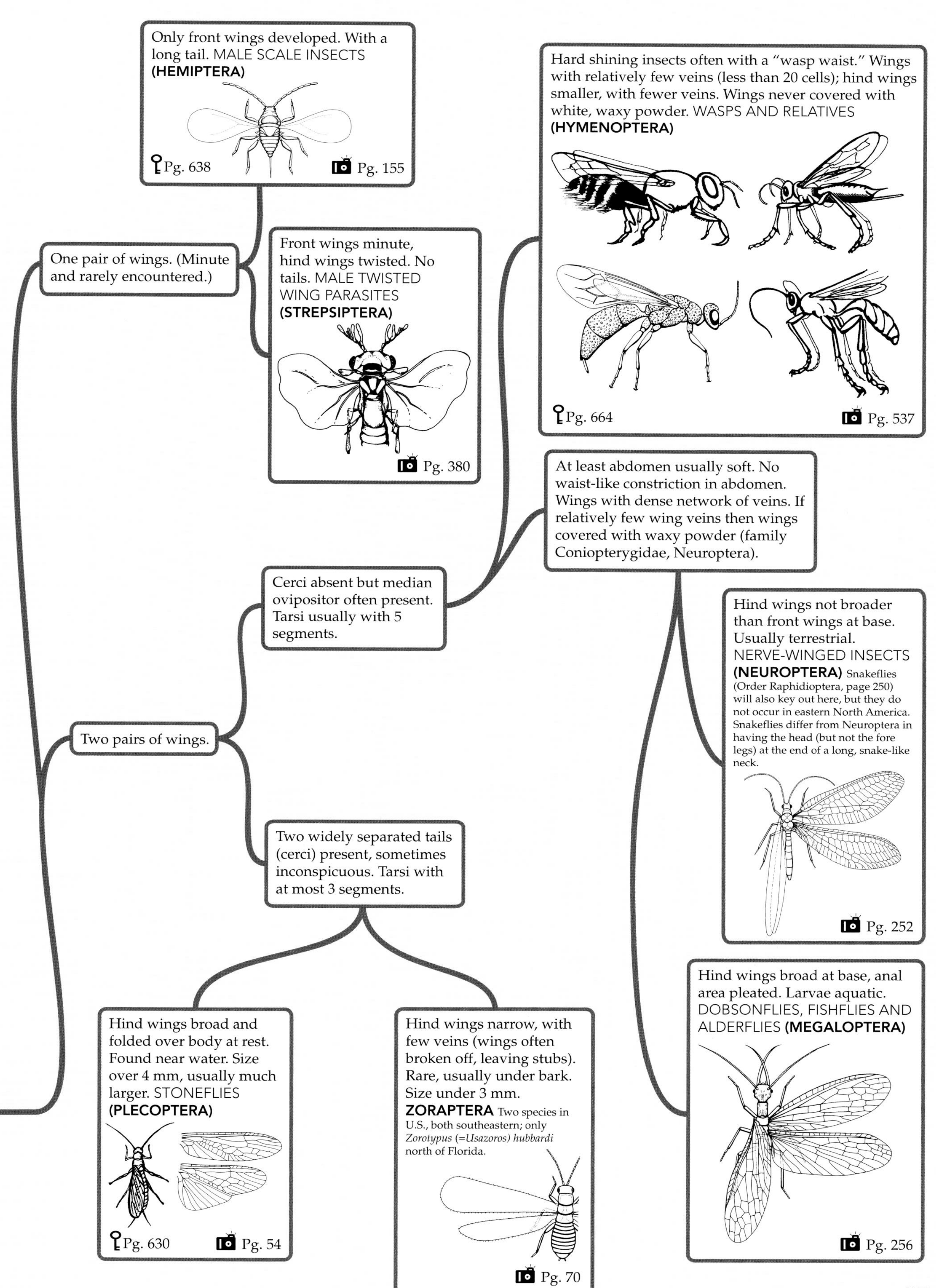

Only front wings developed. With a long tail. MALE SCALE INSECTS (HEMIPTERA)
Pg. 638
Pg. 155
One pair of wings. (Minute and rarely encountered.)
Front wings minute, hind wings twisted. No tails. MALE TWISTED WING PARASITES (STREPSIPTERA)
Pg. 380
Hard shining insects often with a "wasp waist." Wings with relatively few veins (less than 20 cells); hind wings smaller, with fewer veins. Wings never covered with white, waxy powder. WASPS AND RELATIVES (HYMENOPTERA)
Pg. 664
Pg. 537
At least abdomen usually soft. No waist-like constriction in abdomen. Wings with dense network of veins. If relatively few wing veins then wings covered with waxy powder (family Coniopterygidae, Neuroptera).
Cerci absent but median ovipositor often present. Tarsi usually with 5 segments.
Hind wings not broader than front wings at base. Usually terrestrial. NERVE-WINGED INSECTS (NEUROPTERA) Snakeflies (Order Raphidioptera, page 250) will also key out here, but they do not occur in eastern North America. Snakeflies differ from Neuroptera in having the head (but not the fore legs) at the end of a long, snake-like neck.
Pg. 252
Two pairs of wings.
Two widely separated tails (cerci) present, sometimes inconspicuous. Tarsi with at most 3 segments.
Hind wings broad at base, anal area pleated. Larvae aquatic. DOBSONFLIES, FISHFLIES AND ALDERFLIES (MEGALOPTERA)
Pg. 256
Hind wings broad and folded over body at rest. Found near water. Size over 4 mm, usually much larger. STONEFLIES (PLECOPTERA)
Pg. 630
Pg. 54
Hind wings narrow, with few veins (wings often broken off, leaving stubs). Rare, usually under bark. Size under 3 mm. ZORAPTERA Two species in U.S., both southeastern; only Zorotypus (=Usazoros) hubbardi north of Florida.
Pg. 70

KEY TO THE MOST COMMONLY ENCOUNTERED INSECT LARVAE

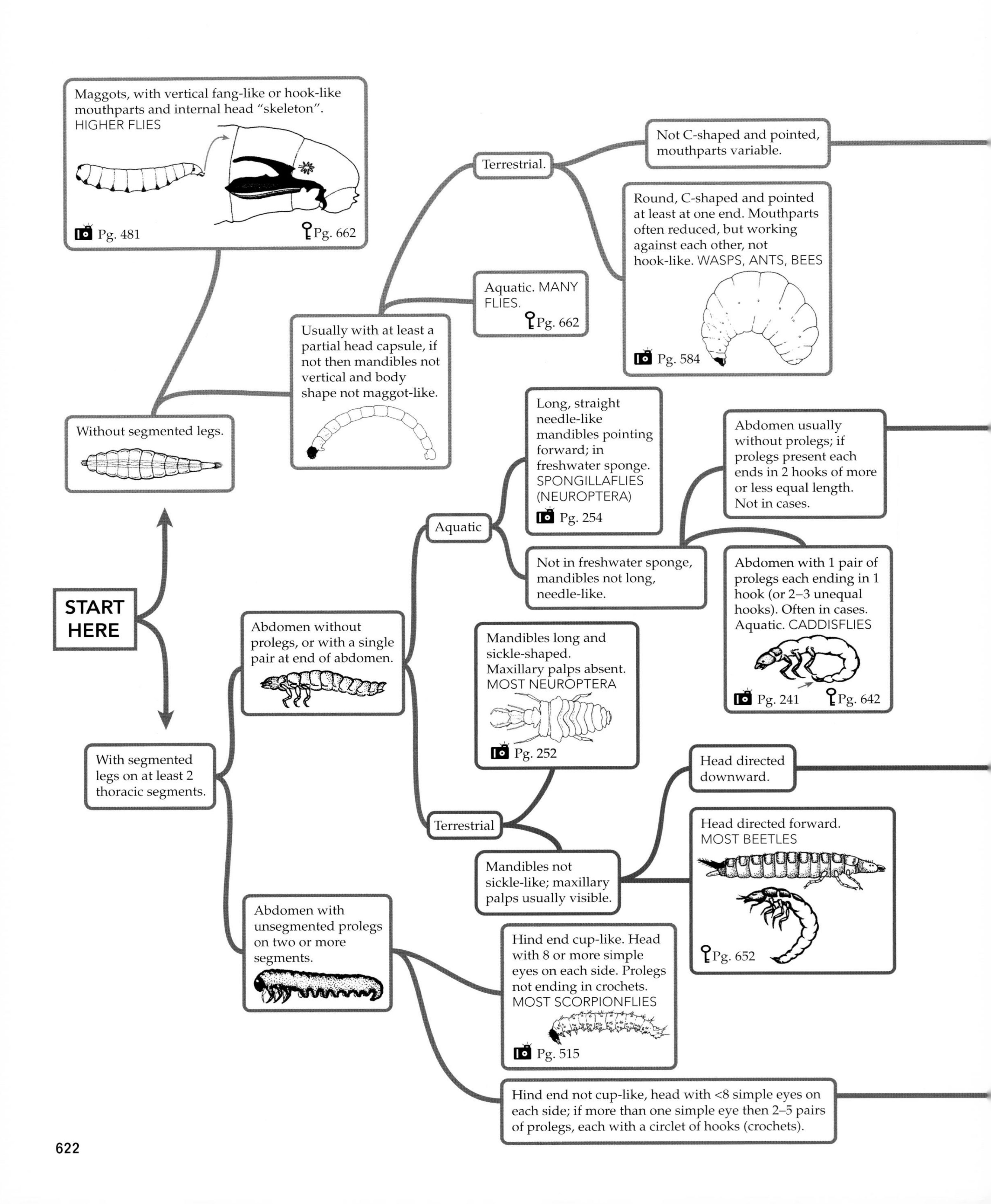

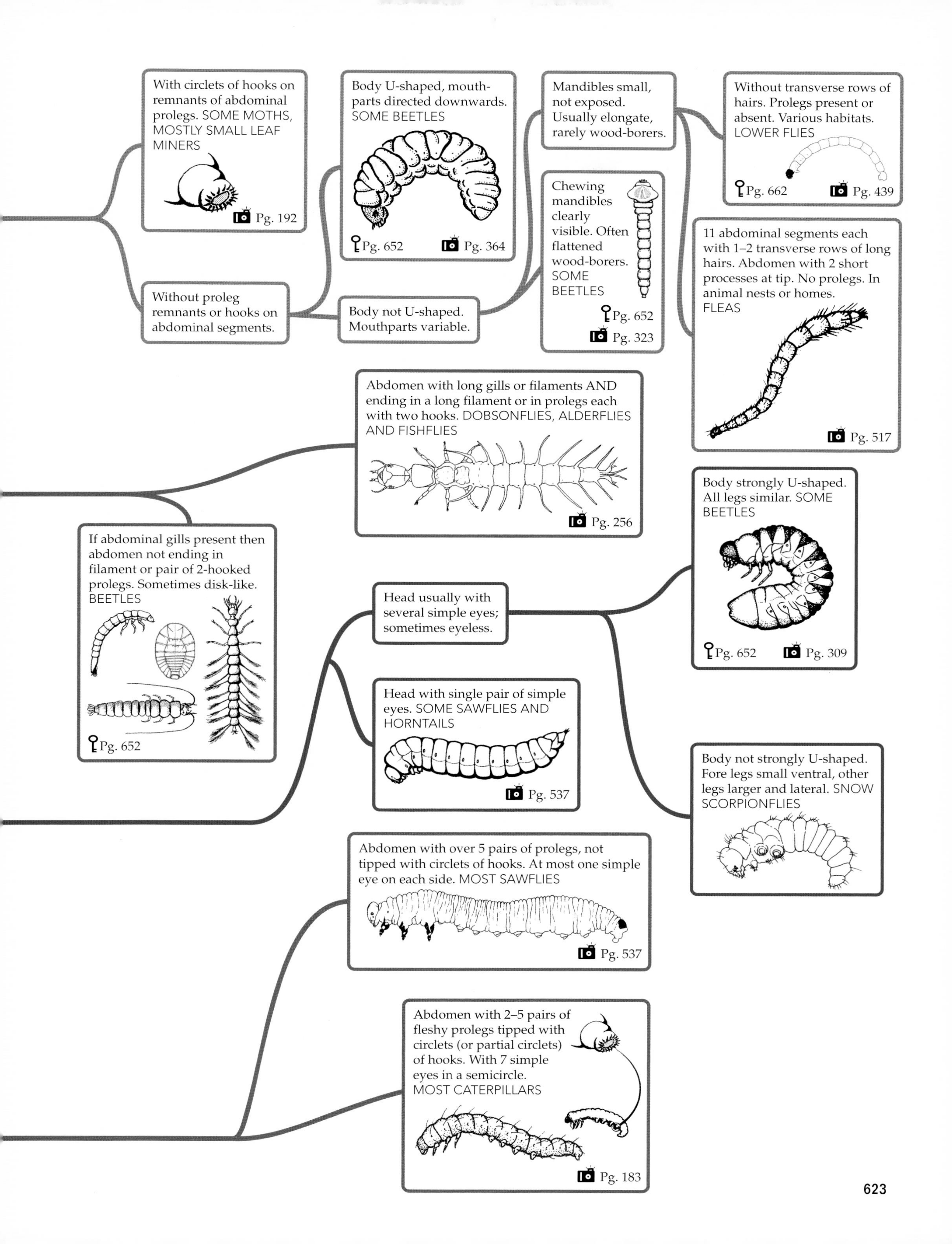

With circlets of hooks on remnants of abdominal prolegs. SOME MOTHS, MOSTLY SMALL LEAF MINERS
Pg. 192
Without proleg remnants or hooks on abdominal segments.
Body U-shaped, mouthparts directed downwards. SOME BEETLES
Pg. 652
Pg. 364
Body not U-shaped. Mouthparts variable.
Mandibles small, not exposed. Usually elongate, rarely wood-borers.
Chewing mandibles clearly visible. Often flattened wood-borers. SOME BEETLES
Pg. 652
Pg. 323
Without transverse rows of hairs. Prolegs present or absent. Various habitats. LOWER FLIES
Pg. 662
Pg. 439
11 abdominal segments each with 1–2 transverse rows of long hairs. Abdomen with 2 short processes at tip. No prolegs. In animal nests or homes. FLEAS
Pg. 517
Abdomen with long gills or filaments AND ending in a long filament or in prolegs each with two hooks. DOBSONFLIES, ALDERFLIES AND FISHFLIES
Pg. 256
If abdominal gills present then abdomen not ending in filament or pair of 2-hooked prolegs. Sometimes disk-like. BEETLES
Pg. 652
Head usually with several simple eyes; sometimes eyeless.
Body strongly U-shaped. All legs similar. SOME BEETLES
Pg. 652
Pg. 309
Head with single pair of simple eyes. SOME SAWFLIES AND HORNTAILS
Pg. 537
Body not strongly U-shaped. Fore legs small ventral, other legs larger and lateral. SNOW SCORPIONFLIES
Abdomen with over 5 pairs of prolegs, not tipped with circlets of hooks. At most one simple eye on each side. MOST SAWFLIES
Pg. 537
Abdomen with 2–5 pairs of fleshy prolegs tipped with circlets (or partial circlets) of hooks. With 7 simple eyes in a semicircle. MOST CATERPILLARS
Pg. 183

EPHEMEROPTERA KEY ONE: Mayfly adults.

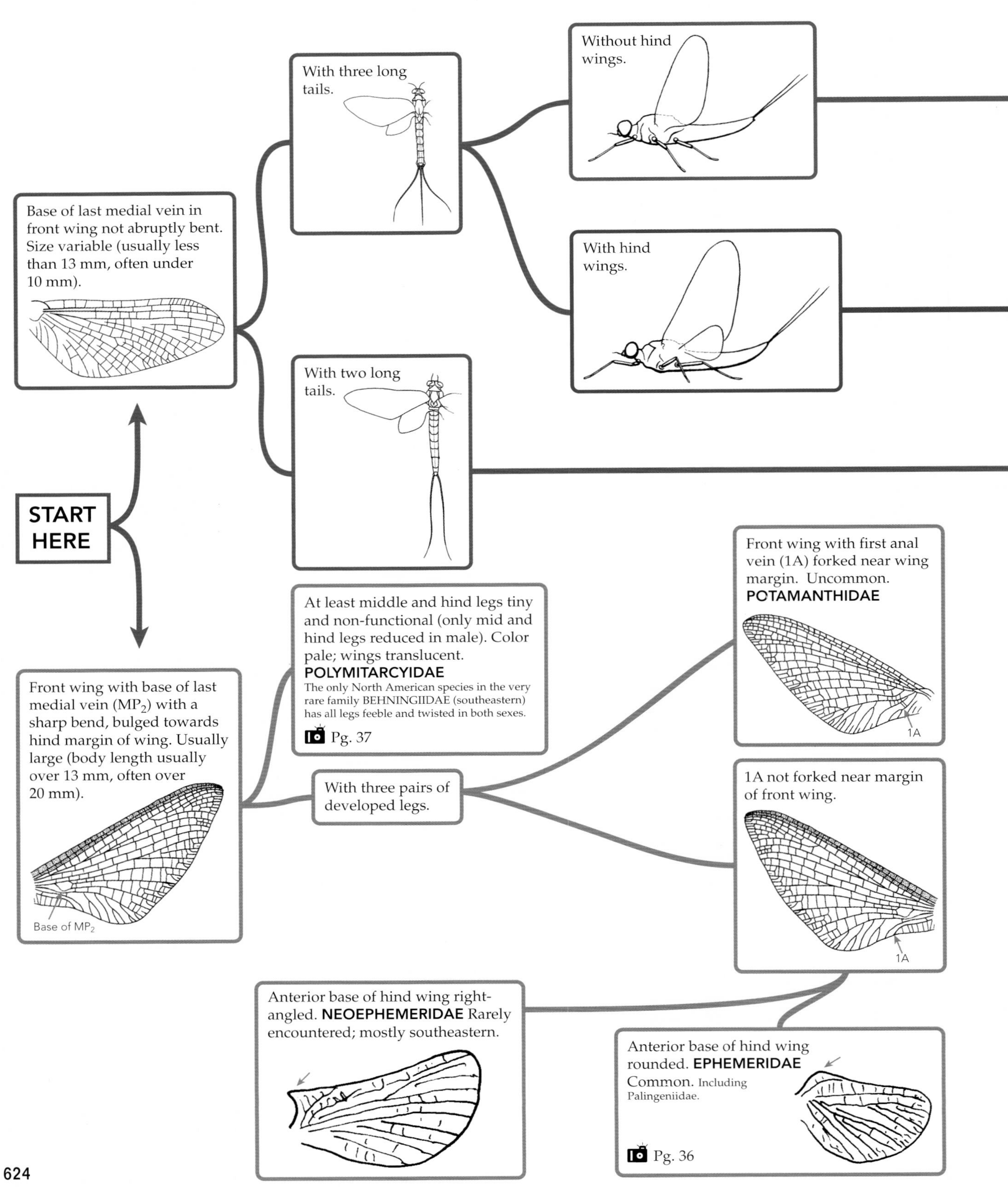

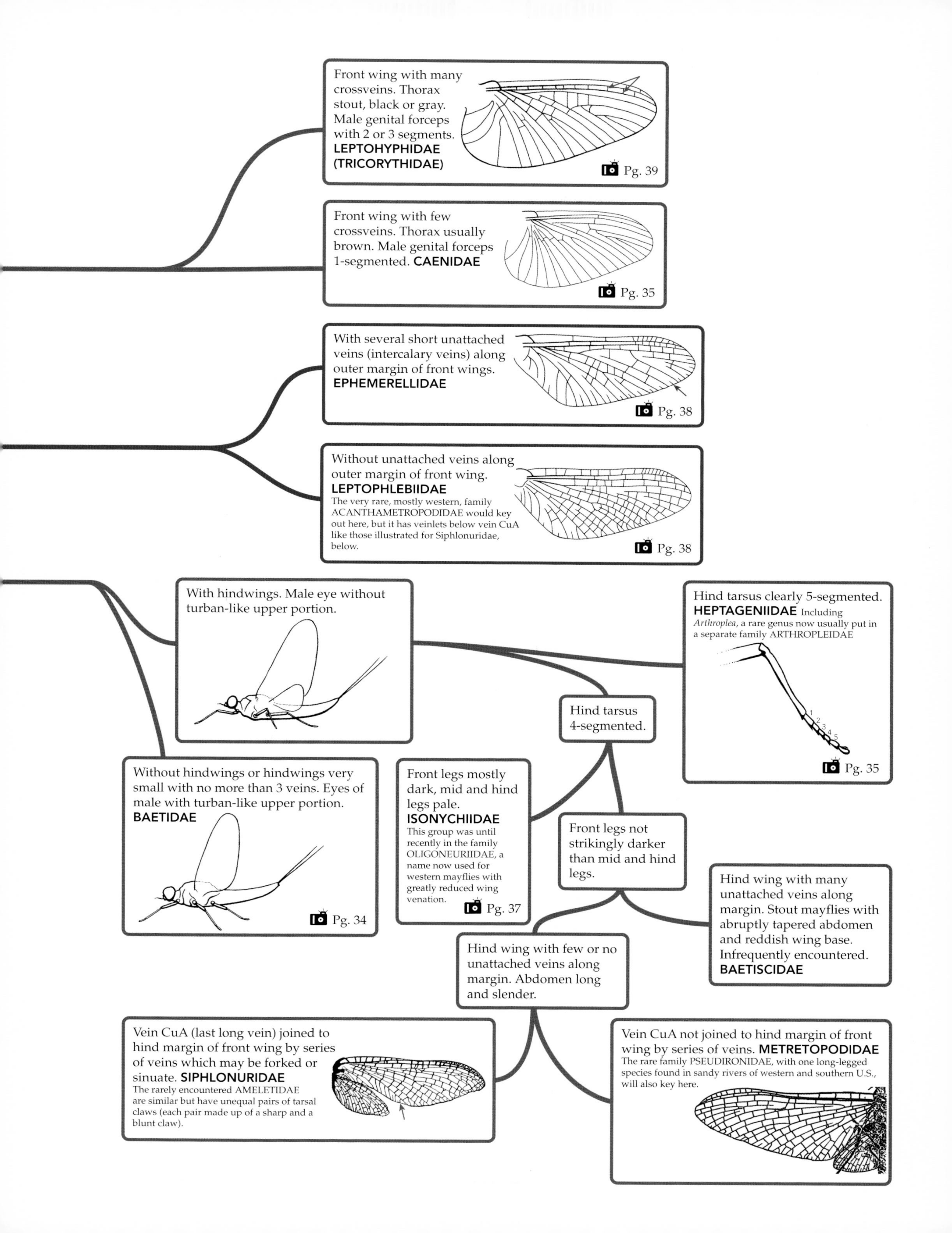

Front wing with many crossveins. Thorax stout, black or gray. Male genital forceps with 2 or 3 segments. LEPTOHYPHIDAE (TRICORYTHIDAE)
Pg. 39
Front wing with few crossveins. Thorax usually brown. Male genital forceps 1-segmented. CAENIDAE
Pg. 35
With several short unattached veins (intercalary veins) along outer margin of front wings. EPHEMERELLIDAE
Pg. 38
Without unattached veins along outer margin of front wing. LEPTOPHLEBIIDAE
The very rare, mostly western, family ACANTHAMETROPODIDAE would key out here, but it has veinlets below vein CuA like those illustrated for Siphlonuridae, below.
Pg. 38
With hindwings. Male eye without turban-like upper portion.
Hind tarsus clearly 5-segmented. HEPTAGENIIDAE Including Arthroplea, a rare genus now usually put in a separate family ARTHROPLEIDAE
1
2
3
4
5
Pg. 35
Hind tarsus 4-segmented.
Without hindwings or hindwings very small with no more than 3 veins. Eyes of male with turban-like upper portion. BAETIDAE
Pg. 34
Front legs mostly dark, mid and hind legs pale. ISONYCHIIDAE
This group was until recently in the family OLIGONEURIIDAE, a name now used for western mayflies with greatly reduced wing venation.
Pg. 37
Front legs not strikingly darker than mid and hind legs.
Hind wing with many unattached veins along margin. Stout mayflies with abruptly tapered abdomen and reddish wing base. Infrequently encountered. BAETISCIDAE
Hind wing with few or no unattached veins along margin. Abdomen long and slender.
Vein CuA (last long vein) joined to hind margin of front wing by series of veins which may be forked or sinuate. SIPHLONURIDAE
The rarely encountered AMELETIDAE are similar but have unequal pairs of tarsal claws (each pair made up of a sharp and a blunt claw).
Vein CuA not joined to hind margin of front wing by series of veins. METRETOPODIDAE
The rare family PSEUDIRONIDAE, with one long-legged species found in sandy rivers of western and southern U.S., will also key here.

EPHEMEROPTERA KEY TWO: Mayfly nymphs.

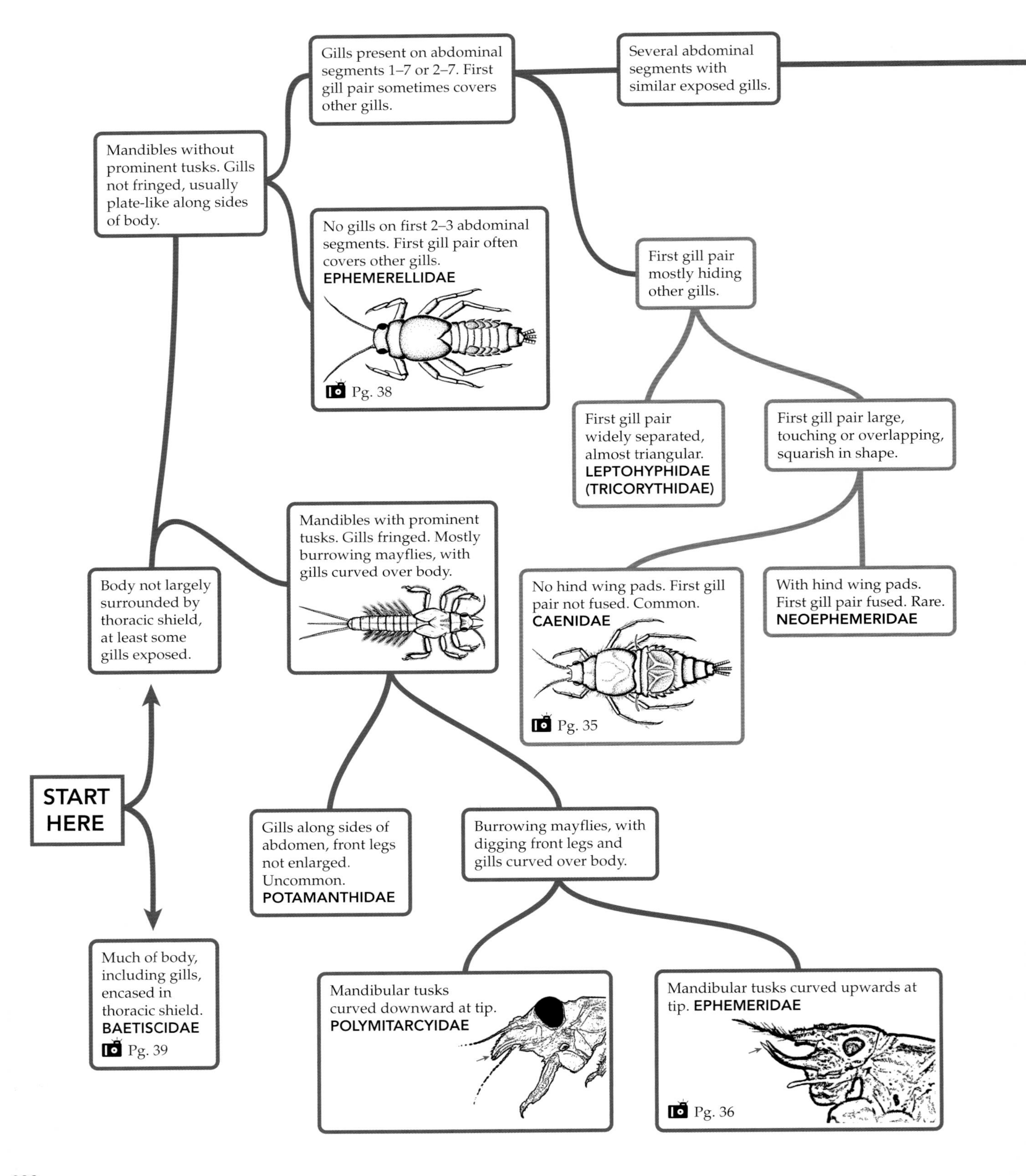

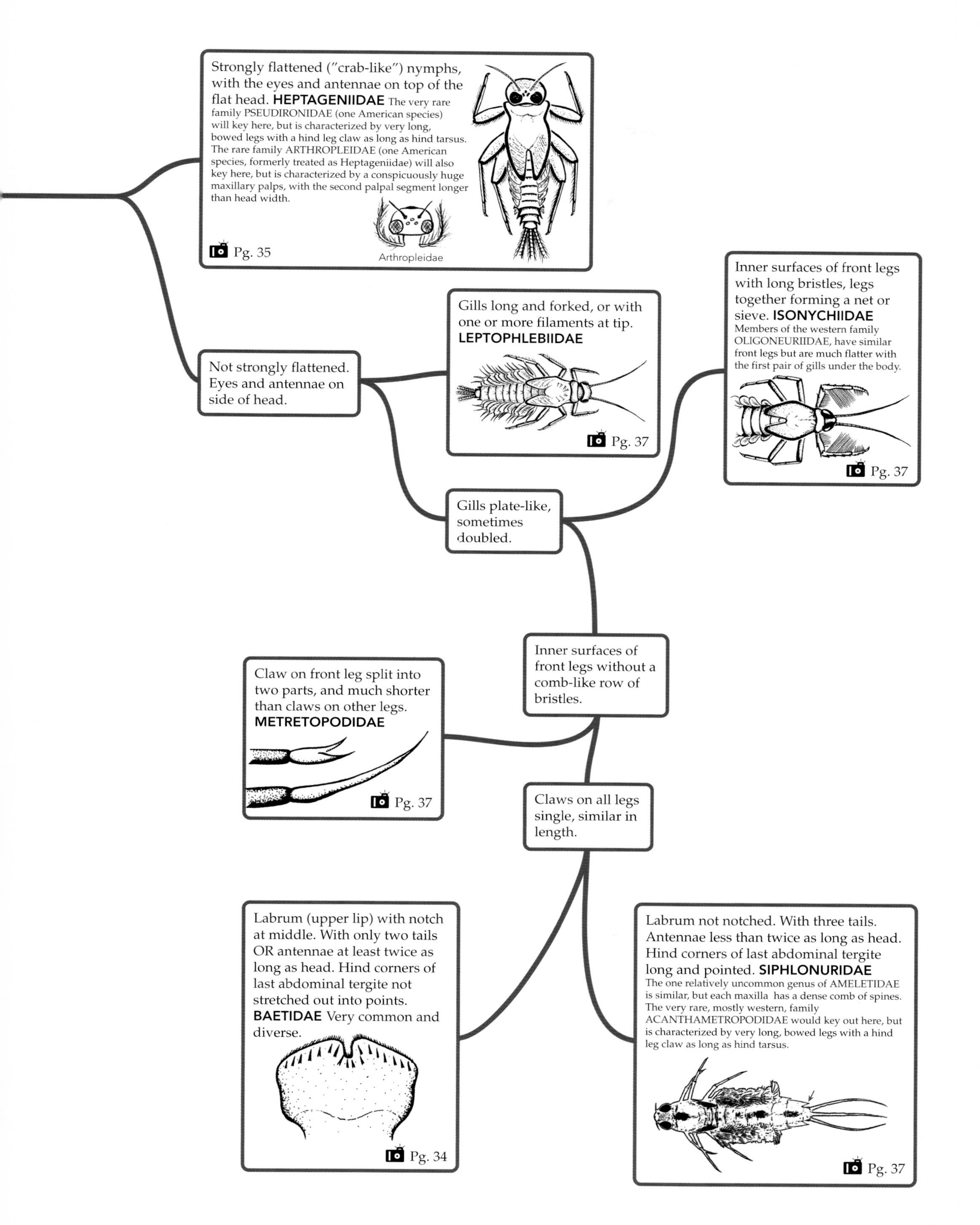

Strongly flattened ("crab-like") nymphs, with the eyes and antennae on top of the flat head. HEPTAGENIIDAE The very rare family PSEUDIRONIDAE (one American species) will key here, but is characterized by very long, bowed legs with a hind leg claw as long as hind tarsus. The rare family ARTHROPLEIDAE (one American species, formerly treated as Heptageniidae) will also key here, but is characterized by a conspicuously huge maxillary palps, with the second palpal segment longer than head width.
Arthropleidae
Pg. 35
Not strongly flattened. Eyes and antennae on side of head.
Gills long and forked, or with one or more filaments at tip. LEPTOPHLEBIIDAE
Pg. 37
Inner surfaces of front legs with long bristles, legs together forming a net or sieve. ISONYCHIIDAE Members of the western family OLIGONEURIIDAE, have similar front legs but are much flatter with the first pair of gills under the body.
Pg. 37
Gills plate-like, sometimes doubled.
Inner surfaces of front legs without a comb-like row of bristles.
Claw on front leg split into two parts, and much shorter than claws on other legs. METRETOPODIDAE
Pg. 37
Claws on all legs single, similar in length.
Labrum (upper lip) with notch at middle. With only two tails OR antennae at least twice as long as head. Hind corners of last abdominal tergite not stretched out into points. BAETIDAE Very common and diverse.
Pg. 34
Labrum not notched. With three tails. Antennae less than twice as long as head. Hind corners of last abdominal tergite long and pointed. SIPHLONURIDAE The one relatively uncommon genus of AMELETIDAE is similar, but each maxilla has a dense comb of spines. The very rare, mostly western, family ACANTHAMETROPODIDAE would key out here, but is characterized by very long, bowed legs with a hind leg claw as long as hind tarsus.
Pg. 37

ODONATA KEY ONE: Dragonfly and damselfly adults.

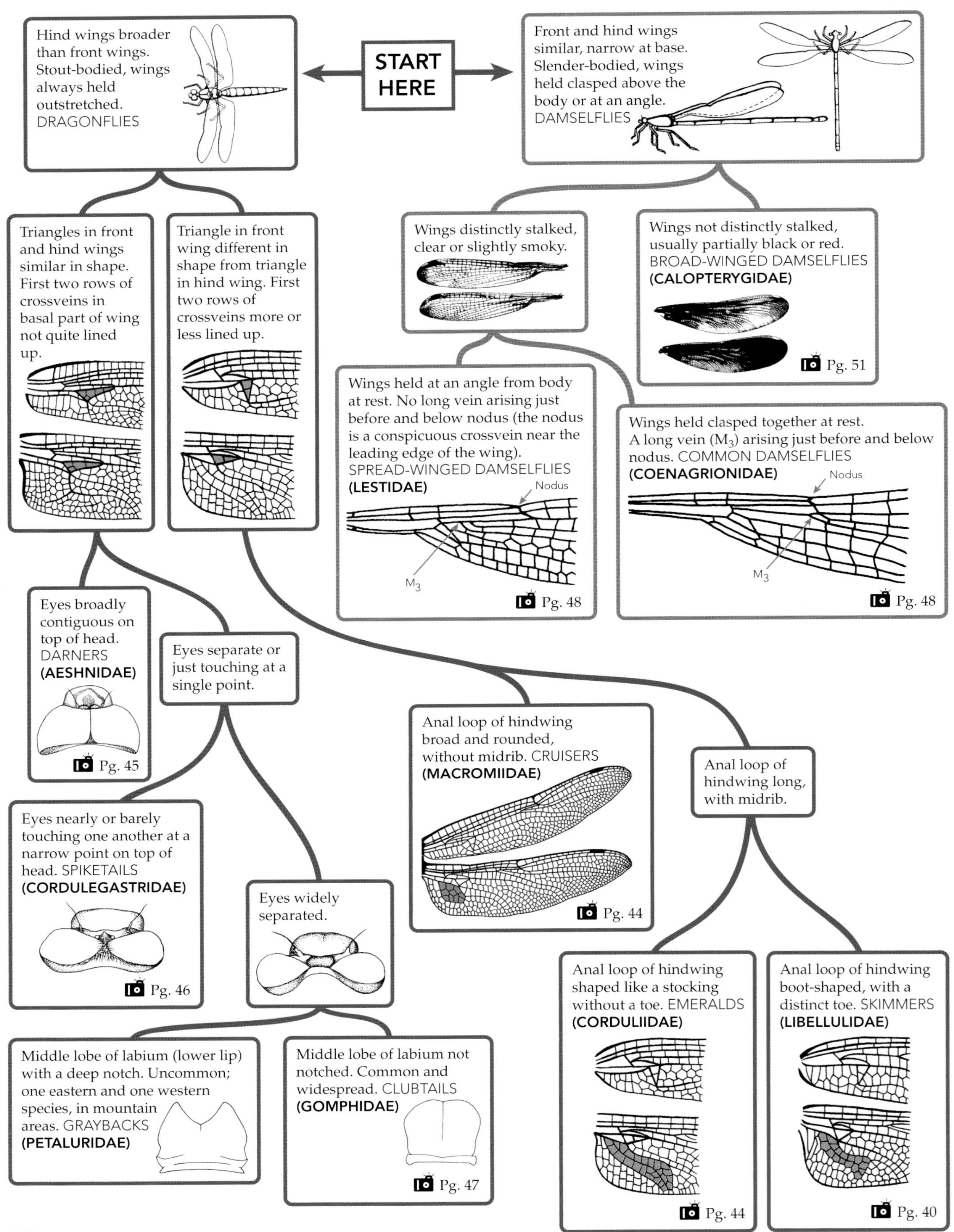

ODONATA KEY TWO: Dragonfly and damselfly nymphs.

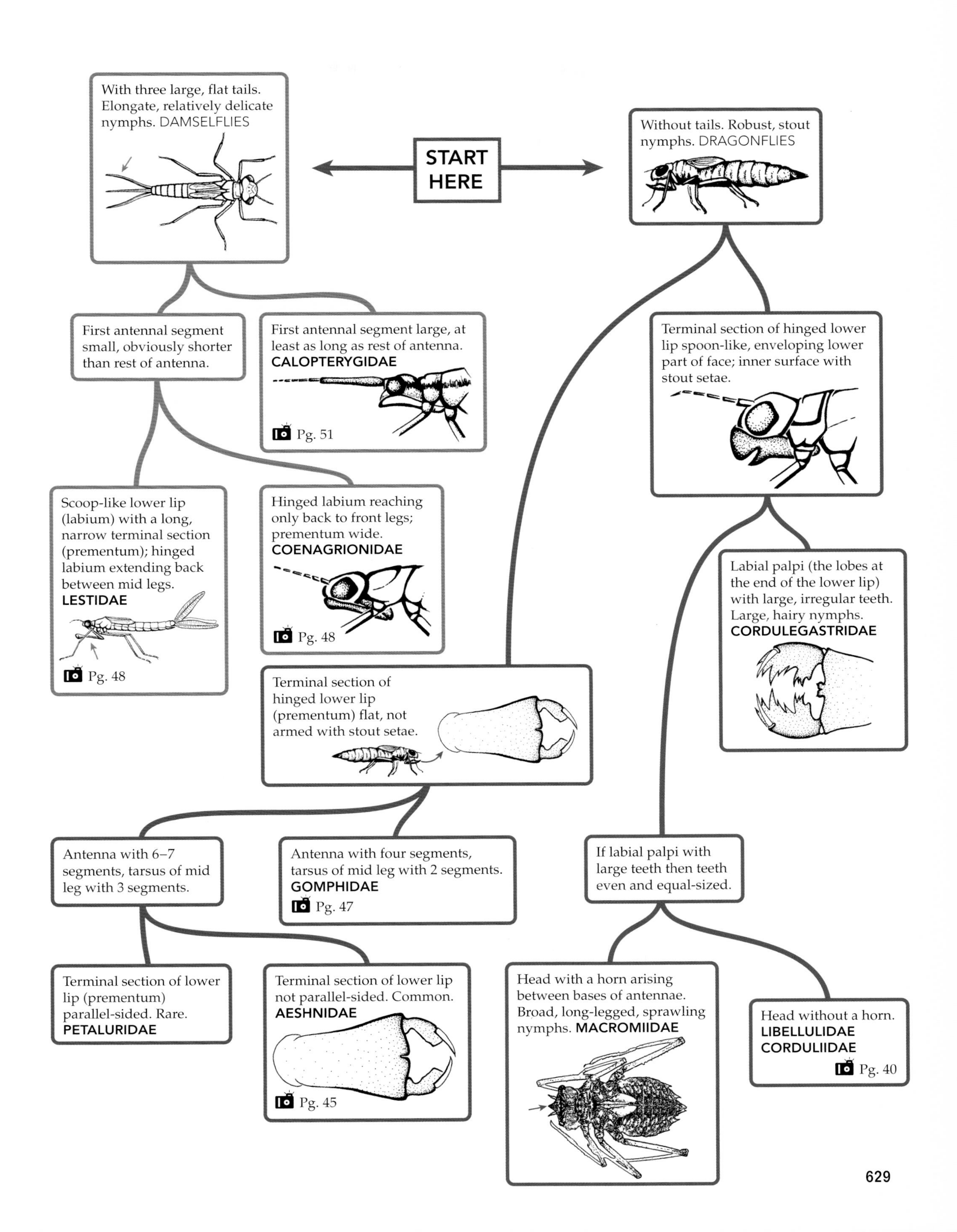

PLECOPTERA KEY ONE: Stonefly adults.

START HERE

Huge stoneflies, 25–65 mm in length but usually over 40 mm. Anal (trailing) area of forewing with rows of crossveins. Labium ending in four equal short lobes. GIANT STONEFLIES **(PTERONARCYIDAE)**

Pg. 54

Small to large stoneflies, usually well under 25 mm but if 20–35 mm then labium ending in 2 large outer lobes and 2 much smaller inner lobes, and anal area of forewing with at most one row of crossveins.

Broad-bodied, cockroach-like stoneflies, 12–18 mm in length. Only one pair of palps visible in dorsal view (looking at the head from above). Two ocelli. Labium ending in four small equal lobes. Basal tarsal segment much shorter than apical segment. Found in mountain streams. ROACH-LIKE STONEFLIES **(PELTOPERLIDAE)**

Pg. 54

Either small (less than 14 mm, usually less than 10 mm) and narrow-bodied, with basal tarsal segment as long as apical segment OR larger with two pairs of palps extending in front of head (thus visible from above), and end of labium with large outer lobes and small inner lobes. Usually 3 ocelli.

First segment of tarsus at least as long as last segment. Labium (lower lip) ending in four lobes (glossae large). Mostly winter, fall and spring stoneflies.

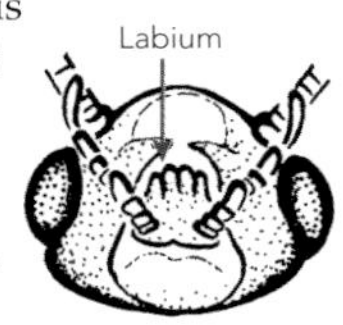

First segment of tarsus obviously shorter than last segment. Labium with two large outer lobes (paraglossae); inner lobes (glossae) very small so the labium seems to end in just two lobes. Mostly summer stoneflies.

Cerci (tails) long, at least 4 segments.

Cerci short and one segmented.

Second tarsal segment as long as first. WINTER STONEFLIES **(TAENIOPTERYGIDAE)**

Pg. 56

Second tarsal segment shorter than first or last segments. SMALL WINTER STONEFLIES **(CAPNIIDAE)**

Pg. 55

Usually no gill remnants on thorax; if gill remnants present they are never branched. Size variable, usually under 15 mm; if larger then paraglossa not bent in at abrupt angle.

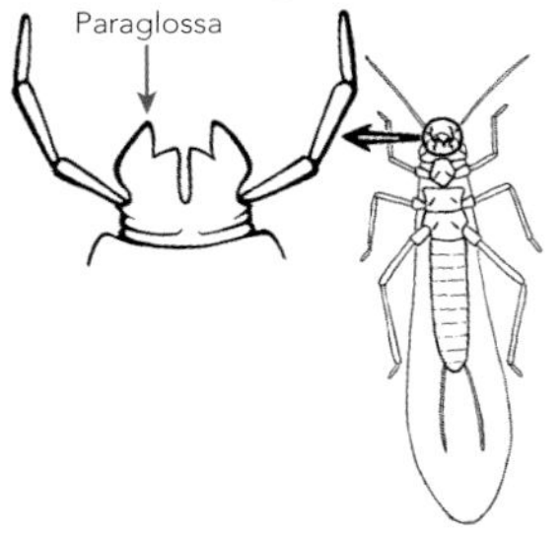

Thorax with remnants of branched gills.* Usually over 20 mm. Paraglossa bent in at abrupt angle. COMMON STONEFLIES **(PERLIDAE)**

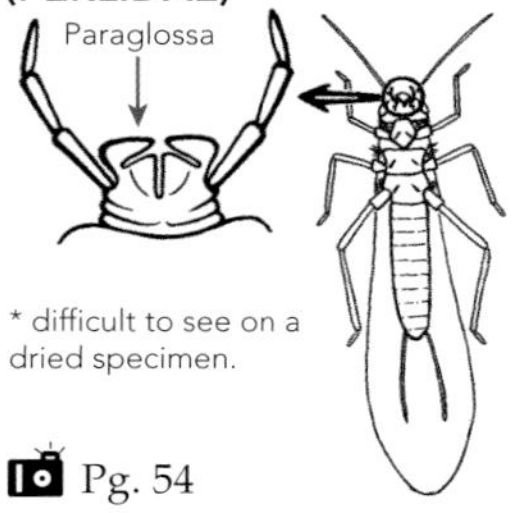

* difficult to see on a dried specimen.

Pg. 54

Front wings wrapped around the slender body giving it a cigar-shape; no "X" pattern in front wing. ROLLED-WINGED STONEFLIES **(LEUCTRIDAE)**

Pg. 57

Front wings, held flat at rest, with a distinctive "X" in the apical wing venation. BROADBACKS **(NEMOURIDAE)**

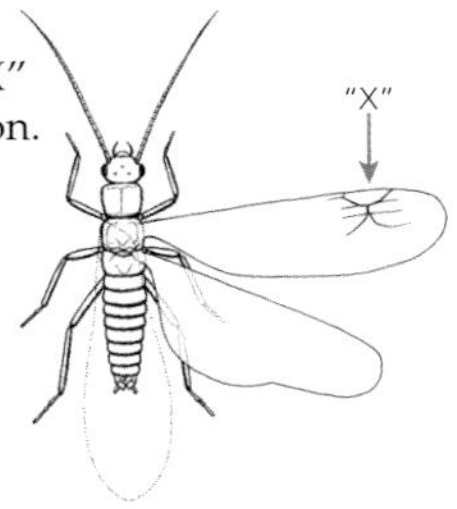

Pg. 56

Pronotum quadrate, with angulate corners. Color variable, sometimes green. Hind wing with 5–10 anal veins in a broad anal area. **PERLODIDAE**

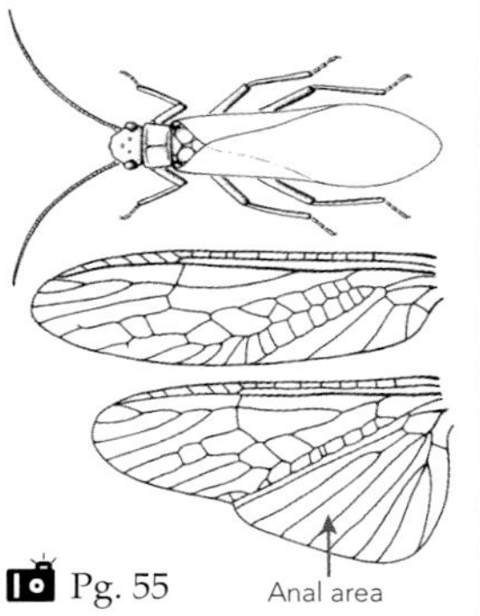

Pg. 55

Pronotum rounded. Color almost always green or yellow. Hind wing usually with 1–4 anal veins in a narrow anal area. GREEN STONEFLIES **(CHLOROPERLIDAE)**

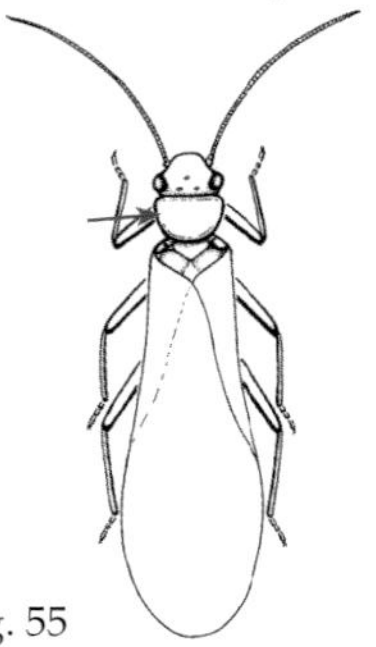

Pg. 55

PLECOPTERA KEY TWO: Stonefly nymphs.

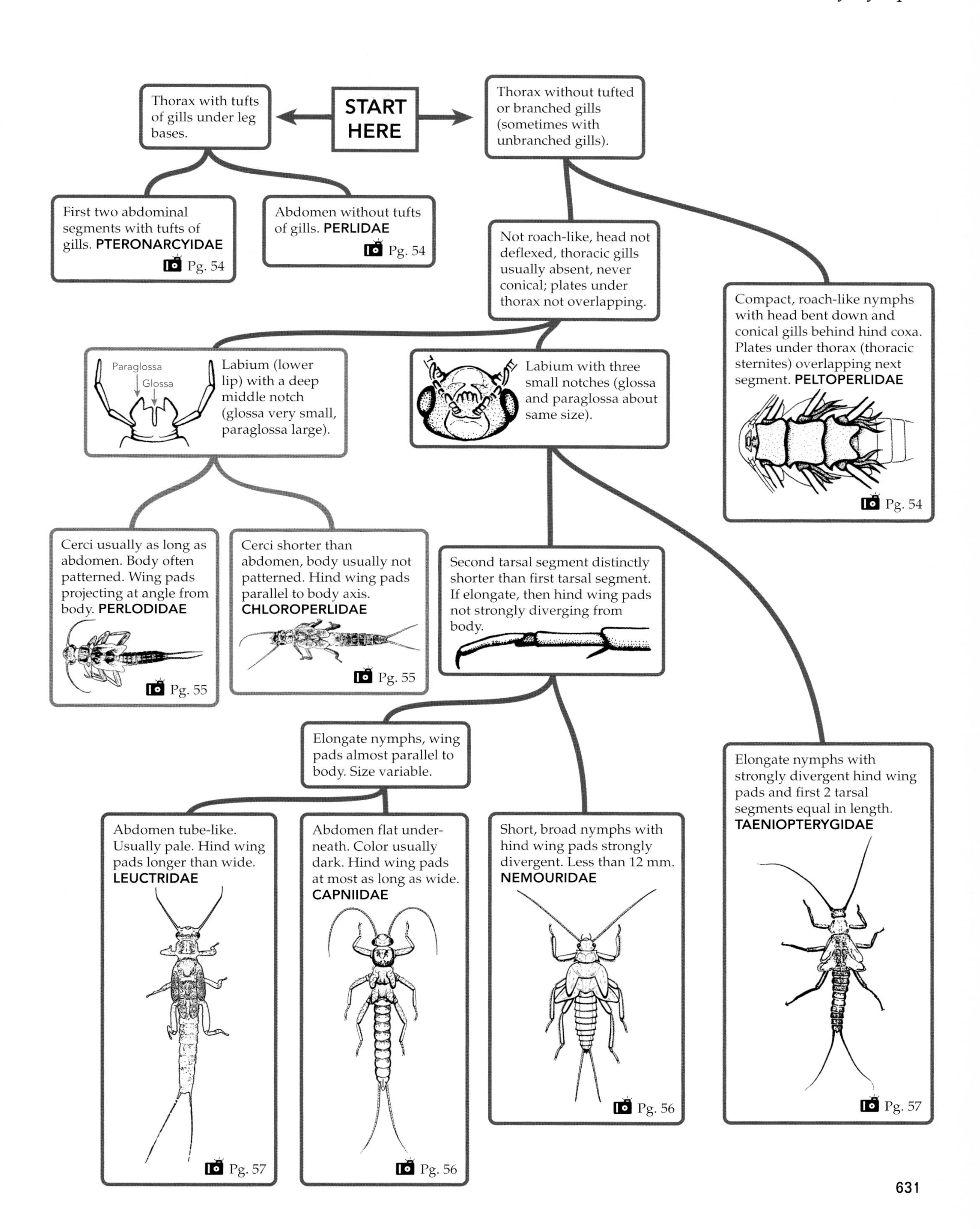

ORTHOPTERA KEY: Grasshoppers and crickets.

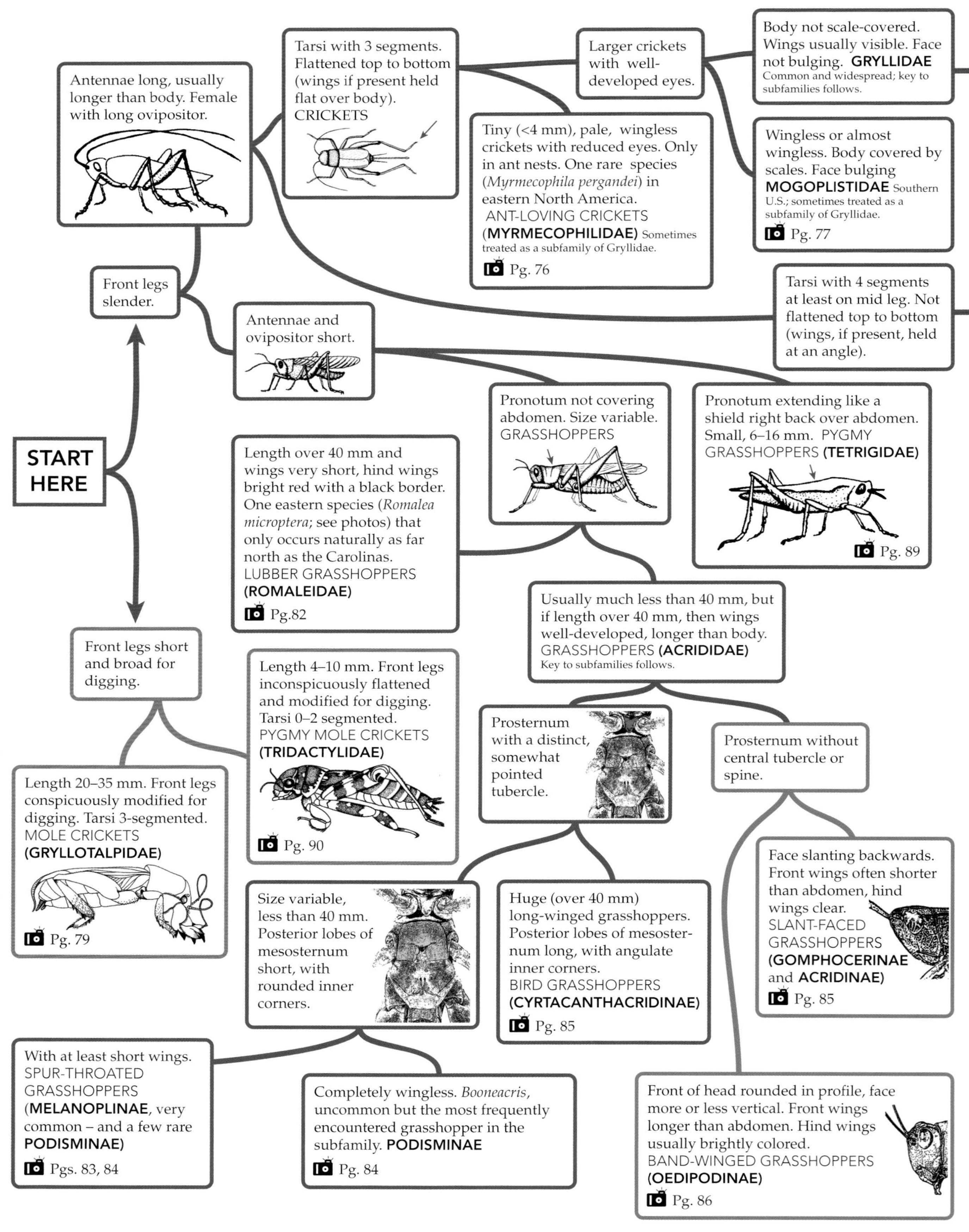

Brown or black, usually stout-bodied and with ocelli. Wings not extending beyond tip of abdomen.

Second (middle) tarsal segment large and depressed (flattened top to bottom). Most northeastern species are uncommon, less than 10 mm, and have the ovipositor flat and strongly curved upwards.

Size over 12 mm. Ovipositor straight, needle-like. More than three pairs of long moveable (arising in sockets) tibial bristles. Uncommon in northeast, not reaching Canada. ENEOPTERINE BUSH CRICKETS **(ENEOPTERINAE)** Pg. 76

Elongate, delicate crickets with no ocelli. Often pale green. Wings longer than abdomen. TREE CRICKETS **(OECANTHINAE)** Pg. 77

Second (middle) tarsal segment small and compressed (flattened side to side). Ovipositor straight or slightly curved. Very common; size variable.

Size less than 9 mm. Ovipositor flat and strongly curved up. Common species either black with red head and pronotum, or brown with only three pairs of hind tibial bristles. BUSH CRICKETS **(TRIGONIDIINAE)** Pg. 76

Size over 14 mm. Hind tibia not appearing bristly; with only short, stout, immobile spines. FIELD CRICKETS AND HOUSE CRICKETS **(GRYLLINAE)** Also the southeastern subfamily BRACHYTRUPINAE (very short ovipositor, ocelli forming a row rather than a triangle). Pg. 76

Size less than 12 mm. Hind tibia appearing bristly, with pairs of long slender bristles in sockets. GROUND CRICKETS **(NEMOBIINAE)** Pg. 76

Body not strongly arched (topline almost straight), with pronotum strongly differentiated from other segments. Wings usually present at least as stubs. Tarsi depressed. KATYDIDS AND OTHER LONG-HORNED GRASSHOPPERS **(TETTIGONIIDAE)**

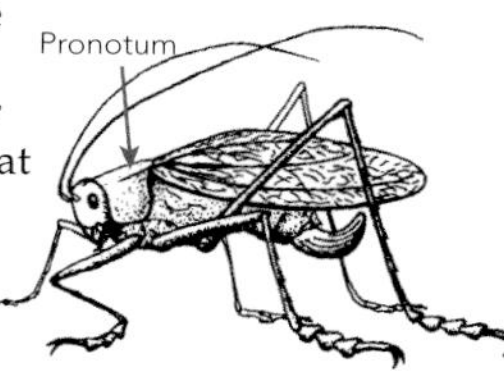

Elongate, wingless predators with conspicuously enlarged and spiny front and middle legs. Introduced species (*Saga pedo*) known from one county in Michigan. ARMORED KATYDIDS **(SAGINAE)**

Front and middle legs neither enlarged nor with stout spines. Wings present or absent.

Topline of body convex (body arched) and uniform in texture. Wings entirely absent. Tarsi compressed.

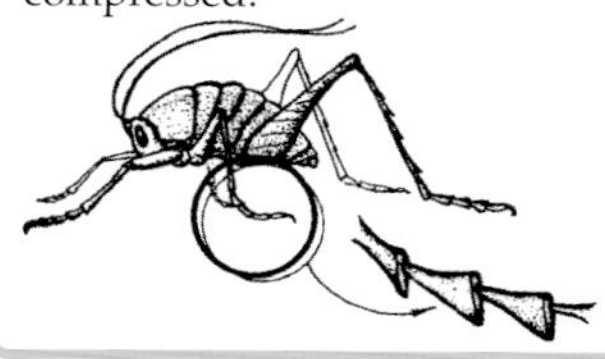

Antennae conspicuously separated, with a gap between the bases. LEAF-ROLLING CRICKETS **(GRYLLACRIDIDAE)** (unlikely to be encountered in the northeast) Pg. 79

Antennae inserted very close together, bases almost touching. CAMEL AND CAVE CRICKETS **(RHAPHIDOPHORIDAE)** Pg. 78

Front wings (if present) elongate-oval, not clamshell-like.

Front wings oval, leaflike, more or less enclosing body like a clamshell. One distinctive species (*Pterophylla camellifolia*) in northeast. TRUE KATYDIDS **(PSEUDOPHYLLINAE)** Pg. 79

Not "cone-headed"; front of head not extending in front of antennal bases.

Front of head strikingly long and extending conelike in front of antennal bases. CONEHEADS **(COPIPHORINAE)** Pg. 82

Front tibia with dorsal spines. Wings usually greatly reduced; native species brown and cricket-like (but see photos of the introduced *Metrioptera roeselii*). Pronotum large and shieldlike. SHIELD-BACKED KATYDIDS **(TETTIGONIINAE)** Pg. 79

Head somewhat angulate when viewed from the side, not rounded and short. Hind wings not projecting beyond front wings. Size and color variable, but usually less than 20 mm. Tympanal openings on fore tibia slit-like.

Without yellow dorsal stripe (if with any dorsal stripe then tympanal opening slit-like). Male without long, tube-like cerci. Size variable.

Head and pronotum with yellow dorsal stripe. Delicate green katydids less than 20 mm long. Male with conspicuously long tube-like cerci. Tympana (ears) on fore tibia exposed, oval. DRUMMING KATYDIDS **(MECONEMATINAE)** one uncommon introduced species, *Meconema thalassinum*.

Front tibia without dorsal spines. Wings usually well-developed, sometimes moderately shortened. Pronotum relatively small. MEADOW KATYDIDS **(CONOCEPHALINAE)** Pg. 80

Head rounded and short. Large (usually over 20 mm), usually green, species with hindwing usually projecting beyond forewing. Tympana (ears) on fore tibia exposed, oval. BUSH KATYDIDS **(PHANEROPTERINAE)** Pg. 79

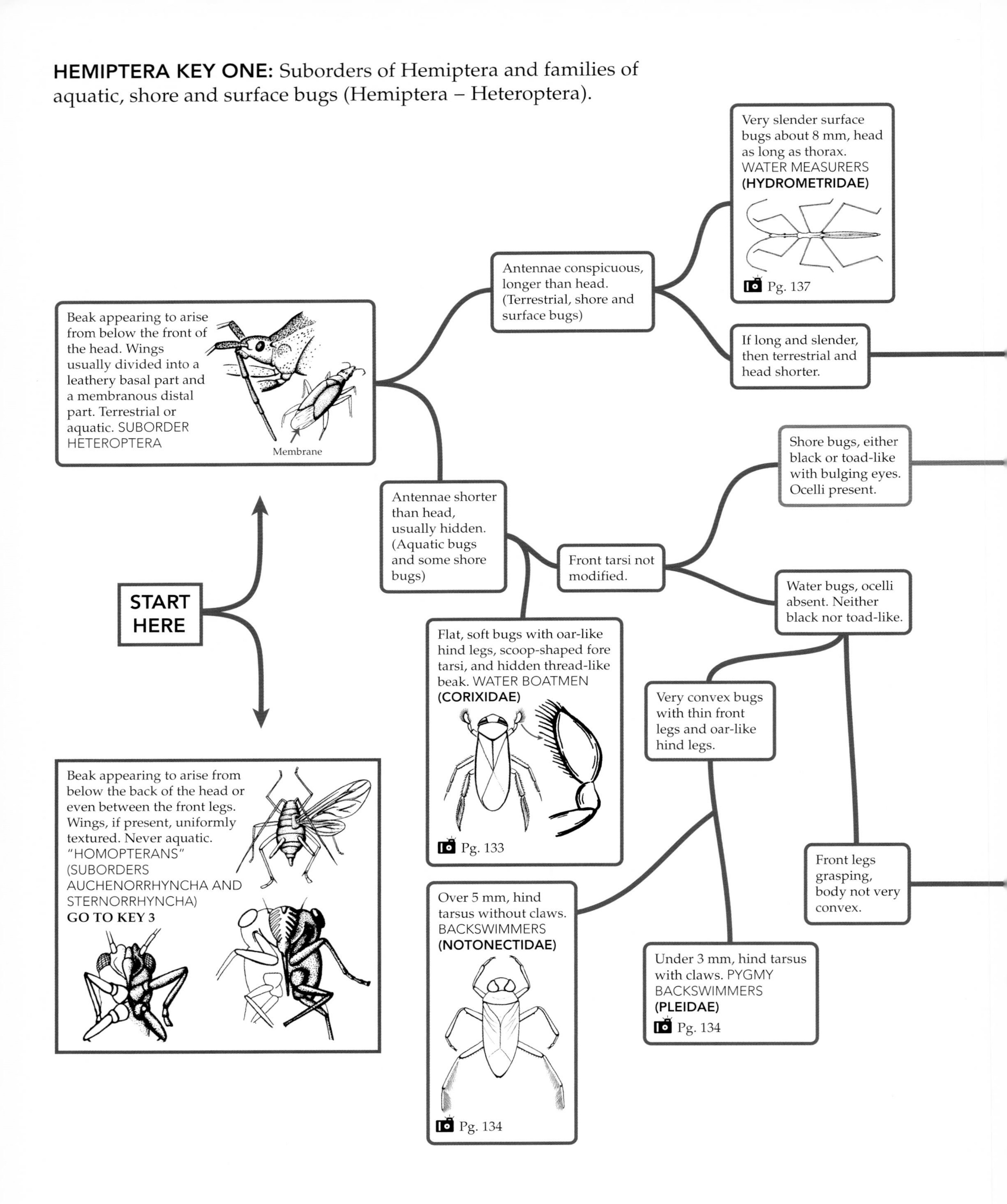
HEMIPTERA KEY ONE: Suborders of Hemiptera and families of aquatic, shore and surface bugs (Hemiptera – Heteroptera).
START HERE
Beak appearing to arise from below the front of the head. Wings usually divided into a leathery basal part and a membranous distal part. Terrestrial or aquatic. SUBORDER HETEROPTERA
Membrane
Beak appearing to arise from below the back of the head or even between the front legs. Wings, if present, uniformly textured. Never aquatic. "HOMOPTERANS" (SUBORDERS AUCHENORRHYNCHA AND STERNORRHYNCHA) GO TO KEY 3
Antennae conspicuous, longer than head. (Terrestrial, shore and surface bugs)
Very slender surface bugs about 8 mm, head as long as thorax. WATER MEASURERS (HYDROMETRIDAE)
Pg. 137
If long and slender, then terrestrial and head shorter.
Antennae shorter than head, usually hidden. (Aquatic bugs and some shore bugs)
Front tarsi not modified.
Shore bugs, either black or toad-like with bulging eyes. Ocelli present.
Water bugs, ocelli absent. Neither black nor toad-like.
Flat, soft bugs with oar-like hind legs, scoop-shaped fore tarsi, and hidden thread-like beak. WATER BOATMEN (CORIXIDAE)
Pg. 133
Very convex bugs with thin front legs and oar-like hind legs.
Front legs grasping, body not very convex.
Over 5 mm, hind tarsus without claws. BACKSWIMMERS (NOTONECTIDAE)
Pg. 134
Under 3 mm, hind tarsus with claws. PYGMY BACKSWIMMERS (PLEIDAE)
Pg. 134

Terrestrial bugs. Not found on the water surface. Tarsal claws apical. If wing membrane veinless (Anthocoridae) then with a cuneus (side of wing notched). **GO TO KEY 2**

Greenish, elongate, usually wingless. If winged, membrane without veins. WATER TREADERS **(MESOVELIIDAE)**

Pg. 137

Claws at ends of legs. Wing membrane white, veinless. Less than 3 mm. VELVET WATER BUGS **(HEBRIDAE)**

Pg. 136

Wing membrane without 4–5 loop-like closed cells; if with any closed cells then terrestrial.

Wing membrane with 4–5 loop-like closed cells (some rare genera lack a wing membrane). Semiaquatic. SHORE BUGS **(SALDIDAE)**

Macroveliidae (2 western North American species) have 6 closed cells in the wing membrane.

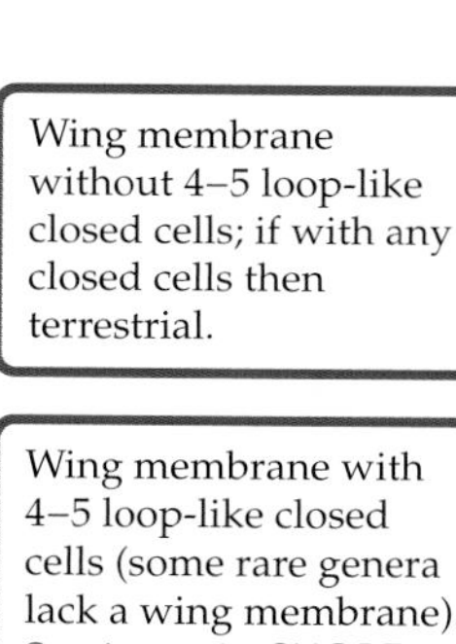

Pg. 135

Shore and surface bugs either with tarsal claws far above end of leg OR wing membrane without veins AND cuneus absent OR wingless and greenish surface bugs.

Body blackish. With or without wings.

Claws inserted before end of leg. Wings, if present, with veins. Size variable.

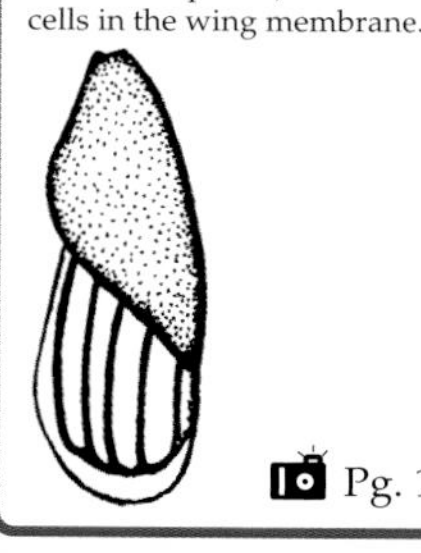

Mid legs closer to hind legs than front legs; body not broadest on front half. WATER STRIDERS **(GERRIDAE)**

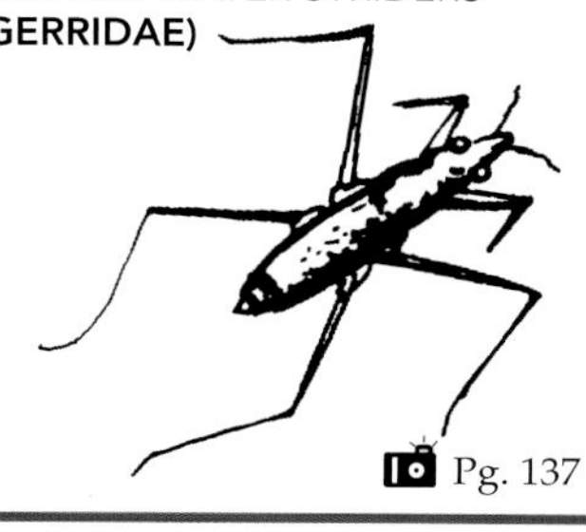

Pg. 137

Mid legs closer to front legs than to hind legs, OR body with a broad-shouldered appearance. RIFFLE BUGS **(VELIIDAE)**

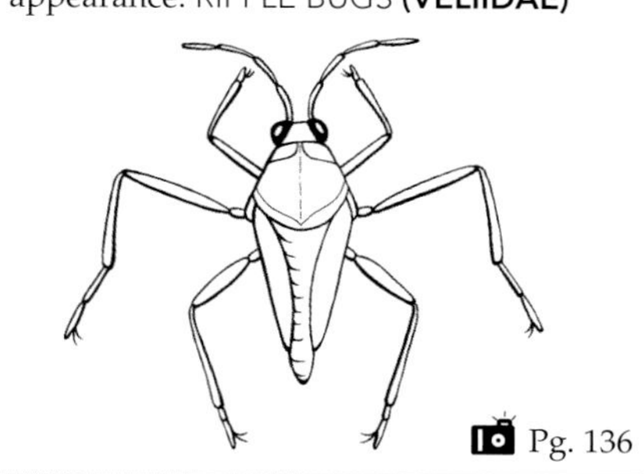

Pg. 136

Antennae visible from above. Eyes not bulging. Velvety black. VELVETY SHORE BUGS **(OCHTERIDAE)**

Pg. 135

Antennae hidden, not visible from above. Eyes bulging. Color grey-brown. TOAD BUGS **(GELASTOCORIDAE)**

Pg. 135

Without a snorkel-like breathing tube.

With a snorkel-like terminal breathing tube (made up of 2 filaments). WATER SCORPIONS **(NEPIDAE)**

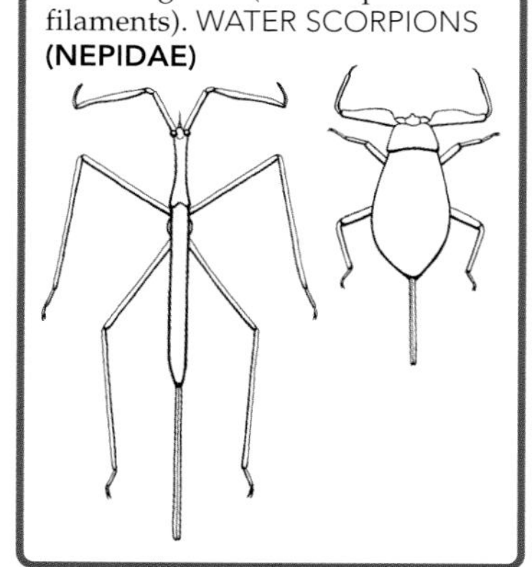

Body length over 20 mm. Wing membrane with distinct veins. Abdomen with short flap-like terminal appendages. GIANT WATER BUGS **(BELOSTOMATIDAE)**

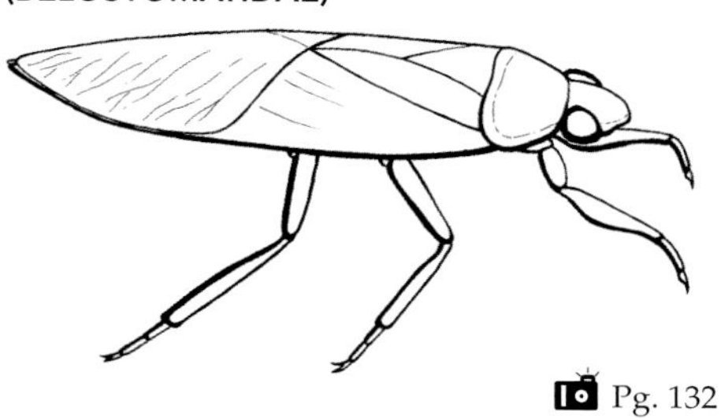

Pg. 132

Body length less than 13 mm. Wing membrane veinless. No terminal tails or flaps. CREEPING WATER BUGS **(NAUCORIDAE)**

Pg. 135

HEMIPTERA KEY TWO: Terrestrial true bugs (Heteroptera).

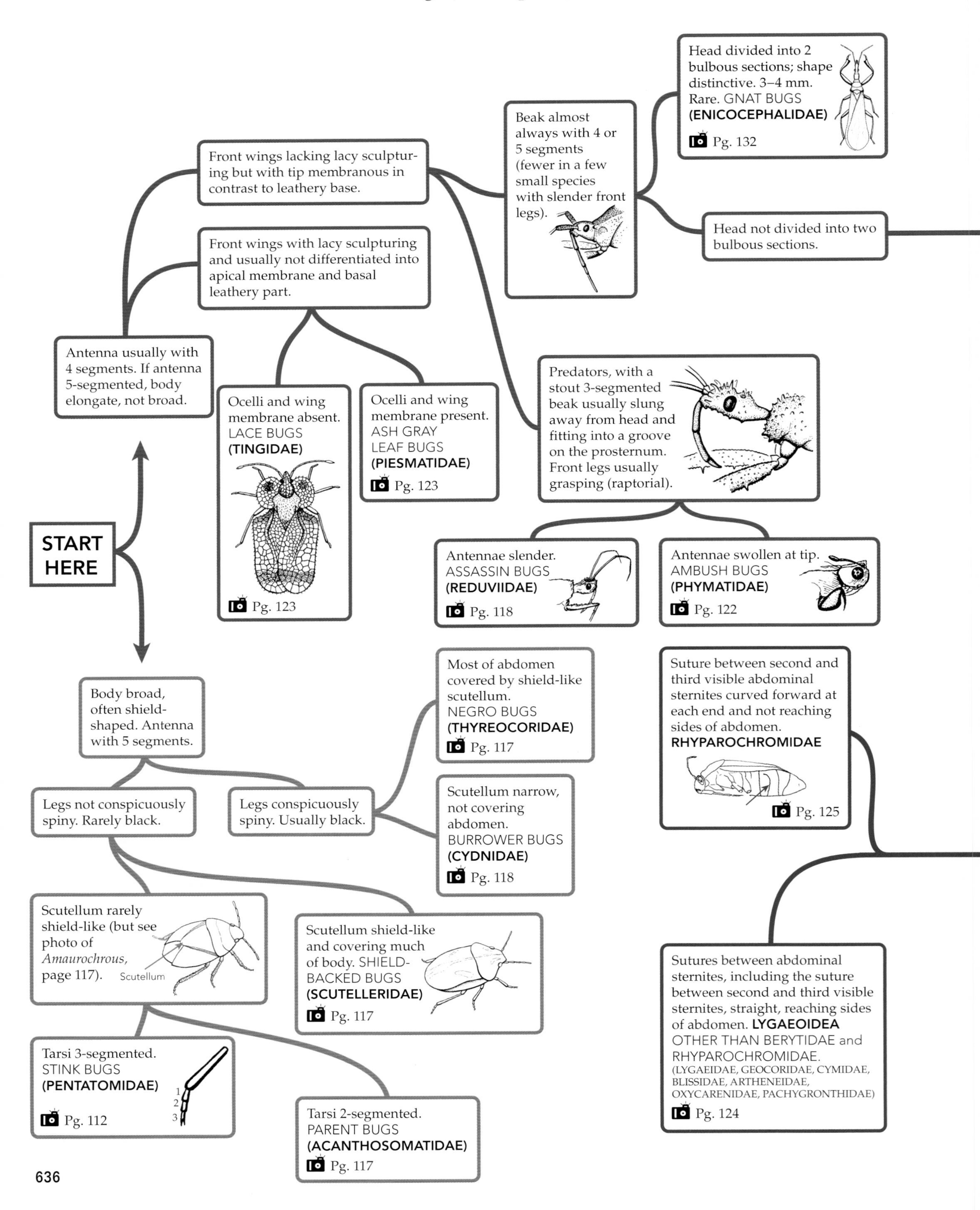

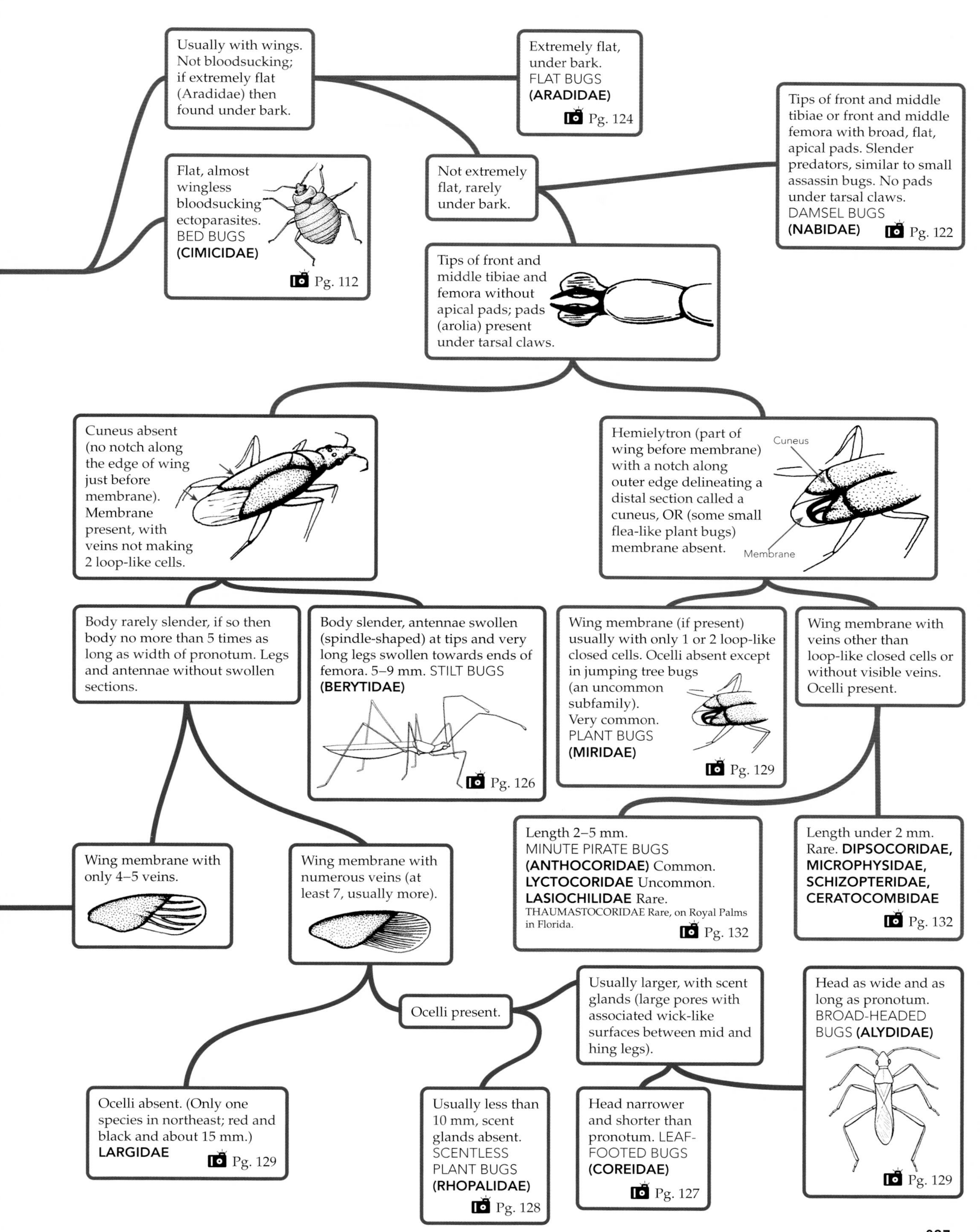

Usually with wings. Not bloodsucking; if extremely flat (Aradidae) then found under bark.
Extremely flat, under bark. FLAT BUGS (ARADIDAE) Pg. 124
Flat, almost wingless bloodsucking ectoparasites. BED BUGS (CIMICIDAE) Pg. 112
Not extremely flat, rarely under bark.
Tips of front and middle tibiae or front and middle femora with broad, flat, apical pads. Slender predators, similar to small assassin bugs. No pads under tarsal claws. DAMSEL BUGS (NABIDAE) Pg. 122
Tips of front and middle tibiae and femora without apical pads; pads (arolia) present under tarsal claws.
Cuneus absent (no notch along the edge of wing just before membrane). Membrane present, with veins not making 2 loop-like cells.
Hemielytron (part of wing before membrane) with a notch along outer edge delineating a distal section called a cuneus, OR (some small flea-like plant bugs) membrane absent.
Cuneus
Membrane
Body rarely slender, if so then body no more than 5 times as long as width of pronotum. Legs and antennae without swollen sections.
Body slender, antennae swollen (spindle-shaped) at tips and very long legs swollen towards ends of femora. 5–9 mm. STILT BUGS (BERYTIDAE) Pg. 126
Wing membrane (if present) usually with only 1 or 2 loop-like closed cells. Ocelli absent except in jumping tree bugs (an uncommon subfamily). Very common. PLANT BUGS (MIRIDAE) Pg. 129
Wing membrane with veins other than loop-like closed cells or without visible veins. Ocelli present.
Wing membrane with only 4–5 veins.
Wing membrane with numerous veins (at least 7, usually more).
Length 2–5 mm. MINUTE PIRATE BUGS (ANTHOCORIDAE) Common. LYCTOCORIDAE Uncommon. LASIOCHILIDAE Rare. THAUMASTOCORIDAE Rare, on Royal Palms in Florida. Pg. 132
Length under 2 mm. Rare. DIPSOCORIDAE, MICROPHYSIDAE, SCHIZOPTERIDAE, CERATOCOMBIDAE Pg. 132
Ocelli present.
Usually larger, with scent glands (large pores with associated wick-like surfaces between mid and hing legs).
Head as wide and as long as pronotum. BROAD-HEADED BUGS (ALYDIDAE) Pg. 129
Ocelli absent. (Only one species in northeast; red and black and about 15 mm.) LARGIDAE Pg. 129
Usually less than 10 mm, scent glands absent. SCENTLESS PLANT BUGS (RHOPALIDAE) Pg. 128
Head narrower and shorter than pronotum. LEAF-FOOTED BUGS (COREIDAE) Pg. 127

HEMIPTERA KEY THREE: "Homopterans" (Suborders Sternorrhyncha and Auchenorrhyncha).

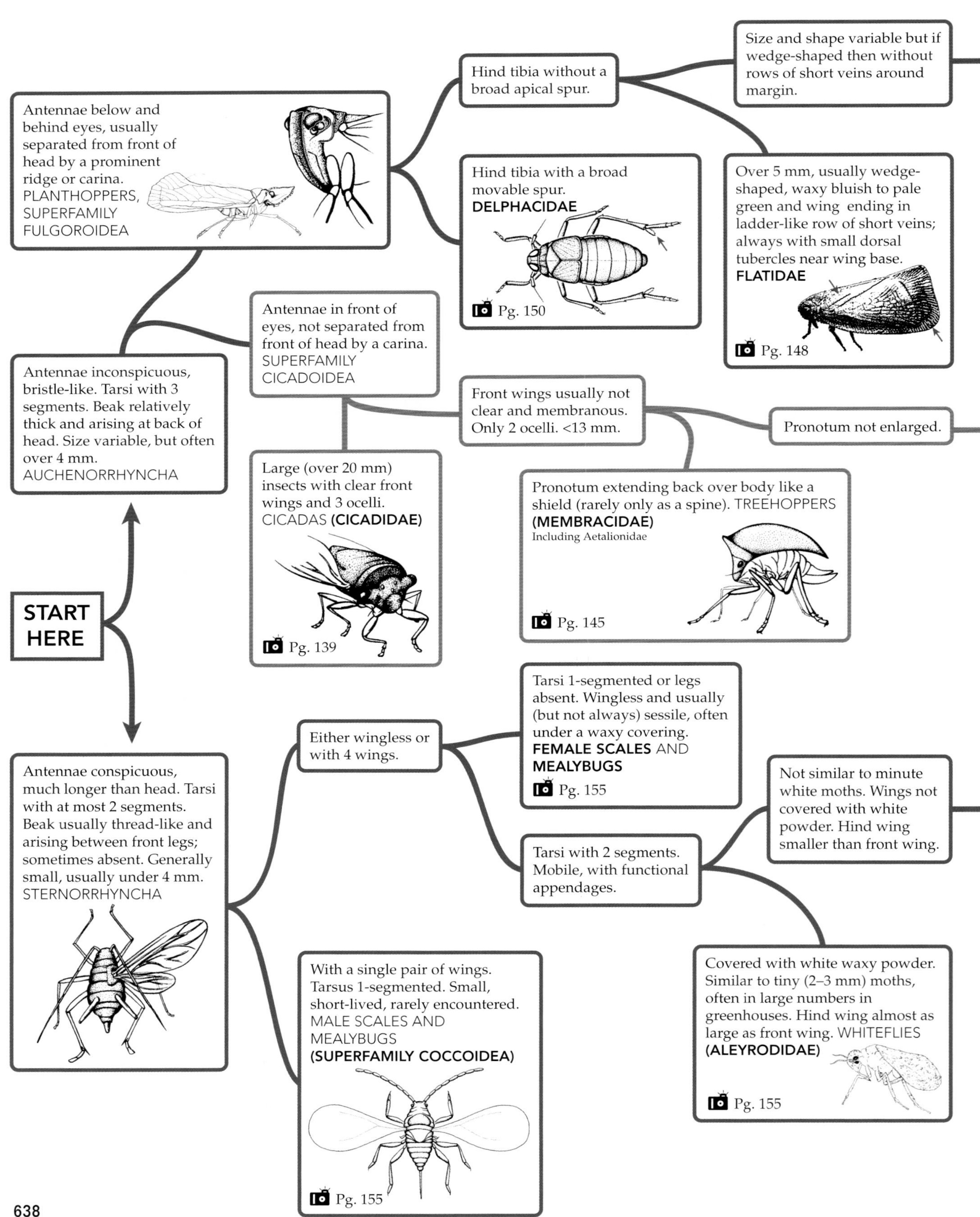

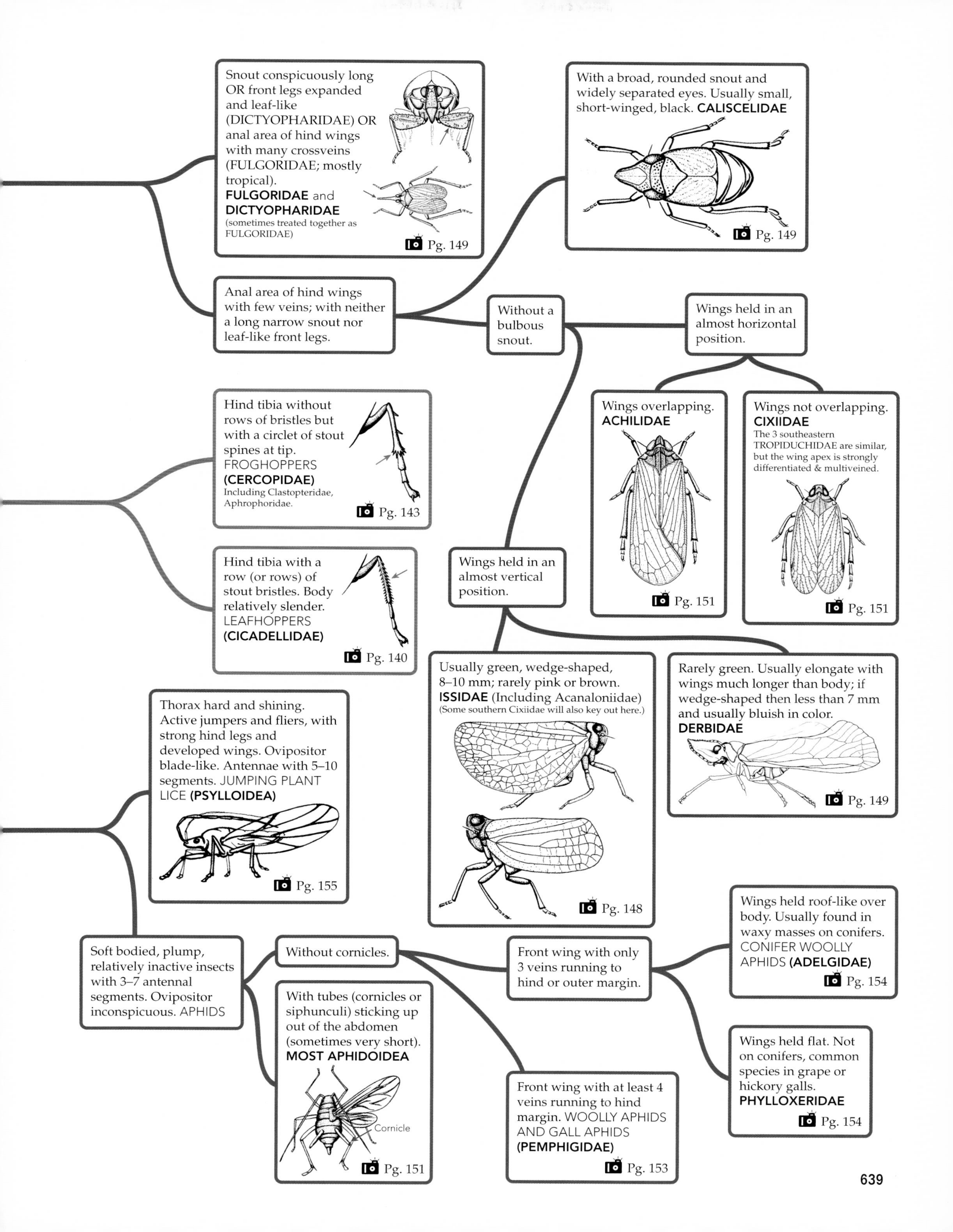
Snout conspicuously long OR front legs expanded and leaf-like (DICTYOPHARIDAE) OR anal area of hind wings with many crossveins (FULGORIDAE; mostly tropical).
FULGORIDAE and DICTYOPHARIDAE
(sometimes treated together as FULGORIDAE)
Pg. 149
With a broad, rounded snout and widely separated eyes. Usually small, short-winged, black. CALISCELIDAE
Pg. 149
Anal area of hind wings with few veins; with neither a long narrow snout nor leaf-like front legs.
Without a bulbous snout.
Wings held in an almost horizontal position.
Wings overlapping.
ACHILIDAE
Pg. 151
Wings not overlapping.
CIXIIDAE
The 3 southeastern TROPIDUCHIDAE are similar, but the wing apex is strongly differentiated & multiveined.
Pg. 151
Hind tibia without rows of bristles but with a circlet of stout spines at tip.
FROGHOPPERS
(CERCOPIDAE)
Including Clastopteridae, Aphrophoridae.
Pg. 143
Hind tibia with a row (or rows) of stout bristles. Body relatively slender.
LEAFHOPPERS
(CICADELLIDAE)
Pg. 140
Wings held in an almost vertical position.
Usually green, wedge-shaped, 8–10 mm; rarely pink or brown.
ISSIDAE (Including Acanaloniidae)
(Some southern Cixiidae will also key out here.)
Pg. 148
Rarely green. Usually elongate with wings much longer than body; if wedge-shaped then less than 7 mm and usually bluish in color.
DERBIDAE
Pg. 149
Thorax hard and shining. Active jumpers and fliers, with strong hind legs and developed wings. Ovipositor blade-like. Antennae with 5–10 segments. JUMPING PLANT LICE (PSYLLOIDEA)
Pg. 155
Soft bodied, plump, relatively inactive insects with 3–7 antennal segments. Ovipositor inconspicuous. APHIDS
Without cornicles.
With tubes (cornicles or siphunculi) sticking up out of the abdomen (sometimes very short).
MOST APHIDOIDEA
Cornicle
Pg. 151
Front wing with only 3 veins running to hind or outer margin.
Wings held roof-like over body. Usually found in waxy masses on conifers.
CONIFER WOOLLY APHIDS (ADELGIDAE)
Pg. 154
Wings held flat. Not on conifers, common species in grape or hickory galls.
PHYLLOXERIDAE
Pg. 154
Front wing with at least 4 veins running to hind margin. WOOLLY APHIDS AND GALL APHIDS
(PEMPHIGIDAE)
Pg. 153

TRICHOPTERA KEY ONE: Caddisfly adults.

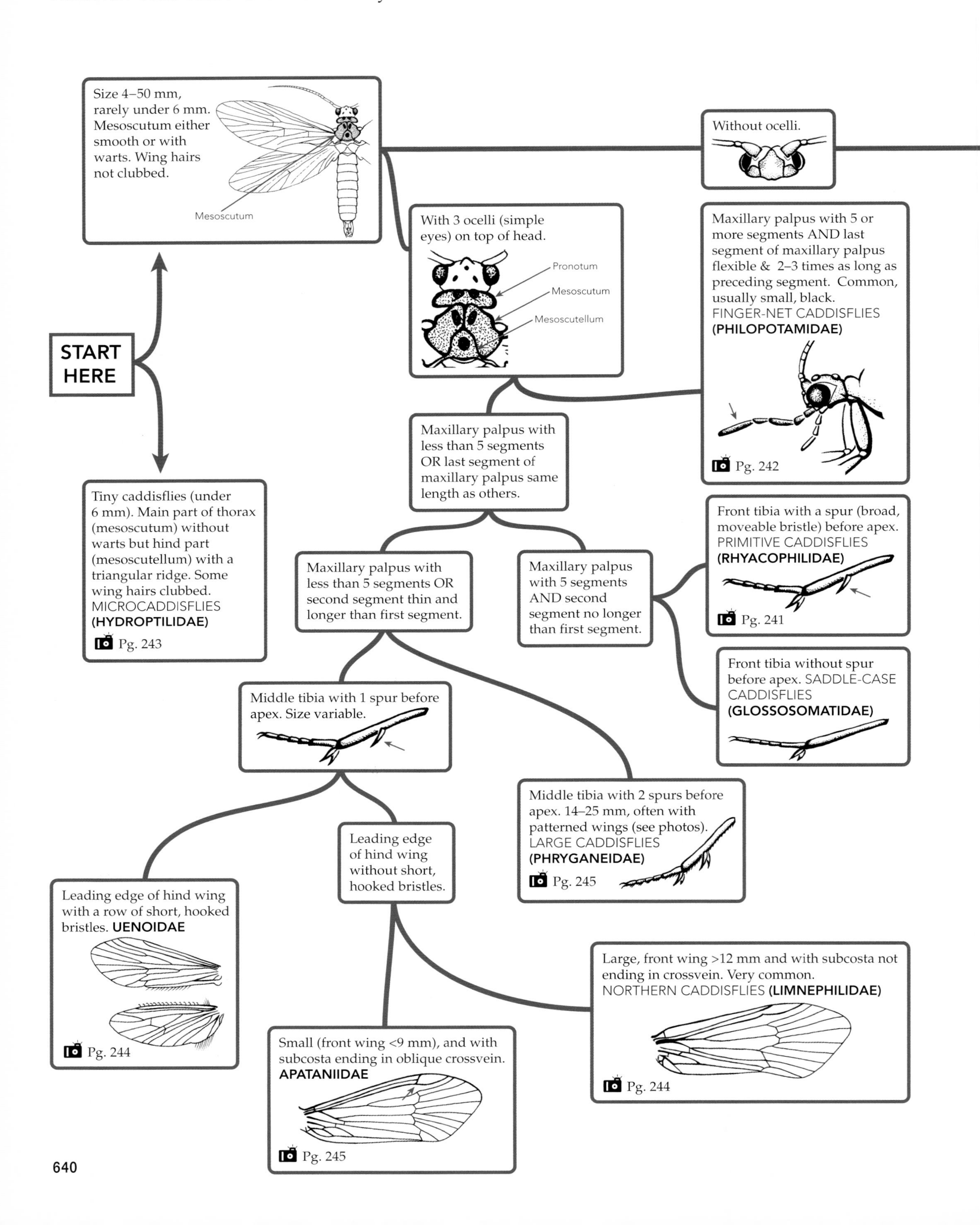

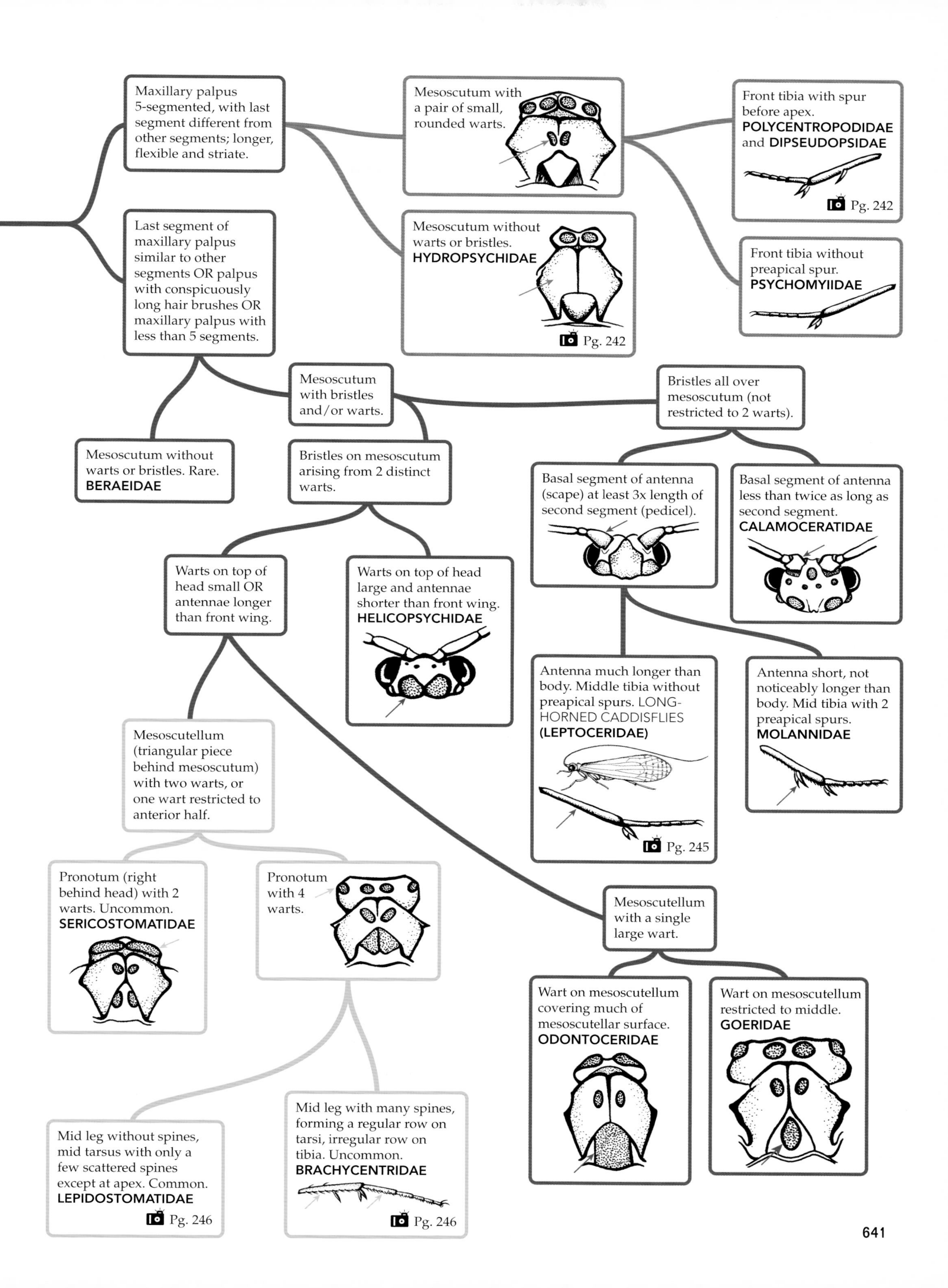
Maxillary palpus 5-segmented, with last segment different from other segments; longer, flexible and striate.
Mesoscutum with a pair of small, rounded warts.
Front tibia with spur before apex. POLYCENTROPODIDAE and DIPSEUDOPSIDAE
Pg. 242
Front tibia without preapical spur. PSYCHOMYIIDAE
Mesoscutum without warts or bristles. HYDROPSYCHIDAE
Pg. 242
Last segment of maxillary palpus similar to other segments OR palpus with conspicuously long hair brushes OR maxillary palpus with less than 5 segments.
Mesoscutum with bristles and/or warts.
Bristles all over mesoscutum (not restricted to 2 warts).
Mesoscutum without warts or bristles. Rare. BERAEIDAE
Bristles on mesoscutum arising from 2 distinct warts.
Basal segment of antenna (scape) at least 3x length of second segment (pedicel).
Basal segment of antenna less than twice as long as second segment. CALAMOCERATIDAE
Warts on top of head small OR antennae longer than front wing.
Warts on top of head large and antennae shorter than front wing. HELICOPSYCHIDAE
Antenna much longer than body. Middle tibia without preapical spurs. LONG-HORNED CADDISFLIES (LEPTOCERIDAE)
Pg. 245
Antenna short, not noticeably longer than body. Mid tibia with 2 preapical spurs. MOLANNIDAE
Mesoscutellum (triangular piece behind mesoscutum) with two warts, or one wart restricted to anterior half.
Pronotum (right behind head) with 2 warts. Uncommon. SERICOSTOMATIDAE
Pronotum with 4 warts.
Mesoscutellum with a single large wart.
Wart on mesoscutellum covering much of mesoscutellar surface. ODONTOCERIDAE
Wart on mesoscutellum restricted to middle. GOERIDAE
Mid leg without spines, mid tarsus with only a few scattered spines except at apex. Common. LEPIDOSTOMATIDAE
Pg. 246
Mid leg with many spines, forming a regular row on tarsi, irregular row on tibia. Uncommon. BRACHYCENTRIDAE
Pg. 246

TRICHOPTERA KEY TWO: Caddisfly larvae.

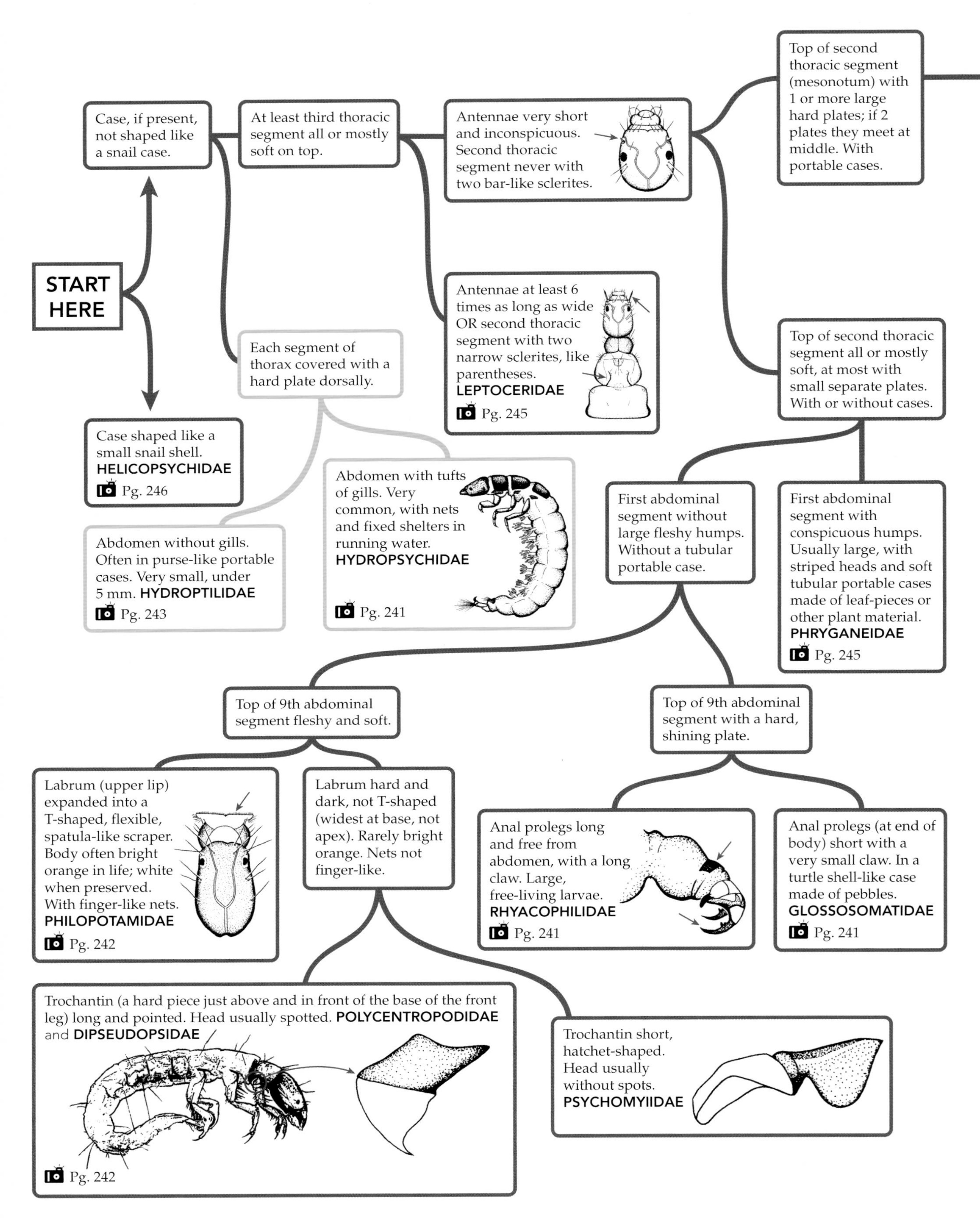

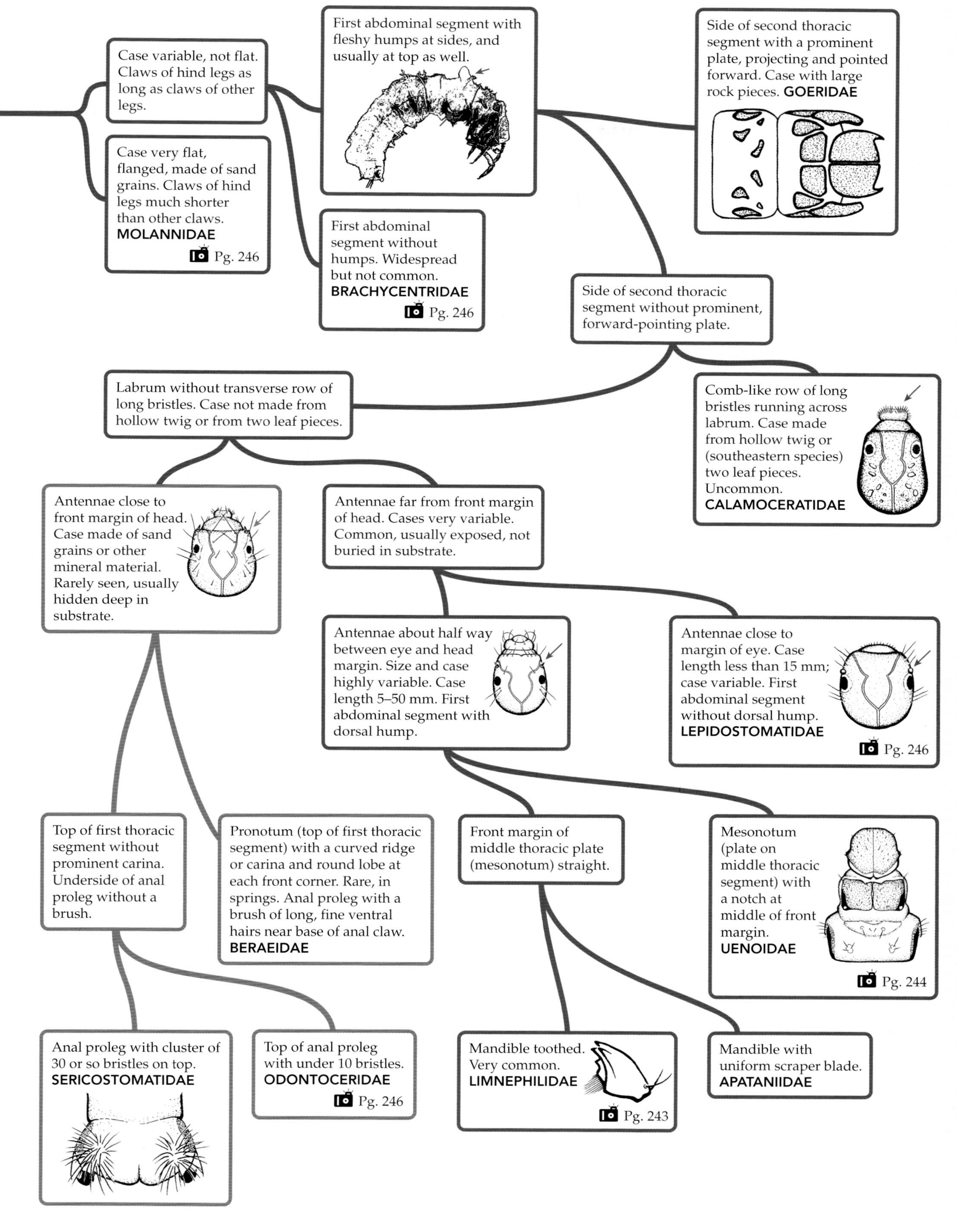

Case variable, not flat. Claws of hind legs as long as claws of other legs.
Case very flat, flanged, made of sand grains. Claws of hind legs much shorter than other claws. MOLANNIDAE
Pg. 246
First abdominal segment with fleshy humps at sides, and usually at top as well.
First abdominal segment without humps. Widespread but not common. BRACHYCENTRIDAE
Pg. 246
Side of second thoracic segment with a prominent plate, projecting and pointed forward. Case with large rock pieces. GOERIDAE
Side of second thoracic segment without prominent, forward-pointing plate.
Labrum without transverse row of long bristles. Case not made from hollow twig or from two leaf pieces.
Comb-like row of long bristles running across labrum. Case made from hollow twig or (southeastern species) two leaf pieces. Uncommon. CALAMOCERATIDAE
Antennae close to front margin of head. Case made of sand grains or other mineral material. Rarely seen, usually hidden deep in substrate.
Antennae far from front margin of head. Cases very variable. Common, usually exposed, not buried in substrate.
Antennae about half way between eye and head margin. Size and case highly variable. Case length 5–50 mm. First abdominal segment with dorsal hump.
Antennae close to margin of eye. Case length less than 15 mm; case variable. First abdominal segment without dorsal hump. LEPIDOSTOMATIDAE
Pg. 246
Top of first thoracic segment without prominent carina. Underside of anal proleg without a brush.
Pronotum (top of first thoracic segment) with a curved ridge or carina and round lobe at each front corner. Rare, in springs. Anal proleg with a brush of long, fine ventral hairs near base of anal claw. BERAEIDAE
Front margin of middle thoracic plate (mesonotum) straight.
Mesonotum (plate on middle thoracic segment) with a notch at middle of front margin. UENOIDAE
Pg. 244
Anal proleg with cluster of 30 or so bristles on top. SERICOSTOMATIDAE
Top of anal proleg with under 10 bristles. ODONTOCERIDAE
Pg. 246
Mandible toothed. Very common. LIMNEPHILIDAE
Pg. 243
Mandible with uniform scraper blade. APATANIIDAE

BEETLE KEY ONE: Suborders of Coleoptera, families of Adephaga, and selected families of Polyphaga. Start here if you are unsure about where to begin.

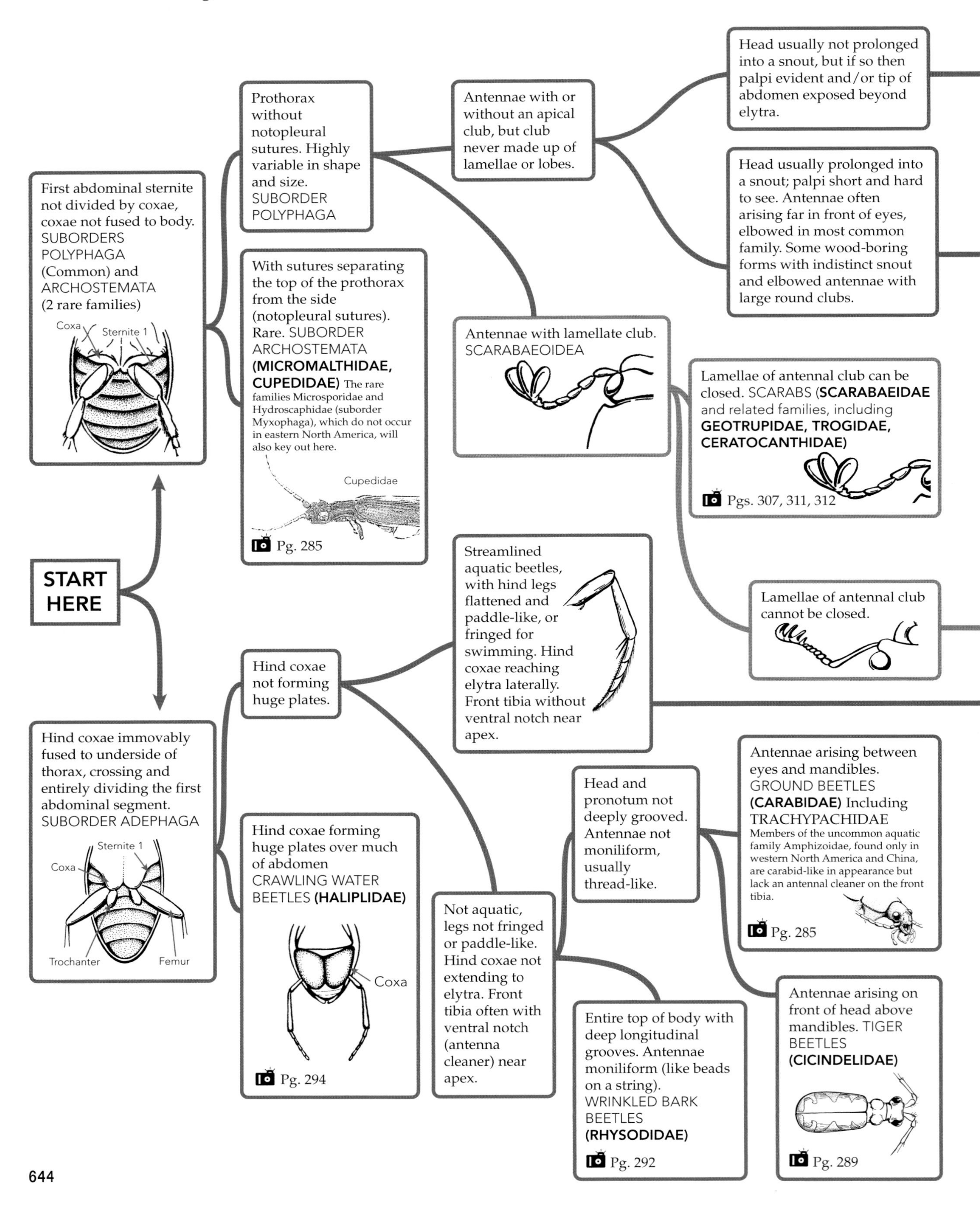

Tarsi variable, but if hind leg with 3rd segment lobed and concealing 4th, then antennae clubbed. **SEE BEETLE KEY 2**

Each leg with 5 tarsomeres, but with 3rd segment of tarsus expanded, lobed and concealing a small 4th segment between the spongy lobes. Antennae not clubbed. If the 2nd or 4th tarsomere is lobed (rather than the 3rd), go to key 2. Scirtidae (key 4, 4th segment lobed) are often misidentified as Chrysomelidae, so count carefully!

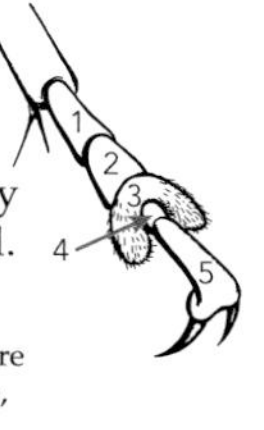

Elytra usually short, exposing tip of abdomen. With a weak, broad snout and stout hind femora. BEAN WEEVILS **(BRUCHIDAE)**

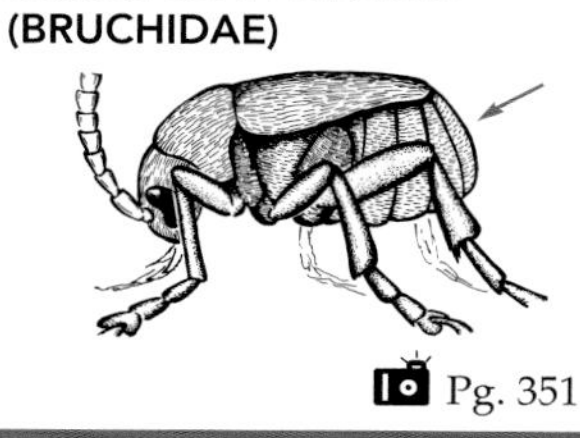

Pg. 351

Eyes notched by antennal bases. Antennae long, usually at least half as long as body. MOST LONG-HORNED BEETLES **(CERAMBYCIDAE)**

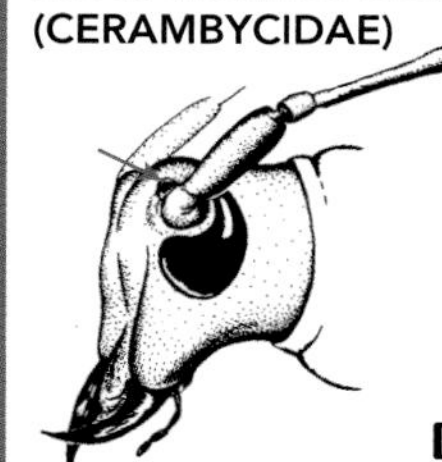

Pg. 360

Elytra not shortened, shape different and without a broad beak.

Eyes not notched by antennal bases. Antennae less than half as long as body.

Scutellum heart-shaped. Antennae (at least ♂) branched or sawtoothed. **PTILODACTYLIDAE** (Infrequently collected).

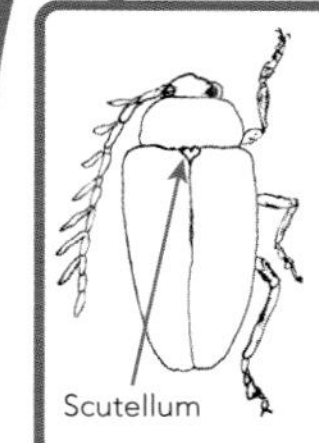

Pg. 297

Scutellum not heartt-shaped. Antennae uniform.

Covered with bristly hairs. Head at least as wide as prothorax, prothorax rounded and narrower than abdomen. Most Cleridae have a distinct antennal club and will come out in key 2. CHECKERED BEETLES **(CLERIDAE)**

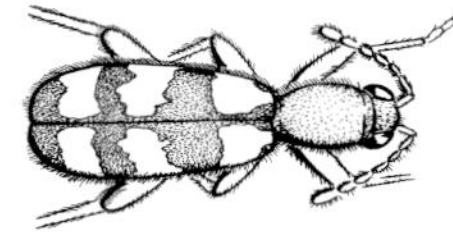

Pg. 321

Elytra without bristly hairs. Head not broad.

Narrow, parallel-sided and very flat. Head constricted behind eyes. Under bark or in stored products. **SOME SILVANIDAE**

Pg. 341

Leaf-eating beetles, usually not flat and parallel-sided; if so then head not constricted behind eyes. Usually brightly colored, often shiny.

Tibiae usually without paired apical spurs. LEAF BEETLES **CHRYSOMELIDAE** (an enormous and variable family, very common). The single North American species of CHELONARIIDAE, a rare southeastern species (distinctively convex, with white scale patches on the elytra), would also key out here. A few other beetles have an expanded third tarsomere and might be mistakenly keyed out here; try key 2 if your beetle is dull-colored or associated with fungi, dead trees or decaying material.

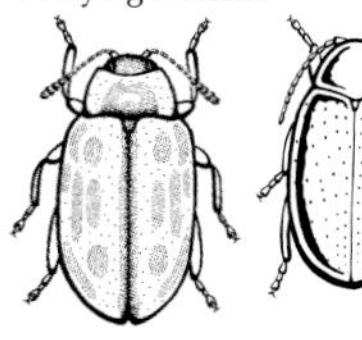

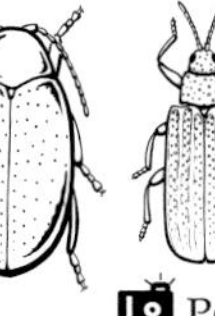

Pg. 351

Tibiae with paired apical spurs.

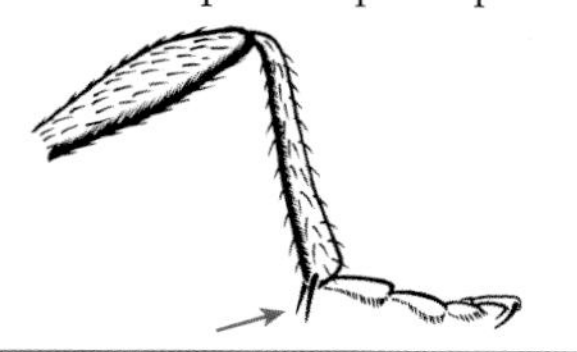

Head with a narrow neck and a constriction behind eye. **MEGALOPODIDAE** (a small group, infrequently collected).

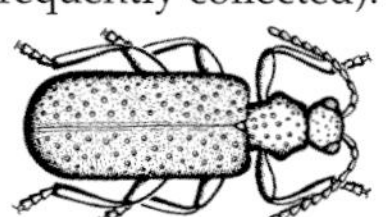

Head not constricted behind eye. **ORSODACNIDAE**

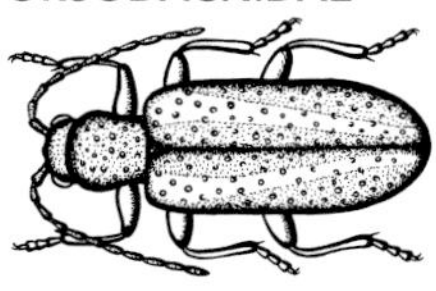

Pg. 351

Antennae elbowed. WEEVILS AND BARK BEETLES **(CURCULIONIDAE)**

Pg. 370

Antennae not elbowed FUNGUS WEEVILS **(ANTHRIBIDAE),** PRIMITIVE WEEVILS **(BRENTIDAE)** and some other small families of Curculionoidea

Pgs. 378, 379

Antennae not elbowed; head with small horn. BESSBUGS **(PASSALIDAE)**

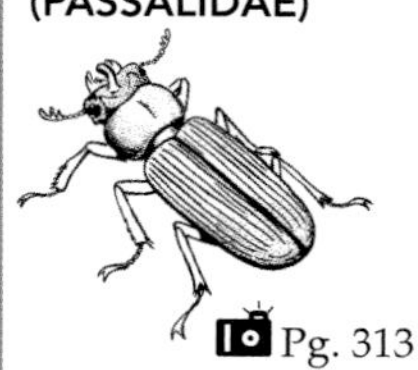

Pg. 313

Antennae elbowed. Males often with big jaws. STAG BEETLES **(LUCANIDAE)**

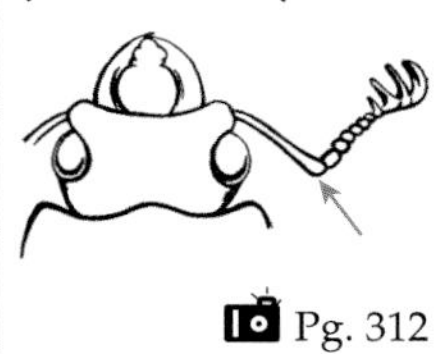

Pg. 312

Eyes entire. Length 2–40 mm. DIVING BEETLES **(DYTISCIDAE)** BURROWING WATER BEETLES (NOTERIDAE) are similar, but small (1–5 mm) with a hidden scutellum AND a front tibial spur (except in some minute southeastern species).

Pgs. 292, 295

Eyes divided. WHIRLIGIGS **(GYRINIDAE)**

Pg. 295

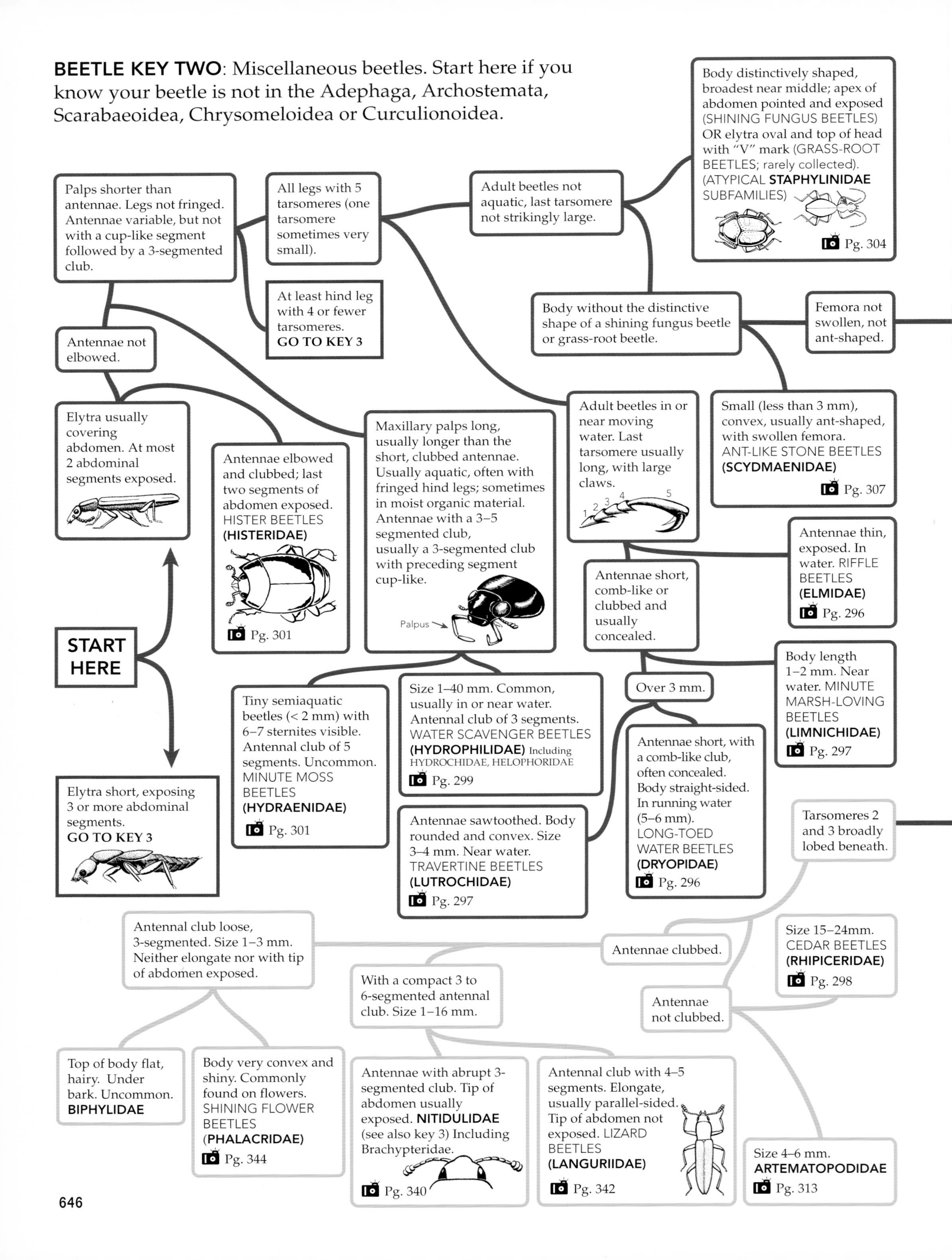
BEETLE KEY TWO: Miscellaneous beetles. Start here if you know your beetle is not in the Adephaga, Archostemata, Scarabaeoidea, Chrysomeloidea or Curculionoidea.
START HERE
Elytra usually covering abdomen. At most 2 abdominal segments exposed.
Elytra short, exposing 3 or more abdominal segments. GO TO KEY 3
Antennae not elbowed.
Antennae elbowed and clubbed; last two segments of abdomen exposed. HISTER BEETLES (HISTERIDAE) Pg. 301
Palps shorter than antennae. Legs not fringed. Antennae variable, but not with a cup-like segment followed by a 3-segmented club.
Maxillary palps long, usually longer than the short, clubbed antennae. Usually aquatic, often with fringed hind legs; sometimes in moist organic material. Antennae with a 3–5 segmented club, usually a 3-segmented club with preceding segment cup-like.
Palpus
Tiny semiaquatic beetles (< 2 mm) with 6–7 sternites visible. Antennal club of 5 segments. Uncommon. MINUTE MOSS BEETLES (HYDRAENIDAE) Pg. 301
Size 1–40 mm. Common, usually in or near water. Antennal club of 3 segments. WATER SCAVENGER BEETLES (HYDROPHILIDAE) Including HYDROCHIDAE, HELOPHORIDAE Pg. 299
All legs with 5 tarsomeres (one tarsomere sometimes very small).
At least hind leg with 4 or fewer tarsomeres. GO TO KEY 3
Adult beetles not aquatic, last tarsomere not strikingly large.
Adult beetles in or near moving water. Last tarsomere usually long, with large claws.
1 2 3 4 5
Body distinctively shaped, broadest near middle; apex of abdomen pointed and exposed (SHINING FUNGUS BEETLES) OR elytra oval and top of head with "V" mark (GRASS-ROOT BEETLES; rarely collected). (ATYPICAL STAPHYLINIDAE SUBFAMILIES) Pg. 304
Body without the distinctive shape of a shining fungus beetle or grass-root beetle.
Femora not swollen, not ant-shaped.
Small (less than 3 mm), convex, usually ant-shaped, with swollen femora. ANT-LIKE STONE BEETLES (SCYDMAENIDAE) Pg. 307
Antennae thin, exposed. In water. RIFFLE BEETLES (ELMIDAE) Pg. 296
Antennae short, comb-like or clubbed and usually concealed.
Body length 1–2 mm. Near water. MINUTE MARSH-LOVING BEETLES (LIMNICHIDAE) Pg. 297
Over 3 mm.
Antennae sawtoothed. Body rounded and convex. Size 3–4 mm. Near water. TRAVERTINE BEETLES (LUTROCHIDAE) Pg. 297
Antennae short, with a comb-like club, often concealed. Body straight-sided. In running water (5–6 mm). LONG-TOED WATER BEETLES (DRYOPIDAE) Pg. 296
Tarsomeres 2 and 3 broadly lobed beneath.
Antennae clubbed.
Antennae not clubbed.
Size 15–24mm. CEDAR BEETLES (RHIPICERIDAE) Pg. 298
Size 4–6 mm. ARTEMATOPODIDAE Pg. 313
Antennal club loose, 3-segmented. Size 1–3 mm. Neither elongate nor with tip of abdomen exposed.
With a compact 3 to 6-segmented antennal club. Size 1–16 mm.
Top of body flat, hairy. Under bark. Uncommon. BIPHYLIDAE
Body very convex and shiny. Commonly found on flowers. SHINING FLOWER BEETLES (PHALACRIDAE) Pg. 344
Antennae with abrupt 3-segmented club. Tip of abdomen usually exposed. NITIDULIDAE (see also key 3) Including Brachypteridae. Pg. 340
Antennal club with 4–5 segments. Elongate, usually parallel-sided. Tip of abdomen not exposed. LIZARD BEETLES (LANGURIIDAE) Pg. 342
646

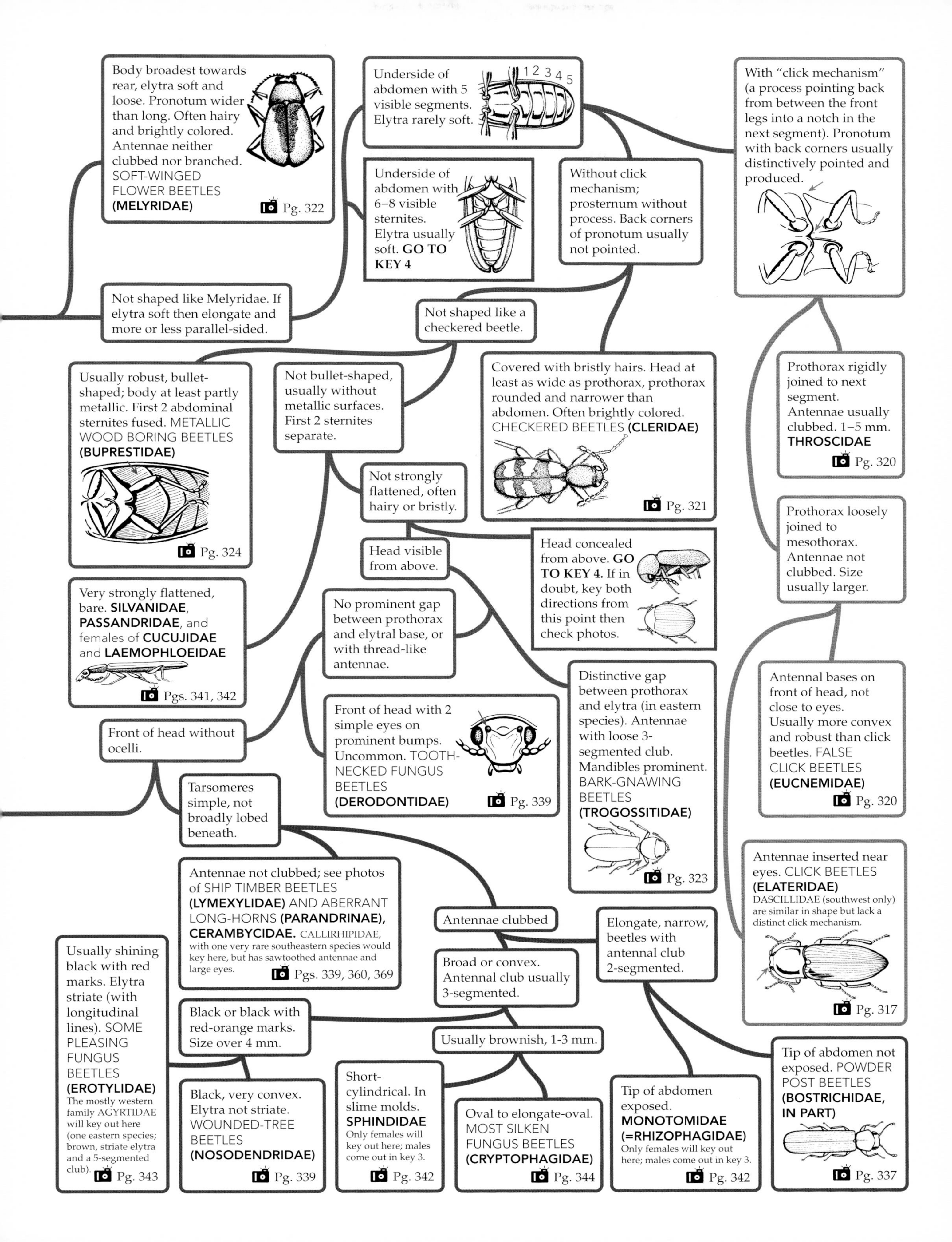
Body broadest towards rear, elytra soft and loose. Pronotum wider than long. Often hairy and brightly colored. Antennae neither clubbed nor branched. SOFT-WINGED FLOWER BEETLES (MELYRIDAE) Pg. 322
Underside of abdomen with 5 visible segments. Elytra rarely soft.
1 2 3 4 5
With "click mechanism" (a process pointing back from between the front legs into a notch in the next segment). Pronotum with back corners usually distinctively pointed and produced.
Underside of abdomen with 6–8 visible sternites. Elytra usually soft. GO TO KEY 4
Without click mechanism; prosternum without process. Back corners of pronotum usually not pointed.
Not shaped like Melyridae. If elytra soft then elongate and more or less parallel-sided.
Not shaped like a checkered beetle.
Usually robust, bullet-shaped; body at least partly metallic. First 2 abdominal sternites fused. METALLIC WOOD BORING BEETLES (BUPRESTIDAE) Pg. 324
Not bullet-shaped, usually without metallic surfaces. First 2 sternites separate.
Covered with bristly hairs. Head at least as wide as prothorax, prothorax rounded and narrower than abdomen. Often brightly colored. CHECKERED BEETLES (CLERIDAE) Pg. 321
Prothorax rigidly joined to next segment. Antennae usually clubbed. 1–5 mm. THROSCIDAE Pg. 320
Not strongly flattened, often hairy or bristly.
Prothorax loosely joined to mesothorax. Antennae not clubbed. Size usually larger.
Head visible from above.
Head concealed from above. GO TO KEY 4. If in doubt, key both directions from this point then check photos.
Very strongly flattened, bare. SILVANIDAE, PASSANDRIDAE, and females of CUCUJIDAE and LAEMOPHLOEIDAE Pgs. 341, 342
No prominent gap between prothorax and elytral base, or with thread-like antennae.
Distinctive gap between prothorax and elytra (in eastern species). Antennae with loose 3-segmented club. Mandibles prominent. BARK-GNAWING BEETLES (TROGOSSITIDAE) Pg. 323
Antennal bases on front of head, not close to eyes. Usually more convex and robust than click beetles. FALSE CLICK BEETLES (EUCNEMIDAE) Pg. 320
Front of head without ocelli.
Front of head with 2 simple eyes on prominent bumps. Uncommon. TOOTH-NECKED FUNGUS BEETLES (DERODONTIDAE) Pg. 339
Tarsomeres simple, not broadly lobed beneath.
Antennae inserted near eyes. CLICK BEETLES (ELATERIDAE) DASCILLIDAE (southwest only) are similar in shape but lack a distinct click mechanism. Pg. 317
Antennae not clubbed; see photos of SHIP TIMBER BEETLES (LYMEXYLIDAE) AND ABERRANT LONG-HORNS (PARANDRINAE), CERAMBYCIDAE. CALLIRHIPIDAE, with one very rare southeastern species would key here, but has sawtoothed antennae and large eyes. Pgs. 339, 360, 369
Antennae clubbed
Elongate, narrow, beetles with antennal club 2-segmented.
Usually shining black with red marks. Elytra striate (with longitudinal lines). SOME PLEASING FUNGUS BEETLES (EROTYLIDAE) The mostly western family AGYRTIDAE will key out here (one eastern species; brown, striate elytra and a 5-segmented club). Pg. 343
Broad or convex. Antennal club usually 3-segmented.
Black or black with red-orange marks. Size over 4 mm.
Usually brownish, 1-3 mm.
Tip of abdomen not exposed. POWDER POST BEETLES (BOSTRICHIDAE, IN PART) Pg. 337
Black, very convex. Elytra not striate. WOUNDED-TREE BEETLES (NOSODENDRIDAE) Pg. 339
Short-cylindrical. In slime molds. SPHINDIDAE Only females will key out here; males come out in key 3. Pg. 342
Oval to elongate-oval. MOST SILKEN FUNGUS BEETLES (CRYPTOPHAGIDAE) Pg. 344
Tip of abdomen exposed. MONOTOMIDAE (=RHIZOPHAGIDAE) Only females will key out here; males come out in key 3. Pg. 342

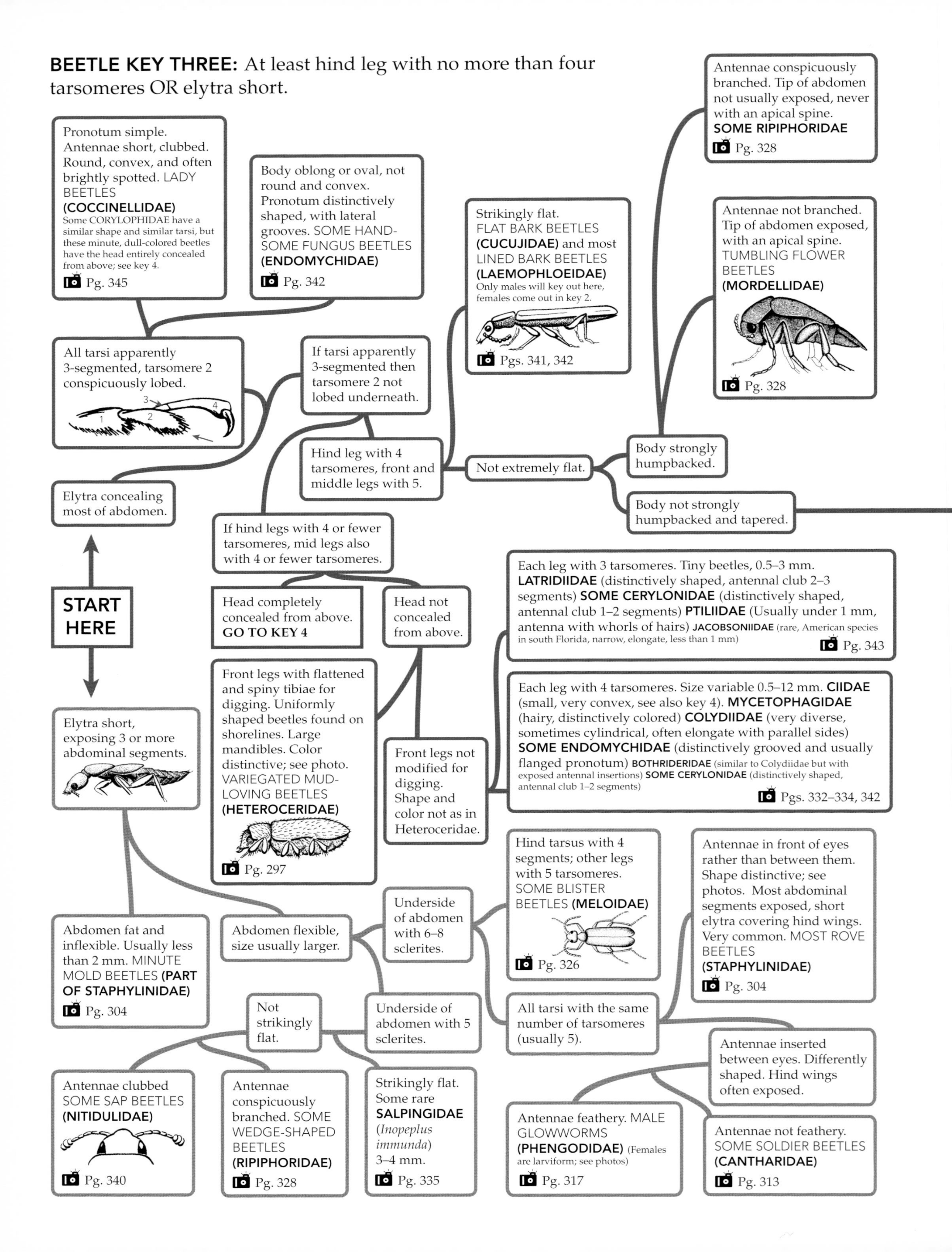

BEETLE KEY THREE: At least hind leg with no more than four tarsomeres OR elytra short.
START HERE
Elytra concealing most of abdomen.
All tarsi apparently 3-segmented, tarsomere 2 conspicuously lobed.
3
1
2
4
Pronotum simple. Antennae short, clubbed. Round, convex, and often brightly spotted. LADY BEETLES (COCCINELLIDAE) Some CORYLOPHIDAE have a similar shape and similar tarsi, but these minute, dull-colored beetles have the head entirely concealed from above; see key 4. Pg. 345
Body oblong or oval, not round and convex. Pronotum distinctively shaped, with lateral grooves. SOME HANDSOME FUNGUS BEETLES (ENDOMYCHIDAE) Pg. 342
If tarsi apparently 3-segmented then tarsomere 2 not lobed underneath.
Hind leg with 4 tarsomeres, front and middle legs with 5.
Strikingly flat. FLAT BARK BEETLES (CUCUJIDAE) and most LINED BARK BEETLES (LAEMOPHLOEIDAE) Only males will key out here, females come out in key 2. Pgs. 341, 342
Not extremely flat.
Body strongly humpbacked.
Antennae conspicuously branched. Tip of abdomen not usually exposed, never with an apical spine. SOME RIPIPHORIDAE Pg. 328
Antennae not branched. Tip of abdomen exposed, with an apical spine. TUMBLING FLOWER BEETLES (MORDELLIDAE) Pg. 328
Body not strongly humpbacked and tapered.
If hind legs with 4 or fewer tarsomeres, mid legs also with 4 or fewer tarsomeres.
Head completely concealed from above. GO TO KEY 4
Head not concealed from above.
Front legs with flattened and spiny tibiae for digging. Uniformly shaped beetles found on shorelines. Large mandibles. Color distinctive; see photo. VARIEGATED MUD-LOVING BEETLES (HETEROCERIDAE) Pg. 297
Front legs not modified for digging. Shape and color not as in Heteroceridae.
Each leg with 3 tarsomeres. Tiny beetles, 0.5–3 mm. LATRIDIIDAE (distinctively shaped, antennal club 2–3 segments) SOME CERYLONIDAE (distinctively shaped, antennal club 1–2 segments) PTILIIDAE (Usually under 1 mm, antenna with whorls of hairs) JACOBSONIIDAE (rare, American species in south Florida, narrow, elongate, less than 1 mm) Pg. 343
Each leg with 4 tarsomeres. Size variable 0.5–12 mm. CIIDAE (small, very convex, see also key 4). MYCETOPHAGIDAE (hairy, distinctively colored) COLYDIIDAE (very diverse, sometimes cylindrical, often elongate with parallel sides) SOME ENDOMYCHIDAE (distinctively grooved and usually flanged pronotum) BOTHRIDERIDAE (similar to Colydiidae but with exposed antennal insertions) SOME CERYLONIDAE (distinctively shaped, antennal club 1–2 segments) Pgs. 332–334, 342
Elytra short, exposing 3 or more abdominal segments.
Abdomen fat and inflexible. Usually less than 2 mm. MINUTE MOLD BEETLES (PART OF STAPHYLINIDAE) Pg. 304
Abdomen flexible, size usually larger.
Underside of abdomen with 6–8 sclerites.
Hind tarsus with 4 segments; other legs with 5 tarsomeres. SOME BLISTER BEETLES (MELOIDAE) Pg. 326
All tarsi with the same number of tarsomeres (usually 5).
Antennae in front of eyes rather than between them. Shape distinctive; see photos. Most abdominal segments exposed, short elytra covering hind wings. Very common. MOST ROVE BEETLES (STAPHYLINIDAE) Pg. 304
Antennae inserted between eyes. Differently shaped. Hind wings often exposed.
Antennae feathery. MALE GLOWWORMS (PHENGODIDAE) (Females are larviform; see photos) Pg. 317
Antennae not feathery. SOME SOLDIER BEETLES (CANTHARIDAE) Pg. 313
Underside of abdomen with 5 sclerites.
Not strikingly flat.
Antennae clubbed SOME SAP BEETLES (NITIDULIDAE) Pg. 340
Antennae conspicuously branched. SOME WEDGE-SHAPED BEETLES (RIPIPHORIDAE) Pg. 328
Strikingly flat. Some rare SALPINGIDAE (Inopeplus immunda) 3–4 mm. Pg. 335

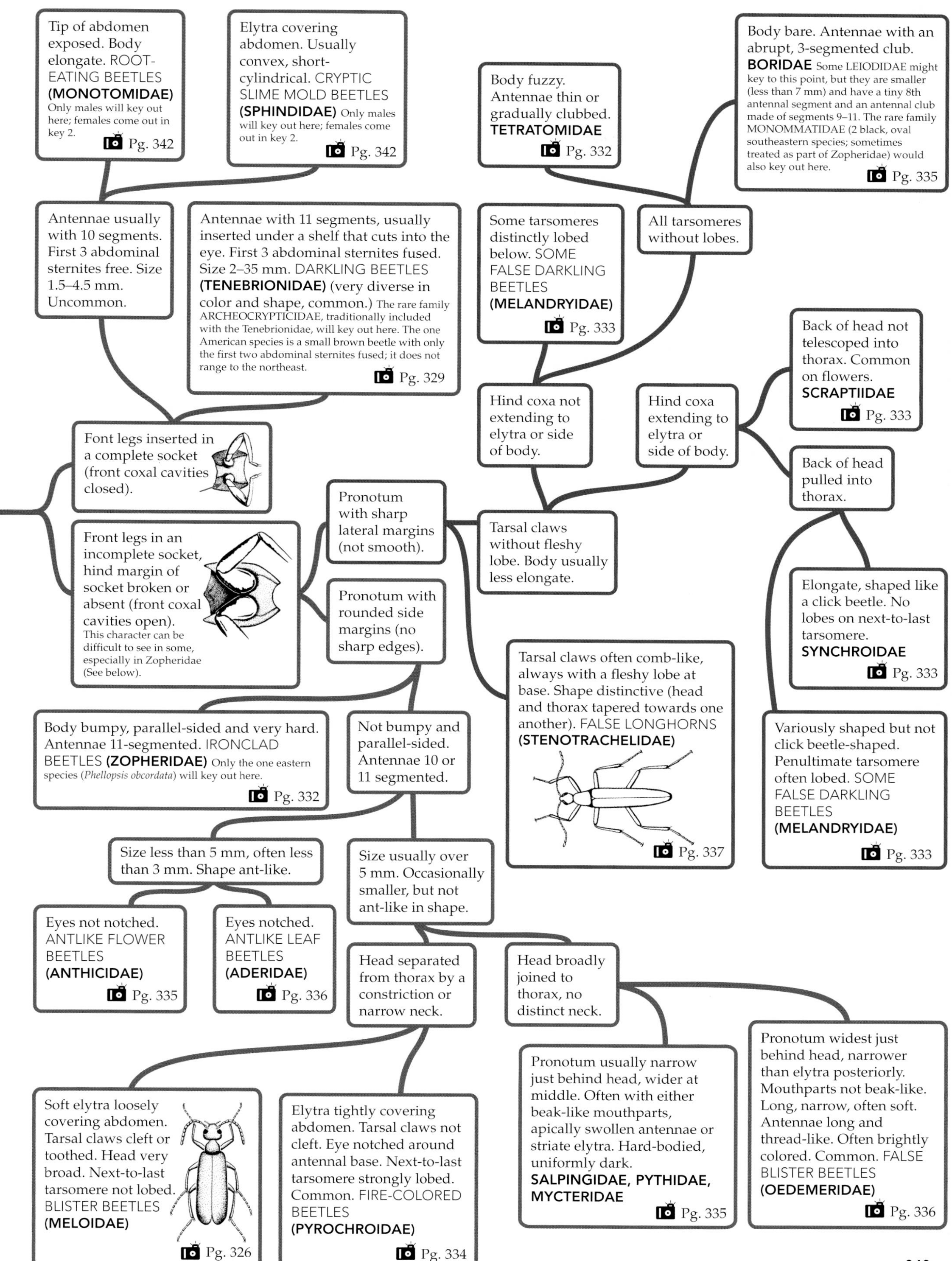

Tip of abdomen exposed. Body elongate. ROOT-EATING BEETLES (MONOTOMIDAE) Only males will key out here; females come out in key 2. Pg. 342
Elytra covering abdomen. Usually convex, short-cylindrical. CRYPTIC SLIME MOLD BEETLES (SPHINDIDAE) Only males will key out here; females come out in key 2. Pg. 342
Body fuzzy. Antennae thin or gradually clubbed. TETRATOMIDAE Pg. 332
Body bare. Antennae with an abrupt, 3-segmented club. BORIDAE Some LEIODIDAE might key to this point, but they are smaller (less than 7 mm) and have a tiny 8th antennal segment and an antennal club made of segments 9–11. The rare family MONOMMATIDAE (2 black, oval southeastern species; sometimes treated as part of Zopheridae) would also key out here. Pg. 335
Antennae usually with 10 segments. First 3 abdominal sternites free. Size 1.5–4.5 mm. Uncommon.
Antennae with 11 segments, usually inserted under a shelf that cuts into the eye. First 3 abdominal sternites fused. Size 2–35 mm. DARKLING BEETLES (TENEBRIONIDAE) (very diverse in color and shape, common.) The rare family ARCHEOCRYPTICIDAE, traditionally included with the Tenebrionidae, will key out here. The one American species is a small brown beetle with only the first two abdominal sternites fused; it does not range to the northeast. Pg. 329
Some tarsomeres distinctly lobed below. SOME FALSE DARKLING BEETLES (MELANDRYIDAE) Pg. 333
All tarsomeres without lobes.
Back of head not telescoped into thorax. Common on flowers. SCRAPTIIDAE Pg. 333
Font legs inserted in a complete socket (front coxal cavities closed).
Hind coxa not extending to elytra or side of body.
Hind coxa extending to elytra or side of body.
Back of head pulled into thorax.
Front legs in an incomplete socket, hind margin of socket broken or absent (front coxal cavities open). This character can be difficult to see in some, especially in Zopheridae (See below).
Pronotum with sharp lateral margins (not smooth).
Tarsal claws without fleshy lobe. Body usually less elongate.
Pronotum with rounded side margins (no sharp edges).
Elongate, shaped like a click beetle. No lobes on next-to-last tarsomere. SYNCHROIDAE Pg. 333
Tarsal claws often comb-like, always with a fleshy lobe at base. Shape distinctive (head and thorax tapered towards one another). FALSE LONGHORNS (STENOTRACHELIDAE) Pg. 337
Body bumpy, parallel-sided and very hard. Antennae 11-segmented. IRONCLAD BEETLES (ZOPHERIDAE) Only the one eastern species (Phellopsis obcordata) will key out here. Pg. 332
Not bumpy and parallel-sided. Antennae 10 or 11 segmented.
Variously shaped but not click beetle-shaped. Penultimate tarsomere often lobed. SOME FALSE DARKLING BEETLES (MELANDRYIDAE) Pg. 333
Size less than 5 mm, often less than 3 mm. Shape ant-like.
Size usually over 5 mm. Occasionally smaller, but not ant-like in shape.
Eyes not notched. ANTLIKE FLOWER BEETLES (ANTHICIDAE) Pg. 335
Eyes notched. ANTLIKE LEAF BEETLES (ADERIDAE) Pg. 336
Head separated from thorax by a constriction or narrow neck.
Head broadly joined to thorax, no distinct neck.
Pronotum widest just behind head, narrower than elytra posteriorly. Mouthparts not beak-like. Long, narrow, often soft. Antennae long and thread-like. Often brightly colored. Common. FALSE BLISTER BEETLES (OEDEMERIDAE) Pg. 336
Pronotum usually narrow just behind head, wider at middle. Often with either beak-like mouthparts, apically swollen antennae or striate elytra. Hard-bodied, uniformly dark. SALPINGIDAE, PYTHIDAE, MYCTERIDAE Pg. 335
Soft elytra loosely covering abdomen. Tarsal claws cleft or toothed. Head very broad. Next-to-last tarsomere not lobed. BLISTER BEETLES (MELOIDAE) Pg. 326
Elytra tightly covering abdomen. Tarsal claws not cleft. Eye notched around antennal base. Next-to-last tarsomere strongly lobed. Common. FIRE-COLORED BEETLES (PYROCHROIDAE) Pg. 334

BEETLE KEY FOUR: From BEETLE KEYS TWO and THREE (beetles with more than 5 abdominal segments visible ventrally, plus selected groups with the head hidden from above).

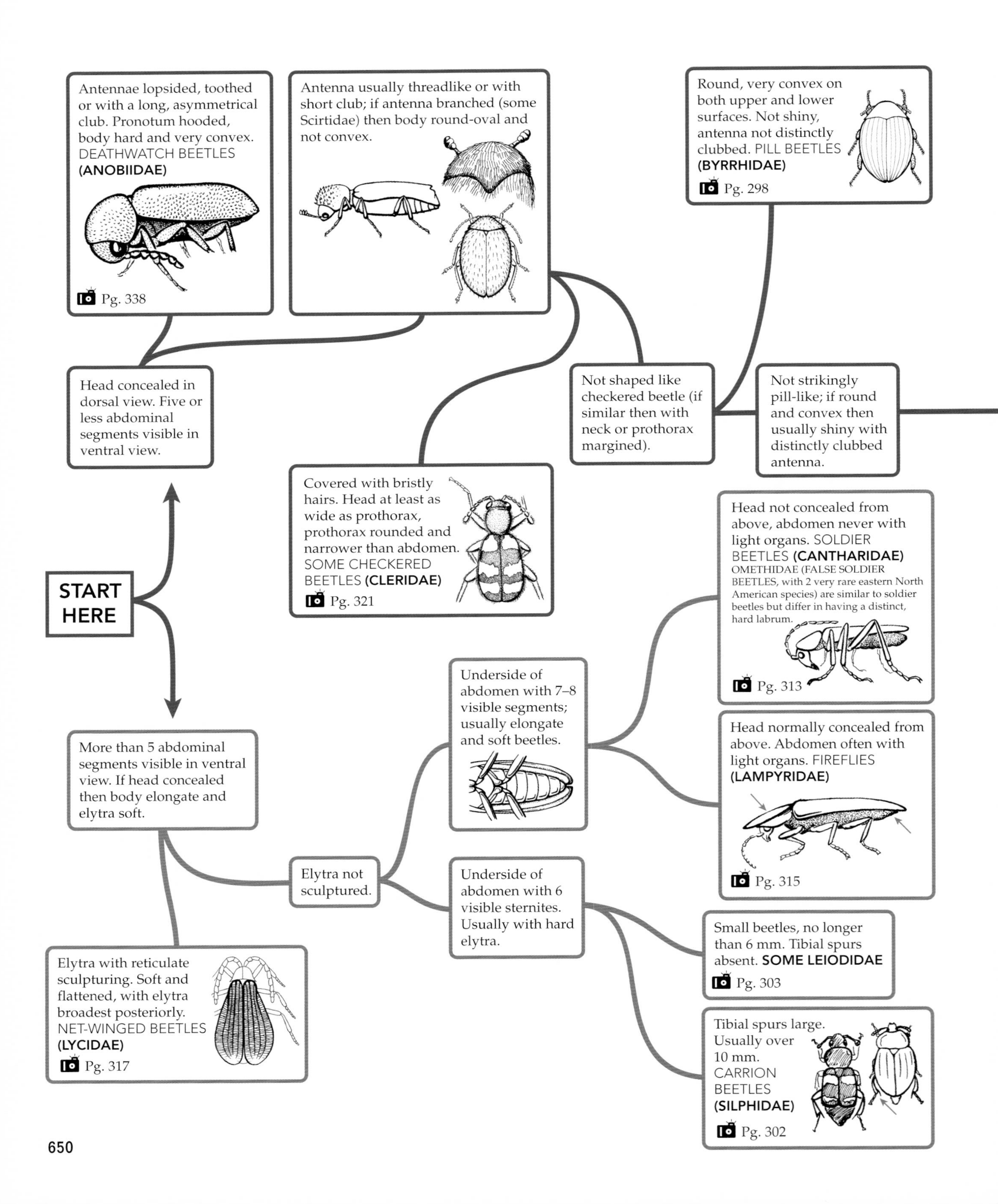

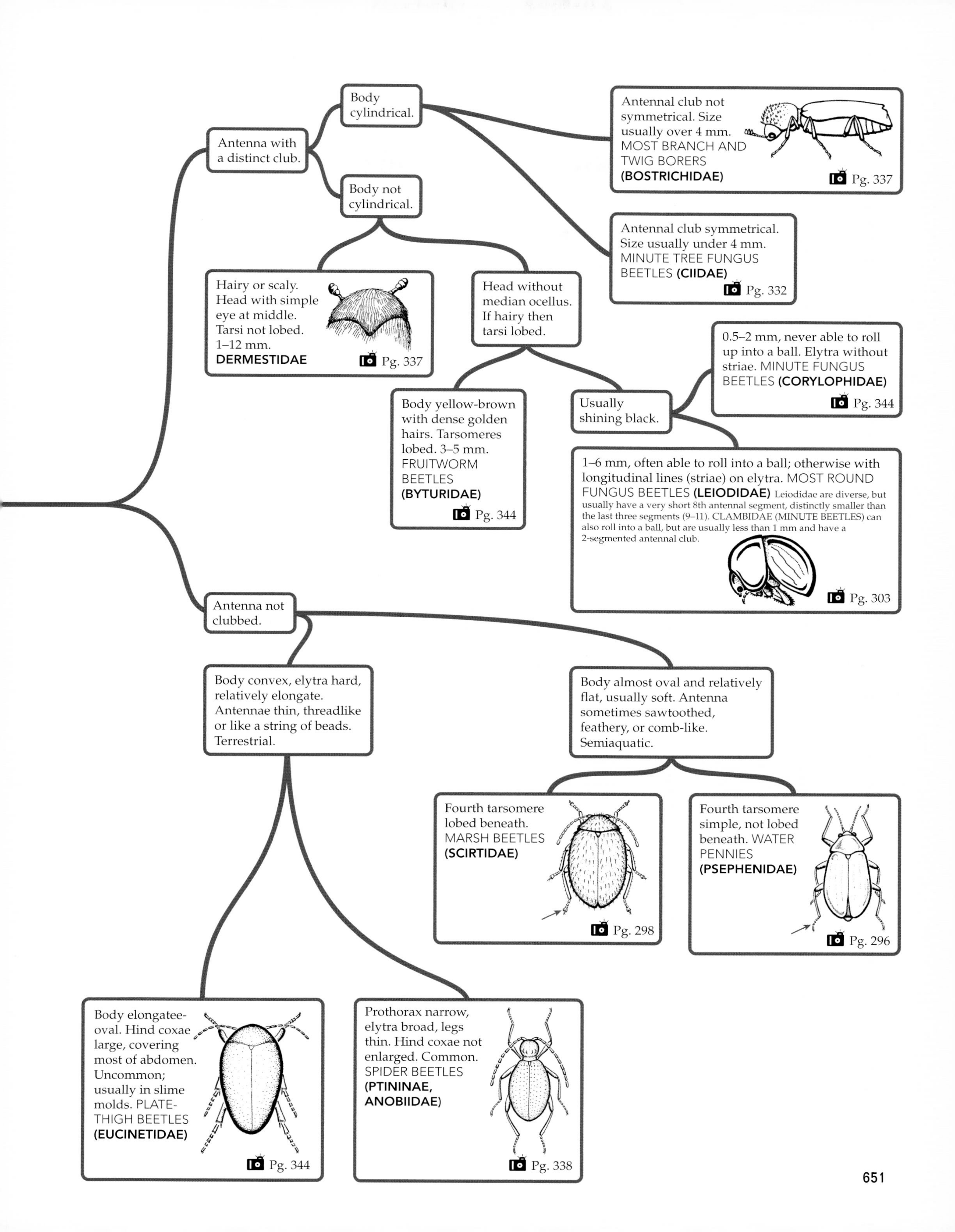

Antenna with a distinct club.
Body cylindrical.
Antennal club not symmetrical. Size usually over 4 mm. MOST BRANCH AND TWIG BORERS (BOSTRICHIDAE) Pg. 337
Antennal club symmetrical. Size usually under 4 mm. MINUTE TREE FUNGUS BEETLES (CIIDAE) Pg. 332
Body not cylindrical.
Hairy or scaly. Head with simple eye at middle. Tarsi not lobed. 1–12 mm. DERMESTIDAE Pg. 337
Head without median ocellus. If hairy then tarsi lobed.
Body yellow-brown with dense golden hairs. Tarsomeres lobed. 3–5 mm. FRUITWORM BEETLES (BYTURIDAE) Pg. 344
Usually shining black.
0.5–2 mm, never able to roll up into a ball. Elytra without striae. MINUTE FUNGUS BEETLES (CORYLOPHIDAE) Pg. 344
1–6 mm, often able to roll into a ball; otherwise with longitudinal lines (striae) on elytra. MOST ROUND FUNGUS BEETLES (LEIODIDAE) Leiodidae are diverse, but usually have a very short 8th antennal segment, distinctly smaller than the last three segments (9–11). CLAMBIDAE (MINUTE BEETLES) can also roll into a ball, but are usually less than 1 mm and have a 2-segmented antennal club. Pg. 303
Antenna not clubbed.
Body convex, elytra hard, relatively elongate. Antennae thin, threadlike or like a string of beads. Terrestrial.
Body elongatee-oval. Hind coxae large, covering most of abdomen. Uncommon; usually in slime molds. PLATE-THIGH BEETLES (EUCINETIDAE) Pg. 344
Prothorax narrow, elytra broad, legs thin. Hind coxae not enlarged. Common. SPIDER BEETLES (PTININAE, ANOBIIDAE) Pg. 338
Body almost oval and relatively flat, usually soft. Antenna sometimes sawtoothed, feathery, or comb-like. Semiaquatic.
Fourth tarsomere lobed beneath. MARSH BEETLES (SCIRTIDAE) Pg. 298
Fourth tarsomere simple, not lobed beneath. WATER PENNIES (PSEPHENIDAE) Pg. 296

KEY TO COLEOPTERA LARVAE: Aquatic beetle larvae and larvae of the most commonly encountered terrestrial families.

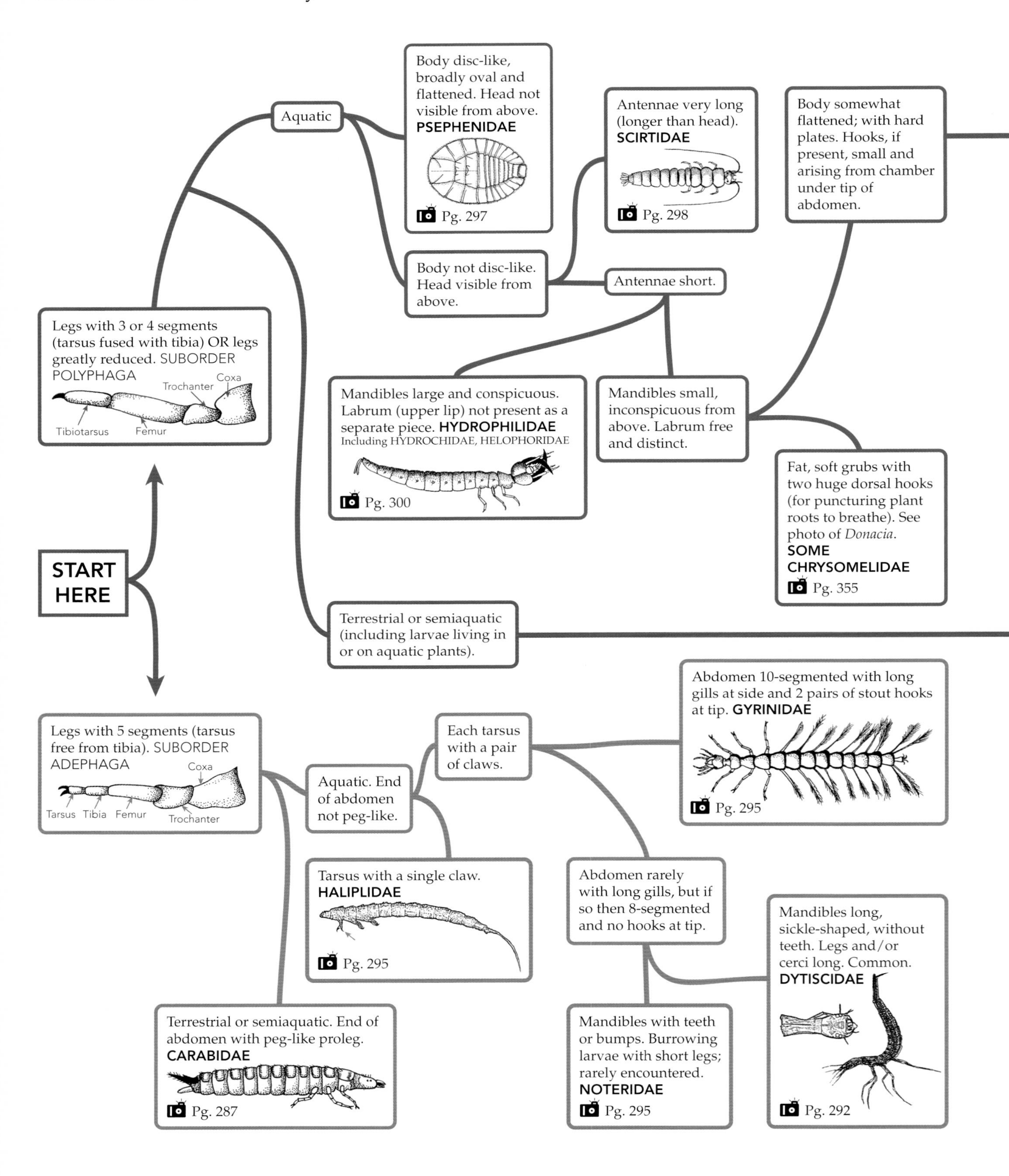

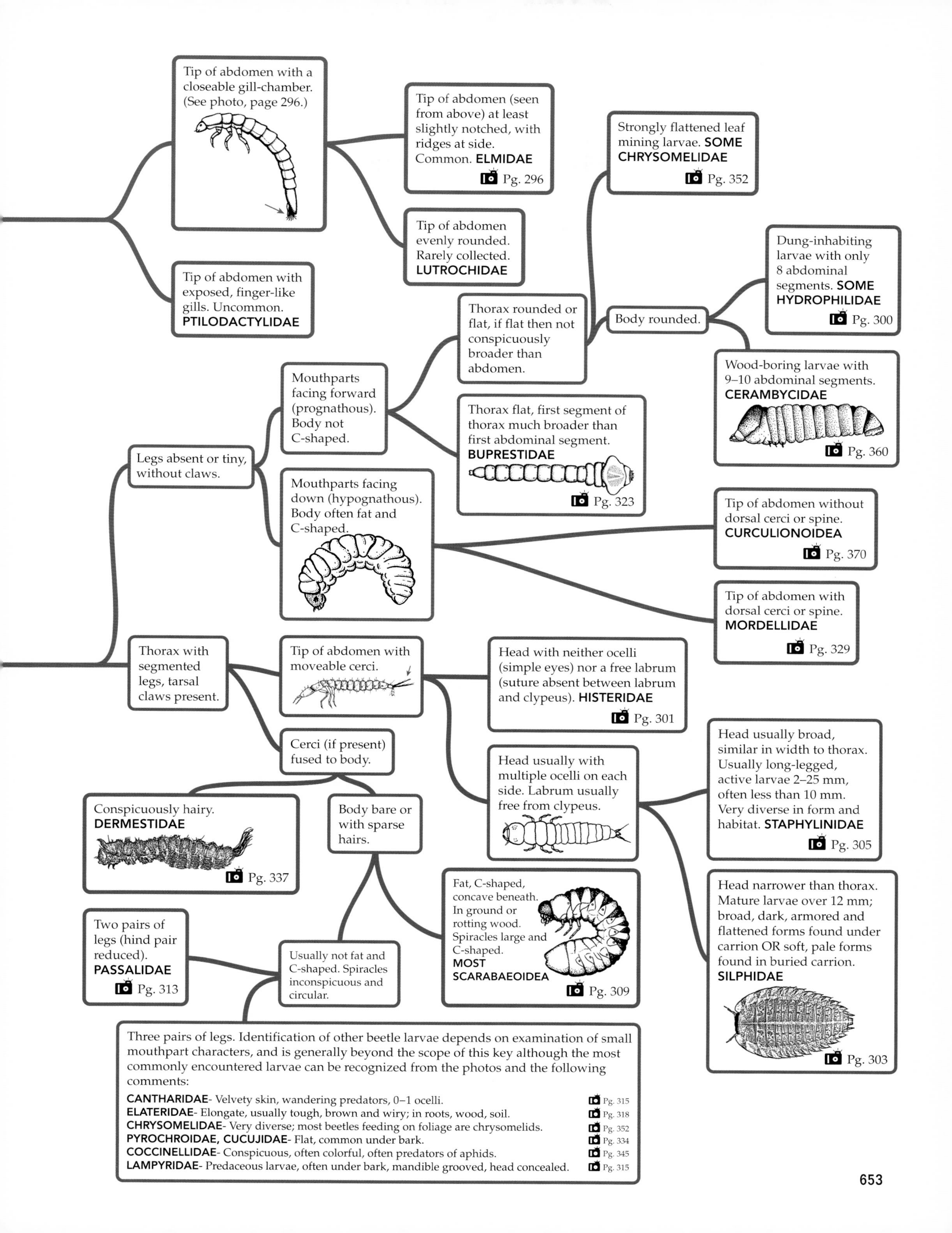
Tip of abdomen with a closeable gill-chamber. (See photo, page 296.)
Tip of abdomen (seen from above) at least slightly notched, with ridges at side. Common. ELMIDAE Pg. 296
Tip of abdomen evenly rounded. Rarely collected. LUTROCHIDAE
Tip of abdomen with exposed, finger-like gills. Uncommon. PTILODACTYLIDAE
Strongly flattened leaf mining larvae. SOME CHRYSOMELIDAE Pg. 352
Dung-inhabiting larvae with only 8 abdominal segments. SOME HYDROPHILIDAE Pg. 300
Body rounded.
Thorax rounded or flat, if flat then not conspicuously broader than abdomen.
Wood-boring larvae with 9–10 abdominal segments. CERAMBYCIDAE Pg. 360
Mouthparts facing forward (prognathous). Body not C-shaped.
Thorax flat, first segment of thorax much broader than first abdominal segment. BUPRESTIDAE Pg. 323
Legs absent or tiny, without claws.
Mouthparts facing down (hypognathous). Body often fat and C-shaped.
Tip of abdomen without dorsal cerci or spine. CURCULIONOIDEA Pg. 370
Tip of abdomen with dorsal cerci or spine. MORDELLIDAE Pg. 329
Thorax with segmented legs, tarsal claws present.
Tip of abdomen with moveable cerci.
Head with neither ocelli (simple eyes) nor a free labrum (suture absent between labrum and clypeus). HISTERIDAE Pg. 301
Cerci (if present) fused to body.
Head usually with multiple ocelli on each side. Labrum usually free from clypeus.
Head usually broad, similar in width to thorax. Usually long-legged, active larvae 2–25 mm, often less than 10 mm. Very diverse in form and habitat. STAPHYLINIDAE Pg. 305
Conspicuously hairy. DERMESTIDAE Pg. 337
Body bare or with sparse hairs.
Fat, C-shaped, concave beneath. In ground or rotting wood. Spiracles large and C-shaped. MOST SCARABAEOIDEA Pg. 309
Head narrower than thorax. Mature larvae over 12 mm; broad, dark, armored and flattened forms found under carrion OR soft, pale forms found in buried carrion. SILPHIDAE Pg. 303
Two pairs of legs (hind pair reduced). PASSALIDAE Pg. 313
Usually not fat and C-shaped. Spiracles inconspicuous and circular.
Three pairs of legs. Identification of other beetle larvae depends on examination of small mouthpart characters, and is generally beyond the scope of this key although the most commonly encountered larvae can be recognized from the photos and the following comments:
CANTHARIDAE- Velvety skin, wandering predators, 0–1 ocelli. Pg. 315
ELATERIDAE- Elongate, usually tough, brown and wiry; in roots, wood, soil. Pg. 318
CHRYSOMELIDAE- Very diverse; most beetles feeding on foliage are chrysomelids. Pg. 352
PYROCHROIDAE, CUCUJIDAE- Flat, common under bark. Pg. 334
COCCINELLIDAE- Conspicuous, often colorful, often predators of aphids. Pg. 345
LAMPYRIDAE- Predaceous larvae, often under bark, mandible grooved, head concealed. Pg. 315

DIPTERA KEY ONE: The main groups of flies and the calyptrate families.

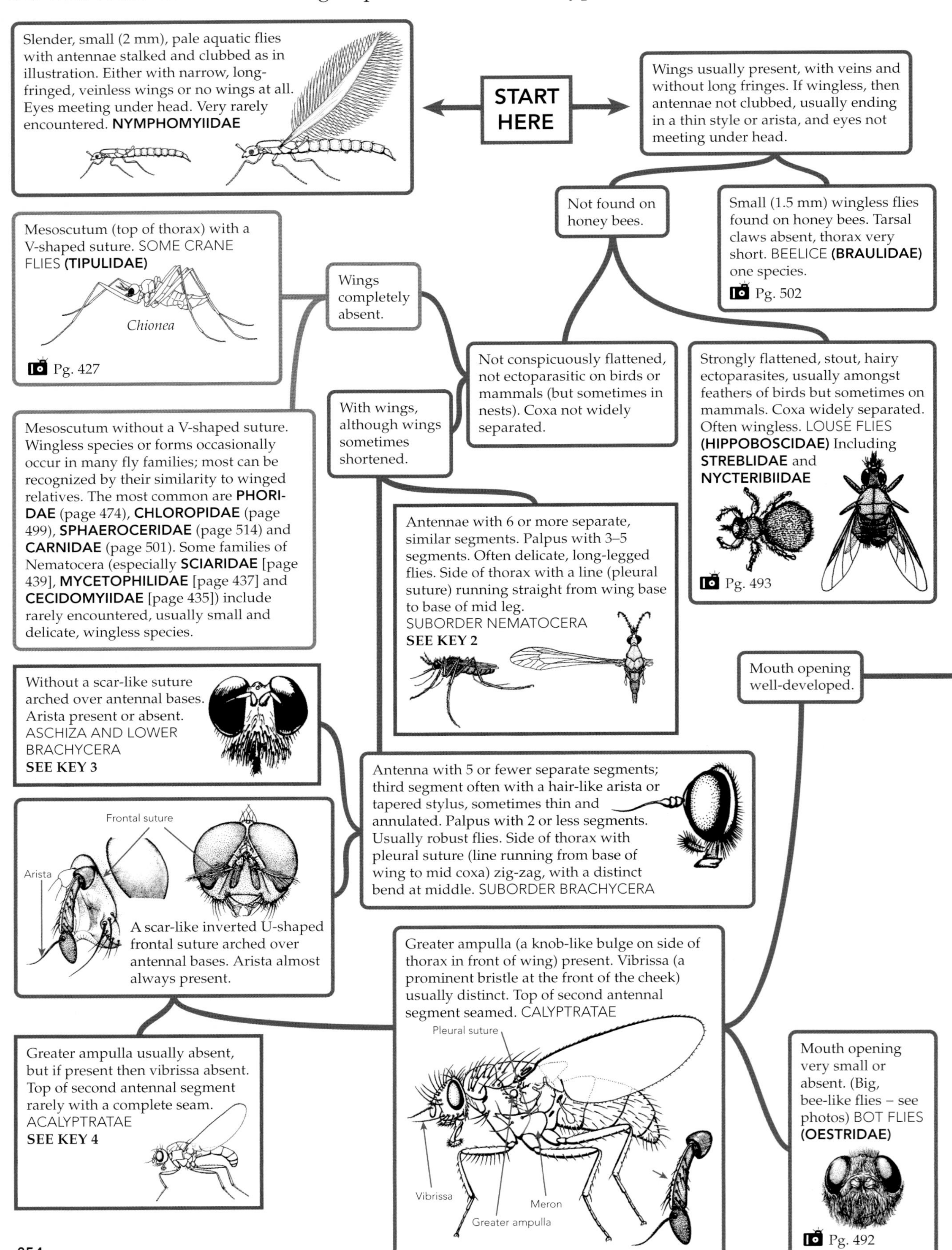

Either metallic in color (BLOW FLIES) or with dense silky yellow hairs (like corn silk) on side of thorax (CLUSTER FLIES). Arista plumose. **CALLIPHORIDAE** Pg. 481

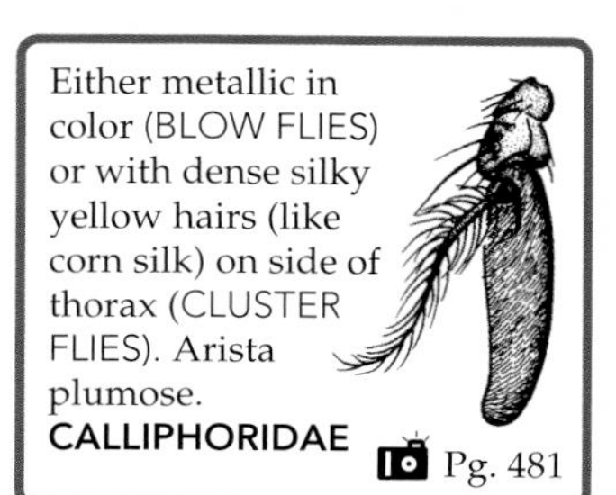

Wing dark, or dark with white apical spot. **MOST RHINOPHORIDAE** (One common black species, *Melanophora roralis*) Pg. 492

Wing clear. Body often striped in black and gray. FLESH FLIES AND SATELLITE FLIES **(SARCOPHAGIDAE)**

Notopleuron (side of thorax just in front of wing base) with 2 bristles. SATELLITE FLIES **(MILTOGRAMMINAE)** Some rarely encountered, atypical, House Fly-like snail parasitoids in the Calliphoridae tribe Angioneurini will also key out here.

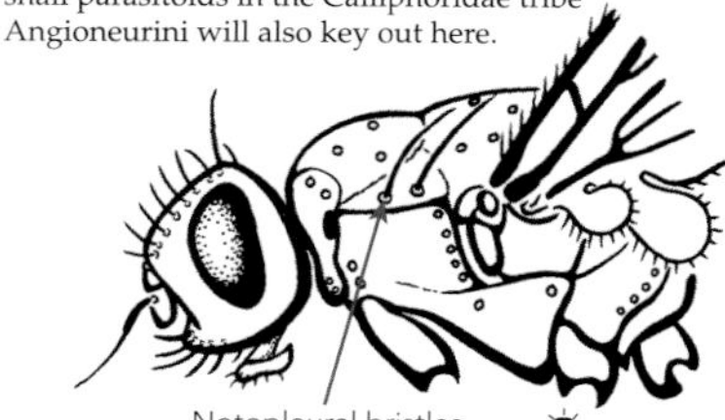

Pg. 483

Subscutellum (back part of thorax, below scutellum) not evenly bulging.

Not metallic. Often gray and black. Arista usually short-haired, or plumose only on basal half.

Arista

Notopleuron with 3–4 bristles. FLESH FLIES **(SARCOPHAGINAE)**

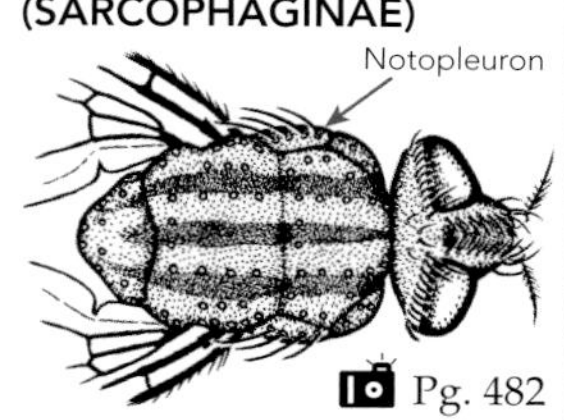

Pg. 482

Meron (the plate just above and behind the mid coxa) with row of bristles.

Subscutellum distinctly and evenly bulging. Usually very bristly, but one subfamily bare-bodied. Very diverse. PARASITIC FLIES **(TACHINIDAE)**

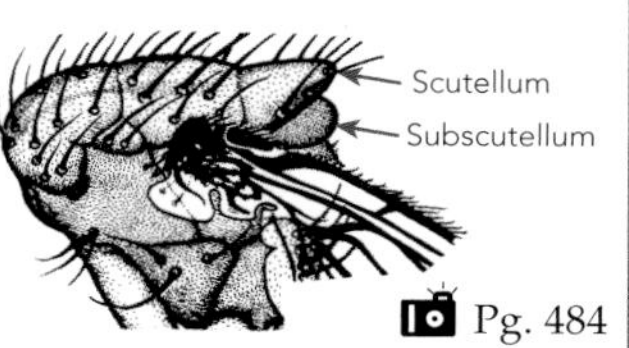

Pg. 484

Front of head with strong inclinate interfrontal bristles. Most commonly found in shoreline algae. **SOME ANTHOMYIIDAE (Fucelliinae)** A few other Anthomyiidae, including one that forms galls on ferns, will also key out here.

Pg. 479

Front of head without interfrontal bristles. Often yellowish and fuzzy or elongate and bare. **SCATHOPHAGIDAE**

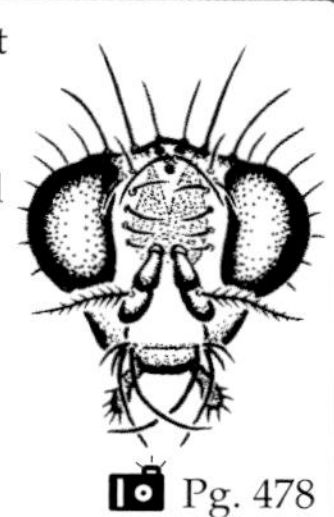

Pg. 478

Meron without a row of bristles. HOUSE FLIES, ROOT MAGGOTS, PILOSE YELLOW DUNG FLIES AND OTHER MUSCOIDEA **(MUSCIDAE, ANTHOMYIIDAE, FANNIIDAE, SCATHOPHAGIDAE)**

Underside of scutellum without erect hairs.

Lower calypter (the flap beneath the wing) much smaller than upper calypter, sometimes linear. Vein A1 usually traceable to wing margin. *Acridomyia*, a rare genus of Anthomyiidae parasitic on grasshoppers, has a short vein A1, but differs from other muscoids in lacking a palpus.

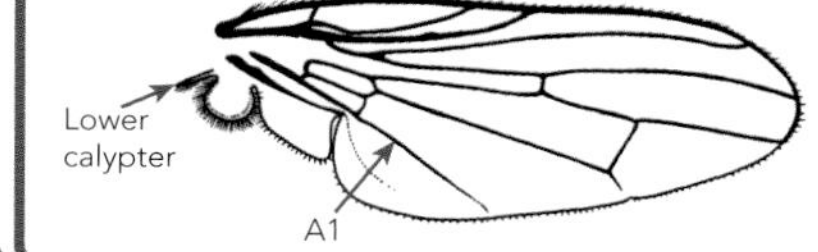

Underside of scutellum with erect hairs. ROOT MAGGOTS AND RELATIVES **(MOST ANTHOMYIIDAE)**

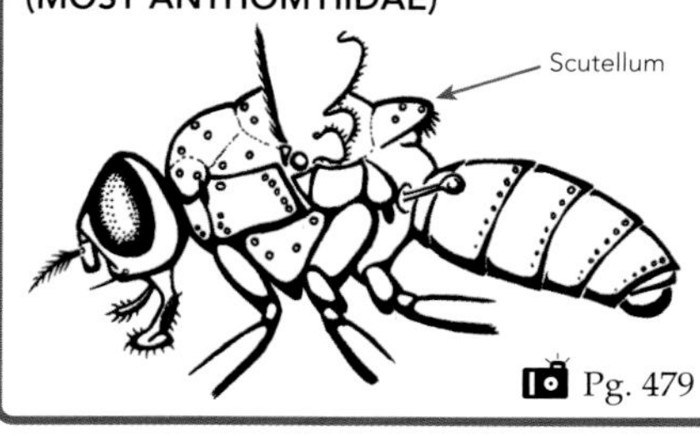

Pg. 479

Last 2 wing veins, even if extended, would not meet before wing apex. Subcosta curved up just before it meets the costa.

Last 2 wing veins short; last vein (A2) curved towards second last vein (A1); if extended these veins would meet far before wing margin. Most of subcosta running straight to costa, not curved up in apical third. LITTLE HOUSE FLIES AND THEIR RELATIVES **(FANNIIDAE)**

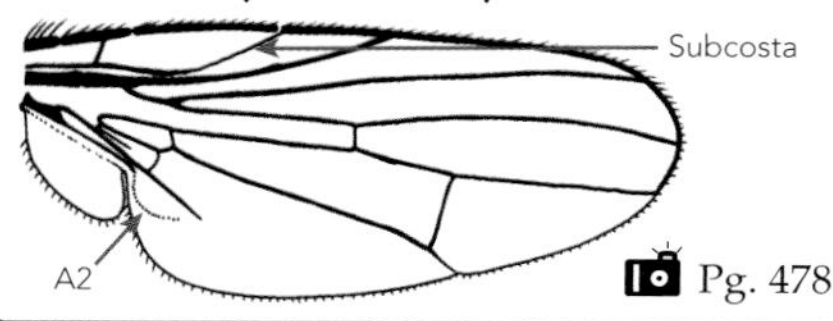

Pg. 478

Lower calypter very broad, much larger than upper calypter, never linear. Vein A1 fading out well before wing margin. **MUSCIDAE**

A1

Lower calypter

Pg. 476

DIPTERA KEY TWO: The long-horned flies (suborder Nematocera). Use this key for flies with 6 or more antennal segments.

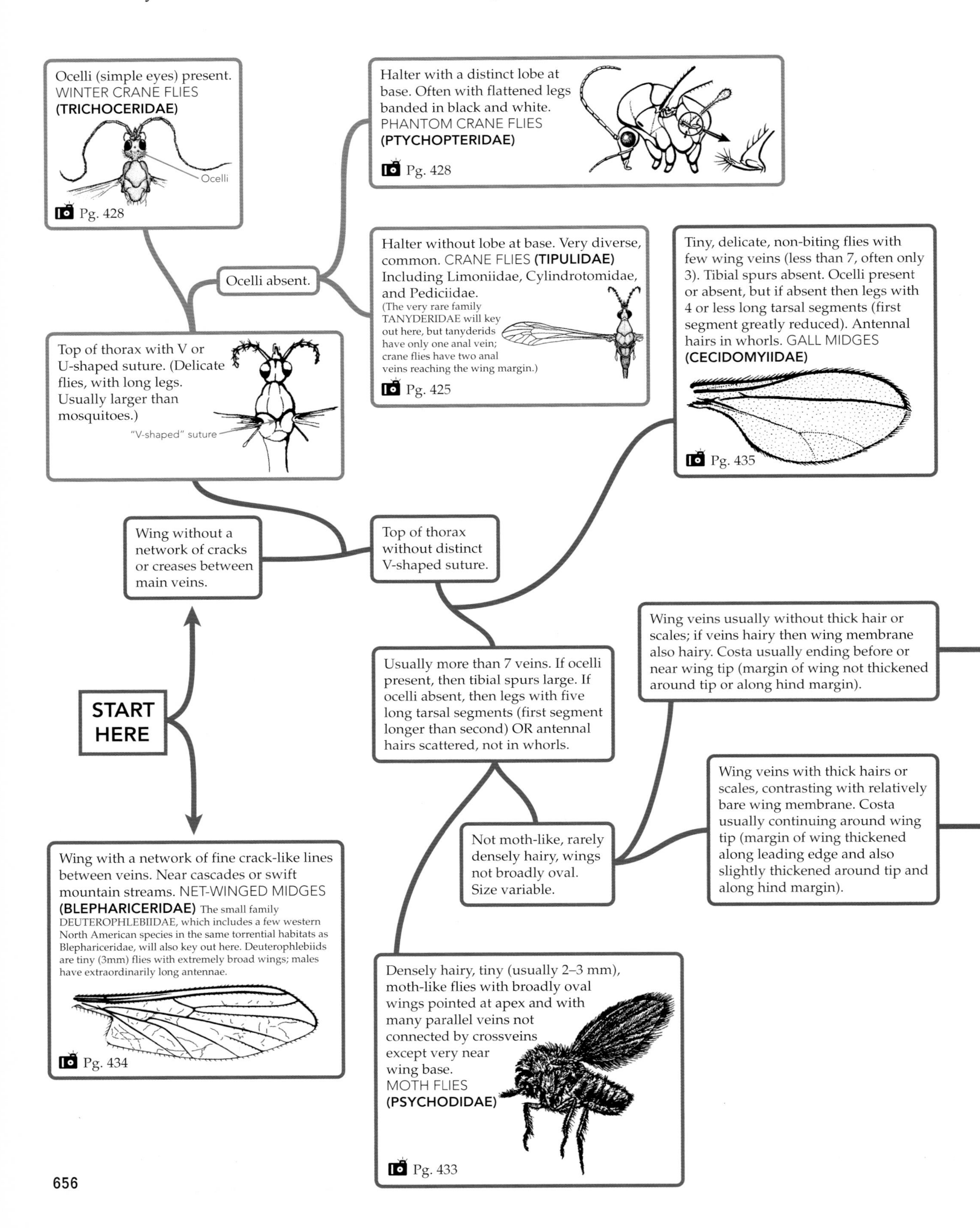

Eyes meeting over antennae, forming an eye bridge. Usually small, brown to black flies. Abdomen broadest at junction with thorax. Coxae of moderate size. DARK-WINGED FUNGUS GNATS **(SCIARIDAE)**

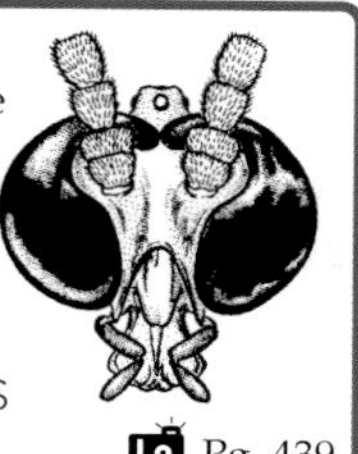

Pg. 439

Eyes rarely meeting over antennae. Coxae often conspicuously long. Size and color variable but often pale or patterned. Abdomen broadest at middle or apex, often narrowed at junction with thorax. FUNGUS GNATS **(MYCETOPHILIDAE)** Including Keroplatidae, Bolitophilidae, Diadocidiidae, Lygistorrhinidae and Ditomyiidae.

Coxa

Tibial spurs

Pg. 437

With large tibial spurs. Three ocelli present.

Tibial spurs small or absent. Ocelli absent.

Small, compact flies. Female with biting mouthparts. Male antenna plumose. Radial veins thickened and pushed towards base and wing margin. BITING MIDGES **(CERATOPOGONIDAE)**

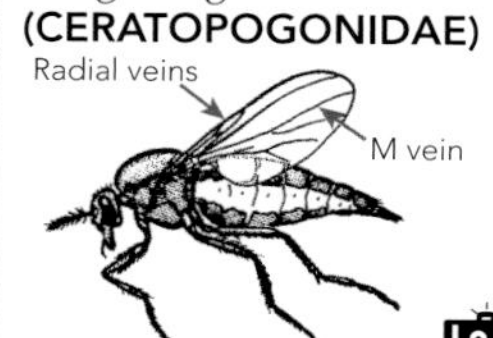

Pg. 428

Thorax strongly developed between scutellum and abdomen, usually with a distinct longitudinal groove or depression. Non-biting midges. Vein M (medial vein) with only one branch reaching wing margin. Ocelli absent. Males almost always with plumose antennae. MIDGES **(CHIRONOMIDAE)**

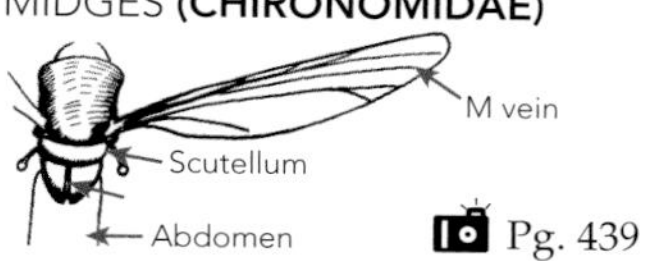

Pg. 439

Thorax without a longitudinal groove behind scutellum. Vein M almost always with two branches reaching wing margin, if with only one branch then ocelli present.

Wing without closed cell or broad basal cell.

Wing with a closed cell at middle or (rarely) with a broad cell at base with 6 veins branching off outer edge. **ANISOPODIDAE** *Rachicerus*, an uncommon xylophagid with unusual comb-like antennae, will also key here, as will the small family PACHYNEURIDAE with one western North American species.

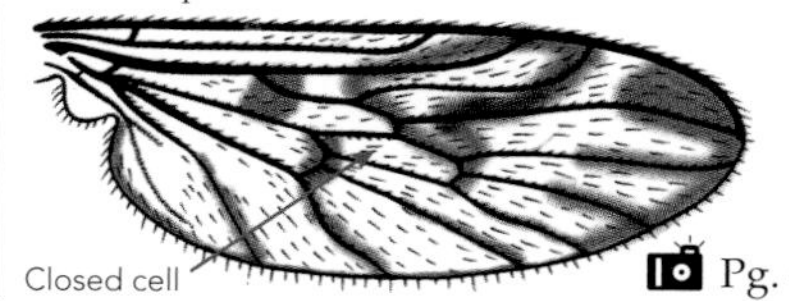

Pg. 434

Non-biting, antenna not plumose. Similar in size and shape to a long-legged and delicate mosquito. Radial veins in middle of wing, stem of R_2 conspicuously arched. Rarely collected, found near water. **DIXIDAE**

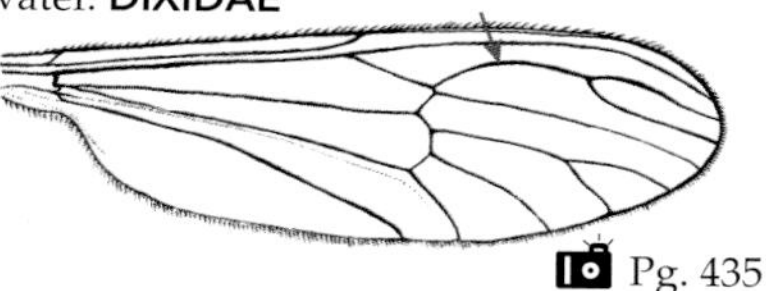

Pg. 435

Antennae slender, at least twice as long as head; usually longer than head and thorax combined. Antennae often feathery or with long hairs.

Antennae thick, without long hairs, shorter than head and thorax combined.

Ocelli present. Elongate flies.

Ocelli absent. Stocky flies.

Small flies, less than 4 mm. Feet without visible pads. MINUTE BLACK SCAVENGER FLIES **(SCATOPSIDAE)** (Flies in the rare family CANTHYLOSCELIDAE (=Synneuridae) are similar but their eyes nearly meet below the antennae and they have 4-segmented palps unlike the one-segmented palps of scatopsids.)

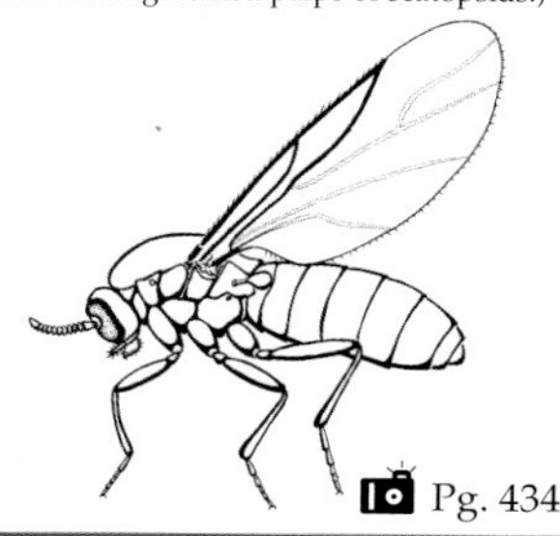

Pg. 434

Wing veins uniform. Reddish in color. Found only near seeps. **THAUMALEIDAE**

Pg. 435

Wing veins strong along leading margin. Very common biting flies. BLACK FLIES **(SIMULIIDAE)**

Pg. 432

Scales present on wing veins, legs and body. Females with long, biting mouthparts. 3–9 mm. MOSQUITOES **(CULICIDAE)**

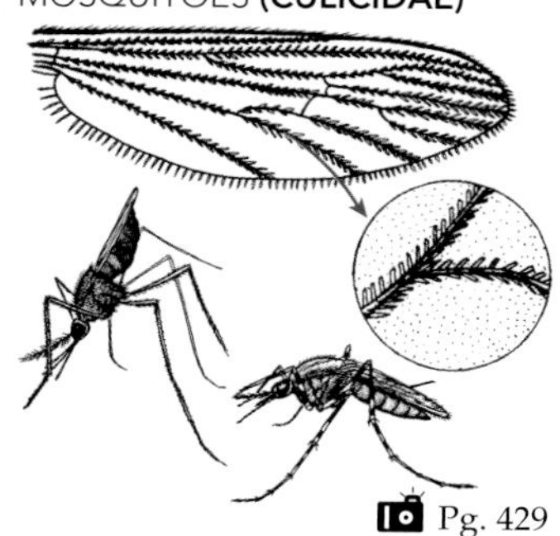

Pg. 429

Wing veins densely hairy; without broad scales. Usually non-biting.

Wing length usually over 2 mm. Mid femur no thicker than other legs. Common. PHANTOM MIDGES **(CHAOBORIDAE)**

Pg. 432

Wing length always less than 2 mm. Mid femur swollen. Uncommon; on or near singing tree frogs. **CORETHRELLIDAE**

Pg. 432

Medium-sized flies (over 5 mm, often much larger). Each foot with 3 conspicuous pads (a central empodium and a pulvillus under each tarsal claw). MARCH FLIES **(BIBIONIDAE)** The one eastern species in the rare family AXYMYIIDAE (with two small shining spots on top of the thorax; see page 439) will also key out here.

Pg. 437

DIPTERA KEY THREE: Lower Brachycera and Aschiza.

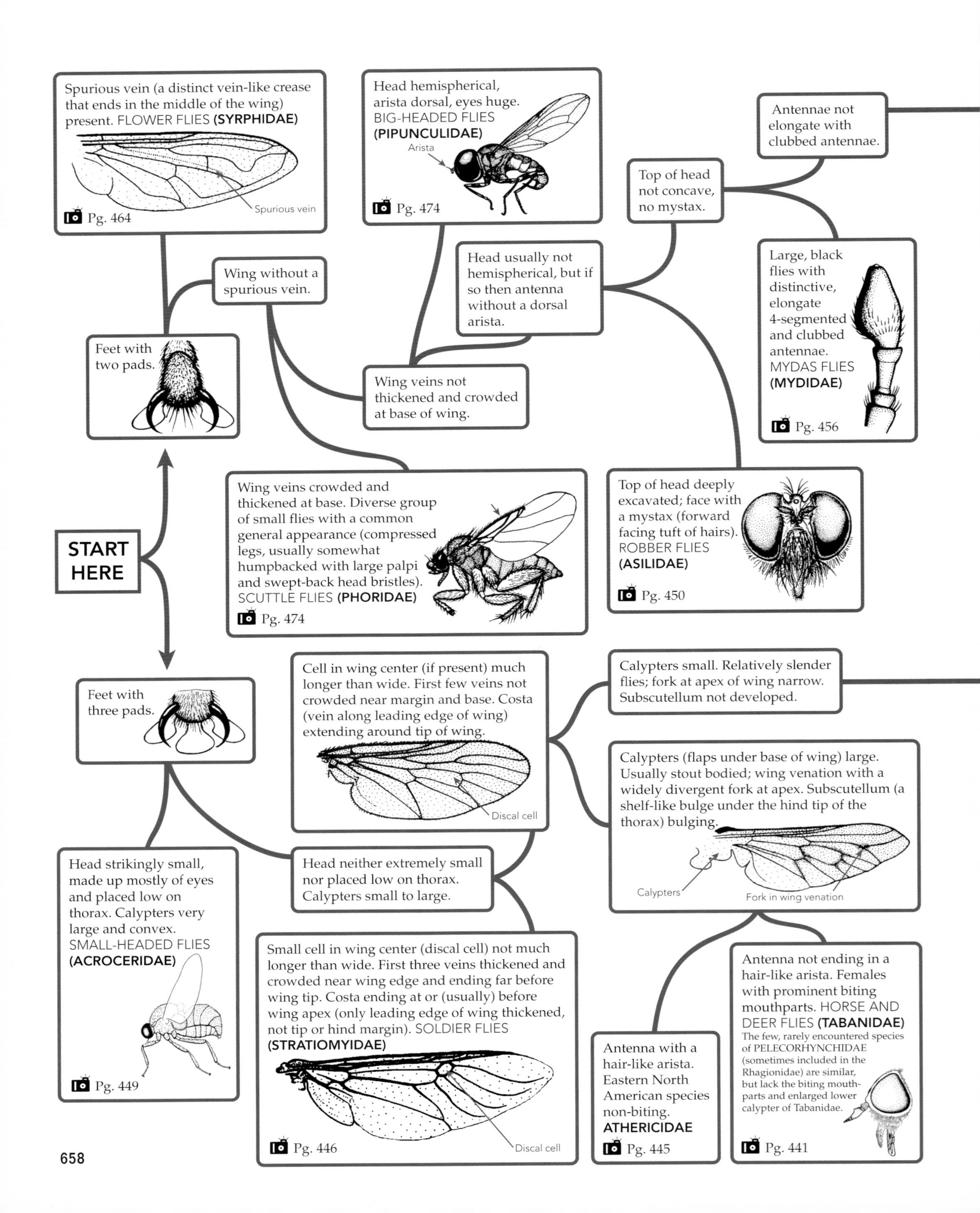

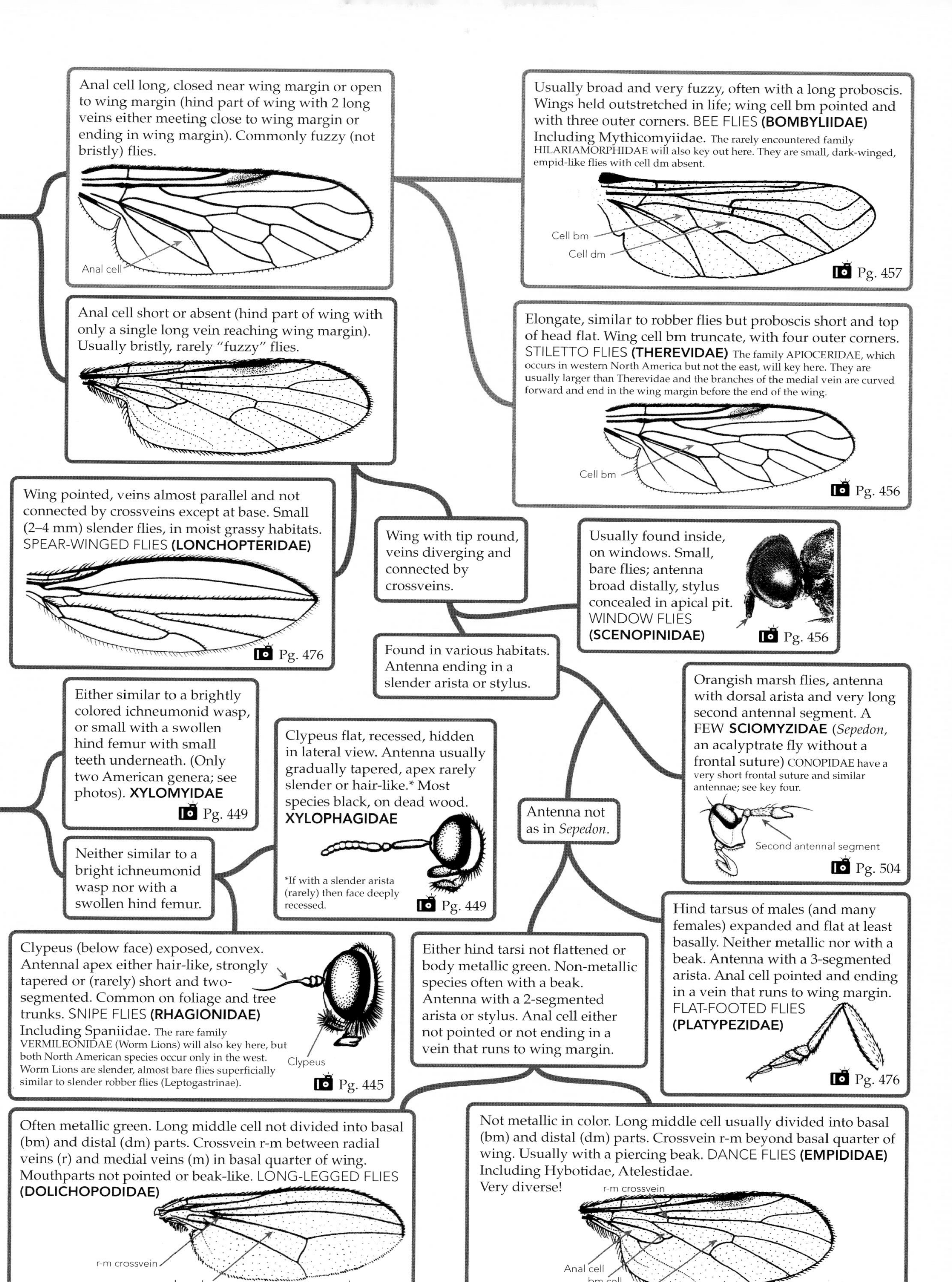
Anal cell long, closed near wing margin or open to wing margin (hind part of wing with 2 long veins either meeting close to wing margin or ending in wing margin). Commonly fuzzy (not bristly) flies.
Anal cell
Usually broad and very fuzzy, often with a long proboscis. Wings held outstretched in life; wing cell bm pointed and with three outer corners. BEE FLIES (BOMBYLIIDAE) Including Mythicomyiidae. The rarely encountered family HILARIAMORPHIDAE will also key out here. They are small, dark-winged, empid-like flies with cell dm absent.
Cell bm
Cell dm
Pg. 457
Anal cell short or absent (hind part of wing with only a single long vein reaching wing margin). Usually bristly, rarely “fuzzy” flies.
Elongate, similar to robber flies but proboscis short and top of head flat. Wing cell bm truncate, with four outer corners. STILETTO FLIES (THEREVIDAE) The family APIOCERIDAE, which occurs in western North America but not the east, will key here. They are usually larger than Therevidae and the branches of the medial vein are curved forward and end in the wing margin before the end of the wing.
Cell bm
Pg. 456
Wing pointed, veins almost parallel and not connected by crossveins except at base. Small (2–4 mm) slender flies, in moist grassy habitats. SPEAR-WINGED FLIES (LONCHOPTERIDAE)
Pg. 476
Wing with tip round, veins diverging and connected by crossveins.
Usually found inside, on windows. Small, bare flies; antenna broad distally, stylus concealed in apical pit. WINDOW FLIES (SCENOPINIDAE)
Pg. 456
Found in various habitats. Antenna ending in a slender arista or stylus.
Orangish marsh flies, antenna with dorsal arista and very long second antennal segment. A FEW SCIOMYZIDAE (Sepedon, an acalyptrate fly without a frontal suture) CONOPIDAE have a very short frontal suture and similar antennae; see key four.
Second antennal segment
Pg. 504
Either similar to a brightly colored ichneumonid wasp, or small with a swollen hind femur with small teeth underneath. (Only two American genera; see photos). XYLOMYIDAE
Pg. 449
Clypeus flat, recessed, hidden in lateral view. Antenna usually gradually tapered, apex rarely slender or hair-like.* Most species black, on dead wood. XYLOPHAGIDAE
*If with a slender arista (rarely) then face deeply recessed.
Pg. 449
Antenna not as in Sepedon.
Neither similar to a bright ichneumonid wasp nor with a swollen hind femur.
Hind tarsus of males (and many females) expanded and flat at least basally. Neither metallic nor with a beak. Antenna with a 3-segmented arista. Anal cell pointed and ending in a vein that runs to wing margin. FLAT-FOOTED FLIES (PLATYPEZIDAE)
Pg. 476
Clypeus (below face) exposed, convex. Antennal apex either hair-like, strongly tapered or (rarely) short and two-segmented. Common on foliage and tree trunks. SNIPE FLIES (RHAGIONIDAE) Including Spaniidae. The rare family VERMILEONIDAE (Worm Lions) will also key here, but both North American species occur only in the west. Worm Lions are slender, almost bare flies superficially similar to slender robber flies (Leptogastrinae).
Clypeus
Pg. 445
Either hind tarsi not flattened or body metallic green. Non-metallic species often with a beak. Antenna with a 2-segmented arista or stylus. Anal cell either not pointed or not ending in a vein that runs to wing margin.
Often metallic green. Long middle cell not divided into basal (bm) and distal (dm) parts. Crossvein r-m between radial veins (r) and medial veins (m) in basal quarter of wing. Mouthparts not pointed or beak-like. LONG-LEGGED FLIES (DOLICHOPODIDAE)
r-m crossvein
bm+dm
Pg. 462
Not metallic in color. Long middle cell usually divided into basal (bm) and distal (dm) parts. Crossvein r-m beyond basal quarter of wing. Usually with a piercing beak. DANCE FLIES (EMPIDIDAE) Including Hybotidae, Atelestidae. Very diverse!
r-m crossvein
Anal cell
bm cell
dm cell
Pg. 460

DIPTERA KEY FOUR: Acalyptrate Diptera.

Arista usually conspicuously plumose (with long hairs above and below); fat part of antenna pointed down. Very common. **DROSOPHILIDAE**
(The rare family DIASTATIDAE is similar, but has spinules along the leading edge of the wing).

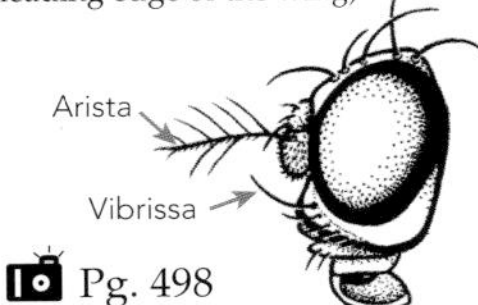

Pg. 498

Arista usually not plumose, fat part of antennae usually directed forward.

Wings clear, without dark patches.

Wings with distinct spots or maculations.

Slender, usually brown or yellowish; never dull gray. Wings with an apical wing spot and often with additional markings; never mottled. Some common species found in grasses. **OPOMYZIDAE**
Pg. 500

Stout, dull gray flies usually with brown markings and banded legs. Wings often mottled but never with an apical wing spot. Uncommon, found on tree wounds. **ODINIIDAE**
Pg. 511

Front femur without single, stout spine. Body usually not strikingly slender, but if so then wings with distinct spots or maculations.

Ocellar bristles minute or absent. Head with a bright orange band across the front. On tree wounds. **AULACIGASTRIDAE**
Pg. 511

Elongate, slender flies almost always with a stout spine on the front femur; wings usually clear but *Ischnomyia* (page 501) has extensively darkened wings. **ANTHOMYZIDAE**
Pg. 501

Ocellar bristles strong. Face without orange band.

Vibrissa or forward-pointing vibrissa-like bristles present at front of cheek.

Arista almost bare and face flat. Often with a large shining triangle around ocelli. Orbital bristles short. GRASS FLIES **(CHLOROPIDAE)**
(Eastern CARNIDAE, page 501, will also key here. Carnids are black, 1–3 mm, with long head bristles including a strong vibrissa.) Also CRYPTOCHAETIDAE (one exotic species, California, with no arista).
Pg. 499

Arista hairy and/or face bulging. Ocelli often raised but not usually surrounded by large shiny triangle. Bristles above eye (orbital bristles) often long. SHORE FLIES **(EPHYDRIDAE)**

Pg. 499

Vibrissa absent. **PSILIDAE** The small family STRONGYLOPHTHALMYIIDAE with one distinctive eastern species (page 503) will also come out here.

Pg. 502

Wing with anal cell (near base of last vein); long middle cell of wing usually divided into basal (bm) and distal (dm) cells (as shown in diagram of fruit fly wing below.)

Wing without anal cell; middle cell (bm + dm) not divided by crossvein.

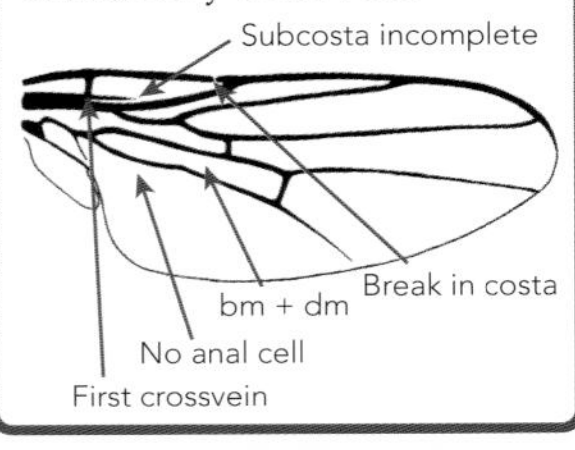

Subcosta not bent, or bent at less than 90°. Wings pictured or clear.

Head broad and hammerhead-shaped; antennae widely separated. One distinctive genus. STALK-EYED FLIES **(DIOPSIDAE)**
Pg. 510

Subcosta bent abruptly up towards front margin of wing but not reaching wing margin. Usually with conspicuous wing markings. FRUIT FLIES **(TEPHRITIDAE)**

Subcosta
dm cell
bm cell
First crossvein
Anal cell

Costa with a break or gap near end of subcosta, sometimes also above first crossvein. Many common families; size variable. Arista with or without long hairs.

Costa entire, without distinct breaks. Arista often with long hairs. Rarely collected flies, under 4 mm. **ASTEIIDAE**
Also PERISCELIDIDAE (second antennal segment with dorsal seam).
Pg. 512

START HERE

First segment of hind tarsus longer and no thicker than second segment.

Subcosta incomplete (not reaching wing margin), often short and inconspicuous

First crossvein (h)
Subcosta incomplete

Eyes not at ends of stout stalks.

Subcosta complete and distinct, running to costa.

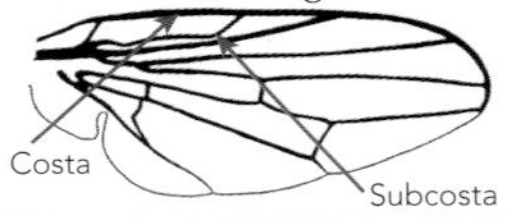

First segment of hind tarsus short and thicker than second. Usually small, dark flies; very common. **SPHAEROCERIDAE**

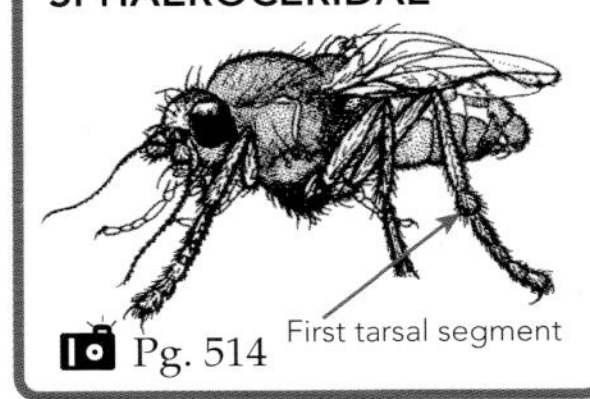

Pg. 514

Ocelli (simple eyes) absent. Large, uncommon flies with pictured wings. **PYRGOTIDAE**
Pg. 508

Proboscis usually short (long in some small, black flies). Arista usually hair-like, longer than thick part of antenna (rarely greatly reduced). Frontal suture well developed.

Ocelli usually present in a triangular pattern of 3 on top of head. If ocelli absent (some Conopidae) then with a conspicuous, forward-pointing proboscis.

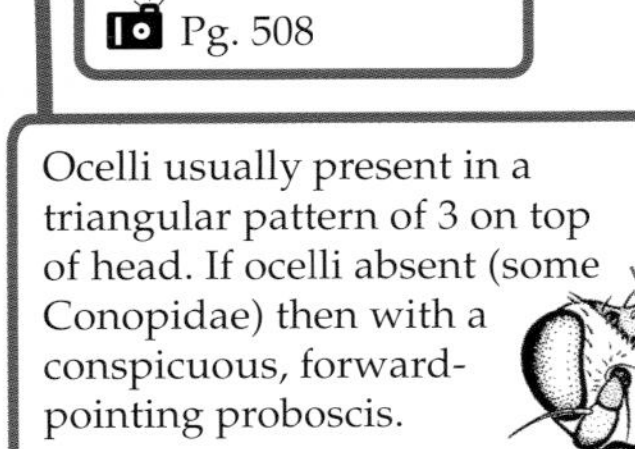

Usually with a long forward-pointing proboscis (bent at base or at base and middle). If antenna with a hair-like arista, then arista shorter than the thick part of antenna. Frontal suture small, sometimes difficult to see. Elongate, bare, often wasp-like flies, head broader than thorax. THICK-HEADED FLIES **(CONOPIDAE)** **and STYLOGASTRIDAE**
Pgs. 507, 508

Postocellar bristles convergent or parallel. Costa broken twice (above subcosta and above the first crossvein). Small, usually black flies; often with a long, bent proboscis. **MILICHIIDAE** Some uncommon flies in other families will also key out here, including some species of tiny gray TETHINIDAE and CANACIDAE (normally on sea shores) and some little yellowish CHYROMYIDAE.
Pg. 501

Postocellar bristles divergent (rarely absent). Costa not broken above first crossvein. Color variable, but often with extensive white or yellow markings. Female with a stiff, tubular abdominal tip. Larvae are usually leaf miners. **MOST AGROMYZIDAE**

Pg. 502

Anal cell with an angulate, often pointed outer corner (usually) OR top of thorax without strong bristles on front half. **ULIDIIDAE (= OTITIDAE)**

Anal cell

Pg. 496

Anal cell with rounded outer corner. Top of thorax with strong bristles (dorsocentrals) on front half. **PALLOPTERIDAE**
Pg. 498

Wings with smoky areas, especially at tip and leading edge; sometimes with other markings. Arista arising near apex of first flagellomere. Mid tibia usually with one or two preapical dorsal bristles (missing in *Heteromeringia*, page 510). Usually yellow to orange with body markings, but sometimes black. **CLUSIIDAE**

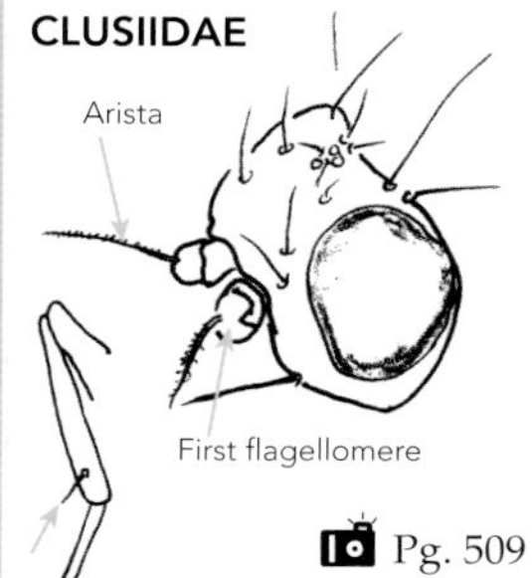

Pg. 509

Wing glassy, clear, rarely with spots or other markings. Arista arising near base of first flagellomere. Mid tibia without preapical dorsal bristles. Usually black, rarely pale. **PIOPHILIDAE** Some AGROMYZIDAE (postocellar bristles divergent like Piophilidae but with strong inclinate frontal bristles) could key out here, along with some rarely collected flies in other families including some species of tiny gray TETHINIDAE and CANACIDAE (normally on sea shores) and little yellowish CHYROMYIDAE (postocellar bristles convergent).
Pg. 511

Strikingly long-legged flies with two long veins converging at wing tip and no bristles between the ocelli. STILT-LEGGED FLIES **(MICROPEZIDAE)** TANYPEZIDAE (only 2 North American species, uncommon) are similar but have larger eyes and have bristles between the ocelli. NERIIDAE (mostly tropical but present in southwestern USA) are similar but with an apical arista.

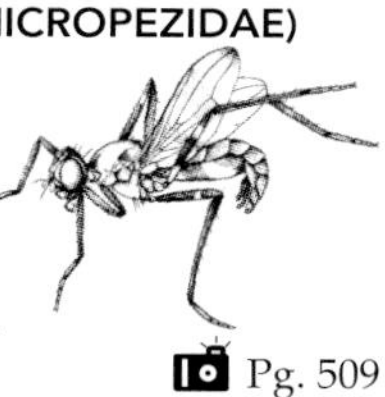

Pg. 509

Neither ant-like nor with stout ventral femoral bristles.

At least one femur with conspicuous strong, short bristles underneath. Often ant-like. **MOST RICHARDIIDAE**, including all eastern species. Uncommon.
Pg. 496

Leading edge of wing with only small setulae. If (rarely) with distinct costal spines then postocellar bristles divergent.

Halter whitish. Wings often patterned.

Halter black, body usually shining dark, legs black. Wings clear. On fallen trees. **LONCHAEIDAE** Pg. 497

Not stilt-legged, veins not converging at wing apex.

Cheek without a strong anterior bristle (vibrissa).

Cheek with one or more enlarged bristles at lower anterior corner (small in some tiny, gray seashore species).

Leading edge of wing spiny. Postocellar bristles convergent.

Costa (edge of wing) not broken at end of subcosta. Without a distinct vibrissa.

Costa broken at end of subcosta. With or without distinct vibrissa (a single outstanding bristle at the front lower corner of the cheek).

Arista bare. Not humpbacked. **HELEOMYZIDAE**
Pg. 512

Arista plumose. Thorax humpbacked. **CURTONOTIDAE**
Pg. 513

Palpus absent or minute. Usually ant-shaped, with a spherical head. Often on dung. With bristle(s) below and behind posterior spiracles. ANT-LIKE SCAVENGER FLIES **(SEPSIDAE)**

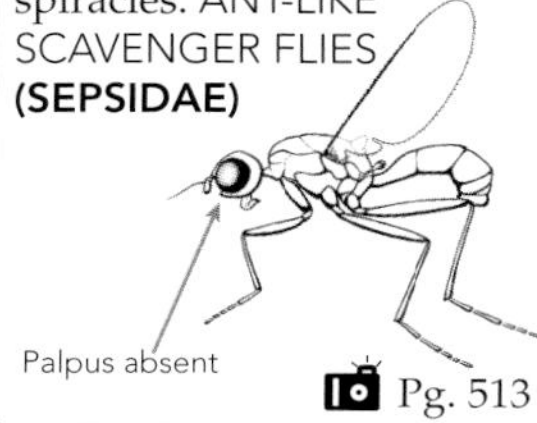

Pg. 513

Not ant-shaped. Palpus present. Posterior spiracle without bristles.

Strongly flattened bristly flies found on seashore algae. **COELOPIDAE** *Orygma* (Sepsidae) are similar seaweed flies, but have very small palps and a high, bare cheek.
Pg. 507

Wings clear or inconspicuously marked. Commonly gray with black spots. **CHAMAEMYIIDAE** Also ACARTOPHTHALMIDAE (uncommon, postocellar bristles divergent, vibrissa present but indistinct).
Pg. 503

Thorax not strongly flattened.

Tibiae without preapical dorsal bristle.

Tibiae with dorsal bristle before apex.

Wing with conspicuous pattern. **PLATYSTOMATIDAE** Also some ULIDIIDAE (end of anal cell usually with an angulate extension, costa not broken at first crossvein).
Pg. 495

Postocellar bristles parallel or absent. 2–21 mm.

Postocellar bristles convergent. Size under 6 mm. **LAUXANIIDAE** RHINOTORIDAE will also key out here: one species (a stout dark fly over 7 mm) occurs in Florida.

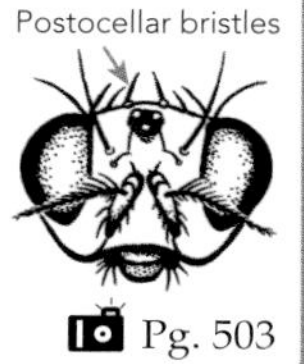

Pg. 503

Clypeus (sclerite below face) bulging. Second antennal segment short. Pale orange flies, 4–12 mm. **DRYOMYZIDAE**

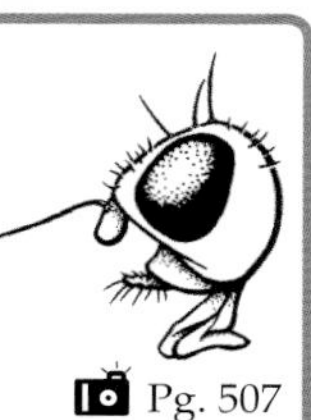

Pg. 507

Clypeus small, not bulging out below bottom of face. Second antennal segment often elongate. MARSH FLIES OR SNAIL-KILLING FLIES **(SCIOMYZIDAE)** (Some common species are superficially similar to DRYOMYZIDAE.)

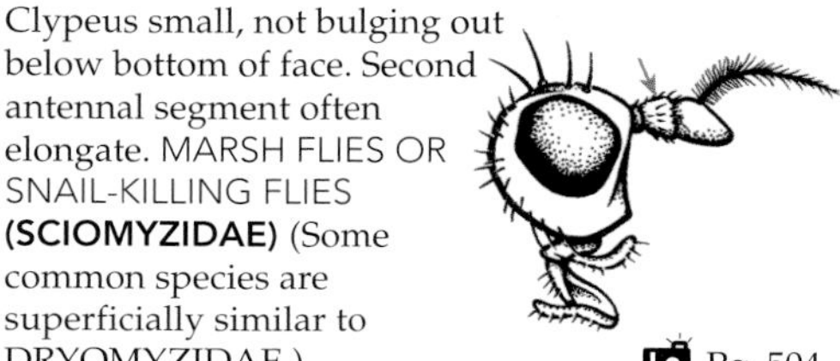

Pg. 504

KEY TO DIPTERA LARVAE: Aquatic families and some commonly encountered terrestrial fly larvae.

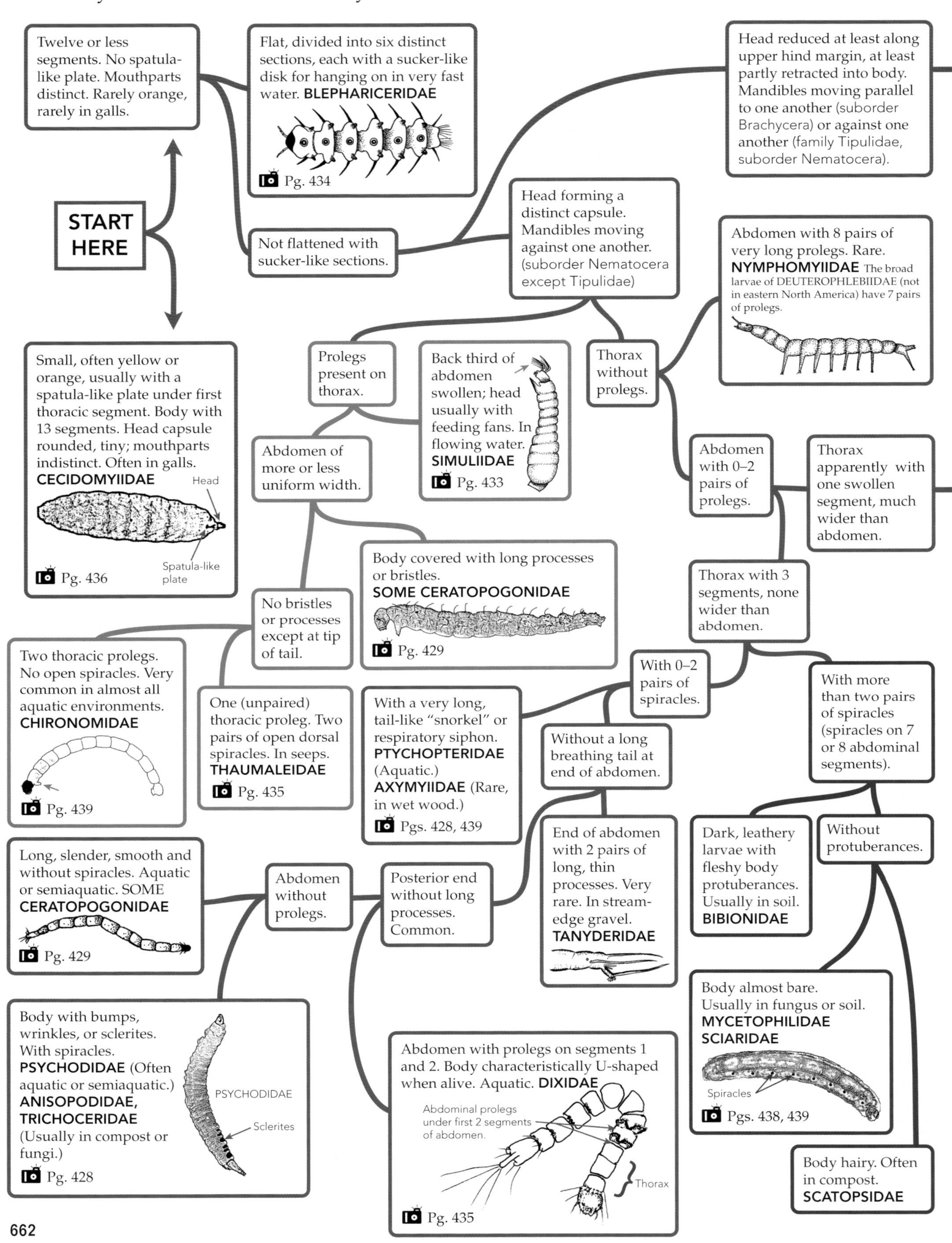

Mandibles moving against each other in an oblique or horizontal plane. Head capsule retractable into thorax, upper back of capsule reduced. Usually fat, soft larvae tapered at front and with lobes at abdominal tip. Usually in water, wet soil or rotting wood. **TIPULIDAE**

Pg. 425

Mandibles moving in a vertical plane. Head capsule usually deeply retracted or reduced to an internal skeleton. If head capsule present then body usually either flattened, spindle-shaped, or with abdomen ending in either a deep pit, distinct prolegs or a single tube.

At least upper front part of head capsule complete and exposed. If greatly reduced then with slender internal rods. ("Lower" Brachycera – Only families with aquatic members are keyed here.)

Head capsule reduced to a characteristically shaped internal skeleton, only tips of fang-like mouthhooks exposed. "Maggots." (Muscomorpha – a large and difficult group; only aquatic maggots are keyed here.)

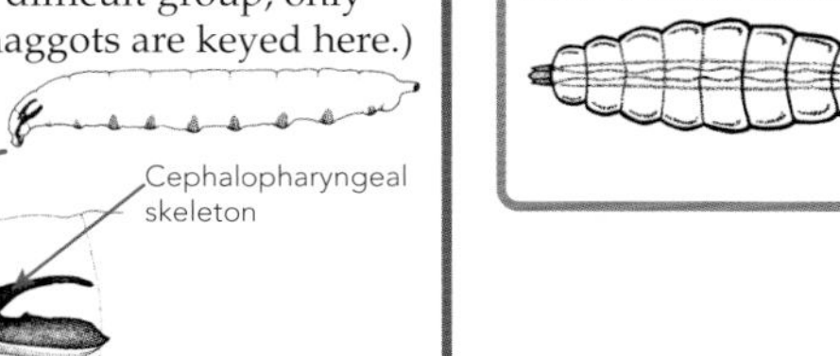

Body flattened top to bottom. Skin textured like sandpaper. **STRATIOMYIDAE**

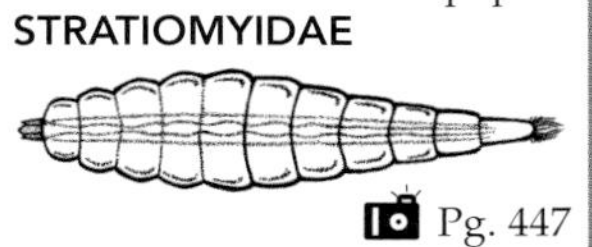

Pg. 447

Body not flat; skin smooth.

Labrum with brush-like tufts of bristles. **CULICIDAE**

Pg. 429

Labrum without brushes, but end of antenna usually with tuft of blade-like bristles. **CHAOBORIDAE**
Also CORETHRELLIDAE; rare, tail with tuft of ventral bristles rather than a fan-like row.

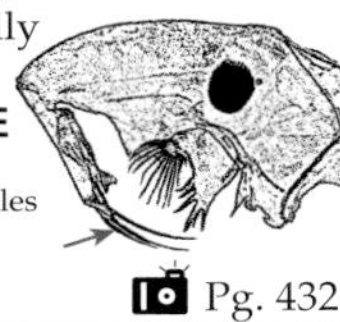

Pg. 432

Posterior spiracles not in deep pit.

Posterior spiracles in a deep pit. Aquatic species in pitcher plants. **SARCOPHAGIDAE**

Pg. 483

With at most one tapered process at end of abdomen. Prolegs present or absent.

With prolegs and tapered processes on every abdominal segment, processes on last segment very long and haired. **ATHERICIDAE** (*Atherix*)

Pg. 445

Abdomen encircled by 7 rings, each with 3 or 4 tubercles. Skin with longitudinal grooves. Tip of abdomen with at most one process. **TABANIDAE**

Pg. 441

Each abdominal segment with at most one proleg or tubercle. Tip with 2 or more lobes.

Body not ending in a tail or tail-like tubes.

Body ending in one or two tail-like tubes or a single stalk-bearing both posterior spiracles.

Anterior spiracular lobes present but simple and small, without multiple lobes. Common in decaying material; the few aquatic or semiaquatic species are rarely encountered, flattened larvae. SCUTTLE FLIES **(PHORIDAE)**

Anterior spiracles absent or with finger-like lobes.

Usually two tails or tip of tail split; spiracles separate. Body tapered. **SOME EPHYDRIDAE**

With a single tail-like or stalk-like breathing tube ending in 2 contiguous spiracles. Front of body blunt in common aquatic species. **SYRPHIDAE**

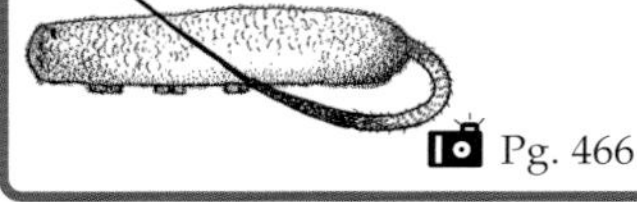

Pg. 466

Abdomen ending in pit surrounded by 4 lobes, 2 spiracles open at base of upper lobes. **DOLICHOPODIDAE**

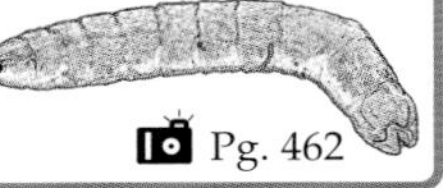

Pg. 462

Without open spiracles OR abdomen not ending in 4 lobes. **EMPIDIDAE**

Body tapered at both ends, usually with bumps, spicules or other vestiture on most of body.

Body smooth except for transverse spinulose rows. Usually broadest at or near tip (posterior).

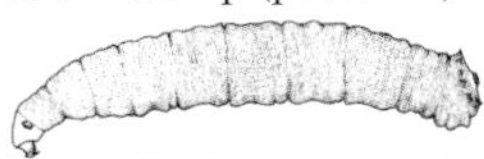

Tip of abdomen without spines or slender tubes, with tubercles surrounding posterior spiracles. Head skeleton with a solid bridge (ventral arch) below base of mouthhooks. **SOME SCIOMYZIDAE**
(size variable, usually over 5 mm)

Pg. 505

Tip (posterior) of abdomen with 2 small spines or breathing tubes not surrounded by tubercles. **SOME EPHYDRIDAE**
(size variable, usually less than 4 mm)

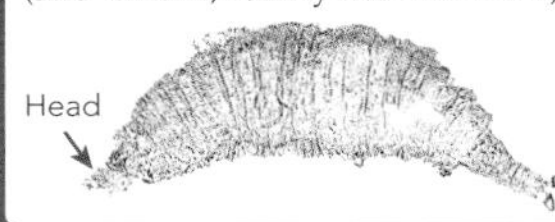

Tip of abdomen without tubercles around the margin. **MUSCIDAE**

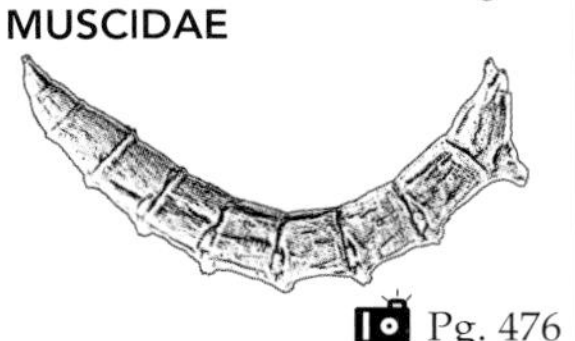

Pg. 476

Tip of abdomen with pairs of tubercles around the margin. **SCATHOPHAGIDAE**

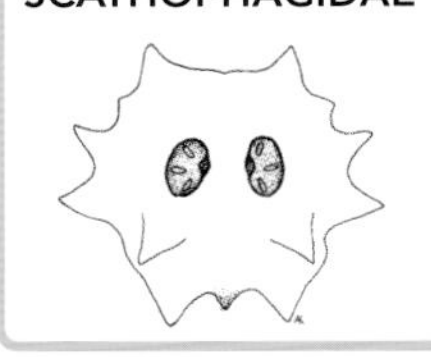

HYMENOPTERA KEY ONE: Sawflies, horntails and miscellaneous families.

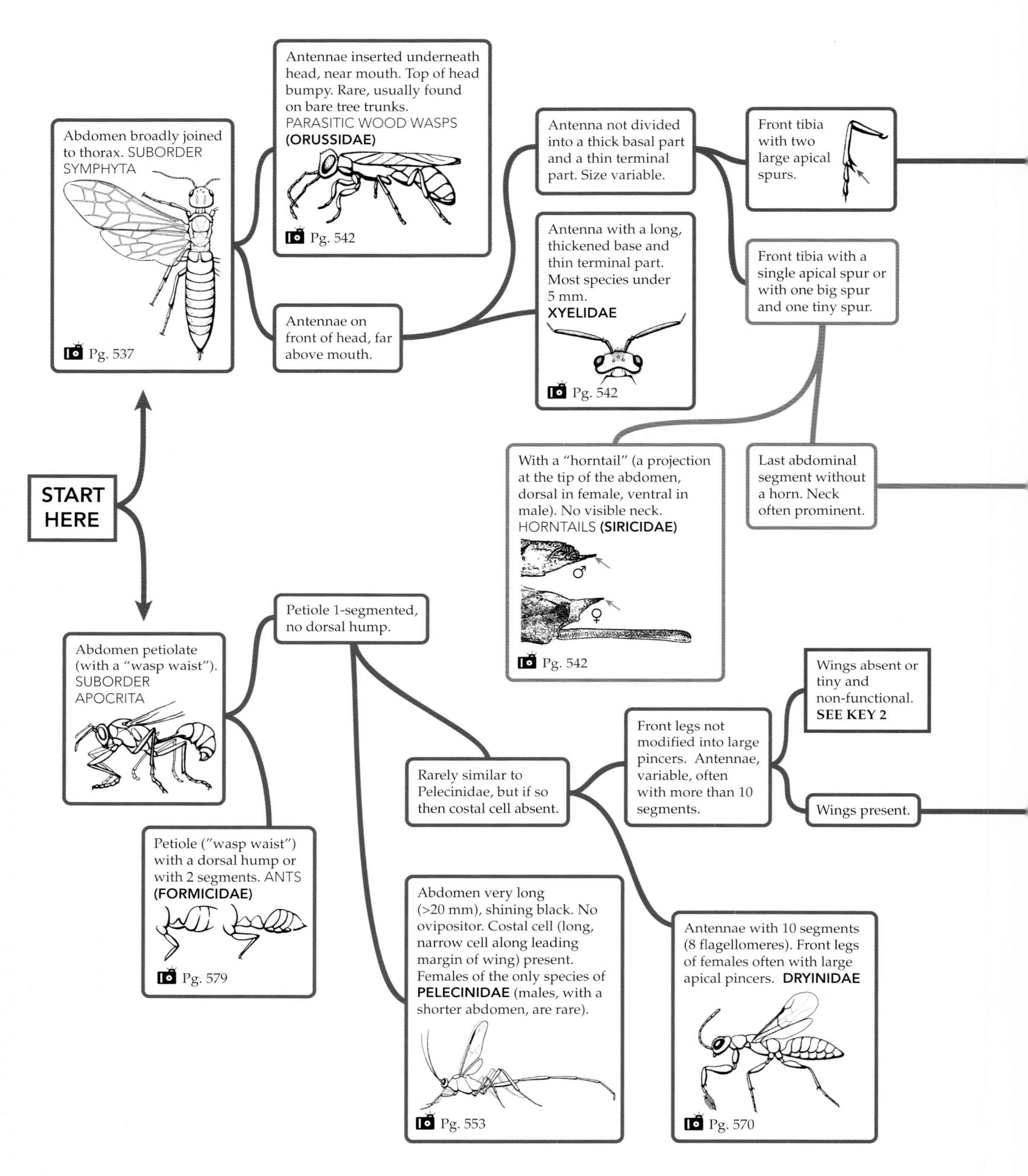

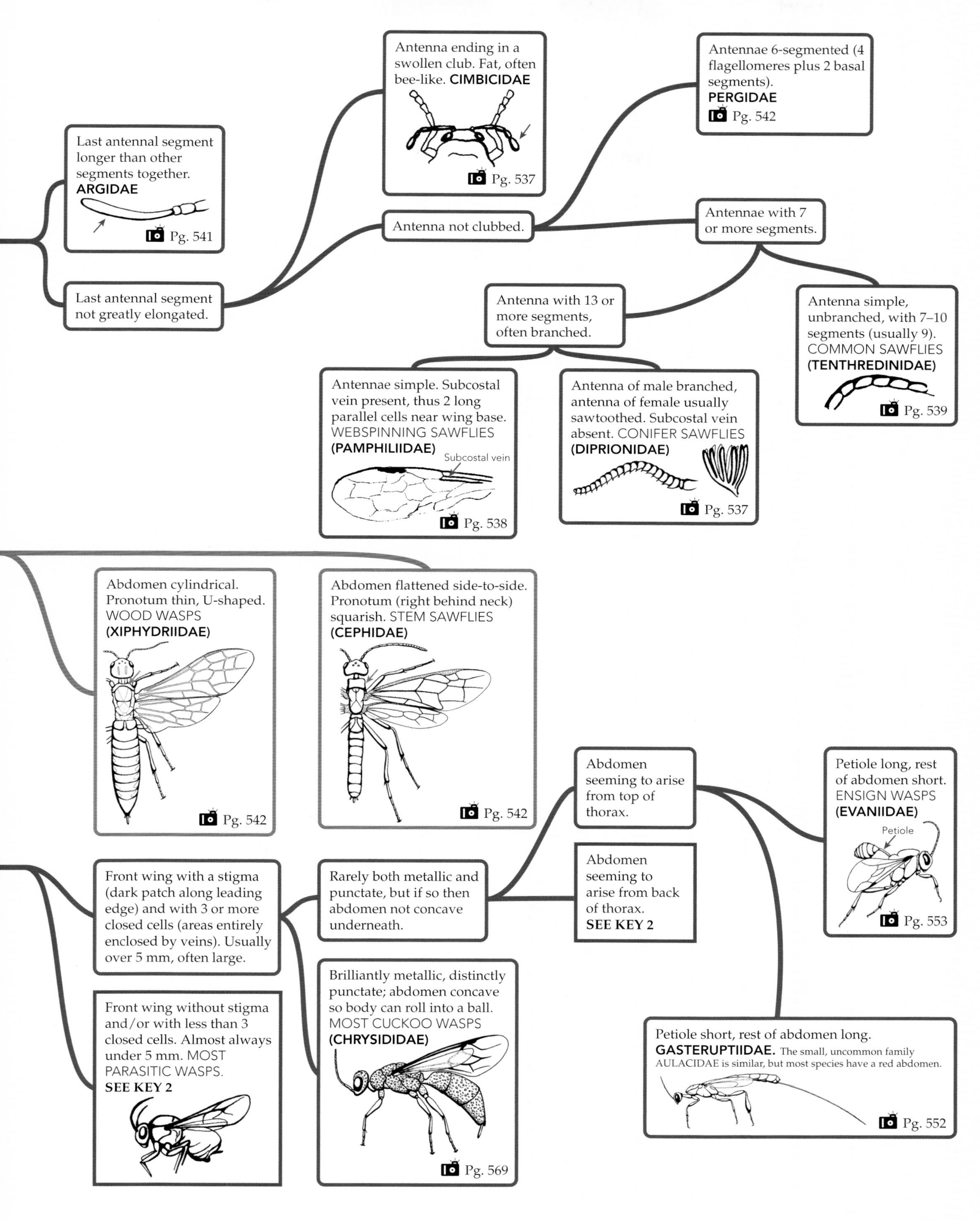
Last antennal segment longer than other segments together. ARGIDAE
Pg. 541
Last antennal segment not greatly elongated.
Antenna ending in a swollen club. Fat, often bee-like. CIMBICIDAE
Pg. 537
Antenna not clubbed.
Antennae 6-segmented (4 flagellomeres plus 2 basal segments). PERGIDAE
Pg. 542
Antennae with 7 or more segments.
Antenna with 13 or more segments, often branched.
Antenna simple, unbranched, with 7–10 segments (usually 9). COMMON SAWFLIES (TENTHREDINIDAE)
Pg. 539
Antennae simple. Subcostal vein present, thus 2 long parallel cells near wing base. WEBSPINNING SAWFLIES (PAMPHILIIDAE)
Subcostal vein
Pg. 538
Antenna of male branched, antenna of female usually sawtoothed. Subcostal vein absent. CONIFER SAWFLIES (DIPRIONIDAE)
Pg. 537
Abdomen cylindrical. Pronotum thin, U-shaped. WOOD WASPS (XIPHYDRIIDAE)
Pg. 542
Abdomen flattened side-to-side. Pronotum (right behind neck) squarish. STEM SAWFLIES (CEPHIDAE)
Pg. 542
Front wing with a stigma (dark patch along leading edge) and with 3 or more closed cells (areas entirely enclosed by veins). Usually over 5 mm, often large.
Rarely both metallic and punctate, but if so then abdomen not concave underneath.
Abdomen seeming to arise from top of thorax.
Abdomen seeming to arise from back of thorax. SEE KEY 2
Petiole long, rest of abdomen short. ENSIGN WASPS (EVANIIDAE)
Petiole
Pg. 553
Front wing without stigma and/or with less than 3 closed cells. Almost always under 5 mm. MOST PARASITIC WASPS. SEE KEY 2
Brilliantly metallic, distinctly punctate; abdomen concave so body can roll into a ball. MOST CUCKOO WASPS (CHRYSIDIDAE)
Pg. 569
Petiole short, rest of abdomen long. GASTERUPTIIDAE. The small, uncommon family AULACIDAE is similar, but most species have a red abdomen.
Pg. 552

HYMENOPTERA KEY TWO: Most Aculeata and Parasitica (from HYMENOPTERA KEY ONE).

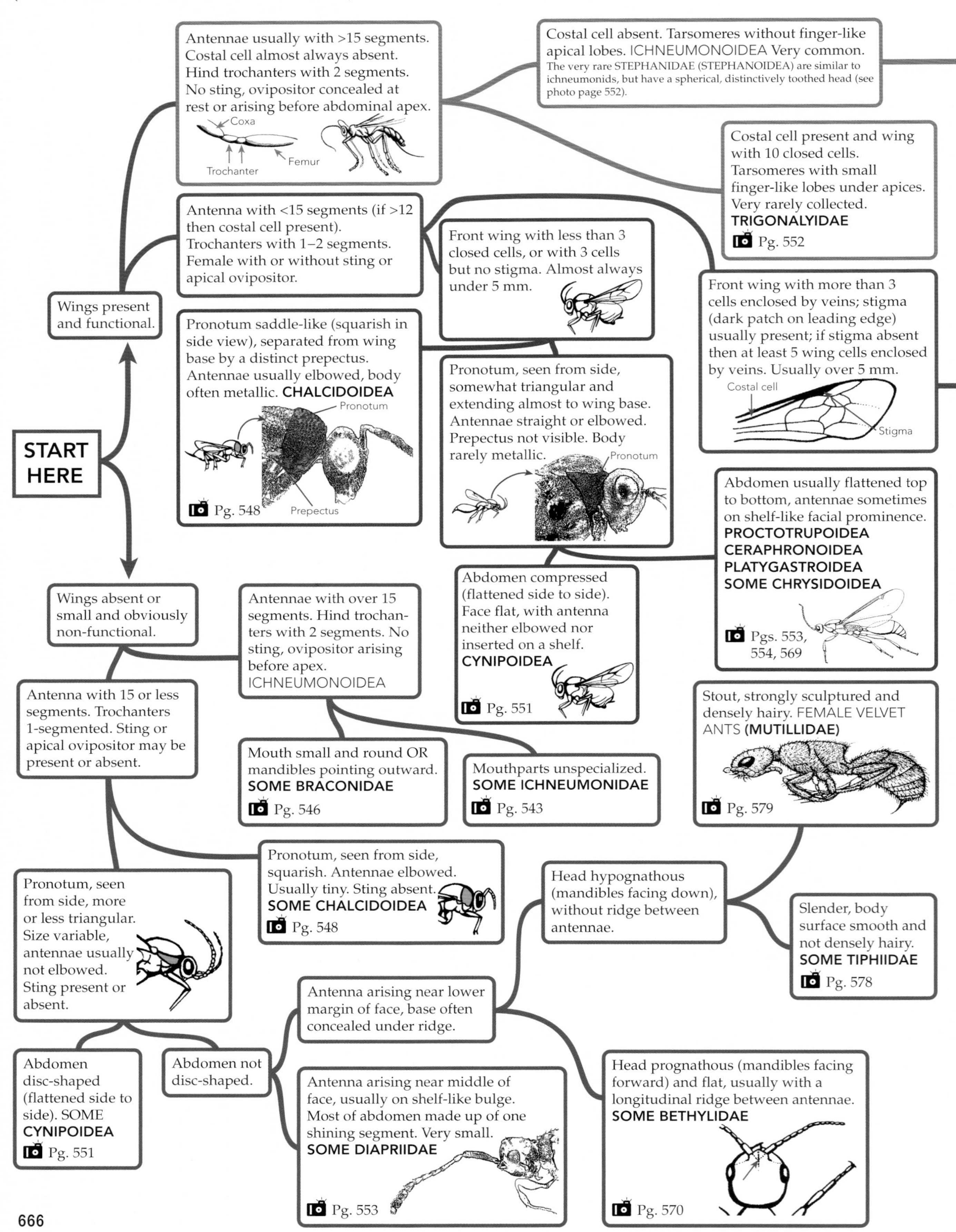

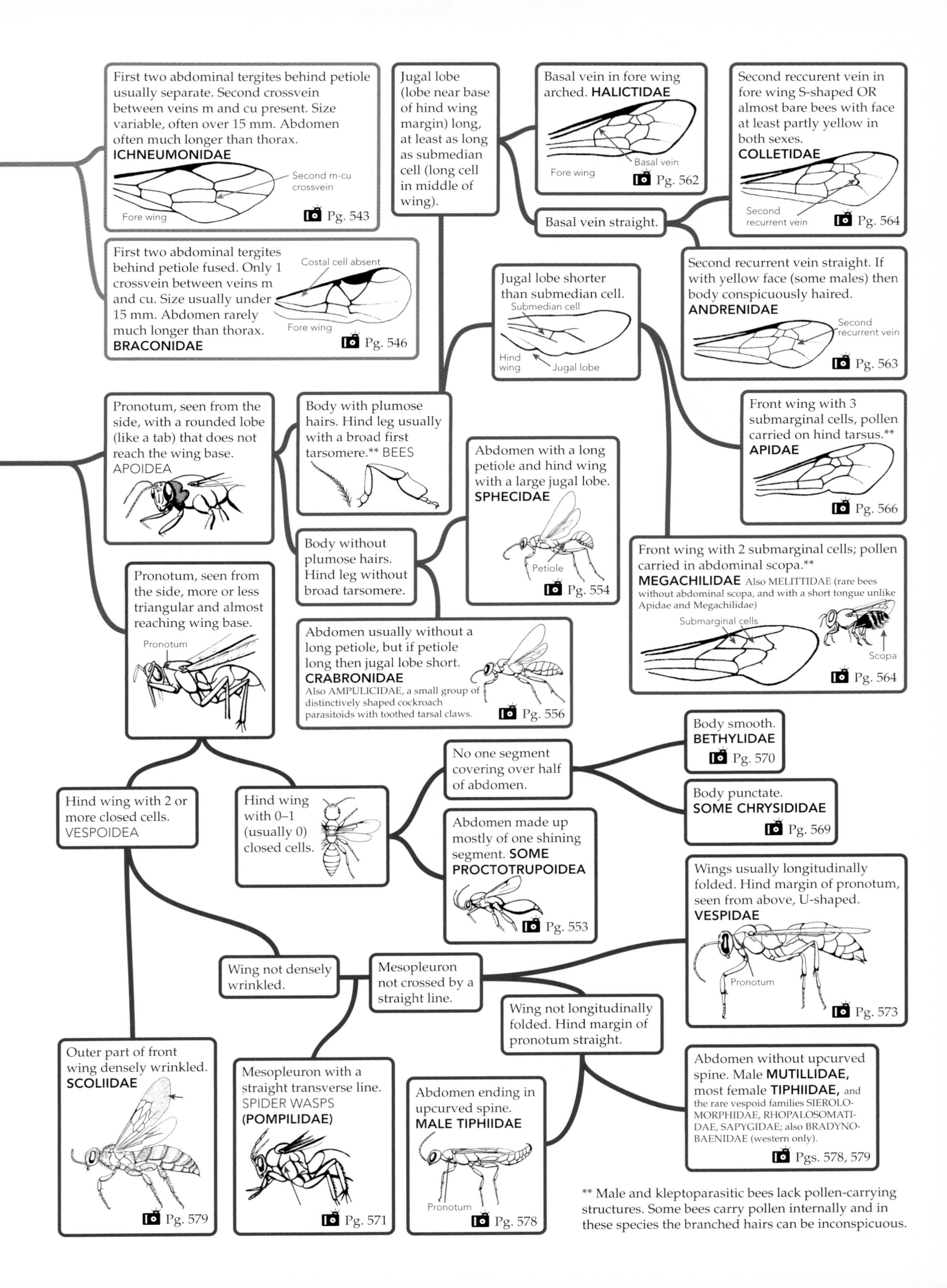
First two abdominal tergites behind petiole usually separate. Second crossvein between veins m and cu present. Size variable, often over 15 mm. Abdomen often much longer than thorax.
ICHNEUMONIDAE
Second m-cu crossvein
Fore wing
Pg. 543
First two abdominal tergites behind petiole fused. Only 1 crossvein between veins m and cu. Size usually under 15 mm. Abdomen rarely much longer than thorax.
BRACONIDAE
Costal cell absent
Fore wing
Pg. 546
Jugal lobe (lobe near base of hind wing margin) long, at least as long as submedian cell (long cell in middle of wing).
Basal vein in fore wing arched. HALICTIDAE
Basal vein
Fore wing
Pg. 562
Basal vein straight.
Second reccurent vein in fore wing S-shaped OR almost bare bees with face at least partly yellow in both sexes.
COLLETIDAE
Second recurrent vein
Pg. 564
Second recurrent vein straight. If with yellow face (some males) then body conspicuously haired.
ANDRENIDAE
Second recurrent vein
Pg. 563
Jugal lobe shorter than submedian cell.
Submedian cell
Hind wing
Jugal lobe
Front wing with 3 submarginal cells, pollen carried on hind tarsus.**
APIDAE
Pg. 566
Front wing with 2 submarginal cells; pollen carried in abdominal scopa.**
MEGACHILIDAE Also MELITTIDAE (rare bees without abdominal scopa, and with a short tongue unlike Apidae and Megachilidae)
Submarginal cells
Scopa
Pg. 564
Pronotum, seen from the side, with a rounded lobe (like a tab) that does not reach the wing base.
APOIDEA
Body with plumose hairs. Hind leg usually with a broad first tarsomere.** BEES
Abdomen with a long petiole and hind wing with a large jugal lobe.
SPHECIDAE
Petiole
Pg. 554
Body without plumose hairs. Hind leg without broad tarsomere.
Abdomen usually without a long petiole, but if petiole long then jugal lobe short.
CRABRONIDAE
Also AMPULICIDAE, a small group of distinctively shaped cockroach parasitoids with toothed tarsal claws.
Pg. 556
Pronotum, seen from the side, more or less triangular and almost reaching wing base.
Pronotum
Body smooth.
BETHYLIDAE
Pg. 570
No one segment covering over half of abdomen.
Body punctate.
SOME CHRYSIDIDAE
Pg. 569
Hind wing with 2 or more closed cells.
VESPOIDEA
Hind wing with 0–1 (usually 0) closed cells.
Abdomen made up mostly of one shining segment. SOME PROCTOTRUPOIDEA
Pg. 553
Wings usually longitudinally folded. Hind margin of pronotum, seen from above, U-shaped.
VESPIDAE
Pronotum
Pg. 573
Wing not densely wrinkled.
Mesopleuron not crossed by a straight line.
Wing not longitudinally folded. Hind margin of pronotum straight.
Outer part of front wing densely wrinkled.
SCOLIIDAE
Pg. 579
Mesopleuron with a straight transverse line.
SPIDER WASPS
(POMPILIDAE)
Pg. 571
Abdomen ending in upcurved spine.
MALE TIPHIIDAE
Pronotum
Pg. 578
Abdomen without upcurved spine. Male MUTILLIDAE, most female TIPHIIDAE, and the rare vespoid families SIEROLOMORPHIDAE, RHOPALOSOMATIDAE, SAPYGIDAE; also BRADYNOBAENIDAE (western only).
Pgs. 578, 579
** Male and kleptoparasitic bees lack pollen-carrying structures. Some bees carry pollen internally and in these species the branched hairs can be inconspicuous.

Selected references

The following list includes only a few selected references for each major group of insect, chosen either because they include keys likely to be of use to naturalists in northeastern North America, or because they are major references followed in organizing this book. More complete reference lists can be found in the recent entomological textbooks that are cited, especially the authoritative recent volume by Grimaldi and Engel. Regularly updated lists of entomological literature are posted on several websites, many of which are themselves excellent sources of information. The Tree of Life (tolweb.org/tree), for example, is an authoritative website that provides linkages to information on the phylogeny and taxonomy of all major insect groups. Other good places to start web searches for entomological information include sites posted by entomological organizations such as the Entomological Society of Canada (www.esc-sec.org), the Entomological Society of America (www.entsoc.org) and the Biological Survey of Canada (www.biology.ualberta.ca/bsc/bschome.htm). The site should be of particular interest owing to the recent addition of a digital journal of insect identification. Updates to this book, along with other linkages and illustrated identification guides to Ontario species of a growing list of taxa, can be found at the University of Guelph Insect Collection Web Site (www.uoguelph.ca/~samarsha). A recently developed website likely to be of interest to users of this book is BugGuide.net (bugguide.net), which includes an impressive compilation of original photographs and associated information posted by an "online community of naturalists." There is a steady growth of excellent new websites posted by experts (amateurs and professionals) on most popular (and many unpopular) taxa, and an apparently exponential growth of websites posted by people who are clearly not experts. Only a few of the former are listed below, as websites are constantly changing and most easily located using Google or other search engines.

General Reference, Insect Classification, and Arthropod Identification

Brues, C.T, A.L. Melander, and F.M. Carpenter. *Classification of Insects: Keys to the Living and Extinct Families of Insects, and to the Living Families of Other Terrestrial Arthropods.* Cambridge, MA: Harvard University, Museum of Comparative Zoology, 1954.

Cranshaw, W. *Garden Insects of North America.* Princeton, NJ: Princeton University Press, 2004.

Danks, H.V. (ed.). "Canada and Its Insect Fauna." *Memoirs of the Entomological Society of Canada* 108 (1979).

Grimaldi, D., and M.S. Engel. *Evolution of the Insects.* Cambridge, UK: Cambridge University Press, 2005.

Gullan, P.J., and P.S. Cranston. *The Insects: An Outline of Entomology.* 3rd edition. Oxford: Blackwell Publishing, 2005.

Merritt, R.W., and K.W. Cummins (eds.). *Aquatic Insects of North America.* 3rd edition. Dubuque, IA: Kendall/Hunt, 1996.

Mullen, G., and L. Durden (eds.). *Medical and Veterinary Entomology.* San Diego, CA: Academic Press, 2002.

Naumann, I.D. et al. (eds.). *Insects of Australia.* 2nd edition. Melbourne: Melbourne University Press: 1991.

Stehr, F. (ed.). *Immature Insects.* Vols. 1 and 2. Dubuque, IA: Kendall/Hunt, 1987 and 1991.

Triplehorn, C.A., and N.F. Johnson. *Borror and Delong's Introduction to the Study of Insects.* 7th edition. Belmont, CA: Brooks/Cole–Thomson Learning , 2005.

Wilson, E.O. *The Insect Societies.* Cambridge, MA: Belknap Press. 1971.

General Reading on Insect Natural History and Diversity

Askew, R.R. *Parasitic Insects*. New York: American Elsevier Publishing, 1971.

Berenbaum, M.R. *Bugs in the System: Insects and Their Impact on Human Affairs*. Reading, MA: Perseus Books, 1995.

Conniff, R. *Spineless Wonders*. New York: Henry Holt, 1996.

Eisner, T.A. *For Love of Insects*. Cambridge, MA: Belknap Press, 2003.

Evans, H.E. *Life on a Little Known Planet*. New York: E.P. Dutton, 1968.

Hocking, B. *Six-legged Science*. Cambridge, MA: Schenkman Publishing, 1971.

Marshall, S.A. *Insects of Algonquin Park.* Whitney, ON: Friends of Algonquin Park, 1997.

Stokes, D.W. *A Guide to Observing Insect Lives*. Boston, Little, Brown, 1983.

Waldbauer, G. *What Good Are Bugs? Insects in the Web of Life.* Cambridge, MA: Harvard University Press, 2003.

Chapter 1: Collembola and Other Primarily Wingless Hexapods

Christiansen, K.A., and P.F. Bellinger. *The Collembola of North America North of the Rio Grande*. Grinnell, IA: Grinnell College, 1980–81.

Hopkin, S.P. *The Biology of Springtails, Insecta: Collembola.* New York: Oxford University Press, 1997.

Tuxen, S.L. *The Protura: A Revision of the Species of the World with Keys for Determination*. Paris: Hermann, 1964.

Wygodzinsky, P. "A Revision of the Silverfish (Lepismatidae, Thysanura) of the United States and the Caribbean Area." *American Museum Novitates* 2481 (1972): 1–26.

Wygodzinsky, P., and K. Schmidt. "Survey of the Microcoryphia (Insecta) of the Northeastern United States." *American Museum Novitates* 2071 (1980): 1–17.

Chapter 2: Ephemeroptera and Odonata

Dunkle, S.W. *Dragonflies Through Binoculars: A Field Guide to Dragonflies of North America.* NY: Oxford University Press, 2000.

Edmunds, G.F., S.L. Jenson, and L. Berner. *The Mayflies of North and Central America.* Minneapolis: University of Minnesota Press, 1976.

Walker, E.M. *General: The Zygoptera – Damselflies.* Vol. 1 of The Odonata of Canada and Alaska. Toronto: University of Toronto Press, 1953.

———. *The Anisoptera – Four Families*. Vol. 2 of The Odonata of Canada and Alaska. Toronto: University of Toronto Press, 1958.

———, and P.S. Corbet. *The Anisoptera – Three Families*. Vol. 3 of The Odonata of Canada and Alaska. Toronto: University of Toronto Press, 1975.

Westfall, M.J. Jr., and M.L. May. *Damselflies of North America.* Gainesville, FL: Scientific Publishers, 1996.

Chapter 3: Plecoptera

Hynes H.B.N. "Biology of Plecoptera." *Annual Review of Entomology* 26 (1976): 135–153.

Stark, B.P., S.W. Szczytko, and C.R. Nelson. *American Stoneflies: A Photographic Guide to the Plecoptera.* Columbus, OH: Caddis Press, 1998.

Stewart K.W., and B.P. Stark *Nymphs of North American Stonefly Genera (Plecoptera).* Entomological Society of America. Ann Arbor MI: Thomas Say Foundation, 1988.

Chapters 4 and 5: Orthopteroids

Helfer, J.R. *How to Know the Grasshoppers, Cockroaches, and Their Allies.* Dubuque, IA: Wm. C. Brown, 1972.

Hoffman, K.M. "Earwigs (Dermaptera) of South Carolina, with a Key to the Eastern North American Species and a Checklist of the North American Fauna." *Proceedings of the Entomological Society of Washington* 89 (1987): 1–14.

Vickery, V.R., and D.K. McE. Kevan. *The Grasshoppers, Crickets and Related Insects of Canada and Adjacent Regions: Ulonata: Dermaptera, Cheleutoptera, Notoptera, Dictuoptera, Grylloptera, and Orthoptera*. Part 14 of The Insects and Arachnids of Canada. Ottawa: Agriculture Canada Publications, 1985.

Walker, T.J., and T.E. Moore. *Singing Insects of North America*. 2003. http://buzz.ifas.ufl.edu/

Weesner, F.M.. *The Termites of the United States: A Handbook*. Loring, VA: National Pest Control Association, 1965.

Chapter 6: Hemipteroids

Beirne, B.P. "The Leafhoppers (Homoptera, Cicadellidae) of Canada and Alaska." *Canadian Entomologist Supplement* 2 (1956): 1–180.

Hamilton, K.G.A. *The Spittlebugs of Canada (Homoptera: Cercopidae). Part 10 of The Insects and Arachnids of Canada.* Ottawa: Agriculture Canada Publications, 1982.

Henry, T.J. "Phylogenetic Analysis of Family Groups Within the Infraorder Pentatomomorpha (Hemiptera, Heteroptera) with Emphasis on the Lygaeoidea." *Annals of the Entomological Society of America* 90 (1997): 275–301.

———, and R.C. Froeschner. *Catalog of the True Bugs of Canada and the Continental United States.* Leiden: E.J. Brill, 1988.

Kim, K. C., H. D. Pratt, and C. J. Stojanovich. *The Sucking Lice of North America: An Illustrated Manual for Identification.* University Park, PA: Pennsylvania State University Press, 1986.

Maw, H.E.L., R.G. Foottit, K.G.A. Hamilton and G.G.E. Scudder. *Checklist of the Hemiptera of Canada and Alaska.* Ottawa: NRC Press, 2000.

McPherson, J.E. *The Pentatomoidea (Hemiptera) of Northeastern North America.* Carbondale, IL: Southern Illinois University Press, 1982.

Mockford, E.L. *North American Psocoptera.* Gainesville, FL: Sandhill Crane Press, 1993.

Slater, J.A., and R..M. Baranowski. *How to Know the True Bugs.* Dubuque, IO: Wm. C. Brown, 1978.

Stannard, L.J. "The Thrips, or Thysanoptera, of Illinois." *Illinois Natural History Survey Bulletin* 29 (1968): 215–552.

Tuff, D.W. "A Key to the Lice of Man and Domestic Animals." *Texas Journal of Science* 28 (1977): 145–59.

Chapter 7: Lepidoptera

Covell, C.V., Jr. *A Field Guide to the Moths of Eastern North America.* Boston,: Houghton Mifflin, 1984.

Handfield, L. *Le guide des Papillons du Québec* (including color illustrations of 1,448 species). Boucherville, QC: Broquet, 1999.

Layberry, R.A., P.W. Hall, and J.D. LaFontaine. *The Butterflies of Canada.* Toronto: University of Toronto Press, 1998.

Opler, P.A. and G.O. Krizek. *Butterflies East of the Great Plains: An Illustrated Natural History.* Baltimore: Johns Hopkins University Press, 1984.

Troubridge, J.T., and J.D. Lafontaine. *Moths of Canada* http://www.cbif.gc.ca/spp_pages/misc_moths/phps/mothindex_e.php

Wagner, D.L. *Caterpillars of Eastern North America: A Guide to Identification and Natural History.* Princeton, NJ: Princeton University Press, 2005.

Chapter 8: Trichoptera

Ross, H.H. *The Caddis Flies, or Trichoptera, of Illinois.* Champaign, IL: Illinois Natural History Survey, 1944. Reprinted in 1972 by Entomological Reprint Specialists.

Wiggins, G.B. *Larvae of the North American Caddisfly Genera (Trichoptera).* 2nd edition. Toronto: University of Toronto Press, 2000.

———, *Caddisflies: The Underwater Architects.* Toronto: University of Toronto Press, 2005.

Chapter 9: Neuroptera and Megaloptera

Oswald, J.D. "Revision and Cladistic Analysis of the World Genera of the Family Hemerobiidae (Insecta: Neuroptera)." *Journal of the New York Entomological Society* 101 (1983): 143–299.

Penny, N.D., P.A. Adams, and L.A. Stange. "Species Catalog of the Neuroptera, Megaloptera, and Rhaphidioptera of America North of Mexico." *Proceedings of the California Academy of Science* 147 (1997): 1–194

Throne, A.L. "The Neuroptera – Suborder Planipennia of Wisconsin, Part 1: Introduction and Chrysopidae." *Michigan Entomologist* 4 (1971): 65–78.

———, "The Neuroptera – Suborder Planipennia of Wisconsin, Part 2: Hemerobiidae, Polystoechotidae, and Sisyridae." *Michigan Entomologist* 4 (1971): 79–87.

Chapter 10: Coleoptera

Arnett, R.H., M. C. Thomas, P.E. Skelley, J. H. Frank (eds.). *American Beetles*. Vols. 1 and 2. Boca Raton, FL: CRC Press, 2002.

Bousquet, Y.A. (ed.). "Checklist of Beetles of Canada and Alaska." *Research Branch, Agriculture Canada Publication* 1861E, 1991. (www.canacoll.org/Coleo/Checklist/checklist.htm)

Downie, N.M., and R.H. Arnett. *The Beetles of Northeastern North America.* 2 vols. Gainesville, FL: Sandhill Crane Press, 1996.

Goulet, H., and Y. Bousquet. *Ground Beetles of Canada.* www.cbif.gc.ca/spp_pages/carabids/phps/index_e.php

White, R.E. *A Field Guide to the Beetles of North America.* Boston: Houghton Mifflin, 1983.

Yanega, D. *Field Guide to Northeastern Longhorned Beetles.* Champaign, IL: Illinois Natural History Survey, 1996.

Chapter 11: Diptera, Mecoptera and Siphonaptera

Benton, A.H. *An Illustrated Key to the Fleas of the Eastern United States.* Fredonia, NY: Marginal Media, 1983.

Ferrar, P. *A Guide to the Breeding Habits and Immature Stages of Diptera Cyclorrhapha.* Leiden: Brill/ Scandinavian Science Press, 1987.

Foote, R.H., F.L. Blanc, and A.L. Norrbom. *Handbook of the Fruit Flies (Diptera: Tephritidae) of America North of Mexico.* Ithaca, NY: Comstock Publishing, 1993.

Gagné, R.J. *The Plant-feeding Gall Midges of North America.* Ithaca, NY: Comstock Publishing, 1989.

Holland, G.P. "The Fleas of Canada, Alaska and Greenland (Siphonaptera)." *Memoirs of the Entomological Society of Canada* 130 (1985).

McAlpine, J.F., B.V. Peterson, G.E. Shewell, H.J. Teskey, J.R. Vockeroth, and D.M. Wood, eds. *Manual of Nearctic Diptera.* Vols. 1—3. Ottawa: Research Branch, Agriculture Canada Monographs 27, 28 and 32, 1981–89.

Oldroyd, H. *The Natural History of Flies.* London: Weidenfeld and Nicolson, 1964.

Stubbs, A., and P. Chandler. *A Dipterist's Handbook.* Hanworth, UK: The Amateur Entomologist's Society, 1978.

Thompson, F.C. *The BioSystematic Database of World Diptera.* 2004. http://www.sel.barc.usda.gov/Diptera/biosys.htm

Thornhill, A.R., and J.B. Johnson. "The Mecoptera of Michigan." *Great Lakes Entomologist* 7 (1974): 33–53.

Webb, D.W., N.D. Penny, and J.C. Martin. "The Mecoptera, or Scorpionflies, of Illinois." *Illinois Natural History Survey Bulletin* 31 (1975): 250–316.

Chapter 12: Hymenoptera

Akre, R.D., A. Greene, J.F. MacDonald, P.J. Landolt, and H.G. Davis. *Yellowjackets of America North of Mexico.* Washington DC: U.S. Department of Agriculture, Agriculture Handbook No. 552, 1980.

Bohart, R.M., and A.S. Menke. *Sphecid Wasps of the World: A Generic Revision.* Berkeley, CA: University of California Press, 1976.

Gauld, I.,and B. Bolton, eds. *The Hymenoptera.* Oxford: Oxford University Press, 1988.

Gibson, G.A.P., J.T. Huber, and J.B. Woolley, eds. *Annotated Keys to the Genera of Nearctic Chalcidoidea (Hymenoptera).* Ottawa: NRC Press, 1997.

Goulet, G.A.P., and J.T. Huber. *Hymenoptera of the World: An Identification Guide to Families.* Ottawa: Agriculture Canada, 1993.

Hanson, P.E., and I.D. Gauld. *The Hymenoptera of Costa Rica.* Oxford: Oxford University Press, 1995.

Hölldobler, B., and E.O. Wilson. *The Ants.* Cambridge, MA: Belknap Press, 1990.

Michener, C.D. *The Bees of the World.* Baltimore: Johns Hopkins University Press, 2000.

O'Neill, K.M. *Solitary Wasps: Behavior and Natural History.* Ithaca, NY: Comstock Publishing, 2000.

Chapter 13: Other Arthropods

Dindal, D. L. (ed.). *Soil Biology Guide.* New York: John Wiley & Sons, 1990.

Levi, H.V, L.R. Levi, and H.S. Zim. *Spiders and Their Kin.* New York: Golden Press, 1968.

Roth, V. *Spider Genera of North America.* 3rd ed. Gainesville, FL: American Arachnological Society, 1993.

Chapter 14: Entomological Techniques

Marshall, S.A., R.S. Anderson, R.E. Roughley, V. Behan-Pelletier, and H.V. Danks. "Terrestrial Arthropod Biodiversity: Planning a Study and Recommended Sampling Techniques." *Bulletin of the Entomological Society of Canada* 26(1), Suppl. 1994.

Martin, J.E.H. *The Insects and Arachnids of Canada, Part 1: Collecting, Preparing and Preserving Insects, Mites, and Spiders.* Ottawa: Agriculture Canada Publication 1643, 1977.

Shaw, J. *John Shaw's Closeups in Nature.* New York: Amphoto Books, 1987.

Acknowledgments

I could not possibly have put together such a wide-ranging book without the generous support of colleagues, friends and family. First and foremost, my sons Alex and Stephen kept me company in the field, ferreted out invisible insects for me to photograph, and generally made my field work a pleasure. My wife, Christine, supported the whole crazy enterprise, and all of my colleagues in the University of Guelph Insect Systematics laboratory helped out in many ways. I am particularly indebted to Matthew Buck and Steven Paiero for sharing their impressively broad entomological expertise, and for their help in obtaining and identifying many of the species herein. Other friends in the entomological community generously helped with the identification of images, and I am much indebted to the following colleagues for their generous advice and assistance with identifications: Ernest Bernard (Collembola); Rob Cannings, Colin Jones and Paul Pratt (Odonata); Shelley Ball and Steven Burian (Ephemeroptera); Riley Nelson (Plecoptera); Dan Johnson and Thomas Walker (Orthoptera); Andy Hamilton, Eric Maw, Geoff Scudder and Michael Schwartz (Hemiptera); Phil Careless and Edwin Mockford (Psocoptera); Jason Dombrosky, Jean-Francois Landry, Ken Stead, Jim Troubridge and David Wagner (Lepidoptera); Ralph Holzenthal, John Morse and Glenn Wiggins (Trichoptera); Norman Penny and Catherine Tauber (Neuroptera); Steve Ashe, Robert Anderson, Bruce Gill, Vasily Grebennikov, Serge Laplante, Laurent LeSage, James Lloyd, Rob Roughley and Henry Stockwell (Coleoptera); Andrew Bennett, Gary Gibson, Henri Goulet, John Huber, Lubomir Masner, Tatiana Romankova, Michael Sharkey, Cory Sheffield, David Smith and Gary Umphrey (Hymenoptera); Rowland Shelley (myriapods); Robert Bennett, Chris Buddle, G.B. Edwards, Robert Holmberg, Heather Proctor and David Walter (arachnids); and Joan Jass and Barbara Klausmeier (isopods). Many of my colleagues in the dipterological community, including Kevin Barber, Art Borkent, Bohdan Bilyj, Brian Brown, Yonghsheng Cui, Neal Evenhuis, Eric Fisher, Ben Foote, Ray Gagné, Steven Gaimari, Jon Gelhaus, Graham Griffiths, Kevin Holston, Lloyd Knutson, Owen Lonsdale, Mehrdad Parchami-Araghi, James O'Hara, Bradley Sinclair, Jeff Skevington, Christian Thompson, Richard Vockeroth, Don Webb, Monty Wood and Chen Young, generously helped by checking images, specimens or facts. Many other colleagues have at some time looked at my images or identified associated specimens, and I am indebted to them all.

Several people generously reviewed specific parts of the draft manuscript, including Ernest Bernard (Apterygotes), Steven Burian (Mayflies), Rob Cannings (Dragonflies), Thomas Walker, Darryl Gwynn and Dan Johnson (Orthoptera), Thomas Henry (Hemiptera), Jim Troubridge and Jean-Francois Landry (Lepidoptera), Glenn Wiggins (Trichoptera), Jeff Skevington (Diptera), Robert Anderson and Serge Laplante (Coleoptera), John Huber (Hymenoptera) and Robb Bennett (non-insect arthropods). Kevin Barber, Joe Shorthouse, Gard Otis, Christine Schisler, Jonathan Schmidt, Mehrdad Parchami-Araghi and George Barron also offered useful comments on parts of the developing manuscript. George Barron provided encouragement and advice throughout the project.

Projects dealing with the identification, biology and distribution of a wide range of arthropod species are inescapably dependent on information only available in major institutional insect collections. This book would have been impossible without daily access to the University of Guelph insect collection, an irreplaceable library of specimens, distributional data and identifications contributed by generations of entomologists since its establishment in 1863.

I am most grateful to the various illustrators who helped with the picture keys, especially Ian Smith, who illustrated the original keys in the manual for my "Natural History of Insects" course, and Monika Musial, who added many of the drawings needed to expand the original keys to cover a wider range of taxa for my "Insects in Relation to Wildlife" course. Some illustrations were also done by Christine Schisler, Rebecca Langstaff and Katie Bethune-Lehman; a few are my own. Dave Cheung redesigned the keys specifically for this book, and helped convert several photographs to illustrations for the keys. Rob McAleer first encouraged me to start this book, as *Bugs of Ontario*, and Lionel Koffler of Firefly Books suggested that I expand *Bugs of Ontario* into something much broader. I am indebted to the professionals at Firefly Books who helped turn those suggestions into reality.

Index of Photographs

Note: Page numbers are followed by a number corresponding to the photograph on the page.

This index identifies insects by genus and species.

Cloeon dipterum, 34.3–4
Clogmia albipunctatus, Large Drain Fly, 433.9

Index of Photographs

Index of Common Family Names

Note: Page numbers in bold type indicate a photograph in the text section.

General Index

Note: Orders include all levels of classification down to the superfamily level. Families are listed separately (in bold type) and include subfamilies. Page numbers in bold type indicate a photograph in the text section.